The Central Nervous System of Vertebrates   Volume 1

«Le système nerveux est au fond tout l'animal;
les autres systèmes ne sont là que pour le servir
ou pour l'entretenir...»

GEORGES CUVIER (1812)

»Hirnanatomie allein getrieben wäre eine sterile Wissenschaft.
Erst in dem Momente, wo man die Frage nach dem Verhältnis
der anatomischen Struktur zu der Funktion aufwirft,
gewinnt sie Leben.«

LUDWIG EDINGER (1908)

"Our primary interest is in the behavior of the living body,
and we study brains because these organs are the chief instruments
which regulate behavior."

CHARLES JUDSON HERRICK (1948)

Springer

Berlin
Heidelberg
New York
Barcelona
Budapest
Hong Kong
London
Milan
Paris
Santa Clara
Singapore
Tokyo

R. Nieuwenhuys
H. J. ten Donkelaar
C. Nicholson

The
Central Nervous System
of Vertebrates

Volume 1

With Chapters
in Cooperation with:

W. J. A. J. Smeets
H. Wicht

With 474 Figures and 16 Tables

 Springer

ISBN 3-540-56013-0 Springer-Verlag Berlin Heidelberg New York Tokyo

Libary of Congress Cataloging-in-Publication-Data
Nieuwenhuys, R., 1927–
    The central nervous system of vertebrates / R. Nieuwenhuys, H.J. ten Donkelaar, C. Nicholson :
    with contributions by J.L. Dubbeldam ... [et al.]. p.   cm.
    Includes bibliographical references.
    ISBN 3-540-56013-0 (hardcover : alk. paper)
    1. Central nervous system – Anatomy. 2. Vertebrates – Anatomy. 3. Anatomy, Comparative.
    4. Neuroanatomy. I. Donkelaar, H.J. ten (Hendrik Jan), 1946–, II. Nicholson, Charles.
    III. Title. [DNLM: 1. Central Nervous System – anatomy & histology. 2. Anatomy, Comparative.
    3. Vertebrates – anatomy & histology. WL
    300 NB82ca 1997]   QM451.N498 1997   573.8'616 – dc21   DNLM/DLC
    for Library of Congress

© Springer-Verlag Berlin Heidelberg 1998
    Printed in Germany

Production: PRO EDIT GmbH, Heidelberg
Typesetting: Mitterweger Werksatz GmbH, Plankstadt
Cover Design: Erich Kirchner, Heidelberg
SPIN: 10087711   27/3136 – 4 3 2 1 0 – Printed on acid-free paper

Preface

The plans for this book began to form more than 30 years ago when the first author was head of the section of comparative neuroanatomy at the Central Institute for Brain Research (now called the Institute for Brain Research) in Amsterdam, The Netherlands. Working there implied that one was part of a tradition since research at the "Brain Institute" – as it was usually known – had focussed exclusively on comparative neuroanatomy for 50 years.

The first director of the institute, Cornelius Ubbo Ariëns Kappers (1877–1946) held this position from the opening of the institute in 1909 until his death and was a disciple and associate of Ludwig Edinger (1855–1918), one of the founders of comparative neuroanatomy. Ariëns Kappers had become, during his long directorship, a neuroanatomist of international fame. His books included the monumental Vergleichende Anatomie des Nervensystems (1920–1921) and, in collaboration with G. Carl Huber and Elizabeth C. Crosby, The Comparative Anatomy of the Nervous System of Vertebrates (1936). His collected reprints, encompassing seven large volumes, his library and his enormous collection of preserved and sectioned brains, were all emphatically present in the institute.

The research of Ariëns Kappers and his numerous associates was almost exclusively based on non-experimental material which, after fixation, was embedded in celloidin, sectioned at 20 mm, stained with the Weigert-Pal method to reveal myelinated fibres, and counterstained with paracarmine for cell bodies. Such material has obvious limitations for analysis of the microstructure and, looking back, one can only marvel that so much could be achieved with material of this kind. As an aside, it is astonishing how much aesthetic appeal this type of material provides, with its blue-black fibre tracts contrasting with the pink masses of the cells.

By the 1960s "Kappers, Huber and Crosby" still remained the only available work on comparative neuroanaomy, and, indeed, it was reprinted in 1967 by the Hafner Publishing Company to try to satisfy the continuing need for such a comprehensive study. But the original work was obviously outdated by then and, furthermore, it had always been limited somewhat as a systematic treatise of comparative neuroanatomy, since it neither attempted to describe the principles of the subject nor did it set the chosen species in any wider zoological context. Finally, the limitations of Ariëns Kappers' histological archive mandated that any new comprehensive survey of vertebrate comparative neuroanatomy would have to be based on new histological material. These factors led the first author to plan the present work.

In the early 1970s Charles Nicholson was invited to join the project. Some years earlier he and Rudolf Nieuwenhuys had collaborated on a study of the structure of the mormyrid cerebellum. When this research was published, it was accompanied by a comprehensive review of what was then known about the function of the electrosensory system in fishes (which was thought to be related to the hypertrophy of the mormyrid brain). It was originally planned that the present work would follow a somewhat similar format, combining a description of structure with a functional commentary, at an introductory level. Just as with other living material, the book grew vigorously during its prolonged period of gestation. The book gradually developed into a reference work with the emphasis on structure and a more limited coverage of function. There were several reasons for this. First, neurophysiological research is even more biased towards mammals than neuroanatomy so that any comparative survey would be extremely fragmentary in its coverage. Second, a true comparative neurophysiology has never been attemped; rather, neurophysiologists have selected preparations which provide unusual opportunities for studying a particular cell or system. In fact there are several groups, including the Myxinoids, the Brachiopterygians and the Dipnoans, which have been completely neglected by neurophysiologists so far.

For all of these reasons the original "design" of the book could not be realized. However, the principle that structure and function are merely facets of a single reality is retained. For instance, the recent combined structural and functional research on the central nervous system of the lamprey, focussing on the control of locomotion, has been treated in detail and accounts in part for the length of Chap. 10.

As the extent of the work and its dominant structural content became more clear, it was evident that the first author could not hope to complete the anatomical surveys in a reasonable timeframe without the assistance of a second comparative anatomist. Consequently, during the initial phase of the preparation of the text (i.e. in the early 1980s) Glenn Northcutt (Scripps Institute of Oceonography, San Diego, Calif., USA) was invited to join the project. During that phase he read early versions of Chaps. 9–17 and provided valuable comments. Although keenly interested, he felt that the format of the work had been determined already so that it would be difficult for him to make a real and creative contribution and, so far as the book was concerned, our paths diverged. Mary Sue Northcutt, however, was kind enough to edit initial versions of several chapters, and we expect that even the final versions will bear traces of her unequalled editorial skill. We are sure that the present work would have benefited from Glenn Northcutt's vast knowledge of comparative neuroanatomy. Some consolation was offered by the fact that in the final phase of the production of this book, his associate, Helmut Wicht, from the Department of Anatomy of the J.W. Goethe University in Frankfurt am Main, Germany, accepted our invitation to update the chapter on Myxinoids. We felt justified in appealing to Helmut Wicht for this task, because it was mainly as a result of his research that the said chapter had become outdated!

Fortunately, another accomplished comparative neuroanatomist, in the person of Hans ten Donkelaar, was willing to join the project and provide the essential help to see this project through. Since completing his thesis research under the direction of Rudolf Nieuwenhuys, Hans ten Donkelaar had focussed his attention on amphibians and reptiles and was ideally suited to cover these topics in the present work. He has also been invaluable in working with the first author on the day-to-day mechanics of the production of material for the book, since both authors worked at the same institution for many years, while the third author, located in the United States, has interacted by means of trans-Atlantic visits which have become too numerous to count anymore.

Finally, as the scope of the book continued to expand, and more to the point, the scientific literature was generated at an ever-increasing pace, it became necessary to find assistance with specific chapters which will be further detailed below.

In 1968 the first author moved to the Department of Anatomy at the University of Nijmegen, The Netherlands, where he then had at his disposal the requisite facilities for the realization of a new book on comparative vertebrate anatomy. Some of the most important supplementary elements were:

1. A number of highly competent technicians, among them Marion Cremers-Cornelisz, Carla de Vocht-Poort and Nelly Driessen-Verijdt, were available to prepare serial sections through the brains and selected parts of the spinal cord of representative species of all vertebrate classes, with Nissl and Klüver-Barrera stains and the silver-impregnation of Bodian, as the standard techniques.

2. The same technicians and their colleagues Roelie de Boer-van Huizen, Henk Joosten and Theo Hafmans also kept pace with the stream of new anatomical techniques, including the various anterograde and retrograde tracer methods, histofluorescence and immunohistochemistry. Electronmicroscopy also found wide application. Theo Hafmans deserves special mention, since he was responsible for much of the photography at the light microscopical level.

3. The presence of an illustration department (founded by Prof. H.J. Lammers) with by no less than ten competent illustrators led by Christiaan van Huijzen and later by Joep de Bekker. Almost all of the half-tone drawings, illustrating the gross anatomy of the brains of 17 species, and the India-ink drawings showing the microscopical structure at representative levels of the brains and spinal cords of 18 species, which form the central core of the present work, were fortunately prepared before stringent budget cuts essentially brought the illustration department to an end. The fact that one of the artists, Mr. J. P. M. Maas, long after his retirement and until the conclusion of the project, continued preparing high quality illustrations for the work, deserves a special mention. Our grateful thanks are also due to Christiaan van Huijzen, who prepared the accompanying posters 1–3. Roel Seidell prepared three sets of drawings for the gross anatomy series. Most of the drawings for Chaps. 18–20 were done by Marlu de Leeuw, who also prepared posters 4 and 5.

The final phase of this project would not have been possible without the generous support of Mr. A. Dousma (Audiovisual Group, University of Nijmegen) and of the Department of Anatomy and Embryology (University of Nijmegen).

During the period 1992–1997, Professor D.F. Swaab, the present director of the Netherlands Institute for Brain Research, Amsterdam, kindly put the facilities of this Institute (where it all began!) at the disposal of the first author. The crucial help of Mr. G. van der Meulen, photographer at that Institute is especially acknowledged.

4. Without adequate secretarial assistance a comprehensive project of this kind could not be accomplished. Four successive, competent and highly devoted secretaries, Trudy van Son-Verstraeten, Margaret Shak Shie, Marion van de Coevering and Inge Eijkhout, have lent their support to the project.

An essential element leading to the present work was the comprehensive programme of comparative neuroanatomical research which was planned and realized in Nijmegen. This encompassed all major groups of vertebrates, except for birds and mammals, and within the framework of this programme, numerous doctoral theses were prepared. As mentioned above, the second author of the present work was one of the first "promovendi" in this programme of research at Nijmegen.

Two other fundamental and important chapters also originated with work begun as Ph.D. projects. These are the chapter on "Cartilaginous Fishes" by Wil Smeets, Department of Anatomy, Free University, Amsterdam, and the chapter on "Holosteans and Teleosts" by Hans Meek and Rudolf Nieuwenhuys, Department of Anatomy, University of Nijmegen.

Thus, the research material for most of the required animal groups were covered in the course of the highly productive investigations and using the excellent facilities at the University of Nijmegen. Two groups, the birds and the mammals, offered problems, however. Work on the central nervous system of birds had never been included in the Nijmegen research programme. We were fortunate, however, to find Jaap Dubbeldam (Professor of Neurobehavioral Morphology, State University Leiden, The Netherlands), an expert in this field, willing to prepare the chapter on the brain and spinal cord of this group.

The problems with the mammals were primarily quantitative in nature. The body of knowledge on the central nervous system of this group of animals has grown to such enormous proportions that comprehensive surveys of current knowledge of the structure of brain and spinal cord of a single species can be provided only by a large team of experts. To take three recent examples: The Rat Nervous System, 2nd edn, (1995), edited by G. Paxinos includes 22 contributions by at least 60 authors and covers 1136 pages. The Cerebral Cortex (1984–1995) edited by A. Peters and E.G. Jones spans 12 volumes with an average of 500 pages per volume and the contributors are too numerous to count. Finally, Comparative Neurobiology in Chiroptera, (1996) by G. Baron, H. Stephan and H. Frahm, is devoted solely to bats, yet comprises three volumes and 1596 pages!

It was tempting to leave out the mammals from the present work, because of the difficulties they presented, but we felt that a book on comparative vertebrate neuroanatomy would be incomplete without this group. It was evident, however, that within the limits set and repeatedly revised upwards over the years, for the present work only a concise and eclectic survey of the structure of the mammalian brain and spinal cord, with emphasis on comparative aspects, could be included. Very fortunately Jan Voogd, head of the Department of Anatomy, Erasmus University Rotterdam, The Netherlands, and Paul van Dongen, a former collaborator at the Department of Anatomy Nijmegen, now scientific editor, Department of Clinical Research at Janssen-Cilag, Tilburg, the Netherlands, were kind enough to cooperate with us on this part of the project and provide the essential, but succinct, coverage that we desired. Paul van Dongen also provided a concise, yet comprehensive, chapter on brain size in vertebrates.

As noted above, and emphasised wherever possible in this work, structure and function belong together and should be studied jointly wherever possible; but additional levels of conceptual integration are required too. From the perspective of the theory of evolution, the functioning brain forms only the intermediate level in a continuum extending between two ultimate frontiers, the abiotic and biotic environment, on one hand, and the genetic reservoir of interbreeding populations, on the other. This means that genetic and comparative developmental neurobiology, neuroethology and neuroecology ultimately should be included in comparative neuroscience. Genetics, ethology and ecology are all flourishing fields of science. The comparative approach has rarely been very central in these field of research; however, it is hoped that this book will pave the way to some extent for these new and highly promising provinces of the neurosciences.

In concluding this preamble the invaluable editing assistance of Ms. S. Bakker, M.Sc., is especially acknowledged, and, finally, we extend our most sincere thanks to the publishing house of Springer-Verlag and their staff – especially Dr. R. Lange, Mr. R.-P. Fischer, Mr. E. Kirchner, Ms. A. Clauss, Ms. S. Sundell and Ms. G. Wiegel, PRO EDIT GmbH, for their kind help during the preparation of this book.

Abcoude/Nijmegen/New York, August 1997   RUDOLF NIEUWENHUYS
                                                HANS J. TEN DONKELAAR
                                                CHARLES NICHOLSON

Purpose and Plan

The purpose of this book is threefold: (1) to set forth the general foundations, concepts and principles of comparative neuroanatomy, (2) to present a comprehensive survey of our current knowledge of the structural organisation of the brain and spinal cord of the various groups of vertebrates, and (3) to provide some ideas for future research.

The plan of the book reflects these goals. The chapters forming the general part attempt to define the context in which the more detailed later chapters were formulated. Today, neuroscience is a burgeoning field in which ideas are changing rapidly. While we hope that the main body of this work will remain useful for many years, we recognise that as time passes it will be difficult for those referring to it to envisage the frame of reference of the authors. One objective of these general chapters is to indicate this frame. A second goal is to provide some definitions so that when the reader later encounters terms or concepts that are unfamiliar, the general section may be referred to for clarification.

The special character of Chap. 6 should be emphasised. Although numerous contributions to the methodology of comparative neuroanatomy have been made during the last century, the synthesis presented in the second part of that chapter is new. The reader should not expect, however, that all of the structural interpretations discussed in the second, specialised part are realised in the light of the methodology outlined in Chap. 6. Rather, the methodology presented is offered as a framework for future research. The same holds true for the final section of Chap. 6. The programme of comparative neuroanatomy outlined there is not an introduction to the voluminous specialised part, in the sense that this part represents the realisation of such a programme. Most of the book, i.e. the presentation of a comprehensive survey of our current knowledge of the vertebrates, is realised in the large, specialised part of the work. Our intention here was to achieve a balanced survey of our current knowledge on the brain and spinal cord of the various groups *and not* to profile the results of our own research. Nevertheless, all of these chapters have been written by authors with a first-hand knowledge of the central nervous system of the pertinent group of animals.

In the preparation of the specialised parts many choices had to be made. The first choice was between arranging the data according to the groups of animals or according to the divisions of the central nervous system, i.e. the spinal cord, the rhombencephalon and so on. We decided to follow the first approach because it would enable us to begin each chapter with some introductory notes on the animals and their lifestyle, before moving on to the central nervous system. Moreover, readers who want to study the central nervous system region by region will find our format easy to use, because each chapter is organized in a similar way and begins with a table of contents, making it simple to extract the relevant comparative information on any chosen brain structure.

Other important choices included the species to be dealt with and the selection of the illustrations. For a discussion of these aspects the reader is referred to Chap. 8. We only note here that the pictorial central theme of all chapters of the specialised part is formed by (1) drawings of entire animals, (2) simple illustrations showing the brain in position, (3) four standard views of the brains, showing their macrostructure, and (4) drawings of sequences of transversely oriented microscopical sections through the spinal cord and brain of a number of species.

The third and final purpose of the book, i.e. to provide a basis for and to indicate some of the prospects for future comparative neuroanatomical research, is evident in all parts of the work. As already mentioned, a conceptual basis and programme of future research is described in Chap. 6. The separate chapters of the specialised section, by recording what is known of the structural organisation of the central nervous system of the various groups of vertebrates, also serve to highlight the many gaps in our knowledge. Finally, the concluding chapter briefly surveys the major fields and "levels" of current comparative neurobiology and indicates many of the goals that lie in the future.

Acknowledgements for the Use of Illustrations

Thanks for the use of illustrations are due to the following sources. Acknowledgments to the authors who were so kind to give their permission to use figures, are made in the figure legends.

Academic Press, London
Gans C (ed) Biology of the Reptilia, Vols 9 and 10 (1979)

Academic Press, San Diego
Paxinos G (ed) The Rat Nervous System, 2nd Ed (1995)

Akademie Verlag, Berlin
Journal für Hirnforschung

American Association for the Advancement of Science, Washington DC
Science

The American Physiological Society, Bethesda
Journal of Neurophysiology

American Society of Zoologists, Lawrence, Kansas
American Zoologist

ANKHO International Inc, Fayetteville, NY
Brain Research Bulletin

Johann Ambrosius Barth, Leipzig
Nova Acta Leopoldina

Brill, Leiden
Netherlands Journal of Zoology

Cambridge University Press, Cambridge, UK
The Behavioral and Brain Science; The Journal of Physiology

Cell Press, Cambridge, MA
Neuron

Clarendon Press, Oxford
Benton MJ (ed) The Phylogeny and Classification of the Tetrapods, Vol 1 (1988)

Current Biology Ltd, London
Current Opinions in Neurobiology

Elsevier Science, Amsterdam
Brain Research; Brain Research Reviews; Progress in Brain Research

Elsevier Science Ireland, Shannon
Neuroscience Letters

Elsevier Science Inc, New York
Journal of Chemical Neuroanatomy

Elsevier Science Ltd, Kidlington, UK
Neuroscience; Progress in Neurobiology; Trends in Neurosciences

Gustav Fischer Verlag, Jena – Stuttgart
Fortschritte der Zoologie; Zoologisches Jahrbuch (Physiologie)

WH Freeman and Co, New York
Carroll RL, Vertebrate Paleontology and Evolution (1988)

S Karger AG, Basel
Acta Anatomica; Brain, Behavior and Evolution

Macmillan Magazines Ltd, London
Nature

The University of Michigan Press, Ann Arbor
Fish Neurobiology (Vol 1, Northcutt RG, Davis RE, eds; Vol 2, Davis RE, Northcutt RG, eds, 1983)

National Institute of Mental Health, Rockville, Maryland
Greenberg N, MacLean PD (eds) Behavior and Neurobiology of Lizards (1978)

The New York Academy of Sciences, New York
Annals of the New York Academy of Sciences

Oxford University Press, New York
Shepherd GM (ed) The Synaptic Organization of the Brain, 3rd ed (1990)

Oxford University Press, Oxford
European Journal of Neuroscience

Plenum Press, New York
Ebbesson SOE (ed) Comparative Neurology of the Telencephalon (1980); Jones EG, Peters A (eds) Cerebral Cortex, Vol 8A (1990)

The Royal Society, London
Philosophical Transactions of the Royal Society (London), Section B

Science Reviews Ltd, Northwood, Middlesex
Scientific Progress, Oxford

Sinauer Associates Inc, Sunderland, MA
Hille B (ed) Ionic Channels of Excitable Membranes, 2nd ed (1992)

Society for Neuroscience, Washington DC
The Journal of Neuroscience

Springer Verlag, Berlin-Heidelberg-New York
Advances in Anatomy, Embryology and Cell Biology; Anatomy and Embryology; Cell and Tissue Research; Experimental Brain Research; Journal of Comparative Physiology A: Studies in Brain Function

Swets Publishing Service, Lisse, the Netherlands
Acta Morphologica Neerlando-Scandinavica

Georg Thieme Verlag, Stuttgart
Hassler R, Stephan H (eds) Evolution of the Forebrain (1966)

John Wiley & Sons, New York
Ulinski PS, Dorsal Ventricular Ridge (1983); Kettenmann H, Grantyn R (eds) Practical Electrophysiological Methods (1992)
The American Journal of Anatomy; The Journal of Comparative Neurology; Journal of Morphology; Journal of Neurobiology

List of Contributors

DUBBELDAM, J. L.
Van der Klaauw Laboratory, University of Leiden,
P. O. Box 9516, 2300 RA Leiden,
The Netherlands

MEEK, J.
Department of Anatomy & Embryology,
University of Nijmegen, P. O. Box 9101,
6500 HB Nijmegen, The Netherlands

NICHOLSON, C.
Department of Physiology & Biophysics, New York
University Medical Center, 550 First Avenue,
New York, 10016 NY, USA

NIEUWENHUYS, R.
Papehof 25, 1391 BD Abcoude,
The Netherlands

SMEETS, W. J. A. J.
Department of Anatomy & Embryology,
Free University, Van der Boechorststraat 7,
1081 BT Amsterdam, The Netherlands

TEN DONKELAAR, H. J.
Department of Anatomy & Embryology,
University of Nijmegen, P.O. Box 9101,
6500 HB Nijmegen, The Netherlands

VAN DONGEN, P. A. M.
Linge 9, 5032 EV Tilburg, The Netherlands

VOOGD, J.
Department of Anatomy, Erasmus University
Rotterdam, P. O. Box 1738, 3000 DR Rotterdam,
The Netherlands

WICHT, H.
Klinikum der J.-W.-Goethe-Universität, Zentrum
der Morphologie, University of Frankfurt,
Theodor-Stern-Kai 7, 60590 Frankfurt am Main,
Germany

Contents

VOLUME 2

II. SPECIALISED PART (Contd.)

VOLUME 3

II. SPECIALISED PART (Contd.)

III. GENERAL CONCLUDING PART

I. GENERAL INTRODUCTORY PART

CHAPTER 1

Structure and Function of the Cellular Elements in the Central Nervous System

C. NICHOLSON

1.1
Introduction

The purpose of this and the general chapters is to clarify the context in which the later specialised sections were formulated. From the conception of this book to its publication, more than 20 years have passed. During this time the nervous system was studied from the twin perspectives of structure and function: structure was defined largely by the available neuroanatomical staining methods, while functional analysis was confined mainly to the recording of the electrical activity of neurons. These perspectives dominate this book, but during the past decade the discipline of molecular biology has made rapid and extraordinary inroads into neuroscience and is changing the way the brain is perceived. In parallel, electrical recording methods are being augmented by techniques such as optical recording of neuronal activity, positron-emission tomography and functional magnetic resonance imaging. In short, neuroscience is undergoing the sort of paradigm shift referred to by Kuhn (1970) in his influential book *The Structure of Scientific Revolutions*. Although not entirely neglecting these more recent perspectives, we have chosen not to excessively emphasise such studies, because to have done so would have compromised the continuity of the book. In some senses, therefore, this work closes a rich and immensely productive period of neuroscience with what we hope is a thoughtful overview that will remain valuable and stimulating as unimagined insights emerge from many new techniques.

In order to minimise the size of this chapter we shall direct the reader's attention wherever possible to published reviews and texts for in-depth discussion of the topics. From this it is clear that these early chapters are not intended as an introduction to current cellular neuroscience but rather as a sketch of the field. In a way, the state of contemporary neuroscience is defined by recent introductory textbooks and monographs. In this chapter we have drawn most heavily on four, which are listed in alphabetical order: Alberts et al. (1994), *Molecular Biology of the Cell*, 3rd edn, as its title indicates, goes far beyond neurobiology. Kandel et al. (1995), *Essentials of Neural Science and Behavior*, is a first edition and more concise than the widely used textbook *Principles of Neuroscience*, 3rd edn., by Kandel et al. (1991). The short book gives a broad and beautifully illustrated picture from molecules to behaviour. While many gaps remain, the holy grail of contemporary neuroscience is indeed to explain behaviour in terms of molecular biology. Peters et al. (1991), *The Fine Structure of the Nervous System*, 3rd edn, has long been the pre-eminent guided tour of the electron microscopy of the central nervous system (CNS); the illustrations represent the finest level of ultrastructural microscopy that can be achieved. *Neurobiology*, by Shepherd (1994), has its origins in electrophysiology and has acquired valuable molecular biological perspectives in the third edition. Many other excellent works could have been selected; indeed, in contrast to the situation of 20 years ago, there is now a plethora of introductory books on neuroscience. In this chapter, one or more of these references is often given after a heading, indicating that much of that section relies on these sources. When additional information is introduced, the specific sources are cited in the text. The figures are not intended to fully illustrate the topics mentioned in the text, but rather to give some impressions of how the cellular level was thought of at the time this book was completed. In a few cases, the figures include a comparison with earlier ideas on the topic, as a reminder that our present views will also become the subject of history in a few years.

1.2
Structural Features of Cellular Elements

This section begins with an overview of the structural elements of the CNS. To do this, the rather specialised nervous tissue is placed in the more general biological context of the epithelium.

1.2.1
CNS as an Epithelium

A striking characteristic of the nervous system is the dense aggregation of cells, which has some of the characteristics of an epithelium (Watson 1976; Palay and Chan-Palay 1977). The aggregate of cells abuts a lumen on one side – the ventricles or central canal – while on the other side, the cells contact a basement membrane that separates the neuronal tissue from blood vessels and perivascular connective tissue. This basement membrane is, from an anatomical perspective, an insignificant structure composed of a thin layer of glycosaminoglycans and related substances (Palay and Chan-Palay 1977; Peters et al. 1991). This epithelial relationship is most clear in anamniotes and in the floor of the third ventricle in higher forms in which ependymal cells stretch from the ventricular surface to the basal lamina on the external surface of the brain. In some of the simpler nervous systems, blood vessels do not penetrate the tissue, but later in evolution the increasing mass of tissue necessitated an invasion of the blood vessels in order to bring metabolic substrates to, and remove waste products from, the nerve cells. The epithelial relationship was preserved, however, when the basement membrane invaded along with the vessels.

1.2.2
Structural Features of Neurons (Peters et al. 1991)

Nerve cells are strikingly different from the other cells that constitute the body; a primary source of this dissimilarity lies in neuronal morphology. Many examples of the diversity of neuronal morphology will be found in subsequent chapters.

Here, Fig. 1.1 simply presents some idea of the development of ideas about one of the best-known neurons of the brain: the Purkinje cell (note that Purkinje used the German spelling of his name until 1850 and only after that employed the Czech form, see Pannese 1994, p. 1).

The term "neuron" was enunciated by Waldeyer in 1891, and by that time it was finally recognised by many, though not all, neuroanatomists that nerve cells were individual entities. Early studies of the brain using inadequate histological techniques had given the impression that the nervous system was a weblike reticulum. Consequently, the question of neuronal classification was moot. The brilliant research of Santiago Ramón y Cajal (summarised in two massive volumes published in the French edition in 1909 and 1911 and finally available in 1995 in an English translation), utilising the staining method originated by Golgi in 1873 (Szentágothai 1975), was needed to establish a different view: the neuron doctrine, providing the conceptual basis upon which modern neuroscience is founded. Actually, this view had been put forward also by His and by Forel shortly before Cajal began his publications. Details of the fascinating history of these ideas can be found in Shepherd (1991).

Cajal concluded that neurons are distinct membrane-bounded entities, each separated from its neighbours by a small region of extracellular space. He also believed that each neuron was functionally polarised. These ideas faced years of opposition but were finally vindicated by the images derived from the electron microscope, many years after Cajal's death.

A critical consequence of the neuron doctrine is that functional continuity between neurons is established by specialised junctions that can transmit signals across the spaces between nerve cells. These junctions are the chemical synapses that do indeed provide a functional polarisation to the nerve cell and to the entire nervous system.

Fig. 1.1a–e. Neuron morphology exemplified by the Purkinje cell. **a Drawing of J.E. Purkinje published in 1837 showing limited resolution attainable in that era (from Szentágothai 1975). b** S. Ramón y Cajal's illustration of a human Purkinje cell impregnated using the Golgi method (from Ramón y Cajal 1911). *a,* the axon; *b,* recurrent collateral; *c,* cavity containing capillaries; *d,* space occupied by basket cells. **c** Composite photomicrograph of Purkinje cell (Golgi impregnation) from a sagittal section through C2 region of the cerebellum of a mormyrid electric fish (from Nieuwenhuys and Nicholson 1969). Note the exceedingly regular straight dendrites compared with those in human beings. **d** Purkinje cell from guinea-pig cerebellum labelled with horseradish peroxidase (from Rapp et al. 1994). **e** Computer reconstruction of Purkinje cell shown in **d** to reveal electrotonic length and provide data for numerical models of electrophysiological properties of the Purkinje cell dendritic tree (from Rapp et al. 1994)

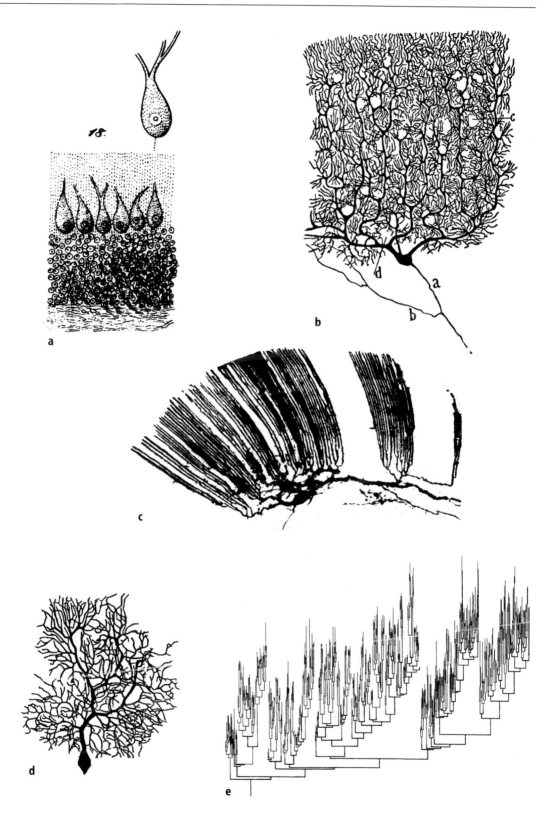

Electrical junctions are also found between cells of the nervous system, which re-introduces some aspects of the reticulum but does not compromise the essential integrity of the nerve cell.

Thus, the view emerged of the neuron as consisting of a receptive surface, the dendrites, a cell body or soma, and a distinct process for the transmission of information in the direction of the next neuron, the axon and its terminal ramifications, the telodendron. The motoneuron exemplifies such a cell. The clarity of this view has been clouded by the discovery of dendro-dendritic as well as axo-axonic synapses, together with an increasing number of cells where the distinct polarisation of the neuron is far from evident, on structural as well as molecular grounds (Craig and Banker 1994). It is now clear that rather than being an archetype, the classical motoneuron may be at one end of the spectrum of neuronal variation.

At the root of the problems of neuronal classification is the lack of knowledge of the relationship between structure and function in the cell (see, for example, Craig and Banker 1994). It has long been a premise that differences in neuronal form implied differences in function, but we still lack so much knowledge of neuronal function that the notion is far from established.

1.2.2.1
Soma (Peters et al. 1991; Alberts et al. 1994)

The soma or cell body is typically a distinct enlargement of the neuron that may range from 5 to 100 μm in diameter. It is characterised by the pres-

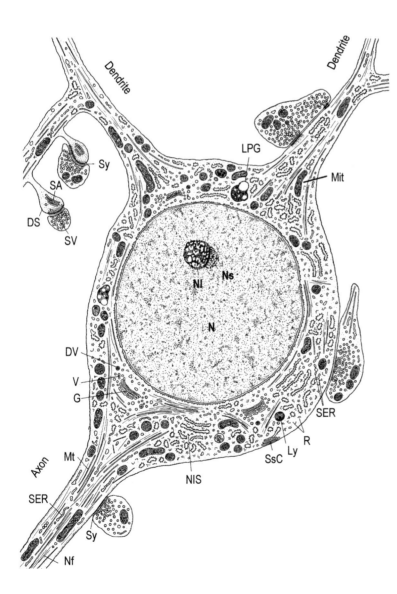

Fig. 1.2. Cytology of neuron cell body and proximal dendrites. This generalised drawing from ultrastructural material shows main features. A large nucleus (*N*) has a number of pores in its membrane, a dense nucleolus (*NI*) and nucleolar satellite (*Ns*). External to the nucleus is found the rough endoplasmic reticulum, also called Nissl substance (*NIS*). Free ribosomes (*R*) are common in the cytoplasm; smooth endoplasmic reticulum (*SER*) is found in the form of tubules or cisternae, and flattened subsurface cisternae (*SsC*) occur immediately beneath the cell membrane. The Golgi apparatus (*G*) consists of a short stack of flattened cisternae; vesicles (*V*) and dense core vesicles (*DV*) are found in the vicinity. Lysosomes (*Ly*) are numerous in the perikaryon and may become lipofuscin pigment granules (*LPG*). Mitochondria (*Mit*) occur throughout the cytoplasm. The axon arises from the axon hillock. Microtubules (*Mt*) stream into the axon from the perikaryon, along with neurofilaments (*Nf*). Dendrites extend from the cell body and contain organelles similar to those in the perikaryon. Dendritic spines (*DS*) project from the dendrite and often contain flattened cisternae known as the spine apparatus (*SA*). Axon terminals form synapses (*Sy*) both on spines (usually excitatory) and on the main dendritic trunk (often inhibitory). The enlarged presynaptic terminal contains synaptic vesicles (*SV*) (from Lentz 1971)

ence of a nucleus, in which resides the genetic specification of the cell. The soma is distinguished from the perikaryon, which refers only to the cytoplasm surrounding the nucleus and not to that within the processes of the cell (Fig. 1.2).

The perikaryon contains the Nissl substance, which can be identified with the rough, or granular, endoplasmic reticulum. Classically, the Nissl substance has been the target of the Nissl stain, which remains an effective method of revealing the location, size and density of cell bodies in the CNS (Pannese 1994). This rough endoplasmic reticulum is studded with ribosomes and polysomes, organelles that are concerned with protein synthesis. In addition, free ribosomes exist within the cytoplasm.

In proximity to the Nissl substance, and not always easily distinguishable from it, is the smooth, or agranular, endoplasmic reticulum. This structure approaches in many places the membrane surface of the neuron in the form of hypolemmal cisternae.

The Golgi apparatus may be considered a special configuration of the agranular reticulum that lies in the middle zone of the perikaryon. It appears as a tangled skein of tortuous anastomosing strands and small vacuoles around a nucleus, but remains a single organelle. At the electron microscopic level it has the appearance of a collection of flattened cisternae. The close packing of the cisternae and absence of ribosomes distinguishes the Golgi apparatus from the agranular reticulum and the rough endoplasmic reticulum. Coated vesicles are associated with the Golgi apparatus, and this structure has a major role in packaging secretory material.

Lysosomes occur in the somata of all neurons. These distinct vacuoles contain acid phosphatase and other hydrolytic enzymes. They remove unwanted material from the interior of the cell. It is possible that, as the cell ages, the lysosomes become transformed into lipofuscin granules, which constitute the "wear-and-tear" pigments of the cell.

Mitochondria are key organelles in the existence of the cell. They have a granular or rodlet-like appearance and are about 0.1 μm in diameter. They can be very elongated on occasion, up to 20 μm. Mitochondria have two membranes, an outer and an inner one, usually separated by a space some 4–8 nm wide. The inner membrane of the mitochondrion forms rows of cristae or folds that are usually oriented at right angles to the long axis of the organelle. The major function of the mitochondrion is to produce the high-energy compound adenosine triphosphate (ATP) by oxidative metabolism (Alberts et al. 1994). The mitochondrion is the principal source of ATP in the cell, although some is produced throughout the cytoplasm by the anaerobic process of glycolysis. An interesting, not to say bizarre, aspect of the ubiquitous mitochondria is that they appear to be descendants of independent prokaryotic micro-organisms, probably the so-called purple bacteria, that lived symbiotically within the eukaryotic cells (Margulis 1970; Yang et al. 1985).

1.2.2.2
Cytoskeleton
(Kandel et al. 1995; Alberts et al. 1994)

An intracellular structure exists, the cytoskeleton (see Fig. 1.2), which consists of a number of thread-like filaments that extend throughout the cell. There are three components of the cytoskeleton, each of which is a polymer consisting of many repeating monomers. The components are: microtubules, neurofilaments and microfilaments.

Microtubules are composed of long, straight protein tubes with a diameter of about 25 nm and 13 protofilaments forming the tubular walls. They can be as long as 100 μm.

Neurofilaments (neurofibrils) are about 10 nm in diameter and are the most abundant fibrillar component of axons. Neurofilaments are built with fibres that twist around each other and they retain silver nitrate, making them an important element in the classic stains used to define the structure of nerve cells.

Microfilaments are 3–5 nm in diameter, contain actin and are wound in a two-strand helix. They are attached to cell membranes via associated proteins linked to the actin. They are also able to interact with the extracellular matrix through the mediation of transmembrane proteins called integrins.

The cytoskeleton is under continuous modification and plays at least two roles. It is a structural determinant for the neuron and a substrate for intracellular transport of material (see Sect. 2.2.10).

1.2.2.3
Nucleus (Peters et al. 1991; Alberts et al. 1994)

The nucleus is typically a large round structure in the centre of the soma. It contains an obvious cytoplasm, the karyoplasm, that contains chromatin. The nucleus usually contains a conspicuous spherical nucleolus. The nucleus has a distinct envelope that delineates it from the rest of the soma. This envelope is formed by an apparent double membrane, and within this narrow space lies the perinuclear cisterna. Communication between the two compartments occurs, however, since the nuclear membrane is fenestrated by numerous circular pores. These pores connect the inner and outer nuclear membranes, and they are continuous on the outside with the cisternae of the endoplasmic reticulum. Although the pores are about 90 nm in diameter, they appear to be filled with material and have considerable structure, so that the actual size of molecules able to pass through the pores is limited to perhaps 15 nm. Time-lapse photomicrography has revealed that at least some nuclei rotate slowly; the reason for this is not known.

The nucleolus is the repository of the genetic material of the cell, the DNA. The DNA has two functions: to confer the genetic material on the new cells after division and to provide the molecular information required for the construction of cellular proteins. In stark contrast to many cells of the body, neurons do not divide in the adult animal, so that the major role of DNA in the mature neuron is to specify proteins.

1.2.2.4
Cell Membrane
(Peters et al. 1991; Alberts et al. 1994)

The membrane is the bounding surface that defines the form of the nerve cell. The membrane consists of a lipid bilayer, composed of phospholipids, glycolipids and cholesterol, some 4–5 nm thick, with the hydrophobic heads of the lipid molecules facing the intracellular and extracellular environments. Originally, it was thought (Davson and Danielli 1970) that the lipids were coated with proteins (Fig. 1.3a), but now the concept is that the lipid is a fluid environment in which float a wide variety of proteins, so that the membrane resembles a "fluid mosaic" (Singer and Nicolson 1972), though the proteins may be more restricted in their movement than was initially assumed (Jacobson et al. 1995). The proteins have many functions; some may be found on one surface or the other of the membrane, while others span the entire bilayer. Amongst these latter proteins are the ionic channels that endow the membrane of the nerve cell with its unique electrical properties. Other proteins are involved in the transduction of a vast variety of chemical signals across the membrane, including cell-cell recognition, antigenicity, and the response to hormones (Fig. 1.3b).

1.2.2.5
Dendrites (Peters et al. 1991)

Dendrites comprise a vast protoplasmic expansion; in the majority of neurons, this is enclosed in a membrane that forms the primary receptive surface for incoming contacts from other neurons (Figs. 1.1,

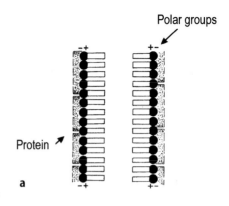

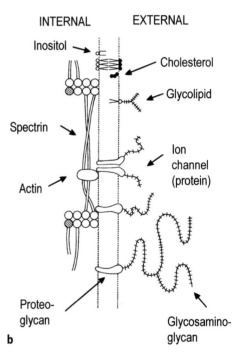

Fig. 1.3a,b. The neuronal membrane then and now. **a** Initial ideas about the lipid bilayer circa 1940, when it was envisaged by Danielli as being composed of symmetrical polar groups, with the outer surface of the membrane coated in protein and facing aqueous medium on the inside and outside of the cell (modified from Davson and Danielli 1970). **b** Simplified view of the membrane in 1994. The lipid bilayer has stood the test of time and its constituent molecules have been identified (e.g. cholesterol and inositol), but the proteins are now seen as being diverse and scattered throughout the membrane. Ion channels are formed from proteins that span the membrane along with glycoproteins and proteoglycans. Each of these proteins has long-chain sugar molecules (e.g. glycosaminoglycan) attached to its outer face and extending into the extracellular space. Glycolipids are also attached to the outer surface. Together these long-chain molecules form the external glycocalyx. Many proteins are anchored in place with an internal membrane cytoskeleton containing an assortment of molecules such as spectrin and actin (modified from Shepherd 1994)

Fig. 1.4. Patterns of dendrites. An unusual picture showing a section through the cerebellar molecular layer parallel to the brain surface stained with the Bodian-Ziesmer reduced-silver method. The Purkinje cell dendrites are perpendicular to the surface so they are cut transversely and appear as *black dots*. The amazing regularity of the dendritic pattern is due to the fact that this is a section through the valvular ridge of a mormyrid electrosensory teleost (*Petrocephalus bovei*), in which the neuronal architecture has extraordinary regularity (see example of mormyrid Purkinje cell in Fig. 1.1c). The dendritic trees of the Purkinje cells spread vertically in the figure and the parallel fibres run horizontally. *Scale bar* represents 10 μm (from Nieuwenhuys and Nicholson 1969)

1.4). It is this feature that distinguishes the dendrite from the axon, although exceptions to this polarisation are now well documented. Shepherd (1994, p. 63) has stated that "dendrites are all those branches of a nerve cell that do not fulfil the criteria for being an axon." The term dendrite connotes the branching pattern of a tree, and this successive bifurcation is usually seen in dendrites, though non-branching receptive extensions exist. Dendrites contain Nissl substance in their initial parts, the endoplasmic reticulum persists throughout, and hypolemmal cisternae are prominent, as are microtubules.

A major feature of the majority of dendrites, which is not shared by axons, is the presence of spines (Fig. 1.2). These consist of a slender neck ending in a terminal bulb with a length of some 1–2 μm. The spine density varies with the position on the tree. On pyramidal cells from the cerebral cortex of the cat there are about 4000 spines, but most are found on the terminal arborisations. For comparison, a Purkinje cell in the rat cerebellum may have as many as 175 000 spines. Spines contain a considerable amount of actin and may show plastic changes with repeated activation or during afferentation or de-afferentation of the cell (Pannese 1994). Dendritic spines are the site of contact of the majority of excitatory chemical synapses. In contrast, inhibitory synapses make direct contact with the main dendritic trunks and branches.

1.2.2.6
Axon (Peters et al. 1991)

The axon begins with a hillock, devoid of Nissl substance, and typically arises from the cell body (Fig. 1.2), but many examples are known where the site of origin is a large proximal dendrite. The axon may proceed for a considerable distance away from the parent cell, possibly emitting collaterals along its path – this has been designated as the Golgi type I cell. Such long axons are usually myelinated. If the axon ramifies within the territory defined by the dendritic arborisation of the neuron, it is called a Golgi type II cell. These axons are often unmyelinated. Most, but not all, neurons have axons.

Axons often terminate in a branching expansion called a telodendron which ends in synaptic boutons, but boutons may also form as enlargements of the axonal process along its course. Synaptic boutons are distinguished by the presence of vesicles containing transmitter substances. Axons may also participate in the formation of tight junctions.

Axons contain microtubules, neurofilaments, rough endoplasmic reticulum and multivesicular bodies. This complex cytoskeleton serves as a substrate for axoplasmic transport.

1.2.2.7
Myelin Sheaths (Peters et al. 1991)

Axons with a diameter of 2 μm or more in both the peripheral nervous system and the CNS are ensheathed in myelin. In the periphery, the sheath is the product of Schwann cells, while in the CNS the sheath is formed by oligodendrocytes. All myelin sheaths are interrupted by nodes of Ranvier, where the axon is free of the sheath for a short distance to permit the regeneration of the action potential. A major function of myelin is to insulate the nerve fibres and permit the more efficient transmission of electrical impulses. This means the same velocity of propagation can be achieved with a substantially smaller myelinated fibre compared with one that is not so enclosed. Interestingly, myelin has never been identified in the invertebrate nervous system, or in that of agnathan vertebrates.

1.2.2.8
Terminals and Synapses (Peters et al. 1991)

Axons end in synaptic terminals or boutons which are separated by a cleft from a specialised region on another cell, or in some cases the same cell as that giving rise to the axon (Fig. 1.2). This complex is called a synapse and is one of the defining characteristics of nerve cells. The term synapse was coined by Sherrington in a textbook of physiology (Foster and Sherrington 1897). The synapse is of seminal importance to the understanding of the nervous system, and its function will be described in more detail later in this chapter.

The pre-synaptic structure can have a variety of shapes but most commonly approximates a sphere with a diameter substantially larger than the axon leading into it. The classical chemical synapse is characterised by the presence of vesicles in the pre-synaptic terminal some 40–50 nm in diameter. These vesicles contain transmitter molecules which are released when they fuse with membrane facing the synaptic cleft. Within the presynaptic terminal at the surface adjacent to the cleft is found a specialised intracellular extension of the membrane – the presynaptic grid, apparently an attachment site for vesicles. After release, the transmitter molecules diffuse across the cleft, typically between 10 and 20 nm in width, and activate specific receptors in the postsynaptic membrane. The release of transmitter is brought about by depolarisation of the presynaptic terminal. The consequences of post-synaptic receptor activation by the transmitter may be either depolarisation (leading to excitation of the target cell) or hyperpolarisation (leading to inhibition of the target).

The shapes and size of the vesicles vary considerably. Clear, coated and dense-core vesicles have all been identified. There has also been considerable discussion of the shape of vesicles in relation to synaptic function. It is now appreciated that vesicle shape, as revealed by the electron microscope, is a function of both the chemical content of the vesicle and the method of fixation. Moreover, the role of a particular transmitter is not determined uniquely by its chemical composition but also depends on the postsynaptic receptor, which recognises the transmitter and the iontophore, the ion-selective channel in the membrane that the receptor opens. Consequently, the shape of a vesicle is only a reliable guide to synaptic function within a specific system and fixation paradigm.

In addition to the chemical synapse, the electrical junction is now well established. Here the pre- and postsynaptic sites are actually joined by a complex of narrow pores that allow both the passage of electric current and the movement of small molecules between the cells.

In the majority of cases, synapses are established between axons and dendrites. As remarked above, however, virtually all other permutations of contacts have now been seen somewhere in the nervous system. In addition, there may be vesicle-filled terminals without any obvious postsynaptic targets. Such structures presumably release their contents into the extracellular or ventricular space and mediate a less specific mode of chemical signalling, termed non-synaptic transmission or volume transmission.

1.2.2.9
Protein Synthesis and Trafficking
(Kandel et al. 1995; Alberts et al. 1994)

One of the major purposes of the various parts of the cell described so far is the production and distribution of proteins. Messenger RNA (mRNA) in the nucleus reads information from the DNA and emerges into the cytoplasm via the nuclear pores. The mRNA gives rise to three classes of protein: cytosolic proteins, nuclear and mitochondrial proteins, and cell membrane proteins and secretory products.

Cytosolic proteins are generated by the association of mRNA with free ribosomes to form polyribosomes or polysomes. Cytosolic proteins build the fibrillar elements that form the cytoskeleton; they also form the numerous enzymes that catalyse metabolic reactions in the cell. Nuclear and mitochondrial proteins are also formed on free polysomes but are then targeted to their specific organelle by a process of post-translational importation, based on receptors on the nuclear and mitochondrial membranes.

Cell membrane proteins and secretory products form on polysomes attached to the endoplasmic reticulum. Indeed, it is the attachment of the ribosomes that gives rise to the rough endoplasmic reticulum (and the absence is a characteristic of smooth or agranular reticulum). The ribosomal mRNA stains intensely with toluidine blue and cresyl violet, and this is the basis of the Nissl stain. Cell membrane and secretory proteins are extensively modified after translation to tailor them to their eventual role. The nuclear membrane, endoplasmic reticulum, Golgi apparatus, plasma membrane, secretory granules and endosomes are all constituted from different proteins. The production begins in the endoplasmic reticulum and continues in the Golgi apparatus, where a variety of vesicles are formed and distributed through the cell. Some of these vesicles form secretory granules or synaptic vesicles.

1.2.2.10
Intracellular Transport of Substances
(Kandel et al. 1995)

The fact that the major synthetic machinery of the nerve cell resides in the cell body, but neurons can have very distant extensions, implies the need for transport of materials. This was directly demonstrated in axons by making local constrictions and observing the accumulation of transported material (Weiss and Hiscoe 1948). Since the early studies were performed, a much more detailed picture has emerged.

Fast axonal transport carries material in both an anterograde (away from the soma) and a retrograde (toward the soma) direction. The material includes proteins and synaptic vesicles. These substances move in the anterograde direction at a speed of some 400 mm per day along microtubules with the aid of a molecular "motor" molecule called kinesin, which is an ATPase. Retrograde transport is mediated by another motor molecule, dynein, at a speed that is half to two thirds that of anterograde transport. Some of the material carried by retrograde transport is composed of material originally moved out by anterograde transport, as part of a scavenging operation; the remainder consists of material entering from the extracellular space, such as nerve growth factor.

Slow anterograde transport moves the cytosol, i.e. cytoskeletal elements and soluble proteins. Two components have been identified, one with an average velocity of 0.2–2.5 mm per day and another one some twice as fast. Subunits of microtubules and neurofilaments are carried by the slower component. The faster component carries a variety of proteins, including actin and enzymes for intermediary metabolism.

Dendrites also depend heavily on intracellular transport, but detailed analysis is difficult because of the inaccessibility. There is evidence that dendrites release substances, including dopamine (Geffen et al. 1976). These substances may be related to the local control of blood flow, but, undoubtedly, they are also involved in volume transmission. These substances are presumably transported within the dendrite, much as occurs in the axon.

1.2.3
Glia and Ependymal Cells
(Kettenmann and Ransom 1995)

The term neuroglia, often translated as "nerve glue", was coined by Virchow in 1846. Succinct and interesting descriptions of the history of the glial cells, with its many disputes and ambiguities, have been provided by Somjen (1988) and Young (1991). Today it is convenient to consider three classes of non-neuronal cell: the macroglia, comprising the astrocytes and oligodendrocytes, the microglia, and the ependyma.

Taken together, these three cell types probably exceed the number of nerve cells in the brain because the so-called glial index, or ratio of glia to neurons, is greater than two for most brain regions (Blinkov and Glezer 1968). The function of these elements remains an enigma, however. Two defining characteristics of glia are that they are electrically inexcitable (do not produce action potentials) and lack synapses, although gap junctions between glial cells are common. This has meant that electrophysiologists paid little attention to them for many years, since neurons were much more rewarding to study. The advent of new techniques has, however, revived the study of glial and ependymal cells.

We also note that the concept of a paraneuron has been introduced and elaborated by Fujita (1979). He envisaged paraneurons as cells which have not been classified as neurons and yet share some neuronal features; thus these cells are not glia either. To date, the concept has not gained wide currency.

1.2.3.1
Astrocytes (Peters et al. 1991)

In classic gold-sublimate impregnations, the light microscope reveals astrocytes as characteristically star-shaped cells; hence the name (Fig. 1.5). In the white matter, astrocytes have numerous fibrils and consequently are designated as fibrous astrocytes. Those in grey matter have fewer fibrils and thus are known as protoplasmic astrocytes. Astrocytes frequently have endfeet that attach to blood vessels or form a glia limitans at the surface of the nervous system (Fig. 1.5).

In the past, astrocytes were identified by their appearance after use of a staining technique that was thought to preferentially reveal that cell type, but today they are usually characterised by an appropriate immunocytochemical technique. The best-known marker for astrocytes is glial fibrillary acidic protein (GFAP), though not all astrocytes stain with this marker. When the brain is injured, resting astrocytes are transformed into a reactive form characterised by increased levels of GFAP (McMillian et al. 1994).

1.2.3.2
Oligodendrocytes (Peters et al. 1991)

Most anatomists agree that oligodendrocytes were identified by del Rio Hortega in a series of publications that began in 1921. They are involved primarily in the production of myelin in the CNS to form a sheath around the axons of neurons. Thus, they are analogous to the Schwann cells of the peripheral nervous system. Not surprisingly, oligodendrocytes predominate in white matter, where they align with

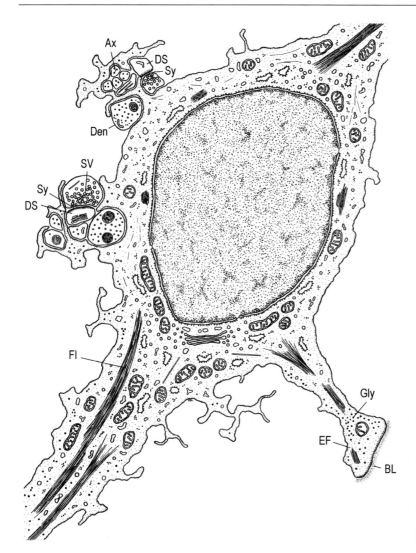

Fig. 1.5. Protoplasmic astrocyte. This star-shaped non-neuronal cell has cytoplasmic processes extending into the neuropil or terminating in endfeet (*EF*) on blood vessels. Such terminations are bounded by a basal lamina (*BL*). Protoplasmic astrocytes occur in the grey matter, contain glycogen granules (*Gly*) and have fewer cytoplasmic fibrils (*FI*) than the fibrous astrocytes of the white matter. The cell has a very irregular outline and the expansions may form very thin sheets. Bundles of small axons (*Ax*) are enveloped by glial processes. Dendritic spines (*DS*) and synaptic terminals, identified by the synapse (*Sy*) and synaptic vesicles (*SV*), in the CNS are also enclosed by glial expansions. Drawn from electronmicrograph (from Lentz 1971)

the nerve fibres, but they also occur in grey matter, since myelin is present there, too, and sometimes are seen also as the satellites of a nerve cell.

1.2.3.3
Microglia (Streit 1995)

Until recently there was considerable doubt as to whether microglia existed as a distinct cell class, in part because of the plastic nature of this type of cell. Production of relatively specific antibodies has largely laid to rest the question of the existence of microglia, however. It is commonly accepted that a major function of microglia is phagocytosis, but that is doubtless a gross simplification. The origin of these cells continues to be unresolved.

1.2.3.4
Ependyma, Tanycytes and Radial Glia (Peters et al. 1991)

Ependyma are the non-neuronal cells that line the ventricles. They are cuboidal and limited in extent. When the cells extend further they are generally referred to as tanycytes. During development, and in some amphibia and reptiles, tanycytes extend all the way to the pial surface, but in adult mammals the distance is too great and the cells end on blood vessels. Many ependyma, and some tanycytes, have cilia that extend into the ventricles where they beat to prevent the formation of a stagnant layer of fluid at the ventricular surface.

Like many other questions of glial classification, the simple scheme given above fails to accommodate all the variants that have been observed. In particular, the radial glia present problems. For example, in the mammalian cerebellum the Bergmann glia (Golgi epithelial cells) extend from the pial surface to the Purkinje cell layer, where they have their cell bodies and terminate. They are easy to recognise by their characteristic morphology in the cerebellar molecular layer and were classified as radially extending astrocytes by Palay and Chan-Palay (1974) and Reichenbach and Robinson (1995). Reichenbach and Robinson (1995) considered that tanycytes belong to the class of ependymoglia or radial glia that have more in common with ependyma than with astrocytes. A paradigm of this type of cell is the retinal Müller cell that is by far the best-characterised cell of its type (Reichenbach and Robinson 1995). In contrast, Privat et al. (1995) included radial astrocytes, Müller cells of the retina, Bergmann glia and tanycytes under the heading of astrocytes.

1.2.3.5
Functions of Glia
(Barres 1991; Kettenmann and Ransom 1995)

While neurons, as a class of cells, can be said to be responsible for electrical signalling in the brain and spinal cord, no such clear characterisation of glia has yet emerged. Instead, we presently have an increasing collection of seemingly unrelated hypotheses about their function, some of which will be listed here.

1. Despite the fact that glia do not generate active electrical responses, they are now known to possess many, if not all, the voltage-gated channels found in neurons (Ritchie 1992). Much of this work has been done on Schwann cells, but it seems likely that channels for K^+, Na^+ Ca^{2+} and Cl^- are to be found on all glia, though the relative abundance differs among cells. One view of the functional significance is that channels are manufactured in glia for use by neurons, but it seems more likely that they play a role, so far unknown, in the glia themselves.

2. It is clear that some types of glia are involved in ensheathing neurons, particularly the axons. In the case of those glia that generate myelin, the function is to improve the efficiency of impulse transmission along the axon, by essentially insulating the axons from the extracellular microenvironment.

3. Since some glia extend to the pial or ventricular surfaces and others have prominent endfeet on blood vessels, it has been suggested that glia are involved in the transport of substances from one location to another. A specific and fairly well documented example of this is that glia are involved in potassium homeostasis in the microenvironment, either by the so-called spatial buffer mechanism or by uptake mechanisms (Newman 1995).

4. Glial cells synthesise and release many neuroactive substances, including neuropeptides, amino acid transmitters, eicosanoids, steroids and growth factors (Martin 1992). One presumed role for this multitude of potential chemical signals is to mediate communication with neurons, and there is increasing support for a symbiotic view of neurons and glia (e.g. Abbott 1991; Chiu and Kriegler 1994).

5. An important role for glia is the uptake of transmitters released from neurons to prevent prolonged action (Schousboe and Westergaard 1995). Glia apparently take up most amino acid transmitters; of particular importance is the removal of glutamate, which otherwise can easily reach cytotoxic levels. In the case of glutamate, the glial cells convert the amino acid to glutamine, which is then transported out of the cell, enters the extracellular space and is taken up by neurons to re-synthesise glutamate (Wiesinger 1995).

6. In another context, one that lies beyond that of the present chapter, it should be noted that glia are strongly implicated in the guidance of neuronal migration during development (Rakic 1995).

Despite these various proposed roles for glia, the consensus is that we are still ignorant of the major functions of these cells. One reason for this may be the tendency to group all these cells together, whereas, at least on anatomical grounds, they are a widely differentiated group of cells and the functional roles may differ significantly between glial cell types. The diversity of glial architectonics was recognised as far back as the studies of Retzius (1894).

1.2.4
Brain Cell Microenvironment (Nicholson 1997)

The neuron is critically dependent on its environment for survival and possibly for the expression of some of its functions. The environment maintains an appropriate concentration of essential metabolic and ionic substrates for the cells. The most immediate aspect of this microenvironment is the extracellular space, but in the wider sense the blood-brain barrier and glia also come into the picture; this whole complex has been termed the brain cell microenvironment (Schmitt and Samson 1969), or brain extracellular microenvironment to distinguish it from the intracellular region (Fig. 1.6).

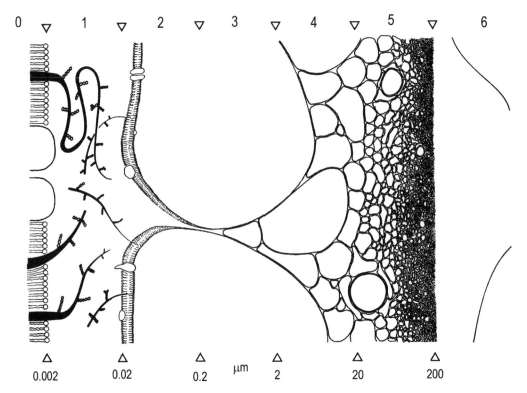

Fig. 1.6. Brain cell microenvironment. This drawing was made on a logarithmic scale in the horizontal axis, which indicates increments by a factor 10 on the lower scale. Beginning on the *left* at 0.002 μm (*zone 0, top* of figure), the lipid bilayer is seen with an ion channel and long-chain glycosaminoglycans protruding into the extracellular space (*zone 1*). At 0.02 μm the membranes of the next cells are encountered (*zone 2*) and as the magnification continues to diminish, the aggregate of cellular processes (including cell bodies, dendrites, axons and protoplasmic extensions of astrocytes) assumes the topology of the soap phase of a soap film (*zones 3–5*). It is this level of resolution where macroscopic diffusion theory becomes valid and the concepts of volume fraction and tortuosity are meaningful (*zones 4–5*). As the magnification continues to decrease, whole blood vessels are encountered (*zone 5*) and finally the bounding surface of the brain (*zone 6*) (from Nicholson 1979)

1.2.4.1
Extracellular Space (Nicholson 1997)

The extracellular space consists of the totality of the narrow spaces that envelop the cellular elements of the brain. The actual width almost certainly varies from place to place, but the average value is of the order of 20 nm. The total fraction of the brain thus delineated has been the subject of much controversy, stemming mainly from different methods used to estimate the space. Presently, there is general agreement that most regions of the nervous system have an extracellular volume fraction of between 15 and 25 %. When molecules diffuse through the extracellular space they frequently encounter the membranes of cells. This effect can be quantified in the so-called tortuosity of the extracellular space, which leads to an apparent diffusion coefficient for most molecules that is about 2.5 times less than that in free solution (Nicholson 1993).

The interstices of the nervous system are bathed predominantly in dilute saline consisting of approximately 150 mM NaCl with small amounts of other ions, notably K$^+$ and Ca^{2+}, together with dissolved oxygen, carbon dioxide and glucose, as well as a number of lesser constituents, including the antioxidant ascorbate (Rice et al. 1995). The extracellular space is the final conduit for the metabolic and ionic substrates that are vital to the neurons. These constituents are brought to the brain via the vascular system and the cerebrospinal fluid-filled spaces. The extracellular space also contains long-chain sugar molecules tethered to the cell membranes and bearing a negative charge (Figs. 1.3b, 1.6), as well as a certain amount of hyaluronate, but the amount and distribution of these large extracellular molecules is uncertain.

1.2.4.2
Non-synaptic or Volume Transmission
(Nieuwenhuys 1985; Fuxe and Agnati 1991;
Agnati et al. 1995)

Non-synaptic or volume transmission signifies the flow of "informational substances" between cells (Schmitt 1984) to provide a mode of chemical signalling that complements conventional synaptic transmission. The extracellular space is thus envisaged as a communication channel (Fig. 1.6), and cells release chemical signals that diffuse to appropriate receptor sites (Nicholson and Rice 1991). These sites may be at another location on the releasing cell, or they may be hundreds of micrometres or more away.

The anatomical evidence for volume transmission will be reviewed in detail in Chap. 2. Although this mode of information transfer is less well established than conventional synaptic transmission, the weight of evidence in its favour is steadily increasing.

1.2.5
Blood-Brain Barrier (Bradbury 1979)

It has now been established that a blood-brain barrier exists in all adult vertebrates so far studied (Cserr and Bundgaard 1984). This barrier effectively isolates the brain from the blood and allows only very specific and controlled access of ions and water-soluble molecules. Lipid-soluble compounds, including dissolved gases, can pass readily across the membranes of the cells forming the blood-brain barrier and enter the brain. The barrier is identified with the capillary endothelial cells surrounding the blood vessels of the brain. These cells are connected by tight junctions, so that the ensemble effectively excludes all water-soluble organic molecules, and ions, from diffusion into the extracellular space. These compounds gain access by active transport and this is the mechanism by which the metabolic substrate glucose enters the brain.

The CNS arises from ectodermal tissue while the blood vessels derive from mesoderm; consequently, the blood-brain barrier maintains the isolation of these two tissues. The reason for the penetration of blood vessels into the brain is undoubtedly related to the need to supply metabolic substrates to the nervous tissue, with oxygen and glucose being the most critical substances.

It has long been held that the blood-brain barrier confers immune privilege on the brain. Recent studies have confirmed this to some degree but at the same time indicated that the brain is not entirely isolated from the rest of the immune system (Cserr and Knopf 1992).

1.3
Overview of Neurochemistry

From one perspective the brain might be regarded as a biochemical machine. Although little space is now devoted to the bioenergetics of the brain in contemporary neuroscience textbooks, a very brief outline is needed because some aspects of comparative structure are best appreciated from that viewpoint.

1.3.1
Glucose, Oxygen and ATP
(Siesjö 1978; Alberts et al. 1994)

The major external energy substrates of the brain are glucose and oxygen; in fact, about 20 % of all oxygen entering the human body under resting conditions is used by the brain. These two substrates are converted within the cells of the brain to ATP; at typical consumption rates the human brain has only a reserve measured in tens of seconds for ATP. Consequently, there is a vital need to supply the substrates in quantity to all parts of the brain, and this places major constraints on the structure of the brain parenchyma.

Glucose and oxygen enter the brain from blood via the blood vessels; carbon dioxide resulting from metabolism leaves via this route. The final transportation step between vessel and cell is accomplished by diffusion. The latter process is ineffective over distances of more than 100 μm, given typical consumption rates of the brain, so the brain capillary system must ensure that every cell is within about 50 μm of a capillary, although the precise density of vessels varies with brain region (Purves 1972). This requirement accounts for the great ramification of the vascular system. The limitations imposed by diffusion also limit cell size to the order of some tens of micrometres in diameter. In *Amphioxus*, where blood vessels do not penetrate regions of the nervous system, the ventricular system is probably the main channel for the movement of metabolic substrates, and the neurons cluster in the subventricular area or even intrude into the ventricle (see Chap. 9).

Glucose can be converted into ATP without the need for oxygen in the process of anaerobic glycolysis, but in mammals the main function of glycolysis is to provide substrates for the far more efficient aerobic process known as the citric acid or Krebs cycle that, utilising oxygen, produces most of the ATP actually used in the brain. Glycolysis takes place in the neuronal cytoplasm but the Krebs cycle is confined to mitochondria, thus emphasising the unique importance of these organelles. The metabolic pathways are, of course, extremely complex with possibilities of alternative substrates and shunts; these are also closely regulated by positive and negative feedback mechanisms operating through the metabolic intermediates themselves.

1.3.2
Na$^+$–K$^+$ Pump (Alberts et al. 1994)

The major part of the ATP produced in the brain is used to power the Na$^+$–K$^+$ pump. It has been estimated that 20%–40% of all brain energy is used in this single transport system, although the estimates are not very certain (Siesjö 1978). In contrast, the synthesis of transmitters and proteins may account for only a few percent of brain energy expenditure (Siesjö 1978).

The pumping mechanism that transports Na$^+$ and K$^+$ across the membrane resides in an enzyme, sodium-potassium ATPase, located in the lipid bilayer. Internal Na$^+$ complexes with a protein and diffuses out across the membrane. At the outside the complex dissociates, thus releasing the Na$^+$ and maintaining the concentration gradient for the complex. The protein now changes its form, a process that does not require energy, so that it can bind K$^+$ and the new complex diffuses back into the cell. At the inner membrane surface the new complex again dissociates to release the K$^+$ and the protein is converted back into a form that can transport Na$^+$. This last conversion, however, requires the use of ATP, i.e. energy. The pump typically moves three Na$^+$ out of the cell and two K$^+$ into it, although the ratio can vary.

The transport of Na$^+$ out of the cell and K$^+$ into the cell underlies the resting potential (see below), in the sense that it establishes and maintains the differences in ionic concentration that are necessary to generate potentials. In addition, such transport enables osmotic forces to be balanced and the volumes of cells to be regulated. In the final analysis, it is the continuous action of the Na$^+$–K$^+$ pump that is largely responsible for maintaining the steady state of the neuron. The Na$^+$-gradient is used also to drive a variety of other active transport mechanisms.

1.4
Basic Physiology of Neurons

Since it is thought that the rationale for much of the form of the brain is related to the function, this outline presents concepts of neuronal electrophysiol-ogy. This rests largely on studies made with microelectrodes. The resting potential, passive electrical properties, active properties and synaptic transmission will be briefly reviewed.

1.4.1
Resting Potential
(Shepherd 1994; Johnston and Wu 1995)

All neurons and glia maintain an intracellular potential that is negative with respect to the outside. A typical value is −70 mV, but the range is probably −40 to −90 mV, depending on the cell. This potential is the result of two factors: (a) potassium is about 30 times more concentrated inside the cell than outside, and (b) the membrane is more permeable to potassium than to any other ion. These conditions create a so-called Nernst potential across the membrane which can be calculated from basic physico-chemical principles. In reality there is some permeability to Na$^+$ and Cl$^-$, and for very precise estimates a modification of the Nernst theory, named after Goldman, Hodgkin and Katz, can be used. In order to maintain general electroneutrality, K$^+$ is balanced by equal amounts of anions, and in order to maintain osmotic neutrality, the paucity of extracellular potassium is compensated by an abundance of sodium outside the cell, so that the gradient of this ion is reversed across the membrane in comparison to potassium. Finally, these ionic gradients would slowly dissipate with time and, as noted below, with active processes, if it were not for the action of the sodium-potassium pump (see above).

The picture of a resting cell is therefore one of a dynamic steady state in which electrical, diffusional and osmotic forces are balanced through the constant expenditure of energy. Consequently, the neuron can be viewed as a dissipative structure, the anatomical and functional state of which is maintained only by the continuous expenditure of energy. This arrangement endows the nervous system with a type of dynamism that is absent from static structures. The implications of such dissipative structures have been discussed from both a thermodynamic and a philosophical perspective by Katchalsky (1971).

1.4.2
Passive Electrical Properties
(Johnston and Wu 1995)

The resting potential can be envisaged as a battery, or source of electromotive force, that is available to do electrical work for the cell. Some of this work may be used in the transport of ions, but much of it is expended in sending signals between neurons. In order to convey such signals, the elements of the nerve cell must have some of the properties of electrical conductors. The extent and form of these properties determine the nature of electrical signalling within the nervous system.

There are three distinct conductive regions associated with a nerve cell: the cytoplasm, the membrane and the extracellular microenvironment. The cytoplasm, which includes the axoplasm, and the extracellular microenvironment are quite similar electrically and consist principally of a dilute saline solution having a specific conductivity of about $0.02 \, S \, cm^{-1}$. This is often stated in the reciprocal form, namely that the specific resistivity is about $50 \, \Omega \, cm$. Note that this refers only to the actual material filling the cell or extracellular space. The total conductivity of a length of axon or volume of extracellular space is determined both by the specific conductivity of the filling material and by the geometry of the enclosing structure.

The membrane of a cell is predominantly a lipid bilayer. Such a layer possesses the characteristics of an insulator, enclosing a conductive salt solution, and this has the desirable consequence of making the axon, dendrite or soma into a simple, insulated cable. Unfortunately, such a pure lipid bilayer would be unable to generate a resting or action potential because this property depends on the existence of channels within the membrane that selectively allow certain ions to permeate the membrane. The existence of such channels inevitably means that the membrane has conductive properties as well as insulating attributes. The result of this is that the membrane has high, but finite, resistance together with capacitative properties. This means that charge can be stored in the immediate vicinity of the membrane as well as traverse it. Conceptually, it may be hard to envisage the coexistence of conductive and capacitative properties; in fact, the conductivity resides in the discrete channels separated by regions of non-conducting membrane. This combination of properties makes dendrites and axons into "leaky cables".

Cable structures of the sort described above mean that when a pulse is injected into one end it diminishes in amplitude with distance due to current loss across the membrane and is simultaneously slowed down by the capacitative properties of the lipid bilayer. As a consequence of the leaky properties of axons, they do not rely on passive electrical properties for signalling when long distances are involved. The passive properties are important, however, in most dendrites and in some short axons. Indeed, as shown by Wilfred Rall

(1977), they are a major mechanism for the integration of synaptic information in dendrites and these properties can be modelled accurately.

1.4.3
Voltage-Gated Channels
(Hille 1992; Johnston and Wu 1995)

The passive membrane properties just described are, by definition, independent of the voltage across the membrane. In particular, the channels responsible for the resting conductivity appear to be voltage insensitive and their identity is rarely discussed. There is increasing evidence that in some cases, such as glia (Ransom and Sontheimer 1995), they are potassium channels of the inward rectifier type (see below) and so do actually exhibit voltage dependency, though in glial cells under normal resting conditions this is not evident. The properties of nerve cells responsible for rapid signalling, however, reside in the voltage-dependent properties, and for this reason they are referred to as voltage-gated channels.

With the advent of the patch clamp together with the ability to clone channels, a major thrust of contemporary neuroscience is to characterise voltage-gated channels at both the electrical and the molecular level. The molecular sequencing has arrived at a point where it has become apparent that the principal subunits of the voltage-gated Na^+, Ca^{2+} and K^+ channels are members of a related gene family (Catterall 1993). Despite this fundamental similarity, the three classes of channels show very different voltage-dependent activation, ion conductance and inactivation properties, which is stimulating research to elucidate the full three-dimensional structure of the channels. The major known channels are:

1. Two functionally distinct Na^+-channels which differ in their inactivation properties. These are referred to as $Na_{(fast)}$ and $Na_{(slow)}$.
2. Potassium channels probably show the greatest diversity and in consequence are used "like the stops on an organ, the diversity of available channels is used to give timbre to the functions played by excitable cells" (Hille 1992, p. 115). Several major classes of potassium channels can be distinguished as follows: delayed rectifier, $K_{(DR)}$, transient (A) current, $K_{(A)}$, slowly inactivating "delay" channel, $K_{(D)}$, muscarine-sensitive or M channel, $K_{(M)}$, inward rectifier current, $K_{(IR)}$, and calcium-gated channels, $K_{(Ca)}$ and $K_{(AHP)}$.
3. Among the calcium channels are the high-threshold or L-type ("long-lasting"), $Ca_{(L)}$, low-threshold or T-type ("transient"), $Ca_{(T)}$, high-

threshold N-type, $Ca_{(N)}$ and the high-threshold P-type ("Purkinje cell"), $Ca_{(P)}$.

4. Much less is known about the anion channels but they also appear to form a fairly diverse group and both voltage-sensitive and Ca^{2+}-sensitive channels have been identified.

5. There is increasing evidence for proton channels, identified by local changes in pH; however, such pH changes may be brought about by the movement of HCO_3^- or other "acid equivalents", which makes the task of unambiguous identification difficult (Chesler 1990).

1.4.3.1
Action Potentials in Axons
(Hille 1992; Johnston and Wu 1995)

When an axon is locally depolarised by a certain amount, voltage-sensitive Na^+ channels [now identified as $Na_{(fast)}$] open, allowing this ion to enter the axon down its electrochemical gradient. The entry of this positive ion further depolarises the axon, leading to a positive feedback cycle. This event is terminated both by an intrinsic molecular mechanism within the Na^+ channel that causes it to inactivate and by the opening of voltage-sensitive K^+ channels [now identified as the delayed rectifier $K_{(DR)}$]. The latter channels allow K^+ ions to leave the cell, since the equilibrium potential for this ion is negative with respect to the depolarised cell. These two events restore the membrane potential to its resting level. This sequence of events is known as the action potential (Ritchie 1995).

The experimental and theoretical basis for the action potential was firmly established by Hodgkin and Huxley in the 1940s and 1950s working on the giant axon of the squid (Hodgkin and Huxley 1952; Hodgkin 1992; Huxley 1995; Fig. 1.7a–d). Subsequently, the basic mechanism has been confirmed in virtually all axons in many vertebrates as well as invertebrates.

The entire action potential has a duration of about 1 ms and an amplitude of about 70 mV in a mammal (as well as in the squid axon) and constitutes a brief electrical impulse. It is followed by a period of similar duration during which no action potential can be initiated; this is the absolute refractory period and corresponds with the inactivation of the Na^+ channel (Fig. 1.7e–f). A relative refractory period follows while K^+ channels remain open, and a larger-than-normal stimulus is required to elicit an action potential. The currents that are associated with the ionic movements complete their circuits through the adjacent axonal membrane and extracellular space and in so doing depolarise the neighbouring axonal region. This, in

turn, generates a further action potential and, consequently, this disturbance propagates along the axon. The velocity of propagation is determined largely by the finite amount of time that it takes to change the charge on the capacity of the membrane. In typical axons conduction velocities range from below 1 m/s to two orders of magnitude more. In myelinated fibres the voltage-dependent channels occur only at the nodes of Ranvier, thus compelling the action potential to jump from node to node (saltatory conduction), which leads to an increase in conduction velocity (Ritchie 1995).

Fig. 1.7a–f. Action potential in an axon. **a** Capillary electrode (*e*) in centre of the giant axon of a squid (*ax*). The *right-hand image* was formed directly by a microscope, the *left-hand one* was produced in a mirror, so the axon is seen from two directions. The giant axon was discovered by J.Z. Young, and physiological experiments with electrodes of the type shown here were carried out by A.L. Hodgkin and A.F. Huxley to measure the membrane and action potential. By introducing a second electrode, in the form of a concentric wire around the capillary tube, a voltage-clamp experiment can be carried out, leading to an understanding of the currents underlying the action potential (from Young 1957). **b** Action potential computed by Hodgkin and Huxley (1952), based on their experimental data. *Broken curve* shows membrane potential across squid axon, amplitude shown on *left* in millivolts (mV), and time scale in milliseconds (ms) at *bottom*. *Continuous line* shows total conductance change across membrane. **c** Actual propagated action potential (*dotted line*) recorded by Cole and Curtis (1939) and actual conductance change (*white area*). Conductance change measured in $mmho\,cm^{-2}$ ($mS\,cm^{-2}$) (from Hodgkin and Huxley 1952). **d** Computed components of propagated action potential. The conductance change due to opening of sodium channels is g_{Na}, the conductance change due to opening of potassium channels is g_K, and the summed conductance of the two channels is g. The *conductance scale* is in $mmho\,cm^{-2}$ ($mS\,cm^{-2}$). The resulting action potential, $-V$, is in millivolts (mV), the time scale in milliseconds (ms) (from Hodgkin and Huxley 1952). **e** Contemporary recording of gating in single Na channels. Patch-clamp records of Na currents (I_M) due to the opening of single Na channels in mouse toe muscles during a voltage step (E_M) from -80 to $-40\,mV$ using a cell-attached patch recording (see Chap. 7). *Upper part* of **e** shows ten consecutive trials; *lower part* shows ensemble average (mean waveform), obtained by adding together 352 repeats of similar recordings, to show how macroscopic currents are formed from individual channel events. In practice the ensemble average is formed by the opening of a population of channels, which manifests itself as a change in g_{Na} (see **d**). Modified from Hille (1992), based on data supplied by J.B. Patlak. **f** Working hypothesis of a voltage-gated Na channel. The channel is a macromolecule that spans the lipid bilayer. The entrance of the channel is adorned by long-chain sugar molecules on its extracellular (*EC*) side, and the channel narrows here to provide a selectivity filter able to discriminate between cations. The channel itself consists of an aqueous pore through the centre of the molecule. The channel opens or closes depending on the trans-membrane electric field experienced by charge groups (the voltage sensor) within the macromolecule. On the intracellular side (*IC*), there is a molecular gate which inactivates the channel for a brief time after opening and serves the function of providing the axon with an absolute refractory period. Anchor proteins on the intracellular side retain the channel at its location in the membrane (from Hille 1992)

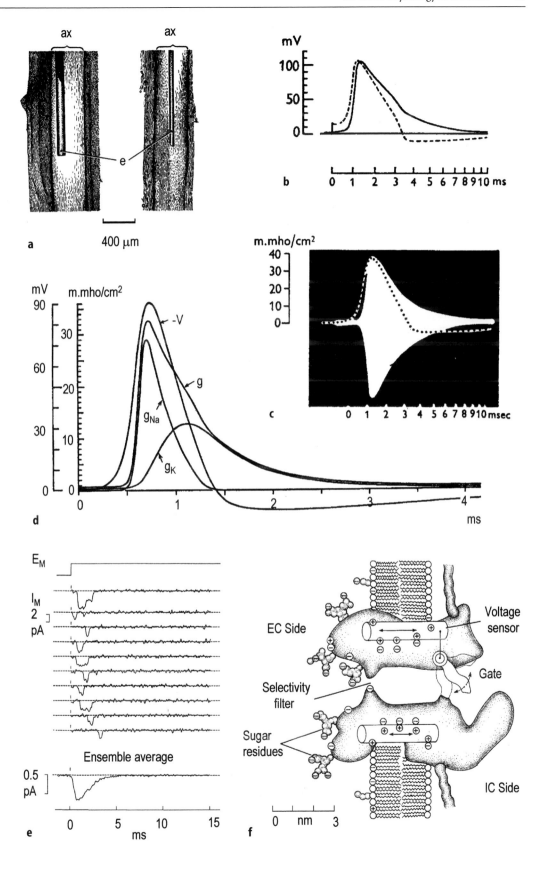

a

400 µm

b

c

d

e

Ensemble average

f

EC Side

IC Side

Voltage sensor

Gate

Selectivity filter

Sugar residues

Action potentials are stereotyped all-or-none phenomena that have an invariable form which is independent of the way in which they are evoked. Consequently, information in axons is not transmitted by the shape of these potentials but rather by the frequency with which they are generated. The immutability of the form of the action potential enables it to travel over long distances, from spinal motoneuron to muscles of the foot, for example without compromising the fidelity of the signal.

1.4.3.2
Action Potentials in Dendrites
(Johnston et al. 1996)

For many years it was thought that dendrites behaved like passive cables and that their integrative properties could be accurately calculated from a knowledge of basic resistive, capacitative and geometrical parameters (Rall 1977). More recently it has been shown that action potentials occur in the dendrites of some cells. It is now established that the dendrites of hippocampal (primarily CA1) and cortical pyramidal cells support Na^+- and Ca^{2+}-mediated action potentials while cerebellar Purkinje cell dendrites support only Ca^{2+} spikes.

The discovery of dendritic action potentials has importance in several areas of neuronal function. It likely permits entirely new modes of dendritic integration, since distant dendritic depolarisation can now influence the soma more directly by generating a propagating action potential (Llinás and Nicholson 1971; see also Chap. 7, Figs. 5, 9). When several action potentials are present at the same time in a dendrite they can interact, possibly by annihilation due to the refractory period. It must be said, though, that it is not clear under what circumstances dendritic spikes occur, and their functional role remains speculative.

Finally, the fact that dendritic action potentials are mediated largely by Ca^{2+} may have profound importance. It is now clear that the entry of Ca^{2+} is essential for many intracellular processes, including the release of substances and cellular plasticity and growth (Llinás 1988).

1.4.5
Ligand-Gated Channels
(Hille 1992; Shepherd 1994)

The primary realm of ligand-gated channels is chemically mediated synaptic transmission. Synaptic transmission is one of the characteristics that distinguishes the nervous system from other tissues. It is the major mechanism by which information is transferred from one neuron to another. Two

modes of transmission exist: chemical and electrical. Of these, the chemical mode is the most prevalent and important in the vertebrate CNS.

Chemical transmission is probably best conceived of as a specialisation of neurohumoral mechanisms that arose in primitive cell groups (and continues to persist in cellular systems) as a means of communication (Lentz 1968). In its original form, such communication probably consisted of the release of a neuroactive substance by one cell and its transduction by neighbouring elements. Specificity was determined by two factors. The first was the transport properties of the medium that contained the cells, primarily diffusion or bulk flow. The other factor determining specificity was the distribution of receptors for the neuroactive substance.

In the nervous system of vertebrates, chemical synaptic transmission takes place at specialised junctions, usually of the order of a micrometre in diameter with a narrow cleft separating the pre- and postsynaptic components (Fig. 1.8a–b). The presynaptic component is commonly an enlarged axonal terminal and the postsynaptic element is a dendrite or dendritic spine, although virtually all combinations of junctional elements have been found in some context. The neurohumoral compound is packaged in vesicles that are found in quantity in the presynaptic terminal. When the terminal is adequately depolarised by an invading action potential, Ca^{2+}-selective channels open in the terminal, and the local entry of this ion leads to the fusion of the vesicles with the terminal membrane, followed by the release of the contents into the synaptic cleft (Matthews 1996). The transmitter rapidly diffuses across the narrow gap separating the cells to reach receptors on the postsynaptic membrane. Activation of these receptors leads to the opening of associated ionic channels, allowing ions to move across the membrane and so change the potential of the postsynaptic cell. The entire process takes about 1 ms.

These facts are now common knowledge, and attention is presently turning to the precise kinetics of Ca^{2+} entry into presynaptic terminals, to the roles of proteins, such as synaptobrevin and syntaxin, and to the process of vesicle docking and membrane recycling within the terminal (Matthews 1996), while on the postsynaptic side, the molecular structure of the receptors is being clarified. It is worth recalling the great contributions of past physiologists, including Du Bois Reymond, Dale, Loewi, Sherrington, Katz and Eccles, who gradually revealed the intricacies of chemical synaptic transmission (see Kuffler et al. 1984). It used to be thought that only a few transmitter substances were

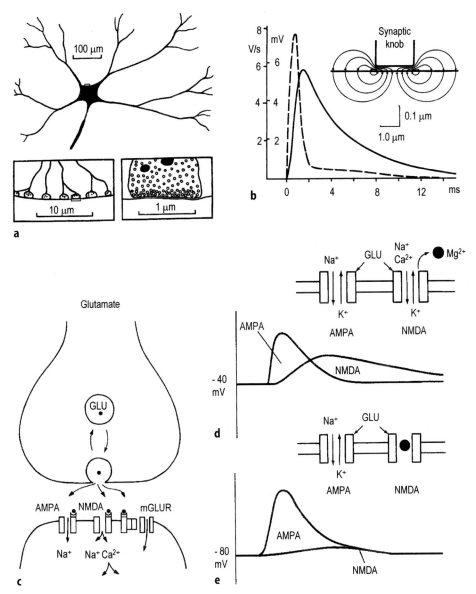

Fig. 1.8a–d. Chemical synaptic transmission. **a,b** Concept of synaptic transmission in 1957. **a** Sketch of motoneuron with small area of soma indicated and enlarged successively in the *insets* to show set of synaptic boutons and vesicles. **b** Average excitatory postsynaptic potential (EPSP) recorded in motoneuron. *Continuous line* is mean of several EPSPs (mV scale on *inside* of *ordinate*) and *broken line* indicates estimate of subsynaptic current needed to generate EPSP, based on plausible values of membrane electrical parameters (V/s scale on *left* of *ordinate*). *Inset* is schematic of synaptic knob and postsynaptic membrane with cleft indicated using two different length scales (from Eccles 1957). **c,d** Glutaminergic synapse, 1994. **c** Simplified schematic of complex events know known to occur in pre- and postsynaptic terminal. Glutamate (*GLU*) is synthesised and stored in vesicles from which it is released, via the mediation of Ca^{2+}, into the synaptic cleft. On arriving at the postsynaptic membrane, glutamate acts at two sites denoted by the names of specific agonists for these receptors, α-amino-3-hydroxy-5-methyl-4-isoxazole-propionic acid (*AMPA*) and N-methyl-D-aspartic acid (*NMDA*). The AMPA receptor opens a channel that allows the move-

ment of Na^+ and K^+, while the NMDA receptor regulates a Ca^{2+}-permeable channel (which also allows Na^+ and K^+ to move through it) that is normally blocked by Mg^{2+} at typical resting potential of -80 mV. When the membrane becomes depolarised (by the activation of the AMPA receptor, for example), the Mg^{2+} block is relieved and Ca^{2+} enters the postsynaptic area to cause further depolarisation and activates second messenger systems. The AMPA and NMDA receptors are examples of ionotropic receptors; metabotropic glutamate receptors also exist, indicated here by *mGLUR*. **d, e** Postsynaptic potentials generated by glutamate and ionic mechanism. At -80 mV (*panel e*) GLU opens AMPA channel but NMDA channel is blocked by Mg^{2+} (*filled black circle*), so almost all of the EPSP is due to current flowing through AMPA channel. At -40 mV (*panel d*), however, the Mg^{2+} block is relieved and current now flows through the NMDA channel. AMPA-mediated EPSP and NMDA-mediated EPSP are shown separately for illustration purposes; in reality, the GLU-mediated EPSP is the sum of these two components (modified from Shepherd 1994). **c, d** and **e** illustrate some of the complexity of present concepts of central synaptic transmission

present in the CNS but we now see an increasingly rich "chemoarchitecture" (Nieuwenhuys 1985).

1.4.5.1
Excitatory Synaptic Transmission
(Johnston and Wu 1995; Kandel et al. 1995)

Identification of transmitter substances has been difficult, owing to the small size of the junctions and their inaccessibility, particularly in the CNS. The best-studied synapse is the neuromuscular junction connecting the axon of the motoneuron with the muscle. While this is a rather specialised example, it has all the basic characteristics of chemical synaptic transmission. The transmitter is acetyl-choline (ACh) and with suitable analysis it has been shown to be released in small packets or quanta (Kuffler et al. 1984). Arrival of the ACh at the postsynaptic site activates receptors that open channels for Na^+ and K^+. Unlike the axonal action potential, both these channels open simultaneously, so that the postsynaptic element depolarises to a level that approximates the arithmetic mean of the sodium and potassium equilibrium potentials, about -15 mV. Transmitter release is brief and its action is terminated by enzymatic destruction mediated by acetylcholinesterase at the neuromuscular junction, but cellular uptake and diffusion are dominant mechanisms at other synapses.

Another synapse that has been extensively studied is the giant synapse of the squid. Here, microelectrodes have been inserted in both pre- and postsynaptic processes and the details of the kinetics and Ca^{2+} activation have been elucidated (Llinás et al. 1981). Despite the enormous number of studies on this synapse, the transmitter has not been conclusively identified, illustrating the difficulty of this type of study. The consensus is that the transmitter is the excitatory amino acid, glutamate.

Glutamate is probably the most ubiquitous transmitter in the vertebrate CNS and has been the subject of intense study. Following the introduction of specific receptor agonists it became possible to classify glutaminergic synapses according to their binding properties. At present, the N-methyl-D-aspartic acid (NMDA) receptor and the α-amino-3-hydroxy-5-methyl-4-isoxazole-proprionic acid (AMPA) receptor (which also binds quisqualate and kainic acid) are recognised and a number of receptors are being cloned (Fig. 1.8c–d). Glutamate (and possibly aspartate) acts at both receptors, but the effects differ. The NMDA receptor behaviour is particularly complex and interesting and may mediate a form of learning known as long-term potentiation.

When the ligand acts on a receptor complex directly associated with a channel that admits ions, the mechanism is referred to as ionotropic. There are other classes of action of glutamate on receptors that act through intermediates associated with metabolism and involving GTP-binding proteins (G proteins). Such receptor actions are termed metabotropic, and the majority of known transmitter actions may actually fall into this latter category. For example, dopamine, serotonin and noradrenaline, adenosine, somatostatin, vasopressin and oxytocin, to name only a few, can all have G protein-mediated metabotropic actions. Metabotropic activation can diminish membrane conductance rather than enhance it and may be excitatory or inhibitory, depending on the specific receptor-effector system.

1.4.5.2
Inhibitory Synaptic Transmission
(Johnston and Wu 1995; Kandel et al. 1995)

Many chemical synapses exert an inhibitory effect on the postsynaptic element by causing channels selective to K^+ or Cl^- to open. Since the equilibrium potentials for these ions are slightly negative with respect to the resting level in the cell, this causes a hyperpolarisation of the membrane and also holds the cell at this level in the face of any competing excitatory influences. One of the best-known inhibitory substances is γ-aminobutyric acid (GABA). Two classes of receptors exist for this amino acid, known as $GABA_A$ and $GABA_B$. The $GABA_A$ receptor activates a Cl^- channel via a fast ionotropic action, while the $GABA_B$ receptor is metabotropic and activates a second messenger cascade to open a K^+ channel. Another inhibitory transmitter is glycine, which acts in the spinal cord via an ionotropic mechanism to open Cl^- channels. It is clear that the actual sign of transmitter action need not be correlated with the substance but depends on the nature of the channels to which the receptor is coupled.

1.4.5.3
Nonvesicular Release of Transmitter

There is evidence that biogenic amine transmitters can be released from neurons by reversal of the normal carrier-mediated re-uptake process (Levi and Raiteri 1993). Further studies have shown that similar mechanisms may release glutamate from glial cells, particularly under pathophysiological conditions (Billups and Attwell 1996). Although the observations supporting these hypotheses are fairly meagre, they do prompt caution in the interpretation of neuroanatomical structure.

1.4.5.4
Nitric Oxide

Among the ever-increasing roll call of chemicals that are able to modulate neuronal activity, it came as a surprise to discover that the toxic gas nitric oxide (NO) was generated and apparently widely used as a signal in the CNS (Vincent and Hope 1992). NO is generated by the enzyme NO synthase and acts on soluble guanylyl cyclase to increase cyclic GMP levels throughout the nervous system. Since NO is a gas, it is not constrained to move in the extracellular space but can pass through cells.

1.4.6
Electrical Synapses and Gap Junctions
(Dermietzel and Spray 1993)

Electrical synapses are specialised regions where information can be transferred between cells by direct electrical continuity. Such junctions also permit the passage of molecules between cells. Historically, there was a period when prominent neuroscientists believed that electrical synapses were a major mechanism for neuronal communication, but it was soon established that this was not the case (Grundfest 1975). Today it is known that membrane channels consisting of six matching subunits in each cell form junctions that allow the passage of molecules as large as 1 kDa between cells. When these molecules are charged, current may be transferred, but it is also possible to move chemical species per se, such as Ca^{2+} and inositol triphosphate (IP_3). The subunits are formed of molecules called connexins. Connexin43 is found predominantly in astrocytes, while connexin32 is found in oligodendrocytes and some neurons. Gap junctions have been implicated in the mediation of waves of changing Ca^{2+} concentration in glioma cell cultures (Charles et al. 1992) and in developmental processes, but the full significance is still unclear. In those neurons where tight electrical synchrony is of paramount importance, such as some cells mediating electric signalling in fish and in the inferior olive, gap junctions do mediate electrical coupling (Bennett et al. 1991). This is unlikely to be their primary role in astrocytes, however, and there they may be involved in the spatial buffering of K^+. A specialised and curious electrical synapse exists in the axon cap of the Mauthner cell in teleosts, in which a true electric-field effect, devoid of junctions, occurs (Faber and Korn 1978). This is thought to be very unusual, however.

1.5
The Broad Picture

In comparison to physics and chemistry, biology is a "softer" science, lacking an extensive theoretical basis described by quantitative laws. Neuroscience is even "softer" than biology and, indeed, has only recently been perceived as a distinct discipline worthy of a name. We lack theories of brain organisation and we lack explanations of memory or consciousness. Even leaving aside those great issues, we lack explanations of what the differences in dendritic pattern between, say, a pyramidal cell and a Purkinje cell signify, not to mention the radically different local circuitry of cerebral cortex and cerebellum.

It would be true to say, therefore, that we are largely ignorant of how the brain works. Despite this, neuroscience is one of the fastest growing areas of science today. The reasons are various, but a major one is the application of concepts of molecular and cellular biology to the nervous system. Today we are beginning to conceive of a nerve cell differently from the way we have done since the 1950s.

The view that became solidified in the period 1950–1960 was itself a product of evolving concepts, driven as always by technological advances. The great era of neurohistology occurring around the turn of the century, exemplified by Ramón y Cajal's monumental two-volume treatise (Ramón y Cajal 1909, 1911), had established the neuron doctrine, i.e. that nerve cells were discrete entities, not elements of a continuous syncytium, with distinct geometries revealed so exquisitely by the Golgi impregnation. This wealth of anatomical information remained largely unassimilated at a functional level until the microelectrode was perfected. It is true that the concept of the synapse was elaborated as an essential element in the reflex arc by Sherrington, using gross electrical recording. Furthermore, the chemical nature of synaptic transmission was established by such pioneers as Loewi (1921) and Dale et al. (1936). These results had already led to an overconfident theoretical interpretation of neuronal circuitry (McCulloch and Pitts 1943). It really needed a microelectrode, however, to complement physiologically the resolution of the histological picture and permit great conceptual advances.

Microelectrodes came into popular use with the study of Ling and Gerard (1949), though there were certainly others with similar ideas. These electrodes permitted the intracellular electrical activity of single cells to be correlated with their anatomy. At this time, Hodgkin, Huxley and others were using the favourable size of the squid giant axon to definitively study the action potential (Hodgkin and Huxley 1952), while Fatt and Katz (1951) explored synaptic transmission. The explosive progress in cellular electrophysiology was summarised in a book by Eccles (1957), who was himself actively contributing to the understanding of excitatory and inhibitory synaptic transmission at the time.

The marriage of cellular anatomy and electrophysiology led to greatly improved descriptions of local neuronal circuits, perhaps nowhere more thoroughly documented than in the cerebellum (Eccles et al. 1967). It is cautionary, however, to reflect that, at least in the cerebellum, the description of cellular physiology did little to clarify the role of the cerebellum in the brain as a whole. Indeed, most concepts of function and understanding of pathways came from an earlier era of ablation and electrical stimulation experiments (Dow and Moruzzi 1958). This situation remained almost unchanged until very recently. Now, however, so-called systems physiology is beginning to move forward, through both the application of new methods, such as functional imaging, and the extension of classical recording utilising sophisticated computer technology. These approaches are being joined with the use of genetically engineered animals to strategically alter the systems under study.

At the microscopic level, all the organelles revealed by electron microscopy, but seemingly lacking function for so many years, are now being understood through molecular biology and the extension of microelectrode techniques to the patch clamp, enabling single channels to be studied physiologically.

Thus we see that the concept of the nerve cell (and, by extension, the glial cell) has gradually come into focus over the years as techniques provided increasing structural and functional resolution. The nerve cell has formed the basis from which neuroscience is now moving in two directions: towards a systems description ending in behaviour and towards a molecular description ending in the genome. This process creates a hierarchy of explanations, but perhaps the most exciting and challenging thing about the hierarchy is that it is not rigid. Insights are constantly being forged by bringing together the discoveries at disparate levels: a theme that will be echoed throughout this work.

References

Abbott NJA (1991) Glial-neuronal interaction. Ann NY Acad Sci 633:1–639

Agnati LF, Zoli M, Stromberg I, Fuxe K (1995) Intercellular communication in the brain: wiring versus volume transmission. Neuroscience 69:711–726

Alberts B, Bray D, Lewis J, Raff M, Roberts K, Watson JD (1994) Molecular biology of the cell, 3rd edn. Garland, New York

Barres BA (1991) New roles for glia. J Neurosci 11:3685–3694

Bennett MVL, Barrio LC, Bargiello TA, Spray DC, Hertzberg E, Saez JC (1991) Gap junctions: new tools, new answers, new questions. Neuron 6:305–320

Billups B, Attwell D (1996) Modulation of non-vesicular glutamate release by pH. Nature 379:171–174

Blinkov SM, Glezer II (1968) The human brain in figures and tables. Basic Books/Plenum, New York

Bradbury MW (1979) The concept of a blood-brain barrier. Wiley, Chichester

Catterall WA (1993) Structure and function of voltage-gated ion channels. Trends Neurosci 16:500–506

Charles AC, Naus CCG, Zhu D, Kidder GM, Dirksen ER, Sanderson MJ (1992) Intercellular calcium signaling via gap junctions in glioma cells. J Cell Biol 118:195–201

Chesler M (1990) The regulation and modulation of pH in the nervous system. Progr Neurobiol 34:401–427

Chiu SY, Kriegler S (1994) Neurotransmitter-mediated signaling between axons and glial cells. Glia 11:191–200

Cole KS, Curtis HJ (1939) Electric impedance of the squid giant axon during activity. J Gen Physiol 22:649–670

Craig AM, Banker G (1994) Neuronal polarity. Annu Rev Neurosci 17:267–310

Cserr HF, Bundgaard M (1984) Blood-brain interfaces in vertebrates: a comparative approach. Am J Physiol 246:R277–R288

Cserr HF, Knopf PM (1992) Cervical lymphatics, the blood-brain barrier and the immunoreactivity of the brain: a new view. Immunol Today 13:507–512

Dale HH, Feldberg W, Vogt M (1936) Release of acetylcholine at voluntary motor nerve endings. J Physiol (Lond) 86:353–380

Davson H, Danielli JF (1970) The permeability of natural membranes, 2nd edn. Hafner, Darien (reprint of 1952 edition published by Cambridge University Press, Cambridge)

Del Río Hortega P (1921) La glía de escasa radiaciones (oligodendroglía). Bol Real Sociedad Espan Historia Nat 21:63–92

Dermietzel R, Spray DC (1993) Gap junctions in the brain: what type, how many and why? Trends Neurosci 16:186–192

Dow RS, Moruzzi G (1958) The physiology and pathology of the cerebellum. University of Minnesota Press, Minneapolis

Eccles JC (1957) The physiology of nerve cells. Johns Hopkins, Baltimore

Eccles JC, Ito M, Szentágothai J (1967) The cerebellum as a neuronal machine. Springer, Berlin Heidelberg New York

Faber DS, Korn HE (1978) Neurobiology of the Mauthner cell. Raven, New York

Fatt P, Katz B (1951) An analysis of the end-plate potential recorded with an intra-cellular electrode. J Physiol (Lond) 115:320–370

Foster M, Sherrington CS (1897) A textbook of physiology, part III: the central nervous system, 7th edn. MacMillan, London

Fujita T (1979) Current views on the paraneuron concept. Trends Neurosci 2:27–30

Fuxe K, Agnati LF (eds) (1991) Volume transmission in the brain. Raven, New York (Advances in neuroscience, vol 1)

Geffen LB, Jessell TM, Cuello AC, Iversen LL (1976) Release of dopamine from dendrites in rat substantia nigra. Nature 260:258–260

Golgi C (1873) Sulla struttura della sostanza grigia del cervello. Gazz Med Ital Lombardia 33:244–246

Grundfest H (1975) History of the synapse as a morphological and functional structure. In: Santini M (ed) Golgi centennial symposium. Raven, New York, pp 39–50

Hille B (1992) Ionic channels of excitable membranes, 2nd edn. Sinauer, Sunderland, MA

Hodgkin AL (1992) Chance and design. Cambridge University Press, Cambridge

Hodgkin AL, Huxley AF (1952) A quantitative description of membrane current and its application to conduction and excitation in nerve. J Physiol (Lond) 117:500–544

Huxley AF (1995) Electrical activity of nerve: the background up to 1952. In: Waxman SG, Kocis JD, Stys PK (eds) The axon. Oxford University Press, New York, pp 3–10

Jacobson K, Sheets ED, Simson R (1995) Revisiting the fluid mosaic model of membranes. Science 268:1441–1442

Johnston D, Wu M-SW (1995) Foundations of cellular neurophysiology. MIT Press, Cambridge, MA

Johnston D, Magee JC, Colbert CM, Christie BR (1996) Active properties of neuronal dendrites. Annu Rev Neurosci 19:165–186

Kandel ER, Schwartz JH, Jessel JM (1991) Principles of neural science, 3rd edn. Appleton and Lange, Norwalk, CT

Kandel ER, Schwartz JH, Jessell TM (1995) Essentials of neural science and behavior. Appleton and Lange, Norwalk, CT

Katchalsky A (1971) Thermodynamics of flow and biological organization. Zygon J Rel Sci 6:99–125

Kettenmann H, Ransom BR (1995) Neuroglia. Oxford University Press, New York

Kuffler SW, Nicholls JG, Martin AR (1984) From neuron to brain, 2nd edn. Sinauer, Sunderland, MA

Kuhn TH (1970) The structure of scientific revolutions, 2nd edn. University of Chicago, Chicago

Lentz TL (1968) Primitive nervous systems. Yale Universtiy, New Haven

Lentz TL (1971) Cell fine structure. Saunders, Philadephia

Levi G, Raiteri M (1993) Carrier-mediated release of neurotransmitters. Trends Neurosci 16:415–419

Ling G, Gerard RW (1949) The normal membrane potential of frog sartorius fibers. J Cell Comp Physiol 34:383–396

Llinás C (1988) The intrinsic electrophysiological properties of mammalian neurons: insights into central nervous system function. Science 242:1654–1664

Llinás R, Nicholson C (1971) Electrophysiological properties of dendrites and somata in alligator Purkinje cells. J Neurophysiol 34:532–551

Llinás R, Steinberg IZ, Walton K (1981) Relationship between presynaptic calcium current and postsynaptic potential in squid giant synapse. Biophys J 33:323–351

Loewi O (1921) Über humorale Übertragbarkeit der Herznervenwirkung. Pflügers Arch 189:239–242

Margulis L (1970) Origin of eukaryotic cells. Yale University Press, New Haven

Martin DL (1992) Synthesis and release of neuroactive substances by glial cells. Glia 5:81–94

Matthews G (1996) Neurotransmitter release. Annu Rev Neurosci 19:219–233

McCulloch WS, Pitts WH (1943) A logical calculus of the ideas immanent in nervous activity. Bull Math Biophys 5:115–133

McMillian MK, Thai L, Hong JS, O'Callaghan JP, Pennypacker KR (1994) Brain injury in a dish: a model for gliosis. Trends Neurosci 17:138–142

Newman EA (1995) Glial cell regulation of extracellular potassium. In: Kettenmann H, Ransom BR (eds) Neuroglia. Oxford University Press, New York, pp 717–731

Nicholson C (1979) Brain cell microenvironment as a communication channel. In: Schmitt FO, Worden FG (eds) The neurosciences. Fourth study program. MIT Press, Cambridge, MA, pp 457–476

Nicholson C (1993) Ion-selective microelectrodes and diffusion measurements as tools to explore the brain cell microenvironment. J Neurosci Methods 48:199–213

Nicholson C (1997) Brain cell microenvironment. In: Adelman G, Smith B (eds) Encyclopedia of neuroscience, CD-ROM, 2nd edn. Elsevier, Amsterdam (in press)

Nicholson, C, Rice ME (1991) Diffusion of ions and transmitters in the brain cell microenvironment. In: Fuxe K, Agnati LF (eds) Volume transmission in the brain. Raven, New York, pp 279–294

Nieuwenhuys R (1985) Chemoarchitecture of the brain. Springer, Berlin Heidelberg New York

Nieuwenhuys R, Nicholson C (1969) Aspects of the histology of the cerebellum of mormyrid fishes. In: Llinás R (ed) Neurobiolgy of cerebellar evolution and development. American Medical Association, Chicago, pp 135–169

Palay SL, Chan-Palay V (1974) Cerebellar cortex, cytology and organization. Springer, Berlin Heidelberg New York

Palay SL, Chan-Palay V (1977) General morphology of neurons and neuroglia. In: Kandel ER (ed) The handbook of physiology, section 1: the nervous system, vol 1. American Physiological Society, Bethesda, MD, pp 5–37

Pannese E (1994) Neurocytology. Thieme, Stuttgart

Peters A, Palay SL, Webster H de F (1991) The fine structure of the nervous system. Neurons and their supporting cells, 3rd edn. Oxford University Press, New York

Privat A, Gimenez-Ribotta M, Ridet J-L (1995) Morphology of astrocytes. In: Kettenman H, Ransom BR (eds) Neuroglia. Oxford University Press, New York, pp 4–22

Purves MJ (1972) The physiology of the cerebral circulation. Cambridge University Press, Cambridge

Rakic P (1995) Radial glial cells: scaffolding for brain construction. In: Kettenman H, Ransom BR (eds) Neuroglia. Oxford University Press, New York, pp 746–762

Ramón y Cajal S (1909) Histologie du système nerveux de l'homme et des vertébrés. Tome I. Malone, Paris (reprint CSIC, Madrid 1955; English translation Oxford University Press, New York, 1995)

Ramón y Cajal S (1911) Histologie du système nerveux de l'homme et des vertébrés. Tome II. Malone, Paris (reprint CSIC, Madrid 1955; English translation Oxford University Press, New York, 1995)

Rall W (1977) Core conductor theory and cable properties of neurons. In: Kandel ER (ed) Handbook of physiology, section 1: the nervous system, vol I. Cellular biology of neurons, part 1. American Physiological Society, Bethesda, MD, pp 39–97

Ransom CB, Sontheimer H (1995) Biophysical and pharmacological characterization of inwardly rectifying K^+ currents in rat spinal cord astrocytes. J Neurophysiol 73:333–346

Rapp M, Segev I, Yarom Y (1994) Physiology, morphology and detailed passive models of guinea-pig cerebellar Purkinje cells. J Physiol (Lond) 474:101–118

Reichenbach A, Robinson SR (1995) Ependymoglia and ependymoglia-like cells. In: Kettenman H, Ransom BR (eds) Neuroglia. Oxford University Press, New York, pp 58–84

Retzius G (1894) Die Neuroglia des Gehirns beim Menschen und Säugethieren. Biol Untersuch 6:1–28

Rice ME, Lee EJK, Choy YY (1995) High levels of ascorbic acid, not glutathione, in the CNS of anoxia-tolerant reptiles contrasted with levels in anoxia-intolerant species. J Neurochem 64:1790–1799

Ritchie JM (1992) Voltage-gated ion channels in Schwann cells and glia. Trends Neurosci 15:345–351

Ritchie JM (1995) Physiology of axons. In: Waxman SG, Kocis JD, Stys PK (eds) The axon. Oxford University Press, New York, pp 68–96

Schmitt FO (1984) Molecular regulators of brain functioning: a new view. Neuroscience 13:991–1001

Schmitt FO, Samson FE (1969) The brain cell microenvironment. Neurosci Res Prog Bull 7:277–417

Schousboe A, Westergaard N (1995) Transport of neuroactive amino acids in astrocytes. In: Kettenmann H, Ransom BR (eds) Neuroglia. Oxford University Press, New York, pp 246–258

Shepherd GM (1991) Foundations of the neuron doctrine. Oxford University Press, New York

Shepherd GM (1994) Neurobiology, 3rd edn. Oxford University Press, New York

Siesjö BK (1978) Brain energy metabolism. Wiley, Chichester

Singer SJ, Nicolson GL (1972) The fluid mosaic model of the structure of cell membranes. Science 175:720–731

Somjen GG (1988) Nervenkitt: notes on the history of the concept of neuroglia. Glia 1:2–9

Streit WJ (1995) Microglial cells. In: Kettenman H, Ransom BR (eds) Neuroglia. Oxford University Press, New York, pp 85–96

Szentágothai J (1975) What the "reazione nera" has given to us. In: Santini M (ed) Golgi centennial symposium. Raven, New York, pp 1–12

Vincent SR, Hope BT (1992) Neurons that say NO. Trends Neurosci 15:108–113

Waldeyer HWG (1891) Über einige neuere Forschungen im Gebiete der Anatomie des Centralnervensystems. Dtsch Med Wochenschr 17:1213–1218, 1244–1246, 1267–1269, 1287–1289, 1331–1332, 1352–1356

Watson WE (1976) Cell biology of brain. Chapman and Hall, London

Weiss P, Hiscoe HB (1948) Experiments on the mechanism of nerve growth. J Exp Zool 107:315–395

Wiesinger H (1995) Glia-specific enzyme systems. In: Kettenmann H, Ransom BR (eds) Neuroglia. Oxford University Press, New York, pp 488–499

Yang D, Oyaizu Y, Olsen H, Woese CR (1985) Mitochondrial origins. Proc Natl Acad Sci U S A 82:4443–4447

Young JZ (1957) The life of mammals. Oxford University Press, New York

Young JZ (1991) The concept of neuroglia. Ann NY Acad Sci 633:1–26

CHAPTER 2

Structure and Organization of Centres

R. NIEUWENHUYS

2.1
Introduction

The preceding chapter dealt with the structure and function of the cellular elements in the central nervous system. In the present chapter the same features of assemblies of neurons will be discussed. The approach will be purely morphological, passing from simple to complex and from generalized to specialized, although some theories about the phylogenetic changes in the vertebrate neuraxis will be considered. Following some notes on grey and white matter, the periventricular grey and the neuropil and fibre zone which covers this grey peripherally in the brain of many different anamniotes will be covered. The neuropil and its differentiation into various zones or sheets will next be considered, and then the organisation of the diffuse or reticular grey will be discussed. Having dealt with the structure and function of simple and generalised grisea, we will then survey the organisation of more complex neuronal assemblies. Attention will be paid to the position, cytoarchitecture and organisation of cell masses or nuclei as well as to their dendritic and axonal ramifications and extrinsic connections. Many grisea in the central nervous system of vertebrates show a laminar organisation, and this interesting feature will be treated in some detail. The last section is devoted to the chemoarchitecture of grisea.

2.2
Grey and White Matter

If we inspect freshly cut sections through the brain and spinal cord of mammals, some regions appear a shiny white, while others are greyish. This subdivision into substantia grisea and substantia alba, or into grey and white matter, is best seen in the spinal cord and in the telencephalon. In the spinal cord the grey matter is located centrally and has a typical H-shaped appearance in transverse sections, surrounded by a mantle of white matter. In the telencephalon the relation of grey and white matter is reversed. Here, a superficially situated sheet of grey matter – termed cortex cerebri – surrounds a central core of white matter. Grey matter consists of perikarya, dendritic trees, the initial axonal segments of neurons and the terminal segments and synaptic endings of axons. White matter is composed of myelinated and unmyelinated nerve fibres, and it is actually the myelin which creates the typical glistening white appearance.

The designations 'grey' and 'white' matter are widely used in the description of central nervous systems of vertebrates. In large areas of the brain, however, – for example, in the reticular formation and the hypothalamus – neurons, dendritic processes and fibres are so interspersed that a subdivision into grey and white compartments is impossible. Note also that in anamniotes and reptiles the white matter contains, besides nerve fibres, extensive networks of dendritic arborisations and axonal ramifications. In acraniates and cyclostomes the axons are devoid of myelin; hence, in these groups there is, strictly speaking, no white matter.

2.3
Periventricular Grey

2.3.1
Conceptual Framework

This discussion is based on the diagrammatic longitudinal section through a part of the wall of the central nervous system of a hypothetical simple vertebrate (Fig. 2.1). In this conceptual model the neuronal perikarya have retained their embryonic position and constitute a narrow, dense, continuous zone of periventricular or central grey. The neurons have a few long, slender, non- or poorly ramifying dendrites which extend peripherally. Such neurons have been designated by Ramón-Moliner (1969) as *leptodendritic* (*leptos* = slender, elegant). The wide peripheral zone is occupied by unmyelinated axons, and it is assumed that synaptic contacts are made at places where axons and dendrites meet. It follows that the peripheral zone is both a fibre layer and a continuous neuropil zone, i.e. a meshwork of dendrites and axons, where each axon makes synaptic contact with dendrites of other neurons and where the dendrites of every neuron are in synaptic rela-

tions with many axons. The axons of the neurons in our conceptual model of the central grey emanate either from the soma or from the proximal part of a dendrite and pass peripherally to contribute to the neuropil and fibre zone.

The configuration in Fig. 2.1 is hypothetical, but many of its features are derived from studies of simple structures like the primordium hippocampi of the lamprey (Johnston 1902, 1912; Heier 1948) and the pallium of brachiopterygians (Nieuwenhuys 1983). In such structures each cell probably contacts different, though overlapping sets of fibres, and this property will shape the potential response patterns of the various elements.

Ramón-Moliner (1967, 1969, 1975) believed that in all vertebrates the periventricular regions of the brain stem and of the diencephalon are populated by leptodendritic neurons. He designated this area of the brain as the leptodendritic core and surmised that this entity "could have an even more primitive significance than that of the surrounding isodendritic core" (Ramón-Moliner 1975, p. 95). The isodendritic core, which corresponds roughly with the reticular formation, will be considered later in this chapter. The leptodendritic core of Ramón-Moliner closely corresponds to 'the core of the neuraxis' or

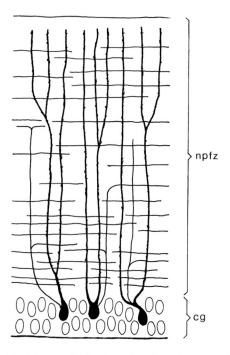

Fig. 2.1. Longitudinal section through a part of the wall of the central nervous system of a hypothetical simple and primitive vertebrate. The neuronal perikarya constitute a narrow zone of central grey (*cg*). The leptodendritic neurons extend their dendrites peripherally into a neuropil and fibre zone (*npfz*). Axons of the neurons depicted also pass peripherally and contribute to the neuropil and fibre zone

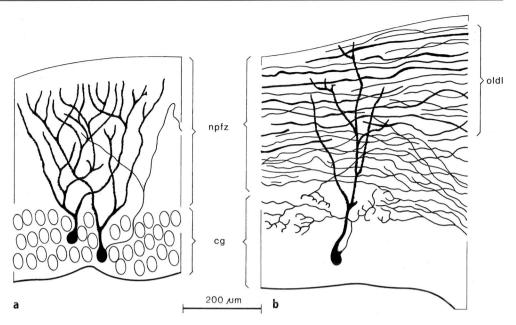

Fig. 2.2a,b. Aspects of the structure of the lateral part of the pallium of the tiger salamander, *Ambystoma tigrinum*. In this and many other parts of the brain of that species a central zone of periventricular grey (*cg*) is covered by a peripheral neuropil and fibre zone (*npfz*). **a** Two typical isodendritic neurons as seen in a transverse section; **b** a similar neuron in a horizontal section. The superficial part of the neuropil and fibre zone contains coarse axons of the tractus olfactorius dorsolateralis (*oldl*). The axon of the cell depicted issues branching collaterals in the superficial zone of the central grey and the same zone also receives terminal branches from axons passing through the neuropil and fibre zone. (modified from Herrick 1927, Fig. 10A, and Herrick 1934, Fig. 1B)

'the greater limbic system' described by one of the present authors (Nieuwenhuys 1985, 1996; Nieuwenhuys et al. 1988, 1989), albeit on quite other grounds than dendroarchitectonic properties (see Chap. 22, Sect. 12.3).

Conditions resembling those depicted in Fig. 2.1, but showing more complexity, are found in many parts of the central nervous system of primitive actinopterygians, dipnoans and amphibians. The following differences are seen: Firstly, the zone of periventricular grey is usually wider. Secondly, the dendritic trees of the neurons are larger and radiate out towards the meningeal surface (Figs. 2.2, 2.3, 2.6, 2.8); neurons of this type were termed isodendritic by Ramón-Moliner (1962, 1967, 1975; *isos* = similar, unchanging, uniform). Thirdly, the periventricular grey may show various degrees of differentiation, such as (a) the formation of grisea with different axonal destinations, (b) the occurrence of different cell types within the various grisea, and (c) the formation of different grisea with a distinct cytoarchitecture. Finally, the neuropil and fibre zone may likewise show differentiation of various kinds, including the formation of separate neuropil zones and distinct fibre paths.

2.3.2
Differentiation of Periventricular Grey

Within a morphologically homogeneous zone of periventricular grey neuronal populations can occur where the axons still contact distinct targets. Two examples illustrate this: (a) In favourable Golgi- or reduced silver preparations of the spinal cord and brain stem of urodeles motoneurons can be distinguished from surrounding neuronal elements because their axons enter the nerve roots (Herrick 1948). (b) Using retrograde tracer techniques it has been shown that in anurans the periventricular grey of the lower rhombencephalon contains a subpopulation of cells which projects to the contralateral half of the cerebellum (van der Linden and ten Donkelaar 1987). This population of cells can be looked upon as a 'foreshadowing' or 'primordium' of the inferior olive as it occurs in cartilaginous fishes and amniotes.

Differences in the size, shape, density and stainability of perikarya allow subdivision of the periventricular grey in many places into separate cytoarchitectonic zones, areas or fields. Examples of such parcellations of the periventricular grey are presented in Figs. 2.18 and 2.22–2.27. When such sets of neurons form clusters they are regarded as periventricular nuclei (Figs. 2.19, 2.20). Golgi prep-

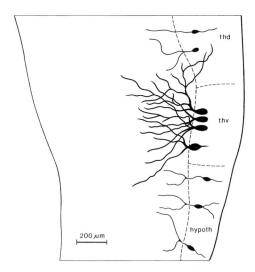

Fig. 2.3. Transverse section through part of the wall of the diencephalon of the lamprey, *Lampetra fluviatilis*. The pars ventralis thalami (*thv*), which contains isodendritic neurons is flanked by the pars dorsalis thalami (*thd*) and the hypothalamus (*hypoth*), both of which are composed of small, leptodendritic neurons. (based on Heier 1948, Fig. 10B)

arations reveal that periventricular areas which differ in cytoarchitectonic appearance may also differ in their dendroarchitecture. Thus, Fig. 2.3 shows that in the diencephalon of the lamprey the isodendritic ventral thalamus is dorsally and ventrally flanked by leptodendritic areas. Periventricular grisea which are well segregated cytoarchitectonically at the level of the cell bodies, however, are often incompletely segregated at the level of their dendritic trees. Such trees can spread widely in the

neuropil and fibre zone, where they intertwine with the dendritic arborisations of the neurons which make up adjacent grisea. This is exemplified in Fig. 2.4, which shows that in the rhombencephalon of the mud puppy, long dendrites of neurons in the superior reticular nucleus enter the domain of the motor trigeminal nucleus.

As regards differentiation within periventricular grisea, comparison of Nissl and Golgi preparations often reveals the presence of neurons of different types within the confines of particular cytoarchitectonic entities. Commonly, the deeper zone of periventricular grisea contains mainly cells with radially oriented dendritic arborisations, whereas in the peripheral zone neurons with tangentially extending dendrites prevail (e.g., the motor nucleus of V in Fig. 2.4). Herrick (1948) noted that throughout the periventricular grey of urodeles, small Golgi type II cells are present whose axons and dendrites spread in the immediate vicinity of the cell body while their axons participate in the formation of the periventricular neuropil (see Sect. 2.4). It seems likely that local circuit neurons also occur in the extensive periventricular grey of the brains of cyclostomes, primitive actinopterygians, dipnoans and crossopterygians, but adequate Golgi analyses are not available.

The periventricular grey has been considered to represent not only a simple, but also a primitive condition (e.g., Herrick 1933a, 1948). Several authors disagreed, however, and thought that in some parts of the brain of urodeles and lepidosirenid lungfishes the periventricular position of the neuronal perikarya is the result of a secondary sim-

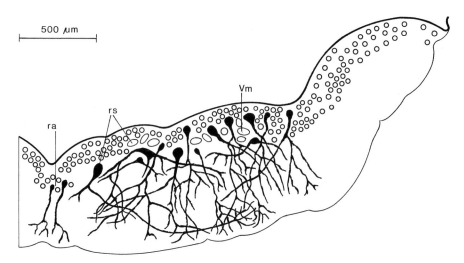

Fig. 2.4. Transverse hemisection through the rostral part of the rhombencephalon of the mud puppy, *Necturus maculatus*, showing neuronal configurations in the basal plate. Note that the dendritic fields of the motor trigeminal nucleus (*Vm*) and the nucleus reticularis superior (*rs*) overlap, and that both of these centres contain neurons of different types. *ra*, Nucleus raphes. (based on Herrick 1930, Figs. 12 and 13)

plification. The two groups mentioned are known to be *paedomorphic* vertebrates because they retain many juvenile characters and lack many adult features (Bemis 1984; Wake 1966; Wake and Roth 1989; Roth et al. 1993). Thus, the fact that the tectum mesencephali of urodeles and lepidosirenid lungfishes, contrary to their homologues in most other vertebrates, show no trace of lamination, but contain only a periventricular cell plate, has been considered a paedomorphic trait, due to suppression of cell migration (Northcutt 1977; Schmidt et al. 1989). Söderberg (1922) studied the development of the telencephalon in urodeles and anurans and also concluded that the ventricular location of cells in the amphibian brain is due to a secondary simplification. She observed that distinct layers of migrated cells are formed in the pallial parts of the cerebral hemispheres during ontogenesis, but that later in development these layers reunite with the periventricular cell zone. Similar observations were made by Holmgren (1922) and Bäckström (1924) in the telencephalon of cartilaginous fishes.

2.4
'Primitive' or 'Primordial' Neuropil Zone and Its Differentiation

2.4.1
Introductory Notes

'Neuropil' is a term given to areas of the central nervous system that are occupied by a feltwork of interlacing dendrites and axonal terminal ramifications. Most of the synaptic junctions occur in the neuropil, and therefore this tissue component forms a continuous synaptic field. Although dendrites and axonal ramifications are both essential components of neuropil zones, one of the components is frequently emphasized and this has led to the usage of the terms dendritic and axonal neuropil (Herrick 1933a, 1934, 1948; Scheibel 1979). The Golgi method remains the best technique for the analysis of the texture of the neuropil – a branch of neuroanatomy designated as neuropil architectonics (Szentágothai and Réthelyi 1973).

In Fig. 2.1 the axonal component of the zone peripheral to the central cell layer consists entirely of longitudinally oriented, unmyelinated fibres making synaptic contact with the peripherally extending dendrites of the periventricular neurons. Here there is no structural or functional segregation of fibres and terminals, hence the term neuropil and fibre zone. In parts of 'real' central nervous systems, however, differentiation is often seen. Thus, in the medial part of the brachiopterygian pallium, secondary olfactory fibres are situated exclusively in the more superficial parts of the neuropil and fibre zone, whereas the deeper regions harbour axons from other sources (Braford and Northcutt 1974; von Bartheld and Meyer 1986). Comparable spatial segregations of fibres of different origins may already have been present in the central nervous system of the earliest vertebrates. A further differentiation from the configuration shown in Fig. 2.1 is the outgrowth of preterminal collaterals. Studies on the spinal cord of larval and adult anamniotes (cf. Ramón y Cajal 1909) have shown three categories of such collaterals in this part of the neuraxis: internal, interstitial and external (Fig. 2.5). The internal collaterals pass to the periventricular grey matter, where they enter into synaptic contact with perikarya and dendrites of the neurons located within this compartment. The interstitial collaterals synapse with the peripherally extending dendrites of motoneurons and other large elements. The terminal segments of the dendrites just mentioned ramify and spread tangentially in the most superficial zone of the cord, where they form a dense plexus marginalis or perimedullaris with the external collaterals (Fig. 2.6). The plexus formed by the three types of collaterals plays a major role in the studies of Herrick (1927, 1930, 1933a,b, 1934, 1939, 1942, 1948) on the structure and differentiation of the neuropil.

2.4.2
Studies of Herrick on Urodeles

Herrick, a noted comparative neuroanatomist, subjected the brains of larval and adult urodeles to meticulous analysis using the Golgi technique. He found that neuropil is a very prominent component of the central nervous system in these animals, but one which varies in extent and complexity from place to place. He conjectured that many of the specialized structures found in the brains of urodeles and amniotes have evolved from a primitive or primordial neuropil zone.

"In the generalized brains here under consideration [i.e. the brains of urodeles] the neuropil is so abundant and so widely spread that it evidently plays a major role in all central adjustments, ..." (Herrick 1948, p. 29).

"Axons or collaterals interweave to form the closely knit fabric which pervades both gray and white substance everywhere. This is the axonic neuropil, within the meshes of which dendrites of neurons ramify widely. These axons are unmyelinated, and every contact with a dendrite or a cell body is a synaptic junction" (Herrick 1948, pp. 31, 32).

"All neuropil is a synaptic field, and, since in these amphibian brains it is an almost continuous fabric spread throughout the brain, its action is fundamentally integrative. But it is more than this. It is germinative tissue, the matrix from which much specialized structure of higher brains has been differentiated" (Herrick 1948, p. 32).

"The diffuse neuropil is probably the primordial form, going back to the earliest evolutionary stages of nervous differentiation (coelenterates)" (Herrick 1948, p. 33).

"From the primordial diffuse neuropil, differentiation advanced in two divergent directions. One of these ... led to the elaboration of the stable architectural framework of nuclei and tracts, the description of which comprises the larger part of current neuroanatomy. This is the heritable structure, which determines the basic pattern of those components of behavior which are common to all members of the species. The second derivative of the primordial neuropil is the apparatus of individually modifiable behavior – conditioning, learning, and ultimately the highly specialized associational tissues of the cerebral cortex" (Herrick 1948, p. 33).

Herrick (1933b, 1934, 1942, 1948) indicated that the neuropil in the urodelan brain is arranged in

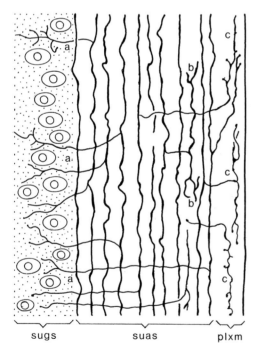

sugs   suas   plxm

Fig. 2.5. Longitudinal section through the lateral funiculus of the spinal cord of an older frog larva showing, apart from ascending and descending axons constituting the substantia alba spinalis (*suas*), three types of collaterals: internal (*a*), interstitial (*b*) and external (*c*). The internal collaterals enter the substantia grisea spinalis or spinal grey matter (*sugs*), the external collaterals participate in the formation of a dense plexus marginalis (*plxm*) or perimedullaris. (modified from Ramón y Cajal 1909, Fig. 224)

three layers, deep or periventricular, intermediate, and superficial. These layers correspond with the areas of distribution of the internal, intermediate and external collaterals observed in the spinal cord of anamniotes (Fig. 5).

2.4.3
Deep Neuropil Zone

The deep neuropil is a dense entanglement of fine internal axonal collaterals, within which the cell bodies of the periventricular grey are closely enmeshed (Fig. 2.7). Some of its constituent axons come from Golgi type II cells, others are slender fibres which enter the grey traversing the zone peripheral to the cell layer (Fig. 2.2), and still others are fine collaterals or terminals of the coarse axons of the long ascending and descending tracts. Herrick (1934, 1948) thought that the periventricular neuropil is present throughout the central nervous system and that in this 'all-pervasive' neuropil the synaptic junctions are made directly on the cell bodies. He supposed that this dense axonal neuropil constitutes a diffuse, continuous synaptic field, which exerts a general tonic or reinforcing influence upon all the neurons embedded in it. Herrick (1948) believed that the deep neuropil persists in some parts of the mammalian brain as subependymal and periventricular systems of fine fibres.

The observations just summarised are highly intriguing and raise several questions: (a) Is a deep neuropil, enveloping all perikarya, present throughout the periventricular grey of the urodeleCNS? (b) Is a comparable neuropil present in other anamniotes? (c) Do the extremely thin fibres which wind their way through the periventricular grey make synaptic contact with the somata of many neuronal perikarya, as the descriptions and illustrations of Herrick suggest? These questions remain unanswered.

Herrick's extensive studies have shown that a periventricular neuropil of the type shown in Fig. 2.7 is present in several parts of the urodele brain and occurs in some telencephalic regions of anurans. Because of the lack of Golgi studies, beyond those of Herrick, the ubiquity of a deep or central neuropil in 'lower' vertebrates cannot be assessed. Moreover, we lack electron-microscopical studies devoted to axosomatic synapses in the periventricular grey. Note, however, that the perikarya of the neurons forming the periventricular layer of the teleostean tectum mesencephali have almost no synapses (Meek 1981).

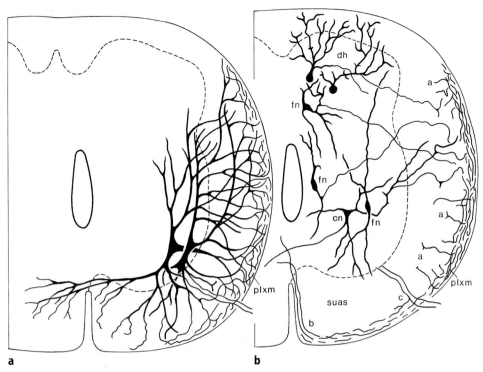

Fig. 2.6a,b. Transverse hemisections through the spinal cord of an older larva of the toad, *Bufo vulgaris*. **a** Two motoneurons. The dendrites of these elements branch out into the substantia alba spinalis (*suas*) and participate in the formation of the plexus marginalis (*plxm*), which constitutes the most superficial part of the lateral funiculus. **b** Funicular neurons (*fn*), commissural neurons (*cn*), dorsal horn cells (*dh*) and the axonal part of the plexus marginalis. The latter receives external collaterals from axons passing through the lateral funiculus (*a*), axons of certain commissural cells (*b*) and collaterals from the axons of motoneurons (*c*). Golgi technique. (modified from Ramón y Cajal 1909, Fig. 223)

2.4.4
Intermediate Neuropil Zone

The intermediate neuropil occupies most of the zone peripheral to the periventricular cell layer and represents the principal synaptic field in the urodele brain. It is formed by the interstitial axonal collaterals synapsing on the intermediate parts of the dendrites of periventricular grey neurons. Herrick (1934, 1948) pointed out that during evolution three

Fig. 2.7.a Detail of the neuropil of the grey substance of the dorsal sector of the nucleus olfactorius anterior of the tiger salamander, *Ambystoma tigrinum*, as observed in a horizontal section through the cerebral hemispheres at the level of the dorsal border of the olfactory bulb. Golgi technique. **b** Outline of the section from which the detail shown in **a** is taken. *bol*, Bulbus olfactorius; *nuola*, nucleus olfactorius anterior; *phip*, primordium hippocampi; *ppir*, primordium piriforme; *vl*, ventriculus lateralis. (from Herrick 1934, Figs. 1, 2)

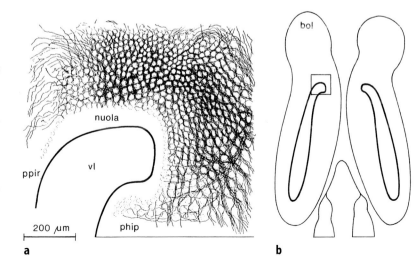

important differentiation processes have occurred within the confines of the intermediate neuropil zone, viz. the formation of fibre tracts, the formation of specific neuropil fields, and the formation of nuclei. Herrick believed that these processes can be uncovered by studying various parts of the urodelan brain and by comparing these with each other and with their homologues in 'higher' vertebrates.

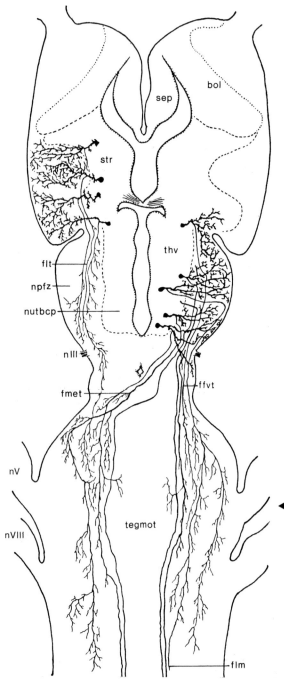

As regards fibre systems, Herrick found that most of the long ascending and descending tracts lie within the intermediate zone of neuropil and that in urodeles this zone shows a range of gradations between a homogeneous web of neuropil and long, well-fasciculated tracts. Many of the long fibres have collateral branches throughout their length, but others are well-isolated conductors between origin and termination. Two examples taken from Herrick's work (1934, 1939) show these differences. Figure 2.8 shows that in larval *Ambystoma* the efferents descending from the striatum and the ventral thalamus issue numerous collaterals over almost their entire length. In contrast, in the adult *Ambystoma* the very coarse myelinated axons of the dorsolateral olfactory tract appear entirely free from functional connections with their surrounding structures over their trajectory shown in Fig. 2.2b. Herrick believed that the structural variations shown by axonal systems in the urodelan brain can be arranged in a series which reflects the phylogenetic development of fibre paths: "In phylogeny the long well-organized tracts seem to have been formed by a concentration of the fibres of the neuropil" (Herrick 1948, p. 32).

The second example of differentiation within the intermediate neuropil in the urodele brain is the presence of local, circumscribed fields of neuropil such as the striatal neuropil, shown in Fig. 2.9. It contains the very thorny dendrites of striatal neurons and a dense feltwork of extremely delicate axons. Some of the latter derive from small intrinsic neurons of the striatal grey, but the largest and most characteristic component comprises the terminal arborisations of the thalamofrontal tract arising from the dorsal thalamus (Herrick 1927, 1933a, 1948). Herrick (1948) observed comparable formations in other parts of the telencephalon, in the diencephalon, in the roof of the midbrain and in the isthmus region.

Fig. 2.8. Horizontal section showing some patterns of connectivity in the brain of a 38-mm larva of the tiger salamander, *Ambystoma tigrinum*. Neurons situated in the ventral part of the striatum (*str*) send their axons via the lateral forebrain bundle (*flt*) to the pars ventralis thalami (*thv*) and the nucleus tuberculi posterioris (*nutbcp*), where they branch in the neuropil and fibre zone (*npfz*) of these structures. Neurons in the two centres last mentioned issue, in their turn, axons which descend via the fasciculus medianus tegmenti (*fmet*) and the fasciculi ventrales tegmenti (*ffvt*) to the tegmentum motoricum (*tegmot*) of the rhombencephalon, where most of them ramify and terminate. Some fibres were observed to enter the fasciculus longitudinalis medialis (*flm*). *bol*, Bulbus olfactorius; *nIII*, nervus oculomotorius; *nV*, nervus trigeminus; *nVIII*, nervus octavus; *sep*, septum. (based on Herrick 1939, Fig. 23)

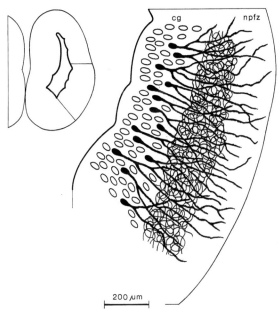

Fig. 2.9. The striatum of the tiger salamander, *Ambystoma tigrinum*. The deep part of the neuropil and fibre zone (*npfz*) is occupied by a dense axonal feltwork. *cg*, Central grey. Golgi technique (modified from Herrick 1927, Fig. 9)

The third differentiation process that Herrick (1934, 1948) believed took place within the intermediate neuropil is the formation of specific nuclei: "When some of these neuropil fields are analyzed it is found that their fibrous connections conform with those of specific 'nuclei' of higher brains. In *Ambystoma* such a field is not a 'nucleus' for it contains no cell bodies; but examination of the corresponding area in anurans and reptiles may show all stages in the differentiation of a true nucleus by migration of cell bodies from the central gray outward into the alba. Such a series of phylogenetic changes can be readily followed in the corpus striatum, the geniculate bodies, the interpeduncular nucleus, and many other places and particularly the tectum opticum and the pallial fields of the cerebral hemispheres" (Herrick 1948, p. 32).

Thus Herrick thought that differentiation processes within the intermediate neuropil zone had led to the development of numerous specific nuclei and their interconnecting fibre tracts. As already mentioned, Herrick thought that the 'stable architectural framework' formed by these structures is concerned with programming of the stereotyped patterns of innate behaviour.

2.4.5
Superficial Neuropil Zone

The third neuropil zone observed in the urodele brain, i.e. the layer of superficial neuropil, is a submeningeal sheet of interlacing dendritic and external axonal terminals. In some places it is absent, but in others it is elaborately organized. Herrick hypothesised that this superficial neuropil zone is a medium for strictly individual adjustments, and that in the telencephalon of 'higher' vertebrates the cerebral cortex develops within this layer.

If we adhere to the definition that the neuropil is a feltwork of interlacing dendrites and axon terminals containing numerous synaptic contacts, but no neuronal perikarya, then extensive neuropil compartments or zones in 'lower' vertebrates occur which are lacking in 'higher' forms. It can be postulated that during phylogeny two processes occur leading to this 'vanishing' of neuropil zones, viz. the reduction in length of dendrites by 'retraction' and the reduction in length of dendrites by migration of their parent perikarya in the direction in which these processes extend.

As regards 'retraction' of dendrites, in the spinal cord of anamniotes and reptiles, motoneurons and other large elements extend long peripherally directed dendrites through large regions of the anterior and lateral funiculi (Fig. 2.6). These long dendrites are the targets of the interstitial and external collaterals which emanate from axons passing longitudinally in the funiculi (Fig. 2.5). Thus, these funiculi not only represent fibre compartments but also contain an extensive neuropil, corresponding to the intermediate and superficial zones of Herrick. The dendrites of the neurons in the avian and mammalian spinal cord are much reduced in length, however, and penetrate only for a short distance into the anterior and lateral funiculi, or are confined entirely to the grey substance. Due to this 'retraction' of dendrites the interstitial and external collaterals lose their targets and disappear, and with them their corresponding neuropil zones. Only the central collaterals remain (see Fig. 2.16). The segregation of grey and white matter has become complete, so the white matter now contains only fibres and lacks synapses. Comparable changes may have taken place in the brain stem, but here the spatial relations between grey and white matter become much more intricate than in the spinal cord. If we compare large interneurons in the spinal cord of anamniotes and reptiles with those in mammals, the main difference may be characterised as the result of a 'retraction' of dendrites from the anterior and lateral funiculi. The 'transformation' of anamniote to mammalian motoneurons, however, is more complex and involves both a 'retraction' and a reorientation of dendrites. The dendrites of mammalian motoneurons are mainly oriented longitudinally and the dendritic trees of functionally related elements are organized in rostro-caudally oriented column-like ensembles (Laruelle 1937; Romanes 1964; Scheibel and Scheibel 1970a, 1973; Scheibel 1979).

The second phenomenon, i.e. the migration of neuronal perikarya in the direction of extension of their main dendrites, has received much attention. In Herrick's opinion, the formation of specific fields of neuropil often precedes the migration of neuronal perikarya and the formation of nuclei. Similar phenomena are prominent in Ebbesson's (1980) parcellation theory, whereas the outward migration of neurons per se has been the subject of several theories, including the chemotactic theory of Ramón y Cajal (1893, 1909) and Ariëns Kappers' (cf. Ariëns Kappers et al. 1936) concept of neurobiotaxis. It is important to note that during the ontogenetic development of the central nervous system of amniotes, migration of neuroblasts out from a periventricular matrix to a more peripheral position is very common. Presumably due to a belief in the correctness of Haeckel's biogenetic rule, which says that ontogenesis is an abbreviated recapitulation of phylogeny, the migration of neuroblasts during ontogenesis has often been equated with the gradual displacement of neurons during phylogeny.

Based on extensive studies of the structure of the brain of urodeles, Herrick concluded that during phylogeny: (a) parts of an initially diffuse neuropil zone gradually change into specialised fields of neuropil to which afferent axons from a limited number of sources contribute; (b) the neurons which extend their dendrites peripherally show a preference for these specialised fields of neuropil; (c) when the dendritic ramifications become confined to a single field of neuropil, their parent cell bodies migrate towards and enter that field of neuropil; and (d) by the latter event a specialised field of neuropil becomes transformed into a specific nucleus.

In some of his writings Herrick (1933a,b, 1934) selected the structural relations in the dorsal thala-

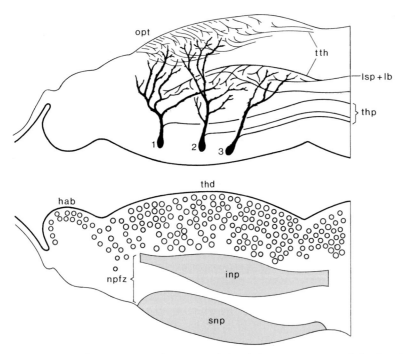

Fig. 2.10. Typical connections of neurons of the thalamus dorsalis (*thd*) of the mud puppy, *Necturus maculosus*, as seen in a horizontal section. The thalamus dorsalis is represented by an undifferentiated zone of periventricular grey. Optic tract fibres (*opt*) and tectothalamic fibres (*tth*) contribute substantially to a superficial neuropil zone (*snp*). Other tectothalamic fibres and axons originating from the spinal cord and the rhombencephalon, and passing to the diencephalon via the lemnisci spinalis (*lsp*) and bulbaris (*lb*), participate in the formation of a deeper, interstitial neuropil (*inp*). Three neurons of the dorsal thalamus are drawn. Their dendrites spread in the neuropil and fibre zone (*npfz*), where they are in synaptic connection with terminals or collaterals of various sensory tracts. The dendrites of neurons *1* and *2* participate in the formation of the intermediate as well as the superficial neuropil zone; the dendrites of neuron *3* confine themselves to the intermediate neuropil zone. The axons of the neurons depicted enter the tractus thalamopeduncularis (*thp*). *hab*, Ganglion habenulae. (Combined from two figures of Herrick: 1933b: Fig. 4 and 1934: Fig. 6)

mus of the mud puppy to illustrate some aspects of the development just sketched (Fig. 2.10). In the mud puppy, and in urodeles in general, the dorsal thalamus comprises a homogeneous and continuous zone of periventricular grey, which is flanked by a wide neuropil and fibre zone. Within the latter, separate intermediate and superficial neuropil zones are differentiated as discrete synaptic fields. The afferents of the superficial zone consist primarily of optic tract fibres, whereas the intermediate zone receives mainly exteroceptive somatosensory fibres via the spinal and bulbar lemnisci. Both zones also receive tectothalamic fibres. The dendrites of the periventricular perikarya penetrate the overlying neuropil and fibre zones. Herrick observed that most of the thalamic neurons extend dendritic branches into both the superficial and the intermediate neuropil zones, but a few confine themselves to one or the other. He stated that this incipient segregation increases in the course of phylogeny and is followed by a migration outward of the perikarya from the central grey toward either the intermediate or the superficial neuropil zone. These processes of segregation and migration result in the general somatosensory thalamic nuclear complex and the lateral geniculate body. Herrick observed that in the frog some thalamic neurons are segregated in specific relation to the superficial 'geniculate' neuropil, and that in reptiles the development of the lateral geniculate body is consummated. He emphasized that the sequence of events, as exemplified in the thalamus, is likely repeated in many other parts of the brain.

2.4.6
The Parcellation Theory of Ebbesson

The parcellation theory, enunciated by Ebbesson (1980, 1984), states that during both ontogenesis and evolution nervous systems become more complex, not by one system invading another, but by differentiation and parcellation that involves the selective loss of connections of the newly formed daughter aggregates and subsystems. "Neurons of a given primordial cluster have a greater variety, but less specific input than the more recent clusters, and as differentiation and parcellation of neural systems evolve, some inputs are lost, i.e. differentiation and parcellation involve losing, *not* gaining a variety of inputs" (Ebbesson 1980, p. 206).

From the foregoing it appears that the idea that diffuse and undifferentiated grisea change through time by the differential loss of connections and the subsequent segregation or parcellation of more homogeneous and discrete neuronal populations had already been clearly expressed by Herrick

(1933a,b, 1948), but Ebbesson (1980) does not make any reference to the pertinent work of Herrick.

2.4.7
Chemotaxis (Cajal) and Neurobiotaxis (Ariëns Kappers)

Regarding the causes of the displacement of neurons and the outgrowth of their processes, Ramón y Cajal (1893, 1909) hypothesised that, in addition to mechanical factors, chemical factors may play an important role in these mechanisms. He believed that neuroblasts are endowed with a 'sensibilité chimiotactique' and are able to carry out amoeboid movements under the influence of attracting and repelling substances, secreted by neuronal and glial elements. He also believed that the multiple connections of each neuron "sont le produit de nombreuses influences chimiotactiques, successives et echelonnées" (Ramón y Cajal 1909, p. 658). It should be emphasized that Cajal's theory of chemotaxis sought only to explain the displacement of neuronal elements and the outgrowth of their processes during *ontogenetic* development; more recent neuroembryological investigators (for reviews see Dodd and Jessell 1988; Sanes 1989) have confirmed the correctness of this theory. We discuss here the chemotaxis theory of Cajal because it is often raised in connection with Ariëns Kappers' theory of neurobiotaxis (Ariëns Kappers 1919, 1922, 1928, 1929, 1932, 1941; Ariëns Kappers et al. 1936 and numerous other publications).

Ariëns Kappers observed that the motor nuclei in the brain stem of various vertebrates may show striking differences in position, and that these differences can be correlated with the location of the principal afferent fibre systems of these nuclei. Thus, in cartilaginous fishes the abducens nucleus is located close to the ventricular surface, in the immediate vicinity of the medial longitudinal fasciculus, which is very large in these animals, but in teleosts the same nucleus occupies a submeningeal position, close to the strongly developed tectobulbar tract. According to Ariëns Kappers, the medial longitudinal fasciculus and the tectobulbar tract both represent optic reflex tracts. Ariëns Kappers believed that during phylogeny, when the fibre systems which are afferent to a particular centre change in size, the main dendrites of the cells in that centre will be directed toward the largest source of afferents, and that this orientation of dendrites may be followed by a migration of the neuronal perikarya in the same direction so that the length of the dendrites becomes considerably reduced. He emphasised that this cell migration is selective and occurs only in the direction of fibre groups which have a functional relation with the shifting cells. Thus the abducens nucleus may shift from one path for visual reflexes (the medial longitudinal fasciculus) to a position close to another path for visual reflexes (the tractus tectobulbaris). The enormous increase in the number of taste fibres occurring in the hindbrain of certain teleosts, however, does not have the slightest influence on the eye muscle nuclei, since there is no functional relation between the two. Yet this increase in number of taste fibres does induce migration of the motoneurons supplying the jaws and gills, the activity of which is strongly influenced by gustatory impulses. Ariëns Kappers designated the presumed phylogenetic shifts of neurons as 'neurobiotaxis'. Extrapolating from comparative neuroanatomical data, Ariëns Kappers, his co-workers (e.g., Black 1917a,b, 1920, 1922, Addens 1933) and others, described numerous examples of neurobiotaxis throughout the vertebrate kingdom and in all parts of the brain. In fact, the changes presumed by Herrick (1933b, 1934, 1948) to occur during the phylogenesis of the thalamus and discussed earlier in the present section were considered (Herrick 1933b, p. 441) as being "in accordance with Ariëns Kappers' law of neurobiotaxis."

There can be no doubt that among different vertebrate species striking differences occur in the spatial relations between certain cell masses and their afferent fibre systems, and that these differences, if arranged in a particular sequence, may suggest the changes that have occurred during phylogenesis. To this extent Ariëns Kappers' concept of neurobiotaxis is plausible; however, he went far beyond this by attempting to formulate a causal explanation of the presumed migrations of neuronal populations. He claimed that the outgrowth of dendrites and shifting of neurons takes place only between 'stimulatively' correlated centres, without making clear what this means in an evolutionary context. "Thus the positions and relations of the dendrites and cell bodies of neurons of the central nervous system are regulated in conformity with that law of psychology which has long been known as the law of association. According to this law, the simultaneousness of excitations or their successive occurrence is the leading factor" (Ariëns Kappers et al. 1936, pp. 77, 78). Moreover, he held that the position of neurons and the orientation of their processes is influenced by electric fields generated by active fibre tracts, and that neurobiotaxis thus may be regarded as a manifestation of the general biological phenomenon known as galvanotaxis. Here again, an explanation of how such a tropism could work during phylogenetic development is lacking. Finally, Ariëns Kappers adduced throughout his writings results of

embryological investigations as evidence supporting the correctness of his essentially phylogenetic theory. Many of these results could not be confirmed by later embryological studies (cf. Hamburger 1952, 1955). Because of the conceptual and factual flaws just indicated, we agree with Kuhlenbeck (1970) that Ariëns Kappers' theory of neurobiotaxis must be abandoned as a significant causal explanation for the evolutionary development of structural relations in the central nervous system.

2.5
Diffuse or Reticular Grey

2.5.1
Introductory Notes

During the ontogenesis of the central nervous system neuroblasts migrate away from the periventricular matrix and either aggregate into clusters (see Sect. 2.6) or spread diffusely over part of the neural wall. In a typical diffuse grey the neuronal perikarya are fusiform or multipolar in shape. Their long, poorly ramifying dendrites extend in all directions, but show a preference for a plane perpendicular to the longitudinal axis of the central nervous system. Longitudinally passing fibres surround areas of diffuse grey and many others pass through it. The latter are randomly arranged. Surrounding and through-passing axons issue numerous collaterals, which also have a preference for the transverse plane. Diffuse or reticular areas obey the so-called stacked-chips architectonic principle: the predominant terminal arborisation of entering axons and the dendrites of the reticular cells are arranged in transversely oriented flat, disc-shaped building blocks (Scheibel and Scheibel 1958; Szentágothai 1983; Szentágothai and Réthelyi 1973). Typical examples of diffusely organized grisea include: the lateral hypothalamus, the reticular formation of the brain stem and the zona intermedia of the spinal cord. Some structural features of these three areas will now be briefly discussed.

2.5.2
Some Typical Reticular Formations

The lateral hypothalamus is a cellular continuum extending from the lateral preoptic area to the beginning of the mesencephalic tegmentum. Because the structure of the lateral preoptic area resembles the lateral hypothalamic area, these two are together often designated as the lateral preoptico-hypothalamic continuum (LPHC). Throughout its length the LPHC is traversed by the fasciculus medialis telencephali or medial forebrain

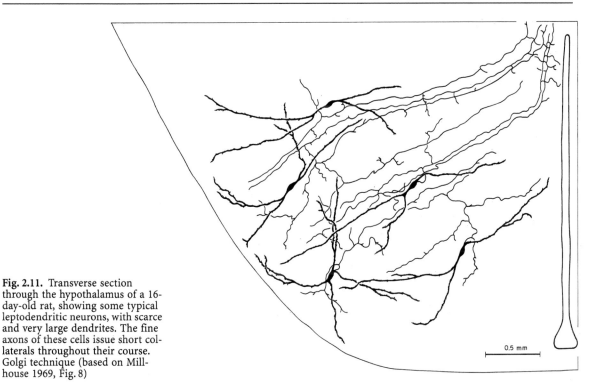

Fig. 2.11. Transverse section through the hypothalamus of a 16-day-old rat, showing some typical leptodendritic neurons, with scarce and very large dendrites. The fine axons of these cells issue short collaterals throughout their course. Golgi technique (based on Millhouse 1969, Fig. 8)

bundle (MFB). The latter is a large system consisting mainly of loosely arranged, thin myelinated and unmyelinated ascending and descending axons, arising from a variety of extrahypothalamic and hypothalamic sites (Nieuwenhuys et al. 1982). The following data on the neurons in the LPHC and their relations to the fibres of the MFB are based on the studies of Millhouse (1967, 1969, 1979), Szentágothai et al. (1968), Szentágothai (1983) and McMullen and Almli (1981).

The neuronal somata in the LPHC are fusiform and taper gradually into two or three dendritic trunks. Their straight and long dendrites are sparse and do not frequently divide (Fig. 2.11). The dendritic trees of most LPHC neurons are oriented perpendicular to the fibres of the MFB, but some ventral elements extend in parallel with that bundle (Figs. 2.12, 2.13). Many of the transversely oriented neurons have the long axis of their somata and the arborisation of their dendrites aligned in a dorsomedial-to-ventrolateral direction (Fig. 2.11). The LPHC elements with their fusiform somata and their 2–5 sparsely ramifying dendrites conform to Ramón-Moliner's (1969, 1975) leptodendritic classification.

The mostly unmyelinated fibres of the MFB make numerous boutons en passant with dendrites of LPHC neurons. The MFB axons issue collaterals throughout their hypothalamic course, which are oriented perpendicular to their parent shafts and are often in parallel with the dendrites of the LPHC neurons they contact (Figs. 2.12, 2.13). Thus, each LPHC neuron receives a substantial input from a few MFB collaterals via repetitive axodendritic synapses and also samples activity from a much greater portion of the bundle via non-repetitive boutons en passant (Millhouse 1969). The axons of the LPHC neurons bifurcate; one axonal division passes caudally with the descending MFB and another division travels rostrally. Both divisions give off numerous collaterals to other LPHC neurons (Millhouse 1969).

The foregoing is not an exhaustive survey of the structural relations in the LPHC. A recent cytoarchitectonic analysis (Geeraedts et al. 1990a,b) has shown that this continuum is not homogeneous, and the Golgi study of McMullen and Almli (1981) has revealed that, apart from the 'typical' LPHC neurons shown in Figs. 2.11 and 2.12, a few other neurons (among them large multipolar elements) do occur in the area under discussion. These variations, however, do not detract from the overall structural principle of interacting transversely oriented dendritic fields and axon collaterals. Concerning the applicability of the 'stacked-chips principle', Szentágothai (1983) remarked that the discs of dendritic and axonal processes are not as flat and regular in the hypothalamus as in the reticular formation or in the spinal intermediate grey, but that this pattern is clearly recognizable.

The reticular formation, or, more precisely, the medial reticular formation, extends throughout the brain stem, consisting of interspersed large triangular or multipolar neurons and bundles of fibres. The neurons have long, straight, radiating dendrites with few branches (Fig. 2.14). They typically belong to Ramón-Moliner's (1962) category of isodendritic neurons. Indeed, the appearance of these neurons is so characteristic that Ramón-Moliner and Nauta (1966) designated the medial reticular formation as the "isodendritic core of the brain stem." Longitudinal sections show that the majority of the dendrites of the reticular cells are oriented in the transverse plane of the brain stem (Fig. 2.15). Due to the great length and the rectilinear course of its dendrites, the medial reticular formation appears in transverse sections as a continuum of overlapping dendritic fields (Fig. 2.14).

Afferents from a variety of sources converge upon the medial reticular formation (Fig. 2.14). The Golgi studies of Scheibel and Scheibel (1958) have shown that the terminal branches and collaterals of the various afferent systems run approximately perpendicular to the long axis of the brain stem, i.e. in the same plane as most dendrites of the reticular cells (Fig. 2.15). These collaterals and terminal branches produce restricted arborisations, which overlap extensively in the medial reticular formation (Fig. 2.14). The efferents of the reticular cells are long, bifurcating axons which take a longitudinal, ascending and descending course and emit numerous collaterals. The patterns of these collaterals vary from simple non-arborising terminals to richly branching plexuses, and a single axon may produce a surprising variety of patterns (Scheibel and Scheibel 1958; Fig. 2.15). From the structure of the medial reticular formation, shown in Fig. 2.14, it is clear that its neurons are under the convergent influence of stimuli from many different sources. As indicated by the Scheibels: "the degree of overlap of the collateral afferent plexuses is so great that it is difficult to see how any specificity of input can be maintained" (Scheibel and Scheibel 1958, p. 34). However, the maximal overlap of afferent fields and dendritic arbors seen in transverse sections (Fig. 2.14) is in marked contrast to the relations observed in sagittal sections. The collaterals from the pyramidal tract afferents, and from a single

pti

fmt

100 μm

Fig. 2.12. Sagittal section through the lateral hypothalamus of a 12-day-old rat, showing some axons of the fasciculus medialis telencephali (*fmt*) and a few typical hypothalamic neurons. The first axons issue numerous short collaterals which remain within the lateral hypothalamus. Other collaterals of the same axons leave the lateral hypothalamus to enter the pedunculus thalami inferior (*pti*). Golgi technique (redrawn from Millhouse 1979, Fig. 6)

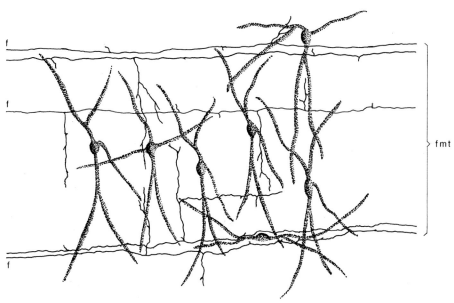

Fig. 2.13. Relations between fibres (*f*) of the fasciculus medialis telencephali (*fmt*) and lateral hypothalamic neurons within its path. By way of the various fmt fibres there is a convergence of neural signals from different sources (e.g., hippocampus, septum, amygdala and olfactory tubercle) upon single neuronal units. (redrawn from Millhouse 1967, Fig. 16)

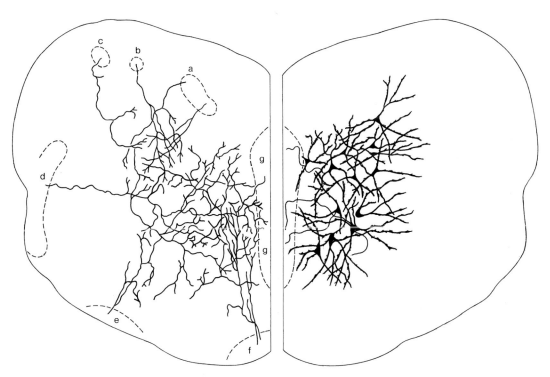

Fig. 2.14. Transverse section through the upper third of the medulla oblongata of a 10-day-old kitten, which illustrates the overlapping of afferents from various sources in the neuropil of the medial reticular formation. The afferents include (*a*) fibres of the descending periventricular system, (*b*) vestibuloreticular fibres, (*c*) fibres from the fasciculus uncinatus, (*d*) fibres from the tractus descendens of the trigeminal nerve, (*e*) spinoreticular fibres, (*f*) pyramidal tract fibres, and (*g*) collaterals of medial reticulospinal tract fibres. Golgi technique (redrawn from Scheibel and Scheibel 1958, Fig. 2). On the *right side*, neurons as observed in the medial reticular formation of a newborn cat have been added (redrawn from Scheibel and Scheibel 1973, Fig. 2)

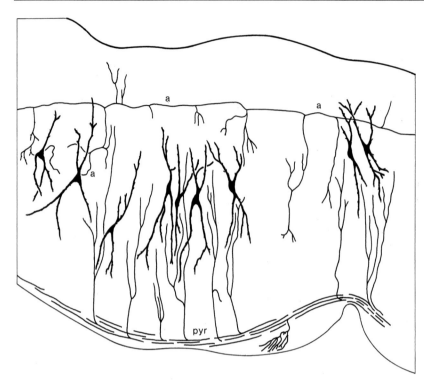

Fig. 2.15. Sagittal section through the rhombencephalon of a 10-day-old rat. Collaterals of the pyramidal tract (*pyr*) and from a single reticular cell axon (*a*) illustrate the tendency toward organisation of the afferent terminals in planes perpendicular to the longitudinal axis of the brain stem. The organisation of reticular dendrites parallels these terminals. Golgi technique. (redrawn from Scheibel and Scheibel 1958, Fig. 3)

reticular cell axon, as shown in Fig. 2.15, illustrate the tendency toward organization of terminals in planes perpendicular to the long axis of the brain stem. Because the reticular dendrites parallel these terminals, the Scheibels thought that "the reticular core might be considered as a series of neuropil segments" (Scheibel and Scheibel 1958, p. 35). According to Szentágothai (1983), the work of the Scheibels shows that the medial reticular formation is organized according to what he called 'the stacked-chips architectonic principle'.

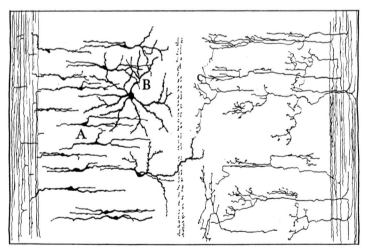
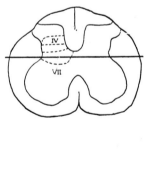

Fig. 2.16. Horizontal section through the base of the dorsal horn of the spinal cord of a 14-day-old cat at S-1, showing some neurons at the *left* and a number of terminal collaterals on the *right*. *A*, Interneurons showing the characteristic transverse dendritic organisation of cells in this region; *B*, single large cell, belonging to the *noyau gris intermediaire* of Cajal, spreading its dendrites in all directions. The collaterals issuing from fibres situated in the deeper parts of the lateral funiculus, which are probably propriospinal in derivation, spread mainly in transverse planes, parallelling the dendrites of the propriospinal neurons. Golgi technique, × 150. (from Scheibel and Scheibel 1968, Fig. 12)

Herrick (1933b, 1934, 1948) regarded certain specialized fields of neuropil as the direct forerunners of specific nuclei. Ramón-Moliner (1969, 1975) and Ramón-Moliner and Nauta (1966) attributed a similar primordial or matrix function to the reticular formation. These authors postulated that "the isodendritic core represents a pool of pluripotential neurons which in the course of phylogeny have remained relatively undifferentiated and in charge of processing afferent signals of very heterogeneous origin" (Ramón-Moliner and Nauta 1966, p. 311). The reticular formation was thought "to constitute a pool of pluripotential neurons which, in the course of evolution, have remained diffusely distributed throughout the brain stem, while retaining generalized morphological features. The usually more circumscript allodendritic and idiodendritic cell groups would have become segregated from this isodendritic pool in the course of phylogeny, probably as a result of being monopolized by specific functions and connexions" (Ramón-Moliner 1969, p. 66). The term 'phylogenetic segregation' was used to denote this assumed course of events (Ramón-Moliner and Nauta 1966).

The third region in the mammalian central nervous system, in which the so-called stacked-chips principle can be clearly recognized, is the zona intermedia of the spinal cord. In this region, which encompasses the laminae V through VIII of Rexed, the dendritic arborisations of many interneurons are disc-shaped; in transverse Golgi sections they show a radiate, isodendritic pattern, but in longitudinal sections these dendritic trees are clearly flattened (Fig. 2.16). The preterminal and terminal axon collaterals which enter the intermediate zone of the spinal cord have, like the dendrites, a preference for the transverse plane (Scheibel and Scheibel 1968; Réthelyi 1976; Szentágothai 1983; Szentágothai and Réthelyi 1973; Fig. 2.17). These collaterals which arise mainly from the axons of propriospinal neurons situated at other levels of the cord run directly through large regions of the intermediate grey, giving one or a few synaptic knobs to many neurons situated in widely different areas. These structural relations indicate (Szentágothai 1964) that the internuncial network of the spinal cord is established by a synaptic system having both tremendous divergence and convergence.

From the foregoing it appears that three different grisea in the mammalian central nervous system, viz. the lateral hypothalamus, the reticular formation of the brain stem and the intermediate zone of the spinal cord, consist essentially of rostrocaudally arranged discs of dendrites and axonal processes. Note that these three structures form part of a single continuum, and the interneuronal interactions within the flattened, chiplike units of this continuum are effected mainly by non-repetitive synapses *en passant* and by repetitive synaptic contacts between axonal branches and dendritic ramifications running in parallel. Specialised synaptic formations are rare in diffuse grisea (Scheibel and

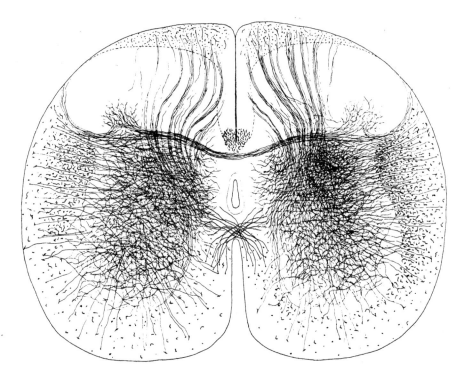

Fig. 2.17. Transverse section through the lumbar spinal cord of a 2-day-old cat, showing the pattern of collaterals. (from van Gehuchten 1891, Fig. 19)

Scheibel 1958; Szentágothai 1964; Szentágothai et al. 1968). The question arises whether the corresponding areas in the central nervous system of other vertebrate groups show a structural organization like that in the diffuse grisea of mammals, but we do not know because adequate Golgi studies are lacking.

2.6
Nuclei

2.6.1
Introductory Notes

Since Nissl developed his staining method more than 100 years ago (Nissl 1885; cf. also Kreuzberg 1984), it has been one of the most powerful tools in neuroanatomy. It is an indispensable first step in the analysis of any part of the CNS. The method reduces the baffling intricacy of nervous tissue to a representation that shows only the neuronal perikarya, leaving the feltwork of dendrites, axons and terminals hidden. This single technique has led to the birth of a separate neuroanatomical discipline, cytoarchitectonics, which aims to subdivide the grey matter of the central nervous system into adjacent provinces, called grisea. The latter include periventricular grisea, diffuse or reticular grisea, cell masses or nuclei, and areas of laminated grey matter, including cortices. The boundaries between these grisea are based on differences in size, shape, density and stainability of their perikarya. Periventricular and diffuse grisea have already been dealt with, and laminated grey matter will be covered later in this chapter; this section will focus on nuclei, defined as *delineable clusters of neuronal somata*. The application of the Nissl method is an essential preliminary to the analysis of a region of the CNS. In order to illustrate this procedure, pairs of photomicrographs of Nissl-stained transverse sections through various brain regions of a number of vertebrates are presented (Figs. 2.18–2.29). In the left panel of each pair the cellular structure is shown as such, but in the right panel the cytoarchitectonic boundaries between grisea are indicated, so the reader can decide whether to agree or not with the delineations drawn. There is another reason for presenting these photomicrographs here. In the second part of the present work, drawings of sequences of transversely oriented microscopical sections through the spinal cord and brain are presented. In the left half of these drawings the cellular pattern is depicted semidiagrammatically (for details on the procedure followed see Chap. 8). From the series of photomicrographs the reader can compare the original material with our final schematization.

With regard to their position, the nuclei in the central nervous system can be placed in three broad categories: periventricular, intermediate and superficial.

2.6.2
Periventricular Nuclei

Periventricular nuclei occur in two different configurations. Either they are embedded in a diffuse central grey, or the central grey is locally entirely broken up in separate cell masses. Thus, in the South American lungfish *Lepidosiren paradoxa*, as well as in the green frog *Rana esculenta*, the motor facial nucleus and the nucleus reticularis medius are embedded in a zone of diffuse central grey (Figs. 2.19, 2.20). The breaking up of the central grey into separate cell masses is very common. In the series presented it can seen in the basal plate of the rhombencephalon of the lamprey (Fig. 2.18), in the rhombencephalic alar plate of *Rana* (Fig. 2.20), in the midbrain of the bowfin *Amia calva* (Fig. 2.22), in the diencephalon of the lamprey (Fig. 2.23), the reedfish *Erpetoichthys calabaricus* (Fig. 2.24), the false featherfin *Xenomystus nigri* (Fig. 2.25), the frog (Fig. 2.26) and the turtle *Testudo hermanni* (Fig. 2.27), and finally in the dorsal ventricular ridge and in the septum of *Testudo* (Fig. 2.29). Between a condition in which the periventricular zone is occupied entirely by cell masses and the presence of a completely homogeneous

Fig. 2.18. Transverse section through the rostral part of the ▶ rhombencephalon of the lamprey, *Lampetra fluviatilis*. The periventricular grey is broken up into a number of distinct separate cell masses. The nucleus octavomotorius intermedius, the nucleus of the radix descendens nervi trigemini and a reticular cell group in the medial part of the tegmentum belong in the category of intermediate nuclei. *dors*, Nucleus dorsalis areae octavolateralis; *gcrh*, griseum centrale rhombencephali; *int*, nucleus intermedius areae octavolateralis; *Mü*, cell of Müller; *nuomi*, nucleus octavomotorius intermedius; *nurdV*, nucleus of the radix descendens nervi trigemini; *rm*, nucleus reticularis medius; *Vm*, nucleus motorius nervi trigemini. Nissl stain

Fig. 2.19. Transverse section through the intermediate part of the rhombencephalon of the lungfish, *Lepidosiren paradoxa*. The alar plate contains two lateral line centres, the nucleus dorsalis and the nucleus intermedius of the area octavolateralis, both of which comprise a periventricular part and an intermediate part. The nucleus vestibularis magnocellularis and the nucleus raphes superior are typical intermediate nuclei. The nucleus reticularis medius and the rostral part of the motor facial nucleus are embedded in the small-celled, diffuse periventricular grey of the basal plate. *dors*, Nucleus dorsalis areae octavolateralis; *int*, nucleus intermedius areae octavolateralis; *lrz*, lateral reticular zone; *ras*, nucleus raphes superior; *rm*, nucleus reticularis medius; *vem*, nucleus vestibularis magnocellularis; *VIImr*, nucleus motorius nervi facialis, pars rostralis. Nissl stain

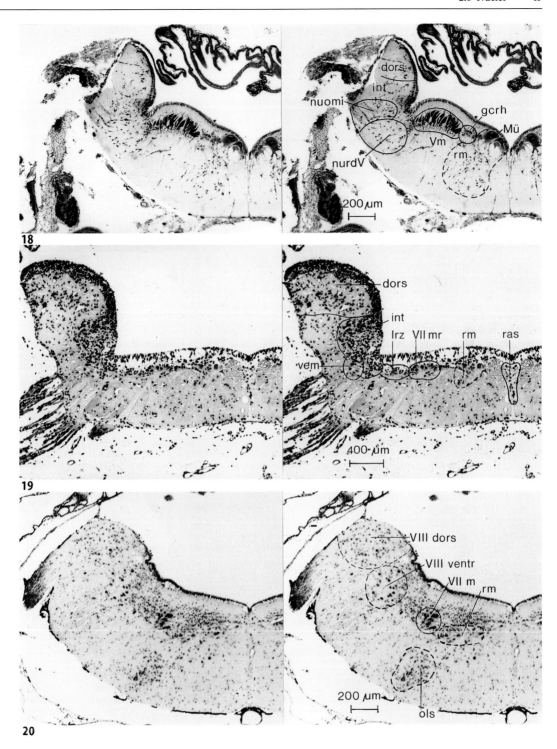

Fig. 2.20. Transverse section through the rostral part of the rhombencephalon of the frog, *Rana esculenta*. The periventricular grey of the alar plate is largely broken up into two conspicuous cell masses, the dorsal and ventral octavus nuclei. The periventricular grey of the basal plate contains two nuclei, the nucleus reticularis medius and the motor facial nucleus. The neuropil and fibre zone of the basal plate contains one distinct nucleus, the superior olive. *ols*, Oliva superior; *rm*, nucleus reticularis medius; *VIIm*, nucleus motorius nervi facialis; *VIIIdors*, nucleus dorsalis nervi octavi; *VIII-ventr*, nucleus ventralis nervi octavi. Nissl stain

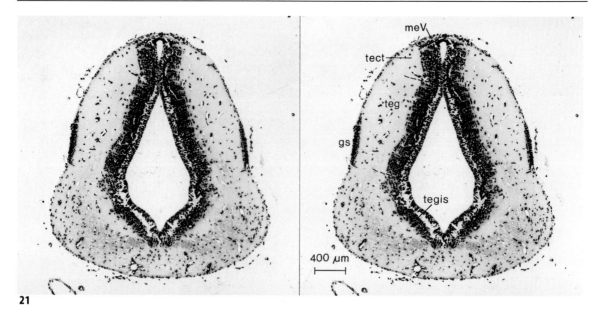

21

Fig. 2.21. Transverse section through the caudal part of the mesencephalon, and through the isthmus region of the lungfish, *Lepidosiren paradoxa*. The ventricular cavity is surrounded by a continuous zone of central grey, which comprises, from *dorsal* to *ventral*, the mesencephalic trigeminal nucleus, the tectal cell layer and the cellular zones of the tegmentum mesencephali and the tegmentum isthmi. The com-

the pact griseum superficiale isthmi et mesencephali belongs, as its name indicates, in the category of superficial nuclei. *gs*, Griseum superficiale isthmi et mesencephali; *meV*, nucleus mesencephalicus nervi trigemini; *tect*, tectum mesencephali; *teg*, tegmentum mesencephali; *tegis*, tegmentum isthmi. Nissl stain

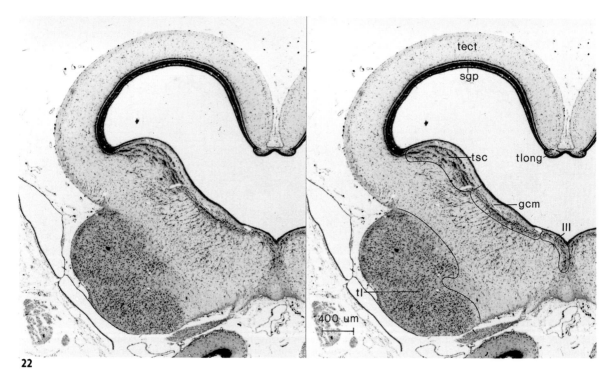

22

Fig. 2.22. Transverse section through the mesencephalon of the bowfin, *Amia calva*. The periventricular grey is broken up into five zones or nuclei, i.e. the nucleus tori longitudinalis, the periventricular grey layer of the tectum, the torus semicircularis, the mesencephalic central grey and the oculomotor nucleus. The huge nucleus of the torus lateralis belongs in

category of superficial nuclei. *gcm*, Griseum centrale mesencephali; *sgp*, stratum griseum periventriculare; *tect*, tectum mesencephali; *tl*, nucleus tori lateralis; *tlong*, torus longitudinalis; *tsc*, torus semicircularis; *III*, nucleus nervi oculomotorii. Nissl stain

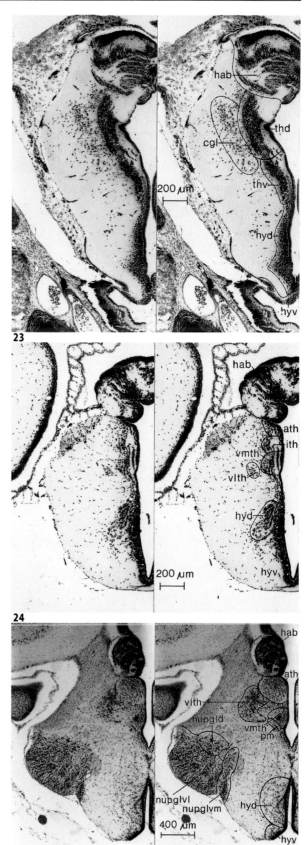

Fig. 2.23. Transverse section through the diencephalon of the lamprey, *Lampetra fluviatilis*. The periventricular grey can be subdivided into five different cell zones: the ganglion habenulae, the dorsal thalamus, the ventral thalamus, the dorsal hypothalamic nucleus and the ventral hypothalamic nucleus. The large lateral geniculate body belongs in the category of intermediate nuclei. Note that the ventral thalamus is clearly heterogeneous. Its ventral part contains a few dark, triangular elements which are clearly larger than those in the remainder of this cell mass. *cgl*, Corpus geniculatum laterale; *hab*, ganglion habenulae; *hyd*, nucleus dorsalis hypothalami; *hyv*, nucleus ventralis hypothalami; *thd*, thalamus dorsalis; *thv*, thalamus ventralis. Nissl stain

Fig. 2.24. Transverse section through the diencephalon of the reedfish, *Erpetoichthys calabaricus*. The diencephalon contains at the level of this section, apart from a number of periventricularly situated cell masses, only a single small cell mass belonging in the category of intermediate nuclei, viz. the ventrolateral thalamic nucleus. *ath*, Nucleus anterior thalami; *hab*, ganglion habenulae; *hyd*, nucleus dorsalis hypothalami; *hyv*, nucleus ventralis hypothalami; *ith*, nucleus intermedius thalami; *vlth*, nucleus ventrolateralis thalami; *vmth*, nucleus ventromedialis thalami. Nissl stain

Fig. 2.25. Transverse section through the diencephalon of the false featherfin, *Xenomystus nigri* (Teleostei). The diencephalon contains at the level of this section not only a number of periventricular cell masses and a large intermediate griseum, the ventrolateral thalamic nucleus, but in addition a large submeningeally situated group of cell masses, the preglomerulosus complex. *ath*, Nucleus anterior thalami; *hab*, ganglion habenulae; *hyd*, nucleus dorsalis hypothalami; *hyv*, nucleus ventralis hypothalami; *nupgld*, nucleus preglomerulosus dorsalis; *nupglvl*, nucleus preglomerulosus ventrolateralis; *nupglvm*, nucleus preglomerulosus ventromedialis; *pm*, nucleus preopticus magnocellularis; *vlth*, nucleus ventrolateralis thalami; *vmth*, nucleus ventromedialis thalami. Nissl stain

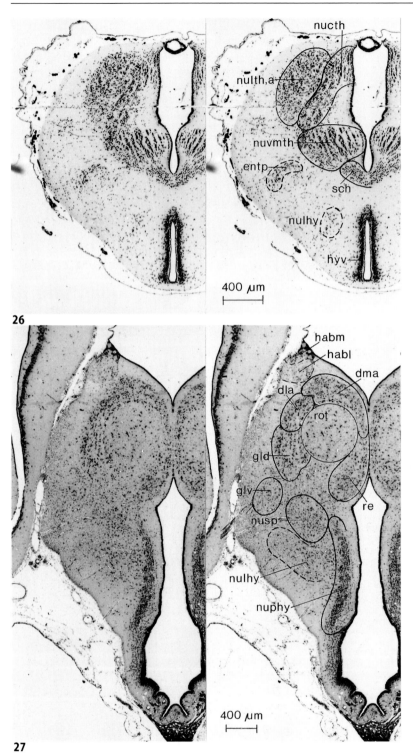

Fig. 2.26. Transverse section through the intermediate part of the diencephalon of the frog, *Rana esculenta*. The periventricular zone is occupied by a number of distinct cell masses, viz. the central thalamic nucleus, the ventromedial thalamic nucleus, the suprachiasmatic nucleus and the ventral hypothalamic nucleus. The posterior entopeduncular nucleus and the lateral hypothalamic nucleus belong in the category of intermediate nuclei. The large lateral thalamic nucleus may be considered either as a periventricular cell mass or as an intermediate nucleus. The interpretation of the cell masses is based on Neary and Northcutt (1983). *entp*, Nucleus entopeduncularis posterior; *hyv*, nucleus ventralis hypothalami; *nucth*, nucleus centralis thalami; *nulhy*, nucleus lateralis hypothalami; *nulth,a*, nucleus lateralis thalami, pars anterior; *nuvmth*, nucleus ventromedialis thalami; *sch*, nucleus suprachiasmaticus. Nissl stain

Fig. 2.27. Transverse section through the diencephalon of the turtle *Testudo hermanni*. The diencephalon contains at the level of this section a number of fairly distinct cell masses. The medial habenular nucleus, the dorsomedial anterior nucleus, the reuniens nucleus and the periventricular hypothalamic nucleus may be considered as belonging to the periventricular zone, whereas the lateral habenular nucleus, the dorsolateral anterior nucleus, the nucleus rotundus, the dorsal and ventral parts of the lateral geniculate nucleus, the suprapeduncular nucleus and the lateral hypothalamic nucleus may be categorised as intermediate nuclei. The interpretation of the cell masses is based on Powers and Reiner (1980). *dla*, Nucleus dorsolateralis anterior; *dma*, nucleus dorsomedialis anterior; *gld*, nucleus geniculatus lateralis, pars dorsalis; *glv*, nucleus geniculatus lateralis, pars ventralis; *habl*, nucleus habenularis lateralis; *habm*, nucleus habenularis medialis; *nulhy*, nucleus lateralis hypothalami; *nuphy*, nucleus periventricularis hypothalami; *nusp*, nucleus suprapeduncularis; *re*, nucleus reuniens; *rot*, nucleus rotundus. Nissl stain

central grey, many gradations occur. Thus, in the midbrain and isthmus region of *Lepidosiren* (Fig. 2.21) the structural differences between the various cellular areas which surround the ventricle are minimal, and the boundary between the tectal and the tegmental grey must be based on an additional, non-cytoarchitectonic criterion such as the ventral extent of the tectal layer of optic fibres. In the urodele telencephalon (Fig. 2.28) a number of periventricular cellular areas can be distinguished, but the cytoarchitectonic differences between areas are slight and the exact boundaries hard to determine.

2.6.3
Intermediate Nuclei

Intermediate nuclei are embedded in the neuropil and fibre zone which externally covers the periventricular grey. Typical examples include the nucleus octavomotorius intermedius and the nucleus of the radix descendens of the trigeminal nerve in the rhombencephalon of the lamprey (Fig. 2.18), the lateral geniculate body in the diencephalon of the same animal (Fig. 2.23), the magnocellular vestibular nucleus in *Lepidosiren* (Fig. 2.19), the superior olive (Fig. 2.20), the posterior entopeduncular and lateral hypothalamic nuclei in *Rana* (Fig. 2.26), the ventrolateral thalamic nucleus in the reedfish *Erpetoichthys calabaricus* (Fig. 2.24), a whole series of nuclei in the diencephalon of *Testudo* (Fig. 2.27) and the medial septal nucleus in the telencephalon of the tiger salamander (Fig. 2.28) and that of the turtle (Fig. 2.29).

In the above examples, the intermediate cell masses are surrounded by the neuropil and fibre zone. Frequently, however, such cell masses lie against the periventricular grey or are even directly continuous with corresponding zones within that grey. An example occurs in the telencephalon of the tiger salamander (Fig. 2.28), where the zone of peripherally migrated 'intermediate' cells of the primordium hippocampi cannot be delineated from the periventricular zone. Examples of the positioning of intermediate nuclei directly against the periventricular grey are the lateral line nuclei in the lamprey (Fig. 2.18) as well as in *Lepidosiren* (Fig. 2.19), the ventrolateral thalamic nucleus in the false featherfin *Xenomystus nigri* (Fig. 2.25), the lateral thalamic nucleus in the frog (Fig. 2.26) and the lateral part of the dorsal ventricular ridge in the turtle (Fig. 2.29).

2.6.4
Superficial Nuclei

Superficial nuclei occupy a submeningeal position and are generally situated peripheral to the large fibre streams that occupy a central position in the wall of the neuraxis. Classical examples include the inferior olive, the pontine nuclei and certain parts of the dorsal thalamus, such as the medial and lateral geniculate bodies, in the mammalian brain. It is remarkable, however, that such superficial nuclei also occur in the brain of most groups of non-mammalian vertebrates. The griseum superficiale isthmi et mesencephali of dipnoans (Fig. 2.21), the huge nucleus tori lateralis in the holostean *Amia* (Fig. 2.22) and the preglomerulosus complex occurring in osteoglossomorphs and other teleost groups (Fig. 2.25) are selected to illustrate this point.

2.6.5
A Note on Cytoarchitectonics

The goal of cytoarchitectonics, i.e. the delineation of grisea, is to reveal units of biological significance (e.g., Olszewski and Baxter 1954). In many parts of the brain, however, it is difficult to draw unambiguous cytoarchitectonic boundaries. In the many neurobiological cytoarchitectonic atlases boundaries are nevertheless shown, but the reasons for locating the boundaries are rarely made explicit (see Swanson 1984).

Apart from the inherent subjectivity of the cytoarchitectonic approach per se, it is doubtful that the grey matter consists solely of adjacent portions and is built entirely as a kind of three-dimensional jigsaw puzzle. Examination of Nissl preparations frequently reveals that particular populations of perikarya which are easily distinguishable by their size, shape, or stainability do not respect the boundaries as drawn in cytoarchitectonic atlases.

2.6.6
Open and Closed Nuclei

The Nissl technique is the first step in the study of the organization of the grey matter. Golgi analysis then enables us to take the configuration of the dendritic trees of the constituent neurons of cell masses into consideration. This dendroarchitectonic approach has led Mannen (1960) to a subdivision of nuclei into two groups: open nuclei (*noyaux ouverts*) and closed nuclei (*noyaux fermés*). In open nuclei the dendrites cross the cytoarchitectural limits of the nucleus and penetrate neighbouring territories. Thus, the actual domain of the nucleus is

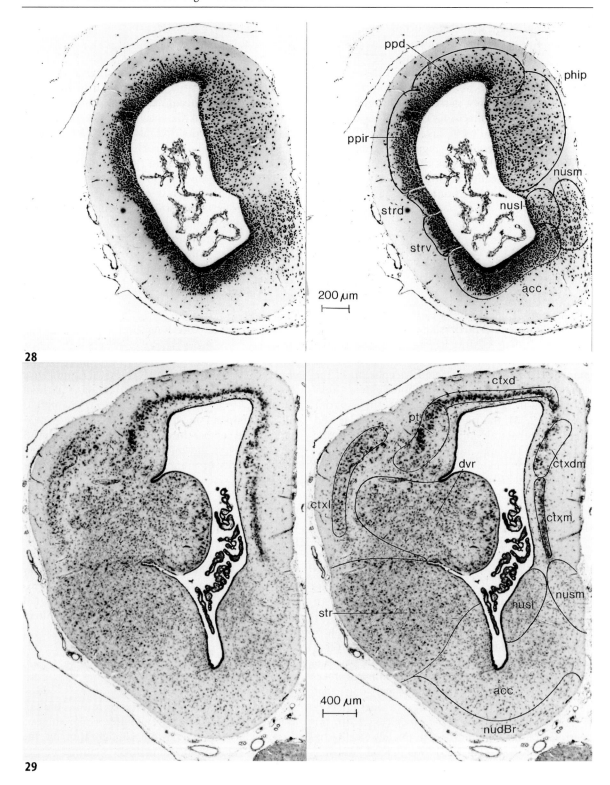

much larger than that observed in Nissl preparations. Most open nuclei are, in their turn, invaded by the dendrites of adjoining nuclei. The dendrites of closed nuclei, on the other hand, do not invade the neighbouring territories significantly, nor is their territory significantly invaded by dendrites of the neighbouring grey matter. Some examples will now be given.

In most mammals the nucleus ventromedialis hypothalami (VMH) is a discrete collection of neuronal somata which is situated in the ventral part of the medial hypothalamic area. It is separated from the surrounding areas by a relatively cell free zone (Fig. 2.30). Golgi studies (Millhouse 1969, 1973a,b, 1978, 1979) have shown that the VMH neurons have

◀ **Fig. 2.28.** Transverse section through the intermediate part of the telencephalon of the tiger salamander, *Ambystoma tigrinum*. The lateral ventricle is surrounded by a continuous zone of periventricular grey which, in accordance with Herrick (1927, 1948), is subdivided into seven areas, the primordium hippocampi, the primordium pallii dorsalis, the primordium piriforme, the dorsal and ventral parts of the striatum, the nucleus accumbens and the lateral septal nucleus. The structural differences between the various areas situated in the lateral hemisphere wall are only slight. In the medial pallial sector, which is occupied by the primordium hippocampi, the compact central grey layer is dispersed by outward migration of the cells throughout almost the entire thickness of the wall. Herrick (1948) regarded this dispersal of cells as a first step toward cortical differentiation. The peripheral zone of the primordium hippocampi could be considered as belonging to the category of intermediate nuclei. The medial septal nucleus, which is located in the medial sector of the subpallium, belongs to the same category. *acc*, Nucleus accumbens; *nusl*, nucleus septi lateralis; *nusm*, nucleus septi medialis; *phip*, primordium hippocampi; *ppd*, primordium pallii dorsalis; *ppir*, primordium piriforme; *strd*, corpus striatum, pars dorsalis; *strv*, corpus striatum, pars ventralis. Nissl stain

Fig. 2.29. Transverse section through the intermediate part of the telencephalon of the turtle, *Testudo hermanni*. Throughout most of the pallium the neuronal perikarya have migrated away from the intermediate vicinity of the ventricular surface to form a number of true cortical fields. The remainder of the pallium is occupied by a large cell mass known as the dorsal ventricular ridge. This cell mass comprises a periventricular component as well as a component belonging to the intermediate zone. In the subpallium the cells are dispersed over the entire width of the hemisphere wall. Cytoarchitectonic boundaries here are hard to draw. However, in our opinion the five cell masses indicated can be recognised with sufficient clarity. The medial septal nucleus may be considered as belonging to the category of intermediate nuclei, whereas the nucleus of the diagonal band of Broca belongs in the superficial zone. The interpretation of the cell masses is based mainly on Powers and Reiner (1980). *acc*, Nucleus accumbens; *ctxd*, cortex dorsalis; *ctxdm*, cortex dorsomedialis; *ctxl*, cortex lateralis; *ctxm*, cortex medialis; *dvr*, dorsal ventricular ridge; *nudBr*, nucleus of the diagonal band of Broca; *nusl*, nucleus septi lateralis; *nusm*, nucleus septi medialis; *pt*, pallial thickening; *str*, corpus striatum. Nissl stain

a few long, straight and infrequently branching spiny dendrites. Most of these processes extend outside the nucleus into adjacent hypothalamic areas, where they overlap extensively with those of adjoining cellular groups (Fig. 2.31). The nucleus is surrounded by a fibre capsule, composed of fibres from various hypothalamic and extrahypothalamic sources. Accordingly, the dendritic tree of each VMH neuron can be subdivided into a central or nuclear portion, an intermediate or capsular portion and a peripheral or extranuclear portion. The nuclear portions of the dendritic trees presumably receive afferents from neighbouring VMH cells, but the capsular and extranuclear portions receive input from a variety of sources. Neurons like those found in the VMH are not unique to this nucleus but have a ubiquitous distribution in the hypothalamus. They belong to a type characterized (Millhouse 1978) as the 'generalized hypothalamic neuron'. The dendritic fields of the various hypothalamic areas constitute a continuum, and it is only by its greater cell density and by its capsule that the VMH can be distinguished from adjacent areas. There is only a slight structural difference between the VMH and the adjacent diffuse, reticular lateral hypothalamic nucleus.

The structural motif seen in the VMH is common in the central nervous system. Mannen (1960) indicated that most motor nuclei in the brain stem consist of radiate neurons extending their dendrites far beyond the cytoarchitectonic boundaries of these centres. Ramón-Moliner and Nauta (1966) surmised that the typical open character of motor and many other nuclei is related to the heterogeneity of their afferent relationships.

The typical feature of closed nuclei is that the dendrites of their constituent neurons do not extend beyond their cytoarchitectural boundaries. Well-known examples include Clarke's column in the lower thoracic and the lumbar spinal cord (Fig. 2.32a) and the inferior olive (Fig. 2.32b–d). The latter consists of neurons with highly branching and curved dendrites, which often turn back toward the soma (Ramón y Cajal 1909; Scheibel and Scheibel 1955; Foster and Peterson 1986). Figures 2.2–2.32c,d show that in the inferior olive the discreteness, i.e. the separation of arborisation in space, while valid for the entire nucleus, does not apply to individual cells.

Mannen (1960) stated that not only Clarke's column and the inferior olive, but all precerebellar nuclei are of the closed type, as are the cerebellar nuclei and many sensory nuclei in the brain stem. Ramón-Moliner and Nauta (1966) designated the closed nuclei, in which dendrites do not extend into adjacent fibre bundles, as 'hodophobic'. They

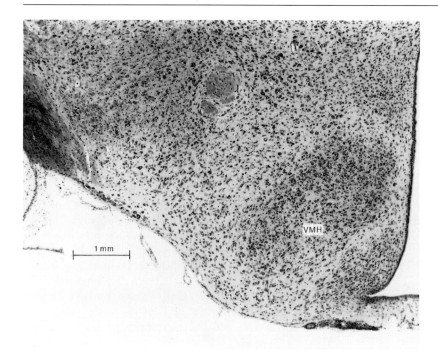

Fig. 2.30. Transverse section through the hypothalamus of the rat to illustrate the position and the cytoarchitecture of the nucleus ventromedialis hypothalami (*VMH*). Klüver-Barrera technique

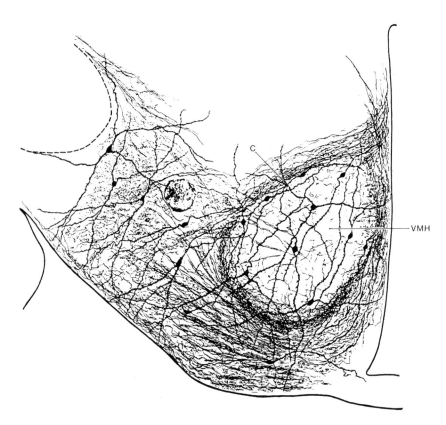

Fig. 2.31. Transverse section through the hypothalamus of a 10-week-old mouse. The nucleus ventromedialis hypothalami (*VMH*) is set off from the rest of the hypothalamus by a fibre capsule (*C*). Because the dendrites of the nucleus ventromedi-alis hypothalami cells reach far beyond this capsule, the cell mass belongs to the so-called open nuclei. Golgi technique. (from Millhouse 1973a, Fig. 1)

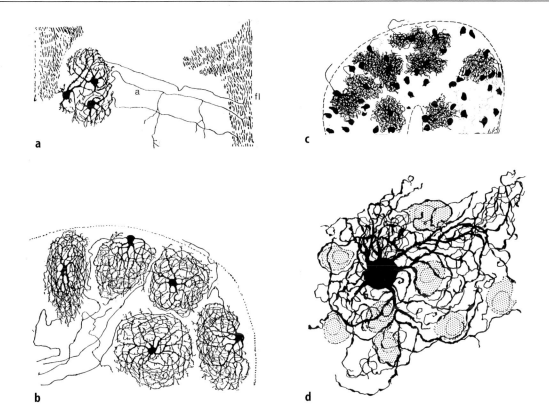

Fig. 2.32a–d. The structure of 'closed' nuclei as revealed by Golgi preparations. **a** Transverse section through part of the lower thoracic cord of a newborn mouse, showing some cells of the column of Clarke and their axons (*a*) which enter the lateral funiculus (*fl*). **b** Cells in the inferior olive of a newborn human infant. **c** Groups of well-impregnated cells in the inferior olive of a human infant; the cell bodies are drawn from a superimposed thionine preparation. **d** Inferior olive cell with well-developed dendritic arbor containing the cell bodies of at least eight adjacent olive cells. (from Ramón y Cajal 1909: Figs. 145A and 415B, and from Scheibel and Scheibel 1955: Figs. 13C and 14D)

believed that the closed nuclei with their allodendritic and idiodendritic cells (cf. Ramón-Moliner 1962) had gradually become segregated from a pool of pluripotential isodendritic neurons, which in the course of evolution remained diffusely distributed throughout the brain stem. In their opinion, closed centres, composed of cells with specialised dendritic trees, appeared during phylogenetic development whenever a given group of neurons became dominated by one of the original afferent systems.

Interestingly, closed-cell aggregates may form part of larger complexes. The *glomérulos* (Lorente de Nó 1922) or barrels (Woolsey and Van der Loos 1970) in the somatosensory cortex of rodents are a good example. These structures, which occupy the fourth layer of the somatosensory cortex, consist of a cylinder of densely packed cells surrounding a core of lower cell density. Golgi preparations reveal that the dendrites of the constituent cells are confined to the cytoarchitectural boundaries of the

barrels (Fig. 2.33). Each barrel receives segregated inputs related to an individual vibrissa, or whisker, on the snout. The vibrissae are arranged in a grid-like pattern in which each has a unique position. Within the somatosensory cortex each barrel also occupies a unique position and the topographical arrangement of these structures closely corresponds to that of the vibrissae (Woolsey and van der Loos 1970; Woolsey 1987).

Mixed configurations of open and closed nuclei also occur. Thus, in the hypoglossal nucleus most dendrites are curved and confined to the cytoarchitectural boundary of the nucleus, but some are straight and extend ventrolaterally into the adjacent reticular formation (Fig. 2.34).

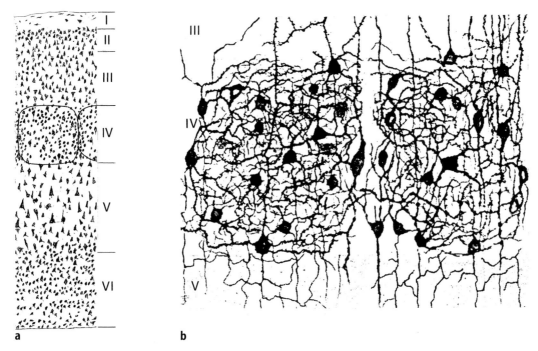

Fig. 2.33a,b. The somatosensory cortex of the mouse. **a** Cytoarchitecture; the fourth layer contains specialised structures consisting of a cylinder of densely packed cells surrounding a region of lower cell density. These structures were designated as glomérulos by Lorente de Nó (1922) and as barrels by Woolsey and van der Loos (1970). **b** Two glomérulos as observed in Golgi preparations. (from Lorente de Nó 1922: Figs. 2 and 8)

2.7
Patterns of Neuronal Interconnections

2.7.1
Introductory Notes

Any centre in the brain or spinal cord contains both neuronal somata and their dendritic trees, the terminal axonal ramifications of other neurons, and, most importantly, the synaptic contacts between neurons engaged in the processing of information within that centre. In many centres the interconnections show spatial order; they are patterned at both the cellular and the supracellular level. In the present section we will focus on this patterning of synaptic contacts.

2.7.2
Distribution of Synaptic Contacts Over the Receptive Surface of Neurons

The patterning of synaptic contacts over the receptive surface of neurons is illustrated by the following 12 examples:

1. The large neuronal elements present in *layer IV of the mammalian spinal cord* have three different sets of dendrites that radiate dorsally, medially and laterally from the soma. Scheibel and Scheibel

(1966, 1968) have shown that these three sets of dendrites receive segregated inputs from three different groups of afferents (Fig. 2.35). The dorsal dendrites receive their presynaptic contacts from the dense terminal plexuses of coarse cutaneous primary afferent fibres. Golgi analysis of sections

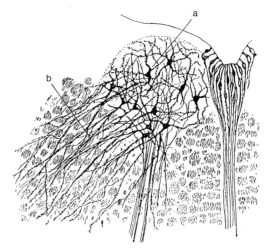

Fig. 2.34. Transverse section through the hypoglossal nucleus of an 8-day-old cat. Most dendrites are curved and remain within the confines of the nucleus (*a*), but some extend ventrolaterally into the adjacent reticular formation (*b*). Golgi method (from Ramón y Cajal 1909: Fig. 292)

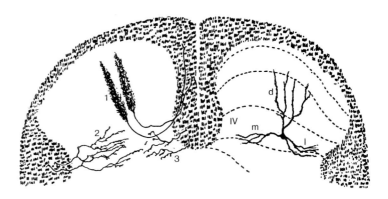

Fig. 2.35. A large secondary sensory neuron in lamina *IV* of the spinal cord of the cat (*right*), provided with dorsal (*d*), medial (*m*) and lateral (*l*) sets of dendrites, and the three principal sets of afferents that innervate its three dendritic domains (*left*): *1*, terminal arborisations of coarse, cutaneous primary afferent fibres; *2*, collaterals from corticospinal fibres; *3*, terminal collaterals emerging from the deepest part of the dorsal funiculus. These collaterals arise presumably from the axons of large lamina IV and lamina V neurons. (from Scheibel and Scheibel 1968, Fig. 5)

cut in different planes reveals that these plexuses consist of narrow, longitudinally arranged neuropil plates. The lateral dendrites are contacted by collaterals and terminals of the lateral corticospinal tract. The medial dendrites receive terminal fibres from the most ventral portion of the dorsal funiculus, an area of intraspinally derived projection elements. Scheibel and Scheibel (1966, 1968) have pointed out that these three sets of dendrites sample input of highly contrasting significance. The dorsal system extracts information from a series of parallel neuropil plates acting as a first level of representation for parallel cutaneous fields upon the surface of the skin. The integrative operations of this system are presumably modulated by cortical (suprasegmental) and spinal (segmental) mechanisms.

2. Another example of segregation of presynaptic terminals upon individual dendrites of single neurons is found in the *medial reticular formation of the brain stem*. The large neurons have long, sparsely branching dendrites which spread in planes oriented perpendicular to the long axis of the brain stem. Within these planes the dendrites radiate out in all directions. Comparison of the pattern of the dendrites with the convergent pattern of collaterals and terminal axonal branches which impinge on them suggests that each major dendrite is under the presynaptic control of one dominant afferent system (Scheibel and Scheibel 1958; Fig. 2.14).

3. Configurations similar to those mentioned under 2 are found in certain parts of the *brains of urodeles*. In the central nervous system of this group most neuronal perikarya are located in a zone of periventricular grey, from where they extend their dendrites into a peripheral neuropil and fibre zone (Fig. 2.2). The latter zone contains, in the rhombencephalic alar plate, several bundles of primary afferent axons (Fig. 2.36). The Golgi studies of Herrick (1948) have shown that in the larval

tiger salamander many alar plate neurons distribute their dendritic branches over three or more such bundles (neurons 1 and 2 in Fig. 2.36). In the diencephalon of the mud puppy, the external layer of the thalamic region is differentiated into separate visual and somatosensory neuropil zones (Fig. 2.10). Most thalamic neurons extend their dendritic branches into both neuropil zones (neurons 1 and 2 in Fig. 2.10). Herrick (1933a,b, 1934, 1948) regarded the 'bi' – or 'polymodal' neurons just discussed as primitive in relation to 'unimodal' neurons, extending their entire dendritic tree within a single domain of afferents (e.g., neuron3 in Figs. 2.10 and 2.36). He conjectured that during evolution monopolisation of neurons by one afferent system precedes the migration of these elements towards the neuropil – or fibre zone from where they receive their afferent impulses, and thus the formation of specific nuclei. Monopolisation of neurons by particular afferent systems has also been suggested as an evolutionary mechanism by which specific nuclei have developed out of a reticular core, consisting of isodendritic elements like those discussed under 2 (Ramón-Moliner 1967, 1975).

4. A good example of a partial segregation of inputs is shown by the *Mauthner cell*. Mauthner cells are a pair of giant neurons found in the reticular formation of many fish and amphibians. They are symmetrically located in the rhombencephalon at the level of entrance of the eighth nerve. Two main dendrites, lateral and ventral, arise from the fusiform soma (Fig. 2.37a,b). The Mauthner cells give rise to a strikingly large pair of myelinated axons which decussate at the level of the cell body and descend through the length of the spinal cord (Fig. 2.37a,c), contacting motoneurons and interneurons. The axon hillock and initial segment are surrounded by a dense, specialized neuropil known as the axon cap. The Mauthner cell is an essential component of the neural network underlying the startle response, i.e. a very rapid escape-and-avoid

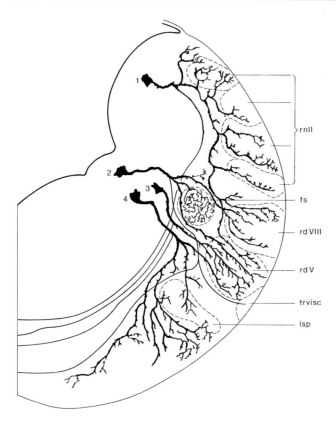

rnll

fs

rd VIII

rd V

trvisc

lsp

Fig. 2.36. Transverse hemisection of the larval rhombencephalon of the tiger salamander, *Ambystoma tigrinum*. Four types of neurons in the sensory alar plate showing preference for one or more sets of primary afferent fibres are depicted. The superficial zone of the alar plate contains, according to Herrick (1914, 1948), seven bundles of primary afferent fibres, i.e. the fasciculus solitarius (*fs*), the radix descendens nervi octavi (*rdVIII*), the radix descendens nervi trigemini (*rdV*), and four bundles of primary afferent lateral line fibres (radices nervi lineae lateralis: *rnll*). Neuron *1* is in synaptic relation with all of the bundles of primary afferent fibres, except for the fasciculus solitarius. Neuron *2* makes its chief connections with general and/or special viscerosensory fibres of the fasciculus solitarius and less intimate connections with octavus and trigeminal fibres. Neuron *3* connects exclusively with root fibres of the trigeminus. Neuron *4* connects with the trigeminus and also with the spinal lemniscus (*lsp*) and the motor zone in the basal plate. The axons of all four cells depicted decussate in the raphe and may ascend in the general bulbar lemniscus. The axon of neuron 2 divides, one branch ascending in the tractus visceralis secundarius (*trvisc*) of the same side and the other crossing to the opposite side. Herrick mentions that some similar neurons connect only with the fasciculus solitarius and have unbranched axons entering the tractus visceralis secundarius only, and adds: "Numberless permutations of the various types of connections here shown have been observed." (modified from Herrick 1948, Fig. 9)

reaction (Faber and Korn 1978; Korn 1987). The morphology and synaptology of the Mauthner cell has been extensively studied in teleosts at the light- (Bartelmez 1915; Bodian 1937) and electron-microscopical levels (Robertson et al. 1963; Nakajima 1974; Nakajima and Kohno 1978). Nakajima (1974) conducted a thorough ultrastructural study of the morphology of synaptic endings on the Mauthner cell of the goldfish. Building on the light-microscopical work of Bodian (1937), he distinguished six categories of terminals (Fig. 2.37e): (a) the large myelinated club endings of posterior eighth nerve fibres, (b) small myelinated club endings, (c) unmyelinated club endings, (d) large-vesicle boutons, (e) small-vesicle boutons, and (f) the spiral club endings. Endings *a* and *b* and a certain proportion of *e* and *f* exhibit both chemical synapses and gap junctions.

Several of the synapse types onto the Mauthner cell are segregated in particular regions of the cell body and dendrites. Thus the large myelinated club endings are located at the distal part of the lateral dendrite, and the unmyelinated club endings establish synaptic contacts on the soma and the axon hillock region, whereas the spiral fibre endings surround and wrap around the axon hillock and the initial segment (Fig. 2.37d). Activation of the poste-

rior eighth nerve and its large myelinated club endings evokes a dual excitatory postsynaptic potential (EPSP) with early and later components produced by electrotonic and chemical synapses, respectively. Some unmyelinated club endings and

Fig. 2.37a–e. The Mauthner cell in the goldfish, *Carassius*. **a** Dorsal view of the brain stem and cervical spinal cord. The position of the Mauthner cells and their axons is indicated. **b** A reconstruction of the Mauthner cells from a series of transverse sections through the brain and cervical spinal cord of a young goldfish. Impregnation according to Bodian. **c** Transverse section through the cervical spinal cord to show the size and position of the Mauthner axons. **d** The synaptic endings and their distribution on the Mauthner cell of the goldfish. **e** The various synaptic endings, as seen with the light microscope (LM, *upper row*) and the electron microscope (EM, *lower row*). *a*, Axon; *ah*, axon hillock; *cb*, cerebellum; *cc*, canalis centralis; *cd*, cap dendrite; *chs*, chemical synaptic junction; *flm*, fasciculus longitudinalis medialis; *fMth*, fibre of Mauthner; *g*, gap junction; *id*, inferior dendrite; *ld*, lateral dendrite; *lmce*, large myelinated club ending; *lobX*, lobus vagi; *lvb*, large-vesicle bouton; *medsp*, medulla spinalis; *Mth*, cell of Mauthner; *nVIII*, nervus octavus; *rm*, nucleus reticularis medius; *sf*, spiral fibres; *sfe*, spiral fibre ending; *smce*, small myelinated club ending; *ss*, symmetrical synapse; *svb*, small-vesicle bouton; *tect*, tectum mesencephali; *tel*, telencephalon; *trvisc*, tractus visceralis secundarius; *uce*, unmyelinated club ending; *vd*, ventral dendrite; *vem*, nucleus vestibularis magnocellularis; *vq*, ventriculus quartus (**d** and **e**, modified from Nakajima and Kohno 1978, Fig. 1)

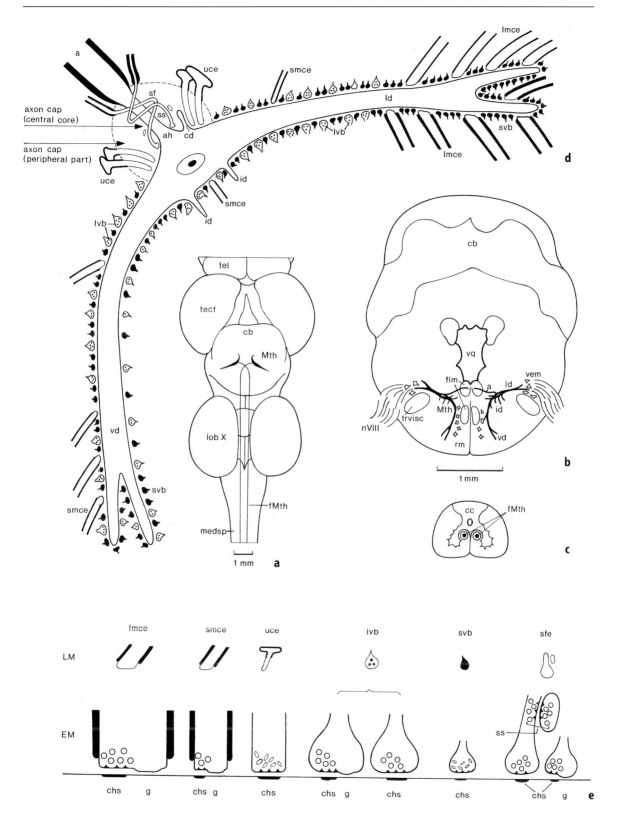

the small vesicle boutons both arise from the same type of inhibitory interneurons, the so-called passive hyperpolarizing potential (PHP)-exhibiting cells (Korn 1987). The spiral fibre endings are also believed to originate from interneurons and to exert an inhibitory action on the initial segment.

5. The *cerebellar Purkinje cells* are representative of neurons in which different zones of the receptive somatodendritic surface receive sharply separated inputs. As can be seen in Fig. 2.38, the Purkinje dendritic tree develops from one side of the soma and extends into the outer, molecular layer of the cerebellar cortex. In mammals and birds the dendritic system of the Purkinje cells is made up of primary, secondary and tertiary branches, compressed into a flattened domain oriented perpendicular to one of the two major cerebellar inputs, the parallel fibre system. The primary and secondary branches have a smooth surface, but the short tertiary branches are densely covered by short, thick spines, hence their designation as spiny branchlets. The Purkinje cells receive their main synaptic input from four sources: (a) Parallel fibres, originating from granule cells in the deepest zone of the three-layered cerebellar cortex, synapse with the spines situated on the tertiary branchlets of the Purkinje cells, and since each fibre makes a single synapse with a Purkinje cell, its excitatory effect on the cell is small. (b) Olivocerebellar fibres, continuing on as climbing fibres, attain the molecular layer and split into a number of branches which climb, ivy-like,

along the dendritic arborisations of the Purkinje cells. Each Purkinje cell receives a single climbing fibre which terminates with serial synapses on the smooth primary and secondary dendrites of the Purkinje cells and so exerts a strong excitatory action, overriding all other inputs, on the Purkinje cell. (c) Stellate cells distributed through the molecular layer receive an excitatory input from the parallel fibers and make inhibitory synapses onto the dendrites of Purkinje cells. (d) Basket cells are found in the deep zone of the cerebellar molecular layer; their axons pass horizontally in the same plane as the dendritic arborisations of the Purkinje cells, and have numerous descending collaterals which form intricate plexuses around the somata of the Purkinje cells. About 20 basket-cell collaterals participate in a single pericellular basket on the axon hillock of the Purkinje cells, where they exert a strong inhibitory action (Eccles et al. 1967). Interestingly, basket cells are found only in birds and mammals (Llinás and Hillman 1969).

6. Another conspicuous neuron exhibiting a prominent proximo-distal segregation of input along its receptive, somatodendritic surface is the *large pyramidal cell*, found in the cornu ammonis of the mammalian hippocampal formation. The cornu ammonis is a simple three-layered cortex consisting of peripheral and central plexiform layers and an intermediate cellular layer. The cellular layer contains, in addition to small elements, the somata of the hippocampal pyramidal cells, which extend

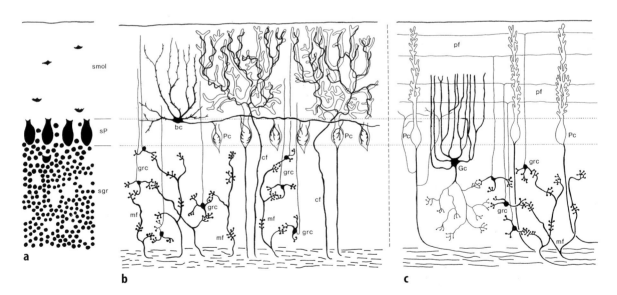

Fig. 2.38a–c. The mammalian cerebellar cortex: **a** cytoarchitecture; **b** neuronal elements as seen in sagittal Golgi sections, and **c** the same in transverse sections. *bc*, Basket cell; *cf*, climbing fibre; *Gc*, Golgi cell; *grc*, granule cell(s); *mf*, mossy fibres; *Pc*, Purkinje cell; *pf*, parallel fibres; *sgr*, stratum granulare; *smol*, stratum moleculare; *sP*, stratum Purkinje. Note that the stellate cells of the molecular layer are omitted from this figure. (**b** and **c**, based on Figs. 3, 27, 36, 91 from Ramón y Cajal 1935)

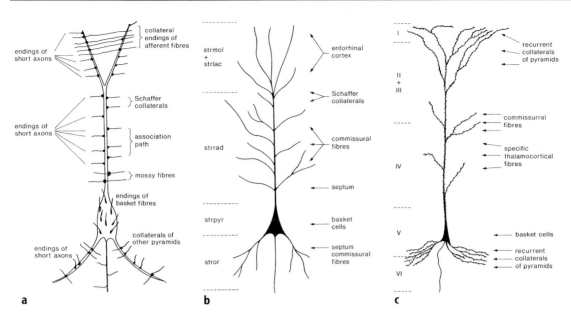

Fig. 2.39a–c. The relative distribution of afferents over cortical pyramidal cells: **a** pyramidal cell of hippocampal field CA3; **b** pyramidal cell of hippocampal field CA1; **c** a large fifth-layer pyramid of the neocortex. **a** is reproduced including its labelling from Fig. 34 of Lorente de Nó (1934). According to our present state of knowledge the apical dendritic stems in field CA3 are considerably shorter than depicted. Moreover, Schaffer collaterals do not occur in this field. **b** is slightly modified from Fig. 355 of Stephan (1975). **c** is reproduced from Scheibel and Scheibel (1970b), Fig. 6b. *strlac*, Stratum lacunosum; *strmol*, stratum moleculare; *stror*, stratum oriens; *strpyr*, stratum pyramidale; *strrad*, stratum radiatum

stout apical dendrites into the peripheral plexiform layer. Some distance from the soma the apical dendrites divide into a number of terminal branches. The apical stem dendrites make up the stratum radiatum, whereas the layer in which the terminal branches of the apical dendrites are situated is known as the stratum lacunosum-moleculare. A number of basal dendrites arise from the soma and fan out in the central plexiform layer or stratum oriens (Fig. 2.39a,b). The axons of the hippocampal pyramidal cells enter a subependymal fibre layer known as the alveus. A certain proportion of these axons descend via the fornix to the lateral septal nucleus, many others pass to the adjacent entorhinal cortex, and still others remain within the hippocampal formation. During their initial course through the stratum oriens, as well as in the alveus, the axons of the pyramidal cells issue numerous collaterals.

Lorente de Nó (1934) was one of the first to note the laminar distribution of the various synaptic afferents over the surface of the hippocampal pyramidal cells. On the basis of Golgi studies, he concluded that collaterals of other pyramidal cells terminate on the basal dendrites, whereas the axons of basket cells, the cell bodies of which are located in the stratum oriens, are the only afferents contacting the soma of the pyramidal cells. The granule cells in the fascia dentata (another part of the hippocampal formation) issue so-called mossy fibres. These end in large bulbous swellings on the apical dendrites near their origin at the pyramidal cell bodies. Schaffer collaterals, the coarse, myelinated branches of the axons of other pyramidal cells, were already known to terminate on the distal end of the apical dendritic shafts, and Lorente de Nó discovered that collateral endings of afferent fibres contact the terminal branches of the apical dendrites. Later experimental and electron-microscopical studies have considerably extended the observations of Lorente de Nó (for review see Stephan 1975). It is now known that septal afferents terminate on the basal dendrites and on the proximal part of the apical dendrites, that commissural fibres end on basal dendrites and on the intermediate part of the shafts of the apical dendrites, and that the extrinsic fibres contacting the terminal branches of the apical dendrites originate from the entorhinal cortex. Electron-microscopical studies (Hamlyn 1963; Blackstad 1967; Gottlieb and Cowan 1972) have shown that the contacts with the pyramidal cells of the mossy fibres, of the Schaffer collaterals and of the entorhinal cortical afferents are of type I, i.e. they have asymmetrical synaptic profiles and spherical vesicles, whereas the axosomatic synapses of the basket cells are type II contacts, with sym-

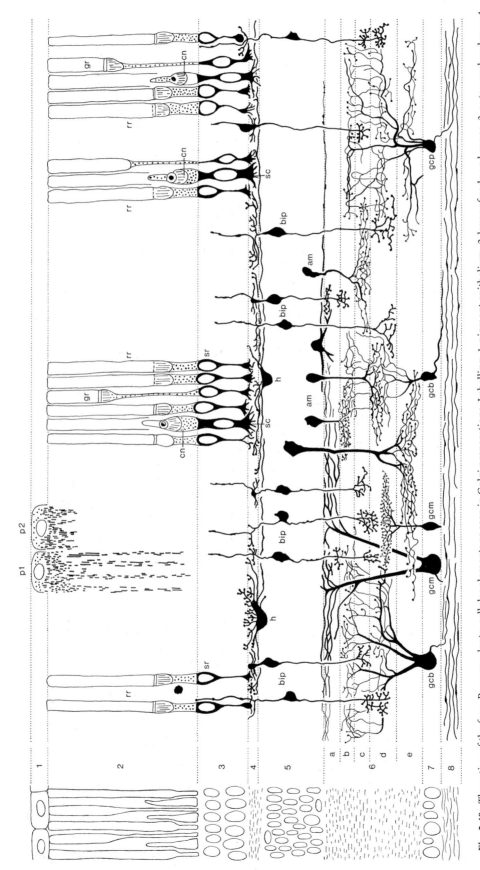

Fig. 2.40. The retina of the frog. Rana esculenta; cellular elements as seen in Golgi preparations. Labelling: *1*, pigment epithelium; *2*, layer of rods and cones; *3*, outer nuclear layer; *4*, outer plexiform layer; *5*, inner nuclear layer; *6*, inner plexiform layer; *a,b*, etc., sublaminae of 6; *7*, ganglion cell layer; *am*, amacrine cells; *bip*, bipolar cells; *cn*, cone; *gcb*, ganglion cell, bistratified; *gcm*, ganglion cell, monostratified; *gcp*, ganglion cell, polystratified; *gr*, green rod; *h*, horizontal cell; *p1*, pigment cell with the pigment migrated into the protoplasmic expansions of the element; *p2*, pigment cell with the pigment largely retracted from the expansions; *rr*, red rod; *sc*, soma of cone; *sr*, soma of rod (based on Ramón y Cajal 1893, Planche II, and Ramón y Cajal 1911, Fig. 218)

metrical membrane profiles and flattened vesicles. As is well known, type I synaptic contacts are considered to be excitatory while type II contacts are considered inhibitory.

7. Experimental data summarised by Scheibel and Scheibel (1970b) indicate that the placement of synaptic contacts along the *large fifth-layer cortical pyramids* of the neocortex show a segregation of input comparable to that seen on the hippocampal pyramids (Fig. 2.39c). The following four categories were described by Scheibel and Scheibel (1970b): (a) Specific thalamocortical afferent systems, such as the geniculocalcarine (visual) radiation, terminate in part directly upon the intermediate third of the shafts of the apical dendrites. (b) Axons originating from a specific locus on one side of the cortex project through the corpus callosum to the corresponding locus in the other side of the cortex. These commissural projections terminate on the oblique branches of the apical dendritic shafts. (c) Recurrent collaterals of pyramidal cells synapse upon the basal dendrites and upon the apical arches of surrounding pyramidal cells. (d) Basket cells produce pericellular envelopes of terminals surrounding the somata of the pyramidal cells which are of type II and hence are probably inhibitory.

8. Segregation of input at the supracellular level

is seen in the deeper layers of complex *retinas*, like that of the frog. In the retina of this animal the inner plexiform layer can be subdivided into five sublaminae. The Golgi studies of Ramón y Cajal (1893) have shown that the retinal ganglion cells distribute their dendrites differentially over these sublaminae. The dendrites of some types of ganglion cells are confined to one sublayer, other cells distribute their dendrites over two sublayers, and still others spread out their dendrites in three sublayers. In the various sublayers of the inner plexiform layer the diverse types of ganglion cells contact the axonal endings of particular sets of bipolar cells and the dendrites of different types of amacrine cells (Fig. 2.40). Thus, the various types of ganglion cells receive input from different sources and, presumably, process information related to various aspects of visual images (Lettvin et al. 1961).

9. Dendritic configurations closely resembling those of the retinal ganglion cells occur in the *tectum mesencephali of many non-mammalian vertebrates.* The tectum of teleost fishes, which has been thoroughly studied (for reviews see Vanegas and Ito 1983; Meek 1983, 1990), is an example.

In most teleosts the tectum is laminated, encompassing from depth to surface (1) the stratum periventriculare (SPV), (2) the stratum album centrale

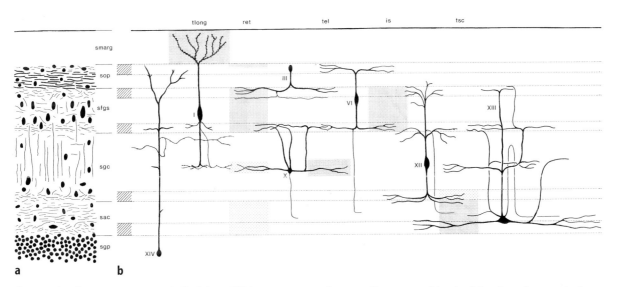

Fig. 2.41a,b. The tectum mesencephali of the goldfish, *Carassius*. **a** Silver-impregnated transverse section through the tectum, showing the various layers. **b** The main tectal neuron types as seen in transverse Golgi preparations. *Cross-hatching* represents plexiform layers. The sites of termination of the more important tectal afferent systems are indicated by *fields of dots*. The sources of these afferents are indicated at the *top*. **b** is based on Meek and Schellart (1978), Figs. 3–19, and the numbering of the cell types is also adopted from these authors. These cell types have been distinguished primarily on the basis of dendritic properties; *I*, large pyramidal neuron extending most of its dendrites into the marginal layer; *III*, monostratified neuron; *VI* and *X*, bistratified neurons; *XII* and *XIII*, multistratified neurons; *XIV*, nonstratified neuron. The neuron types I, III and XIV represent local circuit neurons, whereas the remaining types VI, X, XII and XIII represent efferent neurons. *is*, Nucleus isthmi; *ret*, retina; *sac*, stratum album centrale; *sfgs*, stratum fibrosum et griseum superficiale; *sgc*, stratum griseum centrale; *sgp*, stratum griseum periventriculare; *smarg*, stratum marginale; *sop*, stratum opticum; *tel*, telencephalon; *tlong*, torus longitudinalis; *tsc*, torus semicircularis

(SAC), (3–4) the stratum griseum centrale (SGC), (5) the stratum fibrosum et griseum superficiale (SFGS), (6) the stratum opticum (SOP), and (7) the stratum marginale (SMARG) (Fig. 2.41a). Several of these layers contain one or two plexiform zones, indicated diagrammatically as cross-hatched blocks in Fig. 2.41b. A Golgi analysis of the tectum of the goldfish by Meek and Schellart (1978) revealed 15 different tectal cell types in this species, some of which have been depicted in Fig. 2.41b. The very common neurons of type XIV, which have their perikarya in the SVP, have a long, slender, ascending dendritic stem which reaches the stratum opticum and ramifies there. At various levels the apical dendrite issues short side branches. Meek and Schellart characterised these cells as non-stratified, because their dendrites do not show a preference for particular layers or plexiform zones; in this respect they contrast with most other tectal neurons. Taking this feature as a criterion, Meek and Schellart distinguished monostratified neurons (e.g., type III), bistratified neurons (e.g., types VI and X), and multistratified neurons (e.g., types XII and XIII). The large type I pyramidal neurons, the spiny apical dendritic branches of which ramify profusely in the marginal layer, were placed by Meek and Schellart in a separate category. Four cell types have myelinated axons which leave the tectum (types VI, X, XI and XIII); the remaining types are interneurons.

Experimental studies have shown that the various tectal afferent systems show a laminar distribution. Five of these will be briefly discussed: (a) the retinotectal system, (b) the telencephalotectal system and the afferents from (c) the torus longitudinalis, (d) the torus semicircularis, and (e) the nucleus isthmi. The lamination pattern of these afferent systems is indicated diagrammatically by fields of dots in Fig. 2.41b.

(a) A large contralateral retinotectal tract has been described by many authors for a variety of teleosts. This tract projects massively to the SFGS and the SOP, in particular its superficial part, with some sparse termination in the SGC and the SAC. It is noteworthy that the SFGS also receives afferents from the pretectal region. The latter can be considered a 'visual' region, since it receives bilateral input from the retina.

(b) The telencephalotectal afferents arise from the central part of the area dorsalis telencephali, Dc, and terminate for the major part in the SGC, with a concentration in the intermediate zone of this layer.

(c) Afferents from the torus longitudinalis terminate in a single tectal layer, called the superficial marginal layer. This layer is present only in actinopterygian fishes, as is the torus longitudinalis. The latter is a bilateral, archlike, longitudinally oriented strip of tissue attached to the medial aspect of each tectal half. The torus longitudinalis contains densely packed small granule cells, giving rise to fine unmyelinated axons that pass laterally through the superficial marginal layer. The afferents to the torus longitudinalis originate mostly in the valvula cerebelli, a cerebellar structure also present only in actinopterygians.

(d) Afferents from the torus semicircularis project predominantly to the SAC, with some additional projections to the SPV and the SGC. The torus semicircularis is an important relay centre, situated in the lateral part of the tegmentum of the midbrain. It receives auditory, mechanoreceptive lateral line and, in many teleostean species, electroreceptive lateral line input from sensory centres in the rhombencephalon.

(e) Isthmotectal projections arise from the nucleus isthmi, generally one of the most conspicuous centres in the midbrain tegmentum. The fibres of this projection terminate predominantly in the SFGS. The nucleus isthmi is a 'visual' nucleus, since it receives its input from the 'visual' nucleus pretectalis and from a subtype of the tectal type XIV cell, also considered 'visual'. So the SFGS may be designated as the 'visual' afferent zone, both because of its retinal input and because of afferents from the pretectal region and the nucleus isthmi.

Considering the laminar organization of the tectal afferents and the configuration of the dendritic domains of the tectal neurons, together with data on tectal neurons, Meek (1983) proposed a scheme for the intratectal connectivity in teleosts. This framework includes the following features (Fig. 2.41b): Type I neurons are the tectal afferents for toral (cerebellar) input and they may process some visual (via dendrites in the deep part of the SFGS) and telencephalic input (via dendrites in the SGC). Type III neurons are visual interneurons, since they receive exclusively direct (retinal) and indirect (pretectal, isthmic) visual input. The bulk of type XIV neurons are also visual interneurons. Among the efferent neurons, the bistratified type VI neurons probably transfer visual information beyond the tectum. Type X neurons send telencephalic and toral information (via type I neurons) out of the tectum, integrated with visual information. The large multistratified efferent neurons of type XII and type XIII integrate information from all tectal input systems. For type XIII neurons, however, the deep tectal input is dominant, whereas visual input is more important for type XII than for type XIII neurons.

The considerable degree of connective differentiation shown in Fig. 2.41b suggests that each tectal cell type receives a characteristic sample of the multimodal information available in the different presynaptic zones. Meek (1983) emphasized that his analysis of tectal circuitry was provisional. Meek did survey several electrophysiological studies of the teleostean tectum, but he concluded that few of those results could be correlated with the available morphological data.

10. The last example of a cell type in which different parts of the dendritic system have different synaptic contacts and subserve different functions is the *large mitral cell in the olfactory bulb*. The latter centre has a similar structure in all vertebrates (Nieuwenhuys 1967a). The slender olfactory receptor cells are situated in the nasal mucosa and, contrary to most other sense cells, they have an axon, and so are designated as neurosensory cells. Their axons, which are extremely fine, follow a peripheral course, enter the olfactory bulb, and terminate in spherical neuropil configurations called glomeruli. In these structures the primary olfactory terminals synapse with the most distal dendrites of the peripheral secondary olfactory neurons, i.e. the mitral cells.

In most vertebrates the olfactory bulb is laminated. Passing from peripheral to central, the following layers occur (Fig. 2.42): (a) the stratum fibrosum, composed of interlacing primary olfactory fibres; (b) the stratum glomerulosum; (c) the stratum plexiforme, where dendrites of mitral cells and of small granule elements interact; (d) the stratum mitrale, consisting of the somata of the principal bulbar neurons. The dendrites of these elements differ among the various groups of vertebrates. In amphibians, and in many other non-mammalian vertebrates, the mitral cells have coarse, widely spreading dendrites ending in glomerular tufts (Fig. 2.42b), but in mammals each mitral cell has a single primary dendrite entering a glomerulus and several secondary dendrites branching in the external plexiform layer (Fig. 2.42d). The synapses of the central processes of the olfactory receptor cells and the dendrites of the mitral cells in the glomeruli are asymmetric with round vesicles. Such, so-called type I synapses are generally considered to be excitatory; in the case of the mitral cells this has been confirmed physiologically (Shephard and Greer 1990). (e) The stratum granulare is formed by the densely packed somata of small granule cells. In most vertebrates the stratum granulare is separated from the layer of mitral cells by a distinct plexiform layer, and separate external and internal plexiform layers can be distinguished (Fig. 2.42c). Remark-

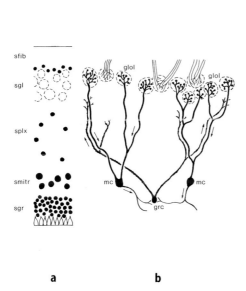
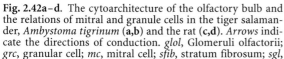
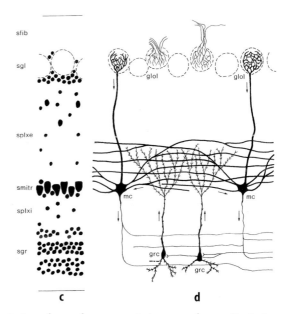

Fig. 2.42a–d. The cytoarchitecture of the olfactory bulb and the relations of mitral and granule cells in the tiger salamander, *Ambystoma tigrinum* (**a,b**) and the rat (**c,d**). *Arrows* indicate the directions of conduction. *glol*, Glomeruli olfactorii; *grc*, granular cell; *mc*, mitral cell; *sfib*, stratum fibrosum; *sgl*, stratum glomerulosum; *sgr*, stratum granulare; *smitr*, stratum mitrale; *splxe*, stratum plexiforme externum; *splxe*, stratum plexiforme internum. [modified from Herrick 1924: Figs. 1, 4 and 10 (**b**), Scheibel and Scheibel 1975: Figs. 1 and 5, and Valverde (1965): Fig. 25 (**d**)]

ably, the small neurons in the bulbar granular layer have no axon, reminiscent of the amacrine cells in the retina. In amphibians the granule cells have a few long, widely spreading, thorny dendrites extending into the plexiform layer (Fig. 2.42b). In mammals the granule cells have several short basal dendrites and a long dendrite that spreads peripherally and ramifies in the external plexiform layer among the secondary dendrites of the mitral cells. The branching distal portions of these dendrites are densely studded with long gemmules (Fig. 2.42d). The axons of the mitral cells traverse the granular layer and turn caudally to enter the olfactory parts of the cerebral hemispheres. During their course through the granular layer they emit numerous collaterals which contact granule cells. The granule cells also receive a heavy input via centrifugal fibres from the cerebral hemispheres. This influence is indirect, since it reaches the granule cells via interneurons situated in the granular layer. In mammals the secondary dendrites of the mitral cells and the spines of the peripheral granule cell dendrites are richly interconnected by dendrodendritic synapses, organized in reciprocally oriented pairs. The mitral-to-granule cell synapses are excitatory, whereas the adjacent granule-to-mitral cell synapses are inhibitory. These peculiar reciprocal synapses constitute extremely short inhi-

bitory pathways from mitral cell to mitral cell (Shepherd and Greer 1990).

From the foregoing it may be concluded that in mammals the somata of the mitral cells receive EPSPs, originating in the glomeruli via their main dendrites, and inhibitory postsynaptic potentials (IPSPs), elicited by granule cells via their secondary dendrites.

The ultrastructural and functional relations between the dendrites of granule and mitral cells in the amphibian olfactory bulb are unknown. However, assuming that these are comparable to those in mammals, it seems likely that functions subserved in amphibians by the most distal and the more proximal segments of the same dendrite are, in mammals, distributed over different dendrites.

11. Neuronal somata are often the target of special *presynaptic configurations*, as for instance the *pericellular baskets* around the somata of hippocampal and neocortical pyramidal cells. The following four configurations may be taken as typical (cf. Ramón y Cajal 1909, 1911; Lorente de Nó 1981). (a) A fibre which crosses a neuronal soma or dendrite leaves a single bouton. (b) A fibre passing along the surface of a soma and/or dendrite establishes multiple synapses by means of boutons en passant. (c) A fibre which ramifies in the vicinity of a soma and/or dendrite contributes boutons de passage as well as

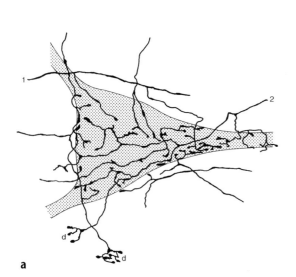

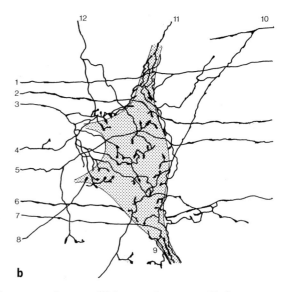

Fig. 2.43a,b. Synaptic boutons on neuronal somata, as seen in Golgi preparations of the CNS of young cats. **a** Large interneuron in the spinal cord of a 15- to 16-day-old cat. Fibres *1* and *2* both have a collateral branch which terminates in a cluster of boutons terminaux for dendrites (*d*) (redrawn from Lorente de Nó 1981: Fig. 3–52. **b** Large neuron in the medial reticular formation of the medulla oblongata from a sagittally cut preparation of an 11-day-old kitten. Sagittally running

fibres *1–5* and *7–8* establish synaptic contact with the neuron via boutons terminaux or boutons en passant. Fibre *6* does not make contact with the neuron. Fibres 3, 4 and 8 give rise to rather elaborate bouton clusters on the soma. Fibres *9* and *11* produce abundant boutons on the primary dendrites. Fibre *10* is a collateral of a sagittal fibre, and fibre *12* approaches from a rostrolateral direction to terminate on the soma. (redrawn from Scheibel and Scheibel 1955, Fig. 1)

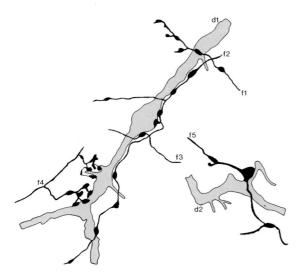

Fig. 2.44. Mode of formation of synapses on dendrites of neurons situated in the spinal ventral horn of a 20-day-old cat. The longer segment (*d1*) begins approximately 100 µm from the soma. The shorter segment (*d2*) belongs to a subdivision of a main dendrite. Fibres *f1* and *f3* cross dendrite d1 and leave a single bouton on it; similarly, fibre *f5* leaves a single large bouton on dendrite d2 at the initial part of a spine. Fibre *f4* forms two dense clusters of synaptic boutons with dendrite d1. Fibre *f2*, finally, passes along dendrite d1 and establishes multiple synapses with dendrite d1. Golgi rapid method. (redrawn from Lorente de Nó 1981, Fig. 3–53)

boutons terminaux throughout a large area. (d) A fibre forms a cluster of boutons terminaux that occupies a small area. Examples of these configurations are illustrated in Figs. 2.43 and 2.44.

Turning now to the special presynaptic configurations around neuronal somata, attention is drawn first to the dense *perisomatic axonal plexuses* found in various parts of the mammalian central nervous system (Fig. 2.45a,b). Comparable structures occur in the telencephalon of urodeles (Fig. 2.7). Such plexuses form typical pericellular baskets (Fig. 2.45c) in the cerebellar cortex, the hippocampus and the neocortex through special types of local circuit neurons.

A second configuration specific to neuronal somata is formed by a coarse fibre which terminates in a grape-like manner with *large boutons* on the soma of one or a few adjacent neurons. Examples are the terminals of IA muscle spindle afferents on the surface of the soma of spinal motoneurons (Bodian 1966; Poritsky 1969; Fig. 2.46) and the secondary vestibular fibres which ascend through the medial longitudinal fasciculus to terminate with a few large boutons on somata in the oculomotor nucleus (Szentágothai 1964; Fig. 2.47).

A third typical perisomatic presynaptic formation is formed by the *end bulbs* and the *calyces* first described by Held (1893, 1897) in the medial

nucleus of the trapezoid body of mammals. End bulbs are clusters of large synaptic boutons, whereas calyces are cuplike perisomatic structures which embrace the postsynaptic soma. Calyces occur both in mammals and in several non-mammalian groups. One example is the calyces found in the tangential nucleus of teleosts (Ramón y Cajal 1908; Bodian 1937; Hinojosa 1973).

The entire spectrum of the endings under discussion, ranging from small end-bulbs to large calyces, has been shown to occur in the anteroventral cochlear nucleus of the cat (Brawer et al. 1974; Brawer and Morest 1975; Lorente de Nó 1981; Fig. 2.48). In this nucleus the Held endings make contact with cells of only one type, the brush or bushy cells. These elements have a large spherical soma but only a limited dendritic tree (Fig. 2.48a). Club endings are depicted in Fig. 2.48c,h,j, whereas Fig. 2.48b,d,e,f,g shows a number of typical calyces. Brush cells contacted by a calyx have a one-to-one relationship with cochlear afferents, since each anterior cochlear branch forms only one terminal calyx and each brush cell receives only one. As can be seen in Fig. 2.48, the calyces have ragged contours. The lacunae in these endings contain synaptic boutons formed by a number of fibres; some of these may be thin collaterals of cochlear branches whereas others are intrinsic (Lorente de Nó 1981). The large synaptic endings found in the anteroventral cochlear nucleus may allow second-order neurons to preserve the time pattern of responses occurring in the primary auditory fibres.

12. The soma is necessarily the sole target of presynaptic endings in those remarkable neurons which lack dendrites. Such *adendritic cells* occur in the rhombencephalic and mesencephalic lateral line centres of electric teleosts. Most of these elements receive large electrical synapses and issue coarse, myelinated, axonal processes (Fig. 2.49).

In mormyrids the large unipolar cells of the nucleus of the electrosensory lateral line lobe (ELLL) receive afferents from so-called knollenorgan receptors. These fibres terminate with large club endings which appear to be purely electrical (Szabo et al. 1983; Bell and Szabo 1986; Fig. 49a). The axons of the unipolar cells project via the lateral lemniscus to the anterior extrolateral nucleus of the torus semicircularis in the mesencephalon, where they again terminate with electrical synapses. Thus, the knollenorgan projection is characterised by electrical synapses and large fibres, features that ensure the greatest fidelity of temporal information. Such information is crucial for electrocommunication between mormyrids (Bell and Szabo 1986). In this group of teleosts the electrosensory system is of the active type and includes a motor

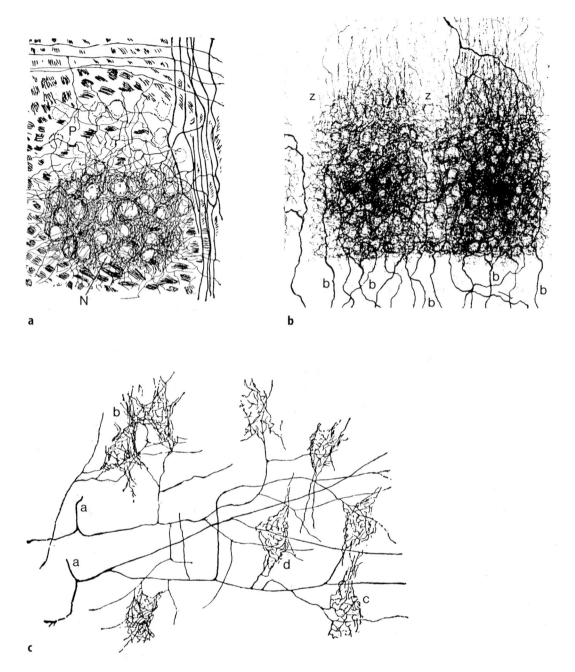

Fig. 2.45a–c. Pericellular axonal plexuses in the mammalian CNS. **a** Motor nucleus of V in an 8-day-old rabbit; the pericellular nests (*N*) consist principally of collaterals of the constituent motor neurons of the nucleus, but sensory collaterals, which constitute a plexus (*P*) situated directly dorsal to the nucleus, also contribute to the formation of the nests (reproduced from Ramón y Cajal 1909, Fig. 385). **b** Sensory cortex of a young mouse; the specific somatosensory efferents from the thalamus (*b*) constitute patches of a dense terminal feltwork around the neuronal somata in the fourth cortical layer; these patches correspond to cylindrically shaped structures, designated as glomérulos by Lorente de Nó (1922) and as barrels by Woolsey and van der Loos (1970). They are separated by zones (*z*) in which the axonal feltwork is clearly less dense (reproduced from Lorente de Nó 1922, Fig. 26). **c** Pericellular ramifications in the layer of medium-sized and large pyramidal cells in the motor cortex of a 25-day-old infant: *a*, axons giving rise to long horizontal branches; *b–d*, pericellular baskets. (from Ramón y Cajal 1911, Fig. 362)

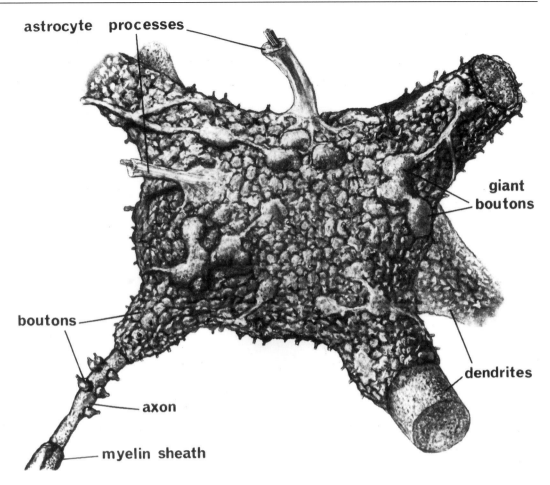

Fig. 2.46. Synaptic terminal boutons on the surface of a motoneuronal perikarya, reconstructed from serial electron micrographs. The giant boutons belong to IA muscle spindle afferents. (slightly modified from Poritsky 1969, Fig. 15)

system that generates the electric organ discharge (EOD). The central circuitry related to the knollenorgan receptors enables a fish to distinguish its own EODs from those of other fish.

The electrosensory system of mormyrids is strikingly similar to that of gymnotid fishes, although these two groups are separated by a large taxonomic distance. In both families the ELLL is a laminated structure. However, whereas in mormyrids the neurons related to the knollenorgan fibres form a separate nucleus beneath the laminae of the lobe, the corresponding elements in gymnotids form a single row in the second layer of the ELLL. These pear-shaped adendritic cells are of two types, C1 and C2. The majority are C1 elements and receive a single giant terminal knob or a calyx covering much of the soma. The C2 elements are contacted by several thin terminals of separate presynaptic fibres approaching the soma in a radial manner (Réthelyi and Szabo 1973b; Fig. 2.49b). As in mormyrids, the axons of the C2 cells form a rapid

electrosensory pathway which terminates in a conspicuous area of the mesencephalon called the magnocellular mesencephalic nucleus. This centre contains scattered large adendritic neurons, which are contacted by about 20 club endings (Rethélyi and Szabo 1973a; Sotelo et al. 1975; Fig. 2.49c). It is believed that some of these club endings are formed by the axons of the rhombencephalic C1 and C2 cells. The cells and the surrounding nucleus are found in the gymnotids *Gymnotus* and *Sternarchus*. Corresponding large, adendritic neurons are present in the sixth layer of the highly remarkable multilaminar torus semicircularis of another gymnotid, *Eigenmannia* (Carr et al. 1981; Carr and Maler 1985, 1986; Fig. 2.54d).

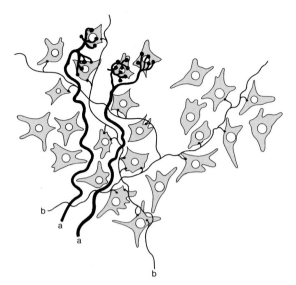

Fig. 2.47. Two different types of presynaptic collaterals, as found in Golgi preparations of the oculomotor nucleus of the cat. The coarse 'a' type collaterals break up into terminal arborisations inside a well-circumscribed territory of the nucleus and establish numerous contacts with the body of one or a few neighbouring motoneurons. The much thinner 'b' type collaterals issue short terminals all over their course, ending in individual terminal knobs. The large 'a' type collaterals belong to secondary vestibular neurons, while the thin 'b' type collaterals originate mainly from neurons situated in the reticular formation or in the interstitial nucleus of Cajal (redrawn from Szentágothai 1964, Fig. 8–7)

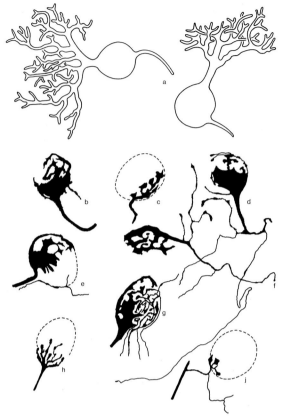

Fig. 2.48. Brush or bushy cells and their axosomatic synapses in the anteroventral cochlear nucleus of cats about 12–15 days old, as seen in Golgi preparations: *a*, two large brush or bushy cells with a large spherical soma and a limited dendritic tree; the anterior branches of cochlear nerve fibres form in the anteroventral cochlear nucleus typical calyces (*b*, *d–g*) and bulbs (*c, h, j*) of Held but also give off collaterals that end by forming single boutons or clusters of boutons of various sizes (*d, f*); in the synaptic apparatus shown in *g* the holes in the calyces are filled by synaptic boutons provided by thin collaterals of other cochlear fibres and by branches of endogenous origin (based on Lorente de Nó 1981, Figs. 3–5 and 3–12)

2.7.3
Functional Significance of the Distribution of Synaptic Contacts Over the Receptive Surface of Neurons

In the present section the functional significance of the inter-neuronal connectivity patterns described in the previous section will be discussed, starting with a summary of basic concepts (see also Chap. 1). The patterns are presented in Fig. 2.50, where the numbers refer to neurons and the letters refer to synaptic domains. These symbols will be used throughout this section.

1. Afferents make synapses on the surface of the somata and the dendrites, which is therefore designated as the receptive surface. The synapses release chemical neurotransmitters that interact with receptors in the postsynaptic membrane and generate a postsynaptic potential (PSP) (Eccles 1964).

2. A synapse can be either excitatory or inhibitory (Eccles 1964; Shepherd 1990). An excitatory synapse generates an EPSP while an inhibitory synapse generates an IPSP. It is likely that all neurons receive both excitatory and inhibitory synapses.

3. The PSPs resulting from the activation of many synapses sum together algebraically (i.e., EPSPs depolarise, while IPSPS hyperpolarise the membrane) after modification by the electrotonic properties of the cell (Shepherd 1990; Johnston and Wu 1995). In most cases this summation is accurately described by regarding the dendrites as passive electrotonic conductors, but there is increasing evidence for active properties in dendrites. By these means the soma and axon hillock of a neuron become depolarised.

4. The zone where the soma of a neuron becomes the axon is of paramount functional importance, because it is at this so-called axon hillock-initial segment zone that the action potentials are generated (Eccles 1957) when this trigger zone reaches a critical level of depolarisation. Most commonly, the axon originates from the soma, but in many neu-

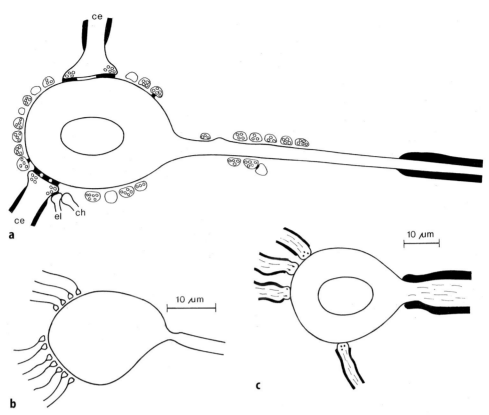

Fig. 2.49a–c. Some adendritic neurons in the brains of electric fishes. **a** Large neuron in the nucleus of the ELLL of a mormyrid fish. Elements of this type receive two types of terminals: club endings (*ce*) that form gap junctions (electrical synapses) with the soma, and small boutons that form chemical (*ch*) or electrical (*el*) synapses with the soma as well as with the axon hillock. Boutons also form gap junctions with the club endings. These boutons are contacted, in turn, by other boutons establishing chemical synapses (redrawn from Bell and Szabo 1986, Fig. 4). **b** An adendritic pear-shaped neuron from the second layer of the lateral line lobe of Gymnotus carapo. This element is approached in a radial fashion by a number of thin presynaptic fibres (based on Réthelyi and Szabo 1973b, Figs. 18 and 19). **c** A large unipolar neuron from the magnocellular mesencephalic nucleus of *Gymnotus carapo*. The coarse axon is covered with myelin right from its origin. The soma of this adendritic element is contacted by a number of presynaptic fibres terminating upon it with club endings (based on Réthelyi and Szabo 1973a, Figs. 11–15)

rons the axon arises from a dendrite (Fig. 2.50: 3a,b; 9a,c).

5. Most neurons receive synaptic input from many other neurons, a phenomenon called convergence. The numbers of synapses on individual neurons can range from hundreds to thousands; on human Purkinje cells the synapses number more than 200 000! (Llinás 1987).

6. The size and number of synaptic contacts made by a single terminal fibre on a neuron may vary considerably (see Figs. 2.37, 2.39a, 2.44, 2.46–2.49). Several factors affect how much a set of synaptic inputs influences a cell. It is often assumed that each synaptic terminal causes an amount of excitation or inhibition proportional to the size of the area of synaptic contact. Most excitatory synapses are found on dendritic spines, however, and the slender geometry of these structures introduces potential subtlety into the generation of the EPSP

(Koch and Poggio 1983; Yuste and Denk 1995). Inhibitory synapses are generally not on spines. It is also assumed that the power of the synaptic action of a given terminal fibre on a neuron depends on the number of synaptic contacts. One of the most powerful synaptic junctions is that between a climbing fibre afferent and the dendrites of a Purkinje cell. Each Purkinje cell receives a single climbing fibre, which establishes as many as 200–300 synaptic contacts with the dendritic tree (Llinás 1987; Fig. 2.38).

7. It follows from 4 that synaptic contacts on the soma of a neuron are able to exert maximal control over the spike-initiating zone of the axon hillock-initial segment. Moreover, in most neurons it is in the soma that the final integration of all PSPs occurs (Fig. 2.50). Consequently, very large synapses or synapse complexes at this location will have a strong influence on the output of that cell. In

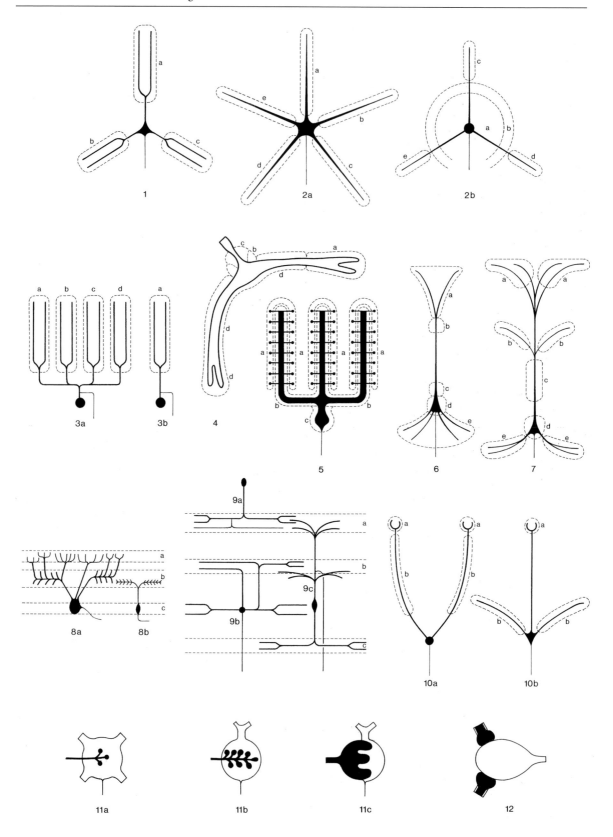

some cases the soma plays the dominant role in transducing synaptic input, while the dendritic input plays only a modulatory role. This holds for the neurons in the anteroventral cochlear nucleus (AVCN) of mammals, the somata of which receive massive synaptic terminals, whereas their dendritic trees are small (Figs. 2.48; 2.50: 11). The output, in the form of unit responses recorded from these so-called brush cells, closely resembles that of the afferent fibres of the auditory nerve (the 'primary-like' response, see Chap. 15 in Shepherd 1994). The influence of somatic input is greatest in neurons without dendrites (Figs. 2.49; 2.50: 12). Such neurons occur in various electrosensory centres of electric fishes (see Chap. 15 of the present work) and the reason is probably to preserve temporal information, as in the mammalian AVCN.

◀ **Fig. 2.50.** The patterns of interneuronal connectivity discussed in this chapter and depicted in Figs. 2.35–2.49. *1*, A large neuron in lamina IV of the spinal cord of the cat, provided with a large dorsal set of dendrites and smaller medial and lateral sets of dendrites. The three sets of dendrites receive segregated inputs from three different groups of afferents (cf. Fig. 2.35). *2a*, A large isodendritic neuron of the rhombencephalic reticular formation, having a number of long, straight dendrites, each of which is under presynaptic control of one predominant afferent system (cf. Fig. 2.14). *2b*, A leptodendritic neuron of the nucleus ventromedialis hypothalami. The central (nuclear) and capsular portions of the dendrites of this neuron receive afferents from different sources and the same holds true for the individual peripheral, extranuclear portions of the dendrites (cf. Fig. 2.31). *3a*, A neuron in the rhombencephalic alar plate of a larval salamander. Its dendrites enter the domains of four functionally different bundles of primary afferent fibres; hence, these elements may be designated as 'polymodal' (cf. Fig. 2.36). *3b*, A 'unimodal' variant of neuron 3a. *4*, A teleostean Mauthner neuron, receiving six different types of synapses, arranged in four different domains (cf. Fig. 2.37). *5*, A mammalian Purkinje cell. The receptive surface of elements of this type receives the three sharply separated, proximodistally arranged inputs shown in Fig. 2.38, together with the stellate cell input to the dendrites. *6*, A hippocampal pyramidal cell, exhibiting likewise a far-going proximodistal segregation of input along its somato-dendritic surface (cf. Fig. 2.39a,b). *7*, A neocortical pyramidal cell, showing a segregation of input comparable to that seen in hippocampal pyramids (cf. Fig. 2.39c). *8*, Retinal ganglion cells in the frog, distributing their dendrites mainly to one (*b*), and two (*a*) subzones of the inner plexiform layer (cf. Fig. 2.40). *9*, Monostratified (*a*), bistratified (*b*) and multistratified (*c*) neuronal elements in the tectum mesencephali of the goldfish (cf. Fig. 2.41). *10*, Mitral cells from the olfactory bulb of an amphibian (*a*) and a mammal (*b*). In the amphibian the main inputs impinge upon the most distal and more proximal segments of the dendrites, but in the mammals these inputs are distributed over different dendrites (cf. Fig. 2.40). *11a*, A large grapelike ending on the soma of a mammalian motoneuron (cf. Figs. 2.46, 2.47). *11b,c*, An end bulb (*b*) and a calyx (*c*) terminating on somata of large brush cells in the anteroventral nucleus of the cat (cf. Fig. 2.48). *12*, Two large club endings terminating on an adendritic neuron in the lateral line lobe of an electric fish (cf. Fig. 2.49).

8. When PSPs are decrementally conducted along the receptive surface, there is the implication that the segregated proximo-distal placement of synapses from various sources (Fig. 2.50: 5, 6, 7) indicates a hierarchical organisation of input.

9. Two concepts about the location of inhibitory synapses may be formulated. The first is that inhibitory synapses are most effective in counteracting excitatory synapses when they are located as close as possible to them. In such cases the inhibition is not only due to local hyperpolarisation but is also a function of the shunting of dendritic impedance by the open inhibitory channels that serves to 'clamp' the membrane close to the equilibrium potential of that particular IPSP (Johnson and Wu 1995). The location of inhibitory synapses close to excitatory ones may explain the fact that within compartment (a) of Mauthner cells (Fig. 2.50: 4) the excitatory large myelinated club endings are intermingled with (presumably) inhibitory small-vesicle boutons (Fig. 2.37), and that within compartment (d) (Fig. 2.50: 4) the (presumably) excitatory large-vesicle boutons and the (presumably) inhibitory small-vesicle boutons are equally dispersed. The second concept is that inhibitory synapses are most effective in preventing the generation of action potentials when located close to the site of impulse generation. Therefore, inhibitory synapses are often concentrated on the somata and proximal dendrites of neurons (Fig. 2.50: 4b, 6d, 7d) or on the axon hillock itself (Figs. 2.49a, 2.50: 4c, 5c).

10. From the perspective of afferent input, dendrites can be categorised as monoreceptive and polyreceptive elements (Scheibel and Scheibel 1970a).

Monoreceptive dendrites have a single type of input, suggesting a relatively uniform functional role for that dendrite, with the final integrative process occurring in the soma. An example is the large reticular neuron, depicted in Fig. 2.50: 2a, which has a number of monoreceptive dendrites. Some authors (e.g., Ramón-Moliner and Nauta 1966) hold that such cells are phylogenetically ancient. More discrete and specific nuclei would have developed from a pool of such primitive reticular elements, during which process the dendritic system (a) would have been gradually monopolised by one type of afferent, and (b) would have been transformed into more concentrated allodendritic or idiodendritic patterns (e.g., Fig. 2.32). Note that the remarkably heterogeneous input of large reticular elements (Fig. 2.14), which may include somatosensory (spinal, trigeminal, vestibular) and cerebellar as well as pyramidal afferents, does not exclude the possibility that these afferents have very specific functions. Heterogeneity should not be equated with lack of specificity. This may be different for the peculiar secondary sensory cells of larval amphibians, depicted in Fig. 2.50: 3a. Herrick (1948) pointed out that these cells receive afferent fibres from tactile, vestibular, lateral line and gustatory roots and so are unable to preserve specificity. Herrick suggested that the organisation of these larval second-order neurons facilitates activation of large masses of musculature by sensory excitation of any kind. He observed that during further ontogenetic development the dendritic systems of the secondary sensory cells become more restricted, so that in the adult stage many dendrites are contacted by only a single bundle of primary afferents. Such a 'unimodal' element is depicted in Fig. 2.50: 3b. Herrick (1948) suggested that this process of local specialisation of structure within the sensory field may have a parallel in phylogenetic history. It is tempting to regard the mammalian secondary sensory spinal neuron (Fig. 2.50: 1), which has three sets of monoreceptive dendrites, as an advanced version of the neuron shown in Fig. 2.50: 3b. Both represent unimodal secondary sensory relay cells, receiving their primary afferents via a set of dendrites, designated as (*a*) in the figure. However, the element in Fig. 2.50: 1 has two sets of basal dendrites (*b* and *c*), with which it probably samples modulatory inputs from spinal (segmental) as well as cortical (suprasegmental) mechanisms (Scheibel and Scheibel 1968). This cell resembles the mammalian mitral cell (Fig. 2.50: 10b), in which the terminal tuft of the primary dendrite (*a*) is contacted by primary olfactory fibres, whereas the output from the cell body is under the control of intrinsic neurons of the olfactory bulb, which impinge upon the secondary dendrites (*b*).

The afferents contacting *polyreceptive dendrites* are often spatially segregated, forming a pattern of hierarchically organised preferential input loci. In laminated structures the polyreceptive dendrites are generally radially oriented, whereas the various input systems and their terminals form layers orthogonal to the dendrites. Frequently, the radially oriented dendrites extend lateral branches into the fibre layers, e.g. in the pyramidal cells in the neocortex (Fig. 2.50: 7), and this configuration is even more pronounced in the retinal ganglion cells (Fig. 2.50: 8a) and in many neurons in the tectum of non-mammalian vertebrates (Fig. 2.50: 9b,c). Thus, dendritic sectors receiving particular inputs may become transformed into large domains. In both the retina and the tectum, some neurons confine their dendritic branches to a single layer and consequently to a single source of afferents (Fig. 2.50: 8b, 9a).

Neurons of the same type in different species, presumably with similar functions, may differ in the organisation of dendritic afferents. This is seen, for instance, between the mitral cells in the olfactory bulb of amphibians, with their polyreceptive dendrites, and the corresponding elements in mammals, with their monoreceptive dendrites (Fig. 2.50: 10a,b). The outgrowth of new dendrites into a domain where afferents of a particular type (represented by the tufts of apical dendrites of the axonless granule cells) are concentrated and the translocation of the synaptic contacts between the granule and mitral cells from the primary toward the secondary dendrites of the latter are the processes which may have transformed the 'amphibian type' into the 'mammalian type' of mitral cells.

The dendrites of some neurons combine *mono-and polyreceptive* features. The neurons from the ventromedial hypothalamic nucleus (Fig. 2.50: 2b) show this hybrid condition. The intranuclear, initial portions of the dendrites receive input from one source *a*, and the same holds for the intermediate capsular segments of the dendrites, which all receive input *b*. However, the extranuclear distal segments of the dendrites get disparate afferents *c,d,e*, thus resembling the entire dendrites of the characteristic element of the reticular formation, depicted in Fig. 2.50: 2a.

11. The rule that the soma and the dendrites serve as purely passive electrotonic conductors, presented above under 3, does not hold for all neurons. Electrophysiological observations indicate that the dendrites of several types of neurons, including Purkinje cells, hippocampal and neocortical pyramids, are able to generate and propagate action potentials (Johnston et al. 1996). The dendrites have voltage-gated channels for Ca^{2+} or Na^+ which may not be uniformly distributed. Such dendritic properties provide an entirely new mode of dendritic integration, since distant dendritic depolarisation can now influence the soma more strongly by generating propagating action potentials, and the entry of Ca^{2+} into dendrites may lead to further local functional changes. This means that in certain types of cells, the size and the distance from the soma are not the only criteria of importance in determining the 'weight' of an excitatory synaptic contact.

2.7.4
Synaptic and Non-synaptic Chemical Neurotransmission

The orthodox view that neurons in the vertebrate CNS interact exclusively by means of synaptic contacts has been challenged during the past decade. There is mounting evidence for the non-synaptic chemical interaction, or volume transmission, in the central nervous system of vertebrates (for reviews see: Schmitt 1984; Nieuwenhuys 1985; Nieuwenhuys et al. 1989; Buma et al. 1989; Fuxe and Agnati 1991; Bach-y-Rita 1993; Golding 1994; Agnati et al. 1995).

A classical chemical synapse is characterised by a presynaptically located cluster of synaptic vesicles, containing the neurotransmitter, electron-dense appositions to the cytoplasmic side of the pre- and postsynaptic membranes, and a widened synaptic cleft (Pappas and Waxman 1972; Gray and Guillery 1966; Gerschenfeld 1973; Heuser et al. 1979; Fig. 2.51a). During classical synaptic transmission the vesicles fuse with the plasma membrane and release their content of neurotransmitter molecules into the synaptic cleft, where they diffuse to the postsynaptic membrane, bind to receptors and produce a PSP. Transmitter action is often terminated by re-uptake into cells or enzymatic degradation.

In *non-synaptic chemical transmission*, vesicles containing a neurotransmitter or another neuroactive substance (such as a neuropeptide) fuse with non-specialised parts of the neurolemma and release their content by exocytosis in the extracellular space. The latter constitutes a continuous system of narrow aqueous voids, which extends throughout the neuraxis, separating the cellular elements and occupying about 20 % of a typical brain volume (Nicholson 1997). It is assumed that, following non-synaptic release of a neuroactive substance, the molecules diffuse through the extracellular space and reach specific receptors in a much wider volume than release into a specialised synaptic cleft would allow (Fig. 2.51b). Some evidence supporting

the existence of non-synaptic chemical neurotransmission in the central nervous system may be summarised as follows:

1. Release of neuromediators from axon terminals devoid of synaptic membrane specialisations has long been known to be the rule in the peripheral autonomic nervous system and is also common in the CNS of invertebrates (Buma and Roubos 1986; Schmidt and Roubos 1987). It is plausible that this mode of intercellular communication, which, in its duration and range, is intermediate between the local, fast synaptic transmission and the global, slow endocrine communication, also occurs in the vertebrate CNS.

2. In the mammalian CNS a complex of centres and pathways, which may be characterised as a caudally extended limbic system, shows an extraordinary richness and density of neuroactive substances, particularly neuropeptides (Nieuwenhuys 1985, 1996; Nieuwenhuys et al. 1988, 1989). Given the presence of a multitude of specific messengers and specific receptors for all of these messengers, it is likely that non-synaptic interneuronal communication prevails here. Such a mode of communication implies a very economical use of the available space in the CNS. Moreover, although the various messengers all travel in the same extracellular space, selectivity of action is maintained through the presence of specific receptors for all of these substances.

3. In many parts of the CNS of vertebrates thin, unmyelinated fibres occur. Immunohistochemical studies have shown that these fibres usually contain monoamines or neuropeptides. The fibres often have numerous small varicosities filled with syn-

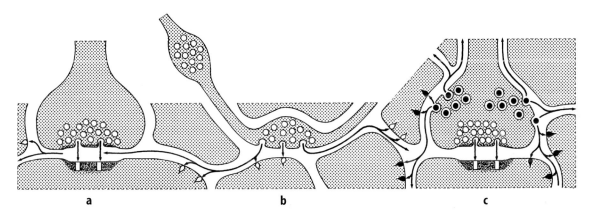

Fig. 2.51a–c. Synaptic and non-synaptic neurotransmission. **a** Classical synapse in which a presynaptic element influences a single postsynaptic element. **b** An axonal varicosity which, by means of non-synaptic secretion of a neuromediator, influences several other neuronal elements. **c** A synaptic terminal containing small, clear vesicles as well as large dense-cored vesicles; the small vesicles discharge their content in the synaptic cleft, whereas the content of the dense-cored vesicles is released at non-synaptic sites of the terminal membrane. The width of the extracellular space has been exaggerated. The *arrows*, which indicate the flow of neuromediator molecules, point to synaptic and non-synaptic membrane receptors (modified from Nieuwenhuys 1985, Fig. 50, and Nieuwenhuys et al. 1989, Fig. 7)

aptic vesicles. Because analysis of serial sections reveals that these varicosities usually do not show any trace of the membrane specialisations typical for synapses, it is assumed that these structures are involved in non-synaptic chemical neurotransmission (Arluison 1981; Baraban and Aghanian 1981; Maxwell et al. 1983; Light et al. 1983; Clements et al. 1985; Mobley and Greengard 1985; Takagi et al. 1986; Dolabela de Lima and Singer 1987; Mori et al. 1987; Ulfhake et al. 1987; Buma et al. 1989; Séguéla et al. 1989, 1990; Christenson et al. 1990; Fig. 2.51b).

4. Many synaptic terminals contain two different types of vesicles: small, clear vesicles containing a classical neurotransmitter and larger, so-called dense-core vesicles containing one or more neuropeptides. Some of the small vesicles are found in the immediate vicinity of the synaptic contact zone, but the dense-core vesicles are generally located away from the active synaptic site. This has led to the hypothesis that these terminals release their classical neurotransmitter synaptically, but they release their neuropeptide(s) from non-specialised parts of their axolemma (Thureson-Klein et al. 1988; Buma et al. 1989; Nieuwenhuys et al. 1989; Fig. 2.51c).

5. Herkenham (1987) analysed a large number of publications on the localisation of neuroactive substances and the receptors to which they bind and concluded that each has a discrete and heterogeneous distribution in the central nervous system, but the two distributions do not regularly correspond to each other. He suggested that this 'mismatch' may reflect the widespread occurrence of non-synaptic interneuronal communication, although there may also be unresolved methodological issues.

6. Direct evidence for this extrasynaptic communication includes the demonstration of (a) non-synaptic exocytosis and (b) the location of receptor complexes on the surface of neurons beyond the confines of postsynaptic specialisations.

The occurrence of non-synaptic exocytosis from dense-core vesicles has been demonstrated in several parts of the CNS of the rat, including the area postrema (Buma et al. 1984), the trigeminal subnucleus caudalis (Zhu et al. 1986; Thureson-Klein et al. 1988), the periaqueductal grey (Buma et al. 1989, 1992), the hypothalamus (Pow and Morris 1988) and the median eminence (Buma and Nieuwenhuys 1987, 1988).

Evidence for receptor complexes outside the confines of postsynaptic specialisations is growing. For example, Dana et al. (1989) demonstrated that in the area tegmentalis ventralis of the rat, binding sites for neurotensin are not exclusively present opposite abutting axon terminals but instead are distributed along the plasma membrane, with few sites associated with synaptic junctions. In another study, Aoki and Pickel (1992) provided morphological evidence that catecholamines may act beyond morphologically identifed synapses.

2.7.5
Summary

It has been shown that dendrites can be categorised in relation to their afferents as monoreceptive and polyreceptive, and that in polyreceptive dendrites the synaptic terminals from different sources often show a distinct proximo-distal segregation. The somata of many different types of neurons are under the influence of strong presynaptic configurations.

In all parts of the vertebrate CNS the principal functional operations are generated by specific circuits formed by neurons and their synaptic connections. However, morphological evidence strongly suggests that non-synaptic communication plays an important role in shaping the functional activities in many of these centres. Such influences most probably have a slow onset and long duration but are not necessarily 'diffuse' or 'aspecific'. It is reasonable to assume that different neurons in a given region possess different receptors. Hence, each neuroactive substance travelling through the extracellular space after its non-synaptic release will address only those neurons which have 'their own' specific receptor.

2.8
Lamination

2.8.1
Introductory Notes

Lamination, i.e. differentiation into layers, is common in the central nervous system of vertebrates. In most anamniotes the wall of the neuraxis contains a central zone of periventricular grey, covered externally by a neuropil and fibre zone. The neuronal somata in the central grey extend their dendritic trees into the neuropil and fibre zone, where they synapse with axons and terminals (Figs. 2.1, 2.2, 2.9). It is important to note that the structure of the cerebellar cortex, which is laminated in most vertebrates (Larsell 1967; Nieuwenhuys 1967b), can be directly derived from the simple, and presumably primitive, configuration just discussed (see Figs. 2.1 and 2.38). The cerebellar molecular layer can be regarded as a specialised neuropil and fibre zone because all axons in it are oriented in the same direction, parallel to the external surface. Similarly, the Purkinje cells are specialised derivatives of cells found in the periventricular grey of many vertebrates. Both types of cell send their dendrites into a peripheral zone of neuropil. However, the dendritic trees of the Purkinje cells are, in contrast to cells in the periventricular grey, flattened and oriented perpendicular to the parallel fibres in the molecular layer. Taking into account, in addition to the cerebellum, the neuronal organization of other, complex laminated structures, such as the retina, the vagal lobe of cyprinids, the electrosensory lateral line lobe of gymnotids, the tectum mesencephali, the olfactory bulb and the hippocampus, it appears that most of these structures share the following three features:

1. One or more plexiform layers, oriented parallel to the surface of the brain
2. One or more cellular layers, likewise oriented parallel to the surface of the brain
3. Radially oriented elements (dendrites, axons, entire neurons), which make functional links between two or more of the tangentially oriented layers

These three basic features will be discussed first. Then the functional significance of lamination will be considered, and finally the organization of the tectum mesencephali will be commented upon.

2.8.2
Basic Features of Laminated Structures

2.8.2.1
One or Several Neuropil Zones, Oriented Parallel to the Surface of the Brain

These plexiform layers are usually composed of axonal and dendritic components but may also contain neuronal somata. The axonal components of the plexiform layers may (a) have their somata beyond the laminated structure and enter the latter as afferent fibres, (b) originate from intrinsic neurons, and (c) arise as collaterals from the axons of efferent neurons. The afferents of all laminated structures are topographically organised and the afferents are either primary sensory fibres or sensory fibres of higher order. In these cases sensory surfaces are mapped onto the plexiform layers. Some examples follow:

1. The electrosensory body surface is mapped no less than four times on the large disc-shaped electrosensory lateral line lobes of the gymnotid teleost *Eigenmannia virescens* (Carr et al. 1982; Fig. 2.52).
2. The gustatory surface of the palatal organ of cyprinid teleosts, like the goldfish, is mapped on

three layers (8, 10 and 13) of the large vagal lobe of these fishes (Morita et al. 1980; Finger 1981a,b; Fig. 2.53).
3. The retina, and consequently the visual world, is topically projected in one or more plexiform layers of the tectum mesencephali (e.g., Vanegas and Ebbesson 1973: teleosts: Fig. 2.41; Lázár and Székely 1969; Lázár et al. 1983: anuran amphibians: Figs. 2.2–2.56b; Butler and Northcutt 1971; Dacey and Ulinski 1986c: reptiles; Hartwig 1970: birds).
4. The retinal surface is mapped via two different systems, the direct retinocollicular (e.g., Dräger and Hubel 1975; Goldberg and Robinson 1978; Stein 1981) and the indirect retino-geniculo-cortico-tectal (e.g., Lund 1964; Rhoades 1981; Cadusseau and Roger 1985), on the superficial zone of the colliculus superior.
5. The somatosensory body surface projects in an orderly fashion, via spinocollicular (Antonetty and Webster 1975; Yeziersky 1988) and trigeminocollicular (Killackey and Erzurumlu 1981; Huerta et al. 1983; Cadusseau and Roger 1985) fibre systems, upon the deeper zone of the superior colliculus.
6. The cochlea, and consequently the auditory environment, is represented in the deeper zone of the colliculus superior (Dräger and Hubel 1975; Cadusseau and Roger 1985) via fibres originating from the inferior colliculus.

It is commonly stated that the various sensory maps are topographically ordered; however, because these maps do not preserve shape or size, but only continuity and connectedness, they are *mathematically topological* instead of topographical.

Retinotectal projection may segregate into several terminal zones. Two types of segregation may be identified (Potter 1976). In one, the fibres run in a single lamina, then turn sharply to descend (in reptiles and birds) or ascend (in mammals) for a variable distance before terminating. The terminal arbors form synaptic laminae in reptiles and birds. The other type, which is present in teleosts and anurans, has fibres which run in 3–4 laminae and form terminal arbors in, or adjacent to, these laminae. In this second type the terminals are branched extensions of the stem fibres, without an intervening fibre segment bending toward or away from the external surface. Morphological and physiological studies on laminated structures have shown that if fibres associated with either the same or different sensory modalities topically project to different synaptic zones, all the projections are in spatial correspondence; i.e. the various 'sensory maps' are all in register. Two examples illustrate this interesting feature.

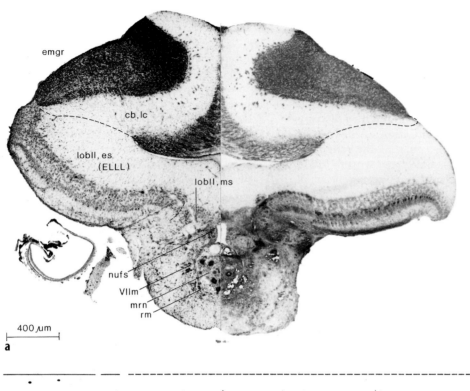

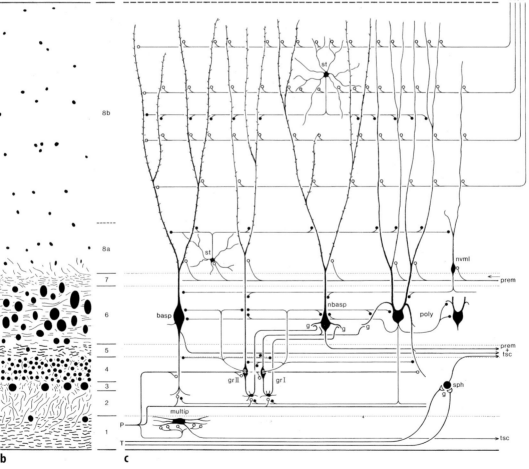

The *torus semicircularis* of the gymnotid teleost *Eigenmannia* is a large multilayered structure (Carr and Maler 1985, 1986; Carr et al. 1981; Fig. 2.54) which is predominatly a second-order station for electrosensory input. Afferent fibres from theELLL, also strongly developed,(Fig. 2.52) run vertically in the torus, projecting a map of the electrosensory surface upon several laminae. Projections related to different types of electroreceptors, termed the T-system and the P-system, terminate differentially. The T-system is involved mainly in electrocommunication and the terminals are found in layer VI, where they contact the adendritic giant cells characteristic of this layer. The P-system, which subserves active electrolocation, synapses in laminae III, V, VII, VIIIC and VIIID. The maps of the electrosensory surface within the various toral laminae are all in register and constitute together a single stack of congruent 'pisciculi' in the torus. Thus, the head region is mapped rostrally, the body caudally, the dorsum medially, and the left and right ventral sides laterally (Carr et al. 1981; Heiligenberg and Bastian 1984).

◀ **Fig. 2.52a-c.** The electrosensory lateral line lobe (*ELLL*) of the gymnotid teleost *Eigenmannia virescens*. **a** Transverse section through the rhombencephalon at the level of the ELLL. The *left half* represents a photomicrograph of a Nissl section, the *right half* that of the adjacent Klüver-Barrera section. **b** A strip from the ELLL, drawn from a Bodian section. The lobe can be subdivided into eight layers (*1–8*). Three of these are fibre layers (1, 5 and 7), two are neuropil zone (2 and 8), and the remaining three are cellular layers (3, 4 and 6). **c** A semidiagrammatic representation of neuronal elements and of the principal extrinsic and intrinsic connections of the ELLL (based on Maler 1976, Maler et al. 1981, and Carr and Maler 1986). The ELLL receives afferents from two types of lateral line receptors, T and P, from the caudal lobe of the cerebellum and from the nucleus praeeminentialis. It contains four types of efferent neurons: the spherical, multipolar, basilar pyramidal and non-basilar pyramidal cells. The remaining elements depicted, i.e. the granule type I cells, the granule type II cells, the polymorph cells and the neurons of the ventral molecular layer are local circuit neurons. *Open circles* indicate putatively excitatory chemical synapses, *filled circles* indicate putatively inhibitory chemical synapses. *basp*, Basilar pyramidal cell; *cb,lc*, cerebellum, caudal lobe; *emgr*, eminentia granularis; *g*, gap junction contact; *grI*, granule cell type I; *grII*, granule cell type II (with long apical dendrite); *lobll,es*, lobus lineae lateralis, electrosensory part (=ELLL); *lobll,ms*, lobus lineae lateralis, mechanosensory part; *mrn*, medullary relay nucleus of electromotor system; *multip*, multipolar cell; *nbasp*, non-basilar pyramidal cell; *nufs*, nucleus fasciculi solitarii; *nvml*, neuron of the ventral molecular layer; *P*, P-type afferent; *poly*, polymorph cell; *prem*, nucleus praeeminentialis; *rm*, nucleus reticular medius; *sph*, spherical cell; *st*, stellate cell; *T*, T-type afferent; *tsc*, torus semicircularis; *VIIm*, nucleus motorius nervi facialis

In addition to the electrosensory projections from the ELLL, there is another electrosensory input to the torus which originates from the posterior part of the eminentia granularis of the cerebellum. This cerebellar electrosensory input is confined to lamina VIIIB. Moreover, the descending nucleus of the trigeminal nerve projects predominantly to laminae VIIIA and VIIIC, with a smaller projection to laminae VIIID and IX, and the tectum mesencephali projects to lamina IX (Carr and Maler 1986). The spatial organisation of these projections is unknown, but it may well be that their terminal fields are in register with the electrosensory 'pisciculi' represented in layers III through VIII. All but two laminae (I and VI) of the torus contain cells which project to the tectum mesencephali. By this topically ordered torotectal projection, the toral 'pisciculi' are transferred to the tectum, where they are in spatial register with the retinotopic map in the tectum, so that a given radial column in the latter structure receives electrosensory and visual input from corresponding portions of the animal's environment (Carr and Maler 1986).

The mammalian homologue of the tectum mesencephali, i.e. the *colliculus superior* (Fig. 2.55), is a second example of a laminated structure in which several superimposed topically ordered sensory projections are in spatial correspondence. The retina, and consequently the visual environment, is represented in an orderly fashion in the superficial zone of the superior colliculus, and comparable representations of the surface of the body and of the auditory environment can be detected in the deeper collicular zone. Physiological experiments have shown that these individual sensory representations are all in close spatial register with one another (Gordon 1973; Dräger and Hubel 1975; Harris et al. 1980). The principal function of the superior colliculus (and of the non-mammalian tectum mesencephali as well) is to participate in the control of orienting responses, i.e. the steering of rapid combined movements of eyes, head and body towards external stimuli. The deep layers of the superior colliculus send descending efferents to a variety of centres in the brain stem and spinal cord involved in the organisation of these movements. Interestingly, the cells of origin of these descending tectal efferents are topically organised, and this deep layer 'premotor' organisation is in register with the various sensory representations in the superior colliculus (Stein 1984; Huerta and Harting 1984).

Intrinsic neurons contributing with their axons to one of the plexiform layers occur in many laminated structures. For example, the horizontal cells in the retina (Fig. 2.40), the periglomerular cells in the olfactory bulb, the type I and type III cells in the

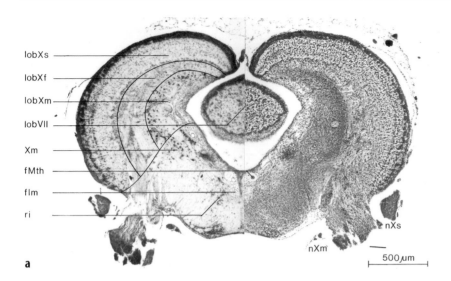

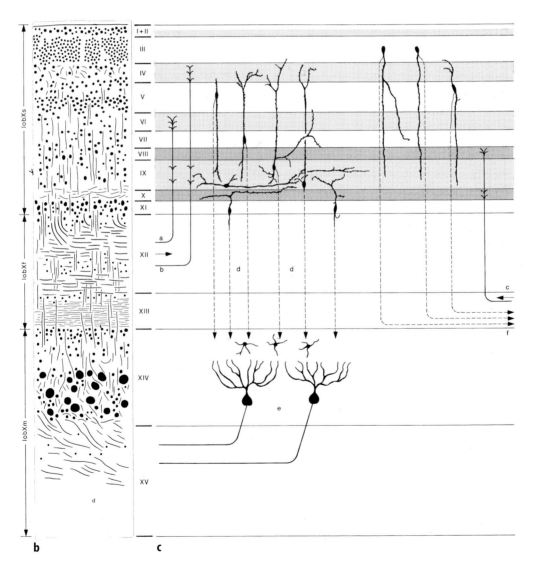

a

b   **c**

tectum mesencephali of teleosts (Fig. 2.41), several types of intrinsic cells in the tectum of the garter snake (Dacey and Ulinski 1986b), and the basket cells in the cerebellum (Fig. 2.38b).

The participation of collaterals of efferent neurons in the formation of plexiform layers has been reported less frequently. However, in the tectum of the garter snake the collaterals of efferent neurons of several types are confined mainly to a single layer (Dacey and Ulinski 1986a), and the recurrent collaterals of large allocortical and neocortical pyramids terminate preferentially on the basal dendrites of cells of the same type (Fig. 2.39).

In several laminated structures the main axons of efferent neurons unite in tangentially oriented sheets, thus further accentuating the layered appearance. Layer 4 of the vagal lobe of the goldfish (Fig. 2.53), layer 5 of the electrosensory lateral line lobe of gymnotids (Fig. 2.52b,c) and layer 7 of the tectum of anurans (Fig. 2.56b) exemplify this.

Before leaving our consideration of the axonal contributions to the plexiform layers in laminated structures, attention should be drawn to a general principle on the texture of these structures enunciated by Maler (1976) and Maler et al. (1981). In studying the ELLL of mormyrids and gymnotid teleosts (Fig. 2.52), these authors found that axons of both afferent and intrinsic elements terminate in a layer-specific fashion, but that these axons, within their layer of termination, end on all possible post-synaptic sites. They thought that this *principle of laminar specificity*, as they termed it, has wide application in laminar structures, noting the organisation of the afferents and the intrinsic axons in the cerebellar cortex (Fig. 2.38) and the specific thalamic nuclei afferents terminating in layer IV of the cerebral cortex. Within that layer such afferents also end on all possible dendritic targets (White 1978; Peters et al. 1979). Maler (1976) made the interesting suggestion that laminar specificity of afferents may be a necessary condition of laminar structures, since by adhering to this principle it may be possible to use relatively simple developmental rules and yet generate a complex circuit.

The dendritic components of the plexiform layers under discussion often consist of segments of radially oriented dendritic trees (such as those of Purkinje cells; Fig. 2.38). Consecutive segments of dendrites often intersect several different neuropil zones [e.g., the apical dendrites of hippocampal pyramids (Fig. 2.39a,b)]. However, it is also frequently observed that neurons in laminated structures prefer particular neuropil zones, by extending part of their dendritic trees tangentially into one or more of these zones. The amacrine cells and the ganglion cells in the retina (Fig. 2.40), many different types of neuron in the tectum of teleosts (Fig. 2.41) and of other non-mammalian vertebrates, the mitral cells in the mammalian olfactory bulb (Fig. 2.42d) and the neocortical pyramids (Fig. 2.39c) are all examples.

Neuronal somata generally do not form part of plexiform layers. Exceptions are those laminated structures in which dense, tangentially arranged feltworks of axon terminals coincide with zones occupied by neuronal somata, for example, the primary visual cortex of the rhesus monkey, in which the dense afferent axonal plexuses in sublayers 4A and 4C coincide with their principal targets, the somata of interneurons.

2.8.2.2
One or More Cellular Layers,
Oriented Parallel to the Surface of the Brain

It appears that sharply defined plexiform layers are often bordered by equally distinct cellular layers, and these cellular layers often contain the somata of identifiable neurons. Thus, in the retina the outer and inner plexiform layers alternate with cellular layers, each consisting of specific types of cells (Fig. 2.40). Comparable situations occur in the mammalian olfactory bulb, in the cornu ammonis and in the cerebellum. In the mammalian olfactory bulb the distinct layer formed by the somata of the large mitral cells separates the external and internal

◀ **Fig. 2.53a–c.** The vagal lobe of the goldfish, *Carassius auratus*. **a** Transverse section through the rhombencephalon at the level of the vagal lobe. The *left half* represents a photomicrograph of a Nissl section, the *right half* that of the adjacent Bodian section. **b** A strip from the vagal lobe, drawn from a Bodian section. The lobe can be subdivided into 15 layers, which are numbered from superficial to deep. Laminae *I–XI* constitute the sensory zone and laminae *XIV* and *XV* the motor zone of the vagal lobe. These two are separated by a fibre zone, comprising laminae *XII* and *XIII*. **c** A semidiagrammatic representation of neuronal elements and of the principal extrinsic and intrinsic connections of the vagal lobe. The neurons are redrawn from Morita et al. (1983), Fig. 5. Afferents from the palatal organ (*a*) reach layers VI and IX, while gill arch input (*b*) terminates in layers II, IV and IX. Diencephalic afferents (*c*) terminate in layers VIII and X. Neurons present in layers V, VII, IX and XI form part of direct vago-vagal reflex loops and terminate in layer XIV (*d*). Axons of large motoneurons situated in layer XIV (*e*) leave the central nervous system via the motor root of the vagal nerve, to innervate the muscle fibres in the palatal organ. Axons of neurons situated in the superficial part of the sensory zone (*f*) leave the vagal lobe and ascend to the nucleus gustatorius secundarius. *flm*, Fasciculus longitudinalis medialis; *fMth*, fibre of Mauthner; *lobVII*, lobus facialis; *lobXf*, lobus vagi, zona fibrosa; *lobXm*, lobus vagi, zona motoria; *lobXs*, lobus vagi, zona sensoria; *nXm*, nervus vagus, pars motoria; *nXs*, nervus vagus, pars sensoria; *ri*, nucleus reticularis inferior; *Xm*, nucleus motorius nervi vagi

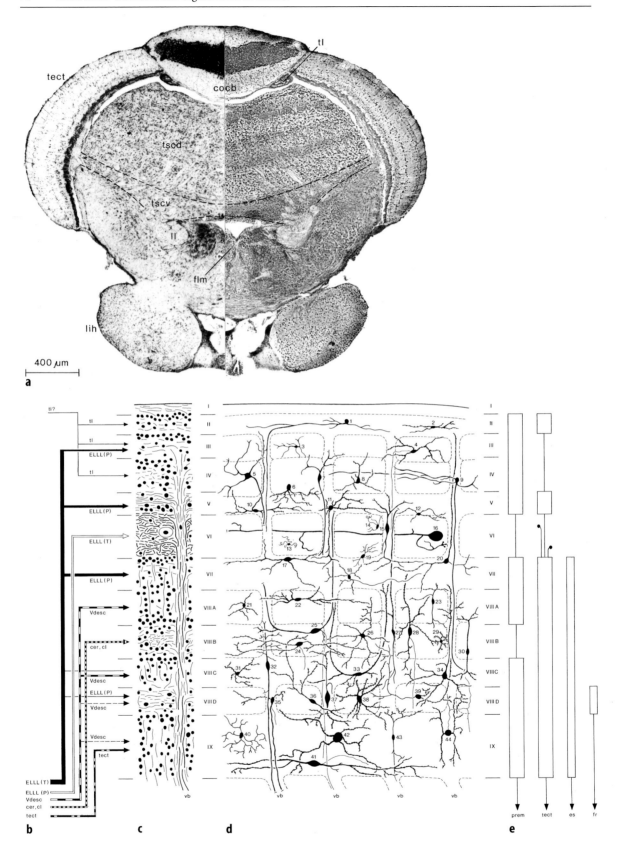

plexiform layers (Fig. 2.42c,d), and the perikarya of the large hippocampal pyramids occupy a corresponding position in the cornu ammonis (Fig. 2.39a,b). In the cerebellum the plexiform zone, i.e. the molecular layer, is bordered by the layer of Purkinje somata, which is one cell deep in birds and mammals. In most vertebrates densely packed granule cells constitute a third cerebellar layer, situated directly beneath the Purkinje somata (Fig. 2.38).

A well-developed molecular layer is also present in the large ELLL of the gymnotid teleost *Eigenmannia*; however, the deeper zone of this structure is more complex than that of the cerebellum with seven layers (Fig. 2.52). Layers 3, 4 and 6 may be designated as cellular layers. Layer 3 contains large adendritic spherical cells, layer 4 consists of the densely packed somata of granule cells, and layer 6 harbours the somata of three types of large cells: the basilar pyramids, the non-basilar pyramids and the polymorphic cells.

In the remarkable, multilayered vagal lobe of cyprinid fishes sharply defined plexiform layers do not alternate with equally distinct cellular layers (Fig. 2.53), but some layers, such as 7, 8 and 13, are predominantly plexiform, whereas others, such as 2, 3 and 6, are mainly cellular. The superficial lamina 14 consists entirely of the densely packed somata of small, unipolar neuronal elements, and so is purely cellular.

A laminar appearance, which is based on gradual differences rather than on a sharp segregation of plexiform and cellular zones, is found in many different neural structures. Thus, in the tectum mesencephali of teleosts only the outermost

SMARG is purely plexiform and only the innermost SGP is purely cellular (Fig. 2.41). In the intervening five layers no fewer than five predominantly plexiform subzones can be distinguished. All, however, contain neuronal somata and several of them contain concentrations of axons. Similar relations are found in the neocortex, where only the most superficial layer may be considered a plexiform layer. The dense axonal plexuses occurring in lamina IV of the somatosensory cortex coincide with laminar concentrations of the somata of local circuit neurons. It is also known that the basilar dendrites of the pyramidal neurons in layer III and V of the neocortex form dense dendritic plexuses which are not devoid of neuronal somata.

In discussing the phenomenon of cellular lamination we have concentrated so far only on the neuronal somata. In many laminated centres neurons occur which are confined to one particular layer. This 'laminar confinement' of neurons is a dominant feature in the torus semicircularis of gymnotids (Fig. 2.54d) and in the mammalian colliculus superior (Fig. 2.55b). Many types of local circuit neurons in the reptilian tectum mesencephali (Dacey and Ulinski 1986b) and in the mammalian neocortex also show this phenomenon (cf., e.g., Lorente de Nó 1922; Lund 1987, 1988; Lund et al. 1979).

2.8.2.3
Radially Oriented Elements (Dendrites, Axons, Entire Neurons), Which Establish Functional Links Between Two or More of the Tangentially Oriented Layers

The third and last basic feature of laminated structures may be designated as 'radial coupling'. Many examples of such coupling of layers have been presented already, e.g. dendritic systems of cerebellar Purkinje cells, hippocampal and neocortical pyramids, bistratified and multistratified neurons in the retina and in the teleostean tectum and the mitral cells in the olfactory bulb (cf. Fig. 2.50). Other examples of neuronal elements effecting radial coupling include (a) the retinal bipolar cells (Fig. 2.40); (b) neurons present in laminae 6–13 of the vagal lobe of cyprinids, which project directly to the motor zone of that structure (Fig. 2.53); (c) granule cells in the ELLL of gymnotids, which form a link between the primary afferent terminals in layers 2 and 4 of that structure, on the one hand, and the basilar and non-basilar pyramids as well as the polymorphous neurons in layer 6, on the other hand (Fig. 2.52); and (d) many types of interneurons in the cerebral cortex, provided with ascending and/or descending axons. In other laminated

◀ **Fig. 2.54a–e.** The torus semicircularis of the gymnotid teleost *Eigenmannia virescens*. **a** Transverse section through the mesencephalon at the level of the torus semicircularis. The *left half* represents a photomicrograph of a Nissl section, the *right half* that of the adjacent Bodian section. **b** Levels of termination of the afferent systems of the dorsal torus (based on Carr and Maler 1986). **c** A strip from the torus semicircularis, pars dorsalis, drawn from a Bodian section. The laminar subdivision as introduced by Carr et al. (1981) is indicated. **d** Cell types of the dorsal torus as determined by the Golgi technique (redrawn from Carr and Maler 1985). **e** Levels of origin of the principal efferent system of the dorsal torus (based on Carr and Maler 1986). *cer,cl*, Cerebellum, caudal lobe; *cocb*, corpus cerebelli; *ELLL (P)*, P-type afferents from the electrosensory lateral line lobe; *ELLL (T)*, T-type afferents from the electrosensory lateral line lobe; *es*, nucleus electrosensorius; *flm*, fasciculus longitudinalis medialis; *fr*, formatio reticularis; *lih*, lobus inferior hypothalami; *ll*, lemniscus lateralis; *prem*, nucleus praeeminentialis; *tect*, tectum mesencephali; *tl*, torus longitudinalis; *tscd*, torus semicircularis, pars dorsalis; *tscv*, torus semicircularis, pars ventralis; *vb*, vertical bundle; *Vdesc*, nucleus descendens nervi trigemini; *1–44*, cell types in the dorsal torus

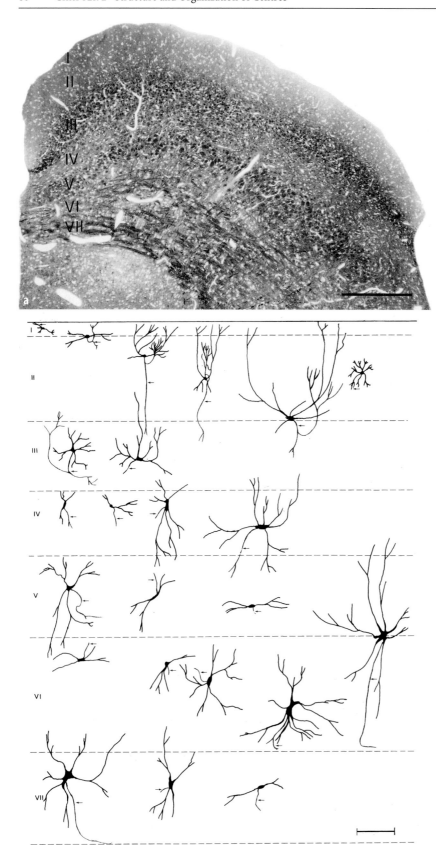

Fig. 2.55a,b. The colliculus superior of the rat. **a** Photomicrograph of a semithin (2 μm) transverse section through the right half of the superior colliculus stained with paraphenylene diamine. *Scale bar* = 0.4 μm. **b** Semidiagrammatic representation of neurons observed in Golgi preparations (redrawn from figures presented by Langer and Lund 1974, Labriola and Laemle 1977, and Tokunaga and Otani 1976). *Roman numbers* indicate collicular layers; *small arrows* point to axons. *Scale bar* = 100 μm (courtesy of Dr. F.J. Albers)

centres such as the torus semicircularis of gymnotids and the colliculus superior the morphological substrate of the radial coupling is less prominent. However, with regard to the torus, Fig. 2.54 shows that the dendritic systems of several cell types intersect two or more sheets of afferents. These dendritic systems probably make synaptic contacts in all of the sheets of afferents intersected, to integrate information from several input channels. The same holds for the superior colliculus (Fig. 2.55). Tokunaga and Otani (1976) found multipolar cells in the intermediate layers (IV, V) of the superior colliculus of rats, with dendrites extending into the superficial layers (I-III). Mooney et al. (1984, 1988) demonstrated dendrites of cells in layers IV-V extending as far as layer II and, reciprocally, dendrites of cells located in layer II extending into layer IV. Moreover, the existence of direct axonal connections between the superficial and deeper collicular layers has been demonstrated. After injecting an anterograde tracer into the superficial grey layer (II) of hamsters, Rhoades et al. (1989) found labelled synaptic contacts in the optic layer as well as in all deeper collicular layers. Axons from cells located in layer II passed through layer III (i.e., the optic layer), making numerous *en passant* contacts. Within layer IV these axons were observed to form networks in which both terminal and *en passant* swellings were observed. Descending further within layer V, terminals of these fibres were similar to those in the optic layer, and another dense terminal field was seen in lamina VI (Rhoades et al. 1989).

2.8.3
Functional Significance of Lamination

In many laminar structures afferent fibres carrying information about spatial senses project in an orderly fashion to one or more laminae, imposing maplike representations of sensory surfaces. When information related to different sensory modalities is transferred to different laminae the representational maps are all in spatial register. Radially oriented elements provide links between corresponding loci of the various maps and integrate the information into an image of the animal's environment which provides accurate guidance for motor activity; i.e. it is transformed into adaptive motor patterns.

A good example of such sensory mapping onto adaptive motor activity is in the vagal lobe of cyprinids (Fig. 2.53), where the spatially ordered representations of the gustatory surface of the palatal organ are linked by radially arranged interneurons to the motor zone of the same lobe (Morita et al. 1983; Morita and Finger 1985; Goehler and Finger

1992). The activity of the motoneurons located in the latter zone leads to adaptive responses of the musculature of the palatal organ, aimed at the selection and movement of food particles (Sibbing 1984; Finger 1988). The role of the tectum mesencephali and its mammalian homologue, the colliculus superior, in transforming sensory information from various sources into motor commands leading to orientation responses has already been discussed. Before reaching the tectum or the colliculus, visual information is already processed by another laminated structure, the retina, and the same thing occurs for other kinds of sensory information reaching the tectum. Thus, in gymnotid teleosts the electrosensory lateral line information, before attaining the tectum, has already passed through two consecutive laminated centres, the rhombencephalic ELLL (Fig. 2.52) and the mesencephalic torus semicircularis (Fig. 2.54). A spatially ordered representation of sensory input also occurs in the sensory areas of the mammalian cortex. For example, the visual cortex is topically organised and contains a complete, ordered map of the visual field. This map is not only found in layer IV, which receives the bulk of the afferents from the lateral geniculate body, but is also demonstrable in other layers that lack a substantial direct input from the lateral geniculate body. Upper layers II and III and lower layers V and VI must therefore acquire their retinotopic order through radially oriented, interlaminar connections. Such radial projections, which have been experimentally demonstrated (e.g., Burkhalter 1989), are probably provided by the local circuit neurons with the ascending and/or descending axons observed in Golgi preparations (Lund 1987, 1988). Maplike representations of the external world as perceived by a sense organ may also be present in cortical areas not in direct receipt of sensory thalamocortical projections. It is interesting that pathways leading out of the primary visual, auditory and somesthetic areas of the neocortex via a number of intercalated association areas ultimately converge upon the hippocampus (for references, see Nieuwenhuys et al. 1988, p. 336), where a map of an animal's spatial environment is formed and stored (O'Keefe and Nadel 1978).

2.8.4
Notes on the Structure and Organization of the Tectum Mesencephali in Teleosts and Amphibians

This section concludes with a brief discussion of aspects of the organisation of the tectum mesencephali in teleosts and amphibians, which exemplifies some general issues.

2.8.4.1
The Teleostean Tectum Mesencephali

The teleostean tectum mesencephali has already been dealt with in Sect. 2.7.2 (Fig. 2.41). We return to it to draw attention to the remarkable nature of its outermost layer. This outermost layer, i.e. the SMARG, is composed mainly of extremely thin (0.1 mm), unmyelinated, mediolaterally running fibres, which closely resemble the cerebellar parallel fibres (Vanegas 1983; Vanegas et al. 1974, 1979, 1984). These marginal fibres originate from small granule cells in the torus longitudinalis (Schroeder 1974; Ito and Kishida 1978), an elongated structure which is bilaterally formed by the most medial parts of both tectal halves (Fig. 2.54a). Like the SMARG, the torus longitudinalis is unique to actinopterygian fishes. The small granule cells in the tori show a conspicuous resemblance to cerebellar granule cells (Ito 1971). The principal sources of afferents to the tori is the valvula cerebelli, likewise a structure unique to actinopterygians (Ito and Kishida 1978). These afferents form synaptic complexes with the dendrites of granule cells that are very similar to cerebellar glomeruli (Vanegas et al. 1984).

The principal postsynaptic elements of the SMARG are the spiny apical dendritic branches of large fusiform neurons, the somata of which are located in the SFGS (type I cells of Meek and Schellart 1978). The marginal fibres make excitatory synapses on the spines of the apical dendritic tufts of these large elements (Vanegas et al. 1979). The synaptic contacts are like those between cerebellar parallel fibres and Purkinje dendrites. However, unlike Purkinje cells, the dendrites of the tectal fusiform neurons are not restricted to a single plane. The tectal dendrites receive both a strong toral input and a limited visual input from the retina. Their axons contribute to a plexiform layer in theSGC, where they make synapses with both intrinsic and efferent elements (Meek and Schellart 1978).

Thus it appears that in actinopterygians a remarkable cerebelloid torus longitudinalis-tectal marginal layer, parallel fibre system is superimposed upon the tectum and that, via equally remarkable large fusiform neurons, this superimposed system is integrated into tectal circuitry. Since this configuration is unique for actinopterygians it may be assumed that the merging of the two components has taken place at an early chondrostean level. Interestingly, a torus longitudinalis and a tectal marginal layer are lacking in polypteriforms, a small group of primitive fishes which appear to have branched off very early from the palaeoniscoid actinopterygian stock. The phenomenon whereby structures within an organism which have developed and differentiated independently unite to form a new structural and functional entity is known as synorganisation (Remane 1966).

2.8.4.2
The Amphibian Tectum Mesencephali

The structure of the tectum mesencephali in the anurans and the urodeles differs considerably. Anurans, like most other vertebrate groups, have a multistratified tectum with alternating cellular and plexiform layers, while urodeles have only two tectal layers, a central cellular and a peripheral neuropil and fibre zone (Herrick 1948; Leghissa 1962). The urodeles share this simple bilaminar tectal configuration with the gymnophiones, another group of amphibians, and with the lepidosirenid lungfishes; furthermore, in these three groups a bilaminar structure is found throughout the neuraxis.

Roth and collaborators have made extensive morphological and histochemical studies (Roth 1986, 1987; Roth et al. 1990; Rettig and Roth 1986; Schmidt et al. 1989) which suggest that the differences between the anuran and the urodelan tectum are less profound than their overall histology suggests. Roth (1986; see Fig. 2.56a) reported that in some salamanders the zone of central grey is divisible into separate deep, intermediate and superficial cellular layers (2, 4, 6), which are separated from each other by two narrow fibre layers, consisting of unmyelinated axons (3, 5). In the peripheral zone of the urodelan tectum, Roth distinguished five layers, two deeper ones consisting mainly of efferent fibres (7, 8) and three more superficial ones (A, B, C) composed of retinal afferents and terminals. On the basis of an experimental neuroanatomical analysis, Rettig and Roth (1986) concluded that, with regard to the laminar distribution of retinal afferents in the tectum, there are no major differences between urodeles and anurans. Furthermore, Schmidt et al. (1989) demonstrated that the urodelan tectum, in spite of its simple two-layered morphology, shows a high degree of neurochemical differentiation, characterised by a distinct laminar distribution of a number of neuropeptides and of acetylcholinesterase (AChE) activity. This distribution is comparable to that found in anurans.

Using the Golgi method, Roth et al. (1990) studied the neurons in the urodelan tectum. Using as criteria (a) the position, shape and size of the soma, (b) the angle of dendritic arborisation, and (c) the structure of the dendritic tree and the site of main dendritic arborisation, the following three main cell types were distinguished:

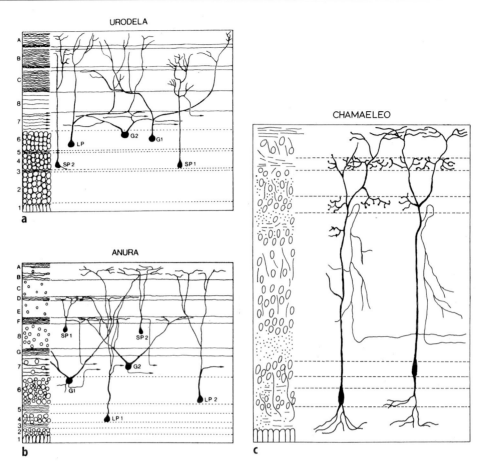

Fig. 2.56a-c. The structure of the tectum mesencephali. **a,b** A comparison between the organisation of the tectum of urodeles and anurans. On the *left side* of the drawings the cell and fibre layers of the tectum are given, according to the nomenclature of Székely and Lázár (1976). *Letters* indicate superficial retinal fibre layers. Three basic types of tectal neurons can be distinguished: (a) neurons with small pear-shaped somata and small dendritic trees making contact with different types of retinal afferents (*SP1, SP2*) – these neurons represent axonless interneurons or efferent elements with short-range axons; (b) neurons with large pear-shaped somata and wide dendritic trees arborising in different layers of retinal afferents (*LP, LP1, LP2*) which represent convergent efferent elements of the tectum and send their axons mostly to the tegmentum or to diencephalic (and telencephalic?) centres; (c) large ganglionic cells with wide and mostly flat dendritic trees (*G1, G2*) which constitute with their axons the crossed (G1) and uncrossed (G2) descending tracts connecting the tectum with the motor zones of the rhombencephalon and the spinal cord (from Roth 1986, Fig. 6). **c** Transverse section through the tectum of the chameleon, *Chamaeleo vulgaris*, showing the laminar organization on the left side and two typical *cellules à crosse* as seen in Golgi preparations on the right side (based on Ramón y Cajal 1909, Fig. 28)

1. Large ganglionic cells with wide, flat, dendritic trees that extend in laminae 7 and 8. The somata are found in lamina 6, directly beneath the efferent layers. The axons originate from one of their dendritic trunks and leave the tectum via laminae 7 and 8.
2. Large piriform cells, which possess a broad dendritic arborisation. In subtype 2a, the somata are located in the uppermost zone of lamina 6, and the dendrites always reach lamina A of retinal afferents; in subtype 2b, the somata are found at various levels in lamina 6, and the dendrites arborise in the deeper laminae (B, C) of retinal afferents; in subtype 2c the somata are located in the deeper zone of lamina 6 and often show several basal dendrites extending into lamina 5; their short apical dendrites break up into a number of branches which ramify mainly in laminae 7 and 8, i.e. the same layers via which their axons leave the tectum.
3. Small piriform neurons, with their somata distributed throughout the central grey. The dendritic trees of these elements are narrow and always extend into laminae A and B. Axons are often difficult to identify; when present, they originate from a dendritic branch and leave the tectum via layer 7 or 8.

Experiments in which HRP was injected in various target areas of tectal efferents revealed a correlation between the morphology of the tectal cells and their projections. Large piriform cells of the subtypes a and c contribute to the crossed and the direct tecto-bulbo-spinal tracts, respectively, whereas the large ganglionic cells contribute to both of these descending pathways. In contrast, the small piriform cells constitute the main origin of the ascending tecto-thalamic tract, the tecto-pretectal projections, and the tecto-isthmic tract and constitute the mass of intrinsic neurons.

Roth (1986, 1987) and Roth et al. (1990) compared their results on the morphology and the efferent projection of tectal neurons in urodeles with corresponding data on the anuran tectum, derived from the studies of Lázár (1984) and Lázár et al. (1983). They concluded that the various types of tectal cells in urodeles are similar to those in anurans (cf. Fig. 2.56b). According to Roth et al., the only major difference in tectal morphology between anurans and urodeles is the degree of migration of the small piriform cells. In anurans these elements are found only in the superficial cellular layers 8, C and E of the tectum, *above* the layers of tectal efferents; in urodeles their somata are found in the three layers of periventricular grey (2, 4, 6), i.e. *below* the layers of tectal efferents. Roth et al. (1990) thought that the lack of cellular migration in the urodelan tectum and in the CNS of urodeles, gymnophiones and lepidosirenid lungfishes in general, is due to a secondary simplification. They pointed out that all these groups are paedomorphic vertebrates because they retain many juvenile characters and lack numerous adult features. Interestingly, many urodele species show behaviour that is at least as complex as that of highly evolved anurans; this is especially true of their visually guided feeding behaviour, where the tectum is critically involved (Roth 1986, 1987). These data caused Roth et al. (1990) to wonder about the functional significance of lamination and migration in the anuran tectum, and the physiological and behavioural consequences of secondary simplification of these morphological characters in urodeles.

The similarities and differences between the urodelan and the anuran tectum mesencephali call for a brief commentary. In a neural centre in which (a) the peripheral zone is occupied by afferent fibres and terminals of a different nature which are arranged in separate laminae, and (b) all constituent neurons have their somata concentrated in a central, periventricular zone, all neurons are potentially able to sample information from any combination of afferents by sending dendrites into particular layers and by making synaptic contacts with the afferents present in these layers. Theoretically, the following five reasons for the periventricular position of the somata may be considered:

1. The centre is in a primitive condition. The neuronal somata are 'still' situated close to their site of origin, i.e. the former ventricular matrix, and populate a persistent embryonic mantle layer.
2. The centre is secondarily simplified due to paedomorphosis.
3. The position of the somata and the initial axonal segments is functionally optimal for the hierarchical integration of the afferent impulses.
4. The position of the somata and the initial axonal segments is functionally optimal, because the neurons receive their input both via the superficial fibre laminae and via deep fibres which invade the grey matter.
5. The position of the somata is optimal for certain trophical or metabolic conditions.

If a neuron situated within such a centre intends to strengthen the influence of the afferent input from one particular source, it can utilise the following three transformations: (a) an increase of the number and/or size of the contacts with the terminals of the axons carrying that particular type of input; (b) a displacement of the site of origin of its axon peripherally toward that part of its receptive surface where the input, the influence of which is to be strengthened, arrives; and (c) the migration of the cell body in the same direction. This migration may lead to a considerable reduction of the total dendritic length of the neuron.

Considering the structural relations in the tectum of urodeles, anurans and other vertebrates, many morphological features of the neurons may be interpreted as the result of the three evolutionary transformations indicated above. Neurons which apparently intensified their contact with particular inputs are present in the tectum of teleosts as monostratified, bistratified and multistratified neurons (Fig. 2.41), and also in reptiles, e.g. in the form of the remarkable *cellules à crosse* described and depicted by Ramón y Cajal (1911; Fig. 2.56c). In anurans the same process may have led to the formation of peculiar appendages to those dendritic branches of the large pear-shaped neurons which extend into the layers of retinal afferents, and also to the formation of the beaded dendritic twigs of the terminal dendrites of the small pear-shaped neurons (Székely and Lázár 1976).

Transposition of the site of origin of the axon with the apparent purpose of strengthening the influence of the input is observed in teleosts, amphibians and reptiles. In the tectum of all these groups there are neurons in which the axon arises from a peripherally directed dendritic shaft, often distant from the soma. Examples are: cell types XII and XIV in the goldfish (Meek and Schellart 1978; Fig. 2.41), the axon-bearing variant of the small piriform cells in urodeles (Fig. 2.56A), the large pyramidal cells and certain large piriform cells in anurans (Fig. 2.56B), and, once again, the reptilian *cellules à crosse* (Fig. 2.56c).

The third and most radical change, i.e. the migration of somata, probably occurred in all tecta in which, in addition to a zone of central grey, one or more layers of more superficially situated cells are present. Such layers can be found in the tectum of most groups of vertebrates, and several examples of neurons found in these layers are depicted in Figs. 2.41 and 2.56b. It is remarkable that none of the cell types occurring in the more superficial tectal layers bear any dendrite descending into the periventricular cell layer. If dendrites of these cells in an earlier evolutionary phase received input via deep afferent axons passing through the periventricular layer, they apparently lost this input before their somata started to migrate peripherally.

We return now to the structural difference between the tectum of anurans and that of urodeles. It is apparent that the three transformations have found a much wider application in the tectum of anurans than in that of the latter group. In this sense, the urodelan tectum is simpler than the anuran tectum. A question which cannot be answered by neuroanatomical data alone is whether the structural condition of the urodelan tectum is genuinely simpler or has become simple through paedomorphosis. Another question which cannot be answered is whether the urodelan tectum is also primitive in the sense of not having an optimal design. If we compare the small piriform cells in the tectum of urodeles with similar elements in the tectum of anurans, it appears that the urodelan elements, with their periventricular somata and their long dendritic shafts, are more primitive than the anuran elements with their migrated somata. The latter seem to answer much better to Ramón y Cajal's (1909) *lois d'économie d'espace et de substance* than the former. These elements conceivably exchange information in a non-synaptic fashion and, as indicated earlier, their periventricular position may be related to certain trophic or metabolic conditions.

2.9
Chemoarchitecture and Chemodifferentiation of Grisea

2.9.1
Introduction

Although the theory that neurons are true secreting cells which act upon one another by the passage of chemical substances was enunciated at the beginning of the 1900s (Scott 1905), it was only half a century later that the significance of humoral transmission for the processing of information in the central nervous system became fully appreciated. It was then established that the transfer of neuronal impulses occurs at morphologically differentiated contact sites, the synapses, and that the influence exerted by a presynaptic element via a neurotransmitter could be either excitatory or inhibitory. A principle first formulated by Dale (1935) for the peripheral nervous system, i.e. that "each neuron releases one and the same neurotransmitter at all its terminals," was thought valid for the central nervous system as well. Once released into the synaptic cleft, the neurotransmitter molecules reach specific receptors embedded in the membrane of the postsynaptic element, and interaction with the receptors opens specific ionic channels. It was held that at all of the synaptic terminals of a given neuron the transmitter opens just one type of ionic channel, characterising either excitatory or inhibitory synapses (e.g., Eccles 1969). Initially, the number of known central neurotransmitters was extremely small, including hardly more than acetylcholine (ACh) and noradrenaline, and it was thought that the brain could manage with one excitatory and one inhibitory neurotransmitter. These early ideas reached their peak around 1960, when several investigators (e.g., Gray 1959, Uchizono 1965) reported that excitatory and inhibitory synapses can be distinguished from each other at the ultrastructural level.

After this period there were two notable developments: the number of neurotransmitters increased steadily, and the views on the modes in which neurons may be influenced widened considerably. By the end of the 1960s, the following CNS transmitters had been identified: ACh, noradrenaline, dopamine, adrenaline and histamine. In the late 1960s and early 1970s it became apparent that, in addition to their metabolic role, certain amino acids such as γ-aminobutyric acid (GABA), glutamic acid, aspartic acid and glycine were neurotransmitters (Graham et al. 1967; Curtis and Johnston 1974).

During the past two decades there has been a dramatic increase in the number of possible neuroactive substances, with the growing recognition that various peptides fulfil the role of neuromediators (e.g., Guillemin 1978, Snyder 1980, Krieger 1983). Present concepts can be summarised as follows:

1. Although there are neurotransmitters that are mainly excitatory in action (e.g., aspartic acid, glutamic acid) and others that are mainly inhibitory (e.g., GABA, glycine), none of the known transmitters can be defined in their own right as either excitatory or inhibitory (and thus as restricted to opening just one type of ionic channel). The effect elicited by a neurotransmitter depends not only on its chemical structure, but also on the nature of the receptor with which it combines. For some neurotransmitters, for example ACh and noradrenaline, both excitatory and inhibitory receptors exist.

2. The binding of some neurotransmitters to receptors does not lead to the opening of ionic channels, (and to the postsynaptic potentials correlated with the ensuing ion displacement), but rather to the activation of much longer lasting intraneuronal processes, which produce enzymes that catalyse the formation of neurotransmitters and neuropeptides. In these processes cyclic nucleotides play an essential biochemical role as intracellular second messengers (Nathanson 1987).

3. In addition to 'classical' synaptic transmission (see Sect. 2.7.4), non-synaptic chemical neurotransmission probably plays an important role in the interneuronal communication in the vertebrate CNS.

4. Certain neurochemicals not directly involved in the process of synaptic transmission are able to influence this process in various ways. These 'neuromodulators' (see Florey 1967; Weight 1979; Kupfermann 1979; Rotsztejn 1980) may be active at the presynaptic side, e.g. by affecting the amount of transmitter released and the time course of transmitter release, as well as on the postsynaptic side, e.g., by regulating the sensitivity of the receptors.

5. We know, due largely to the investigations of Tomas Hökfelt and his associates, that in many neurons more than one neuroactive substance is present (for review see Lundberg and Hökfelt 1983; Hökfelt 1987). 'Multiple neuromediator' neurons commonly synthesise a classical neurotransmitter (e.g., ACh, a monoamine or an amino acid) in combination with one or more neuropeptides. The functions of the neuromediators that are co-released are not well understood. Theoretically, the potential for releasing several messengers greatly increases the possible number of different chemical signals that a neuron can use in communicating with other elements, particularly when the elements innervated have multiple receptors (Iversen 1983; Swanson 1983, 1991; Hökfelt et al. 1984a). Some ideas about the significance of co-localisation of a classical neurotransmitter and a neuropeptide are (see Lundberg and Hökfelt 1983; Hökfelt et al. 1984a): (a) The neuropeptide may be responsible for a slow onset and long duration of the response, while the classical neurotransmitter causes rapid effects of short duration. (b) The neuropeptide may play an auxiliary role as co-messenger in support of the classical neurotransmitter. Such synergistic actions of neuropeptides may be effected in different ways, for instance by blocking presynaptic autoreceptors for the classical neurotransmitters, or by cooperating with the classical transmitter at postsynaptic sites.

Thus, investigations during the past 25 years have shown that in the CNS (a) a multitude of chemical messengers are involved in interneuronal communication; (b) apart from classical synaptic transmission, non-synaptic or paracrine transmission most probably plays an important role; (c) chemical interneuronal signalling includes both the classical, rapid excitatory and inhibitory transmission and a variety of slower, neuromodulatory processes; and (d) many neurons contain, in addition to a classical neurotransmitter, one or several neuropeptides.

It was against the background of the developments just outlined that 'chemical neuroanatomy' (Emson 1983) emerged. Assuming Dale's principle, the initial question was: What is the neurotransmitter of this particular neuron or set of neurons? In this way it was hoped to determine the 'chemical fingerprint' of neurons. Dahlström and Fuxe (1964, 1965) pioneered this area by charting the distribution of monoaminergic cells and fibres in the CNS of the rat, using the formaldehyde-induced fluorescence technique developed by Falck and Hillarp (cf. Falck et al. 1962), and Shute and Lewis (1967; Lewis and Shute 1967), who used AChE staining as a potential marker for cholinergic neurons in combination with lesions. The introduction of the powerful technique of immunohistochemistry (Coons 1958; Sternberger 1979) for the localisation of neurotransmitters and their synthesising enzymes, as well as of neuropeptides, has provided an enormous impetus to the field of chemical neuroanatomy. These techniques are very flexible and can be used in double-label experiments to demonstrate the co-existence of neuroactive substances in single neurons, as well as in combination with tract-tracing techniques such as horseradish perox-idase histochemistry and *Phaseolus vulgaris* leu-coagglutinin to determine the chemical identity of axonal projections (see Chap. 7). The technique of in situ hybridisation has been successfully employed for the identification and localisation of neurons that express messenger RNAs encoding for proteins like the catecholamine enzymes and the precursors of neuroactive peptides.

Ultrastructural studies making use of the power-ful histochemical techniques just mentioned have yielded results which are of great importance for our understanding of the mechanisms of interneu-ronal chemical communication; however, in most studies using these techniques, the labelled neuro-transmitters, enzymes, neuropeptides or messenger RNAs are employed as markers to identify and cha-racterise particular groups of neurons. During the past 25 years, a great number of mapping studies have appeared describing a variety of neu-romediator-specific neuron populations,and their fibre networks and terminal plexuses in many ver-tebrates. It seems that the various neuromediator-specific neuronal populations do not obey simple structural and functional rules. However, some general concepts can be formulated concerning: (a) the relations between neuromediator-specific neu-ron populations and terminal plexuses on the one hand, and the cytoarchitectonic units of classical neuroanatomy on the other; (b) the comparative anatomy of neuromediator-specific neuron popula-tions; and (c) the functional significance of neuromediator-specific neuronal populations (see Nieuwenhuys 1985 for more detail).

Figures 2.57–2.61 are a series of drawings derived from representative mapping studies. These figures provide an overview of the distribution of some neuromediator-specific neuronal populations and their related fibre networks and terminal plex-uses in the brains of a number of vertebrates.

2.9.2
The Relation Between Neuromediator-Specific Neuron Populations and Terminal Plexuses, and the Cytoarchitectonic Units of Classical Neuroanatomy: Chemoarchitecture Versus Cytoarchitecture

2.9.2.1
Neuromediator-specific Neuronal Populations Vary Considerably with Regard to (a) the Area Occupied by Their Perikarya and (b) the Spread of their Fibres and Terminals

Cells containing GABA, somatostatin and enkepha-lin are found in many different areas of the brain, extending throughout most parts (e.g., Fig. 2.58). In contrast, cells containing dopamine, noradrenaline, serotonin, ACTH or angiotensin II occur in a lim-ited number of cell groups which are generally con-fined to a relatively small area of the brain. Finally, the histamine-containing neurons are concentrated in a single cluster situated in the caudal hypothala-mus.

With regard to the spread of fibres, it has been found that noradrenergic and serotoninergic fibres are distributed to virtually all areas of the CNS (e.g., Figs. 2.57, 2.59). Fibres containing adrenaline, vasoactive intestinal polypeptide or oxytocin, on the other hand, project to only a very limited num-ber of centres. The remaining neuromediator-specific neuron populations in the vertebrate brain occupy an intermediate position between these two extremes.

2.9.2.2
Although in a Number of Grisea Most Neurons Contain One Neuromediator, Most Grisea Are Composed of Subpopulations, Each Containing a Different Neuromediator

In mapping studies using an antibody against a particular neuromediator or its synthesising enzyme, it is rarely observed that most neurons in a given cell mass are labelled (e.g., the cholinergic neurons in motor nuclei). Commonly, the antibody labels only a limited number of cells within a par-ticular griseum (Figs. 2.57, 2.58, 2.62). Studies in which antibodies against different neuromediators are used reveal that in many grisea, cells containing different neuromediators occur.

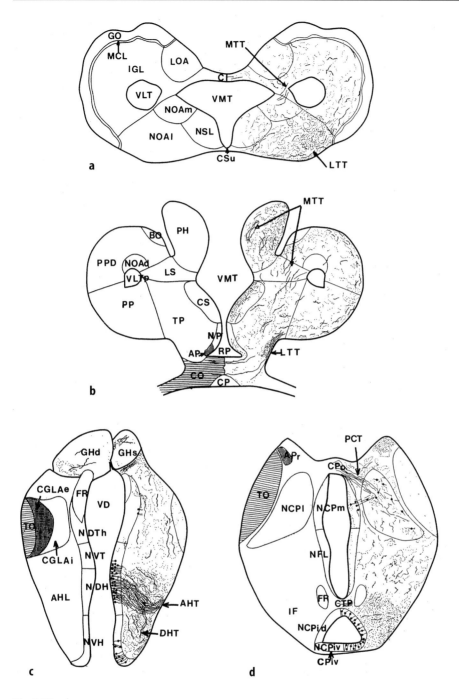

Fig. 2.57a–d

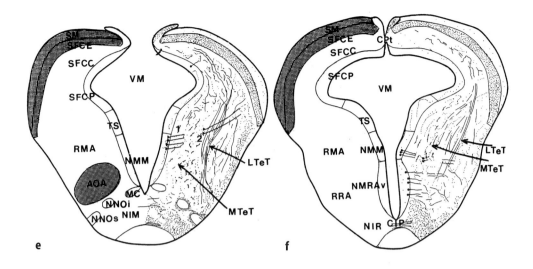

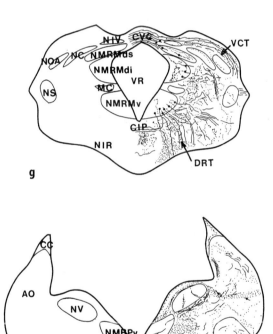

Fig. 2.57e–h

◀ **Fig. 2.57a–h.** Camera lucida drawings of transverse sections of the brain of the lamprey, *Lampetra fluviatilis*, showing the distribution of serotonin-immunoreactive cell bodies (*large dots*) and fibres and terminals (*fine stipples*). The retinofugal system, visualised by tritiated proline autoradiography, is indicated on the *left* of the drawings, the fibres of the optic chiasm and marginal optic tract by *parallel lines*, and the projection zones by *cross-hatching*; the cell bodies of the retinopetal fibres in the mesencephalic tegmentum are indicated in part **e** by *filled triangles*. **a** Olfactory bulb and rostral telencephalon; **b** central telencephalon; **c** diencephalon; **d–f** mesencephalon; **g,h** rhombencephalon. Abbreviations: *AHL*, area hypothalamica lateralis; *AHT*, 5HT fibres in ascending hypothalamic tract; *AO*, area octavolateralis; *AP*, area preoptica; *APr*, area pretectalis; *BO*, bulbus olfactorius; *CC*, crista cerebellaris; *CGLae*, corpus geniculatun laterale anterior, pars externa; *CGLai*, corpus geniculatun laterale anterior, pars interna; *CI*, commissura interbulbaris; *CIP*, commissura interpeduncularis; *CO*, chiasma opticum; *CP*, commissura postoptica; *CPiv*, commissura postinfundibularis ventralis; *CPo*, commissura posterior; *CPt*, commissura post-tectalis; *CS*, corpus striatum; *CSu*, commissura supraoptica; *CTP*, commissura tuberculi posterioris; *CVC*, commissura vestibulocerebellaris; *DHT*, 5HT fibres in descending hypothalamic tract; *DRT*, 5HT fibres in descending rhombencephalic tract; *FR*, fasciculus retroflexus; *GHd*, ganglion habenula dexter; *GHs*, ganglion habenula sinister; *GO*, glomeruli olfactorii; *IF*, infundibulum; *IGL*, internal granular layer; *LOA*, lobus olfactorius accessorius; *LTeT*, 5HT fibres in lateral tegmental tract; *LTT*, 5HT fibres in lateral telencephalic tract; *MC*, Müller cell; *MCL*, mitral cell layer; *MTeT*, 5HT fibres in medial tegmental tract; *MTT*, 5HT fibres in medial telencephalic tract; *NC*, nucleus cerebelli; *NCPid*, nucleus commissurae postinfundibularis pars dorsalis; *NCPiv*, nucleus commissurae postinfundibularis pars ventralis; *NCPl*, nucleus commissurae posterioris lateralis; *NCPm*, nucleus commissurae posterioris medialis; *NDH*, nucleus dorsalis hypothalami; *NDTh*, nucleus dorsalis thalami pars subhabenularis; *NFL*, nucleus fasciculi longitudinalis medialis; *NIM*, nucleus interpeduncularis mesencephali; *NIR*, nucleus interpeduncularis rhombencephali; *NIV*, nucleus motorius nervi trochlearis; *NMM*, nucleus tegmenti mesencephali; *NMRAv, NMRds, NMRMv, NMRPv*, parts of nucleus isthmi rhombencephali; *NNOi*, nucleus nervi oculomotorii intermedius; *NNOs*, nucleus nervi oculomotorii superficialis; *NOA*, nucleus octavomotorius anterior; *NOAd*, nucleus olfactorius anterior dorsalis; *NOAl*, nucleus olfactorius anterior lateralis; *NOAm*, nucleus olfactorius anterior medialis; *NP*, nucleus preopticus; *NS*, nucleus sensibilis nervi trigemini; *NSL*, nucleus septi lateralis; *NVH*, nucleus ventralis hypothalami; *NVT*, nucleus ventralis thalami; *N V*, nucleus motorius nervi trigemini; *N VII*, nervus facialis; *PCT*, 5HT fibres in commissural tract; *PH*, primordium hippocampi; *PP*, primordium piriforme; *PPD*, primordium pallii dorsalis; *RMA*, reticular mesencephalic area; *RP*, recessus preopticus; *RRA*, rhombencephalic reticular area; *SFCC*, stratum fibrosum et cellulare centrale; *SFCE*, stratum fibrosum et cellulare externum; *SFCP*, stratum fibrosum et cellulare periventriculare; *SM*, stratum marginale; *TO*, tractus opticus; *TP*, telencephalic peduncle; *TS*, torus semicircularis; *VCT*, 5HT fibres in vestibulocerebellar tract; *VD*, ventriculus diencephali; *VLTa*, ventriculus lateralis telencephali anterior; *VLTp*, ventriculus lateralis telencephali posterior; *VM*, ventriculus mesencephali; *VMT*, ventriculus medius telencephali; *VR*, ventriculus rhombencephali; *1,2*, groups of retinopetal cells. (adapted from Pierre et al. 1992)

2.9.2.3
The Neuromediator Profiles of the Various Cell Masses Present in the Brain of a Given Species or Group May Differ Considerably

Nieuwenhuys (1985) compiled data concerning the total number of neuromediators and the neuromediator profiles in a large number of cell masses of the brain of mammals (mainly the rat). It appeared that the numbers of neuromediator-specific subpopulations and the neuromediator profiles differ considerably from cell mass to cell mass. Some nuclei in the mammalian brain, such as the nucleus centralis amygdalae, the bed nucleus of the stria terminalis and the periaqueductal grey, contain an extraordinarily large number of neuromediator-specific subpopulations and may be aptly designated as 'multineuromediator complexes'.

2.9.2.4
Neuromediator-specific Neurons Occurring Within the Confines of Given Grisea Are Often Not Equally Dispersed in Their Grisea

The neurons containing a particular neuromediator may be localised to one or more parts of a given centre. Thus, numerous neurons in the inferior reticular nucleus of the lamprey use glutamate as their principal neurotransmitter (Brodin et al. 1989). However, some neurons confined to the medial part of that nucleus also contain a cholecystokinin-like peptide (Brodin et al. 1988; Ohta et al. 1988; Fig. 2.62).

Transmitter specifications of cells present within a given griseum may lead to a chemoarchitectonic subdivision which may or may not correspond to a cytoarchitectonic subdivision. For example, in the golden hamster the suprachiasmatic nucleus can be divided on cytoarchitectonic grounds into dorsomedial and ventrolateral subdivisions. Somatostatin- and vasopressin-containing neurons are localised within the dorsomedial subdivision, whereas vasoactive intestinal polypeptide-immunoreactive neurons are concentrated in the ventrolateral subdivision (Card and Moore 1984; Fig. 2.63). In man, a similar differential distribution of vasopressin- and vasoactive intestinal polypeptide-containing neurons was found in the suprachiasmatic nucleus (Stopa et al. 1984) but is has not been established whether there is a corresponding cytoarchitectonic subdivision.

Fig. 2.58a–f. Line drawings of transverse sections of the brain of the turtle, *Pseudemys scripta*, illustrating the location and relative density of leucine-enkephalin-immunoreactive perikarya (*triangles*), as shown on the *left side* of each drawing, and fibres and terminals (*dots*), as shown on the *right side* of each drawing. **a** Telencephalon; **b** diencephalon; **c,d** mesencephalon; **e** cerebellum and rostral rhombencephalon; **f** caudal rhombencephalon. Abbreviations: *AP*, area pretectalis; *BOR*, basal optic root; *CA*, nucleus centralis amygdalae; *Cb*, cerebellum; *CbL*, nucleus cerebellaris lateralis; *CbM*, nucleus cerebellaris lateralis; *cd*, cortex dorsalis; *cdm*, cortex dorsomedialis; *cm*, cortex medialis; *CN*, core nucleus of the DVR; *CO*, chiasma opticum; *Co*, cochlear nuclei; *CP*, commissura posterior; *cp*, cortex pyriformis; *cpv*, cortex pyriformis, pars ventralis; *d*, area d; *DLA*, nucleus dorsolateralis anterior; *DMA*, nucleus dorsomedialis anterior; *DVR*, dorsal ventricular ridge of the telencephalon; *FLM*, fasciculus longitudinalis medialis; *FRL*, formatio reticularis lateralis mesencephali; *GCL*, granule cell layer; *GLv*, nucleus geniculatus lateralis pars ventralis; *GP*, globus pallidus; *HL*, nucleus habenularis lateralis; *IP*, nucleus interpeduncularis; *La*, nucleus laminaris of the torus semicircularis; *LM*, nucleus lentiformis mesencephali; *LoC*, locus coeruleus; *M*, nucleus mamillaris; *MA*, nucleus medialis amygdalae; *ME*, median eminence; *ML*, molecular layer; *NPd*, nucleus pretectalis dorsalis; *NPv*, nucleus pretectalis ventralis; *nBOR*, nucleus of the basal optic root; *nDB*, nucleus fasciculus diagonalis Brocae; *nDCP*, nucleus dorsalis commissurae posterioris; *nPH*, nucleus periventricularis hypothalami; *nSL*, nucleus septalis lateralis; *nSM*, nucleus septalis medialis; *nSO*, nucleus supraopticus; *nSP*, nucleus suprapeduncularis; *nSPM*, nucleus suprapeduncularis medialis; *nTOL*, nucleus tracti olfactorii lateralis; *nTS*, nucleus tracti solitarii; *nVH*, nucleus ventromedialis hypothalami; *n III*, nucleus nervi oculomotorius; *n VII*, nucleus nervi facialis; *PA*, paleostriatum augmentatum; *Pb*, nucleus parabrachialis; *PD*, peduncularis dorsalis fasciculi prosencephali lateralis; *PH*, primordium hippocampi; *Pr V*, nucleus princeps nervi trigemini; *PT*, pallial thickening; *PV*, pedunculus ventralis fasciculi prosencephali lateralis; *R*, nucleus rotundus; *Rai*, nucleus raphes inferior; *Ras*, nucleus raphes superior; *Re*, nucleus reuniens; *Rm*, nucleus reticularis medialis; *Rs*, nucleus reticularis superior; *Ru*, nucleus ruber; *SGC*, stratum griseum centrale; *SGF*, stratum griseum et fibrosum superficiale; *SN*, substantia nigra; *SO*, stratum opticum; *Tel*, telencephalon; *TO*, tractus opticus; *TOm*, tractus opticus, pars medialis; *TTd*, nucleus descendens nervi trigemini; *TT*, tractus tectothalamicus (adapted from Reiner 1987)

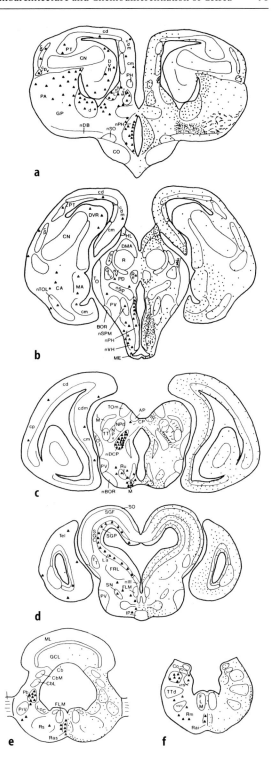

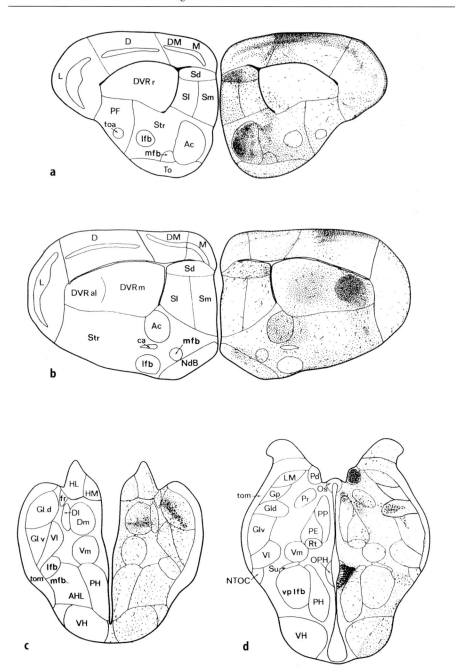

Fig. 2.59a–d.

Fig. 2.59a–g. Camera lucida drawings of transverse sections of the brain of the viper *Vipera aspis*, showing the distribution of serotonin-immunoreactive cell bodies (*large dots*) and fibres and terminals (*fine stippled areas*). **a,b** Telencephalon; **c,d** diencephalon; **e** mesencephalon; **f** cerebellum and rostral rhombencephalon; **g** caudal rhombencephalon. Abbreviations: *Ac*, nucleus accumbens; *AHL*, lateral hypothalamic area; *ca*, commissura anterior; *Cer*, cerebellum; *Co*, nucleus cochlearis; *D*, dorsal cortex; *Dl*, nucleus dorsolateralis thalami; *DM*, dorsomedial cortex; *Dm*, n. dorsomedialis thalami; *DVRal*, dorsal ventricular ridge, pars angulolateralis; *DVRm*, dorsal ventricular ridge, pars angulomedialis; *DVRr*, dorsal ventricular ridge, pars rostralis; *flm*, fasciculus longitudinalis medialis; *fpd*, fasciculus predorsalis; *fr*, fasciculus retroflexus; *GC*, substantia grisea centralis; *Gld*, nucleus geniculatus lateralis pars dorsalis; *Glv*, nucleus geniculatus lateralis pars ventralis; *Gp*, nucleus geniculatus pretectalis; *gr*, granular layer of the cerebellum; *HL*, nucleus habenularis lateralis; *HM*, nucleus habenularis medialis; *Ip*, nucleus interpeduncularis; *L*, lateral cortex; *lfb*, lateral forebrain bundle; *LM*, nucleus lentiformis mesencephali; *ls*, lemniscus spinalis; *M*, medial cortex; *mfb*, medial forebrain bundle; *MV*, nucleus motorius nervi trigemini; *N IV*, nucleus nervi trochlearis; *N Vds*, nucleus descendens nervi trigemni; *N VI*, nucleus nervi abducentis; *N VII*, nucleus nervi facialis; *N VIII*, nucleus vestibulocochlearis; *NCel*, nucleus cerebellaris lateralis; *NCem*, nucleus cerebellaris medialis; *NdB*, nucleus of the diagonal band; *Nm V*, nucleus mesencephalicus nervi trigemni; *NTOC*, thalamic centrifugal optic nucleus; *OPH*, periventricular hypothalamic organ; *Os*, subcommissural organ; *Pd*, nucleus posterodorsalis; *PE*, nucleus lentiformis, pars extensa; *PF*, nucleus perifascicularis; *PH*, nucleus periventricularis hypothalami; *PP*, nucleus lentiformis, pars plicata; *Pr*, nucleus pretectalis; *Pr V*, nucleus princeps nervi trigemini; *Pu*, Purkinje cell layer of the cerebellum; *Rai*, nucleus raphe inferior; *Ras m*, nucleus raphe superior, pars medialis; *RID*, nucleus reticularis inferior, pars dorsalis; *RIV*, nucleus reticularis inferior, pars ventralis; *rm V*, motor root of the trigeminal nerve; *rs V*, sensory root of the trigemnial nerve; *RSL*, nucleus reticularis superior, pars lateralis; *RSM*, nucleus reticularis superior, pars medialis; *Rt*, nucleus rotundus; *sac*, stratum album centrale (optic tectum); *Sd*, nucleus septalis dorsalis; *sgfs*, stratum griseum et fibrosum superficiale (optic tectum); *Sl*, nucleus septalis lateralis; *Sm*, nucleus septalis medialis; *so*, stratum opticum (optic tectum); *Str*, corpus striatum; *Su*, nucleus suprapeduncularis; *sz*, stratum zonale (optic tectum); *To*, tuberculum olfactorium; *toa*, accessory olfactory tract; *tom*, tractus opticus marginalis; *ToS*, torus semicircularis; *ttd*, tractus descendens nervi trigemini; *Ve*, nucleus vestibularis; *VH*, nucleus ventralis hypothalami; *Vl*, nucleus ventrolateralis thalami; *Vm*, nucleus ventromedialis thalami; *vp lfb*, ventral peduncle of the lateral forebrain bundle. (adapted from Challet et al. 1991)

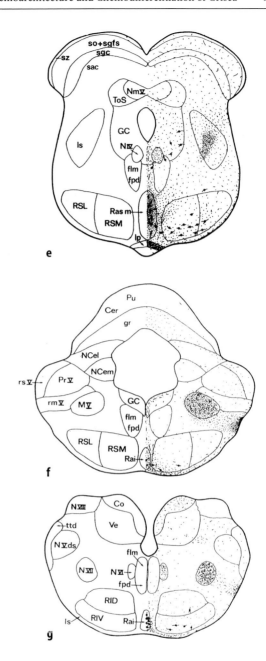

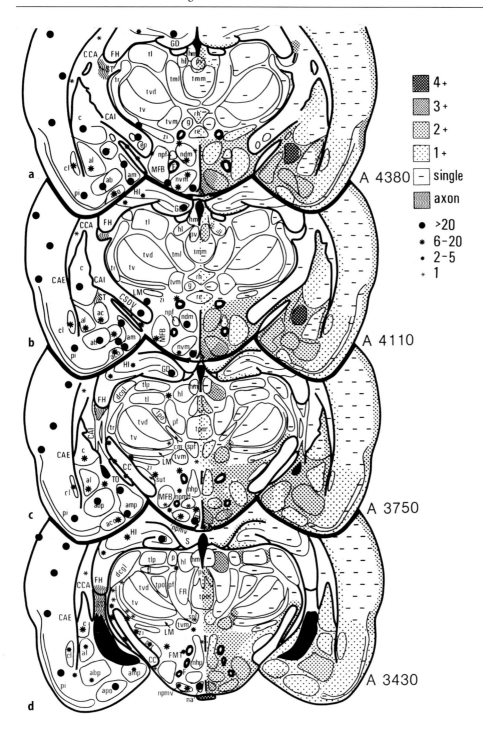

2.9.2.5
The Question of Whether or Not the Neuromediator-specific Subpopulations Present Within a Griseum Correspond to Particular Morphological Classes Is Often Left Unanswered

Information of this kind is confined mainly to identifiable cells. Thus, it is known that the highly characteristic mitral cells in the mammalian olfactory bulb use glutamate or aspartate as their neurotransmitter, and that the Purkinje and basket cells in the cerebellar cortex contain GABA, whereas the small cerebellar granule cells are glutamatergic. However, for cell populations which cannot be as easily classified on morphological grounds, histochemical results concerning neuromediator specification are often presented without any reference to the structural characteristics of the cells found. This holds in particular for studies dealing with the distribution of a particular neuromediator throughout the central nervous system.

◀ **Fig. 2.60.** Line drawings of a series of sections through the diencephalon and caudal telencephalon of the rat, illustrating the distribution and density of somatostatin-like immunoreactive perikarya (*left*) and fibres and terminals (*right*). The number of immunoreactive perikarya per section were indicated by four different symbols, as shown at the *upper right* hand side. The variations in density of fibres and terminals were estimated subjectively and indicated as: 4+, very high; 3+, high density; 2+, medium density; 1+, low density; – single fibres and no immunoreactivity. Abbreviations: *ab*, nucleus amygdaloideus basalis; *abp*, nucleus amygdaloideus basalis posterior; *aco*, nucleus amygdaloideus corticalis; *al*, nucleus amygdaloideus lateralis; *am*, nucleus amygdaloideus medialis; *amp*, nucleus amygdaloideus medialis posterior; *apo*, nucleus amygdaloideus posterior; *c*, nucleus caudatus; *CAE*, capsula externa; *CAI*, capsula interna; *CC*, crus cerebri; *CCA*, corpus callosum; *cl*, claustrum; *cm*, nucleus centromedianus thalami; *CSDV*, commissura supraoptica dorsalis, pars ventralis (Meynert); *dcgl*, nucleus dorsalis corporis geniculati lateralis; *ep*, nucleus entopeduncularis; *F*, fornix; *FH*, fimbria hipocampi; *FMT*, fasciculus mamillothalamicus; *g*, nucleus gelatinosus thalami; *GD*, gyrus dentatus; *HI*, hippocampus; *hl*, nucleus habenulae lateralis; *hm*, nucleus habenulae medialis; *LM*, lemniscus medialis; *MFB*, fasciculus medialis prosencephali; *na*, nucleus arcuatus; *ndm*, nucleus dorsomedialis hypothalami; *nhp*, nucleus hypothalamicus posterior; *npf*, nucleus perifornicalis; *npmd*, nucleus premamillaris dorsalis; *npmv*, nucleus premamillaris ventralis; *nvm*, nucleus ventromedialis hypothalami; *p*, nucleus pretectalis; *pf*, nucleus parafascicularis; *pi*, cortex piriformis; *pv*, nucleus periventricularis thalami; *re*, nucleus reuniens; *rh*, nucleus rhomboideus; *S*, subiculum; *spf*, nucleus subparafascicularis; *ST*, stria terminalis; *sut*, nucleus subthalamicus; *tl*, nucleus lateralis thalami; *tlp*, nucleus lateralis posterior thalami; *tml*, nucleus medialis thalami, pars lateralis; *tmm*, nucleus medialis thalami, pars medialis; *TO*, tractus opticus; *tpm*, nucleus posteromedianus thalami; *tpo*, nucleus posterior thalami; *tr*, nucleus reticularis thalami; *tv*, nucleus ventralis thalami; *tvd*, nucleus ventralis thalami, pars dorsalis; *tvm*, nucleus ventralis medialis thalami, pars magnocellularis; *zi*, zona incerta (adapted from Johansson et al. 1984)

2.9.2.6
Morphologically Homogeneous Sets of Neurons May Be Chemically Heterogeneous with Respect to Neuromediators

Two examples of this heterogeneity are given here.

1. Periglomerular cells in the mammalian olfactory bulb comprise two chemical subsets, one containing GABA, the other dopamine (Mugnaini et al. 1984a,b).
2. Immunohistochemical studies have revealed that several classes of neocortical neurons are chemically heterogeneous. Double-labeling experiments with antibodies raised against glutamate (Glu) and aspartate (Asp) have shown that the immunopositive pyramidal neurons can be subdivided into three populations: elements immunopositive only for Glu, elements immunopositive only for Asp, and elements immunopositive for both Glu and Asp (Giuffrida and Rustioni 1989a,b). As regards nonpyramidal cells, in the visual cortex of the rat a subpopulation of bipolar cells can be labelled with antibodies to vasoactive intestinal polypeptide (Peters and Harriman 1988), whereas in the neocortex of the monkey subpopulations of chandelier cells and of double bouquet cells have been shown to be immunoreactive for corticotropin-releasing factor (Lewis and Lund 1990) and tachykinin (DeFelipe et al. 1990), respectively. Another subpopulation of double bouquet cells has been shown to be immunoreactive for somatotropin release-inhibiting factor (de Lima and Morrison 1989).

2.9.2.7
Subpopulations of Neuromediator-specific Neurons Frequently Extend Beyond the Morphologically Defined Boundaries of Grisea

One example is that of serotoninergic neurons, which are concentrated in the raphe nuclei; however, in all groups of gnathostomes a considerable number of such neurons are scattered in the mesencephalic and rhombencephalic tegmentum, well beyond the raphe (cf., e.g., Stuesse and Cruce 1991, 1992: cartilaginous fishes; Wolters et al. 1986: a lizard; Challet et al. 1991: a snake: Fig. 2.59,e–g; Steinbush and Nieuwenhuys 1983: rat).

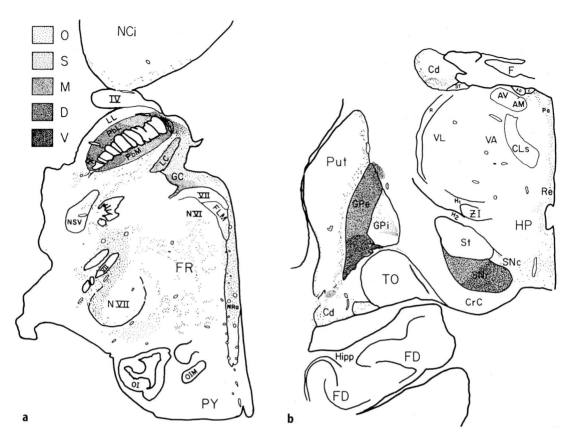

Fig. 2.61a,b. Schematic drawings of transverse sections through the rostral rhombencephalon (**a**) and the rostral mesencephalon, diencephalon and caudal telencephalon (**b**) of the monkey *Cercopithecus aethiops*, showing the distribution and density of enkephalin-immunoreactive fibres and terminals. *Density scales* indicate the relative density of fibres and terminals, which have been assigned values ranging from occasional (*O*), sparse (*S*), moderate (*M*), dense (*D*) and very dense (*V*). Abbreviations: *AD*, nucleus anterior dorsalis thalami; *AM*, nucleus anterior medialis thalami; *AV*, nucleus anterior ventralis thalami; *C*, nucleus centralis; *Cd*, nucleus caudatus; *CLs*, nucleus centralis lateralis superior thalami; *CrC*, crus cerebri; *F*, fornix; *FD*, fascia dentata; *FLM*, fasciculus longitudinalis medialis; *FR*, formatio reticularis; *GC*, substantia grisea centralis; *GPe*, globus pallidus, pars externa; *GPi*, globus pallidus, pars interna; *Hipp*, hippocampus; *HP*, area hypothalamica posterior; H_1, forel H_1; H_2, forel H_2; *LC*, locus coeruleus; *LL*, lemniscus lateralis; *NCi*, nucleus colliculi inferioris; *NRa*, nucleus raphes; *NSV*, nucleus tractus spinalis nervi trigemini; *N VI*, nucleus originis nervi abducentis; *N VII*, nucleus originis nervi facialis; *OI*, nucleus olivaris inferior; *OIM*, nucleus olivaris inferior medialis; *Pa*, nucleus paraventricularis thalami; *PbL*, nucleus parabrachialis lateralis; *PbM*, nucleus parabrachialis medialis; *Put*, putamen; *PY*, tractus pyramidalis; *Re*, nucleus reuniens thalami; *SNc*, substantia nigra, pars compacta; *SNr*, substantia nigra, pars reticulata; *St*, nucleus subthalamicus; *TO*, tractus opticus; *VA*, nucleus ventralis anterior thalami; *VL*, nucleus ventralis lateralis thalami; *ZI*, zona incerta (adapted from Haber and Elde 1982)

2.9.2.8
Numerous Populations or Subpopulations of Neuromediator-Specific Cells Are Not Organized Along the Lines of Defined Cytoarchitectonic Subdivision of the CNS

1. In the preoptico-hypothalamic region many groups of peptidergic neurons are not related to any cytoarchitectonic boundaries.
2. Although the (mainly noradrenergic) locus coeruleus and the (mainly dopaminergic) compact part of the substantia nigra clearly correspond to cytoarchitectonic units, most catecho-

laminergic cell groups have no special structural characteristics in Nissl material; they represent chemoarchitectonic rather than cytoarchitectonic units. Using the formaldehyde-induced fluorescence technique, Dahlström and Fuxe (1964) distinguished 12 groups of catecholaminergic neurons in the brain of the rat, which they labelled A1–A12. Hökfelt and his colleagues (1984b,c) more recently reanalysed the distribution of catecholaminergic neurons in the brain of the rat with immunohistochemical techniques. They confirmed the existence of the cell masses described by Dahlström and Fuxe (1964) and

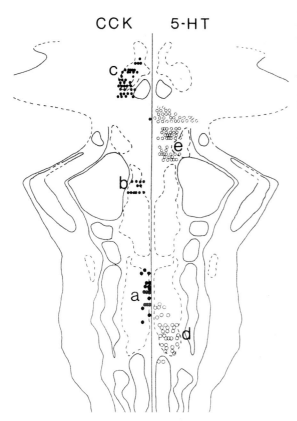

Fig. 2.62. Distribution of cholecystokinin- (*CCK*) and serotonin (*5-HT*)-immunoreactive cell bodies in the brain stem of the lamprey *Lampetra fluviatilis* projected on a topological map. *Filled circles* represent cholecystokin-positive cells; the serotoninergic cells are indicated by *open circles*. The *symbols* represent the actual number of cell bodies as detected in every third section of a serially sectioned brain stem. Symbols: *a*, nucleus reticularis posterior; *b*, nucleus reticularis medius; *c*, nucleus reticularis mesencephali; *d*, caudal group of serotonin-containing cells; *e*, isthmic group of serotonin-containing cells (adapted from Brodin et al. 1988)

The 'non-conformity' phenomena mentioned in this paragraph and the preceding paragraph certainly belong to the most intriguing general results of (immuno-)histochemical studies on the CNS. The 'centres' of classical neuroanatomy, delineated on cytoarchitectonic grounds, are commonly regarded not only as morphological entities, but also as functional units; however, the same holds true for the 'systems' of neuromediator-specific neurons detected by (immuno-)histochemical techniques (see below).

2.9.2.9
The Terminal Plexuses Present Within Many Grisea Are Composed of Subpopulations of Axonal Ramifications, Each Containing a Different Neuromediator. The Neuromediator Profiles of the Terminal Fields in Various Grisea of a Given Species May Differ Considerably

Nieuwenhuys (1985) collected data from the literature concerning the total number of known neuromediators and the neuromediator profiles of the fibres and terminals present within the confines of a large number of grisea in the mammalian CNS. In some nuclei, such as the nucleus ruber, the nuclei lemnisci lateralis and the medial geniculate body, the number of known neuromediator-specific terminal plexuses is limited (<5), but in others, such as the nucleus centralis amygdalae, the nucleus parabrachialis lateralis and the nucleus of the solitary tract, 12 or more different sets of neuromediator-specific terminal ramifications were recorded. The data collected showed that the neuromediator profiles differed from cell mass to cell mass (cf. Nieuwenhuys 1985, Table 5 on p.139). However, some striking similarities became apparent. Thus it was found that (according to what was known in 1985) the terminal plexuses in the nucleus of the solitary tract and in the lateral parabrachial nucleus both comprised 12 subplexuses, each of which contained a different neuropeptide. Moreover, it appeared that the neuropeptides present in the subplexuses of these two functionally related cell masses were exactly the same. It would be highly interesting to have a comparable compilation of data on non-mammalian vertebrates, as well as an update of the mammalian record.

identified four additional catecholaminergic cell groups. Thus, Hökfelt and his colleagues (1984b,c) distinguished nine groups of dopaminergic cell bodies (A8–A16) in the rostral part of the brain (midbrain, hypothalamus and olfactory bulb) and seven groups of noradrenergic cell groups, all of which are situated in the rhombencephalon. Adrenergic neurons are confined to three small cell groups in the caudal rhombencephalon. By analogy with the nomenclature of Dahlström and Fuxe (1964), these cell groups have been labelled C1–C3 (Hökfelt et al. 1974). It should be re-emphasised that of the 19 catecholaminergic cell groups just discussed, only a few correspond to well-defined cytoarchitectonic entities (A6: locus coeruleus; A9: substantia nigra, pars compacta; A16: periglomerular layer of olfactory bulb).

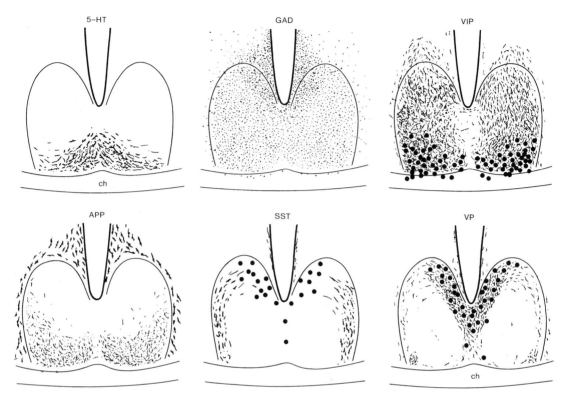

Fig. 2.63. Frontal sections through the nucleus suprachiasmaticus of the golden hamster to illustrate the distribution of serotonin (*5-HT*)-, glutamic acid decarboxylase (*GAD*)-, vasoactive intestinal polypeptide (*VIP*)-, avian pancreatic polypeptide (*APP*)-, somatostatin (*SST*)- and vasopressin (*VP*)-like immunoreactivity within that centre. *ch*, Chiasma opticum (modified from Card and Moore 1984)

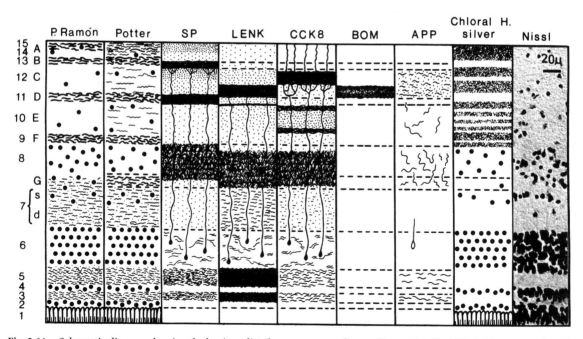

Fig. 2.64. Schematic diagram showing the laminar distribution of five different neuropeptides in the tectum mesencephali of the frog *Rana pipiens* correlated with structural features revealed by Nissl and choral-hydrate silver preparations. *Numbers* designate tectal layers according to P. Cajal as reported by Ramón y Cajal (1911); *letters* designate tectal layers according to Potter (1969). *APP*, Avian pancreatic polypeptide; *BOM*, bombesin; *CCK8*, cholecystokinin octapeptide; *d*, deep portion of layer 7; *LENK*, leucine-enkephalin; *s*, superficial portion of layer 7; *SP*, substance P. (reproduced from Kuljis and Karten 1982)

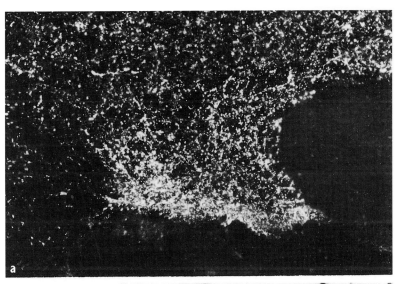

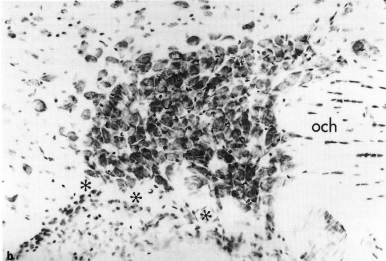

Fig. 2.65a,b. Distribution of substance P-immunoreactive fibres in the nucleus supraopticus of the rat. Darkfield photomicrograph of an immunoperoxidase section (**a**) and an adjoining Nissl-stained section (**b**) taken at roughly the middle of the rostro-caudal extent of the nucleus. Substance P-immunoreactive projections are most prominently distributed to the cell-poor dendritic zone of the nucleus (*asterisks* in **b**). *och*, optic chiasm (reproduced from Bittencourt et al. 1991)

2.9.2.10
Some Neuromediator-specific Terminal Plexuses Conform Exactly to Cytoarchitectonic Units, but Others Extend Beyond the Boundaries of Classical Nuclear Groups or Show No Relation Whatsoever to Known Cytoarchitectonic Entities

'Conformity' is shown, for example, by the serotoninergic plexus in the stratum fibrosum et cellulare externum of the tectum mesencephali of the lamprey (Fig. 2.57: e,f) and in the motor nuclei of layers V and VII of the viper (Fig. 59 f,g), by several plexuses of peptidergic fibres in various layers and sublayers of the anuran tectum mesencephali (Fig. 2.64), and by the plexuses of enkephalinergic fibres in the parabrachial nuclei, the pars reticulata of the substantia nigra and the pars externa of the globus pallidus of the monkey (Fig. 2.61a,b).

'Non-conformity' is shown in many parts of the brain by monoaminergic and peptidergic fibre networks. It may be exemplified by the serotoninergic network in the brains of the lamprey (Fig. 2.57) and the viper (Fig. 2.59), the enkephalinergic fibres in the brain of the turtle (Fig. 2.58), and the somatostatinergic fibres in the brain of the rat (Fig. 2.60).

Plexuses of transmitter-specific fibres often extend beyond the boundaries of cell masses to encompass the adjacent non-cellular regions into which dendrites of the nuclei extend. This phenomenon is seen in the hypothalamus (Figs. 2.57c,d; 2.58b; 2.60; 2.65), but it also occurs in the basal telencephalon of chondrichthyans, cladistians and dipnoans (Fig. 2.74).

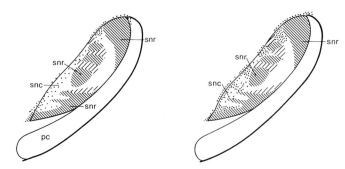

Fig. 2.66. Frontal sections through the substantia nigra of the squirrel monkey illustrate the distribution of substance P- and enkephalin-positive fibres. *Hatched areas* represent networks of fine immunoreactive fibres; *dots* illustrate the more scattered and coarse fibres. *pc*, Pedunculus cerebri; *snc*, substantia nigra, pars compacta; *snr*, substantia nigra, pars reticulata. (modified from Inagaki and Parent 1984)

2.9.2.11
Neuromediator-specific Subpopulations of Fibres and Terminals Occurring Within the Confines of Grisea Often Show a Spatial Segregation Within These Grisea. Such Spatially Segregated Terminal Fields May (a) Conform to Other Neuromediator-Specific Terminal Fields, (b) Be Complementary to Other Terminal Fields, (c) Conform to Cytoarchitectonic Subunits (Areas, Layers, Subfields), and (d) Conform to Sets of Neuromediator-specific Perikarya

Phenomenon *a* (congruity of different neuromediator-specified terminal fields) has been observed in the primate substantia nigra (perfect congruity of substance P- and enkephalin-containing fibres: Fig. 2.66), and in the mammalian striatum, i.e. the caudate-putamen complex. Histochemical and particularly immunohistochemical studies have revealed a remarkable heterogeneity within this complex. The first evidence for this chemoarchitectural heterogeneity came from studies in which a staining technique for the enzyme AChE was applied. These studies showed that in the caudate-putamen complex, 300- to 600-μm-wide zones of low AChE activity stand out against an otherwise AChE-rich background. Graybiel and Ragsdale (1978, 1983), who first identified these zones, designated them as striatal bodies or striosomes (Fig. 2.67). Throughout most of the caudate-putamen complex the striosomes and the matrix in which they are embedded have appeared to represent chemoarchitectonically distinct tissue compartments, which are related to the intrinsic structure of the complex, as well as to the organisation of its afferent and efferent connections. Apart from a low AChE concentration, the striosomes show a high enkephalin, substance P, GABA and neurotensin immunoreactivity, and, in addition to a high AChE concentration, the complementary matrix

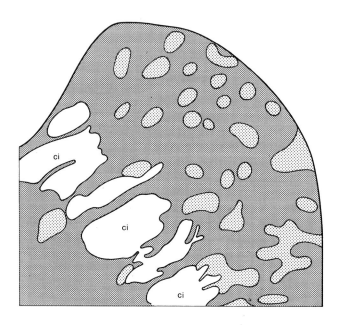

Fig. 2.67. Transverse section through the caudate nucleus of the rhesus monkey, stained for AChE; striosomes poor in AChE embedded in an AChE-rich matrix. *ci*, Capsula interna. (based on Fig. 11 in Graybiel and Ragsdale 1983)

compartment shows a dense plexus of so-matostatin-containing fibres (Graybiel and Ragsdale 1983; Gerfen 1992).

Phenomenon *b* (complementarity of different neuromediator-specified terminal fields) is also very common. It is exemplified by the differences between (a) the striosome and matrix compartments in the caudate-putamen complex, (b) certain layers and sublayers in the tectum of the frog, as for instance the substance P-, L-enkephalin- and CCK8-positive zones in laminae 11 and 12 (Fig. 2.64), and (c) the mutual complementarity of substance P- and neurotensin-immunoreactive fibres and of substance P- and enkephalin-immunoreactive fibres in the parabrachial region of the rat (Fig. 2.68).

Phenomenon *c* (congruity of neuromediator-specified terminal fields and cytoarchitectonic subunits) may be exemplified (a) by the dense plexus of dopaminergic fibres present in the pars lateralis ventralis of the area dorsalis telencephali of the teleost *Gasterosteus aculeatus* (Fig. 2.69b), and (b) by the fact that in primates the vast majority of the striosomes coincide with cell clusters (Goldman-Rakic 1982; Selemon et al. 1994).

Phenomenon *d* (congruity of neuromediator-specified terminal fields and sets of neuromediator-specific perikarya) can be observed in, among many other structures, (a) the suprachiasmatic nucleus (avian pancreatic polypeptide-containing fibres with vasoactive intestinal polypeptide-immunoreactive cell bodies, Fig. 2.63), (b) the striosomes (enkephalin-, GABA- and neurotensin-immunoreactive fibres with substance P- and dynorphin B-immunoreactive cell bodies), and, most strikingly (c) the posterior magnocellular subnucleus of the paraventricular nucleus of the rat, where dopamine β-hydroxylase-immunoreactive (most probably noradrenergic) fibres specifically address the vasopressinergic neurons located in the posterolateral part of that nucleus (Fig. 2.70).

2.9.3
The Comparative Anatomy of Neuromediator-Specific Neuron Populations and Terminal Plexuses

2.9.3.1
Neuromediator-Specific Neuronal Populations and Terminal Plexuses May, with Regard to Their Extension, Vary Considerably Among Different (Groups of) Vertebrates; However, the Similarities Are Much More Striking Than the Differences

This is the overall impression of the present author, obtained from study of the chemical neuroanatomical literature from 1980–1996, and the tenor of the

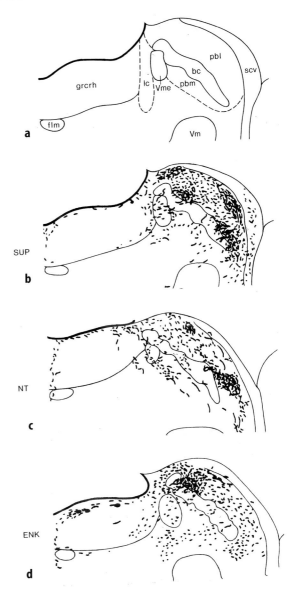

Fig. 2.68a–d. Frontal sections through the parabrachial region of the rat: **a** subdivisions of the region; **b–d** distribution of substance P (*SUP*)-, neurotensin (*NT*)-, and enkephalin (*ENK*)-immunoreactive fibres. *bc*, Brachium conjunctivum; *flm*, fasciculus longitudinalis medialis; *grcrh*, griseum centrale rhombencephali; *lc*, locus coeruleus; *pbl*, nucleus parabrachialis lateralis; *bpm*, nucleus parabrachialis medialis; *scv*, tractus spinocerebellaris ventralis; *Vm*, nucleus motorius nervi trigemini; *Vme*, nucleus mesencephalicus nervi trigemini (modified from Milner et al. 1984)

pertinent sections in the various chapters constituting the specialised part of the present work. As far as the catecholamines are concerned, this impression is reinforced by the recent comparative survey *Phylogeny and Development of Catecholamine Systems in the CNS of Vertebrates*, edited by Smeets and Reiner (1994).

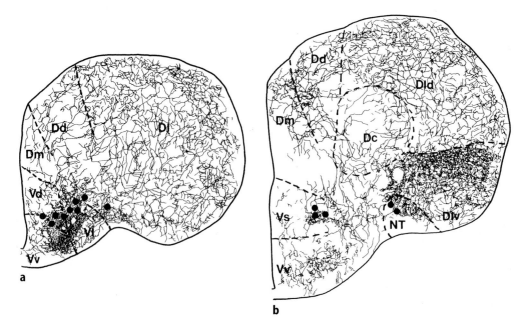

Fig. 2.69a,b. Camera lucida drawings of transverse sections through the rostral (**a**) and intermediate ports (**b**) of the right half of the telencephalon of the stickleback *Gasterosteus aculeatus* showing the distribution of dopamine-immunoreactive structures. *Black dots* denote (groups of) neuronal somata.

Dc, Area dorsalis telencephali, pars centralis; *Dd,* idem, pars dorsalis; *Dld,* idem, pars lateralis dorsalis; *Dlv,* idem, pars lateralis ventralis; *Dm,* idem, pars medialis; *NT,* nucleus taeniae; *Vs,* area ventralis telencephali, pars supracommissuralis; *Vv,* idem, pars ventralis (from Ekström et al. 1990)

2.9.3.2
Because of the Preponderance of Similarities Mentioned Above, the Presence and Distribution of Particular Neuromediators, or of Their Related Enzymes or mRNAs, Can Characterise Neuronal Populations and May thus Be Valuable for Defining Homologies

This notion is directly related to the methodology of establishing homologies in the CNS. As pointed out in Chap. 6, four criteria are of particular importance to the determination of homologies of neuronal groups, namely (a) similarity in relative or topological position, (b) similarity in fibre connections, (c) similarity in special qualities, as for instance the presence of particular cell types or of particular chemical constituents, and (d) continuity of similarity through intermediate species. It is important to note that there is a hierarchy in these criteria, so that similarity in topological position is the first criterion, whereas similarity in fibre connections and in special qualities are additional criteria. Note also that the first criterion is simplex, whereas the second and third criteria are multiplex: A given cell group occupies only one position but has several connections, whereas the number of special qualities, particularly chemical characteristics, is potentially unlimited. Quality and quantity both play a role in the application of the second and third criteria. That is to say, experience may teach us that the presence of a particular afferent or efferent connection, or of a particular chemical substance, is crucial for the identification of a given structure. Then, the pertinent hodological or chemical property becomes a *defining character*. A given character may be defining within a particular group but not within another. With regard to quantity, the probability that two cell groups in the brains of two different vertebrates are homologous increases with the number of special qualities shared by these cell groups.

2.9.3.3
What Has Been Said in the Above Paragraph About Neuromediators also Holds True for Receptor Molecules and for Other Chemical Constituents of the Brain and Spinal Cord

Within the domain of chemical neuroanatomy attention should be directed not only to neuromediators, but also to the distribution of receptor complexes. Although techniques for the localisation of these complexes are available, and these techniques have been successfully applied to the brain of different mammalian species, the paucity of information on the distribution of receptors in the brains of non-mammalian vertebrates prevents us from making a detailed comparative analysis.

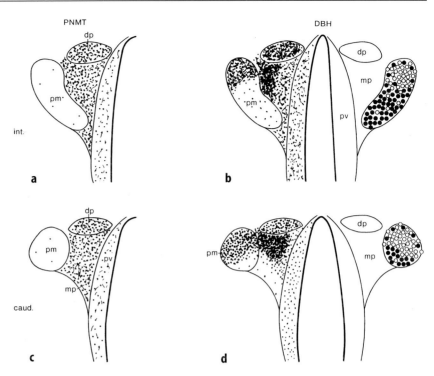

Fig. 2.70a–d. Frontal sections through the intermediate (**a,b**) and caudal part (**c,d**) of the nucleus paraventricularis of the rat illustrate the distribution of PNMT-stained (**a,c**) and DBH-stained fibres (**b,d**). The distribution of oxytocin-stained (*small filled circles*) and vasopressin-stained (*small open circles*) cells in the posterior magnocellular part of the nucleus is shown on the right side of **b** and **d**. The DBH-stained fibres in the posterior magnocellular part of the para-ventricular nucleus (**b,d**) are most probably noradrenergic, because they do not stain with an anti-PNMT serum (**a,c**). Abbreviations indicate the following parts of the paraventricular nucleus: *DBH*, dopamine-β hydroxylase; *dp*, dorsal parvocellular; *mp*, medial parvocellular; *pm*, posterior magnocellular; *PNMT*, phenylethanolamine-*N*-methyltransferase; *pv*, periventricular parvocellular (modified from Swanson et al. 1981)

During the period preceding the current proliferation of immunohistochemical studies on the distribution of neuromediators and neuromediator-related substances, several comparative histochemical studies appeared (e.g., Baker-Cohen 1968a,b; Northcutt and Braford 1980), in which the distribution of enzymes in the brains of different groups of vertebrates were mapped. The enzymes studied included degradative enzymes of neuromediators (e.g., monoamine oxidase, AChE) as well as enzymes not directly related to the synthesis or degradation of neuromediators (e.g., acid phosphatase, succinate dehydrogenase, thiamine pyrophosphatase). These enzyme-histochemical studies yielded the following general results: (a) the distributions of these enzymes are non-uniform in the brain; (b) the enzyme histochemical properties of neuronal populations considered homologous on positional and hodological grounds are generally similar among vertebrates, and (c) enzyme histochemical data may help solve homology problems.

2.9.3.4
Many Neuromediator-Specific Neuronal Groups Can Be Readily Identified on Positional Grounds

Examples of this statement are: (a) the noradrenergic locus coeruleus or A6 group, situated in the most rostral part of the rhombencephalic alar plate; (b) the dopaminergic complex constituted by the A8–A10 groups situated in the tegmentum of the midbrain; (c) and (d) the histaminergic cell cluster and the adrenergic complex C1–C3, situated in the caudobasal hypothalamus and the caudal rhombencephalon, respectively.

2.9.4
The Functional Significance of Neuromediator-Specific Neuronal Populations

Generalisations concerning the function(s) exerted by the various neuromediators are difficult. Most authors who have studied the issue place their findings in the perspective of what is known concerning the functional significance of the various centres, pathways and areas of termination in which that neuromediator is found. Because most of the neuromediator-specified neuronal networks extend over areas bearing different 'functional labels', sentences like: "The present study demonstrated a wide distribution of substance X-containing structures in the central nervous system, suggesting that substance X may be involved in a variety of functions" are frequently encountered in the literature. On the other hand, the notion that a certain population of transmitter-specified neurons may subserve limited functions is also found, often only implicit and unspecified, e.g. by denoting that population as 'the substance X *system*'.

Considering the question of functional homogeneity or functional heterogeneity separately for each neuromediator, it seems highly unlikely that all of the actions of the populations of cholinergic or dopaminergic neurons will ever be integrated into a single functional concept. This does not exclude the possibility that certain subsets of these populations do fulfil a clearly definable functional role, and thus deserve the designation 'system'. The dopaminergic tegmental projection to the striatum in amniotes is a striking case in point (Smeets and Reiner 1994).

The fact that a clearly recognizable noradrenergic locus coeruleus, giving rise to a widely distributed fibre network, is present in all vertebrates strongly suggests that this entity fulfils a basic regulatory role throughout the vertebrate kingdom. The same holds true for the histaminergic 'system'. It is remarkable that in all vertebrates studied so far (Brodin et al. 1990: lamprey; Inagaki et al. 1991: the teleost *Trachurus*; Barroso et al. 1993: the urodele *Triturus*; Airaksinen and Panula 1990: the anuran *Xenopus*; Inagaki et al. 1990: the turtle *Chinemys*; Inagaki et al. 1988, Airaksinen and Panula 1988, Panula et al. 1989, Airaksinen et al. 1989: various mammalian species) the histaminergic neurons form a single cluster in the caudobasal hypothalamus. Only in the lamprey is another group of histaminergic cell bodies present in the border area between the mesencephalon and rhombencephalon. As regards serotonin, it is known that the mainly serotoninergic caudal raphe nuclei project strongly and extensively to the spinal cord. There is evidence suggesting that in mammals this 'raphespinal system' exerts a general level-setting influence on sensory and motor parts of the spinal cord. Impulses travelling along the fibres of this descending projection presumably enhance the level of activity of motoneurons and suppress pain transmission in the spinal dorsal horn (Anderson and Proudfit 1981; Basbaum and Fields 1984; Holstege 1991). These influences would enable an animal in

life-threatening situations to combine maximal motor activity with neglect of painful stimuli. It is conceivable that the 'raphespinal system' exerts this basic function in all vertebrates.

With respect to the neuroactive amino acids, almost all neurons containing GABA or glycine presumably exert an inhibitory action on other neurons, and all members of the neuronal populations containing glutamate and aspartate are excitatory. However, the neurons of these four populations exert their actions in a variety of functional contexts, and attempts to find higher-order functional homogeneities behind these functional heterogeneities may be expected to remain unsuccessful.

As regards the neuropeptides, it is important to note that long before their role as central chemical messengers was discovered many of them were known as neurohormones, peripheral hormones or hypophysiotropic factors. Most members of these three categories of substances exert either a single function or a number of different actions which culminate in a single goal. Because it seems unlikely that the occurrence of these peptides in the periphery as well as in the central nervous system is a mere coincidence, it has been repeatedly suggested that the peripheral and central elements producing one peptide are different components of one functional system. To give a few examples: (a) There is evidence that the intestinal hormone cholecystokinin and the central neuromediator cholecystokinin both inhibit feeding behaviour (for references, see Iversen 1983). (b) It is reasonable to assume that the hypophysiotropic hormone LHRH and certain portions of the central LHRH neuronal network both play a role in the regulation of sexual behaviour and reproduction (Witkin et al. 1982; King et al. 1984). (c) Both peripherally and centrally produced angiotensin II may be implicated in the maintenance of fluid homeostasis. Iversen (1983) has suggested that such parallel actions of peptides in periphery and brain, leading to similar end results, may tell us something about the general integration of brain and body. However, in the examples given, there is at best a coupling between the actions of groups of cells producing a given hormone or hypophysiotropic factor and the possible activities of *part* of the central neuronal network using the same substance as a neuromediator. Moreover, for many other peptides which are known to be active in both the periphery and the brain, there is no known relationship between their peripheral and central actions. For instance, it is hard to understand what the actions of somatostatin or thyrotropin-releasing hormone on the hypophysis could have to do with their activities as central neuromediators.

In the mammalian brain a complex of highly interconnected centres, extending from the medial hemisphere wall to the obex, is characterised by the presence of an extraordinary amount and diversity of neuropeptides (Nieuwenhuys 1985, 1996; Nieuwenhuys et al. 1989). Nieuwenhuys feels justified to denote this complex as a system because it is specifically involved in the regulation of processes directly aimed at the survival of the individual and of the species (eating, drinking, reproduction and agonistic behaviour), and this complex has been designated 'the greater limbic system' by Nieuwenhuys (1996; cf. Chap. 22, Sect. 12), because it encompasses, in addition to a number of classical telencephalic, diencephalic and mesencephalic limbic centres (cf. MacLean 1952; Nauta 1958, 1973), certain rhombencephalic areas which up to now were not considered as belonging to the limbic domain. The wealth of neuropeptides indicates that the way in which the greater limbic system operates is distinctly different from other parts of the brain (Nieuwenhuys 1985, 1996; Herbert 1993).

Important gains in our *neurobiological* insight have generally not come from mapping studies in which the localisation and distribution of a given neuromediator throughout the CNS is presented, but rather from investigations in which morphological, physiological and immunohistochemical data pertaining to the microcircuitry of a particular centre or complex are integrated. As far as mammals are concerned, interesting results of such integrated studies related to numerous different structures can be found in Shepherd's (1990) *The Synaptic Organization of the Brain*. With respect to non-mammalian forms, the work of Grillner and his associates (summarised in Grillner et al. 1991, 1995 and Wallén 1994) on the neuronal network generating locomotor behaviour in the lamprey, and that of Maler and Mugnaini (1994) on the ELLL of the gymnotiform fish *Apteronotus leptorhynchus* are valuable studies.

References

Addens JL (1933) The motor nuclei and roots of the cranial and first spinal nerves of vertebrates. Part I. Introduction. Z Anat Entw Gesch 101:307–410

Agnati LF, Zoli M, Strömberg I, Fuxe K (1995) Intercellular communication in the brain: wiring versus volume transmission. Neuroscience 69:711–726

Airaksinen MS, Panula P (1988) The histaminergic system in the guinea pig central nervous system: an immunocytochemical mapping study using an antiserum against histamine. J Comp Neurol 273:1–24

Airaksinen MS, Panula P (1990) Comparative neuroanatomy of the histaminergic system in the brain of the frog *Xenopus laevis*. J Comp Neurol 292:413–423

Airaksinen MS, Flugge G, Fuchs E, Panula P (1989) The histaminergic system in the tree shrew brain. J Comp Neurol 286:298–310

Anderson EG, Proudfit HK (1981) The functional role of the bulbospinal serotonergic nervous system. In: Jacobs BL, Gelperin A (eds) Serotonin neurotransmission and behavior. MIT Press, London, pp 307–338

Antonetty CM, Webster KE (1975) The organization of the spinotectal projection. An experimental study in the rat. J Comp Neurol 163:449–466

Aoki C, Pickel VM (1992) Ultrastructural relation between β-adrenergic receptors and catecholaminergic neurons. Brain Res Bull 29:257–263

Ariëns Kappers CU (1919) Phenomena of neurobiotaxis as demonstrated by the position of the motor nuclei of the oblongata. J Nerv Ment Dis 50:1–16

Ariëns Kappers CU (1922) Phenomena of neurobiotaxis in the optic system. Libro en honor de D Santiago Ramón y Cajal. Jiménez y Molina, Madrid, pp 267–313

Ariëns Kappers CU (1928) Three lectures on neurobiotaxis and other subjects, delivered at the University of Copenhagen. Levin and Munksgaard, Copenhagen

Ariëns Kappers CU (1929) The evolution of the nervous system in invertebrates vertebrates and man. Bohn, Haarlem

Ariëns Kappers CU (1932) Principles of development of the nervous system (neurobiotaxis). In: Penfield WG (ed) Cytology and cellular pathology of the nervous system, vol 1. Hoeber, New York, pp 45–89

Ariëns Kappers CU (1941) Neurobiotaxic influences in the arrangement of midbrain and 'tween-brain centres. Proc Ned Akad Wet 44:130–139

Ariëns Kappers CU, Huber GC, Crosby EC (1936) The comparative anatomy of the nervous system of vertebrates, including man, vol 1. MacMillan, New York

Arluison M (1981) Les fibres nerveuses serotonergiques du striatum chez le rat et leurs caractéristiques ultrastructurales. J Physiol (Paris) 77:45–51

Bäckström K (1924) Contributions to the forebrain morphology in selachians. Acta Zool 5:123–240

Bach-y-Rita P (1993) Neurotransmission in the brain by diffusion through the extracellular fluid: a review. Neuroreport 4:343–350

Baker-Cohen K (1968a) Comparative enzyme histochemical observations on submammalian brains. Part I. Striatal structures in reptiles and birds. Ergeb Anat Entw Gesch 40:7–41

Baker-Cohen K (1968b) Comparative enzyme histochemical observations on submammalian brains. Part II. Basal structures of the brainstem in reptiles and birds. Ergeb Anat Entw Gesch 40:42–69

Baraban JM, Aghajanian GK (1981) Noradrenergic innervation of serotonergic neurons in the dorsal raphe: Demonstration by electron microscopic autoradiography. Brain Res 204:1–11

Barroso C, Franzoni FM, Fasolo A, Panula P (1993) Organization of histamine-containing neurons in the brain of the crested newt, *Triturus carnifex*. Cell Tissue Res 272:147–154

Bartelmez GW (1915) Mauthner's cell and the nucleus motorius tegmenti. J Comp Neurol 25:87–128

Basbaum AI, Fields HL (1984) Endogenous pain control systems: brainstem spinal pathways and endorphin circuitry. Annu Rev Neurosci 7:309–338

Bell CC, Szabo T (1986) Electroreception in mormyrid fish. Central anatomy. In: Bullock TH, Heiligenberg W (eds) Electroreception. Wiley, New York, pp 375–421

Bemis WE (1984) Paedomorphosis and the evolution of the Dipnoi. Paleobiology 10:293–307

Bittencourt JC, Benoit R, Sawchenko PE (1991) Distribution and origins of substance P-immunoreactive projections to the paraventricular and supraoptic nuclei: partial overlap with ascending catecholaminergic projections. J Chem Neuroanat 4:63–78

Black D (1917a) The motor nuclei of the cerebral nerves in phylogeny. A study of the phenomena of neurobiotaxis. I. Cyclostomi and pisces. J Comp Neurol 27:467–564

Black D (1917b) The motor nuclei of the cerebral nerves in phylogeny. A study of the phenomena of neurobiotaxis. II. Amphibia. J Comp Neurol 28:379–427

Black D (1920) The motor nuclei of the cerebral nerves in phylogeny. A study of the phenomena of neurobiotaxis. III. Reptilia. J Comp Neurol 32:61–98

Black D (1922) The motor nuclei of the cerebral nerves in phylogeny. A study of the phenomena of neurobiotaxis. IV. Aves. J Comp Neurol 34:233–275

Blackstad TW (1967) Cortical gray matter – a correlation of light and electron microscopic data. In: Hydén H (ed) The neuron. Elsevier, Amsterdam, pp 49–118

Bodian D (1937) The structure of the vertebrate synapse. A study of the axon endings on Mauthner's cell and neighboring centers in the goldfish. J Comp Neurol 68:117–159

Bodian D (1966) Synaptic types on spinal motoneurons: an electron microscopic study. Bull Johns Hopkins Hosp 119:16–45

Braford MR Jr, Northcutt RG (1974) Olfactory bulb projections in the bichir, *Polypterus*. J Comp Neurol 156:165–178

Brawer JR, Morest DK (1975) Relations between auditory nerve endings and cell types in the cat's anteroventral cochlear nucleus seen with the Golgi method and Nomarski optics. J Comp Neurol 160:491–506

Brawer JR, Morest DK, Kane EC (1974) The neuronal architecture of the cochlear nucleus of the cat. J Comp Neurol 155:251–300

Brodin L, Buchanan JT, Hökfelt T, Grillner S, Rehfeld JF, Frey P, Verhofstad AAJ, Dockray GJ, Walsh JH (1988) Immunohistochemical studies of cholecystokininlike peptides and their relation to 5-HT, CGRP, and bombesin immunoreactivities in the brainstem and spinal cord of lampreys. J Comp Neurol 271:1–18

Brodin L, Ohta Y, Hökfelt T, Grillner S (1989) Further evidence for excitatory amino acid transmission in lamprey reticulospinal neurons: selective retrograde labeling with (3H)D-aspartate. J Comp Neurol 271:225–233

Brodin L, Hökfelt T, Grillner S, Panula P (1990) Distribution of histaminergic neurons in the brain of the lamprey *Lampetra fluviatilis* as revealed by histamine-immunohistochemistry. J Comp Neurol 292:435–442

Buma P, Nieuwenhuys R (1987) Ultrastructural demonstration of oxytocin and vasopressin release sites in the neural lobe and median eminence of the rat by tannic acid and immunogold methods. Neurosci Lett 74:151–157

Buma P, Nieuwenhuys R (1988) Ultrastructural characterization of exocytotic release sites in different layers of the median eminence of the rat. Cell Tissue Res 252:107–114

Buma P, Roubos EW (1986) Ultrastructural demonstration of nonsynaptic release sites in the central nervous system of the snail *Lymnaea stagnalis*, the insect *Periplaneta americana*, and the rat. Neuroscience 17:867–879

Buma P, Roubos EW, Buys RM (1984) Ultrastructural demonstration of exocytosis of neural, neuroendocrine and endocrine secretions with an in vitro tannic acid (TARI-) method. Histochemistry 80:247–256

Buma P, Veening J, Nieuwenhuys R (1989) Ultrastructural characterization of adrenocorticotrope hormone (ACTH) immunoreactive fibres in the mesencephalic central grey substance of the rat. Eur J Neurosci 1:659–672

Buma P, Veening J, Hafmans T, Joosten H, Nieuwenhuys R (1992) Ultrastructure of the periaqueductal grey matter of the rat: an electron microscopical and horseradish peroxidase study. J Comp Neurol 319:519–535

Burkhalter A (1989) Intrinsic connections of rat primary visual cortex: laminar organization of axonal projections. J Comp Neurol 279:171–186

Butler AB, Northcutt RG (1971) Retinal projections in *Iguana iguana* and *Anolis carolinensis*. Brain Res 26:1–13

Cadusseau J, Roger M (1985) Afferent projections to the superior colliculus in the rat with special attention to the deep layers. J Hirnforsch 26:667–681

Card JP, Moore RY (1984) The suprachiasmatic nucleus of the golden hamster: immunohistochemical analysis of the cell and fiber distribution. Neuroscience 13:415–431

Carr CE, Maler L (1985) A Golgi study of the cell types of the dorsal torus semicircularis of the electric fish *Eigenmannia*: functional and morphological diversity in the midbrain. J Comp Neurol 235:207–240

Carr CE, Maler L (1986) Electroreception in gymnotiform fish. Central anatomy and physiology. In: Bullock TH, Heiligenberg W (eds) Electroreception. Wiley, New York, pp 319–373

Carr CE, Maler L, Heiligenberg W, Sas E (1981) Laminar organization of the afferent and efferent systems of the torus semicircularis of gymnotiform fish: morphological substrates for parallel processing in the electrosensory system. J Comp Neurol 203:649–670

Carr CE, Maler L, Sas E (1982) Peripheral organization and central projections of the electrosensory nerves in gymnotiform fish. J Comp Neurol 211:1390–153

Challet E, Pierre J, Repérant J, Ward R, Miceli D (1991) The serotoninergic system of the brain of the viper, *Vipera aspis*. An immunohistochemical study. J Chem Neuroanat 4:233–248

Christenson J, Cullheim S, Grillner S, Hökfelt T (1990) 5-Hydroxytryptamine immunoreactive varicosities in the lamprey spinal cord have no synaptic specializations – an ultrastructural study. Brain Res 512:201–209

Clements JR, Beitz AJ, Fletcher TF, Mullett MA (1985) Immunocytochemical localization of serotonin in the rat periaqueductal gray: a quantitative light and electron microscopic study. J Comp Neurol 236:60–70

Coons AH (1958) Fluorescent antibody methods. In: Danielli JF (ed) General cytochemical methods. Academic, New York, pp 399–422

Curtis DR, Johnson GAT (1974) Amino-acid transmitters in the mammalian central nervous system. Ergeb Physiol 69:97–188

Dacey DM, Ulinski PS (1986a) Optic tectum of the eastern garter snake, *Thamnophis sirtalis*. II. Morphology of efferent cells. J Comp Neurol 245:198–237

Dacey DM, Ulinski PS (1986b) Optic tectum of the eastern garter snake, Thamnophis sirtalis. III. Morphology of intrinsic neurons. J Comp Neurol 245:283–300

Dacey DM, Ulinski PS (1986c) Optic tectum of the eastern garter snake, *Thamnophis sirtalis*. IV. Morphology of afferents from the retina. J Comp Neurol 245:301–318

Dahlström A, Fuxe K (1964) Evidence for the existence of monoamine-containing neurons in the central nervous system. I. Demonstration of monoamines in the cell bodies of brain stem neurons. Acta Physiol Scand 62 [Suppl 232]:1–55

Dahlström A, Fuxe K (1965) Evidence for the existence of monoamine-containing neurons in the central nervous system. II. Experimentally induced changes in the intraneuronal amine levels of vulvospinal neuron systems. Acta Physiol Scand 64 [Suppl 247]:1–36

Dale HH (1935) Pharmacology and nerve-endings. Proc R Soc Lond B 29:319–322

Dana C, Vial M, Leonard K, Beauregard A, Kitabgi P, Vincent J-P, Rostène W, Beaudet A (1989) Electron microscopic localization of neurotensin binding sites in the midbrain tegmentum of the rat. I. Ventral tegmental area and interfascicular nucleus. J Neurosci 9:2247–2257

DeFelipe J, Hendry SHC, Hashikawa T, Molinari M, Jones EG (1990) A microcolumnar structure of monkey cerebral cortex revealed by immunocytochemical studies of double bouquet cell axons. Neuroscience 37:655–673

de Lima AD, Morrison JH (1989) Ultrastructural analysis of somatostatin-immunoreactive neurons and synapses in the temporal and occipital cortex of the macaque monkey. J Comp Neurol 283:212–227

Dodd J, Jessell TM (1988) Axon guidance and the patterning of neuronal projections in vertebrates. Science 242:692–699

Dolabela de Lima A, Singer W (1987) The serotoninergic fibers in the dorsal lateral geniculate nucleus of the cat: distribution and synaptic connections demonstrated with immunohistochemistry. J Comp Neurol 258:339–351

Dräger UC, Hubel DH (1975) Responses to visual stimulation and relationship between visual, auditory, and somatosensory input in mouse superior colliculus. J Neurophysiol 38:690–713

Ebbesson SOE (1980) The parcellation theory and its relation to interspecific variability in brain organization evolutionary and ontogenetic development, and neuronal plasticity. Cell Tissue Res 213:179–212

Ebbesson SOE (1984) Structure and connections of the optic tectum in elasmobranchs. In: Vanegas H (ed) Comparative neurology of the optic tectum. Plenum, New York, pp 33–46

Eccles JC (1957) The physiology of nerve cells. Johns Hopkins, Baltimore

Eccles JC (1964) The physiology of synapses. Academic, New York

Eccles JC (1969) The inhibitory pathways of the central nervous system. University Press, Liverpool

Eccles JC, Ito M, Szentágothai J (1967) The cerebellum as a neuronal machine. Springer, Berlin Heidelberg New York

Ekström P, Honkanan T, Steinbusch WM (1990) Distribution of dopamine-immunoreactive neuronal perikarya and fibres in the brain of a teleost, Gasterosteus aculeatus L. Comparison with tyrosine hydroxylase- and dopamine-β-hydroxylase-immunoreactive neurons. J Chem Neuroanat 3:233–260

Emson PC (ed) (1983) Chemical neuroanatomy. Raven, New York

Faber DS, Korn H (1978) Electrophysiology of the Mauthner cell: basic properties, synaptic mechanisms, and associated networks. In: Faber DS, Korn H (eds) Neurobiology of the mauthner cell. Raven, New York, pp 47–131

Falck B, Hillarp NA, Thieme G, Torp A (1962) Fluorescence of catecholamines and related compounds condensed with formaldehyde. J Histochem Cytochem 10:384–354

Finger TE (1981a) Enkephalin-like immunoreactivity in the gustatory lobes and visceral nuclei in the brains of goldfish and catfish. Neuroscience 6:2747–2758

Finger TE (1981b) Laminar and columnar organization of the vagal lobe in goldfish: possible neural substrate for sorting food from gravel. Soc Neurosci Abstr 7:665

Finger TE (1988) Sensorimotor mapping and oropharyngeal reflexes in goldfish, Carassius auratus. Brain Behav Evol 31:17–24

Florey E (1967) Neurotransmitters and modulators in the animal kingdom. Fed Proc 26:1164–1178

Foster RE, Peterson BE (1986) The inferior olivary complex of guinea pig: cytoarchitecture and cellular morphology. Brain Res Bull 17:785–800

Fuxe K, Agnati L (1991) Two principal modes of electrochemical communication in the brain: volume versus wiring transmission. In: Fuxe K, Agnata LF (eds) Volume transmission in the brain: novel mechanisms for neural transmission. Raven, New York, pp 1–9

Geeraedts LMG, Nieuwenhuys R, Veening JG (1990a) The medial forebrain bundle of the rat. III. Cytoarchitecture of the rostral (telencephalic) part of the MFB bed nucleus. J Comp Neurol 294:507–536

Geeraedts LMG, Nieuwenhuys R, Veening JG (1990b) The medial forebrain bundle of the rat. IV. Cytoarchitecture of the caudal (lateral hypothalamic) part of MFB bed nucleus. J Comp Neurol 294:537–568

Gerfen CR (1992) The neostriatal mosaic: multiple levels of compartmental organization in the basal ganglia. Annu Rev Neurosci 15:285–320

Gerschenfeld HD (1973) Chemical transmission in invertebrate central nervous system and neuromuscular junction. Physiol Rev 53:1–119

Giuffrida R, Rustioni A (1989a) Glutamate and aspartate immunoreactivity in cortico-cortical neurons of the sensorimotor cortex of rats. Exp Brain Res 74:41–46

Giuffrida R, Rustioni A (1989b) Glutamate and aspartate immunoreactivity in corticospinal neurons of rats. J Comp Neurol 288:154–164

Goehler LE, Finger TE (1992) Functional organization of vagal reflex systems in the brain stem of the goldfish, Carassius auratus. J Comp Neurol 319:463–478

Goldberg ME, Robinson DL (1978) Visual system: superior colliculus. In: Masterson B (ed) Handbook of behavioral neurobiology. Plenum, New York, pp 119–165

Golding DW (1994) A pattern confirmed and refined – synaptic, nonsynaptic and parasynaptic exocytosis. Bioessays 16:503–508

Goldman-Rakic PS (1982) Cytoarchitectonic heterogeneity of the primate neostriatum: subdivision into island and matrix cellular compartments. J Comp Neurol 205:398–413

Gordon BG (1973) Receptive fields in deep layers of cat superior colliculus. J Neurophysiol 36:157–178

Gottlieb DI, Cowan WM (1972) On the distribution of axonal terminals containing spheroidal and flattened synaptic vesicles in the hippocampus and dentate gyrus of the rat and cat. Z Zellforsch 129:413–429

Graham LT, Shank RP, Werman R, Aprison MH (1967) Distribution of some synaptic transmitter suspects in cat spinal cord: glutamic acid aspartic acid g-aminobutyric acid, glycine and glutamine. J Neurochem 14:465–472

Gray EG (1959) Axo-somatic and axodendritic synapses of the cerebral cortex: an electron microscopic study. J Anat 93:320–433

Gray EG, Guillery RW (1966) Synaptic morphology in the normal and degenerating nervous system. Int Rev Cytol 19:111–181

Graybiel AM, Ragsdale CW Jr (1978) Histochemically distinct compartments in the striatum of human monkey and cat demonstrated by acetylthiocholinesterase staining. Proc Natl Acad Sci USA 75:5723–5726

Graybiel AM, Ragsdale CW Jr (1983) Biochemical anatomy of he striatum. In: Emson PC (ed) Chemical neuroanatomy. Raven, New York, pp 427–504

Grillner S, Wallen P, Brodin L, Lansner A (1991) Neuronal network generating locomotor behavior in lamprey: circuitry, transmitters, membrane properties, and simulation. Annu Rev Neurosci 14:169–199

Grillner S, Deliagina T, Ekeberg Ö, El Manira A, Hill RH, Lansner A, Orlovsky GN, Wallén P (1995) Neural networks that co-ordinate locomotion and body orientation in lamprey. Trends Neurosci 18:270–279

Guillemin R (1978) Peptides in the brain: the new endocrinology of the neurons. Science 202:390–402

Haber S, Elde R (1982) The distribution of enkephalin immunoreactive fibers and terminals in the monkey acentral nervous system: an immunohistochemical study. Neuroscience 7:1049–1095

Hamburger V (1952) Development of the nervous system. Ann NY Acad Sci 55:117–132

Hamburger V (1955) Trends in experimental neuroembryology. In: Waelsch H (ed) Biochemistry of the developing nervous system. Academic, New York, pp 52–73

Hamlyn LH (1963) An electron microscope study of pyramidal neurons in the ammon's horn of the rabbit. J Anat 97:189–201

Harris LR, Blakemore C, Donaghy M (1980) Integration of visual and auditory space in mammalian superior colliculus. Nature 288:56–59

Hartwig H-G (1970) Das visuelle System von Zonotrichia leucophrys gambelii. Z Zellforsch Mikrosk Anat 106:556–583

Heier P (1948) Fundamental principles in the structure of the brain. A study of the brain of Petromyzon fluviatilis. Acta Anat [Suppl] VI:1–213

Heiligenberg W, Bastian J (1984) The electric sense of weakly electric fish. Annu Rev Physiol 46:561–583

Held H (1893) Die centrale Gehörleitung. Arch Anat Physiol Anat Abt 201–248

Held H (1897) Beiträge zur Structur der Nervenzellen und ihrer Fortsätze. Arch Anat Physiol 204–294

Herbert J (1993) Peptides in the limbic system: neurochemical codes for co-ordinated adaptive responses to behavioural and physiological demand. Progr Neurobiol 41:723–791

Herkenham M (1987) Mismatches between neurotransmitter and receptor localizations in brain: observations and implications. Neuroscience 23:1–38

Herrick CJ (1914) The medulla oblongata of larval amblystoma. J Comp Neurol 24:343–427

Herrick CJ (1924) The amphibian forebrain. II. The olfactory bulb of amblystoma. J Comp Neurol 37:373–396

Herrick CJ (1927) The amphibian forebrain. IV. The cerebral hemispheres of amblystoma. J Comp Neurol 43:231–325

Herrick CJ (1930) The medulla oblongata of necturus. J Comp Neurol 50:1–96

Herrick CJ (1933a) The amphibian forebrain. VI. Necturus. J Comp Neurol 58:1–288

Herrick CJ (1933b) The evolution of cerebral localization patterns. Science 78:439–444

Herrick CJ (1934) The amphibian forebrain. IX. Neuropil and other interstitial nervous tissue. J Comp Neurol 59:93–116

Herrick CJ (1939) Cerebral fiber tracts of *Amblystoma tigrinum* in midlarval stages. J Comp Neurol 71:511–612

Herrick CJ (1942) Optic and postoptic systems in the brain of *Amblystoma tigrinum*. J Comp Neurol 77:191–353

Herrick CJ (1948) The brain of the tiger salamander. University of Chicago Press, Chicago

Heuser JE, Reeze TS, Dennis MJ, Jan Y, Jan L, Evens L (1979) Synaptic vesicle exocytosis captured by quick freezing and correlated by quantal transmitter release. J Cell Biol 81:275–300

Hinojosa R (1973) Synaptic ultrastructure in the tangential nucleus of the goldfish (*Carassius auratus*). Am J Anat 137:159–186

Hökfelt T (1987) Neuronal mapping by transmitter histochemistry with special reference to coexistence of multiple synaptic messengers. In: Adelman G (ed) Encyclopedia of neurosciences, vol 2. Birkhäuser, Boston, pp 821–824

Hökfelt T, Fuxe K, Goldstein M, Johansson O (1974) Immunohistochemical evidence for the existence of adrenaline neurons in the rat brain. Brain Res 66:235–251

Hökfelt T, Johansson O, Goldstein M (1984a) Chemical anatomy of the brain. Science 225:1326–1334

Hökfelt T, Johansson O, Goldstein M (1984b) Central catecholamine neurons as revealed by immunohistochemistry with special reference to adrenaline neurons. In: Björklund A, Hökfelt T (eds) Handbook of chemical neuroanatomy, vol 2. Classical transmitters in the CNS, part I. Amsterdam, Elsevier, pp 157–276

Hökfelt T, Martensson R, Björklund A, Kleinau S, Goldstein M (1984c) Distributional maps of tyrosine-hydroxylase-immunoreactive neurons in the rat brain. In: Björklund A, Hökfelt T (eds) Handbook of chemical neuroanatomy, vol 2, Classical neurotransmitters in the CNS, part I. Amsterdam, Elsevier, pp 277–379

Holmgren N (1922) Points of view concerning forebrain morphology in lower vertebrates. J Comp Neurol 34:391–440

Holstege G (1991) Descending motor pathways and the spinal motor system: limbic and non-limbic components. Prog Brain Res 87:307–421

Huerta MF, Harting JK (1984) Connectional organization of the superior colliculus. Trends Neurosci 7:286–289

Huerta MF, Frankfurter A, Harting JK (1983) Studies of the principal sensory and spinal trigeminal nuclei of the rat: projections to the superior colliculus, inferior olive, and cerebellum. J Comp Neurol 220:147–167

Inagaki N, Yamatodani A, Ando-Yamamoto M, Tohyama M, Watanabe T, Wada H (1988) Organization of histaminergic fibers in rat brain. J Comp Neurol 273:283–300

Inagaki S, Parent A (1984) Distribution of substance P and enkephalin-like immunoreactivity in the substantia nigra of rat, cat and monkey. Brain Res Bull 13:319–329

Inagaki N, Panula P, Yamatodani A, Wada H (1990) Organization of the histaminergic system in the brain of the turtle *Chinemys reevesii*. J Comp Neurol 297:132–144

Inagaki N, Panula P, Yamatodani A, Wada H (1991) Organization of the histaminergic system in the brain of the teleost *Trachurus trachurus*. J Comp Neurol 310:94–102

Ito H (1971) Fine structure of the carp *Torus longitudinalis*. J Morphol 135:153–164

Ito H, Kishida R (1978) Afferent and efferent fiber connections of the carp *Torus longitudinalis*. J Comp Neurol 181:465–476

Iversen LL (1983) Neuropeptides – what next? Trends Neurosci 6:293–294

Johansson O, Hökfelt T, Elde RP (1984) Immunohistochemical distribution of somatostatin-like immunoreactivity in the central nervous system of the adult rat. Neuroscience 13:265–339

Johnston D, Wu M-SW (1995) Foundations of cellular neurophysiology. MIT Press, Cambridge, MA

Johnston D, Magee JC, Colbert CM, Christie (1996) Active properties of neuronal dendrites. Annu Rev Neurosci 19:165–186

Johnston JB (1902) The brain of *Petromyzon*. J Comp Neurol 12:2–86

Johnston JB (1912) The telencephalon in cyclostomes. J Comp Neurol 22:341–404

Killackey HP, Erzurumlu RS (1981) Trigeminal projections to the superior colliculus of the rat. J Comp Neurol 220:147–242

King JC, Anthony ELP, Gustafson AW, Damassa DA (1984) Luteinizing hormone-releasing hormone (LH-RH) cells and their projections in the forebrain of the bat *Myotis lucifugus lucifugus*. Brain Res 298:289–301

Koch C, Poggio T (1983) A theoetical analysis of the electrical properties of spines. Proc R Soc Lond B 218:455–477

Korn H (1987) The Mauthner cell. In: Adelman G (ed) Encyclopedia of neuroscience, vol II. Birkhäuser, Boston, pp 617–619

Kreuzberg GW (1984) 100 Years of Nissl staining. Trends Neurosci 7:236–237

Krieger DT (1983) Brain peptides: what, where, and why? Science 222:975–985

Kuhlenbeck H (1970) The central nervous system of vertebrates, vol 3, part I: structural elements: biology of nervous tissue. Karger, Basel

Kuljis RO, Karten HJ (1982) Laminar organization of peptide-like immunoreactivity in the anuran optic tectum. J Comp Neurol 212:2188–201

Kupfermann I (1979) Modulatory actions of neurotransmitters. Annu Rev Neurosci 2:447–465

Labriola AR, Laemle LK (1977) Cellular morphology in the visual layer of developing rat superior colliculus. Exp Neurol 5:247–268

Langer TP, Lund RD (1974) The upper layers of the superior colliculus of the rat: a Golgi study. J Comp Neurol 158:405–436

Larsell O (1967) The comparative anatomy and histology of the cerebellum from myxinoids through birds. University of Minnesota Press, Minneapolis

Laruelle L (1937) La structure de la moelle épinière en coupes longitudinales. Rev Neurol 67:695–725

Lázár G (1984) Structure and connections of the frog optic tectum. In: Vanegas H (ed) Comparative neurology of the optic tectum. Plenum, New York, pp 185–210

Lázár G, Székely G (1969) Distribution of optic terminals in the different optic centres of the frog. Brain Res 16:1–14

Lázár G, Tóth P, Csank E, Kicliter E (1983) Morphology and location of tectal projection neurons in frogs: a study with HRP and cobalt filling. J Comp Neurol 215:108–120

Leghissa S (1962) L'evoluzione del tetto ottico nei bassi vertebrati (I). Arch Ital Anat Embriol 67:343–413

Lettvin JY, Maturana HR, McCulloch WS, Pitts WH (1961) Two remarks on the visual systems of the frog. In: Rosenblith WA (ed) Sensory communication. MIT Press, Cambridge, pp 757–776

Lewis DA, Lund JS (1990) Heterogeneity of chandelier neurons in monkey neocortex: corticotropin-releasing factor-

and parvalbumin-immunoreactive populations. J Comp Neurol 293:599–615

Lewis PR, Shute CCD (1967) The cholinergic limbic system: projections to hippocampal formation, medial cortex, nuclei of the ascending cholinergic reticular system, and the subfornical organ and supra-optic crest. Brain 90:521–540

Light AR, Kavookjian AM, Petrtusz P (1983) The ultrastructure and synaptic connections of serotonin-immunoreactive terminals in spinal lamina I and II. Somatosens Res 1:33–50

Llinás RR (1987) Cerebellum, network physiology. In: Adelman G (ed) Encyclopedia of neuroscience, vol I. Birkhäuser, Boston, pp 01–203

Llinás R, Hillman DE (1969) Physiological and morphological organization of the cerebellar circuits in various vertebrates. In: Llinás R (ed) Neurobiology of cerebellar evoution and development. American Medical Association, Chicago, pp 43–73

Lorente de Nó R (1922) La corteza cerebral del ratón. Trabajos Cajal Madrid 20:41–80 (con 26 grabados)

Lorente de Nó R (1934) Studies on the structure of the cerebral cortex. II. Continuation of the study of the ammonic system. J Phys Neurol 46:113–177

Lorente de Nó R (1981) The primary acoustic nuclei. Raven, New York

Lund JS (1987) Local circuit neurons of macaque monkey striate cortex: I. Neurons of laminae 4C and 5A. J Comp Neurol 257:60–92

Lund JS (1988) Anatomical organization of macaque monkey striate visual cortex. Annu Rev Neurosci 11:253–288

Lund JS, Henry GH, MacQueen CL, Harvey AR (1979) Anatomical organization of the primary visual cortex (Area 17) of the cat. A comparison with Area 17 of the macaque monkey. J Comp Neurol 184:599–618

Lund RD (1964) Terminal distribution in the superior colliculus of fibers originating in the visual cortex. Nature 204:1283–1285

Lundberg JM, Hökfelt T (1983) Coexistence of peptides and classical neurotransmitters. Trends Neurosci 6:325–333

MacLean P (1952) Some psychiatric implications of physiological studies on frontotemporal portions of limbic system (visceral brain). Electroencephalogr Clin Neurophysiol 4:407–418

Maler L (1976) Tow types of laminar neural structure; examples from weakly electric fish. Exp Brain Res [Suppl] 1:571–579

Maler L, Mugnaini E (1994) Correlating gamma-aminobutyric acidergic circuits and sensory function in the electrosensory lateral line lobe of a gymnotiform fish. J Comp Neurol 345:224–252

Maler L, Sas EKB, Rogers J (1981) The cytology of the posterior lateral line lobe of high-frequency weakly electric fish (Gymnotidae): dendritic differentiation and synaptic specificity in a simple cortex. J Comp Neurol 195:87–139

Mannen H (1960) 'Noyau fermé' et 'noyau ouvert'. Arch Ital Biol 98:333–350

Maxwell DJ, Leranth CS, Verhofstad AA (1983) Fine structure of serotonin-containing axons in the marginal zone of the rat spinal cord. Brain Res 266:253–259

McMullen NT, Almli CR (1981) Cell types within the medial forebrain bundle: a Golgi study of preoptic and hypothalamic neurons in the rat. Am J Anat 161:323–340

Meek J (1981) A Golgi-electron microscopic study of goldfish optic tectum. II. Quantitative aspects of synaptic organization. J Comp Neurol 199:175–190

Meek J (1983) Functional anatomy of the tectum mesencephali of the goldfish. An explorative analysis of the functional implications of the laminar structural organization of the tectum. Brain Res Rev 6:247–297

Meek J (1990) Tectal morphology: connections, neurones and synapses. In: Douglas RH, Djamgoz MBA (eds) The visual system of fish. Chapman and Hall, London, pp 239–277

Meek J, Schellart NAM (1978) A Golgi study of goldfish optic tectum. J Comp Neurol 182:89–122

Millhouse OE (1967) The medial forebrain bundle: a Golgi analysis. PhD dissertation, University of California, Los Angeles

Millhouse OE (1969) A Golgi study of the descending medial forebrain bundle. Brain Res 15:341–363

Millhouse OE (1973a) The organization of the ventromedial hypothalamic nucleus. Brain Res 55:71–87

Millhouse OE (1973b) Certain ventromedial hypothalamic afferents. Brain Res 55:89–105

Millhouse OE (1978) Cytological observations on the ventromedial hypothalamic nucleus. Cell Tissue Res 191:473–491

Millhouse OE (1979) A Golgi anatomy of the rodent hypothalamus. In: Morgane PJ, Panksepp J (eds) Handbook of the hypothalamus, vol 1. Anatomy of the hypothalamus. Dekker, New York-Basel, pp 221–265

Milner TA, Joh TH, Miller RJ, Pickel VM (1984) Substance P, neurotensin, enkephalin, and catecholamine-synthesizing enzymes: Light microscopic localizations compared with autoradiographic label in solitary efferents to the rat parabrachial region. J Comp Neurol 226:434–447

Mobley P, Greengard P (1985) Evidence for widespread effects of noradrenaline on axon terminals in the rat frontal cortex. Proc Natl Acad Sci U S A 82:945–947

Mooney RD, Klein BG, Jacquin MF, Rhoades RW (1984) Dendrites of deep layer, somatosensory superior collicular neurons extend into the superficial laminae. Brain Res 324:361–365

Mooney RD, Nikoletseas MM, Hess PR, Allen Z, Lewin AC, Rhoades RW (1988) The projection from the superficial to the deep layers of the superior colliculus: an intracellular horseradish peroxidase injection study in the hamster. J Neurosci 8:1384–1399

Mori S, Matsuura T, Takino T, Sano Y (1987) Light and electron microscopic immunohistochemical studies of serotonin nerve fibers in the substantia nigra of the rat, cat and monkey. Anat Embryol (Berl) 176:13–18

Morita Y, Finger TE (1985) Topographic and laminar organization of the vagal gustatory system in the goldfish, Carassius auratus. J Comp Neurol 238:187–201

Morita Y, Ito H, Masai H (1980) Central gustatory paths in the crucian carp Carassius carassius. J Comp Neurol 191:119–132

Morita Y, Murakami T, Ito H (1983) Cytoarchitecture and topographic projections of the gustatory centres in a teleost, Carassius carassius. J Comp Neurol 218:378–394

Mugnaini E, Oertel WH, Wouterlood FF (1984a) Immunocytochemical localization of GABA neurons and dopamine neurons in the rat main and accessory olfactory bulbs. Neurosci Lett 47:221–226

Mugnaini E, Wouterlood FG, Dahl A-L, Oertel WH (1984b) Immunocytochemical identification of GABAergic neurons in the main olfactory bulb of the rat. Arch Ital Biol 1122:83–112

Nakajima Y (1974) Fine structure of the synaptic endings on the Mauthner cell of the goldfish. J Comp Neurol 156:375–402

Nakajima Y, Kohno K (1978) Fine structure of the Mauthner cell: synaptic topography and comparative study. In: Faber DS, Korn H (eds) Neurobiology of the Mauthner cell. Raven, New York, pp 133–166

Nathanson JA (1987) Cyclic nucleotides. In: Adelman G (ed) Encyclopedia of neuroscience, vol 1. Birkhäuser, Basel, pp 294–296

Nauta WJH (1958) Hippocampal projections and related neural pathways to the midbrain in the cat. Brain 81:319–340

Nauta WJH (1973) Connections of the frontal lobe with the limbic system. In: Laitinen LV, Livingston KE (eds) Surgical approaches in psychiatry. Medical and Technical, Lancaster, pp 303–314

Neary TJ, Northcutt RG (1983) Nuclear organization of the bullfrog diencephalon. J Comp Neurol 213:262–278

Nicholson C (1997) Brain cell microenvironment. In: Adelman G, Smith B (eds) Encyclopedia of neuroscience, 2nd edn. Elsevier, Amsterdam (in press)

Nieuwenhuys R (1967a) Comparative anatomy of olfactory centres and tracts. Prog Brain Res 23:1–64

Nieuwenhuys R (1967b) Comparative anatomy of the cerebellum. Prog Brain Res 25:1–93

Nieuwenhuys R (1983) The central nervous system of the brachiopterygian fish, *Erpetoichthys calabaricus*. J Hirnforsch 24:501–533

Nieuwenhuys R (1985) Chemoarchitecture of the brain. Springer, Berlin Heidelberg New York

Nieuwenhuys R (1996) The greater limbic system, the emotional motor system and the brain. Prog Brain Res 107:551–580

Nieuwenhuys R, Geeraedts LMG, Veening JG (1982) The medial forebrain bundle of the rat. I. General introduction. J Comp Neurol 206:49–81

Nieuwenhuys R, Voogd J, van Huijzen C(1988) The human central nervous sytem. A synopsis and atlas, 3rd rev edn. Springer, Berlin Heidelberg New York

Nieuwenhuys R, Veening JG, van Domburg P (1989) Core and paracores; some new chemoarchitectural entities in the mammalian neuraxis. Acta Morphol Neerl Scand 26:131–163

Nissl F (1885) Über die Untersuchungsmethoden der Grosshirnrinde. Neurol Zbl 4:500–501

Northcutt RG (1977) Retinofugal projections in the lepidosirenid lungfishes. J Comp Neurol 174:553–573

Northcutt RG, Braford MR (1980) New observations on the organization and evolution of the telencephalon in actinopterygian fishes. In: Ebbesson SOE (ed) Comparative neurology of the telencephalon. Plenum, New York, pp 41–98

O'Keefe J, Nadel L (1978) The hippocampus as a cognitive map. Clarendon, Oxford

Ohta Y, Brodin L, Grillner S, Hökfelt T, Walsh J (1988) Possible target neurons of the reticulospinal cholecystokinin (CCK) projection to the lamprey spinal cord: immunohistochemistry combined with intracellular staining with Lucifer yellow. Brain Res 445:400–403

Olszewski J, Baxter D (1954) Cytoarchitecture of the human brain stem. Karger, Basel

Panula P, Pirvola U, Auvnen S, Airaksinen MS (1989) Histamine immunoreactive nerve fibers and terminals in the rat brain. Neuroscience 28:585–610

Pappas GD, Waxman SG (1972) Synaptic fine structure – morphological correlates of chemical and electrotonic transmission. In: Pappas GD, Purpura DP (eds) Structure and function of synapses. Raven, New York, pp 1–43

Peters A, Harriman KM (1988) Enigmatic bipolar cell of rat visual cortex. J Comp Neurol 267:409–432

Peters A, Charmian C, Proskauer M, Feldman L, Kimerer L (1979) The projection of the lateral geniculate nucleus to area 17 of the rat cerebral cortex. V. Degenerating axon terminals synapsing with Golgi impregnated neurons. J Neurocytol 8:331–357

Pierre J, Réperant J, Ward R, Vesselkin NP, Rio JP, Miceli D, Kratskin I (1992) The serotoninergic system of the brain of the lamprey, Lampetra fluviatilis: an evolutionary perspective. J Chem Neuroanat 5:195–219

Poritsky R (1969) Two and three dimensional ultrastructure of boutons and glial cells on the motoneuronal surface in the cat spinal cord. J Comp Neurol 135:423–452

Potter HD (1969) Structural characteristics of cell and fiber populations in the tectum of the frog (*Rana catesbeiana*). J Comp Neurol 136:203–232

Potter HD (1976) Axonal and synaptic lamination in the optic tectum. Exp Brain Res [Suppl] 1:506–511

Pow DV, Morris JF (1988) Exocytosis of dense-cored vesicles from synaptic and non-synaptic release sites in the hypothalamus of the rat. J Anat 158:214–216

Powers AS, Reiner A (1980) A stereotaxic atlas of the forebrain and midbrain of the Eastern painted turtle (Chrysemys picta picta). J Hirnforsch 21:125–159

Ramón y Cajal S (1893) La retiné des vertébrés. Cellule 9:119–257

Ramón y Cajal S (1908) Sur un noyau spécial du nerf vestibulaire des poissons et des oiseaux. Trab Lab Invest Biol Univ Madrid 6:1–20

Ramón y Cajal S (1909) Histologie du système nerveux de l'homme et des vertébrés. Tome I. Instituto Ramon y Cajal, Madrid

Ramón y Cajal S (1911) Histologie du système nerveux de l'homme et des vertébrés. Tome II. Instituto Ramon y Cajal, Madrid

Ramón y Cajal S (1935) Die Neuronenlehre. In: Bumke O, Foerster O (eds) Handbuch der Neurologie, vol 1. Anatomie, Springer, Berlin Heidelberg New York, pp 887–994

Ramón-Moliner E (1962) An attempt at classifying nerve cells on the basis of their dendritic patterns. J Comp Neurol 119:211–227

Ramón-Moliner E (1967) La différentiation morphologique des neurones. Arch Ital Biol 105:149–188

Ramón-Moliner E (1969) The leptodendritic neuron: its distribution and significance. In: Petras JM, Noback CR (eds) Comparative and evolutionary aspects of the vertebrate central nervous sytem. Ann NY Acad Sci 167:65–70

Ramón-Moliner E (1975) Specialized and generalized dendritic patterns. In: Santini M (ed) Golgi centennial symposium: perspectives in neurobiology. Raven, New York, pp 87–100

Ramón-Moliner E, Nauta WJH (1966) The isodendritic core of the brain stem. J Comp Neurol 126:311–335

Reiner A (1987) The distribution of proenkephalin-derived peptides in the central nervous system of turtles. J Comp Neurol 259:65–91

Remane A (1966) Phylogenetische Entwicklungsregeln von Organen. In: Hassler R, Stephan H (eds) Evolution of the forebrain. Thieme, Stuttgart, pp 1–8

Réthelyi M (1976) Central core in the spinal grey matter. Acta Morphol Acad Sci Hung 24:64–70

Réthelyi M, Szabo T (1973a) A particular nucleus in the mesencephalon of a weakly electric fish, *Gymnotus carapo*, Gymnotidae. I. Light microscopic structure. Exp Brain Res 17:229–241

Réthelyi M, Szabo T (1973b) Neurohistological analysis of the lateral lobe in a weakly electric fish, *Gymnotus carapo* (Gymnotidae, Pisces). Exp Brain Res 18:323–339

Rettig G, Roth G (1986) Retinofugal projections in salamanders of the family Plethodontidae. Cell Tissue Res 243:385–396

Rhoades RW (1981) Expansion of the ipsilateral visual corticotectal projection in hamsters subjected to partial lesions of the visual cortex during infancy: anatomical experiments. J Comp Neurol 197:425–445

Rhoades RW, Mooney RD, Rohrer WH, Nikoletseas MM, Fish SE (1989) Organization of the projection from the superficial to the deep layers of the hamster's superior colliculus as demonstrated by the anterograde transport of *Phaseolus vulgaris* leucoagglutinin. J Comp Neurol 283:54–70

Robertson JD, Bodenheimer TS, Stage DE (1963) The ultrastructure of Mauthner cell synapses and nodes in goldfish brains. J Cell Biol 19:159–199

Romanes GJ (1964) The motor pools of the spinal cord. In: Eccles JC, Schade JP (eds) Organization of the spinal cord. Elsevier, Amsterdam, pp 93–119

Roth G (1986) Neural mechanisms of prey recognition: an example in amphibians. In: Feder ME, Lauder GV (eds) Predator-prey relationships. University of Chicago Press, Chicago, pp 42–68

Roth G (1987) Visual behavior in salamanders. Springer, Berlin Heidelberg New York

Roth G, Naujoks-Manteuffel C, Grunwald W (1990) Cytoarchitecture of the tectum mesencephali in salamanders: a Golgi and HRP study. J Comp Neurol 291:27–42

Roth G, Nishikawa KC, Naujoks-Manteuffel C, Schmidt A, Wake DB (1993) Paedomorphosis and simplification in the nervous system of salamanders. Brain Behav Evol 42:137–170

Rotsztejn WH (1980) Neuromodulation in neuroendocrinology. Trends Neurosci 3:67–70

Sanes JR (1989) Extracellular matrix molecules that influence neural development. Annu Rev Neurosci 12:491–516

Scheibel AB (1979) Development of axonal and dendritic neuropil as a function of evolving behavior. In: Schmitt FO, Worden FG (eds) The neurosciences fourth study program. MIT Press, Cambridge

Scheibel ME, Scheibel AB (1955) The inferior olive. A Golgi study. J Comp Neurol 102:77–131

Scheibel ME, Scheibel AB (1958) Structural substrates for integrative patterns in the brain stem reticular core. In: Jasper HH, Proctor LD, Knighton RS, Noshay WC, Costello RT (eds) Reticular formation of the brain. Henry Ford Hospital International Symposium. Little Brown, Boston, pp 31–68

Scheibel ME, Scheibel AB (1966) Terminal axonal patterns in cat spinal cord. I. The lateral corticospinal tract. Brain Res 2:333–350

Scheibel ME, Scheibel AB (1968) Terminal axonal patterns in cat spinal cord. II. The dorsal horn. Brain Res 9:32–58

Scheibel ME, Scheibel AB (1970a) Organization of spinal motoneuron dendrites in bundles. Exp Neurol 28:106–112

Scheibel ME, Scheibel AB (1970b) Elementary processes in selected thalamic and cortical subsystems – the structural substrates. In: Schmitt FO (ed) The neurosciences second study program. Rockefeller University Press, New York, pp 443–457

Scheibel ME, Scheibel AB (1973) Dendrite bundles as sites for central programs: an hypothesis. Int J Neurosci 6:195–202

Scheibel ME, Scheibel AB (1975) Dendrite bundles, central programs and the olfactory bulb. Brain Res 95:407–421

Schmidt A, Roth G, Ernst M (1989) Distribution of substance P-like, leucin-enkephalin-like, and bombesin-like immunoreactivity and acetylcholinesterase activity in the visual system of salamanders. J Comp Neurol 288:123–135

Schmidt ED, Roubos EW (1987) Morphological basis for nonsynaptic communication within the central nervous system by exocytotic release of secretory material from the egg-laying stimulating neuroendocrine caudodorsal cells of Lymnaea stagnalis. Neuroscience 20:247–257

Schmitt FO (1984) Molecular regulators of brain function: a new view. Neuroscience 13:991–1001

Schroeder DM (1974) Some afferent and efferent connections of the optic tectum in a teleost, Ictalurus. Anat Rec 178:458

Scott FH (1905) On the metabolism and action of nerve cells. Brain 28:506–526

Séguéla P, Watkins KC, Descarries L (1989) Ultrastructural relationships of serotonin axon terminals in the cerebral cortex of the adult rat. J Comp Neurol 289:129–142

Séguéla P, Watkins KC, Geffard M, Descarries L (1990) Noradrenaline axon terminals in adult rat neocortex: an immunocytochemical analysis in serial thin sections. Neuroscience 35:249–264

Selemon LD, Gottlieb JP, Goldman-Rakic PS (1994) Islands and striosomes in the neostriatum of the rhesus monkey: non-equivalent compartments. Neuroscience 58:183–192

Shepherd GM (1990) The synaptic organization of the brain, 3rd edn. Oxford University Press, New York

Shepherd GM (1994) Neurobiology, 3rd edn. Oxford University Press, New York

Shepherd GM, Greer CA (1990) Olfactory bulb. In: Shepherd GM (ed) The synaptic organization of the brain. Oxford Universtiy Press, New York, pp 133–169

Shute CCD, Lewis PR (1967) The ascending cholinergic reticular system: neocortical, olfactory and subcortical projections. Brain 90:497–520

Sibbing FA (1984) Food handling and mastication in the carp (Gyprinus carpis). PhD thesis, Wageningen, Netherlands

Smeets WJAJ, Reiner A (1994) Catecholamines in the CNS of vertebrates: current concepts of evolution and functional significance. In: Smeets WJAJ, Reiner A (eds) Phylogeny and development of catecholamine systems in the CNS of vertebrates. Cambridge Universtiy Press, pp 463–488

Snyder SH (1980) Brain peptides as neurotransmitters. Science 209:976–983

Söderberg G (1922) Contributions to the forebrain morphology in amphibians. Acta Zool 3:65–121

Sotelo C, Réthelyi M, Szabo T (1975) Morphological correlates of electrotonic coupling in the magnocellular mesencephalic nucleus of the weakly electric fish Gymnotus carapo. J Neurocytol 4:587–607

Stein BE (1981) Organization of the rodent superior colliculus: some comparisons with other mammals. Behav Brain Res 3:175–188

Stein BE (1984) Development of the superior colliculus. Annu Rev Neurosci 7:95–125

Steinbusch HWM, Nieuwenhuys R (1983) The raphe nuclei of the rat brainstem: a cytoarchitectonic and immunohistochemical study. In: Emson PC (ed) Clinical neuroanatomy. Raven, New York, pp 131–207

Stephan H (1975) Allocortex. Handbuch der mikroskopischen Anatomie des Menschen, vol 4: Nervensystem, part 9. Springer, Berlin Heidelberg New York

Sternberger LA (1979) Immunocytochemistry. Wiley, New York

Stopa EG, King JC, Lydic R, Schoene WC (1984) Human brain contains vasopressin and vasoactive intestinal polypeptide neuronal subpopulations in the suprachiasmatic region. Brain Res 297:159–163

Stuesse SL, Cruce WLR (1991) Immunohistochemical localization of serotoninergic, enkephalinergic, and catecholaminergic cells in the brainstem of a cartilaginous fish, Hydrolagus colliei. J Comp Neurol 309:535–548

Stuesse SL, Cruce WLR (1992) Distribution of tyrosine hydroxylase, serotonin, and leu-enkephalin immunoreactive cells in the brainstem of a shark, Squalus acanthias. Brain Behav Evol 39:77–92

Swanson LW (1983) Neuropeptides – new vistas on synaptic transmission. Trends Neurosci 6:294–295

Swanson LW (1984) Review of 'The rat brain in stereotaxis coordinates' by G Paxinos and C Watson. Trends Neurosci 7:53

Swanson LW (1991) Biochemical switching in hypothalamic circuits mediating responses to stress. Prog Brain Res 87:181–200

Swanson LW, Sawchenko PE, Bérod A, Hartman BK, Helle KB, Vanorden DE (1981) An immunohistochemical study of the oprganization of catecholaminergic cells and terminal fields in the paraventricular and supraoptic nuclei of the hypothalamus. J Comp Neurol 196:271–285

Szabo T, Ravaille M, Libouban S, Enger PS (1983) The mormyrid rhombencephalon. I. Light and EM investigations on the structure and connections of the lateral line lobe nucleus with HRP labeling. Brain Res 26:1–19

Székely G, Lázár G (1976) Cellular and synaptic architecture of the optic tectum. In: Llinás R, Precht W (eds) Frog neurobiology. Springer, Berlin Heidelberg New York, pp 407–434

Szentágothai J (1964) Pathways and synaptic articulation patterns connecting vestibular receptors and oculomotor nuclei. In: Bender MB (ed) The oculomotor system. Hoeber Medical Div, Harper and Row, New York, pp 205–223

Szentágothai J (1983) The modular architectonic principle of neural centers. Rev Physiol Biochem Pharmacol 98:11–61

Szentágothai J, Flerkó B, Mess B, Halász B (1968) Hypothalamic control of the anterior pituitary, 3rd rev enlarged edn. Akadémiai Kiadó, Budapest

Szentágothai J, Réthelyi M (1973) Cyto- and neuropil architecture of the spinal cord. In: Desmedt JE (ed) Human reflexes, pathophysiology of motor systems, methodology of human reflexes. New developments in electromyography and clinical neurophysiology, vol 3. Karger, Basel, pp 20–37

Takagi H, Morishima Y, Matsuyama T, Hayashi H, Watanabe T, Wada H (1986) Histaminergic axons in the neostriatum and cerebral cortex of the rat: a correlated light and electron microscopic immunocytochemical study using histidine decarboxylase as a marker. Brain Res 364:114–123

Thureson-Klein AK, Klein RL, Zhu P-C, Kong J-K (1988) Differential release of transmitters and neuropeptides co-stored in central and periperal neurons. In: Zimmermann H (ed) Cellular and molecular basis of synaptic transmission. Springer, Berlin Heidelberg New York, pp 137–151

Tokunaga A, Otani K (1976) Dendritic patterns of neurons in the rat superior colliculus. Exp Neurol 52:189–205

Uchizono K (1965) Characteristics of excitatory and inhibitory synapses in the central nervous system of the cat. Nature 207:642–643

Ulfhake B, Arvidsson U, Cullheim S, Hökfelt T, Brodin E, Verhofstad A, Visser T (1987) An ultrastructural study of 5-hydroxytryptamine-, thyrotropin-releasing hormone- and substance P-immunoreactive axonal boutons in the motor nucleus of spinal cord segments L7–S1 in the adult cat. Neuroscience 23:917–929

Valverde F (1965) Studies on the piriform lobe. Harvard Universtiy Press, Cambridge, Massachusetts

Van der Linden JAM, ten Donkelaar HJ (1987) Observations on the development of cerebellar afferents in Xenopus laevis. Anat Embryol (Berl) 176:431–439

Van Gehuchten A (1891) La structure des centres nerveux. La moelle épinière et le cervelet. Cellule 7:1–44

Vanegas H (1983) Organization and physiology of the teleostean optic tectum. In: Davis RE, Northcutt RG (eds) Fish neurobiology, vol 2: higher brain areas and functions. University of Michigan Press, Ann Arbor, pp 43–90

Vanegas H, Ebbesson SOE (1973) Retinal projections in the perch-like teleost Eugerres plumieri. J Comp Neurol 151:331–358

Vanegas H, Ito H (1983) Morphological aspects of the teleostean visual system: a review. Brain Res Rev 6:117–137

Vanegas H, Laufer M, Amat J (1974) The optic tectum of a perciform teleost. I. General configuration and cytoarchitecture. J Comp Neurol 154:43–60

Vanegas H, Williams B, Freeman JA (1979) Responses to stimulation of marginal fibers in the teleostean optic tectum. Exp Brain Res 34:335–349

Vanegas H, Ebbesson SOE, Laufer M (1984) Morphological aspects of the teleostean optic tectum. In: Vanegas H (ed) Comparative neurology of the optic tectum. Plenum, New York, pp 93–120

Von Bartheld CS, Meyer DL (1986) Central connections of the olfactory bulb in the bichir, Polypterus palmas, reexamined. Cell Tissue Res 244:527–535

Wake DB (1966) Comparative osteology and evolution of the lungless salamanders, family plethodontidae. Mem South Calif Acad Sci 4:1–111

Wake DB, Roth G (1989) Paedomorphosis: new evidence for its importance in salamander evolution. Am Zool 29:134A

Wallén P (1994) Sensorimotor integration in the lamprey locomotor system. Eur J Morphol 32:168–175

Weight FF (1979) Modulation of synaptic excitability. Fed Proc 38:2078–2079

White EL (1978) Identified neurons in mouse Sm I cortex, which are postsynaptic to thalamocortical axon terminals: a combined Golgi-electron microscopic and degeneration study. J Comp Neurol 181:627–662

Witkin JW, Paden ChM, Silverman AJ (1982) The luteinizing hormone-releasing hormone (LHRH) systems in the rat brain. Neuroendocrinol 35:429–438

Wolters JG, ten Donkelaar HJ, Verhofstad AAJ (1986) Distribution of some peptides (substance P, [leu]enkephalin, [met]enkephalin) in the brain stem and spinal cord of a lizard, Varanus exanthematicus. Neuroscience 18:917–946

Woolsey TA (1987) Barrels, vibrissae, and topographic representations. In: Adelman G (ed) Encyclopedia of neuroscience, vol I. Birkhäuser, Boston, pp 111–113

Woolsey TA, van der Loos H (1970) The structural organization of layer IV in the somatosensory region (SI) of mouse cerebral cortex. The description of a cortical field composed of discrete cytoarchitectonic units. Brain Res 17:205–242

Yeziersky RP (1988) Spinomesencephalic tract: projections from the lumbosacral spinal cord of the rat, cat, and monkey. J Comp Neurol 267:131–146

Yuste R, Denk W (1995) Dendritic spines as basic functional units of neuronal integration. Nature 375:682–684

Zhu PC, Thureson-Klein A, Klein RL (1986) Exocytosis from large dense cored vesicles outside the active synaptic zones of terminals within the trigeminal subnucleus caudalis: a possible mechanism for neuropeptide release. Neuroscience 19:43–54

CHAPTER 3

Structure and Organisation of Fibre Systems

R. NIEUWENHUYS

3.1
Introduction

The science of hodology, from the Greek *hodos*, meaning road or path, has two principal aims, (a) the determination of the origin(s), course and site(s) of termination of fibre pathways, and (b) the study of the structure and composition of these pathways. During the past 150 years, an enormous body of data has been accumulated on the fibre pathways in the central nervous system of many vertebrate species, but a general hodology has not been developed. The present chapter will attempt to fill this gap. A brief overview of the evolutionary development of fibre compartments (Sect. 3.2) will be followed by some notes on the structure and the synaptic contacts of individual fibres (Sect. 3.3). Then, the origin, course and termination of fibre pathways (Sect. 3.4) and their structure and composition (Sect. 3.5) will be discussed. The final sec-

tions will present an outline of the three subdisciplines of hodology (Sect. 3.6) and some functional notes (Sect. 3.7). It should be emphasised that the drawings and photomicrographs in this chapter not only illustrate the text, but are central to its message.

3.2
Fibre Compartments

The previous chapter presented a conceptual model of the organisation of the wall of the nervous system of a hypothetical simple vertebrate (Chap. 2, Fig. 2.1). The essentials of this model are repeated here (Fig. 3.1a), as the basis for an exploration of the origin and the evolutionary development of the large fibre compartments in the CNS. The model has a central and a peripheral zone. The central zone contains densely crowded neuronal cell bodies which extend their dendrites throughout the peripheral zone. The axons of central or periventricular neurons also run outward and contribute to the peripheral zone, where many take a longitudinal course. Throughout the peripheral zone, axons synapse with dendrites of periventricular neurons. Consequently, the peripheral zone is both a fibre zone and a continuous neuropil zone, where numerous axodendritic contacts are made. These contacts are made either directly by the unmyelinated axonal shafts or by short collateral branches, termed interstitial collaterals by Ramón y Cajal (1909). Longer, internal collaterals enter the central grey zone to synapse with the proximal dendrites and somata of the intrinsic neurons. One may hypothesise that the following events took place during evolution: (a) The dendrites of the periventricular cells shortened and retracted from the superficial part of the peripheral zone. (b) The lack of recipient structures caused synaptic contacts to vanish from the superficial part of the peripheral zone. (c) Internal collaterals increased in number, and some of the interstitial collaterals were transformed into internal ones. (d) Long, through-conducting myelinated fibres aggregated in the

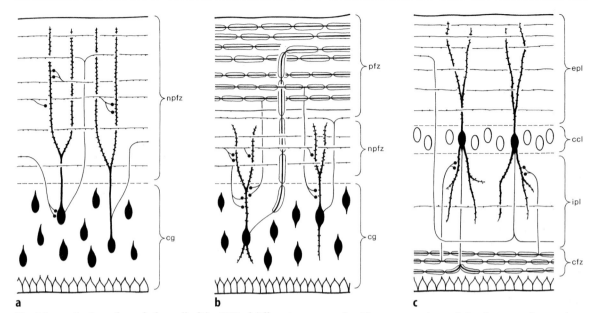

Fig. 3.1a–c. Sections through the wall of the CNS of different vertebrates, to demonstrate the genesis of fibre compartments. **a** Longitudinal section of a part of the wall of the neuraxis of a hypothetical simple and primitive agnathe-like vertebrate. The neuronal perikarya, which themselves constitute a zone of central grey (*cg*), send their dendrites peripherally into a neuropil and fibre zone (*npfz*). The fibres of that zone are all unmyelinated and establish synaptic contacts with the dendrites throughout their extent. **b** The structural relations in the spinal cord and lower brain stem of birds and mammals. The neurons located in the central grey have retracted their dendrites from the external part of the peripheral zone, which itself has become transformed into a fibre zone (*pfz*). **c** Transverse section through the reptilian pallium (or the mammalian hippocampus). The neuronal perikarya have migrated peripherally and form a distinct cortical cell layer (*ccl*), flanked by external and internal plexiform layers (*epl, ipl*). Deep to the ipl, a periventricular or central fibre zone (*cfz*) has developed

peripheral zone and formed a true fibre compartment, within which axons with specific origins and destination have formed into more compact pathways (Fig. 3.1b).

'Stages' in the development just sketched can be observed in the central nervous system of different vertebrate groups. In amphioxus and the cyclostomes, myelinated fibres are absent; here the entire peripheral zone of the neuraxis is potentially a combined fibre and neuropil compartment. Myelinated fibres do occur in the central nervous system of all gnathostomes. However, because in fish, amphibians and reptiles many neurons still extend their dendrites far into the periphery, the pure fibre compartment, or true white matter remains small. In birds and mammals, finally, the dendritic trees of most neurons tend to be much more restricted and the fibre compartment has accordingly increased in size. It is not the case, however, that in these forms the grey matter and the white matter are sharply differentiated. In most places, the central grey matter contains neurons whose distal dendritic segments project to, and receive synaptic contacts from, within the white matter. So even in birds and mammals the compartments of pure grey and white

matter are generally separated by a mixed neuropil and fibre zone.

The tendency of long, through-conducting fibres to concentrate in the superficial zone of the neuraxis was first noticed by Flatau (1897) on the basis of Marchi experiments on the mammalian spinal cord.

The developmental events leading to the formation of a peripherally located fibre compartment are seen most clearly in the spinal cord and the lower brain stem, although superficial fibre zones are seen locally in more rostral parts of the brain. In the telencephalon of amniotes, however, a fibre compartment of an altogether different nature develops. In the pallial part of the cerebral hemispheres of this group, the neurons migrate peripherally during ontogenesis to form one or more cortical layers. The axons of most cortical projection neurons form a deep, central, fibre layer in the subventricular zone (Fig. 3.1c). In reptiles and birds this layer is narrow, but in mammals it is much wider. In whales and primates these deep fibres together constitute an enormous fibre masss known as the centrum semiovale. This 'centrum' contains numerous projection fibres which ascend to, or descend from, the cerebral cortex. However, the bulk of its fibres interconnect various parts of the cortex. In primates these cortico-cortical projections together constitute by far the largest pathway of the brain.

3.3
Notes on Individual Fibres

The nerve fibres in the vertebrate CNS may be unmyelinated or myelinated. In acraniates (Ruiz and Anadón 1989) and cyclostomes (Bertolini 1964; Smith et al. 1970; Bullock et al. 1984) only unmyelinated axons are found, with diameters ranging from 0.1 to 100 μm. Both groups have conspicuously large axons originating from identifiable giant neurons. In acraniates these large fibres, with diameters of 10–30 μm, emanate from the Rohde cells, which are large multipolar neurons that lie across the slitlike central canal of the spinal cord (Rohde 1888a,b; Retzius 1891; Franz 1923; Bone 1960). The Rohde cells form a rostral group of 12

and a caudal group of 14 cells. They are missing in the mid-region of the cord. The most rostral Rohde cell, which is the largest neuron in the acraniate CNS, sends its axon caudally along the midline of the spinal cord, below the central canal. The axons of the remaining rostral Rohde cells form a descending bundle in the dorsolateral part of the cord, while those of the caudal group ascend in the ventromedial part (Fig. 3.2).

In lampreys seven to eight large Müller cells, forming part of the reticular formation of the brain stem, possess conspicuous axons (25–50 μm in diameter, up to 100 μm in adult sea lampreys) which descend via the medial longitudinal fasciculus to the ipsilateral anterior funiculus of the spinal cord (Tretjakoff 1909a,b; Stefanelli 1934; Rovainen 1967, 1978; Rovainen et al. 1973; Fig. 3.3). A pair of large Mauthner cells is also present in the lamprey. Curiously, the axons of these elements, after decussating, go to the dorsal part of the spinal lateral funiculus instead of the anterior funiculus, as in the

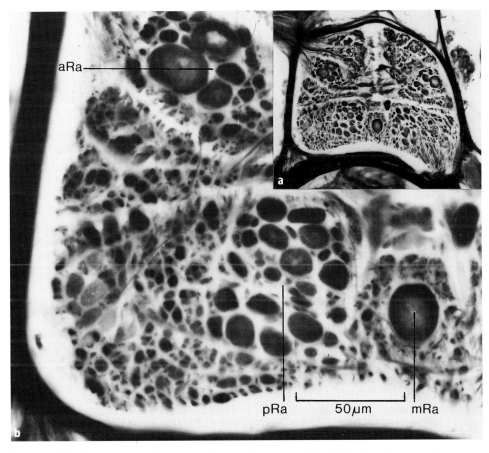

Fig. 3.2a,b. The spinal cord of the lancelet, *Branchiostoma lanceolatum.* **a** transverse section through the intermediate part of the cord; **b** the left ventral quadrant of the cord at higher magnification. Note that the axons, which lack a myelin sheath, vary considerably in size. Most conspicuous are the axons of the giant cells of Rohde, some with a diameter of 20 μm. *aRa,* Anterior Rohde axons; *mRa,* median Rohde axon; *pRa,* posterior Rohde axons

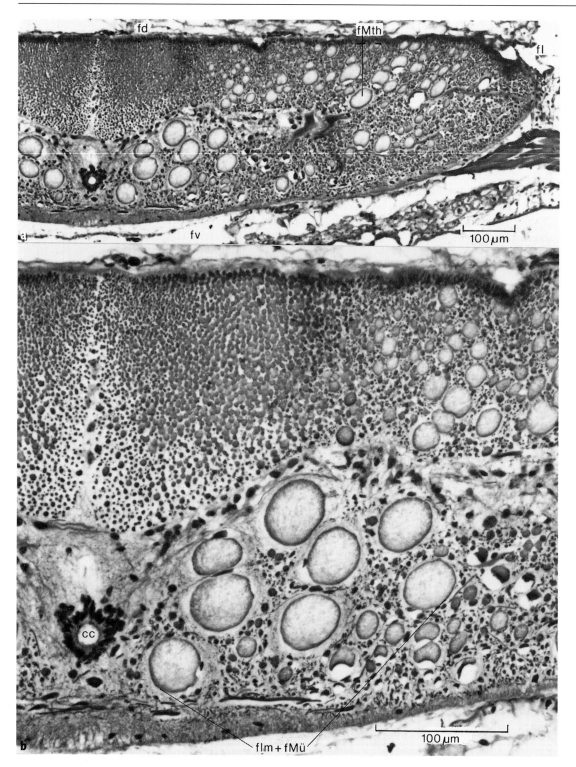

Fig. 3.3a,b. The spinal cord of the lamprey, *Lampetra fluviatilis.* **a** transverse section through the rostral part of the cord; **b** medial part of the cord at higher magnification. The ventral funiculus contains an accumulation of large Müller fibres, the largest with a diameter of 40 μm. Large axons, including the Mauthner axon, also occur in the lateral funiculus. The dorsal funiculus is formed by densely packed, thin axons. All fibres in the CNS of the lamprey are unmyelinated. *cc*, Central canal; *fd*, funiculus dorsalis; *fl*, funiculus lateralis; *flm*, fasciculus longitudinalis medialis; *fMth*, fibre of Mauthner; *fMü*, fibres of Müller; *fv*, funiculus ventralis

case of gnathostomes (Fig. 3.3a). The diameter of the axons of other large reticulospinal neurons may rival that of the Müller axons. Large reticulospinal neurons with coarse descending axons have also been observed in the brain stem of hagfishes (Bone 1963), but typical Müller and Mauthner neurons are lacking.

The CNS system of gnathostomes contains both myelinated and unmyelinated axons. The former possess a sheath of myelin for at least a part of their course. In the central nervous system the myelin sheath is the product of oligodendrocytes, glia cells which spiral their flattened processes around the axon. Adjacent glia processes leave stretches of the axon free of myelin. These interruptions in the myelin sheath, including their naked axon segments, are known as the nodes of Ranvier. Myelination and node formation represent important steps in evolutionary development, enabling the axon to conduct impulses rapidly from node to node in a jumping, or 'saltatory' fashion. Unmyelinated fibres conduct impulses much more slowly and in a continuous manner. The 'advantages' of myelinated fibres over unmyelinated ones are well illustrated by the following passage, quoted from a recent review (Hildebrand et al. 1993, pp. 322–323):

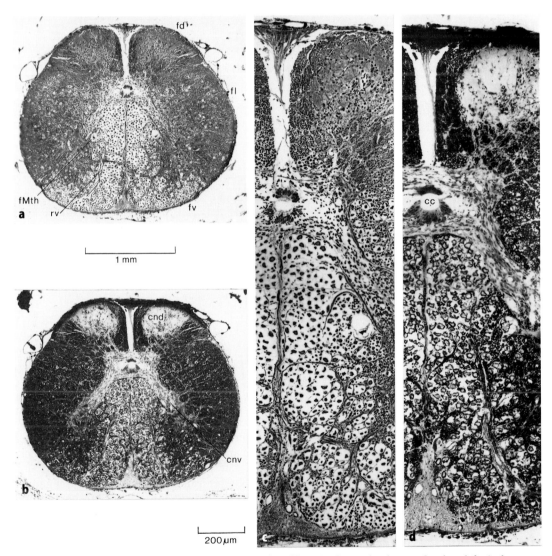

Fig. 3.4a–d. The spinal cord of the holostean fish, *Amia calva*. **a** Transverse section through the rostral portion of the cord, Bodian impregnation; **b** adjacent section, Klüver-Barrera stain; **c,d** the medial parts of **a** and **b** at a higher magnification. The ventral funiculus contains predominantly large fibres; in the lateral funiculus a mixture of thick and thin fibres is observed, whereas the dorsal funiculus comprises a large, dorsal compartment consisting of thin fibres and a small, ventral compartment containing some 25 medium-sized fibres. *cc*, Canalis centralis; *cnd*, cornu dorsale; *cnv*, cornu ventrale; *fd*, funiculus dorsalis, *fMth*, fibre of Mauthner; *fv*, funiculus ventralis; *rv*, radix ventralis

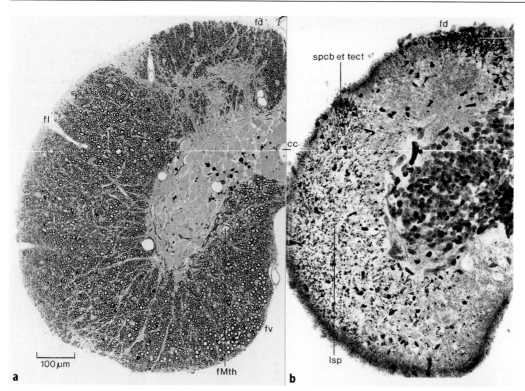

Fig. 3.5a,b. The spinal cord of urodeles. **a** Transverse hemi-section through the cervical cord of the tiger salamander, *Ambystoma tigrinum*, semithin (2 μm) epon section stained with paraphenylenediamine. Note that the superficial zone of the lateral funiculus, which is occupied by the marginal plexus, is devoid of myelinated fibres. **b** Transverse hemisection through the cervical spinal cord of an axolotl (*Ambystoma mexicanum*), which had been hemisected a few seg-ments lower 10 days previously; Nauta technique. Degenerating fibres constitute three bundles, the compact funiculus dorsalis and tractus spinocerebellaris et tectalis, and the more diffuse lemniscus spinalis. *cc*, Canalis centralis; *fd*, funiculus dorsalis; *fl*, funiculus lateralis; *fMth*, fibre of Mauthner; *lsp*, lemniscus spinalis; *fv*, funiculus ventralis; *rd*, radix dorsalis; *spcb et tect*, tractus spinocerebellaris et tectalis

"The velocity of signal conduction by a myelin-ated axon increases directly with the diameter of the fibre, so that a doubling of the diameter gives a doubling of the rate. In contrast, the conduction velocity of an unmyelinated axon increases with the square root of the axon diameter. For the rate to be doubled the calibre of the axon must be four times larger ... . This fundamental difference is clearly illustrated by the fact that an unmyelinated squid giant axon with a diameter of some 500 μm and a mammalian myelinated axon with an outer diam-eter of about 4 μm, both conduct about 20 m/s ... . For a given length, the squid axon occupies 15 000 × more space compared to the mammalian axon. Other calculations indicate that the squid giant axon consumes 5000 × more energy than a myelinated frog axon with a diameter of 12 μm, although the latter conducts more rapidly ... . Thus, in addition to a high conduction velocity, the evolu-tion of the myelinated axon has resulted in a remarkable saving of space and energy."

The myelinated fibres in the central nervous sys-tem of gnathostomes vary considerably in size. In birds and mammals these fibres have external di-ameters (i.e. diameters including the myelin sheaths) of 1–25 μm, though myelinated axons as thin as 0.1–0.2 μm in diameter occur (e.g., Bishop and Smith 1964; Chung and Coggeshall 1983b). The lower brain stem and spinal cord of fishes and reptiles contain myelinated fibres of up to 40 μm in diameter (Figs. 3.4, 3.16). Giant Mauthner neurons occur in most groups of fish and in amphibians. The axon of these elements is myelinated and may have a diameter of 100 μm or more (Fig. 3.4), though in many teleosts and amphibians they are only 20–40 μm in diameter (Figs. 3.5a,c, 3.7d,h). It is noteworthy that in Mauthner fibres the glia lamellae forming the myelin sheath are often wound in a loose spiral (Figs. 3.5a, 3.6b,d, 3.7b–j), and that typical nodes of Ranvier are lacking in this sheath (Leghissa 1956; Celio et al. 1979).

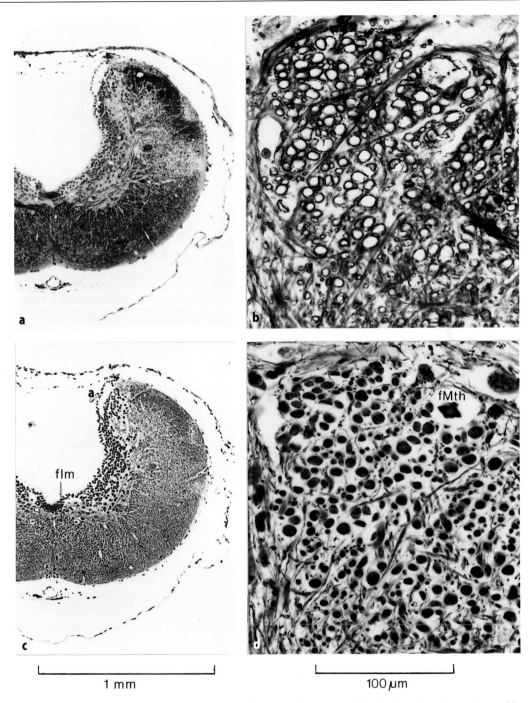

Fig. 3.6a–d. Hemisection through the caudal part of the rhombencephalon of the tiger salamander, *Ambystoma tigrinum*. **a** Klüver and Barrera stain; **b** Bodian impregnation; **c,d** the paramedian parts of the hemisections shown in **a** and **b**, at a higher magnification. *flm*, Fasciculus longitudinalis medialis; *fMth*, fibre of Mauthner

The unmyelinated fibres in the CNS of gnathostomes range in diameter from 0.1 to 3 μm (Figs. 3.7 f–j, 3.12 g–j, 3.20c,d). However, because the vast majority of the myelinated axons in the CNS are very thin (less than 0.5 μm in diameter), the size spectra of myelinated and unmyelinated axons overlap considerably. It should be emphasised that, contrary to current belief, unmyelinated fibres are abundant in gnathostome vertebrates and presumably far outnumber the myelinated fibres.

The axonal process conducts impulses away from the neuronal cell body to its telodendron, where the

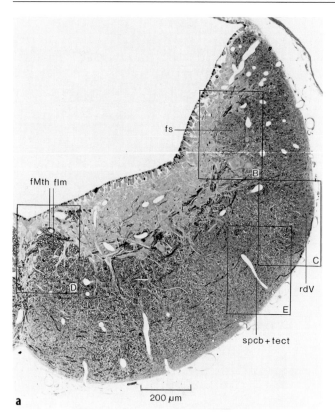

a

200 μm

Fig. 3.7a–i. a Semithin (2 μm) epon section stained with paraphenylenediamine through the intermediate part of the rhombencephalon of the tiger salamander, *Ambystoma tigrinum*. Details of the hemisection shown in **a** are presented in **b–e: b** fasciculus solitarius; **c** radix descendens of V; **d** fasciculus longitudinalis medialis; **e** tractus spinocerebellaris et tectalis. The ultrastructure of these fibre systems is shown in **f–i: f** fasciculus solitarius; **g** radix descendens of V; **h** fasciculus longitudinalis medialis; **i** spinocerebellar and spinotectal tracts. Note the presence of thin, unmyelinated fibres in all four tracts (*arrows*). *d*, Dendrites; *flm*, fasciculus longitudinalis medialis; *fMth*, fibre of Mauthner; *fs*, fasciculus solitarius; *rdV*, radix descendens of V; *spcb + tect*, tractus spinocerebellaris et tectalis; *t*, axon terminal

information coded in these impulses is transmitted to other neuronal elements, muscles or glands. In most central projection neurons the transfer of information is not confined to the distal endings of the axon, but also occurs at other places. The 'devices' for this *en-passant* transfer of information are quite variable, as the following examples will show.

1. The axons of most projection neurons emit one or more collateral branches, often at regular intervals. Thus, the fibres of the mammalian medial forebrain bundle issue a large number of transversely oriented collaterals which run parallel to and synapse with the dendrites of hypothalamic neurons (Millhouse 1969; Fig. 3.8b). A similar relationship has been found between the collaterals of the ascending axons of large neurons in the reticular formation in the mammalian brain stem and the transversely oriented dendritic trees of other, more rostrally situated neurons (Scheibel and Scheibel 1958). Coarse afferent axons passing through the inferior olive make numerous short collaterals which terminate with bouton clusters on the somata of olivary cells (Scheibel and Scheibel 1955; Fig. 3.8c).

2. The axons of neurons involved in the distribution of signals related to coordinated movements or muscle synergies often have multiple axon collater-

als, the arborisations of which rival those of the end of the axon in density and extent. Thus, the axons of large neurons in the tectum mesencephali of the catfish, *Ictalurus*, which are presumably involved in eye movements, send out as many as 19 ramifying collaterals during their descent through the brain stem (Sharma et al. 1985). Tectobulbar axons with similar multicollateral axons also occur in turtles (Sereno 1985; Sereno and Ulinski 1985) and in the cat (Grantyn and Grantyn 1982).

Head movement signals detected by the semicircular canals are mediated through vestibular afferents and vestibulospinal pathways that link each of the three semicircular canals to a particular set of neck muscles. These pathways mediate canal-specific compensatory head movement synergies. Shinoda et al. (1992a,b) studied the morphology of single medial vestibulospinal tract axons involved in these reflex movements. These axons, over one to three cervical segments, give rise to as many as nine collateral branches, which innervate the longitudinally arranged columns of motoneurons of synergic groups of neck muscles at different levels (Fig. 3.8d).

3. In fish it is well established that the giant Mauthner cell mediates the triggering of a fast contralateral body bend, known as the startle reflex, in

response to unexpected stimuli or in avoidance of predator attack (Faber and Korn 1978; Korn 1987). The crossed Mauthner axon runs almost the length of the spinal cord, contacting the motoneurons and interneurons involved in this reflex via short (5–50 μm) collaterals, arising from the axon spaced at 100- to 300-μm intervals (Leghissa 1956; Zottoli 1978). In teleosts these collaterals are ensheathed by myelin to their tip (Celio et al. 1979; Fig. 8e). Greeff and Yasargil (1980) have demonstrated that the impulses are conducted along Mauthner axons in a saltatory mode, although typical nodes of Ranvier have not been observed in these nerve fibres. Yasargil et al. (1982) suggested that the unmyelinated tips of the short collaterals function as nodal equivalents for the saltatory propagation.

4. The axolemma of normal nodes of Ranvier may also be involved in synaptic processes (Bodian and Tailor 1963: primate spinal cord; Sotelo and Palay 1970: lateral vestibular nucleus of rat; Waxman 1972: oculomotor nucleus of a teleost). In such nodes the axon often gives rise to a bulbous swelling which constitutes a miniature collateral and contains characteristic presynaptic vesicles (Fig. 3.9).

Nicol and Walmsley (1991; see also Walmsley 1991) observed that, in the cat, the collaterals of Ia afferents, which enter Clarke's column in the spinal cord, show a wide range of geometries and myelination patterns, including some nodes with a single synaptic bouton and others with two or more synaptic boutons connected by fine unmyelinated lengths of the collateral (Fig. 3.8 f).

5. The unmyelinated large Müller axons in the spinal cord of the lamprey (Fig. 3.3) are unbranched and make *en-passant* efferent synapses with the dendrites of spinal neurons. These specialisations are characterised by thickenings in the pre- and postsynaptic membranes and by focal clusters of round, clear vesicles near the axonal membrane (Bertolini 1964; Smith et al. 1970; Kershew and Christensen 1980). Small gap junctions are often situated near these chemical ones, thus forming 'mixed' synapses (Rovainen 1974a,b, 1978; Christensen 1976). Afferent axo-axonic synapses have occasionally been observed on Müller axons (Wickelgren 1977; Kershew and Christensen 1980).

Ruiz and Anadón (1989) reported that in amphioxus the giant median Rohde axon (Fig. 3.2) establishes minute efferent *en-passant* synapses with their axonal processes; they also found a few reciprocal synapses between Rohde and other unidentified elements.

6. The abundant thin, unmyelinated, axons in the CNS of all vertebrates often have small varicosities. These varicosities may occur either over only cer-

tain trajectories (e.g. the thalamotelencephalic axons observed in a turtle by Heller and Ulinski 1987) or throughout their course [e.g. some efferents of the ventral pallidum of the rat (Swanson et al. 1984) and some calcitonin gene-related peptide-containing axons in the cat neocortex (Conti et al. 1993), Fig. 3.8g]. Such varicosities usually contain small, clear and/or large dense-core vesicles. Analysis of serial sections revealed that the varicosities sometimes participate in synapses, but that in other instances synaptic membrane specialisations are absent, suggesting non-synaptic chemical neurotransmission (Buma 1989; Buma et al. 1992; Veening et al. 1991). These examples indicate that axons, with regard to *en-passant* transfer of information, range from entirely 'closed' myelinated fibres (Fig. 3.8a) to 'open-line' unmyelinated fibres (Fig. 3.8g), with intermediary forms.

3.4
Origin(s), Course, Extent and Site(s) of Termination of Fibre Systems

3.4.1
Introductory Notes

To introduce the aims and problems of hodology, a simple fibre system is diagrammed in Fig. 3.10. This pathway shows the following features: (a) it originates from a single centre; (b) its fibres emanate from cells of one type; (c) its fibres are entirely 'closed' and all pass in the same direction; (d) all fibres follow the same trajectory; (e) it terminates in a single centre; (f) the terminal ramifications of all of its fibres are the same.

Analysis of the origin, course and termination of a given tract requires an extensive experimental analysis. The origin(s) of a fibre system is determined using retrograde tracing techniques. To establish whether a centre that projects to more than one other centre and whether it does so by means of the axons of different neurons or by different branches of axons of the same cells requires multiple retrograde labelling techniques. Anterograde labelling techniques determine the course and site(s) of termination of a tract. For a detailed analysis of the terminal and collateral ramifications of axons, combined light- and electron-microscopical studies on intra-axonally labelled single fibres are often necessary. Current techniques are discussed in Chap. 7.

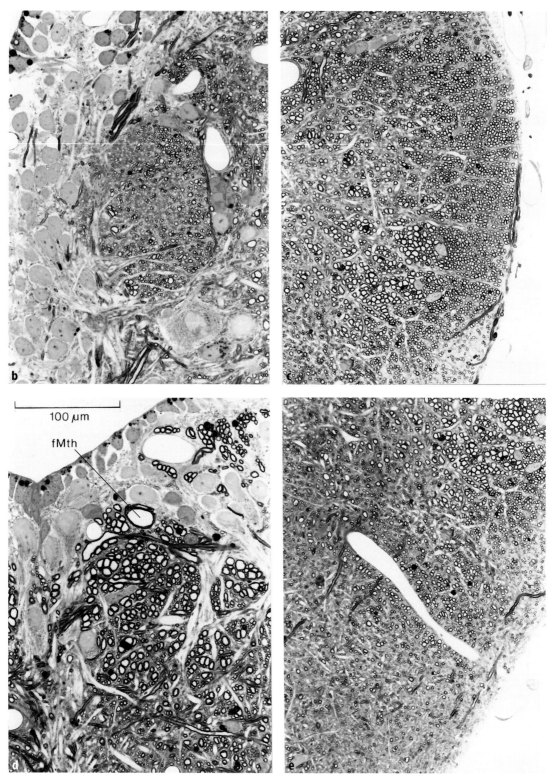

Fig. 3.7b–e

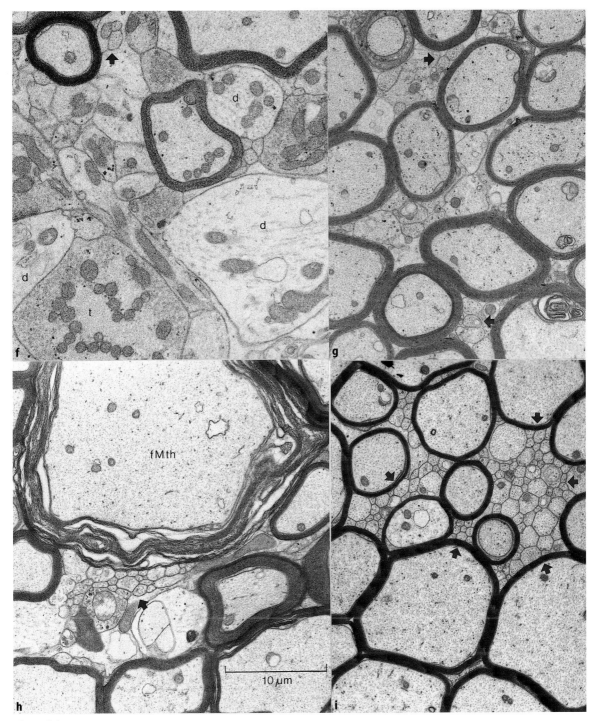

Fig. 3.7f–i.

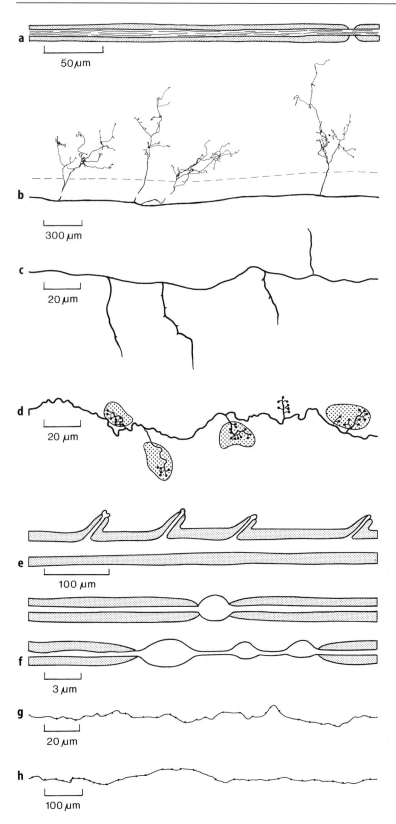

Fig. 3.8a–h. Various types of axons in the vertebrate CNS. **a** Coarse myelinated axon. **b** Reconstruction of a medial vestibulospinal tract axon in the upper cervical cord of the cat, giving rise to several collaterals, injected iontophoretically with HRP (redrawn from Shinoda et al. 1992a). **c** Fibre of the medial forebrain bundle of an 8-day-old mouse, with a number of collaterals which run parallel to the dendrites of lateral hypothalamic neurons. Rapid Golgi technique (redrawn from Millhouse 1969). **d** A heavy afferent fibre to the inferior olive of a 10-day-old kitten, issuing several branches which terminate with bouton clusters on cell bodies in the olive. Golgi method (redrawn from Scheibel and Scheibel 1955). **e** Mauthner axon in the spinal cord of the tench, *Tinca tinca* (based on Yasargil et al. 1982). **f** Two Ia-afferent collaterals obtained from a serial EM study in Clarke's column of the cat spinal cord (simplified from Walmsley 1991). **g** Thin, varicose fibre passing through the rostral hypothalamus of the rat, observed in a *Phaseolus vulgaris*-leucoagglutinin preparation. **h** Camera lucida drawing of a CGRP-containing fibre in the cat neocortex (reproduced from Conti et al. 1993)

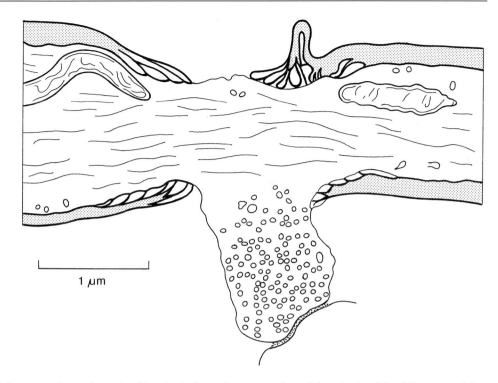

Fig. 3.9. A synaptic bouton arising at the node of Ranvier in the oculomotor nucleus of the spiny boxfish, *Chilomycterus* (simplified from Waxman 1972)

3.4.2
Origin(s), Polarity and Site(s) of Termination of Fibre Systems

Fibre systems originating from a single cell type, within a single centre, consisting entirely of 'closed' axons, and terminating with one type of telodendron in a single centre, like the one shown in Fig. 3.10, do exist but are rare. One example is the tractus telencephalotegmentalis in osteoglossomorph teleosts. This pathway originates from large neurons in a specific zone of the area dorsalis telencephali, designated as Dcd, and descends as a compact, myelinated bundle via the diencephalon (Fig. 3.11) to the tegmentum of the midbrain. Within the tegmentum all of its fibres enter the nucleus lateralis valvulae and terminate with characteristic grapelike endings around adendritic neurons (Nieuwenhuys and Verrijdt 1983).

Experimental studies have revealed that many ascending and descending fibre systems originate from more than one cell type and project to more than one area of termination, as for instance in the mammalian spinothalamic and corticospinal tracts.

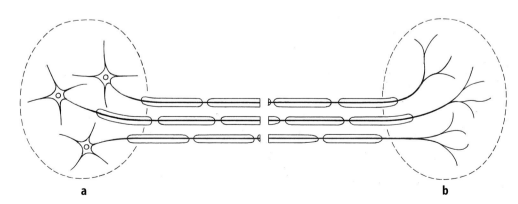

Fig. 3.10. Simple fibre tract projecting from nucleus **a** to nucleus **b**

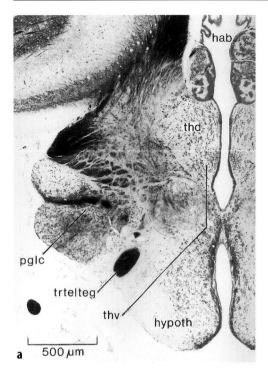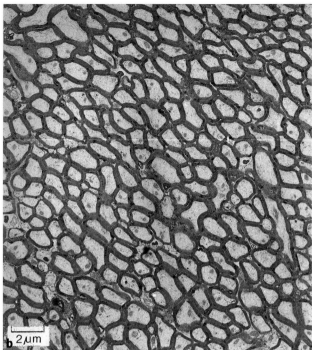

Fig. 3.11a,b. The diencephalon of the teleost *Xenomystis nigri*. **a** Transverse hemisection through the intermediate part of the diencephalon, showing the large, compact tractus telencephalo-tegmentalis; **b** Ultrastructure of the tractus telencephalo-tegmentalis. *hab*, Nucleus habenulae; *hypoth*, hypothalamus; *pglc*, preglomerulosus complex; *thd*, thalamus dorsalis; *thv*, thalamus ventralis; *trtelteg*, tractus telencephalotegmentalis

The spinothalamic tract conveys information about pain and temperature from the spinal cord to the dorsal thalamus. Its cells of origin are distributed over no fewer than six (I, IV, V, VI, VII, VIII) of the ten layers into which Rexed (1954, 1964) subdivided the spinal grey matter (e.g., Comans and Snow 1981). Within the dorsal thalamus the spinothalamic tract projects heavily to the ventral posterolateral nucleus as well as to certain intralaminar nuclei (Mehler 1969; Boivie 1979; Mantyh 1983).

In primates the corticospinal tract originates from the giant neurons of Betz in area 4, and from a much larger number of smaller pyramidal cells in lamina V of the pericentral cortex. The terminal ramifications of its fibres distribute to laminae V, VI and VII of Rexed. However, in animals who execute independent finger movements the corticospinal tract projects directly to motor anterior horn cells situated in lamina IX. In primates these direct cortico-motoneuronal connections are made by the coarse axons of the giant Betz cells; hence these axons constitute, with regard to origin, termination and function, a recognisable subset (Kuypers 1987).

All fibres in the two systems just discussed pass in one direction, but in many other tracts fibres of opposite polarity have been found. For example, the optic tract, which is composed mainly of centripetal fibres from different types of retinal ganglion cells, usually contains numerous centrifugal fibres.

The CNS of all vertebrates has fibre assemblies, composed of axons with very different origins and terminations. The reason why these heterogeneous fibre aggregations are treated as hodological units is simply that they appear, at least over a certain stretch, as delimitable entities. The fasciculus longitudinalis medialis and the fasciculus basalis telencephali, or basal forebrain bundle, are examples.

The fasciculus longitudinalis medialis is a conspicuous bundle of coarse fibres which occupies a periventricular and paramedian position throughout the brain stem in all vertebrates (Figs. 3.6, 3.7a,d, 3.19a,c, 3.23). Caudally, the bundle continues into the anterior funiculus of the spinal cord (Figs. 3.3–3.5, 3.16). Within the fasciculus longitudinalis medialis, fibres originating from many different centres, including the nucleus interstitialis, the tectum, the reticular formation and the vestibular nuclear complex, descend to the spinal cord. However, the bundle also contains ascending fibres which connect the vestibular nuclear complex with the oculomotor nuclei.

The basal forebrain bundle forms the great longitudinal conduction pathway between the telencephalon and caudal parts of the brain. Like the medial longitudinal fasciculus (mlf), the basal fore-brain bundle is composed of fibre components with very different origins and destinations, but unlike the mlf, this bundle consists mainly of thin, loosely arranged fibres. In most vertebrates the basal fore-brain bundle consists of separate medial and lateral bundles with both ascending and descending fibres. Fig. 3.12 shows the structure and composition of the lateral forebrain bundle of the tiger salamander at mid-diencephalic levels where Herrick (1927, 1933, 1948) distinguished about nine descending and one ascending components. In mammals, the medial forebrain bundle appears to be the most complex fibre system of the brain, comprising more than 50 different components (Nieuwenhuys et al. 1982; Figs. 3.21, 3.22).

From a comparative perspective it is important to realise that fibre systems in different vertebrates, which are considered to be homologous on account of similarities in origin and course, may still show marked differences in the sites of termination of some components. In all vertebrates the lateral funiculus carries numerous fibres from the spinal cord to the brain. The course and termination of these fibres has been determined in lampreys (Ronan 1983), hagfishes (Ronan 1983), cartilagi-nous fishes (Hayle 1973; Ebbeson and Hodde 1981), teleosts (Hayle 1973; Oka et al. 1986), lungfishes (Northcutt 1984), salamanders (Nieuwenhuys and Cornelisz 1971), anurans (Ebbeson 1976), reptiles (Ebbesson 1967, 1969; Pritz and Northcutt 1980), birds (Karten 1963) and mammals (e.g., Mehler 1969; Willis and Coggeshall 1978). In all cases, numerous spinal ascending fibres terminated in the rhombencephalic reticular formation, and in all gnathostomes a spinocerebellar projection was found. The site of termination was variable, how-ever. A spinotectal projection has been found in most vertebrates, but not in lampreys, teleosts and anuran amphibians. In the teleost *Oncorhynchus* a projection to the torus semicircularis occurs (Oka et al. 1986), but a direct spinothalamic projection exists only in the nurse shark, *Ginglymostoma* (Ebbeson and Hodde 1981), and in amniotes.

3.4.3
Course, Trajectory and Extent of Fibre Systems

Fibre systems generally follow a fixed and stereo-typed course through theCNS, but there are excep-tions, as the following examples show.

1. The paired Mauthner cells occupy a stereo-typed position at the level of entrance of the eighth nerve (Chap. 2, Fig. 2.37) and have a coarse axon that crosses the midline slightly caudal to the level of the soma before it descends to the spinal cord. In all gnathostomes possessing these giant neurons, their axons run caudally in the anterior funiculus of the spinal cord (Figs. 3.4, 3.6), but in lampreys these axons descend in the dorsal part of the lateral funiculus (Rovainen 1967; Rovainen et al. 1973; Fig. 3.4).

2. The pyramidal or corticospinal tract is a fibre bundle on the ventral surface of the medulla oblon-gata, where it produces a paramedian surface eleva-tion called the pyramid. It consists entirely of de-scending fibres from pyramidal neurons in the fifth layer of the neocortex. Note that the tract is named after the medullary elevation and *not* after its cells of origin. A pyramidal tract is present only in mam-mals; in the majority, the bulk of the corticospinal fibres cross the median plane in the most caudal part of the medulla oblongata and descend in the contra-lateral posterolateral funiculus of the spinal cord to lumbar or sacral levels. However, marked deviations from this common pattern occur in some animals, as seen in the following examples (for a comprehensive comparative survey of the pyramidal tract and other pathways see Verhaart 1970).

(a) The level of decussation may differ, ranging from the upper part of the pons in the monotreme *Tachyglossus* (Goldby 1939; Fig. 3.13b), over the high medulla oblongata in Chiroptera (Fuse 1926) and Pholidota (Chang 1944), to the most common position in the low medulla oblongata, as in carni-vores, primates and most other mammalian groups.

(b) No pyramidal crossing is present in the mole (Hatschek 1907), the giant ant-eater (Verhaart 1963a, 1967) and the hyracoid *Procavia* (Verhaart 1967).

(c) In the lower part of the pyramidal tract of the monotreme *Tachyglossus*, it is remarkable that this bundle decussates completely in the upper pons, then passes very lateral to a position superficial to the descending root of the trigeminal nerve (Fig. 3.13c). At the level of the spinomedullary junc-tion the tract occupies a similar position close to the dorsal horn (Fig. 3.13d). It descends in the most dorsal part of the lateral funiculus, to lower lumbar levels (Goldby 1939; Fig. 3.13e,f). These data show that in *Tachyglossus* the pyramidal tract does not occupy a superficial paramedian position in the lower medulla oblongata; consequently, pyramids are absent in this species.

(d) In those mammals in which the corticospinal tracts reach the most caudal levels of the brain stem via typical pyramids there may be considerable dif-ferences in the spinal trajectory and caudal extent of the tract. Thus, in several marsupials (Martin and Fisher 1968: opossum; Martin et al. 1970: *Tri-chosusus*; Martin et al. 1972: potoro) these fibres decussate to the deepest part of the dorsal funicu-

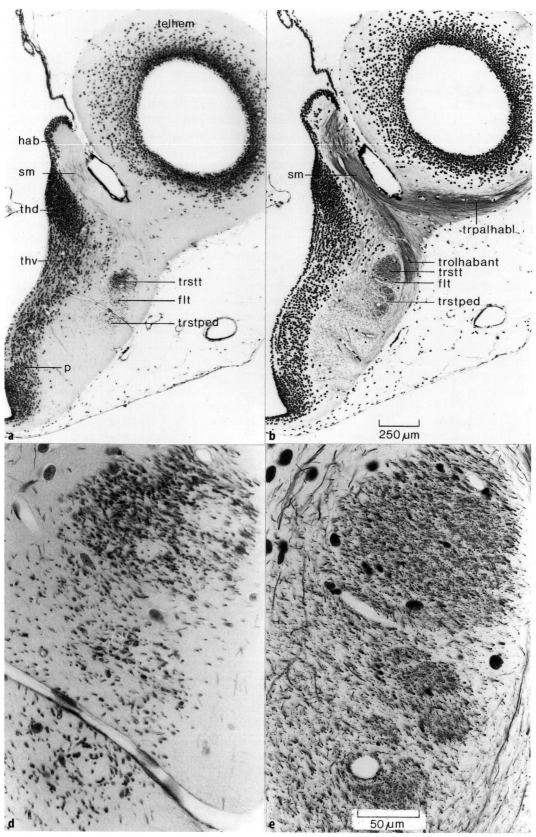

Fig. 3.12a, b, d, e.

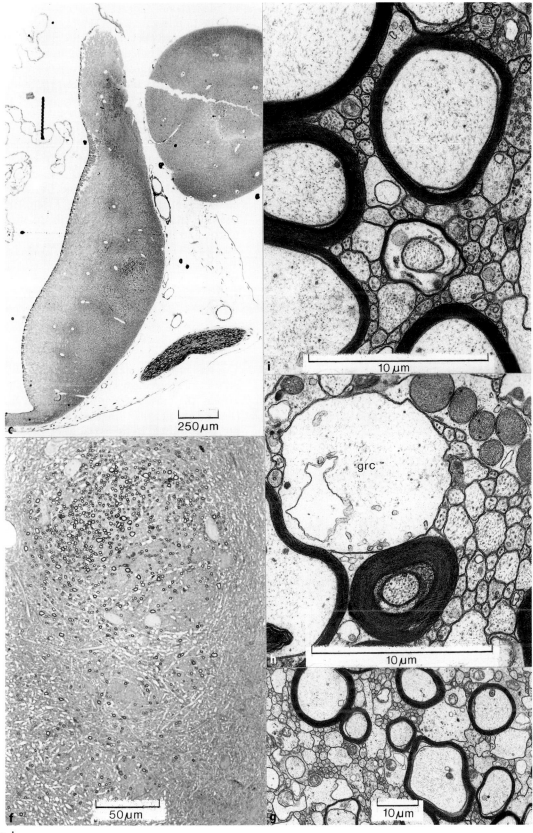

Fig. 3.12c–i.

◄ **Fig. 3.12a–i.** Transverse hemisections through the diencephalon and the caudal part of the telencephalon of the tiger salamander, *Ambystoma tigrinum*, to show the structure of the lateral forebrain bundle. **a** Klüver-Barrera stain; **b** impregnation according to Bodian; **c** semithin (2 μm) epon section stained with paraphenylenediamine; **d–f** the parts of the sections shown in **a–c** containing the lateral forebrain bundle at a higher magnification; **g–i** the ultrastructure of the dorsal portion of the lateral forebrain bundle as seen in transverse sections at approximately the same level as that of **a,b** and **c**. Note the presence of numerous unmyelinated fibres of different size, and of a typical growth cone profile in **h**. *flt*, Fasciculus lateralis telencephali; *gc*, growth cone; *hab*, ganglion habenulae; *p*, nucleus preopticus; *sm*, stria medullaris; *telhem*, telencephalic hemisphere; *thd*, thalamus dorsalis; *thv*, thalamus ventralis; *trolhabant*, tractus olfactohabenularis anterior; *trpalhabl*, tractus palliohabenularis lateralis; *trstped*, tractus striopeduncularis; *trstt*, tractus striotegmentalis

lus, in which they descend to mid- or low thoracic levels (Fig. 3.14a). A similar course is observed in rodents, although here the pyramidal tract descends throughout the lumbar intumescence (e.g., Linowiecki 1914; Goldby and Kacker 1963).

In all primates a large crossed corticospinal tract is found, which descends in the posterolateral funiculus of the spinal cord to sacral levels. In the great apes and in man, however, a smaller, uncrossed anterior corticospinal tract is also present (Fig. 3.14b). The fibres of this anterior tract cross the midline individually at various spinal levels. This anthropoid anterior corticospinal tract varies considerably in both size and caudal extent. Normally, it does not extend below the upper thoracic cord, but Schoen (1964) described a human case in which its fibres could be traced to sacral segments. In rare human cases the fibres of the pyramidal tract of one or both sides do not decussate at all, and descend as a huge anterior corticospinal tract (Verhaart and Kramer 1952; Luhan 1959).

In the hyracoid *Procavia* the pyramidal tract descends entirely uncrossed in the ventral funiculus of the spinal cord into low lumbar levels. Just as in anthropoids, the fibres decussate individually before terminating in the deeper zone of the dorsal horn (Verhaart 1967; Fig. 3.14c).

In the elephant, the pyramidal tract shows a normal course throughout the brain stem, with a decussation at the level of the spinomedullary junction (Verhaart 1962). In the cord, the corticospinal fibres form two bundles, a compact bundle in the most dorsal part of the ventral funiculus and another, more diffuse, bundle bordering the anterior medial fissure (Verhaart and Kramer 1958; Fig. 3.14d). The dorsal bundle is probably composed of crossed fibres and the ventral bundle of uncrossed fibres. In material stained with the Häggqvist method, the ventral bundle was followed

to the second thoracic segment and the dorsal bundle to mid-thoracic levels (Verhaart 1963b).

The marked variability of the pyramidal tract is probably unequalled and has been discussed in some detail because of its considerable general biological significance. Paraphrasing Goldby (1939, p. 522), it is suggested that the earliest mammals possessed a common tendency to develop a pyramidal tract, to bring spinal processes related to motor activity and the processing of somatosensory information under the direct control of the newly acquired neocortex, and that this tendency has been realised independently and in different ways in divergent primitive groups. A similar, though less dramatic variability occurs in the medial lemniscus and in other 'phylogenetically young' pathways (Verhaart 1970).

3.5
Size, Structure and Composition of Fibre Systems

3.5.1
Size of Fibre Systems

The size of fibre systems and the number of fibres present may vary considerably. The single median axon of the most anterior Rohde cell in amphioxus (Fig. 3.2), which coordinates the contractions of the pterygial muscle in the atrial floor (Bone 1960), as well as the axons of the giant Mauthner cells in fish, which trigger the startle reflex, constitute fibre systems in their own right. Fibre systems consisting of fewer than a hundred axons are common in the central nervous system of anamniotes. The bundles of axons of the anterior and posterior Rohde cells in amphioxus (about ten fibres; Fig. 3.2), the bundle of descending Müller axons in the ventral funiculus of the lamprey (7–8 fibres; Fig. 3.3) and the vestibulospinal fasciculus medianus of Stieda in cartilaginous fishes (25–50 fibres; Smeets et al. 1983) are representative. In most anamniotes the medial longitudinal fasciculus encompasses at mid-rhombencephalic levels 100–300 fibres (e.g., Figs. 3.6, 3.7d), but the number of fibres in the majority of the pathways in the CNS of gnathostomes ranges from a thousand (e.g. the tractus descendens of V in the tiger salamander; Fig. 3.7c) to a million (e.g. the human pyramidal tract) or more.

Comparable fibre systems in different species may vary considerably in size and fibre content. Thus it is known that the tractus visceralis secundarius is much larger in highly gustatory cyprinids than in other teleosts, and that in teleosts with a strongly developed electrosensory lateral line system (gymnotids, mormyrids) the lateral lemniscus is exceedingly large. In certain Chiroptera, and par-

Fig. 3.13a–f. The pyramidal tract of the spiny anteater *Tachyglossus* as revealed by the Marchi technique following ablation of part of the motor cortex (from Goldby 1939). The original labelling of the drawings is retained. Levels of the sections: **a** caudal midbrain and rostral pons; **b** mid-pons; **c** caudal medulla oblongata; **d** spinomedullary junction; **e** spinal cord, 7th cervical segment; **f** spinal cord, 24th segment. The letter *D* placed after any abbreviation signifies that the structure shows degenerative changes. *C.-Sp.Tr.*, Corticospinal tract; *Cer.Ped.*, cerbral peduncle; *Dec.*, decussation; *Ext.Arc.F.*, external arcuate fibres; *Inf.C.Q.*, inferior corpus quadrigeminum; *Inf.Ol.*, inferior olive; *L.Lat.*, lemniscus lateralis; *L.Med.*, lemniscus medialis; *N.C.*, nucleus cuneatus; *N.G.*, nucleus gracilis; *N.V.*, nucleus of the spinal root

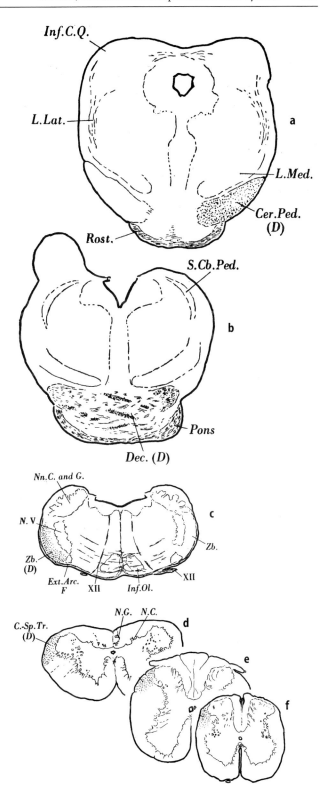

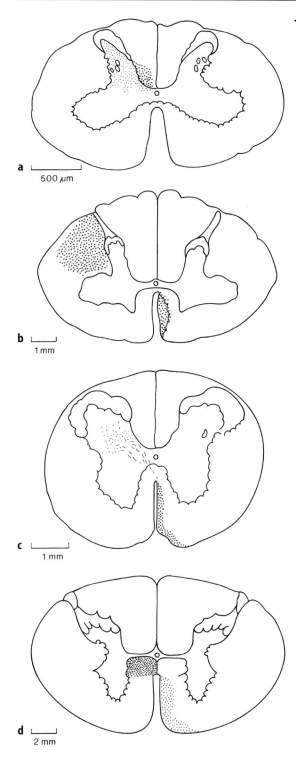

a

500 μm

b

1 mm

c

1 mm

d

2 mm

◀ **Fig. 3.14a–d.** Variations in the spinal course of the pyramidal tract, as observed in transverse sections. Crossed components are indicated to the *left*, ipsilateral components to the *right* of the trigeminal nerve. **a** Section through C-4 of a decorticate specimen of the Tasmanian potoroo, *Potorous,* Nauta technique (redrawn from Martin et al. 1972, Fig. 7). **b** Section through C-5 of a human case with extensive softening of the right cerebral hemisphere following a cerebrovascular accident, Häggqvist technique (redrawn from Schoen 1964, Fig. 5). **c** Section through C-1 of a specimen of the hyracoid, *Procavia,* with a large ablation of the right cerebral cortex, Nauta technique (redrawn from Verhaart 1967, Fig. 5). **d** Section through a high cervical segment of the cord of an elephant. The pyramidal tract has been located on the basis of its fibre pattern as observed in normal Häggqvist material. (based on Verhaart 1970, Fig. 132)

ticularly in cetaceans, the lateral lemniscus attains amazing proportions, which is obviously related to the development of the supersonic orientation system in these animals (Verhaart 1970). However, the acuity of a sense organ is not always directly related to the number of fibres found in the fibre systems associated with it. Linke and Roth (1990) reported that plethodontid salamanders, despite their good visual abilities, have an astonishingly low number of retinal ganglion cells and optic nerve fibres. The five species investigated by Linke and Roth (1990) had total numbers of optic nerve fibres ranging from 26 000 to about 50 000 and their numbers are the lowest found among vertebrates with an elaborated visual system. Other salamanders possess 50 000–80 000 optic nerve fibres, and anurans have about ten times more retinal ganglion cells and optic fibres than urodeles (cf. Table 3.2).

The pyramidal tract is one of the few central pathways which has been analysed quantitatively in many species. Estimates of the total number of fibres have been based on the systematic sampling of transverse light-microscopical sections where the fibres were impregnated with silver (Lassek and co-workers) or stained with the Häggqvist method (Verhaart and co-workers). Table 3.1 shows that the number of fibres in the pyramid varies considerably among species. In the older literature (e.g., Ariëns Kappers et al. 1936) it was held that the size of the pyramidal tract reflected the position of an animal in the evolutionary scale, but this perspective is hard to reconcile with current views on the phylogenetic development of the various mammalian groups. On the basis of a thorough analysis of data derived from the literature on 21 different species of mammals, Towe (1973) concluded that the size of the pyramidal system in mammals is determined primarily by the size of the body and the size of the brain. Heffner and Masterton (1975) did not find any correlation between among mammals the extent of manipulative capability and the size of the pyramidal tract.

Table 3.1. Total number of fibres in the medullary pyramid, observed in light-microscopical preparations of a number of mammals

Animal	Number of fibres	Source
Mouse	32 000	Lassek and Rasmussen (1940)
Opossum	48 000	Towe (1973)
Rat	76 000	Lassek and Rasmussen (1940)
Ferret	86 000	Verhaart and Noorduyn (1961)
Rabbit	102 000	Lassek and Rasmussen (1940)
Cat	186 000	Lassek and Rasmussen (1940)
Macaque	195 000	Verhaart (1948a)
Gibbon	315 000	Verhaart (1948b)
Fur seal	748 000	Lassek and Karlsberg (1956)
Chimpanzee	807 000	Lassek and Wheatley (1945)
Man	>1 000 000	Lassek and Rasmussen (1939) Verhaart (1947, 1950)

3.5.2
Compactness and Discreteness of Fibre Systems

The central nervous system of vertebrates contains pathways which are compact and distinct fibre assemblies. The pyramidal tract at lower medullary levels (Figs. 3.19a, 3.20a), the interstitiospinal tract in the spinal cord of lizards (Fig. 3.15), the tractus telencephalotegmentalis in mormyriform teleosts (Fig. 3.11), the fasciculus solitarius in most tetrapods (e.g. the tiger salamander, Fig. 3.7a,b) and the fornix in mammals (e.g., the rat, Fig. 3.22a,b) are examples. Such discrete pathways are presumably the exception rather than the rule. Verhaart and his collaborators have carried out many hodological studies on the central nervous system of mammals, including the cat (e.g., van Beusekom 1955; Busch 1961, 1964), the monkey (e.g., Verhaart 1954, 1955) and man (e.g. Sie 1956). In these studies, normal and lesioned material from animals and pathological material derived from human clinical cases was analysed following Häggqvist staining. The studies of Verhaart and co-workers revealed that most fibre tracts forming part of the white matter of the mammalian brain stem and spinal cord overlap considerably (Fig. 3.16), and that fibres of some pathways, such as the reticulospinal and rubrospinal projections, are diffusely scattered over large areas of the spinal white matter and so do not form homogeneous tracts. It is likely that a similar situation occurs in non-mammalian vertebrates, but experimental data are lacking.

3.5.3
Fibre Spectrum and Fibre Pattern

It has long been known that the white matter of the vertebrate brain and spinal cord is composed of fibres of many diameters. Flechsig (1876) subdivided the myelinated fibres in the human spinal cord into four categories: coarse fibres, 10–15 μm in diameter; medium-sized fibres with a diameter of 7–9 μm, thin fibres measuring 5–6 μm in diameter, and very thin fibres with a diameter of 2–4 μm. Flechsig also noticed that, while the thickness of the myelin sheath does not always have a fixed relation to the diameter of the entire fibre, in general the coarsest fibres possess the coarsest axis cylinders. Ramón y Cajal (1909) and Bok (1928) also emphasised the considerable variability in size of the fibres in the white matter. Both authors noted that the spinal white matter contained large numbers of thin, unmyelinated axons among the myelinated fibres. The considerable variability in size of the fibres in the white matter of the CNS is illustrated in Figs. 3.5a, 3.7a–e, 3.12f, 3.16 and 3.19.

The first investigator who subjected the fibres in the white matter to a systematic quantitative analysis was the Swedish histologist Häggqvist (1936). He studied the third thoracic segment of the spinal cord of a 13-year-old girl, using a modification of Alzheimer-Mann's methyl blue-eosin staining technique, which was later named after him. The white matter of this segment was subdivided into 14 areas, some of which roughly corresponded to areas known to be occupied by specific fibre pathways, as for instance 11: dorsal spinocerebellar tract, 10: lateral pyramidal tract, and 14: cuneate fasciculus (Fig. 3.17a). In each of these 14 areas, 1200 fibres were measured; hence, the total number of fibres measured amounted to about 17 000. Häggqvist's (1936) main results were:

1. Fibres of very different diameters are intermingled in all parts of the spinal white matter.
2. The coarsest myelinated fibres measure 21 μm, the thinnest less than 1 μm in diameter.
3. Each of the 14 areas of white matter has its own characteristic fibre spectrum (Fig. 3.17b,c).
4. Very fine myelinated fibres with a diameter of less than 3 μm constitute by far the largest fibre contingent in all parts of the spinal white matter (Fig. 3.17). These fine myelinated fibres had been overlooked by previous investigators, including Flechsig (1876), Ramón y Cajal (1909) and Bok (1928).

Even although Häggqvist (1936) confined his analysis to a single specimen and a single level, the study was extremely laborious and time consuming. Most

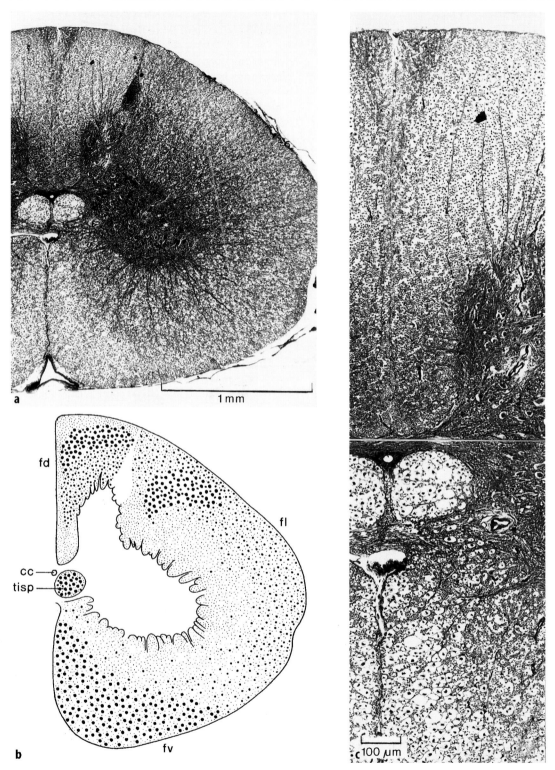

Fig. 3.15a–c. The spinal cord of the tegu lizard, *Tupinambis nigropunctatus*. **a** Transverse hemisection through the sixth spinal segment, Häggqvist stain; **b** diagrammatic representation of the same hemisection (from Kusuma et al. 1979, Fig. 15); **c** the medial part of the hemisection shown in **a** at a higher magnification. Note that the coarse-fibred, most dorsal part of the funiculus ventralis is separated from the remainder of that funiculus by the commissura accessoria. This isolated bundle contains the interstitiospinal tract. *cc*, Canalis centralis; *fd*, funiculus dorsalis; *fl*, funiculus lateralis; *fv*, funiculus ventralis; *tisp*, tractus interstitiospinalis

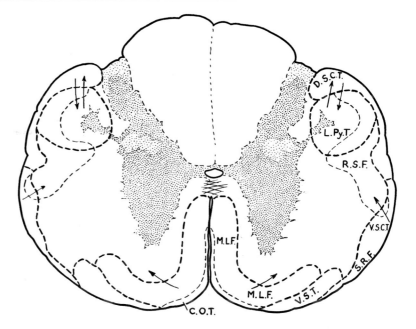

Fig. 3.16. Transverse section through the first cervical segment of the spinal cord of the cat. The position of several fibre systems has been based on experimental Häggqvist material. *Arrows* indicate intermingling of fibres of adjacent tracts. Rubrospinal fibres are scattered over a large area of the lateral funiculus and intermingle extensively with the lateral pyramidal and ventral spinocerebellar fibres. (from van Beusekom 1955, plate VI). *C.O.T.*, Cervico-olivary tract; *D.S.C.T.*, dorsal spinocerebellar tract; *L.Py.T.*, lateral pyramidal tract; *M.L.F.*, medial longitudinal fascicle; *R.S.F.*, rubrospinal fibres; *S.R.F.*, spinoreticular fibres; *V.S.C.T.*, ventral spinocerebellar tract; *V.S.T.*, vestibulospinal tract

later investigators followed a more global approach, focusing on fibre patterns rather than on fibre spectra. The fibre pattern of a fibre system is the *aspect* of these structures in reduced silver, Häggqvist or semi-thin material, resulting from their fibre composition. The fibre spectrum may be regarded as a quantitative specification of the fibre pattern without the spatial arrangement of the fibres. The fibre pattern is usually documented by representative photomicrographs (cf., e.g., Figs. 3.7 a–e, 3.12 f, 3.16, 3.18, 3.19, 3.22), accompanied by descriptions using the terms 'coarse', 'medium-sized' and 'small' fibres. Sometimes the fibre patterns are diagrammed (e.g., Fig. 3.16 d).

Lassek and Rasmussen (1940) studied the fibre pattern of the pyramidal tract at a level just above its decussation in reduced-silver preparations of a series of eight mammals (Fig. 3.18). They found that:

1. In all species the small axons predominate.
2. Man possesses the largest axons in the series. They are between 10 and 25 μm and few in number.
3. The reduced-silver picture of the pyramidal tract of the dog is very similar to that of man, although it seems to have more intermediate-sized fibres.
4. The cat differs from man and dog in possessing no large axons (see, however, Biedenbach et al. 1986).
5. Although the cow weighs ten to twenty times more than man, the axons in the pyramidal tract

of this species are extremely small; in general, this tract resembled the pyramidal tract in the rat, opossum, rabbit, and mouse. In all of these animals the axons are small and closely packed. The mouse has the smallest fibres of the group.
6. No accumulation of either small or large fibres was found in one particular part of the pyramidal tract in the groups examined. These fibres were equally mixed throughout.

The fibre pattern of numerous pathways in the mammalian brain stem and the spinal cord, as observed in Häggqvist material, has been documented by Verhaart and his collaborators (e.g., van Beusekom 1955 and Busch 1961: cat; de Graaf 1967: cetaceans; Verhaart 1954: macaque; Sie 1956: man; and especially Verhaart 1970: representatives of almost all mammalian orders). These studies have shown that most pathways in the mammalian brain stem and spinal cord can be recognised by their characteristic fibre patterns, but that the boundaries between these pathways are not sharp, because the fibres of neighbouring pathways intermingle to a considerable extent. To give an impression of the potential of the Häggqvist technique, Fig. 3.19 presents a transverse section through the lower medulla oblongata of the prosimian *Tupaia* and four detailed pictures at a higher magnification. It will be seen that in the reticular formation small bundles composed of fibres of different calibres occur (Fig. 3.19 b), that the medial longitudinal fascicle contains many large fibres (Fig. 3.19 c), that the pyramidal tract consists exclusively of very fine

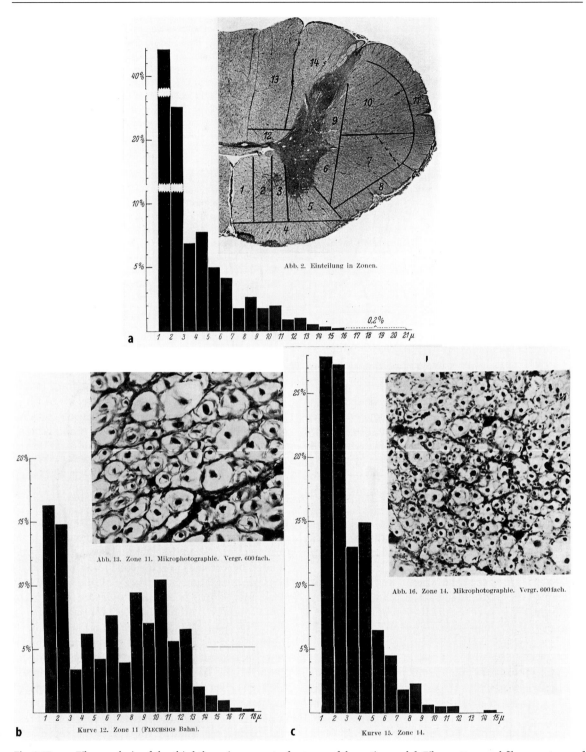

Abb. 2. Einteilung in Zonen.

Abb. 13. Zone 11. Mikrophotographie. Vergr. 600 fach.

Abb. 16. Zone 14. Mikrophotographie. Vergr. 600 fach.

b   Kurve 12. Zone 11 (FLECHSIGS Bahn).

c   Kurve 15. Zone 14.

Fig. 3.17a–c. Fibre analysis of the third thoracic segment of the human spinal cord (three illustrations from Häggqvist's classical study, 1936). **a** Subdivision in zones and fibre spectrum of the entire cord. **b** Fibre pattern and fibre spectrum of zone 11, i.e. the dorsal spinocerebellar tract. **c** Fibre pattern and fibre spectrum of the fasciculus cuneatus

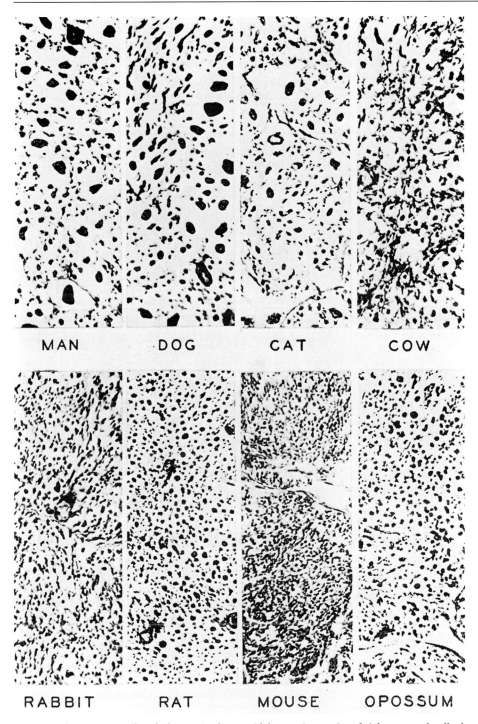

Fig. 3.18. Photomicrographs of silver-stained pyramidal tracts in a series of eight mammals, all taken at a level just above the decussation of the tract (reproduced from Lassek and Rasmussen 1940, Fig. 1). × 720

fibres (Fig. 3.19d; Verhaart 1966), and that the medial lemniscus is composed of fine fibres which are larger than those of the pyramidal tract (Fig. 3.19e).

With regard to non-mammalian vertebrates, van den Akker (1970) documented the fibre pattern of pathways in the spinal cord of the pigeon. The fibre composition of the white matter of the brain stem

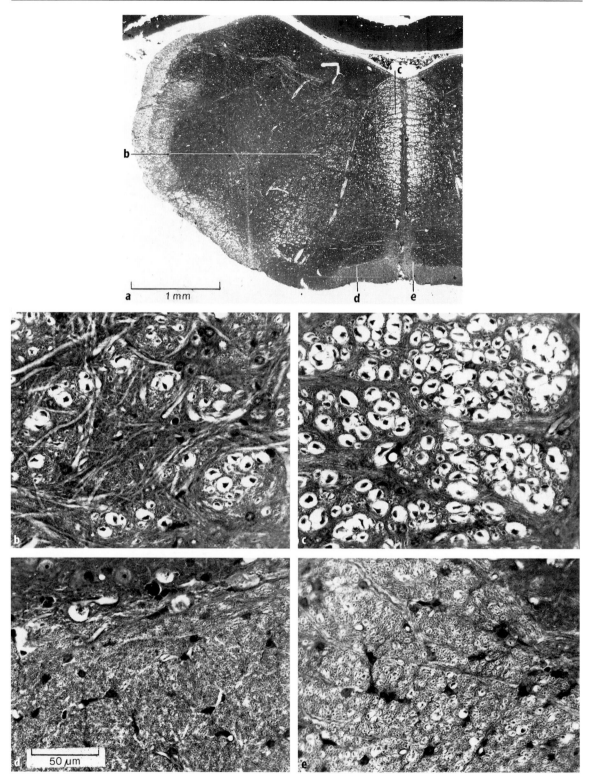

Fig. 3.19. a Hemisection through the medulla oblongata of the tree shrew, *Tupaia glis*, Häggqvist technique; **b–e** details of this section at a higher magnification. The approximate position of the details is indicated in **a**. **b** Medial reticular formation; **c** fasciculus longitudinalis medialis; **d** pyramidal tract; **e** lemniscus medialis

Table 3.2. Approximate numbers of fibres in the optic nerves of some vertebrates

Animal	Total no.	Myelinated	Unmyelinated	(%)	Source
Lampetra	35 000	–	35 000	(100)	Öhman (1977)
Eugerres	200 000	195 000	5 000	(2.5)	Tapp (1974)
Hydromantes	27 000	1 000	26 000	(96)	Linke and Roth (1990)
Salamandra	53 000	4 000	49 000	(93)	Fritzsch, cited in Roth (1987)
Xenopus	58 000	6 000	52 000	(89)	Dunlop and Beazley (1984)
Bufo	330 000	10 000	320 000	(97)	Maturana (1959)
Rana	485 000	15 000	470 000	(97)	Maturana (1959)
Pseudemys	396 000	315 000	81 000	(20)	Geri et al. (1982)
Vipera	65 000	54 000	11 000	(17)	Ward et al. (1987)
Pigeon	2 400 000	1 800 000	600 000	(30)	Binggeli and Paule (1969)
Duck	1 500 000	1 500 000	–	(–)	O'Flaherti (1971)
Opossum	100 000	98 000	2 000	(2)	Kirby et al. (1982)
Mouse	66 000	65 000	1 000	(1.25)	Honjin et al. (1977)
Rat	120 000	120 000	–	(–)	Hughes (1977)
Rabbit	400 000	392 000	8 000	(2)	Vaney and Hughes (1976)
Cat	160 000	160 000	–	(–)	Williams and Chalupa (1983)
Monkey	1 410 000	1 410 000	–	(–)	Ogden and Miller (1966)

and spinal cord of some reptiles was studied by ten Donkelaar and Nieuwenhuys (1979) and Kusuma et al. (1979). Comparable studies of anamniotes have not been made.

Coggeshall and his colleagues carried out an ultrastructural analysis of the white matter of the second sacral segment of the rat spinal cord (Langford and Coggeshall 1981; Chung et al. 1979; Chung and Coggeshall 1983a) and the cat (Chung and Coggeshall 1979; Chung et al. 1985). They found that unmyelinated axons are the predominant population in the spinal white matter of the rat. There were approximately 6000 myelinated axons and 8000 unmyelinated axons in the dorsal funiculi and approximately 1500 myelinated and 4500 unmyelinated axons in the tract of Lissauer. On average, there were 55 000 myelinated and 110 000 unmyelinated axons in the lateral funiculus and 26 000 myelinated and 9000 unmyelinated axons in the ventral funiculus. This yielded a total of 88 500 myelinated axons and 131 500 unmyelinated axons in the white matter of the S-2 level of one side of the rat spinal cord. Unmyelinated fibres were concentrated in the dorsolateral part of the lateral funiculus. In the white matter of the S-2 segment of the cat spinal cord, slightly more than 500 000 fibres and unmyelinated axons, ranging from 0.1 to 1.0 µm in diameter, were present in all parts of the white matter, and the unmyelinated axons made up 45% of the total. As in the rat, the unmyelinated fibres were concentrated in Lissauer's tract and in the dorsal part of the lateral funiculus. The fibre composition of the human posterior columns was also studied by Briner et al. (1988), who found that more than 25% of the component axons are unmyelinated and that many of these unmyelinated fibres represent primary afferent axons.

It appears that only two central pathways, viz. the primary optic projection and the pyramidal tract, have been subjected to a detailed morphometrical analysis at the ultrastructural level. With regard to the primary optic projection, the optic nerves of many vertebrate species have been studied, including the lamprey *Lampetra fluviatilis* (Öhman 1977), the teleost *Eugerres plumieri* (Tapp 1974), a number of plethodontid salamanders including *Hydromantes italicus* and *Batrachoseps attenuatus* (Linke and Roth 1990), the salamander *Salamandra salamandra* (Fritzsch, cited in Roth 1987), the clawed frog, *Xenopus laevis* (Dunlop and Beazley 1984), the toad *Bufo americans* and the frog *Rana pipiens* (Maturana 1959, 1960), the turtle *Pseudemis scripta elegans* (Geri et al. 1982), the viper *Vipera aspis* (Ward et al. 1987), the pigeon *Columba livia* (Binggeli and Paule 1969), the mallard duck, *Anas platyrhynchos* (O'Flaherty 1971), the North American opossum, *Didelphis virginiana* (Kirby et al. 1982), the mouse *Mus wagneri* (Honjin et al. 1977), the pigmented rat (Hughes 1977), the rabbit (Vaney and Hughes 1976), and the cat (Stone and Campion 1978; Williams and Chalupa 1983). The principal results of these studies are:

1. The total number of fibres in the optic nerve varies considerably (Table 3.2).

2. In the optic nerve of the lamprey, as in all parts of the CNS of cyclostomes, myelinated fibres are absent. In most gnathostomes most optic nerve fibres possess a myelin sheath. However, in amphibians only 3%–11% of all fibres are myelinated, and in the miniaturised and highly paedomorphic plethodontid salamander *Batrachoseps* very few myelinated axons are present in the optic nerve (Linke and Roth 1990).

Table 3.3. Range of fibre diameters, location of the modes in the fibre diameter spectra and mean fibre diameter of: (1) myelinated axons including the myelin sheath, (2) myelinated axons excluding the myelin sheath, and (3) unmyelinated axons in the optic nerve of some vertebrates

Animal	Range (μm)	Mode(s) at (μm)		Mean (μm)	Source
1. Myelinated axons including myelin sheath					
Teleost	0.3 – 9.2	1.0			Tapp (1974)
Duck	0.3 – 6.0	1.15			O'Flaherty (1971)
Opossum	0.3 – 6.9	1.13		1.55	Kirby et al. (1982)
Rat	0.4 – 5	1.0		1.00	Hughes (1977)
Mouse	0.3 – 4.2	0.8		0.96	Honjin et al. (1977)
Rabbit	0.3 – 7	0.75		1	Vaney and Hughes (1976)
Monkey	0.4 – 6	1.2			Ogden and Miller (1966)
2. Myelinated axons excluding myelin sheath					
Teleost	0.2 – 7.6	0.7			Tapp (1974)
Turtle	0.35 – 4.5	0.87			Geri et al. (1982)
Opossum	0.12 – 6.1	1		0.78	Kirby et al. (1982)
Cat	0.3 – 8	0.9	2	1.7	Williams and Chaluga (1983)
3. Unmyelinated axons					
Lamprey	0.3 – 3.6				Öhman (1977)
Clawed frog	0.1 – 1.0	0.46			Dunlop and Beazley (1984)
Turtle	0.13 – 0.91	0.42			Geri et al. (1982)
Opossum	0.18 – 1.58	0.57		0.59	Kirby et al. (1982)

3. Table 3.3 presents data derived from the literature concerning (a) the range of fibre diameters, (b) the location of the mode(s) or peak(s) in the fibre diameter spectra, and (c) the mean fibre diameter of myelinated axons including the myelin sheath (external diameter), myelinated axons excluding the myelin sheath (internal diameter), and unmyelinated axons in the optic nerve of a number of vertebrate species. It is clear that the myelinated as well as the unmyelinated fibres vary considerably in size. In almost all species the distribution of the diameters of the fibres is unimodal and skewed toward the large-diameter fibres, with a peak at about 1 μm for the myelinated fibres and at about 0.5 μm for the unmyelinated fibres. Williams and Chalupa (1983), however, found that the axon calibre in the optic nerve of the cat is distributed in a bimodal manner. The class of finest axons, with diameters ranging from 0.3 to 1.5 μm, forms a prominent peak centered at 0.9 μm, and the intermediate calibre class, between 1.5 and 3.5 μm, has a mode at approximately 2 μm. A third group of exceptionally large axons, with diameters between 3.6 and 8.0 μm, forms the extensive tail of the diameter distribution. The proportions of large, intermediate, and small axons were estimated to be 5 %, 45 % and 50 %, respectively. Williams and Chalupa (1983) suggested that these groups of axons correspond to the three principal morphological classes of retinal ganglion cells, α, β, and γ, which are distinguished by their soma size and dendritic pattern, and which in turn probably correspond to the Y, X

and W functional types of ganglion cells, a categorisation based on receptive field properties and conduction velocities (Rodieck 1979; Lennie 1980).

4. Landau et al. (1968) have shown that compound action potentials of nerve tracts may be graphically reconstructed from the histograms of their fibre spectra. In several combined quantitative morphological and electrophysiological studies on the optic nerve (e.g., O'Flaherty 1971; Tapp 1974) a good correlation was found between the theoretical compound action potentials derived from the fibre-size spectrum and the electrophysiological recordings.

5. Analysis of samples taken from several locations in a cross-section of the optic nerve has revealed that the axons of different diameters are not distributed uniformly throughout the nerve (Tapp 1974: teleost; Kirby et al. 1982: opossum; Honjin et al. 1977: mouse; Vaney and Hughes 1976: rabbit; Williams and Chalupa 1983: cat; Reese and Ho 1988: monkey), and similar observations have been made on the optic tract (Cavalcante et al. 1992: opossum; Reese and Guillery 1987: monkey). In the optic nerve of the opossum and the monkey there is a preponderance of fine fibres centrally, while coarse fibres are more numerous in the periphery. Fine fibres occur throughout the optic tract of these animals. Coarse fibres are lacking in the most dorsal part of the tract, but further ventrally, coarse fibres gradually appear and increase steadily in proportion. The largest fibres are found in the most ventral parts of the tract near the pial surface. In

the mammalian primary visual system there is a rearrangement of fibres as they pass from the optic nerve to the optic tract. In the initial part of the optic nerve a simple retinotopic organisation occurs, but central to the optic chiasm there is a gradual segregation of functionally distinct optic axon classes anticipating their differential termination in the various central targets of the primary optic projection (Guillery et al. 1982; Reese and Guillery 1987; Reese and Cowey 1990).

With regard to the pyramidal tract, Leenen et al. (1985) determined the fibre composition and the fibre spectrum of this pathway of the rat in five specimens at two levels: the pyramis medullae (Fig. 3.20a,c) and the second cervical segment. At both levels the pyramidal tract contained a large population of unmyelinated fibres. At the level of the pyramis medullae the number of unmyelinated fibres (140 000 ± 7000) exceeded the number of myelinated fibres (103 000 ± 6000). In contrast, at

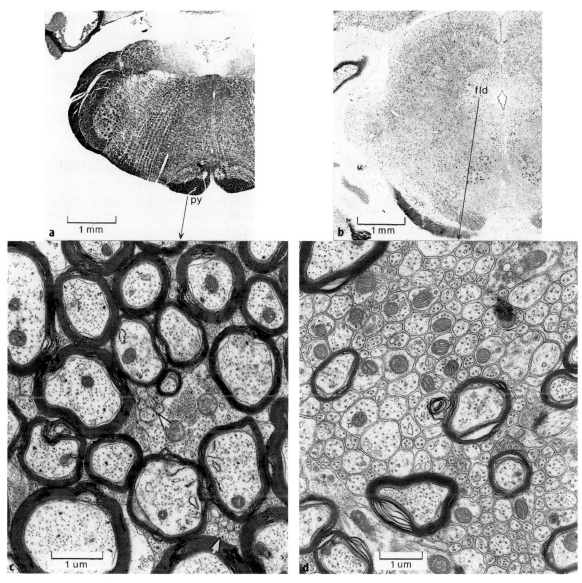

Fig. 3.20a–d. Transverse sections through the medulla oblongata (**a**) and the mesencephalon (**b**) of the rat, to show the positions of the pyramidal tract (*py*) and the fasciculus longitudinalis dorsalis (*fld*). Klüver-Barrera stain. **c,d** Ultra-structure of py and fld. Note that in py, groups of very thin, unmyelinated fibres are present (*arrows*), and that the fld consists mainly of unmyelinated axons

the level of the second cervical segments the numbers of fibres of the two axon populations studied were not significantly different, viz. $43\,000 \pm 2000$ myelinated and $35\,000 \pm 8000$ unmyelinated fibres. However, these numbers mean a significant decrease of myelinated axons (48 %) compared with the pyramis medullae level and an even larger decrease (75 %) in the number of unmyelinated fibres. Diameter distributions showed a similar, monomodal shape for all axon and myelin profiles. For unmyelinated axons the diameter ranged from 0.05 to 1.21 μm, with a mean of 0.18 ± 0.03 μm. For myelinated axons a mean diameter of 0.72 ± 0.12 μm was found (range 0.13–4.92 μm), whereas their mean diameter measured with myelin sheath was 1.08 ± 0.13 μm (range 0.25–6.03 μm). The average thickness of the myelin sheaths was 0.2 μm and was strongly correlated to axon diameter. Harding and Towe (1985) confirmed the presence of large numbers of thin, unmyelinated fibres in the pyramidal tract of the rat at the level of the pyramis.

The pyramidal tract of the cat was analysed at the level of the medullary pyramids by Biedenbach et al. (1986). They found that the number of axons per pyramidal tract averaged 415 000, of which 88 % were myelinated and 12 % were unmyelinated. Of the myelinated fibres, 90 % fell in the diameter range 0.5–4.5 μm. Fibres larger than 9 μm diameter accounted for 1 % of the total; the largest were 20–23 μm. The median diameter of the myelinated axons was 1.60 μm. Unmyelinated fibres averaged 0.18 μm in diameter (range 0.05–0.6 μm). Size distribution was relatively uniform throughout the pyramidal cross-section, with all sizes represented in all regions. The more medial regions, however, had a higher proportion of small fibres than the more lateral regions.

Ralston et al. (1987) studied the pyramidal tract at the level of the medullary pyramid in the rhesus monkey. They reported that most of the profiles which could be interpreted as being non-myelinated axons when viewed in cross-section are actually astroglial processes when examined in longitudinal section, and concluded that non-myelinated axons constitute less than 1 % of the pyramidal tract axons in the primate studied.

Although information concerning the fibre spectrum of other pathways in the mammalian CNS is scant, it is known that the medial forebrain bundle (Sipe and Moore 1977; Nieuwenhuys et al. 1978) and the dorsal longitudinal fasciculus of Schütz (Buma et al. 1992; Fig. 3.20b,d) contain large amounts of thin, unmyelinated fibres, and that such fibres most probably prevail throughout the limbic circuitry of the mammalian brain (Nieuwenhuys et al. 1989; Nieuwenhuys 1996; Holstege 1991). Next to

nothing is known concerning the occurrence of thin, unmyelinated fibres in the CNS of non-mammalian vertebrates. A preliminary ultrastructural study revealed that in the tiger salamander such fibres are present in several pathways, including the medial longitudinal fasciculus (Fig. 3.7 h), the complex formed by the spinocerebellar and spinotectal tracts (Fig. 3.7i), the tractus descendens of the trigeminal nerve (Fig. 3.7g) and the lateral forebrain bundle (Fig. 3.12g–i). Within the last-mentioned bundle numerous growth cone-like structures were found (Fig. 3.12 h). Urodele amphibians are known to be paedomorphic, and it seems likely that the abundance of thin, unmyelinated fibres in the primary optic projection (Linke and Roth 1990, see above), the lateral forebrain bundle and other pathways is related to this condition. It remains an intriguing question whether thin, unmyelinated axons form a sizeable component of particular fibre pathways in other non-mammalian gnathostomes.

3.5.4
Closed and Open Fibre Systems

With regard to communication with their surrounding structures, fibre tracts range from entirely closed to entirely open systems, with many intermediate forms. Closed systems consist of myelinated fibres which traverse the trajectory between their origin and their site of termination without making any *en-passant* functional contacts. Entirely closed systems are presumably rare; examples are the tractus telencephalotegmentalis in the brain of osteoglossomorph teleosts (Fig. 3.11) and certain pathways related to the electromotor system in fish possessing electric organs.

The CNS of anamniote gnathostomes contains numerous pathways consisting mainly of unmyelinated fibres, while in cyclostomes, myelinated fibres are lacking entirely. It is generally assumed (e.g. Herrick 1948; Heier 1948) that unmyelinated pathways are in functional contact throughout their extent, via either their main axons or their collateral branches, with all of the grisea that they pass through, and hence represent open systems (Fig. 3.1a).

Studies using modern tracing techniques (cf. Chap. 7) have revealed the presence of numerous 'new' fibre systems. All of these systems appeared to consist of thin fibres, which had remained unnoticed in lesion-degeneration studies (Holstege 1991). Most of these newly discovered fibre systems do not form compact bundles, but rather consist of loose-textured fibre streams, representing local concentrations in an otherwise mostly diffuse fibre network. These fasciculated and non-fasciculated thin axons are often varicose, either over certain trajectories or throughout their course, and it has already been mentioned that the axonal varicosities may be involved in synaptic as well as in non-synaptic chemical neurotransmission. Histochemical and immunohistochemical studies have revealed that monoaminergic and peptidergic neuronal groups form prominent sources of thin, varicose axons.

In the mammalian brain there is an enormous structural and functional difference between the lateral, cognitive-motor domain and the medial, limbic domain (see Chap. 22, Sect. 12). The 'classical' sensory and motor systems within the lateral domain are composed mainly of discrete centres that are interconnected by well-myelinated, through-conducting, compact fibre bundles. By contrast, in the medial domain, which is concerned with the generation of behavioural patterns directly related to homeostasis and reproduction, ill-defined pathways and diffuse networks of thin, varicose, open fibres prevail (Nieuwenhuys 1994, 1996). The wealth of neuropeptides present in the fibres forming these systems and networks indicates that the way in which this limbic domain operates is distinctly different from other parts of the brain (Nieuwenhuys 1985; Herbert 1993). The diffuseness of the limbic fibre streams does not exclude specificity in the connections (cf., e.g., Roeling et al. 1994; Larsen et al. 1994); however, this feature hampers the elucidation of the circuitry underlying vital behaviours such as eating, drinking, aggression and sexualconduct.

Open systems consisting of thin, varicose fibres have also been observed in limbic regions of the reptilian brain (e.g., Russchen and Jonker 1988), but our knowledge of such connections in non-mammalian vertebrates is scant.

3.5.5
Simple and Composite Fibre Systems

A *projection* may be defined as a set of fibres having one main source and one main site of termination. A *tract* may be defined as a projection which manifests itself as a fibre concentration over at least a part of its course. Such concentrations may be recognised by their fibre pattern in normal material and as accumulations of degenerated or labelled fibres in experimental material. The criteria 'discreteness' and 'compactness' have deliberately been kept out of the definition, because the work of Verhaart and his collaborators (e.g., Verhaart 1954, 1955, 1970; van Beusekom 1955; Busch 1961; Schoen 1964; cf. Sect. 3.5.2) has taught us that most fibre

assemblies in the white matter overlap considerably. For this reason, Verhaart (1955, p. 498) defined a tract as "an area of white matter in which fibres of a certain nature prevail."

Several fibre assemblies in the CNS do not conform to the definition of a tract just given, because they are composed of sets of axons with different origins and different terminations, and often of different polarity. Such a composite fibre assembly is commonly termed a *fasciculus*, or bundle, although some typical tracts are, unfortunately, also designated with that term. The tractus habenulo-interpeduncularis, which is also known as the fasciculus retroflexus, is an example.

The question as to whether the various components of a fasciculus should be regarded as projections or as separate tracts depends on the degree of concentration of their constituent fibres. In the conspicuous, mainly coarse-fibred, medial longitudinal fasciculus (Figs. 3.6, 3.7a,d, 3.19a,c) most components occupy a fixed position and show up in experimental tracing or degeneration studies (Fig. 3.23) as distinct fibre concentrations. For this reason, they may be designated as separate tracts, as is common usage: tractus interstitiospinalis, tractus tectospinalis, tractus vestibulomesencephalicus cruciatus, and so on (Sie 1956; Busch 1961). With more than 50 different components, the loose-textured, mainly thin-fibred fasciculus medialis telencephali is probably the most complex composite bundle of the mammalian brain (Nieuwenhuys et al. 1982; Figs. 3.21, 3.22a,c). Following a meticulous analysis of this bundle in normal material of the rat, Gurdjian (1925, p. 101) stated: 'We would have to analyze the complex more completely than our present preparations permit and note exactly the position of each component and describe the same as an independent fibre tract'. Veening et al. (1982) studied the topography and fibre distribution of some 21 different components of the medial forebrain bundle of the rat, in brains prepared for autoradiography following injections of tritiated amino acids into structures known to contribute fibres to the bundle. This analysis showed that most of the labelled components occupy specific positions within the bundle, but that in general their constituent fibres are too diffusely spread to be designated as tracts. Another reason for avoiding the term 'tract' in the description of the components of the medial forebrain bundle is that most of these components consist of open fibres, which may be expected to make *en-passant* contact with many grisea, rather than with a single one.

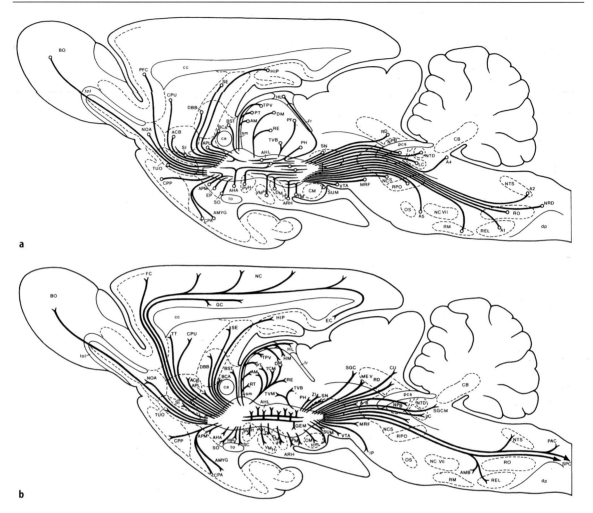

Fig. 3.21a,b. Semidiagrammatic representations of the input (**a**) and the output (**b**) of the medial forebrain bundle of the rat, based on the literature (from Nieuwenhuys et al. 1982)

3.5.6
Topographical Organisation of Fibre Systems

A tract is *topographically organised* if the spatially ordered fashion in which particular parts of its site of origin project to particular parts of its area of termination is reflected in the spatial organisation of its fibres. A topographically organised tract may be *somatotopically organised* if the spatial arrangement of its fibres is either directly or indirectly related to the regional or segmental structure of the body.

Experimental investigations and studies of human clinical cases have shown that numerous fibre systems in the central nervous systems of mammals and man are topographically organised and that some of them also show a somatotopic organisation. The somatotopic organisation of the primary afferent fibres in the posterior funiculus, of the long ascending fibres in the anterolateral funiculus and of the corticospinal fibres in the posterolateral funiculus of the spinal cord, as diagrammatically represented in Fig. 3.24, is well known and does not need further comment here. Our knowledge of the topical organisation of fibre tracts in the CNS of non-mammalian vertebrates is very limited. In all groups of tetrapods (urodeles: Nieuwenhuys and Cornelisz 1971; anurans: Woodburne 1939; Joseph and Whitlock 1968; reptiles: Goldby and Robinson 1962; Ebbeson 1967, 1969; birds: Friedländer 1898; Münzer and Wiener 1898, 1910; van den Akker 1970), however, the dorsal funiculus contains long ascending primary afferent fibres showing a somatotopic arrangement comparable to that in mammals.

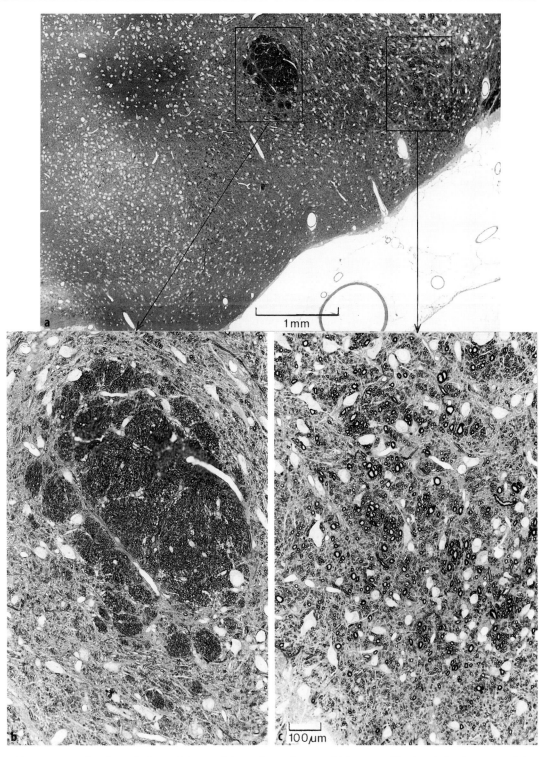

Fig. 3.22. a Transverse section through the caudal part of the hypothalamus of the rat. Semithin (2 μm) epon section stained with paraphenylenediamine. **b,c** Parts of the section shown in **a** at a higher magnification: **b** fornix; **c** part of the loose-textured medial forebrain bundle

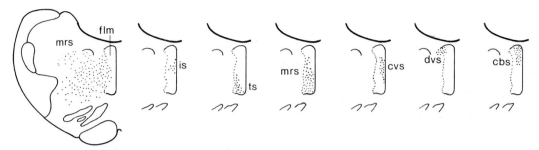

Fig. 3.23. Composition of the fasciculus longitudinalis medialis of the cat at a lower medullary level. *cbs*, Crossed bulbospinal tract; *cvs*, crossed vestibulospinal tract; *dvs*, direct vestibulospinal tract; *flm*, fasciculus longitudinalis medialis; *is*, interstitiospinal tract; *mrs*, medial reticulospinal tract; *ts*, tectospinal tract (redrawn from Busch 1961, Fig. 12)

3.5.7
Chemodifferentiation of Fibre Systems

Neuromediators and their synthesising enzymes can be visualised in neurons and their processes by immunohistochemical techniques. During the past 25 years an enormous number of so-called mapping studies in a wide range of vertebrates have appeared, describing a variety of neuromediator-specific neuron populations. From a survey of this literature we can draw the following conclusions concerning the nature and the disposition of neuromediator-specific fibres:

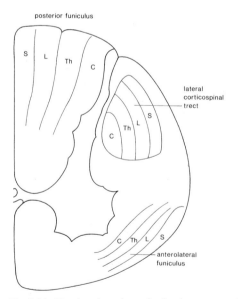

Fig. 3.24. Hemisection through the human cervical spinal cord, showing the somatotopic arrangement of the fibres in the posterior funiculus, the lateral corticospinal tract and the spinothalamic fibres in the anterolateral funiculus. Note that the boundaries between the various sectors are less distinct than indicated here. *C*, Cervical; *L*, lumbar; *S*, sacral; *Th*, thoracic

1. Monoaminergic and peptidergic fibres are generally thin and varicose.
2. There is a paucity of information concerning the nature of the neuromediators present in many of the more compact, myelinated fibre systems of classical neuroanatomy. This may be because these systems use classical neurotransmitters, like acetylcholine or amino acids, which are produced locally in the terminals and so cannot be demonstrated in the axons. Techniques aimed at visualising the synthesising enzyme for a neuromediator may yield negative results in the axons if the concentration of that neuromediator in the terminals is maintained by re-uptake after its release, rather than by continuous synthesis; again, this mechanism is known to be important in neurons containing classical neurotransmitters. Finally, even the axonal levels of neuromediators produced in the soma, such as the neuropeptides, may sometimes be too low to be detected with immunohistochemical techniques.
3. Composite fibre bundles, such as the mammalian ventral amygdalofugal projection, stria terminalis, medial forebrain bundle and fasciculus longitudinalis dorsalis, contain a multiplicity of neuromediators. Because these bundles are composed mainly of thin, varicose fibres which have the potential for influencing other neurons throughout their extent, rather than only in their area of final termination, they may be characterised as 'open multineuromediator channels' (Nieuwenhuys 1985).
4. In fibre systems in which classical neurotransmitters as well as neuropeptides are present, it is likely that both types of substances co-exist within the same axons. This situation may be expected in fibre systems from centres which contain projection neurons of the 'multiple-neuromediator' type. Thus, in raphespinal fibres, serotonin may co-exist with substance P and thyrotropin-releasing hormone, as well as with

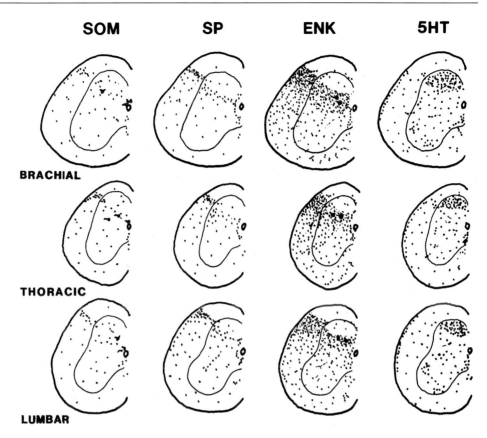

Fig. 3.25. Plots of representative transverse sections showing the distribution of somatostatin (*SOM*)-, substance-P (*SP*)-, enkephalin (*ENK*)- and serotonin (*5HT*)-immunoreactive structures in the frog spinal cord (reproduced from Adli et al. 1988)

other neuropeptides (Hökfelt et al. 1978; Mantyh and Hunt 1984; Basbaum and Fields 1984).

5. With regard to the relations between the fibre systems of classical hodology and the various sets of neuromediator-specific fibres, four configurations have been observed:

(a) A set of neuromediator-specific fibres may constitute or contribute to a classical hodological entity (tract or fascicle).

(b) Sets of neuromediator-specific fibres may constitute diffuse networks or diffuse ascending or descending projections bearing no distinct relation to any classical hodological entity. For example, the various sets of peptidergic fibres which pass through the spinal cord are mostly diffusely distributed over the white matter (Fig. 3.25) or its agnathan equivalent (Fig. 3.26), and the serotoninergic fibres, which descend from the rhombencephalon to the cord, show a similar characteristic.

(c) Combinations of conditions mentioned under (a) and (b) occur. Thus in all verte-

brates throughout most of the brain, noradrenergic and serotoninergic fibres constitute diffuse networks which bear no relation whatsoever to conventional pathways. However, these fibres do enter some of these pathways for what has been called 'epiphytic guidance' to more remote targets (Azmitia and Segal 1978).

(d) In certain regions of the brain stem and spinal cord which are not or not clearly related to the trajectories of classical pathways, sets of monoaminergic fibres attain such a high concentration that they may be designated as special neuromediator-specific bundles. The large 'longitudinal catecholamine bundle' described by Jones and Friedman (1983; Fig. 3.27) is an example. Note, however, that the trajectory of this bundle also contains a variety of peptidergic fibres (Nieuwenhuys et al. 1989). Thus, it may be that the longitudinal catecholamine bundle is not an entity in itself, but rather forms part of an open multineuromediator channel.

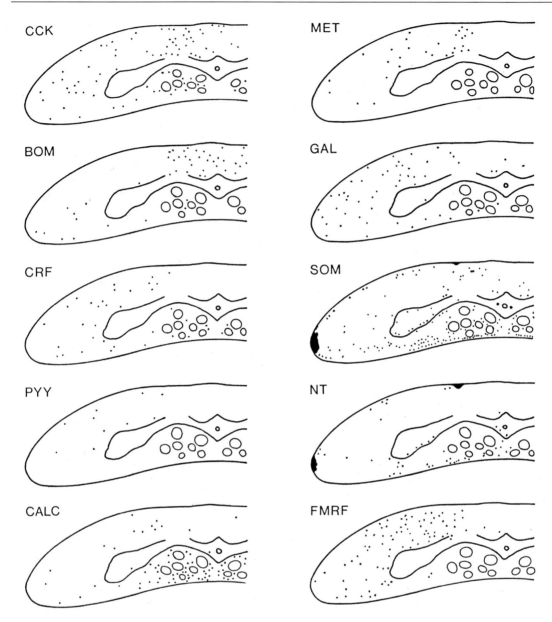

Fig. 3.26. Distribution of fibres with neuropeptide-like immunoreactivity in the lamprey spinal cord. Individual fibres are indicated by *dots*. The antisera were raised against cholecystokinin (*CCK*), metorphamide (*MET*), bombesin (*BOM*), galanin (*GAL*), corticotropin-releasing factor (*CRF*), somatostatin (*SOM*), peptide YY (*PYY*), neurotensin (*NT*), calcitonin (*CALC*) and Phe-Met-Arg-Phe-amide (*FMRF*). The contours of the large reticulospinal Müller axons are drawn in the ventromedial region (redrawn from Buchanan et al. 1987, cf. also Brodin et al. 1988)

3.6
Hodology

3.6.1
Introductory Note

We have seen that fibre systems are complex neural entities defined by the following features: (1.) site(s) of origin; (2.) course; (3.) site(s) of termination; (4.) polarity; (5.) texture, i.e. more compact or more diffuse; (6.) size, expressed in total number of fibres; (7.) nature of fibres: myelinated/unmyelinated; (8.) extent of fibre communication: open/closed; (9.) fibre pattern; (10.) fibre spectrum; (11.) composition: nature and number of components; (12.) neuromediator profile. There is, however, no fibre system for which all of these characteristics are fully known.

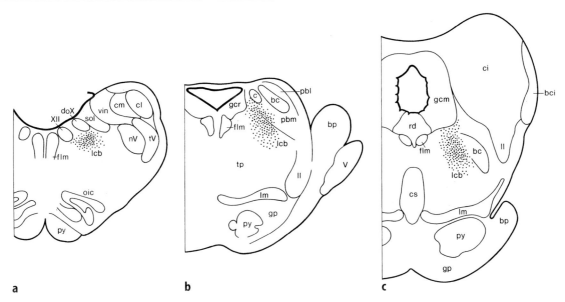

Fig. 3.27a–c. Frontal sections through the right half of the brain stem of the cat to illustrate the position of the longitudinal catecholamine bundle. **a** Caudal rhombencephalon; **b** rostral rhombencephalon; **c** caudal mesencephalon. *bc,* Brachium conjunctivum; *bci,* brachium colliculi inferioris; *bp,* brachium pontis; *c,* locus coeruleus; *ci,* colliculus inferior; *cl,* nucleus cuneatus lateralis; *cm,* nucleus cuneatus medialis; *cs,* nucleus centralis superior; *doX,* nucleus dorsalis nervi vagi; *flm,* fasciculus longitudinalis medialis; *gcm,* griseum centrale mesencephali; *gcr,* griseum centrale rhombencephali; *gp,* gri-seum pontis; *lcb,* longitudinal catecholamine bundle; *ll,* lemniscus lateralis; *lm,* lemniscus medialis; *nR,* nucleus spinalis nervi trigemini; *oic,* oliva inferior complex; *pbl,* nucleus parabrachialis lateralis; *pbm,* nucleus parabrachialis medialis; *py,* tractus pyramidalis; *rd,* nucleus raphes dorsalis; *sol,* nucleus solitarius; *tp,* tegmentum pontis; *tV,* tractus spinalis nervi trigemini; *vin,* nucleus vestibularis inferior; *V,* nervus trigeminus; *XII,* nucleus nervi hypoglossi (based on Fig. 6 of Jones and Friedman 1983)

3.6.2
General Hodology

It is the task of general hodology to provide a semantic framework for the description and characterisation of fibre systems. In the preceding pages an attempt has been made to define a number of general hodological concepts. It may be good to reiterate the definitions, to add a number of new ones, and to attach a brief commentary to some of them.

Fibre system: Any set of fibres showing a predominant, common, orientation, to which a common structural, functional or chemical label, or combination of labels, can be attached. This very wide definition reflects the use of the term in the literature. In a morphological context the term is often used as a synonym for tract, fasciculus, pathway, or projection. The term 'system' is generally used to denote a set of structures which together subserve one function or a set of coherent functions, as, e.g. optic system, extrapyramidal motor system, limbic system. It is common usage to denote the total population of cells in a given central nervous system containing a given neuromediator, substance X, as the substance-X system. If we attribute a func-tional connotation to the term system, this is allowed only if all substance X-containing cells can be brought under one common functional denominator.

Projection: A set of fibres having one main source and one main site of termination. The names of the sites of origin and termination are generally included in the names of projections.

Tractus (plural: tractus) or tract: A projection which manifests itself as a fibre concentration over at least a part of its course. It has already been mentioned that Verhaart (1955, p. 498) defined a tract as "an area of white matter in which fibres of a certain nature prevail." Because by 'nature' is meant: one particular origin and one particular site of termination, this definition is close to ours. Riley (1960, p. 566) defined a tract as "a group of fibres composed of axons subserving a similar or corresponding function."

Fasciculus, fascicle or bundle: A concentration of fibres showing a predominant common orientation, composed of sets of fibres with different origins and termination; components which maintain their individuality throughout the bundle are considered as and named tracts. Riley (1960 p. 566) defined a fasciculus as "a microscopically determinable group of fibres." However, we prefer to reserve the term for fibre assemblies with a composite character.

Pathway or path: "A pathway or path is composed of a chain of correlated neurones such as the cerebro-cerebellar pathway" (Riley 1960, p. 566). In practice, authors do not adhere closely to this definition. The terms 'pathway' and 'path' are often used as synonyms for tract. Conversely, the term projection is often used in the sense reserved here for pathway.

Open multineuromediator channel: A bundle composed of a large number of sets of thin, varicose fibres, each of which contains a different neuromediator or set of neuromediators.

Fibre pattern: the fibre pattern of a zone of white matter or fibre system is the aspect of these structures, stemming from their fibre composition, i.e. the proportions of fibres of various calibres present in them. The fibre pattern is a visual image, which can be perceived and documented by photomicrographs (cf. Figs. 3.5a, 3.6, 3.17, 3.19, 3.22) but cannot be fully described or quantified.

Fibre spectrum: "Percentage frequency distribution of fibre sizes". This definition was borrowed from van Crevel (1958, p. V) who used it, however, to define the term 'fibre pattern'. We maintain that fibre pattern and fibre spectrum are closely related, but not identical concepts. One stands for a visual image, the other for a histogram based on the counting and measuring of fibres.

Fibre class: A group of fibres which can be related to one of the modes in a histogram of fibre diameters, showing a bi-, tri- or polymodal distribution.

3.6.3
Special Hodology

It is the task of special hodology to analyse, describe and name the fibre systems in the CNS. The fibre systems in a variety of vertebrates are described in the specialised part of the present work. With regard to nomenclature, because it is impossible to include all known characters of a fibre system in its name, authors have had to make a selection. Main site of origin and main site of termination are the most frequently used characteristics for naming fibre systems: tractus corticopontinus, tractus spinothalamicus, and so on. A variety of other characteristics or combinations of features, however, have been used for purposes of nomenclature in hodology. For instance: course (fasciculus retroflexus), course + topography (fasciculus longitudinalis medialis), shape [fornix (=vault)], nature of fibres (stria medullaris thalami), function (tractus opticus), function + topography (tractus olfactorius lateralis), function + rank (tractus visceralis secundarius), and even isolation (fasciculus solitarius).

3.6.4
Comparative Hodology

It is the task of comparative hodology to compare and to homologise the fibre systems in related species. Projections in general and tracts in particular are homologised and named on the basis of their sites of origin and termination, which means that their homology is derived from those of the grisea with which they are connected. Fasciculi are usually homologised primarily on the basis of their positional relations. A prominent component of a fasciculus which clearly interconnects two grisea is usually considered a separate tract and named accordingly. In light of these simple, and generally accepted, rules, it may seem strange that fibre connections are able to provide important clues to the identification of cell masses, and yet this is true. In cases in which the topological relations of a given cell mass are obscure, a fibre tract which clearly projects to, or originates from, that cell mass and whose opposite pole is connected with an identified griseum may lead to a hypothesis regarding the homology of that unknown cell mass. The hypothesis will gain in credibility if several such 'guiding connections' are present and all point to the same interpretation, or if the hodological findings are corroborated by (immuno)histochemical data. It is emphasised, however, that similarity in topological position is the first and principal criterion for the recognition of the homology of grisea. The homology of fibre connections is inferred from that of the grisea they interconnect, and not the other way around. The view (e.g., Ariëns Kappers et al. 1936, p. 1255; Campbell and Hodos 1970) that similarity in fibre connections is the primary criterion for the establishment of homologies is incorrect and may lead to dramatically erroneous results, particularly if the positional relations are neglected (cf. Chap. 6, Sect. 6.2.6.3).

3.7
Functional Aspects of Fibre Systems

The overall functional significance of fibre systems is clear: whereas in the neural centres or grisea information is processed and integrated, fibre connections transfer information from centre to centre. The signal which transmits the information is the action potential. The amplitude and duration of action potentials are stereotyped and do not decay as they travel the length of the axon. The strength of the stimulus is encoded in the frequency of the action potentials.

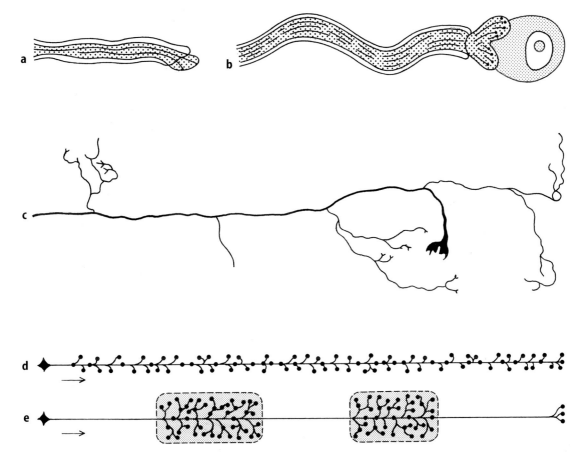

Fig. 3.28. a A large myelinated vestibular fibre in the rhomb-encephalon of the goldfish, ending in a club-shaped terminal, which is known to contact the lateral dendrite of the Mauthner cell; Protargol method (redrawn from Bodian 1937, Fig. 9). **b** Another large myelinated vestibular fibre in the goldfish forming a calyciform ending embracing part of a tangential nucleus cell, Protargol method (redrawn from Bodian 1937, Fig. 8). **c** Ending of a cochlear fibre in the ven-tral cochlear nucleus of a cat several days old, issuing different types of terminal ramifications and one calyx of Held; Golgi rapid method (based on Lorente de Nó 1981, Figs. 3–9). **d** Semidiagrammatic representation of an open-line fibre with output throughout its length. **e** Open-line fibre with output in selected centres (**d** and **e** reproduced from Nieuwenhuys 1996, Fig. 5)

The terminal segments of axons may show a variety of configurations. Some axons form a single terminal knob or calyx immediately after the end of their myelin ensheathment(Fig. 3.28a,b). More frequently, the axon splits up into a set of terminal branches, or telodendron. The various branches may be uniform, but in certain centres, as for instance the mammalian cochlear nuclei, they produce different types of endings, which establish synaptic contacts of varying size (and hence, varying strength) with their respective target elements.

Although axons are specialised primarily for the conduction of nerve impulses, the literature contains a few references to integrative functions of axons (for review see Waxman 1972). Thus it has been observed that at branching points of spinal sensory fibres a filtering of information may occur.

Almost all fibre systems are composed of fibres with different diameters, and the question arises as to what the functional significance of this variation may be. Because of the close relation between axon diameters and conduction velocity, the diversity of fibre diameters adds a temporal pattern to the spatial dimensions of the influence exerted by that tract on its centre(s) of termination. Two examples, both of which are related to the fibre pattern of the pyramidal tract, demonstrate this aspect.

1. The investigations of Leenen et al. (1985) showed that the pyramidal tract of the rat contains numerous thin, unmyelinated fibres (Fig. 3.20c). At the level of the pyramis medullae the number of unmyelinated fibres ($140\,000 \pm 7000$) is much larger than the number of myelinated axons ($103\,000 \pm 6000$). In contrast, at the level of the second cer-

ical segment the numbers of fibres of the two axon populations studied are not significantly different, viz. 43 000 ± 2000 myelinated and 35 000 ± 8000 unmyelinated fibres. However, these numbers mean a significant decrease of myelinated axons (48 %) compared with the pyramis medullae level and a much larger decrease (75 %) in the numbers of unmyelinated fibres. This differential decrease probably means that the relay stations in the lower brain stem are preferentially supplied with unmyelinated, corticofugal fibres. Differences in velocity of impulse propagation for the two fibre populations, together with a differential number of relay stations for these fibre populations, may well ensure the spatiotemporal patterning needed for synchronous activation of target areas at different levels in the central nervous system (Leenen et al. 1985).

2. The very coarse fibres present in the primate pyramidal tract constitute a fast subsystem which impinges directly on the motoneurons innervating the distal extremity muscles.

The functional potentialities of thin, unmyelinated, open fibres are multifarious, as appears in the following survey:

1. These fibres are involved in inter-neuronal communication not only at their end, but also throughout their extent (Fig. 3.28d), or at least at several different levels along their course (Fig. 3.28e).
2. Since these fibres and their terminals usually contain more than one neuromediator (often a classical neurotransmitter in combination with one or several neuropeptides), the following variations in their synaptic action have been proposed:

(a) Even if all neuromediators present are released in equal proportions at all terminals of the fibre, stimulation of that fibre causes different effects in different centres, depending on their complement of postsynaptic receptors (Swanson 1989, 1991).

(b) Peripheral autonomic fibres which innervate sweat glands contain both the classical neurotransmitter acetylcholine (ACh) and the neuropeptide vasoactive intestinal polypeptide (VIP). Lundberg et al. (1979) found that on stimulation of these fibres a single impulse preferentially induces a response which is known to be due to the release of ACh, but that upon stimulation with higher frequencies there is an increased functional effect caused by VIP. Lundberg and Hökfelt (1983) suggested that analogous synergistic actions of a classical neurotransmitter and a neuropeptide released by one and the same neuron could also occur in the central nervous system.

c) A cascade of processes evolve in peptidergic neurons to express the genetic information into biologically active neuropeptides. These processes control the nature as well as the quantities of neuropeptides synthesised in a given peptidergic neuron (de Wied 1987). External factors may influence these intracellular synthetic processes. Thus it is known that the composition of the 'cocktail' of neuropeptides produced by a peptidergic neuron may be strongly influenced by neural as well as hormonal signals, a phenomenon designated by Swanson (1989, 1991) as biochemical switching.

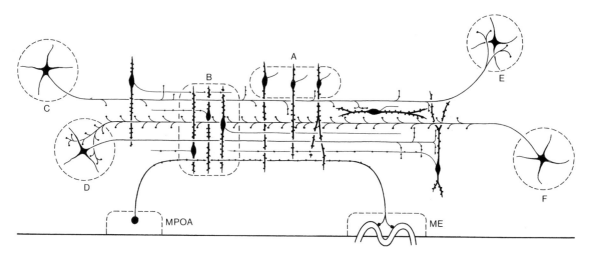

Fig. 3.29. Horizontal projection of the mammalian medial forebrain bundle. *A–F*, Centres discussed in the text; *ME*, median eminence; *MPOA*, medial preoptic area

3. Finally, the possibility should be re-iterated that the neuropeptides released by thin, varicose fibres are involved in non-synaptic rather than synaptic transmission; where this occurs the neural pathway is conveyed in a communication channel of an entirely different nature, i.e. the extracellular space (see also Chap. 2).

The organisation of a bundle composed mainly or exclusively of thin, varicose fibres differs widely from that of the compact, well-myelinated fibre systems of classical neuroanatomy. Figure 3.29 presents a highly simplified diagram of the mammalian medial forebrain bundle. This bundle is known to be composed of numerous long and short ascending and descending components (cf. Fig. 3.21), only a few of which have been included in the diagram. Each of these components is presumably characterised by its own neuromediator profile, and most of them may contain one or several neuropeptides. The medial forebrain bundle is surrounded by grisea, containing neurons which extend their dendrites mainly into it (Fig. 3.29A), and some neuronal groups are embedded in the bundle (Fig. 3.29B; cf. also Chap. 2, Figs. 2.12, 2.13.). Some of the components of the bundle project from one centre to another (Fig. 3.29C–E, F–D); however, the fibres forming these components have the potential for influencing many other centres adjacent to, or within, the bundle. The dendritic trees of the cells in these centres are oriented perpendicular to the bundle and so are optimally placed to sample information from it. Interestingly, several sets of peptidergic fibres originating from the medial preoptic area pass laterally, join the medial forebrain bundle, and pass medially again, to terminate in the median eminence. It is likely that this peculiar detour enables these fibres to make their specific contribution to the complex multipeptidergic language by which the medial forebrain bundle communicates with the cell groups lying within and around it. It should be appreciated that the information flow in the anatomically fixed circuitry depicted in Fig. 3.29 may be strongly influenced by such processes as differential release of neuromediators as a result of stimuli of different strengths, biochemical switching due to hormonal influences, and the non-synaptic release of neuromediators.

References

Adli DSH, Rosenthal BM, Yuen GL, Ho RH, Cruce WLR (1988) Immunohistochemical localisation of substance P, somatostatin, enkephalin, and serotonin in the spinal cord of the Northern leopard frog, *Rana pipiens*. J Comp Neurol 275:106–116

Ariëns Kappers CU, Huber GC, Crosby EC (1936) The comparative anatomy of the nervous system of vertebrates, including man, vol 1. MacMillan, New York

Azmitia EC, Segal M (1978) An autoradiographic analysis of the differential ascending projections of the dorsal and median raphe nuclei in the rat. J Comp Neurol 179:641–668

Basbaum AI, Fields HL (1984) Endogenous pain control systems: brainstem spinal pathways and endorphin circuitry. Annu Rev Neurosci 7:309–338

Bertolini B (1964) Ultrastructure of the spinal cord of the lamprey. J Ultrastruct Res 11:1–24

Biedenbach MA, DeVito HL, Brown AC (1986) Pyramidal tract of the cat: axon size and morphology. Exp Brain Res 61:303–310

Binggeli RL, Paule WJ (1969) The pigeon retina: quantitative aspects of the optic nerve and ganglion cell layer. J Comp Neurol 137:1–18

Bishop GM, Smith JM (1964) The sizes of nerve fibers supplying cerebral cortex. Exp Neurol 9:483–501

Bodian D (1937) The structure of the vertebrate synapse. A study of the axon endings of Mauthner's cell and neighboring centers in the goldfish. J Comp Neurol 68:117–159

Bodian D, Taylor N (1963) Synapse arising at central node of Ranvier, and note on fixation of the central nervous system. Science 139:330–332

Boivie J (1979) An anatomical reinvestigation of the termination of the spinothalamic tract in the monkey. J Comp Neurol 186:343–370

Bok ST (1928) Das Rückenmark. In: Von Möllendorff W (ed) Handbuch der mikroskopischen Anatomie des Menschen, vol 4. Springer, Berlin Heidelberg New York, pp 478–578

Bone Q (1960) The central nervous system in amphioxus. J Comp Neurol 115:27–64

Bone G (1963) The central nervous system. In: Brodal A, Fänge R (eds) The biology of myxine. Universitetsforlaget, Oslo, pp 50–91

Briner RP, Carlton SM, Coggeshall RE, Chung K (1988) Evidence for unmyelinated sensory fibres in the posterior columns in man. Brain 111:999–1007

Brodin L, Buchanan JT, Hökfelt T, Grillner S, Rehfeld JF, Frey P, Verhofstad AAJ, Dockray GJ, Walsh JH (1988) Immunohistochemical studies of cholecystokinin-like peptides and their relation to 5-HT, CGRP, and bombesin immunoreactivities in the brainstem and spinal cord of lampreys. J Comp Neurol 271:1–18

Buchanan JT, Brodin L, Hökfelt T, van Dongen PAM, Grillner S (1987) Survey of neuropeptide-like immunoreactivity in the lamprey spinal cord. Brain Res 408:299–302

Bullock TH, Moore JK, Fields RD (1984) Evolution of myelin sheaths: both lamprey and hagfish lack myelin. Neurosci Lett 48:145–148

Buma P (1989) Synaptic and nonsynaptic release of neuromediators in the central nervous system. Acta Morphol Neerl-Scand 26:81–113

Buma P, Veening J, Hafmans T, Joosten H (1992) Ultrastructure of the periaqueductal grey matter of the rat: an electron microscopical and horseradish peroxidase study. J Comp Neurol 319:519–535

Busch HFM (1961) An anatomical analysis of the white matter in the brain stem of the cat. Thesis, Leiden

Busch HFM (1964) Anatomical aspects of the anterior and lateral funiculi at the spinobulbar junction. Prog Brain Res 11:223–237

Campbell CBG, Hodos W (1970) The concept of homology and the evolution of the nervous system. Brain Behav Evol 3:353–367

Cavalcante LA, Allodi S, Reese BE (1992) Fiber order in the opossum's optic tract. Anat Embryol 186:589–600

Celio MR, Gray EG, Yasargil GM (1979) Ultrastructure of the Mauthner axon collateral and its synapse in the goldfish spinal cord. J Neurocytol 8:19–29

Chang HT (1944) High level decussation of the pyramids in the pangolin, *Manis pentadactyla dalmanii*. J Comp Neurol 81:333–338

Christensen BN (1976) Morphological correlates of synaptic transmission in lamprey spinal cord. J Neurophysiol 39:197–212

Chung K, Coggeshall RE (1979) Primary afferent axons in the tract of Lissauer in the cat. J Comp Neurol 186:451–464

Chung K, Coggeshall RE (1983) Numbers of axons in lateral and ventral funiculi of rat sacral spinal cord. J Comp Neurol 214:72–78

Chung K, Coggeshall RE (1983b) Propriospinal fibers in the rat. J Comp Neurol 217:47–53

Chung K, Sharma J, Coggeshall RE (1985) Numbers of myelinated and unmyelinated axons in the dorsal, lateral, and ventral funiculi of the white matter of the S2 segment of cat spinal cord. J Comp Neurol 234:117–121

Chung K, Langford LA, Applebaum AE, Coggeshall RE (1979) Primary afferent fibers in the tract of the Lissauer in the rat. J Comp Neurol 184:587–598

Comans PE, Snow PJ (1981) Rostrocaudal and laminar distribution of spinothalamic neurons in the high cervical spinal cord of the cat. Brain Res 223:123–127

Conti F, DeBiasi S, Minelli A, Manzoni T, Sternini C (1993) Calcitonin gene-related peptide (CGRP) in the cat neocortex: evidence for a sparse but widespread network of immunoreactive fibers. Cerebral Cortex 4:97–105

de Graaf AS (1967) Anatomical aspects of the cetacean brain stem. Thesis, Leiden

De Wied D (1987) Neuropeptides and behavior. In: Adelman G (ed) Encyclopedia of neuroscience, vol II. Birkhäuser, Boston, pp 839–841

Dunlop SA, Beazley LD (1984) A morphometric study of the retinal ganglion cell layer and optic nerve from metamorphosis in Xenopus laevis. Vision Res 5:417–427

Ebbesson SOE (1967) Ascending axon degeneration following hemisection of the spinal cord in the Tegu lizard (Tupinambis nigropunctatus). Brain Res 5:178–206

Ebbesson SOE (1969) Brain stem afferents from the spinal cord in a sample of reptilian and amphibian species. Ann NY Acad Sci 167:80–101

Ebbesson SOE (1976) Morphology of the spinal cord. In: Llinás R, Precht W (eds) Frog neurobiology. Springer, Berlin Heidelberg New York, pp 697–707

Ebbesson SOE, Hodde KC (1981) Ascending spinal systems in the nurse shark, Ginglymostoma cirratum. Cell Tissue Res 216:313–331

Faber DS, Korn H (1978) Neurobiology of the Mauthner cell. Raven, New York

Flatau E (1897) Das Gesetz der excentrischen Lagerung der langen Bahnen im Rückenmark. Z Klin Med 63:55–152

Flechsig P (1876) Die Leitungsbahnen in Gehirn und Rückenmark des Menschen auf Grund entwicklungsgeschichtlicher Untersuchungen dargestellt. Engelmann, Leipzig

Franz V (1923) Haut, Sinnesorgane und Nervensystem der Akranier. Jen Z Naturwiss 59:401–526

Friedländer A (1898) Untersuchungen über das Rückenmark und das Kleinhirn der Vögel. Neurol Centralbl 17:351–359; 397–409

Fuse G (1926) Vergleichend-anatomische Beiträge zur Kenntnis über die sog. obere, zweite oder proximale Pyramidenkreuzung bei Edentaten, sowie bei einigen fliegenden Säugern. Arb Anat Inst Sendai 12:47–92

Geri GA, Kimsey RA, Dvorak CA (1982) Quantitative electron microscopic analysis of the optic nerve of the turtle, Pseudemys. J Comp Neurol 207:99–103

Goldby F (1939) An experimental investigation of the motor cortex and pyramidal tract of Echidna aculeata. J Anat 73:509–524

Goldby F, Kacker GN (1963) A survey of the pyramidal system on the coypu rat, Myocastor coypus. J Anat 97:517–531

Goldby F, Robinson LR (1962) The central connexions of dorsal spinal nerve roots and the ascending tracts in the spinal cord of Lacerta viridis. J Anat 96:153–170

Grantyn A, Grantyn R (1982) Axonal patterns and sites of termination of cat superior colliculus neurons projecting in the tecto-bulbo-spinal tract. Exp Brain Res 46:243–256

Greeff NG, Yasargil GM (1980) Experimental evidence for saltatory propagation of the Mauthner axon impulse in the tench spinal cord. Brain Res 193:47–57

Guillery RW, Polley EHH, Torrealba F (1982) The arrangement of axons according to fiber diameter in the optic tract of the cat. J Neurosci 2:714–721

Gurdjian ES (1925) Olfactory connections in the albino rat, with special reference to the stria medullaris and anterior commissure. J Comp Neurol 38:127–163

Häggqvist G (1936) Analyse der Fasenverteilung in einem Rückenmarkquerschnitt (Th 3). Z Mikrosk Anat Forsch 39:1–34

Harding GW, Towe AL (1985) Fiber analysis of the pyramidal tract of the laboratory rat. Exp Neurol 87:503–518

Hatschek R (1907) Zur vergleichenden Anatomie des Nucleus ruber tegmenti. Arb Neurol Inst Univ Wien 15:89–136

Hayle TH (1973) A comparative study of spinal projections to the brain (except cerebellum) in three classes of poikilothermic vertebrates. J Comp Neurol 149:463–476

Heffner R, Masterton B (1975) Variation in form of the pyramidal tract and its relationship to digital dexterity. Brain Behav Evol 12:161–200

Heier P (1948) Fundamental principles in the structure of the brain. A study of the brain of Petromyzon fluviatilis. Acta Anat [Suppl] VI:1–213

Heller SB, Ulinski PS (1987) Morphology of geniculocortical axons in turtles of the genera Pseudemys and Crysemys. Anat Embryol (Berl) 175:505–515

Herbert J (1993) Peptides in the limbic system: neurochemical codes for co-ordinated adaptive responses to behavioural and physiological demand. Prog Neurobiol 41:723–791

Herrick CJ (1927) The amphibian forebrain. IV. The cerebral hemispheres of Amblystoma. J Comp Neurol 43:231–325

Herrick CJ (1933) The amphibian forebrain. VI. Necturus. J Comp Neurol 58:1–288

Herrick CJ (1948) The brain of the tiger salamander. University of Chicago Press, Chicago

Hildebrand C, Remahl S, Persson H, Bjartmar C (1993) Myelinated nerve fibres in the CNS. Prog Neurobiol 40:319–384

Hökfelt T, Ljungdahl A, Steinbusch H, Verhofstad A, Nilsson G, Brodin E, Pernow B, Goldstein M (1978) Immunohistochemical evidence of substance-P like immunoreactivity in some 5-hydroxytryptamine containing neurons in the rat central nervous system. Neuroscience 3:517–538

Holstege G (1991) An anatomical review of the descending motor pathways and the spinal motor system. Limbic and non-limbic components. Prog Brain Res 87:307–421

Honjin R, Sakato S, Yamashita T (1977) Electron microscopy of the mouse optic nerve: a quantitative study of the total optic nerve fibers. Arch Histol Jpn 40:321–332

Hughes A (1977) The pigmented-rat optic nerve: fibre count and fibre diameter spectrum. J Comp Neurol 176:263–268

Jones BE, Friedman L (1983) Atlas of catecholamine perikarya, varicosities and pathways in the brainstem of the cat. J Comp Neurol 215:382–396

Joseph BS, Whitlock DG (1968) Central projections of selected spinal cord roots in anuran amphibians. Anat Rec 160:279–288

Karten HJ (1963) Ascending pathways from the spinal cord in the pigeon (Columba livia). Proc 16th Int Congr Cool Wash 2:23

Kershaw P, Christensen BN (1980) A quantitative analysis of ultrastructural changes induced by electrical stimulation of identified spinal cord axons in the larval lamprey. J Neurocytol 9:119–138

Kirby MA, Clift-Forsberg L, Wilson PD, Rapisardi SC (1982) Quantitative analysis of the optic nerve of the North American opossum (Dedelphis virginiana): an electron microscopic study. J Comp Neurol 211:318–327

Korn H (1987) The Mauthner cell. In: Adelman G (ed) Encyclopedia of neuroscience, vol II. Birkhäuser, Boston, pp 617–619

Kusuma A, ten Donkelaar HJ, Nieuwenhuys R (1979) Intrinsic organisation of the spinal cord. In: Gans C (ed) Biology of the reptilia, vol 10: neurology B. Academic, London, pp 59–109

Kuypers HGJM (1987) Pyramidal tract. In: Adelman G (ed) Encyclopedia of neuroscience, vol II. Birkhäuser, Boston, pp 1018–1020

Landau WM, Clare MH, Bishop GH (1968) Reconstruction of myelinated nerve tract action potentials: an arithmetical method. Exp Neurol 22:480–490

Langford LA, Coggeshall RE (1981) Unmyelinated axons in the posterior funiculi. Science 211:176–177

Larsen PJ, Hay-Schmidt A, Mikkelsen JD (1994) Efferent connections from the lateral hypothalamic region and the lateral preoptic area to the hypothalamic paraventricular nucleus of the rat. J Comp Neurol 342:299–319

Lassek AM, Rasmussen GL (1939) The human pyramidal tract. A fiber and numerical analysis. Arch Neurol Psychiatry (Chicago) 42:872–876

Lassek AM, Karlsberg P (1956) The pyramidal tract of an aquatic carnivore (seal). J Comp Neurol 106:425–431

Lassek AM, Rasmussen GL (1940) A comparative fiber and numerical analysis of the pyramidal tract. J Comp Neurol 72:417–428

Lassek AM, Wheatley MD (1945) The pyramidal tract. An enumeration of the large motor cells of area 4 and the axons in the pyramids of the chimpanzee. J Comp Neurol 82–299:302

Leenen LPH, Meek J, Posthuma PR, Nieuwenhuys R (1985) A detailed morphometrical analysis of the pyramidal tract of the rat. Brain Res 359:65–80

Leghissa S (1956) Contribution ultérieure à une meilleure connaissance de l'appareil de Mauthner chez les poissons et observations sur la morphologie de la fibre. In: Ariëns Kappers J (ed) Progress in neurobiology. Amsterdam, Elsevier, pp 45–62

Lennie P (1980) Parallel visual pathways: a review. Vision Res 20:561–594

Linke R, Roth G (1990) Optic nerves in plethodontid salamanders (amphibia, urodela): neuroglia, fiber spectrum and myelination. Anat Embryol (Berl) 181:37–48

Linowiecki AJ (1914) The comparative anatomy of the pyramidal tract. J Comp Neurol 24:509–530

Lorente de Nó R (1981) The primary acoustic nuclei. Raven, New York

Luhan JA (1959) Long survival after unilateral stab wound of medulla with unusual pyramidal tract distribution. Arch Neurol (Chicago) 1:427–434

Lundberg JM, Hökfelt T (1983) Coexistence of peptides and classical neurotransmitters. Trends Neurosci 6:325–333

Lundberg JM, Hökfelt T, Schultzberg M, Uvnäs-Wallenstein K, Köhler C, Said SI (1979) Occurrence of vasoactive intestinal polypeptide (VIP)-like immunoreactivity in certain cholinergic neurons of the cat: evidence from combined immunohistochemistry and acetylcholinesterase staining. Neuroscience 4:1359–1559

Mantyh PW (1983) The spinothalamic tract in the primate: a re-examination using SGA-HRP. Neuroscience 9:847–862

Mantyh PW, Hunt P (1984) Evidence for cholecystokinin-like immunoreactive neurons in the rat medulla oblongata which project to the spinal cord. Brain Res 291:49–54

Martin GF, Fisher AM (1968) A further evaluation of the origin, course and termination of the opossum corticospinal tract. J Neurol Sci 7:177–189

Martin GF, Megirian D, Roebuck A (1970) The corticospinal tract of the marsupial phalanger (Trichosus vulpecula). J Comp Neurol 139:245–258

Martin GF, Megirian D, Conner JB (1972) The origin, course and termination of the corticospinal tracts of the Tasmanian potoroo (Potorous apicalis). J Anat 111:263–281

Maturana HR (1959) Number of fibres in the optic nerve and the number of ganglion cells in the retina of anurans. Nature 183:1406–1407

Maturana HR (1960) The fine anatomy of the optic nerve of anurans. An electron microscope study. J Biophys Biochem Cytol 7:107–120

Mehler WR (1969) Some neurological species differences – a posteriori. Ann NY Acad Sci 167:424–468

Millhouse OE (1969) A Golgi study of the descending medial forebrain bundle. Brain Res 15:341–363

Münzer E, Wiener H (1898) Beiträge zur Anatomie und Physiologie des Centralnervensystems der Taube. Monatsschr Psychiatr Neurol 3:379–406

Münzer E, Wiener H (1910) Experimentelle Beiträge zur Lehre von den endogenen Faersystemen des Rückenmarkes. Monatsschr Psychiatr Neurol 28:1–25

Nicol JM, Walmsley B (1991) A serial section electron microscope study of an identified Ia afferent collateral in the cat spinal cord. J Comp Neurol 413:247–277

Nieuwenhuys R (1985) Chemoarchitechture of the brain. Springer, Berlin Heidelberg New York

Nieuwenhuys R (1994) The neocortex: an overview of its evolutionary development, structural organization and synaptology. Anat Embryol 190:307–337

Nieuwenhuys R (1996) The greater limbic system, the emotional motor system and the brain. Prog Brain Res 107:551–580

Nieuwenhuys R, Cornelisz M (1971) Ascending projections from the spinal cord in the axolotl (Ambystoma mexicanum). Anat Rec 169:388

Nieuwenhuys R, Verrijdt PWY (1983) Structure and connections of the telencephalon of the teleost fish Xenomystis nigri. II. The area dorsalis. Acta Morphol Neerl Scand 21:330

Nieuwenhuys R, Pouwels E, Veening JG (1978) Structure and composition of the medial forebrain bundle. Neurosci Lett [Suppl] 1:190

Nieuwenhuys R, Geeraedts LMG, Veening JG (1982) The medial forebrain bundle of the rat. I. General introduction. J Comp Neurol 206:49–81

Nieuwenhuys R, Veening JG, van Domburg P (1989) Core and paracores: some new chemoarchitectural entities in the mammalian neuraxis. Acta Morphol Neerl Scand 26:131–163

Northcutt RG (1984) Evolution of the vertebrate central nervous system: patterns and processes. Am Zool 24:701–716

O'Flaherty JJ (1971) The optic nerve of the Mallard duck: fiber-diameter frequency distribution and physiological properties. J Comp Neurol 143:17–24

Ogden TE, Miller RF (1966) Studies of the optic nerve of the rhesus monkey: nerve fiber spectrum and physiological properties. Vision Res 6:485–506

Öhman P (1977) Fine structure of the optic nerve of Lampetra fluviatilis (Cyclostomi). Vision Res 17:719–722

Oka Y, Satou M, Ueda K (1986) Ascending pathways from the spinal cord in the him salmon (Landlocked red salmon, Oncorhynchus nerka). J Comp Neurol 254:104–112

Pritz MB, Northcutt RG (1980) Anatomical evidence for an ascending somatosensory pathway to the telencephalon in crocodiles, Caiman crocodilus. Exp Brain Res 40:342–345

Ralston DD, Milroy AM, Ralston III HJ (1987) Non-myelinated axons are rare in the medullary pyramids of the macaque monkey. Neurosci Lett 73:215–219

Ramón y Cajal S (1909) Histologie du système nerveux de l'homme et des vertébrés. Tome I. Maloine, Paris

Reese BE, Guillery RW (1987) Distribution of axons according to diameter in the monkey's optic tract. J Comp Neurol 260:453–460

Reese BE, Ho K-Y (1988) Axon diameter distributions across the monkey's optic nerve. Neuroscience 27:205–214

Reese BE, Cowey A (1990) Fibre organisation of the monkey's optic tract: I. Segregation of functionally distinct optic axons. J Comp Neurol 295:385–400

Retzius G (1891) Zur Kenntniss der Centralnervensystems von Amphioxus lanceolatus. Biol Untersuch 2:29–46

Rexed B (1954) A cytoarchitectonic atlas of the spinal cord in the cat. J Comp Neurol 100:297–351

Rexed B (1964) Some aspects of the cytoarchitectonics and synaptology of the spinal cord. Prog Brain Res 11:58–92

Riley HA (1960) An atlas of the basal ganglia, brain stem and spinal cord. Hafner, New York

Rodieck RW (1979) Visual pathways. Annu Rev Neurosci 2:193–225

Roeling TAP, Veening JG, Kruk MR, Peters JPW, Vermelis MEJ, Nieuwenhuys R (1994) Efferent connections of the hypothalamic 'aggression area' in the rat. Neuroscience 59:1001–1024

Rohde E (1888a) Histologische Untersuchungen über das Nervensystem von Amphioxus. Zool Anz 11:190–196

Rohde E (1888b) Histologische Untersuchungen über das Nervensystem von *Amphioxus lanceolatus*. Schneiders Zool Beitr 2:169–211

Ronan MC (1983) Ascending and descending spinal projections in petromyzontid and myxinoid agnathans. PhD dissertation, University of Michigan, Ann Arbor

Roth G (1987) Visual behavior in salamanders. Springer, Berlin Heidelberg New York

Rovainen CM (1967) Physiological and anatomical studies on large neurons of central nervous system of the sea lamprey (*Petromyzon marinus*). I. Müller and Mauthner cells. J Neurophysiol 30:1000–1023

Rovainen CM (1974a) Synaptic interactions of identified cells in the spinal cord of the sea lamprey. J Comp Neurol 154:189–206

Rovainen CM (1974b) Synaptic interactions of reticulospinal neurons and nerve cell sin the spinal cord of the sea lamprey. J Comp Neurol 154:207–224

Rovainen CM (1978) Müller cells, 'Mauthner 'cells, and other identified reticulospinal meurons in the lamprey. In: Faber DS, Korn H (eds) Neurobiology of the Mauthner cell. Raven, New York, pp 245–269

Rovainen CM, Johnson PA, Roach EA, Mankovsky JA (1973) Projections of individual axons in lamprey spinal cord determined by tracings through serial sections. J Comp Neurol 149:193–202

Ruiz MS, Anadón R (1989) Some observations on the fine structure of the ROHDE cells of the spinal cord of the amphioxus, *Branchiostoma lanceolatum* (Cephalochordata). J Hirnforsch 6:671–677

Russchen FT, Jonker AJ (1988) Efferent connections of the striatum and the nucleus accumbens in the lizard *Gekko gecko*. J Comp Neurol 276:61–80

Scheibel ME, Scheibel AB (1955) The inferior olive. A Golgi study. J Comp Neurol 102:77–131

Scheibel ME, Scheibel AB (1958) Structural substrates for integrative patterns in the brain stem reticular core. In: Jasper HH, Proctor LD, Knighton RS, Noshay WC, Costello RT (eds) Reticular formation of the brain. Henry Ford Hospital International Symposium. Little Brown, Boston, pp 31–68

Schoen JHR (1964) Comparative aspects of the descending fibre systems in the spinal cord. Prog Brain Res 11:203–222

Sereno MI (1985) Tectoreticular pathways in the turtle *Pseudemys scripta*. I. Morphology of tectoreticular axons. J Comp Neurol 223:48–90

Sereno MI, Ulinski PS (1985) Tectoreticular pathways in the turtle *Pseudemys scripta*. II. Morphology of tectoreticular cells. J Comp Neurol 223:91–114

Sharma SC, Dunn-Meynell AA, Bodylack MA (1985) A note on a tectal neuron projecting via the tectobulbar tract in teleosts. Neurosci Lett 59:265–270

Shinoda Y, Ohgaki T, Sugiuchi Y, Futami T (1992a) Morphology of single medial vestibulospinal tract axons in the upper cervical spinal cord of the cat. J Comp Neurol 316:151–172

Shinoda Y, Ohgaki T, Yuriko S, Futami T, Kakei S (1992b) Functional synergies of neck muscles innervated by single medial vestibulospinal axons. Ann NY Acad Sci 656:507–518

Sie PG (1956) Localization of fibre systems within the white matter of the medulla oblongata and the cervical cord in man. Thesis, Leiden

Sipe JC, Moore RY (1977) The lateral hypothalamic area. An ultrastructural analysis. Cell Tissue Res 179:177–196

Smeets WJAJ, Nieuwenhuys R, Roberts BL (1983) The central nervous system of cartilaginous fishes: structure and functional correlations. Springer, Berlin Heidelberg New York

Smith DS, Järlfors U, Beránek R (1970) The organization of synaptic axoplasm in the lamprey (*Petromyzon marinus*) central nervous system. J Cell Biol 46:199–219

Sotelo C, Palay SL (1970) The fine structure of the lateral vestibular nucleus in the rat. II. Synaptic organization. Brain Res 18:93–115

Stefanelli A (1934) I centri tegmentali dell' encefalo dei Petromizonti. Arch Zool Ital 20:117–202

Stone J, Campion JE (1978) Estimate of the number of myelinated axons in the cat's optic nerve. J Comp Neurol 180:799–806

Swanson LW (1989) The neural basis of motivated behavior. Acta Morphol Neerl Scand 26:165–176

Swanson LW (1991) Biochemical switching in hypothalamic circuits mediating responses to stress. Prog Brain Res 87:181–200

Swanson LW, Mogenson GJ, Gerfen CR, Robinson P (1984) Evidence for a projection from the lateral preoptic area and substantia innominata to the 'mesencephalic locomotor region' in the rat. Brain Res 295:161–178

Tapp RL (1974) Axon numbers and distribution, myelin thickness, and the reconstruction of the compound action potential in the optic nerve of the teleost: *Eugerres plumieri*. J Comp Neurol 153:267–274

ten Donkelaar HJ, Nieuwenhuys R (1979) The brainstem. In: Gans C (ed) Biology of the reptilia, vol 10: neurology B. Academic, London, pp 133–200

Towe AL (1973) Relative numbers of pyramidal tract neurons in mammals of different sizes. Brain Behav Evol 7:1–17

Tretjakoff D (1909a) Das Nevensystem von Amnmcoetes. I. Das Rückenmark. Arch Mikrosk Anat 73:607–680

Tretjakoff D (1909b) Das Nevensystem von Amnmcoetes. II. Gehirn. Arch Mikrosk Anat 74:636–779

van Beusekom GT (1955) Fibre analysis of the anterior and lateral funiculi of the cord in the cat. Thesis, Leiden

van Crevel H (1958) The rate of secondary degeneration in the central nervous system. Thesis, Leiden

van den Akker LM (1970) An anatomical outline of the spinal cord of the pigeon. Thesis, Leiden

Vaney DI, Hughes A (1976) The rabbit optic nerve: fibre diameter spectrum, fibre count, and comparison with a retinal ganglion cell count. J Comp Neurol 170:241–252

Veening JG, Swanson LW, Cowan WM, Nieuwenhuys R (1982) The medial forebrain bundle of the rat: II. An autoradiographic study of the topography of the major descending and ascending components. J Comp Neurol 206:82–108

Veening J, Buma P, ter Horst GJ, Roeling TAP, Luiten PGM, Nieuwenhuys R (1991) Hypothalamic projections to the PAG in the rat: Topographical, immuno-electronmicroscopical and functional aspects. In Depaulis A, Bandler R (eds) The midbrain periaqueductal gray matter. Plenum, New York, pp 387–415

Verhaart WJC (1947) On thick and thin fibers in the pyramidal tract. Acta Psychiatry Neurol 22:271–281

Verhaart WJC (1948a) The pes pedunculi and pyramid. J Comp Neurol 88:139–155

Verhaart WJC (1948b) The pes pedunculi and pyramid in hylobates. J Comp Neurol 89:71–78

Verhaart WJC (1950) Hypertrophy of pes pedunculi and pyramid as result of degeneration of contralateral corticofugal fiber tracts. J Comp Neurol 92:1–16

Verhaart WJC (1954) Fiber tracts and fiber patterns in the anterior and the lateral funiculus of the cord in *Macaca ira*. Acta Anat (Basel) 20:330–373

Verhaart WJC (1955) The rubrospinal tract in the cat, the monkey and the ape, its location and fibre content. Monatsschr Psychiatr Neurol 129:487–500

Verhaart WJC (1962) Anatomy of the brain stem of the elephant. J Hirnforsch 5:455–524

Verhaart WJC (1963a) The brain stem of the anteater, *Myrmecophaga jubata* L. J Hirnforsch 6:205–221

Verhaart WJC (1963b) Pyramidal tract in the cord of the elephant. J Comp Neurol 121:45–49

Verhaart WJC (1966) The pyramidal tract of Tupaia, compared to that in other primates. J Comp Neurol 126:43–50

Verhaart WJC (1967) The non-crossing of the pyramidal tract in *Procavia capensis* (Storr) and other instances of absence of the pyramidal crossing. J Comp Neurol 131:387–392

Verhaart WJC (1970) Comparative aspects of the mammalian brain stem and the cord, vol I, II. van Gorcum, Assen

Verhaart WJC, Kramer W (1952) The uncrossed pyramidal tract. Acta Psychiatr Scand 27:181–200

Verhaart WJC, Kramer W (1958) Pyramidal crossing in the elephant. Acta Morphol Neerl Scand 2:174–182

Verhaart WJC, Noorduyn NJA (1961) The cerebral peduncle and the pyramid. Acta Anat (Basel) 45:315–343

Walmsley B (1991) Central synaptic transmission: studies at the connection between primary afferent fibres and dorsal spinocerebellar tract (DSCT) neurones in Clarke's column of the spinal cord. Prog Neurobiol 36:391–423

Ward R, Repérant J, Rio J-P, Peyrichoux J (1987) Étude quantitative du nerf optique chez la Vipère aspic (*Vipera aspis*). CR Acad Sci Paris t 304 (Série III) 12:331–336

Waxman SG (1972) Regional differentiation of the axon: a review with special reference to the concept of the multiplex neuron. Brain Res 47:269–288

Wickelgren WO (1977) Physiological and anatomical characteristics of reticulospinal neurones in lamprey. J Physiol (Lond) 270:89–114

Williams RW, Chalupa LM (1983) An analysis of axon caliber within the optic nerve of the cat: Evidence of size groupings and regional organization. J Neurosci 8:1554–1564

Willis WD, Coggeshall RE (1978) Sensory mechanisms of the spinal cord. Plenum, New York

Woodburne RT (1939) Certain phylogenetic anatomical relations of localizing significance for the mammalian central nervous system. J Comp Neurol 71:215–257

Yasargil GM, Greeff NG, Luescher HR, Akert K, Sandri C (1982) The structural correlate of saltatory conduction along the Mauthner axon in the tench (*Tinca tinca* L.): identification of nodal equivalents at the axon collaterals. J Comp Neurol 212:417–424

Zottoli SJ (1978) Comparative morphology of the Mauthner cell in fish and amphibians. In: Faber D, Korn H (eds) Neurobiology of the Mauthner cell. Raven, New York, pp 13–45

CHAPTER 4

Morphogenesis and General Structure

R. NIEUWENHUYS

4.1
Introductory Note

In this chapter the morphogenesis and the general morphology of the CNS of vertebrates will be discussed. As regards morphogenesis, emphasis will be laid on those processes and features that are of direct importance for the interpretation of the parts of the adult brain and spinal cord.

4.2
Early Development

In all chordates the CNS develops from the neural plate, a thickened and elongated paramedian zone of the external germ layer or ectoderm (Fig. 4.1). The ectoderm along the lateral edges of the neural plate forms bilaterally a bandlike strip, the primordial neural crest, which separates the primordial neural ectoderm from the primordial general body ectoderm. As the neural plate grows, its lateral edges become raised to form the neural folds, whereas its midline region is depressed to form the neural groove (Fig. 4.2). During further development the neural groove deepens and the neural folds meet dorsally and eventually fuse to form the neural tube. It should be noted that as the edges of the neural groove approach each other they carry with them the adjoining primordial general body ectoderm, and that, when the process of fusion occurs, not only the neural ectoderm but also the body ectoderm fuses in the median plane. By this double fusion the neural ectoderm detaches completely from the body ectoderm to form a submerged neural tube. With the closure of the neural tube the cells of the bilateral primordia of the neural crest separate off and move into the space between the dorsal part of the neural tube and the overlying ectoderm; some of these cells become incorporated into the neural tube, however.

The closure of the neural groove does not occur synchronously over its entire length. The neural folds meet first in the hindbrain or upper cervical region and closure then proceeds rostrally and caudally. The temporary openings at either end of the closing neural tube are known as the anterior and posterior neuropores.

The formation of the neural tube does not always come about in the way described above. In the cephalochordate *Branchiostoma*, the lateral borders of the neural grooves separate from the general body ectoderm and the free margins of the latter overgrow the anlage of the CNS and meet dorsally in the median plane. Beneath this now continuous ectodermal layer, the neural groove transforms itself secondarily into a tube (Hatschek 1882). It is known that in birds and mammals the most caudal part of the spinal cord develops from a solid cord of

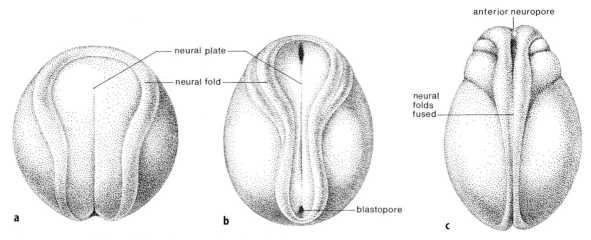

Fig. 4.1a–c. Dorsal views of frog embryos to show the early development of the central nervous system: **a** early neurula; **b** middle neurula; **c** late neurula with the neural tube largely closed (based on Balinsky 1970, Fig. 129)

cells, which is transformed into a hollow tube by cavitation (Criley 1969; Costanzo et al. 1982; Müller and O'Rahilly 1987).

In lampreys and teleosts the entire CNS has been reported to develop from a solid cord of neurectoderm which secondarily attains the form of a tube by cavitation (von Kupffer 1906). However, the difference between this massive formation and the

rolling up of the neural tube occurring in other vertebrates is probably more apparent than real. Reichenbach et al. (1990) recently observed that the formation of the neural tube in teleosts is similar to that in other vertebrates. The only difference appeared to be that in these animals during neurulation the neurectoderm is tightly folded, forming a very narrow neural groove. Judging from

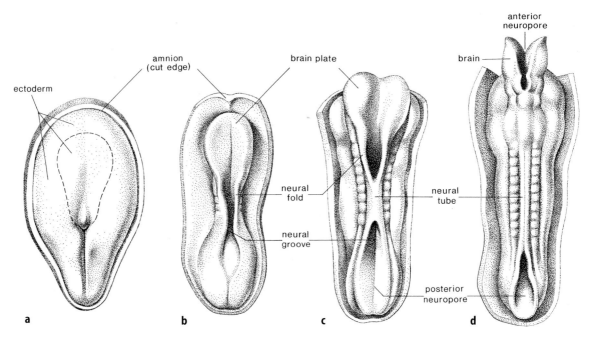

Fig. 4.2a–d. Dorsal aspects of reconstructions of human embryos: **a** 16-day presomite stage; the approximate extension of the neural plate is marked by a *dashed line.* **b** Early somite and neural groove stage of a 20-day embryo. **c** Seven-somite embryo of about the 22nd day. **d** Ten-somite embryo of about the 23rd day (adapted from Noback and Demarest 1975, Fig. 4–1)

some figures in von Kupffer (1906, Figs. 35–39) a similar tight folding, leading to an extremely narrow cleft at the place of the future ventricular cavity, presumably occurs also in the lamprey.

The early neural tube is divisible into a floor plate, a roof plate and bilateral lateral plates, which together enclose a slitlike, fluid-filled ventricular cavity (Fig. 4.3). The floor plate and roof plate are thin and consist of a single layer of epithelial cells, but the lateral plates soon thicken, and it is from these structures that the neuronally differentiated parts of brain and spinal cord come about. A longitudinal ventricular groove develops throughout most of the extent of the neural tube. This groove, the sulcus limitans of His (1893a,b) divides the lateral plates on each side into a ventral basal plate and a dorsal alar plate. This separation indicates a fundamental functional difference, because the primary sensory centres will develop in the alar plate, but the primary motor centres in the basal plate. It

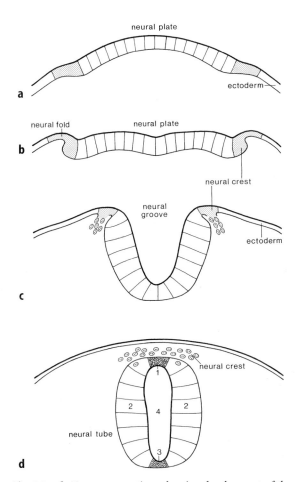

Fig. 4.3a–d. Transverse sections showing development of the neural tube at the level of the spinal cord. **a** Early neural plate stage; **b** late neural plate stage with formation of neural folds; **c** neural groove stage; **d** early neural tube stage. *1*, Roof plate; *2*, lateral plates; *3*, floor plate; *4*, ventricular cavity

should be mentioned parenthetically that in the *Nomina Embryologica* associated with the *Nomina Anatomica* (fourth edition) by the International Anatomical Nomenclature Committee (Tokyo 1977), the terms floor plate, basal plate, alar plate and roof plate have been changed into ventral lamina, ventrolateral lamina, dorsolateral lamina and dorsal lamina, respectively. We do not regard these modifications as improvements and will continue to use the old terms.

4.3
Formation of Brain Regions

According to the classical description of von Kupffer (1906), paraphrased since in numerous textbooks, the following processes lead to the differentiation of the major divisions of the brain (Figs. 4.4, 4.5):

1. Early in development the rostral part of the *neural plate* becomes wider than the caudal part. The former gives rise to the brain, whereas the spinal cord develops from the latter.
2. Prior to the closure of the anterior neuropore the rostral part of the primordial brain shows a transversely oriented fold, the *plica encephali ventralis*. This fold, situated at the level of the rostral end of the notochord, is considered to mark the boundary between two principal regions, the archencephalon and the deuterencephalon. The archencephalon is typically prechordal in position, whereas the deuterencephalon and its caudal continuation, the spinal cord, arise dorsal to the notochord.
3. Soon after formation of the neural tube, its rostral, primordial brain part shows three rostrocaudally arranged dilatations. These so-called *primary brain vesicles* are known as the primary forebrain or prosencephalon, the midbrain or mesencephalon and the primary hindbrain or rhombencephalon. The prosencephalon develops from the archencephalon, whereas the mesencephalon and the rhombencephalon are derivatives of the deuterencephalon. The boundaries between the three primary brain vesicles are marked by not only constrictions, but also by early-developing transversely oriented fibre bundles, known as commissures. Thus the commissura posterior develops dorsally in the border zone between prosencephalon and mesencephalon, whereas another dorsal commissure, the commissura cerebellaris, marks the boundary between the mesencephalon and the rhombencephalon. The narrowing between the two brain parts just mentioned is known as the isthmus, or isthmus rhombencephali. Most authors use these

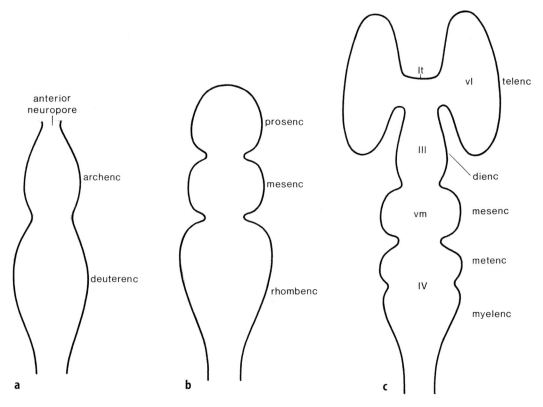

Fig. 4.4a–c. Development of the principal brain divisions: **a** two-vesicle stage; **b** three-vesicle stage; **c** five-vesicle stage. *lt,* Lamina terminalis; *vl,* ventriculus lateralis; *vm,* ventriculus mesencephali; *III,* ventriculus tertius; *IV,* ventriculus quartus

terms to denote just the constriction, but others (e.g., His 1893b, Fig. 4.43) regarded the area surrounding this constriction as a brain segment in its own right. Before the neural tube is entirely closed up, small bulges, the optic vesicles, extend laterally from each side of the developing forebrain. The three primary brain vesicles and the early primordia of the eyes were first observed by M. Malpighi (Diss epistolica de formatione pulli in ovo, London, 1673, quoted from Müller and O'Rahilly 1979) in chick embryos. An exaggerated view of the primary brain vesicles is often suggested in the literature. In most vertebrates these structures present themselves as fusiform widenings, incompletely separated by shallow constrictions. Streeter (1933, p. 474) stated: "The subdivision of the embryonic brain into three primary brain vesicles is an arbitrary expedient rather than a natural phenomenon." Bergquist and Källén (1954) regarded the pros-mesencephalic and mes-rhombencephalic limits as natural boundaries, because they correspond to interneuromeric borders. (The phenomenon of neuromery will be discussed below.) Finally, it should be mentioned that, according to O'Rahilly

and Gardner (1979), the three major subdivisions of the human brain do not begin as vesicles but rather as enlargements on the inner side of the still wide-open neural folds.

4. Ultimately, the prosencephalon and the rhombencephalon show a *differentiation* into two parts. The rhombencephalon becomes subdivided into a rostral metencephalon and a caudal myelencephalon, the latter continuous with the spinal cord. The prosencephalon becomes divided into the rostral telencephalon or endbrain and the caudal diencephalon or between-brain. These subdivisions and the midbrain and the spinal cord, which remain undivided, make up the six major regions of the vertebrate CNS. The boundary between the telencephalon and the diencephalon is usually defined as a plane passing through the decussation of optic fibres ventrally and the velum transversum dorsally. The latter is a ventrally directed fold of the membranous prosencephalic roof, which is clearly present in all vertebrates (Johnston 1909).

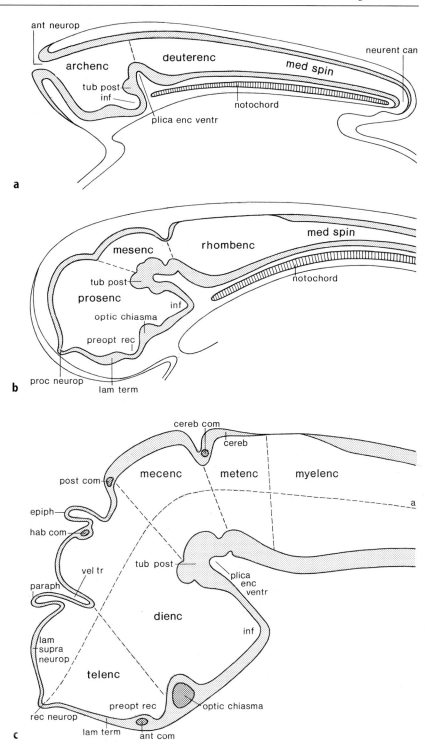

Fig. 4.5a–c. Median sections through the embryonic CNS: **a** two-vesicle stage; **b** three-vesicle stage; **c** five-vesicle stage. *a,* Longitudinal axis of the brain, according to the interpretation of von Kupffer; *ant com,* anterior commissure; *ant neurop,* anterior neuropore; *archenc,* archencephalon; *cereb,* cerebellum; *cereb com,* cerebellar commissure; *deuterenc,* deuterencephalon; *epiph,* epiphysis; *hab com,* habenular commissure; *inf,* infundibulum; *lam term,* lamina terminalis; *lam supra-neurop,* lamina supraneuroporica; *med spin,* medulla spinalis; *mesenc,* mesencephalon; *metenc,* metencephalon; *myelenc,* myelencephalon; *neurent can,* neurenteric canal; *paraph,* paraphysis; *plica enc ventr,* plica encephali ventralis; *post com,* posterior commmissure; *rec neurop,* recessus neuroporicus; *rhombenc,* rhomencephalon; *tub post,* tuberculum posterius; *vel tr,* velum transversum (modified from von Kupffer 1906, Figs. 11, 12 and 13)

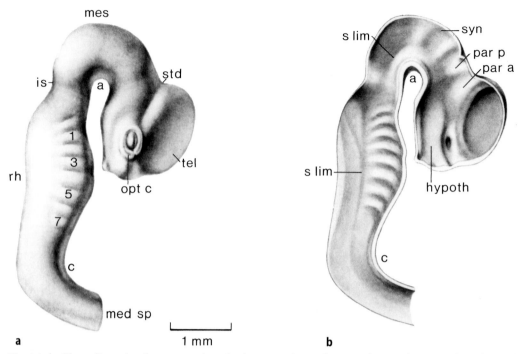

Fig. 4.6a,b. Three-dimensional reconstruction of a rhesus monkey embryo, aged 30 ± 1 days p.c.: **a** lateral view; **b** medial view. See Figs. 4–7 for abbreviations (reproduced with permission from Gribnau and Geysberts 1985, Fig. 4)

As a result of unequal growth of its different regions, three flexuresappear in the developing brain (Figs. 4.6, 4.7): The *cephalic flexure*, which is associated with the formation of the plica encephali ventralis, becomes manifest before the closure of the neural tube. The *cervical flexure*, which like the cephalic flexure is concave ventrally, appears at the junction of the hindbrain and spinal cord. The *pontine flexure*, which differs from the other two in that its convexity is directed ventrally, manifests itself in the middle of the rhombencephalon. In mammals it eventually attains such a depth that the morphologically dorsal sides of the rhombencephalic parts situated in front of and behind the flexure approach each other (Figs. 4.6, 4.7). At these developmental stages the pontine flexure may be considered to mark the boundary between the metencephalon and the myelencephalon. The cerebellum develops from the dorsal part of the metencephalon, whereas its ventral part in mammals gives rise to the pons. Small pontine nuclei are present in birds (Brodal et al. 1950), but these do not effect an externally visible protuberance. Hence, it may be stated that a pons is lacking in all non-mammalian vertebrate groups, and it is this feature which renders it difficult to demarcate the metencephalon from the myelencephalon in these groups. Therefore, it would be good to avoid the use of these terms in non-mammalians and to use the term 'rhombencepha-

lon' for the undivided structure. Unfortunately, this is rarely done. The term medulla oblongata is commonly used to denote the entire rhombencephalon in non-mammalians. However, this is incorrect because the term medulla oblongata is synonymous with myelencephalon and not with rhombencephalon.

The *five brain vesicles* were first described by von Baer (1828, p. 107) in the developing avian brain: "Ich nenne daher die fünf hier aufgezählten Bläschen nach der Reihe von der ersten zur letzten: das Vorderhirn, Zwischenhirn, Mittelhirn, Hinterhirn, und Nachhirn. Sie bilden fünf morphologi-

Fig. 4.7a,b. Three dimensional reconstruction of the brain of ▶ a rhesus monkey, aged 40 ± 1 days p.c.; **a** lateral view; **b** medial view. *a*, Cephalic flexure; *b*, pontine flexure; *b olf*, bulbus olfactorius; *c*, cervical flexure; *c ant*, commissura anterior; *cer hem*, cerebral hemisphere; *ch opt*, chiasma opticum; *c mam*, corpus mammillare; *cp*, cerebellar plate; *epiph*, epiphysis; *epith*, epithalamus; *fi*, foramen interventriculare; *hypoth*, hypothalamus; *is*, isthmus; *l term*, lamina terminalis; *med sp*, medulla spinalis; *mes*, mesencephalon; *met*, metencephalon; *my*, myelencephalon; *n I*, nervus olfactorius; *n II*, nervus opticus; *n V*, nervus trigeminus; *n VIII*, nervus octavus; *opt c*, optic cup; *par a*, parencephalon anterius; *par p*, parencephalon posterius; *rh*, rhombencephalon; *sdd*, sulcus diencephalicus dorsalis; *s lim*, sulcus limitans; *sml*, sulcus mesencephalicus lateralis; *std*, sulcus telodiencephalicus; *sth*, subthalamus; *syn*, synencephalon; *tel*, telencephalon; *tl*, temporal lobe; *thd*, thalamus dorsalis; *thV*, thalamus ventralis; *vel tr*, velum transversum; *1,3 etc.*, rhombomeres (reproduced with permission from Gribnau and Geysberts 1985, Fig. 12)

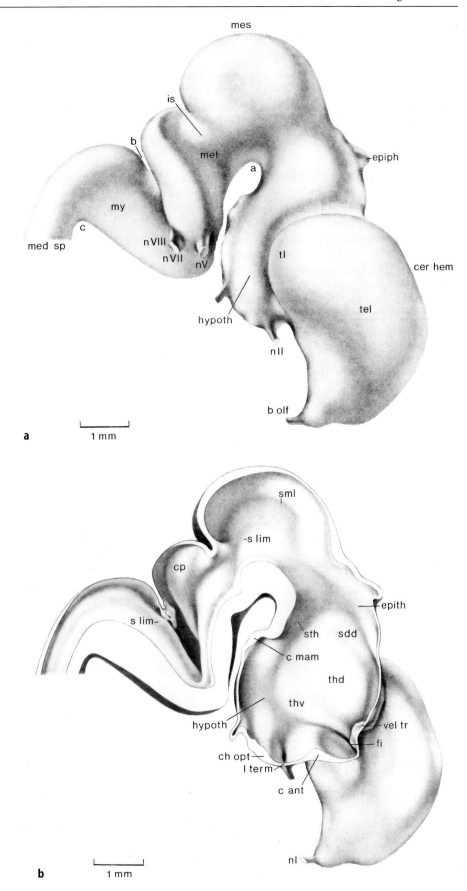

a

1 mm

b

1 mm

sche Elemente des Hirnes, die in Anfange der zwei-ten Periode der Entwickelung noch blosse Bläschen sind." Later in the nineteenth century, the Greek terms still in use were substituted for von Baer's divisions. His (1893b) considered the subdivision proposed by von Baer applicable to the human brain, with the reservation already mentioned that he interpretated the isthmus rhombencephali, situated between the mesencephalon and the metencephalon, as a separate entity. Several later investigators, among them Streeter (1933), Bergquist and Källén (1954) and Bergquist (1964), have pointed out that the five-vesicle concept of von Baer does not provide a satisfactory basis for the subdivision of the vertebrate brain. According to Bergquist and Källén (1954) the tel-diencephalic limit is artificial, because it does not coincide with an interneuromerical boundary, and the validity of the subdivision of the rhombencephalon into a metencephalon and a myelencephalon has also been questioned by these authors. As already mentioned with regard to the subdivision of the rhombencephalon, our views concur with those of Bergquist and Källén, but,

contrary to those authors, we consider a subdivision of the forebrain into a telencephalon and a diencephalon both significant and workable throughout the vertebrate kingdom.

4.4
Morphogenetic Events: Evagination, Invagination, Eversion, Folding and Others

As already discussed in Sect. 4.2, the early neural tube consists of four longitudinal plates or zones, the floor plate, the bilateral lateral plates, and the roof plate, the last being derived from the dorsal midline fusion of the lateral plates (Fig. 4.3). In the present section the various local morphogenetic events by which the neural tube becomes transformed into the adult CNS will be sketched briefly.

Divergence of Lateral Plates. In the early neural tube the lateral plates are vertically oriented and the roof plate is narrow throughout its extent. However, during further development the lateral plates may

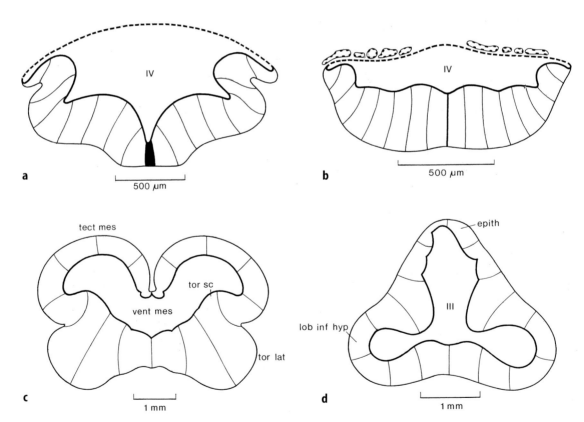

Fig. 4.8a–d. Transverse sections through (**a**) the rhombencephalon of a human embryo of 9.1 mm (based on Fig. 8 of His 1891); (**b**) an older embryo of the shark, *Scyliorhinus* (based on Fig. 99 of von Kupffer 1906); (**c**) the mesencephalon of the holostean fish *Amia*; (**d**) the diencephalon of the chondros-tean fish *Scaphirhynchus*. *epith*, Epithalamus; *lob inf hyp*, lobus inferior hypothalami; *tect mes*, tectum mesencephali; *tor lat*, torus lateralis; *tor sc*, torus semicircularis; *III*, ventriculus tertius; *IV*, ventriculus quartus

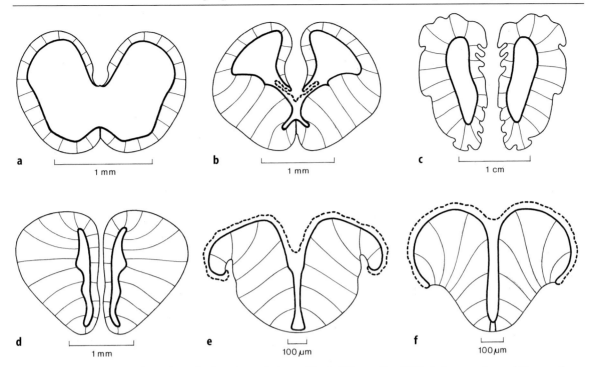

Fig. 4.9a–f. Transverse sections through the telencephalon of: **a** 13-day Chinese hamster embryo; **b** 15-day Chinese hamster embryo; **c** blue whale fetus, 80 cm (outline based on Fig. 7 of Hammelbo 1972); **d** pigeon embryo of 23 mm; **e** larva of the brachiopterygian fish *Polypterus*, 24 mm; **f**, 22-mm larva of the teleost *Gasterosteus*

diverge, with consequent widening of the roof plate. Such an 'opening' of the lateral plates is the dominant morphogenetic event in the rhombencephalon of all vertebrates, where it leads to the formation of the fourth ventricle (Fig. 4.8a,b). The troughlike cavity observed after removal of the thin roof is known as the fossa rhomboidea.

Evagination of the Lateral Plates. Evagination or bulging out of parts of the lateral plates can be observed in different parts of the brain in all vertebrates. Due to this process, extensions of the ventricular cavity surrounded by nervous tissue come about. In the prosencephalic region three subsequent evagination processes occur. The optic evaginations start to develop very early, before closure of the anterior neuropore, and rapidly enlarge to form the optic vesicles, from which the eyes and the optic nerves will develop. Somewhat later, a bilateral evagination of the rostral part of the prosencephalon leads to the formation of the telencephalic hemispheres (Fig. 4.9a–d), and by a secondary evagination of both hemispheric walls the olfactory bulbs are formed. By the hemispheric evagination the dorsal (pallial) parts of the lateral plates are brought into such a position as to cap the ventral (basal) parts (Fig. 4.9a,b). This phenomenon has been described by Holmgren (1922) and others as

the bending inward, or *inversion* of the pallium. The large lobi inferiores hypothalami occurring in chondrichthyan and chondrostean fishes (Fig. 4.8c) are the product of a fourth bilateral evagination in the prosencephalic region.

In all non-mammalian vertebrates with large eyes, the tectum mesencephali is evaginated into conspicuous bilateral lobes (Fig. 4.8d), and evagination is also a common phenomenon during development of the cerebellum. In all vertebrates the most rostral and most dorsal parts of the rhombencephalic lateral walls fuse dorsally in the midline to form a single cerebellar plate. In some groups (amphibians, chelonians: Fig. 4.10a) the cerebellum maintains a simple platelike configuration throughout development, but in others (chondrichthyans, teleosts: Fig. 4.10c, crossopterygians, birds: Fig. 4.10b, mammals) the cerebellar anlage evaginates dorsally, a process leading to the formation of a cerebellar ventricular cavity.

Median Evaginations of the Prosencephalic Roof Plate. The thin prosencephalic roof plate extends from the rostral edge of the commissura posterior to the dorsal end of the lamina terminalis. The point of fusion of these two structures is marked during early development in most vertebrates by a small ventricular recess, the recessus neuroporicus

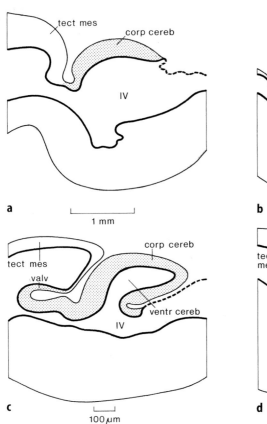

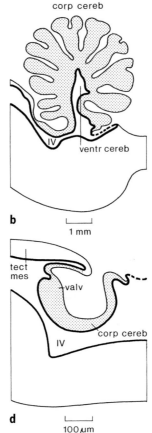

Fig. 4.10a–d. Paramedian sections through the caudal part of the brain of embryos of various vertebrates to show the structural variety of the cerebellum. **a** Alligator embryo (based on Fig. 169A of Larsell 1967); **b** 14-day chick embryo (modified from Streeter 1933, Fig. 5); **c** trout embryo, 16 mm; **d** 18-mm larva of the brachiopterygian fish *Polypterus. corp cereb*, Corpus cerebelli; *tect mes*, tectum mesencephali; *valv*, valvula cerebelli; *ventr cereb*, ventriculus cerebelli; *IV*, ventriculus quartus

(Figs. 4.5b,c, 4.11). A transverse fold known as the velum transversum marks the boundary between the diencephalic and the telencephalic parts of the prosencephalic roof plate (Figs. 4.5c, 4.11, 4.12b). In most groups of fishes the diencephalic roof expands dorsally to form a large saccus dorsalis (Fig. 4.11). This membranous structure is particularly strongly developed in chondrostean and holostean fishes. In the latter group the saccus dorsalis also forms a large, paired lateral diverticle with anterior and posterior parts, the latter extending backwards along the brain stem as far as to the level of the vagal region (Jarvik 1980).

The tela diencephali is bridged by a commissure, the *commissura habenulae*, which divides this structure into a larger rostral part and a smaller caudal part. The caudal part gives rise to two midline structures, the pineal and parapineal organs, which both manifest themselves during ontogenesis initially as small tubular or saccular evaginations (Figs. 4.11, 4.12a). The following brief survey of the development and overall structure of these organs, which together form the pineal complex, is based

mainly on the reviews of J. Ariëns Kappers (1965) and Tsuneiki (1986, 1987).

The *pineal organ* consists in most vertebrates of a tubelike stalk and a terminal vesicle. Its walls contain photoreceptor-like sensory cells or pinealocytes, neurons and supporting elements. Electron-microscopical investigations (Vigh-Teichmann et al. 1983: the chondrichthyan, *Raja*; Ekström 1987: the rainbow trout; Tabata 1982: the blind cave fish, *Astyanax*; Hetherington 1981, the salamander, *Ensatina*) have shown that the sensory cells enter into synaptic contact with the pineal neurons. The axons of the latter pass along the pineal stalk to epithalamic regions of the brain. Some of these fibres end there, but others may pass to the pretectum, the midbrain tegmentum, the thalamus and the hypothalamus.

In lampreys and in anurans the pineal organ develops as an eyelike organ. In these groups the organ is a flattened vesicle, the ventral or basal wall of which differentiates into a 'retina', whereas the dorsal wall forms a lenselike pellucida. Electrophysiological investigations (e.g., Tamotsu and Morita

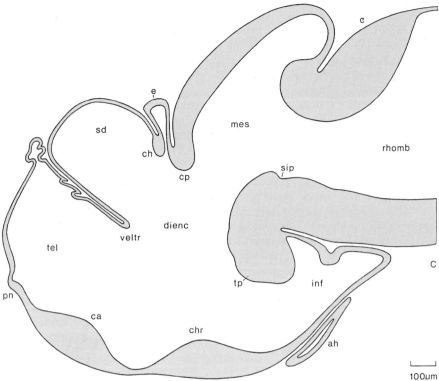

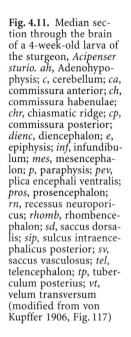

Fig. 4.11. Median section through the brain of a 4-week-old larva of the sturgeon, *Acipenser sturio. ah*, Adenohypophysis; *c*, cerebellum; *ca*, commissura anterior; *ch*, commissura habenulae; *chr*, chiasmatic ridge; *cp*, commissura posterior; *dienc*, diencephalon; *e*, epiphysis; *inf*, infundibulum; *mes*, mesencephalon; *p*, paraphysis; *pev*, plica encephali ventralis; *pros*, prosencephalon; *rn*, recessus neuroporicus; *rhomb*, rhombencephalon; *sd*, saccus dorsalis; *sip*, sulcus intraencephalicus posterior; *sv*, saccus vasculosus; *tel*, telencephalon; *tp*, tuberculum posterius; *vt*, velum transversum (modified from von Kupffer 1906, Fig. 117)

1986: lamprey; Uchida et al. 1992: lamprey; Korf et al. 1981: clawed toad) have shown that the pineal organ is directly sensitive to light.

In lizards and snakes the pinealocytes assume clear characteristics of secretory cells. In birds the pineal organ is generally well developed. Within its walls, which are often strongly folded, both photoreceptor cells and secretory cells occur. In mammals the pinealocytes appear to have entirely transformed into endocrine cells, which form the bulk of the solid pineal organ or epiphysis (Welsh and Storch 1976).

The *parapineal organ* develops from an evagination of the diencephalic roof, situated directly in front of the anlage of the pineal organ (Fig. 4.12a). A small parapineal organ is found in several vertebrate groups, including holosteans, teleosts and anurans (cf. Tsuneki 1986), but in three groups, viz. lampreys, lacertilians and rhynchocephalians, the parapineal anlage develops into a conspicuous parietal eye connected with the brain through a parietal nerve (Fig. 4.12b).

A pineal complex is vestigial or entirely lacking in myxinoids, crocodilians and some mammalian groups. It is noteworthy that, according to Jarvik (1980), the anuran impaired eye, which is also known as the frontal organ, is a parapineal rather than a pineal derivative.

A third unpaired organ, the *paraphysis*, develops as a local thickening of the tela telencephali just in front of the anterior lamina of the velum transversum. The anlage of the organ first forms a single saclike or finger-shaped evagination (Figs. 4.5c, 4.11, 4.12b). During further development its wall may become folded (the dipnoan *Neoceratodus*, reptilians: cf. Tsuneki 1986). In amphibians it takes the shape of a compound racemose, tubular gland (J. Ariëns Kappers 1950, 1956). Its general structure and cytological characteristics strongly suggest a secretory function. A paraphysis is lacking in cyclostomes and in some teleosts.

Median Evaginations of the Diencephalic Floor. As will be set forth in Sect 4.6.6, the thin floor of the diencephalon is formed not by the floor plate, but rather by the lateral plates of the two sides, which at this level are continuous with each other across the median plane. A ventrocaudal evagination of the diencephalic floor leads to the formation of the infundibulum, and a second evagination, taking its departure from the caudal infundibular wall, forms the anlage of the saccus vasculosus (Fig. 4.11). The floor of the infundibulum participates in all vertebrates in the formation of the neural lobe of the hypophysis, or pituitary gland.

A saccus vasculosus is present in all gnathostome fishes, except for the lungfish, *Lepidosiren* and some teleosts (Tsuneki 1986). It is a membranous organ which surrounds an extension of the third ventricle. The walls of the saccus vasculosus, which are often folded, consist of cubic or cylindric cells. Most of these cells are provided on the ventricular side with 10–20 hairlike processes, terminating with small knoblike swellings. These highly typical elements were designated by Dammerman (1910) as 'Krönchenzellen', i.e. coronet cells. According to Dammerman (1910), the coronet cells produce an axonal process that arises from their basal side. Later authors, among them Altner and Zimmermann (1972), have denied, however, that the coronet cells are provided with axons. According to their observations, the axonal processes present in the wall of the saccus vasculosus originate from bipolar liquor-contacting neurons which are situated between the coronet cells. The saccus vasculosus is surrounded by a dense plexus of blood vessels to which the organ owes its name. Its function is unknown.

Eversion of the Lateral Plates. Eversion or rolling outward differs fundamentally from the evagination described above. During this process the side walls recurve laterally and the roof plate is transformed in a greatly expanded tela. An eversion is observed in the telencephalon of the brachiopterygians, *Polypterus* and *Erpetoichthys* (Figs. 4.9e, 4.13b,c) and in all actinopterygians. In the latter, the eversion is combined with a marked thickening of the dorsal part of the lateral telencephalic walls (Fig. 4.9 f). Figure 4.13a–c shows that eversion of the telencephalic lateral plates is the antithesis of the pallial inversion in evaginated forebrains.

Eversion also occurs in other parts of the brain. Thus, the cerebellar plate of rhynchocephalian and lacertilian reptiles is clearly recurved, and the same holds true for the dorsal edges of the rhombencephalic lateral plates during certain developmental stages (Fig. 4.8a). In the shark *Hexanchus* the lateral parts of the rhombencephalic walls are permanently everted (Ariëns Kappers 1908).

Combined Evagination of Lateral Plates and Tela. The hemispheric evagination develops very dissimilarly in the various groups of vertebrates. As pointed out by Källén (1951b,e), two general types of evagination may be distinguished, namely *true evagination* when only nervous substance is included, and *pseudoevagination* when the process involves both the tela and the nervous part of the telencephalic walls. Pseudoevagination occurs in the telencephalon of almost all groups of gnatho-

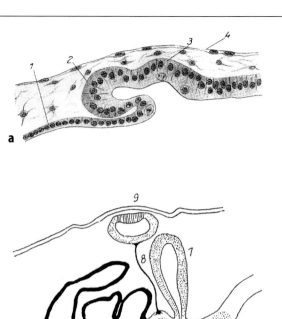

Fig. 4.12a,b. Median sections through dorsal part of the diencephalon of embryonic reptiles, showing development of the pineal complex. **a** The lizard *Lacerta vivipera*, 4-mm-embryo; *1*, diencephalic roof; *2*, anlage of the parapineal organ, which will become the parietal eye; *3*, anlage of the pineal organ, which will differentiate into the epiphysis; *4*, ectoderm. **b** An older embryo of the blindworm *Anguis fragilis*; *1*, paraphysis; *2*, velum transversum; *3*, saccus vasculosus; *4*, commissura habenulae; *5*, pretectum; *6*, commissura posterior; *7*, epiphysis; *8*, parietal nerve; *9*, parietal eye (reproduced from Becarri 1943, Figs. 241, 242)

stome fishes and in urodele amphibians. It leads to the formation of a more or less extensive membranous medial hemisphere wall or septum ependymale. In Fig. 13c and f the results of dorsally and rostrally directed pseudoevagination, respectively, are depicted.

Invagination of Nervous and Membranous Parts of the Neural Tube. Invagination or bulging inward of neuronally differentiated parts of the neural tube can be observed in the cerebellum and in the telencephalon of some groups of vertebrates. In brachiopterygian fishes the entire corpus cerebelli invaginates into the mesencephalic and rhombencephalic ventricular cavities (Fig. 4.10d), and by a similar process the valvula cerebelli of actinopterygian fishes expands rostrally underneath the tectum mesencephali (Fig. 4.10c). In all mammals the hippocampal formation invaginates into the lateral ventricle of the cerebral hemisphere. Local inva-

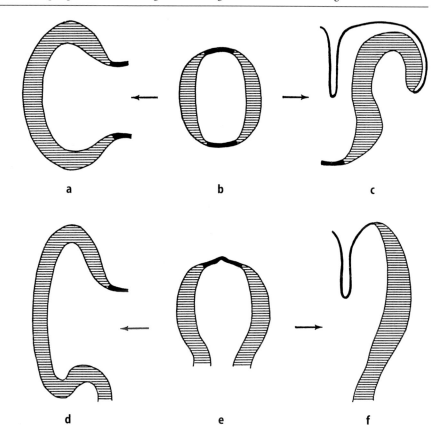

Fig. 4.13a–f. Chief morphogenetic events operative in the telencephalon of anamniotes: **a**, **b** and **c** in transverse section, **d**, **e** and **f** in horizontal section. **b** Early embryonic tube or vesicle stage of the forebrain, common to all vertebrates; **a** lateral evagination; **c** eversion of the lateral plate, concomitant with dorsally directed pseudoevagination; **e** corresponding to **b**, showing the vesicle stage; **d** a rostrally and caudally evaginated hemisphere; **f** a rostrally directed pseudoevagination. *Hatched parts,* lateral plates and their derivatives; *solid black,* membranous structures of the forebrain (reproduced from Nieuwenhuys 1965, Fig. 1)

ginations of the rhombencephalic and prosencephalic parts of the roof plate mark the beginning of the formation of the plexus chorioidei posterior and anterior.

Local Widening of the Wall. The development of certain centres or nuclear complexes may lead to the formation of internal or external protrusions. Thus, the development of the secondary octaval or octavolateral centres in the tegmentum of the midbrain calls forth the development of the tori semicirculares (Fig. 4.8c), intraventricular protrusions which attain amazing proportions in gymnotid and mormyrid teleosts. The dorsal thalami of the dipnoan *Neoceratodus* and the crossopterygian *Latimeria* protrude extensively into the third ventricle, and in the mammalian telencephalon a pair of large ridgelike protuberances form the primordia of the strio-amygdaloid complex (Fig. 4.9b). In the avian telencephalon thickening of the entire lateral hemispheric walls reduces the ventricular cavities to narrow slits (Fig. 4.9d).

As examples of external protrusions the mesencephalic torus lateralis, occurring in actinopterygian, and particularly in holostean fishes (Fig. 4.8d), and the rhombencephalic pons may be mentioned. The latter occurs only in mammals, and its size is strongly positively correlated with that of the neocortex.

The development of decussating and commissural fibre masses effects local or more extensive widenings of the floor plate, the roof plate and the thin median strip in front of the isthmus region, where the lateral plates of the two sides are directly continuous (cf. Sect. 4.5.6.1). Throughout the rhombencephalon the floor plate widens to form a continuous ependymal septum known as the raphe. Comparable formations occur in the tegmentum mesencephali and the tuberculum posterius by the continuous lateral plates, and at the levels of the cerebellum and the tectum mesencephali by the roof plate.

Coarctation of the Ventricular Walls. Coarctation or fusion of ventricular walls is a common phenomenon which may occur in all parts of the brain. The following examples may be cited:

1. The large facial lobes of silurid fishes fuse across the fourth ventricle.
2. The granular layer of the holostean and teleostean cerebellum develops from bilateral protrusions, the *Seitenwülste* of Schaper (1894a,b). Late in development these two protrusions fuse in the

median plane, thereby reducing the cerebellar ventricle to a narrow canal.

3. In gymnotid teleosts and in anuran amphibians the medial surfaces of the right and left tori semicirculares fuse almost completely.

4. The medial surfaces of the thalami on each side of the third ventricle are partially fused in many mammals. This place of fusion is called the massa intermedia or adhaesio interthalamica.

5. The forebrain of embryonic hagfishes contains a well-developed, open ventricular system. However, thickening and inwardly directed growth of the brain walls lead to a reduction, and ultimately to an almost complete obliteration of the ventricular system (von Kupffer 1906; Conel 1929, 1931; Wicht and Northcutt 1992). In the mammalian telencephalon less extensive co-arctations between the ganglionic eminences and the thin-walled septal and pallial portions of the cerebral hemispheres lead to some reduction of the lateral ventricles (Westergaard 1969a,b, 1971; Sturrock 1979).

It should be emphasised that reduction of the size of a ventricular cavity is not always necessarily the result of coarctation. Smart (1972) presented evidence suggesting that in the mammalian spinal cord the decrease in area of the central canal results from continued loss of cells from the ependymal layer and not from fusion of opposing sides, as is commonly believed. A similar 'contraction of the lumen' may well play a role in the diminution of the ventricular cavity in other parts of the brain, as for instance in the telencephalon of batoid chondrichthyans.

Surface Enlargement, Gyrification and Folding. Surface enlargement in the brain is always closely connected with cortex formation and, hence, confined mainly to the cerebellum and the pallial parts of the telencephalic hemispheres. In the avian and mammalian cerebellum the lobi, lobuli and folia, and their separating sulci are, with rare exceptions, oriented perpendicular to the longitudinal axis of the brain and parallel to the course of the parallel fibres in the outermost layer of the cerebellar cortex (Fig. 4.10b). In birds and in mammals the folding of the cerebellar surface is positively correlated with the size of the animals. The same holds true for the cerebellum of cartilaginous fishes, but in larger sharks and rays the entire cerebellar wall and not only the external surface is thrown into transverse folds.

The valvula of the teleostean cerebellum is also folded in some groups, such as the cyprinids and the silurids (Franz 1911a). In mormyrid teleosts the valvula attains amazing dimensions. In this group of fish the valvula grows out of the mesencephalic ventricle and becomes a superficially situated structure, which covers all other parts of the brain (Franz 1911b). Its external surface is considerably enlarged by the formation of fine, transversely oriented ridges in the formation of which only the molecular and Purkinje layers participate (Nieuwenhuys and Nicholson 1969a,b).

The neocortical surface of many mammals is folded (Fig. 4.9c). The pattern of gyri and sulci is relatively constant within the various mammalian orders, but different orders may display highly different folding patterns. The degree of cortical folding correlates with brain size. Small brains are lissencephalic, i.e. their neocortical surface presents no fissures, and in gyrencephalic brains the complexity of folding increases with the size of the brain. Functional specialisation may influence the pattern of gyrification. Many of the cerebral sulci coincide with the borders of cytoarchitectural and functional areas (Zilles et al. 1989).

Surface enlargement may also be observed at the ventricular side of the telencephalic hemispheres. Thus in man the rostral part of the hippocampal formation, or pes hippocampi, which protrudes into the temporal horn of the lateral ventricle, is clearly folded, and the same holds true for the surface of the dorsal ventricular ridge of the rhynchocephalian *Sphenodon*. In the latter species the folding is clearly correlated with a remarkable cortex-like differentiation of the subjacent grey matter.

The morphogenesis of the CNS may be epitomised as follows: neural plate → neural groove → neural tube → modified neural tube. From the foregoing it appears that many different morphogenetic processes are involved in the transformation of the simple neural tube into the adult brain and that most parts of the brain are the product of a combination of several or even many of these processes. To give a single example: In the mammalian telencephalon evagination, invagination, local widening of the lateral walls, local widening of the roof plate and coarctation are all operative during development, and in many mammalian species gyrification can be added to this series.

4.5
Neuromery

4.5.1
Older Studies

During a certain phase of its early development the CNS shows transversely oriented ring-shaped bulges which are called neuromeres. These transitory formations were first described by von Baer (1828)

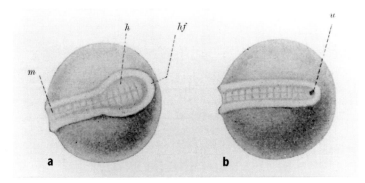

Fig. 4.14a,b. Two views of an egg of *Sala-mandra atra*. *h*, Brain plate; *hf*, anterior brain fold; *m*, medullary plate; *u*, blastopore (reproduced from von Kupffer 1906, Fig. 174a)

in the rhombencephalic region of 3- to 4-day-old chick embryos. Since that time numerous studies have been devoted to these remarkable signs of segmentation. An extensive review of the older literature on this subject (up to about 1950) is beyond the scope of the present work. Suffice it to mention a few points.

1. The existence of two types of segmentation, the primary neuromery of the open neural plate (Fig. 4.14) and the secondary neuromery of the closing and closed neural tube (Figs. 4.15, 4.16) was emphasised by von Kupffer (1906).
2. According to Orr (1887, p. 335) true neuromeres share the following features:
   a) Each neuromere is separated from its neighbours by an external vertical constriction, which corresponds to an internal, sharp dorso-ventral ridge. The constrictions are exactly opposite on each side of the brain.
   b) The elongated cells are oriented radial to the inner curved surface of the neuromere.
   c) On the plane between the apex of the internal ridge and the pit of the external incisure, the cells of adjoining neuromeres are crowded together, though the cells of one neuromere do not extend into another neuromere.
3. A much-debated question was whether or not neuromery occurs in all parts of the CNS. Orr (1887; Figs. 4.14, 4.15, 4.16) was of the opinion that true neuromeres, meeting the criteria mentioned under 2, manifest themselves only in the rhombencephalon, and Streeter (1933) held that the rostral part of the brain is non-segmented (Fig. 4.18); however, most investigators agreed that the entire neuraxis shows signs of segmentation during a certain period of its development (for nomenclature of the neuromeres found in the various parts of the CNS see Table 4.1).
4. It was generally agreed that each of the three primary brain vesicles is composed of one or sev-eral neuromeres, but the numbers of proso-meres, mesomeres and rhombomeres found were in the range of 1–5, 1–2 and 5–8, respectively. However, most investigators identified 11 neuromeres, one telencephalic, two diencephalic, two mesencephalic and six rhombencephalic (e.g., von Kupffer 1906: many species ranging from sharks to man; Hill 1900: *Salmo, Gallus*; Locy 1895: *Squalus*; Kamon 1905: *Gallus*). The two diencephalic neuromeres were designated as the parencephalon and the synencephalon, whereas the first rhombomere was generally agreed to give rise to the cerebellum (Table 4.2).
5. Regular association of branchiomotor nuclei with certain neuromeres was recognised. Thus, the motor nuclei of the Vth, VIIth and IXth nerves were found to develop within the second, the fourth and the sixth rhombomere, respectively (Table 4.2). Neal (1918) emphasised, however, that in the shark *Squalus acanthias* the origin of the motor fibres of the branchial nerves is polymetameric and that the central connections of these nerves overlap (Fig. 4.17).
6. During the first half of the twentieth century the morphological significance of the neuromery phenomenon was played down by numerous authors, among whom Neal (1918), Streeter (1933) and de Lange (1936) may be especially mentioned. It was held that:
   a) Segmentation of the vertebrate body was originally limited to the middle germ layer, and where segmental arrangement occurs in derivatives of the ectoderm it is of secondary origin and is an adaptation to the mesoblast.
   b) The myelomeres are the passive result of the mechanical pressure of the adjacent meso-dermic somites.
   c) The rhombomeres have probably arisen in adaptation to the branchiomeric segmentation.

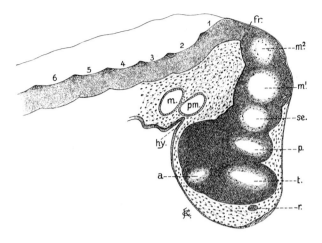

Fig. 4.15. Sagittal section through the brain of a 10-mm larva of the shark *Squalus acanthias*. *a*, Optic stalk; *fr*, fissura rhombo-mesencephalica; *hy*, hypophysis; *m*, mandibular cavity; m^1, m^2, mesomeres; *p*, parencephalon; *pm*, premandibular cavity; *r*, wall of olfactory groove; *se*, synencephalon; *t*, telencephalon; *1,2* etc., rhombomeres (reproduced from von Kupffer 1906, Fig. 89)

d) In front of the rhombencephalon clear signs of segmentation are lacking.

e) True neuromerism, i.e. a primary segmentation of the CNS, probably does not exist.

The argument may be illustrated by a quotation from Streeter (1933, p. 474, cf. also Fig. 4.18): "Instead of a rigid metameric system, the neural tube shows itself responsive at all levels to its environment. Where the environment is truly segmented the tube takes on that character in some degree. At levels where the environment is branchiomeric there we find the neural tube responding with suitable cranial nerves. At levels still further forward, aside of certain special-sense organs, there appear to be no environmental demands, at least in the early stages, and there the neural tube devotes itself to its own requirements in the way of centers of correlation and control, the subdivision of which bears no resemblance to true segmentation."

In the light of these views, it is remarkable that during the second half of the century the following developments pertaining to the neuromery phenomenon have taken place: (a) A serious attempt has been made to create a vertebrate neuromorphology in which neuromeres were proposed to be the basic building blocks of the CNS; (b) the existence of neuromeres throughout the neuraxis was conclusively demonstrated; and (c) several sets of genes that are involved in the establishment or maintenance of the neuromeres were discovered.

4.5.2
The Work of Bergquist and Källén[1]

In 1932, the Swedish neuroembryologist Bergquist published a remarkable monograph on the development of the diencephalon in anamniotes, which was based on very extensive material and on numerous carefully prepared graphical reconstructions. In this study Bergquist arrived at several conclusions which were not in harmony with the then prevailing ideas on the subdivision of the diencephalon. Thus he rejected the well-known subdivision of the diencephalon into four longitudinally arranged zones and denied that ventricular sulci mark the bound-

---

[1] The work of these authors was embedded into that of the Swedish or Holmgren School of comparative neuroembryology. Nils Holmgren (1877–1954) was professor of zoology in Stockholm. He contributed himself to comparative neuroembryology (Holmgren 1922: forebrain of lower vertebrates; Holmgren 1925: forebrain of higher vertebrates; Holmgren 1946: brain of Myxine), and from his laboratory a series of able theses appeared, including those of Palmgren (1921: development of midbrain and cerebellum), Söderberg (1922: development of forebrain in amphibians), Rendahl (1924: development of chick diencephalon), Bergquist (1932: development of diencephalon in anamniotes) and Rudebeck (1945: telencephalon of lungfishes). At the time of their appearance these studies received relatively little attention, and sometimes undeserved negative criticism (cf., e.g., Herrick's remarks on the thesis of Bergquist, mentioned in the text. However, it should be emphasised that several of these studies are of an unequalled precision and thoroughness. Major achievements of the Swedish school include contributions to the interpretation of the actinopterygian telencephalon (Holmgren 1922), the diencephalon of anamniotes, particularly the, up to that time enigmatic, diencephalon of actinopterygians (Bergquist 1932), the avian diencephalon, particularly the recognition of its three constituent neuromeres: synencephalon, parencephalon posterius and parencephalon anterius (Rendahl 1924), the avian telencephalon (Källén 1962) and the subdivision of the vertebrate brain into migration areas (e.g., Bergquist and Källén 1954).

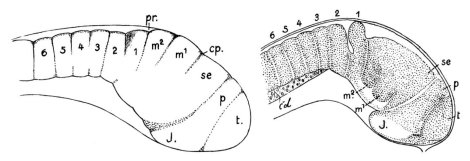

Fig. 4.16. Neuromeres in the brain of larvae of the trout *Salmo purpuratus*: *left*, medial view of the bisected brain of a 20-day-old larva; *right*, sagittal section through the brain of a 22-day-old larva. *cp*, Commissura posterior; *I*, infundibulum; *pr*, plica rhombo-mesencephalica; *remaining abbreviations*: see Fig. 4.15 (reproduced from von Kupffer 1906, Fig. 138a,b, who borrowed these drawings from Hill 1900)

aries of grisea in that brain part. He pointed out that in the developing diencephalon of all amniotes eight centres of proliferation are present, and that these *Grundgebiete* constitute the fundamental morphological units on which the homologisation of diencephalic cell masses should be based. The work of Bergquist was criticised by Herrick (1933), as may appear from the following quotations:

"This new nomenclature is the expression of a pure morphology uncontaminated by functional or other complicating factors. It follows logically from the premises assumed, but it is embarrassing to both the anatomists and the physiologists because it bears no obvious relation to the structural forms with which they are practically concerned. For these people the scheme can have only an academic interest, and this seems to be an unfortunate outcome for morphology." (Herrick 1933, p. 224)

"Most recent students of amphibian brains have recognised that the ventricular sulci are the most obvious and helpful guides in a descriptive anatomical analysis of the internal organisation of these puzzling objects. These sulci mark the boundaries of the larger fields which have distinctive characteristics of histological structure and fibrous connections. The areas so bounded are, accordingly, convenient units of both anatomical and physiological

description. This, of course, is not morphology. But are these forms which have physiological meaning of no significance to morphology, and can they safely be ignored by morphology? These are real forms and morphology as the science of form is concerned with them and their genetic relationships. Will the descriptive account of the origin, proliferation and migration of cellular elements, and this alone, yield an adequate or an intelligible scheme of cerebral organisation, for morphology or anything else? Nerve fibers are quite as important as cell bodies in cerebral organisation and as elements in cerebral forms. By what right does the morphologist ignore them in his study of form?" (Herrick 1993, p. 243)

Some 20 years later, in a long series of papers published in cooperation with his countryman Källén, Bergquist extended his observations on the development of brain centres to cover all classes of vertebrates and all parts of the CNS (Bergquist 1952a–c, 1953, 1954a,b, 1957, 1964; Källén 1951a–e, 1952, 1953a,b, 1954, 1955, 1962; Bergquist and Källén 1953a,b, 1954, 1955). In these studies the *Grundgebiete* or, as they were now called, the migration areas, and the centres developing from these entities, were clearly placed within the context of neuromery.

Table 4.1. Correlation between the brain regions and the designations of the neural bulges (from Vaage 1969)

Encephalon (brain tube)			Myelon (spinal cord)
Encephalomeres			
Prosencephalon P	Mesencephalon M	Rhombencephalon R	
Prosomeres	Mesomeres	Rhombomeres	Myelomeres
Neuromeres			

Table 4.2. Encephalomeres, their relations to adult brain parts and to the motor nuclei of branchial nerves

P1		telencephalon
P2	parencephalon }	diencephalon
P3	synencephalon	
–	*commissura posterior*	
M1		
M2		
–	*fissura rhombo-mesencephalica*	
R1		cerebellum
R2	n. motorius V	
R3		
R4	n. motorius VII	
R5		
R6	n. motorius IX	

The main results of the studies of Bergquist and Källén just enumerated may be summarised as follows. Neuromeres are present during a certain developmental period in all vertebrates (Bergquist 1952a; Källén 1954; Bergquist and Källén 1954). They coincide topographically with zones of high mitotic rate and, hence, with centres of proliferation. In fact, the neuromeric bulges are caused directly by these local intensifications of proliferation (Källén 1952, 1953a, 1962).

During early development three consecutive waves of segmentation can be observed. The first or *proneuromeric* wave disappears before the next or *neuromeric* wave begins, and the neuromeres disappear in their turn completely before a third, the *postneuromeric* wave manifests itself. However, some overlap in time between these three neuromeric waves is found in certain species (Fig. 4.22). There is no one-to-one relationship between proneuromeres, neuromeres and postneuromeres. Two neuromeres often correspond to one proneuromere and a few postneuromeres, or transverse bands as they are also called, often correspond to one neuromere (Bergquist 1964; Källén 1954; Bergquist and Källén 1954; Fig. 4.19).

Simultaneously with, or shortly after, the start of the third or postneuromeric wave, longitudinal zones of high mitotic activity are formed. By interference of the transverse bands and the longitudinal zones a network develops made up of squares or areas with a high proliferative activity. These proliferative centres, which roughly correspond to Bergquist's (1932) *Grundgebiete*, are designated as migration areas (Bergquist 1952b; Bergquist and Källén 1953a,b, 1954; Figs. 4.19–4.21).

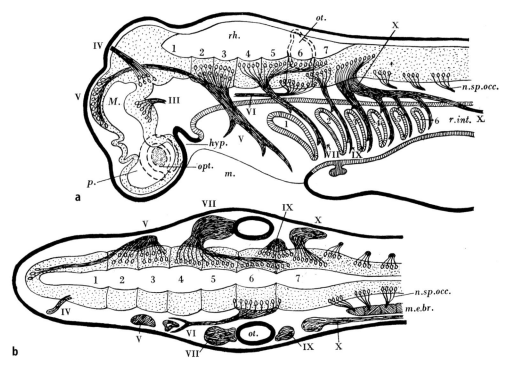

Fig. 4.17a,b. Relations between neuromeres and the origin of cranial nerves in a 13-mm embryo of the shark *Squalus acanthias*; graphical reconstruction of peripheral structures, combined with a sagittal section (**a**), and a horizontal section (**b**). *hyp*, Hypophysis; *M*, midbrain vesicle; *n. sp. occ.*, nervi spinooccipitales; *opt*, optic vesicle; *ot*, otic capsule; *p*, telencephalon; *r. int. X*, ramus intestinalis nervi vagi; *rh*, rhombencephalon; *III, IV*, etc. cranial nerves and their ganglia; *1,2*, etc., rhombomeres, also branchial pouches (reproduced from de Lange 1936, who modified these figures from Neal 1918, Figs. 1, 2)

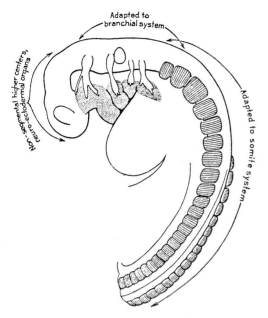

Adapted to branchial system

Non-segmental higher centers, neuro-ectodermal organs

Adapted to somite system

Fig. 4.18. Outline drawing of the CNS of a 25-somite human embryo (reproduced from Streeter 1933, Fig. 6). Streeter provided this illustration with the following caption: "The somites and the condensed tissues of the branchial region are the only body parts that seem to be calling for help from the nervous system and the form of the neural tube is nicely fitted to their service. The character of segmentation shown by them is correspondingly reflected in the neural tube. Otherwise, the environment at this time offers the brain no inducements to segmentation."

The pattern of migration areas shows a remarkable consistency throughout the vertebrate kingdom. Hence, they constitute the fundamental morphological units of the CNS and provide the only sound basis for the homologisation of brain centres (Bergquist 1954b, 1957; Källén 1951b; Bergquist and Källén 1954; Figs. 4.19, 4.20).

The migration areas represent the first signs of nuclear differentiation. They are formed by cells migrating outward from the matrix layer. These cells may remain in direct contact with the matrix layer, or later with the ventricular surface, and thus form an area of periventricular grey, or they may move further outward and contribute to a layer of migrated cells. These layers show subdivisions corresponding to the migration areas themselves (Bergquist 1954b; Källén 1951b; Bergquist and Källén 1954; Fig. 4.23).

In some parts of the brains of amniotes the migration of cells has been shown to be intermittent, taking place in two, three or sometimes even in four successive waves, each of these waves leading to the formation of a new migration layer (Bergquist 1952b, 1954a,b).

In the spinal cord the neuromeres disappear completely and are not followed by postneuromeres or transverse bands (Bergquist and Källén 1954).

Proneuromeres, neuromeres and postneuromeres manifest themselves throughout the brain, but some of the longitudinal proliferation zones do not. In the rhombencephalon four of these zones are found, designated as the columnae dorsalis (D), dorsolateralis (DL), ventrolateralis (VL) and ventralis (V). As indicated in Figs. 4.19–4.21, the dorsal column is confined to the rhombencephalon, whereas the ventral column does not extend beyond the mesencephalon (Källén 1955; Bergquist and Källén 1953a,b, 1954).

In the rhombencephalon six neuromeres (d–i) correspond in a one-to-one fashion to six transverse bands (8–13). The cerebellum develops from the dorsal part of transverse band 8. The four longitudinal migration zones correspond roughly to the functional columns of Herrick (1899, 1905) and Johnston (1902c, 1905; Bergquist 1953; Bergquist and Källén 1953a,b, 1954, 1955; Bengmark et al. 1953; Hugosson 1955).

The mesencephalon arises from two transverse bands (6 and 7) which correspond to a single neuromere (c). The three longitudinal migration zones DL, VL and V, which extend into the mesencephalon, give rise to the area tecti optici, the area dorsalis mesencephali and the area ventralis mesencephali, respectively (Bengmark et al. 1953; Bergquist and Källén 1954).

In the prosencephalon five transverse bands are present (1–5). The domain occupied by transverse bands 1–4 corresponds to that of one prosomere: a, whereas transverse band 5 corresponds to another prosomere: b. The two prosomeres a and b correspond in their turn to a single proneuromere: A. This interpretation is presented in Bergquist and Källén (1955). In several previous publications (e.g., Bergquist and Källén 1953a, 1954) the authors claimed to have observed three separate neuromeres within the domain of prosomere a; however, they admitted now to having misinterpreted the optic and hemispheric evaginations as neuromeres. Two of the four longitudinal proliferation zones, DL and VL, extend into the prosencephalon. The following seven migration areas belong to zone DL: area commissurae posterioris, area caudalis thalami, area medialis thalami (a misnomer; it should be area intermedia thalami, R.N.), area rostralis thalami + area dorsalis telencephali and area optica + area ventralis telencephali. Five migration areas fall within the domain of VL: area fasciculi longitudinalis medialis, area tuberculi posterioris and the areae caudalis-, intermedia- and rostralis hypothalami (Figs. 4.19–4.21). The ganglion habenulae is a derivative of the area caudalis thalami (Bergquist

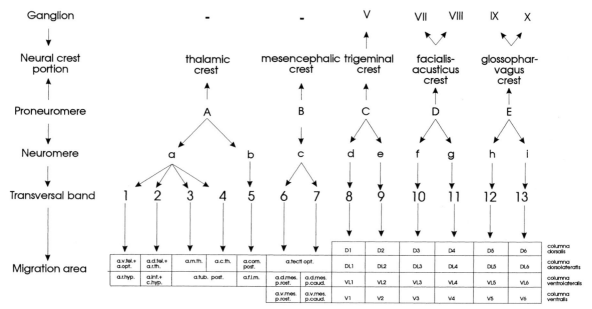

Fig. 4.19. Correspondence between the transverse components at different stages of brain development. If one starts reading from the pro-neuromeres, it is possible to find the corresponding neuromeres and transverse bands with migration areas formed from them (following the *arrows downwards*) and the corresponding neural crest portions and cranial ganglia (following the *arrows upwards*). *a.bulb.,* Area bulbaris; *a.c.hyp.,* area caudalis hypothalami; *a.c.th.,* area caudalis thalami; *a.com.post.,* area commissurae posterioris; *a.d.mes.p.caud.,* area dorsalis mesencephali pars caudalis; *a.d.mes.p.rost.,* area dorsalis mesencephali pars rostralis; *a.d.tel.,* area dorsalis telencephali; *a.f.l.m.,* area fasciculi longitudinalis medialis; *a.int.hyp.,* area intermedia hypothalami; *a.m.th.,* area medialis (should have been intermedia) thalami; *a.opt.,* area optica (preoptic region); *a.r.hyp.,* area rostralis hypothalami; *a.r.th.,* area rostralis thalami; *a, tect(i) opt.,* rea tecti optici; *a.tub.post.,* area tuberculi posterioris; *a.v.mes.p.caud.,* area ventralis mesencephali pars caudalis; *a.v.mes.p.rost.,* area ventralis mesencephali pars rostralis; *a.v.tel.,* area ventralis telencephali; *bursa R.,* bursa Rathkeis; *bursa Sees.,* bursa Seesselis; *ch.opt.,* chiasma opticum; *chord.,* chorda dorsalis; *d.c.,* dorsal mesencephalic column (Palmgren); *ductul.opt.,* ductulus opticus; *evag.hem.,* evagination of hemisphere; *fiss.cerebelli,* fissura cerebellaris; *fiss.O.I,II etc.,* fissura interneuromerica O, I etc., *fov.hyp.,* fovea hypothalamica; *hab.,* habenula; *hypothal.,* hypothalamus; *n.III,* nervus oculomotorius; *neur VII, VIII* etc., neuromere VII, VIII etc.; *prim.ep.,* primordial epiphysis cerebri; *sin.VII, VIII* etc., sinuatio neuromerica VII, VIII etc.; *tela chor.IV,* tela chorioidea ventriculi quarti; *transv.band 1,2* etc., transverse band 1,2 etc.; *I, II* etc., neuromere I, II etc., also cranial nerves I, II etc. (modified from Bergquist and Källén 1954, Fig. 9)

1954a,b; Källén 1951a,b; Bergquist and Källén 1953a,b, 1954).

The boundaries between the mesencephalon and the rhombencephalon and between the mesencephalon and the prosencephalon clearly correspond with interneuromeric boundaries as well as with boundaries between transverse bands. The metencephalon may be interpreted as a derivative of neuromere d and transverse band 8, although the transverse arrangement of the cell material giving rise to the subcerebellar part of this entity is visible during only a very short period. A natural boundary between the telencephalon and the diencephalon cannot be drawn. Transverse bands 1 and 2 extend from the telencephalic into the diencephalic domain. The telencephalon cannot be defined in terms of the hemispherical evagination, because the size of the evaginated parts varies in different species (Källén 1951b; Bergquist and Källén 1954).

Early in development the ventricular grooves are situated in the centre of areas of high mitotic activity. During later development they disappear as a consequence of decreasing mitotic activity and are then replaced by furrows of an entirely different nature, namely the limiting sulci between periventricular cell masses (Källén 1951b; Bergquist and Källén 1954).

The various portions of the cephalic part of the neural crest correspond in a one-to-one fashion to the proneuromeres. Each portion develops shortly after the appearance of the corrsponding proneuromere (Källén 1953b; Fig. 4.19).

After considering the studies of Bergquist and Källén, one cannot help but admire the enormous amount of work achieved by these investigators and their technical assistants. A definitive answer to the question: Is it really true that in the brains of all vertebrates, ranging from the lamprey to man, all grisea derive from 41 strictly homologous migration areas?, would be possible only on the basis of a research programme comparable in extent as well as in depth to that of the two Swedish investigators,

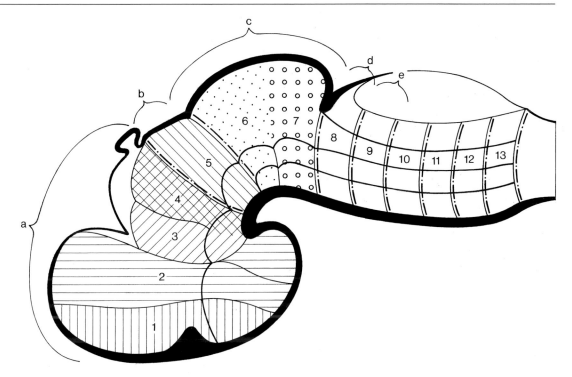

Fig. 4.20. Topography of the neuromeres and of the different migration areas in a vertebrate brain. The median section surface of the brain is *black*. Areas that form transverse bands have been marked *1,2* etc. *Dots* and *dashes* mark the projec- tions of the interneuromerical fissures. The neuromeres are marked *a, b* etc. For abbreviations see Fig. 4.19 (modified from Bergquist and Källén 1954, Fig. 5)

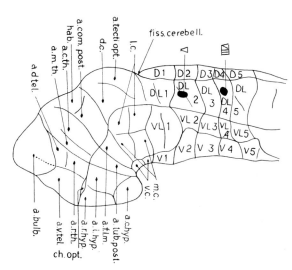

Fig. 4.21. Drawing of a wax-plate model of the left half of the brain of a 12-mm-long larva of the lamprey *Lampetra fluviatilis*, showing the 'external surface' of the ventricular grey matter. The various migration areas are labelled. The positions of some cranial nerves are marked. For abbreviations see Fig. 4.19 (reproduced from Bergquist and Källén 1953a, Fig. 1)

and such a program will most probably never be carried out. Nevertheless, several authors have challenged the correctness of some of the conclusions of Bergquist and Källén in light of their own results. Thus Kuhlenbeck (1956, p. 27) asserted categorically that the transversely oriented neuromerical structures which appear during early development represent transitory phenomena which "keinerlei Beziehungen zum definitiven Bauplan aufweisen," and Vaage (1969) denied that a reduction and disappearance of neuromeres during interneuromeric phases occurs. The reason why Bergquist's (1932, 1953, 1954a,b) parcellation of the diencephalon into eight *Grundgebiete* or migration areas has found no acceptance is most likely that these entities can be recognised only in early developmental stages, whereas the subdivisons of this brain part into four longitudinal zones, as advocated by Herrick (1910, 1917, 1935), Kuhlenbeck (1929a, 1931, 1936), Ariëns Kappers et al. (1936) and others, presents itself clearly in later developmental phases as well as in the adult stage of most vertebrate groups. Finally, it should be noted that the opinion of Bergquist and Källén that the tel-diencephalic limit is artificial and has no morphological importance is at variance with that of most other investigators.

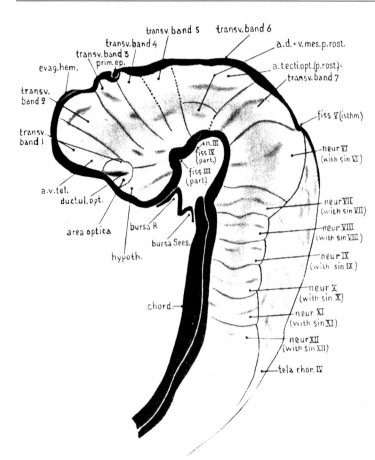

Fig. 4.22. Medial view of the bisected brain of an 8.5-mm-long embryo of the turtle *Lepidochelys olivacea*, based on a graphical reconstruction. The ridges between the transverse bands (*rostrally*) and the neuromeres (*caudally*) are marked. For abbreviations see Fig. 4.19 (modified from Bergquist 1952b, Fig. 2)

4.5.3
Vaage's Investigations on the Chick

Vaage's monograph: "The segmentation of the primitive neural tube in chick embryos (*Gallus domesticus*)", which appeared in 1969, is the most detailed and the most comprehensive study on neuromery in one particular species ever published. Vaage examined living embryos as well as carefully dissected fixed embryos, and studied an extensive material of serially sectioned specimens. He critically compared the results of his own work with those of previous authors. His main results and conclusions may be summarised as follows (cf. Figs. 4.24–4.27). During early morphogenesis the entire neural tube is differentiated into neuromeres. Since these neuromeres are observed in living, dissected, and sectioned specimens it seems evident that they are intravitally existing structures and not artefacts.

During early development two primary encephalomeres, called archencephalon and deuterencephalon, can be clearly observed. Shortly after its formation, the deuterencephalon subdivides into a lesser rostral encephalomere, the mesencephalon, and a larger caudal one, the rhombencepalon. The cerebral tube is then subdivided into three encephalomeres or brain vesicles: the prosencephalon, mesencephalon and rhombencephalon (Table 4.1). New encephalomeres are formed by successive subdivisions of the preceding neuromeres. In this way eight prosomeres, two mesomeres and eight rhombomeres finally develop. The view of Bergquist and Källén (1954) that during ontogenesis three independent waves of segmentation – proneuromeric, neuromeric and postneuromeric – pass over the rostral part of the neural tube could not be confirmed.

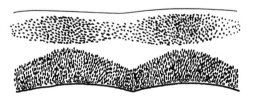

Fig. 4.23. Formation of migration layers. The ventricular grey is divided into two migration areas and the migration layer is subdivided accordingly (modified from Källén 1955, Fig. 2)

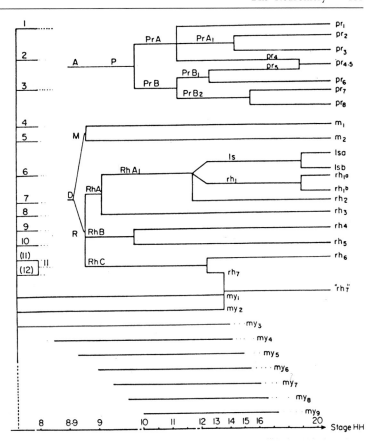

Fig. 4.24. Diagrams showing the subdivision of the archencephalon (*A*), deuterencephalon (*D*) and the spinal cord at different stages of development of the chick. The relationship between the neural tube and the 'primary neuromeres' of Hill (1900, Fig. 1.12) are shown to the *left*. *C*, Cerebellum; *ch*, chorda dorsalis; *F.m-r*, mesrhombencephalic boundary and furrow; *F.my-sp*, myelo-spinal boundary and furrow; *F.p-m*, pros-mesencephalic boundary and furrow; *F.r-sp*, rhombo-spinal boundary and furrow; *F.t-d₁*, primary teldiencephalic boundary and furrow; *F.t-d₂*, secondary tel-diencephalic boundary and furrow; $G_{V}, G_{VII-VIII...}$, ganglion of nerves V, VII–VIII,...; *HH*, stage according to the staging of Hamburger and Hamilton (1951); *Is, Is$_a$, Is$_b$*, rhombencephalic segments Is, Is$_a$, Is$_b$; *I.s.c.1.*, intersomitic cleft 1; *M*, mesencephalon; *my*, myelomere; my_1, $my_{2...}$, myelomere 1, 2,...; m_1, m_2, mesomere 1,2; *Na*, anterior neuropore; *N.C.1.*, first cervical nerve; *Ot.v.*, otic vesicle; *P*, prosencephalon; *PrA, PrA₁*, prosomere A, A$_{1...}$; pr_1, $pr_{2...}$, prosomere 1,2...; *R*, rhombencephalon; *RhA, RhB...*, rhombomere A,B...; rh_1, rh_2, rhombomere 1,2; '*rh7*', 'rhombomere 7'; *Sp.c.*, spinal cord; *T*, telencephalon (according to Vaage 1969)

The developmental pattern of the neural tube is principally the same in all vertebrates. Differences in the numbers of encephalomeres found by various investigators can be explained as due to the examination of different developmental stages.

The cerebral tube differentiates initially in a caudorostral direction, the spinal cord in a rostrocaudal direction. During later development the prosencephalon, the mesencephalon and the rhombencephalon develop independently of each other.

The myelomeres are serially homologous structures, which develop concomitantly and in numerical correspondence with the somites. One portion of the neural crest and one spinal nerve correspond to each myelomere.

The encephalomeres are individually homologous structures in different vertebrates.

In the head region there is no consistent relationship between neuromerism and metamerism (segmentation of mesoderm) and branchiomerism (formation of branchial slits). A corresponding number of encephalomeres exists throughout the vertebrate kingdom, while the head somites and the pharyngeal arches vary.

During early development the wall of the rhombencephalon is faintly subdivided into three rhombomeres, RhA, RhB, RhC. At approximately stage HH-9–10, RhA and RhB subdivide into RhA1, rh3-rh5, whereby five neuromeres (RhA1, rh3, rh4, rh5, RhC) are formed (Fig. 4.25a). RHA1 and RHC subdivide between stages HH-12 and HH-13, the former into Is and rh1, the latter into rh6 and rh7 (Fig. 4.25b). The cerebellar anlage arises from the dorsal part of rh1. Relatively late in development a subdivision of the segments Is and rh1 can be observed (Fig. 4.24).

In the anlage of the mesencephalon two neuromeres, m1 and m2, are formed at stage HH-9 (Fig. 4.25a). The rostral mesomere enlarges and forms the optic tectum. The caudal mesomere undergoes reduction and forms the boundary segment between the mesencephalon and the rhombencephalon. The reduced mesomere m2 has been overlooked by several authors. Bergquist and Källén (1954) misinterpreted m2 by incorporating it in the rhombencephalon.

The prosencephalic vesicle divides at stage HH-10 into two prosomeres, termed PrA and PrB (Fig. 4.25a). At stage HH-11–12 these two prosome-

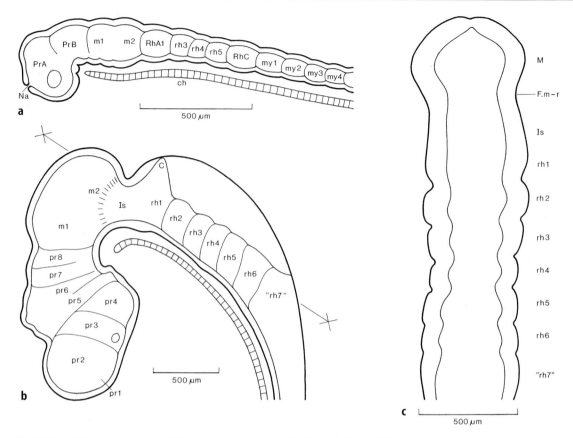

Fig. 4.25a–c. Neuromeric segmentation in the rostral part of the neural tube of chick embryos: **a** lateral view at stage HH-11 (i.e. about 2 days of incubation); **b** lateral view at stage HH-17 (about 2.5 days of incubation); **c** 'horizontal' section through the caudal part of the brain of an embryo at the same stage as that depicted in **b**; *line x-x* indicates the approximate level of this section. For abbreviations see Fig. 4.24 (based on Figs. 39 and 50 of Vaage 1969)

res subdivide into five (pr1, PrA1, pr4, PrB1, PrB2) and slightly later into six prosomeres (pr1, PrA1, pr4, pr5, pr6, PrB2). As indicated in Fig. 4.25b, subdivision of PrA1 into pr2 and pr3 and of PrB2 into pr7 and pr8 leads to a stage in which eight prosomeres (pr1–8) are present.

In amniotes the early embryonic rhombo-spinal boundary is not identical to the later myelo-spinal boundary (Fig. 4.26). The rhombo-spinal boundary is situated medial to the most rostral somite. From stage HH-13 onward the most caudal rhombomere, rh7, undergoes reduction, simultaneous with the reduction of the rostral myelomeres. Subsequently, rh7 merges with the rostral myelomeres to form an enlarged rhombomere, designated as 'rh7' (Figs. 4.24, 4.25b,c). Different numbers of myelomeres merge with the caudalmost rhombomere: none in anamniotes, 4–7 in reptiles and birds, and 3–4 in mammals. This rhombo-spinal merging process is paralleled by the evolutionary development of the spinal accessory and the hypoglossal nerves.

The myelo-spinal boundary is situated between the secondarily enlarged rhombencephalon and the spinal cord.

The boundary between the mesencephalon and the rhombencephalon is marked in early developmental stages by a distinct fissure. Later, a transverse ventricular sulcus situated at the level of the reduced m2 segment marks the boundary between the mesencephalon and the rhombencephalon. This boundary extends between the oculomotor and the trochlear nucleus.

The boundary between the diencephalon and the mesencephalon is identical to that between the prosencephalon and the mesencephalon, as well as to that between the archencephalon and the deuterencephalon. Although, according to several authors (e.g., Herrick 1948; Kuhlenbeck 1956), this boundary is situated directly rostral to the posterior commissure, Vaage (1969) places it with Neal (1898), von Kupffer (1906) and Palmgren (1921) caudal to this commissure.

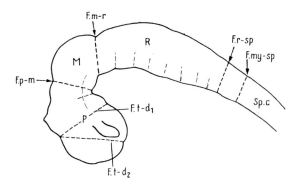

Fig. 4.26. Lateral view of the neural tube of the chick showing the location of the bordering furrows. For abbreviations see Fig. 4.24 (from Vaage 1969)

The tel-diencephalic boundary is an intraprosencephalic border interposed between prosomeres. The boundary between the prosomeres A and B is the first to be formed during the subdivision of the prosencephalon (Fig. 4.25a). The boundary between these two structures is externally marked by a dorsoventral furrow, F.t.-d1 (Fig. 4.26). This boundary disappears during further neurogenesis and another develops, marked by the external furrow F.t.-d2, within the lateral wall of prosomere PrA, separating pr1–2 from the rest of PrA (Figs. 4.25b, 4.26). Both external furrows correspond with a ventricular ridge.

The cranial nerves III-XII do not correspond numerically to the metameres, branchiomeres or neuromeres in the head region, but they appear to bear a fixed relation to certain neuromeres throughout the vertebrate kingdom, while other neuromeres are devoid of peripheral nerves. In the

chick the oculomotor nerve emerges from the rostral mesomere m1 and the nerves IV, V, VI, VII,VIII, IX and X are connected with rhombomeres Is, rh2, rh3, rh4, rh6 and 'rh7', respectively (Fig. 4.27).

The results of a comparison of the work of Vaage with those of other authors will be presented later, in Sect. 4.5.6.

4.5.4
Neuromery in the Mammalian Brain

Brief mention will be made here of the main results of Coggeshall (1964: rat), Keyser (1972: chinese hamster). Gribnau and Geysberts (1985: rhesus monkey), Lakke et al. (1988: rat) and O'Rahilly and collaborators (O'Rahilly and Gardner 1979; O'Rahilly et al. 1989; Müller and O'Rahilly 1987, 1989: man).

Gribnau and Geysberts (1985, Fig. 4.28a,b) observed in rhesus monkey embryos seven rhombomeres (Rh1–7), two isthmic neuromeres (I1,2), two mesomeres (M1,2) and four prosomeres, the synencephalon (IV), the parencephalon posterius (III), the parencephalon anterius (II) and the telencephalic neuromere (I). They noticed that three structures, i.e. the commissura posterior, the fasciculus retroflexus and the zona limitans intrathalamica, coincide with the boundaries between M1 and IV, IV and III and III and II, respectively, and may be considered as landmarks permanently indicating the location of these boundaries (Fig. 4.28d). The zona limitans intrathalamica is a cell-poor zone which marks the boundary between the thalamus ventralis and the thalamus dorsalis. With regard to the prosencephalic neuromeres, the results of Gribnau and Geysberts (1985) resemble in several

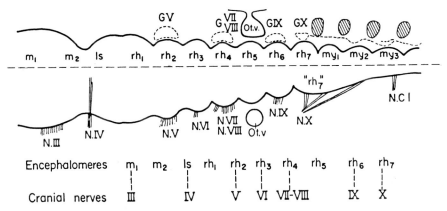

Fig. 4.27. Longitudinal section of the lateral wall of mesencephela and rhombencephala of chick embryos at stages HH-13 (*upper half*) and HH-24 (*lower half*). The anlage of the ganglia to the trigeminal (*GV*), octavo-facial (*GVII-VIII*), glossopharyngeal (*GIX*) and vagus (*GX*) nerves are outlined

in the *upper half*. The oculomotor (*NIII*), trochlear (*NIV*), trigeminal (*NV*), abducens (*NVI*), trigeminal (*NV*), abducens (*NVI*), octavo-facial (*NVII-VIII*), glossopharyngeal (*NIX*) and vagus (*NX*) are outlined in the *lower half*. For other abbreviations see Fig. 4.24 (from Vaage 1969)

points those of Vaage (1969: chick), Coggeshall (1964: rat) and Keyser (1972: Chinese hamster). All of these authors delineated diencephalic neuromeres corresponding in size and position with the synencephalon and the parencephalon posterius of Gribnau and Geysberts (1985). However, the number of neuromeres observed by these authors in front of the parencephalon posterius varies from one to four, and Vaage (1969) subdivided the synencephalon into two separate neuromeres (Fig. 4.28b,c). Lakke et al. (1988) studied the development of the ventricular surface morphology in the prosencephalon of the rat. Just like Coggeshall (1964), they arrived at the conclusion that in this species three prosomeres are present, the most rostral of which comprises the primordia of the telencephalon, the hypothalamus and the parencephalon anterius.

The neuromeric pattern in the brain of human embryos was studied by O'Rahilly and collaborators (O'Rahilly and Gardner 1979; O'Rahilly et al. 1989; Müller and O'Rahilly 1987, 1989). Their main conclusions may be summarised as follows (Fig. 4.29):

1. In the brain of human embryos 13 neuromeres can be discerned, seven rhombencephalic, one isthmic, two mesencephalic and three prosencephalic.
2. The cerebellar anlage originates from the first rhombomere but later encroaches upon the isthmic neuromere.
3. The posterior commissure and the fasciculus retroflexus mark the caudal and rostral boundary, respectively, of the synencephalon, i.e. the most caudal prosomere. The floor area of this prosomere gives rise to the mammillary bodies and the infundibulum.
4. The optic evagination arises from the middle prosomere. The floor area of this prosomere forms the bed of the optic chiasm.
5. The external groove which marks the boundary between the evaginated cerebral hemisphere and the unevaginated portion of the prosencephalon corresponds with a ventricular ridge, the torus hemisphericus. The diencephalo-telencephalic limit passes from the velum transversum, via the torus hemisphericus, to the preoptic recess. The preoptic region belongs to the telencephalon.

4.5.5
Recent Studies on Neuromery

Recent advances in our knowledge and understanding of the neuromery phenomenon will be discussed under the following headings: (a) segmental patterns of differentiation; (b) segmented cell lineage restrictions, (c) neuromeres and the formation of fibre tracts, (d) relations between neuromeres and the centres giving rise to motor roots of spinal and cranial nerves, and (e) segmental expression patterns of putative regulatory genes.

4.5.5.1
Segmental Patterns of Differentiation

Källén (1952, 1953a, 1962) observed that each neuromere contains a centre of cellular proliferation. The question arises whether particular processes of neuronal differentiation also take place within the confines of the neuromeres. This question can be answered in the affirmative for acraniates and agnathans. With regard to acraniates, Bone (1960) observed in the spinal cord of amphioxus six different types of cells, among them primary sensory and internuncial, as well as somatomotor elements, occurring segmentally, singly in each segment. The single, segmental somatomotor neurons (designated by Bone as SMi cells) are large and lie embedded in fan-shaped groups of smaller somatomotor elements (Bone's Smii cells). In accordance with the views of Whiting (1948, 1957), Bone (1960) considered it likely that the SMi cells represent 'archesomatomotor cells', around which the SMii cells later develop. As for agnathans, Whiting (1948, 1957) studied the early development of the CNS in larval lampreys. In the spinal cord of these animals he observed very early developing sensory, internuncial and motor neurons forming a sensorimotor arc in each segment. Whiting surmised that these neurons, together with other early-developing distinctive neurons, reflect the ancestral segmental chordate pattern, which is why he designated them as 'arche-neurons'. Whiting considered it likely that these arche-neurons serve as precursor or pioneer neurons for the later development of groups of cells performing the same function around them.

Several recent studies lend support to Whiting's views. Thus Kimmel and collaborators (Kimmel et al. 1988; Hanneman and Westerfield 1989; Hanneman et al. 1988; Westerfield et al. 1986), using three different methods for the demonstration of early-differentiating neurons (acetylcholinesterase staining, Nomarski interference contrast optics, and staining with the zn-1 monoclonal antibody), showed conclusively that in the early larval spinal cord of the zebrafish, *Brachydanio rerio*, three different kinds of individually identifiable primary motoneurons that repeat in register with the segmental series of myotomes are present. These primary motoneurons form clusters, a single motoneuron of each of the three types being present in each cluster.

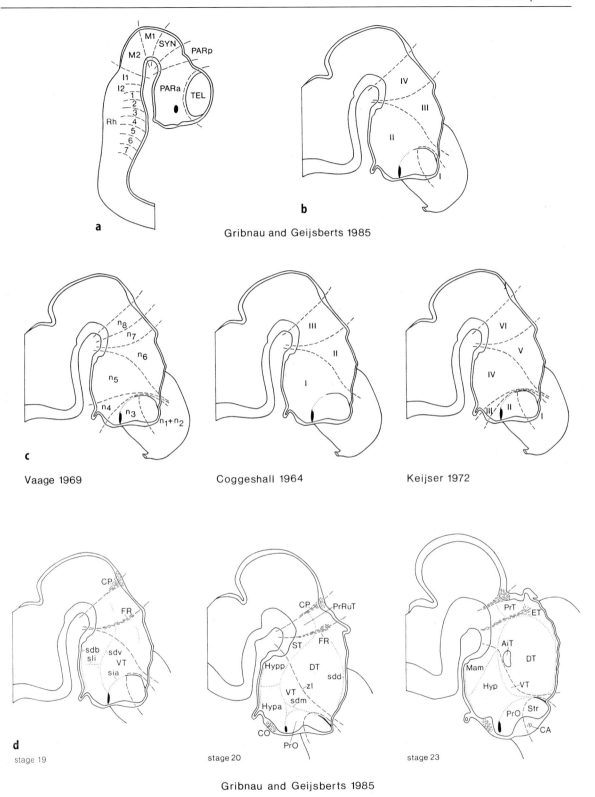

a

b

Gribnau and Geijsberts 1985

c

Vaage 1969

Coggeshall 1964

Keijser 1972

d

stage 19

stage 20

stage 23

Gribnau and Geijsberts 1985

◄ **Fig. 4.28a–d.** Neuromery in the mammalian brain. **a** Brain of a rhesus monkey embryo in stage 15 (30 ± 1 day); **b** prosencephalon of rhesus monkey embryo in stage 19 (36–42 days); **c** neuromeric patterns in the prosencephalon as suggested by some authors, transferred to the outline of **b**; **d** transformations taking place in the developing rhesus monkey diencephalon. The three primary, transversely oriented neuromeres are superseded by longitudinal zones, which in their turn become distorted, mainly because of the disproportionate outgrowth of the dorsal thalamus. The original interneuromeric boundaries remain visible, however, because they are marked by the commissura posterior, the fasciculus retroflexus and the zona limitans intrathalamica. *AiT,* Adhesio interthalamica; *CA,* commissura anterior; *CO,* chiasma opticum; *CP,* commissura posterior; *DT,* dorsal thalamus; *ET,* epithalamus; *FR,* fasciculus retroflexus; *Hyp a,* anterior hypothalamic region; *Hyp p,* posterior hypothalamic region; I_1, I_2., isthmic neuromeres; M_1, M_2., mesencephalic neuromeres; n_1, n_2 etc., neuromeres; *PAR a,* parencephalon anterius; *PAR p,* parencephalon posterius; *PrO,* preoptic region; *PrRuT,* prerubral tegmentum; *PrT,* pretectal region; Rh_1-Rh_7, rhombomeres 1–7; *sdb,* sulcus diencephalicus basalis; *sdd,* sulcus diencephalicus dorsalis; *sdm,* sulcus diencephalicus medius; *sdv,* sulcus diencephalicus ventralis; *sia,* sulcus intraencephalicus anterior; *sli,* sulcus lateralis infundibuli; *ST,* subthalamus; *Str,* striatum; *SYN,* synencephalon; *TEL,* telencephalon; *VT,* ventral thalamus; *zl,* zona limitans intrathalamica; *I, II,* etc., neuromeres (reproduced with permission from Gribnau and Geysberts 1985, Figs. 17–19)

Kimmel and his colleagues (Mendelson 1985, 1986a,b; Metcalfe et al. 1986; Hanneman et al. 1988; Hanneman and Westerfield 1989; Wilson et al. 1990;

Ross et al. 1992; Kimmel 1993) also observed that in the early anlage of the zebrafish brain about 10 neuromeres can be recognised. The three most rostral of these correspond to the telencephalon, the diencephalon and the mesencephalon, whereas the remaining seven occupy the major part of the embryonic hindbrain. With the aid of AChE staining it was demonstrated that in the centre of the basal plate region of all of these encephalomeres single clusters of early-developing neurons arise. The number of neurons that initially form each cluster varied according to the brain region. The largest clusters appeared to be present in the forebrain and midbrain, the diencephalic cluster contained about 11 neurons, and the telencephalic and mesencephalic clusters each contained about five. In contrast, only about two neurons were observed to develop in each rhombomere cluster (Fig. 4.30). Appearance of new neurons was found to occur either by expansion of the early clusters or by formation of new clusters.

In several of the rhombomeres the earliest neurons are specific, identified reticulospinal neurons. For instance, two bilateral pairs of cells always appear to develop first in the centre of the fourth rhombomere. The lateral cell of each pair is the young Mauthner neuron, whereas the medial cell is

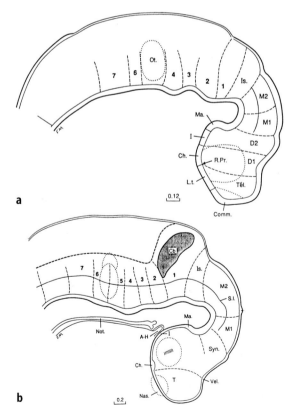

a

b

Fig. 4.29a,b. Median reconstruction of human embryos: **a** stage 13 (28–32 days); **b** stage 14 (31–35 days). *A.H.,* Adenohypophysis; *Cb,* cerebellum; *Ch.,* chiasma opticum; *Comm.,* commissura anterior; *D1,2,* diencephalic neuromeres; *I,* infundibulum; *Is,* isthmus; *L.t.,* lamina terminalis; *M1,2,* mesomeres; *Ma,* mammillary region; *Not.,* notochord; *Ot.,* otic vesicle; *R.Pr.,* recessus preopticus; *S.l.,* sulcus limitans; *Syn.,* synencephalon; *Tél., T,* telencephalon; *Vel.,* velum transversum; *1,2* etc., rhombomeres. (reproduced with permission from O'Rahilly et al. 1989)

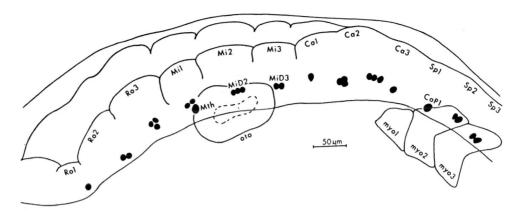

Fig. 4.30. Segmental patterning in the hindbrain of an early embryo of the zebrafish, *Brachydanio rerio*. Single cells or small clusters of cells are present in each hindbrain segment. *Ca1–3,* Caudal rhombomeres; *CaP1,* caudal primary moto-neuron in Sp1; *Mi1–3,* middle rhombomeres; *MiD2–3,* mid- dle dorsal reticulospinal neurons; *Mth,* Manther neuron; *myo1–3,* the first three myotomes; *oto,* otic vesicle; *Ro1–3,* rostral rhombomeres; *Sp1–3,* spinal segments (reproduced from Hanneman et al. 1988)

the ipsilaterally projecting medial reticulospinal neuron MiM1 (Metcalfe et al. 1986; Fig. 4.31).

The individually identifiable reticulospinal neurons occupy characteristic locations in specific rhombomeres, and generally, a cell in one segment shares certain morphological features with particular elements occupying corresponding positions in adjacent segments. On the basis of these shared fea- tures the early-developing reticulospinal neurons were classified into a number of families, each representing a set of segmental homologues. One of these families of segmental homologues includes three neurons, Mauthner, MiD2 cm and MiD3 cm, the axons of which cross the median plane within their own segment, and then descend to the spinal cord by the medial longitudinal fasciculus (MLF;

Fig. 4.31. Segmentally distributed reticulo-spinal neurons in the rhombencephalon of a 5-day-old larva of the zebrafish. Neurons that have crossing projections to the spinal cord are shown on the *left* and those with ipsilateral projections are shown on the *right*. In the brain, all of them are present bilaterally. Most of the types represent single identified cells on each side of the median plane; others (e.g., *RoL1* neurons) are present in clusters. Designation of segments as in Fig. 4.30; except for the Mauthner cell, the neurons are named according to their positions and axonal projections. *Bar,* 25 μm (reproduced with permission from Metcalfe et al. 1986)

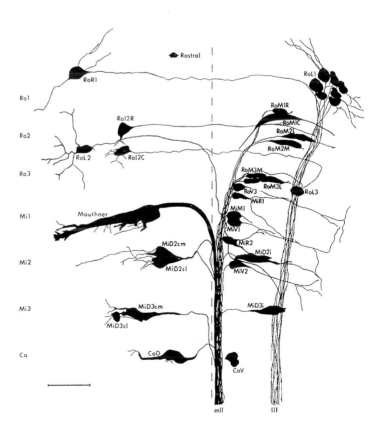

Fig. 4.31). It has been demonstrated that even in the adult brain, the three segmentally homologous reticulospinal neurons just discussed exhibit strikingly similar morphologies (Lee and Eaton 1991). In other families of reticulospinal neurons as well, the iterated cells are nearly exact copies of one another, and the same holds true for the segmentally arranged spinal primary somatomotor neurons already discussed.

After the first neurons begin to differentiate in each rhombomere others are added in the same segmental pattern, including other reticulospinal neurons (Mendelson 1986b) and neurons of other types (Trevarrow et al. 1990).

Segmentally arranged neurons have not been observed in the early embryonic spinal cord of the chick. It has been suggested that during the course of vertebrate evolution, as brain centres for movement control became progressively more dominant over local spinal reflex circuits, intrinsic spinal cord segmentation disappeared (Keynes et al. 1990). Repetitive segmental patterns of differentiation have been observed, however, in the rhombencephalon of the chick. As in teleosts the first neurons to appear and extend axons in this brain part are those of the reticular formation. Although in the chick the reticular neurons are small and numerous and cannot be individually identified, it has been demonstrated that they appear in an alternate manner within the confines of the rhombomeres; at stages HH-11–12 reticular neurons appear in the even-numbered rhombomeres, whereas about a stage later they appear in the odd-numbered rhombomeres (Lumsden and Keynes 1989). As will be discussed below, the formation and disposition of motoneurons in the brain of the chick also conforms to the rhombomere pattern.

A segment-related pattern of early neuronal differentiation has also been observed in the rostral part of the chick brain. Using AChE as a marker for early-differentiating neuroblasts, Puelles et al. (1987b) studied the neurogenesis in the forebrain and midbrain in this animal. They found that the initial appearance of neuroblasts evolves in parallel with the neuromeric segmentation of these regions. Early neuroblasts were found to appear as separate, distinct groups within specific matrix territories at the centre of transversely oriented neuromeric segments. Grossly, however, these groups constitute together a longitudinal zone extending from the tegmentum of the midbrain to the post-chiasmatic region (Fig. 4.32). Puelles et al. (1987b) noticed that the primitive prosencephalon subdivides early into a diencephalic and a secondary prosencephalic segment. Whereas the diencephalic segment further subdivides into synencephalic and parencephalic

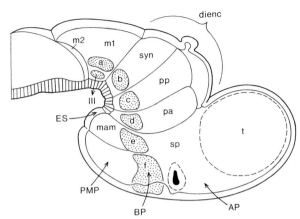

Fig. 4.32. Neuromeres in relation to early neuronal differentiation in the rostral part of the brain of the chick, as observed by Puelles et al. (1987b). According to these authors, the early embryonic prosencephalon subdivides into a diencephalic and a secondary prosencephalic segment. Three neuromeres develop within the diencephalic domain, but the secondary prosencephalon does not show any neuromeric subdivision. Distinct groups of early-differentiating neuroblasts manifest themselves at the centre of the neuromeres. Grossly, these groups constitute together a longitudinally oriented basal plate zone. Further explanation in the text. *AP*, Alar plate; *BP*, basal plate; *dienc*, diencephalon; *ES*, epichordal strip; *m1,2*, mesomeres; *mam*, mamillary region; *PMP*, paramedian plate; *pa*, parencephalon anterius; *pp*, parencephalon posterius; *sp*, secondary prosencephalon; *syn*, synencephalon; *t*, telencephalon; *tub*, tuberal region (modified from Puelles et al. 1987b, Fig. 20)

neuromeres, and the latter splits again into a parencephalon anterius and posterius, the secondary prosencephalon shows, according to Puelles et al. (1987b), no clear-cut neuromeric subdivision. As indicated in Fig. 4.32, this secondary prosencephalon encompasses the telencephalic hemispheres and the preoptic region, as well as the hypothalamus.

It is remarkable that the groups of early-differentiating neurons fit into a transverse neuromeric as well as into a longitudinal zonal pattern. In this respect they resemble the migration areas of Bergquist and Källén (1954). It is conceivable that centres of proliferation would give rise to centres of early differentiation. However, it should be emphasised that nothing is known about the nature and connections of the neurons developing from the early-differentiating neuroblasts observed by Puelles et al. (1987b) and that, hence, it would be precocious to equate them to the clusters of highly characteristic, serially homologous elements observed in the rhombencephalon and spinal cord.

4.5.5.2
Segmental Cell Lineage Restrictions

The presumption of Orr (1887, p. 335) that "the cells of one neuromere do not extend into another neuromere" has been confirmed by the recent experimental studies of Fraser et al. (1990) and Figdor and Stern (1993) on chick embryos. Fraser et al. (1990) marked single neuroepithelial cells in the rhombencephalon and were able to identify the descendants of these cells for up to six generations. When a cell was marked early, before the appearance of the rhombomere boundaries, its labelled descendants were sometimes dispersed into two rhombomeres. However, when a cell was marked later, after appearance of the boundaries, its clone was still dispersed widely within the rhombomere of origin but was now strictly confined within its boundaries.

Figdor and Stern (1993) studied the development of the diencephalon of the chick. According to their observations, four distinct neuromeres, designated by them as D1-D4, are formed in this brain part. Experiments in which either single neuroepithelial cells or small groups of such elements were marked revealed that also in the diencephalon the progeny of the labelled cells fails to cross interneuromeric borders. Injections were often observed to produce progeny that filled the entire neuromere with labelled cells, challenging both boundaries.

The experiments just discussed demonstrate that rhombencephalic as well as diencephalic neuromeres represent discrete units of polyclonal lineage restriction. The cells within each unit are designated to form only one particular part of the brain by the absence of mingling of cells of adjoining segments.

The cellular mechanisms that restrict exchange of cells between neuromeres are not known. It has been suggested that interneuromeric sheets of tightly coherent border cells may form mechanical barriers to the movements of cells between one neuromere and the next, or that the local concentration of particular compounds may impede cell movements at neuromere boundaries (cf. Lumsden 1990; Fraser 1993).

4.5.5.3
Neuromeres and the Formation of Fibre Tracts

In the embryonic central nervous system the fibres growing out from the clusters of early-differentiating cells constitute together a discrete and stereotyped scaffold of transversely and longitudinally oriented axonal bundles. Studies on the brain stem of the zebrafish (Mendelson 1986b, Metcalfe et al. 1986; Kimmel 1993, review) and the chick (Lumsden and Keynes 1989) have shown that the axons of early-differentiating reticular neurons pass medially and then turn sharply caudally to enter (or to pioneer!) the MLF. Some of these axons enter the ipsilateral MLF, others continue to grow across the floor plate before turning into the contralateral bundle. Analogous observations have been made on the more rostral parts of the brain of the zebrafish (Chitnis and Kuwada 1990; Wilson et al. 1990; Ross et al. 1992) and the chick (Figdor and Stern 1993). Interestingly, it has been established that many pathways in the scaffold are situated at the borders of the expression domains of putative regulatory genes (see Sect. 4.5.5.5) and that several of the transversely oriented pathways therein coincide with neuromere boundary zones. Thus Lumsden and Keynes (1989) observed that in the early embryonic rhombencephalon of the chick transversely running axons concentrate at the borders of most rhombomeres, and Figdor and Stern (1993) noticed that in the diencephalon of the same animal the boundaries between the neuromeres D1 and D4 are marked in a very early developmental stage by conspicuous fascicles of axons. Unlike the rhombencephalon, in which these early tracts become progressively more difficult to distinguish with further development, the early fascicles in the diencephalon appear to provide the paths followed by some of the major tracts found in the mature brain. According to the observations of Figdor and Stern (1993), the boundaries between D1/D2, D2/D3, D3/D4 and D4/midbrain are marked in the adult avian brain by the mammillothalamic tract, the fasciculus retroflexus, the rostral border of the posterior commissure and the caudal border of the posterior commissure, respectively. As already mentioned (Sect. 4.5.4), Gribnau and Geysberts (1985) noticed that in the mammalian brain the commissura posterior and the fasciculus retroflexus permanently indicate the location of interneuromeric boundaries (Fig. 4.28d).

4.5.5.4
Neuromeres and the Centres Giving Rise to Motor Roots of Spinal and Cranial Nerves

Throughout the vertebrate kingdom there is a simple one-to-one relationship between myotomes and spinal nerves. Neuromeres developing in register with the myotomes have been observed in many different vertebrate groups. However, groups of clearly segmentally arranged motoneurons have been observed so far only in *Amphioxus* and in early larval lampreys and teleosts (cf. Sect. 4.5.1). It has been experimentally established that in urodeles (Lehmann 1927; Detwiler 1934) and in the chick (Keynes and Stern 1984, 1985) the spinal cord is not intrinsically segmented with respect to motor axon outgrowth.

Data concerning the relationship between rhombomeres and the development of cranial nerve nuclei and their roots are scant in the older literature. It has already been mentioned that according to Neal (1918), in embryos of the shark *Squalus acanthias* the differentiation of the rhombomeres occurs independently of connection with cranial nerve roots and nuclei (Fig. 4.17).

Lumsden and Keynes (1989; cf. also Lumsden 1990; Keynes and Lumsden 1990; Keynes et al. 1990) recently assessed the relation between the pattern of neurogenesis of the cranial branchiomotor nerves and that of the rhombomeres in the chick by the local application of lipid-soluble carbocyanine dyes (DiI and DiO) to individual cranial nerve roots; the fluorescent dye diffuses in the axon membranes to the parent cell bodies, allowing the early motor nuclei to be identified. Their results are summarised below (cf. Fig. 4.33).

The somata of individual cranial nerves have a precise relationship to specific rhombomeres. The motor nerve roots of the first three branchial arches (V, VII and IX) derive from early cell masses that are confined within rhombomeres 2, 4 and 6, respectively. Subsequently, each nerve root is augmented by neurons developing in the caudally adjoining rhombomere. Due to this extension, each consecutive nerve becomes related to a specific consecutive pairing of rhombomeres, each pair lying in register with an adjacent branchial arch. By then the motor nuclei of V, VII and IX occupy serially adjacent positions along the rostrocaudal axis, later forming more or less continuous columns of branchiomotor neurons. By contrast, the somatomotor nuclei remain discontinuous throughout their development. These nuclei also have a constant relationship with specific rhombomeres. Neurons of the fourth nerve are situated in rhombomere 1; those of the sixth span rhombomeres 5 and 6, whereas the hypoglossal nucleus lies in rhombomere 8.

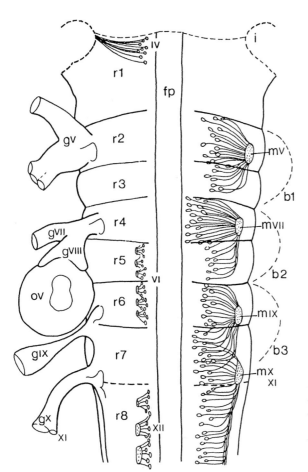

Fig. 4.33. The rhombencephalon of a 3-day-old chick embryo, based on neurofilament-stained and tracer-injected preparations. The relationship of cranial sensory ganglia (*gV–gX*), branchiomotor nuclei, somatomotor nuclei (*IV–XII*) and the combined roots of the sensory and branchial motor nerves (*mV–mXI*) to the rhombomeres (*r1–r8*) and branchial arches (*b1–b3*) is shown. *ov*, Otic vesicle; *fp*, floor plate; *i*, isthmus rhombencephali (reproduced from Lumsden and Keynes 1989 Fig. 4)

4.5.5.5
Segmental Expression of Putative Regulatory Genes

During the past decade considerable progress has been made towards understanding the molecular basis of pattern formation in the CNS. Clues as to how regional identity is specified have come from studies on the development of the fruit fly, *Drosophila melanogaster*. Genes which specify segment identity in this animal are called homeotic selector or HOM genes. Many of these genes have been found to include a sequence of 180 base pairs, called the homeobox, encoding a DNA-binding motif: the homeodomain (Gehring 1987). In *Drosophila* the homeobox-containing HOM genes are clustered in two so-called HOM complexes in the genome, termed the antennapedia complex (ANT-C) and the bithorax complex (BX-C). The

ANT-C is responsible for the pattern of differences between anterior segments, while the BX-C contains genes specifying the differences between posterior segments. The spatial order of the homeobox genes in the ANT-C and BX-C clusters is the same as the order in which the genes are expressed along the anteroposterior body axis. An important feature of this gene expression is that it occurs in sharply defined, segment-restricted patterns. It is believed that all these segment-specifying genes both in the ANT-C and in the BX-C have arisen from a common ancestral homeobox-containing gene. Remarkably, in several vertebrates, including the zebrafish, the frog, the mouse and man, numerous homeobox-containing genes are found, and among these a special group has been detected that, by virtue of a set of striking parallels with the insect HOM genes, may also be called HOM genes. These

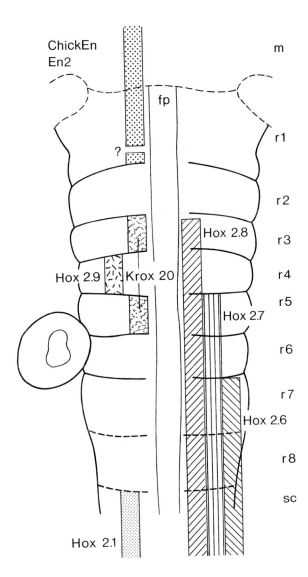

Fig. 4.34. The domains of expression of some genes in the rhombencephalon. For explanation see text (modified from Wilkinson et al. 1989b)

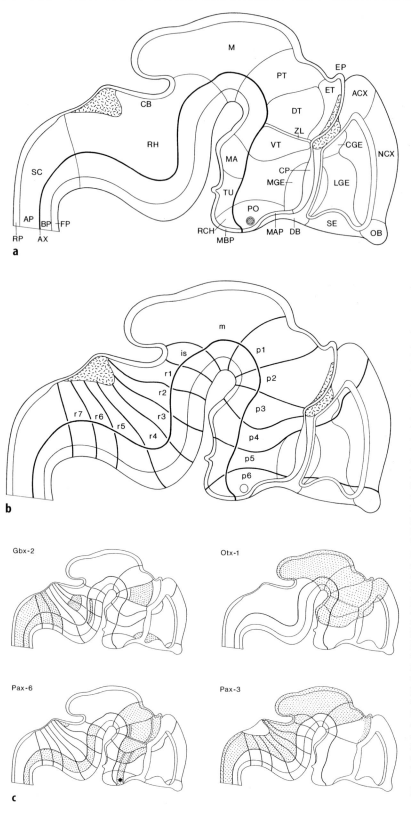

Fig. 4.35a–c. The interpretation of the parts of the brain of a 12.5-day-old mouse embryo according to Bulfone et al. (1993) and Puelles and Rubenstein (1993). The brain is shown in medial view. The medial wall of the telencephalon is opened to show the ganglionic eminences. The site of the optic stalk is indicated by a *hatched circle* in **a**, by an *open circle* in **b**, and by a *black circle* in **c**. The longitudinal and transverse delineations are based largely on the expression patterns of a number of genes. **a** Four longitudinal zones, floor plate (*FP*), basal plate (*BP*), alar plate (*AP*) and roof plate (*RP*), extend from the spinal cord to the rostral limit of the prosencephalon; the *thick line* demarcates the alar and basal plates and defines the longitudinal axis of the brain; the boundaries of the main divisions of the brain and the anlagen of a number of cell masses are indicated. **b** Transverse, neuromeric subdivision of the brain; the diencephalon contains three neuromeres, *p1–p3*; the secondary prosencephalon is likewise composed of three neuromeres, *p4–p6*, all of which extend from the hypothalamus into the telencephalic hemispheres. **c** The expression patterns of some of the genes on which the longitudinal and transverse delineations indicated in **a** and **b** are based. *ACX*, Archicortex; *CB*, cerebellum, *CGE*, caudal ganglionic eminence; *CP*, commissural plate; *DB*, nucleus of the diagonal band; *DT*, dorsal thalamus; *EP*, epiphysis; *ET*, epithalamus; *is*, isthmic neuromere; *LGE*, lateral ganglionic eminence; *M*, mesencephalon; *m*, mesomere; *MA*, mammillary area; *MAP*, median alar plate; *MBP*, median basal plate; *MGE*, medial ganglionic eminence; *NCX*, neocortex; *OB*, olfactory bulb; *p1–6*, prosomeres; *PO*, preoptic area; *PT*, pretectum; *RCH*, retrochiasmatic area; *RH*, rhombencephalon; *r1–7*, rhombomeres; *SC*, spinal cord; *SE*, septum; *TU*, tuberal hypothalamus; *VT*, ventral thalamus; *ZL*, zona limitans inter-thalamica

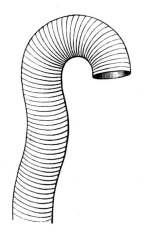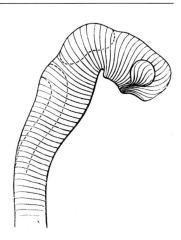

Fig. 4.36. The curvature of the early embryonic brain: **a** simple tube model; **b** brain of a human embryo, reportedly 3 weeks old (reproduced from His 1888, Figs. 12, 13)

are grouped into a small number of tightly linked clusters or HOM complexes. In the mouse at least four such clusters, named Hox-1, Hox-2, Hox-3 and Hox-5, have been discovered. Related genes in the *Drosophila* and murine clusters have extended sequence identity and are organised in the same order along the chromosome. Moreover, there is a simple correlation between the position of a gene in the *Drosophila* and mouse complexes and its pattern of expression along the anteroposterior axis of the embryo (Lewis 1989; Wilkinson and Krumlauf 1990; Krumlauf et al. 1993).

Regional differentiation of the vertebrate central nervous system apparently employs some of the same mechanisms that mediate *Drosophila* development. In situ hybridisation studies have shown that in the mouse each of the Hox genes is expressed in a domain which extends from the caudal end of the neural tube to an anterior limit in the spinal cord or hindbrain, and that in the latter the expression boundaries of these genes coincide precisely with the boundaries of rhombomeres (Wilkinson et al. 1989b). Using the presence of the otocyst adjacent to rhombomeres r5 and r6 as a reference, it was found that the boundaries of expression of the Hox-2.6, -2.7 and -2.8 genes are at the anterior limits of r7, r5 and r3 in the 9.5-day-old embryo, and expression thus occurs at two-segment intervals. By contrast, Hox-2.9 expression appeared to be restricted to a single rhombomere, r4 (Fig. 4.34). These results suggest that Hox genes could play a part in specifying segment phenotype in the neural tube. Further evidence for segmentation of the hindbrain was found by analysis of the expression of the so-called zinc-finger-encoding gene Krox-20, which in the mouse embryo was found to be restricted to two alternate rhombomeres, r3 and r5 (Wilkinson et al. 1989a). The mouse and chick homologues of the

Drosophila segment polarity engrailed gene En-2 (Davis et al. 1988) and ChickEn (Gardner et al. 1988) are expressed more rostrally, in a domain that includes the caudal midbrain and rostralmost hindbrain, probably terminating at the r1-r2 boundary (Fig. 4.34).

Kimmel (1993) reviewed data indicating that in the zebrafish several genes express in patterns that are more or less clearly related to the neuromeres. Thus, the zebrafish paired-box gene pax[zf-a] comes to expression in two separate rostral and caudal domains along the neuraxis. The rostral domain encompasses the dorsal part of the future diencephalon, whereas the caudal domain extends throughout most of the rhombencephalon anlage but is specifically absent in most of the first rhombomere. Remarkably, the zebrafish homologue of the zinc-finger gene Krox-20 appeared to be expressed in the fish in exactly the same rhombomere precursor regions (r3 and r5) as in the mouse. In relation to this observation Kimmel (1993, p. 712) stated: "The conclusion seems inescapable that, as in the mouse, this gene is part of a regulatory network that establishes the pattern of subdivision of the early hindbrain primordium. One would now also suspect that the regulatory network itself must be highly conserved among all vertebrates."

Finally, reference should be made to the recent work of Bulfone et al. (1993) and Puelles and Rubenstein (1993). Bulfone et al. (1993) examined the expression patterns of four genes, denoted by them as potential regulators of development, in the brains of 12.5-day-old mouse embryos. Three of these genes, Dlx-1, Dlx-2 and Gbx-2 are known to encode homeodomain-containing proteins, whereas the fourth, Wnt-3, encodes a putative secreted differentiation factor. They found that most of these genes are expressed in spatially restricted longitudinal or transverse domains. Combination of their findings with data derived from classical neuroembryology led them to propose the following model of embryonic brain organisation (Fig. 4.35).

In the embryonic spinal cord four longitudinal zones are present; from dorsal to ventral they are roof plate, alar plate, basal plate and floor plate. These zones extend rostrally throughout the brain. The boundary between alar plate and basal plate defines the longitudinal axis of the brain. It follows the curvatures of the neural tube and crosses the median plane in the postoptic region. The floor plate extends to the postoptic region, whereas the commissural plate, i.e. the future site of the hippocampal commissure, corpus callosum and anterior commissure, represents the most rostral part of the roof plate. The most rostral part of the basal plate is formed by the basal region of the hypothalamus. The entire telencephalon is a derivative of the alar plate. The left and right alar and basal plates are directly continuous with each other across the median plane. The median basal plate forms the rostral limit of the retrochiasmatic area, whereas the lamina terminalis represents the median alar plate (Fig. 4.35a).

Neuromeric subdivisions are oriented perpendicular to the longitudinal axis of the brain (Fig. 4.35b). Seven rhombomeres, one isthmic neuromere and one mesomere are distinguished. The embryonic prosencephalon is divided into two proneuromeric transverse domains, the diencephalon and the secondary prosencephalon. Each of these regions is subdivided into three hypothetical neuromeres. The diencephalic neuromeres p1, p2 and p3, correspond to the synencephalon, the parencephalon posterius and the parencephalon anterius. The dorsal thalamus develops from the wall of the parencephalon posterius, whereas the ventral thalamus is a derivative of the parencephalon anterius. These two thalamic structures are separated from each other by the cell-free zona limitans intrathalamica. The three neuromeres p4, p5 and p6, which develop from the wall of the secondary prosencephalon, all extend from the hypothalamic region into the telencephalic hemispheres. The principal components of these three hypothetical prosomeres are:

p4: Mammillary area, posterior entopeduncular area, supraoptic/paraventricular area, eminentia thalami, caudal ganglionic eminence and archicortex

p5: Infundibulum, tuberal region, anterior entopeduncular area, medial ganglionic eminence and neocortex

p6: Retrochiasmatic area, suprachiasmatic area, preoptic area, nucleus of the diagonal band, septum and olfactory bulb

In a subsequent study, Puelles and Rubenstein (1993) transferred the mapping of 45 different genes, compiled from the literature, to the model just discussed (Fig. 4.35c). They found that the domains of expression of these genes consistently fall within the postulated areas and respect their hypothesised transverse and longitudinal boundaries.

With regard to the model of the embryonic brain just discussed, the following remarks are in order:

1. The idea that appreciation of the strong curvature of the neural tube is of paramount importance for a sound interpretation of the parts of the embryonic brain is correct and has already been emphasised by His (1888; Fig. 4.36).

2. The neuromeric subdivision of the rhombencephalon, mesencephalon and diencephalon tallies essentially with models proposed by most other investigators.

3. Transversely oriented bandlike formations extending from the hypothalamus into the telencephalic domain have also been observed by Bergquist and Källén (1954; Fig. 4.20). However, whereas according to the authors last mentioned the boundary between the pallium and the subpallium coincides with the boundary between two transverse bands or migration areas, in the model of Bulfone et al. (1993) and Puelles and Rubenstein (1993), the boundary between the basal ganglion and the pallium is oriented perpendicular to the hypothetical prosomeres (4–6). According to the present author's opinion, the boundary between the archicortex and the neocortex and that between the septum and the primordial basal ganglia are oriented parallel to the pallio-subpallial boundary and certainly not perpendicular to it, as suggested by the model.

4.5.6
Neuromery: Summary and Concluding Remarks

The entire neural tube is, during a certain phase of its development, differentiated into neuromeres. Neuromeres may be characterised as follows:

1. They present themselves as bulges separated from each other by external constrictions and internal ventricular ridges.

2. Their constituent matrix cells and early neuroblasts are oriented radial to the ventricular surface of the neuromere.

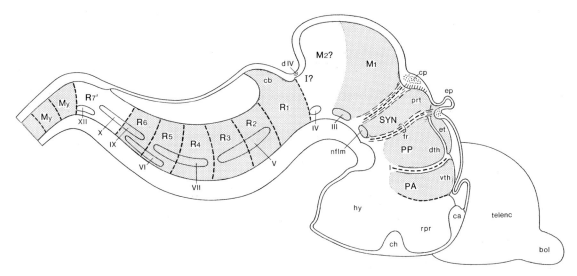

Fig. 4.37. Neuromeres in the vertebrate brain; regions where the neuromere pattern can be clearly recognised are *stippled*; regions where uncertainties exist are left *white*. *bol*, Bulbus olfactorius; *ca*, commissura anterior; *cb*, cerebellum; *ch*, chiasma opticum; *cp*, commissura posterior; *dth*, dorsal thalamus; *ep*, epiphysis; *et*, epithalamus; *fr*, fasciculus retroflexus; *hy*, hypothalamus; *I*, isthmal neuromere(s); *M1,2*, mesomeres; *my*, myelomeres; *nflm*, nucleus (interstitialis) fasciculi longitudinalis medialis; *PA*, parencephalon anterius; *PP*, parencephalon posterius; *prt*, area pretectalis; *R1–6*, rhombomeres; *R7'*, enlarged rhombomere according to Vaage (1969); *SYN*, synencephalon; *telenc*, telencephalon; *vt*, velum transversum; *vth*, ventral thalamus; *zl*, zona limitans interthalamica; *III–XII*, motor cranial nerve nuclei

3. They represent centres of proliferation, early differentiation and migration.
4. They represent units of cell lineage restriction.
5. Their boundaries may be marked by early-developing fibre systems.
6. Many of them have fixed relations with peripheral structures.
7. Their boundaries often correlate with the boundaries of the domains of expression of putative regulatory genes.

Repetitive patterns of differentiation leading to the formation of characteristic individual cells or groups of cells reveal that in cephalochordates, cyclostomes and teleosts the spinal cord segments or myelomeres are serially homologous. The same probably holds true for a number of rhombomeres in all vertebrates.

In some parts of the brain, fixed numbers of neuromeres are found which most probably are individually homologous in all vertebrates. However, because these brain parts alternate with smaller or larger regions in which the neuromere pattern is uncertain, a model of the brain comprising a fixed number of neuromeres which is valid throughout the vertebrate kingdom cannot be presented as yet. The (relative) certainties and the uncertainties pertaining to the neuromere patterns in the various regions of the brain may be summarised as follows (Fig. 4.37):

In all vertebrates the bulk of the hindbrain is most probably composed of six consecutive, individually homologous rhombomeres. The first of these rhombomeres (rh1) is usually larger than the remaining ones. The dorsal part of the rh1 gives rise to the cerebellum. It may be mentioned parenthetically that physiological studies as well as experiments with certain cell-specific marker molecules have revealed that the Purkinje cells in the cerebellum are organised in transverse and parasagittal compartments, and that some of these compartments correspond to the domains of expression of certain homeobox genes (cf., e.g., Oberdick et al. 1993). The pattern according to which the avian rhombomeres 2–6 contribute to individual cranial nerve nuclei (Fig. 4.33) may well be identical to that in other vertebrates.

The total number of rhombomeres is probably larger than six in all vertebrates, but for many groups reliable data are lacking. Hanneman et al. (1988; Fig. 4.30) found nine rhombomeres in the zebrafish, whereas in the chick (Vaage 1969) and in primates (Gribnau and Geysberts 1985: Fig. 4.28a; O'Rahilly et al. 1989; Fig. 4.29) seven rhombomeres were counted. According to Vaage (1969; Fig. 4.27), in amniotes the most caudal rhombomere (rh7) merges with a variable number (3–7) of myelomeres. The enlarged rhombomere rh7' thus formed participates in the formation of the motor nuclei of X, XII and of the cranial part of XI.

The mesencephalon is, according to most authors, composed of two mesomeres, m1 and m2, but the only well-established facts seem to be that (a) the boundary between m1 and the most caudal prosomere is marked by the posterior commissure, and (b) the oculomotor nuclei originate from m1. The mesencephalic domain attributed to m2 varies considerably in size (cf. Figs. 4.25b, 4.28a, 4.29b, 4.32), and the situation is further complicated by the fact that, according to several authors, one (O'Rahilly et al. 1989; Bulfone et al. 1993) or two (Vaage 1969; Gribnau and Geysberts 1985) isthmal neuromeres are intercalated between the mesomeres and the rhombomeres. The trochlear nucleus arises, according to Vaage (1969: Fig. 4.27), from an isthmal neuromere, but according to Lumsden and Keynes (1989: Fig. 4.33), from the first rhombomere.

In most vertebrates the changes that take place in three diencephalic neuromeres, the synencephalon, the parencephalon posterius and the parencephalon anterius, can be traced throughout development. As indicated in Fig. 4.37, the anterior and posterior boundaries of the synencephalon are marked by the fasciculus retroflexus and the posterior commissure, respectively, whereas the anterior and posterior parencephela are largely separated from each other by a cell-free zona limitans. The dorsal part of the synencephalon is occupied by the pretectal area, whereas its ventral part contains the nucleus (interstitialis) of the medial longitudinal fasciculus. The epithalamus and the dorsal thalamus are derivatives of the parencephalon posterius, whereas the ventral thalamus arises from the parencephalon anterius.

Great uncertainty exists with regard to the number and configuration of the neuromeres in the most rostral part of the neural tube, i.e. the region occupied by the hypothalamus, the preoptic area and the telencephalic hemispheres. Two early bilateral evaginations – the optic and the hemispheric – greatly disturb the segmental pattern in this region. Numerous previous workers (e.g. Hill 1900; Fig. 4.16; von Kupffer 1906; Fig. 4.15) considered the entire telencephalon to be the product of a single prosomere, a view which can also be found in the more recent literature (e.g., Gribnau and Geysberts 1985: Fig. 4.28a). Vaage (1969: Fig. 4.25b) held that two prosomeres give rise to the telencephalon, but according to others, the primordia of the hypothalamus and the telencephalon constitute together a continuum derived from a single (Coggeshall 1964: Fig. 4.28c), two (Bergquist and Källén 1954, 1955: Fig. 4.20), or three prosomeres (Bulfone et al. 1993: Fig. 4.35b). Alternatively, the hypothalamus has been considered a product of the parencephalon

anterius (Keyser 1972: Fig. 4.28c; Gribnau and Geysberts 1985: Fig. 4.28b). It must be concluded that the neuromeric organisation of the most rostral part of the neural tube is unclear and offers no reliable landmarks for a 'natural' morphological subdivision of this region. However, the generally accepted subdivision of the prosencephalon into telencephalon and diencephalon should nevertheless be maintained. As mentioned earlier (Sect. 4.3), Johnston's (1909) proposal to define the tel-diencephalic boundary as a plane passing through the velum transversum dorsally and the optic chiasma ventrally offers a satisfactory practical solution to this problem.

4.6
Longitudinal Zones

4.6.1
Investigations of His

According to the classical studies of His (1888, 1891, 1893a,b), which were carried out mainly on human embryonic material, the lateral walls of the CNS consist throughout their extent of two longitudinal zones or plates: the dorsally situated alar plate and the ventral basal plate. His pointed out that the former contains the primary sensory centres, whereas the primary motor centres are found in the latter. Moreover, he remarked upon the fact that the boundary between the alar and basal plates is marked on the ventricular side by a furrow, for which he coined the term 'sulcus limitans'. The exact position of the interface between the two plates was not indicated by His.

4.6.2
Nerve Components:
Gaskell and the 'American School'

By extrapolating certain results of an analysis of the spinal roots in the dog, Gaskell (1886, 1889) made an important contribution to the clarification of the structure of the brain stem. He found that the spinal dorsal roots are composed of fibres of two kinds, somatic sensory and splanchnic sensory, and that ventral roots likewise contain two classes of fibres, splanchnic motor and somatic motor. According to Gaskell, the fibres of these four categories are centrally connected with specific zones of the spinal grey matter. Gaskell conjectured that the four classes of fibres which make up the spinal nerves and their roots are also represented in the cranial nerves. Employing data from the literature, he separated the various cranial nerves into their components and arrived at the conclusion that the groups of cells with which these components are connected are continuations of the corresponding cell zones found at the spinal level (cf. the tabular survey on p. 76 of Gaskell's 1886 paper and Figs. 1 and 2 on plate XVIII of his 1889 paper).

Although Gaskell did not substantiate his nerve-component theory, his ideas had a great impact on a number of American neuroanatomists, among whom Strong, Johnston and Herrick should be especially mentioned. These investigators considered Gaskell's theory a most important clue to a description of the brain in functional terms or, as Herrick (1943) put it: "a physiological analysis by an anatomical method." With this central idea in mind, Osborn (1889) and Strong (1895) studied the cranial nerves and their central and peripheral relations in amphibia. Herrick (1899, 1905, 1906, 1907) analysed the nerve components in various teleosts, whereas Johnston (1902a,b, 1905) selected the sturgeon and the lamprey for similar analyses. As far as the central relations of the cranial nerves and the cerebral zonal pattern in general are concerned, the work of these authors and their numerous followers (together they came to be called the American School of comparative neurologists) can be summarised as follows.

In the brain stem the component parts of the cranial nerves converge into four longitudinal columns, which are rostral continuations of zones of spinal grey. Most of these columns, however, contain specialisatons ("neomorphs in the head", Herrick 1899, p. 175) which do not occur at the spinal level. These columns are, from dorsal to ventral, the somatic sensory, the visceral sensory, the visceral motor and the somatic motor.

The *somatic sensory column* includes a special somatic sensory zone and a general somatic sensory zone. The special somatic sensory zone contains the vestibular, acoustic and lateral line nerve centres. The general somatic sensory zone is constituted by the mesencephalic, princeps and descending sensory nuclei of V. The latter is directly continuous with the dorsal horn of the spinal cord.

The *visceral sensory column* is constituted by the centre of termination of the fasciculus solitarius or fasciculus communis, which latter consists of the afferent components of VII, IX and X. This column, the visceral or vagal lobe of lower vertebrates, is morphologically not divisible into separate nuclei, yet it receives fibres of two different types, namely special visceral afferent fibres, which carry impulses from the taste buds, and general visceral afferent fibres which innervate the viscera. The last-mentioned fibres are also represented at the spinal level, although their exact site of termination in the cord is still a matter of conjecture. It was originally believed that the spinal visceral afferent fibres terminated in Clarke's column (Gaskell 1886, 1889; Johnston 1902c).

The *visceral motor column* contains the cells of origin of (a) preganglionic fibres of the general visceral system and (b) special visceral motor or branchiomotor fibres. The former terminate in autonomic ganglia, where they synapse with elements supplying the glands and musculature of the viscera including the heart and blood vessels. Such general visceral efferent fibers contribute to roots III, VII, IX and X. The special visceral motor fibres innervate, via V, VII, IX and X, the striated muscles of the gill arches and their derivatives. The general visceral efferent elements in the brain stem are considered equivalent to the cells in the lateral horn of the spinal cord.

The *somatic motor column* is regarded as the direct forward continuation of the ventral horn region of the spinal cord. It is constituted by the nuclei of III, IV and VI, which innervate the external eye muscles. In higher vertebrates, the somatic motor column also contains the nucleus of XII, which supplies the musculature of the tongue.

Although the members of the American School of comparative neurologists confined their study of the 'functional columns' to lower vertebrates, they were convinced that their analyses had revealed a basic plan of structure, prevailing throughout the vertebrate kingdom. Herrick (1913) was the first who attempted to apply this zonal subdivision to the human nervous system.

According to the original concept of Gaskell (1886, 1889), the longitudinal columns in the brain stem were thought to consist exclusively of primary afferent and efferent centres. However, various later authors have deliberated about to what extent centres of secondary or higher order also fit into the longitudinal zonal pattern. As regards the rhombencephalon, Johnston (1902c) felt that this part of the brain was readily reducible to the four columns, although he made an exception for the inferior olive. Herrick (1913) was of the opinion that grisea of higher order, which are directly related to the primary afferent and efferent centres, develop and elaborate within the four longitudinal columns. Similar ideas have been expressed by several other authors. Thus Bartelmez (1915) and Tuge (1932), who studied a teleost and a turtle, respectively, incorporated the nuclei motorii tegmenti within the somatic motor column. These nuclei are centres of higher order, comparable to the medial reticular formation of mammals. Herrick (1913), like Johnston (1902c), did not include all of the rhombencephalic cell masses into the four longitudinal columns. He pointed out that a band of associational tissue is present between the primary sensory and motor areas. In his later work, Herrick (1930, 1948) referred to this strip of tissue as the reticular formation or the intermediate zone. It is important to note that in his extensive studies of the urodele brain (1914, 1930, 1948) Herrick played down the significance of the longitudinal zonal pattern to some extent. He noted that in the rhombencephalon of these animals the primary afferent fibres bifurcate and constitute clearly defined, longitudinally arranged nerve components. However, the neurons which form the terminal nuclei of these root fibres appear not to be segregated in relation to the several incoming sensory nerves. Thus Herrick observed that the dendrites of many neurons engaged terminals from sensory fibres of very different sources, e.g. from the skin, from the taste buds and from the internal ear. From these observations Herrick concluded that the sensory grey in the rhombencephalon of urodeles cannot be subdivided into longitudinal functional columns. Instead, he advocated a subdivision of the rhombencephalon into a dorsal, sensory zone, a ventral, motor zone and, between these, the intermediate zone.

It is appropriate at this juncture to mention briefly the views of the American School on the remaining parts of the brain stem, i.e. the cerebellum and the midbrain. As regards the cerebellum, somewhat different opinions have been expressed. Johnston (1902a,b) held that this part of the brain was a derivative of the 'acusticum', i.e. the special somatic sensory zone. Larsell (1967) was also of the opinion that the cerebellum is reducible to the rhombencephalic zonal pattern. However, according to the latter author, the cerebellum is of dual origin: the vestibulo-lateral lobe (lobus flocculonodularis of higher forms) is a direct continuation of the special somatic afferent zone, whereas the corpus cerebelli is derived from the general somatic sensory zone (Fig. 4.38). In one of his earlier papers, Herrick (1906) suggested that the somatic sensory zone included the cerebellum, but later (Herrick 1948) he expressed the opinion that the corpus cerebelli represents a suprasegmental structure which should be excluded from the longitudinal zones.

Although the members of the American School clearly appreciated that the mesencephalon contains direct forward extensions of some functional zones, they were reluctant to reduce this part of the brain in its entirety to the four longitudinal columns observed at the rhombencephalic level (cf. Johnston 1902c, pp. 98–99). Herrick (1948) emphasised that the mesencephalon belongs to the higher parts of the brain, which differ in their organisation radically from the rhombencephalon. Yet he maintained that the three zones of the rhombencephalon: the dorsal, intermediate and ventral zones, can also be recognised at the midbrain level.

Before closing this brief survey of the views of the American School, we should look at the way in which the members of this school delimited the longitudinal zones or columns from each other. Herrick (1913) attributed great significance to the sulcus limitans of His: "This sulcus limitans is the most fundamental landmark of the brain, for it separates on each side a ventrolateral motor plate ...from a dorsolateral sensory plate..." (Herrick 1913, p. 285). In the same treatise Herrick indicates

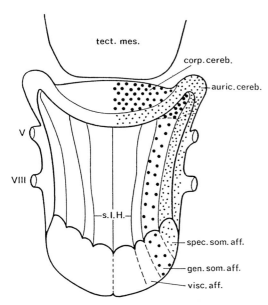

Fig. 4.38. Diagram showing the afferent zones in the rhombencephalon and the way in which these structures, according to Larsell (1967), participate in the formation of the cerebellum. *auric. cereb.*, Auricula cerebelli; *corp.cereb.*, corpus cerebelli; *gen.som.aff.*, general somatic afferent zone; *s.l.H.*, sulcus limitans of His; *spec.som.aff.*, special somatic afferent zone; *tect.mes.*, tectum mesencephali; *visc.aff.*, visceral afferent zone (redrawn from Jansen 1969)

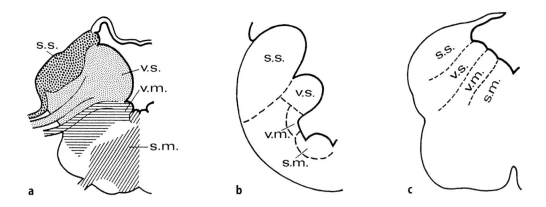

Fig. 4.39a–c. Transverse sections through the rhombencephalon to illustrate the arrangement of the longitudinal columns. a Teleost (redrawn from Herrick 1913, Fig. 883); b sturgeon (redrawn from Johnston 1902c, Fig. 1); c man; delineation of the functional columns as shown in many textbooks of neuroanatomy. *s.m.*, Somatic motor column; *s.s.*, somatic sensory column; *v.m.*, visceral motor column; *v.s.*, visceral sensory column

that in lower vertebrates "... the functionally defined longitudinal columns appear anatomically as well-defined ridges in the wall of the fourth ventricle ..." (Herrick 1913, p. 284). Figure 4.39a shows these relations depicted by Herrick (1913) in a bony fish. Johnston does not discuss the delimitation of the functional columns. However, in two of his papers(1901, 1902c) he presents some diagrams in which the position of these columns is indicated. One of these, showing the columns in the rhombencephalon of the sturgeon, is represented in Fig. 4.39b. It will be noted that in this picture the boundaries of some of the columns do not coincide with ventricular sulci. In more recent years it has become customary to indicate the boundaries of the functional columns as more or less radially oriented curves, which originate from the deepest point of ventricular grooves (cf. Figs. 4.39c, 4.38). It appears that this way of indicating the interfaces of the columns is not based on any direct observations.

4.6.3
Longitudinal Zones in the Brain Stem: Kuhlenbeck and His Followers

Attention may be directed now to the work of Kuhlenbeck (1926, 1927, 1973) and his collaborators Saito (1930) and Gerlach (1933, 1946). Kuhlenbeck (1926, 1927) agreed with the members of the American School that the brain stem fundamentally consists of four longitudinally arranged zones or columns. However, he considered the designations of the columns as somatic sensory, visceral sensory, visceral motor and somatic motor to be unwarranted generalisations. Moreover, he regarded it as

impossible, in some cases, to make a sharp distinction between somatic and visceral centres. For these reasons, Kuhlenbeck preferred a purely morphological terminology; consequently, he indicated the columns from dorsal to ventral as *Dorsalgebiet*, *Intermediodorsalgebiet*, *Intermedioventralgebiet*, and *Ventralgebiet*. The Latin equivalents of these terms: area dorsalis, area intermedioventralis, etc. have been employed by my colleagues and me in our topological analyses of the brain stem (see Sect. 4.6.5). Kuhlenbeck was of the opinion that the four longitudinal zones mentioned extend throughout the deuterencephalon (i.e. the rhombencephalon and mesencephalon). He emphasised that the boundaries of these zones are generally indicated by three distinct ventricular sulci: the sulcus intermedius dorsalis, the sulcus limitans and the sulcus intermedius ventralis (cf. Fig. 4.40b). The scheme of Kuhlenbeck (1926, 1927) was adopted by Saito (1930) in an analysis of the morphological pattern of the brain of a lamprey, and by Gerlach (1933, 1946) in similar studies on a lungfish and a shark. The latter author constructed diagrams of the structural plan (*Bauplan*) of the mesencephalon and the rhombencephalon of the two species. Those for the lungfish *Protopterus* (Gerlach 1933) are represented in Fig. 4.41. A striking feature of these diagrams is that the interfaces between the alar and basal plates and those between the various rhombencephalic columns are indicated as radially oriented lines which, on the ventricular side, take their origin from the ventricular sulci. It is noteworthy that in Kuhlenbeck's diagram (Fig. 4.40b) the boundaries of the columns likewise coincide with ventricular sulci but do not show the rigidly radial orientation indicated by Gerlach.

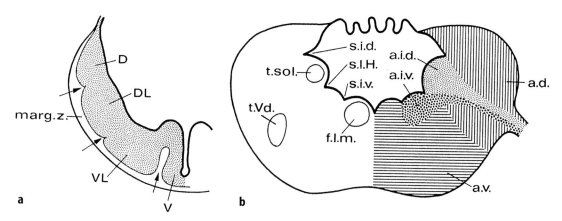

Fig. 4.40a,b. Transverse sections through the rhombencephalon to illustrate the arrangement of the longitudinal columns. **a** Young embryo of the cartilaginous fish *Torpedo*, showing that 'incisures' in the mantle layer (*arrows*) mark the boundaries of the columns (based on Hugosson 1955, Fig. 5); **b** shark (redrawn from Kuhlenbeck 1927, Fig. 92). *a.d.*, Area dorsalis; *a.i.d.*, area intermediodorsalis; *a.i.v.*, area intermedioventralis; *a.v.*, area ventralis; *D*, area dorsalis; *DL*, area dorsolateralis; *f.l.m.*, fasciculus longitudinalis medialis; marg.z., marginal zone; s.i.d., sulcus intermedius dorsalis; s.l.m., sulcus limitans of His; *s.i.v.*, sulcus intermedius ventralis; *t.sol.*, tractus solitarius; *t.V.d.*, tractus descendens nervi trigemini; *V*, area ventralis; *VL*, area ventrolateralis

4.6.4
Zonal Pattern Development: Bergquist and Källén

The embryological studies of Bergquist and Källén and their collaborators (Bergquist 1932; Bergquist and Källén 1953a,b, 1954, 1955; Bengmark et al. 1953; Hugosson 1955, 1957) have already been discussed extensively in Sect. 4.5.2. Hence, I will confine myself here mainly to those aspects of the work of these investigators which have a direct bearing on the zonal pattern of the vertebrate brain. It may be recalled that Bergquist and Källén (1953a,b, 1954) found that early in development, centres appear throughout the brain in which the proliferation of cells is more active than in the intervening parts of the wall. These centres of proliferation give rise to aggregates of migrating cells from which the grisea of the adult brain are directly derived. Extensive comparative studies revealed that the proliferation centres are arranged in a pattern that shows a remarkable consistency throughout the vertebrate kingdom. Consequently, these centres were considered to be fundamental morphological units. Bergquist (1932) designated them as *Grundgebiete*, but Källén (1951b) later introduced the term 'migration areas'. In the rhombencephalon the migration areas manifest themselves as four longitudinal cell cords or columns (Figs. 4.19, 4.20, 4.40a). The most dorsal of these ceases at the level of the isthmus, but the remaining three extend throughout the midbrain (Bengmark et al. 1953; Bergquist and Källén 1954; Hugosson 1955, 1957). These longitudinal columns

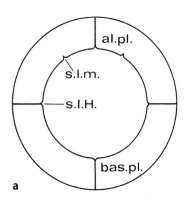

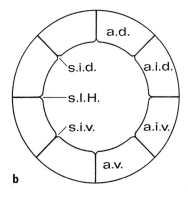

Fig. 4.41a,b. The *Bauplan* of the mesencephalon (**a**) and rhombencephalon (**b**) of the lungfish *Protopterus*. *al.pl.*, Alar plate; *bas.pl.*, basal plate; *s.l.m.*, sulcus lateralis mesencephali; for other abbreviations see Fig. 4.40 (redrawn from Gerlach 1933, Figs. 33, 34)

correspond partly to the four functional zones of the American School. However, since the Swedish authors regarded the embryonic columns as purely morphological entities, they preferred a neutral nomenclature: columnae ventralis, ventrolateralis, dorsolateralis and dorsalis. The most important differences between these embryonic columns and the so-called functional zones appear to be the following: (a) The columna ventralis corresponds largely to the somatic motor zone. However, Bengmark et al. (1953) found that the visceral motor nucleus of Edinger and Westphal also derives from this column. (b) The columna dorsolateralis is not equivalent to the visceral sensory column, since the following somatic sensory centres and regions also develop from it: the vestibular nuclear complex (Hugosson 1955, 1957; cf., however, Vraa-Jensen 1956), the main sensory nucleus of V, and the tectum mesencephali (Bengmark et al. 1953).

It is important to note that in their study of the embryonic cell columns Bergquist and Källén and their co-workers made extensive use of graphic and three-dimensional reconstructions. In the latter the outer layer of the brain wall, i.e. the marginal zone, was deleted in order to reveal the 'external surface' of the mantle layer. This external surface exhibits distinct longitudinal furrows (arrows in Fig. 4.40a), and it is these 'internal sulci' which mark the boundaries between the embryonic columns. The ventricular sulci were considered by Bergquist (1932) and Källén (1951a,b) to be of little importance for the study of the morphological pattern of the brain.

4.6.5
Topological Analysis of the Brain Stem

From the foregoing review it appears that various investigators, operating along different lines of research, have concluded that the grisea in the brain stem are arranged in four longitudinal zones or columns. According to the original concept of Gaskell, only the primary afferent and efferent nuclei display this zonal arrangement, but Herrick, Johnston and others later found that grisea of secondary or higher order also fit into the longitudinal zones. Still later, Kuhlenbeck and his co-workers advanced the idea that the entire grey matter of the brain stem is reducible to a system of four longitudinal zones. This concept was corroborated by the embryological observations of Bergquist and Källén and their co-workers, who found that the grisea in the brain stem can all be attributed to four columns of neuroblasts that manifest themselves early in development as migration areas.

As regards the delimitation of the longitudinal zones, several authors, among them Herrick and Kuhlenbeck, have emphasised that their boundaries correspond to ventricular grooves. Bergquist and Källén, on the other hand, attributed little significance to the ventricular sulci, and instead emphasised the importance of distinct 'internal sulci' in the embryonic brain, i.e. incisures in the mantle layer that mark the boundaries of the columns. It should be stressed that the exact location of the interfaces between the various zones so far have not been defined in the literature, although it has become customary to indicate them in cross-sections as more or less radially oriented curves (Figs. 4.38, 4.39c).

Although the so-called four-column concept has been found in all textbooks since the beginning of this century, study of the relevant literature reveals that many important questions regarding the structural and functional organisation of the brain stem are still open. This is simply because cross-sections do not reveal the rostrocaudal extent of sulci and of cell zones. Questions still awaiting a definitive answer include: (a) To what extent is the brain stem divisible into a motor basal plate and a sensory alar plate? (b) To what extent are the centres contained within the basal plate and the alar plate arranged in a longitudinal zonal pattern? (c) To what extent are the boundaries of such zones marked by ventricular sulci? (d) To what extent do nuclei, falling under common functional denominators, fit into a longitudinal zonal pattern? (e) Does the mesencephalon show a longitudinal zonal pattern, and if so, to what extent does this pattern correspond to that of the rhombencephalon?

In order to tackle these questions, I (Nieuwenhuys 1972, 1974) developed a procedure which makes it possible to survey the entire ventricular surface of the brain stem, with its sulci, and the underlying cell masses in a single reconstruction. This procedure involves essentially two steps: (a) Cell masses are projected upon the ventricular surface; i.e. roughly brought back to their site of origin. (b) The ventricular surface, with the outlines of the nuclei and the position of the sulci marked upon it, is flattened out; i.e. is subjected to a topological transformation. With the aid of this procedure topological analyses of the brain stems of the following species have been carried out: the lamprey *Lampetra fluviatilis* (Nieuwenhuys 1972), the cartilaginous fishes *Scyliorhinus canicula, Squalus acanthias, Raja clavata* and *Hydrolagus collei* (Smeets and Nieuwenhuys 1976; Smeets et al. 1983), the brachiopterygian *Erpetoichthys calabaricus* (Nieuwenhuys and Oey 1983), the chondrostean *Scaphirhynchus platorynchus* (Nieuwenhuys, unpublished); the holosteans *Lepisosteus osseus* (Nieuwenhuys and

Pouwels 1983) and *Amia calva* (Heijdra and Nieu-wenhuys 1994); the dipnoan *Lepidosiren paradoxa* (Thors and Nieuwenhuys 1979), the crossopteryg-ian *Latimeria chalumnae* (Kremers and Nieuwen-huys 1979), the urodele *Ambystoma mexicanum* (Opdam and Nieuwenhuys 1976), the anurans *Rana esculenta* and *Rana catesbeiana* (Opdam et al. 1976) and *Xenopus laevis* (Nikundiwe and Nieuwenhuys 1983), and the turtle *Testudo hermanni* (Cruce and Nieuwenhuys 1974).

The principal results of these analyses may be summarised as follows (Fig. 4.42):

1. In all species studied a sulcus limitans is present in the rhombencephalon, dividing this brain part into a basal plate and an alar plate.

2. The designation of the alar plate as 'sensory' and the basal plate as 'motor' is correct insofar as all primary afferent centres are situated within the former and all primary efferent centres within the latter.

3. In the rhombencephalon of most species ex-amined the grey matter is arranged in four longi-tudinal zones and in many places three sulci, i.e. the sulcus intermedius ventralis, sulcus limitans and sulcus intermedius dorsalis, mark the boundaries between these morphological enti-ties.

4. The four rhombencephalic zones may be desig-nated as area ventralis, area intermedioventralis, area intermediodorsalis and area dorsalis. These areas coincide largely, but not entirely, with the

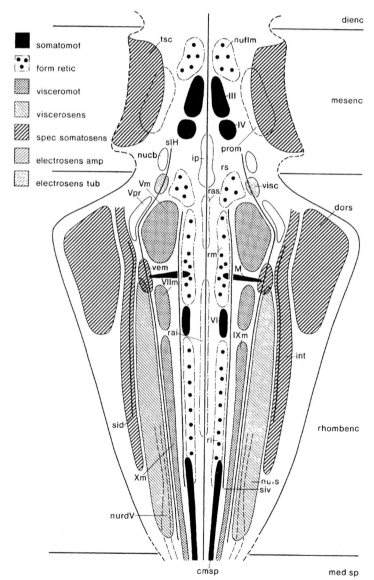

Fig. 4.42. Topological chart showing the rel-ative positions of ventricular sulci and cell masses in the brain stem of anamniotes. *cmsp,* Columna motoria spinalis; *dienc,* diencephalon; *dors,* nucleus dorsalis areae octavalateralis; *int,* nucleus intermedius areae octavolateralis; *ip,* nucleus interpedun-culus; *M,* cell of Mauthner; *med sp,* medulla spinalis; *mesenc,* mesencephalon; *nucb,* nucleus cerebelli; *nuflm,* nucleus of the flm; *nufs,* nucleus fasciculi solitarii; *nurdV,* nucleus of the radix descendens nervi trige-mini; *prom,* nucleus porfundus mesence-phali; *rai,* nucleus raphes inferior; *ras,* nucleus raphes superior; *rhombenc,* rhomb-encephalon; *ri,* nucleus reticularis inferior; *rm,* nucleus rericularis medius; *rs,* nucleus reticularis superior; *sid,* sulcus intermedius dorsalis; *SlH,* sulcus limitans of His; *siv,* sul-cus intermedius ventralis; *tsc,* torus semicir-cularis; *vem,* nucleus vestibularis magnocel-lularis; *visc,* nucleus visceralis secundarius; *III,* nucleus motorius nervi oculomotorii; *IV,* nucleus motorius nervi trochlearis; *Vm,* nucleus motorius nervi trigemini; *Vpr,* nucleus sensorius princeps nervi trigemini; *VI,* nucleus nervi abducentis; *VIIm,* nucleus motorius nervi facialis; *IXm,* nucleus moto-ris nervi glossopharyngei; *Xm,* nucleus motoris nervi vagi (modified from Nieuwen-huys and Meek 1985)

functional columns of Herrick and Johnston (Fig. 4.39). The most obvious incongruity is that the area intermediodorsalis contains, in addition to the viscerosensory nucleus of the fasciculus solitarius, two general somatosensory centres, i.e. the nuclei of the radix descendens and the principal nucleus of the trigeminal nerve, and one or more vestibular, i.e. special somatosensory centres. Thus in all fishes the conspicuous nucleus vestibularis magnocellularis lies in the medial part of the area intermediodorsalis, often even encroaching upon the adjacent area intermedioventralis.

5. Certain centres of secondary or higher order fit, with regard to both position and function, into the longitudinal zonal pattern. For example, the medial reticular formation or nucleus motorius tegmenti forms part of the somatomotor column and the nucleus visceralis secundarius, which is situated directly lateral to the sulcus limitans, represents the rostral extreme of the viscerosensory column.

6. The basal plate contains a number of evidently non-motor relay centres (not shown in Fig. 4.42). Two precerebellar nuclei, the inferior olive and the nucleus funiculi lateralis, are situated in the superficial part of the ventral and intermedioventral area, respectively. The superficial part of the intermedioventral zone harbours, in addition, the superior olive and the nucleus of the lemniscus lateralis, sensory relay centres which both belong to the octavolateral system.

4.6.6
Longitudinal Zones in the Rostral Part of the Brain

4.6.6.1
Overall Pattern

Wilhelm His (1888) subdivided the early neural tube into four longitudinal zones: floor plate, basal plate, alar plate and roof plate. He pointed out that the basal plate contains the motor centres, whereas the sensory centres are found in the alar plate, and that the boundary between these two plates is demarcated by a ventricular groove, the sulcus limitans (His 1893b). Later investigations have led to a subdivision of both the basal and alar plate into two longitudinal zones. We have seen that the four zones or columns can be clearly distinguished in the rhombencephalon (Figs. 4.38–4.42), and the question arises as to whether, and to what extent, these entities manifest themselves in the rostral part of the brain as well. This and related questions pertaining to the course of the longitudinal axis of

the brain, the location of the rostral end of the brain, and the transformation of the neural plate into the neural tube have been much debated in the literature. The ensuing survey will show that widely different answers have been given, and that we are still far removed from a consensus concerning the basic subdivision of the rostral part of the vertebrate neuraxis.

His (1893a,b) indicated that in human embryos the sulcus limitans can be traced throughout the brain and terminates in the recessus (pre)opticus. He subdivided the brain into six ringlike regions, each comprising a basal plate and an alar plate derivative (Fig. 4.43). According to His, the floor plate terminates at the infundibular recess and the median structure extending from this recess to the rostral limit of the chiasma opticum forms part of the lamina terminalis (Fig. 44a,d).

Von Kupffer (1906) remained unable to trace the sulcus limitans into the forebrain and doubted that the six ringlike structures distinguished by His (1893b; Fig. 4.43) represent 'primitive brain segments'. Von Kupffer attributed to the brain a longitudinal axis which ends in the recessus neuroporicus (Fig. 4.5c). According to him, this recess marks the boundary between the roof and the anterior wall of the prosencephalon. This anterior wall or lamina terminalis represents the permanent median closure of the brain.

In 1909 Johnston published an extensive study on the morphology of the forebrain vesicle in vertebrates. The principal results of this study may be summarised as follows:

1. The anterior boundary of the neural plate is formed by a transversely oriented terminal ridge. The optic chiasm is formed in this terminal ridge and therefore occupies the anterior border of the brain floor. These relations indicate that the chiasma region must be taken from the lamina terminalis and added to the brain floor.

2. The sulcus limitans of His marks the boundary between the alar plate and the basal plate. Throughout the CNS the alar plate is sensory, the basal plate motor. In those brain regions in which the primary sensory or motor centres are reduced or wanting, owing to reduction or absence of the peripheral organs, correlating mechanisms constitute the whole of the zone concerned. The longitudinal functional zones constitute the most fundamental division of the brain; hence, the sulcus limitans is the most important landmark in the brain. The two sulci converge at the anterior end of the brain to meet in the recessus preopticus. The meeting point marks the anterior end of the central axis of the

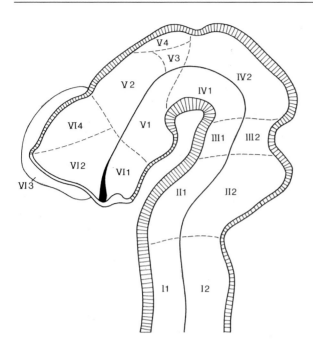

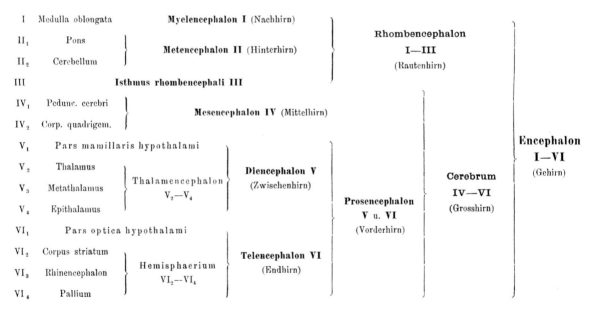

Fig. 4.43. The subdivision of the human brain as proposed by His (1893b). The diagram, showing a medial view of the brain of an approximately 1-month-old human embryo, is based on Fig. 1 of the paper mentioned

brain. (As mentioned, His was also of the opinion that the sulcus limitans entered the preoptic recess, but according to him, the left and right sulci approached each other only after the completion of a fusion of both basal plates in the median plane.)

3. The bending down of the sulcus limitans is caused by the absence of motor centres in front of the third nerve and by a strong development

of sensory and correlating centres in the rostral part of the alar plate.

4. Contrary to von Kupffer (1906), Johnston did not attribute any morphological significance to the recessus neuroporicus.

In a series of embryological studies, Kingsbury (1920, 1922, 1930) presented a new view on the spatial relations of the longitudinal zones in the rostral

Fig. 4.44a-f. The zonal pattern of the vertebrate brain (*upper figures*) and the projection of that pattern upon the neural plate (*lower figures*), according to the interpretations of (**a**) His (1893b); (**b**) Kingsbury (1922) and(**c**) Bergquist and Källén (1954). *ap,* Alar plate; *bp,* basal plate; *cd,* columna dorsalis; *cdl,* columna dorsolateralis; *cv,* columna ventralis; *cvl,* columna ventrolateralis; *f,* floor plate; *i,* infundibulum; *m,* recessus mamillaris; *n,* notochord; *p,* recessus preopticus; *r,* roof plate (**a** and **b** are modified from Kingsbury 1922, Fig. 1)

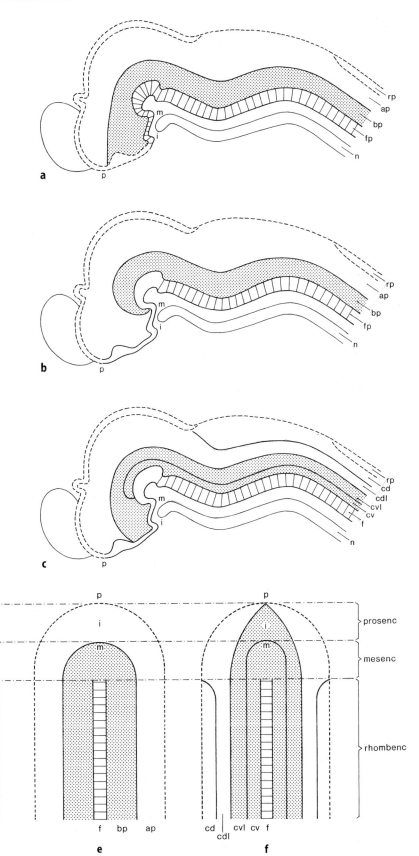

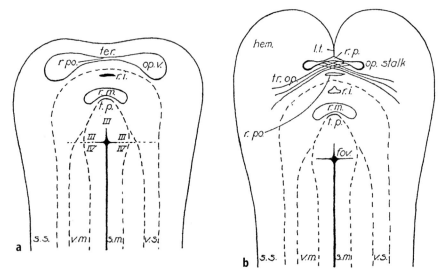

Fig. 4.45a,b. Two diagrams of the arrangements of the functional columns in the anterior portion of the neural tube, prepared by Johnston (1923, Fig. 1) based on the work of Kingsbury (1922). **a** Open neural plate; **b** neural tube represented as laid open along the dorsal seam. Further explanation in the text. *hem*, Hemisphere; *l.t.*, lamina terminalis; *op.stalk*, optic stalk; *op.v.*, optic vesicle; *r.i.*, recessus infundibuli; *r.m.*, recessus mamillaris; *r.p.*, recessus preopticus; *r.p.o.*, recessus postopticus; *s.m.*, somatic motor column; *s.s.*, somatic sensory column; *t.p.*, tuberculum posterius; *ter.*, terminal ridge; *tr.op.*, tractus opticus; *v.m.*, visceral motor column; *v.s.*, visceral sensory column; *III,IV*, position of the nuclei of cranial nerves

part of the vertebrate brain. He pointed out that the floor plate of the developing brain and spinal cord was usually described as a zone devoid of neuroblasts, consisting of neuroglial elements, forming a median ependymal plate, but that this formation and its exact rostral extent had received little attention thus far. He studied these relations in representatives of three vertebrate classes (the chondrichthyan *Squalus*, the chick and the calf) and observed that "the differentiated floor-plate, characterized by the presence of the ependymal layer only and the neuroglial processes (neuroglia fibers), often grouped together in parallel radial bundles, terminates at the fovea isthmi and cephalad of this point, in the floor of the midbrain, the characteristic arrangement of the floor-plate is lacking, while ependymal, mantle, and marginal layers appear. In other words, the differentiation shown is that characteristic of the lateral wall of the neural tube. Cephalad of the midbrain the medial floor thins in the hypothalamic region, to thicken again at the chiasma" (Kingsbury 1920, p. 117). With Johnston (1909), Kingsbury (1920, 1922) was of the opinion that the optic chiasma marks the rostral edge of the neural plate. According to him, the lateral plates of the two sides are continuous across the median plane in front of the rostral end of the floor plate. He conjectured that the basal plate terminates with the midbrain and that the

boundary between the motor and sensory zones terminates near the mammillary recess. The findings and views of Kingsbury (1920, 1922) are represented diagrammatically in Fig. 4.44b,e.

The views of Kingsbury were soon accepted by Johnston. In a remarkable paper, published in 1923 – the last contribution of this noted scientist to comparative neuroanatomy – he changed his mind with regard to the extent of the floor plate and attempted to harmonise the 'four-column doctrine' with Kingsbury's findings. Johnston (1923, Fig. 1) summarised his new interpretation in two diagrams, reproduced here as Fig. 4.45a,b, one showing the arrangement of the functional columns in the anterior portion of the open neural plate (a), the other (b) representing the neural tube as laid open along the dorsal seam. It will be seen that the anterior end of the motor basal lamina is marked by the tuberculum posterius, and that the mammillary and infundibular recesses lie in the region where the visceral sensory zones are confluent across the median plane. The pits which evaginate to form the optic vesicles are located within the somatosensory zones and are connected with one another across the median plane by the primitive optic groove, which later becomes the postoptic recess. The primitive optic groove is bounded in front by the terminal ridge which, as stated earlier (Johnston 1909), forms the anterior border of the neural plate.

In *b* the relationships are the same as in *a*, except that the region of the future hemispheres has expanded rostrally, whereas the terminal ridge has been converted into the optic chiasma, indicated by the crossing fibres of the optic tracts. The lamina terminalis is the rostral end of the dorsal seam which meets the terminal ridge. Johnston mentioned that he had restudied the sulcus limitans in a wide range of comparative material and had been "quite unable to find any clear evidence that the sulcus in the hindbrain and cord regions, which is undoubtedly the sulcus limitans of His, continues farther forward than the mamillary recess in any form studied" (Johnston 1923, p. 153), and somewhat further on he even stated (Johnston 1923, p. 156): "The materials at my disposal do not make clear that it can be traced forward with certainty beyond the level of the trigeminus." Johnston (1923) included the hypothalamus in the viscerosensory column because, according to his observations, this region receives strong projections from gustatory centres in the most rostral part of the rhombencephalon.

It has already been noted that Bergquist and Källén (1953a, 1954) also observed a longitudinal zonal pattern in the vertebrate brain (cf. Sect. 4.5.2 and Figs. 4.19–4.21). These authors did not study the course and the extent of ventricular sulci, or the localisation of functional zones, but rather the patterns of early cellular proliferation and migration. They distinguished some forty migration areas which, according to their observations, show a remarkable consistency throughout the vertebrate kingdom. Bergquist and Källén noticed that in the rhombencephalon these migration areas are arranged in four longitudinal zones, i.e. dorsal, dorsolateral, ventrolateral and ventral. The dorsal zone was observed to end at the rostral margin of the rhombencephalon, while the ventral zone was traced to the rostral end of the midbrain. The remaining two zones appeared to extend to the rostral end of the brain, which Bergquist and Källén (1953a, 1954), with Kingsbury (1920, 1922), located in the region of the optic chiasma. Bergquist and Källén's conception of the zonal organisation of the CNS is represented diagrammatically in Fig. 4.44c,f. They emphasised that the four longitudinal zones of migration areas (cf. here Figs. 4.19, 4.20) in the rhombencephalon roughly correspond to the functional columns of Herrick and Johnston, but that more rostrally they should be looked upon as purely morphological entities. Reference to Figs. 4.19 and 4.20 shows that the area commissurae posterioris, the thalamus and the entire telencephalon belong to the dorsolateral zone, whereas the area tuberculi posterioris and the hypothalamus contribute the rostral part of the ventrolateral zone.

Before we close this survey of the literature on the localisation of the rostral end of the neural tube and the presence of longitudinal zones in the rostral part of the brain we should mention the recent studies of Puelles and collaborators (Puelles et al. 1987a,b; Puelles and Rubenstein 1993; Bulfone et al. 1993). Puelles et al. (1987a) pointed out that different assumptions about the mode of closure of the rostral neuropore have led to different concepts concerning which structure in the adult brain marks or represents the rostral margin of the early embryonic neural plate. Thus His (1893a,b), Kingsbury (1920, 1922) and von Kupffer (1906) located this margin in the infundibular, preoptic and neuroporic recesses, respectively. Puelles et al. (1987a) inserted a piece of black nylon thread through the rostral neuropore of chick embryos and fixed this thread to its ventral lip. These operations were carried out at a number of intermediate stages during the process of closure of the rostral neuropore. The embryos were killed at a later stage, by which time the neuropore had disappeared. In the cleared specimens the threads were always situated at the upper border of the lamina terminalis, irrespective of the stage at which the marker was inserted. According to Puelles et al. (1987a), these experiments warrant the following conclusions (for orientation the reader is referred to Fig. 4.5): (a) The lamina terminalis develops from the most rostral part of the neural plate and is not the product of fusion of the free edges of the neural groove, as His (1893a,b) supposed. (b) The topological rostral end of the brain lies at the upper limit of the lamina terminalis and is marked by the neuroporic recess, as proposed by von Kupffer (1906).

Puelles et al. (1987b) studied the neurogenesis in the forebrain and midbrain of the chick, using AChE as a marker for early-differentiating neuroblasts. As already discussed in Sect. 4.5.5.1, these authors found that in the centre of the mesencephalic and prosencephalic neuromeres groups of early neuroblasts appear which constitute together a longitudinally oriented array extending from the tegmentum of the midbrain to the post-chiasmatic region (Fig. 4.32). The principal results and conclusions of Puelles et al. (1987b) pertaining to the longitudinal zonal organisation of the embryonic brain may be summarised as follows (cf. Figs. 4.32, 4.35):

1. The series of precociously differentiating cell groups is situated in the morphologically dorsal part of the basal plate and may be designated together as the basal plate zone.

2. These 'pioneering' cell groups represent migration areas as defined by Bergquist and Källén
(1954); most of them had already been
observed by Rendahl (1924) in his study on the
development of the diencephalon of the chick.

3. Many cells expressing enzymes of the catecholamine biosynthetic pathway are distributed in a
longitudinal sector of the neural tube wall,
coinciding with the basal plate zone (cf. also
Puelles and Medina 1994).

4. The basal plate band extends caudally into the
mesencephalon but does not continue into the
rhombencephalon, due to the presence of an
interposed undifferentiated m2 neuromere.

5. The 'pioneering' cell groups may be designated
from caudal to rostral as (a) area tegmentalis,
(b) the anlage of the nucleus of the medial longitudinal fasciculus, (c) area tuberculi posterioris, (d) area retromamillaris, (e) area mamillaris lateralis, and (f) area retrochiasmatica.

6. The basal plate band is one of four longitudinal
zones, which are, in dorsoventral order: the alar
plate zone, the basal plate zone, the paramedian
zone and the epichordal zone.

7. The alar plate and paramedian zones are characterised by the fact that their constituent neurons mature at a slower pace than those in the
basal plate zone.

8. Rostrally, the paramedian, basal plate and alar
plate zones are continuous across the median
plane. The paramedian zones are continuous in
the hypothalamic floor, the basal plates in the
prospective retrochiasmatic area, and the alar
plates in the lamina terminalis.

9. The epichordal zone is an AChE-positive, presumably glial, strip which runs along the midline of the rhombencephalon, mesencephalon
and diencephalon. It disappears just behind the
mammillary recess. The epichordal zone is so
named because it is co-extensive with the notochord. Puelles et al. (1987b) coined a new name
for this formation, in order to avoid confusion
with the floor plate of Kingsbury (1920, 1922,
1930). This structure terminates at the fovea
isthmi and is defined on the basis of differentiation markers appearing later in development
than the AChE activity in the epichordal strip,
such as the formation of an ependymal palis-
sade (Kingsbury 1922) and accumulation of
glycogen (Vaage 1969).

10. The development of the optic vesicle and the
telencephalic hemisphere has nothing to do
with the formation of neuromeres. These
paired structures clearly grow out from circular
and not ringlike prosencephalic fields at widely
different developmental stages. The optic vesi-

cles and telencephalic hemispheres should be
looked upon as specialised alar plate derivatives, with a status comparable to that of the
cerebellum and the tectum mesencephali.

Finally, it should be mentioned that in some recent
studies already alluded to in Sect. 4.5.5.5, Bulfone
et al. (1993) and Puelles and Rubenstein (1993)
demonstrated that their longitudinal zonal model is
consistent with the expression patterns of several
putative regulatory genes (Fig. 4.35c).

Following this survey of the literature concerning the presence of longitudinal zones in the rostral
part of the brain in general, brief mention will be
made of zonal subdivisions of the various parts of
this region, i.e. the mesencephalon, the diencephalon and the telencephalon.

4.6.6.2
Longitudinal Zones in the Mesencephalon

Most authors agree that the two principal longitudinal zones of the rhombencephalon, i.e. the primarily motor basal plate and the primarily sensory alar
plate, extend into the midbrain, and, according to
some, the boundary between these two entities is
marked by a sulcus limitans also at mesencephalic
levels in the embryonic (e.g., Gribnau and Geysberts 1985: Figs. 4.6b, 4.7b; O'Rahilly et al. 1989:
Fig. 4.29) and in the adult brain (e.g., Holmgren and
van der Horst 1925; Gerlach 1933, 1946; Heier
1948). It is also generally agreed that only two of the
four functional rhombencephalic columns, i.e. the
somatic motor and the somatic sensory, extend into
the midbrain, with the reservation that one evidently general visceromotor cell mass, the nucleus
Edinger-Westphal, is located within its confines.

From an embryological point of view, it is important to note that three of the four longitudinal zones
which Bergquist and Källén (1953a, 1954: Figs. 19,
20) distinguished in the developing rhombencephalon, i.e. the ventral, ventrolateral and dorsolateral
zones, were traced throughout the midbrain. As
already discussed in Sect. 4.6.4, the embryonic longitudinal zones delineated by Bergquist and Källén
are composed of series of early-differentiating and
migrating cell groups.

Nieuwenhuys and collaborators (for references
see Sect. 4.6.5) subjected the brain stem, including
the mesencephalon, of a considerable number of
representative vertebrate species to a topological
analysis. Their findings may be summarised as follows (cf. Fig. 4.42):

1. In the clawed toad *Xenopus* the sulcus limitans
extends into the caudal part of the midbrain, but
in all other species examined (ranging from the

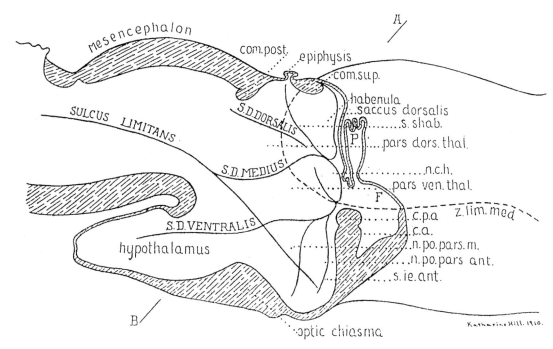

Fig. 4.46. The ventricular sulci and main divisions of the diencephalon of the axolotl, according to Herrick (1910, Fig. 22). *Line A–B* indicates the plane of the section shown in Fig. 4.47a. *c.a.*, Commissura anterior; *c.p.a.*, commissura pallii anterior; *com. post.*, commissura posterior; *com. sup.*, commissura superior; *n.c.h.*, nucleus of commissura hippocampi; *n.p.o. pars ant.*, nucleus preopticus pars anterior; *n.po. pars m.*, nucleus preopticus, pars magnocellularis; *pars dors. thal.*, pars dorsalis thalami; *pars ven. thal.*, pars ventralis thalami; *SD*, sulcus diencephalicus; *s. ie. ant.*, sulcus interencephalicus anterior; *s.shab.*, sulcus subhabenularis; *z. lim. med.*, zona limitans medialis; *P*, paraphysis; *F*, foramen interventriculare

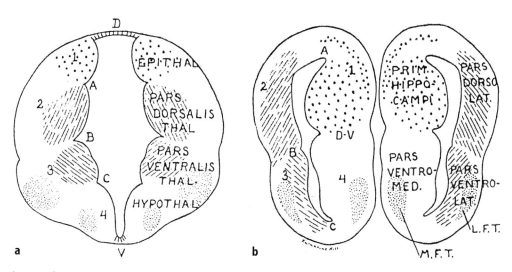

Fig. 4.47a,b. Cross-sections through the diencephalon (**a**) and the telencephalon (**b**) of an amphibian to illustrate the way in which, according to Herrick (1910, Figs. 83, 84), the cerebral hemispheres have been formed by the lateral evagination of the walls of the neural tube. Numbers *1–4* and letters *A*, *B* and *C* mark corresponding structures in **a** and **b**. *A*, Sulcus diencephalicus dorsalis and dorsal angle of hemisphere; *B*, sulcus diencephalicus medius and zona limitans lateralis; *C*, sulcus diencephalicus ventralis and ventral angle of hemisphere; *D*, roof plate; *D–V*, zona limitans and fissura limitans hippocampi; *L.F.T.*, lateral forebrain tract; *M.F.T.*, medial forebrain tract; *V*, floor plate

lamprey to the turtle) this sulcus was not traced beyond the isthmus rhombencephali.

2. Several authors, including Holmgren and van der Horst (1925), Gerlach (1933), Ariëns Kappers et al. (1936) and Kuhlenbeck and Niimi (1969), have misinterpreted mesencephalic ventricular grooves which are obviously not continuous with the rhombencephalic sulcus limitans as being part of that sulcus.

3. The general arrangement of the cell masses in the mesencephalon does not exhibit a clear morphological zonal pattern, although this part of the neuraxis may be divided functionally into primarily motor and sensory zones. The motor zone, which includes the nuclei of III and IV, the nucleus of Edinger-Westphal, the nucleus of the fasciculus longitudinalis medialis and the nucleus ruber, occupies the medial part of the tegmentum mesencephali. The sensory zone comprises the lateral part of the tegmentum and the tectum. The lateral part of the tegmentum contains a large special somatosensory relay centre, the torus semicircularis. In most groups of non-mammalian vertebrates this torus manifests itself as an intraventricular protrusion. In mammals it becomes transformed into the colliculus inferior (cf. Puelles et al. 1994: Fig. 4.14). The tectum mesencephali is to be considered as a general and special somatosensory correlation centre of higher order. However, in all gnathostomes it also contains a primary somatosensory cell group: the nucleus mesencephalicus of V.

4.6.6.3
Longitudinal Zones in the Diencephalon

During the late nineteenth century and the early twentieth century, the diencephalon was usually subdivided into three moieties: epithalamus, thalamus and hypothalamus (e.g., His 1893b: Fig. 4.43). In 1910 the noted comparative neuroanatomist Charles Judson Herrick published a paper entitled: "The Morphology of the Forebrain in Amphibia and Reptilia", in which he proposed a subdivision of the diencephalon into four longitudinal zones: epithalamus, thalamus dorsalis, thalamus ventralis and hypothalamus. Herrick placed these zones in the perspective of the fundamental longitudinal divisions of the early neural tube as defined by His (1893a,b). He emphasised that these four diencephalic zones represent not only structural but also functional entities. Moreover, he expressed the opinion that the four diencephalic zones are rostrally directly continuous with four functionally related telencephalic regions.

The ensuing synopsis of Herrick's (1910) study will be presented largely in his own words: "The absence of peripheral motor nerves rostrally of the midbrain involves the great reduction of the ventral lamina [i.e. the basal plate, R.N.] in this region. Accordingly, the sulcus limitans disappears in the diencephalon. In the human embryo (His) it appears to end in the preoptic recess, which would imply that the chiasma ridge and hypothalamus belong in the ventral lamina. And such indeed has been clearly shown to be the case, ..." (Herrick 1910, p. 466). "From the absence of the ventral lamina rostral to the chiasma ridge, it follows that the remaining parts of the diencephalon and the telencephalon belong to the dorsal lamina [i.e. the alar plate, R.N.], ..." (Herrick 1910, pp. 466–467). It is noteworthy that in some later studies on the urodelan brain (Herrick 1935, 1948) Herrick does not make reference to the continuation of the sulcus limitans into the diencephalic domain.

Referring to the relations in the amphibian diencephalon (cf. Figs. 4.46, 4.47a) Herrick (1910, p. 477) stated: "Besides the unpaired membranous roof plate and floor plate, there are four ridges on each side, the epithalamus, the pars dorsalis thalami, the pars ventralis thalami and the hypothalamus separated by the dorsal, median and ventral sulci of the diencephalon. The mode of development in early embryos shows that the eight massive columns are not produced by a passive plication of the walls due

Fig. 4.48a–d. Longitudinal zones in the diencephalon, as revealed by the studies of Kuhlenebeck. **a** 'Generalised' urodele amphibian (Kuhlenbeck 1956, Fig. 11): *crosses*, epithalamus; *circles*, thalamus dorsalis; *black dots*, pretectal component neighbourhoods of dorsal thalamus; *vertical hatching*, thalamus ventralis; *oblique hatching*, hypothalamus; *X*, basal plate of deuterencephalon. **b** Chick embryo of 120-h incubation (Kuhlenbeck 1973, Fig. 221): *crosses*, epithalamus; *circles*, dorsal thalamus; *horizontal hatching*, tectum mesencephali; *vv*, torus semicircularis; *xx*, tegmental cell cord; *vertical hatching*, ventral thalamus; *oblique hatching*, hypothalamus. **c** Rabbit embryo, 15 mm (Kuhlenbeck 1973, Fig. 238): *crosses*, epithalamus; *circles*, dorsal thalamus; *vertical hatching*, thalamus ventralis; *oblique hatching //*, hypothalamus; *horizontal hatching*, mesencephalic alar plate; *oblique hatching*, lamina terminalis; *xx*, tegmental cell cord. **d** Human embryo, 10–11 mm long (Kuhlenbeck 1973 Fig. 240): *circles*, epithalamus; *vertical hatching*, thalamus dorsalis; *horizontal hatching*, thalamus ventralis; *oblique hatching*, hypothalamus; *xx*, basal plate. *B1, B2,* Telencephalic basal ganglionic cell masses; *CA, ca,* commissura anterior; *ce,* cerebellar plate; *CH,* chiasmatic ridge; *CL,* commissura pallii; *CN, hc,* habenular commissure; *CP, cp,* commissura posterior, commissural plate (in **d**); *ep,* epiphysis; *gh,* ganglionic hills; *in,* infundibulum; *me,* mesencephalon; *op,* optic chiasma; *sa,* sulcus intraencephalicus anterior; *sd,* sulcus diencephalicus dorsalis; *SF, sf, si,* sulcus lateralis infundibuli; *SL, sl,* sulcus limitans; *sm,* sulcus diencephalicus medius; *SR, slm,* sulcus lateralis mesencephali; *sv,* sulcus diencephalicus ventralis; *SY,* groove of synencephalic neuromere; *sy,* synencephalon; *te,* telencephalon; *vt,* velum transversum

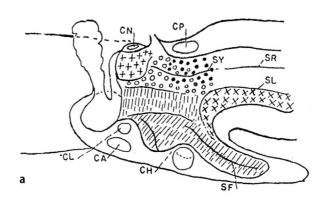

a

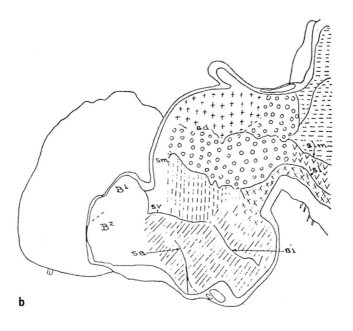

b

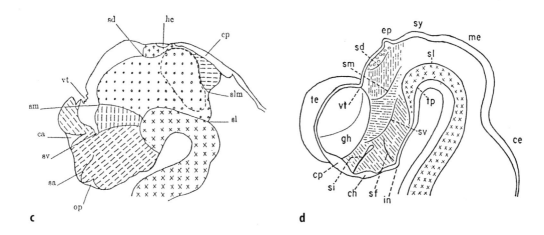

c   d

to extrinsic forces, but that each column is a center of more active proliferation of neuroblasts than the intervening sulci, and has, therefore, doubtless been differentiated under the influence of definite functional requirements."

"The roof plate and the floor plate converge into the lamina terminalis, where of course they end. The four massive columns on each side converge into the interventricular foramen, and in larvae with wide foramina and adult urodeles they may be followed through the foramina into the evaginated hemispheres. Bearing in mind the fact that during development the roof plate and floor plate retain permanently their primitive attachments to the lamina terminalis, and that it is only the massive lateral columns which are evaginated into the hemispheres, it clearly follows that these columns of the diencephalon are continued into the hemispheres in the form shown by the accompanying diagram [reproduced here in Fig. 4.47b, R.N.], the zona limitans lateralis representing the locus of the sulcus medius and the zona limitans medialis the line of union of the dorsal and ventral columns in the lateral evaginations rostral to the fusion of the roof plate and the floor plate in the lamina terminalis" (Herrick 1910, p. 477).

The pars dorsalis thalami is devoted to somatic sensory correlations, whereas the pars ventralis thalami and its telencephalic continuation (i.e. the striatum) represent motor correlation centres. The sulcus diencephalicus medius marks the boundary between sensory and motor correlation areas and is therefore a secondary extension of the sulcus limitans, though not a part of it (Fig. 4.46). Herrick (1910, p. 469) emphasised that "this division of the thalamus by the sulcus limitans and sulcus medius into dorsal afferent and ventral efferent parts is fundamental and characteristic of the vertebrate phylum as a whole." According to his observations, it is very evident in generalised fishes and in both embryonic and adult amphibians and reptiles. It is also easily recognised in mammalian embryos, though often secondarily obscured in adult mammals.

Two large composite bundles of projection fibres, i.e. the medial and lateral forebrain bundles, stress the functional continuity of the ventromedial and ventrolateral parts of the telencephalic hemispheres with, respectively, the hypothalamus and the pars ventralis thalami (Fig. 4.47).

A subdivision of the diencephalon into four longitudinal zones was also advocated by Kuhlenbeck. He investigated embryological material of many different vertebrate groups, including cyclostomes, selachians, teleosts, dipnoans, urodeles and anurans (Kuhlenbeck 1929a), reptiles (Kuhlenbeck 1931), birds (Kuhlenbeck 1936, 1937, 1939) and

mammals (Kuhlenbeck and Miller 1942, 1949). The results of these studies, which have been repeatedly reviewed by Kuhlenbeck (1954, 1956, 1973, 1977; see also the sections prepared by the same author in a paper by Christ 1969), may be summarised as follows (cf. Fig. 4.48):

1. Four longitudinal cell columns, epithalamic, thalamus dorsalis, thalamus ventralis and hypothalamus, represent "die Grundbestandteile des Zwischenhirnbauplans" (Kuhlenbeck 1929a, 1931, 1936), i.e. the fundamental units of the structural plan of the vertebrate diencephalon.

2. The fundamental zonal pattern of the diencephalon is readily recognisable in adult anamniotes as well as in certain ontogenetic 'key stages' in amniotes, but during later development the primitive original pattern in amniotes breaks up into many separate diencephalic centres.

3. In embryonic and adult anamniotes and in embryonic amniotes the fundamental diencephalic zones are often separated from each other by three ventricular sulci, sulcus diencephalicus dorsalis, medius and ventralis. A fourth ventricular groove, the sulcus infundibuli, marks the boundary between the dorsal and ventral parts of the caudal hypothalamus. If the sulci are sufficiently developed they afford great help in morphological analysis and interpretation of cellular structures; however, the spatial relationships of the cellular pattern within the diencephalic wall constitute the primary criterion, while the sulci are an auxiliary criterion for the identification of the zonal organisation. In some vertebrate forms or in some developmental stages the longitudinal cell columns are separated from each other by limiting zones poor in cells and characterised by growing fibre connections. Thus the developing mammalian dorsal and ventral thalami are separated from each other by a zona limitans intrathalamica, within which the fasciculus thalamicus (or field H1 of Forel) develops.

4. The most rostral part of the basal plate extends into the caudal portion of the diencephalon. It is represented by the tegmental cell cord, which corresponds in mammals and man to the pre-rubral tegmentum described by Papez (1940). This rostral extreme of the basal plate arches around the tuberculum posterius; it is bordered by the sulcus limitans and terminates in the primary infundibular or mamillary recess.

5. Since the four longitudinal diencephalic zones develop rostral to the sulcus limitans, they must be interpreted as derivatives of the alar plates.

6. The four longitudinal diencephalic zones are to be considered as purely structural entities which

are independent of the functional somatic and visceral zones of the brain stem.

7. The nucleus subthalamicus, the entopeduncular nucleus and the globus pallidus differentiate from the hypothalamic zone.
8. The preoptic region represents an intrinsic part of the diencephalon and should not be regarded as a part of the telencephalon impar.

A subdivision of the diencephalon into four longitudinally oriented zones has been widely accepted as applicable in all vertebrates (cf., e.g., Ariëns Kappers et al. 1936; Northcutt 1987). Kahle (1956, 1958) and Müller and O'Rahilly (1988), who both studied human embryonic material, added a fifth, subthalamic zone, interposed between the ventral thalamus and the hypothalamus. However, the ontogenetic studies of numerous authors, including Bergquist (1932, 1952a–c, 1953, 1954a,b, 1964), Bergquist and Källén (1954), Keyser (1972), Gribnau and Geysberts (1985), Figdor and Stern (1993), Bulfone et al. (1993) and particularly Puelles and Rubenstein (1993), have convincingly shown that the dorsal and ventral thalami are both neuromeric derivatives, and hence oriented perpendicular to the longitudinal axis of the brain, and that these spatial relations become instantly clear if the curvature of the neural tube at the junction of the mesencephalon with the prosencephalon is taken into account (Figs. 4.6, 4.7, 4.20, 4.28, 4.35, 4.37, 4.48). The epithalamus stems from the same neuromere as the dorsal thalamus. The status of the hypothalamus is less clear, but according to several authors (Bergquist 1932; Bergquist and Källén 1954, 1955; Puelles et al. 1987b, Puelles and Rubenstein 1993; Bulfone et al. 1993), three neuromeres or transverse bands participate in the formation of this brain part. In light of these findings, the thesis of Kuhlenbeck (1954, 1956, 1973) that neuromeres are transient phenomena which have nothing to do with the definitive structural plan of the diencephalon is untenable. The four diencephalic zones distinguished by him are not longitudinally arranged and do not represent fundamental morphological units. This is not to deny that the subdivision of the diencephalon as proposed by Herrick and Kuhlenbeck is convenient and simple, and that the four entities distinguished clearly belong to different functional domains. Opinions differ about where the sulcus limitans ends anteriorly, and the epithalamus, thalamus dorsalis, thalamus ventralis and hypothalamus certainly do not represent rostral continuations of the well-known functional rhombencephalic columns. However, Herrick (1910, cf. also Herrick 1948, pp. 56–61) claimed with some justification that the sensory and motor zones of the brain extend rostrally via the diencephalon into the telencephalon. Moreover, Herrick's (1910) analysis brought to light some basic functional relations between telencephalic and diencephalic regions of the vertebrate brain.

4.6.6.4
Longitudinal Zones in the Telencephalon

In all vertebrates the telencephalon is composed of three principal parts, the telencephalon medium or impar, the telencephalic hemispheres and the olfactory bulbs. The telencephalon medium, which is generally small, maintains the early embryonic tubelike condition. It corresponds roughly to the preoptic region and is caudally directly continuous with the diencephalon. Hemispheric formation occurs embryonically by two fundamentally different processes. In most vertebrates the hemispheres form by evagination or bulging-out of the lateral walls of the forebrain vesicles. In actinopterygian fishes, in contrast, the hemispheres form by an outward bending or eversion of the dorsal portions of the lateral walls of the early embryonic forebrain vesicle. In some groups of fishes (Holocephali, Dipnoi, Crossopterygii) some parts of the walls of the hemispheres show a condition intermediate between evagination and eversion. The olfactory bulbs generally come about by secondary evagination of the rostral parts of the hemispheres.

Numerous investigators have indicated that the walls of the telencephalic hemispheres can be subdivided into a number of longitudinal zones. Thus, it has already been noted that Herrick (1910, Fig. 4.47) distinguished four quadrants or sectors in the amphibian hemispheres: pars ventromedialis (septum), pars ventrolateralis (striatum), pars dorsolateralis (pyriform lobe), and pars dorsomedialis (primordium hippocampi). In a later study Herrick (1922) indicated that these sectors can also be distinguished in the brains of fishes, although in these forms the homologues of the amphibian dorsolateral and ventrolateral quadrants cannot be sharply delineated from each other.

Holmgren (1922, Fig. 4.49) indicated that in the telencephalon of all Anamnia a dorsally situated pallium and a ventrally situated subpallium can be distinguished. The boundary between these two parts is marked by a ventricular groove, the sulcus limitans pallii lateralis, and by a cell-free zona limitans lateralis. According to Holmgren, the pallium in both evaginated and everted hemispheres can be subdivided into three longitudinal zones: pyriform pallium, general pallium and hippocampal pallium. In the subpallium of both hemisphere types he also distinguished three zones. In evaginated forebrains

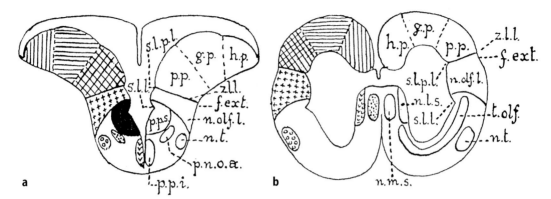

Fig. 4.49a, b. Longitudinal zons in telencephela of the everted (**a**) and inverted type (**b**), according to Holmgren (1922, Figs. 40, 41). *f. ext.,* Fovea externa; *g.p.* general pallium; *f.p.,* hippocampal pallium; *n.l.s.m* nucleua lateralis septi; *n.m.s.,* nucleus medialis septi; *n.olf.I.,* nucelus olfactorius lateralis; *n.t.,* nucleus taeniae; *p.n.o.a.,* nucleus olfactoris anterior, pars precommissuralis; *p.p.,* pyriform pallium; *p.p.i.,* corpus precommissurale, pars inferior; *p.p.s.,* corpus precommissurale, pars superior; *s.l.l.,* sulcus limitans lateralis; *s.l.p.l.,* sulcus limitans pallii lateralis; *t.olf.,* tuberculum olfactorium; *z.l.l.,* zona limitans lateralis

these zones are represented by the nucleus olfactorius lateralis, the tuberculum olfactorium and the septum, the last containing a periventricular lateral septal nucleus and a migrated medial septal nucleus. The homologues of these subpallial cell masses in the everted brains were designated by Holmgren with names that are no longer in use. However, it will be clear that, according to him, the various pallial and subpallial cell masses in evaginated and everted forebrains, though differing markedly in topographical position, correspond with regard to their relative or topological position.

Kuhlenbeck (1929b, 1938, 1973, 1977) sought to answer the question of which fundamental units or *Grundbestandteile* can be distinguished in the telencephalon of vertebrates. To this end he studied larval or adult material of a number anamniotes and embryonic material of some reptilian, avian and mammalian species. His main results may be summarised as follows (Fig. 4.50):

1. In the telencephalon of all vertebrates a dorsal area, D, and a basal area, B, can be distinguished.
2. The structural plan of the hemispheric lobes in all gnathostomes is characterised by a longitudinal zonal system. Within the dorsal area three such zones, D1, D2 and D3, can be distinguished, whereas in the basal area four zones, B1, B2, B3 and B4, are delineable. Zone B4 is restricted to a telencephalic configuration which includes paired, rostrally directed evagination of the hemispheres (Fig. 4.50b,d–f). The seven zones represent the *Grundbestandteile* of the telencephalic hemispheres. As such they are comparable to the longitudinal zones in the brain stem and in the diencephalon.

3. The boundaries between the seven longitudinal zones are marked by ventricular sulci.
4. The longitudinal zones form the basis for the homologisation of telencephalic nuclei: "Zur Homologisierung bestimmter, in sich cytoarchitektonisch einheitlicher Zellmassen ist es erforderlich, die Lage dieser Zellmassen im Bauplan zu erkennen. Nur auf Grund dieses Kriteriums und zwar ohne jede Rücksicht auf etwaige Strukturunterschiede oder Faserverbindungen ist eine Homologie feststellbar. Zellmassen des Endhirns, welche einem in Bezug auf den Bauplan gleichen Wandabschnitt entstammen, sind homolog" (Kuhlenbeck 1929b, p. 49).
5. The seven *Grundbestandteile* and their delimiting sulci can be readily recognised in the telencephalon of all gnathostome anamniotes, as well as in certain embryonic 'key stages' of amniotes.
6. In sauropsides D1 unites with the basal area to constitute the intraventricular protrusion, which previous investigators denoted as the striatal complex.
7. In birds the seven *Grundbestandteile* give rise to the following formations (Fig. 4.50d–f): D3 constitutes the hippocampal formation. D2 splits up into a medial part and a lateral part. The medial part becomes cortical and represents the parahippocampal cortex, whereas the lateral part forms the non-cortical hyperstriatum accessorium and contributes, with D1, to a lateral corticoid lamina. D1, besides taking part in the formation of the corticoid lamina just mentioned, forms the matrix for the following nuclei: D1a (hyperstriatum dorsale + ventrale), D1b (neostriatum), D1c (ectostriatum), D1d (archistriatum) and D1e (nucleus basalis). B1 and

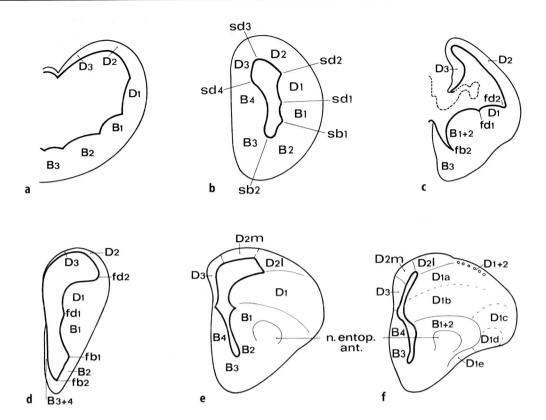

Fig. 4.50a–f. Longitudinal zones and their delimiting ventricular sulci in the telencephalon according to Kuhlenbeck. **a** Embryo of the snake, *tropidonotus natrix* (Kuhlenbeck 1938 Fig. 1C); **b** newly hatched lizard *Lacerta agilis* (Kuhlenbeck 1929b, Fig. 19); **c** pig embryo, length about 40 mm (Kuhlenbeck 1929b, Fig. 26); **d, e, f:** chick embryos, 6, 8 and 10 days old, respectively (Kuhlenbeck 1929 b, Fig. 2; Kuhlenbeck 1973, Fig. 296; Kuhlenbeck 1938, Fig. 3)

B2 fuse their cell masses and give rise to the paleostriatum augmentatum. B3 forms the nuclei of the septum and B4 forms the nucleus basimedialis superior.

8. In mammals the neocortex develops from the lateral part of D2, whereas D1, B1 and B2 participate in the formation of the nucleus caudatus, the putamen and the amygdaloid complex.

The work of the Swedish comparative neuroembryologists Bergquist and Källén has already been discussed in Sect. 4.5.2. These authors found that in all vertebrates cellular areas of high activity with respect to both cell multiplication and migration develop in the wall of the early neural tube. These migration areas appeared to show a remarkable consistency throughout the vertebrate kingdom and were thus considered to represent the fundamental morphological units of the central nervous system (Bergquist and Källén 1954; Fig. 4.20). The boundaries between the migration areas are not marked by ventricular sulci, but rather by 'incisures' in the external surface of the mantle layer. (Such 'inci-

sures' are indicated by arrows in Fig. 4.51a). During further development the cells contributing to the migration areas may remain in contact with the matrix zone, or they may move further outward to form a layer of migrated cells (Fig. 4.23). Proliferation and migration appeared to be intermittent processes, and in many parts of the brain the first active period is followed after a while by a new migration wave. In some parts of the brains of amniotes Bergquist and Källén recorded three or even four successive migrations, each giving rise to the formation of a new migration layer. The separate migration layers with a given migration area (denoted as *I, II*, etc. in Fig. 4.51) may split up into separate external and internal zones (designated as *e* and *i* in Fig. 4.51), or they may be subdivided into cell aggregates forming the primordia of the nuclei in the mature brain. Bergquist and Källén (Bergquist 1932, Källén 1951b, Bergquist and Källén 1954) emphasised that analysis of the ontogenetic development of brain nuclei forms the only sound basis for their homologisation. "The morphological position of a certain nucleus can thus be deter-

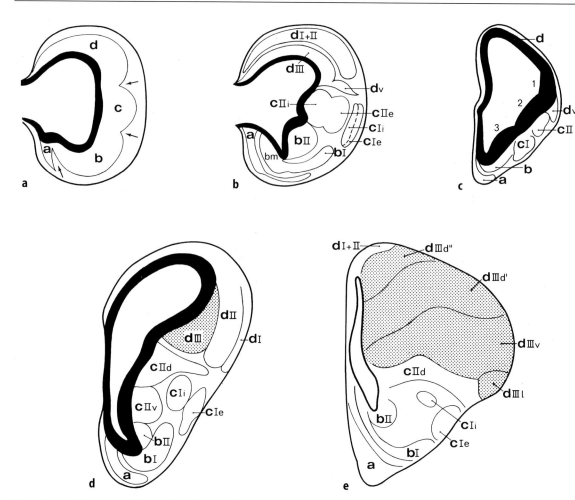

Fig. 4.51a–e. The mode of formation of telencephalic nuclei from four cell columns **a, b, c, d,** according to Källén. **a,b,** Schematic summary of the formation of cell masses in the amniote hemispheres; **a** young stage, showing the four cell columns; **b** older stage, showing the presence of migration layers and their subdivisions (Källén 1955, Fig. 8). **c** Human embryo, 17 mm; the ventricular ridges are numbered (Källén 1951c, Fig. 8). **d,e** Chick embryos, 6.5 and 14 days old, respectively (Källén 1962, Figs. 5, 9). *Arrows* in **a** indicate 'incisures' in the mantle layer. Further explanation in text

mined by answers to the following questions: 1) from which migration area does the nucleus develop, 2) within which migration layer is it formed and 3) within which part of the migration layer does it develop?" (Bergquist and Källén 1954, p. 641).

Källén (1951a–e; 1955, 1962) devoted a series of papers to the development of the telencephalon, studying representative species of all vertebrate groups, except for the myxinoids and the crossopterygians. His main findings and conclusions may be summarised as follows:

1. In all vertebrates studied two migration areas, i.e. the area dorsalis telencephali and the area ventralis telencephali, are formed in the forebrain. The former gives rise to the pallium, the latter to the subpallium.

2. The subpallial migration area splits up into three longitudinal cell columns a, b and c. Together with the area dorsalis telencephali, d, these three columns constitute the four fundamental units from which all telencephalic cell masses develop (Fig. 4.51a).

3. The area dorsalis telencephali (*d*) gives rise to the various pallial and, where present, cortical formations. In reptiles, however, *d* – more specifically, the lateral part of its third migration layer, dIII – differentiates into the large dorsal ventricular ridge, which therefore should be considered a pallial derivative. In birds the dIII formation expands enormously, giving rise to the formations designated by Huber and Crosby (1929, Fig. 4.51d,e) as hyperstriatum dorsale (dIIId'), hyperstriatum ventrale (dIIId"), neostriatum (dIIIv) and ectostriatum (dIIIl). In mammals *d*

participates in the formation of the amygdaloid complex and a special derivative of dIII, designated as *dv* in Fig. 4.51b,c, represents the anlage of the claustrum.

4. The dorsalmost subpallial column (*c*) in Anamnia forms the nucleus olfactorius lateralis (cf. Fig. 4.49). Within this column two migration layers, cI and cII, are formed in all amniote groups, both subdividing into an external and an internal part (Fig. 4.51b). In birds cII splits up into a dorsal part cIId and a ventral part cIIV. The cIId cell mass forms, together with cIe, the paleostriatum augmentatum. The paleostriatum primitivum arises from cIi, and cIIV and bII both contribute to the nucleus accumbens (Fig. 4.51d,e). In mammals cIe participates in the formation of some amygdaloid nuclei, whereas cIi gives rise to the globus pallidus, the nucleus basalis and the nucleus centralis amygdalae. The mammalian nucleus caudatus and putamen are derivatives of cIIi and cIIe, respectively.

5. The middle subpallial column (*b*) in Anamnia represents the tuberculum olfactorium. In amniotes there are two successive proliferations in this column, forming bI and bII (Fig. 4.51b-e). The medial part of bI (*bm*) enters the septum forming the lateral septal nucleus. The lateral part of b1 gives rise to the tuberculum olfactorium. In all amniotes bII represents or contributes substantially to the nucleus accumbens. In mammals the bII formation forms, in addition, the caput nuclei caudati and the bed nucleus of the stria terminalis.

6. The ventral subpallial column (*a*) in all evaginated forebrains becomes partly incorporated in the septum. Its medial part forms the nucleus medialis septi, whereas the nucleus of the diagonal band develops from its lateral part. However, due to the rostral hemispheric evagination, parts of columns *b* and *c* also come to lie in the septum. It has already been mentioned that the septal part of *b* (*bm*) forms the nucleus lateralis septi. In mammals the medial part of column *c* also contributes to the formation of the nucleus lateralis septi. Among lower vertebrates the mode of formation of the medial hemisphere walls differs considerably from group to group and these differences greatly influence the size, extent and composition of the septal nuclei.

Northcutt, finally, divided the reptilian telencephalon into six longitudinal phylogenetic or prototypic columns, three pallial ones (PI–PIII) and three subpallial ones (BI–III). He based this subdivision on the ontogenetic evidence of Källén (1951d) and on his own experimental hodological and architectonic analysis of adult nuclear groups (Northcutt 1967, 1970). He speculated about which structural and connectional changes within the various prototypic columns may have occurred in the telencephalon of the various amniote texa. In a later paper, Northcutt (1981) expressed the opinion that the cerebral hemispheres of all vertebrates are formed by homologous roof (pallial) and floor (subpallial) regions, and that a comparable number of pallial and subpallial subdivisions occur in all vertebrate radiations. They are primitive characters, being homologous among all vertebrates. Northcutt emphasised, however, that numerous differentiations that have occurred within the prototypic columns are less widely distributed among vertebrate radiations and are probably derived characteristics. Similarities between such characteristics observed in different groups will often be homoplastic rather than homologous.

From the foregoing survey of the literature, it appears that a subdivision of the vertebrate telencephalic hemispheres into a dorsal pallium and a basal subpallium is generally accepted, and that, according to most authors, both of these regions are divisible into three longitudinal zones or columns. The six zones thus distinguished occupy corresponding topological positions in all vertebrates and are to be considered as fundamental units, providing a basis for the homologistion of individual cell masses. The most spectacular result so far of tracing the ontogenesis of these zones has been the discovery that the dorsal ventricular ridge of reptiles and the greatly expanded dorsolateral quadrant of the avian hemispheres are not occupied by striatal formations, as Edinger (1908), Ariëns Kappers (1921), Huber and Crosby (1929) and other previous workers believed, but rather represent pallial derivatives. The thorough neuroembryological investigations of Källén (1962) have clearly shown that the cell masses in the avian hemispheres, known as the hyperstriatum dorsale, hyperstriatum ventrale, neostriatum and ectostriatum, are all pallial formations, and that only one of the 'striata' stacked in the lateral wall of the avian hemispheres, i.e. the palaeostriatum augmentatum, corresponds to the mammalian caudate-putamen complex. Interestingly, this embryological interpretation of the parts of the avian telencephalon was later fully confirmed by the results of histochemcial, immunohistochemical and experimental hodological studies (cf. Chap. 6, Sect. 6.2.6.4). It should be emphasised, however, that the telencephalic zones or columns described by the various authors, although grossly corresponding, are by no means identical. For example, Holmgren (1922) compared the cell masses situated along the greatly expanded ependy-

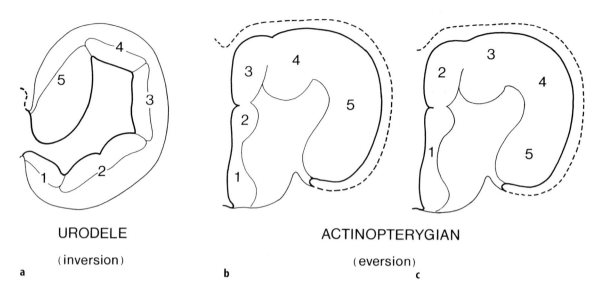

URODELE

(inversion)

a

b

ACTINOPTERYGIAN

(eversion)

c

Fig. 4.52a–c. Location of the principal cell masses in the telencephalic hemispheres of urodeles (**a**) and actinopterygians, according to Holmgren (**b**) and Kuhlenbeck (**c**). For further explanation see text

mal surface of the everted teleostean forebrain in the way indicated in Fig. 4.52a,b with the periventricular grisea in the evaginated urodelan telencephalon. Kuhlenbeck (1929b, 1973, 1977) and Miller (1940) located the pallio-subpallial boundary in the teleostean forebrain further dorsal than Holmgren did (Fig. 4.52c), a decision which greatly influenced the interpretation of all periventricular grisea in that forebrain, as indicated diagrammatically by the numbers in Fig. 4.52. Källén (1962) emphasised that his migration areas (Fig. 4.51) are not fully comparable to the ventricular ridges separated by ventricular furrows which Kuhlenbeck (1929b, 1938, 1973, 1977; Fig. 4.50) considered the fundamental units in the vertebrate forebrain. The following two examples may illustrate this point. (a) Kuhlenbeck (1929b; Fig. 4.50c) reported that ridge D1 contributes in mammals to the caudate-putamen complex, but the corresponding migration zone, i.e. the lateral part of *d*, according to Källén (1951c; Fig. 4.51c) does not participate in the formation of this complex. (b) The avian palaeostriatum augmentatum is, according to Kuhlenbeck (1938; Fig. 4.50d,e,f), a product of the fused ridges B1 and B2; Källén (1962; Fig. 4.51d,e) reported, however, that this cell mass develops exclusively from migration zone *c*, which roughly corresponds to ridge B1. It is also worth recalling that whereas Herrick (1910), Holmgren (1922), Kuhlenbeck (1929b, 1973, 1977) and Northcutt (1967) considered the septum and its nuclei the product of a single zone or column, according to Källén (1951b,d,e) all three of the subpallial migration zones participate in the formation of the septal area. Finally, it should be emphasised that Kuhlenbeck (1929b, 1973, 1977) and Källén (1951b–d) considered the telencephalic zones as purely morphological entities, but that according to Herrick (1910) these zones, just like the diencephalic ones, represent both structural and functional units.

One fundamental question remains: Are the longitudinal telencephalic zones discussed above really longitudinally oriented? That is to say: Do these columns run parallel to the longitudinal axis of the brain? For finding a reply to these questions the following points, which have been extensively discussed in previous sections of this chapter, are relevant:

1. Neuromeres represent transversely oriented bulges of the neuraxis.
2. The neural tube is strongly curved during development.
3. According to several previous workers, the sulcus limitans indicates the course of the axis of the brain; however we know now that this groove rostrally cannot be followed beyond the isthmus region.
4. Bergquist and Källén (1954) distinguished four longitudinal migration columns in the brain. Two of these, the columnae dorsolateralis and ventrolateralis, appeared to extend throughout the length of the brain. The boundary between these zones terminates immediately behind the optic chiasma (Fig. 4.20).
5. Puelles and collaborators (Puelles et al. 1987b; Puelles and Rubenstein 1993; Bulfone et al. 1993)

constructed a longitudinal axis of the brain on the basis of phenomena of early differentiation and of the expression patterns of a number of putative regulatory genes. This axis terminates, just like the zonal boundary of Bergquist and Källén, immediately behind the optic chiasma (Fig. 4.35).

According to Bergquist and Källén (1954), the area ventralis telencephali or subpallium and the area dorsalis telencephali or pallium form part of transverse bands 1 and 2 (Fig. 4.20). These bands represent postneuromeric structures which, as the designation 'transverse' indicates, are oriented perpendicular to the longitudinal axis of the brain. Because the subpallial migration zones a, b and c, delineated by Källén (1951b–d), run parallel to the pallio-subpallial boundary, they have to be considered likewise as transversely oriented structures. The same holds true for the three 'theoretical' neuromeres, which according to Bulfone et al. (1993) participate in the formation of the mammalian telencephalic hemispheres (Fig. 4.35). However, as already pointed out in Sect. 4.5.5.5, the authors mentioned allocated the main regions of the adult mammalian telencephalon in a strange and inconsistent way to the domains of their theoretical neuromeres. Hence, all that can be said at present concerning the direction of the six basic telencephalic zones or columns is that, according to the extensive comparative embryological investigations of Bergquist and Källén, these entities are oriented perpendicular to the longitudinal axis of the brain.

4.6.7
Longitudinal Zones: Summary and Conclusions

All parts of the CNS have been subdivided into two or more longitudinal zones. Ventricular sulci have been considered to mark the boundaries between these longitudinal zones. The longitudinal zones or columns distinguished have often been considered to represent both structural and functional entities.

According to His, the entire CNS can be subdivided into a dorsal, primarily sensory alar plate and a ventral, primarily motor basal plate, and the boundary between these two plates is marked throughout the neuraxis by a ventricular groove, the sulcus limitans. The subdivision proposed by His manifests itself clearly in the spinal cord and the rhombencephalon; however, attempts to localise the alar and basal plates and to trace the sulcus limitans in the more rostral parts of the brain have led to widely different results. All that can be said at present is that Bergquist and Källén (1954) and Bulfone et al. (1993) have adduced neuroembryological

and immunohistochemical evidence indicating that the basal and alar plates extend as purely morphological entities throughout the brain, and that the boundary between these two plates terminates rostrally in the postoptic recess.

In the rhombencephalon the basal plate and the alar plate can both be subdivided into two morphological zones or areas. Thus from ventral to dorsal a ventral, an intermedioventral, an intermediodorsal and a dorsal area can be distinguished. The boundaries between these zones are generally marked by ventricular sulci. Analysis of the spatial relations of the central components of the various cranial nerves has led to a subdivision of the rhombencephalon into four functional zones: somatomotor, visceromotor, viscerosensory and somatosensory. Topological analyses of the rhombencephala of many different vertebrates have revealed that the morphological and the functional zones in that brain part coincide largely but not entirely.

The sulcus limitans does not extend into the midbrain; hence, a subdivision of this brain part into a basal plate and an alar plate cannot be based on the localisation of this landmark. However, the mesencephalic cell masses are more or less clearly arranged into three longitudinal zones, medial tegmental, lateral tegmental and tectal. The medial zone contains mainly motor and motor-coordinating centres and hence can be considered a rostral extension of the somatomotor zone. The rhombencephalic somatosensory zone continues rostrally into the lateral tegmental and tectal zones. In all non-mammalian vertebrates the former contains a large special somatosensory relay centre, the torus semicircularis, whereas the latter represents mainly a general and special somatosensory centre of higher order.

Herrick and Kuhlenbeck subdivided the diencephalon into four longitudinal zones: epithalamus, thalamus dorsalis, thalamus ventralis and hypothalamus. Kuhlenbeck looked upon these zones as purely morphological units, but according to Herrick they represent structural as well as functional entities. There is now ample evidence that most of the four diencephalic zones of Herrick and Kuhlenbeck are not longitudinally oriented and do not represent fundamental morphological units. Ontogenetic studies have convincingly shown that the dorsal and ventral thalami are both neuromeric derivatives, and thus are oriented perpendicular rather than parallel to the longitudinal axis of the brain. The epithalamus develops from the same neuromere as the dorsal thalamus. The hypothalamus is the only diencephalic region which is more or less longitudinally oriented.

Within the telencephalic hemispheres of all vertebrates a dorsal pallium and a basal subpallium can be distinguished, and according to many authors, the cell masses contained within both of these regions are arranged in three zones, all of which are oriented parallel to the pallio-subpallial boundary. However, because, according to the thorough embryological studies of Källén and Bergquist, the subpallium and the pallium originate from post-neuromeric structures, designated as transverse bands 1 and 2 (Figs. 4.19, 4.20), the six telencephalic zones or columns are also oriented perpendicular to the axis of the brain.

4.7
Plan and Basic Subdivision of the Vertebrate Brain

4.7.1
Principles of Comparative Embryology

In 1828, Karl Ernst von Baer (1792–1876), the founder of embryology, published the first and principal volume of his master-work: "Über Entwicklungsgeschichte der Thiere. Beobachtung und Reflexion". In this treatise he presented a detailed account of the development of the chick, followed by a discussion of vertebrate ontogenesis in general. He formulated four general laws of development, which are of lasting significance and hence worth quoting in full (von Baer 1828, p. 224):

1. "Daß das Gemeinsame einer größeren Thiergruppe sich früher im Embryo bildet, als das Besondere." (That the general characteristics of a larger group appear earlier in the embryo than do the special characteristics.)
2. "Aus dem Allgemeinsten der Formverhältnisse bildet sich das weniger Allgemeine und so fort, bis endlich das Speciallste auftritt." (Less general structural relations develop from the more general, and so on, until finally the most special appear.)
3. "Jeder Embryo einer bestimmten Thierform, anstatt die andern bestimmten Formen zu durchlaufen, scheidet sich vielmehr von ihnen." (Each embryo of any given form, instead of passing through the adult stages of other forms, departs more and more from them.)
4. "Im Grunde ist also nie der Embryo einer höheren Thierform einer anderen Thierform gleich, sondern nur seinem Embryo." (Fundamentally, the embryo of a higher animal form never resembles the adult of another animal form, but only its embryo.)

One important consequence of von Baer's thesis that the embryo develops from the general to the specific is that the state in which each organ or structural component first appears must represent the general state of that organ or component within the group. Embryology will therefore be of paramount importance for the detection of the generalised type, the fundamental structural pattern or *Bauplan* upon which the species belonging to a particular group is built. The discovery of the type or *Bauplan* of a given group is an important aim of comparative morphological research. Conversely, such a type or *Bauplan* may provide the basis for more detailed analyses of organs or structural components of a particular group. The type was defined by von Baer (1828, p. 208) as "das Lagerungsverhältniss der Theile" (the spatial relationships of the parts).

Within a phylogenetic context the type or morphotype is the set of characters believed to have been present in the common ancestor of a given monophyletic group, based on a determination of shared primitive characters of the stem taxa (Northcutt 1985). This approach, which is known as cladistic analysis, will be discussed in Chap. 6. Suffice it to mention here that this approach requires the determination of the polarity (that is: primitive versus derived condition) of the characters being examined, and that the laws of von Baer presented above play a prominent role in the determination of this polarity. It is important to note that the term character "may be applied to any recognizable attribute of an organism" (Carroll 1988, p. 5), but that in morphology this term refers particularly to the spatial relationships of parts.

4.7.2
The General Plan of the Vertebrate Brain

What has been said above about the importance of embryology for comparative anatomy in general is fully applicable to the study of the CNS. The preceding sections show that numerous attempts have been made to derive the essence of the structural organisation of the vertebrate neuraxis from embryological studies. The presence of brain vesicles, neuromeres or transverse bands and longitudinal zones appeared to be the most important basic features detected.

If we survey the data discussed, the following conclusions may be drawn:

1. Even with respect to the significance of the three basic features just mentioned considerable differences of opinion exist.

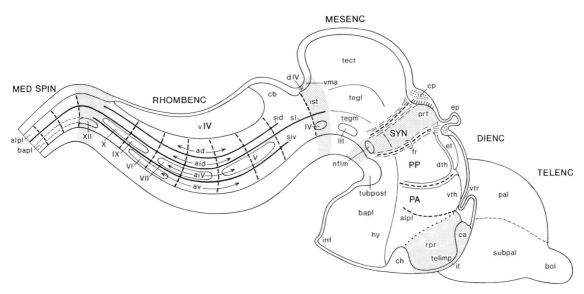

Fig. 4.53. The principal landmarks in the vertebrate brain as revealed by ontogenetic studies. The boundaries of the most important real and apparent longitudinal zones are superimposed upon the pattern of neuromeres (cf. Fig. 4.37). *ad*, Area dorsalis; *aid*, area intermediodorsalis; *aiv*, area intermedioventralis; *alpl*, alar plate; *av*, area ventralis; *bapl*, basal plate; *bol*, bulbus olfactorius; *cb*, cerebellum; *ch*, chiasma opticum; *cp*, commissura posterius; *ep*, epiphysis; *et*, epithalamus; *hy*, hypothalamus; *inf*, infundibulum; *ist*, isthmus; *lt*, lamina terminalis; *MED SPIN*, medulla spinalis; *nflm*, nucleus of the fasciculus longitudinalis medialis; *PA*, parencephalon anterius; *pal*, pallium; *PP*, parencephalon posterius; *prt*, area pretectalis; *sid*, sulcus intermedius dorsalis; *siv*, sulcus intermedius ventralis; *sl*, sulcus limitans; *subpal*, subpallium; *SYN*, synencephalon; *tect*, tectum; *tegkl*, tegmentum laterale; *tem*, tegmentum mediale; *telimp*, telencephalon impar; *tubpost*, tuberculum posterius; *Vth*, ventral thalamus; *vtr*, velum transversum; *vIV*, ventriculus quartus; *III–XII*, motor cranial nerve nuclei; transitional regions are marked with *dots*

2. So far, the research programs of only two groups, that of Kuhlenbeck and his allies and that of Bergquist and Källén and their allies, have included species belonging to almost all vertebrate groups.

3. The research programs mentioned under 2 have led to widely divergent results with regard to the essential features of the vertebrate brain; moreover, these programs were carried out before the advent of cladistics as a rigorous and logically consistent methodology for the determination of morphotypes and the unravelling of phylogenetic relationships. It follows that the determination of the morphotypes of, say, the vertebrate, the gnathostome, the tetrapod and the amniote brain remains an important aim of future research. However, this does not absolve us from the obligation to present a survey of the basic subdivision of the brain, as used in the various chapters of the specialised section of the present work. This subdivision rests first and foremost on the results of comparative embryological research and therefore approaches a morphotype. However, functional factors and the practical consideration of anatomical and didactic convenience have also been taken into account. In Fig. 4.53 the entities, boundaries, border

zones and landmarks to be discussed have been gathered together in the framework of the brain of an entirely hypothetical vertebrate embryo.

4.7.3
Subdivision of the Vertebrate Brain

Although the *subdivision* of the vertebrate brain into three primary brain vesicles, prosencephalon, mesencephalon and rhombencephalon, is widely used, its significance has been challenged by several investigators, among them Streeter (1933) and Bergquist (1966). In the present author's opinion, these vesicles, and their derivatives in adult brains, can be readily recognised in most groups although they do not feature as conspicuously as in many textbook diagrams (including Fig. 4.4). In many brains the derivatives of the three vesicles cannot be sharply delimited from each other; rather, they appear to be separated from or connected with each other by transitional zones. Moreover, their boundaries are generally more distinct on the dorsal side of the neuraxis than on the ventral side.

There is no sharp *boundary* between the spinal cord and the rhombencephalon, and there is evidence indicating that this boundary has shifted caudally in the course of phylogeny. The spinorhombencephalic boundary is situated between the exit of the last vagal and the first spinal root. In gnathostome anamniotes a varying number of so-called spino-occipital nerves (Fürbringer 1897) emerge from the area intervening between the two roots mentioned. These spino-occipital nerves are considered as originally being spinal nerves, which have been included secondarily in the head. During this assimilation these nerves lose their sensory, dorsal roots and become entirely somatomotor in nature. In amniotes the spino-occipital nerves unite in a single nerve root, the hypoglossal, which innervates the tongue muscles and the remaining hypobranchial musculature.

In Fig. 4.53 the rhombencephalon is depicted as being composed of *seven neuromeres* or transverse bands. Although nothing can be observed of these nerve segments on gross inspection of adult brains, several recent embryological studies (cf. Sect. 4.5) have shown their presence convincingly. Most motor cranial nerve nuclei have appeared to be derivatives of two adjacent rhombomeres, and the reticular neurons in the rhombencephalon have also been shown to be segmentally arranged, even in adult forms (cf., e.g., Lee et al. 1993).

The rhombencephalon can also be subdivided into *four longitudinal zones*, the boundaries of which are marked by ventricular sulci: s. intermedius dorsalis, s. limitans and s. intermedius ventralis. These longitudinal zones are considered here as purely morphological entities and designated accordingly as area dorsalis, area intermediodorsalis, area intermedioventralis and area ventralis. They correspond largely, but not entirely, with the well-known functional zones of Herrick and Johnston.

The *cerebellum* is to be considered a dorsal outgrowth of the first rhombomere.

In non-mammalian vertebrates the *rhombencephalon* cannot be subdivided into a rostral metencephalon and a caudal myelencephalon, since macroscopic or microscopic landmarks upon which such a subdivision can be based are lacking.

The rostral boundary of the rhombencephalon is marked by a constriction known as the *isthmus*. It is worthy of note that several authors, among them Vaage (1969; Fig. 4.25) and O'Rahilly et al. (1989; Fig. 4.29), considered the neural wall surrounding this narrowing to be a derivative of a separate isthmic neuromere, and that His (1893b; Fig. 4.43) interpreted this area as one of the six main divisions of the brain. I consider the area surrounding the isthmus a transitory region, containing a number of characteristic cell masses, including the locus coeruleus, the nucleus visceralis secundarius, the nucleus isthmi and the interpeduncular nucleus.

According to von Kupffer (1906; Fig. 4.5c), the boundary between the rhombencephalon and mesencephalon is dorsally marked by the cerebellar commissure. The exit of the trochlear nerve from the velum medullare anterius, or its decussation within that structure, represents a more reliable landmark of that boundary. However, an exception should be made here for the actinopterygians. In this group the valvula cerebelli develops from the anterior medullary velum (Fig. 4.10). This development, which shows amazing differences among the various actinopterygian groups, strongly influences the central course and the site and mode of decussation of the fibres of the trochlear nerve (van der Horst 1918).

The *sulcus limitans* is a most important structural and functional landmark in the spinal cord and the rhombencephalon, because it marks the boundary between the sensory alar plate and the motor basal plate. Rostrally, this ventricular groove cannot be traced beyond the isthmus region. However, there is evidence suggesting that the alar and basal plated extend, at least as morphological entities, throughout the brain, and that the interface between these two entities terminates rostrally in the postoptic recess (Fig. 4.53).

The *mesencephalon* is divisible into three longitudinal zones, medial tegmental, lateral tegmental and tectal. Functionally, the medial tegmental zone is closely associated with the rhombencephalic somatomotor zone, whereas the lateral tegmental and tectal zones represent rostral continuations of the somatosensory zone.

A transitional zone, known as the *synencephalon*, connects the mesencephalon with the prosencephalon. There is convincing evidence that this zone is the product of a single neuromere. Its rostral and caudal boundaries are marked by the fasciculus retroflexus and the posterior commissure, respectively. The dorsal part of the synencephalon is occupied by the area pretectalis, a complex of visual centres, whereas its ventral part contains the large-celled nucleus of the fasciculus longitudinalis medialis (flm).

Embryological studies have shown that the synencephalon develops from the most caudal prosomere; thus, according to morphological standards, this brain part should be allocated to the diencephalon. However, because (a) the pretectum is structurally and functionally closely related to the tectum, and (b) the nucleus of the flm represents the most rostral part of the reticular formation of the brain stem, it is convenient and common practice to treat the synencephalon as forming part of the midbrain.

It is practical and practicable to subdivide the *prosencephalon* into a rostral telencephalon and a caudal diencephalon. No anatomist will deny the correctness of this thesis, but the theoretical foundations upon which this subdivision rests or should rest have been endlessly debated in the literature. Because it is extremely difficult to decipher the vestiges of neuromery in the most rostral parts of the neuraxis, the problem revolves around the following interlocking questions: (a) Does the telencephalon, as its name implies, represent the morphologically most rostral end of the brain? (b) Where is the rostral end of the neuraxis located? (c) Does the telencephalon represent a complete ring or segment of the brain? (d) Do the basal plate and the alar plate extend into the telencephalon, or is this brain part only a derivative of the alar plate? The sometimes widely diverging answers given to these questions have been extensively discussed in the preceding sections of this chapter and need not be reconsidered here. Suffice it to mention that, in my opinion, the telencephalon and the hypothalamus are probably derived from the same set of neuromeres, and that the basal plate is not involved in the formation of the telencephalon.

As regards the location of the telodiencephalic boundary, most neuroanatomists are, with Johnston (1909), of the opinion that this boundary can best be defined as a plane passing from the velum transversum to the optic chiasma and that, hence, the preoptic region belongs to the telencephalon. The ideas behind this concept are, briefly, that (a) the lamina terminalis represents the rostral closure of the neuraxis; (b) the telencephalon arises from the rostral part of the neural tube; (c) the area of the neural wall involved in the evagination leading to the formation of the telencephalic hemispheres is situated at some distance behind the lamina terminalis. It follows that (d) the lamina terminalis and the regio preoptica, located immediately behind that structure, represent together the telencephalon impar or telencephalon medium, i.e. the unevaginated rostralmost portion of the endbrain. From a purely morphological point of view, this interpretation is correct; however, because the preoptic region forms a structural as well as a functional continuum with the hypothalamus, it may be convenient to treat these two entities as forming a single complex and to allocate this preoptico-hypothalamic continuum to the diencephalon.

Two neuromeres, the *parencephalon posterius* and the *parencephalon anterius*, are probably involved in the formation of the morphologically dorsal part of the diencephalon *sensu strictiori* (that is, exclusive of the synencephalon). The epithalamus and the dorsal thalamus are derived from the parencephalon posterius, whereas the ventral thalamus arises from the parencephalon anterius. The

morphological status of the hypothalamus is uncertain. It seems likely that this brain part originates from the neuromeres which also give rise to the telencephalon, but the parencephala, particularly the parencephalon anterius, may also be involved.

The portion of the floor of the brain which, due to the strong cephalic flexure, protrudes into the third ventricle is known as the tuberculum posterius. Clear morphological landmarks are rare in this part of the brain; hence, this protrusion has been allocated to different domains. Herrick (1910, 1948) considered it as belonging to the mesencephalon, but the extensive embryological studies of Bergquist (1932) have shown that the tuberculum posterius and the adjacent parts of the lateral wall of the brain form part of the diencephalon. According to Bergquist and Källén (1954; Figs. 4.19, 4.20), the transverse bands which give rise to the dorsal and ventral parts of the thalamus merge ventrally into the area tuberculi posterioris.

Herrick (1910) divided the *diencephalon* into four longitudinal zones: epithalamus, dorsal thalamus, ventral thalamus and hypothalamus. These zones are, according to Herrick, separated form each other by ventricular grooves, the sulcus diencephalicus dorsalis, -medius and -ventralis (Figs. 4.46, 4.54). He emphasised that these sulci define regions which are functionally as well as morphologically distinct. This zonal subdivision of

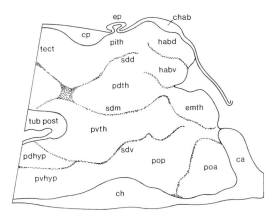

Fig. 4.54. The diencephalic zonal pattern in the adult alpine newt, *Triturus alpestris*. The boundaries of the periventricular cell masses are marked by ventricular sulci. *ca,* Commissura anterior; *ch,* chiasma opticum; *chab,* commissura habenulae; *cp,* commissura posterior; *emth,* eminentia thalami; *ep,* epiphysis; *habd,* nucleus habenulae, pars dorsalis; *habv,* idem, pars ventralis; *pdhyp,* pars dorsalis hypothalami; *pdth,* pars dorsalis thalami; *pith,* pars intercalaris thalami; *poa,* nucleus preopticus, pars anterior; *pop,* idem, pars posterior; *pvhyp,* pars ventralis hypothalami; *pvth,* pars ventralis thalami; *sdd,* sulcus diencephalicus dorsalis; *sdm,* idem, medius; *sdv,* idem, ventralis; *tect,* tectum; *tubpost,* tuberculum posterius (modi<fied from Wicht and Himstedt 1988)

the diencephalon has since been adopted by the great majority of all vertebrates. Ariëns Kappers et al. (1936, p. 867) considered it applicable to all vertebrates, with the possible exception of certain cyclostomes. Although we now know that two of Herrick's zones, i.e. the dorsal thalamus and the ventral thalamus, are direct derivatives of neuromeres, and that the topographically horizontal orientation of these entities is due to the strong curvature of the rostral part of the brain, it cannot be denied that Herrick's subdivision offers a convenient starting point for the structural and functional analysis of the diencephalon. It is for this practical reason that this partitioning has been used throughout the specialised part of the present work.

The *telencephalon* of all vertebrates is divisible into four major regions: olfactory bulb, pallium, subpallium or basis, and telencephalon impar.

References

Altner H, Zimmermann H (1972) The saccus vasculosus. In: Bourne G (ed) Structure and function of nervous tissue, vol 5. Academic, New York, pp 293–328

Ariëns Kappers CU (1908) Eversion and inversion of the dorsolateral wall in different parts of the brain. J Comp Neurol 18:433–436

Ariëns Kappers CU (1921) Die vergleichende Anatomie des Nervensystems der Wirbeltiere und des Menschen, II. Abschnitt. Bohn, Haarlem

Ariëns Kappers J (1950) The development and structure of the paraphysis cerebri in urodeles with experiments on its function in Ambystoma mexicanum. J Comp Neurol 92:93–127

Ariëns Kappers J (1956) On the development, structure and function of the paraphysis cerebri. In: Ariëns Kappers J (ed) Progress in neurobiology. Elsevier, Amsterdam, pp 130–145

Ariëns Kappers J (1965) Survey of the innervation of the epiphysis cerebri and the accessory pineal organs of vertebrates. Prog Brain Res 10:87–153

Ariëns Kappers CU, Huber GC, Crosby EC (1936) The comparative anatomy of the nervous system of vertebrates, including man. MacMillan, New York

Baer KE von (1828) Über die Entwickelungsgeschichte der Thiere, Beobachtung und Reflexion. Bornträger, Königsberg

Balinsky BJ (1970) An introduction to embryology. Saunders, Philadelphia

Bartelmez GW (1915) Mauthner's cell and the nucleus motorius tegmenti. J Comp Neurol 25:87–128

Becarri N (1943) Neurologia. Comparata Sansoni, Florence

Bengmark S, Hugosson R, Källén B (1953) Studien über Kernanlagen im Mesencephalon sowie im Rostralteil des Rhombencephalon von Mus musculus. Z Anat Entw Gesch 117:73–91

Bergquist H (1932) Zur Morphologie des Zwischenhirns bei niederen Wirbeltieren. Acta Zool 13:57–303

Bergquist H (1952a) The formation of neuromeres in homo. Acta Soc Med Ups 57:23–32

Bergquist H (1952b) Transversal bands and migration areas in Lepidochelys olivacea. Kgl Fysiogr Sällsk Lund Handl N F 63:13

Bergquist H (1952c) Studies on the cerebral tube in vertebrates. The neuromers. Acta Zool (Stockh) 33:117–187

Bergquist H (1953) On the development of diencephalic nuclei and certain mesencephalic relations in Lepidochelys olivacea and other reptiles. Acta Zool (Stockh) 34:155–190

Bergquist H (1954a) Morphogenesis of diencephalic nuclei in homo. Kgl Fysiogr Sällsk Lund Handl N F 64:20

Bergquist H (1954b) Ontogenesis of diencephalic nuclei in vertebrates. Kgl Fysiogr Sällsk Lund Handl N F 65:6

Bergquist H (1957) Comments on the architype of the vertebrate brain. Kgl Fysiogr Sällsk Lund Handl 27:153–160

Bergquist H (1964) Die Entwicklung des Diencephalons im Lichte neuer Forschung. Prog Brain Res 5:223–229

Bergquist H, Källén B (1953a) Studies on the topography of the migration areas in the vertebrate brain. Acta Anat (Basel) 17:353–369

Bergquist H, Källén B (1953b) On the development of neuromeres to migration areas in the vertebrate cerebral tube. Acta Anat (Basel) 18:65–73

Bergquist H, Källén B (1954) Notes on the early histogenesis and morphogenesis of the central nervous system in vertebrates. J Comp Neurol 100:627–659

Bergquist H, Källén B (1955) The archencephalic neuromery in Ambystoma punctatum: an experimental study. Acta Anat (Basel) 24:208–214

Bone Q (1960) The central nervous system in amphioxus. J Comp Neurol 115:27–64

Brodal A, Jansen J, Kristiansen K (1950) Experimental demonstration of a pontine homologue in birds. J Comp Neurol 92:23–70

Bulfone A, Puelles L, Porteus MH, Frohman MA, Martin GR, Rubinstein JL (1993) Spatially restricted expression of Dix-1, Dix-2 (Tes-1), Gbx-2, and Wnt-3 in the embryonic day 12.5 mouse forebrain defines potential transverse and longitudinal boundaries. J Neurosci 13:3155–3172

Carroll RL (1988) Vertebrate paleontology and evolution. Freeman, New York

Chitnis AB, Kuwada JY (1990) Axonogenesis in the brain of zebrafish embryos. J Neurosci 10:1892–1905

Christ JF (1969) Derivation and boundaries of the hypothalamus, with atlas of hypothalamic grisea. In: Haymaker W, Anderson E, Nauta WJH (eds) The hypothalamus. Thomas, Springfield, Illinois, pp 13–60

Coggeshall RE (1964) A study of diencephalic development in the albino rat. J Comp Neurol 122:241–269

Conel J (1929) The development of the brain of Bdellostoma stouti. I. Internal growth changes. J Comp Neurol 47:343–403

Conel J (1931) The development of the brain of Bdellostoma stouti. II. External growth changes. J Comp Neurol 47:343–403

Costanzo R, Watterson RL, Schoenwolf GC (1982) Evidence that secondary neurulation occurs autonomously in the chick embryo. J Exp Zool 219:233–240

Criley BB (1969) Analysis of the embryonic sources and mechanisms of development of posterior levels of chick neural tubes. J Morphol 128:465–501

Cruce WLR, Nieuwenhuys R (1974) The cell masses in the brain stem of the turtle, Testudo hermanni; a topographical and topological analysis. J Comp Neurol 156:277–306

Dammerman KW (1910) Der Saccus vasculosus der Fische ein Tieforgan. Z Wiss Zool 96:654–726

Davis CA, Noble-Topham SE, Rossant J, Joyner AC (1988) Expression of the homeobox-containing gene En-2 delineates a specific region of the developing mouse brain. Genes Dev 2:361–371

De Lange D (1936) The head problem in chordates. J Anat 70:515–547

Detweiler SR (1934) An experimental study of spinal nerve segmentation in Amblystoma with reference to the plurisegmental contribution to the brachial plexus. J Exp Zool 67:395–441

Edinger L (1908) Vorlesungen über den Bau der nervösen Zentralorgane. II. Vergleichende Anatomie des Gehirns, 7th edn. Vogel, Leipzig

Ekström P (1987) Photoreceptors and CSF-contacting neurons in the pineal organ of a teleost fish have direct axonal connections with the brain: an HRP-electron-microscopic study. J Neurosci 7:987–995

Figdor MC, Stern CD (1993) Segmental organization of embryonic diencephalon. Nature 363:630–634

Franz V (1911a) Das Kleinhirn der Knochenfische. Zool Jahrb 32:401–464

Franz V (1911b) Das Mormyridenhirn. Zool Jahrb 32:465–492

Fraser S, Keynes R, Lumsden A (1990) Segmentation in the chick embryo hindbrain is defined by cell lineage restrictions. Nature 344:431–435

Fraser SE (1993) Segmentation moves to the fore. Fate mapping and cell lineage studies reveal neuromeres in the avian forebrain, suggesting that similar cellular and molecular mechanisms operate in both forebrain and hindbrain. Curr Biol 3:787–789

Fürbringer M (1897) Über die spino-occipitalen Nerven der Selachier und Holocephalen und ihre vergleichende Morphologie. Festschr Gegenbaur 3:349–788

Gardner CA, Darnell DK, Poole SJ, Ordahl CP, Barald KF (1988) Expression of an engrailed-like gene during development of the early chick nervous system. J Neurosci Res 21:426–437

Gaskell WH (1886) On the structure, distribution and function of the nerves which innervate the visceral and vascular systems. J Physiol 7:1–81

Gaskell WH (1889) On the relation between the structure, function, distribution and origin of the cranial nerves; together with a theory of the origin of the nervous system of vertebrata. J Physiol 10:153–211

Gehring W (1987) Homeoboxes in the study of development. Science 236:1245–1252

Gerlach J (1933) Über das Gehirn von Protopterus annectens. Anat Anz 75:305–448

Gerlach J (1946) Beiträge zur vergleichenden Morphologie des Selachierhirnes. Anat Anz 96:79–165

Gribnau AAM, Geijsberts LGM (1985) Morphogenesis of the brain in staged rhesus monkey embryos. Adv Anat Embryol Cell Biol 91:1–69

Hamburger V, Hamilton H (1951) A series of normal stages in the development of the chick embryo. J Morphol 88:49–91

Hammelbo T (1972) On the development of the cerebral fissures in Cetacea. Acta Anat (Basel) 82:606–618

Hanneman E, Westerfield M (1989) Early expression of acetylcholinesterase activity in functionally distinct neurons of the zebrafish. J Comp Neurol 284:350–361

Hanneman E, Trevarrow B, Metcalfe WK, Kimmel CB, Westerfield M (1988) Segmental pattern of development of the hindbrain and spinal cord of the zebrafish embryo. Development 103:49–48

Hatschek B (1882) Studien über die Entwicklung von Amphioxus. Arb Zool Inst Univ Wien 4:1–88

Heier P (1948) Fundamental principles in the structure of the brain. Acta Anat [Suppl] VI:213

Heijdra YF, Nieuwenhuys R (1994) Topological analysis of the brainstem of the bowfin, Amia calva. J Comp Neurol 339:12–26

Herrick CJ (1899) The cranial and first spinal nerves of Menidia: a contribution upon the nerve components of the bony fishes. J Comp Neurol 9:153–1455

Herrick CJ (1905) The central gustatory paths in the brains of bony fishes. J Comp Neurol 15:375–456

Herrick CJ (1906) On the centers for taste and touch in the medulla oblongata of fishes. J Comp Neurol 16:403–439

Herrick CJ (1907) A study of the vagal lobes and funicular nuclei of the brain of the codfish. J Comp Neurol 17:67–78

Herrick CJ (1910) The morphology of the forebrain in amphibia and reptilia. J Comp Neurol 20:413–547

Herrick CJ (1913) Anatomy of the brain. In: The reference handbook of the medical sciences, vol 2, 3rd edn. Wood, New York, pp 274–342

Herrick CJ (1914) The medulla oblongata of larval Amblystoma. J Comp Neurol 24:343–427

Herrick CJ (1917) The internal structure of the midbrain and thalamus of Necturus. J Comp Neurol 28:215–348

Herrick CJ (1922) Functional factors in the morphology of the forebrain of fishes. Libro en honor de D Santiago Ramón y Dajal. Jiméney Molina, Madrid, pp 142–204

Herrick CJ (1930) The medula oblongata of Necturus. J Comp Neurol 50:1–96

Herrick CJ (1933) Morphogenesis of the brain. J Morphol 54:233–258

Herrick CJ (1935) A topographic analysis of the thalamus and midbrain of Amblystoma. J Comp Neurol 62:239–261

Herrick CJ (1943) The cranial nerves. A review of fifty years. Denison Univ Bull J Sci Lab 38:41–51

Herrick CJ (1948) The brain of the tiger salamander. University of Chicago Press, Chicago

Hetherington TE (1981) Morphology of the pineal organ in the salamander Ensatina eschscholtzi. J Morphol 169:191–206

Hill C (1900) Developmental history of primary segments of the vertebrate head. Zool Jahrb 13:393–446

His W (1888) Zur Geschichte des Gehirns sowie der centralen und periferischen Nervenbahnen beim menschlichen Embryo. Abh Math Phys Kl Kgl Sächs Ges Wiss 14:339–393

His W (1891) Die Entwickelung des menschlichen Rautenhirns, vom Ende des ersten bis zum Beginn des dritten Monats. I. Verlängertes Mark. Abh Math Phys Kl Kgl Sächs Ges Wiss 17:1–75

His W (1893a) Über das frontale Ende des Gehirnrohres. Arch Anat Physiol Anat Abt 157–172

His W (1893b) Vorschläge zur Eintheilung des Gehirns. Arch Anat Physiol Anat Abt 172–180

Holmgren N (1922) Points of view concerning forebrain morphology in lower vertebrates. J Comp Neurol 34:391–459

Holmgren N (1925) Points of view concerning forebrain morphology in higher vertebrates. Acta Zool 6:413–477

Holmgren N (1946) On two embryos of Myxine glutinosa. Acta Zool 27:1–90

Holmgren N, van der Horst CJ (1925) Contribution to the morphology of the brain of Ceratodus. Acta Zool 6:59–165

Huber GC, Crosby EC (1929) The nuclei and fiber paths of the avian diencephalon, with consideration of telencephalic and certain mesencephalic centers and connections. J Comp Neurol 48::1–186

Hugosson R (1955) Studien über die Entwicklung der longitudinalen Zellsäulen und der Anlagen der Gehirnnervenkerne in der Medulla oblongata bei verschiedenen Vertebraten. Z Anat Entw Gesch 118:543–566

Hugosson R (1957) Morphologic and experimental studies on the development and significance of the rhombencephalic longitudinal cell columns. Thesis, Lund

International Anatomical Nomenclature Committee (1977) Nomina anatomica, 4th edn. Excerpta Medica, Amsterdam

Jansen J (1969) On cerebellar evolution and organization from the point of view of a morphologist. In: Llinás R (ed) Neurobiology of cerebellar evolution and development. American Medical Association, Chicago, pp 881–893

Jarvik E (1980) Basic structure and evolution of vertebrates, vol 1. Academic, New York

Johnston JB (1901) Das Gehirn und die Cranialnerven der Anamnier. Ergebn Anat Entw Gesch 11:973–1112

Johnston JB (1902a) The brain of Acipenser. Zool Jahrb 15:59–260

Johnston JB (1902b) The brain of Petromyzon. J Neurol 12:1–86

Johnston JB (1902c) An attempt to define the primitive functional divisions of the central nervous system. J Comp Neurol 12:87–106

Johnston JB (1905) The cranial nerve components of Petromyzon. Morphol Jahrb 34:149–203

Johnston JB (1909) The morphology of the forebrain vesicle in vertebrates. J Comp Neurol 19:457–539

Johnston JB (1923) Further contributions to the study of the evolution of the forebrain. J Comp Neurol 36:143–192

Kahle W (1956) Zur Entwicklung des menschlichen Zwischenhirnes: Studien ueber die Matrix-phasen und die örtlichen Reifungsunterschiede im embryonalen menschlichen Gehirn: II. Mitteilung. Dtsch Z Nervenheilkd 175:259–318

Kahle W (1958) Über die längszonale Gliederung des menschlichen Zwischenhirnes. In: Currie SB, Martini L (eds) Pathophysiologica diencephalica. Springer, Vienna, pp 134–142

Källén B (1951a) Contributions to the ontogeny of the nuclei and the ventricular sulci in the vertebrate forebrain. Kgl Fysiogr Sällsk Lund Handl N F 62:5–34

Källén B (1951b) Embryological studies on the nuclei and their homologization in the vertebrate forebrain. Kgl Fysiogr Sallsk Lund Handl N F 62:3–34

Källén B (1951c) The nuclear development in the mammalian forebrain with special regard to the subpallium. Kgl Fysiogr Sällsk Lund Handl N F 61:3–43

Källén B (1951d) On the ontogeny of the reptilian forebrain. Nuclear structures and ventricular sulci. J Comp Neurol 95:307–347

Källén B (1951e) Some remarks on the ontogeny of the telencephalon in some lower vertebrates. Acta Anat (Basel) XI:537–548

Källén B (1952) Notes on the proliferation processes in the neuromeres in vertebrate embryos. Acta Soc Med Ups 57:111–118

Källén B (1953a) On the significance of the neuromeres and similar structures in vertebrate embryos. J Embryol Exp Morphol 1 (4):387–392

Källén B (1953b) Notes on the development of the neural crest in the head of musculus. J Embryol Exp Morphol 1 (4):393–398

Källén B (1954) On the segmentation of the central nervous system. Kgl Fysiogr Sällsk Lund Handl N F 64 (18):3–10

Källén B (1955) Neuromery in living and fixed chick embryos. Kgl Fysiogr Sällsk Lund Handl 25:1–6

Källén B (1962) Embryogenesis of brain nuclei in the chick telencephalon. Ergeb Anat Entw Gesch 36:62–82

Kamon K (1905) Zur Entwicklungsgeschichte des Gehirns des Hünchens. Anat Hefte (Wiesbaden) 30:560–650

Keynes R, Lumsden A (1990) Segmentation and the origin of regional diversity in the vertebrate central nervous system. Neuron 2:1–9

Keynes R, Cook G, Davies J, Lumsden A, Norris W, Stern C (1990) Segmentation and the development of the vertebrate nervous system. J Physiol (Paris) 84:27–32

Keynes RJ, Stern CD (1984) Segmentation in the vertebrate nervous system. Nature 310:786–789

Keynes RJ, Stern CD (1985) Segmentation and neural development in vertebrates. Trends Neurosci 8:220–223

Keyser AJM (1972) The development of the diencephalon of the Chinese hamster. Acta Anat (Basel) [Suppl] 59/1 (83):1–181

Kimmel CB (1993) Patterning the brain of the zebrafish embryo. Annu Rev Neurosci 16:707–732

Kimmel CB, Sepich DS, Trevarrow B (1988) Development of segmentation in zebrafish. Development 104:197–207

Kingsbury BF (1920) The extent of the floor-plate of His and its significance. J Comp Neurol 32:113–135

Kingsbury BF (1922) The fundamental plan of the vertebrate brain. J Comp Neurol 34:461–491

Kingsbury BF (1930) The developmental significance of the floor-plate of the brain and spinal cord. J Comp Neurol 50:177–207

Korf H-W, Liesner R, Meissl H, Kirk A (1981) Pineal complex of the clawed toad, Xenopus laevis Daud: structure and function. Cell Tissue Res 216:113–130

Kremers JWM, Nieuwenhuys R (1979) A topological analysis of the brainstem of the crossopterygian fish, Latimeria chalumnae. J Comp Neurol 187:613–637

Krumlauf R, Marshall H, Studer M, Nonchev S, Sham MH, Lumsden A (1993) Hox Homeobox genes and regionalisation of the nervous system. J Neurobiol 24:1328–1340

Kuhlenbeck H (1926) Betrachtungen über den funktionellen Bauplan des Zentralnervensystems. Folia Anat Jpn 4:111–135

Kuhlenbeck H (1927) Vorlesungen über des Zentralnervensystem der Wirbeltiere. Fischer, Jena

Kuhlenbeck H (1929a) Über die Grundbestandteile des Zwischenhirnbauplans der Anamnier. Morphol Jahrb 63:50–95

Kuhlenbeck H (1929b) Die Grundbestandteile des Endhirns im Lichte der Bauplanlehre. Anat Anz 67:1–51

Kuhlenbeck H (1931) Über die Grundbestandteile des Zwischenhirns bei Reptilien. Morphol Jahrb 66:244–317

Kuhlenbeck H (1936) Ueber die Grundbestandteile des Zwischenhirnbauplans der Vögel. Morphol Jahrb 77:61–109

Kuhlenbeck H (1937) The ontogenetic development of the diencephalic centers in a bird's brain (chick) and comparison with the reptilian and mammalian diencephalon. J Comp Neurol 66:23–75

Kuhlenbeck H (1938) The ontogenetic development and phylogenetic significance of the cortex telencephali in the chick. J Comp Neurol 69:273–301

Kuhlenbeck H (1939) The development and structure of the pretectal cell masses in the chick. J Comp Neurol 71:361–387

Kuhlenbeck H (1954) The human diencephalon. A summary of development, structure, function and pathology. Confin Neurol [Suppl] 14:1–230

Kuhlenbeck H (1956) Die Formbestandteile der Regio praetectalis des Anamnier-Gehirns und ihre Beziehungen zum Hirnbauplan. Folia Anat Jpn 28:23–44

Kuhlenbeck H (1973) The central nervous system of vertebrates, vol 3, part II: overall morphologic pattern. Karger, Basel

Kuhlenbeck H (1977) The central nervous system of vertebrates, vol 5, part I: derivatives of the prosencephalon: diencephalon and telencephalon. Karger, Basel

Kuhlenbeck H, Miller RN (1942) The pretectal region of the rabbit's brain. J Comp Neurol 76:323–365

Kuhlenbeck H, Miller RN (1949) The pretectal region of the human brain. J Comp Neurol 91:369–407

Kuhlenbeck H, Niimi K (1969) Further observations on the morphology of the brain in the Holocephalian Elasmobranchs Chimaera and Callorhynchus. J Hirnforsch 11:267–314

Lakke EAJF, van der Veeken JGPM, Mentink MMT, Marani E (1988) A SEM study on the development of the ventricular surface morphology in the diencephalon of the rat. Anat Embryol (Berl) 179:73–80

Larsell O (1967) The comparative anatomy and histology of the cerebellum from Myxinoids through birds. University of Minnesota Press, Minneapolis

Lee KKL, Eaton RC (1991) Identifiable reticulospinal neurons of the adult zebrafish, Brachydanio rerio. J Comp Neurol 304:34–52

Lee KKL, Eaton RC, Zottoli SJ (1993) Segmental arrangement of hindbrain neurons in the goldfish, Carassius auratus. J Comp Neurol 329:539–556

Lehmann F (1927) Further studies on the morphogenetic role of the somites in the development of the nervous system of amphibians. J Exp Zool 49:93–131

Lewis J (1989) Genes and segmentation. Nature 341:382–383

Locy WA (1895) Contributions to the structure and development of the vertebrate head. J Morphol 11:497–544

Lumsden A (1990) The cellular basis of segmentation in the developing hindbrain. Trends Neurosci 13:329–335

Lumsden A, Keynes R (1989) Segmental patterns of neuronal development in the chick hindbrain. Nature 337:424–428

Mendelson B (1985) Soma position is correlated with time of development in three types of identified reticulospinal neurons. Dev Biol 112:489–493

Mendelson B (1986a) Development of reticulospinal neurons of the zebrafish. I. Time of origin. J Comp Neurol 251:160–171

Mendelson B (1986b) Development of reticulospinal neurons of the zebrafish. II. Early axonal outgrowth and cell body position. J Comp Neurol 251:172–184

Metcalfe WK, Mendelson B, Kimmel CB (1986) Segmental homologies among reticulospinal neurons in the hindbrain of the zebrafish larva. J Comp Neurol 251:147–159

Miller RN (1940) The telencephalic zonal system of the teleost Corydoras paliatus. J Comp Neurol 72:149–176

Müller F, O'Rahilly R (1987) The development of the human brain, the closure of the caudal neuropore, and the beginning of secondary neurulation at stage 12. Anat Embryol (Berl) 176:413–430

Müller F, O'Rahilly R (1988) The development of the human brain, including the longitudinal zoning in the diencephalon at stage 15. Anat Embryol (Berl) 179:55–71

Müller F, O'Rahilly R (1989) The human brain at stage 17, including the appearance of the future olfactory bulb and the first amygdaloid nuclei. Anat Embryol (Berl) 180:353–369

Neal HV (1898) The segmentation of the nervous system in Squalus acanthiasis. Bull Mus Comp Zool Harv 31:293–315

Neal HV (1918) Neuromeres and metameres. J Morphol 31:293–315

Nieuwenhuys R (1965) The forebrain of the crossopterygian Latimeria chalumnae Smith. J Morphol 117:1–24

Nieuwenhuys R (1972) Topological analysis of the brain stem of the lamprey Lampetra fluviatilis. J Comp Neurol 145:165–177

Nieuwenhuys R (1974) Topological analysis of the brain stem: a general introduction. J Comp Neurol 156:255–276

Nieuwenhuys R, Nicholson C (1969a) A survey of the general morphology, the fiber connections, and the possible functional significance of the gigantocerebellum of mormyrid fishes. In: Llinás R (ed) Neurobiology of cerebellar evolution and development. 1st Int Symp Inst Biomed Res Am Med Ass, Chicago, pp 107–134

Nieuwenhuys R, Nicholson C (1969b) Aspects of the histology of the cerebellum of mormyrid fishes. In: Llinás R (ed) Neurobiology of cerebellar evolution and development. 1st Int Symp Inst Biomed Res Am Med Ass, Chicago, pp 135–169

Nieuwenhuys R, Meek J (1985) Constructional principles of the brain stem in anamniotes, with emphasis on actinopterygian fishes. In: Duncker H-R, Fleischer G (eds) Functional morphology in vertebrates. Fisher, Stuttgart, pp 515–528

Nieuwenhuys R, Oey PL (1983) Topological analysis of the brainstem of the reedfish, Erpetoichthys calabaricus. J Comp Neurol 213:220–232

Nieuwenhuys R, Pouwels E (1983) The brainstem of actinopterygian fishes. In: Northcutt R, Davis (eds) Fish neurobiology, vol 1: brainstem and sense organs. University of Michigan Press, Ann Arbor, MI, pp 25–87

Nikundiwe AM, Nieuwenhuys R (1983) The cell masses in the brainstem of the South African clawed frog Xenopus laevis: a topographical and topological analysis. J Comp Neurol 213:199–219

Noback CR, Demarest RJ (1975) The human nervous system. McGraw-Hill, New York

Northcutt RG (1967) Architectonic studies of the telencephalon of Iguana iguana. J Comp Neurol 130:109–147

Northcutt RG (1970) The telencephalon of the western painted turtle, Chrysemys picta belli. University of Illinois Press, Urbana (Illinois biological monographs, no 43)

Northcutt RG (1981) Evolution of the telencephalon in nonmammals. Annu Rev Neurosci 4:301–350

Northcutt RG (1985) The brain and sense organs of the earliest craniates: reconstruction of a morphotype. In: Foreman RE, Gorbman A, Dodds JM, Olson R (eds) Evolutionary biology of primitive fishes. Plenum, New York, pp 81–112

Northcutt RG (1987) Evolution of the vertebrate brain. In: Adelman G (ed) Encyclopedia of neuroscience, vol 1. Birkhäuser, Basel, pp 415–418

Oberdick J, Schilling K, Smeyne RJ, Corbin JG, Bocchiaro C, Morgan JI (1993) Control of segment-like patterns of gene expression in the mouse cerebellum. Neuron 10:1007–1018

Opdam P, Nieuwenhuys R (1976) Topological analysis of the brainstem of the axolotl Ambystoma mexicanum. J Comp Neurol 165:285–306

Opdam P, Kemali M, Nieuwenhuys R (1976) Topological analysis of the brainstem of the frogs Rana esculenta and Rana catesbeiana. J Comp Neurol 165:307–332

O'Rahilly R, Gardner E (1979) The initial development of the human brain. Acta Anat (Basel) 104:123–133

O'Rahilly R, Müller F, Bossy J (1989) Atlas des stades de développement de l'encéphale chez l'embryon humain etudié par des reconstructions graphiques du plan médian. Arch Anat Histol Embryol 72:3–34

Orr HA (1887) Contributions to the embryology of the lizard. J Morphol 1:311–372

Osborn HF (1889) A contribution to the internal structure of the amphibian brain. J Morphol 2:51–96

Palmgren A (1921) Embryological and morphological studies on the mid-brain and cerebellum of vertebrates. Acta Zool 2:1–94

Papez JW (1940) The embryologic development of the hypothalamic area in mammals. Proc Assoc Res Nerv Ment Dis 20:31–51

Puelles L, Medina L (1994) Development of neurons expressing tyrosine hydroxylase and dopamine in the chicken brain: a comparative segmental analysis. In: Smeets WJAJ, Reiner A (eds) Phylogeny and development of catecholamine systems in the CNS of vertebrates. Cambridge University Press, Cambridge, pp 381–404

Puelles L, Rubenstein JL (1993) Expression patterns of homeobox and other putative regulatory genes in the embryonic mouse forebrain suggest a neuromeric organization. Trends Neurosci 16:472–479

Puelles L, Domenech-Ratto G, Martinez de la Torre M (1987a) Location of the rostral end of the longitudinal brain axis: review of an old topic in the light of marking experiments on the closing rostral neuropore. J Morphol 194:163–171

Puelles L, Amat JA, Martinez-de-la-Torre (1987b) Segment-related, mosaic neurogenetic pattern in the forebrain and mesencephalon of early chick embryos: I. Topography of AChE-positive neuroblasts up to stage HH18. J Comp Neurol 266:247–268

Puelles L, Robles C, Martinez de la Torre M, Martinez S (1994) New subdivision schema for the avian torus semicircularis: neurochemical maps in the chick. J Comp Neurol 340:98–125

Reichenbach A, Schaaf P, Schneider H (1990) Primary neurulation in teleosts – evidence for epithelial genesis of central nervous tissue as in other vertebrates. J Hirnforsch 31:153–158

Rendahl H (1924) Embryologische und Morphologische Studien über das Zwischenhirn beim Huhn. Acta Zool 5:241–344

Ross LS, Parrett T, Stephen S, Easter S jr (1992) Axonogenesis and morphogenesis in the embryonic zebrafish brain. J Neurosci 12:467–482

Rudebeck B (1945) Contributions to forebrain morphology in Dipnoi. Acta Zool 26:9–156

Saito T (1930) Über das Gehirn des japanischen Flussneunauges (Entosphenus japonicus Martens). Folia Anat Jpn 8:189–263

Schaper A (1894a) Die morphologische und histologische Entwicklung des Kleinhirns der Teleostier. Morphol Jahrb 21:625–708

Schaper A (1894b) Die morphologische und histologische Entwicklung des Kleinhirns der Teleostier. Anat Anz 9:489–501

Smart IHM (1972) Proliferative characteristics of the ependymal layer during the early development of the spinal cord in the mouse. J Anat 111:365–380

Smeets WJAJ, Nieuwenhuys R (1976) Topological analysis of the brainstem of the sharks Squalus acanthias and Scyliorhinus canicola. J Comp Neurol 165:333–368

Smeets WJAJ, Nieuwenhuys R, Roberts BL (1983) The central nervous system of cartilaginous fishes. Springer, Berlin Heidelberg New York

Söderberg G (1922) Contributions to the forebrain morphology in amphibians. Acta Zool 3:65–121

Streeter GL (1933) The status of metamerism in the central nervous system of chick embryos. J Comp Neurol 57:455–476

Strong OS (1895) The cranial nerves of Amphibia. J Morphol 10:101–231

Sturrock RR (1979) A comparison of the processes of ventricular coarctation and choroid and ependymal fusion in the mouse brain. J Anat 129:235–242

Tabata M (1982) Persistence of pineal photosensory function in blind cave fish, Astyanax Mexicanus. Comp Biochem Physiol 73A:125–127

Tamotsu S, Morita Y (1986) Photoreception in pineal organs of larval and adult lampreys, Lampetra japonica. J Comp Physiol A 159:1–5

Thors F, Nieuwenhuys R (1979) Topological analysis of the brainstem of the lungfish Lepidosiren paradoxa. J Comp Neurol 187:589–612

Trevarrow B, Marks DL, Kimmel CB (1990) Organization of hindbrain segments in the zebrafish embryo. Neuron 4:669–679

Tsuneki K (1986) A survey of occurrence of about seventeen circumventricular organs in brains of various vertebrates with special reference to lower groups. J Hirnforsch 27:441–470

Tsuneki K (1987) A histological survey on the development of circumventricular organs in various vertebrates. Zool Sci 4:497–521

Tuge H (1932) Somatic motor mechanisms in the midbrain and medulla oblongata of Chrysemys elegans (Wied). J Comp Neurol 55:185–271

Uchida K, Nakamura T, Morita Y (1992) Signal transmission from pineal photoreceptors to luminosity-type ganglion cells in the lamprey, Lampetra japonica. Neuroscience 47:241–247

Vaage S (1969) The segmentation of the primitive neural tube in chick embryos (Gallus domesticus). Ergeb Anat Entw Gesch 41:1–88

van der Horst CJ (1918) Die motorische Kerne und Bahnen in dem Gehirn der Fische, ihr taxonomischer Wert und ihre neurobiotactische Bedeutung. Tijdschr Ned Dierk Vereen 16:168–270

Vigh-Teichmann I, Vigh B, Manzano e Silva MJ, Aros B (1983) The pineal organ of Raja clavata: opsin immunoreactivity and ultrastructure. Cell Tissue Res 228:139–148

Von Kupffer C (1906) Die Morphogenie des Zentralnervensystems. In: Hertwig O (ed) Handbuch der Vergleichenden und Experimentellen Entwickkungslehre der Wirbeltiere, vol 2, part 3. Fischer, Jena, pp 1–272

Vraa-Jensen G (1956) On the correlation between the function and structure of nerve cells. Acta Psychiatr Neurol Scand [Suppl] 109:1–88

Welsh U, Storch V (1976) Comparative animal cytology and histology. Sidgwick and Jackson, London

Westerfield M, McMurray JV, Eisen JS (1986) Identified motoneurons and their innervation of axial muscles in the zebrafish. J Neurosci 6:2267–2277

Westergaard E (1969a) The cerebral ventricles of the golden hamster during growth. Acta Anat (Basel) 72:533–548

Westergaard E (1969b) The cerebral ventricles of the rat during growth. Acta Anat (Basel) 74:405–423

Westergaard E (1971) The lateral cerebral ventricles of human foetuses with a crown-rump length of 26–178. Acta Anat (Basel) 79:409–422

Whiting HP (1948) Nervous structure of the spinal cord of the young larval brook-lamprey. Q J Microsc Sci 89:359–384

Whiting HP (1957) Mauthner neurones in young larval lampreys (Lampetra sp). Q J Microsc Sci 98:163–178

Wicht H, Himstedt W (1988) Topologic and connectional analysis of the dorsal thalamus of Triturus alpestris Amphibia, Urodela, Salamandridae. J Comp Neurol 267:545–561

Wicht H, Northcutt RG (1992) The forebrain of the pacific hagfish: a cladistic reconstruction of the ancestral craniate forebrain. Brain Behav Evol 40:25–64

Wilkinson DG, Krumlauf R (1990) Molecular approaches to the segmentation of the hindbrain. Trends Neurosci 13:335–339

Wilkinson DG, Bhatt S, Cook M, Boncinelli E, Krumlauf R (1989a) Segmental expression of Hox-2 homeobox-containing genes in the developing mouse hindbrain. Nature 341:405–409

Wilkinson DG, Bhatt S, Chavrier P, Bravo R, Charney P (1989b) Segment specific expression of a zinc finger gene in the developing nervous system of the mouse. Nature 337:461–464

Wilson SW, Ross LS, Parrett T, Easter SS jr (1990) The development of a simple scaffold of axon tracts in the brain of the embryonic zebrafish, Brachydanio rerio. Development 108:121–145

Zilles K, Armstrong E, Moser KH, Schleicher A, Stephan H (1989) Gyrification in the cerebral cortex of primates. Brain Behav Evol 34:143–150

CHAPTER 5

Histogenesis

R. NIEUWENHUYS

5.1
Early Histogenesis of the Neural Tube

5.1.1
Neuroepithelium

The neural plate and early neural tube are formed by a single layer of columnar cells, the neuroepithelium (Fig. 5.1a). As this layer thickens it gradually acquires the configuration of a pseudostratified epithelium; that is to say, its nuclei become arranged in more and more layers, but all elements remain in contact with the external and internal surface (Figs. 5.1b, 5.2, 5.3a). Mitotic figures are found exclusively along the ventricular surface (Fig. 5.1). His (1889) believed that these mitoses belong to cells which form a ventricular layer of germinal cells (*Keimzellen*), and that the more peripherally located cells represent spongioblasts, primordial glial elements forming a syncytial meshwork (*Markgerüst*). Neuroblasts produced by the germinal cells were supposed to migrate peripherally in the intercellular spaces of this meshwork.

His' concept of neurogenesis was challenged by F.C. Sauer. On the basis of thorough cytological studies on chick, rat and pig embryos (Sauer 1935a,b, 1936), Sauer concluded that: (a) "... the neural tube is composed of discrete cells and is not at any time a syncytium" (Sauer 1935b, p. 20); (b) "... the radially arranged columnar cells, designated by His (1889) as spongioblasts, and the rounded cells in stages of mitosis near the lumen, which His named germinal cells, are not two types of cells, but are the interkinetic and mitotic stages of the same cell" (Sauer 1935a, p. 393; Fig. 5.4). Thus, according to Sauer, the wall of the early neural tube consists of a single type of epithelial cell in various stages of the mitotic cycle. The resting cells reside in the superficial part of the wall. The nuclei of the elements that are going to divide are translocated toward the ventricular surface. At the end of this migration phase, the peripheral process of the cell loses its contact with the external surface and retracts, and the cell rounds up and divides into two daughter cells. The neuroepithelial cells are attached to each other at the surface bordering the lumen by a terminal bar net; according to Sauer (1935a), this accounts for the typical ventricular position of the mitotic figures. Each daughter cell produces a new peripheral process, and their nuclei move away from the ventricle.

Sauer's observations have since been confirmed by numerous studies using different techniques, among them spectrophotometric nuclear analysis (Sauer and Chittenden 1959), autoradiography (Sauer and Walker 1959; Fujita 1963, 1966), and electron microscopy (Fisher and Jacobson 1970; Hinds and Ruffett 1971; Hinds and Hinds 1974; Aström and Webster 1991; Knyihar-Csillik et al. 1995).

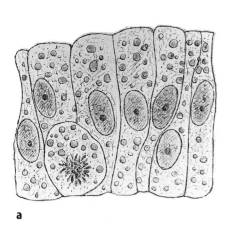

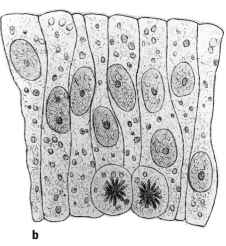

Fig. 5.1a,b. Neuroepithelium in early embryos of the shark *Pristiurus*. **a** Encephalic part of the neural plate in an embryo of 1.75 mm total length. **b** Encephalic part of the neural tube just after its closure in an embryo of 2.5 mm total length. All cells contain droplets of yolk (reproduced from His 1889, Figs. 35, 36)

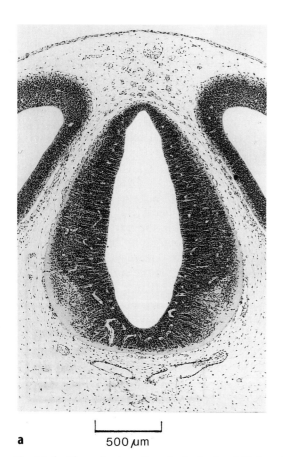

a   500 μm

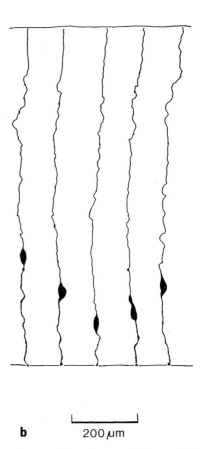

b   200 μm

Fig. 5.2a,b. The palisade pattern in the brain of 12-day-old Chinese hamster embryos. **a** Transverse section through the diencephalon; throughout most of the wall the matrix cells and early neuroblasts show a distinct radial arrangement. Hematoxylin and eosin stain. **b** Neuroepithelial cells in the wall of the pallium, drawn from a rapid Golgi preparation

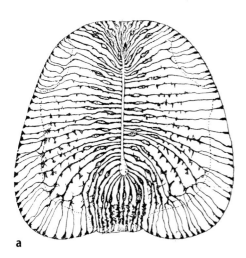

Fig. 5.3a,b. Radially arranged cellular elements in the developing spinal cord as seen in Golgi preparations. **a** Neuroepithelial cells in the spinal cord of a 5-day-old chick embryo. **b** Ependymal cells (*a*), astroblasts (*b*) and young astrocytes (*c*) in the spinal cord of a newborn mouse (after Ramón y Cajal 1909, Figs. 259, 261)

5.1.2
Formation of Embryonic Layers and Zones

At a certain developmental stage the nuclei of the elongated neuroepithelial cells withdraw from the most superficial zone of the neural tube and the wall becomes divisible into an outer anuclear and an inner nuclear zone (Fig. 5.5: 3–4). The outer zone, or marginal layer, consists exclusively of the external cytoplasmic processes of the neuroepithe-

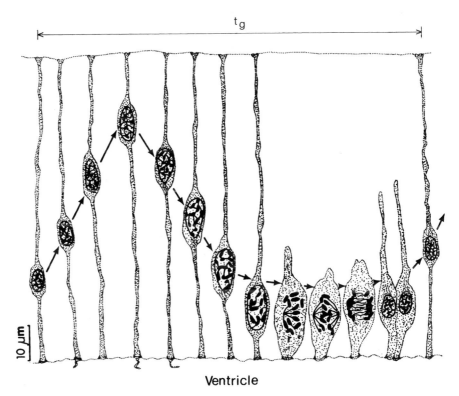

Fig. 5.4. A neuroepithelial cell in the early neural tube of the chick during the various phases of the mitotic cycle (modified from Jacobson 1991, Fig. 2.2)

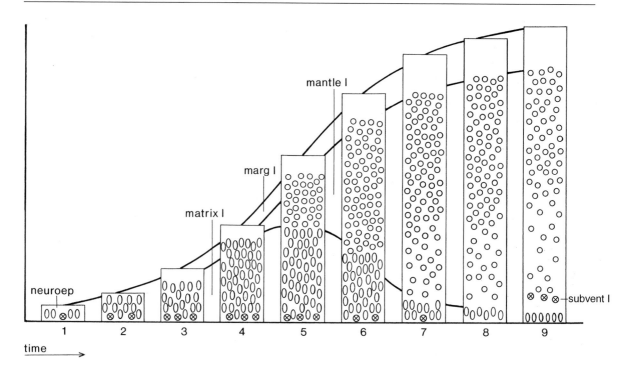

Fig. 5.5. Ontogenesis of the CNS. The histogenesis of the wall of the neural tube is subdivided into nine phases. The following developmental events are indicated: Transformation of monolayered neuroepithelium into a pseudo-stratified epithelium ($1{\to}4$); increase ($2{\to}4$), culmination (5), decrease ($5{\to}7$) and exhaustion (8) of matrix layer; appearance (3) and development ($3{\to}9$) of marginal layer; appearance (5) and expansion ($5{\to}9$) of marginal layer; appearance of subventricular layer (9). *mantle l*, Mantle layer; *marg l*, marginal layer; *matrix l*, matrix layer; *neuroep*, monolayered neuroepithelium; *subvent l*, subventricular layer (modified from Keyser 1972, Fig. 33a)

lial cells for a certain period, but it is soon invaded by the axonal processes of maturing neuroblasts. The inner zone is termed the matrix layer (Kahle 1951; Fujita 1963, 1966). It contains the densely crowded nuclei of a morphologically homogeneous cell population, all elements of which participate in the proliferation process. The matrix cells are the precursors of all neuronal and macroglial elements of the CNS.

The matrix layer can be subdivided into three zones, the M or mitotic, the I or intermediate, and the S or synthetic zone (Fig. 5.6). Fujita (1963, 1966) characterised the translocation of the nuclei of the matrix cells during a generation cycle as an elevator movement. He showed that the nuclei of the matrix cells at the time of DNA synthesis (ts) are located in the superficial half of the matrix layer (S-zone). When the nuclei have finished DNA synthesis, they descend during a postsynthetic or premitotic period, t2, through the I-zone to enter the M-zone. The matrix cells divide there and, after mitotic time, tm, both nuclei of the daughter cells pass to the I-zone, where they spend the postmitotic and presynthetic period, t1. Finally, they enter the S-zone once again, where a new generation cycle begins.

During a certain period the matrix layer represents a purely proliferative compartment. Mother cells produce more mother cells, with the surface area as well as the thickness of the tube increasing steadily (Fig. 5.5: 2–4). This period of symmetrical division of germinal cells is followed by a period of asymmetrical division, in which one of the daughter cells resulting from each mitosis differs from both its mother cell and its sister cell, in that it withdraws from the mitotic cycle and migrates out from the matrix layer (Fig. 5.6: n). These postmitotic elements or neuroblasts will then form a third compartment, the mantle layer, which is situated between the matrix and marginal layers (Fig. 5.5: 5). The elements giving rise to one postmitotic and one proliferative daughter cell are denoted as stem cells. In a later stage, both daughter cells resulting from a mitosis differ from their mother cell and become postmitotic elements. The cells undergoing such a final mitosis are termed differentiated cells (Jacobson 1991).

When stem cells appear in the matrix layer, the period of pure proliferation has come to an end. As more and more dividing neuroepithelial cells switch to the presumptive stem cell mode and begin

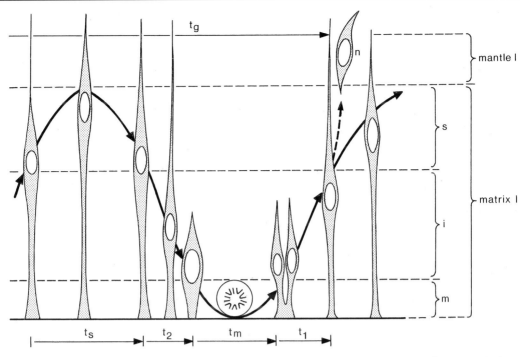

Fig. 5.6. Representation of an 'elevator movement' of the matrix cell. *I*, Intermediate zone; *M*, mitotic zone; *mantle l*, mantle layer; *matrix l*, matrix layer; *n*, neuroblast just differentiated from the matrix cell; *S*, s-zone, i.e. zone of DNA synthesis; *tg*, generation time of the matrix cell; *tm*, mitotic time; *ts*, DNA-synthetic time; t_1, postmitotic resting time; t_2, premitotic resting time (modified from Fujita 1966, Fig. 1)

to generate postmitotic daughter cells, the mantle layer increases rapidly in thickness. During this phase, proliferation and stem cells coexist in the matrix layer. At some later point in time matrix cells start to produce two postmitotic cells, and gradually more and more matrix elements switch to this developmental mode. Within the mantle layer, which further expands *pari passu* with the progressive depletion of the matrix layer (Fig. 5.5: 5–8), the grisea of the CNS originate by migration and aggregation of neuroblasts. The definitive neuronal aggregates as observed in the adult brain often differentiate from larger transitional complexes of neuroblasts, as for instance the migration layers described by Bergquist and Källén (1954; see below).

The processes described above, although uniformly programmed, do not take place at the same time in different parts of the neuraxis. They are temporo-spatially patterned. This is expressed in several ways. Firstly, the proliferative activity in the early neural tube shows local maxima that coincide with the formation of neuromeres and, in a later phase, of migration areas (Bergquist and Källén 1954; see below). Secondly, the waxing and waning of the matrix layer shows, with regard to both amplitude and duration, profound differences among the various parts of the CNS (Kahle 1951;

Fig. 5.7). Thirdly, using autoradiographic techniques, Altman and Bayer established (in a long series of papers, summarised in Bayer and Altman 1995a,b), that in the rat the times of origin, migratory pathways and settling patterns of neuronal populations may vary considerably in different parts of the nervous system, but that functionally related centres (e.g. dorsal thalamus and neocortex) develop synchronously. In several grisea, including the cerebellar cortex, the hippocampus and the olfactory bulb, the time of origin of the various constituent cell types was determined. The pertinent findings appeared to be in harmony with the rule that projection neurons are generally produced before local circuit neurons (Jacobson 1991). Bayer and Altman (1995b, p. 1093) postulated that "... the precise order of neurogenesis is synchronized in germinal systems so that neurons can meet and interact at the right place at the right time to produce a properly 'wired' CNS," and that "... this 'preestablished harmony' operates throughout development as neural circuits are laid down."

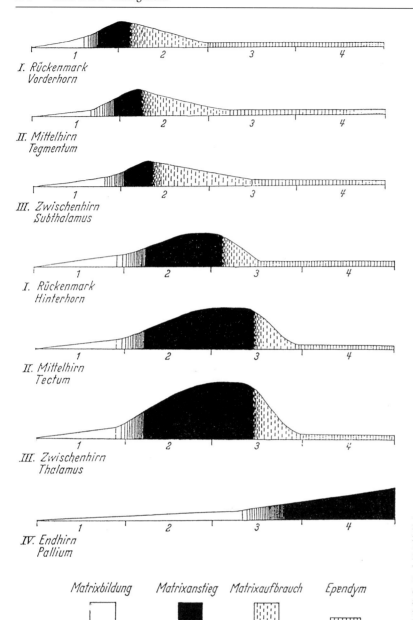

I. Rückenmark
Vorderhorn

II. Mittelhirn
Tegmentum

III. Zwischenhirn
Subthalamus

I. Rückenmark
Hinterhorn

II. Mittelhirn
Tectum

III. Zwischenhirn
Thalamus

IV. Endhirn
Pallium

Matrixbildung   Matrixanstieg   Matrixaufbrauch   Ependym

Fig. 5.7. Development of the matrix layer in various parts of the human brain during the first 4 intrauterine months. This development is subdivided into four phases: formation (*Matrixbildung*), increase (*Matrixanstieg*), decrease (*Matrixaufbrauch*) and complete depletion and formation of ependyma (*Ependym*) (after Kahle 1951, Figs. 10, 11)

5.1.3
Development of Glia and Blood Vessels

As regards gliogenesis, Fujita (1965, 1966) held that the germinal cells which constitute the matrix layer first give rise to neuroblasts and that only later, after the completion of neurogenesis, do the remaining germinal cells transform into glioblasts or differentiate into ependymal cells. Studies using Golgi and electron-microscopic techniques have shown, however, that radial glial cells already exist during late stages of neurogenesis (Rakic 1971,

1972; Schmechel and Rakic 1979). As the name implies, radial glial cells are a specialised nonneuronal cell class characterised by an elongated fibre that radially spans the entire wall of the neuraxis from the ventricular to the meningeal surface (Rakic 1987b; Figs. 5.8, 5.9a). More recently, immunocytochemical staining for the specific glial fibrillary acid protein (GFAP) (Choi and Lapham 1978; Levitt and Rakic 1980; Levitt et al. 1981) provided more direct evidence for the presence of radial glial cells during early stages of neurogenesis. These results indicate that, contrary to the views of Fujita,

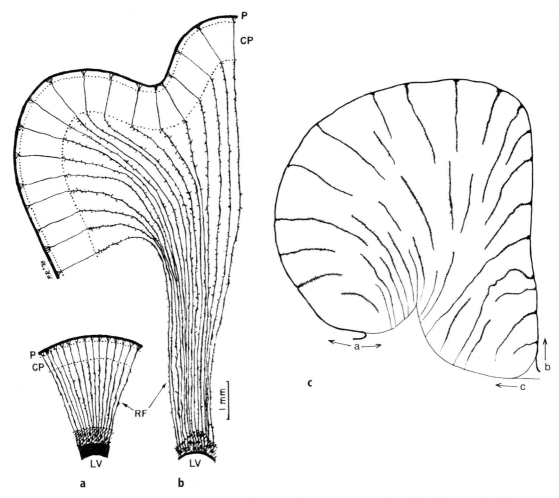

Fig. 5.8a–c. Distribution of radial glial cells as observed in Golgi preparations. **a** Dorsal portion of the occipital lobe of an 81-day-old rhesus monkey fetus. **b** Section from the corresponding region from a 138-day-old fetus. Due to the increased thickness of the telencephalic wall and the marked tangential expansion of the superficially situated cortical plate (*CP*) during the intervening period, radial glial fibres (*RF*) that traverse the wall of the evaginated hemisphere from the lateral ventricle (*LV*) to the pial surface (*P*) become several times longer and considerably curved. **c** Transverse section through the adult telencephalon of the teleost fish *Gasterosteus aculeatus*. Due to an eversion of the lateral walls of the telencephalon, the ependymal surface has greatly expanded. Radial glial fibres converge from this large surface (*a–b*) to the much smaller meningeal surface (*a–c*) (**a** and **b** after Schmechel and Rakic 1979, Fig. 1; **c** after Nieuwenhuys 1966, Fig. 2)

the precursor cell population of the matrix layer contains at least two cell types.

In most non-mammalian species, radial glial cells persist in many regions throughout the life span, but in mammals they represent a transient class of cells, which gradually disappears from most regions. The classical Golgi studies of von Lenhossék (1895) and Ramón y Cajal (1909; Fig. 5.3b), and more recent investigations using antibodies against GFAP (Choi 1981; Choi et al. 1983; Choi and Kim 1984; Hirano and Goldman 1988; Voigt 1989) have shown that in mammals most radial glial cells transform into astrocytes. The immunohistochemical studies just mentioned revealed that at least

part of the oligodendrocytes are also derived from radial glial cells.

As the epithelial wall of the developing neural tube steadily becomes thicker, blood vessels penetrate into it. These structures exhibit the same overall orientation as the radial glial fibres (Figs. 5.9b, 5.10). It is important to note that the encephalic blood vessels are clearly labelled as invaders by their ensheathing connective tissue and the uninterrupted basal lamina. Even the smallest capillaries are always separated from the (epithelial) neural tissue proper by a basal lamina (Palay 1967).

This brief overview of the early histogenesis of the neural tube is derived almost entirely from

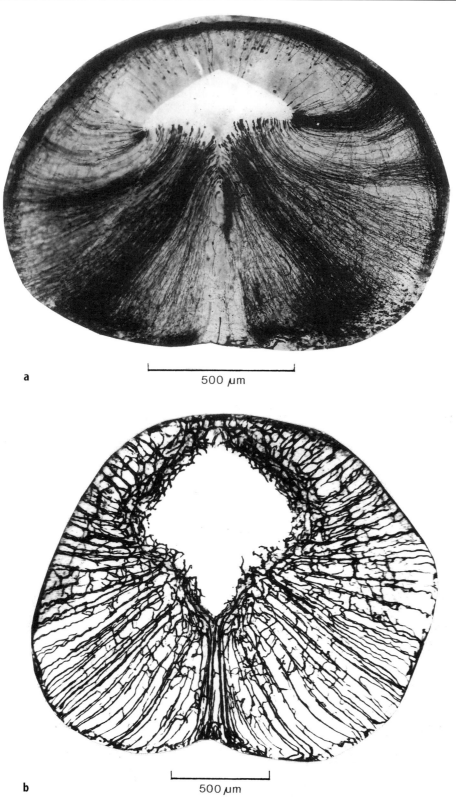

a   500 μm

b   500 μm

Fig. 5.9. a Transverse section through the mesencephalon of a 16-day-old Chinese hamster embryo, showing the distribution of radial glial fibres (*ventral* and *lateral parts*) and neuroepithelial cells and young neuroblasts (*dorsal side*). Rapid Golgi technique. **b** Transverse section through the mesencephalon of a 16-day-old rat embryo, showing the vascular pattern following intracardial injection with India ink (courtesy of Dr. P. Kucera, Lausanne)

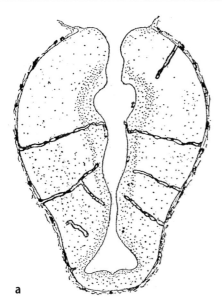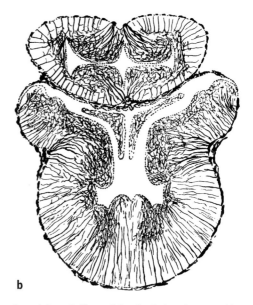

a   **b**

Fig. 5.10a,b. Vascular pattern in (**a**) the caudal diencephalon and rostral mesencephalon of the lamprey *Petromyzon marinus* (after Sterzi 1907, Fig. 73) and (**b**) the rhombence-phalon and caudal cerebellum of the shark *Squalus acanthias* (after Sterzi 1909, Fig. 369)

studies on mammals. Comparable studies on non-mammalian vertebrates are scant. However, there is no reason to believe that the histogenesis of the CNS in non-mammalian vertebrates differs substantially from that in mammals. The three classical layers, matrix, mantle layer and marginal layer, can be clearly observed in all parts of the lateral plates of the neural tube in all vertebrates. An international group of neuroembryologists (Boulder Committee 1969) has proposed to replace the names of the three layers just mentioned by the 'geographical' terms ventricular zone, intermediate zone and marginal zone. I do not consider these changes improvements and suggest that the old terms be maintained.

5.1.4
Proliferative Compartments

Apart from the ubiquitous ventricular matrix, two other proliferative compartments, i.e. the *subventricular zone* and the *external germinal layer*, have been observed in the developing brain. The literature on the subventricular zone, which includes the studies of Allen (1912), Kershmann (1938), Globus and Kuhlenbeck (1944), Smart (1961), Privat and Leblond (1972), Altman (1966), Fujita (1966), Blakemore (1969), Blakemore and Jolly (1972), Sturrock and Smart (1980), and many others, may be summarised as follows:

The subventricular zone, sometimes also denoted as the subependymal layer or cell plate, has

been described only in the lateral and basal walls of the mammalian telencephalon. This layer develops at the junction of the matrix zone and the mantle layer. Most authors agree that this layer appears relatively late, at a moment when neurogenesis is largely or entirely completed (Fig. 5.6: 9). However, according to Rakic (1974), the subventricular zone can be recognised in the pallial wall of the rhesus monkey at as early as 45 days of gestation (the total duration of gestation is about 165 days in this animal). It persists after birth and, in a vestigial manner, into adult life and even senescence. Contrary to the ventricular matrix cells, the subventricular elements divide in situ and do not exhibit interkinetic nuclear migration.

The subventricular layer gives rise to special classes of neurons and to all types of macroglial elements, with the possible exception of ependymal cells. According to several authors, including Altman (1966), Fujita (1966) and Privat and Leblond (1972), the subventricular layer is involved mainly or exclusively in the production of glial cells.

A particularly strongly developed subventricular zone can be observed in the mammalian striatal primordia. Smart (1972a,b, 1973, 1976, 1981, 1985; cf. also Smart and Sturrock 1979) attributed to this striatal subventricular zone a special phylogenetic status, on the basis of the following precepts and considerations. (a) The efficiency of the central nervous system is partly a function of the number of nerve cells available to it, and its evolution will therefore be associated with improvements in cell

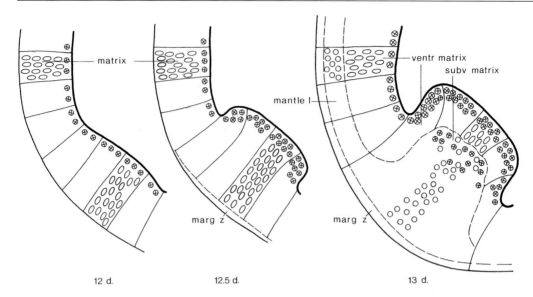

Fig. 5.11. Transverse sections through the telencephalon, showing the development of the striatum and the adjacent part of the pallium in embryos of the Chinese hamster, *Cricetulus griseus*. In *12-* and *12.5-day-old* embryos a typical palisade pattern is found throughout the wall of the telencephalon. By the *13th day* of development the typical palisade pattern has vanished in the striatal anlage and a subventricular matrix has formed in that anlage

production. (b) Because neurons do not divide, but originate from a population of matrix cells, the production of the maximum number of nerve cells in a given number of generations will be maximised by a delay of differentiation, that is, by investing more cell cycles in the production of matrix cells. (c) Given the pseudostratified nature of the matrix layer, proliferation (without a corresponding increase in area of the ventricular surface) will lead to a considerable thickening of this layer. (d) The nuclei of the pseudostratified cells in the matrix layer migrate to the ventricular surface on entering mitosis. This produces a situation in which the ventricular surface becomes saturated with mitotic figures in areas of high pseudostratification or high mitotic rate. (e) Escape from such 'choking' of the matrix without slowing down the rate of cell production may be achieved by increasing the surface area of the matrix layer or by abandoning the interkinetic nuclear migration. (f) the formation of a non-migratory proliferative compartment will enable a developing structure to increase in volume instead of in area.

In areas of the mouse CNS in which particularly large numbers of neurons are being produced (spinal dorsal horn, dorsal thalamus, neopallium, striatal primordia), Smart found a high degree of pseudostratification, associated with crowding of mitotic figures at the ventricular surface. In these areas there was a relatively late onset of differentiation and the occurrence of mitotic figures which were located away from the ventricular surface. He observed that the phenomena just mentioned are particularly prominent in the ventricular elevations that give rise to the corpus striatum. In these elevations a zone of proliferating cells is formed at the external side of the ventricular matrix (Fig. 5.11). This subventricular zone gradually becomes a prominent feature of the striatal primordia and, by the 14th day of gestation (E14), takes over from the ventricular matrix as the major site of mitotic activity. The zone increases in volume up to E16 and then declines progressively. The thin subventricular zone observed in late embryonic and postnatal stages, which is concerned mainly with the production of macroglial elements, is to be considered a remnant of this conspicuous neuron-producing layer.

Golgi studies of the developing mouse forebrain revealed that in earlier stages radial glial cells provided with long peripherally extending processes are present throughout the telencephalon. However, in the ventricular elevations the processes of these cells soon lose contact with the meningeal surface and by about E12 do not extend beyond the matrix layer. Because it is widely held that the fibres of the radial glial cells play an important role in guiding migrating neuroblasts throughout the CNS (see below), it must be concluded that this guiding system is not operative during most of the development of the mammalian basal ganglia.

In the reptilian telencephalon there are well-developed ridges which protrude from the lateral wall into the ventricular cavity. Studies of embryonic reptilian material revealed that in these ridges mitotic figures occur exclusively at the ventricular surface and that, hence, a subventricular layer is entirely absent.

From these observations Smart concluded that (a) the formation of a subventricular layer composed of proliferating elements which divide without entering the elevator movement seems to be a feature which evolved during the reptile-mammal transition; (b) the appearance of this layer represents an escape from the constraints of interkinetic nuclear migration; and (c) the genesis of the mammalian basal ganglia is much more complex than that of the adjacent pallium and should be considered one of the major evolutionary achievements in the conglomerate of changes which occurred during the transition from the reptilian to the mammalian grade of organisation of the telencephalon.

The Boulder Committee (1970) suggested that the subventricular zone, together with the ventricular, intermediate and marginal zones, belongs to the 'fundamental' or 'cardinal' zones of the embryonic vertebrate CNS. Given the fact that the subventricular zone has so far been observed only in the mammalian telencephalon, much credit cannot be given to this suggestion.

The third and final proliferative compartment to be discussed, i.e. the *external germinal layer*, is confined to the cerebellum. This layer, also referred to as the external granular layer, develops from the ventricular matrix in the rostral part of the rhombic lip. The rhombic lip is a thickened germinal zone in the rhombencephalic alar plate, situated directly adjacent to the attachment of the roof of the fourth ventricle. From this zone, the layer spreads by tangential migration of its elements over the entire external surface of the cerebellar anlage. This transitory germinal zone gives rise to the cerebellar granule, stellate and, where present, basket cells. Following its formation, the layer increases in thickness as a result of proliferation. This proliferative phase gradually diminishes with the onset of the phase of neurogenesis. During this phase the immature neurons, destined to become the local circuit neurons mentioned, migrate inward following a straight radial path. The radially arranged processes of the early-developing Bergmann glial cells are believed to act as guidelines for the migrating elements (Mugnaini and Forstrønen 1967; Rakic 1971).

The external germinal layer, the development and fate of which has been extensively studied in birds and mammals (for review see Jacobson 1991), has also been observed in amphibians (Urey et al. 1987) and teleosts (Schaper 1894a,b; Pouwels 1978), and may be expected to occur in all gnathostomes.

From this survey it may be concluded that all neuronal and macroglial elements in the vertebrate nervous system are derived from a single proliferative compartment, the ventricular matrix, either directly or indirectly via one of the two secondary matrices, i.e. the subventricular zone and the external germinative layer.

5.2 Natural Coordinate System of the Neuraxis

Our knowledge concerning the morphogenesis and histogenesis of the central nervous system warrants the thesis that this organ contains a built-in natural coordinate system (Nieuwenhuys 1972, 1974; Smart 1978, 1982, 1985; Smart and McSherry 1986). This thesis may be elucidated as follows:

During the early morphogenesis of the CNS the neural plate is transformed into the neural tube. On the neural tube the bilateral lateral plates, which are interconnected dorsally and ventrally by the roof plate and the floor plate, respectively, can be distinguished. The structures just mentioned surround a central, fluid-filled ventricular cavity. The natural coordinate system includes: (a) two natural surfaces, i.e. the ventricular and meningeal surfaces of the neural tube, and (b) a system of more or less radially oriented curves or vectors which connect these two surfaces.

The vectors manifest themselves in the direction and orientation of:

1. The epithelial cells, which constitute the neural plate and early neural tube (Fig. 5.1)
2. The matrix cells, which span the width of the lateral plates during early development (Figs. 5.2b, 5.3a, 5.4)
3. The radial glial cells, which are present during early neurogenesis in all vertebrates and throughout development in most anamniotes (Figs. 5.3b, 5.8, 5.9a)
4. The overall arrangement of the cellular elements in the wall of the developing CNS (Fig. 5.2a)
5. Many blood vessels (Figs. 5.9b, 5.10)

It is important to note that the processes of the elements mentioned under *b* and *c* are reported to play an important role as guidelines during the migration of neuroblasts or their precursors (see below). This implies that many developing neurons shift in a simple way within the natural coordinate system. The neuroblasts and their precursors are products of mitoses which, as discussed, take place in the immediate vicinity of the ventricular surface; consequently, this surface can be considered the 'starting' or 'zero' plane of neurogenesis. The natural coordinate system, which may be considerably

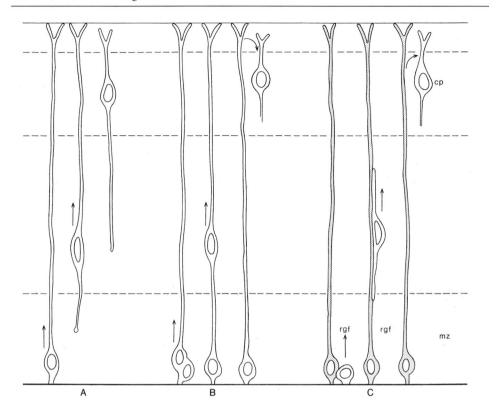

Fig. 5.12. Migration of neuroblasts from the pallial matrix zone (*mz*) to the cortical plate (*cp*); *rgf*, radial glial fibres. Mechanisms proposed by (*A*) Morest (1970), (*B*) Berry and Rogers (1965), and (*C*) Rakic (1972) (modified from McConnell 1988, Fig. 3)

deformed during ontogenesis (Fig. 5.8; Chap. 4, Figs. 4.8, 4.9), reveals the general morphology and the basic organisation of the CNS.

5.3
Patterns of Migration

5.3.1
Introductory Note

In anamniotes, most postmitotic cells settle just peripheral to the matrix compartment. Thus, *pari passu* with the reduction of the latter compartment, these elements form a periventricular zone of grey matter, within the confines of which more or less individualised cell masses may develop (cf. Chap. 2). However, particularly in amniotes, there may be a considerable distance between the site where a given neuron was generated and the position which it occupies in the mature brain. The intervening migration process involves the dislocation of postmitotic cells from the other elements in the matrix zone and the active movement of the neuroblasts through the surrounding cellular environment to their final destination, before they settle to establish their specific synaptic connections

(Rakic 1987a, 1990). In terms of orientation and directionality of cell movement, three main types of migration: radial, deflected radial, and tangential can be distinguished. These types or modes of migration will now be briefly considered.

5.3.2
Radial Migration

During a radial migration the neuroblasts move simply from the ventricular surface to the meningeal surface. The following mechanisms have been suggested to be operative during this displacement:

1. Morest (1970) studied the early development of neurons in the telencephalon of opossum pouch young in Golgi material. He concluded that: (a) neuroblasts maintain central and peripheral primitive processes that extend to the ventricular and meningeal surfaces, respectively; (b) not the cells per se, but rather their nuclei and perikaryal regions move through their peripheral primitive process outward; and (c) this translocation is accompanied or followed by a loss of the central primitive process, and the cell later detaches itself also from the meningeal surface (Fig. 5.12a).

Berry and Rogers (1965), who studied the development of the cortex in the rat, suggested a similar mechanism. They observed in their Golgi preparations 'binucleate' elements maintaining their attachment to both the ventricular and meningeal surfaces, and suggested that following the nuclear division one of the daughter nuclei travels peripherally through the primitive process and that the cytoplasmic division occurs only later, after one nucleus has attained its destination in the anlage of the cortex, i.e. the so-called cortical plate (Fig. 5.12b).

Morest and collaborators concluded from later Golgi studies that also in the tectum opticum (Domesick and Morest 1977a,b) and the rhombencephalic acoustic nuclei of the chick (Book and Morest 1990), neuroblasts migrate by translocation of their perikarya radially from the matrix zone into the matrix layer. They believed that this mode of initial radial migration (which may be followed by a later tangential migration; see below) applies to neuroblasts in all parts of the brain.

2. On the basis of Golgi and electron-microscopic studies on the cerebral (Rakic 1972, 1988; cf. also Schmechel and Rakic 1979; Sidman and Rakic 1973) and cerebellar cortices (Rakic 1971) of the monkey, Rakic proposed that radially oriented glial fibres provide contact guidance paths for migrating neuroblasts. In the pallium the neuroblasts generated in the ventricular matrix zone were supposed to travel to the anlage of the cerebral cortex along transient glial elements, the conspicuous peripheral processes of which span the full thickness of the pallial wall (Figs. 5.8a, 5.12c). In the developing cerebellum postmitotic elements from the external granular layer (cf. Sect. 5.1.4) were thought to migrate inward, through the molecular and Purkinje cell layers, to the (internal) granular layer along Bergmann fibres, i.e. the radially oriented peripheral processes of the Golgi epithelial cells.

The following pieces of evidence support a role for the radial glia in guiding young neurons:

1. The development of the radial glial cells precedes the onset of migration and these elements remain present as long as migration lasts. This coincidence is particularly striking in the adult avian telencephalon. It is known that in adult canaries newly generated neurons are inserted into the synaptic network of the hyperstriatum ventrale, pars caudalis, a centre of the song-control system (Goldman and Nottebohm 1983; Burd and Nottebohm 1985). These neurons originate from a discrete matrix zone adjacent to the lateral ventricle and differentiate into mature neurons after having migrated over a distance of about 5 mm. It is remarkable that, precisely in the region through which these young neurons migrate, persistent long radial glial fibres are present (Alvarez-Buylla 1990; Alvarez and Nottebohm 1988; Alvarez-Buylla et al. 1987). Conversely, in the anlage of the mammalian striatum, where radial migration of neuroblasts does not occur, the system of long radial glial fibres disappears early in development (Smart and Sturrock 1979).

2. Electron-microscopic analysis shows that during their migration neuroblasts are closely apposed to neighbouring glial fibres.

It should be emphasised that perikaryal translocation (Fig. 5.12a,b) and neuroblast migration along radial glial fibres (Fig. 5.12c) are not mutually exclusive. It might well be that the former mode of displacement is operative during early development, whereas the latter process prevails during later development, particularly in regions where considerable distances have to be bridged.

During the past decade, neuronal and glial migration patterns have been studied extensively with a variety of techniques, including:

- Infection of cells in the matrix zone with recombinant retroviruses, followed by histochemical identification of their progeny (e.g., Gray et al. 1988: avian tectum; Walsh and Cepko 1992, 1993: rat cortex; Leber and Sanes 1995: chick spinal cord)
- Labelling and tracing sets of migrating cells that share a neurotransmitter (e.g., Phelps et al. 1993; Phelps and Vaughn 1995: rat spinal cord)
- Mapping of regional differences with quail/chick homotopic transplants (Senut and Alvarado-Mallart 1987; Martínez et al. 1992: avian tectum)
- Monitoring of living cells in the embryonic murine telencephalon, following their labelling in the matrix zone with the lipophilic dye DiI (Fishel et al. 1993)
- Analysis of the pattern of cortical cell dispersion using transgenic mice in which roughly half of the brain cells are coloured by a transgene (Tan and Breen 1993)
- The use of a DM-20 cRNA probe to follow, by in situ hybridisation, the oligodendrocyte lineage during embryonic development (Timsit et al. 1995: mouse)

The available space does not permit a full discussion of the results obtained with these techniques. A few points may be mentioned, however(for reviews see Hatten 1993; Grove 1993; Price 1993). During early development the neuroepithelial elements which form the matrix zone show a remarkable mobility in the plane of that zone, exceeding the passive movement needed to make way for new daughter cells. This 'fluidity' of the matrix zone is hard to reconcile with the widely accepted concept that the proliferating neuroepithelial cells are firmly attached to each other at the ventricular surface (cf. Sect. 5.1.1).

The common observation of radial arrays of marked or labelled postmitotic elements in the developing mantle zone indicates that during early neurogenesis neuron migration is predominantly radial. Radial migration may be followed by a later tangential displacement. There is evidence suggesting that the majority of neurons in the cerebral cortex and the tectum move only radially, but that some show, in addition, a tangential displacement. The range of tangential migration may vary considerably. Thus the large multipolar neurons in the

stratum griseum centrale of the avian tectum are known to have a limited range of migration; cortical neurons may translocate across considerable distances, whereas certain neuron populations in the rhombencephalon travel over amazing distances before reaching their destination (see below).

Neurons, as well as glial elements, are probably involved in the tangential dispersal just indicated. Timsit et al. (1995) presented evidence that in the mouse oligodendrocytes spread from a limited ventral zone tangentially over the entire brain and spinal cord.

5.3.3
Deflected Radial Migration

During morphogenesis the walls of the neural tube may be considerably deformed. The vectors inter-

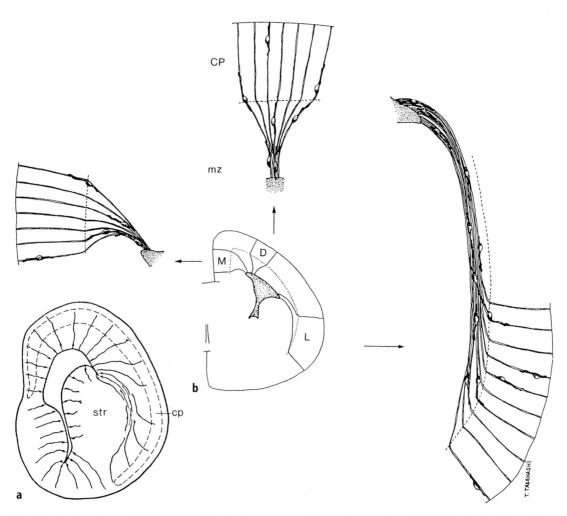

Fig. 5.13a,b. Development of the telencephalon in the mouse. **a** The distribution of Golgi-stained processes of periventricular cells at E17 (modified from Smart and Sturrock 1979, Fig. 8). **b** Patterns of radial glial alignment and neuro- blast migration in medial (*M*), dorsal (*D*), and lateral (*L*) hemispheric regions. The somata of radial glia are not illustrated (reproduced from Misson et al. 1991, Fig. 1). *cp, CP,* Cortical plate; *mz,* matrix zone; *str,* striatal anlage

connecting the ventricular and the meningeal surfaces of the neural tube (cf. Sect. 5.2) are important
indicators of the deformations that have taken
place. In curved parts of the wall these vectors may
diverge from the smaller ventricular surface toward
the larger ventricular surface (Fig. 5.8a,b) or, conversely, converge from the larger ventricular surface
toward the smaller meningeal surface (Fig. 5.8c; cf.
also Chap. 4, Figs. 4.8, 4.9). In these instances the
direction of the vectors and, hence, also the direction of migration in accordance with these vectors,
may be designated as radial. However, in some
parts of the CNS the walls become distorted to such
an extent that the course of the vectors, and that of
corresponding migrations, can no longer be
adequately denoted as radial. Two examples may
elucidate this situation.

Smart and Sturrock (1979) studied the development of the telencephalon of the mouse in Golgi
preparations. They observed that early in development the telencephalic ependymal gliocytes show a
radial orientation. However, later in development
the ependymoglial fibres in the striatal anlage lose
contact with the meningeal surface, and those emanating from the lateral part of the pallial ventricular
surface begin to deflect ventrally in association
with the appearance of the cortical plate. Thus, the
ependymal processes will pursue curving courses
which circumscribe the striatal primordium
(Fig. 5.13a). Given the close association between the
ependymal glial fibres and the migration of neuroblasts from the ventricular matrix toward the cortical plate (Rakic 1972, 1988), it may be expected

that the glial processes not only reveal the formal
changes in the lateral hemisphere wall, but also indicate the route along which migrating neuroblasts
attain the ventrolateral part of the cortical plate,
which forms the anlage of the prepiriform cortex.
This presumption has been substantiated by the
work of Misson et al. (1991), who studied the migration of neuroblasts in relation to the radial glial fibre
system in the murine neopallium. These authors
combined retroviral labelling of cortical neuroblasts
with staining of the radial glial fibres with a monoclonal antibody selective for cells of astroglial lineage. They found that during embryonic days 16–18
the leading processes and the somata of migrating
neuroblasts are close to and aligned with adjacent
glial fibres, despite substantial variations in the patterns of alignment of the fibre fascicles. As regards
the alignment of the glial fibres and of the neuroblasts apposed to them, Misson et al. (1991) distinguished three patterns (Fig. 5.13b), namely *radial
divergent*, observed in the dorsal pallial region, *curving divergent*, occurring in the medial pallial region,
and *curving convergent, then divergent*, prevailing in
the most lateral pallial region. It is important to note
that the glial fibres emanating from the lateral angle
of the lateral ventricle pass over a considerable distance *tangentially* through the well of the cerebral
hemisphere, and that the same holds true for the
neuroblasts migrating along these fibres. However,
it will be clear that the course of these glial fibres
and the direction of this migration, though *topographically tangential*, is *topologically vectorial*. In
order to distinguish this mode of displacement

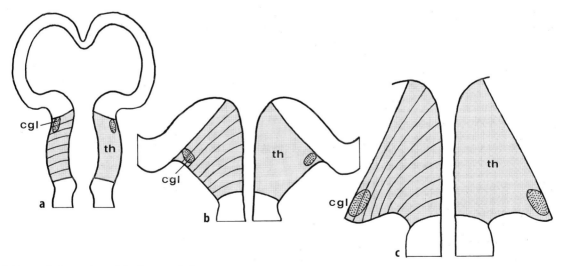

Fig. 5.14a–c. Deformation of the human thalamus (*th*) during ontogenesis. Horizontal sections through the diencephalon in an early embryonic stage (**a**), a later stage (**b**) and the
adult stage (**c**). Note the remarkable translocation of the corpus geniculatum laterale (*cgl*). The vectorial pattern is conjectural (based on Schwalbe 1880 and Hochstetter 1919;
reproduced form Nieuwenhuys 1994)

from the true tangential migration, to be discussed in the next section, it is designated here as *deflected radial migration*. Bayer and Altman (1995b) traced the migration of neuroblasts in the developing neocortex of the rat with the aid of thymidine radiography. They correctly concluded that neuroblasts bound for the ventrolateral part of the cortex migrate tangentially, forming a 'lateral cortical stream'. However, they failed to appreciate the deflected radial or vectorial character of this migration.

The second example of distortion of the wall of the CNS concerns the human diencephalon at the level of the dorsal thalamus. The investigations of Schwalbe (1880) and Hochstetter (1919) showed that during ontogenesis the human dorsal thalamus is deformed in such a fashion that its initially lateral surface becomes its caudal surface. This apparent vanishing of the free lateral wall of the thalamus is accompanied by a remarkable shift of the anlage of the lateral geniculate body (Fig. 5.14), but it may be expected to affect the positional relations of all other parts of the thalamus as well. A similar, though less dramatic, deformation of the thalamus occurs during development in all mammals.

The vectorial pattern indicated in Fig. 5.14 is conjectural and not based on observed histological features, like the orientation of cellular streams or the course of glial fibres or blood vessels. However, if the assumption that during ontogenesis the neuroblasts in the mammalian thalamus shift mainly according to this vectorial pattern is correct, it follows that cellular migration in this brain part cannot be adequately studied in transverse sections (as is commonly done).

Before leaving radial, or better, vectorial migration, we should recall that the descendents of

marked neuroepithelial cells are dispersed widely in avian rhombencephalic (Fraser et al. 1990) and diencephalic neuromeres (Figdor and Stern 1993) but do not cross the (vectorially oriented) interneuromeric borders (cf. Chap. 4, Sect. 4.5.5.2). Interestingly, Fishell et al. (1993) observed that labelled neuroepithelial cells move extensively in the plane of the pallial matrix but do not cross the boundary between this matrix and that of the adjacent anlage of the striatum.

5.3.4
Tangential Migration

It has been known since the time of W. His (1831–1904) that neuroblasts in the CNS may migrate tangentially over considerable distances. In a remarkable memoir, devoted to the early development of the human rhombencephalon, His (1890a) reported that the 'rhombic lip', i.e. the zone directly adjacent to the line of attachment of the attenuated roof of the fourth ventricle, is the seat of abundant cell proliferation and has to be regarded as the germ centre for the cells which form the inferior olive, the pontine grey and other cell masses in the brain stem. He noticed that the neuroblasts bound for these nuclei form a large stream passing tangentially from the alar plate into the basal plate (Fig. 5.15).

The observations of His (1890a,b) have been confirmed by numerous authors, including Essick (1907, 1912: rabbit, pig, man), Hayashi (1924: man), Harkmark (1954a,b: chick), Taber Pierce (1966: mouse) and Altman and Bayer (1987a–d: rat). Essick (1907, 1912) observed that the nuclei pontis are formed by a rostrally directed migration

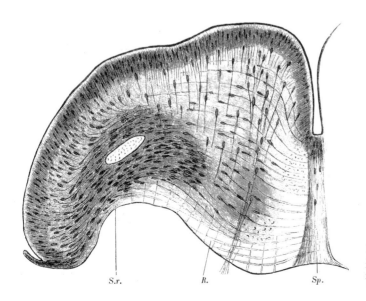

Fig. 5.15. Transverse section through the medulla oblongata of a human embryo, total length 13.6 mm, with neuroblasts schematically blocked in (reproduced from His 1890a, Fig. 16)

S.r.   *R.*   *Sp.*

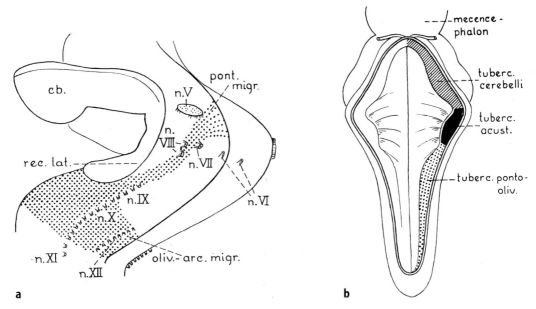

Fig. 5.16a,b. Long tangential migrations in the human rhombencephalon. **a** The olivo-arcuate and pontine migrations in man. Ventrolateral view based on a reconstruction of the rhombencephalon of a 23-mm human embryo by Essick (1912). **b** The main subdivisions of the most dorsal part of the rhombencephalic alar plate. The caudal subdivision represents the rhombic lip, which gives rise to the pontine and olivary nuclei (reproduced from Harkmark 1954b, Figs. 76, 79)

through a restricted pathway, the corpus pontobulbare, and that a superficial migration over the lateral and ventral surface of the medulla oblongata gives rise to the nuclei arcuati and the inferior olive (Fig. 5.16a). According to Hayashi (1924), different sectors of the rhombic lip give rise to different structures (Fig. 5.16b), a finding which was later fully confirmed by the autoradiographic analyses of Taber Pierce (1966) and Altman and Bayer (1987a–d). The authors last mentioned found that the neurons originating from the rhombic lip, destined for the various precerebellar nuclei, have different times of origin and different migratory routes. They distinguished an intramural migratory stream providing the cells of the inferior olive, a posterior subpial migratory stream giving rise to the contralateral external cuneate and lateral reticular nuclei, and an anterior subpial migratory stream along which the cells forming the nucleus reticularis tegmenti pontis and the pontine nuclei attain their destination. Harkmark (1954a,b) established that in the chick lesions of particular parts of the rhombic lip lead to defects in the cell strands and in the resultant inferior olivary and pontine nuclei.

As regards the mechanism of the tangential migrations just discussed, Ono and Kawamura (1989, 1990) and Rakic (1990) observed that neurons generated in the rhombic lip assume a bipolar shape and migrate closely apposed to the neuronal surface provided by fibre tracts that run parallel to the brain surface. They believed that the migrating elements are guided by these axons toward their destination. Bourrat and Sotelo (1988, 1990) provided evidence suggesting that the neuroblasts generated by the rhombic lip first develop an axon, and that the soma subsequently moves down this process.

Tangential migration of neuroblasts also occurs in the spinal cord, as may appear from the following two examples. Leber and Sanes (1995) injected recombinant retroviruses into the lumen of the spinal cord of chick embryos. They studied the migration of cells out of the matrix zone by infecting embryos at a fixed stage and varying the time of analysis. Isolated clusters of cells were interpreted to be single clones. At first, most clones consisted of radial arrays of cells, suggesting that the initial migration is predominantly radial (Fig. 5.17a). In many clones, however, neurons turned orthogonally, to migrate ventrally or dorsally, parallel to the surface of the spinal cord or to the medial edge of the motoneuron pool (Fig. 5.17b). The majority migrated from dorsal to ventral. The ventrally migrating cells originated at all dorsoventral levels of the spinal cord, but were most likely to originate in the dorsal region. Most of these cells were unipolar in shape, with leading processes and axons extended in the direction of apparent migration (Fig. 5.17b). Comparable observations have been made by Phelps and Vaughn (1995), who studied

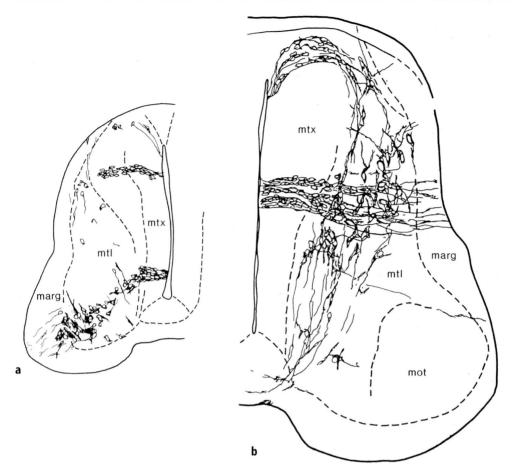

Fig. 5.17a,b. Patterns of migration in the embryonic chick spinal cord, as revealed by injections of recombinant retroviruses into the central canal. **a** Three streams of cells, to be considered as separate clones, are visible. Some cells are migrating in the mantle layer. Virus was injected at stage 8 (according to Hamburger and Hamilton 1951), and the animal was killed at stage 22. **b** Radial migration from two separate clones, combined with complex dorsal-to-ventral tangential migration. Some of the ventrally migrating cells give rise to commissural axons. Virus was injected at stage 9, and the animal was killed at stage 29 (reproduced from Leber and Sanes 1995, Figs. 2, 4)

the development of a group of early-differentiating cholinergic neurons in the cervical spinal cord of the rat. They found at the time of its first appearance, that this cell group surrounds the most ventral part of the matrix zone, but that it later migrates dorsally, to become the group of cholinergic dorsal horn cells seen in older animals. Electron-microscopical analysis revealed that during their translocation the immature cholinergic neurons are directly apposed to a system of decussating axons. These axons, which together form the so-called circumferential bundle (cf. Holley 1982; Holley et al. 1982), pass ventrally along the external surface of the matrix zone. On the basis of these observations, Phelps and Vaughn (1995) considered it likely that the cholinergic neurons use the axons of the circumferential bundle for guidance during their dorsal migration.

Tangential migration is not confined to the rhombencephalon and the spinal cord, but also occurs in the more rostral parts of the brain. It has already been mentioned that in the avian optic tectum and in the mammalian cerebral cortex the most prominent pattern of migration is vectorial, but that in both of these structures cells may migrate tangentially over appreciable distances (see Sect. 5.3.2).

5.4
Formation of Cell Masses

Although the neuroblasts destined to form a given cell mass may travel over great distances through the wall of the neural tube before attaining their ultimate position, it is reasonable to assume that most cell masses in the vertebrate brain result from

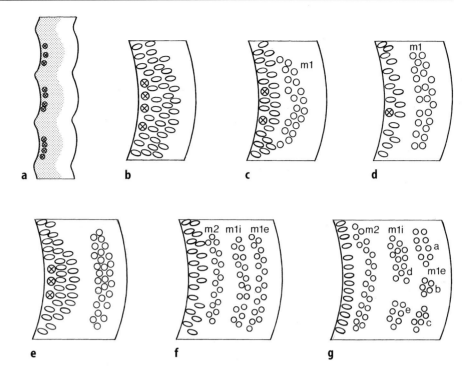

Fig. 5.18a–g. Development of brain nuclei according to Bergquist and Källén. **a** Postneuromeres or transverse bands. **b** Local proliferative activity leads to the formation of a migration area. **c,d,** Neuroblasts produced within a migration area form a separate migration layer: *m1*. **e** A second wave of mitotic activity leads to the production of numerous new neuroblasts. **f** The new neuroblasts aggregate in another migration layer, *m2*, and the first migration layer splits up into separate external and internal zones, *m1e*, *m1i*. **g** Layers *m1e* and *m1i* rearrange themselves and form a number of cell concentrations, *a–e*, which represent definitive brain nuclei (based on Källén 1951b; Bergquist 1953, 1954; and Bergquist and Källén 1954)

relatively short, straight or radial migration. The assumption of the prevalence of such simple migrations is implicit in the comparative studies on the ontogenesis of brain nuclei to be discussed now.

Kuhlenbeck (1929a,b, 1937) held that the migrations leading to the formation of cell masses in the brain start from particular fundamental components (*Grundbestandteile*), i.e. ventricular areas bounded by internal sulci (Chap. 4, Figs. 4.40, 4.41, 4.48, 4.50), whereas according to Bergquist and Källén (Bergquist 1932, 1953, 1954; Källén 1951a–d, 1952, 1955, 1962, 1965; Bergquist and Källén 1954), these migrations lead to the formation of so-called migration areas. The latter manifest themselves as cushion-like thickenings of the embryonic mantle layer, produced by local intensifications of the proliferative activity in the ventricular matrix (Chap. 4, Figs. 4.19–4.22, 4.40, 4.51).

According to the observations of Bergquist and Källén, the following events may take place within a given migration area (Fig. 5.18):

1. Radial migration is a continuous process, and the cells migrate over only short distances. The resultant cell mass remains in direct contact with, or at least close to, the ventricular surface (Fig. 5.18b). This is the prevalent situation observed in most parts of the brains of anamniotes (Chap. 2, Figs. 2.18–2.24, 2.26, 2.28).

2. Migrating cells detach themselves from the ventricular zone and form a separate migration layer (Fig. 5.18c,d).

3. New consecutive waves of local proliferative activity lead to the formation of additional migration layers (Fig. 5.18e,f). In certain areas of the diencephalon and telencephalon of amniotes four consecutive waves of proliferation and four resultant migration layers were observed (cf., e.g., Chap. 4, Fig. 4.51d).

4. The migration layers split up into two, or sometimes three, sublayers (Fig. 5.18 f; Chap. 4, Fig. 4.51b,d).

5. Each of the layers or sublayers within a migration area may fractionate into two or more separate cell masses (Fig. 5.18g).

Comparable phenomena were observed by Kuhlenbeck (1929a,b, 1973) in some of his 'fundamental components'. Thus, telencephalic component D1, which can be delimited in all gnathostomes, differ-

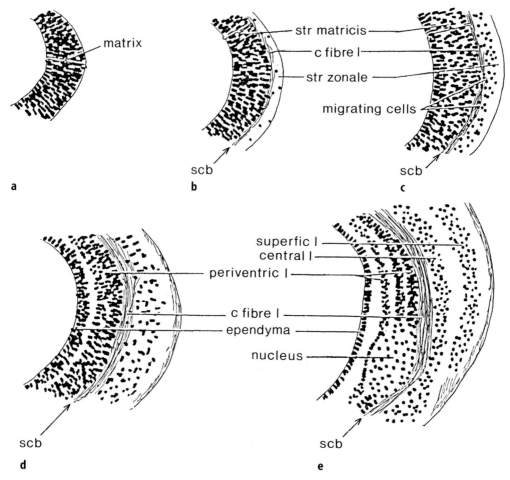

Fig. 5.19a–e. Development of layers in the reptilian brain. *central l*, Central layer; *c fibre l*, central fibre layer; *periventric l*, periventricular layer; *scb*, subcentral boundary; *str matricis*, stratum matricis; *str zonale*, stratum zonale; *superfic l*, superficial layer (reproduced from Senn 1968b, Fig. 6)

entiates in birds, according to Kuhlenbeck, into five separate cell masses (Chap. 4, Fig. 4.50).

The formation of layers also plays a prominent role in the work of Senn (1968a,b, 1970, 1972, 1974), who studied the development of the CNS in a number of amphibian and reptilian species, mainly in non-experimental, silver-impregnated material. Curiously enough, the detailed description and subdivision of the optic tectum of the adult chameleon, published in 1986 by P. Ramón, was the basis of Senn's embryological explorations. By studying the ontogenesis of the reptilian optic tectum, Senn found that the 14 concentric tectal zones described by P. Ramón develop from three fundamental layers: periventricular, central and superficial. Subsequent studies revealed that these layers can also be distinguished in all other parts of the CNS. "... the brain (and the spinal cord) is built according to a layer principle, a fundamental stratification, which is the same for all parts" (Senn 1970, p. 522). Senn's principal results may be summarised as follows:

1. The three fundamental layers develop according to a fixed pattern. Initially, only a cellular matrix is present (Fig. 5.19a), but soon a stratum zonale, situated peripheral to the matrix, can be distinguished (Fig. 5.19b). Tangential fibres concentrate in the border zone between the matrix and the stratum zonale (Fig. 5.19b–e). These fibres constitute a central white layer, which maintains itself throughout development. The interface between the central white layer and the matrix zone is termed the subcentral boundary (Fig. 5.20). Cells leave the matrix compartment, migrate through the central fibre layer and invade the stratum zonale to form a loose cellular layer (Fig. 5.19c,d). This layer of migrated cells subsequently divides into two layers, one

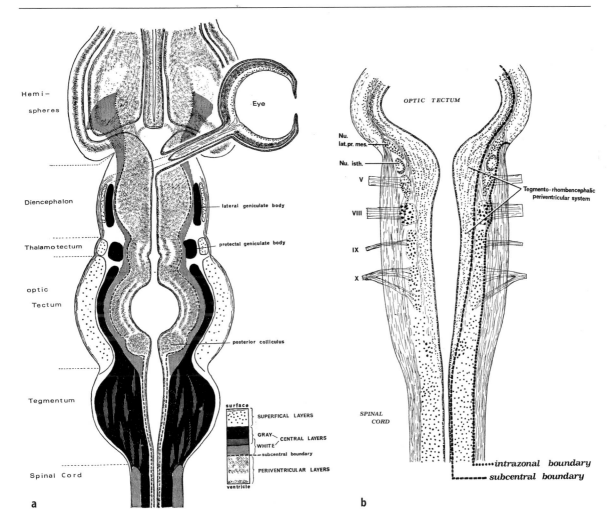

Fig. 5.20a,b. Horizontal sections through the brain of a lizard (**a**) and the brain stem of a frog (**b**) to show the disposition of the three principal layers. Note that in the rhombencephalon of the frog a periventricular layer is present which is lacking in the lizard. *Nu. isth.*, Nucleus isthmi; *Nu. lat. pr. mes.*, nucleus lateralis profundus mesencephali (reproduced from Senn 1970, Fig. 9; 1972, Fig. 8)

becoming the central grey layer, the other representing the anlage of the superficial grey layer.

2. Within all three of the principal cellular layers a further differentiation into sublaminae, or the formation of discrete cell masses or nuclei may occur (Fig. 5.19e).
3. The further development of the three principal layers shows profound local differences.
4. In some regions, as for instance the reptilian rhombencephalon and the amphibian and reptilian spinal cord (Fig. 5.20), almost all cells produced in the matrix layer pass the central fibre layer and migrate into the stratum zonale. Where this occurs, the remaining periventricular layer ultimately consists only of the ventricle-lining ependymal cells. In other regions numerous cells differentiate within the compartment situated

between the ventricular surface and the subcentral boundary, and the matrix is gradually transformed into a periventricular layer (Fig. 5.19d,e). This layer is strongly developed in the mesencephalon, the pretectum and the thalamus. The (amphibian and reptilian) telencephalic hemispheres consist almost entirely of a periventricular layer (Fig. 5.20a). In the tectum and pretectum the periventricular layer is split up into several sublayers (Fig. 5.19e, 5.20a). The rhombencephalon of anurans contains, unlike the corresponding region in reptiles, a well-developed periventricular region (Fig. 5.20).

5. The central layer is strongly developed in the spinal cord and the rhombencephalon, but it is also present in the tectum, the pretectum and the thalamus (Fig. 5.20a). The tegmental and cere-

bellar grey matter is almost exclusively derived from the central grey layer.

6. The superficial layer is found only in the tectum, the pretectum and the thalamic region (Fig. 5.20a).

7. The pretectal region shows a striking resemblance to the tectum. These centres together constitute a single complex.

8. The embryonic development of the hypothalamus resembles that of the telencephalic hemispheres.

9. The subdivision into concentric layers is one of three possible morphological subdivisions of the CNS. The other two are the subdivision into longitudinal, dorsoventrally arranged zones, according to His and Herrick, and the rostrocaudal subdivision into brain parts, according to von Kupffer.

10. Homologous cell masses may be situated in different layers. Thus, the nucleus interstitialis of the fasciculus longitudinalis medialis is situated in the central grey, but its amphibian homologue is embedded in the periventricular grey, and a similar difference exists between the reptilian and amphibian nucleus motorius tegmenti (i.e., the rhombencephalic medial reticular formation).

It is a pity that the validity of the stratification principle, as enunciated by Senn, has not been systematically tested in other vertebrates, and that no attempts have been made to harmonise his results with those of Bergquist and Källén.

In Chap. 2 of the present work, the cell masses in the vertebrate brain have been provisionally placed in three categories: periventricular, intermediate and superficial (cf. Chap. 2, Figs. 2.18–2.24). It should be emphasised that this subdivision is exclusively based on comparison of adult material and, hence, has no direct relation with the subdivision as advocated by Senn.

The investigations of Kuhlenbeck, Bergquist and Källén, as well as Senn, discussed above (and in the previous chapter) were carried out on a considerable number of different species. These investigations represent contributions to *comparative* neuroembryology, with the central aims being to discover fundamental features of the brain of vertebrates and to present a morphological interpretation of the parts of the brains investigated, that is, to establish their homologies. The material studied by these authors consisted exclusively of serial sections of non-experimental material, treated with routine techniques such as hematoxylin and eosin or cresyl violet staining, or silver impregnation according to Cajal or Bodian. Graphical reconstructions were often made to elucidate the spatial relationships of the structures studied. Bergquist's (1932) monograph on the diencephalon of anamniotes exemplifies this type of research. Containing detailed analyses and interpretations of the development of the diencephalon and adjoining regions of 12 species, ranging from the lamprey to the frog, illuminated by more than 40 carefully prepared graphical reconstructions, it is a monument of comparative neuroembryological research.

During the past decades, numerous experimental neuroembryological studies have appeared and developmental neurobiology has become a flourishing branch of the neurosciences (see Jacobson 1991 for an excellent review). Although practically all of these studies are devoted to a single species and are not primarily intended as contributions to comparative neuroembryology, many of them nevertheless have a bearing on the central questions of that classical discipline. It is not feasible to present here a detailed survey of all pertinent data. However, the principal results of some recent studies on the development of the brain of rodents will be dealt with in the next section.

5.5
Some Investigations on the Development of the Rodent Brain

5.5.1
Thymidine Radiography Studies of Altman and Bayer

During the past 25 years, Joseph Altman and Shirley Bayer have published a series of more than 60 papers and monographs on the development of the rat brain. Their work centres on two techniques of [^{3}H] thymidine autoradiography, namely (a) long-survival [^{3}H] thymidine autoradiography after multiple injections, to date the time of origin of neurons in adult brains, and (b) short- and sequential-survival [^{3}H] thymidine autoradiography, to determine the neuroepithelial sources of various populations of neurons and to follow their paths of migration and their order of settling. The following survey of the principal results and deductions of Altman and Bayer, which is based primarily on two recent reviews published by these authors (Bayer and Altman 1995a,b), will be presented largely in their own words.

"In spite of the tremendous variety of mammals and their considerable variations in habitat and environment, the mammalian central nervous system (CNS) contains universally shared core components. Mammals also share developmental patterns." Rodents, especially rats, are well suited to serve "as model species for understanding mammalian neural development" (Bayer and Altman 1995b, p. 1079).

"There is growing evidence that, notwithstanding its apparently homogeneous cellular composition, the neuroepithelium (or proliferative germinal matrix) contains a blueprint (Bauplan) of the anatomy of the central nervous system" (Bayer and Altman 1995b, p. 1080).

"Throughout the spinal cord, hindbrain, floor of the midbrain, and diencephalon, it is possible to subdivide the neuroepithelium into small patches by such landmarks as evaginations or invaginations, thickenings or thinnings, and regional differences in label uptake within 2 h after injection of [³H] thymidine" (Bayer and Altman 1995b, p. 1082).

Particular attention has been paid to the development of the diencephalon. Earlier studies (Altman and Bayer 1978a,b, 1979) led to the conclusion that "... the major nuclear divisions of the mature diencephalon are relatable to identifiable proliferative divisions of the third ventricle epithelium of the early embryo" (Altman and Bayer 1979, p. 518). These proliferative divisions were found to manifest themselves as dorsoventrally arranged neuroepithelial lobules (Fig. 5.21). In later studies (Altman and Bayer 1988a–c, 1989a–c) the transformation of the thalamic neuroepithelium was subjected to a detailed analysis. It was found that the thalamic

lobules become further partitioned into sublobules, characterised as regional concavities and convexities, and that these sublobules represent putative germinal sources of specific thalamic nuclei. Reinvestigation of the hypothalamus (Altman and Bayer 1986) led to a subdivision into five discrete anteroposterior segments and to the identification of several patches of the neuroepithelium from which neurons could be traced outward to form specific hypothalamic nuclei. It was suggested that the lobulation of the ventricular surface of the developing thalamus is necessary to accommodate the particularly numerous mitotic cells in this region (Altman and Bayer 1988a). In the hypothalamic neuroepithelium and in many other brain regions, the neuroepithelial patches, as identified with the aid of thymidine radiography, were often not demarcated by furrows and bulges. The general pattern of the development of nuclear structures in the diencephalon has been summarised in a diagram (Bayer and Altman 1995b; Fig. 2), which is reproduced here as Fig. 5.22. It will be seen that at an early point in time (top diagram, Fig. 522), spatially segregated neuroepithelial patches (shaded bars) give rise to specific populations of neurons (shaded ellipses). At a later time (bottom diagram, Fig. 5.22) the diencephalic neuroepithelium is transformed in such a fashion that successive neuronal populations are being generated at the same sites.

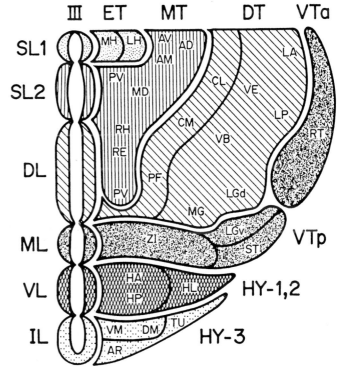

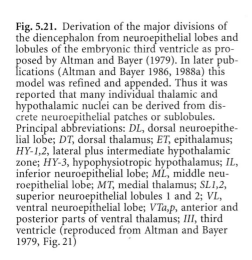

Fig. 5.21. Derivation of the major divisions of the diencephalon from neuroepithelial lobes and lobules of the embryonic third ventricle as proposed by Altman and Bayer (1979). In later publications (Altman and Bayer 1986, 1988a) this model was refined and appended. Thus it was reported that many individual thalamic and hypothalamic nuclei can be derived from discrete neuroepithelial patches or sublobules. Principal abbreviations: *DL*, dorsal neuroepithelial lobe; *DT*, dorsal thalamus; *ET*, epithalamus; *HY-1,2*, lateral plus intermediate hypothalamic zone; *HY-3*, hypophysiotropic hypothalamus; *IL*, inferior neuroepithelial lobe; *ML*, middle neuroepithelial lobe; *MT*, medial thalamus; *SL1,2*, superior neuroepithelial lobules 1 and 2; *VL*, ventral neuroepithelial lobe; *VTa,p*, anterior and posterior parts of ventral thalamus; *III*, third ventricle (reproduced from Altman and Bayer 1979, Fig. 21)

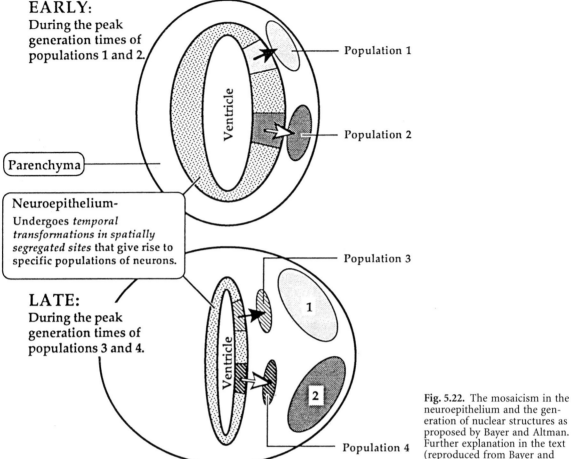

EARLY:
During the peak generation times of populations 1 and 2.

Population 1

Population 2

Parenchyma

Neuroepithelium-
Undergoes *temporal transformations in spatially segregated sites* that give rise to specific populations of neurons.

LATE:
During the peak generation times of populations 3 and 4.

Population 3

Population 4

Fig. 5.22. The mosaicism in the neuroepithelium and the generation of nuclear structures as proposed by Bayer and Altman. Further explanation in the text (reproduced from Bayer and Altman 1995b, Fig. 2)

"One of the most consistent findings is that each population of neurons in the central nervous system is generated according to a precisely timed schedule (neurogenetic timetable) with little variation between different animals of the same strain" (Bayer and Altman 1995b, p. 1084; cf. also Bayer and Altman 1995a, in which the neurogenetic timetables of all major cell groups in the CNS of the rat are provided).

The theory that a pre-existing network of radial glial fibres guide migrating neurons throughout the CNS to their destinations is rejected. According to the authors (Bayer and Altman 1995b), in many regions of the brain migrating neurons follow complex trajectories unrelated to the orientation of radial glial fibres.

It is hypothesised that "this chronology of neuron production is a necessary prerequisite for the proper anatomical and functional development of the CNS" (Bayer and Altman 1995b, p. 1084).

A few comments on the work of Altman and

Bayer are in order. With regard to proliferation and the formation of cell masses, the findings of Altman and Bayer (e.g., Bayer and Altman 1995b, 1987) show a striking resemblance to the migration areas of Bergquist and Källén (e.g., 1954; cf. Figs. 5.18 and 5.22). [It may be recalled that these migration areas result from an intersection of (post)neuromeres with longitudinal zones.] However, Altman and Bayer (1988a) deny a relationship between the migration areas and their 'neuroepithelial patches'. In fact, they attribute little value to the concept of neuromeric organisation, for the following two reasons: (a) no consensus has ever been reached with regard to the number or identity of the neuromeres, and (b) neuromeric theory is conceptually tied to the segmental organisation of the spinal cord and the rhombencephalon but does not account for the organisation of the diencephalon in horizontally oriented dorsoventral tiers. As has been shown in Chap. 4, Sect. 4.5, the first point is correct, but the second is not. Altman and Bayer

(1988a) also stated that the techniques used by Bergquist and Källén were inadequate to study the dynamics of proliferation and migration.

It is my opinion that the results of Altman and Bayer could be adequately compared with those of Bergquist and Källén only with the help of graphical reconstructions, but figures of that type have never been published by the authors first mentioned. This is a serious limitation of the work of Altman and Bayer. They indicate that the matrix layer contains a blueprint or *Bauplan* of the CNS, consisting of a mosaic of 'proliferative patches', 'cytogenetic sectors' or 'neuroepithelial zones', but they fail to show this mosaic to the reader. One reason for their reluctance to produce such figures is hinted at near the end of their extensive monograph on the development of the hypothalamus, where they state (Altman and Bayer 1986, p. 163): "The examination of hypothalamic neurogenesis suggests that there is a mosaic of discrete cell lines in the ventral third ventricle neuroepithelium, each characterized by a special chronology of precursor cell production and spatial distribution. Much further work is needed to translate this approach into an accurate map of the neuroepithelial zones of the germinal matrix of the hypothalamus." A second and more fundamental difficulty in the production of such maps may well be that the 'discrete cell lines' or 'proliferative patches' shift in position during development, and hence do not form together a static mosaic, but rather a dynamic kaleidoscopic picture. If so, the spatial organisation of these entities does not represent a *Bauplan,* i.e. "a set of invariant configurational relationships common to a group of organisms" (Kuhlenbeck 1967, p. 233).

If it is true that most 'neuroepithelial zones' or 'proliferative patches' cannot be delimited from each other by morphological landmarks, then for comparative studies of the entities material treated with thymidine radiography must be available for all species to be studied.

Several of the non-radial migration patterns observed by Bayer and Altman (1995b) may well represent cases of deflected radial or vectorial migration, as described in Sect. 5.3.3.

5.5.2
Homeobox and Other Genes in the Embryonic Mouse Forebrain

Bulfone et al. (1993) investigated the patterns of expression of three homeobox genes (Dlx-1, Dlx-2 and Gbx-2) and one member of the Wnt gene family (Wnt-3), all potential regulators of development in the CNS of the mouse at embryonic day 12.5. As has already been discussed in Chap. 4,

Sect. 4.5.5.5, they found that each of these genes is expressed in spatially restricted transverse and longitudinal domains within the forebrain. They concluded that these patterns of expression support the existence of longitudinal zones (alar plate, basal plate), extending throughout the brain (Chap. 4, Fig. 4.35a), and are also consistent with neuromeric models of the forebrain. They postulated the presence of six neuromeres (p1–p6) in the prosencephalon (Chap. 4, Fig. 4.35b).

In a subsequent study, Puelles and Rubenstein (1993) collected data from the literature about some 45 homeobox and other genes that are expressed in spatially restricted domains in the embryonic forebrain. Mapping of the expression patterns of these genes onto their E12.5 model consistently resulted in maps that fall within the hypothesised domains and respected the hypothesised boundaries (Chap. 4, Fig. 4.35c). They expressed the expectation that the information gathered will facilitate attempts to understand the genetic hierarchy that controls differentiation in specific neuroepithelial domains.

5.5.3
Homeobox Genes in the Embryonic Rat Forebrain

Alvarez-Bolado, Rosenfeld and Swanson recently published an extensive study on the spatiotemporal expression patterns of four POU-III homeodomain transcription factor genes, Brn-1, Brn-2, Brn-4 and Tst-1, in the developing rat forebrain vesicle, in embryos ranging in age from embryonic day 10 to 17 (Alvarez-Bolado et al. 1995). POU-III mRNA expression was studied in serial sections (Fig. 5.23), and the results were plotted on carefully prepared maps, which greatly facilitate comparison of the distribution patterns of the four genes at the various ages investigated. An extensive appendix was devoted to the preparation of these maps (Alvarez-Bolado and Swanson 1995).

It was found that the ventricular and subventricular matrices, as well as the mantle layer in the wall of the forebrain, each show a unique pattern of regionalised POU-III expression. Expression patterns showed a clear tendency to correlate with major morphological features in the forebrain, such as internal or external sulci associated with proliferative zones (Fig. 5.24). Thus, clear differences in hybridisation intensity were observed between the pretectal region, the dorsal thalamus, the ventral thalamus, the hypothalamus, the basal ganglia and the cerebral cortex (Fig. 5.24). Transverse bands of hybridisation extending from the roof to the floor of the forebrain, corresponding to proposed neuromeres, were not observed with these four probes.

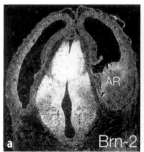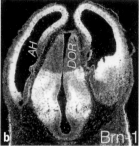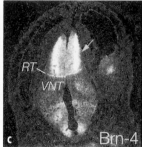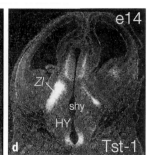

Fig. 5.23a–d. The basic regionalisation of the diencephalon of a 14-day-old rat embryo, as shown in four consecutive sections, hybridised with probes for four POU-III transcription factor mRNAs (indicated in the *lower right corners* of the individual figures). The primordium of the dorsal thalamus (*DOR*) is clearly delimited from the rest of the diencephalon, especially in **b** and **c**. The primordium of the reticular nucleus (*RT*) is represented by a thin band of cells just ventral to the presumed dorsal thalamus. The presumed zona incerta (*ZI*) shows intense expression of *Tst-1* at this stage. *Brn-4* and *Tst-1* probes label a dorsoventral band of cells that lies in the presumed dorsal thalamus, between the matrix and the mantle layer. *AH*, Ammon's horn; *AOB*, accessory olfactory bulb; *AR*, amygdala ridge; *CTX*, cerebral cortex (*l, m*: lateral, medial parts); *DOR*, dorsal thalamus (*c, r*: caudodorsal, rostroventral parts); *EPI*, epithalamus; *ets*, external sulcus; *HIP*, hippocampus; *HY*, hypothalamus; *INF*, infundibulum; *MAM*, mammillary level, hypothalamus; *OB*, olfactory bulb; *PIN*, pineal gland; *PR*, pallial ridge; *PRT*, pretectal region; *RT*, nucleus reticularis thalami; *shb*, sulcus subhabenularis; *shy*, sulcus hypothalamicus; *smi*, sulcus diencephalicus medius; *SPR*, septal ridge; *sps*, sulcus pallidostrialalis; *SR*, striatal ridge; *TEM*, thalamic eminence; *vmh*, nucleus ventromedialis hypothalami; *VNT*, ventral thalamus; *ZI*, zona incerta (reproduced from Alvarez-Bolado et al. 1995, Fig. 17)

Alvarez-Bolado et al. (1995) criticised the prosomeric model of Puelles and his colleagues (Bulfore et al. 1993; Puelles and Rubenstein 1993), discussed above (Sect. 5.5.2) and depicted in Chap. 4, Fig. 4.35. They pointed out that none of the prosomeric borders clearly extend to the ventral surface of the brain, and that Puelles and his colleagues did not assign specific adult nuclei to ventral regions of the pretectal, dorsal/epithalamic, and ventral thalamic columns or neuromeres along the base of the diencephalon. In the present author's opinion this criticism is correct. The borders of the three regions indicated cannot be traced to the ventral surface of the brain (cf. Chap. 4, Figs. 4.37, 4.53). However, is this a valid reason for rejecting a neuromeric model of the diencephalon altogether? The pretectum, the dorsal thalamus and the ventral thalamus are almost certainly direct derivatives of three prosomeres, the synencephalon, the parencephalon posterius and the parencephalon anterius, and the findings of Alvarez-Bolado and his colleagues are in harmony with this interpretation, if the curvature of the neural tube is taken into consideration. In the rhombencephalon neuromery manifests itself most clearly, though not in a straightforward one-to-one way, in the generation of the cranial motor nuclei, situated in the ventral basal plate, and much less in the sensory alar plate, as clearly indicated in Alvarez-Bolado and colleagues' (1995) Fig. 22. However, this feature does not invalidate a neuromeric model of the rhombencephalon.

Surveying the literature on the expression of genes in the developing brain, Alvarez-Bolado et al.

(1995, p. 281) noticed that "... spatiotemporal maps based on serial section analysis at different ages, necessary for determining the extent to which expression is found throughout various adjacent regions and how these patterns may change, are very uncommon," and that, hence, "... frequent claims that a gene is involved in 'brain segmentation' should be viewed cautiously." "An understanding of the true regionalization of the neural tube – its fundamental plan – will emerge only after comparing the expression patterns of (probably) hundreds of genes in physically accurate, rather than hypothetical or assumed, maps, as well as clarifying mechanisms regulating their coordinate and sequential expression, whether or not gene programs are repeated in a metameric way."

Alvarez-Bolado and his colleagues (1995, pp. 281–282) conclude their study by outlining a vast program for future research and offering an almost eschatological perspective: "A cardinal feature of the CNS as an organ is its division into hundreds of more or less discrete cell groups that are interconnected very precisely by axons to form the biological circuit that coordinates and directs behaviour and metabolism. The developmental genetic code responsible for assembling this macro-circuitry (in the absence of experiential refinements of synaptology after birth) from the neural plate region of the trilaminar embryonic disc is poorly understood. However, a fundamental part of the solution includes knowledge of the complete set of genes regulated during neural plate and tube morphogenesis, along with their spatiotemporal pat-

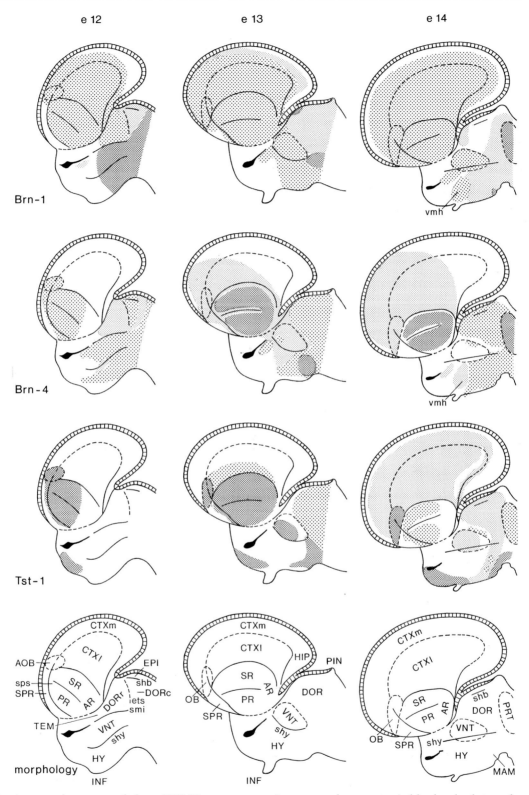

Fig. 5.24. Spatiotemporal patterns of three POU-III transcription factor mRNAs (*Brn-1, Brn-p4, Tst-1*) in the developing rat forebrain. The distribution of the POU-III mRNAs has been mapped onto flat representations of the embryonic forebrain at the *12th*, *13th* and *14th days* of embryonic develop-ment. *Bottom row* shows anatomical landmarks that can be recognised easily in the transverse sections upon which the grapical reconstructions are based. For abbreviations see Fig. 23 (based on Alvarez-Bolado et al. 1995, Fig. 2).

terns of expression. Ultimately, then, it should be possible to compare directly a chromosomal map of the brain developmental gene network with an anatomical map of adult brain circuitry."

5.5.4
Comments on the Studies Discussed

The studies of Altman and Bayer, Puelles and his collaborators, and Alvarez-Bolado and his colleagues, just summarised, have been selected to exemplify current neuroembryological research. These studies have the following features in common:

1. They were carried out on a group of closely related species, i.e. the rodents.
2. In all of them the intention is expressed to contribute to more than just the neuroembryology of rodents. Bayer and Altman (1995a, p. 1078) recommended rodents, especially rats, "as model species for understanding mammalian neural development." Puelles and Rubenstein (1993; cf. also Rubenstein et al. 1994) indicated that the model they described is intended as an anatomical foundation that can guide future studies of forebrain development, and as a framework for studying the evolution of the forebrain. Alvarez-Bolado et al. (1995), lastly, intended to contribute to the determination of the fundamental plan of neural tube regionalisation.
3. In all three studies it is indicated that at least some of the boundaries of the marked or labeled regions are coincident with morphological features, such as ventricular or external sulci.

However, there are also notable methodological and interpretational differences between the three studies selected. As regards *methodology*, Alvarez-Bolado and colleagues took great pains to present clear and accurate pictorial surveys of their findings. They mapped their results (i.e. the distribution of POU-III transcription factor mRNAs) onto flat representations of the embryonic forebrain, at different ages, with 'deevaginated' cerebral hemispheres (Fig. 5.24). They emphasise that careful mapping is a *conditio sine qua non* for understanding the true regionalisation of the CNS. Moreover, they presented bilateral flattened maps to show the regionalisation of the early neural tube (Alvarez-Bolado et al. 1995, Fig. 22). Puelles and his colleagues used a diagrammatic median section through the brain of a 12.5-day-old mouse embryo as a general reference for plotting their results (Chap. 4, Fig. 4.35); they also presented an orthogonalised topological map of an embryo of the same age for similar purposes (Bulfone et al. 1993, Fig. 7; Puelles and Rubenstein 1993, Fig. 4B). Altman and

Bayer, on the other hand, illustrated their publications with hundreds of photomicrographs, but they prepared no reconstructions whatsoever. This serious limitation greatly hampers comparison of their results with those of others, as far as regionalisation is concerned.

As regards *interpretation*, it is highly remarkable that the differences of opinion between the three groups selected concerning the essential regionalisation of the rodent brain closely resemble those concerning the basic subdivision of the vertebrate brain during the period of classical comparative neuroembryology. On the basis of the expression patterns of a large number of homeobox and other genes, Puelles and his colleagues present a prosomeric model of the forebrain, according to which the diencephalon and the secondary prosencephalon are both composed of three neuromeres. Alvarez-Bolado and his colleagues, who mapped the expression patterns of four other homeobox genes, did not detect any trace of segmentation in the forebrain. Rather, they observed clear differences in hybridisation intensity of the POU-III transcription factor mRNAs studied between such entities as the cerebral cortex, basal ganglia, dorsal thalamus, ventral thalamus and hypothalamus. They tended to consider the three diencephalic regions just mentioned, with Herrick (1910) and others, as dorsoventrally arranged, horizontally oriented entities, and concluded their discussion of regionalisation as follows (Alvarez-Bolado et al. 1995, p. 281): "It would not be surprising if one conclusion of current molecular analyses of brain development is that the 'true' early regions of the neural tube resemble the subdivisions classically recognised on the basis of gross morphology, connectivity, and cytoarchitecture."

Finally, Altman and Bayer (1988a, p. 372) stated that the embryonic brain displays neither a transverse (neuromeric) nor a longitudinal organisation. They concluded from their thymidine radiography experiments that the neuroepithelium is an irregular mosaic of 'cytogenetic sectors' or 'proliferative patches' destined to give rise to the various cell masses in the CNS.

5.6
Development of Fibre Systems

5.6.1
Introduction

During the first half of this century, numerous publications appeared on the ontogenesis of fibre systems. Prominent among these are the studies of His (1904: man), Tello (1923: chick; 1934: mouse),

Windle (1932a,b: cat), Windle and Austin (1935: chick), Windle and Baxter (1936: rat), Angulo y González (1939: rat), and those of Coghill and Herrick on the salamanders *Ambystoma punctatum* and *Ambystoma triginum* (Coghill 1926, 1928, 1929, 1930, 1931; Herrick 1937, 1938a–c; Herrick and Coghill 1915). All of these studies, except for those of His and Coghill, were based on silver-impregnated serial sections of non-experimental material.

It is important to note that the work of Coghill and Herrick, as well as that of Windle and his associates, was aimed at determining the relation between the early differentiation of the CNS and the development of behaviour. Both groups studied larval or embryonic specimens which were physiologically tested immediately before preservation. Accordingly, Coghill and Herrick used a 'behavioural staging' in their analysis of the development of the CNS of *Ambystoma*: non-motile stage, early flexure stage, coil stage, early swimming stage, and so on. With regard to the interpretation of the earliest activities of vertebrate embryos, Coghill and Herrick (e.g., Coghill 1929; Herrick 1948) and Windle and his associates (e.g., Windle et al. 1934; Windle 1950) arrived at diametrically opposed concepts. According to the authors first mentioned, reactive behaviour starts as a total pattern, which means that the animal, when first activated from the sensory zone, reacts with a total response of all of the neuromuscular mechanism which is capable of functioning. This total pattern of activity expands as more and more of the neuromuscular mechanism becomes functional. After its completion the apparatuses of local reflexes ('partial patterns') gradually become individuated within the larger frame of the total pattern. As early as 1915, Herrick and Coghill emphasised that the typical monosynaptic reflex arc in the mammalian spinal cord represents the end rather than the beginning of the developmental sequence and is not to be regarded as a primitive form. Windle and his associates, on the other hand, held that in birds and mammals the first responses to stimulation are localised, simple reflexes, which only later become integrated into behavioural patterns (for review see Hooker 1952).

More recently, the origin and early development of fibre systems have been analysed with a variety of techniques, including electron microscopy (Singer et al. 1979: the newt, *Triturus*; Nordlander and Singer 1982a,b; Nordlander 1987: *Xenopus*; Kevetter and Lasek 1982: *Xenopus*; Nakao and Ishikawa 1987: the lamprey; Knyiher-Csillik et al. 1995: rhesus monkey), horseradish peroxidase labelling (Metcalfe et al. 1986: zebrafish; Nordlander 1987: *Xeno-pus*; Lee et al. 1993: goldfish), immunohistochemistry (e.g. Easter et al. 1993: mouse: labelling with an antibody to neuron-specific class III β-tubulin), or combinations (e.g., Bernardt et al. 1990; Kuwada et al. 1990a: zebrafish: horseradish peroxidase, DiI and lucifer yellow labelling, as well as labelling with a monoclonal antibody against acetylated tubulin; Ross et al. 1992: zebrafish: AChE histochemistry; immunohistochemistry with antibodies to the cell surface marker HNK-1 and anti-acetylated tubulin antibodies, horseradish peroxidase labelling and electron microscopy).

The ensuing brief survey of hodogenesis is based on both the older and the more recent studies mentioned, which have been selected from a vast literature. Following some general remarks on tract formation, the main axonal bundles observed in early embryonic and larval vertebrates will be discussed. This survey is illustrated by some representative figures reproduced from the work of Herrick (1937, 1938a, Fig. 5.27), Windle and Austin (1935, Fig. 5.28), and Windle (1935, Fig. 5.29), and by a composite drawing of the fibre systems in the CNS of an early zebrafish embryo (Fig. 5.26), based on the recent studies by Bernardt et al. (1990), Kuwada et al. (1990a,b), Wilson et al. (1990), Ross et al. (1992), and Kimmel (1993).

5.6.2
Formation of Axonal Bundles

During early development the axons of differentiating neuroblasts pass to the marginal zone of the neural tube, where they assemble in distinct bundles, situated between the most peripheral parts of the neuroepithelial cells. During further development, these superficially situated bundles increase in size, and finally merge to become the primordial white matter of the neuraxis (e.g., Kevetter and Lasek 1982).

Singer and his associates discovered that in the marginal zone of the developing spinal cord of the newt *Triturus viridescens* and *Xenopus* (Singer et al. 1979), prior to the appearance of axons, longitudinally arranged, tunnel-like spaces appear between the peripheral processes of the neuroepithelial cells, and that these channels are subsequently invaded by growing axons. Similar channels were observed between the radial, processes of the ependymal cells in the spinal cord in regenerating newt (Nordlander and Singer 1978) and lizard tails (Egar and Singer 1972). Comparable phenomena have been observed during normal development in other parts of the CNS, for instance, in the optic nerve of the mouse (Silver and Robb 1979).

Singer et al. (1979) proposed that the preformed

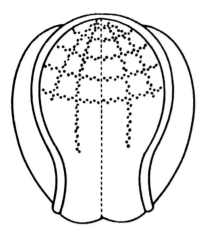

Fig. 5.25. The pattern of hypothetical substrate pathway precursor fields projected upon the neural plate (reproduced from Katz et al. 1980)

intracellular channels guide growing axons during development and that the primitive neuroepithelium and its derivative primitive glia contain a developmental genetic 'blueprint' of the primary nervous pathways.

Katz and Lasek (1978, 1979, 1981) found that in *Xenopus* tadpoles, optic nerve and Mauthner cell axons transplanted to unnatural places in the CNS behave as if they are following discrete, reproducible pathways in the substrate. On the basis of these experiments, Katz et al. (1980) and Katz and Lasek (1981) postulated the presence of substrate pathways in the developing CNS, which they defined as "sets of similar guidance cues aligned in continuous discrete routes" (Katz et al. 1980, p.825). Their experiments suggested the presence of two stereotyped longitudinal routes, i.e. a ventral substrate pathway extending in the ventral marginal zone from the caudal diencephalon to the spinal cord, running the length of the basal plate, and thus called the *basal substrate pathway*, and a dorsal, *alar substrate pathway*, extending in the alar plate of the developing hindbrain and spinal cord.

In order to obtain some idea of how the pattern of substrate pathways is foreshadowed in the neural plate stage, Katz et al. (1980) sketched the routes of a number of major tracts, such as the radix descendens of V, the anterior commissure, and the habenulo-interpeduncular tract, on C.-O. Jacobson's (1959) neural plate fate map of the axolotl CNS. The resultant simple, meridional pattern is shown in Fig. 5.25. They proposed that this pattern is topologically preserved as the neural plate is transformed into the neural tube, and that it manifests itself by then as the three-dimensional organisation of the early marginal zone formed by the

peripheral processes of developing neuroepithelial and radial glial elements.

In relation to the blueprint hypothesis of Singer et al. (1979) and the substrate pathway concept of Katz et al. (1980), it is worthy of note that Knyihar-Csillik et al. (1995) recently found that in the prospective lateral column of the spinal cord of rhesus monkey embryos, end feet of radial glial elements establish subpial compartments of extracellular space, which somewhat later are filled by outgrowing axons.

A detailed discussion of how developing axons find their way through the developing CNS, attain their destination(s), and establish appropriate synaptic contacts is beyond the scope of the present work. However, it may be mentioned that, in general, small numbers of precociously differentiating neurons provide the axons which pioneer the various pathways. These cells project a growth cone which navigates through the developing brain by following a stereotyped and precise route. These primary or pioneer neurons are believed to be responsible for establishing the first functional circuits of the CNS. Other axons, developing from vastly more numerous, later-maturing neurons, join the pioneer axon(s) and, when the wiring is accomplished, the pioneer neurons undergo cell death, leading to degeneration of their axons (Wilson et al. 1990; Knyihar-Csillik et al. 1995). It has been established that the axon growth cones of pioneer fibres are generally larger and more complex than those of 'follower' axons (Reh and Constantine-Paton 1985; Nordlander 1987; LaMantia and Rakic 1990; Williams et al. 1991; Kim et al. 1991).

During the past decade, new insights have been obtained into the molecular mechanisms which guide axonal growth cones. Several molecules, among them neural cell adhesion molecule (N-CAM) and laminin, are expressed on the neural epithelial cells that surround pioneering axons and provide permissive substrates that promote axon extension but do not impart specific directional information. A group of proteins named netrins has been shown to exert chemotropic guidance to growth cones, whereas another protein, collapsin (thus named because of its ability in vitro to cause the collapse of certain growth cones), has appeared to exert a repulsive influence on growth cones. Interestingly, the adhesion molecules mentioned, the chemoattractive netrins, and the repulsive collapsin have been shown to be involved in the control of growth cone guidance in vertebrates and invertebrates. The data on growth cone guidance just discussed are derived from the review articles of Dodd and Jessell (1988) and Goodman (1994), to which the reader is referred for details.

In the brain stem and the diencephalon the early-developing dorsoventral axons are confined mainly to interneuromeric boundaries (cf. Chap. 4, Sect 4.5.5.3). Because the boundaries of many neuromeres in these regions coincide with the boundary regions of the expression domains of putative regulatory genes, the hypothesis has been advanced that the transcription factors encoded by these genes regulate the expression of molecules involved in growth cone guidance (Wilson et al. 1993).

5.6.3
Early Pathways in the Vertebrate Brain and Spinal Cord

The earliest axonal tracts in the central nervous system of vertebrates develop from a limited number of loci or clusters of precociously differentiating neurons. In the rhombencephalon and spinal cord these clusters are clearly segmentally arranged (see Chap. 4, Sect. 4.5.5.1). Coghill (1928) distinguished three areas of early differentiation in the rostral part of the brain of *Ambystoma* embryos: one in the cerebral hemisphere, a second close behind the chiasma ridge, and a third along the boundary of the diencephalon and mesencephalon. These areas correspond to the dorsorostral, ventrorostral and ventrocaudal clusters of AChE-positive cells observed by Ross et al. (1992) in early zebrafish embryos (Fig. 5.26). Comparable clusters have been observed in the brains of embryos of other vertebrates.

The earliest tracts in the vertebrate neuraxis are organised in a stereotyped orthogonal pattern. This first set of tracts serves as a 'scaffold' on which

most later tracts form (Easter and Taylor 1989; Wilson et al. 1990; Easter et al. 1993). It encompasses two longitudinal fibre streams, designated here as the dorsolateral and ventrolateral bundles, a primitive telencephalic and a primitive diencephalic axonal bundle, i.e. the telencephalic and dorsoventral diencephalic tracts, respectively, of Chitnis and Kuwada (1990) and Ross et al. (1992), and the posterior commissure. The three tracts last mentioned are oriented perpendicular to the rostral part of the ventrolateral bundle (Fig. 5.26). These early tracts will now be discussed.

The *dorsolateral bundle* (DLB) extends from the isthmus region caudally to the end of the spinal cord. It corresponds to the dorsolateral sensory tract or primitive sensory tract of Herrick and Coghill (1915), the lateral longitudinal fascicle of Windle (1932b), Metcalfe et al. (1986) and Kimmel (1993), the dorsolateral fascicle of Nordlander (1987) and the dorsolateral tract of Nakao and Ishikawa (1987).

Initially, the DLB is composed exclusively of the longitudinally running main axonal processes of the embryonic Rohon-Beard cells, large transient elements which have been observed in representatives of all vertebrate groups (for review see Nieuwenhuys 1964). In later developmental stages axons of spinal ganglion cells and, at the level of the rhombencephalon, descending fibres of the trigeminal nerve are added to the bundle (Figs. 5.26, 5.27b, 5.28b). In all gnathostomes these fibres of peripheral origin gradually entirely replace the Rohon-Beard axons. The most rostral part of the DLB is formed by axons descending from the

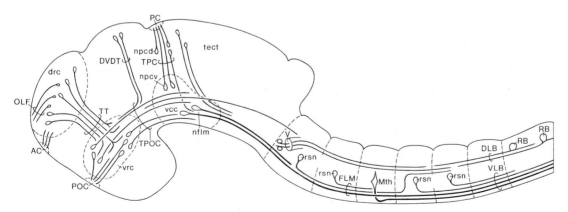

Fig. 5.26. Axonal tracts in the brain of a 30-h embryo of the zebrafish, *Brachydanio rerio*. *AC*, Anterior commissure; *DLB*, dorsolateral bundle; *DVDT*, dorsoventral diencephalic tract; *drc*, dorsorostral cluster of early-differentiating cells; *FLM*, fasciculus longitudinalis medialis; *Mth*, Mauthner neuron; *nflm*, nucleus of flm; *npcd*, dorsal nucleus of the posterior commissure; *npcv*, idem, ventral ; *OLF*, olfactory nerve; *PC*, posterior commissure; *POC*, postoptic commissure; *RB*, Rohon-Beard cells; *rsn*, reticulospinal neurons; *tect*, tectum mesencephali; *TPC*, tract of the postoptic commissure; *TT*, telencephalic tract; *V*, trigeminal nerve; *vcc*, ventrocaudal cluster of early-differentiating cells; *VLB*, ventrolateral bundle; *vrc*, ventrorostral cluster of early-differentiating cells (compiled from data in Wilson et al. 1990, Chitnis and Kuwada 1990, and Ross et al. 1992)

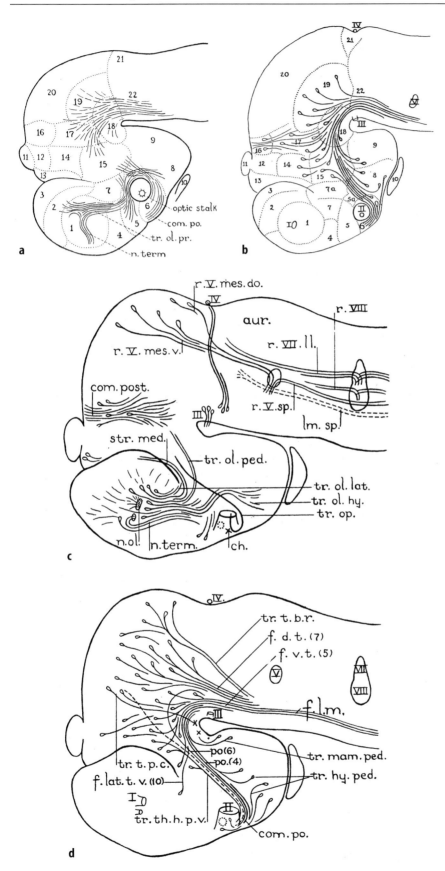

◄ **Fig. 5.27a–d.** Arrangement of nerve fibres in some developmental stages of the salamanders *Ambystoma punctatum* (**a,b**) and *A. tigrinum* (**c,d**), drawn by Herrick (1937, 1938a). **a** Coil stage; *numbers* refer to cerebral areas, as distinguished by Coghill (1930, 1931). Fibres from the rostral end of the nasal placode enter the primary olfactory region (*1*) and extend a primary olfactory tract (*tr.ol.pr*) into the secondary olfactory (*4*) and preoptic (*5*) regions. According to Coghill (1930, 1931), these fibres represent the nervus terminalis (*n.term.*). In the region of the future chiasma ridge a definite strand of nerve fibres crosses the median plane in the postoptic commissure (*com.po.*). Fibres covering the ventral thalamus (*15*), the ventral nucleus of the posterior commissure (*17*), the dorsal tegmentum (*19*) and the mesencephalic motor region (*18*) converge toward the tegmentum of the rhombencephalon (reproduced from Herrick 1937, Fig. 1). **b** Fibres of the postoptic commissure, posterior commissure (*16*) and tegmental fascicles of the early swimming stage. Note that fibres originating from the postoptic region (*6*), the ventral (*8*) and dorsal (*9*) parts of the hypothalamus, the ventral thalamus (*15*), and the ventral nucleus of the posterior commissure (*17*) converge upon the mesencephalic motor region (*18*), and that the latter and several of the 'higher' centres mentioned contribute to the primordial fasciculus longitudinalis medialis. *Roman numerals* refer to cranial nerve roots (reproduced from Herrick 1938a, Fig. 5). **c** Diagrammatic projection of the olfactory connections, posterior commissure, optic tract and some root fibres of cranial nerves III, IV, V, VII (i.e. the anterior lateral line) and VIII, as observed in a 10-mm larva of *A. tigrinum*, the CNS of which is somewhat further developed than that of the specimen of *A. punctatum*, upon which the diagram shown in **b** is based. **d** Diagrammatic projection of the components of the postoptic commissure, fasciculus longitudinalis medialis, and of several fascicles converging upon the mesencephalic motor tegmentum. *Broken line* indicates the course of fibres of the postoptic commissure, which connect with the rostral end of the tectum (**c** and **d** reproduced from Herrick 1938a, Figs. 19, 20). *aur*, Auricula cerebelli; *ch*, chiasma opticum; *comp.po.*, commissura postoptica; *com.post.*, commissura posterior; *f.d.t.*, fasciculi dorsales tegmenti; *f.l.m.*, fasciculus longitudinalis medialis; *f.lat.t.v.*, ventral fascicles of fasciculus lateralis telencephali; *f.v.t.*, fasciculi ventrales tegmenti; *lm.sp.*, lemniscus spinalis; *n.ol.*, nervus olfactorius; *n.term.*, nervus terminalis; *po.(4)(6)*, fibres from the postoptic commissure entering tegmental fascicles; *r.V.mes.do.*, dorsal division of mesencephalic root of V nerve; *r.V.mes.v.*, idem, ventral division; *str.med.*, medullaris thalami; *tr.hy.ped.*, tractus hypothalamus-peduncularis; *tr.mam.ped.*, tractus mamillo-peduncularis; *tr.ol.hy.*, tractus olfacto-hypothalamicus; *tr.ol. lat.*, tractus olfactorius lateralis; *tr.ol.ped.*, tractus olfacto-peduncularis; *tr.op.*, tractum opticus; *tr.t.b.r.*, tractus tectobulbaris tectus; *tr.t.p.c.*, tractus tectopeduncularis cruciatus; *tr.th.h.p.v.*, tractus thalamo-hypothalamicus, pars ventralis

mesencephalic trigeminal nucleus (Coghill 1930; Windle 1932a; Herrick 1938a; Angulo y Gonzalez 1939; Easter et al. 1993; Fig. 5.27c).

The DLB is not composed exclusively of primary afferent axons. Metcalfe et al. (1986) observed in zebrafish embryos that the axon of a rostral rhombencephalic reticular neuron, designated as RoL2 (see Chap. 4, Fig. 4.31), which initially descends in the ventrolateral bundle, switches abruptly to the DLB in the middle of rhombomere 5, with which it passes to the spinal cord. In the lamprey the axon of the Mauthner cell takes a similar course. In the spinal cord of zebrafish embryos axons of spinal fun-

icular and commissural neurons have been observed to join the DLB. Some of these fibres appeared to terminate in the border zone of the mesencephalon and diencephalon (Bernhardt et al. 1990; Kuwada et al. 1990a). In amphibians axons of spinal neurons ascend ventral to the DLB (Nordlander 1987) and enter the brain as the primordial spinal lemniscus (Herrick 1938a; Fig. 5.27b). It is also noteworthy that, apart from sensory trigeminal nerve fibres, primary afferent axons of the eighth (Figs. 5.27c, 5.28) and, where present, of the lateral line nerves (Fig. 5.27c) form distinct longitudinal bundles in the most superficial zone of the rhombencephalic alar plate in early developmental stages.

The *ventrolateral bundle* (VLB) extends throughout the brain and spinal cord. At spinal levels it is composed initially of the axons of primitive motoneurons, which in amphibians constitute the earliest descending tract of the CNS (Herrick and Coghill 1915; Coghill 1929). Descending axons of intersegmental interneurons are soon added to this primitive motor tract (Nordlander 1987), and somewhat later the bundle is invaded by axons originating from the primordial rhombencephalic reticular formation and from a group of early-differentiating cells situated at the level of the synencephalon (i.e. the so-called ventrocaudal cluster in the brain of the zebrafish: Fig. 5.26). The segment of the VLB extending through the brain stem constitutes the primordial fasciculus longitudinalis medialis (FLM). Numerous authors, among them His (1904), Tello (1923), Windle (1935, Fig. 5.29), Windle and Austin (1935, Fig. 5.28) and Herrick (1937, 1938a, Fig. 5.27a,b,d), have emphasised that the FLM is one of the earliest differentiating fibre tracts of the brain. Coghill (1931) and Herrick (1937) considered it likely that in *Ambystoma* larvae the long descending axons of the FLM influence the lower apparatus of swimming even before they receive any nervous excitations from sensory centres. Windle (1935), on the other hand, held that in embryonic cats the rostral component of the FLM forms a path, shared by both optic and olfactory systems in early behavioural mechanisms.

The most rostral part of the VLB is formed by a set of composite fascicles and diffusely arranged fibres extending from the tegmentum of the midbrain to the postoptic commissure. With Wilson et al. (1990), Ross et al. (1992) and Easter et al. (1993), I will designate this fibre complex as the *tract of the postoptic commissure* (TPOC). The postoptic commissure is situated in the caudal part of the chiasma ridge and runs between the two sides of the rostral diencephalon. Several authors (Tello 1923; Coghill 1930; Herrick 1937, 1938a, Fig. 5.27a,b; Windle 1932a, 1935, Fig. 5.29; Windle and Austin 1935,

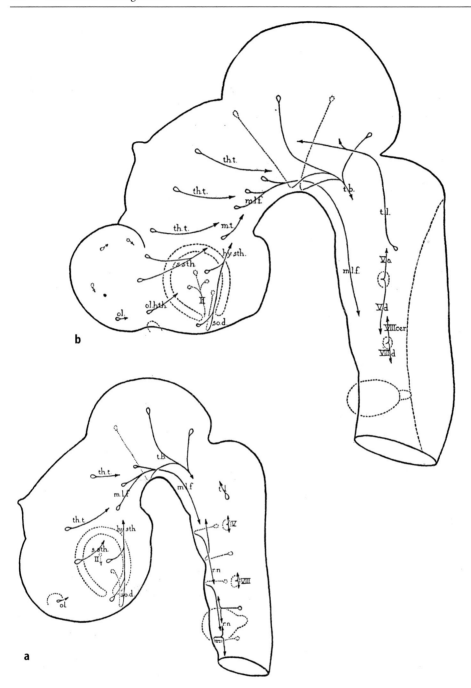

Fig. 5.28a,b. Fibre tracts in the brain of 60-h-(**a**) and 73-h-(**b**) old chick embryos as observed in pyridine silver preparations. The positions of eye and auditory placode are indicated by *broken lines*. The primary motor neurons have been omitted; fibres which cross from the opposite side are also indicated by *broken lines*. *Bifurcations* mean that tracts pass one way or the other, not that fibres divide. In **b** reticular pathways and most secondary afferent neurons of the rhombencephalon have been omitted. *hy.sth.,* Hypothalamosubthalamic tracts; *m.l.f.,* medial longitudinal fascicle; *m.t.,* mamillotegmental tract; *ol.,* higher-order olfactory neurons; *ol.hth.,* olfacto-hypothalamic tract; *r.n.,* reticular neurons of rhombencephalon; *so.d.,* dorsal supra-optic decussation; *s.sth.,* subthalamic tract; *t.b.,* tecto-bulbar and tecto-spinal tracts; *th.t.,* thalamo-tegmental or thalamo-bulbar tract; *t.l.,* trigeminal lemniscus; *II,* optic nerve; *V a.,* ascending trigeminal primary tract; *V d.,* descending trigeminal primary tract; *VIII cer.,* primary vestibulo-cerebellar tract; *VIII d.,* descending vestibular primary tract (reproduced from Windle and Austin 1935, Figs. 3, 4)

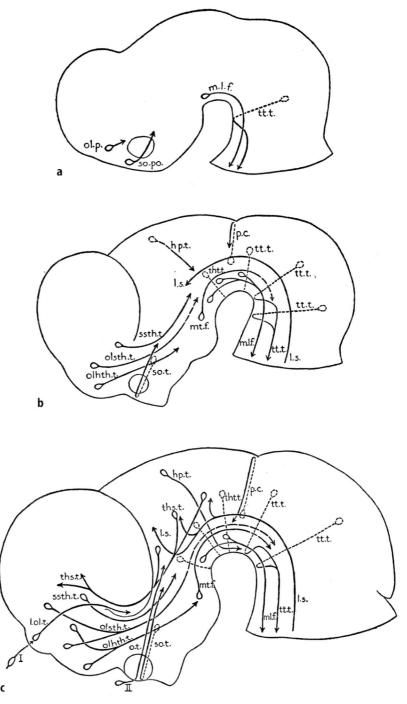

Fig. 5.29a–c. Outlines of the brains of 7-mm, 10-mm and 15-mm cat embryos on which are projected the principal fibre tracts, as observed in pyridine silver preparations. Omitted are the terminal nerve and various structures below the diencephalon, including the primary motor nerve nuclei of the midbrain. Crossed neurons are *dotted*, and ultimate incorporation into descending tracts is not shown. *I*, Olfactory nerve; *II*, optic nerve; *hp.t.*, habenulo-peduncular tract; *l.ol.t.*, lateral olfactory tract; *l.s.*, lemniscus system; *m.l.f.*, me-dial longitudinal fascicle; *mt.f.*, mamillo-tegmental fascicle; *olhth.t.*, olfacto-hypothalamic tract; *olsth.t.*, olfacto-subthalamic tract; *ol.p.*, olfactory pathway neurons (indifferent); *o.t.*, optic tract; *p.c.*, posterior commissure; *so.po.*, supraoptic system: preoptic component; *so.t.*, supraoptic system: hypothalamic component; *ssth.t.*, strio-subthalamic tract; *ths.t.*, thalamo-strial or thalamo-cortical tract; *tht.t*, thalamo-tegmental neurons (crossed); *tt.t*, tecto-tegmental neurons (crossed) (reproduced from Windle 1935, Figs. 2, 3, 4)

Fig. 5.28; Wilson et al. 1990) have remarked upon the very early development of this commissure.

Herrick (1937, 1938a) is the only author who presents a detailed description of the composition of the early embryonic postoptic commissure. He observed in his silver-impregnated material of *Ambystoma* larvae that neurons situated in the chiasma ridge itself and in the ventral parts of the thalamus and hypothalamus contribute to this commissure (Fig. 5.27b,d). Herrick (1937, p. 417) emphasised that the postoptic commissure develops much earlier than the optic systems: "There is local differentiation of nerve fibers of the postoptic commissure system in the chiasma ridge beginning more than 2.5 days before any retinal fibers enter the brain, and this intrinsic system of conductors seems to be sufficiently mature to be capable of nervous functions before the entrance of any fibers of retinal origin." Easter and Taylor (1989) have shown that also in *Xenopus* the postoptic commissure and its associated tracts are present before the optic axons reach the brain, and similar findings have been reported by Wilson et al. (1990) for the zebrafish.

The postoptic commissure is contiguous with the TPOC, which runs caudally through the ventral diencephalon and enters the anterior tegmentum. Here most of its fascicles form a single complex with those of the rostral FLM, but some terminate more dorsally in the tegmentum of the midbrain (Fig. 5.26).

Elements of the TPOC have already been described by His (1904, p. 152). He observed in 4- to 5-week-old human embryos "ein vom Riechhirn bis zum Mittelhirn verfolgbarer Zug von Längsfasern," which he designated as the tractus hypothalamicus.

According to Herrick (1938a; Fig. 5.27b,d), the fibre assembly termed here TPOC is, in early *Ambystoma* larvae, composed largely of crossed and ipsilateral fibres originating from the chiasma ridge area and from the ventral part of the hypothalamus, to which fibres from the dorsal part of the hypothalamus and the tuberculum posterius are added. Some of the ten tegmental fascicles described by Herrick (1936, 1948) in the adult brain of *Ambystoma* also belong to the TPOC complex. This holds in particular for the dorsal fascicle (7), originating from the dorsal thalamus, rostral tectum and dorsal tegmentum, and for the ventral fascicle (5), which arises mainly from the ventral thalamus and the tuberculum posterius (Fig. 5.27d).

The hypothalamo-subthalamic and thalamo-tegmental fibres, as well as the supraoptic decussation, observed by Windle and Austin (1935) in chick embryos (Fig. 5.28) all belong to the TPOC complex, and the same holds true for the supra-

optic system, the strio-subthalamic tract and the mamillo-tegmental fascicle observed by Windle (1935) in cat embryos (Fig. 5.29).

In the zebrafish the TPOC consists principally of direct and crossed fibres which originate from the ventrorostral cluster of early-differentiating cells situated in the presumptive hypothalamus. These fibres assemble in several caudally directed bundles. Some of these are contiguous with the FLM, but others terminate somewhat more dorsally in the tegmentum of the midbrain (Fig. 5.26).

Three morphologically transversely oriented bundles, i.e. the telencephalic tract, the dorsoventral diencephalic tract and the tract of the posterior commissure, complete the simple and sterotyped set of axonal tracts, observed by Wilson et al. (1990), Chitnis and Kuwada (1990) and Ross et al. (1992) in the brain of early zebrafish embryos (Fig. 5.26).

The *telencephalic tract* (TT; by the authors mentioned, also designated as supraoptic tract, SOT) consists of axons which project from the dorsorostral to the ventrorostral cluster of early-differentiating cells. The dorsorostral cluster includes the part of the telencephalon that receives input from the olfactory epithelium by way of the olfactory nerve. Therefore, this cluster represents the anlage of the olfactory bulb and probably other parts of the telencephalon as well. Axonal tracts comparable to the TT have been observed in *Ambystoma* (Herrick 1937, 1938a, Fig. 5.27a,c), the chick (Windle and Austin 1935: tractus olfacto-hypothalamicus: Fig. 5.28b) and the cat (Windle 1935: olfacto-hypothalamic tract; olfactosubthalamic tract: Fig. 5.29b,c) but were not readily identified in the mouse (Easter et al. 1993).

Coghill (1930, p. 322) observed an early-developing axonal bundle in *Ambystoma* embryos which he interpreted as the nervus terminalis: "Running through the primary olfactory center and into the secondary is a slender fiber tract which lies immediately against the external limiting membrane. It arises from unipolar cells in the anteroventral part of the olfactory epithelium and is directed toward the postoptic region. It is regarded as the n. terminalis." His observations were confirmed by Herrick (1937, 1938a, Fig. 5.27a,c).

The *dorsoventral diencephalic tract* (DVDT) has so far been clearly observed only in the zebrafish (Wilson et al. 1990; Wilson and Easter 1991; Chitnis and Kuwada 1990). It originates from a small group of cells situated in the wall of the primordial epiphysis. Its fibres pass ventrally and join the TPOC, where they take a rostral course (Fig. 5.26). Wilson et al. (1990) pointed out that this pineal projection develops long before retinal axons enter the brain.

The *tract of the posterior commissure* (TPC) contains axons of two groups of early-differentiating cells, one forming part of the so-called ventrocaudal cluster, the other situated more dorsal in the synencephalon. These two groups of cells have been designated in Fig. 5.26 as the dorsal and ventral nuclei of the posterior commissure. Axons originating from both groups have been observed to decussate in the posterior commissure. Their site of termination is unknown. Individual axons of the cells of the ventral nucleus have been observed to join the TPOC and to extend to the posterior region of the tegmentum. Numerous authors, among them His (1904: man), Coghill (1926, 1930: *Ambystoma*, Fig. 5.27c), Windle and Austin (1935: chick) and Windle (1935: cat, Fig. 5.29b,c), noticed that the posterior commissure appears early in ontogenesis.

Although the number of species in which the early development of fibre tracts has been investigated is limited, the very early appearance of the DLB, VLB, TPC, DVDT and TT strongly suggests that these bundles form part of the basic plan of the vertebrate brain and spinal cord and, hence, either represent or foreshadow fibre systems present in all species belonging to this group. For this reason, these fibre systems have been added in Fig. 5.30 to the pictorial survey of landmarks in the vertebrate brain presented in Fig. 4.53.

At spinal levels, the DLB and VLB have been observed not only in vertebrates, but also in larval

(Bone 1959) and adult amphioxus (Bone 1960; Chap. 9, Figs. 9.7, 9.8). In all vertebrates the descending trigeminal tract represents the encephalic part of the DLB. The FLM, i.e. the brain-stem segment of the VLB, represents the principal encephalospinal pathway in all non-mammalian vertebrates. In the poorly developed brain of amphioxus a possible trace of an encephalic part of the DLB is formed by the axons of the so-called B-cell group, but an FLM homologue is entirely lacking. In larval specimens of this species the VLB, i.e. Bone's (1959) tract III, is composed only of axons of primary somatic motor neurons. Lacalli et al. (1994) suggested that the fibres in amphioxus larvae which descend from the conspicuous lamellar body (see Chap. 9) correspond to the vertebrate DVDT, and that the most rostral nerve of that form may well be the counterpart of the terminal and/or olfactory nerves in vertebrates.

A postoptic commissure is present in all vertebrates, including the lampreys (Heier 1948) and the myxinoids (Jansen 1930; Wicht and Northcutt 1992, 1993). It is remarkable that this early-developing "commissure diencéphalique primitive" (Tello 1923, p. 93) is situated in the immediate vicinity of the rostral end of the neuraxis (Fig. 5.30).

As we have seen, the TPOC represents the most rostral portion of the VLB. It extends from the postoptic commissure to the tegmentum of the midbrain. Its composition in early zebrafish embryos is

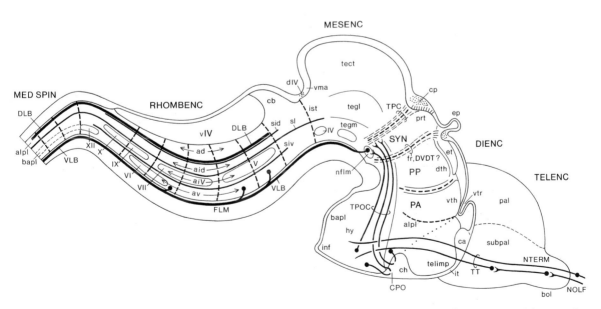

Fig. 5.30. Some early-developing axon bundles, projected upon a diagram showing the most important landmarks in the vertebrate brain. *CPO,* Commissura postoptica; *DLB,* dorsolateral bundle; *DVDT,* dorsoventral diencephalic tract; *FLM,* fasciculus longitudinalis medialis; *NOLF,* nervus olfac-torius; *NTERM,* nervus terminalis; *TPC,* tract of the posterior commissure; *TPOC,* tract of the postoptic commissure; *TT,* telencephalic tract; *VLB,* ventrolateral bundle; for remaining abbreviations see Chap. 4, Fig. 4.53

shown in Fig. 5.26. However, it should be noted that its further development and its relation to the fibre connections of the hypothalamus in adult teleosts are unknown. *Ambystoma* is so far the only species in which the development of what is called here TPOC has been analysed in some detail. This analysis (Herrick 1936, 1937, 1938a, 1948), carried out on non-experimental silver-impregnated material, suggested that the cells situated in the region of the optic stalk and in the hypothalamus, which in the early embryo project to the motor tegmentum (directly as well as via the postoptic commissure), are also present in the adult form (Fig. 5.27). As regards the functional significance of the TPOC, Windle (1935) and Herrick (1938a) assumed that this early-developing bundle forms part of a projection conveying olfactory impulses to the efferent centres in the brain stem and spinal cord.

The TPC is a conspicuous bundle in all vertebrates, which is generally considered to be related to the visual system, particularly to the mediation of pupillary light reflexes. However, the fact that this bundle and its commissure are well developed in the myxinoids, forms with tiny, rudimentary eyes, indicates that this is certainly not the whole story. The TPC, which represents an important landmark in the brain, indicating the diencephalo-mesencephalic boundary, has recently gained further status by coinciding with the boundaries of the expression domains of several putative regulatory genes (cf. Chap. 4, Sect. 4.5.5.5 and Figdor and Stern 1993; Puelles and Rubenstein 1993). Thus, in the zebrafish embryo, the TPC marks the caudal boundary of the prosencephalic expression domain of the paired-box gene *pax[zf-a]* (Krauss et al. 1991).

According to Wilson et al. (1990), the DVDT observed in early zebrafish embryos is not comparable to projections from either the pineal organ or the habenular nucleus that have been described in adult anamniotes. However, it cannot be excluded that the fibres of the somewhat later developing fasciculus retroflexus follow the same dorsoventral trajectory. Neither Wilson et al. (1990) nor Ross et al. (1992) indicate an alternative route for this tract. The tract of the habenular commissure described by the latter authors most probably corresponds to the stria medullaris.

The TT, finally, is probably composed of both nervus terminalis and olfactory fibres. Experimental studies (Northcutt and Puzdrowski 1988: the lamprey *Ichthyomyzon unicuspis*; von Bartheld and Meyer 1988: the lamprey *Lampetra planeri*, the lungfish, *Protopterus dolloi*, and the bichir, *Polypterus palmas*) have shown that in a number of adult anamniotes the central projections of the nervus terminalis follow a route closely corresponding to that observed by Coghill (1930, 1931) and Herrick (1937, 1938a, Fig. 5.27a,c) in *Ambystoma*. The direct connections between the anlage of the olfactory bulb and the presumptive hypothalamus observed by Herrick (1937, 1938a, Fig. 5.27a,c) in *Ambystoma* larvae and by Chitnis and Kuwada (1990), Wilson et al. (1990) and Ross et al. (1992) in embryos of the zebrafish (Fig. 5.26) have no well-documented equivalents in any adult vertebrate. During later development these pioneering fibres are possibly replaced by an indirect projection, which is synaptically interrupted in the septal region or its equivalents. Experimental neuroanatomical studies have shown the presence of bulbo-septal projections in teleosts, urodeles and anurans (cf. Chaps. 15, 18 and 19).

Acknowledgement. I thank Dr. I.H.M. Smart for reading the manuscript of this chapter and for many helpful suggestions.

References

Allen E (1912) The cessation of mitosis in the central nervous system of the albino rat. J Comp Neurol 22:547–568

Altman J (1966) Proliferation and migration of undifferentiated precursos cells in the rat during postnatal gliogenesis. Exp Neurol 16:263–278

Altman J, Bayer SA (1978a) Development of the diencephalon in the rat. I. Autoradiographic study of the time of origin and settling patterns of neurons of the hypothalamus. J Comp Neurol 182:945–972

Altman J, Bayer SA (1978b) Development of the diencephalon in the rat. II. Correlation of the embryonic development of the hypothalamus with the time of origin of its neurons. J Comp Neurol 182:973–994

Altman J, Bayer SA (1979) Development of the diencephalon in the rat. VI. Reevaluation of the embryonic development of the thalamus on the basis of thymidine-radiographic datings. J Comp Neurol 188:501–524

Altman J, Bayer SA (1986) Development of the rat hypothalamus. Adv Anat Embryol Cell Biol 100:1–178

Altman J, Bayer SA (1987a) Development of the precerebellar nuclei in the rat. I. The precerebellar neuroepithelium of the rhombencephalon. J Comp Neurol 257:477–489

Altman J, Bayer SA (1987b) Development of the precerebellar nuclei in the rat. II. The intramural olivary migratory stream and the neurogenetic organization of the inferior olive. J Comp Neurol 257:490–512

Altman J, Bayer SA (1987c) Development of the precerebellar nuclei in the rat. III. The posterior precerebellar extramural migratory stream and the lateral reticular and external cuneate nuclei. J Comp Neurol 257:513–528

Altman J, Bayer SA (1987d) Development of the precerebellar nuclei in the rat. IV. The anterior precerebellar extramural migratory stream and the nucleus reticularis tegmenti pontis and the basal pontine gray. J Comp Neurol 257:529–552

Altman J, Bayer SA (1988a) Development of the rat thalamus. I. Mosaic organization of the thalamic neuroepithelium. J Comp Neurol 275:346–377

Altman J, Bayer SA (1988b) Development of the rat thalamus. II. Time and site of origin and settling pattern of neurons derived from the anterior lobule of the thalamic neuroepithelium. J Comp Neurol 275:378–405

Altman J, Bayer SA (1988c) Development of the rat thalamus. III. Time and site of origin and settling pattern of neurons of the retinal nucleus. J Comp Neurol 275:406–428

Altman J, Bayer SA (1989a) Development of the rat thalamus. IV. The intermediate lobule of the thalamic neuroepithelium, and the time and site of origin and settling pattern of neurons of the ventral nuclear complex. J Comp Neurol 284:534–566

Altman J, Bayer SA (1989b) Development of the rat thalamus. V. The posterior lobule of the thalamic neuroepithelium and the time and site of origin and settling pattern of neurons of the medial geniculate body. J Comp Neurol 284:567–580

Altman J, Bayer SA (1989c) Development of the rat thalamus. VI. The posterior lobule of the thalamic neuroepithelium and the time and site of origin and settling pattern of neurons of the lateral geniculate and lateral posterior nuclei. J Comp Neurol 284:581–601

Alvarez-Blado G, Swanson LW (1995) Appendix: on mapping patterns in the embryonic forebrain. J Comp Neurol 355:287–295

Alvarez-Bolado G, Rosenfeld MG, Swanson LW (1995) Model of forebrain regionalization based on spatiotemporal patterns of POU-III homeobox gene expression, birthdates, and morphological features. J Comp Neurol 355:237–295

Alvarez-Buylla A (1990) Mechanisms of neurogenesis in adult avian brain. Experientia 46:948–955

Alvarez-Buylla A, Nottebohm F (1988) Migration of young neurons in adult avian brain. Nature 335:353–354

Alvarez-Buylla A, Buskirk DR, Nottebohm F (1987) Monoclonal antibody reveals radial glia in adult avian brain. J Comp Neurol 264:159–170

Angulo y González AW (1939) Histogenesis of the monopolar neuroblast and the ventral longitudinal path in the albino rat. J Comp Neurol 71:325–360

Aström E, Webster H de F (1991) The early development of the neopallial wall and area choroidea in fetal rats. Springer, Berlin Heidelberg New York

Bayer SA, Altman J (1987) Directions in neurogenetic gradients and patterns of anatomical connections in the telencephalon. Prog Neurobiol 29:57–106

Bayer SA, Altman J (1995a) Neurogenesis and neuronal migration. In: Paxinos G (ed) The rat nervous system, 2nd edn. Academic, San Diego, pp 1041–1078

Bayer SA, Altman J (1995b) Principles of neurogenesis, neuronal migration and neural circuit formation. In: Paxinos G (ed) The rat nervous system, 2nd edn. Academic, San Diego, pp 1079–1098

Bergquist H (1932) Zur Morphologie des Zwischenhirns bei niederen Wirbeltieren. Acta Zool (Stockh) 13:57–303

Bergquist H (1953) On the development of diencephalic nuclei and certain mesencephalic relations in Lepidochelys olivacea and other reptiles. Acta Zool (Stockh) 34:155–190

Bergquist H (1954) Ontogenesis of diencephalic nuclei in vertebrates. A comparative study. Lunds Univ Arsskr Avd 65:1–34

Bergquist H, Källén B (1954) Notes on the early histogenesis and morphogenesis of the central nervous system in vertebrates. J Comp Neurol 100:627–659

Bernhardt RR, Chitnis AB, Lindamer L, Kuwada JY (1990) Identification of spinal neurons in the embryonic and larval zebrafish. J Comp Neurol 302:603–616

Berry M, Rogers AW (1965) The migration of neuroblasts in the developing cerebral cortex. J Anat 99:491–709

Blakemore WF (1969) The ultrastructure of the subependymal plate in the rat. J Anat 104:423–433

Blakemore WF, Jolly RD (1972) The subependymal plate and associated ependyma in the dog. An ultrastructural study. J Neurocytol 1:69–84

Bone Q (1959) The central nervous system in larval acraniates. Q J Microsc Sci 100:509–527

Bone Q (1960) The central nervous system in amphioxus. J Comp Neurol 115:27–64

Book KJ, Morest DK (1990) Migration of neuroblasts by perikaryal translocation: role of cellular elongation and axonal outgrowth in the acoustic nuclei of the chick embryo medulla. J Comp Neurol 297:55–76

Boulder Committee (1969) Embryonic vertebrate central nervous system: revised terminology Anat Rec 166:257–262

Bourrat F, Sotelo C (1988) Migratory pathways and neuritic differentiation of inferior olivary neurons in rat embryo. Axonal tracing study using the in vitro slab technique. Brain Res 467:19–37

Bourrat F, Sotelo C (1990) Early development of the rat precerebellar system: migratory routes, selective aggregation and neuritic differentiation of the inferior olive and lateral reticular nucleus neurons. An overview. Arch Ital Biol 128:151–170

Bulfone A, Puelles L, Porteus MH, Frohman MA, Martin GR, Rubenstein JLR (1993) Spatially restricted expression of Dlx-1, Dlx-2 (Tes-1), Gbx-2, and Wnt-3 in the embryonic day 12.5 mouse forebrain defines potential transverse and longitudinal segmental boundaries. J Neurosci 13:3155–3172

Burd GD, Nottebohm F (1985) Ultrastructural characterization of synaptic terminals formed on newly generated neurons in a song control nucleus of the adult canary forebrain. J Comp Neurol 240:143–152

Chitnis AB, Kuwada JY (1990) Axogenesis in the brain of zebrafish embryos. J Neurosci 10:1892–1905

Choi BH (1981) Radial glia of developing human fetal spinal cord: Golgi, immunohistochemical and electron microscopic study. Dev Brain Res 1:249–267

Choi BH, Kim RC (1984) Expression of glial fibrillary acidic protein in immature oligodendroglia. Science 223:407–409

Choi BH, Lapham LW (1978) Radial glia in the human fetal cerebrum: a combined Golgi, immunofluorescence and electron microscope study. Brain Res 148:295–311

Choi BH, Kim RC, Lapham L (1983) Do radial glia give rise to both astroglia and oligodendroglial cells? Dev Brain Res 8:119–130

Coghill GE (1926) Correlated anatomical and physiological studies of the growth of the nervous system in amphibia. VI. The mechanism of integration in Amblystoma punctatum. J Comp Neurol 41:95–152

Coghill GE (1928) Correlated anatomical and physiological studies of the growth of the nervous system of amphibia. J Comp Neurol 45:227–247

Coghill GE (1929) Anatomy and the problem of behaviour. Cambridge Unversity Press, Cambridge

Coghill GE (1930) Correlated anatomical and physiological studies of the growth of the nervous system in amphibia. IX. The mechanism of association of Amblystoma punctatum. J Comp Neurol 51:311–375

Coghill GE (1931) Correlated anatomical and physiological studies of the growth of the nervous system in amphibia. X. Corollaries of the anatomical and physiological study of amblystoma from the age of earliest movement to swimming. J Comp Neurol 53:147–168

Dodd J, Jessell TM (1988) Axon guidance and the patterning of neuronal projections in vertebrates. Science 242:692–699

Domesick VB, Morest DK (1977a) Migration and differentiation of ganglion cells in the optic tectum of the chick embryo. Neuroscience 2:459–476

Domesick VB, Morest DK (1977b) Migration and differentiation shepherd's crook cells in the optic tectum of the chick embryo. Neuroscience 2:477–492

Easter SS Jr, Taylor JSH (1989) The development of the Xenopus retinofugal pathway: optic fibers join a pre-existing tract. Development 107:553–573

Easter SS Jr, Ross LS, Frankfurter A (1993) Initial tract formation in the mouse brain. J Neurosci 13:285–299

Egar M, Singer M (1972) The role of ependyma in spinal cord regeneration in the urodele, Triturus. Exp Neurol 37:422–430

Essick CR (1907) The corpus ponto-bulbare – a hitherto undescribed nuclear mass in the human hindbrain. Am J Anat 7:119–135

Essick CR (1912) The development of the nuclei pontis and the nucleus arcuatus in man. Am J Anat 13:25–54

Figdor MC, Stern CD (1993) Segmental organization of embryonic diencephalon. Nature 363:630–634

Fishell G, Mason CA, Hatten ME (1993) Dispersion of neural progenitors within the germinal zones of the forebrain. Nature 362:636–638

Fisher S, Jacobson M (1970) Ultrastructural changes during early development of retinal ganglion cells in *Xenopus*. Z Zellforsch Mikroskop Anat 104:165–177

Fraser S, Keynes R, Lumsden A (1990) Segmentation in the chick embryo hindbrain is defined by cell lineage restrictions. Nature 344:431–435

Fujita S (1963) The matrix cell and cytogenesis in the developing central nervous system. J Comp Neurol 120:37–42

Fujita S (1965) An autoradiographic study on the origin and fate of the subpial glioblasts in the embryonic chick spinal cord. J Comp Neurol 124:51–60

Fujita S (1966) Application of light and electron microscopic autoradiography to the study of cytogenesis of the forebrain. In: Hassler R, Stephan H (eds) Evolution of the forebrain. Thieme, Stuttgart, pp 180–196

Globus JH, Kuhlenbeck H (1944) The subependymal cell plate (matrix) and its relationship to brain tumors of the ependymal type. J Neuropathol Exp Neurol 3:1–35

Goldman SA, Nottebohm F (1983) Neuronal production, migration, and differentiation in a vocal control nucleus of the adult female canary brain. Proc Natl Acad Sci U S A 80:2390–2394

Goodman CS (1994) The likeness of being: phylogenetically conserved molecular mechanisms of growth cone guidance. Cell 78:353–356

Gray GE, Glover JC, Owens GE, Majors J, Sanes JR (1988) Radial arrangement of clonally related cells in the chicken optic tectum: lineage analysis with a recombinant retrovirus. Proc Natl Acad Sci U S A 85:7356–7360

Grove EA (1993) Fixed or fluid? Curr Biol 3:470–473

Hamburger V, Hamilton H (1951) A series of normal stages in the development of the chick embryo. J Morphol 88:49–91

Harkmark W (1954a) Cell migrations from the rhombic lip to the inferior olive, the nucleus raphe and the pons. A morphological and experimental investigation on chick embryos. J Comp Neurol 100:115–210

Harkmark W (1954b) The rhombic lip and its derivatives in relation to the theory of neurobiotaxis. In: Jansen J, Brodal A (eds) Aspects of cerebellar anatomy. Grundt Tanum, Oslo, pp 264–184

Hatten ME (1993) The role of migration in central nervous system neuronal development. Curr Opin Neurobiol 3:38–44

Hayashi M (1924) Einige wichtige Tatsachen aus der ontogenetischen Entwicklung des menschlichen Kleinhirns. Dtsch Z Nervenheilkd 81:74–82

Heier P (1948) Fundamental principles in the structures of the brain. A study of the brain of *Petromyzon fluviatilis*. Acta Anat (Basel) [Suppl] 8:1–213

Herrick CJ (1910) The morphology of the forebrain in amphibia and reptilia. J Comp Neurol 20:413–547

Herrick CJ (1936) Conduction pathways in the cerebral peduncle of Amblystoma. J Comp Neurol 63:293–352

Herrick CJ (1937) Development of the brain of Amblystoma in early functional stages. J Comp Neurol 67:381–422

Herrick CJ (1938a) Development of the cerebrum of Amblystoma during early swimming stages. J Comp Neurol 68:203–241

Herrick CJ (1938b) Development of the brain of *Amblystoma punctatum* from early swimming to feeding stages. J Comp Neurol 69:13–30

Herrick CJ (1938c) The brains of *Amblystoma punctatum* and *A. tigrinum* in early feeding stages. J Comp Neurol 69:371–426

Herrick CJ (1948) The brain of the tiger salamander. University of Chicago Press, Chicago

Herrick CJ, Coghill GE (1915) The development of reflex mechanisms in Amblystoma. J Comp Neurol 25:65–85

Hinds JW, Hinds PL (1974) Early ganglion cell differentiation in the mouse retina: an electron microscopic analysis utilizing serial sections. Dev Biol 37:381–416

Hinds JW, Ruffett TL (1971) Cell proliferation in the neural tube: an electron microscopic and Golgi analysis in the mouse cerebral vesicle. Z Zellforsch 115:226–264

Hirano M, Goldman JE (1988) Gliogenesis in rat spinal cord: evidence of origin of astrocytes and oligodendrocytes from radial precursors. J Neurosci Res 21:155–167

His W (1889) Die Neuroblasten und deren Entstehung im embryonalen Mark. Arch Anat Physiol Anat Abt 249–300

His W (1890a) Die Formentwicklung des menschlichen Vorderhirns vom Ende des ersten bis zum Beginn des dritten Monats. Abh Kön Sächs Ges Wiss Math Phys Kl 15:675–735

His W (1890b) Die Entwicklung des menschlichen Rautenhirns vom Ende des ersten bis zum Beginn des dritten Montas. I. Verlängertes Mark. Abh Kön Sächs Ges Wiss Math Phys Kl 17:1–74

His W (1904) Die Entwickelung des menschlichen Gehirns während der ersten Monate. Hirzel, Leipzig

Hochstetter F (1919) Beiträge zur Entwicklungsgeschichte des menschlichen Gehirns. Deuticke, Vienna

Holley JA (1982) Early development of the circumferential axonal pathway in mouse and chick spinal cord. J Comp Neurol 205:371–382

Holley JA, Nornes Ho, Morita M (1982) Guidance of neuritic growth in the transverse plane of embryonic mouse spinal cord. J Comp Neurol 205:360–370

Hooker D (1952) The prenatal origin of behavior. University of Kansas Press, Lawrence, Kansas

Jacobson CO (1959) The localization of the presumptive cerebral regions in the neural plate of the axolotl larva. J Embryol Exp Morphol 7:1–21

Jacobson M (1991) Developmental neurobiology, 3rd edn. Plenum, New York

Jansen J (1930) The brain of myxine glutinosa. J Comp Neurol 49:359–507

Kahle W (1951) Studien über die Matrixphasen und die örtliche Reifungsunterschiede im embryonalen menschlichen Gehirn. Dtsch Z Nervenheilkd 166:273–302

Källén B (1951a) The nuclear development in the mammalian forebrain with special regard to the subpallium. Kgl Fysiogr Sällsk Lund Handl NF 61 (9):1–43

Källén B (1951b) Contributions to the ontogeny of the nuclei and the ventricular sulci in the vertebrate forebrain. Kgl Fysiogr Sällsk Lund Handl NF 62 (3):1–50

Källén B (1951c) Embryological studies on the nuclei and their homologization in the vertebrate forebrain. Kgl Fysiogr Sällsk Lund Handl NF 62 (5):1–36

Källén B (1951d) On the ontogeny of the reptilian forebrain. J Comp Neurol 95:307–348

Källén B (1952) Notes on the proliferation processes in the neuromeres in vertebrate embryos. Acta Soc Med Ups 57:111–118

Källén B (1955) Notes on the mode of formation of brain nuclei during ontogenesis. CR Ass Anatomistes 42e Réunion, pp 1747–1756

Källén B (1962) Embryogenesis of brain nuclei in the chick telencephalon. Ergeb Anat Entw Gesch 36:61–82

Källén B (1965) Early morphogenesis and pattern formation in the central nervous system. In: Dellaan RL, Ursprung H (eds) Organogenesis. Holt, Rinehart and Winston, New York, pp 107–128

Katz MJ, Lasek RJ (1978) Eyes transplanted to tadpole tails send axons rostrally in two spinal cord tracts. Science 199:202–203

Katz MJ, Lasek RJ (1979) Substrate pathways which guide growing axons in *Xenopus* embryos. J Comp Neurol 193:817–832

Katz MJ, Lasek RJ (1981) Substrate pathways demonstrated by transplanted Mauthner axons. J Comp Neurol 195:627–641

Katz MJ, Lasek RJ, Nauta HJW (1980) Ontogeny of substrate pathways and the origin of the neural circuit pattern. Neuroscience 5:821–833

Kershman J (1938) The medulloblast and the medulloblastoma. A study of human embryos. Arch Neurol Psychiatry 40:937–967

Kevetter GA, Lasek RJ (1982) Development of the marginal zone in the rhombencephalon of *Xenopus laevis*. Dev Brain Res 4:195–208

Keyser AJM (1972) The development of the diencephalon of the chinese hamster. Acta Anat (Basel) [Suppl] 59/1 (83):1–181

Kim GJ, Shatz CJ, McConnell SK (1991) Morphology of pioneer and follower growth cones in the developing cerebral cortex. J Neurobiol 2:629–642

Kimmel CB (1993) Patterning the brain of the zebrafish embryo. Annu Rev Neurosci 16:707–732

Knyihar-Csillik E, Csillik B, Rakic P (1995) Structure of the embryonic primate spinal cord at the closure of the first reflex arc. Anat Embryol (Berl) 191:319–540

Krauss S, Johansen T, Korzh V, Fjose A (1991) Expression pattern of zebrafish pax genes suggests a role in early brain regionalization. Nature 353:267–270

Kuhlenbeck H (1929a) Über die Grundbestandteile des Zwischenhirnbauplans der Anamnier. Morphol Jahrb 63:50–95

Kuhlenbeck H (1929b) Die Grundbestandteile des Endhirns im Lichte der Bauplanlehre. Anat Anz 67:1–51

Kuhlenbeck H (1937) The ontogenetic development of the diencephalic centers in a bird's brain (chick) and comparison with the reptilian and mammalian diencephalon. J Comp Neurol 66:23–75

Kuhlenbeck H (1967) The central nervous system of vertebrates, vol 1. Karger, Basel

Kuhlenbeck H (1973) The central nervous system of vertebrates, vol 3, part 2: overall morphologic pattern. Karger, Basel

Kuwada JY, Bernhardt R, Chitnis AB (1990a) Pathfinding by identified growth cones in the spinal cord of zebrafish embryos. J Neurosci 10:1299–1308

Kuwada JY, Bernhardt RR, Nguyen N (1990b) Development of spinal neurons and tracts in the zebrafish embryo. J Comp Neurol 302:617–628

Lacalli TC, Holland ND, West JE (1994) Landmarks in the anterior central nervous system of amphioxus larvae. Philos Trans R Soc Lond B 344:165–185

LaMantia A-S, Rakic P (1990) Axons overproduction and elimination in the corpus callosum of the developing rhesus monkey. J Neurosci 10:2156–2175

Leber SM, Sanes JR (1995) Migratory paths of neurons and glia in the embryonic chick spinal cord. J Neurosci 15:1236–1248

Lee RKK, Eaton RC, Zottoli SJ (1993) Segmental arrangement of reticulospinal neurons in the goldfish hindbrain. J Comp Neurol 329:539–556

Levitt P, Rakic P (1980) Immunoperoxidase localization of glial fibrillary acidic protein in radial glial cells and astrocytes of the developing rhesus monkey brain. J Comp Neurol 193:815–840

Levitt P, Cooper MC, Rakic P (1981) Coexistence of neural and glial precursor cells in the cerebral ventricular zone of the fetal monkey: an ultrastructural immunoperoxidase study. J Neurosci 1:27–39

Martínez S, Puelles L, Alvarado-Mallart M (1992) Tangential neuronal migration in the avian tectum: cell type identification and mapping of regional differences with quail/chick homotopic transplants. Dev Brain Res 6:153–163

McConnell SK (1988) Development and decision-making in the mammalian cerebral cortex. Brain Res Rev 13:1–23

Metcalfe WK, Mendelson B, Kimmel CB (1986) Segmental homologies among reticulospinal neurons in the hindbrain of the zebrafish larva. J Comp Neurol 251:147–159

Misson J-P, Austin CP, Takahashi T, Cepko CL, Caviness VS Jr (1991) The alignment of migrating neural cells in relation to the murine neopallial radial glial fiber system. Cerebral Cortex 1:221–229

Morest DK (1970) A study of neurogenesis in the forebrain of the opposum pouch young. Z Anat Entw Gesch 130:265–305

Morest DK (1970) A study of neurogenesis in the forebrain of opossum pouch young. Z Anat Entw Gesch 130:265–305

Mugnaini E, Forstrønen PF (1967) Ultrastructural studies on the cerebellar histogenesis. I. Differentiation of granule cells and development of glomeruli in the chick embryo. Z Zellforsch Mikrosk Anat 77:115–143

Nakao T, Ishizawa A (1987) Development of the spinal nerves in the lamprey: I. Rohon-Beard cells and interneurons. J Comp Neurol 256:342–355

Nieuwenhuys R (1964) Comparative anatomy of the spinal cord. Prog Brain Res 11:1–57

Nieuwenhuys R (1966) The interpretation of the cell masses in the teleostean forebrain. In: Hassler R, Stephan H (eds) Evolution of the forebrain. Thieme, Stuttgart, pp 32–39

Nieuwenhuys R (1972) Topological analysis of the brainstem of the lamprey *Lampetra fluviatilis*. J Comp Neurol 145:165–178

Nieuwenhuys R (1974) Topological analysis of the brainstem: a general introduction. J Comp Neurol 156:255–276

Nieuwenhuys R (1994) Comparative neuroanatomy: place, principles, practice and programme. Eur J Morphol 32:142–155

Nordlander RH (1987) Axonal growth cones in the developing amphibian spinal cord. J Comp Neurol 263:485–496

Nordlander RH, Singer M (1978) The role of ependyma in regeneration of the spinal cord in the urodele amphibian tail. J Comp Neurol 180:349–374

Nordlander RH, Singer M (1982a) Spaces precede axons in *Xenopus* embryonic spinal cord. Exp Neurol 75:221–228

Nordlander RH, Singer M (1982b) Morphology and position of growth cones in the developing *Xenopus* spinal cord. Dev Brain Res 4:181–193

Northcutt RG, Puzdrowski RL (1988) Projections of the olfactory bulb and nervus terminalis in the silver lamprey. Brain Behav Evol 32:96–107

Ono K, Kawamura K (1989) Migration of immature neurons along tangentially oriented fibers in the subpial part of the fetal mouse medulla oblongata. Exp Brain Res 78:290–300

Ono K, Kawamura K (1990) Mode of neuronal migration of the ponyine stream in fetal mice. Anat Embryol (Berl) 182:11–19

Palay SL (1967) Principles of cellular organization in the nervous system. In: Quarton GC, Melnechuk T, Schmitt FO (eds) The neurosciences. Rockefeller University Press, New York, pp 24–31

Phelps PE, Vaughn JE (1995) Commissural fibers may guide cholinergic neuronal migration in developing rat cervical spinal cord. J Comp Neurol 355:38–50

Phelps PE, Barber RP, Vaughn JE (1993) Embryonic development of rat sympathetic preganglionic neurons – possible migratory substrates. J Comp Neurol 330:1–14

Pouwels E (1978) On the development of the cerebellum in the trout, *Salmo gairdneri*. I. Patterns of cell migration. Anat Embryol (Berl) 152:291–308

Price J (1993) Organizing the cerebrum. Nature 362:590–591

Privat A, Leblond CP (1972) The subependymal layer and neighboring region in the brain of the young rat. J Comp Neurol 146:277–302

Puelles L, Rubenstein JL (1993) Expression patterns of homeobox and other putative regulatory genes in the embryonic mouse forebrain suggest a neuromeric organization. Trends Neurosci 16:472–479

Rakic P (1971) Neuron-glia relationship during granule cell migration in developing cerebellar cortex. A Golgi and electronmicroscopic study in *Macacus rhesus*. J Comp Neurol 141:283–312

Rakic P (1972) Mode of cell migration to the superficial layers of fetal monkey neocortex. J Comp Neurol 145:61–84

Rakic P (1974) Neurons in rhesus monkey visual cortex: systematic relation between time of origin and eventual disposition. Science 183:425–427

Rakic P (1987a) Neuronal migration. In: Adelman G (ed) Encyclopedia of neuroscience, vol II. Birkhäuser, Boston, pp 825–827

Rakic P (1987b) Radial glial cells. In: Adelman G (ed) Encyclopedia of neuroscience, vol II. Birkhäuser, Boston, pp 1022–1024

Rakic P (1988) Specification of cerebral cortical areas. Science 241:170–176

Rakic P (1990) Principles of neural cell migration. Experientia 46:883–891

Ramón P (1896) Estructura del encéfalo del camaleón. Rev Trimest Micrograf 1:46–82

Ramón y Cajal S (1909) Histologie du système nerveux de l'homme et des vertébrés, vol I (reprint 1955, Instituto Ramón y Cajal, Madrid)

Reh TA, Constantine-Paton M (1985) Growth cone-target interactions in the frog retinotectal pathway. J Neurosci Res 13:89–100

Ross LS, Parrett T, Easter SS Jr (1992) Axonogenesis and morphogenesis in the embryonic zebrafish brain. J Neurosci 12:467–485

Rubenstein JLR, Martinez S, Shimamura K, Puelles L (1994) The embryonic vertebrate forebrain: the prosomeric model. Science 266:578–580

Sauer FC (1935a) Mitosis in the neural tube. J Comp Neurol 62:377–407

Sauer FC (1935b) Cellular structure of the neural tube. J Comp Neurol 63:13–23

Sauer FC (1936) The interkinetic migration of embryonic epithelial nuclei. J Morphol 60:1–11

Sauer ME, Chittenden AC (1959) Desoxyribonucleic acid content of cell nuclei in the neural tube of the chick embryo; evidence for intermitotic migration of nuclei. Exp Cell Res 16:1–6

Sauer ME, Walker BE (1959) Radiographic study of interkinetic nuclear migration in the neural tube. Proc Soc Exp Biol Med 101:557–560

Schaper A (1894a) Die morphologische und histologische Entwicklung des Kleinhirns der Teleostier. Anat Anz 9:489–501

Schaper A (1894b) Die morphologische und histologische Entwicklung des Kleinhirns der Teleostier. Morphol Jahrb 21:625–708

Schmechel DE, Rakic P (1979) A Golgi study of radial glial cells in developing monkey telencephalon: morphogenesis and transformation into astrocytes. Anat Embryol (Berl) 156:115–152

Schwalbe G (1880) Beiträge zur Entwicklungsgeschichte des Zwischenhirns. Sitz Ber Jen Ges Med Naturwiss 20:2–7

Senn DG (1968a) Bau und Ontogenese von Zwischen- und Mittelhirn bei Lacerta sicula (Rafinesque). Acta Anat (Basel) [Suppl] 55:1–155

Senn DG (1968b) Der Bau des Reptiliengehirns im Licht neuer Ergebnisse. Verh Naturf Ges Basel 79:25–43

Senn DG (1970) The stratification in the reptilian central nervous system. Acta Anat (Basel) 75:521–552

Senn DG (1972) Development of tegmental and rhombencephalic structures in a frog (Rana temporatia L.). Acta Anat (Basel) 82:525–548

Senn DG (1974) Notes on the amphibian and reptilian thalamus. Acta Anat (Basel) 87:555–596

Senut MC, Alvarado-Mallart RM (1987) Cytodifferentiation of quail tectal primordium transplanted homotopically into the chick embryo. Dev Brain Res 32:187–205

Sidman RL, Rakic P (1973) Neuronal migration, with special reference to developing human brain: a review. Brain Res 62:1–35

Silver J, Robb RM (1979) Studies on the development of the eye cup and optic nerve in normal mice and in mutants with congenital optic nerve aplasia. Devl Biol 68:175–190

Singer M, Nordlander RH, Egar M (1979) Axonal guidance during embryogenesis and regeneration in the spinal cord of the newt: the blueprint hypothesis of neuronal pathway patterning. J Comp Neurol 185:1–22

Smart IHM (1961) The subependymal layer of the mouse brain and its cellular production as shown by radioautography after thymidine-H^3 injection. J Comp Neurol 116:325–349

Smart IHM (1972a) Proliferative characteristics of the ependymal layer during the early development of the mouse diencephalon, as revealed by recording the number, location, and plane of cleavage of mitotic figures. J Anat 113:109–129

Smart IHM (1972b) Proliferative characteristics of the ependymal layer during the early development of the spinal cord in the mouse. J Anat 111:365–380

Smart IHM (1973) Proliferative characteristics of the ependymal layer during the early development of the mouse diencephalon as revealed by recording the number, location and plane of cleavage of mitotic figures. J Anat 116:67–91

Smart IHM (1976) A pilot study of cell production by the ganglionic eminences of the developing mouse brain. J Anat 121:71–84

Smart IHM (1978) Cortical histogenesis and the 'glial coordinate system' of Nieuwenhuys. J Anat 126:419–420

Smart IHM (1981) Proliferative characteristics of the ependymal layer during the early development of the mouse diencephalon as revealed by recording the number, location, and plane of cleavage of mitotic figures. J Anat 113:109–129

Smart IHM (1982) Radial unit analysis of hippocampal histogenesis in the mouse. J Anat 134:1–31

Smart IHM (1985) Differential growth of the cell production systems in the lateral wall of the developing mouse telencephalon. J Anat 141:219–229

Smart IHM, McSherry GM (1986) Gyrus formation in the cerebral cortex of the ferret. II. Description of the internal histological changes. J Anat 147:27–43

Smart IHM, Sturrock RR (1979) Ontogeny of the neostriatum. In: Divac I, Oberg RGE (eds) The neostriatum. Pergamon, Oxford, pp 127–146

Sterzi G (1907) Il sistema nervoso centrale dei vertebrati, vol I: ciclostomi. Draghi, Padova

Sterzi G (1909) Il sistema nervoso centrale dei vertebrati, vol 2. Pesci libro I: selaci part I: anatomia. Draghi, Padova

Sturrock RR, Smart IHM (1980) A morphological study of the mouse subependymal layer from embryonic life to old age. J Anat 130:391–415

Taber Pierce E (1966) Histogenesis of the nuclei griseum pontis, corporis pontobulbaris and reticularis tegmenti pontis (Bechterew) in the mouse. An autoradiographic study. J Comp Neurol 126:219–240

Tan SS, Breen S (1993) Radial mosaicism and tangential cell dispersion both contribute to mouse neocortical development. Nature 362:638–640

Tello JF (1923) Les différenciation neuronales dans l'embryon du poulet, pendant les premiers jours de l'incubation. Trabajos Lab Invest Biol Madrid 21:1–93

Tello JF (1934) Les différenciation neurofibrillaires dans le prosencéphale de la souris de 4 à 15 millimètres. Trabajos Lab Invest Biol Madrid 29:339–395

Timsit S, Martinez S, Allinquant B, Peyron F, Puelles L, Zalc B (1995) Oligodendrocytes originate in a restricted zone of the embryonic ventral neural tube defined by DM-20 mRNA expression. J Neurosci 15:1012–1024

Urey NJ, Gona AG, Hauser KF (1987) Autoradiographic studies of cerebellar histogenesis in the premetamorphic bullfrog tadpole: I. Generation of the external granular layer. J Comp Neurol 266:234–246

Voigt T (1989) Development of glial cells in the cerebral wall of ferrets: direct tracing of their transformation from radial glia into astrocytes. J Comp Neurol 289:74–88

Von Bartheld CS, Meyer DL (1988) Central projections of the nervus terminalis in lampreys, lungfishes, and bichirs. Brain Behav Evol 32:151–159

Von Lenhossék M (1895) Der feinere Bau des Nervensystems im Lichte neuester Forschungen, 2nd edn. Fischer, Berlin

Walsh C, Cepko CL (1992) Widespread dispersion of neuronal clones across functional regions of the cerebral cortex. Science 355:434–440

Walsh C, Cepko CL (1993) Clonal dispersion in proliferative layers of developing cerebral cortex. Nature 362:632–635

Wicht H, Northcutt RG (1992) The forebrain of the pacific hagfish: I. A cladistic reconstruction of the ancestral craniate forebrain. Brain Behav Evol 40:25–64

Wicht H, Northcutt RG (1993) Secondary olfactory projections and pallial topography in the pacific hagfish, *Eptatretus stouti*. J Comp Neurol 337:529–542

Williams RW, Borodkin M, Rakic P (1991) Growth cone distribution patterns in the optic nerve of fetal monkeys: implications for mechanisms of axon guidance. J Neurosci 11:1081–1094

Wilson SW, Easter SS Jr (1991) Stereotyped pathway selection by growth cones of early epiphysial neurons in the embryonic zebrafish. Development 112:723–746

Wilson SW, Ross LS, Parrett T, Easter SS Jr (1990) The development of a simple scaffold of axon tracts in the brain of the embryonic zebrafish, *Brachydanio rerio*. Development 108:121–145

Wilson SW, Placzek M, Furley AJ (1993) Border disputes: do boundaries play a role in growth-cone guidance? Trends Neurosci 16:316–323

Windle WF (1932a) The neurofibrillar structure of the 7-mm cat embryo. J Comp Neurol 55:99–138

Windle WF (1932b) The neurofibrillar structure of the five-and-one-half-millimeter cat embryo. J Comp Neurol 55:315–331

Windle WF (1935) Neurofibrillar development of cat embryos: extent of development in the telencephalon and diencephalon up to 15 mm. J Comp Neurol 63:139–171

Windle WF (1950) Reflexes of mammalian embryos and fetuses. In: Weiss P (ed) Genetic neurology. Universtiy of Chicago Press, Chicago, pp 214–222

Windle WF (1935) Neurofibrillar development in the central nervous system of chick embryos up to 5 days' incubation. J Comp Neurol 63:431–463

Windle WF, Baxter RE (1936) The first neurofibrillar development in albino rat embryos. J Comp Neurol 63:173–199

Windle WF, Orr DW, Minear WL (1934) The origin and development of reflexes in the cat during the third fetal week. Physiol Zool 7:600–617

CHAPTER 6

Comparative Neuroanatomy: Place, Principles and Programme[1]

R. NIEUWENHUYS

6.1
The Place of Comparative Neuroanatomy

Comparative neuroanatomy or neuromorphology is a biological subdiscipline, the central aim of which is: the elucidation and interpretation of the form, structure and ultrastructure of the nervous system (cf. Sect. 6.3.1).

Comparative neuroanatomy encompasses at least five fields of research, i.e. comparative neuroembryology, formal or pure comparative neuroanatomy, phylogenetic comparative neuroanatomy, functional comparative anatomy and evolutionary morphology. The place of neuromorphology among other special and general (sub)disciplines is indicated in Table 6.1. All elements in this table should be thought of as being interconnected by bidirectional communication channels. This is a most important point; the various sciences and disciplines constitute together an organism and not the contents of a chest of drawers.

A distinction has been made between special (neuro)biological sciences and the general biological sciences: taxonomy, phylogenetics and evolutionary biology. Taxonomy is the science of classification of organisms. Phylogenetics is the science which orders the result of taxonomy in patterns (dendrograms) expressing presumed consanguinities. Evolutionary biology is the science which attempts to explain the mechanisms that produced the phylogenetic diversification of organisms.

Two general or universal sciences, mathematics and philosophy, have been added in the diagram. Mathematics is the most concise, exact and consistent language; hence, it should be attempted to express the results of all special sciences, including the biological sciences, as much as possible in that language (d'Arcy Thompson 1961). The addition of philosophy requires little explanation; no special science can flourish without an on-going reflection upon its aims and principles, and upon the nature of its results.

Table 6.1. The place of neuromorphology among other special and general (sub)disciplines

Evolutionary biology	
Phylogenetics	
Taxonomy	

Neurosciences	Special biological sciences
neuromorphology	morphology
neurophysiology	physiology
neuroethology	ethology
neuroecology	ecology
neurochemistry	biochemistry
neurogenetics	genetics

Mathematics
Philosophy

---

[1] A preliminary version of this chapter has been published in Nieuwenhuys (1994).

6.2
Principles of Comparative Neuroanatomy

6.2.1
The Fundamental Structure of the CNS

The CNS is a transformed bilaterally symmetrical tube, which contains a built-in, natural coordinate system.

The CNS of vertebrates is, throughout its development, bilaterally symmetrical and rostrocaudally polarised, with a brain anlage in front and a spinal cord anlage behind. From a very early developmental phase onward, the brain anlage is divisible into a rostrocaudally arranged series of swellings or vesicles, prosencephalon, mesencephalon and rhombencephalon.

The embryonic neural tube in all vertebrates is initially composed of an undifferentiated pseudostratified neuroepithelium along its entire length. The neuroepithelium is bounded by two surfaces, the ventricular and the meningeal.

All neuroepithelial cells, except for those in mitosis, are radially oriented and extend from the ventricular surface to the meningeal surface. These neuroepithelial cells are polarised, in that their ventricular parts differ clearly from their meningeal parts.

The radial orientation shown by the neuroepithelial cells manifests itself during further development in other structures, such as ependymal gliocytes, blood vessels and many axonal and dendritic processes. This predominantly radial orientation of

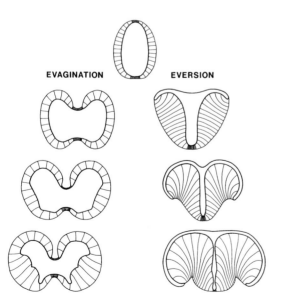

Fig. 6.1. Transverse sections illustrating how evaginated as well as everted telencephala develop from an early embryonic tube-shaped anlage.

its constituent structural elements is a fundamental feature of the developing CNS, indicating a specific spatial relationship between its two natural surfaces, the ventricular and the meningeal. The two natural surfaces and a system of curves or radii, indicating the specific spatial relationship between these two, constitute the natural coordinate system of the CNS (Chap. 5, Sect. 5.2).

During development the neural tube and its walls may be bent and deformed in many different ways (Chap. 4, Sect. 4.4). A consequence of these deformations may be that the radii, although they remain to indicate the specific spatial relationships between the ventricular and meningeal surfaces, will pursue a non-radial course over certain stretches of their trajectories (Chap. 5, Sect. 5.4.3). This being so, the term radius has to be replaced by vector, and the adjective radial by vectorial.

It is of particular importance that mitoses in the CNS generally occur at the ventricular surface; hence, this surface not only is the basis of the natural coordinate system, but also represents the starting or zero plane of neurogenesis as well as gliogenesis. That the natural coordinate system reveals the general morphology of the wall of the CNS may be elucidated by the following two examples:

First, in most vertebrates the lateral walls of the tube-shaped early embryonic anlage of the telencephalon evaginate, a process leading to the formation of hollow cerebral hemispheres (Fig. 6.1, left panel). However, in actinopterygians the dorsal parts of the telencephalic walls diverge, with a consequent widening of the initially narrow dorsal closure of the ventricle. This eversion produces solid hemispheres with a very extensive ventricular surface. The transformation of the natural coordinate system, as revealed by the radial glial cells, tells the story of this remarkable morphogenetic process (Fig. 6.1, right panel; Nieuwenhuys 1962, 1964).

Second, in the early embryonic mammalian telencephalon the matrix cells and the ependymal gliocytes show a radial orientation. However, later in development the ependymoglial fibres in the striatal anlage lose contact with the meningeal surface, and those emanating from the lateral part of the pallial ventricular surface begin to deflect ventrally in association with the appearance of the cortical plate. Thus, the ependymal processes will pursue curving courses which circumscribe the striatal primordium (Chap. 5, Fig. 5.13a). Not only do these fibres reveal the formal changes in the lateral hemisphere walls, they also indicate the remarkable route along which migrating neuroblasts attain the ventral part of the cortical plate, which forms the anlage of the prepiriform cortex (Smart and Sturrock 1979; Misson et al. 1991; Chap. 5, Fig. 5.13b).

6.2.2
Cartesian and Natural Coordinate Systems

Any morphological approach to the CNS of verte-brates should take the natural coordinate system, and not the artificial, orthogonal, Cartesian coordinate system, as its point of departure.

In anatomy it is customary to employ for descriptive purposes an orthogonal, Cartesian coordinate system. In this system three axes or directions, i.e. rostral-caudal or anterior-posterior, dorsal-ventral and medial-lateral, are distinguished. This coordinate system with the related terminology is also frequently used in the naming of structures within the CNS: lateral pallium, nucleus dorsomedialis anterior, nucleus septi medialis, tractus olfactorius lateralis and so on. Such names are appropriate for the description of structures identified in one particular species, but they may be totally inadequate to designate correspond-

ing parts or entities in the brains of different vertebrates, as may appear from the following four examples.

Six structures, i.e. the commissura anterior, the commissura habenulae, the commissura posterior, the tuberculum posterius, the infundibulum, and the chiasma opticum can be easily recognised in the diencephalon of all vertebrates and hence may be considered landmarks in that brain part. Figure 6.2 clearly shows that these structures do not occupy a fixed topographical position with respect to each other.

A cell mass, the anlage of which is situated in the dorsal most part of the lateral plate, directly adjacent to the site of attachment of the roof plate, will occupy a dorsomedial position in an evaginated telencephalon, but a ventrolateral position in an everted telencephalon (Fig. 6.1).

In all gnathostomes the principal secondary olfactory projection terminates in a discrete part of the pallial wall. In amphibians this target area of the

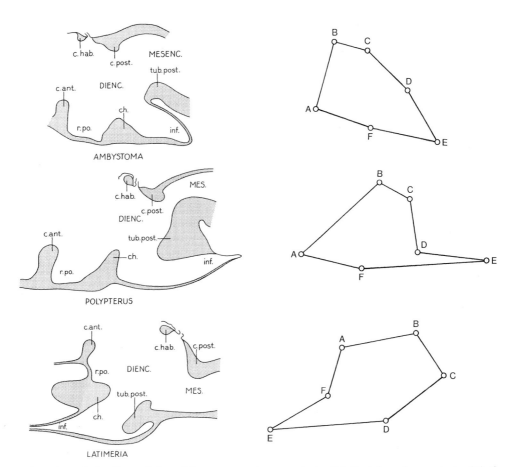

Fig. 6.2. Median sections through the diencephalon of three anamniotes, the tiger salamander, *Ambystoma tigrinum,* the bichir *Polypterus ornatipinnis,* and the coelacanth *Latimeria chalumnae,* showing the position of six landmarks: the commissura anterior (*A*), the commissura habenulae (*B*), the commissura posterior (*C*), the tuberculum posterius (*D*), the infundibulum (*E*) and the chiasma opticum (*F*). In the *right panel* the centres of these landmarks are interconnected and presented as hexagons

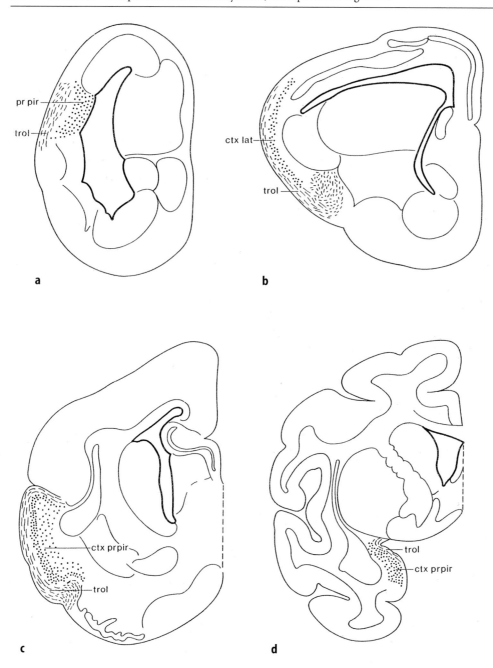

Fig. 6.3a–d. Transverse hemisections through the telencephalic hemispheres of: (**a**) the tiger salamander, *Ambystoma tigrinum*; (**b**) the tegu lizard, *Tupinambis nigropunctatus*; (**c**) the opossum, *Didelphis virginiana*; (**d**) the rhesus monkey, *Macacus rhesus*. The position of the tractus olfactorius lateralis and its principal target is indicated. *ctx lat*, Cortex lateralis; *ctx prpir*, cortex prepiriformix; *pr pir*, primodium piriforme; *trol*, tractus olfactorius lateralis

principal olfactory projection is constituted by an area of periventricular grey, known as the primordium piriforme, which is situated in the dorsolateral hemisphere wall (Fig. 6.3a); in reptiles it is represented by a sheet of migrated cells: the cortex lateralis (Fig. 6.3b). The position of the mammalian equivalent of the amphibian piriform primordium and the reptilian lateral cortex is strongly influenced by the expansion of the neocortex. Thus, in the primitive opossum. the olfactory or prepiriform cortex occupies a ventrolateral position (Fig. 6.3c), but in the rhesus monkey (and in primates in general) it is displaced ventromedially by the greatly expanded neocortex (Fig. 6.3d).

The lateral geniculate body is a derivative of the rostral part of the thalamus; hence, early in development its anlage occupies a rostral or rostrolateral position in the diencephalon of all vertebrates, and the same holds true for this centre in the adult stage of most non-mammalian vertebrates. However, due to a remarkable deformation of the thalamus, the human lateral geniculate body shifts during ontogenesis within the diencephalon from a rostral to a caudolateral position (Chap. 5, Sect. 6.3.3; Chap. 5, Fig. 5.14).

6.2.3
Significance of Topology

The central nervous systems of all vertebrates are topologically equivalent.

Topology is the geometry of distortion, the branch of mathematics, which investigates the properties that remain unchanged when geometric configurations are subjected to one-to-one continuous transformations. In these transformations it is allowed to change distances, to bend, stretch or twist the configurations. However, it is forbidden to 'cut' or to 'tear'. Thus a triangle drawn on a sheet of rubber could be imagined as stretched into other triangles, circles, or any other closed curve, and all of these figures are considered topologically equivalent. Yet, a triangle or a circle cannot be continuously transformed into a straight line.

Edinger (1908a,b) suggested that in the course of vertebrate evolution a new brain, or neencephalon has been added to a primitive, old brain, or palaeencephalon. The ideas of MacLean (1970), although developed some 60 years later, show a striking resemblance to those of Edinger. According to MacLean, the brain of higher primates is composed of three neural formations: reptilian brain, paleomammalian brain, neomammalian brain, that reflect, as their names imply, ancestral relationships to reptiles, early mammals, and late mammals. In spite of these concepts, it is generally assumed that the brains of all vertebrates are morphologically and topologically equivalent (cf. Johnston 1923, p. 186: "New structures have not appeared, ..."), which means that, essentially, they can be transferred into each other by means of one-to-one continuous topological transformations (Fig. 6.4). This does not exclude of course that certain apparently 'new' structures, as for instance the torus longitudinalis in actinopterygians, the dorsal ventricular ridge in reptiles and birds and the 'neo'cortex in mammals can be derived only from 'general primordia' or field homologues in other vertebrates. On the other hand, reduction of certain peripheral fields and regression of special sense organs may lead to corresponding reductions and regressions in the CNS. Thus, cervical and lumbar enlargements of the spinal cord are lacking in limbless amphibians and reptiles. In the microsmatic lizard *Anolis*, the olfactory bulbs are tiny and the lateral or piriform cortex is largely replaced by a membranous structure (Fig. 6.5). In the anosmatic cetaceans olfactory bulbs are entirely lacking (although their anlagen are present in early embryonic stages; Buhl and Oelschläger 1988).

As long as the elements of the natural coordinate system (see above, Sect. 6.2.2) are clearly recognisable and a sufficient number of other landmarks or invariants are present (see Sect. 6.2.6.2), the topological transformation of the brains of two related species into each other will not present major problems. However, orientation (and therewith interpretation) may become extremely difficult if one or more of the elements of the natural coordinate system become lost. Thus, in the brain of myxinoids the ventricular cavity is largely obliterated, parts of the meningeal surface may fuse, and the radial ependymoglia is replaced in an early developmental phase by astroglia. Hence, the topological relations in this brain are hard to judge and the interpretation of its parts is fraught with difficulties (Wicht and Northcutt 1992; cf. Sect. 6.2.6.4).

It is important to note that topological analysis as such may reveal morphological correspondence between two or more structures, but not whether the corresponding structures are homologous or homoplastic (see below, Sect. 6.2.6).

6.2.4
Significance of Taxonomy

A critical selection of the species to be studied is an indispensable first step in all comparative (neuro)biological studies. This means that the specific aims of such studies should be in harmony with the taxonomic position of the chosen species.

Earlier generations of comparative neuroanatomists attempted to reconstruct the phylogenetic history of the vertebrate brain by comparing the brains of arbitrarily chosen species belonging to the various classes of vertebrates: a cyclostome, a fish, an amphibian and so on. However, we now know that conclusions drawn from the analysis of such a 'series' of vertebrates are of little value, because vertebrates have not evolved linearly. During vertebrate phylogeny four distinct major radiations (Agnatha, Chondrichthyes, Osteichthyes, Tetrapoda) have evolved, and each of these radiations has given rise to several new groups. In fact, none of the extant vertebrate groups is ancestral to any other group. In principle, it is possible to recon-

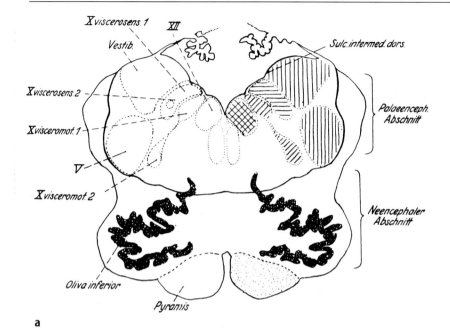

a

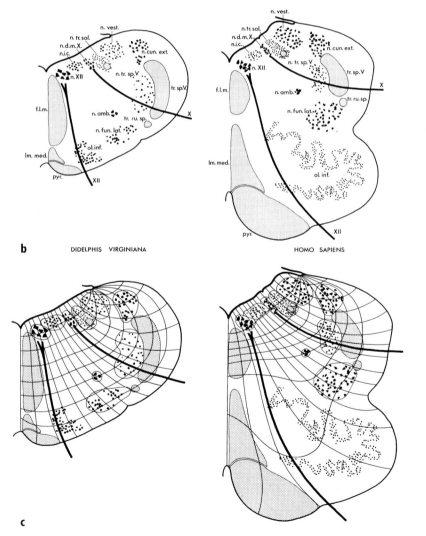

b   DIDELPHIS VIRGINIANA   HOMO SAPIENS

c

Fig. 6.4. a,b Two interpretations of the structural differences between the medulla oblongata of a marsupial and of man as observed in transverse sections: **a** superimposed sections of the kangaroo, *Macropus* (*heavy outline*) and of man (*thin outline*), intended to show that in the course of evolution a ventral, neencephalic part has been added to a dorsal, palaeencephalic (reproduced from Haller von Hallerstein 1934); **b** hemisections of the opossum, *Didelphis virginiana* and of man. **c** Topological analysis of the two sections shown in **b**, indicating that the section of the opossum can be transferred into that of man by means of a one-to-one topological transformation. This transformation shows that no new parts have been added, but rather that some provinces have considerably enlarged. *f.l.m.*, Fasciculus longitudinalis medialis; *lm.med.*, lemniscus medialis; *n.amb.*, nucleus ambiguus; *n.cu-n.ext.*, nucleus cuneatus externus; *n.d.m.x.*, nucleus dorsalis motorius vagi; *n.fun.lat.*, nucleus funiculi lateralis; *n.i.c.*, nucleus intercalatus; *n.tr.sol.*, nucleus tractus solitarii; *n. tr.sp.V.*, nucleus tractus spinalis nervi trigemini; *n.vest.*, nuclei vertibulares; *ol.inf.*, oliva inferior; *pyr.*, tractus pyramidalis; *tr.ru.sp.*, tractus rubrospinalis; *tr.sp.V.*, tractus spinalis nervi trigemini; *X, XII*, cranial nerves part

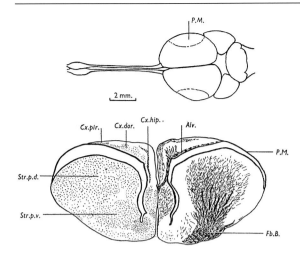

Fig. 6.5. *Above*: Dorsal view of the rostral part of the brain of the microsmatic lizard, *Anolis garmani*. *Below*: Transverse section through the rostral part of the telencephalic hemispheres of *Anolis lineatopus*. Note that the piriform or lateral cortex is largely reduced to a pallial membrane. *Alv.*, alveus; *Cx.Dor.*, dorsal cortex; *Cx. hip.*, hippocampal cortex; *Cx.pir.*, piriform cortex; *Bv.B.*, forebrain bundles; *P.M.*, pallial membrane; *Str.p.d.*, striatum, pars dorsalis; *Str.p.v.*, striatum, pars ventralis (reproduced from Goldby and Gamble 1957, Figs. 2, 4)

struct the phylogeny of the CNS of vertebrates and to produce morphotypes of the brains of the common ancestral groups of craniates, jawed vertebrates, tetrapods, anapsid amniotes, etc.; however, for such a reconstruction a comprehensive cladistic analysis (see below, Sect. 6.3.4) based on vast numbers of species would be required. So far, such comprehensive analyses have been carried out at an overall level for only a few parts of the brain, such as the dorsal thalamus of jawed vertebrates (Butler 1994a) and the dorsal pallium of amniotes (Butler 1994b). The level of these analyses is characterised as 'overall' because in them only grisea and their major fibre connections are taken into consideration. Detailed analyses of the kinds of morphological changes that neural systems (i.e. sets of interconnected neurons) have undergone, intentionally directed at disclosing the developmental and evolutionary mechanisms responsible for these changes, can be successfully carried out only on a small number of carefully selected, closely related species. The same holds true for other neurobiological disciplines, for instance, neuroethology. "Neuroethological studies of closely related species that have undergone adaptive radiation are the key to understanding the principles of evolution in the sensory and neural systems controlling behavior" (Roth 1987, p. VI).

The necessity of selection appears immediately from the facts that the vertebrate kingdom encom-

passes approximately 50 000 species and that many groups display an amazing diversity. The fact that in several groups, particularly the cartilaginous fish, the teleosts, the birds and the mammals, this diversity is clearly reflected in the relative size, external appearance and internal structure of the brain underscores the need for a critical selection of the species to be used in comparative neurobiological research.

6.2.5
Central (Neuro)morphological Concepts

6.2.5.1
Introduction

Comparative neuroanatomy derives all of its principles and basic concepts from its mother science: morphology. The foundations of morphology were laid during the first half of the nineteenth century, but the interpretation of its results and the formulation of most of its basic concepts have been strongly influenced by the Darwinian revolution of 1859. Hence, in what follows the basic concepts which comparative neuroanatomy shares with morphology will be treated in their historical context.

6.2.5.2
Homology and Analogy

The concept of homology expresses the existence of typical and specific correspondences between the parts of members of natural groups of living beings. The recognition and the systematic search for homologies marks the beginning of morphology as a separate scientific discipline. The foundations of this discipline were laid by three men: J.W. von Goethe (1749–1832), Étienne Geoffroy St-Hilaire (1772–1844) and Richard Owen (1804–1892).

It is of paramount importance that homology is essentially a purely empirical concept, and that, hence, the validation of the criteria to be used for determining homologies is based entirely on comparative experience. "In der Homologie und ihren verschiedenen Formen liegt aber nur der Ausdruck der vergleichenden Erfahrung" (Gegenbaur 1898, p. 25).

It was recognised from the very beginning of morphology that *the position of parts with respect to each other* forms the prime and principal criterion for the establishment of homologies. "Dagegen ist das beständigste der Platz" (von Goethe 1795). "Le principe des connexions est invariable: un organe est plutôt diminué, effacé, anéanti que transposé" (Geoffroy St-Hilaire 1818, p. 405). "These relationships are mainly, if not wholly, determined by the relative position and connection of parts..." (Owen 1849, p. 6).

From the beginning, similarity in function was excluded as a criterion for the establishment of homologies. Owen (1843, p. 379) defined homologue as: "The same organ in different animals under every variety of form and function." "An organ is homologous with another because of what it *is*, not because of what it *does*" (de Beer 1971, p. 3).

The terms 'analogy' and 'analogue' are used to denote similarity in function. Thus, Owen (1843, p. 374) defined analogue as "...a part or organ in one animal which has the same function as another part or organ in a different animal." It should be noted that, according to Owen's definitions, structures can be both homologous and analogous. Several later authors (e.g., Jacobshagen 1925, Bock 1963) considered this overlap unfeasible and proposed definitions in which analogy features as the opposite of homology (see below, Sect. 6.2.5.3). I recommend, with Gegenbaur (1898), Boyden (1943, 1969) and many others, that the term analogy be used in Owen's sense.

It stands to reason that if all recognisable structures or entities were to exert a single, discrete function, and if throughout the animal kingdom the link between particular structures and functions were absolute and invariable, there would be no need for the separate terms homology and analogy; 'correspondence' would do. However, even very early observations showed that morphologically corresponding structures may well exert quite different functions. Thus, in 1555 Belon published an *Histoire de la nature des Oyseaux*, in which he 'homologised' the wing of a bird and the arm of a man, and showed that the skeletal parts of these two extremities correspond one to one. During the first half of the nineteenth century, numerous other cases of homologous-but-not-analogous parts were found, the classic example of which is that the ear ossicles (incus and malleus) of mammals and the bones of jaw articulation (quadrate and articular, respectively) of reptiles and other gnathostomes are homologous (Reichert 1837). After the enunciation and acceptance of the theory of evolution, change of function of organs (Dohrn 1875) and substitution of an organ with a particular function by another organ exerting the same function (Kleinenberg 1886) were proclaimed important evolutionary principles (cf. Sewertzoff 1931, pp. 182–183).

In annelid worms, arthropods and vertebrates a serial repetition of body parts along the rostrocaudal axis of the body can be observed. Each such division of the body is termed a segment, metamere or somite. Owen (1843, p. 7) pointed out that in ver-

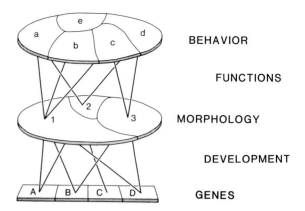

Fig. 6.6. Genes, morphology and behaviour are shown as three major levels of biological organisation, with the levels of development and function interposed between them. Causal relations between the three major levels are shown as *solid lines.* (reproduced from Striedter and Northcutt 1991, Fig. 3)

tebrates the endoskeletal components of these segments are for the most part composed of elements or components similar in number and arrangement. He termed this kind of repetition in the segments of the same skeleton *serial homology* (Fig. 6.8b). This term is still in use to designate the specific relation between corresponding (skeletal as well as non-skeletal) parts of segments in an individual, as for example the spinal dorsal root ganglia.

Ontogenetic investigations may be of great importance for the determination of the relative position of parts or structures, and thus for the establishment of homologies. However, this does not mean that structures which have developed from corresponding anlagen are always strictly homologous. A particular anlage or primordium may give rise to a single structure in one species but differentiate into several structures in another, related species. The term *field homology* has been introduced to indicate the relationship between a particular anlage and its derivatives in different species. Smith (1967, p. 102) defined the meaning of this term as follows: "Derivation of structures, however similar or dissimilar, from a common anlage, or in other words from the same ontogenetic source of the same or different segments, of any two or more compared individuals or groups of individuals."

Darwin's (1859) theory of evolution, enunciated in *"On the Origin of Species by Means of Natural Selection, ..."*, offered a natural explanation of the astonishing diversity of living forms, as well as of the numerous typical correspondences between the parts of members of natural groups of organisms, known since Owen (1843) as *homologies.* Accord-

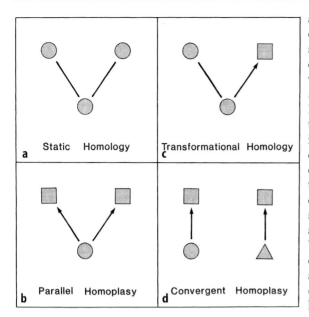

a Static Homology

c Transformational Homology

b Parallel Homoplasy

d Convergent Homoplasy

Fig. 6.7a–d. The definitions of static homology, transformational homology, parallel homoplasy, and convergent homoplasy as used by Wiley (1981), Striedter and Northcutt (1991) and the present author. In each of the four cases, an ancestral character is shown together with two derived characters. The same symbols denote the same characters, and different symbols denote different characters. Phylogenetic transformations are indicated by *arrows*, and retention of characters without transformation is indicated by *solid lines*. (reproduced from Striedter and Northcutt 1991, Fig. 1)

ingly, numerous post-*Origin* authors re-defined the concept of homology in phylogenetic terms. Lankester (1870) proposed replacing homology by homogeny (a term which has not caught on) and explained the meaning of this term as follows (Lankester 1870, p. 36): "Structures which are genetically related, in so far as they have a single representative in a common ancestor, may be called homogenous." Gegenbaur (1898, p. 24) defined homology as follows: "Homologie im engeren Sinne bezeichnet das Verhältnis zwischen zwei Organen gleicher Abstammung, die somit aus der gleichen Anlage hervorgegangen, gleiches morphologisches Verhalten darbieten."

Other well-known 'phylogenetic' homology definitions include those of Simpson, Ghiselin and Wiley. Simpson (1961, p. 78) stated simply: "Homology is resemblance due to inheritance from a common ancestry." Because resemblance is by no means a typical feature of homologous structures, Campbell (1987) preferred the definition of Ghiselin (1966, p. 29), which states that "structures or other entities are homologous when they could, in principle, be traced back through a genealogical series to a stipulated common ancestral precursor irrespective of morphological similarity." Northcutt (1984)

and Striedter and Northcutt (1991) have pointed out that the above definition is still not entirely satisfactory because it fails to exclude from the category of true homologies cases of parallelism, in which similar characters evolve *by independent transformations from the same ancestral character*. In their opinion, Wiley's (1981, pp. 121–122) definition of homology clearly excludes cases of parallelism from true homology. Wiley suggested that "a character of two or more taxa is homologous if this character is found in the common ancestor of these taxa, or, two characters (or a linear sequence of characters) are homologues if one is directly (or sequentially) derived from the other(s)." Striedter and Northcutt (1991) noted that in this definition Wiley implicitly makes a distinction between two different types of homology: (a) static homology and (b) transformational homology. The authors explain the difference between these types of homology as follows (Striedter and Northcutt 1991, p. 179; Fig. 6.7a,b): "In cases of static homology, an ancestral character is retained in two or more descendant taxa without any transformation. (...) In cases of transformational homology, the ancestral character is retained in only one of the descendant taxa and is transformed into a different character (or series of characters) by a single transformation (or linear sequence of transformations) in the other taxon (or series of taxa)."

In the phylogenetic homology definitions of Lankester, Gegenbaur, Simpson, Ghiselin and Wiley presented above (and in those of numerous other authors), shared common ancestry is the essential feature. It is noteworthy that several adherents to the theory of evolution, among them Zangerl (1948) and Remane (1954, 1956), have challenged the correctness of including phylogeny in the definition of homology. The latter stated: "Nicht die Phylogenie entscheidet über die Homologie, sondern die Homologie über die Phylogenie" (Remane 1954, pp. 171–172). During the first decades of the twentieth century a number of workers, among them Naef (1919), Jacobshagen (1925, 1927) and Lubosch (1925), went one step further by claiming that phylogenetic explanations are entirely beyond the scope of morphology. They advocated a pure, strictly formal morphology, not contaminated by evolutionary or functional speculations. Accordingly, Jacobshagen (1925, p. 81) defined the concept of homology as follows: "Organe, die in einem Bauplan oder dessen Grundformteilen denselben Bestandteil verkörpern, nennen wir, unbekümmert um etwaige Form- und Funktionsunterschiede, homolog," whereas Lubosch (1925, p. 33) stated succinctly: "Homolog sind Organe in gleichen Lagebeziehungen."

The concept of homology has appeared to be applicable also in other domains of biology, particularly in biochemistry, genetics and ethology. Thus, proteins (particularly enzymes) and (parts of) DNA and RNA molecules have been homologised on the basis of statistically significant similarities in their amino acid and nucleotide sequences, respectively. Genes of different animals have also been homologised (cf., e.g., Patterson 1988) and several authors, among them Tinbergen (1951) and Baerends (1958), have shown that the concept of homology can also be applied to species-specific behaviours. The genome and the species-specific behaviours are strongly, though indirectly coupled, as follows: genome → development of morphological characters → phenotype → behaviours. It has been suggested that these causal interrelationships imply that homologous behaviours depend on homologous structures, and that homologous structures in turn will follow similar processes of differentiation, depending on the influence of homologous genes. Striedter and Northcutt (1991), who extensively reviewed the pertinent literature, felt that genes, developmental processes, morphological structures, physiological functions and behaviours all constitute different levels of biological organisation. In their opinion, this hierarchical nature of biological organisation leads towards a hierarchical concept of homology that precludes any attempts to reduce homologies at one level of organisation to those of another level. Striedter and Northcutt emphasise the following two points: (a) The various levels of biological organisation are causally interrelated, but the causal relationships between characters at different levels are complex. They cite numerous examples showing that there is no simple one-to-one correspondence between characters at different levels of organisation (Fig. 6.6). (b) Changes in the causal relationships between characters at different levels may occur during the course of evolution, but these changes leave the homology of these characters at their own level of organisation entirely unaffected.

6.2.5.3
Homoplasy

The concept of homoplasy was introduced by Lankester in 1870. After having proposed that the term homology be replaced, in light of the theory of evolution, by the term homogeny, Lankester (1870, p. 41) characterised homoplasy as follows: "Homoplasy includes all cases of close resemblance of form which are not traceable to homogeny, all details of agreement not homogenous, in structures which are broadly homogenous, as well as in structures

having no genetic affinity." According to Hodos (1987), homoplastic similarities occur when organisms with differing genealogies respond in a similar way to similar adaptive pressures from the environment. Two forms of homoplasy, namely parallelism or parallel homoplasy, and convergence or convergent homoplasy, can be distinguished (Fig. 6.7c,d). Wiley (1981, p. 12) defined *parallel homoplasy* as the independent evolution of similar characters from the same ancestral character, and convergent homoplasy as the independent evolution of similar characters from different ancestral characters.

Some authors have designated the phenomenon homoplasy by a different name. Thus, Gegenbaur (1898, p. 25) used the term 'homomorphic' and Remane (1956) denoted *Anpassungsähnlichkeit* as *Analogie*. Jacobshagen's (1925) concept of analogy also comes close to what has been denoted above as homoplasy. In fact, the concept of analogy, as defined by that author, is to be considered as a non-evolutionary version of the concept of homoplasy. It has already been mentioned that Jacobshagen (1925, 1927) advocated a pure, strictly formal morphology. On that account, he rejected the common usage of denoting organs subserving the same function as analogous: "Die übliche Behauptung, analog seien 'physiologisch gleichbedeutende' Organe, ist ein blamables Dogma im Munde eines Morphologen. Was weiß er von physiologisch-gleicher Bedeutung, er, der Mann der Form?". Jacobshagen (1925, p. 198) arrived at the following definition: "Organe übereinstimmender oder ähnlichen Baues, die nicht denselben Bestandteil des Bauplanes verkörpern und somit, trotz ihrer Ähnlichkeit, einen ganz verschiedenen morphologischen Wert besitzen, nennt man analog."

6.2.5.4
Structural Plan or Morphotype

It is customary in morphology to unite the obvious and unquestionable structural features (i.e. the evidently homologous organs or parts including their spatial relationships) shared by the members of a natural groups in formulas or diagrams, denoted as plans, structural plans *(Baupläne)*, types or archetypes (Fig. 6.8). In the prephylogenetic era these plans represented abstractions or idealisations expressing the *formal* relationships between the members of natural groups. The mental process by which the members of species, as found in nature, can be transformed into a plan or type, and vice versa, was denoted as metamorphosis. It is important to note, however, that during the first half of the nineteenth century the conceptual status

Die Endhirnbaupläne bei den Amphibien können demnach in der folgenden Weise ausgedrückt werden:

Anuren .... $O, D_1, D_2, D_3, B_1, B_2, B_3, B_4.$

Urodelen ...... $O, D_1, D_2, D_3, B_1, B_2, B_{3+4}.$

Gymnophionen . $O, D_1, D_2, D_3, B_1, \dfrac{B_2}{B_s}, B_3, B_4.$

Die Bauplanformel des Reptilienendhirns lautet somit:

$$O, \frac{}{D_3(s)}, \frac{}{D_2a(s)}, \frac{}{D_{(2+1)}b(s)}, \frac{}{D_{(2+1)}c(s)}, D_1, \frac{B_1, B_2}{B_s}, B_3, B_4.$$

Für das Endhirn der Vögel lautet demnach die Bauplanformel:

$$O, \frac{}{D_3(s)}, \frac{}{D_2a(s)}, D_2b, D_{(2+1)}c, D_1, \frac{B_{1+2}}{B_s}, B_{3+4}.$$

a

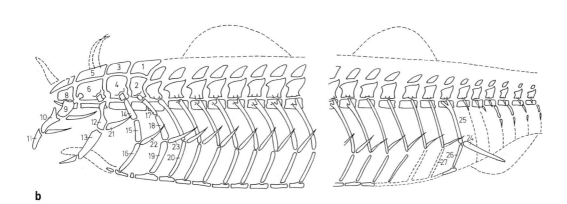

b

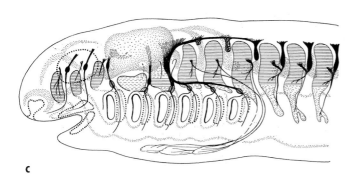

c

Fig. 6.8a–c. Structural plans (*Baupläne*) of parts of the vertebrate body. **a** Formulas expressing the arrangement of cell masses (O, D_1, D_2, etc.) in the telencephalon of amphibians, reptiles and birds. The *symbols* placed under the *horizontal lines* represent migrated nuclei. In order to gain some insight into the spatial arrangement of the various cell masses, these formulas should be compared with the diagrammatic cross-sections reproduced in Chap. 4, Fig. 4.50. (formulas and the accompanying texts reproduced from Kuhlenbeck 1929). **b** The archetype of the vertebrate skeleton after Owen. **c** The metameric organisation of the head of cartilaginous fishes after Goodrich (**b, c** reproduced from Starck 1977, Fig. 5)

of the types and their members changed dramatically. Under the influence of the German *Naturphilosophie* (J.W. von Goethe, 1749–1832; L. Oken, 1779–1851). The types were looked upon as forming part of a transcendental reality, a realm of Platonic ideas. The members of natural groups were interpreted as the (varying) appearances or adumbrations of eternal and invariant ideas. For this reason the early nineteenth-century morphologists are commonly designated as 'typologists', 'transcendentalists', 'idealistic morphologists', or even 'nature philosophers'. Remane (1956, p. 13) pointed out that the influence of the philosophical concepts outlined above was confined to the explanation of the results of the early morphologists, but did not affect their methodology and scientific achievements as such. On that account, Remane recommended characterising the epoch of prephylogenetic morphology as that of *pure morphology* rather than that of idealistic morphology.

After the advent of the theory of evolution the types, morphotypes or *Baupläne* retained their central position in morphology; however, they were now reinterpreted as representing ancestral forms (Russell 1916; Lubosch 1931; Starck 1965). Most evolutionary morphologists, among them Haeckel (1866) and Gegenbaur (1878), interpreted the homologous resemblances of animals, as assembled in a *Bauplan*, as being due to inheritance, but their differences as due to adaptation. [Remarkably, a comparable difference features prominently in the antithesis between *Planmäßigkeit* and *Planstörung*, as presented in Jacobshagen's (1925) reformed, non-evolutionary morphology.] However, Darwin himself (cf. Gould 1982, footnote 37) pointed out that a sharp distinction between ancestral and adaptive characters cannot be made, and Cain (1964) argued and adduced some evidence suggesting that the major plans upon which the members of the different phyla and classes of animals are built are adaptive for broad functional specialisations, and are retained merely because of that. That the *Bauplan* or morphotype concept is nevertheless very useful in the realm of phylogenetic morphology (and classification) may be explained as follows: (a) The stream of adaptations shows numerous ramifications. (b) It is an extreme, though acceptable simplification (cf. Sect. 6.3.1 and Fig. 6.21) to position at each bifurcation of the stream a hypothetical stem form, representing the common ancestor of all downstream individuals. (c) The methodology known as cladistics (see Sect. 6.3.3) enables us to design images, i.e. morphotypes, of these hypothetical common ancestors.

6.2.5.5
Concluding Remarks

Homology is the central concept of all biological comparisons, and perhaps even for all biology (Wake 1994). Homologies define groups in the inter-nested pattern of classification, and this pattern forms *the* basis (and *an* end point) of all comparative biological research. The defining character of homoplasy is that the resemblances denoted as such are non-homologous.

The central morphological concepts of homology, homoplasy and structural plan are fully valid and applicable in neuromorphology. However, the *criteria* according to which homologies are established in the CNS have to be critically tested, all the more so because in this system, contrary to most other organ systems, many microscopic structures are compared.

6.2.6
Criteria for the Determination of Homology in the CNS

6.2.6.1
Introduction

Comparative experience has taught that the following four criteria: (a) similarity in position, (b) similarity in fibre connections, (c) similarity in special quality, and (d) continuity of similarity through intermediate species, are of particular importance for the establishment of the homology of cell masses in the CNS, and that similarity in position has to be considered as the principal criterion in this context. These criteria will now be discussed.

6.2.6.2
Principal Criterion: Similarity in Position

In light of what has been stated above in Sect. 6.2.3 and Sect. 6.2.5.2, it will be clear that the *relative or topological position*, rather than the topographical position, should be taken into consideration here. It is surprising that in the neuroanatomical literature, even in articles dealing specifically with the establishment of homologies, similarity in topographical position is recommended again and again as an important clue to homology (e.g., Campbell and Hodos 1970; Campbell 1976; Butler 1994a). The fallacy of this 'criterion' has already been elucidated in Sect. 6.2.3; hence, I confine myself here to a single additional example, illustrated in Fig. 6.9.

Two nuclei can be clearly recognised in the subpallial part of the medial wall of the telencephalic hemispheres of amphibians and other tetrapods,

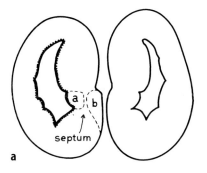 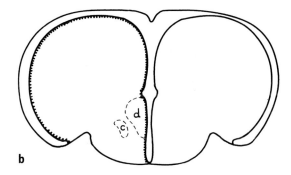

Fig. 6.9. Transverse sections through the telencephalic hemispheres of a urodele amphibian and of a teleost. The position of some subpallial cell masses (*a-d*) is indicated. Further explanation in the text

i.e. a periventricularly situated lateral septal nucleus (a) and a submeningeal medial septal nucleus (b) (Fig. 6.9a). In the dorsal part of the subpallium of the everted teleostean telencephalon likewise two nuclei are present, a migrated nucleus (c) and a periventricular nucleus (d) (Fig. 6.9b). Using similarity in topographical position as a criterion, Droogleever Fortuyn (1961) and Crosby et al. (1966) homologised the teleostean nuclei c and d with the lateral and medial septal nuclei, respectively, of tetrapods. It will be clear that if the unevaginated condition of the teleostean telencephalon is taken into consideration, and if it were allowed to compare the brain of teleosts directly with those of tetrapods, cell mass d would be homologous to the lateral septal nucleus (a), and cell mass c to the medial septal nucleus (b). Even within different evaginated telencephala, nuclei may differ considerably in topographical position (Fig. 6.11).

The *Bauplan* or *morphotype* of the brain, i.e. the set of invariant configurational relationships common to all species included in a comparative study, is of paramount importance for the determination of the topological position, and therewith for the establishment of the homologies, of neural cell masses (Kuhlenbeck 1929, 1933, 1967; Nieuwenhuys and Bodenheimer 1966). This means that, for the determination of the topological position of neural cell masses, in addition to the mutual cytoarchitectonic boundaries, the spatial relationships to such landmarks as ventricular sulci (e.g. the sulcus limitans), cell-free zones (e.g. the zona limitans intrathalamica), commissures (e.g. the optic chiasma), and fibre bundles (e.g. the fasciculus retroflexus), should be taken into account. For details the reader is referred to Chap. 4, Sects. 4.7.2 and 4.7.3, and to Chap. 4, Fig. 4.53 and Chap. 5, Fig. 5.30.

Cell masses situated at some distance from the ventricular surface should be homologised on the basis of their *primary position*, rather than on the basis of their secondary or definitive position. This aspect has been emphasised by several members of the 'Swedish' or 'Holmgren' school of comparative neuroembryology (cf. Chap. 4, Sect. 4.5.2), as may appear from the following quotations:

"If ... it can be proved that two nuclei are formed from the same portion of the neuroblastic layer, thus from the same portion of a (secondary) segments, they are homologous" (Palmgren 1921, p. 3). "The only basis for the homologization of brain nuclei ought to be the ontogenetic development of the nuclei from the regions where the first morphologically visible cell migration takes place, the so-called migration areas (roughly corresponding to Bergquist's (1932) *Grundgebiete*)" (Källén 1951, p. 30). "If it can be proved that two nuclei in different species develop from the same anlage and in a similar way, they must be looked upon as homologous. The longer the developments of two nuclei are similar, the stricter is the homology between the nuclei" (Källén 1951, p. 6).

The idea that the cells destined to form a given cell mass are generally generated in a discrete field or compartment of the ventricular matrix is implicit in the statements just quoted. It tallies with the thesis of Bayer and Altman (1995) that the matrix contains a blueprint of the anatomy of the CNS, although there is no consensus about the exact spatial relationships within the blueprint for either mammals (cf. Chap. 5, Sect. 5.5) or vertebrates in general (cf. Chap. 5, Sect. 5.4). In spite of these uncertainties, tracing nuclei back to their *primary topological position*, i.e. to the position once occupied by their matrix formation at the ventricular surface, may be essential for the determination of the homology. It stands to reason that, as indicated by Palmgren, Källén and others, the exact site or origin of cell masses can be assessed only by embryological research. However, given the fact that most cell masses in the vertebrate brain result from relatively short, radial migrations, an acceptable approximation of their primary position can

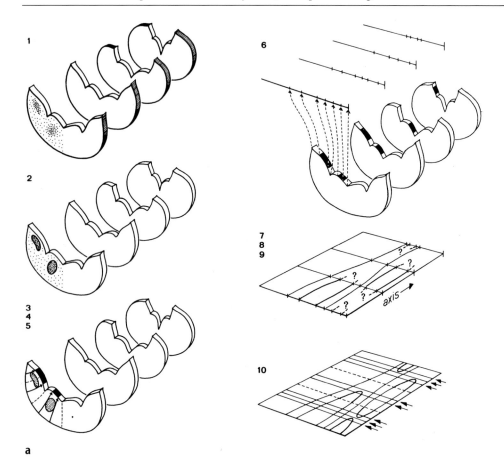

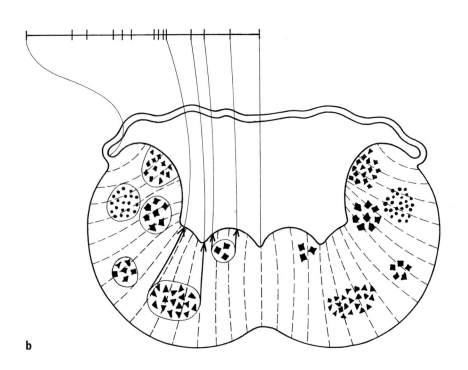

◀ **Fig. 6.10a,b.** Topological reconstruction of the brain stem. **a** Steps involved in the preparation of such a reconstruction: *1,* selection of sections; *2,* drawing of sections; *3,* introduction of projection curves derived from the natural coordinate system (cf. Sect. 6.2.2); *4,* drawing of tangent curves; *5,* projection of cell masses; *6,* transformation of the curvilinear profile of the ventricular surface into a straight line; *7,* introduction of an orthogonal system of coordinates; *8,* transfer of the lines representing the ependymal surfaces of the sections to the coordinate system; *9,* connection of corresponding points; *10,* collection of additional data concerning the beginning and the end of ventricular sulci and cell masses, and completion of the reconstruction. **b** A more detailed picture showing how data derived from a transverse section are transferred to a straight line (**a** is reproduced from Nieuwenhuys 1974, to which the reader is referred for details)

be obtained by projecting them back to the ventricular surface with the aid of the vectors of the natural coordinate system. This type of projection is used in the method known as topological analysis and has been applied to the brain stem of numerous non-mammalian vertebrates (Nieuwenhuys 1972, 1974; cf. Chap. 4, Sect. 4.6.5; Fig. 6.10) and to the telencephalon of teleosts (Braford 1995). In more complex structures, such as the thick-walled portions of the diencephalon and telencephalon of amniotes, ontogenetic studies are indispensable for a reliable determination of the primary position of their constituent cell masses. For example, the origin of the mammalian prepiriform cortex (and thus its primary position) has long puzzled neuromorphologists. Rose (1935) believed that this cortical field, which in adult mammals is far removed from the lateral ventricle, shares its matrix formation with the corpus striatum, which is why he assigned this cortex to what he called the cortex striaticus or the cortex semiparietinus. However, the investigations of Smart and Sturrock (1979) and Misson et al. (1991) have shown that the matrix formation of the prepiriform cortex is situated dorsal to that of the striatum, and that the neuroblasts produced there attain their destination by a highly remarkable long, deflected radial migration (cf. Chap. 5, Sect. 6.3.3 and Chap. 5, Fig. 5.13). Thus, as shown in Fig. 6.11, the amphibian primordium piriform, the reptilian lateral cortex and the mammalian prepiriform cortex all can be traced back to sectors of ventricular matrix occupying corresponding topological positions.

The derivatives of particular matrix areas can often be homologised only as fields. Some sectors of the ventricular matrix give rise to many different cell masses. It has already been discussed (Chap. 5, Sect. 6.4) that certain *Grundgebiete* in the brains of amniotes give rise to several layers of migrated cells, that some of these layers may split up into two or even three sublayers, and that all of these layers

and sublayers may fractionate into two or more separate cell masses. The ontogenetic events in such a *Grundgebiet* are diagrammatically illustrated in Fig. 6.12. If we imagine now that the configurations depicted in Fig. 6.12b–f represent not only ontogenetic stages leading to the configuration shown in Fig. 6.12g but, in addition, the configurations as encountered in adult species related to the bearer of the condition in g, then the cell mass m1e in species f would be homologous as a field to the cell masses *a*, *b* and *c* in species g, the cell layer m1 in species d would be homologous as a field to layers m1e and m1i in species f, and so on, and finally, the single periventricular zone of species b would be homologous as a field to the sum of all layers and cell masses observed in each of the species c–g.

It is clear that in situations comparable to the one just discussed the study of ontogenetic material will be necessary for the establishment of the homology of the various cell masses. However, it may be added that if such material is not available, analysis of the fibre connections of these cell masses may well yield important clues with regard to their homology.

6.2.6.3
First Auxiliary Criterion:
Similarity in Fibre Connections

The value of fibre connections as clues to the homology of cell masses has been the subject of considerable controversy; in fact, the spectrum of opinions concerning this issue ranges from "no value whatsoever" to "by far the most important criterion". In what follows the views of a number of authors on the value and validity of connectional evidence for the institution of homologies will be summed up first. Then a brief commentary will be attached to some of these views, and finally I will intimate my own opinion on this vexed issue.

During the first half of this century, Charles Judson Herrick was the undisputed leader of the 'American School of Comparative Neuroanatomy'. This school pursued the creation of a functional neuroanatomy, in which all structures should bear functional names (cf. Chap. 4, Sect. 4.6.2; Sect. 6.3.5). Within the framework of this research program, analysis of fibre connections was considered to be essential for the disclosure of the functional significance of cell masses and other grisea. Herrick himself spent over 50 years carrying out a meticulous analysis of the fibre connections in the brain of the tiger salamander, *Ambystoma tigrinum*, and some related species (summarised in Herrick 1948). This analysis was based on a huge amount of Golgi material and on non-experimental, silver-impreg-

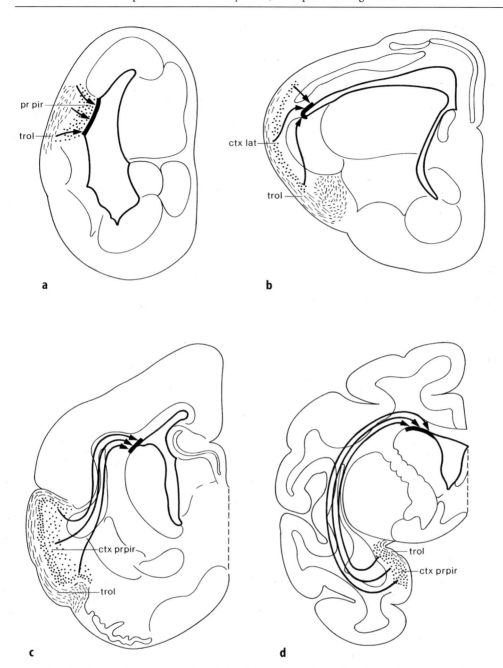

Fig. 6.11a–d. The same transverse sections through the telencephalon of (**a**) the tiger salamander, (**b**) the tegu lizard, (**c**) the opossum and (**d**) the rhesus monkey as shown in Fig. 6.3. The end stations of the tractus olfactorius lateralis are projected back to their sites of origin at the ventricular surface

nated series of larval, juvenile and adult specimens. His primary interest in this study of the salamander brain was to inquire into "the origins of the structural features and physiological capacities of the human brain and the general principles in accordance with which these have been developed in the course of vertebrate evolution" (Herrick 1948, p. 4). In 1909 Herrick published a short paper on 'The

Criteria of Homology in the Peripheral Nervous System'. He concluded this study as follows (Herrick 1909, p. 209): "The principles outlined above for guidance in determining homologies in the peripheral nervous system can be applied, *mutatis mutandis* to tracts within the central nervous system. It will not be necessary to make the application here in detail. The treatment of the grey nuclei and

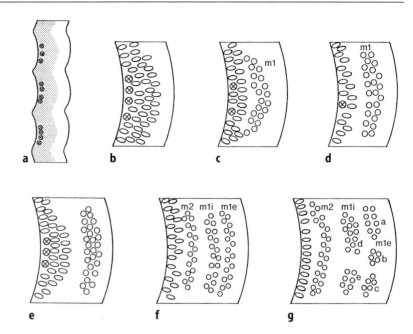

Fig. 6.12a–g. Development of brain nuclei according to Bergquist and Källén. **a** Postneuromeres or transverse bands. **b** Local proliferative activity leads to the formation of a migration area. **c, d** Neuroblasts produced within a migration area form a separate migration layer, *m1*. **e** A second wave of mitotic activity leads to the production of numerous new neuroblasts. **f** The new neuroblasts aggregate in another migration layer, *m2*, while the first migration layer splits up into separate external and internal zones: *m1e, m1i.* **g** The layers *m1e* and *m1i* rearrange themselves and form a number of cell concentrations: *a-e*, which represent definitive brain nuclei (based on Källén 1951, Bergquist 1953, 1954, and Bergquist and Källén 1954)

correlation centers will also be controlled by similar rules, the aim being to homologize only such structures as are genetically related and to use functional connections wherever possible as guides to homology."

In the introductory part of his summarising treatise on the brain of the tiger salamander, Herrick (1948, p. 18) emphasised that the mammalian names which he applied to structures in the brain of that species rarely implied exact homology: "... these areas are to be regarded as primordia from which the designated mammalian structures have been differentiated. The relationships here implied have been established by several independent lines of evidence: 1) The relative positions and fibrous connections of cellular masses and the terminal connections of tracts. In so far as these arrangements conform with the mammalian pattern, they may be regarded as homologous. 2) Embryological evidence. The early neural tubes of amphibians and mammals are similar, and subsequent development of both has been recorded. ... 3) The relationships of supposed primordia of mammalian structures may be tested by the comparative method. In an arrangement of animal types which approximates the phylogenetic sequence from the most generalized amphibians to man, there are many instances of progressive differentiation of amphibian primordia by successive increments up to the definitive human form."

During the period 1920–1960, Kuhlenbeck published a long series of comparative neuroanatomical studies aimed at disclosing the morphological pattern of the diencephalon and telencephalon (cf. Chap. 4, Sect. 4.6.6 for references and details). Within the framework of his 'rigorous formanalytic approach,' Kuhlenbeck homologised cell masses exclusively on the basis of their relative positions, excluding fibre connections categorically as clues to homology, because of their relation to functional systems: "Zur Homologisierung bestimmter, in sich cytoarchitektonisch einheitlicher Zellmassen ist es erforderlich, die Lage dieser Zellmassen im Bauplan zu erkennen. Nur auf Grund dieses Kriteriums und zwar ohne jede Rücksicht auf etwaige Strukturunterschiede oder Faserverbindungen ist eine Homologie feststellbar" (Kuhlenbeck 1929, p. 49). "Grisea, welche gleichen Wandabschnitten eines Grundbestandteils entstammen, sind homolog, unabhängig von ihrer Struktur, von ihren sekundären Lagebeziehungen und von ihren Faserverbindungen, d.h. ihrer Einbeziehung in funktionelle Systeme" (Kuhlenbeck 1933, p. 309).

During about the same period the comparative neuroembryological studies of Holmgren and his followers appeared (cf. Chap. 4, Sects. 4.5.2, 4.6.4). These workers studied the development of various parts of the brain in serial sections from closely graded series of embryos of a considerable number of species. With the aid of this material they sought to establish the relationship between the cell masses in the adult brain and the first signs of differentiation in the wall of the neural tube, i.e. the formation of *Grundgebiete* or 'migration areas'. As discussed in the previous section, the members of the 'Holmgren School' considered the primary position of the anlagen of neural cell masses decisive for their homology; similarity in fibre connections was considered to be a less important criterion: "The preponderant number of conclusions about homology of various nuclei and portions of the vertebrate brain are inferences founded on similar fibre connections of the nuclei. When a nucleus in a lower animal has been proved to have the same or similar connections as a nucleus in a higher animal, the conclusion drawn therefrom was that these nuclei are homologous. Very often little or no regard has been paid to the morphological position of the nuclei, and still less to their ontogenetic formation. In the very idea of homology there is a claim that the homologous portions should be of the same genetic origin. In order to decide if two nuclei, or portions of the brain in two separate types of animals are homologous or not, it is, above all, necessary to investigate, if possible, the embryological origin of these nuclei. If later it can be shown that the nuclei in question have the same fibre connections, then a valuable support is naturally given to the conclusion. Fibre connections as decisive factors must be a secondary consideration. In those cases when two brain portions plainly prove to be of the same genetic origin, the nuclei are homologous even if the fibre connections are not precisely the same. The latter are subject to variations in the animal series. Connections existing in a lower animal may be lacking in a higher, while, on the other hand, here new connections may have arisen. The fibre connections of a nucleus may well be comprehended as an expression of its function in one case or the other. But the function of the nuclei has nothing to do with their homology, that being a purely morphological conception" (Palmgren 1921, pp. 2–3).

Holmgren (1922, p. 393) discussed the homologies of the various pallial formations in the vertebrate telencephalon. After an exposé of Edinger's subdivision of the pallium in palaeencephalic and neencephalic parts, which was based on the presence or absence of olfactory fibres, respectively, he continued: "I cannot accept this view, since it supposes that a brain nucleus would not be able to change its connections without changing its morphological value. It is a well-known matter of fact that the connections of a nerve-nucleus can change in different closely related species (for instance, the nucleus rotundus in bony fishes). But none will for that reason declare that this nucleus is not a homologous one."

A few years later, in an extensive paper devoted to the development of the reptilian and mammalian telencephalon, Holmgren (1925, p. 454) stated: "The connections of a morphological brain unit may shift, but the unit does not therefore change in

another unit. Consequently, I do not from a comparative anatomical point of view lay stress upon fibre connections as decisive arguments by homologizing brain areas. They may be of great importance as supplementary arguments, but not as decisive".

The work of Kuhlenbeck, as well as that of the 'Swedish school', was criticised by Herrick. Concerning Kuhlenbeck's (1921a–c, 1922) earlier form-analytic studies on the telencephalon, he remarked trenchantly (Herrick 1922, pp. 199–200): "Kuhlenbeck in a series of papers has developed a very original scheme of forebrain morphology which is carried through the vertebrate series from cyclostomes to man... . This scheme is based chiefly on topographic relations with total neglect or misapprehension of the fibrous connections of the parts and a failure to grasp essential fundamental morphologic relations in lower forms."

In 1932, Bergquist, one of Holmgren's students, published a thorough and able study on the development of cell masses in the diencephalon of anamniotes. Herrick's (1933) commentary to that study has already been quoted extensively in Chap. 4, Sect. 4.5.2. Suffice it to repeat here some sentences from that commentary (Herrick 1933, p. 243): "Will the descriptive account of the origin, proliferation and migration of cellular elements, and this alone, yield adequate or an intelligible scheme of cerebral organization, for morphology or anything else? Nerve fibres are quite as important as cell bodies in cerebral organization and as elements in cerebral forms. By what right does the morphologist ignore them in his study of form?"

The great work on the comparative anatomy of the nervous system of vertebrates of Ariëns Kappers, Huber and Crosby (1936) does not contain a chapter on the principles of neuromorphology. However, in the section dealing with the telencephalon of myxinoids it is stated that, when it comes to the establishment of homologies, fibre connections are "always more significant guides than positional relations" (Ariëns Kappers et al. 1936, p. 1255).

Nieuwenhuys and Bodenheimer (1966) compared the diencephalon of the primitive brachiopterygian fish *Polypterus* with that of other anamniotes with the aim of testing the validity of the three criteria, i.e. ventricular grooves, nuclear boundaries and fibre connections, which are usually employed as a clue to identification of structures in that brain part. On the basis of the following arguments, Nieuwenhuys and Bodenheimer (1966, p. 436) concluded that fibre systems are of little value for the establishment of homologies in the anamniote diencephalon: "1) Fiber systems are defined by their origin and termination. Since these origins and terminations are actually nuclear masses, we are defining fibers by the nuclei they interconnect. If we then turn around and homologize nuclear masses on the basis of the fibers to which they are connected we must be aware that such circular reasoning may lead to erroneous conclusions. If, however, we define a fiber tract by a characteristic topographical position of one segment in the tract, then we cannot be certain whether tracts with corresponding topographical positions at that segment actually go to nuclei with any similarity of morphological position. For instance, fibers in the lateral part of the telencephalic peduncle of amphibians have different origins and terminations from fibers in a similar place in actinopterygians... . 2) A more practical restriction of fiber tracts as a basis for homologization is that many cell masses have no well-defined fiber projection, but only diffuse connections, as e.g. the periventricular thalamic regions and the nucleus medianus tuberculi posterioris of *Polypterus*." It should be emphasised that the study by Nieuwenhuys and Bodenheimer was based exclusively on non-experimental material.

The neuroanatomist C.B.G. Campbell and the ethologist W. Hodos devoted a series of papers to the concept of homology in neurobiology (Campbell and Hodos 1970; Campbell 1976, 1982, 1987; Hodos 1974, 1976). With regard to the value of fibre systems between cell masses as a criterion for homology, Campbell and Hodos (1970, p. 362) stated: "The objections of Nieuwenhuys and Bodenheimer (1966) to the use of fiber connections between cell masses as a criterion of homology seems to be based on: 1) an artificial dichotomy between nuclear groups and their fiber connections, and 2) failure to take into account the full potential of the Nauta technique and its variations ... to determine the origin and termination of fibers in thecentral nervous system. We quite agree that defining a fiber tract by a characteristic topographic position without knowledge of the source and destination of the fibers within it may lead to an erroneous conclusion as to what that tract is. However, it is possible, using these modern experimental methods, to determine the source and destination of fibers quite accurately." They argued that the CNS is composed of "a number of subsystems having characteristic relationships with each other and normally functioning together in an integrated manner so that the regulatory and other responses of the system are adaptive (or at least not nonadaptive) for the organism. These subsystems are composed of populations of neurons having characteristic relations with other subsystems. Such relations between and within cell populations are transacted by the various cell processes. The char-

acteristic of interconnectedness is the essence of a nervous system. Therefore, we feel that the afferent and efferent connections of a given cell population must be taken into account when determining to which specific subsystem the cell population may belong" (Campbell and Hodos 1970, p. 362). This same point is reemphasised at the end of the article (Campbell and Hodos 1970, p. 365): "We will not see this path [i.e. the path of evolution] unless we recognize that cell groups in the CNS and their fiber connections constitute functional subsystems, each with an evolutionary history that has been shaped by various selective pressures, and are not merely isolated morphological entities." In accordance with these statements, Campbell and Hodos considered similarity in fibre connections an important criterion for the establishment of homologies. They felt, however, that this criterion should be used in combination with other evidence: "In summary, the following types of data seem to us to be the most useful in order to establish homologies in the central nervous system: 1) experimentally determined fiber connections 2) topology 3) topography 4) the position of reliably occurring sulci 5) embryology 6) morphology of individual neurons 7) histochemistry 8) electrophysiology 9) behavioral changes resulting from stimulation, lesions, etc. The greater the degree of concordance among the characters, the stronger becomes the justification for drawing the inference that structures in two different species may have been derived from corresponding structures in a common ancestor" (Campbell and Hodos 1970, p. 364).

Similar thoughts are expressed in the other papers cited above. The suggestion to include the role of neural structures in the physiological or behavioural functions of organisms as a criterion for their homology is repeated later by Campbell (1987). Hodos (1974, p. 22) asserted that "one cannot deny the existence of homologous behavior since we know that behavior is a character of organisms that is responsive to the pressure of natural selection and survival." However, since behavior does not exist independently of structure, in his opinion "the concept of behavioral homology is totally dependent on the concept of structural homology" (Hodos 1976, p. 165). Accordingly, he offered the following definition of behavioural homology: "Behaviors are considered homologous to the extent that they can be related to specific structures that can, in principle, be traced back through a genealogical series to a stipulated ancestral precursor irrespective of morphological similarity" (Hodos 1976, p. 156).

Some comments on this variegated set of opinions follow. Herrick was perfectly right in stating that fibre connections as elements in cerebral form cannot be ignored by morphologists. It follows, then, that Kuhlenbeck's categorical exclusion of fibre connections as clues to the homology of cell masses was unjustified.

Campbell and Hodos correctly stated that the dichotomy between nuclear groups and their fibre connections is artificial. However, it is a matter of fact that, due to the dominance of techniques in neuroanatomy, cytoarchitectonics, myeloarchitectonics and experimental hodology have become separate subdisciplines.

It is hard to disagree with Campbell and Hodos' statement that the origin and destination of fibre connections can generally be determined more reliably and more accurately with experimental techniques than with non-experimental techniques. This holds, a fortiori, for the modern tracing techniques, which by now have entirely replaced Nauta's axon degeneration technique (cf. Chap. 7).

Again, Campbell and Hodos' contentions that the characteristic of interconnectedness is the essence of a nervous system, and that it is a major task of comparative neuroanatomy to track the functional subsystems of the brain and to study the evolutionary history of these subsystems are undisputable (cf. Sect. 6.3.5). It should be kept in mind, however, that homology is a morphological concept. To quote de Beer (1971, p. 3) once again: "An organ is homologous with another because of what it *is*, not because of what it *does*." Cell masses, fibre connections and other delimitable entities in the CNS can be placed into two different contexts: (a) the purely formal context and (b) the functional and behavioural context. The second context is irrelevant for the establishment of morphological homologies. The concept of homology has been fruitfully applied to patterns of innate behaviour. However, these patterns belong to another level of organisation of an organism than its morphology; hence behavioural and morphological homologies cannot be lumped together (Striedter and Northcutt 1991; cf. Sect. 6.2.5.2; Fig. 6.6). It follows that data derived from electrophysiology, or behavioural changes resulting from stimulation or lesion experiments, are not relevant for the establishment of morphological homologies, and that a judgement of the validity of fibre connections as clues to the homology of cell masses has to be based on the morphological and not on the functional properties of these structures.

It stands to reason that the use of fibre connections for the homologisation of neural cell masses should be preceded by the homologisation of these fibre connections themselves. A comparison with the muscular system is relevant here. Comparative

experience has taught us that for the homologisation of muscles neither their relation to bones nor their relation to one another, but rather their innervation is the most useful criterion (cf., e.g., Appleton 1928; Jones 1979). This, however, "... implies the introduction of another parallel, but independent conception – that of the homologization of the nerves themselves" (Appleton 1928, p. 366). The requirement of an independent conception of, or methodology for, their homologisation cannot be fulfilled for the fibre connections in the CNS. These structures are homologised on the basis of their sites of origin and termination. So, Nieuwenhuys and Bodenheimer correctly stated that a *petitio principii* or a vicious circle results if, in turn, we base the homologies of these sites of origin and termination upon the homology of fibre connections.

Nevertheless, there are situations in which fibre connections may provide important clues to the identification of cell masses, as the following two examples may show. Firstly, as stated already in Chap. 3, Sect. 3.6.4, in cases where the topological relations of a given cell mass are obscure, a fibre tract which clearly projects to or originates from that cell mass, whereas its opposite pole is connected with a griseum, the identity of which is obvious and undisputed, may allow us to propose a hypothesis regarding the homology of that enigmatic cell mass.

Secondly, experimental hodological studies may assist in or even lead to the detection and identification of cell masses which are cytoarchitectonically not sufficiently individualised. For example, following injection of retrograde tracers into the cerebellum of anurans, a cluster of labelled cells can be observed in the periventricular grey of the contralateral lower rhombencephalon, just ventral to the hypoglossal nucleus (Cochran and Hackett 1977; van der Linden and ten Donkelaar 1987). Given the facts that (a) in amniotes a cell mass occupying a corresponding topological position known as the inferior olive is present, and that (b) this cell mass in the group mentioned projects to the contralateral cerebellum, the anuran cell mass can be looked upon as the primordium of the amniote inferior olive.

The presupposition that fibre systems are invariant with regard to their sites of origin and termination is implicit to their use as criteria for the homologisation of cell masses. Hence, the question arises whether this presupposition is correct. With Northcutt (1984, p. 70), I am of the opinion that: "Many, if not most, neural pathways appear to be very stable phylogenetically, and the majority of these pathways appear to have arisen with the origin of vertebrates or, shortly after, with the origin of jawed vertebrates." The exceptions to this rule are not negligible, however. Cell masses may show considerable differences in their fibre connections, even in closely related species, as is exemplified by Striedter's (1992) experimental study on the connections of the lateral preglomerular nucleus in ostariophysean teleosts. Fibre systems originating in different groups of vertebrates from obviously homologous cell masses may terminate in obviously non-homologous cell masses. Thus, the ascending pathway which in all amniotes originates from the principal trigeminal sensory nucleus terminates in mammals in a diencephalic centre, i.e. the ventral posteromedial thalamic nucleus. However, in birds this pathway terminates in the evidently telencephalic nucleus basalis.

If we survey the facts and opinions discussed above, the following conclusions seem to be warranted:

1. A neural cell mass may change its connections without losing its morphological identity.
2. The homology of a cell mass can be confirmed by connectional evidence but should never be rejected solely on the basis of such evidence.
3. Similarity in fibre connections is an important auxiliary criterion for the establishment of the homology of cell masses.

6.2.6.4
Second Auxiliary Criterion: Similarity in Special Quality

This criterion is derived from the work of Remane (1956). It refers to highly specific morphological (e.g. the presence of a particular cell type) or histochemical features (e.g. the presence of a particular neurotransmitter, neuromodulator or enzyme). The use of histochemical and particularly immuno-histochemical techniques in neuroanatomy has enormously increased during the past several decades. This development has meant that the similarities in special quality revealed by these techniques are currently frequently used for the establishment or affirmation of homologies. On this account, it may be good to elucidate the strength and limitations of this criterion with a number of examples.

The Identification of the Avian and Reptilian Homologues of the Mammalian Caudate-Putamen Complex. The mammalian caudate-putamen complex or neostriatum is a conspicuous cell mass which, at rostral telencephalic levels, occupies a central position in the cerebral hemispheres (Fig. 6.13a). It was long thought that the array of dorso-ventrally arranged centres which occupy the thickened wall of the avian hemispheres represent the homologue of the mammalian caudate-putamen complex, hence they all received names ending in the suffix '-striatum' (Fig. 6.13b). In reptiles the ventrolateral walls of the cerebral hemispheres are likewise thickened and protrude into the lateral ventricles. In the older literature (summarised by Ariëns Kap-

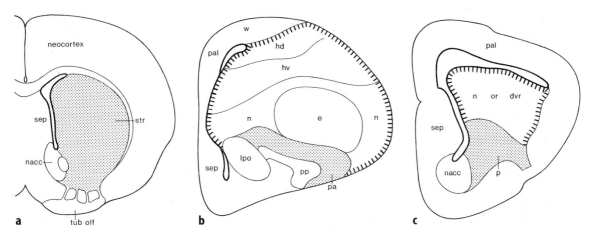

Fig. 6.13a–c. Transverse hemisections through the rostral part of the telencephalon of the rat (**a**), the pigeon (**b**) and a lizard (**c**). *Dotted area* represents the striatum according to current insights, based mainly on histochemical and immunohistochemical studies. *Dashes* in **b** and **c** indicate the dorsal extension of the striatum according to previous descriptions.

dvr, Dorsal ventricular ridge; *e,* ectostriatum; *hd,* hyperstriatum dorsale; *hv,* hyperstriatum ventrale; *lpo,* lobus perolfactorius; *p,* palaeostriatum; *pa,* palaeostriatum augmentatum; *pal,* pallium; *pp,* palaeostriatum primitivum; *sep,* septum; *str,* striatum; *tub olf,* tuberculum olfactorium; *w,* 'Wulst'

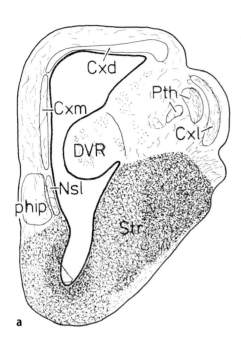
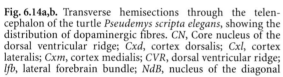
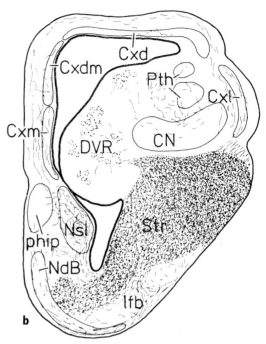

Fig. 6.14a,b. Transverse hemisections through the telencephalon of the turtle *Pseudemys scripta elegans*, showing the distribution of dopaminergic fibres. *CN*, Core nucleus of the dorsal ventricular ridge; *Cxd*, cortex dorsalis; *Cxl*, cortex lateralis; *Cxm*, cortex medialis; *CVR*, dorsal ventricular ridge; *lfb*, lateral forebrain bundle; *NdB*, nucleus of the diagonal band of Broca; *Nsl*, nucleus septi lateralis; *Nsm*, nucleus septi medialis; *Pth*, pallial thickening; *phip*, primordium hippocampi; *Str*, striatum. It is important to note that the ventromedial extension of the field of dopamine fibres largely coincides with the unlabelled nucleus accumbens (reproduced from Smeets et al. 1987)

pers et al. 1936) these intraventricular protrusions were interpreted in toto as the corpus striatum. They are composed of a smaller, ventral part, which has received a variety of names, including paleostriatum, nucleus basalis and ventrolateral area, and a dorsal part, known as the neostriatum or the dorsal ventricular ridge (Fig. 6.13c).

The results of histochemical and immunohistochemical studies have dramatically changed our views on the extent of the avian and reptilian striatum. Taking as criteria (a) a high succinic dehydrogenase activity (Baker-Cohen 1968), (b) an intensely positive reaction for AChE (Karten 1969; Nauta and Karten 1970; Karten and Dubbeldam 1973), and (c) the presence of a rich catecholamine plexus (Juorio and Vogt 1967; Parent and Olivier 1970; Karten and Dubbeldam 1973), which in the light of immunohistochemical studies may be interpreted as consisting of dopaminergic fibres and terminals (Smeets 1988, 1994; Smeets et al. 1987 Fig. 6.14; Reiner 1994; Reiner et al. 1994; Wynne and Güntürkün 1995), it appeared that only the paleostriatum of reptiles and the paleostriatum augmentatum of birds are homologous to the caudate-putamen complex of mammals.

The strongly positive AChE reaction reflects the presence of intrinsic striatal cholinergic neurons (Butcher and Woolf 1982; Fibiger 1982; Hoogland and Vermeulen-van der Zee 1990; Henselmans and Wouterlood 1994). The striatal dopamine, on the other hand, is extrinsic and contained in terminals of striatal afferent fibres originating from the substantia nigra (Brauth and Kitt 1980; Fallon and Moore 1978). The nigrostriatal dopaminergic projection is reciprocated by a striatonigral projection, numerous fibres of which contain substance P (Reiner et al. 1984; Anderson and Reiner 1991; Smeets 1991).

Finally, note that the nucleus accumbens, a cell mass situated in the rostral, basomedial part of the telencephalon, closely resembles the striatum structurally as well as chemoarchitectonically (Figs. 16.3, 6.14; Chap. 2, Fig. 2.59a). In the avian brain the nucleus accumbens forms part of the lobus parolfactorius (Kitt and Brauth 1981; Fig. 6.13b). The nucleus accumbens shows, like the striatum proper, strong AChE activity and a dense plexus of dopaminergic fibres. The latter arise mainly from a separate set of mesencephalic dopaminergic neurons, designated as A10 and located in the area tegmentalis of Tsai.

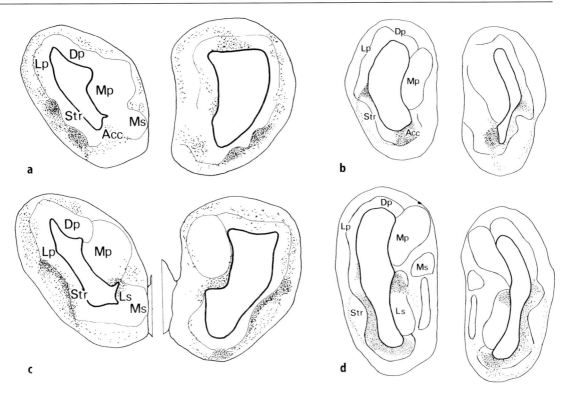

Fig. 6.15a–d. Transverse sections through the rostral part of the telencephalon of the newt, *Pleurodeles waltlii* (**a**,*c*) and the frog *Rana ridibunda* (**b**,**d**), showing the distribution of immunoreactive fibres. Tyrosine hydroxylase-immunoreactive fibres are shown on the *left*; dopamine-immunoreactive fibres are shown on the *right*. *Acc*, Nucleus accumbens; *Dp*, dorsal pallium; *Lp*, lateral pallium; *Ls*, lateral septum; *Mp*, medial pallium; *Ms*, medial septum; *Str*, striatum (reproduced from González and Smeets 1991)

The Localisation of the Striatum and Nucleus accumbens in Amphibians. It is generally agreed that in amphibians the striatum or striatal primordium is formed by a zone of periventricular grey situated in the ventrolateral part of the telencephalic hemispheres, and that the nucleus accumbens consists of an adjacent area of periventricular grey which, in the rostral part of the hemispheres, arches around the ventral angle of the lateral ventricle (Herrick 1948; Hoffman 1963; Kicliter and Ebbesson 1976; Chap. 2, Fig. 2.28). This interpretation rests almost exclusively on topological grounds; however, it is corroborated by the fact that in anurans the striatum and the nucleus accumbens both show high AChE activity (Northcutt 1974). Immunohistochemical studies have shown that the centres just mentioned and the adjacent neuropil also contain a plexus of dopaminergic fibres. Interestingly, in urodeles the strongest dopamine innervation is found in the striatum, while in anurans the nucleus accumbens contains the densest plexus of dopamine-immunoreactive fibres. In the latter group the striatum is innervated only moderately by dopamine-immunoreactive fibres, and in the most rostral part of the telencephalon this innervation is almost lacking (González and Smeets 1991; González et al. 1994; Fig. 6.15). These findings do not necessitate changes in the morphological interpretation of the cell masses mentioned. The fact that the special quality 'presence of a dense plexus of dopaminergic fibres' does not hold for certain parts of the anuran striatum does not challenge the morphological status of that structure. Rather, it indicates that this particular special quality, when applied to the identification of the amphibian striatum and nucleus accumbens, does not have the status of a defining character. However, the differences in dopaminergic innervation, localisation and size of the cell groups giving rise to this innervation do indicate that the organisation of the *functional complex*, formed by striatum, nucleus accumbens and the dopaminergic neurons in the caudal diencephalon and the tegmentum of the midbrain, in urodeles differs markedly from that in anurans (González and Smeets 1991; González et al. 1994).

The Localisation of the Striatum in Some Groups of Fish. Many uncertainties exist concerning the interpretation of the parts of the telencephalon of fishes.

Northcutt and colleagues made an interesting attempt to define the major subdivisions of the telencephalon and to determine the possible homologues of these subdivisions in other vertebrates with the aid of immunohistochemical techniques. The distributions of several neuropeptides, among them enkephalin (ENK) and substance P (SP), a neurotransmitter (serotonin: 5HT), and a neurotransmitter-related enzyme that is involved in catecholamine synthesis (tyrosine hydroxylase: TH) were examined in the shark *Squalus acanthias* (Northcutt et al. 1988), the lungfish *Protopterus annectens* (Reiner and Northcutt 1987), and the Senegal bichir, *Polypterus senegalus* (Reiner and Northcutt 1992). What follows will be confined to the localisation of the corpus striatum. Northcutt et al. (1988) pointed out that in *Squalus*, a periventricular area situated in the basal part of the hemisphere, named area periventricularis ventrolateralis (APVL), has histochemical features that are most similar to those of the striatum of amniotes: ENK-positive and SP-positive neurons and fibres and TH-positive and 5HT-positive fibres (Fig. 6.16A 1–4). Furthermore, a prominent SP-positive fibre bundle was traced from the APVL to the tuberculum posterius and tegmentum mesencephali, and this bundle appeared to be reciprocated by ascending TH-positive, probably dopaminergic fibres. Given the fact that the APVL occupies a topological position corresponding to that of the striatum of tetrapods, the interpretation presented by Northcutt et al. (1988) is most probably correct. Northcutt et al. (1988) also thought that a conspicuous zone of migrated cells, known as area superficialis basalis (ASB), exhibits histochemical similarities to the globus pallidus of amniotes. In my opinion, this cell zone also belongs to the striatum. Histochemically, it hardly differs from the APVL and it has been shown to contain a rather dense plexus of dopamine-immunoreactive fibres in the sandy ray, *Raja radiata* (Meredith and Smeets 1987).

In *Polypterus* the dorsal part of the subpallium contains a periventricular cell group, named dorsal nucleus of the ventral telencephalic area (Vd), which is also typically present in all actinopterygians (Nieuwenhuys 1963, 1966; Nieuwenhuys and Meek 1990a). According to Reiner and Northcutt (1992), the following features of the labelling patterns obtained suggested that Vd is the homologue of the striatum of cartilaginous fishes, dipnoans and amniotes (Fig. 6.16B1–4): (a) an abundance of ENK-positive and SP-positive fibres; (b) an abundance of TH-positive fibres, possibly dopaminergic and of posterior tubercle/tegmental origin; (c) an abundance of 5HT fibres; (d) the presence of an SP-positive fibre bundle that descends from basal tel-

encephalic levels to terminate in the posterior tubercle/tegmentum. I concur with this interpretation, although some of the histochemical features (presence of plexuses of enkephalinergic and serotoninergic fibres) are by no means specific for Vd (cf. Fig. 6.16B1–4). Reiner and Northcutt (1992) also pointed out that the holostean and teleostean cell mass Vd, which on positional and structural grounds may be homologissed with Vd in *Polypterus*, also shares a number of histochemical features with the latter.

Reiner and Northcutt (1987) divided the subpallium of the lungfish *Protopterus* into a medial (Sm) and a lateral part (Sl). They interpreted Sl as the homologue of the striatum and the nucleus accumbens of amniotes, on the basis of the following immunohistochemical characteristics: (a) the presence of SP-positive neurons and fibres, ENK-positive neurons and fibres, and TH-positive and 5HT-positive fibres; (b) Sl appears to give rise to a descending SP-positive striatonigral pathway and appears to receive a, presumably dopaminergic, reciprocal nigrostriatal input (Fig. 6.16C1–4). This interpretation is most probably correct. However, I believe that the nucleus intercalatus of Reiner and Northcutt (1987) and the adjacent rather narrow zone of subpallial periventricular grey specifically represent the striatum of tetrapods (cf. Nieuwenhuys and Hickey 1965; Nieuwenhuys 1969; Nieuwenhuys and Meek 1990b; Chap. 16). This area corresponds positionally and structurally to the amphibian striatum (Chap. 2, Fig. 2.28). Moreover, it is specifically related to the large striatotegmental tract.

Fig. 6.16. Transverse (hemi)sections through the mid- ▶ telencephalon of the shark *Squalus acanthias* (*A1–4*), the Senegal bichir, *Polypterus senegalus* (*B1–4*), and the African lungfish, *Protopterus annectens* (*C1–4*), illustrating the distribution of perikarya (*triangles*) and fibres/terminals (*dots*) containing enkephalin (*ENK*), substance P (*SP*), tyrosine hydroxylase (*TH*) and serotonin (*5-HT*). High-contrast photomicrographs of Nissl-stained sections are shown in the *left hand column*. A, Nucleus A (*Squalus*); *APVL*, area periventricularis ventrolateralis; *ASB*, area superficialis basalis; *DP*, dorsal pallium; *DPs*, dorsal pallium, pars superficialis; *IFB*, interstitial nucleus of basal forebrain bundle; *IN*, intercalated nucleus; *LP*, lateral pallium; *LPl*, lateral pallium, pars lateralis; *MP*, medial pallium; *P*, nucleus P (*Squalus*); *P1-d*, dorsal part of the first pallial zone; *P1-v*, ventral part of the first pallial zone; *P2*, second pallial zone; *P3*, third pallial zone; *pd*, pars dorsalis of medial pallium; *pi*, pars intermedia of the medial pallium; *pv*, pars ventralis of the medial pallium; *S*, septal nucleus; *Sl*, lateral subpallium; *Sm*, medial subpallium; *Vd*, dorsal nucleus of the ventral telencephalic area; *Vl*, lateral nucleus of ventral telencephalic area; *Vn*, another nucleus of ventral telencephalic area; *Vv*, ventral nucleus of ventral telencephalic area (reproduced from Northcutt et al. 1988, Fig. A1–4; Reiner and Northcutt 1992, Fig. B1–4; Reiner and Northcutt 1987, Fig. C1–4)

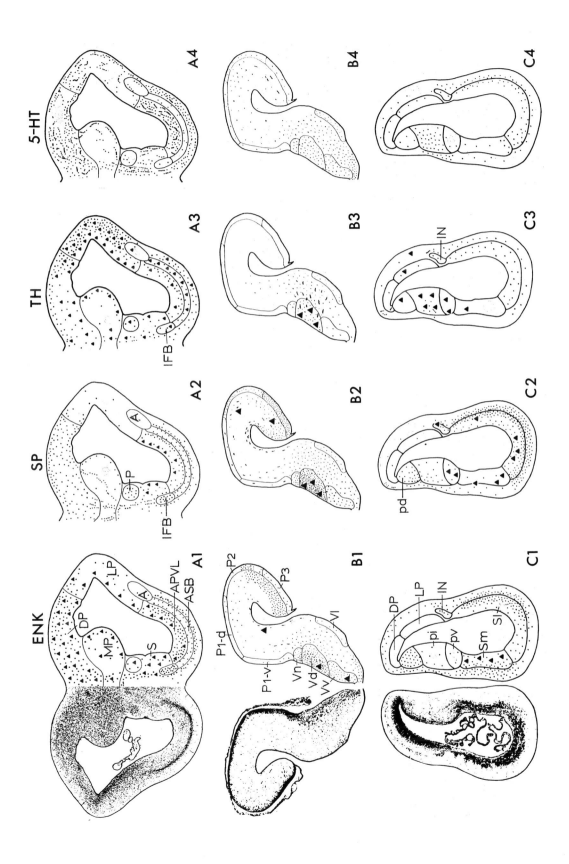

From the foregoing it appears that in the telencephalon of chondrichthyans, cladistians and dipnoans the distribution of biogenic amines and neuropeptides can aid considerably in defining cellular aggregates and their morphological identity.

The Localisation of the Striatum in Teleosts. The teleostean subpallial area Vd (Chap. 2, Fig. 2.69a), on account of its topological position, can be homologised with the similarly designated cell group in *Polypterus* (Fig. 6.16B). Because the latter cell group is homologous to the striatum in sharks, dipnoans and tetrapods, the teleostean area Vd also represents a striatal primordium. Reiner and Northcutt (1992) have adduced immunohistochemical evidence in support of this homology, such as (a) presence of dopaminergic/TH-positive and 5HT-positive fibres and (b) abundance of SP-positive and ENK-positive perikarya and fibres. Meek (1994) pointed out, however, that among teleosts there is a large interspecific variability with regard to the dopaminergic innervation of the forebrain, and that Vd does not belong to the most heavily dopamine-innervated regions (Chap. 2, Fig. 2.69). Another intriguing observation mentioned by Meek (1994) is the absence in teleosts of dopaminergic midbrain neurons comparable to those of the substantia nigra and/or ventral tegmental area. In my opinion, these observations do not challenge the fact that in the teleostean telencephalon an entity is present which is morphologically equivalent to the striatum of other fish and tetrapods; however, they do indicate that a functional complex comprising a set of telencephalic centres and a set of mesencephalic centres which are reciprocally connected by a strong substance P-containing descending projection and an equally strong ascending dopaminergic projection may have changed considerably in the course of actinopterygian evolution.

The Interpretation of the Dorsomedial Wall of the Telencephalic Hemispheres of Lungfishes. The dorsomedial part of the dipnoan cerebral hemispheres has been interpreted in two divergent ways. One group of investigators, among them Elliot Smith (1908), Holmgren (1922), Holmgren and van der Horst (1925), Rudebeck (1945), Nieuwenhuys and Hickey (1965) and Nieuwenhuys (1969), thought that this part of the dipnoan telencephalon is subpallial and represents the dorsal part of the septum, but another group, including Schnitzlein and Crosby (1967), Clairambault and Capanna (1973), Northcutt (1986a, 1995) and Reiner and Northcutt (1987), held that it is of a pallial nature and represents the medial pallium or primordium hippocampi of amphibians. The authors last mentioned

believed that the area in question is divisible into three dorsoventrally arranged subareas (Fig. 6.16c1–4). In order to resolve this question, von Bartheld et al. (1990) compared morphological, hodological and histochemical parameters of the African lungfish *Protopterus* and the Australian lungfish *Neoceratodus* with those of the amphibian species *Ambystoma* and *Xenopus*. They concluded that the dorsomedial telencephalon of lungfishes represents a subpallial, but not a medial pallial structure. This conclusion was based on striking

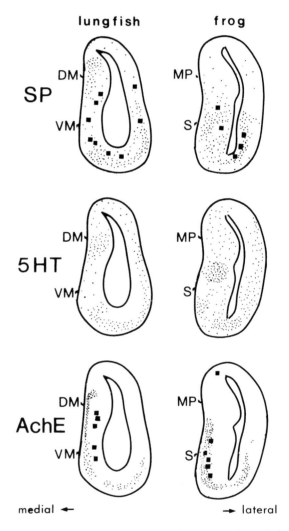

Fig. 6.17. Transverse sections through the right telencephalic hemispheres of a lepidosirenid lungfish (*left*) and a frog (*right*), to show the striking topographical differences in the distribution of some chemical markers, which support the view that the dorsomedial part of the dipnoan hemisphere corresponds to the dorsal part of the septal region in frogs. *Solid squares,* Immunoreactive cell bodies; *dots,* fibres and terminals; *AChE,* acetylcholinesterase; *DM,* dorsomedial telencephalon; *5HT,* serotonin (5-hydroxytryptamine); *MP,* medial pallium; *S,* septum; *SP,* substance P; *VM,* ventromedial telencephalon (reproduced from von Bartheld et al. 1990)

differences in the patterns of distribution of a number of chemical markers in the medial wall of the telencephalon of lungfishes as compared with amphibians (Fig. 6.17).

The Interpretation of the Central Nucleus of the Prosencephalon of Myxinoids. The brain of myxinoid cyclostomes differs considerably from that of other vertebrates. During ontogenesis many cells migrate away from the ventricular surface, and late in development an extreme reduction of the ventricular system takes place, with the consequent loss of ventricular landmarks (Conel 1929, 1931). In the adult stage the prosencephalon consists of a solid, rostrocaudally compressed body, within which numerous more or less distinct cell masses can be distinguished. The superficial zone of the telencephalon proper is occupied by a beltlike laminated structure, which is generally considered to represent the pallium (Fig. 6.18). The core of the solid prosencephalon is occupied by a conspicuous cell mass, given by Wicht and Northcutt (1992, 1994) the neutral name nucleus centralis prosencephali (Nc). It comprises three parts, dorsal (d), medial (m) and ventrolateral (vl), the last of which is by far the largest (Fig. 6.18). This deeply buried cell mass has puzzled many investigators, and widely divergent names and interpretations have been given to it, as may appear from the following survey.

Holm (1901) termed the central prosencephalic cell mass 'nucleus rotundus' and considered it comparable to the nucleus rotundus of other fishes, which apparently referred to a diencephalic cell mass of teleosts, now called nucleus glomerulosus. Edinger (1906) emphasised that this cell mass is situated caudal to the epithelial vestiges of the lateral ventricle and concluded that, on account of its position, it represents the *Thalamusganglien* of other vertebrates.

Holmgren (1919) advanced the theory that in the forebrain of myxinoids during ontogenesis a 'hyperinversion' occurs by which the homologue of the primordium hippocampi of lampreys, as delineated by Johnston (1912), rolls inward and attains a central position in the forebrain. Holmgren emphasised, however, that he considered Johnston's hippocampal primordium of the lamprey, as well as the central nucleus of myxinoids, a diencephalic structure. The ontogenetic studies of Conel (1929, 1931) have conclusively shown that Holmgren's (1919) hyperinversion theory is incorrect.

Jansen (1930) concurred with Holmgren (1919) that the central nucleus is homologous to the primordium hippocampi of lampreys, which he, however, regarded as a telencephalic structure. He

stated that his findings concerning the fibre connections of the central nucleus are in harmony with this interpretation, and that this interpretation is also supported by the embryological development of myxinoids. The latter statement is surprising, because Jansen (1930) was acquainted with the work of Conel (1929, 1931; cf. Wicht and Northcutt 1992). Jansen (1930, p. 475) concluded: "The term primordium hippocampi is, therefore, retained. It refers to a region which I regard as being the matrix of the future hippocampal and possibly general cortex."

Ariëns Kappers et al. (1936, p. 1255) briefly reviewed the various interpretations given to the central nucleus up to that time. They subscribed to Jansen's interpretation, adding an important parenthetic clause: "Recently Jansen called attention to its connections – always more significant guides than positional relations – and showed that, judged by such criteria, the region is quite probably to be regarded as the forerunner of the hippocampus of higher forms."

In 1946 Holmgren published a voluminous study on the development of the brain of the Atlantic hagfish, *Myxine glutinosa*, in which he made no mention of his hyperinversion theory of 1919, and in which he confirmed Conel's (1929, 1931) observations concerning the fate of the ventricular system. In a paper devoted to the comparative anatomy of *Myxine*, Crosby and Schnitzlein (1974) labelled the central nucleus as thalamus dorsalis, without presenting any explanation for this interpretation.

During the period 1974–1983, a number of histochemical studies on the brain of cyclostomes appeared (Wächtler 1974, 1975, 1983; Kusunoki et al. 1981). In all of these publications it is repeated that the central nucleus shows high AChE activity and resembles in this respect the striatum of the lamprey (Chap. 2, Fig. 2.57b: CS) and of gnathostomes, rather than the primordium hippocampi of the latter.

Wicht and Northcutt (1992, 1994) recently subjected the forebrain of the Pacific hagfish, *Eptatretus stouti*, to a detailed morphological and immunohistochemical analysis. Taking the results of the ontogenetic studies of Conel (1929, 1931) into consideration, they concluded that the central prosencephalic nucleus (a name coined by them) represents a diencephalic structure. Wicht and Northcutt conjectured that the central nucleus originates on the extreme lateral wall of the forebrain vesicle and, developing in a mediorostral direction, goes on to protrude into the ventricular space. The lateral ventricles become closed as the ventricular surfaces of the olfactory bulbs and the anterior (ventricular) surfaces of the central prosencephalic nuclei meet.

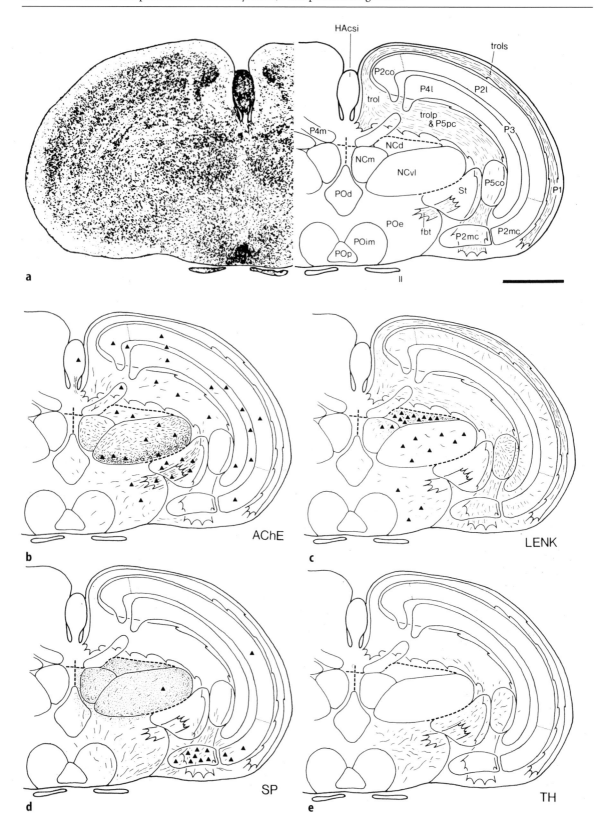

◀ **Fig. 6.18a–e.** Transverse sections through the prosencephalon of the Pacific hagfish, *Eptatretus stouti*, at the level of entrance of the optic nerve. **a** High-contrast photomicrograph of a cresyl-violet-stained section (*left*) and a mirror-image line drawing (*right*), showing the cytoarchitecture. The shading (*fine lines*) highlights the major fibre pathways (*areas with irregular outlines*) in this frame only. In **b–e**, *fine lines* and *dots* represent immuno-labelled fibres and terminals, respectively; *triangles* represent labelled perikarya. The *heavy dashed line* in **a–e** indicates vestigial parts of the forebrain ventricular system. **b** Distribution of acetylcholinesterase; **c** distribution of leu-enkephalin; **d** distribution of substance P; **e** distribution of tyrosine hydroxylase*fbt*, Fasciculus basalis telencephali; *HAcsi*, corpus sinistrum habenulae; *NC*, nucleus centralis prosencephali; *NCd*, NC, pars dorsalis; *NCm*, NC, pars medialis; *NCvl*, NC, pars ventrolateralis; *St*, striatum. *Scale bar* = 0.5 mm . (reproduced from Wicht and Northcutt 1994)

The remnants of the lateral ventricles, already observed by Edinger (1906), and indicated by dashed lines in Fig. 6.18, bears witness to this developmental process. Having established that the central nucleus is topologically a rostral extension of the diencephalon, Wicht and Northcutt (1994) attempted to establish the homology of this and of other enigmatic cell masses in the forebrain of the hagfish by studying its chemoarchitecture. It is important to note that Wicht and Northcutt (1992) interpreted a part of the nucleus olfactorius anterior of previous authors as being the striatum, mainly because they found that the basal forebrain bundle entertains a special relation with this area (Fig. 6.18).

Within the framework of their chemoarchitectonic analysis, Wicht and Northcutt (1994) localised AChE, enkephalin, substance P, tyrosine hydroxylase and a-melanocyte-stimulating hormone by means of (immuno)histochemistry (Fig. 6.18). They concluded that there are very few chemoarchitectural similarities shared by hagfishes and other craniates, and consequently a comparative chemoarchitectural analysis proved to be of limited value in revealing homologies among cell groups of hagfishes, lampreys and gnathostomes. With regard to the central nucleus, Wicht and Northcutt (1994) found that the chemical profile of its dorsal subnucleus differs considerably from that of the medial and ventrolateral subnuclei. The medial and ventrolateral subnuclei appeared to have the highest AChE content of any forebrain area and these nuclei also showed a relatively large number of leucine-enkephalin-labelled cells and fibres, as well as dense substance P-positive terminal formations (Fig. 6.18). Wicht and Northcutt (1994) admit that these are all histochemical properties characterising the striatum of other craniates. However, they reject this homology on the following topological, connectional and histochemical grounds (Wicht

and Northcutt 1994, p. 156): "The striatum in craniates typically occupies the ventrolateral hemispheric wall ventrolateral to the lateral ventricles, whereas the medial and ventrolateral subnuclei of the central prosencephalic nucleus are located posterior to the (vestigial) lateral ventricles in hagfishes (Wicht and Northcutt 1992). Furthermore, there is no connection between these nuclei and the basal forebrain bundle; there seem to be no major ascending thalamic projections, and there are practically no catecholaminergic inputs to these nuclei, all characters one would expect to find in a typical striatum."

Wicht and Northcutt (1994) also consider the possibility that the medial and ventrolateral subnuclei of the central prosencephalic nucleus together represent the medial pallium or hippocampal formation, as suggested by Jansen (1930), but they reject this homology for the following reasons (Wicht and Northcutt 1994, p. 156): "There is compelling embryological and topological evidence that these nuclei are diencephalic in nature (Edinger 1906; Holmgren 1946; Wicht and Northcutt 1992) and the *combined* presence of high amounts of AChE, LENK, SP and 5HT makes this hypothesis even more unlikely."

Finally, it may be mentioned that Wicht and Northcutt (1994) reported that their identification of the myxinoid striatum, as presented in 1992, is to a degree supported by histochemical data, such as the presence of large neurons which stain intensively for AChE and of a TH-positive, probably dopaminergic, innervation (Fig. 6.18).

These six examples indicate the value of chemoarchitectonic analyses for the solution of questions of homology. In general, it may be stated that: (a) in cases of obscure topological relations histochemical data, and particularly the *combined* presence of certain neurochemicals, may give important clues as to the homology of grisea; (b) the homology of a griseum, established on topological and hodological grounds, may be confirmed by histochemical data; and (c) a homology can never be definitively rejected solely on the basis of histochemical data.

6.2.6.5
Third Auxiliary Criterion: Continuity of Similarity Through Intermediate Species

This criterion is, like the preceding one, derived from the work of Remane (1956). In morphology it is used to detect homologies of structures which differ in position and special quality, as well as to exclude homologies of structures of similar form and position. For example, the skull of a hyena and that of a gorilla both bear a median crest of similar shape and position. However, because taxonomically intermediate forms are lacking, these structures are to be considered not homologous, but rather parallel homoplastic.

In comparative neuroanatomy the 'continuance criterion' is of particular importance for the discrimination between homologous and parallel homoplastic structures. Analyses based on large numbers of species, carefully selected on the basis of their mutual taxonomic relationships, have revealed that a considerable number of neural structures which were formerly considered to be homologous are in fact homoplastic (Northcutt 1981, 1984; Sects. 6.2.4 and 6.3.3). For example, in many anamniotic vertebrates including lampreys, cartilaginous fishes, chondrosteans, crossopterygians, lungfishes, apodans and salamanders, an electrosensory system is present. In all of these groups the fibres which connect the peripheral electroreceptors with the CNS enter the brain via a separate dorsal root of the anterior lateral line nerve and terminate in a cell mass known as the dorsal octavolateral nucleus. As its name indicates, this nucleus is situated in the most dorsal part of the rhombencephalic alar plate. A few teleost groups (Ictaluridae, Notopteridae, Gymnotidae, Mormyridae) are also electroreceptive. In these fish the electrosensory fibres also enter the rhombencephalon and terminate in the so-called electroreceptive lateral line lobe (ELLL), a structure which differs slightly in its topological position from the dorsal octavolateral nucleus in other anamniotes. The fact that electroreceptors and a dorsal octavolateral nucleus are entirely lacking in holosteans, i.e. the group of actinopterygian fishes standing taxonomically between the chondrosteans and the teleosts, renders it likely that the dorsal octavolateral nucleus found in most anamniote groups and the teleostean ELLL are homoplastic rather than homologous. Similarly, the taxonomic relationships within the large group of teleosts strongly suggest that the ELLL of notopterids and mormyrids is homoplastic with that of ictalurids and gymnotids (McCormick 1982; Northcutt 1984; Chap. 15).

6.2.6.6
Summary and Concluding Remarks

Cell masses in the CNS which occupy a corresponding relative or topological position are homologous or, in some cases, parallel homoplastic. For the establishment of the homology of cell masses it is often necessary to take their primary topological position, that is, the position of their anlagen at the embryonic ventricular surface, rather than their

secondary position in the adult stage into consideration.

Topological correspondence is to be considered the principal homology criterion because of its sine-qua-non character. Neural cell masses which do not occupy corresponding topological positions are non-homologous. A cell mass in the CNS of a given vertebrate is often not strictly homologous to a single cell mass in another vertebrate, but rather homologous as a field to two or more cell masses in that other vertebrate, or vice versa.

Similarity in fibre connections and similarity in special qualities are important auxiliary criteria for the establishment of neural cell masses. The strength of these criteria is that they are *multiplex;* cell masses in different vertebrates may have several or even many fibre connections in common, and the number of special qualities – particularly (immuno)histochemical characteristics – shared by such cell masses is potentially unlimited. Nevertheless, these criteria are to be ranked as auxiliary criteria, because homologies can never be rejected on the basis of absence of the similarities specified.

The criterion of continuity of similarity through intermediate species is of particular importance for distinguishing homologous cell masses from homoplastic ones.

6.3
The Programme of Comparative Neuroanatomy

6.3.1
The Central Aim of Comparative Neuroanatomy

The central aim of comparative neuroanatomy is the elucidation and interpretation of the form, structure and ultrastructure of the CNS and its parts. The ensuing elementary survey of biological relationships and interactions may serve to bring the scope of this definition into focus.

Living animals are embedded in and form part of a complex environment. The environment may be defined as the totality of extrinsic organisms, things, factors and conditions affecting an organism. A distinction can be made between the abiotic and the biotic environment of an organism. The abiotic environment includes such essential factors as water, air and solar radiation. The biotic environment encompasses the conspecifics with which a population or community is formed, heterospecifics – among them potential preys and predators – and the entire vegetable environment.

Animals entertain mutual and reciprocal relations with their environment. Much of the interaction with their abiotic and biotic environment manifests itself in patterns of motivated or goal-oriented behaviour. These behavioural patterns may be subdivided into two broad classes, one related primarily to survival of the individual, the other to survival of the species as a whole. Exploratory, ingestive, thermoregulatory, and agonistic behaviours are fundamental to the survival of an individual, whereas reproductive (sexual and parental) and social behaviours are necessary for survival of the species (Swanson 1989). All of these behaviours, except for exploratory behaviour, are specifically directed to one or a few components of the environment.

The organs and parts of a living animal entertain structural and functional relations with each other, and it is these mutual relations which render it possible for the animal to function as an organism, as a whole. A living animal is adapted to its environment, which means that the intrinsic harmony of its organs and parts continues into an extrinsic harmony.

Discussing the relations between animals and their environments, Sewertzoff (1931, pp. 124–133) made a distinction between ectosomatic organs, i.e. organs which entertain a direct functional or biological relation to conditions of the external world (*Bedingungen der Außenwelt*), and endosomatic organs, which lack such direct relations to the environment. The ectosomatic organs of vertebrates include the skin and dermal skeleton, the fins of fishes and extremities of tetrapods, the simple and special sense organs and the digestive system including its glands. Prominent among the endosomatic organs are the heart, the endocrine glands and the CNS. Sewertzoff emphasised that the endosomatic organs are 'adapted', just like the ectosomatic ones, but that their adaptations are directly related not to the conditions of the external world, but rather to the structure and function of ectosomatic organs and other endosomatic organs.

The CNS fulfils a unique role in the organism, in that it steers and governs all other parts of the body. Cuvier, the noted French anatomist, emphasised this unique role by observing as early as 1812: "– le système nerveux est au fond tout l'animal; les autres systèmes ne sont là que pour le servir où pour l'entretenir." (Cuvier 1812, p. 76)

An essential feature of the CNS as an organ is its division into numerous more or less discrete cell groups that are interconnected by axons to form a circuit system of tremendous intricacy. Within this circuit, systems subserving particular sensory and motor functions can be distinguished, and considerable progress has been made during the past several decades in clarifying the neural circuitry underlying the various motivated behaviours, at least in the rat (cf., e.g., Swanson 1987, 1989). These behavioural systems appear to be hierarchically organised, and thus may be envisioned as central extensions of the network of functional interrelationships outlined above. For example, the central neural mechanism subserving defensive behaviour encompasses the telencephalic prefrontal cortex, hippocampus, amygdala and septal region, the diencephalic medial hypothalamus, the mesencephalic periaqueductal grey and the dorsomedial part of the rhombencephalic reticular formation (Swanson 1989).

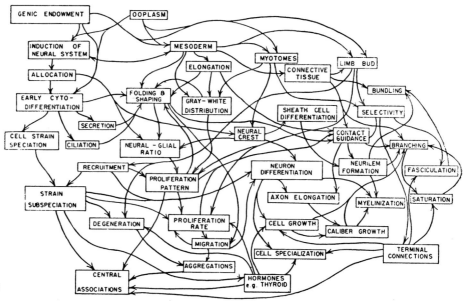

Fig. 6.19. Causal relations in neurogenesis (reproduced from Weiss 1955, Fig. 14)

The adult CNS is the end product of a developmental process during which (a) the neural plate is induced from the adjacent mesoderm, and (b) this apparently simple plate of neuroectoderm is transformed into an ultracomplex information-processing system. Experimental neuroembryological studies (cf. Weiss 1955, Jacobson 1991 for reviews) have revealed that an intricate network of causal factors and relations is involved in neurogenesis (Fig. 6.19). Major aspects of the development of the CNS include (a) the patterned generation of neurons and glial elements, (b) the formation of specific nuclei or areas, each with their own complement of specific neuronal phenotypes, (c) the formation of specific neural pathways, and (d) the establishment of specific synaptic connections.

The ontogenesis of the CNS occurs under genetic control, and many processes taking place in the adult CNS are also influenced by the genome. Throughout development there is interaction between the genome and the internal state of the developing cells, and the latter is in turn influenced by neighbouring elements. This means that the development of the CNS is not fully determined by the genetic code; epigenetic cellular interactions also play an important role in shaping the organ.

Our knowledge concerning the modes in which the genome influences the developing and the adult CNS, though still limited, has increased significantly during the past decade. As regards ontogenesis, it appears that an interrelated group of so-called homeotic genes display precise spatiotemporal patterns of expression in the developing CNS and, hence, may well be involved in establishing the identity of particular brain regions. Homeotic genes are expressed in various combinations in rhombomeres (Chap. 4, Sect. 4.5.5.5), whereas the expression patterns of such genes in the forebrain have been reported to correlate with the boundaries of major domains, such as the thalamus, the striatum and the cerebral cortex (Chap. 5, Sect. 5.5.3). It seems likely that homeotic genes fulfil different functions in different phases of development. Thus, in the early neuroepithelium they may help define specific regions, but during later development they may be involved in specifying particular neuronal phenotypes. The proteins encoded by homeotic genes are transcription factors that regulate the expression of large numbers of target genes, some belonging themselves to the homeotic category. Indeed, evidence is accumulating that an intricate hierarchical network of regulator genes is responsible for brain development (Alvarez-Boledo et al. 1995). Probably, each transcription regulator is involved in the differentiation of numerous neurons, and multiple transcription regulators participate in the specification of each individual neuron.

The establishment of specific synaptic connections is a complex process involving numerous developmental steps. It is known that late in development the as yet immature synapses are subjected to an activity-dependent modelling or 'fine tuning'. A group of genes denoted as 'immediate-early genes' are involved in this maturation process. The same genes operate again during learning, which may be considered as an ongoing developmental process (Greenspan and Tully 1994). It is generally assumed that during learning the 'weight' of certain synapses, and therewith the flow of information through the circuitry in which the pertinent neurons are embedded, is permanently changed.

Finally, it may be mentioned that the genome is directly involved in the regulation of the functional state of the neuronal networks which form the morphological substrate of the various motivated or goal-oriented behaviours. These networks contain numerous peptidergic neurons, most of which produce several (up to ten) different neuropeptides. The expression of the genes encoding for these peptides is regulated by transcription factors, the activity of which is strongly influenced by circulating hormones. There is evidence that by such hormonal influences the ratios of the various neuropeptides released from the terminals of the pertinent neurons may be selectively altered. By this 'biochemical switching' (Swanson 1989, 1991) the flow of information through the circuitry associated with motivated behaviours may be considerably changed. It may be concluded that the endocrinon exerts its influence on behaviour by way of two interacting networks, a genomic one and a neuronal one (Greenspan and Tully 1994).

The relations discussed so far are diagrammatically illustrated in Fig. 6.20. It may be concluded that living organisms are composed of highly interactive systems, acting at many levels, extending from the genomic 'level' to the 'level' of the environment.

Living organisms form part of a community, a population and ultimately, a species. A species is a group of interbreeding natural populations that is reproductively isolated from other such groups (Mayr and Ashlock 1991, p. 26). The members of a species, though closely similar in many respects, are not identical. They differ with respect to both their phenotypes and their genotypes. The species is a fundamental biological unit characterised and determined by a common gene pool.

In some comparative biological studies numerous specimens of a species are examined (for instance to analyse the influence of domestication, or to determine the average relative brain weight of

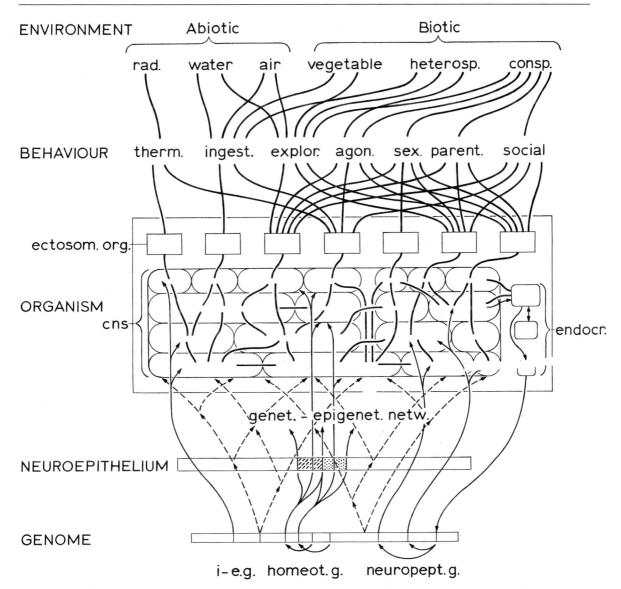

Fig. 6.20. A living organism envisioned as a continuous network of relations extending from its genome to its environment. Three types of genes, immediate-early (*i-e.g.*), homeotic (*homeot.g.*) and neuropeptide (*neuropept.g.*), exemplify the genome. Two homeotic genes show a segment-restricted pattern of expression in the embryonic neuroepithelium. The organism is composed of ectosomatic and endosomatic organs; the central nervous system (*cns*) and the endocrine glands (*endocr*) belong to the latter category. Under the heading *BEHAVIOUR* the following classes of motivated or goal-oriented behaviour are indicated: *thermo*regulatory, *ingest*ive, *explor*atory, *agon*istic, *sex*ual, *parent*al and *social*. The abiotic environment encompasses, apart from water and air, solar radiation (*rad*), and the biotic environment includes, apart from plants, conspecific (*consp.*) and heterospecific (*heterosp.*) animals

that species), but most of these studies are directed at the 'supraspecies level', which means that attributes or characters of specimens belonging to two or more different species are compared (Fig. 6.21). Prior to 1859, the justification for such comparisons was sought in affinities to be traced back either to the Creator or to a realm of platonic ideas (van der Klaauw 1947; cf. Sect. 6.2.5.4). The theory of evolution offers a natural explanation for the similarities observed in taxonomically related species. These similarities are based on consanguinity. In its most simple form: Species B and C have descended from a common ancestral species A. The diagram depicted centrally in Fig. 6.21 indicates these phylogenetic relationships. Bifurcations of this type form the elementary units in 'family trees' or cladograms of natural groups. They reflect in an extremely simplified way the process of speciation, that is, the process by which a single complex of individuals, interconnected by genealogical relationships, is

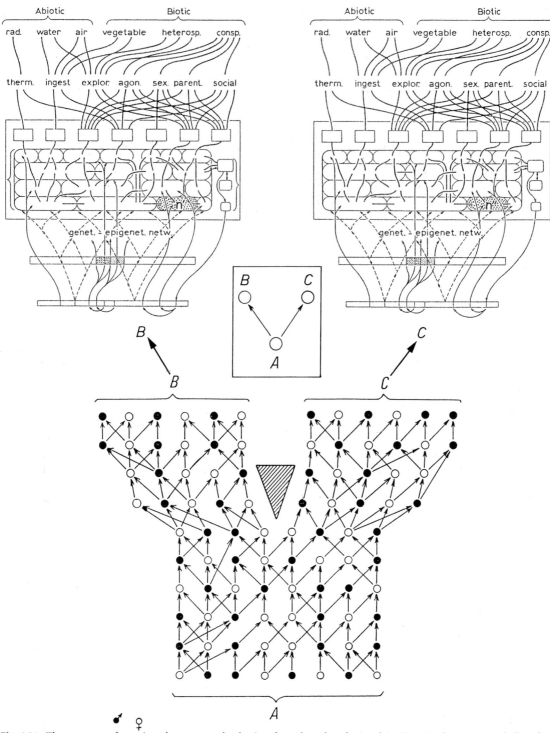

Fig. 6.21. The process of species cleavage as the basis of diversification. *Lower half* of the figure (reproduced from Hennig 1966, Fig. 4) shows the genealogical and therewith the phylogenetic relationships between the individuals of species *B* and *C* with the individuals of the common stem species *A*. The diagrams forming the *upper half* of the figure are identical to that depicted in Fig. 6.20, but now symbolise the networks of relations at the species level. The gene pools of species *B* and *C* form the link between the lower and the upper parts of the figure. A cell mass *n'* in species *B* and its homologue *n"* in species *C* are *stippled*

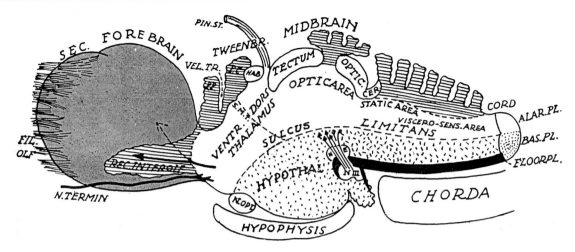

Fig. 6.22. The brain of a hypothetical lamprey-like fish. (reproduced from Ariëns Kappers 1929, Fig. 45)

split up into two separate complexes (Hennig 1966; Fig. 6.21, lower part). This means that, if the sets of interactions and interrelationships depicted in the upper half of Fig. 6.21 are taken as representing species B and C, the gene pools of these two species are the products of the interactions shown in the lower part of that figure.

According to the current neo-Darwinian theory, an intricate web of the following factors is considered to be involved in the process of speciation and in evolution in general: (a) the continuing changes in the external and internal abiotic and biotic environment of populations, (b) the existing variety in the genetic make-up of individuals belonging to a particular species, (c) the continuous production of new variations through random mutation, (d) the sexual reproduction resulting in individuals with new combinations of genes, (e) the enormous reproductive capacity of species, (f) the process of natural selection acting on the different capacities of individuals of species to survive, and (g) the effect of all these processes culminating in an uneven, non-random contribution of the potential parents to next generations (Osse 1983).

As indicated at the outset, the elementary survey of current biology just presented was necessary to bring the scope of our definition of the central aim of comparative neuroanatomy, i.e. *the elucidation and interpretation of the form, structure and ultrastructure of the CNS and its parts*, into focus. It will be clear from our considerations that the entire biotic world represents a continuous spatiotemporal network of relations and interactions. This means that entities within this network (populations, organisms, central nervous systems, neural cell masses, neurons, synapses, ion channels) can be interpreted only in terms of their intrinsic and extrinsic rela-

tions and interactions. For instance, the interpretation of cell mass n^1 in the CNS of species B (Fig. 6.21) includes:

1. Qualitative and quantitative analysis of its constituent neurons and their intrinsic relations
2. Determination of its topological and hodological relationships
3. Determination of the functional, behavioural and ecological networks in which this cell mass is embedded
4. Analysis of its ontogenetic history
5. Determination of the genomic influences involved in its formation
6. Placing of all of the desiderata enumerated under 1–5 in a comparative perspective

This means, in its most simple form, that cell mass n^1 is compared with its homologue n^{11} in a closely related species C (Fig. 6.21). If cell mass n^1 differs typically (standing for: above the level of 'noise') in any of its attributes from cell mass n^{11}, it may be assumed in light of the theory of evolution that cell masses n^1 and n^{11} not only are different, but *have become* different from their common ancestral cell mass n. Moreover, it may be assumed that the change is either directly or indirectly related to an adaptation to a new situation, and that the mechanism involved is the tandem dualism of genetic variability and natural selection (Mayr 1969). Reference to Fig. 6.21 shows that evolutionary change, often denoted simply as the result of interactions between genetic and environmental factors, is in reality an ultracomplex process, and that genetic mutations that effect an evolutionary change in adult characters, morphological or behavioural, do so indirectly by influencing the process of development. "Evolutionary change must in a sense be fun-

neled through the developmental mechanism"
(Sperry 1958, p. 128).

From the foregoing it may be clear that the complete interpretation of a CNS or any of its parts is not realisable. For this reason the term elucidation, meant as anything conducive to interpretation, has been included in our definition of the central aim of comparative neuroanatomy. Somewhat arbitrarily, within the field of comparative neuroanatomy the following domains of elucidation, or subdisciplines, may be distinguished: pure neuromorphology, phylogenetic neuromorphology, comparative neuroembryology and genetic neuromorphology, functional neuromorphology and evolutionary morphology. These subdisciplines will now be discussed.

6.3.2
Pure Neuromorphology

The establishment of homologies is the central aim of pure neuromorphology. It is the specific contribution of neuromorphology to neurobiology. The establishment of homologies is not an esoteric goal in itself, but rather an activity of eminent practical value: Structures need to be named, and their names should express their homology with corresponding structures in other, related species (Owen 1866, p. xii). In comparative neuroanatomy (and in biology in general: cf. Remane 1956, p. 59) we are far removed from this ideal situation. It has already been indicated in Sect. 6.2.2 that the prevailing habit of labelling neural structures on the basis of their topographical position (e.g. nucleus dorsomedialis anterior) is inadequate to express morphological correspondence, let alone homology. Homologous structures may occupy quite different topographical positions even in closely related species, and the same holds true for the same structure at different ontogenetic stages of the same species. Thus, in discussing the development of the human brain, His (1888, p. 350) stated: "Die Adjectiva dorsal und ventral oder medial und lateral lassen uns bei den stattfindenden Verschiebungen der einzelnen Zonenabschnitte im Stich." A logically consistent topological nomenclature could be based on a procedure involving the following three steps: First, the various axial curvatures are disposed of and the brain is transformed into a straight and stretched organ: the neuraxis; then, the results of the various special morphogenetic events such as evagination or eversion are disposed of and the early embryonic tubelike condition is restored; finally, the tube is opened along its dorsomedian seam and the walls are folded laterally. The topographical terms referring to the orthogonal, Cartesian coordinate system

could be adequately used for such stretched and unfolded models of the brains of all vertebrate species. However, the introduction of such a new terminology would have great practical drawbacks. Provisional, so-called neutral names should be introduced only if the homology of a given structure cannot be established with reasonable certainty.

There is nothing to be said against the use of function-denoting names in comparative neuroanatomy, on the condition that all structures thus named are strictly homologous (cf. Sect. 6.3.4). Bulbus olfactorius is a fully adequate and unambiguous term. The name tectum opticum is acceptable, because in most vertebrates retinofugal, optic fibres constitute by far the largest input to the structure denoted as such. However, in all vertebrates it receives, apart from visual fibres, projections related to other sensory modalities (somatosensory, acoustic, lateral line). In mormyrid and gymnotid teleosts the electrosensory lateral line projections originating from the torus semicircularis may be more massive than those from the retina, and in myxinoids, cave teleosts such as *Astyanax mexicanus jordani*, the urodele *Proteus anguineus* and the Gymnophiones the eyes and the retinotectal projections are strongly reduced. For all of these reasons the term tectum mesencephali is preferable to tectum opticum. Finally, it may be mentioned that the original name of the nuclear complex now generally denoted as thalamus was thalamus opticus. The adjective opticus was dropped because it appeared to be untenable in light of experimental hodological findings.

Comparison of homologous structures stands central in (neuro)morphology. Wherever possible, such comparisons should be placed in a phylogenetic (Sect. 6.3.3) or evolutionary context (Sect. 6.3.6). However, if the material studied does not justify such contexts, it is scientifically fully correct to arrange the structures compared in grades, i.e. according to attributes derived from these structures themselves, such as size or degree of complexity. Such comparisons should be specified as purely morphological, and the criterion or criteria used should always be made explicit.

6.3.3
Phylogenetic Neuromorphology

There is substantial, though largely circumstantial, evidence that biological entities are diverse because they have become diverse. Hence, it should be attempted to place the results of any neuromorphological study in a historical context. This goal has been pursued from the beginning of comparative neuroanatomical research. Within its confines it has been attempted (a) to determine the nature of the brain in the earliest vertebrates, (b) to reconstruct the phylogenetic history of the vertebrate brain, and (c) to derive generalisations – trends, rules, principles or laws – from that history. Cyclostomes represent the most primitive vertebrates, and because the brain of hagfishes shows many highly aberrant features, it was generally assumed that the brain of the earliest vertebrates must have been lamprey-like (Johnston 1902; Ariëns Kappers 1929, Fig. 6.22;Herrick 1948) and that the brain of the lamprey offers an optimal starting point for the study of the evolutionary development of the vertebrate brain (Nieuwenhuys 1977). Northcutt (1985, p. 81) criticised this approach by emphasising that both living taxa of cyclostomes "have a long evolutionary history during which many changes have obviously occurred."

The works by Ariëns Kappers and his allies (Ariëns Kappers 1920/1921, 1947; Ariëns Kappers 1929; Ariëns Kappers et al. 1936) may be considered attempts at reconstructing the phylogenetic history of the CNS of vertebrates. It is important to note that in these works this history is presented as a progressive process, thought to be driven by an undetermined intrinsic force. "In the nervous system, as in all living substance, there is an active striving, an inherent tendency, to supplement the activities and to elaborate the various impulses, which is recognizable chiefly through the end results, but which may be felt consciously as an indeterminate realization of the force of thought and will. The results of this entelechic tendency ... are seen in the progressive development of the brain in accordance with a general plan, in the progressive differentiation and adjustment of its constituents, and in their mutual general relations" (Ariëns Kappers et al. 1936, p. XII).

Edinger's 'palaeencephalon-neencephalon' theory, Herrick's 'invasion' theory, Ariëns Kappers' theory of neurobiotaxis, and Ebbesson's 'parcellation' theory may exemplify attempts at inferring phylogenetic generalisations from observed differences between the brains of vertebrates. According to Edinger (1908a,b), the brain of vertebrates can be subdivided into two fundamental functional units, the palaeencephalon and the neencephalon (Fig. 6.23). The palaeencephalon, which extends from the olfactory bulb to the spinal cord, is present, with all its character subdivisions, from cyclostomes to man. It is, as its name implies, the oldest part of the CNS, and cyclostomes and most fishes possess only this part. The activities which depend on this entity are common to all vertebrates and include locomotion, ingestion of food and reproduction. The palaeencephalon is the seat of all reflex mechanisms and of innate instinctive behaviour. Contrary to the palaeencephalon, the second principal unit, i.e. the neencephalon, shows a marked progressive evolutionary development. Beginning from modest vestiges in the chondrichthyan and amphibian pallium, it expands steadily to culminate in the human neocortex. The appearance and elaboration of this new entity involves marked changes in the behavioural repertoire, to the effect that the (palaeencephalic) reflexes and instincts become increasingly subordinated to (neencephalic) associative and intelligent actions.

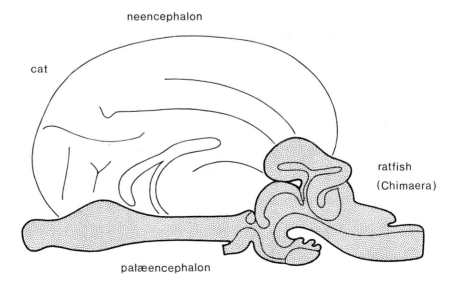

Fig. 6.23. A cat brain superimposed upon the brain of the holocephalian fish *Chimaera monstrosa*, to show the increase resulting from the addition of the neencephalon to the palaeencephalon. (redrawn from Edinger 1908b, Fig. 2)

"Nun muss als ein Hauptgewinn der bisherigen Studien ausgesprochen werden, dass der ganze Mechanismus vom Rückenmarkende bis zum Riechnerven bei allen hohen und niederen Vertebraten im Prinzipe überall ganz gleichartig angeordnet ist, dass also für die einfachsten Funktionen durch die ganze Reihe hindurch gleichartige Unterlagen bestehen. Erst mit dem Auftreten des Grosshirnes ändert sich das Verhältnis total. Jetzt, schon von den Selachiern, sicher von den Amphibien ab tritt ein Apparat auf, der sich über jenes Palaeencephalon schaltet. In kontinuierlicher Reihe nimmt er durch die Amphibien bis zu den Säugern immer mehr zu und in der Reihe der letzteren wächst er zu dem enormen Gebilde an, das beim Menschen Träger aller höheren seelischen Funktionen ist" (Edinger 1908a, pp. VIII-IX). "Nimmt man nun die Ausbildung des Neencephalons etwa zum Masse, so kann man in der Tat von höher und niedriger stehenden Gehirnen sprechen. Denn so gleichartig das Palaeencephalon und der beschriebene Gundapparat selbst bei den Säugern bleiben, so mannichfaltig gestaltet sich die Ausbildung der Pallialteile des Neencephalon" (Edinger 1908a, p. IX).

Flaws in Edinger's theory include (a) the 'linear' treatment of the vertebrate kingdom, (b) vagueness about the localisation and delineation of the neencephalic components in the brain of non-mammalian vertebrates, and (c) failure to include neocortical dependencies such as the pyramidal tracts and the cerebellar hemispheres into the neencephalon. Several other theories and concepts, such as Herrick's (1948, p. 64) concept of the expanding intermediate zone, MacLean's (1970) triune brain concept and Jerison's (1973) ideas concerning the progressive enlargement of the brain, are all related to Edinger's neencephalisation theory.

The invasion hypothesis of Herrick (1921, 1948), sometimes also designated as the 'smell-brain' theory, was eloquently summarised by that author as follows:

"The regional differentiation of the anatomically distinct centers of the entire endbrain behind the olfactory bulbs, therefore, primitively arose as a result of the invasion of the original secondary olfactory area by diverse non-olfactory systems, and the entire history of the subsequent evolutionary differentiation of this part of the brain can be written in terms of the interaction of these two systems of conduction fibers – those descending from the olfactory bulb and those ascending from the betweenbrain. ... The details of this dramatic history cannot be recounted here. In a general view of the process it may be said that in cyclostomes the entire endbrain and a large part of the betweenbrain are dominated by the olfactory system, the non-olfactory components entering this territory from the midbrain being relatively small and incompletely known. As we ascend the vertebrates scale the non-olfactory systems assume progressively greater importance. In urodeles a considerable part of the thalamus is devoted exclusively to non-olfactory correlations, but no part of the cerebral hemispheres is wholly free from olfactory connections. In reptiles the ascending systems are greatly enlarged and a portion of the corpus striatum complex appears to be devoted exclusively to them. Here there is well-defined cerebral cortex, most of which is clearly dominated by its olfactory connections (hippocampus and pyriform lobe), though in another part (the general cortex) somatic systems predominate. ... In mammals somatic systems with no admixture of olfactory elements come to dominate the architecture and functions of the cerebral hemispheres, until in man, whose olfactory organs are greatly reduced, the olfactory centers are crowded down into relatively obscure

crannies of the hemisphere by the overgrown somatic systems" (Herrick 1921, pp. 449-450).

Similar thoughts were expressed by Ariëns Kappers et al. (1936). Invasion of regions or cell groups by new projections has also been supposed to occur in other parts of the CNS. Thus, Noback and Shriver (1969) claimed that, from a certain phylogenetic stage onward, spinal lemniscal fibres, which originally did not extend beyond the reticular formation of the brain stem, started to invade the dorsal thalamus, and that, similarly but in the opposite direction, neocortical efferents, originally terminating only in the reticular formation, developed branches which descended to the spinal cord and thus established a direct corticospinal connection. The correctness of the application of the invasion hypothesis to the evolution of the telencephalon has been challenged by Northcutt (1981, 1984), who pointed out that in all of the numerous gnathostomes the secondary olfactory connections were studied with experimental tracing techniques; these projections appeared to spread over a relatively small telencephalic area. However, the theory has recently regained its validity with the findings that in the lamprey the secondary olfactory fibres extend over most of the telencephalon (Northcutt and Puzdrowski 1988; Polenova and Vesselkin 1993), and that in myxinoids practically the entire telencephalon is dominated by secondary olfactory projections (Wicht and Northcutt 1993; Fig. 6.26).

The theory of neurobiotaxis, put forth by Ariëns Kappers (e.g. 1929), states that during phylogeny the outgrowth of dendrites and the migration of cell bodies belonging to certain centres are influenced by electric, stimulative forces, selectively exerted by other, functionally related, centres. For a discussion and evaluation of this theory the reader is referred to Chap. 2, Sect. 2.4.7.

The parcellation theory, enunciated by Ebbesson (1980, 1984), proposed the opposite of the invasion hypothesis discussed above. According to the parcellation theory, the major trend in brain evolution is not that fibres establish new connections by invading territories they did not innervate previously, but rather that neuronal populations split up in such a fashion that the brains of the oldest and most primitive vertebrates contained mainly diffuse grisea giving rise to widely overlapping connections, and that in the course of evolution brains possessing more discrete cell masses and more restricted fibre connections gradually evolved. A comment on this theory is given below.

A rigorous and potentially very powerful methodology for determining how brains (or parts thereof) have changed during evolution has been

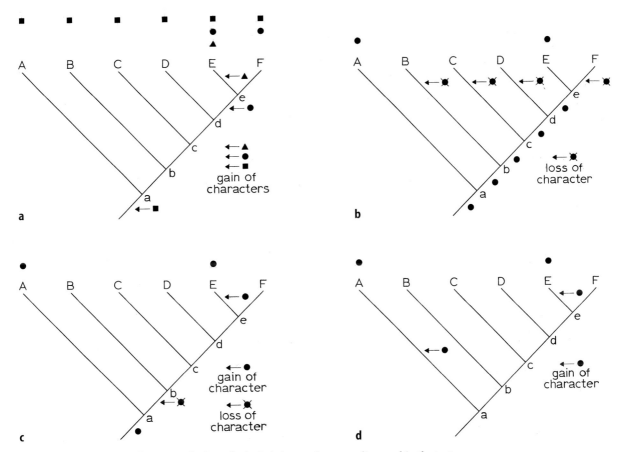

Fig. 6.24a–d. Cladograms illustrating the hypothetical phylogenetic events discussed in the text

developed by Glenn Northcutt and his associates (e.g., Northcutt 1984, 1986a,b, 1995; Wulliman and Northcutt 1990; Wulliman and Roth 1994; Striedter 1991, 1992; Striedter and Northcutt 1991; Butler 1994a,b, 1995). This methodology, which will be called here *neurocladistics*, because it is derived from the field of phylogenetic systematics or cladistics (Hennig 1966; Wiley 1981; Forey 1990), may be outlined as follows (Fig. 6.24). A set of characters is drawn from a particular neural complex in a smaller or larger group of interrelated vertebrate taxa. Knowledge concerning the phyletic interrelationships of the taxa involved in the inquiry is expressed in a cladogram, i.e. a branching scheme involving only bifurcations. Each end segment of the cladogram represents one or several monophyletic groups of extant species from which the selected characters are derived. These end segments are the terminal taxa of the cladistic analysis (A–F in Fig. 6.24a). The branching points are occupied by hypothetical ancestral forms (a–e), which represent the last common ancestors of all groups which are situated distal to them on the cladogram. Any

assembly of taxa which includes all descendants of one last common ancestor is called a monophyletic group (e.g. C–F in Fig. 6.24a, but not A–C, which represents a so-called paraphyletic group because it does not include all descendants of the ancestor a). Two adjacent terminal taxa (or monophyletic groups of terminal taxa) which are connected by a single branching point are referred to as sister groups or adelphotaxa (e.g. E and F in Fig. 6.24a).

The characters selected are, in accordance with their presence or absence, distributed over the terminal taxa of the cladogram. On the basis of the distribution of the characters, it is determined which of them are probably primitive or ancestral (plesiomorphic) and which are probably specialised or derived (apomorphic).

A character that appears to be present in only a single terminal taxon (character 'triangle' in Fig. 6.24a) is called autapomorphic for that taxon, i.e. one assumes that this character has evolved independently in that taxon after it became separated from its sister group. Characteristics that are present in sister groups (character 'circle' in

Fig. 6.24a) but not in the neighbouring taxa (the so-called out-groups of the sister groups) are called synapomorphies of those sister groups. In this case, one assumes that the characters were present in the last common ancestor of the sister groups, but not in the last common ancestor of the sister group and the out-groups. Characteristics that are present in sister groups and in the out-groups (e.g. character 'square' in Fig. 6.24a) are called symplesiomorphic characters of the sister groups. It is noteworthy that both symplesiomorphic ('square') and synapomorphic ('circle') characters of the adelphotaxa E and F are *homologous* characters. The terms '(syn)apomorphic' and '(sym)plesiomorphic' convey information about the so-called polarity of the homologous characters. Synapomorphic, derived characters are indicative of monophyletic groups, while symplesiomorphies are not. It is also noteworthy that both terms are valid only with respect to a defined taxonomical level in the cladogram. Thus, in the example above, the character 'square' is a symplesiomorphy of the monophyletic group D-F, but it may well be a synapomorphy of the monophyletic group A-F, if it has evolved in their common ancestor a.

Quite frequently, the distribution of characters across the terminal taxa of the cladogram does not allow the relatively simple conclusions regarding the characters of the common ancestors described above (character 'circle' in Fig. 6.24b,c,d). Under these conditions, it is necessary to apply the principle of parsimony in order to decide between different hypotheses. In the case shown in Fig. 6.24b, one assumes that the character 'circle' is plesiomorphic for the monophyletic group A-F and that it has been present in all ancestors a–e. This also implies that the character has a steady phylogenetic history and is a homology between A and E. Under this assumption, four independent losses of the character (arrows) are required to explain the actual distribution. Alternatively, one might assume that the character was lost in the last common ancestor of B-F and re-evolved in taxon E (Fig. 6.24c). This assumption requires two evolutionary events (one loss and one gain). Another possibility is the assumption that the character was absent in the last common ancestor a and evolved independently in the taxa A and E (Fig. 6.24d). Again, two evolutionary events are required to explain the actual distribution.

Obviously, using the principle of parsimony, the last two hypotheses will be favoured over the first one, because they require fewer evolutionary transformations. By virtue of the same principle, it is not possible to decide between the last two hypotheses, as both require two events. Nevertheless, the last two hypotheses both imply that the character 'circle' has no steady history in the monophyletic group A-F. Such characters are called non-homologous or homoplastic, irrespective of the (profound) similarities that may exist between them.

Within a taxon studied, a particular character may be present in one group and absent in another, or a character may exist in a variable state in the taxon under study. In order to reconstruct the phylogenetic history it is necessary to determine the direction of change or polarity (i.e. primitive versus derived condition) of the characters or character states studied. Two criteria are used in comparative neuroanatomy to determine the polarity of transformation, i.e. out-group analysis and ontogenetic transformation.

In out-group analysis, the following operational rules are used (Hennig 1966; Watrous and Wheeler 1981): If a particular character is present in some members of a monophyletic group but absent in other members, the condition that is also found in the sister group is the primitive or plesiomorphic one. Similarly, for a given character with two or more states within a monophyletic group, the state also occurring outside this group is the primitive state. If the character studied contains only two states, the alternative state is the derived one.

In the analysis of development aimed at determination of the polarity of characters or character states the laws of von Baer (1828; cf. Chap. 4, Sect. 4.7) play a prominent role. According to these laws, development within a natural group is a progress from the general to the special. Early embryos of animals belonging to such a group closely resemble each other, and the degree of resemblance decreases as development proceeds. Thus, two or more taxa will follow the same developmental course to the stage of their divergence. It follows that characters or character sets observed during early development reflect the plesiomorphic condition, whereas those observed during later stages will reflect progressively more apomorphic conditions.

Neurocladistics is a new and significant branch of comparative neuroanatomy. Numerous data, dumped in the literature as isolated and dead facts, can be brought to life within this conceptual framework. It is felt appropriate at this juncture to mention briefly the results of some neurocladistic studies.

Pathways ascending from the spinal cord to the brain have been studied with the aid of experimental techniques in a large number of vertebrate species. Northcutt (1984; Fig. 6.25) subjected the results of these studies to a cladistic analysis, on the basis of which he arrived at the following conclu-

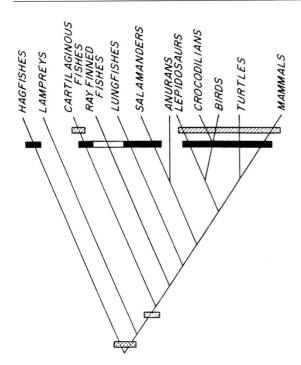

Fig. 6.25. A cladogram showing the distribution of some ascending pathways: spinoreticular (*stippling*); spinocerebellar (*random dashes*); spinotectal (*solid bars*); uncertain spinotectal (*open bar*); spinothalamic (*hatching*) (reproduced from Northcutt 1984, Fig. 4)

sions: Spinoreticular pathways occur in all vertebrate species examined and probably represent a shared primitive character that arose with the origin of vertebrates. Spinocerebellar pathways occur in all jawed vertebrates and probably represent a shared primitive character of Gnathostomes. Spinotectal pathways occur in most vertebrates but they do not appear to exist in lampreys, ray-finned fishes, and anuran amphibians. Given the distribution of this pathway, it is more probable that a spinotectal pathway was present in ancestral vertebrates and has been independently lost (three events) than that ancestral vertebrates did not possess such a pathway and it has evolved independently (a minimum of four events – three gains and one loss, three gains and two losses, or five gains). A spinothalamic pathway is known in some elasmobranchs and amniotes. Given this distribution, it is most probable that this pathway evolved independently in the two groups (two events). Alternately, if it is hypothesised that a spinothalamic pathway was present in the earliest vertebrates, one must argue that such a pathway has been lost five to six times independently.

Ascending pathways connecting the torus semicircularis (or colliculus inferior) with the telencephalon are synaptically interrupted in the diencephalon. A torotelencephalic pathway through the dorsal thalamus is probably a primitive character for gnathostomes, but a torotelencephalic pathway through the lateral preglomerular nucleus (which is a derivative of the tuberculum posterius) is probably a derived character for actinopterygian fishes (Striedter 1991).

It has already been mentioned that, in the past, several types of phylogenetic changes were described for the brain. Thus Herrick (1921, 1948) claimed that during evolution the axons or axon collaterals of particular neuronal populations 'invade' cell groups they did not previously innervate; Ebbesson (1980, 1984), on the other hand, argued that the principal trend in brain evolution is that neuronal populations split up in such a fashion that the daughter aggregates selectively lose some of the 'old' connections. In order to find out how the fibre connections of the teleostean lateral preglomerular nucleus may have changed in the course of phylogeny, Striedter (1992) examined the connections of this cell mass in four closely related ostariophysean teleosts with the fluorescent tracer DiI, and subjected the results obtained to a cladistic analysis. He summarised the data obtained as follows: "Individual cell groups may vary in size and complexity, may disappear or appear 'de novo', may change the strength of their connections, and may gain or lose connections. Furthermore, changes in the size of a cell group may be uncoupled from changes in its number of subdivisions, and both of these types of changes may occur independently of changes in connections. Although increases in the size and complexity of the cell groups outnumber the decreases in this system, and losses of connections outnumber the gains by a narrow margin, the results of the present study clearly suggest that no single type of phylogenetic change predominates, as was proposed by Herrick (1948), Ebbesson (1980, 1984) and others. Instead, many different types of phylogenetic changes interact to produce a remarkable variety of species differences. Furthermore, the individual cell groups within this system seem to change largely independently of one another. For example, decreases in the size of one cell group are frequently accompanied by increases in the size of one of its targets, and vice versa" (Striedter 1992, pp. 355–356).

The dorsal thalamus of gnathostomes comprises two basic divisions – the lemnothalamus and the collothalamus – that receive their principal input from lemniscal pathways and from the midbrain roof, respectively. Analysis of the evolution of the dorsal thalamus reveals that the collothalamic, i.e. the midbrain-sensory relay nuclei, are homologous to each other in all vertebrate radiations as discrete

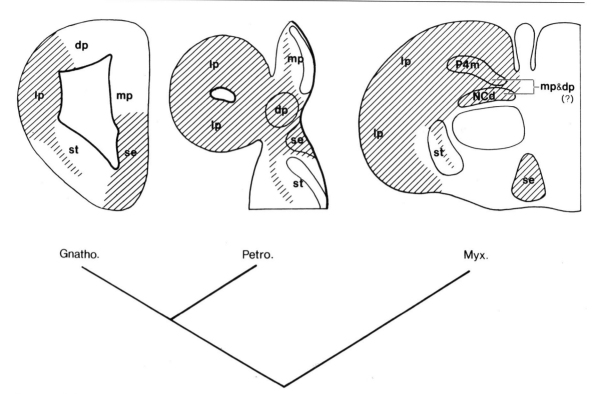

Fig. 6.26. Cladogram showing the genealogical interrelationships of the three major recent craniate taxa, the gnathostomes, the petromyzontids and the myxinoids, and diagrammatic cross-sections through a telencephalic hemisphere of a hagfish (*right*), a lamprey (*middle*) and a urodele amphibian (*left*). The extent of secondary olfactory projections is indicated by *hatching*. *dp*, Dorsal pallium; *lp*, lateral pallium; *mp*, medial pallium; *NCd*, dorsal subnucleus of the central prosencephalic nucleus; *P4m*, medial subdivision of the fourth pallial layer; *se*, septum; *st*, striatum (reproduced from Wicht and Northcutt 1993, Fig. 8)

nuclei, and that the dorsal lateral geniculate nucleus of mammals and the dorsal lateral optic nucleus of other amniote groups are likewise homologous as discrete nuclei. Most other thalamic nuclear groups in mammals are collectively homologous as a field to nuclei or nuclear groups in other vertebrate radiations (Butler 1994a).

Wicht and Northcutt (1993) compared the secondary olfactory projections in hagfishes with those in lampreys and gnathostomes. It appeared that hagfishes have more extensive secondary olfactory projections than lampreys, and that lampreys have more extensive projections than gnathostomes. Taking the cladistic relations of these three taxa into consideration (Fig. 6.26), they concluded that: (a) extensive secondary olfactory systems are a plesiomorphic character for craniate brains, and (b) extensive non-olfactory telencephalic centres arose with the origin of gnathostomes.

These examples may suffice to show that a neurocladistic analysis is nothing but the distribution of characters and their polarities over a dendrogram, which latter expresses a hypothesis concerning the phylogenetic relations within a given taxon.

Characters and character states which, on the basis of their distribution and in light of the principle of parsimony, were most probably present in the hypothetical common ancestor of a given taxon are incorporated in the *morphotype* of that taxon. This set of shared primitive characters forms the point of departure for a reconstruction of the phylogenetic development of the organ or other entity from which the studied characters are derived. This reconstruction is based on the sets of shared derived characters (synapomorphies) encountered at each bifurcation in the dendogram, until the end segments, and the autapomorphies positioned there, are attained.

The purpose of cladism is the reconstruction of phylogeny. It will be appreciated that the dendrograms over which the neural characters and character states are distributed in order to reconstruct the phylogeny of the CNS and its parts are based themselves on nothing but the distribution of characters and character states. In order to avoid circular reasoning it is required that the hypothesis of phylogeny, i.e. the dendrogram which forms the basis of a neurocladistic study, be based on characters

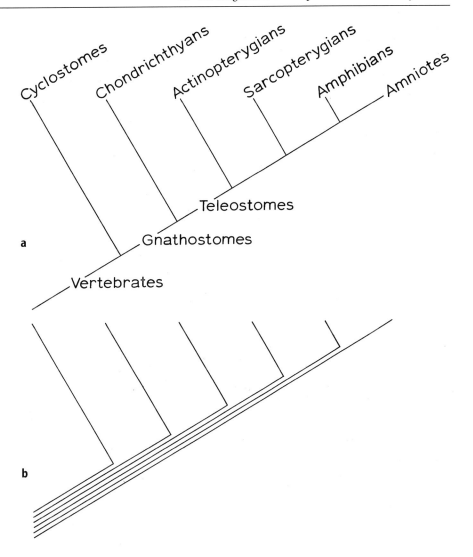

Fig. 6.27a,b. Relations between phylogenesis and ontogenesis. **a** Simplified cladogram showing the phylogenetic relationship of the major vertebrate group; **b** the ontogenetic development of the same groups according to von Baer (1828)

that are independent, to as much an extent as possible, of the characters being analysed (Butler 1994a,b). However, because the CNS is connected to all other parts of the body, it can be argued that neural characters are of particular importance for the determination of phylogenetic affinities. Therefore, in all neurocladistic studies the question should be posed whether the distribution of the characters studied suggests alternatives for the hypothesis of phylogeny started from. This holds in particular if the taxon selected for a comparative neuromorphological study includes species or groups of which the relationships are disputed. (For an interesting, though unsuccessful attempt to determine the interrelationships of living sarcopterygians on the basis of neural characters, the reader is referred to Northcutt 1986a.)

The variety and variability of the kinds of changes playing a role in brain phylogeny doubtless reflect the 'patchwork' or 'mosaic' course of evolu-

tion. Attempts to gain insight into the constraints, the mechanisms and the biological significance of these phylogenetic changes will have to be undertaken in light of (onto)genetic (Sect. 6.3.4), functional, ethological and ecological information (Sect. 6.3.5). Such attempts are beyond the scope of phylogenetic neuromorphology and belong in the domain of evolutionary morphology (Sect. 6.3.6).

6.3.4
Comparative Neuroembryology and Genetic Neuromorphology

The only comprehensive general treatise on the comparative neuroembryology of vertebrates that has appeared so far is von Kupffer's (1906) excellent contribution to Hertwig's *Handbuch*. In this contribution the CNS is subdivided into three primary and five secondary brain vesicles and the significance of neural segmentation is emphasised (cf. Chap. 4).

During the period of 'classical' comparative neuroembryology, which ended around 1970, numerous studies appeared which were based almost exclusively on serial sections of non-experimental material, stained with simple routine techniques. Reference to Chaps. 4 and 5 shows that the central and unresolved question during this period was: Is the CNS of vertebrates fundamentally divisible into longitudinal zones (as was held by Herrick and Kuhlenbeck; cf. Chap. 4, Sect. 4.6), or into rostrocaudally arranged transverse bands or neuromeres (as was emphasised by several members of the Swedish school; cf. Chap. 4, Sect. 4.5.2), or are the basic units the product of a combination of a longitudinal and a transverse subdivision (as was advocated by Bergquist and Källén; cf. Chap. 4, Secs. 4.5.2, 4.6.7 and 4.7).

Since 1970, developmental neurobiology has focussed almost entirely on a few groups (rodents, primates) or species (chick, clawed toad, zebrafish). Hence, so far there has been no modern period of comparative neuroembryology. Remarkably, however, the general issues discussed in current neuroembryological publications revolve around the same questions as those discussed in the classical period (cf. Chap. 5, Sect. 5.5).

It may be expected that comparative studies on the structure and genomic organisation of the genes involved in the development of the CNS, and on the spatiotemporal patterns of expression of these genes, will (a) help resolve the long-standing uncertainties referred to above, and (b) increase our insight into the phylogenetic development of the CNS and its genetic basis. This prognosis is based on the following notions:

1. All phylogenetic changes have to be funneled through ontogenesis, or, to quote Garstangs' (1922) maxim: Ontogenesis does not recapitulate phylogeny [as Haeckel, 1866, asserted], but creates it.
2. All phylogenetic changes are correlated with genomic changes.
3. The relations between ontogenesis and phylogenesis are complex. Apart from Haeckel's 'biogenetic law', numerous other 'laws' or 'rules' pretending to govern this relationship have been formulated, but most of these have too many exceptions to be useful for a 'translation' of ontogenetic into phylogenetic processes (for reviews see Remane 1956; Gould 1977; Northcutt 1990). An exception should be made for the laws of von Baer (1828; cf. Chap. 4, Sect. 4.7.1), which provide a sound basis for drawing parallels between the ontogenesis and phylogenesis of the various vertebrate groups (Løvtrop 1987; Fig. 6.27), particularly in the CNS.

4. In the past decade, dramatic progress has been made towards the identification of genes involved in vertebrate neurogenesis. It appears that the products of numerous so-called homeobox genes act as regulatory transcription factors in the control of spatial patterning in the embryonic CNS. The homeobox genes form a diverse multigene family. Their distinguishing feature, the homeobox, is a DNA sequence of approximately 180 base pairs that codes for a protein domain of approximately 60 amino acids, the homeodomain. Proteins containing a homeodomain bind to and regulate the transcription rate of a multitude of target genes involved in morphogenesis. Homeodomain proteins are grouped into different families or classes, including *Antp, engrailed, even-skipped* and *POU* classes.

Homeoboxes were originally identified in *Drosophila* in homeotic selector and segmentation genes. In this species the *Antp*-class genes are grouped together in the homeotic complex, *HOM-C*. In the mouse and man, the genes of this class are in four complexes, *Hox-a* through *Hox-d*, and it seems likely that four comparable complexes are present in all gnathostomes. The data just summarised have been derived from the following reviews: Gehring (1987), Krumlauf (1992), Carroll (1995), Holland and Garcia-Fernàndez (1996), to which the reader is referred for details.

Molecular genetic techniques render it possible to assess the activity patterns of genes in histological sections. As extensively discussed in Chap. 4, Sect. 4.5.5.5 and Chap. 5, Sects. 5.5.2 and 5.5.3, members of the *Hox* gene family are expressed in segment-specific patterns in the developing hindbrain, whereas numerous other genes, among them members of the *POU* class, are expressed in spatially restricted forebrain and midbrain domain. Given the highly conserved character of the homeobox genes, it may be expected that their patterns of expression can be used as molecular landmarks for homology. Hence, comparison of their expression limits in different species may help define the fundamental pattern of brain regionalisation. In light of the laws of von Baer (cf. Chap. 5, Sect. 57.1), it will be important to include in such comparative studies very early developmental stages (neural plate, early neural tube) and to employ accurate mapping techniques (cf. Alvarez-Bolado and Swanson 1995).

5. Comparative studies on the genomic organisation and diversity of *Hox* and other developmentally relevant genes are starting to provide

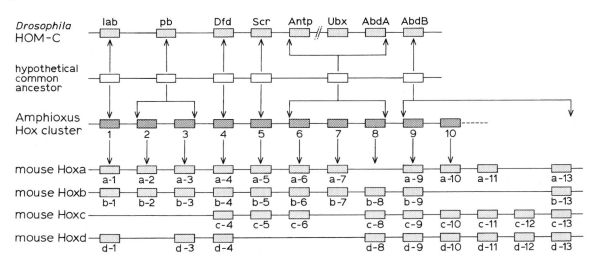

Fig. 6.28. Suggested evolutionary relationships between the complex of homeotic selector genes, *HOM-C* in *Drosophila* and the *Hox* gene clusters in amphioxus and mammals, together with the deduced complement of *Hox* genes in the hypothesised common ancestor of chordates and arthropods. (adapted from Holland and Garcia-Fernández 1996, Fig. 2)

insight into the genetic basis for phylogenetic development of the vertebrate brain (Fig. 6.28). In mammals there are 39 *Hox* genes, forming four separate clusters, Hox a–d. Structurally, the genes in these four clusters correspond to each other in such a fashion that 13 paralogous subgroups of genes: 1–13 can be distinguished (McGinnis and Krumlauf 1992).

Garcia-Fernàndez and Holland (1994) recently established that in the genome of the cephalochordate amphioxus a single *Hox* gene cluster is present. This cluster has similar genomic organisation to the four mammalian *Hox* clusters and contains homologues of at least the first ten paralogous groups of mammalian *Hox* genes in a collinear array. Data summarised by Holland (1992), Holland and Garcia-Fernàndez (1996) and Carroll (1995) suggest that in the genome of several nonmammalian gnathostomes (teleosts, *Xenopus*, chick) four *Hox* gene clusters are present, having a 1:1 correspondence with the four mammalian clusters.

Taking data concerning the organisation of the *Hox*-like genes in *Drosophila* and other invertebrates into consideration, Garcia-Fernàndez and Holland (1994; cf. also Holland et al. 1994) deduced the following picture of the phylogenesis of the *Hox* genes (Fig. 6.28):

1. There was a considerable increase in the number of *Hox* genes early in chordate evolution.
2. The presence of at least two AbdB-related genes in amphioxus indicates that tandem duplication of an ancestral AbdB-like gene preceded the evolution of typical vertebrate features.

3. *Hox* cluster duplications occurred after the divergence of the cephalochordate and vertebrate lineages.
4. There may have been two phases of cluster duplication, one close to the origin of vertebrates, and one close to the origin of gnathostomes.

Carroll (1995) collected data from the literature indicating that not only the *Hox* genes, but also other important developmental gene families such as *Wnt* and *Dlx* genes, are generally several-fold larger in teleosts and mammals than in primitive chordates and invertebrates. However, the complement of *Hox*, *Wnt* and *Dlx* genes is comparable in teleosts and mammals.

Because *Hox*, *Dlx*, *Wnt* and other families of regulatory genes are expressed in the embryonic brain, the data reviewed above suggest that the origin of new genes (by tandem duplication within a single cluster as well as by cluster duplication) may well have been important in enabling the early evolution of the vertebrate brain, but that the subsequent evolution and diversification of the brain of gnathostomes is mainly the product of regulatory changes within larger, but essentially fixed sets of developmental genes (Carroll 1995; Holland and Graham 1995; Holland and Garcia-Fernàndez 1996).

The great potential of gene duplication in the evolution of increasing complexity has been stressed by Holland and his colleagues (Holland 1992; Holland et al. 1994; Holland and Garcia-Fernàndez 1996), in line with the classical work of Ohno (1970). These authors speculated that by this type of mutation redundant genes are created. Some of these will have been lost (cf. the open places in the complement of mammalian *Hox* genes in Fig. 6.28), but others will eventually have been co-opted to new roles, facilitating the evolution of new developmental pathways.

Further comparative analyses of the organisation and expression of *Hox* and other regulatory genes will certainly contribute to our understanding of the genetic basis for evolutionary changes in the development of the brain.

6.3.5
Functional Neuromorphology

Biological structures are not simply formal or historical entities but, first and foremost, parts of living organisms in which they exert or participate in one or more functions. Hence, in a meaningful (comparative) (neuro)anatomy functional aspects cannot be left out of consideration.

Kasper et al. (1994) opened a recent publication on the cerebral cortex with the sentence "Belief in correlation between structure and function is one of the articles of faith of neuroscience." It is no exaggeration to say that this article of faith has always played a prominent role in comparative neuroanatomy. Indeed, Edinger (1908b, p. ix) stated: "Hirnanatomie allein getrieben wäre eine sterile Wissenschaft. Erst in dem Momente, wo man die Frage nach dem Verhältnis der anatomischen Struktur zu der Funktion aufwirft, gewinnt sie

Leben." J.B. Johnston, another founding father of comparative neuroanatomy, went one step further by asserting (Johnston 1902, p. 88): "The time has come, as I think every comparative neurologist feels, that we must break away from the old cumbersome nomenclature and descriptions imposed by human anatomy, and create a new and simple mode of describing the brain, which shall express rather than obscure the functions of its several parts." Somewhat earlier, Gaskell (1886) had advanced the concept that the peripheral and the CNS can be subdivided into units or components which have *at the same time a structural and functional significance.* As detailed in Chap. 4, Sect. 4.6.2, Gaskell analysed the various cranial nerves into their components and concluded that these components are centrally connected to longitudinally arranged cellular zones or columns. Application of Gaskell's approach to the brain of various anamniotes led Johnston and Herrick to the well-known concept that in the rhombencephalon four longitudinal columns, somatosensory, viscerosensory, visceromotor and somatomotor, can be distinguished (Fig. 6.29). In light of this result, the authors mentioned considered the question of the extent to which centres in more rostral parts of the brain also fit into the longitudinal pattern detected and/or can be defined according to elementary functional relationships.

Johnston (1902, p. 98) stated: "The attempt to reduce the mid, 'tween, and forebrain to terms of these four functional divisions leads us into a field

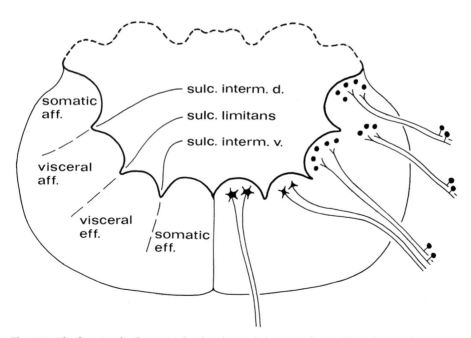

Fig. 6.29. The functional columns in the rhombencephalon according to Herrick and Johnston

where few data for certain conclusions are at present to be found," and further (Johnston 1902, p. 105): "There are certain parts of the central grey matter in the rostral region of the brain (for example, the nucleus of the posterior commissure and the corpus striatum) which hold such relations as to defy attempts to interpret them as modified representatives of any structure in the spinal cord or hind brain. The same is to be said of the dorsal part of the brain at its extreme rostral end." However, he continued his search for rostral continuations of the rhombencephalic functional columns. In studies on the selachian and actinopterygian telencephalon (Johnston 1911a,b) he claimed that distinct equivalents of the somatosensory and viscerosensory columns were present in the dorsal part of the telencephalon of these groups. The eversion of the actinopterygian telencephalon was stated to be due to a local hypertrophy of the viscerosensory column. These views of Johnston have found no acceptance in the literature. Johnston's claim that an integral functional nomenclature could be designed for the brain rested upon the supposition that the relations between structure and function within this organ remain constant throughout the vertebrate kingdom. "I can find no ground for supposing that morphological entities in the brain have different functions in different classes of vertebrates" (Johnston 1911b, p. 526).

Herrick (1903) was also of the opinion that the goal to be pursued by comparative neuroanatomy was a complete functional subdivision of the CNS, and that analysis of the nerve components was conducive to this goal. Discussing the morphological subdivision of the brain, Herrick (1908, p. 403) argued: "The real unit of the nervous system is unquestionably the functional system of neurones." With regard to the parts of the brain situated rostral to the rhombencephalon, he declared that the mesencephalon and diencephalon are not natural regions. He recommended the use of the term ophthalmencephalon in analogy to rhinencephalon: "Functionally and genetically, the retinae, optic nerves, chiasma and tracts and the optic thalamus (*sensu stricto*) should be associated with the optic tectum of the mid-brain to form an *ophthalmencephalon* whose boundaries cross freely those of the classic encephalic regions" (Herrick 1908, p. 401). Somewhat later, Herrick (1922) proposed the subdivision of that anamniote telencephalon behind the olfactory bulb into medial, dorsal and lateral olfactory areas, an olfacto-somatic and a somatic area. This subdivision has not caught on, except for actinopterygian fishes (cf. Ariëns Kappers et al. 1936). In his later writings (e.g. 1930, 1943) Herrick was more critical about the possibility for the creation of an integral functional subdivision of the brain. In his monograph on the brain of the tiger salamander (Herrick 1948), he distinguished three longitudinal zones performing three general classes of functions: a dorsal receptive or sensory zone, a ventral emissive or motor zone and, between these and infiltrating them, an intermediate zone of correlation and integration. He pointed out that the intermediate zone shows a progressive evolutionary development and comprises in man more than half of the total weight of the brain.

Johnston and Herrick attempted to realise a complete functional subdivision of the vertebrate brain on the basis of an analysis of the central relations of the cranial nerves and their components in non-experimental material. Although this attempt remained unsuccessful, their work – and that of the American school in general – brought into focus the fact that the study of the morphological substrates of the various functions of the CNS is one of the most salient tasks of comparative neuroanatomy.

During the past few decades a number of highly reliable experimental techniques for establishing the origin, course and termination of neural connections have been developed (cf. Chap. 7). With the aid of these techniques the central representations of the various sense organs have been traced 'antegradely' beyond their primary recipient centres, and the same holds true for the central representation of effectors, which have been traced 'retrogradely' centralward. In this way the various sensory and motor systems, including their successive fibre systems and relay centres, have been and are being charted. In the chapters devoted to the various groups of vertebrates ample consideration will be given to the results of these experimental hodological studies.

Some general comments on the results and prospects of functional comparative anatomy follow. Wherever thoroughly analysed, the fibre connections in the vertebrate CNS have been found to be far more complexly organised than was previously assumed. As discussed in Chap. 3, at least 12 different characters can be attributed to these entities, and there is hardly a fibre system in the vertebrate brain for which all of these are fully known. Clearly, much hodological work remains to be done within the realm of functional comparative neuroanatomy.

There are very few sensory and motor relay centres in the brain receiving their afferents solely from the next lower centre and projecting solely to the next higher centre in the cascade; there is preference rather than exclusivity. The centres forming the functional array are more tightly and intensely connected to each other than to other centres. A related aspect is that many functional systems are 'unipolar', by which is meant that they are structurally and functionally discrete at only one end of the array. Thus, the mammalian visual system is discrete at its retinal side, but it loses its identity in the neocortex at the transition of the visual association areas and the multimodal association area, and the same feature presents itself at several subcortical levels of the visual system, for instance, in the deeper layers of the superior colliculus. These features put serious constraints on the possibility of relating discrete functions to discrete structures. Brodal (1981, p. 690) remarked upon the fact that "... as research progresses it becomes increasingly difficult to separate functionally different regions in the brain The borderlines between 'functional systems' become more and more diffuse."

As stated in Chap. 2, Sect. 2.9.4, important gains in our 'understanding' of the brain have come from correlative electrophysiological, structural and ultrastructural studies. The number of centres in the brain of non-mammalian vertebrates which have been analysed at this level of resolution is very small indeed, and true comparative studies of this type have remained limited so far to a single centre, the cerebellum (Llinás 1969).

Comparative studies have revealed numerous parallels in the organisation of functional systems and complexes. Many centres in the brains of different vertebrates are homologous as well as analogous. However, there are numerous exceptions to this rule (cf., e.g., ten Donkelaar 1982, Striedter 1991). Johnston's (1911b) assertion that morphological entities in the brain have the same functions in different classes of vertebrates has been disproved.

Wherever possible, the structures and structural variations observed in the brains of different vertebrates should be placed in the context of behaviour and environment. The ethological and ecological significance of many centres and functional complexes, – for instance, the motor trigeminal nucleus in lampreys, the vagal lobes in cyprinids and the infrared system in certain snakes – is perfectly clear, but more often than not the biological significance of neural structures is unknown. 'Biological' neuromorphology obviously has a great future. Homoplastic similarities have usually been developed in response to similar environmental factors. The number of ways in which vertebrate nervous systems respond to such factors is presumably limited. Thus, as emphasised by Northcutt (1981, p. 343): "An examination of homoplastic characters will reveal the 'rules' of neural adaptation and how the nervous system evolves." If two closely related species show two obviously different states with regard to a particular neural character (or set of related characters), a comparative ethological analysis may well yield clues as to the biological significance of these different character states. Techniques have recently become available which render it possible to label directly the neural populations involved in particular types of behaviour. These techniques are potentially very useful for neuroethological studies, the more so because they can be combined with immunohistochemistry and tract-tracing (cf. Chap. 7, Sect. 7.4.5). However, it should be cautioned that what has been said above on the delineability of function systems holds a fortiori for behavioural systems.

6.3.6
Evolutionary Neuromorphology

It is the task of evolutionary (neuro)biology (a) to reflect on the question of which specific factors and processes might have caused the structural, functional and other changes through which organisms (viewed as living, active and purposeful agents) have passed, and (b) to develop explanatory scenarios. If we survey the enormous (though still highly incomplete) body of knowledge on the phylogenetic development of the vertebrate brain, two predominant phenomena present themselves compellingly: (a) diversification, and (b) increase in complexity.

The current neo-Darwinian theory of evolution gives a satisfactory general explanation for the first of these phenomena: the multiple changes leading to diversification are to be interpreted as adaptations to new environmental situations, and the mechanism involved is the tandem dualism of genetic variability and natural selection (Mayr 1969). Within and beyond neurobiology, most 'explanations' given to particular evolutionary changes are mere repetitions of, or paraphrases on, the general principles just mentioned (cf., e.g., Chap. 31 in Butler and Hodos 1996). Attempts should be made, of course, to infer *which specific factors or processes* have led to particular changes. As has already been indicated, combined neuroanatomical, neurophysiological, ethological and ecological studies on a limited number of closely related species showing some discrete differences may well provide clues as to which specific selective pressures have been operative in the group to which these species belong (Bock 1980; Roth 1987; Burggren and Bemis 1990). On the other hand, in the brains of all vertebrates examined so far, numerous structures are present, the functional significance and, hence, the possible adaptive value of which are entirely unknown. Finally, it seems likely that many brain characters do not reflect adaptations to a single specific ecological situation.

The phenomenon 'increase in complexity' can be clearly observed in the brains of many different vertebrate (and invertebrate!) radiations, but its 'evolutionary' status is controversial. According to Remane (1956, 1966), organs and organ systems may show two different evolutionary trends: *Anpassungsreihen*, or trends of adaptation, and *Vervollkomnungsreihen*, or trends of improvement. Remane distinguished many different trends of improvement, two of which, i.e. differentiation and synorganistion, in his opinion, play a dominant role in the CNS. Differentiation requires little comment: it is the phenomenon whereby organs tend to become more complex in structure and function during evolution. The concept of synorganistion transcends differentiation. It refers to the phenomenon whereby structures which have developed and differentiated quite independently may unite and become integrated into new apparatuses or functional systems. Remane (1966) suggested that during evolution of the CNS, new functional entities are formed by the increasing linking of groups of neurons. Thus, he regarded differentiation and synorganisation as manifestations of improvement or progress, and he claimed a special status for these phenomena within the theory of evolution.

Most adherents of the modern neo-Darwinian theory reject the concept of evolutionary progress. They claim that all extant species are basically equal, because they all have successfully tackled the problems of survival. Moreover, they claim that the idea of progress stems from the pre-evolutionary view that all living organisms can be arranged in one single linear and hierarchical 'ladder of life', or *scala naturae*, and finally they assert that evolutionary progress implies the presence of an intrinsic driving force apart from the external drive of selective pressure (cf. e.g. Benton 1987; Dawkins 1992; Maynard Smith and Szathmáry 1995; Butler and Hodos 1996).

Personally, I do not see why an increase in complexity of the brain, which manifests itself so obviously in many taxa, could not in principle be satisfactorily explained as a functional adaptation to a particular mode of life. I fully agree with Bullock (1984, p. 473) when he states: "The brain has diversified and advanced in evolution more than any other organ," and: "This evolution was not in one line or ladder but in many radiations of descendant stocks. Nevertheless, it is a mistake to deny lower and higher forms, inasmuch as most members of later-appearing classes are more advanced than earlier classes, have more distinguished parts and processes, and larger repertoires of behavior," and finally (Bullock 1993, p. 89): "Careful use of words such as 'higher', 'advanced' and 'bet-

ter' is obligatory in discussions of evolution, but they need not be avoided. The time is past for reflex hypersensitivity to any reference that might be interpreted as implying a scala naturae." Evolutionary progress does exist. The challenge is to develop plausible explanatory scenarios for this intriguing phenomenon. It is of particular importance in this context: (a) to distinguish evolutionary grades from phylogenetic clades (cf. e.g. Schaeffer 1965; Gould 1976), and (b) to fully appreciate the considerable constraints which may be laid on the genetic-epigenetic pathways that connect gene mutations with morphological changes. Through these constraints random processes may well be changed into pseudo-orthogenetic, directed processes. To give an example, such constraints may well have led to a considerable expansion of the neocortex in a number of mammalian taxa.

6.3.7
Summary and Concluding Remarks

The biotic world represents a continuous spatio-temporal network of relations and interactions. It is the task of biology to explore this network. Within the domain of neuromorphology or comparative neuroanatomy the relations of the CNS and its parts are systematically explored.

A cytoarchitectonic analysis is a good starting point for all kinds of neuromorphological research. Such an analysis involves two phases: making an inventory and interpretation. During the first, the various neuronal aggregates (cf. Chap. 2) are delineated; during the second phase the cell masses are named, i.e. they are homologised with cell masses in the CNS of other, related species. Topological relations are of paramount importance for the establishment of homologies, which may also require embryological research. Once established, homologies pave the way for the exploration of other relations: phylogenetic, hodological and functional.

A systematic cladistic analysis is the method of choice for placing the results of a comparative neuroanatomical study in a phylogenetic context. Experimental hodological techniques determine the fibre connections of the various neuronal aggregates in a CNS with great precision. Cladistic analysis of the results of comparative hodological studies will reveal which pathways have remained constant, and what changes in the 'wiring' of the CNS have occurred in the course of phylogeny.

Within the domain of neuromorphology, functional relations can be analysed in two different ways: (a) with the aid of experimental hodological techniques, the arrays of centres and pathways primarily related to particular sensory and motor functions can be traced, and (b) active centres and pathways can be visualised, for instance, with the immediate-early gene technique (Chap. 7, Sect. 7.4.5). The functional complexes thus delineated can again be subjected to a cladistic analysis.

It is to be emphasised that in the branch of neuromorphology, which is aimed at the subdivision of the CNS into functional complexes, the actual function of these complexes, i.e. the processing of information, is not considered.

In contrast, analysis of the intrinsic organisation of neuronal aggregates is an indispensable prerequisite for physiological studies aimed at an understanding of the way in which information is processed within such entities. The Golgi method, electron microscopy and immunohistochemistry, and particularly techniques in which these approaches are combined, form essential tools in this analysis.

It is the task of evolutionary neurobiology to address the question as to which factors and processes may have caused the structural and functional changes through which the CNS has passed in the course of phylogeny. The following two approaches may provide answers to this question:

1. Correlative ecological, ethological neurophysiological and neuroanatomical studies, carried out on closely related species showing some discrete differences
2. Comparative studies on the structure and organisation of the genes involved in the development of the CNS and on the spatiotemporal pattern of expression of these genes

Ontogenetic studies of this type will not only provide an insight into the genetic basis of evolutionary changes, but also contribute to our understanding of how these changes are channelled through, and structured by, the pathways and programs of development. Since both of the approaches just outlined involve multidisciplinary research on highly complex relations and interactions, progress in evolutionary neurobiology is likely to be slow without new research paradigms.

Acknowledgement. I thank Dr. Helmut Wicht for stimulating discussions and for his help in phrasing Sect. 6.3.3., and Prof. Peter Holland for critically reading Sect. 6.3.4.

References

Alvarez-Bolado G, Swanson LW (1995) Appendix: on mapping patterns in the embryonic forebrain. J Comp Neurol 355:287–295

Alvarez-Bolado G, Rosenfeld MG, Swanson LW (1995) Model of forebrain regionalization based on spatiotemporal patterns of POU-III homeobox gene expression, birthdates, and morphological features. J Comp Neurol 355:237–295

Anderson KD, Reiner A (1991) Striatonigral projection neurons: a retrograde labeling study of the precentages that contain substance P or enkephalin in pigeons. J Comp Neurol 303:658–673

Appleton AB (1928) The muscles and nerves of the post-axial region of the tetrapod thigh, part I. J Anat 62:364–401

Ariëns Kappers CU (1920–1921) Die vergleichende Anatomie des Nervensystems der Wirbeltiere und des Menschen. Bohn, Haarlem

Ariëns Kappers CU (1929) The evolution of the nervous system in invertebrates, vertebrates and man. Bohn, Haarlem

Ariëns Kappers CU (1947) Anatomie comparée du système nerveux. Bohn, Haarlem

Ariëns Kappers CU, Huber GC, Crosby EC (1936) The comparative anatomy of the nervous system of vertebrates. MacMillan, New York

Baerends GP (1958) Comparative methods and the concept of homology in the study of behaviour. Arch Neerl Zool 13 [Suppl 1]:401–417

Baker-Cohen KF (1968) Comparative enzyme histochemical observations in submammalian brains. I. Striatal structures in reptiles and birds. Ergeb Anat Entw Gesch 10:7–41

Bayer SA, Altman J (1995) Principles of neurogenesis, neuronal migration, and neural circuit formation. In: Paxinos G (ed) The rat nervous system. Academic, San Diego, pp 1079–1098

Benton MJ (1987) Progress and competition in macroevolution. Biol Rev 62:305–338

Bergquist H (1932) Zur Morphologie des Zwischenhirns bei niederen Wirbeltieren. Acta Zool (Stockl) 13:57–303

Bergquist H (1953) On the development of diencephalic nuclei and certain mesencephalic relations in Lepidochelys olivacea and other reptiles. Acta Zool (Stockh) 34:155–190

Bergquist H (1954) Ontogenesis of diencephalic nuclei in vertebrates. Kgl Fysiogr Sällsk Lund Handl N F 65 (6):1–34

Bergquist H, Källén B (1954) Notes on the early histogenesis and morphogenesis of the central nervous system in vertebrates. J Comp Neurol 100:627–659

Bock WJ (1963) Evolution and phylogeny in morphologically uniform groups. Am Naturalist 97:265–285

Bock WJ (1980) The definition and recognition of biological adaptation. Am Zool 20:217–227

Boyden A (1943) Homology and analogy: a century after the definitions of 'homologue' and 'analogue' of Richard Owen. Q Rev Biol 18:228–241

Boyden A (1969) Homology and analogy. Science 164:455–164

Braford MR Jr (1995) Comparative aspects of forebrain organization in the ray-finned fishes: Touchstones or not? Brain Behav Evol 46:259–274

Brauth SC, Kitt CA (1980) The paleostriatal system of Caiman crocodilus. J Comp Neurol 189:437–465

Brodal A (1981) Neurological anatomy. Oxford University Press, New York

Buhl EH, Oelschläger HA (1988) Morphogenesis of the brain in the harbour porpoise. J Comp Neurol 277:109–125

Bullock TH (1984) Comparative neuroscience holds promise for quiet revolutions. Science 225:473–478

Bullock TH (1993) How are more complex brains different? Brain Behav Evol 41:88–96

Burgren WW, Bemis WE (1990) Studying physiological evolution: paradigms and pitfalls. In: Nitecki MH (ed) Evolutionary innovations. University of Chicago Press, Chicago, pp 191–228

Butcher LL, Woolf NJ (1982) Cholinergic and serotonergic systems in the brain and spinal cord: anatomic organization, role in intercellular communication processes and interactive mechanisms. Prog Brain Res 55:3–40

Butler AB (1994a) The evolution of the dorsal thalamus of jawed vertebrates. Brain Res Rev 19:29–65

Butler AB (1994b) The evolution of the dorsal pallium in the telencephalon of amniotes: cladistic analysis and a new hypothesis. Brain Res Rev 19:66–101

Butler AB (1995) The dorsal thalamus of jawed vertebrates: a comparative viewpoint. Brain Behav Evol 46:209–223

Butler AB, Hodos W (1996) Comparative vertebrate neuroanatomy: evolution and adaptation. Wiley-Liss, New York

Cain AJ (1964) The perfection of animals. Viewpoints Biol 3:36–63

Campbell CBG (1976) Morphological homology and the nervous system. In: Masterton RB, Hodos W, Jerison H (eds) Evolution, brain, and behavior: persistent problems. Erlbaum, Hillsdale, NJ, pp 143–151

Campbell CBG (1982) Some questions and problems related to homology. In: Armstrong E, Falk D (eds) Primate brain evolution. Methods and concepts. Plenum, New York, pp 1–11

Campbell CBG (1987) Homology. In: Adelman G (ed) Encyclopedia of neuroscience, vol I. Birkhäuser, Boston, pp 501–502

Campbell CBG, Hodos W (1970) The concept of homology and the evolution of the nervous system. Brain Behav Evol 3:353–367

Carroll SB (1995) Homeotic genes and the evolution of arthropods and chordates. Nature 376:479–485

Clairambault P, Capanna E (1973) Suggestions for a revision of the cytoarchitectonics of the telencephalon of Protopterus, Protopterus annectens (Owen). Boll Zool 40:149–171

Cochran SL, Hackett JT (1977) The climbing fiber afferent system of the frog. Brain Res 121:362–367

Conel J (1929) The development of the brain of Bdellostoma stouti. I. External growth changes. J Comp Neurol 47:343–403

Conel J (1931) The development of the brain of Bdellostoma stouti. II. Internal growth changes. J Comp Neurol 52:365–499

Crosby E, Schnitzlein HB (1974) The comparative anatomy of the telencephalon of the hagfish, Myxine glutinosa. J Hirnforsch 15:211–236

Crosby EC, DeJonge BR, Schneider RC (1966) Evidence for some of the trends in the phylogenetic development of the vertebrate telencephalon. In: Hassler R, Stephan H (eds) Evolution of the forebrain. phylogenesis and ontogenesis of the forebrain. Thieme, Stuttgart, pp 117–135

Cuvier G (1812) Sur un nouveau rapprochement à établir entre les classes qui composent le règne animal. Ann Mus Hist Nat 19:73–84

Dawkins R (1992) Progress. In: Keller EF, Lloyd EA (eds) Keywords in evolutionary biology. Harvard University Press, Cambridge, pp 263–272

de Beer GR (1971) Homology, an unsolved problem. Oxford University Press, London

Dohrn A (1875) Der Ursprung der Wirbelthiere und das Prinzip des Funktionswechsels. Engelmann, Leipzig

Droogleever Fortuyn J (1961) Topographical relations in the telencephalon of the sunfish Eupomotis gibbosus. J Comp Neurol 116:249–264

Ebbesson SOE (1980) The parcellation theory and its relation to interspecific variability in brain organization, evolutionary and ontogenetic development, and neuronal plasticity. Cell Tissue Res 213:179–212

Ebbesson SOE (1984) Evolution and ontogeny of neural circuits. Behav Brain Sci 7:321–366

Edinger L (1906) Über das Gehirn von Myxine glutinosa. In: Abhandlungen der königlich preußischen Akademie der Wissenschaften vom Jahre 1906. Königliche Akademie der Wissenschaften, Berlin, pp 1–36

Edinger L (1908a) The relations of comparative anatomy to comparative psychology. J Comp Neurol 18:437–457

Edinger L (1908b) Vorlesungen über den Bau der Nervösen Zentralorgane des Menschen und der Tiere. II. Vergleichende Anatomie des Gehirns. Vogel, Leipzig

Elliot Smith G (1908) The cerebral cortex in Lepidosiren, with comparative notes on the interpretation of certain features of the forebrain in other vertebrates. Anat Anz 33:513–540

Fallon JH, Moore RY (1978) Catecholamine innervation of the basal forebrain. IV. Topography of the dopamine projection to the neostriatum. J Comp Neurol 180:545–580

Fibiger HC (1982) The organization and some projections of cholinergic neurons of the mammalian forebrain. Brain Res Rev 4:327–388

Forey PL (1990) Cladistics. In: Brigs DEG, Crowther PR (eds) Palaeobiology, a synthesis. Blackwell, London, pp 430–434

Garcia-Fernàndez J, Holland PWH (1994) Archetypal organization of the amphioxus Hox gene cluster. Nature 370:563–566

Garstang W (1922) The theory of recapitulation: a critical restatement of the biogenic law. Zool J Linn Soc Lond 35:81–101

Gaskell WH (1886) On the structure, distribution and function of the nerves which innervate the visceral and vascular systems. J Physiol 7:1–81

Gegenbaur C (1878) Grundriß der vergleichenden Anatomie. Engelmann, Leipzig

Gegenbaur C (1898) Vergleichende Anatomie der Wirbelthiere, vol I. Engelmann, Leipzig

Gehring WJ (1987) Homeoboxes in the study of development. Science 236:1245–1252

Geoffroy Saint-Hilaire E (1818) Philosophie anatomique. I. Des organes respiratoires sous le rapport de la détermination et de l'identité de leurs pièces osseuses. Méquignon-Marvis, Paris

Ghiselin MT (1966) An application of the theory of definitions to systematic principles. Syst Zool 15:127–130

Goldby F, Gamble HJ (1957) The reptilian cerebral hemispheres. Biol Rev 32:383–420

González A, Smeets WJAJ (1991) Comparative analysis of dopamine and tyrosine hydroxylase immunoreactivities in the brain of two amphibians, the anuran Rana ridibunda and the urodele Pleurodeles waltlii. J Comp Neurol 303:457–477

González A, Muñoz M, Muñoz A, Marin O, Smeets WJAJ (1994) On the basal ganglia of amphibians: dopaminergic mesostriatal projections. Eur J Morphol 32:271–274

Gould SJ (1976) Grades and clades revisited. In: Masterton RB, Hodos W, Jerison H (eds) Evolution, brain, and behavior: persistent problems. Erlbaum, Hillsdale, NJ, pp 115–122

Gould SJ (1977) Ontogeny and phylogeny. Harvard University Press, Cambridge

Gould SJ (1982) Darwinism and the expansion of evolutionary theory. Science 216:380–387

Greenspan RJ, Tully T (1994) Group report: How do genes set up behavior? In: Greenspan RJ, Kyriacou CP (eds) Flexibility and constraint in behavioral systems. Wiley, Chichester, pp 65–80

Haeckel E (1866) Generelle Morphologie der Organismen: allgemeine Grundzüge der organischen Formen-Wissenschaft, mechanisch begründet durch die von Charles Darwin reformirte Descendenz-Theorie, vol 1, 2. Reimer, Berlin

Haller von Hallerstein V (1934) Äußere Gliederung des Zentralnervensystems. In: Bolk L, Göppert E, Kallius E, Lubosch W (eds) Handbuch der vergleichenden Anatomie der Wirbeltiere, vol 2, part 1. Urban und Schwarzenberg, Berlin, pp 1–318

Hennig W (1966) Phylogenetic Systematics (translated by DD Davis, R Zangerl). University of Illinois Press, Urbana

Henselmans JML, Wouterlood FG (1994) Light and electron microscopic characterization of cholinergic and dopaminergic structures in the striatal complex and the dorsal ventricular ridge of the lizard Gekko gecko. J Comp Neurol 345:69–83

Herrick CJ (1903) The doctrine of nerve components and some of its applications. J Comp Neurol 13:301–312

Herrick CJ (1908) The morphological subdivision of the brain. J Comp Neurol Psychol 18:393–408

Herrick CJ (1909) The criteria of homology in the peripheral nervous system. J Comp Neurol 19:203–209

Herrick CJ (1921) A sketch of the origin of the cerebral hemispheres. J Comp Neurol 32:429–454

Herrick CJ (1922) Functional factors in the morphology of the forebrain of fishes. In: Libro en Honor de DS Ramón y Cajal. Jiménez y Molina, Madrid, pp 143–202

Herrick CJ (1930) Localization of function in the nervous system. Proc Nat Acad Sci U S A 10:643–650

Herrick CJ (1933) Morphogenesis of the brain. J Morphol 54:233–258

Herrick CJ (1943) The cranial nerves: a review of fifty years. Denison Univ Bull J Sci Lab 38:41–51

Herrick CJ (1948) The brain of the tiger salamander Ambystoma tigrinum. University of Chicago Press, Chicago

His W (1888) Zur Geschichte des Gehirns sowie der centralen und periferischen Nervenbahnen beim menschlichen Embryo. Abh Math-Phys Kl Kgl Sächs Ges Wiss 14:339–393

Hodos W (1974) Comparative study of brain-behavior relationships. In: Goodman IJ, Schein MW (eds) Birds: brain and behavior. Academic, New York, pp 15–25

Hodos W (1976) The concept of homology and the evolution of behavior. In: Masterton RB, Hodos W, Jerison H (eds) Evolution, brain, and behavior: persistent problems. Erlbaum, Hillsdale, NJ, pp 153–167

Hodos W (1987) Homoplasy. In: Adelman G (ed) Encyclopedia of neurosciences, vol I. Birkhäuser, Boston, p 502

Hoffman HH (1963) The olfactory bulb, accessory olfactory bulb and hemisphere of some anurans. J Comp Neurol 120:317–368

Holland PWH (1992) Homeobox genes in vertebrate evolution. Bioessays 14:267–273

Holland PWH, Garcia-Fernàndez J (1996) Hox genes and chordate evolution. Dev Biol 173:382–395

Holland PWH, Graham A (1995) Evolution of regional identity in the vertebrate nervous system. Perspect Dev Neurobiol 3:17–27

Holland PWH, Garcia-Fernández J, Holland LZ, Williams NA, Holland ND (1994) The molecular control of spatial patterning in amphioxus. J Mar Biol Ass UK 74:49–60

Holm F (1901) The finer anatomy of the nervous system of Myxine glutinosa. Morphol Jahrb 29:365–401

Holmgren N (1919) Zur Anatomie des Gehirnes von Myxine. Kungliga Svenska Vetenskapsakademiens Handlingar 60:1–96

Holmgren N (1922) Points of view concerning forebrain morphology in lower vertebrates. J Comp Neurol 34:491–459

Holmgren N (1925) Points of view concerning forebrain morphology in higher vertebrates. Acta Zool 13:413–477

Holmgren N (1946) On two embryos of Myxine glutinosa. Acta Zool 27:1–90

Holmgren N, van der Horst CJ (1925) Contribution to the morphology of the brain of Ceratodus. Acta Zool 6:59–165

Hoogland PV, Vermeulen-VanderZee E (1990) Distribution of choline acetyltranferase immunoreactivity in the telencephalon of the lizard Gekko gecko. Brain Behav Evol 36:378–390

Jacobshagen E (1925) Allgemeine vergleichende Formenlehre der Tiere. Klinkhardt, Leipzig

Jacobshagen E (1927) Zur Reform der allgemeinen vergleichenden Formenlehre der Tiere. Fischer, Jena

Jacobson M (1991) Developmental neurobiology. Plenum, New York

Jansen J (1930) The brain of Myxine glutinosa. J Comp Neurol 49:359–507

Jerison HJ (1973) Evolution of the brain and intelligence. Academic, New York, London

Johnston JB (1902) An attempt to define the primitive functional divisions of the central nervous system. J Comp Neurol 12:87–106

Johnston JB (1911a) The telencephalon of selachians. J Comp Neurol 21:1–113

Johnston JB (1911b) The telencephalon of ganoids and teleosts. J Comp Neurol 21:489–581

Johnston JB (1912) The telencephalon of cyclostomes. J Comp Neurol 22:341–404

Johnston JB (1923) Further contributions of the study of the evolution of the forebrain. J Comp Neurol 36:143–192

Jones CL (1979) The morphogenesis of the thigh of the mouse with special reference to tetrapod muscle homologies. J Morphol 162:275–310

Jurio AV, Vogt M (1967) Monoamines and their metabolites in the avian brain. J Physiol (Lond) 189:489–518

Källén B (1951) Embryological studies on the nuclei and their homologization in the vertebrate forebrain. Lunds Universitets Årsskrift 62:3–34

Karten HJ (1969) The organization of the avian telencephalon and some speculations on the phylogeny of the amniote telencephalon. Ann NY Acad Sci 167:164–179

Karten HJ, Dubbeldam JL (1973) The organization and projections of the paleostriatal complex in the pigeon (Columba livia). J Comp Neurol 148:61–90

Kasper EM, Larkman AU, Lübke J, Blakemore C (1994) Pyramidal neurons in layer 5 of the rat visual cortex. I. Correlation among cell morphology, intrinsic electrophysiological properties, and axon targets. J Comp Neurol 339:459–474

Kicliter E, Ebbesson SOE (1976) Nonolfactory cortex; organization of the 'nonolfactory' telencephalon. In: Llinás R, Precht W (eds) Frog neurobiology. A handbook. Springer, Berlin Heidelberg New York, pp 946–972

Kitt CA, Brauth SE (1981) Projections of the paleostriatum upon the midbrain tegmentum in the pigeon. Neuroscience 6:1551–1566

Kleinenberg N (1886) Die Entstehung des Annelid aus der Larve von Lopadorhynchus. Z Wiss Zool 44:1–227

Krumlauf R (1992) Evolution of the vertebrate Hox homeobox genes. Bioessays 14:245–252

Kuhlenbeck H (1921a) Zur Morphologie des Urodelenvorderhirns. Jen Z Nat 57:463–490

Kuhlenbeck H (1921b) Zur Histologie des Anurenpalliums. Anat Anz 54:280–285

Kuhlenbeck H (1921c) Die Regionen des Anurenvorderhirns. Anat Anz 54:304–316

Kuhlenbeck H (1922) Über den Ursprung der Grosshirnrinde. Anat Anz 55:337–365

Kuhlenbeck H (1929) Die Grundbestandteile des Endhirns im Lichte der Bauplanlehre. Anat Anz 67:1–51

Kuhlenbeck H (1933) Bemerkungen über die theoretischen Grundlagen der Hirnmorphologie. Anat Anz 75:305–309

Kuhlenbeck H (1967) The central nervous system of vertebrates, vol 1: propaedeutics to comparative neurology. Karger, Basel

Kusunoki T, Katoda T, Kishida R (1981) Chemoarchitectonics of the forebrain of the hagfish, Eptatretus burgeri. J Hirnforsch 22:285–293

Løvtrup S (1987) The theoretical basis of evolutionary thought. Ann Sci Nat Zool 8:219–236

Lankester ER (1870) On the use of the term homology in modern zoology, and the distinction between homogenetic and homoplastic agreements. Ann Mag Nat Hist 6:34–44

Llinás R (1969) Neurobiology of cerebellar evolution and development. American Medical Association, Chicago

Lubosch W (1925) Grundriss der wissenschaftlichen Anatomie. Thieme, Leipzig

Lubosch W (1931) Geschichte der vergleichenden Anatomie. In: Bolk L, Göppert E, Kallius E, Lubosch W (eds) Handbuch der vergleichenden Anatomie der Wirbeltiere, vol 1. Urban and Schwarzenberg, Berlin, pp 3–76

MacLean P (1970) The triune brain, emotion and scientific bias. In: Schmitt FO (ed) The neuroscience second study program. Rockefeller University Press, New York, pp 336–-349

Maynard Smith J, Szathmáry E (1995) The major transitions in evolution. Freeman, Oxford

Mayr E (1969) Grundgedanken der Evolutionsbiologie. Naturwissenschaften 56:392–397

Mayr E, Ashlock PD (1991) Principles of systematic zoology. McGraw-Hill, New York

McCormick CA (1982) The organization of the octavolateralis area in actinopterygian fishes: a new interpretation. J Morphol 171:159–181

McGinnis W, Krumlauf R (1992) Homeobox genes and axial patterning. Cell 68:283–302

Meek J (1994) Catecholamines in the brains of Osteichthyes (bony fishes). In: Smeets WJAJ, Reiner A (eds) Phylogeny and development of catecholamine systems in the CNS of vertebrates. Cambridge University Press, Cambridge, pp 49–76

Meredith GE, Smeets WJAJ (1987) Immunocytochemical analysis of the dopamine system in the forebrain and midbrain of Raja radiata: evidence for a substantia nigra and ventral tegmental area in cartilaginous fish. J Comp Neurol 265:530–548

Misson J-P, Austin CP, Takahashi T, Cepko CL, Caviness VS Jr (1991) The alignment of migrating neural cells in relation to the murine neopallial radial glial fiber system. Cerebral Cortex 1:221–229

Naef A (1919) Idealistische Morphologie und Phylogenetik. Fischer, Jena

Nauta WJH, Karten HJ (1970) A general profile of the vertebrate brain, with sidelights on the ancestry of cerebral cortex. Schmitt FO (ed) The neurosciences: second study program. Rockefeller University Press, New York, pp 7–26

Nieuwenhuys R (1962) The morphogenesis and the general structure of the actinopterygian forebrain. Acta Morphol Neerl Scand 5:65–78

Nieuwenhuys R (1963) The comparative anatomy of the actinopterygian forebrain. J Hirnforsch 6:13–192

Nieuwenhuys R (1964) Further studies on the general structure of the actinopterygian forebrain. Acta Morphol Neerl Scand 6:65–79

Nieuwenhuys R (1966) The interpretation of the cell masses in the teleostean forebrain. In: Hassler R, Stephan H (ed) Evolution of the forebrain. Phylogenesis and ontogenesis of the forebrain. Thieme, Stuttgart, pp 32–39

Nieuwenhuys R (1969) A survey of the structure of the forebrain in higher bony fishes (Osteichthyes). Ann N Y Acad Sci 167:31–64

Nieuwenhuys R (1972) Topological analysis of the brainstem of the lamprey Lampetra fluviatilis. J Comp Neurol 145:165–178

Nieuwenhuys R (1974) Topological analysis of the brainstem: a general introduction. J Comp Neurol 156:255–276

Nieuwenhuys R (1977) The brain of the lamprey in a comparative perspective. Ann N Y Acad Sci 229:97–145

Nieuwenhuys R (1994) Comparative neuroanatomy: place, principles, practice and programme. Eur J Morphol 32:142–155

Nieuwenhuys R, Bodenheimer TS (1966) The diencephalon of the primitive bony fish Polypterus in the light of the problem of homology. J Morphol 118:415–450

Nieuwenhuys R, Hickey M (1965) A survey of the forebrain of the Australian lungfish Neoceratodus forsteri. J Hirnforsch 7:433–452

Nieuwenhuys R, Meek J (1990a) The telencephalon of actinopterygian fishes. In: Jones, EG, Peters A (eds) Cerebral cortex, vol 8A. Plenum, New York, pp 31–73

Nieuwenhuys R, Meek J (1990b) The telencephalon of sarcopterygian fishes. In: Jones, EG, Peters A (eds) Cerebral cortex, vol 8A. Plenum, New York, pp 75–106

Noback CR, Shriver JE (1969) Encephalization and the lemniscal systems during phylogeny. Ann N Y Acad Sci 167:118–128

Northcutt RG (1974) Some histochemical observations on the telencephalon of the bullfrog, Rana catesbeiana. J Comp Neurol 157:379–390

Northcutt RG (1981) Evolution of the telencephalon in nonmammals. Annu Rev Neurosci 4:301–350

Northcutt RG (1984) Evolution of the vertebrate central nervous system: patterns and processes. Am Zool 24:701–716

Northcutt RG (1985) The brain and sense organs of the earliest craniates: reconstruction of a morphotype. In: Foreman RE, Gorbman A, Dodds JM, Olson R (eds) Evolutionary biology of primitive fishes. Plenum, New York, pp 81–112

Northcutt RG (1986a) Lungfish neural characters and their bearing on sarcopterygian phylogeny. J Morphol Suppl 1:277–297

Northcutt RG (1986b) Speculations on pattern and cause. In: Cohen MJ, Strumwasser F (eds) Comparative neurobiology: modes of communication in the nervous system. Wiley, Chichester, pp 351–378

Northcutt RG (1990) Ontogeny and phylogeny: a re-evaluation of conceptual relationships and some applications. Brain Behav Evol 36:116–140

Northcutt RG (1995) The forebrain of gnathostomes: in search of a morphotype. Brain Behav Evol 46:278–318

Northcutt RG, Puzdrowski RL (1988) Projections of the olfactory bulb and nervus terminalis in the silver lamprey. Brain Behav Evol 32:96–107

Northcutt RG, Reiner A, Karten HJ (1988) An immunohistochemical study of the telencephalon of the spiny dogfish, Squalus acanthias. J Comp Neurol 227:250–267

Ohno S (1970) Evolution by gene duplication. Springer, Berlin Heidelberg New York

Osse JWM (1983) Morphology and evolution. Acta Morphol Neerl Scand 21:49–67

Owen R (1843) Lectures on the comparative anatomy and physiology of the invertebrate animals. Longmans-Brown-Green and Longmans, London

Owen R (1849) On the nature of limbs. Van Voorst, London

Owen R (1866) On the anatomy of vertebrates. Fishes and reptiles. Longmans Green, London

Palmgren A (1921) Embryological and morphological studies on the mid-brain and cerebellum of vertebrates. Acta Zool (Stockh) 1:1–94

Parent A, Olivier A (1970) Comparative histochemical study of the corpus striatum. J Hirnforsch 12:73–81

Patterson C (1988) Homology in classical and molecular biology. Mol Biol Evol 5:603–625

Polenova OA, Vesselkin NP (1993) Olfactory and nonolfactory projections in the river lamprey (Lampetra fluviatilis) telencephalon. J Hirnforsch 34:261–279

Reichert C (1837) Über die Visceralbogen der Wirbeltiere im allgemeinen und deren Metamorphosen bei den Vögeln und Säugetieren. Arch Anat Physiol: 120–222

Reiner A (1994) Catecholaminergic innervation of the basal ganglia in mammals: anatomy and function. In: Smeets WJAJ, Reiner A (eds) Phylogeny and development of catecholamine systems in the CNS of vertebrates. Cambridge University Press, Cambridge, pp 247–272

Reiner A, Northcutt RG (1987) An immunohistochemical study of the telencephalon of the African lungfish. J Comp Neurol 256:463–481

Reiner A, Northcutt RG (1992) An immunohistochemical study of the telencephalon of the Senegal bichir (Polypterus senegalus). J Comp Neurol 319:359–386

Reiner A, Brauth SE, Karten HJ (1984) Evolution of the amniote basal ganglia. Trends Neurosci 7:320–325

Reiner A, Karle EJ, Anderson KD, Medina L (1994) Catecholaminergic perikarya and fibers in the avian nervous system. In: Smeets WJAJ, Reiner A (eds) Phylogeny and development of catecholamine systems in the CNS of vertebrates. Cambridge University Press, Cambridge, pp 135–181

Remane A (1954) Morphologie als Homologienforschung. Verh Dtsch Zool 1954:159–183

Remane A (1956) Die Grundlagen des natürlichen Systems, der vergleichenden Anatomie und der Phylogenetik. Geest and Portig, Leipzig

Remane A (1966) Phylogenetische entwicklungsregeln von Organen. In: Hassler R, Stephan H (eds) Evolution of the forebrain. Thieme, Stuttgart, pp 1–8

Rose M (1935) Cytoarchitektonik und Myeloarchitektonik der Groszhirnrinde. In: Bumke O, Foerster O (eds) Handbuch der Neurologie, vol 1, Anatomie. Springer, Berlin, pp 588–778

Roth G (1987) Visual behavior in salamanders. Springer, Berlin Heidelberg New York

Rudebeck B (1945) Contribution to forebrain morphology in Dipnoi. Acta Zool 26:9–156

Russell ES (1916) Form and function. Murray, London

Schaeffer B (1965) The role of experimentation in the origin of higher levels of organization. Syst Zool 14:318–336

Schnitzlein HN, Crosby EC (1967) The telencephalon of the lungfish, Protopterus. J Hirnforsch 9:105–149

Sewertzoff AN (1931) Morphologische Gesetzmäßigkeiten der Evolution. Fischer, Jena

Simpson GG (1961) Principles of animal taxonomy. Columbia University Press, New York (Columbia biological series, vol 20)

Smart JHM, Sturrock B (1979) Ontogeny of the neostriatum. In: Divac I, Oberg RGE (eds) The neostriatum. Pergamon, Oxford, pp 127–146

Smeets WJAJ (1988) The monoaminergic systems of reptiles investigated with specific antibodies against serotonin, dopamine, and noradrenaline. In: Schwerdtfeger WK, Smeets WJAJ (eds) The forebrain of reptiles. Current concepts of structure and function. Karger, Basel, pp 97–109

Smeets WJAJ (1991) Comparative aspects of the distribution of substance P and dopamine immunoreactivity in the substantia nigra of amniotes. Brain Behav Evol 37:179–188

Smeets WJAJ (1994) Catecholamine systems in the CNS of reptiles: structure and functional correlations. In: Smeets WJAJ, Reiner A (eds) Phylogeny and development of catecholamine systems in the CNS of vertebrates. Cambridge Universtiy Press, Cambridge, pp 103–133

Smeets WJAJ, Jonker AJ, Hoogland PV (1987) Distribution of dopamine in the forebrain and midbrain of the red-eared turtle, Pseudemys scripta elegans, reinvestigated using antibodies against dopamine. Brain Behav Evol 30:121–142

Smith HM (1967) Biological similarities and homologies. Syst Zool 16:101–102

Smith LM, Ebner FF, Colonnier M (1980) The thalamocortical projection in Pseudemys turtles: a quantitative electron microscope study. J Comp Neurol 190:445–462

Sperry RW (1958) Developmental basis of behavior. In: Roe A, Simpson GG (eds) Behavior and evolution. Yale University Press, New Haven, pp 128–139

Starck D (1965) Vergleichende Anatomie der Wirbeltiere von Gegenbaur bis heute. Verh Dtsch Zool 28:51–67

Starck D (1977) Tendenzen und Strömungen in der vergleichenden Anatomie der Wirbeltiere im 19. und 20. Jahrhundert. Natur Museum 107:93–102

Striedter GF (1991) Auditory, electrosensory and mechanosensory lateral line pathways through the forebrain in channel catfishes. J Comp Neurol 312:311–331

Striedter GF (1992) Phylogenetic changes in the connections of the lateral preglomerular nucleus in ostariophysan teleosts: a plurastic view of brain evolution. Brain Behav Evol 39:329–357

Striedter GF, Northcutt RG (1991) Biological hierarchies and the concept of homology. Brain Behav Evol 38:177–189

Swanson LW (1987) The hypothalamus. In: Björklund A, Swanson LW (eds) Handbook of chemical neuroanatomy, vol 5, integrated systems of the CNS, part I. Elsevier, Amsterdam, pp 125–277

Swanson LW (1989) The neural basis of motivated behavior. Acta Morphol Neerl Scand 26:165–176

Swanson LW (1991) Biochemical switching in hypothalamic circuits mediating responses to stress. Prog Brain Res 87:181–200

Swertzoff AN (1931) Morphologische Gesetzmässigkeiten der Evolution. Fisher, Jena

ten Donkelaar HJ (1982) Organization of descending pathways to the spinal cord in amphibians and reptiles. Prog Brain Res 57:25–67

Thompson d'Arcy W (1961) On growth and form. Abridged edition by JT Boinner. Cambridge University Press, Cambridge

Tinbergen N (1951) The study of instinct. Clarendon, Oxford

van der Klaauw CJ (1947) Inleiding. In: Ihle JEW (ed) Leerboek der vergelijkende ontleedkunde van de vertebraten. Oosthoek, Utrecht, pp 1–10

van der Linden JAM, ten Donkelaar HJ (1987) Observations on the development of cerebellar afferents in Xenopus laevis. Anat Embryol (Berl) 176:431–439

von Baer KE (1828) Über die Entwickelungsgeschichte der Thiere, Beobachtung und Reflexion. Bornträger, Königsberg

von Bartheld CS, Collin SP, Meyer DL (1990) Dorsomedial telencephalon of lungfishes: a pallial or subpallial structure? Criteria based on histology, connectivity, and histochemistry. J Comp Neurol 294:14–29

von Goethe JW (1795) Erster Entwurf einer allgemeinen Einleitung in die vergleichende Anatomie, ausgehend von der Osteologie. In: von Goethe JW (ed) Schriften zur Anatomie, Zoologie und Physiognomik (1962). Deutscher Taschenbuch Gesamtausgabe, vol 37, Munich, pp 69–122

von Kupffer C (1906) Die Morphogenie des Zentralnervensystems. In: Hertwig O (ed) Handbuch der Vergleichenden und Experimentellen Entwickkungslehre der Wirbeltiere, vol 2, part 3. Fischer, Jena, pp 1–272

Wächtler K (1974) The distribution of acetylcholinesterase in the cyclostome brain. I. Lampetra planeri (L). Cell Tissue Res 152:259–270

Wächtler K (1975) The distribution of acetylcholinesterase in the cyclostome brain. II. Myxine glutinosa. Cell Tissue Res 159:109–120

Wächtler K (1983) The acetylcholine-system of cyclostomes with special references to the telencephalon. J Hirnforsch 24:63–70

Wake DB (1994) Review of: Hall BK (ed) Homology, the hierarchical basis of comparative biology. Academic, San Diego. Science 265:268–269

Watrous LE, Wheeler QD (1981) The outgroup comparison method of character analysis. Syst Zool 30:1–11

Weiss P (1955) Special vertebrate organogenesis. I. Nervous system (Neurogenesis). In: Willier BH, Weiss PA, Hamburger V (eds) Analysis of development. Saunders, London, pp 346–401

Wicht H, Northcutt RG (1992) The forebrain of the Pacific hagfish: a cladistic reconstruction of the ancestral craniate forebrain. Brain Behav Evol 40:25–64

Wicht H, Northcutt RG (1993) Secondary olfactory projections and pallial topography in the pacific hagfish, Eptatretus stouti. J Comp Neurol 337:529–542

Wicht H, Northcutt RG (1994) An immunohistochemical study of the telencephalon and the diencephalon in a myxinoid jawless fish, the Pacific hagfish, Eptatretus stouti. Brain Behav Evol 43:140–161

Wiley EO (1981) Phylogenetics. Wiley, New York

Wulliman MF, Northcutt RG (1990) Visual and electrosensory circuits of the diencephalon in mormyrids: an evolutionary perspective. J Comp Neurol 297:537–552

Wullimann MF, Roth G (1994) Descending telencephalic information reaches longitudinal torus and cerebellum via the dorsal preglomerular nucleus in the teleost fish, Pantodon buchholzi: a case of neural preaptation? Brain Behav Evol 44:338–352

Wynne B, Güntürkün O (1995) Dopaminergic innervation of the telencephalon of the pigeon (Columba livia): a study with antibodies against tyrosine hydroxylase and dopamine. J Comp Neurol 357:446–464

Zangerl R (1948) The methods of comparative anatomy and its contribution to the study of evolution. Evolution 2:351–374

CHAPTER 7

Notes on Techniques

H. J. TEN DONKELAAR AND C. NICHOLSON

7.1
Introduction

The early students of the structure of the vertebrate CNS (see Conn 1948; Clarke and O'Malley 1968; Spillane 1981; Shepherd 1991) had to rely on the quick analysis of either fresh or, at best, partially preserved biological material. Thus, Reil's introduction of alcohol for fixation in 1809 was an important step. Formalin fixation was introduced much later (Blum 1893). Both Remak's (1836) description of axons and their sheaths, and Purkyně's 1838 paper (Purkyně 1838) paper on the cerebellar neurons named after him (see Chap. 1, Fig. 1.1), were based on unstained embryonic material. The first stains to be used, predominantly carmine (von Gerlach 1858), gave rather unsatisfactory pictures. Nevertheless, Deiters (1865) was able to differentiate between dendrites and axons. The introduction of hematoxylin, the Nissl technique, the Weigert technique for selective staining of myelin sheaths, and Golgi's (1873) method for selectively impregnating nerve cells with silver nitrate opened up vast new possibilities for studying the detailed structure of the CNS.

Although the Marchi technique (Marchi and Algeri 1885) revealed degenerating myelin sheaths, and so enabled fibre connections to be traced, neuroanatomists had to wait for the much later developed silver impregnation techniques (Nauta and Gygax 1954; Fink and Heimer 1967) to visualise degenerating unmyelinated axons and axon terminals. The use of tracers based on axonal transport for studies of neuronal connectivity (Kristensson and Olsson 1971; Cowan et al. 1972; LaVail and LaVail 1972), however, stimulated a renaissance in neuroanatomy. The simultaneous development of fluorescence microscopy (see Fuxe et al. 1970) and immunohistochemical techniques (e.g., Sternberger et al. 1970) for the demonstration and mapping of neurotransmitters and their enzymes, neuropeptides and proteins greatly improved our knowledge of chemical neuroanatomy. The technique of in situ hybridisation (Valentino et al. 1992; Wilkinson 1992) is rapidly becoming an essential tool in this field. The combination of various techniques and the use of electron microscopy has made it possible to study microcircuits, synaptic interactions, and the transmitters involved.

Electrophysiological techniques have been applied to the nervous system for a long time, starting around 1786 with Galvani's observations on muscle contractions in a frog, published in 1791. Subsequent studies by Bell, Magendie, Müller, Marshall Hall and many others (see Brazier 1959, 1961; Spillane 1981) laid the basis for twentieth-century neurophysiology founded by Sherrington (e.g., Sherrington 1906). With improving neurophysiological instrumentation, e.g., the development of the Horsley and Clarke (1908) stereotaxic system, the development of amplifiers and cathode-ray oscillographs, microelectrodes and refined stimulation techniques, it became possible to record the action potentials of single fibres or cells (single-unit recording). The recording of antidromic potentials, which occurs when the axon is stimulated at some distance from the cell body, leading to backward propagation of action potentials into the soma, has been especially useful for correlation with anatomical data. The further application of electrophysiological and tracing techniques to in vitro preparations such as brain slices or isolated brains has been particularly advantageous.

This chapter provides an overview of the various techniques used in the special section (Chaps. 9–22) as well as a bibliography of brain atlases arranged by chapter.

7.2
Tissue-Staining Techniques

Much of the organisation of the CNS can be studied by using specific tissue-staining techniques. *Nissl's* (1885, 1894) *technique*, involving the use of basic aniline dyes to stain sections fixed in alcohol, remains the routine technique for investigations of the cytoarchitecture of the CNS. For the staining of nerve fibres, various silver-impregnation techniques have been developed. In 1904, Ramón y Cajal and Bielschowsky, independently of each other, developed the first silver methods that proved suitable for routine use. *Bodian's* (1936) *modification* of Bielschowsky's technique is still widely used to analyse the predominantly unmyelinated fibres in the anamniote CNS. Rager's modification of the Bodian technique is very suitable for developing fibre tracts (Rager et al. 1979). For the staining of peripheral nerves Sudan Black B has been advocated (Filipski and Wilson 1984; Nishikawa 1987).

Many features of the main fibre connections of the CNS of amniotes have been elucidated by staining cell bodies and myelin sheaths. Ever since Weigert discovered that myelin treated in potassium dichromate stains bright red with acid fuchsin (Weigert 1884; Pal 1887), the staining of myelin sheaths around the axons of neurons has been of great value in neuroanatomical studies. More recent modifications of the *Weigert-Pal technique* (see Lowe and Cox 1990) include the procedures developed by Loyez (1920), Weil (1928) and Woelcke (1942). These techniques, however, involve chromate treatment, which is incompatible with the use of a cresylecht violet counterstain for the mapping of the cell bodies. The discovery that Luxol

Fast Blue stains the myelin without requiring previous chromation of the sections was therefore of great value (Klüver and Barrera 1953). The *Klüver-Barrera technique* combines a myelin stain with a cresyl violet staining of cell bodies and is the method of choice for the simultaneous visualisation of neurons and myelinated fibres. With *Häggqvist's* (1936) *modification* of the Alzheimer-Mann methylene blue-eosin stain, the axons stain blue and the myelin sheaths of the individual fibres stain red. It therefore can be used to analyse the calibre spectrum of fibre systems.

Golgi (1873) made a major advance in the application of silver staining. *Golgi's technique* and its modifications (see Ramón-Moliner 1970; Scheibel and Scheibel 1970; Valverde 1970; Millhouse 1981) only randomly impregnate cells. Ramón y Cajal, van Gehuchten, Retzius, and von Lenhossek, among others, adopted Golgi's impregnation method. Golgi's (1873) original ('slow') method was to harden fresh tissue in potassium dichromate, followed by immersion in weak silver nitrate. In a later modification (Golgi 1875) osmium tetroxide and potassium dichromate were used for hardening the tissue ('rapid Golgi'). Cox's (1891) modification (Golgi-Cox) fixes the tissue initially in a potassium dichromate/mercuric chloride solution, followed by hardening in dichromate. The value of the Golgi technique is that it reveals only a small number (5–10%) of the neurons in an area, but it impregnates these neurons with their dendrites in full detail, although less so their axons. The small number of cells impregnated permits the cutting of thick sections (80–300 μm), which makes it possible to obtain a three-dimensional view of neurons with their processes (see Chap. 1, Fig. 1.1 for some examples of the Golgi method).

Intracellular staining is a valuable tool for studying the morphology of a neuron (see Kater and Nicholson 1973; Kitai and Bishop 1981). Introduced into a neuron by intracellular injection, such stains provide a means for correlating a neuron's morphology with its physiological properties and permit unequivocal identification of the type of cell recorded in a physiological experiment. Intracellular stains can be roughly divided into two classes. The first are fluorescent dyes such as *Procion Yellow* (Stretton and Kravitz 1968) and *Lucifer Yellow* (Stewart 1978) and have the advantage that filled processes are easily detected when viewed with near-ultraviolet light and the appropriate filters. Fluorescent dyes have the disadvantage of fading with prolonged viewing (Stewart 1978). The second major class of stains are light absorbing. The chief examples are *horseradish peroxidase* (HRP; see Kitai and Bishop 1981; Kitai et al. 1989), *cobalt chloride* (Pitman et al. 1972), *nickel-lysine* (Fredman 1987) and *biocytin* (Horikawa and Armstrong 1988; Kita and Armstrong 1991). These stains are typically invisible when injected into neurons. The tissue is subsequently treated to form a dark reaction product which can be viewed with conventional optics. Since the reaction product is electron dense, it can be used for combined light- and electron-microscopic studies (e.g., Cullheim and Kellerth 1976; Mesulam 1982; Izzo 1991).

7.3
Tract-Tracing Techniques

7.3.1
The Classical Degeneration Techniques

Much of our present-day knowledge of the fibre connections in the vertebrate brain has been obtained by analysis of the degeneration which occurs following lesions in the nervous system. Waller (1850) first described the process of *antero-grade degeneration* in the peripheral nervous sys-

tem. After cutting the nerve supply to the tongue in frogs, he observed the changes in teased, unstained preparations. However, the first person to realise that anterograde degeneration could be used as a tracing technique was Türck, a Viennese neurologist, who in 1849 described his findings in autopsy material of human beings who had suffered spinal cord compression (see Clarke and O'Malley 1968; Leonard 1979). Ramón y Cajal (1928) found that transected peripheral axons seemed to dissolve and then coalesce into droplets which were highly sensitive to silver impregnation (degeneration argyrophilia).

The changes in a neuron which follow transection of its axon are shown in Fig. 7.1. Effects confined to the parts of a damaged neuron distal to the lesion are called *anterograde*, those affecting the soma and axon proximal to the lesion *retrograde*. Changes in undamaged neurons connected synaptically with the damaged one are *transneuronal* or *trans-synaptic* changes (see Cowan 1970; Brodal 1981). If an axon is transected, its peripheral parts, including its terminal ramifications and boutons, and the myelin sheath degenerate. *Wallerian degen-*

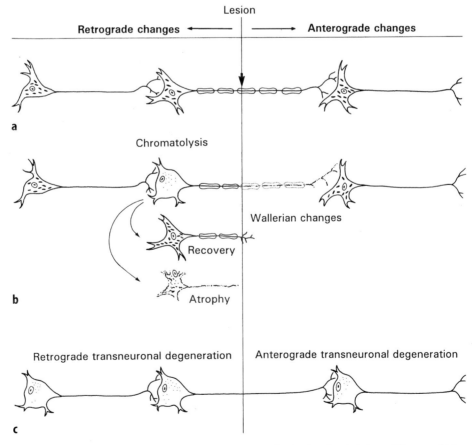

Fig. 7.1a–c. The changes in a neuron which follow transection of its axon. See text for further details (after Williams and Warwick 1975 and Brodal 1981)

eration can be shown by *Marchi staining* of the degenerating myelin (Marchi and Algeri 1885), but, although widely used, the Marchi technique has two limitations: it cannot identify the terminal field of tracts, because axons lose their myelin sheath before they terminate, and unmyelinated fibres, of course, cannot be traced at all. Nonidez (1939) introduced a more sensitive silver-impregnation technique.

Glees (1946) and Nauta (1950; Nauta and Gygax 1951, 1954), with modifications by others (see Heimer 1970a; de Olmos et al. 1981), developed techniques for staining terminal structures and unmyelinated fibres. Ebbesson, in particular, applied the *selective silver impregnation* of degenerating axons and their synaptic terminals to non-mammalian species (see Ebbesson 1970). Anterograde degeneration techniques have two major disadvantages. The techniques are useful mainly when a single tract or homogeneous group of neurons can be selectively damaged. If fibres from other sources travel through the injured region one *cannot* unequivocally identify the origin of a particular pathway. A second problem is survival time. Degeneration takes time, particularly in cold-blooded vertebrates, and different fibre systems degenerate at different rates. Furthermore, the terminal arborisations degenerate more rapidly than the axon itself. Among the numerous modifications of the *Nauta technique* the two procedures described by Fink and Heimer (1967) appear to have been most widely used. With their techniques the terminal boutons are usually clearly impregnated, while the terminal fibres are less conspicuous. Therefore, the field of termination of a degenerating fibre bundle can be more precisely outlined than with the Nauta technique. Electron-microscopic studies of degenerating boutons (see Heimer 1970b) opened up new possibilities for establishing the synaptic relationship of fibre systems in great detail. One can determine not only whether a contact between two structures is a synapse, but also whether there are synaptic contacts with dendrites, spines or the soma itself. Although rather laborious, anterograde degeneration of fibres and terminals can be combined with Golgi impregnation (Blackstad 1965; Somogyi et al. 1979).

The *retrograde changes* that occur in the perikaryon of a cell after injury have been of great value for studying the site of origin of connections of the nervous system. The typical retrograde changes include: (a) an obvious dissolution of the large Nissl bodies (*chromatolysis*) – in certain cells a residual amount of Nissl substance remains, largely confined to a narrow peripheral zone (Lieberman 1971); (b) swelling of the perikaryon; and (c) a peripheral displacement (eccentricity) of the nucleus. These changes were described as *primäre Reizung* by Nissl (1892, 1894), or as *réaction à distance* by Marinesco (1898). Early changes after severance of the axon can vary considerably in intensity. Usually, they are clearer in very young animals than in adults (von Gudden 1870; Brodal 1939, 1940). In mammals, the acute retrograde cell changes are most pronounced between 1 and 3 weeks after axotomy. In cold-blooded vertebrates such as reptiles, however, this process takes at least 3 weeks (Robinson 1969; ten Donkelaar 1976). It should be emphasised that it is extremely difficult to judge retrograde changes in small neurons (see also Beran and Martin 1971). A second limitation of the retrograde cell degeneration techniques lies in the apparent resistance of a neuron to severance of its axon if collateral branches are present proximal to the site of the injury. These collaterals are thought to be capable of protecting the cell against the harmful effect of losing a substantial portion of its cytoplasm (Ramón y Cajal 1928; Cragg 1970; Lieberman 1971).

7.3.2
Modern Tract-Tracing Techniques

The limitations inherent to the anterograde and retrograde degeneration techniques were mostly solved with the introduction of the axonal transport techniques. In tracing fibre connections, both central and peripheral, two new techniques have proved particularly valuable: the autoradiographic identification of labelled substances (e.g., [3H] leucine, [3H] proline), absorbed by neurons and transported anterogradely by their axons (Fig. 7.3a), and the retrograde transport (Fig. 7.3b) of identifiable 'marker' substances (e.g., horseradish peroxidase, fluorescent dyes). Nowadays, the *autoradiographic technique*, introduced by Lasek et al. (1968) and Hendrickson (1969) and popularised by Cowan et al. (1972), has been replaced by anterograde transport techniques making use of lectins or dextran amines.

The axonal transport techniques depend on the physiological activity of neurons. Weiss and Hiscoe (1948) first demonstrated the phenomenon of *axonal transport* (see Fig. 7.2). Following ligation of a peripheral nerve, 'damming' of axoplasm was found proximal to the constriction. Neurons continuously produce protein and other material and transport it at a rate of about 1 mm/day to their terminals. Lubínska (1964) demonstrated the presence of retrograde transport in axons, whereas Ochs and Burger's (1958) studies with labelled aminoacids led to the discovery of the fast phase of anterograde

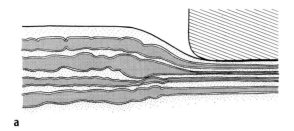

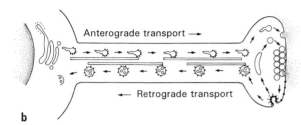

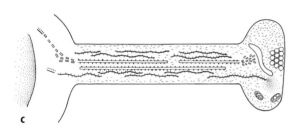

Fig. 7.2a–c. Axonal transport mechanisms: **a** Weiss and Hiscoe's (1948) first demonstration of the axonal transport phenomenon; **b** the mechanism for fast anterograde and retrograde transport; **c** the mechanism for slow anterograde transport (**b** and **c** based on Lasek and Katz 1987, as well as on Hammerschlag et al. 1994)

axonal transport (see Ochs 1982). Current views on the mechanisms of axonal transport are shown in Fig. 7.2. At least five components can be distinguished (see Grafstein and Forman 1980; Lasek and Katz 1987; Vallee and Bloom 1991; Hammerschlag et al. 1994): (a) fast anterograde transport of small vesiculotubular structures, neurotransmitters, membrane proteins and lipids; (b) transport of mitochondria; (c) fast retrograde transport of lysosomal vesicles and enzymes; (d) slow anterograde transport of microfilaments and metabolic enzymes; (e) slow anterograde transport of neurofilaments and microtubules (see also Chap. 1).

Anterograde tracing techniques making use of plant lectins and bacterial toxins (see Sawchenko and Gerfen 1985) and of the more recently introduced fluorescent or biotinylated *dextran amines* (Glover et al. 1986; Nance and Burns 1990; Schmued et al. 1990; Veenman et al. 1992) have considerable

advantages over the previously used autoradiographic techniques. *Wheat germ agglutinin* (WGA; from *Triticum vulgaris*) alone or conjugated with HRP has been particularly thoroughly studied (see Trojanowski et al. 1982). WGA is known to bind *N*-acetyl-D-glucosamine and sialic acid, which are common constituents of glycoconjugates found on neuronal cell membranes. These sugars appear capable of acting as 'receptors' for exogenously applied WGA or WGA-HRP (Sawchenko and Gerfen 1985). WGA-HRP is now used mainly as a sensitive probe for retrograde tracing (see Trojanowski et al. 1982; Llewellyn-Smith et al. 1992). *Cholera toxin B* subunit (CTb), either unconjugated or conjugated with HRP, is also a sensitive retrograde tracer (Trojanowski et al. 1982; Wan et al. 1982). CTb can be conjugated with fluorescein isothiocyanate (FITC) or with tetramethylrhodamine isothiocyanate (TRITC) and used for double-labelling studies (Dederen et al. 1994). *Phaseolus vulgaris leucoagglutinin* (PHA-L; from red kidney beans), delivered iontophoretically into the CNS through fine-tipped micropipettes and localised immunohistochemically, labels discretely, directly, and quite completely cell bodies and their dendrites at their site of injection, along with their axons and terminal specialisations (Gerfen and Sawchenko 1984; Gerfen et al. 1989; Groenewegen and Wouterlood 1990). The PHA-L technique is easily adaptable for use in combination with other techniques (see Záborszky and Heimer 1989). Nonetheless, PHA-L has two significant disadvantages: (a) a multistep immunohistochemical procedure must be used to visualise the tracer, and (b) PHA-L labelling can be capricious in its efficacy. Effective alternatives are biocytin (see King et al. 1989; Kita and Armstrong 1991) and biotinylated dextran amine (BDA; Veenman et al. 1992). Biotinylated tracers have the advantage of visualisation with a simple avidin-biotinylated HRP complex (ABC) procedure followed by a standard or metal-enhanced diaminobenzidine reaction (Hsu et al. 1981). BDA has been shown to be an effective anterograde pathway tracer which can easily be used in combination with immunohistochemical labelling or other tracers (Veenman et al. 1992; Dolleman-van der Weel et al. 1994) and for electron microscopy (Wouterlood and Jorritsma-Byham 1993).

The introduction of *horseradish peroxidase* (HRP) as a retrograde tracer (Kristensson and Olson 1971; LaVail and LaVail 1972) has greatly accelerated and improved the description of connectivity in the vertebrate nervous system. HRP was first used for retrograde mapping of long-distance neuronal projections, but it has proved to be very versatile. It can be used as an effective ante-

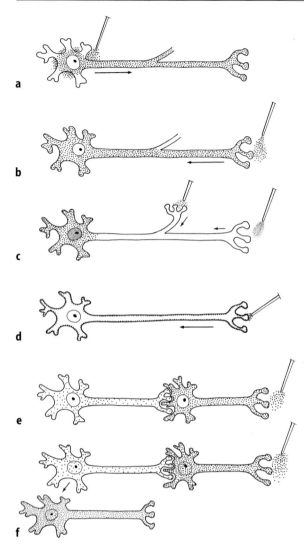

a

b

c

d

e

f

Fig. 7.3a–f. Modern tract-tracing techniques. **a** Anterograde transport from uptake sites in neuronal soma (e.g., isotopically labelled amino acid, lectins, BDA); **b** retrograde transport from terminal uptake sites (e.g., HRP); **c** multiple labelling technique in which one fluorescent tracer (e.g., Nuclear Yellow) is taken up by terminals, another one (e.g., Fast Blue) by a collateral; **d** retrograde labelling of membranes by externally applied lipophilic dyes (e.g., DiI); **e** retrograde transneuronal transport of WGA-HRP; **f** retrograde transneuronal transport of neurotropic viruses (e.g., herpes simplex virus type 1). The viruses are replicated in recipient neurons after transneuronal transfer (based on Williams and Warwick 1975 and Kuypers and Ugolini 1990)

times (depending on the system and species used), perfusion, and cutting, sections can be stained in various ways: the incubation procedure using 3,3'-diaminobenzidine tetrahydrochloride (DAB), described by Graham and Karnovsky (1966), Adams' (1981) heavy metal intensification of the DAB procedure, and Mesulam's (1978) technique, using tetramethylbenzidine (TMB) as chromogen, are the most widely used. Limitations, however, are the difficulty of combining HRP with other tracers, and its limited use in in vitro studies. Therefore, fluorescent molecules with different emission spectra were introduced by Kuypers and co-workers (see Kuypers et al. 1980; Kuypers and Huisman 1984). Retrograde *fluorescent tracers* such as Fast Blue (FB) and Nuclear Yellow (NY) can be applied similarly to HRP and are particularly useful for studying the existence of axonal branching (see Fig. 7.3c). In the multiple retrograde fluorescent tracer technique, more than one fluorescent substance (e.g., FB and NY) can be used as tracers, which, after retrograde transport through divergent axon collaterals, reach the same parent cell body where they can be visualised independently by means of fluorescence microscopy (FB in the cytoplasm, NY in the nucleus of the cell body).

Fluorescent dextran amines and lipophilic *carbocyanine tracers* such as DiI and DiO, originally developed to study cell lineage (Gimlich and Braun 1985) and as markers in cell cultures (Honig and Hume 1986), respectively, are now widely used as neuronal tracers. Fluorescein- and rhodamine-conjugated dextran amines (FDA and RDA) are effective retrograde and anterograde tracers (see Nance and Burns 1990), particularly for the developing nervous system (Glover et al. 1986; Manns and Fritzsch 1991; Fritzsch 1993). The fluorescent carbocyanine dyes DiI and DiO are excellent retrograde and anterograde tracers in the developing nervous system, both in vivo and in vitro (Honig and Hume 1989). These dyes are retained in neurons placed in culture and are non-toxic. The major mechanism of translocation for these molecules is lateral diffusion in the membrane (Fig. 7.3d), rather than fast axonal transport (Honig and Hume 1986). Godement and co-workers showed that carbocyanine dyes can be used to label axonal projections in *fixed* tissue (Godement et al. 1987). DiI tracing can be used in combination with immunohistochemistry (Holmqvist et al. 1992) and is also effective in in vitro preparations.

Tracing chains of neurons requires the use of *transneuronal tracers*, which are transferred between connected neurons (Fig. 7.3e,f). Transneuronal labelling can be obtained with *tetanus toxin* fragments (e.g., Schwab and Thoenen 1976), WGA

rograde tracer, it is a superb intracellular marker (Bishop and King 1982), and it can be localised at the electron-microscopic level (see Mesulam 1982). HRP has been widely applied for studies of neuronal connectivity throughout vertebrates. It can be used in various ways, e.g., by injection, in a slow-release gel (Griffin et al. 1979), or recrystallised on a fine tungsten needle. After appropriate survival

(Ruda and Coulter 1982) and the WGA-HRP conjugate (e.g., Harrison et al. 1984), which bind to specific receptors or neuronal membranes (see Sawchenko and Gerfen 1985). However, such transneuronal labelling is relatively weak and can be detected in only some of the synaptically connected neurons. Novel developments include the use of live neurotropic viruses such as *herpes simplex* virus type 1 and *herpes virus suis* (pseudorabies) for transneuronal tracing (Ugolini et al. 1989; Kuypers and Ugolini 1990; Strack and Loewy 1990; Blessing et al. 1994). The viruses are replicated in recipient neurons after transneuronal transfer. This replication, which is a unique characteristic of viruses, produces strong transneuronal labelling (see Fig. 7.3f). Herpes viruses represent powerful tools for demonstrating neuronal connections across synapses, for example, between peripheral nerves and neurons in the brain (Kuypers and Ugolini 1990).

Functional mapping of neuronal circuits is possible with activity markers. The *[¹⁴C]-2-deoxyglucose* (2-DG) *technique* introduced by Sokoloff et al. (1977) relies on the uptake of a non-metabolisable form of [¹⁴C]-labelled glucose that then becomes trapped within the cell. The degree of 2-DG label relates to the intensity of metabolic activity, which is particularly large at active synapses. The regions with high metabolic activity are subsequently visualised by autoradiography. The chief disadvantages of this technique are its limited resolution, the long processing time required for autoradiography, and the difficulty in distinguishing between synaptic excitation and inhibition, both of which can enhance metabolic activity and 2-DG uptake. Other activity markers are sulforhodamine, which has been used in an in vitro turtle brain stem-cerebellum preparation (Keifer et al. 1992), and the immediate-early genes and their proteins (see Sect. 7.4.5).

7.4
Fluorescence Histochemical and Immunohistochemical Techniques

7.4.1
Fluorescence Histochemical Techniques

Since Falck and Hillarp (see Falck et al. 1962) developed a fluorescent histochemical technique to visualise monoamine-containing cells in the brain, various refinements of the histofluorescence technique (see Fuxe et al. 1970; Moore 1981) have been introduced for the intracellular demonstration of biogenic amines. Such techniques, permitting the characterisation of neurons on the basis of the pro-

duction of monoamines, appeared extraordinarily useful (see Moore and Bloom 1978; Björklund and Hökfelt 1984; Parent et al. 1984).

Falck and Hillarp's technique, based on the condensation of a biogenic amine with formaldehyde in a gas-phase reaction, overcame the earlier problems of the water solubility and extreme variability of the monoamines. The use of the gas-phase reaction on dried or freeze-dried tissue prevented the diffusion of the water-soluble biogenic amines and produced an intense fluorophore that could readily be identified in the fluorescence microscope (see Moore 1981). Subsequent improvements such as the vibratome-formaldehyde technique (Hökfelt and Ljungdahl 1972) and the glyoxylic acid technique (Björklund et al. 1972; Lindvall and Björklund 1974) have led to further advances. There is general agreement (see Björklund et al. 1975; Lindvall and Björklund 1978; Moore 1981) that the fluorescence histochemical techniques demonstrate only dopamine, noradrenaline, adrenaline and serotonin in neurons in the CNS. The original technique of Falck and Hillarp was sufficiently sensitive to demonstrate the cell bodies and terminal plexuses of axons containing catecholamines, but it did not demonstrate the preterminal axons. The newer techniques are sufficiently sensitive to demonstrate the *entire* neuron.

However, the fluorescence histochemical techniques have three major limitations (see Moore 1981): (a) they are applicable only to monoamine neuron systems; (b) because unstained sections have to be analysed with dark-field illumination, and only fluorescent material is clearly visible, interpretation of the exact localisation of fluorescent structure may be difficult; (c) the apparent absence of an innervation in fluorescent histochemical material does not necessarily mean that it does not exist. Nowadays, the fluorescence histochemical techniques have been replaced by immunohistochemical techniques for the demonstration of neurotransmitters, neuropeptides and proteins (see Sect. 7.4.2).

7.4.2
Immunohistochemistry of Neurotransmitters, Neuropeptides and Proteins

Immunocytochemical techniques have replaced the fluorescence histochemical techniques, since they proved to be very sensitive for the light- and electron-microscopic identification of chemically distinct neurons within the nervous system. Two major types of immunocytochemical techniques can be distinguished on the basis of whether the marker used for identification of the site of the antigen-antibody reaction is attached directly to the antibody or indirectly to an anti-intermediary immunoglobulin. The direct labelling techniques, such as the direct immunofluorescence technique developed by Coons and Kaplan (1950), were introduced first. The antibody itself is conjugated with a label such as a fluorescent molecule to be visualised in the microscope. Direct labelling techniques greatly reduce the reactivity of the antibodies and are unable to demonstrate cellular constituents that are present in low concentrations in tissue. The more sensitive indirect immunocytochemical labelling techniques make use of an anti-immunoglobulin G intermediary to couple the antibodies to the marker compound. Two standard immunocytochemical techniques have proved successful for light as well as electron microscopy, i.e. the *peroxidase anti-peroxidase* technique (see Sternberger 1979) and the *avidin-biotin-peroxidase* technique (Hsu et al. 1981). The primary antibody is not directly labelled; rather, a secondary antibody binds to the primary one and it is this secondary reagent that is labelled or binds to a label. The advantages of such indirect labelling over direct labelling are: (a) it acts as an amplification step – more than one molecule of the secondary antibody will bind to the primary antibody; (b) it avoids the necessity of subjecting valuable primary antiserum to a chemical reaction (see Sternberger 1979; Cuello 1983). The reaction product formed in peroxidase reactions can be intensified with metal ions such as cobalt chloride or nickel ammonium sulphate.

With *antibodies directed against neurotransmitters* (or their synthesising enzymes) or against neuropeptides, it is possible to characterise those neurons expressing the transmitter or peptide, to analyse their synaptic input (see Sect. 7.4.3), and to study their efferent projections. Immunohistochemical techniques can also be used to detect axonal tracers such as PHA-L and BDA (see Sect. 7.3.2) and proteins such as calcium-binding proteins (Celio 1990) or those formed by immediate-early gene expression such as Fos or Jun (see Sect. 7.4.5).

The most widely used immunohistochemical techniques are those that employ antibodies against catecholamine-synthesising enzymes such as *tyrosine hydroxylase* (Goldstein et al. 1972, 1973; Hökfelt et al. 1973, 1984b) and, more recently, antibodies developed for the visualisation of the catecholamines (dopamine, noradrenaline, adrenaline) themselves (Geffard et al. 1984; Hökfelt et al. 1984a; Steinbusch and Tilders 1987). The distribution of the catecholaminergic system has been extensively studied throughout vertebrates and recently summarised (Smeets and Reiner 1994). Similarly, the localisation of serotonin in the CNS, mostly using the immunohistochemical technique developed by

Steinbusch and co-workers (Steinbusch et al. 1978; Steinbusch and Tilders 1987), is well-known for many vertebrates. Descriptions of the techniques involved and the underlying problems can be found in various handbooks (Björklund and Hökfelt 1983; Cuello 1983, 1993; Steinbusch 1987) and in a large number of manuals. The distribution in the CNS of monoamines, other neurotransmitters such as GABA and glycine, and of various peptides has been extensively described in the various volumes of the *Handbook of Chemical Neuroanatomy* (Elsevier, Amsterdam), particularly for mammals.

7.4.3
Combinations with Tract-Tracing

Although the use of retrograde tracers in combination with techniques for visualising transmitters or their synthesising enzymes dates back to the mid-1970s (Ljungdahl et al. 1975) and many different combinations have been applied for the *tracing of transmitter-specific pathways* (see Skirboll et al. 1989; Záborsky and Heimer 1989), two methods for the combination of tracer studies with transmitter (or peptide) identification have become widely used: (a) retrograde fluorescent tracers (e.g., Fast Blue, Fluoro-Gold) in combination with immunohistochemistry based on fluorescent markers (see Skirboll et al. 1984, 1989), and (b) combinations of HRP, PHA-L or BDA tracing with immunohistochemistry (Záborsky et al. 1985; Záborsky and Heimer 1989). Immunostaining permits visualisation of transmitters or peptides, and fluorescent immunomarkers such as FITC or TRITC can be conveniently combined with fluorescence immunohistochemistry through the use of filter combinations (Skirboll et al. 1984, 1989). The combination of immunohistochemistry with retrograde tracing permits the visualisation of neurotransmitters in the same neuron that is labelled from a particular part of the CNS. Disadvantages of the combination of fluorescent tracers with immunohistochemistry are the fading of the tracers and the unsuitability for EM studies. Therefore, combinations of HRP, PHA-L or BDA tracing with immunohistochemistry have largely replaced the combination of immunofluorescence with fluorescence tracing. Various combinations are possible (see Záborsky and Heimer 1989; Smith and Bolam 1992) and especially useful for analysis of neuronal microcircuits and synaptic interactions (see Sect. 7.5.).

7.4.4
In Situ Hybridisation Histochemistry

The technique of *in situ hybridisation* is rapidly becoming an essential tool in the field of chemical neuroanatomy. In situ hybridisation originated in the field of molecular genetics (John et al. 1969; Pardue and Gall 1969) for the detection of ribosomal nucleotide sequences in cells using labelled ribosomal RNA. Subsequently, this technique has been used extensively for the cellular localisation of specific sequences of both DNA and mRNA within tissue sections (see Young 1990; Valentino et al. 1992). With the growing list of characterised mRNAs encoding proteins and neuropeptide precursors, and the availability of convenient kits for producing isotopically or non-isotopically labelled probes, the utility of in situ hybridisation as a neuroanatomical technique has increased enormously (see Young 1990; Emson 1993). The use of non-radioactive probes has several advantages over radiolabelled probes, notably the ability to obtain single cell resolution of the signal, and its speed and safety. Moreover, non-radioactive probes offer the possibility of carrying out in situ hybridisation in a whole mount (see Wilkinson 1992). Analysis of the spatial expression of genes is of particular importance for developmental neurobiology.

In situ hybridisation histochemistry making use of non-radioactive, biotinylated probes can also be used in combination with axonal tracing (see Chronwall et al. 1989) and immunocytochemistry (Emson 1993; Augood et al. 1994; Wahle 1994).

7.4.5
Immediate-Early Genes and Their Proteins

Immediate-early genes (IEGs) or proto-oncogenes are a group of genes responsive to trans-synaptic stimulation and membrane electrical activity in neural cells. In contrast to the late-response genes, IEG transcription is activated rapidly and transiently within minutes of stimulation (see Sheng and Greenberg 1990; Morgan and Curran 1991, 1995; Dubner and Ruda 1992; Sagar and Sharp 1993; Curran and Morgan 1995). The most commonly studied of these genes are the *c-fos* and *c-jun* genes encoding the peptides Fos and Jun that act in the form of dimers, as 'third messengers' or transcription factors to control the expression of late-response genes (see Morgan and Curran 1991). *C-fos* is induced rapidly and transiently in neurons, whereas *c-jun* is expressed over long periods (Morgan and Curran 1991, 1995; Jenkins et al. 1993; Curran and Morgan 1995). The duration of the reaction may vary from a few hours (e.g., Fos expression

after a sensory stimulus) to several days (e.g., Jun after axotomy). In particular, the *fos* proto-oncogene or *c-fos* has provided a useful marker of neuronal activity with which the effects of behavioural, pharmacological, electrical and physiological stimuli can be traced in the nervous system. Among the stimuli found to induce the expression of *c-fos* are direct electrical or chemical (e.g., kainic acid) stimulation of the nervous system, damage by lesions or ischaemia, and nociceptive stimulation (see Morgan and Curran 1991; Dubner and Ruda 1992; Sagar and Sharp 1993). *C-fos* has also been implicated in the control of growth and differentiation of neurons, as well as in regeneration and cell death (e.g., de Felipe et al. 1993; Curran and Morgan 1995). Fos immunostaining may provide a method to map circuits in the CNS involved in a particular type of behaviour and can be used in combination with immunohistochemistry and tract-tracing (Sagar and Sharp 1993). This type of *functional mapping* has been used extensively for the neuroendocrine system, especially the hypothalamo-pituitary adrenal axis (Sagar et al. 1988; Sharp et al. 1991) and increasingly for the motor system (e.g., Sagar et al. 1988; Wan et al. 1992).

7.5
Techniques for Analysis of Neuronal Microcircuits and Synaptic Interactions

Various techniques have been suggested for analysis of *neuronal microcircuits* and synaptic interactions of neuronal elements. A prerequisite for such an analysis is the capability of visualising the relationships between the neuronal elements, e.g., the terminals from one group of neurons on cell bodies or dendrites of another population. For this visualisation, combinations of tracing and labelling procedures are available, including: (a) anterograde degeneration studies of fibres and axon terminals combined with Golgi impregnation (Blackstad 1965, 1981; Somogyi 1978; Somogyi et al. 1979); (b) anterograde tracing with lectins such as PHA-L combined with immunohistochemistry (e.g., Wouterlood 1988; Smith and Bolam 1992); (c) combinations of anterograde and retrograde tracing with immunohistochemistry (e.g., Záborsky and Heimer 1989; Bolam and Ingham 1990; Smith and Bolam 1992); and (d) the use of intracellular Lucifer Yellow injections in fixed brain slices combined with retrograde and anterograde tracing, immunohistochemistry and ultrastructural studies (Buhl and Lübke 1989; Buhl 1992, 1993; Wouterlood et al. 1990, 1993; Meredith and Arbuthnott 1993).

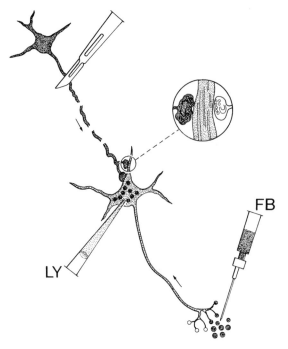

FB

LY

Fig. 7.4. The tracing of successive links in a neuronal circuit. See text for further explanation(based on Wouterlood et al. 1993 and Heimer 1995)

The fourth, most promising technique allows the visualisation of at least three elements in one preparation (see Fig. 7.4): (a) fibre input from a particular centre to (b) a morphologically identified and intracellularly labelled neuron, and (c) the output of that neuron to another area. An anterograde tracer, e.g. PHA-L or BDA, can be injected to label a certain input to a particular neuron which has been retrogradely labelled from one of its output areas by a fluorescent tracer such as Fast Blue. After suitable survival time, tissue slices can be prepared in which the retrogradely labelled neurons can be identified and, with the aid of a specialised microscope (see Buhl and Lübke 1989; Meredith and Arbuthnott 1993), injected with an intracellular marker, e.g. Lucifer Yellow. Synaptic contacts between the neurons investigated can be studied at the electron-microscopic level. Transmitter immunohistochemistry can be added to characterise the participating neurons in terms of their chemical messengers (see Wouterlood et al. 1993). Various applications can be found in a recent volume of the *IBRO Handbook Series Methods in the Neurosciences* (Meredith and Arbuthnott 1993).

Another promising technique in this field uses *confocal microscopy* to analyse the three-dimensional (3-D) structure of functionally identified neurons and their input (e.g., Wallén et al. 1988; Wallén 1993). Confocal microscopy has also

proved able to reveal the morphology of presynaptic terminals at the neuromuscular junction (e.g., Lichtman et al. 1989). The latter approach relies on high-quality video camera technology (Inoué 1986). *Fluorescent probes* can be used also to stain living nerve terminals (Magrassi et al. 1987), and the development and plasticity of neuromuscular connectivity has been extensively studied using this method (e.g., Rich and Lichtman 1989a,b; Balice-Gordon and Lichtman 1990, 1993; van Mier and Lichtman 1994).

7.6
Electrophysiological Techniques

As anatomists revealed the seemingly endless varieties and intricacies of neuronal structure and interconnections, electrophysiological techniques evolved to intercept and analyse the signals moving between nerve cells. As early as 1850, von Helmholtz recorded the conducted action potential from the exterior of muscle, and this era (see Brazier 1959, 1961) culminated in the detailed studies of Erlanger and Gasser (1937) and their collaborators on conduction in peripheral nerves. Berger (1929) detected electrical signals measuring $10-100\,\mu V$ from the closed human scalp, and this began the use of electroencephalography (EEG), still a widely employed clinical tool. It was the invention of the microelectrode, however, which was to prove the single most significant advance in understanding the electrical behaviour of brain cells. Early microelectrodes were made by pulling out a heated glass Pyrex tube by hand and are usually credited to Ling and Gerard (1949), although recently Brown and Flaming (1986) have argued that the first microelectrode recordings on vertebrate cells were actually made by Graham and Gerard (1946). The significant fact is that these electrodes remain patent even down to a tip diameter of less than $1\,\mu m$. They can then be filled with a conducting saline solution and connected to an amplifier via a chloride-coated silver wire in the saline. This simple concept allows both the 3-D mapping of the extracellular potentials of the brain and the intracellular recording from single neurons.

Microelectrode techniques evolved in many ways, and the books by Lavallée et al. (1969) and Brown and Flaming (1986) contain useful descriptions. Automated electrode pullers were developed which are now controlled by microprocessors. The tip geometry can be carefully manipulated by multi-stage pulling and heating and subsequently modified by bevelling, and the filling solution is often introduced using capillary glass with a fine glass fibre in it. But the basic concept remains that of a slender glass electrode, sharp enough to penetrate the membrane of a cell with the minimum of damage.

A very important development occurred when Sakmann and Neher (1983) modified the microelectrode tip and made various other innovations which enabled a patch of cell membrane to be studied, so that individual channels could be analysed. Other significant developments of microelectrode technology included the ion-selective microelectrode, which introduced a liquid membrane into the tip permitting recording of ionic concentrations, and the carbon-fibre-filled microelectrode that permitted voltammetric determination in vivo of the concentration of electroactive transmitters, such as dopamine, noradrenaline and serotonin.

Numerous methods of optical recording of nervous activity are now beginning to appear. Some are based on the use of voltage-sensitive dyes to record membrane potentials; other methods depend on changes in the light scattering properties of cells or tissue in response to nervous activity, while still others rely on the use of intracellular compounds that alter their optical properties in response to calcium or other ions.

7.6.1
EEG, MEG and Evoked Potentials

The *electroencephalogram* (EEG) is recorded from the surface of the scalp with an array of chlorided silver disks lightly attached with conducting paste. The electrodes are connected to sensitive AC-coupled amplifiers and the output is often registered on a chart recorder. The signals, with an amplitude of a few tens of microvolts, show complex patterns that vary with the state of alertness of the subject and many other parameters. The EEG is sensitive to many types of input, including sound and visual stimulation, which lead to evoked potentials on the scalp. Abnormalities in the EEG are an effective tool in the diagnosis and localisation of epilepsy. Because of its relative simplicity and low cost, the EEG is a very widely used diagnostic tool. Its application to the study of human brain physiology is detailed in many books and papers, for example, Bodis-Wolner (1982) and Karrer et al. (1984).

The EEG is generated when a population of active neurons sets up small currents in the extracellular space of the brain (see Sect. 7.6.2), and these currents, when they have the correct distribution, can create tiny potential differences on the scalp (Nuñez 1981). There are two fundamental limitations of the EEG: (a) the scalp potentials are greatly attenuated by the impedance of the skull and scalp itself, and (b) because recordings are confined to a surface, consisting of the outside of the head, different distributions of current generators, corresponding to different populations of active cells, can give rise to similar potential distributions. In fact, it is possible for populations of neurons to be active and yet not produce any potential at the surface of the head (the so-called closed field, to be described below).

Because of the limitations of the EEG there has been an effort to record the magnetic fields generated by the intrinsic currents of the brain, rather than the surface potentials. This method is known as *magnetoencephalography* (MEG). The magnetic signals are not attenuated by the skull and scalp; however, there are still configurations of current generators that cannot be detected by this method. The magnetic fields generated by the brain are much smaller than the earth's magnetic field and elaborate technology is required to detect them. The instrumentation relies on arrays of superconducting quantum interference devices (SQUIDS) that are cooled in liquid helium and operated in elaborately shielded rooms and the technique presently is confined to research environments. The techniques are described by Williamson et al. (1983, 1989).

In a research context, both EEG and MEG have been used primarily with human subjects (e.g., Näätänen et al. 1994); however, a few very interesting animal studies have been done. In particular, Bullock and co-workers have long drawn attention to the fact that there are remarkable differences in the EEG spectrum among species and these should reflect differences in underlying brain organisation (Bullock and Basar 1988). Some MEG studies have been carried out in animal models to clarify the origin of the signals (Okada and Nicholson 1988; Kyuhou and Okada 1993).

With both EEG and MEG it is common to evoke the signals by means of visual or auditory stimuli or, where appropriate, electrical stimulation of a nerve (Bodis-Wollner 1982). Potentials can also be evoked by *magnetic stimulation* to induce currents directly in the brain (Cracco et al. 1993). This method has the advantage that it can be used without the need for any electrode contact, making it useful in human studies and as a diagnostic tool.

7.6.2
Extracellular Field Potentials

All neurons are embedded in an extracellular microenvironment that consists of the diminutive interstitial spaces between cells, comprising the extracellular space. Although narrow (perhaps

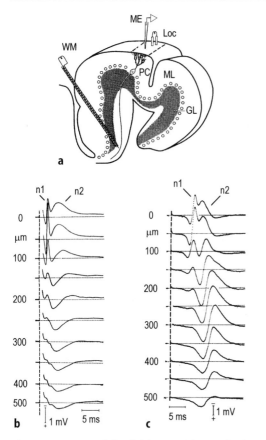

Fig. 7.5a–c. Extracellular field potentials reveal information about synaptic transmission and dendritic physiology in alligator cerebellum. **a** Schematic of alligator cerebellum with sagittal section. A local surface-stimulating electrode (*Loc*) excites a beam of parallel fibres in the molecular layer (*ML*), which is seen in the superficial field potentials in **b** and **c** as an early fast negative potential (*n1*). The parallel fibres synaptically activate the Purkinje cells (*PC*) and this produces a later superficial negative field potential (*n2* in **b** and **c**). Also in **a**, granule cell layer (*GL*) and location of a white matter (*WM*)-stimulating electrode that can be used to antidromically activate Purkinje cells. **b** Passive dendritic potentials. Field potentials measured at indicated depths below the surface. There is a superficial postsynaptic negative potential *n2*, which reverses with depth to a positive wave revealing that the dendrites behave like passive cable structures. **c** Active dendritic properties. Another experiment with the same paradigm as in **b**. The postsynaptic negative wave *n2* persists with increasing latency down to 450 µm, which is approximately the cell body layer of the Purkinje cells. This indicates that the excitatory postsynaptic potential generates an action potential in the Purkinje cell dendrites which propagates to the soma. (Later studies showed that this action potential was generated by a calcium current; see also Fig. 7.8). This figure illustrates the fact that field potential analysis is a powerful method for analysing neuronal circuitry (adapted from Nicholson and Llinás 1971).

opened by the transmitters released at the synapses, and equal and opposite currents flow across the neighbouring passive membranes. The exact distribution of the currents depends on the intrinsic electrical parameters of the neuron and its geometry. Inward or outward synaptic currents across the membrane lead to currents in the extracellular space and, by virtue of the resistance of these narrow extracellular spaces, potential differences are generated in the tissue, known as *extracellular field potentials* (Fig. 7.5). These field potentials relate in both their temporal pattern and their spatial distribution to the underlying neuronal activity; however, the relationship is quite complex. The EEG, described above, is a special case of a continuous 3-D field potential within the brain that can be recorded at a restricted location on the scalp.

The earliest technique for studying nervous activity relied on using large electrodes to measure evoked responses from excised tissue. The concepts developed to understand extracellular potentials around excised muscle (Bishop and Gilson 1929) and nerve fibres (Lorente de Nó 1947b) were applied to ensembles of neurons (Bishop and O'Leary 1942; Lorente de Nó 1947a). Lorente de Nó (1947a) identified three types of ensemble geometry (Hubbard et al. 1969). These were the *open field*, the *closed field* and the *mixed field*. In extracellular potential recording it is usually necessary that a group of neurons be active at the same time and in the same way in order to generate sufficient voltage to be recorded by the electrode. In the open field, all the neurons are aligned parallel to each other. By mapping the distribution of the potential in the tissue, information about the location of the current generators in the neurons can be obtained (Fig. 7.5). In general, experience has shown (Eccles 1951) that the potentials recorded from the CNS are produced by excitatory postsynaptic potentials (EPSPs). One of the best examples of this type of analysis is the work of Eccles, Llinás and Sasaki, summarised by Eccles et al. (1967), in understanding the interaction between parallel fibres and Purkinje cells in the cerebellum. This analysis was later extended to demonstrate the existence of dendritic action potentials (Nicholson and Llinás 1971; Fig. 7.5).

The second type of extracellular field potential is the closed field (Lorente de Nó 1947a; Baker and Precht 1972). Here the neurons are arranged in a spherical geometry and, when they are excited synchronously, the rather surprising observation is made that no extracellular potentials are recorded outside the territory of the cells (this can be demonstrated theoretically as well). Thus, the only way to detect such populations by their extracellu-

averaging 20 nm in width), these spaces are numerous and comprise a volume fraction of about 20 % of the total brain tissue (Nicholson and Rice 1991). When neurons are activated by synaptic input, currents flow in or out of the membrane channels

lar field potentials is to insert a recording electrode within the spherical ensemble.

The third type of extracellular field potential is the mixed field, which combines both of the first two. In reality, this is probably the most commonly encountered, and its presence implies that some proportion of the neurons in an ensemble are active but not contributing to the field potential outside the population of active cells.

The generation of extracellular field potentials can be analysed in terms of basic electrostatic theory (Nicholson 1973; Llinás and Nicholson 1974). Such an analysis shows that by combining potentials from different locations in a suitable way the distribution of the underlying sources and sinks of current (i.e. synaptic sites) can be computed and localised. This is the so-called *current-source density analysis* (Nicholson and Freeman 1975; Freeman and Nicholson 1975; Mitzdorf 1985) which, when suitable measurements can be made, provides a powerful method for studying neuronal populations and continues to be applied with increasing sophistication (e.g., Tenke et al. 1993; Ketchum and Haberly 1993; Plenz and Aertsen 1993).

7.6.3
Single and Multiple Unit Recording

Using either a glass microelectrode or an insulated metal electrode (Geddes 1972) it is possible to record *single unit* activity from a neuron (Fig. 7.6 a-e). This consists of a sequence of impulses that represent action potentials in the cell (Segundo 1970; Perkel 1970). Since action potentials usually arise in the cell body or axon hillock and travel down the axon, unit activity is recorded from axons and cell bodies, rather than from the dendrites of cells. The recording electrode usually touches the axon or the cell body, or is very close to it, and senses the rapid change in extracellular potential corresponding to the passage of the nerve impulse. Electrodes for recording unit activity usually have fairly low impedance and are AC-coupled to the amplifier so that a high gain may be used. Many of the best recordings are made with metal electrodes, which are often made from tungsten and more recently from carbon fibres (Armstrong-James and Millar 1979). Other variants are platinum and steel. The metal is often etched to a very fine tip and always insulated with glass, epoxy or another plastic (Geddes 1972). Elaborate methods have been developed to produce electrodes suitable for recording from very tiny fibres, such as C-fibres (Merrill and Ainsworth 1972).

It is often possible, using a single electrode, to record the *multiple unit activity* from more than

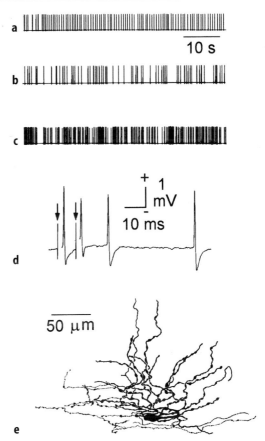

Fig. 7.6a–e. Unit activity recorded from stellate cell in turtle cerebellum. **a-c** Spontaneous action potentials recorded from vicinity of stellate cell similar to the one shown in **e.** The *time scale* in **a** applies to panels **a-c** and the amplitude in each case is about 2 mV positive. **d** Some of the individual spikes in more detail; the *downward arrows* indicate where Loc stimulation (see Fig. 7.5) was applied and caused the cell to fire, following synaptic activation by the parallel fibres. **e** A stellate cell that was penetrated with an HRP-containing micropipette after spikes were recorded from its vicinity and then filled with the stain. This experiment shows how information is transmitted in the form of spike trains and how electrophysiology and cell staining may be combined (modified from Chan and Nicholson 1986)

one cell. Single cells can then be distinguished on the basis of amplitude and wave shape (Gerstein 1970). Alternatively, an array of recording electrodes may be used (Sugihara et al. 1993). An ingenious extension of the multielectrode system was employed by Meister et al. (1991), who placed an isolated retina on a substrate containing an array of embedded electrodes and simultaneously recorded up to 100 ganglion cells.

When it is desired to mark the position of the recording electrode in the tissue, a dye-filled glass micropipette and dyes such as Fast Green or Alcian Blue, released by iontophoresis or a steel electrode can be used and a Prussian Blue precipitate formed by passing current to release iron into the tissue, followed by suitable post mortem treatment (Nicholson and Kater 1973). Since such *extracellular marking techniques* (except for Prussian Blue) often fail, investigators generally rely on stereotaxic placement for electrode location.

Physiological methods are also valuable for tracing fibre connections. Most often, this is done by electrically stimulating the fibre pathway near the site of their axon terminals. Action potentials then invade the cells of origin, and these can be located by mapping the evoked potentials with a microelectrode. This stimulation technique is known as *antidromic stimulation.* In one study the method was used to construct a detailed map of the olivocerebellar projection by recording antidromic responses in the inferior olive (Armstrong et al. 1974). In another set of experiments the relations of the substantia nigra to the superior colliculus were determined (e.g., Hikosaka and Wurtz 1983).

7.6.4
Ion-Selective and Voltammetric Microelectrodes

Microelectrodes can be modified to sense ions (ion-selective microelectrodes) or made so that they are sensitive to certain electroactive compounds, such as dopamine (voltammetric microelectrodes). *Ion-selective microelectrodes* (ISMs; Ammann 1986; Nicholson 1993) contain a liquid membrane composed of a hydrophobic solvent and an ion carrier. Liquid membranes are available for H^+ (pH), K^+, Na^+, Ca^{2+} and Cl^-. The concentration is determined by measuring a modified Nernst potential across the liquid membrane when the tip is exposed to an appropriate ion. A second barrel of the ISM records local potential, and this can be subtracted from that across the ion-selective barrel and also used to map extracellular field potentials. Electrodes of this type have revealed that the ionic concentrations (K^+ in particular) in the brain extracellular microenvironment fluctuate when cells are active (Syková 1992; Nicholson 1993). During pathophysiological events such as spreading depression, seizure or epilepsy, the variations are very large and may contribute to the pathology (Hansen 1985; Nicholson 1993).

A variation of the ISM method, based on introducing and sensing the ion tetramethylammonium (TMA^+) has been used to measure the diffusion characteristics of the brain extracellular microenvironment (Nicholson and Phillips 1981; Nicholson 1993). The analysis of the diffusion characteristics has revealed that the typical *extracellular volume fraction* of the brain extracellular microenvironment is 20 %; i.e. this amount of brain tissue is extracellular space. Diffusion analysis also reveals that the *tortuosity* is about 1.6; tortuosity is a measure of the hindrance imposed by cellular obstructions. These two measures give some information about the way cells are packed together in the brain and also have implications for the movement of chemical signals. In general, a small molecule diffusing in the extracellular space has an apparent diffusion coefficient that is reduced by the square of the tortuosity compared with free diffusion in water (i.e. a factor of about 2.6). These relations hold for most regions of the CNS examined so far in several different species. Again, during various forms of activity and pathology, the parameters can change. In particular, the volume fraction can fall to as little as 5 % during severe ischaemia (Syková et al. 1994).

Voltammetric microelectrodes (Adams and Marsden 1982; Nicholson and Rice 1988) are presently made by pulling a microelectrode with a small carbon fibre in it (about 8 µm in diameter). The fibre electrode is connected to a current-measuring system and inserted into tissue, and the potential difference between the electrode tip and tissue varied through a range of voltages which includes the oxidation and reduction peak potentials of the electroactive substance of interest. As the potential passes through this peak potential, the current flow on the carbon fibre reaches a maximum value which is related to the concentration of the substance. In practice, quite complex voltage sequences are used to help discriminate among different substances. The most frequently studied electroactive substance is dopamine, but the method can also detect noradrenaline, serotonin and ascorbate (Kawagoe et al. 1993).

7.6.5
Intracellular Recording

Some of the most spectacular early successes with *intracellular recording* were achieved not with a microelectrode, but rather with quite large capillary electrodes inserted in the giant axon of the squid by Hodgkin, Huxley, Cole and others (see Chap. 1). The major technical advance here was the introduction of the voltage clamp by Marmont and Cole (Cole 1968), which used very fast electronic current feedback circuitry to 'clamp' the membrane potential at any desired potential, while the required current was measured. From these studies came the detailed theory of the action potential (Hodgkin and Huxley 1952; Hodgkin 1967).

Apart from continued studies on the giant axon, most subsequent intracellular work was done with the glass micropipette. Although Graham and Gerard (1946) and Ling and Gerard (1949) developed the glass micropipette for intracellular

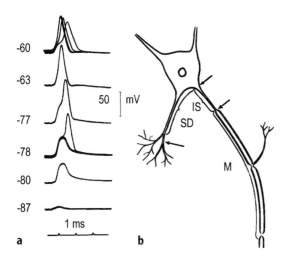

Fig. 7.7a,b. Intracellular recording of antidromic invasion of a motoneuron from cat spinal cord. **a** Recordings from cell body of motoneuron showing arrival of antidromic impulse produced by exciting the axon at a distant location. The intracellular electrode was connected to a 'bridge' amplifier that allowed the electrode to also pass current and polarise the cell body. As the cell was increasingly hyperpolarised (indicated by the *numbers* on the *left*, which are the intracellular potential in millivolts), the invading action potential broke down into smaller components. This was interpreted to mean that the antidromic invasion was stopping at various locations. With normal resting potential, the action potential reflected the successive invasion of the medullated axon, the initial segment of the axon and the soma-dendritic expansion. As the invasion failure occurred more and more distant from the recording site, the remnants of the action potential were conducted electronically to the soma and so appeared as attenuated and slowed waveforms. **b** Schematic of a motoneuron showing soma-dendritic expansion (*SD*) (note that the dendrites are very schematised in this figure), the initial segment of the axon (*IS*) which lacks myelin, and the medullated, or myelinated axon (*M*). The *three arrows* indicate the most likely locations of the blockage of the antidromic invasion (modified from Eccles 1957)

recording, many of the early discoveries with this technique about central neurons (Fig. 7.7) and synaptic transmission were made by Eccles and co-workers (Eccles 1957, 1964) and Katz and Miledi, working on the neuromuscular junction (Katz 1962). Thereafter, the technique was applied throughout the nervous system. By combining intra- and extracellular data, great advances were made in understanding neuronal circuitry in the spinal cord, cerebellum, and olfactory bulb, to mention only some areas (Eccles et al. 1967; Hubbard et al. 1969; Shepherd 1990). Most recently, the basis of spinal locomotion in the lamprey has been greatly clarified by Grillner and others (see Chap. 10).

Over the years a number of advances have been made in microelectrode fabrication. Electrode pullers have been improved with the use of microprocessors and a strategically applied air-jet to cool the pulled glass more quickly (Brown and Flaming

1986). The ease with which electrodes can be filled with a chosen electrolyte has been greatly increased by the introduction of a glass fibre into the borosilicate capillary stock used to fabricate electrodes. In some instances, bevelling the tip of the electrode aids penetration of cells (Brown and Flaming 1986).

Apart from the availability of sharp micropipettes with tips less than 1 µm in diameter, intracellular recording has also benefited from improvements in electronics. A major advance was the use of a so-called bridge amplifier, based initially on the Wheatstone Bridge circuit but now achieved with operational amplifier circuits, which permits a single microelectrode to simultaneously record the intracellular potential and pass current to depolarise or hyperpolarise a cell. Continued advances in electrode design and electronic sophistication have led to the single-electrode voltage clamp, and much of this technology is now routinely available in systems for patch recording.

7.6.6
Patch Recording

The *patch clamp* was developed by Neher and Sakmann (1976) and represents a major advance in single-cell recording. The idea is elegantly simple but was not easy to perfect, though it is a now routine technique in many laboratories. A fire-polished micropipette with a tip diameter of 1 µm or more is pressed against a patch of cell membrane and is induced to form a seal with the membrane, effectively electrically isolating that area of membrane and the ion channels within it. The channels can then be studied with voltage clamp techniques or other recording paradigms. The technique has been greatly elaborated (Fig. 7.8) and now permits patches to be removed from cells with either membrane face contacting the electrolyte in the electrode, or the method may be used to study the channels in situ (Sakmann and Neher 1983). In another useful variant, the high-resistance seal is retained and the patch removed, allowing very low impedance recording of intracellular potentials; this is particularly valuable for recording from small cells.

Initially, it was thought that the patch clamp could be applied only to the relatively clean membranes obtained in tissue cultures and acutely isolated cells. Now the technique has advanced to the point that it can be used in brain slices (Edwards et al. 1989; Blanton et al. 1989; Edwards 1995), and it is becoming the method of choice for intracellular recording, supplanting the 'sharp' microelectrode in many instances. The patch clamp is also an essential adjunct to techniques designed to manip-

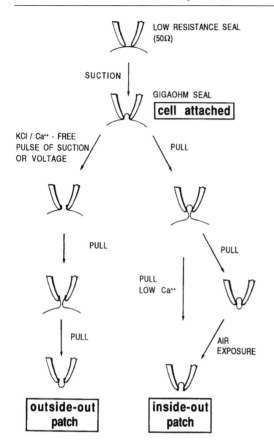

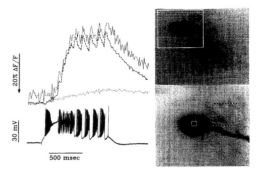

Fig. 7.9. Optical recording of calcium transients from a Purkinje cell in a slice of guinea-pig cerebellum. An intracellular microelectrode in the soma of the cell injects the calcium-sensitive dye Fura-2 and also records the membrane potential. The cell is visualised with a 25 x water-immersion lens and a cooled CCD camera. Somatic and dendritic regions were selected from the images (indicated on *right* of figure). By spatially averaging the fluorescent signal (excited at 380 nm) over the region indicated in the *dotted box* and comparing the results with those obtained in the soma, it was possible to show that the calcium transients were generated in the dendrites. The optical signals are shown in the *upper waveform* on the *left*; *scale* indicates percentage change in fluorescence. The *lower potential recording* shows typical dendritic calcium spikes, together with sodium-mediated action potentials, as seen with a somatic recording electrode. Note that the calcium spikes were reduced in amplitude because of their distant site of origin. This figure shows the ability of optical techniques to localise function in nerve cells (from Ross and Lasser-Ross 1992)

Fig. 7.8. Various types of patch-recording configurations. Beginning with a simple mechanical contact of the patch electrode to the cell (*upper panel*, 50 Ω input resistance), a slight suction applied to the electrode results in a gigaohm seal. Further manipulation provided opens the membrane (*left panel*), and this configuration is appropriate for recording from small cells in slices. With a clean cell membrane from cultured cells, further manipulations result in a cell-free patch that can be either an outside-out patch (*left panels*) or an inside-out-patch (*right panels*). The patch-clamp technique has provided unprecedented access to small populations of channels in membranes. By modifying the channels through molecular biological techniques, molecular subunits can be added or deleted and the resulting changes in electrical activity studied (from Hamill 1992)

ulate the composition of ion channels in the membrane through the application of the techniques of molecular biology.

7.6.7
Optical Recording

An important development in recording brain activity is the concept of using *voltage-sensitive dyes* to record membrane potential (Blasdel 1989; Salzberg 1989; Kim et al. 1989; Grinvald et al. 1994). Such dyes enter the membrane environment by various means and change their fluorescence or absorbance in relation to the local membrane potential.

The resulting changes in light intensity can be recorded with arrays of photodiodes or a camera system and so reveal the activity of cell populations. In the course of research with voltage-sensitive dyes it was found that *intrinsic optical signals* occur in the brain even when voltage-sensitive dyes are not present (Lieke et al. 1989; Frostig et al. 1990). These approaches have been used successfully to study ocular dominance columns and other functional characteristics in the visual cortex (Blasdel 1989; Lieke et al. 1989; Haglund et al. 1993; Arieli et al. 1995).

Another optical technique that has produced some very interesting data is the measurement of changes in intracellular Ca^{2+} and Na^+ (Fig. 7.9; Tank et al. 1988; Sugimori and Llinás 1990; Miyakawa et al. 1992; Ross and Lasser-Ross 1992; Midtgaard 1994), after injecting single nerve cells with *fluorescent probes* (such as the Ca^{2+} sensor, *Fura-2*) that respond to these ions. At present, techniques employing fluorescent probes are restricted to the cell cultures, brain slices or isolated spinal cord preparations (see O'Donovan et al. 1993, 1994) described in the next section.

7.7
In Vitro Preparations

As the various techniques described in Sect. 7.6 have developed in sensitivity and sophistication, a need for new preparations has naturally arisen. While the most desirable preparation has always been the intact animal, the more and more delicate recording methods required preparations that could be isolated from the vibrations arising out of respiration and heartbeat. Isolated nerves, including the giant axon of the squid, provided an invaluable starting point for these more precise measurements. The use of the squid axon led to the employment of numerous invertebrate preparations to study both isolated and intact neuronal circuits (Bullock and Horridge 1965). Various forms of tissue culture allowed the use of isolated cells from the mammalian CNS. In order to have available isolated mammalian circuits, the brain slice was pioneered and has now become a standard preparation for many investigations (Dingeldine 1984). Finally, it is possible to isolate whole brain regions, certain areas, such as the retina, and to use complete, isolated brains of the lamprey, the frog and the turtle (Jahnsen 1990) and even of mammals (Mühlethaler et al. 1993).

7.7.1
Cell Culture

Banker and Goslin (1991) summarised the field of *neuronal cell culture* up to that time. The first example of tissue culture applied to neuronal tissue was by Harrison (1907), but there was limited subsequent use of the approach until the late 1960s because of both technical problems in maintaining sterile cultures and difficulties peculiar to neuronal cultures. One of the latter is the inability of immature nerve cells to divide readily after removal from the brain.

Cultures can be divided into two different classes: The first is *primary cultures*, which consist of cells removed from the living CNS and then cultured in various ways. The drawbacks here are the heterogeneity of the cells and, because of the inability of neurons to divide, the inevitable demise of the population. An alternative strategy is to use *clonal cell lines*, which are immortalised cells of a single type that reliably divide and can be induced to differentiate; the issue here is the extent to which such cells resemble any of those in the CNS.

Primary neuronal cultures can be subdivided into various technical approaches. Probably the most widely used are *dissociated cell cultures* prepared from suspensions of individual cells dissociated from the CNS. These cells are plated onto a suitable substrate and can differentiate, extend processes, form networks and make synaptic contacts. Although individual cells may have many of the characteristics of the cells in the CNS, the networks generally differ from those found in situ. Such cultures can be maintained for months and can develop characteristic electrical activity. Short-term cultures of this type, lasting a few hours and referred to as *acutely dissociated cells*, are frequently used for patch-clamp studies on channels. Most primary cultures contain a variety of cells including both neurons and glia, but it is possible to choose the culture conditions so that only glia survive. This has led to studies of astrocytes and of oligodendrocytes and Schwann cells in culture. Unfortunately, the channels in cultured astrocytes can change (Barres et al. 1990). This problem has been circumvented by rapidly removing cells along with many of their processes using the *tissue print method* (Barres 1992).

Clonal cell lines are typically derived from tumour cells; they are homogeneous and divide. They can be induced to differentiate, i.e. send out processes, by various treatments. They are usually electrically active and can synthesise transmitters. They also have the advantage than many techniques of molecular biology can be used to modify their behaviour. The main disadvantage is that the cells do not represent a recognisable neuron type, and their particular properties, channel populations, transporters etc., may therefore not correspond to those found in the CNS. One of the best-known clonal cell lines is the PC12 cell (Greene and Tischler 1976), derived from a tumour of the adrenal medulla, which has several characteristics of the chromaffin cell, including the ability to synthesise and release catecholamines.

Instead of culturing dissociated cells, small pieces of tissue can be excised and cultured. These are known as *explant* or *organotypic cultures*. Generally, immature tissue works best and, again, the cultures may remain viable for months. Such cultures retain some of the local circuitry and geometry of the brain, and different brain regions can be co-cultured to study interactions. A drawback of the explant is that it retains its thickness and density, which makes visual identification of single cells difficult. An advance in this area was made by Gähwiler (1981, 1988), who adapted the *roller-tube technique* for work with CNS explants; this makes it possible to 'thin out' the explants. Easy access to neurons not only facilitates the use of standard electrophysiological techniques and intracellular labelling, but also makes these cultures amenable to patch-clamp techniques (Gähwiler and Knöpfel

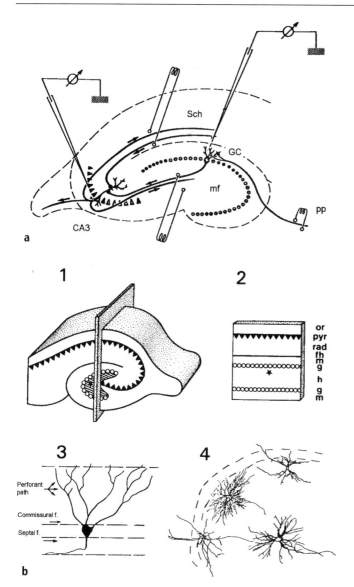

Fig. 7.10a,b. Preparation and recording paradigm for hippocampal slices. **a** Schematic drawing of cross-sectioned hippocampal slice with typical stimulation and recording sites. Pyramidal cells of CA3 can be activated both orthodromically and antidromically, depending on which fibre tract is chosen for stimulation. *GC*, Granule cell layer; *Sch*, Schaffer collaterals; *pp*, perforant path; *mf*, mossy fibres. **b1** Laminar structure of typical rodent hippocampus. **b2** Section through layers, perpendicular to surface. **b3** Some of the afferent pathways to a granule cell: *Perforant path*, perforant pathway; *Comm. f.*, commissural fibres, *Sept. f.*, septal fibres. **b4.** Some cells of the hilar region (*asterisk* in **b2**) stained with intracellular HRP injection. Studies from the laboratory of Misgeld have shown the majority of these cells to be inhibitory. This figure illustrates the potential of this popular slice preparation for studying local neuronal circuitry and, again, the value of combining electrophysiology and intracellular staining (from Misgeld 1992)

1990). The successful demonstration of functional synaptic connections between co-culture explants enables the study of synaptic transmission between remote brain areas. An organotypic spinal cord-dorsal root ganglion-skeletal muscle co-culture of embryonic rats is a useful model for morphological and physiological studies of the spinal reflex arc (Spenger et al. 1991; Streit et al. 1991).

Organotypic cultures are also used in developmental studies on target attraction and target recognition. Gähwiler's technique has been used in studies on the development of thalamocortical connections (e.g., Bolz et al. 1992, 1993). Other successful co-culture techniques are Lumsden and Davies' (1983) collagen gel approach and the use of a por-

ous and transparent membrane as a support for the explants (Romijn et al. 1988; Blakemore and Molnár 1990; Stoppini et al. 1991). In the latter technique, the explants are maintained at the interface between air and a culture medium. In organotypic cultures the lipophilic carbocyanine dyes such as DiI are highly useful (e.g., Blakemore and Molnár 1990; Bolz et al. 1990). The approach of Lumsden and Davies has been used extensively in studies on the development of connections from the mammalian cortex (e.g., Heffner et al. 1990; Joosten et al. 1991; O'Leary and Koester 1993).

7.7.2
Brain Slices

The *brain slice* was introduced by McIlwain for neurochemical studies (McIlwain et al. 1951); intracellular resting potentials were recorded by Li and McIlwain (1957) and more extensive neuronal responses by Yamamoto and McIlwain (1966). Initial experiences with slices as an electrophysiological preparation were somewhat disappointing, however, for two reasons. Firstly, the original device for making slices, the McIlwain chopper, did not produce slices viable for electrophysiology except from the hippocampus, and indeed was not always satisfactory for neurochemistry (Garthwaite et al. 1979). Alternative techniques such as slicing by hand, which is not often used, and the vibrating disposable blade slicer led to much improved slice quality. A parallel advance occurred in the design of the chambers in which the slices resided during the physiological experiments. There are two basic designs (Alger et al. 1984). The first is the interface chamber, wherein the lower surface of the slice is bathed in flowing artificial cerebrospinal fluid (ACSF) while the upper surface is exposed to a humidified mixture of 95 % oxygen and 5 % carbon dioxide, which was also used to saturate the ACSF. In the second type of chamber, the slice is submerged in flowing ACSF. Over the years there has been discussion of the relative merits of the two chamber designs, and there is no clear advantage in one over the other in terms of slice viability, although a specific experimental requirement may favour one design. Typical experiments consist of rapidly removing the brain from the skull, cutting slices at 300–400 µm thickness and incubating them for an hour in ACSF to allow recovery from the initial ischaemia and mechanical trauma during preparation (depolarisation of tissue). Slices typically can be used for 4–15 h without significant deterioration.

Almost every region of the CNS has been sliced (Dingeldine 1984), but the region used most has been the hippocampus of the rat (Fig. 7.10). The major advantage of the slice is that it permits whole cells to be studied in an appropriate microenvironment. In many cases, when slices are cut with appropriate orientation, the most important neuronal circuitry is preserved; this is particularly true in slices from the hippocampus.

The vast numbers of studies with slices attest to their utility. There are drawbacks which relate mainly to the problem of diffusion of substrates. In the intact animal, a blood vessel carrying oxygen and glucose is probably within 50 µm of any point in the tissue, but in the slice all substrates must diffuse in from the slice surface. Even when both surfaces are accessible, slice thickness is limited to about 400 µm for this reason. Another consequence of this limitation is that slices must be bathed in 95 % oxygen (the other 5 % is usually carbon dioxide to provide pH buffering) in order to ensure that the centre of the slice is adequately oxygenated; this means that the outer regions of the slice are bathed in excess oxygen. Finally, typical ACSF is a simplified medium that doubtless omits several substances present in brain CSF.

7.7.3
Isolated CNS Preparations

When the primary need is to have authentic brain circuitry available, particularly circuits that involve different brain regions, an isolated CNS preparation is desirable. Many of these have been developed, and their success often depends on special features of the preparation that enable it to survive outside the body, or on elaborate support strategies that reduce the consequences of a lack of blood supply.

Among preparations that lend themselves to isolation, the retina is one of the more popular (Ames and Masland 1990). A major reason for the success of this preparation is that the whole structure is less thick than a typical slice (the average retina is about 200 µm thick) while containing complex circuitry that can be readily excited by an easily controlled stimulus (light). Another useful preparation is the spinal cord. This has been taken from the frog (see Grantyn 1992) or the chick embryo (O'Donovan and Landmesser 1987; O'Donovan 1989), and the spinal cord of young rats can be maintained in vitro long enough for tracing and electrophysiological studies (Chmykhova et al. 1991; Karamian et al. 1991; Kerkut and Bagust 1995).

A factor that enhances viability of isolated CNS preparations is a decreased need for oxygen and metabolic substrates, and this can be achieved in some cases by cooling mammalian preparations but is often found naturally in nonmammalian species. Of particular note here are the lamprey (Brodin and Grillner 1990), frogs (Cochran et al. 1987; Straka and Dieringer 1993; Luksch et al. 1996), and freshwater turtles (Hounsgaard and Nicholson 1990). The lamprey spinal cord has no intrinsic blood vessels and is oxygenated directly from the cerebrospinal fluid (CSF). The brain stem has intrinsic blood vessels but is probably also oxygenated to a large extent from the CSF (Brodin and Grillner 1990). Rovainen (1967a,b) took advantage of these favourable conditions by developing an in vitro preparation of the lamprey nervous system. The spinal cord and brain stem can be maintained in vitro for periods of 2–3 days at temperatures around 7–10 °C. This in vitro preparation of the lamprey CNS has been used extensively as an experimentally amenable model of synaptic transmission and integrative functions, particularly of the motor system (see Brodin and Grillner 1990 and Chap. 10).

The turtle brain has an unusual resistance to anoxia (Lutz et al. 1985; Hounsgaard and Nicholson 1990). In vitro preparations have been used to investigate several regions of the turtle CNS. The isolated olfactory bulb was the first in vitro preparation from the CNS of turtles (Mori and Shepherd 1979). In vitro preparations of the telencephalon (e.g., Connors and Kriegstein 1986; Kriegstein and Connors 1986; Larson-Prior et al. 1991), of the cerebellum (e.g., Chan and Nicholson 1986; Hounsgaard and Midtgaard 1988), and of the isolated cerebellum – brain stem – spinal cord (Keifer and Houk 1989; Keifer et al. 1992; Sarrafizadeh and Houk 1994) are excellent models for combined anatomical and physiological studies.

For most mammalian tissues it is not possible to maintain adequate physiological integrity without perfusing the vascular system with some form of blood substitute. Usually, this takes the form of oxygenated ACSF. Examples of such preparations are the rat hypothalamus (Bourque 1990) and the isolated guinea-pig brain (Llinás et al. 1981; Mühlethaler et al. 1993). Isolated neonatal brains or brain stem-spinal cord preparations can be kept alive without perfusion of the vascular system. An in vitro brain stem-spinal cord preparation for studies of motor systems for mammalian respiration and locomotion is available (e.g., Smith and Feldman 1987). An isolated CNS preparation of the newly born South American opossum, *Monodelphis domestica*, has been used for studies on the development and regeneration of synaptic interactions (e.g., Nicholls et al. 1990; Saunders et al. 1992; Woodward et al. 1993, Møllgård et al. 1994).

Acknowledgements. The comments of Drs. John Nicholls, Ivan Urbán, Jan van Gisbergen and Jan Veening have been gratefully received. The artwork by Marlu de Leeuw and the secretarial assistance of Inge Eijkhout are gratefully acknowledged.

References

Adams JC (1981) Heavy metal intensification of DAB-based HRP reaction product. J Histochem Cytochem 29:775
Adams RN, Marsden CA (1982) Electrochemical methods for monoamine measurements in vitro and in vivo. In: Iversen LL, Iversen SD, Snyder SH (eds) Handbook of psychopharmacology, vol 15. Plenum, New York, pp 1–74
Alger BE, Dhanjal SS, Dingeldine R, Garthwaite J, Henderson G, King GL, Lipton P, North A, Schartzkroin TA, Sears M,

Segal M, Whittingham TS, Williams J (1984) Brain slice methods. In: Dingeldine R (ed) Brain slices. Plenum, New York, pp 381–437
Ames A, Masland RH (1990) The rabbit retina in vivo. In: Jahnsen H (ed) Preparations of vertebrate central nervous system in vitro. Wiley, Chichester, pp 183–202
Ammann D (1986) Ion-selective microelectrodes. Springer, Berlin Heidelberg New York
Arieli A, Shoham D, Hildesheim R, Grinvald A (1995) Coherent spatiotemporal patterns of ongoing activity revealed by real-time optical imaging coupled with single-unit recording in the cat visual cortex. J Neurophysiol 73:2072–2093
Armstrong DM, Harvey RJ, Schild RF (1974) Topographical localization in the olivocerebellar projection: an electrophysiological study in the cat. J Comp Neurol 154:287–302
Armstrong-James M, Millar J (1979) Carbon fibre microelectrodes. J Neurosci Methods 1:279–287
Augood SJ, McGowan EM, Finsen BR, Heppelman B, Emson PC (1994) Non-radioactive in situ hybridization using alkaline phosphatase-labelled oligonucleotides. In: Wisden W, Morris BJ (eds) In situ hybridization protocols for the brain. Academic, London, pp 81–97
Baker R, Precht W (1972) Electrophysiological properties of trochlear motoneurons as revealed by IV nerve stimulation. Exp Brain Res 14:127–157
Balice-Gordon RJ, Lichtman JW (1990) In vivo visualization of the growth of pre- and postsynaptic elements of neuromuscular functions in the mouse. J Neurosci 10:894–908
Balice-Gordon RJ, Lichtman JW (1993) In vivo observations of pre- and postsynaptic changes during the transition from multiple to single innervation at developing neuromuscular junctions. J Neurosci 13:834–855
Banker G, Goslin K (1991) Culturing nerve cells. MIT Press, Cambridge, MA
Barres BA (1992) The tissue print method. In: Kettenmann H, Grantyn R (eds) Practical electrophysiological methods. A guide for in vitro studies in vertebrate neurobiology. Wiley-Liss, New York, pp 96–100
Barres BA, Chun LLY, Corey DP (1990) Ion channels in vertebrate glia. Annu Rev Neurosci 13:441–474
Beran RL, Martin GF (1971) Reticulospinal fibers of the opossum, Didelphis virginiana. I. Origin. J Comp Neurol 141:453–466
Berger H (1929) Über das Elektrenkephalogramm des Menschen. Arch Psychiat Nervenkr 87:527–570
Bielschowsky M (1904) Die Silberimprägnation der Neurofibrillen. J Psychol Neurol (Lpz) 3:169–188
Bishop GA, King JS (1982) Intracellular horseradish peroxidase injections for tracing neural connections. In: Mesulam M-M (ed) Tracing neural connections with horseradish peroxidase. Wiley, Chichester, pp 185–247 (IBRO handbook series: methods in the neurosciences, vol 1)
Bishop GH, Gilson AS (1929) Action potentials from skeletal muscle. Am J Physiol 89:135–151
Bishop GH, O'Leary JL (1942) Factors determining the form of the potential record in the vicinity of the synapses of the dorsal nucleus of the lateral geniculate body. J Cell Comp Physiol 31:315–331
Björklund A, Hökfelt T (eds) (1983) Methods in chemical neuroanatomy. Elsevier, Amsterdam (Handbook of chemical neuroanatomy, vol 1)
Björklund A, Hökfelt T (eds) (1984) Classical transmitters in the CNS. Elsevier, Amsterdam (Handbook of chemical neuroanatomy, vol 2, part I)
Björklund A, Lindvall O, Svensson LÅ (1972) Mechanisms of fluorophore formation in the histochemical glyoxylic acid method for monoamines. Histochemie 32:113–131
Björklund A, Falck B, Lindvall O (1975) Microspectrofluorometric analysis of cellular monoamines after formaldehyde or glyoxylic acid condensation. In: Bradley PB (ed) Methods in brain research. Wiley, London, pp 249–294
Blackstad TW (1965) Mapping of experimental axon degeneration by electron microscopy of Golgi preparations. Z Zellforsch 67:819–834

Blackstad TW (1981) Tract tracing by electron microscopy of Golgi preparations. In: Heimer L, RoBards MJ (eds) Neuroanatomical tract-tracing methods. Plenum, New York, pp 407–440

Blakemore C, Molnár Z (1990) Factors involved in the establishment of specific interconnections between thalamus and cerebral cortex. Cold Spring Harbor Symp Quant Biol 55:491–504

Blanton MG, Lo Turco JJ, Kriegstein A (1989) Whole cell recording from neurons in slices of reptilian and mammalian cerebral cortex. J Neurosci Methods 30:203–210

Blasdel GG (1989) Visualization of neuronal activity in monkey striate cortex. Annu Rev Physiol 51:561–581

Blessing WW, Ding Z-Q, Li Y-W, Gieroba ZJ, Wilson AJ, Hallsworth PG, Wesselingh SL (1994) Transneuronal labelling of CNS neurons with herpes simplex virus. Prog Neurobiol 44:37–53

Blum F (1893) Der Formaldehyd als Härtungsmittel. Z Wiss Mikrosk 10:314–315

Bodian D (1936) A new method for staining nerve fibres and nerve endings in mounted paraffin sections. Anat Rec 65:89–97

Bodis-Wollner I (1982) Evoked potentials. Ann NY Acad Sci Vol 388

Bolam JP, Ingham CA (1990) Combined morphological and histochemical techniques for the study of neuronal microcircuits. In: Björklund A, Hökfelt T, Wouterlood FG, van den Pol AN (eds) Analysis of neuronal microcircuits and synaptic interactions. Elsevier, Amsterdam, pp 125–198 (Handbook of chemical neuroanatomy, vol 8)

Bolz J, Novak N, Götz M, Bonhoeffer T (1990) Formation of target-specific neuronal projections in organotypic slice cultures from rat visual cortex. Nature 346:359–362

Bolz J, Novak N, Staiger V (1992) Formation of specific afferent connections in organotypic slice cultures from rat visual cortex co-cultured with lateral geniculate nucleus. J Neurosci 12:3054–3070

Bolz J, Götz M, Hübener M, Novak N (1993) Reconstructing cortical connections in a dish. Trends Neurosci 16:310–316

Bourque CW (1990) The isolated and perfused mammalian hypothalamus. In: Jahnsen H (ed) Preparations of vertebrate central nervous system in vitro. Wiley, Chichester, pp 203–232

Brazier MAB (1959) The historical development of neurophysiology. In: Field J, Magoun HW, Hall VE (eds) Handbook of physiology, vol 1: neurophysiology. American Physiological Society, Washington DC, pp 1–58

Brazier MAB (1961) A history of the electrical activity of the brain. Pitman, London

Brodal A (1939) Experimentelle Untersuchungen über retrograde Zellveränderungen in der unteren Olive nach Läsionen des Kleinhirns. Z Ges Neurol Psychiatr 166:624–704

Brodal A (1940) Modification of Gudden method for study of cerebral localization. Arch Neurol Psychiatr 43:46–58

Brodal A (1981) Neurological anatomy in relation to clinical medicine, 3rd edn. Oxford University Press, New York

Brodin L, Grillner S (1990) The lamprey CNS in vitro, an experimentally amenable model for synaptic transmission and integrative functions. In: Jahnsen H (ed) Preparations of vertebrate central nervous system in vitro. Wiley, Chichester, pp 103–153

Brown KT, Flaming DG (1986) Advanced micropipette techniques for cell physiology. Wiley, Chichester

Buhl EH (1992) Intracellular lucifer yellow injection in fixed brain slices. In: Bolam JP (ed) Experimental neuroanatomy. A practical approach. IRL Press, Oxford, pp 187–212

Buhl EH (1993) Intracellular injection in fixed brain slices: a highly versatile tool to examine neuronal geometry in combination with other neuroanatomical techniques. In: Meredith GE, Arbuthnott GW (eds) Morphological investigations of single neurons in vitro. Wiley, Chichester, pp 27–46 (IBRO handbook series: methods in the neurosciences, vol 16)

Buhl EH, Lübke J (1989) Intracellular lucifer yellow injection in fixed brain slices combined with retrograde tracing, light and electron microscopy. Neuroscience 28:3–16

Bullock TH, Basar E (1988) Comparison of ongoing compound field potentials in the brains of invertebrates and vertebrates. Brain Res Rev 13:57–75

Bullock TH, Horridge GA (1965) Structure and function in the nervous systems of invertebrates. Freeman, San Francisco

Celio MR (1990) Calbindin D-28k and parvalbumin in the rat nervous system. Neuroscience 35:375–475

Chan CY, Nicholson C (1986) Modulation of neuronal activity by applied electric fields in the isolated turtle cerebellum. J Physiol (Lond) 371:89–114

Chmykhova NM, Karamian OA, Kozhanov VM (1991) Sensorimotor connections in the lumbar spinal cord of the young rat: a morphological study. Neuroscience 43:569–576

Chronwall BM, Lewis ME, Schwaber JS, O'Donohue TL (1989) In situ hybridization combined with retrograde fluorescent tract tracing. In: Heimer L, Záborsky L (eds) Neuroanatomical tract-tracing methods 2. Plenum, New York, pp 265–297

Clarke E, O'Malley CD (1968) The human brain and spinal cord. University of California Press, Berkeley (2nd, revised and enlarged edition: Norman Publishing, San Francisco, 1996)

Cochran SL, Kasik P, Precht W (1987) Pharmacological aspects of excitatory synaptic transmission to second-order neurons in the frog. Synapse 1:102–123

Cole KS (1968) Membranes, ions and impulses. University of California Press, Berkeley

Conn HJ (1948) The history of staining. Biotech Publications, Geneva

Connors BW, Kriegstein AR (1986) Cellular physiology of the turtle visual cortex: distinctive properties of pyramidal and stellate neurons. J Neurosci 6:164–177

Coons AH, Kaplan MH (1950) Localization of antigens in tissue cells. II. Improvement in a method for the detection of antigen by means of fluorescent antibody. J Exp Med 91:1–14

Cowan WM (1970) Anterograde and retrograde transneuronal degeneration in the central and peripheral nervous system. In: Nauta WJH, Ebbesson SOE (eds) Contemporary research methods in neuroanatomy. Springer, Berlin Heidelberg New York, pp 217–249

Cowan WM, Gottlieb DL, Hendrickson AE, Price JL, Woolsey TL (1972) The autoradiographic demonstration of axonal connections in the central nervous system. Brain Res 37:21–51

Cox W (1891) Impregnation des centralen Nervensystems mit Quecksilbersalzen. Arch Mikrosk Anat 37:16–21

Cracco RQ, Amassian VE, Maccabee PJ, Cracco JB, Rudell A, Eberle L (1993) Insights into cerebral function revealed by magnetic coil stimulation. Adv Neurology 63:43–50

Cragg BG (1970) What is the signal for chromatolysis? Brain Res 23:1–21

Cuello AC (ed) (1983) Immunohistochemistry. Wiley, Chichester (IBRO handbook series: methods in the neurosciences, vol 3)

Cuello AC (ed) (1993) Immunohistochemistry II. Wiley, Chichester (IBRO handbook series: methods in the neurosciences, vol 14)

Cullheim S, Kellerth JO (1976) Combined light and electron microscopical tracing of neurones, including axons and synaptic terminals, after intracellular injection of horseradish peroxidase. Neurosci Lett 2:307–313

Curran T, Morgan JI (1995) Fos: an immediate-early transcription factor in neurons. J Neurobiol 26:403–412

Dederen PJWC, Gribnau AAM, Curfs MHJM (1994) Retrograde neuronal tracing with cholera toxin B subunit: comparison of three different visualization methods. Histochem J 26:856–862

de Felipe C, Jenkins R, O'Shea R, Williams TSC, Hunt SP (1993) The role of immediate-early genes in the regeneration of the central nervous system. In: Seil FJ (ed) Neural regeneration. Raven, New York, pp 263–271 (Advances in neurology, vol 59)

de Olmos J, Ebbesson SOE, Heimer L (1981) Silver methods of the impregnation of degenerating axoplasm. In: Heimer L, RoBards MJ (eds) Neuroanatomical tract-tracing methods. Plenum, New York, pp 117–170

Deiters OFK (1865) Untersuchungen über Gehirn und Rückenmark des Menschen und der Säugetiere (M Schultze ed). Vieweg, Braunschweig

Dingeldine R (1984) Brain slices. Plenum, New York

Dolleman-van der Weel MJ, Wouterlood FG, Witter MP (1994) Multiple anterograde tracing, combining Phaseolus vulgaris leucoagglutinin with rhodamine- and biotin-conjugated dextran amine. J Neurosci Methods 51:9–21

Dubner R, Ruda MA (1992) Activity-dependent neuronal plasticity following tissue injury and inflammation. Trends Neurosci 15:96–102

Ebbesson SOE (1970) The selective silver-impregnation of degenerating axons and their synaptic endings in non-mammalian species. In: Nauta WJH, Ebbesson SOE (eds) Contemporary research methods in neuroanatomy. Springer, Berlin Heidelberg New York, pp 132–161

Eccles JC (1951) Interpretation of action potentials evoked in the cerebral cortex. EEG Clin Neurophysiol 3:449–464

Eccles JC (1957) The physiology of nerve cells. John Hopkins Univ Press, Baltimore

Eccles JC (1964) The physiology of synapses. Springer, Berlin Heidelberg New York

Eccles JC, Ito M, Szentágothai J (1967) The cerebellum as a neuronal machine. Springer, Berlin Heidelberg New York

Edwards FA (1995) Patch-clamping in brain slices: Synaptic transmission from ATP to long-term potentiation. J Neurosci Methods 59:59–65

Edwards FA, Konnerth A, Sakmann B, Takahashi T (1989) A thin slice preparation for patch clamp recordings from neurones of the mammalian central nervous system. Pflügers Arch 414:600–612

Emson PC (1993) In situ hybridization as a methodological tool for the neuroscientist. Trends Neurosci 16:9–16

Erlanger J, Gasser HS (1937) Electrical signs of nervous activity. University of Pennsylvania Press, Philadelphia

Falck B, Hillarp NÅ, Thieme G, Thorp A (1962) Fluorescence of catecholamines and related compounds condensed with formaldehyde. J Histochem Cytochem 10:348–354

Filipski GT, Wilson MVH (1984) Sudan black B as a nerve stain for whole cleared fishes. Copeia 1:204–208

Fink RP, Heimer L (1967) Two methods for selective silver impregnation of degenerating axons and their synaptic endings in the central nervous system. Brain Res 4:369–374

Fredman SM (1987) Intracellular staining of neurons with nickel-lysine. J Neurosci Methods 20:181–194

Freeman JA, Nicholson C (1975) Experimental optimization of current source-density technique for anuran cerebellum. J Neurophysiol 38:369–382

Fritzsch B (1993) Fast axonal diffusion of 3000 molecular weight dextran amines. J Neurosci Methods 50:95–103

Frostig RD, Lieke EE, Ts'o DY, Grinvald A (1990) Cortical functional architecture and local coupling between neuronal activity and the microcirculation revealed by in vivo high-resolution optical imaging of intrinsic signals. Proc Natl Acad Sci U S A 87:6082–6086

Fuxe K, Hökfelt T, Jonsson G, Ungerstedt U (1970) Fluorescence microscopy in neuroanatomy. In: Nauta WJH, Ebbesson SOE (eds) Contemporary research methods in neuroanatomy. Springer, Berlin Heidelberg New York, pp 275–314

Gähwiler BH (1981) Organotypic monolayer cultures of nervous tissue. J Neurosci Methods 4:329–342

Gähwiler BH (1988) Organotypic cultures of neural tissue. Trends Neurosci 11:484–489

Gähwiler BH, Knöpfel T (1990) Cultures of brain slices. In: Jahnsen H (ed) Preparations of vertebrate central nervous system in vitro. Wiley, Chichester, pp 77–100

Galvani A (1791) De viribus electricitatis in motu musculari. Commentarius De Bononiensi Scientarium et Artium Instituto atque Academia Commentarii 7:363–418 (English translation by JF Fulton and MD Stanton 1953: Burndy Library Publ No 10, Norwalk, Connecticut)

Garthwaite J, Woodhams PL, Collins MJ, Balazs R (1979) On the preparation of brain slices: morphology and cyclic nucleotides. Brain Res 173:373–377

Geddes LA (1972) Electrodes and the measurement of bioelectric events. Wiley, New York

Geffard M, Buys RM, Seguela P, Pool CW, Le Moal M (1984) First demonstration of highly specific and sensitive antibodies against dopamine. Brain Res 294:161–165

Gerfen CR, Sawchenko PE (1984) An anterograde neuroanatomical staining method that shows the detailed morphology of neurons, their axons and terminals: immunohistochemical localization of an axonally transported plant lectin, Phaseolus vulgaris leucoagglutinin (PHA-L). Brain Res 290:219–238

Gerfen CR, Sawchenko PE, Carlsen J (1989) The PHA-L anterograde tracing method. In: Heimer L, Záborsky L (eds) Neuroanatomical tract-tracing methods 2. Plenum, New York, pp 19–47

Gerstein GL (1970) Functional association of neurons: detection and interpretation. In: Schmitt FO, Quarton GC, Melnechuk T, Adelman G (eds) The neurosciences. Second study program. Rockefeller University Press, New York, pp 648–661

Gimlich RL, Braun J (1985) Improved fluorescent compounds for tracing cell lineage. Dev Biol 109:509–514

Glees P (1946) Terminal degeneration within the central nervous system as studied by a new silver method. J Neuropathol Exp Neurol 5:54–59

Glover JC, Petursdottir G, Jansen KS (1986) Fluorescent dextran-amines used as axonal tracers in the nervous system of the chicken embryo. J Neurosci Methods 18:243–254

Godement P, Vanselow J, Thanos S, Bonhoeffer F (1987) A study in developing visual systems with a new method of staining neurons and their processes in fixed tissue. Development 101:697–713

Goldstein M, Fuxe K, Hökfelt T (1972) Characterization and tissue localization of catecholamine synthesizing enzymes. Pharmacol Rev 24:293–309

Goldstein M, Anagnoste B, Freedman LS, Roffman M, Ebstein RP, Park DH, Fuxe K, Hökfelt T (1973) Characterization, localization and regulation of catecholamine synthesizing enzymes. In: Usdin E, Snyder S (eds) Frontiers in catecholamine research. Pergamon, New York, pp 69–78

Golgi C (1873) Sulla struttura della sostanza grigia del cervello. Gazz Med Ital Lombardia 33:244–246

Golgi C (1875) Sui gliomi del cervello. Riv Sper Freniatria Med Leg 1:66–78

Golgi C (1879) Di una nuova reazione apparentemente nera delle cellule nervose cerebrali ottenuta col bichloruro di mercurio. Arch Sci Med 3:1–7

Grafstein B, Forman BS (1980) Intracellular transport in neurons. Physiol Rev 60:1167–1283

Graham J, Gerard RW (1946) Membrane potentials and excitation of impaled single muscle fibers. J Cell Comp Physiol 28:99–117

Graham RC, Karnovsky MI (1966) Glomerular permeability. Ultrastructural cytochemical studies using peroxidase as protein tracers. J Exp Med 124:1123–1134

Grantyn R (1992) Pair recording: quantitative reconstruction of a reflex pathway. In: Kettenmann H, Grantyn R (eds) Practical electrophysiological methods. A guide for in vitro studies in vertebrate neurobiology. Wiley-Liss, New York, pp 323–329

Greene LA, Tischler AS (1976) Establishment of a noradrenergic clonal line of rat adrenal pheochromocytoma cells which respond to nerve growth factor. Proc Natl Acad Sci U S A 73:2424–2428

Griffin G, Watkins LR, Mayer DJ (1979) HRP pellets and slow-release gels: two new techniques for greater localization and sensitivity. Brain Res 168:595–601

Grinvald A, Lieke EE, Frostig RD, Hildesheim R (1994) Cortical point-spread function and long-range lateral interactions revealed by real-time optical imaging of macaque monkey primary visual cortex. J Neurosci 14:2545–2568

Groenewegen HJ, Wouterlood FG (1990) Light and electron microscopic tracing of neuronal connections with Phaseolus vulgaris leucoagglutinin (PHA-L), and combinations with other neuroanatomical techniques. In: Björklund A, Hökfelt T, Wouterlood FG, van den Pol AN (eds) Analysis of neuronal microcircuits and synaptic interactions. Elsevier, Amsterdam, pp 47–124 (Handbook of chemical neuroanatomy, vol 8)

Häggqvist G (1936) Analyse der Faserverteilung in einem Rückenmarkquerschnitt (Th 3). Z Mikrosk Anat Forsch 39:1–34

Haglund MM, Ojemann GA, Blasdel GG (1993) Optical imaging of bipolar cortical stimulation. J Neurosurg 78:785–793

Hamill OP (1992) Cell-free patch clamp. In: Kettenmann H, Grantyn R (eds) Practical electrophysiological methods. Wiley-Liss, New York, pp 284–288

Hammerschlag R, Cyr JL, Brady ST (1994) Axonal transport and the neuronal cytoskeleton. In: Siegel GL, Agranoff BW, Albers RW, Molinoff PB (eds) Basic neurochemistry. Raven, New York, pp 545–571

Hansen AJ (1985) Effect of anoxia on ion distribution in the brain. Physiol Rev 65:101–148

Harrison PJ, Hultborn H, Jankowska E, Katz R, Storai B, Zytnicki D (1984) Labelling of interneurones by retrograde transsynaptic transport of horseradish peroxidase from motoneurones in rats and cats. Neurosci Lett 45:15–19

Harrison RG (1907) Observations on the living developing nerve fiber. Anat Rec 1:116–118

Heffner CD, Lumsden AGS, O'Leary DDM (1990) Target control of collateral extension and directional axon growth in the mammalian brain. Science 247:217–220

Heimer L (1970a) Selective silver-impregnation of degenerating axons and their terminals. In: Nauta WJH, Ebbesson SOE (eds) Contemporary research methods in neuroanatomy. Springer, Berlin Heidelberg New York, pp 106–113

Heimer L (1970b) Bridging the gap between light and electron microscopy in the experimental tracing of fiber connections. In: Nauta WJH, Ebbesson SOE (eds) Contemporary research methods in neuroanatomy. Springer, Berlin Heidelberg New York, pp 162–172

Heimer L (1995) The human brain and spinal cord, 2nd edn. Springer, Berlin Heidelberg New York

Hendrickson AE (1969) Electron microscopic radioautography: identification of origin of synaptic terminals in normal nervous tissue. Science 165:194–196

Hikosaka O, Wurtz RH (1983) Visual and oculomotor functions of monkey substantia nigra pars reticulata. IV. Relation of substantia nigra to superior colliculus. J Neurophysiol 49:1285–1301

Hodgkin AL (1967) The conduction of the nervous impulse. Liverpool University Press, Liverpool

Hodgkin AL, Huxley AF (1952) A quantitative description of membrane current and its application to conduction and excitation in nerve. J Physiol (Lond) 117:500–544

Hökfelt T, Ljungdahl Å (1972) Modification of the Falck-Hillarp formaldehyde fluorescence method using the vibratome: simple, rapid and sensitive localization of catecholamines in sections of unfixed, or formalin fixed brain tissue. Histochemie 29:325–339

Hökfelt T, Fuxe K, Goldstein M, Joh T (1973) Immunohistochemical studies of three catecholamine-synthesizing enzymes: aspects and methodology. Histochemie 33:251–254

Hökfelt T, Johansson O, Goldstein M (1984a) Central catecholamine neurons as revealed by immunohistochemistry with special reference to adrenaline neurons. In: Björklund A, Hökfelt T (eds) Classical transmitters in the CNS, part I.

Elsevier, Amsterdam, pp 157–276 (Handbook of chemical neuroanatomy, vol 2)

Hökfelt T, Martensson R, Björklund A, Kleinau S, Goldstein M (1984b) Distribution maps of tyrosine-hydroxylase-immunoreactive neurons in the rat brain. In: Björklund A, Hökfelt T (eds) Classical transmitters in the CNS, part I. Elsevier, Amsterdam, pp 277–379 (Handbook of chemical neuroanatomy, vol 2)

Holmqvist BI, Östholm T, Ekström P (1992) DiI tracing in combination with immunocytochemistry for analysis of connectivities and chemoarchitectonics of specific neural systems in a teleost, the Atlantic salmon. J Neurosci Methods 42:45–63

Honig MC, Hume RI (1986) Fluorescent carbocyanine dyes allow living neurons of identified origin to be studied in long-term cultures. J Cell Biol 103:171–187

Honig MC, Hume RI (1989) DiI and DiO: versatile fluorescent dyes for neuronal labelling and pathway tracing. Trends Neurosci 12:333–341

Horikawa K, Armstrong WE (1988) A versatile means of intracellular labeling: injection of biocytin and its detection with avidin conjugates. J Neurosci Methods 25:1–11

Horsley V, Clarke RH (1908) The structure and functions of the cerebellum examined by a new methods. Brain 31:45–124

Hounsgaard J, Midtgaard J (1988) Intrinsic determinants of firing pattern in Purkinje cells of the turtle cerebellum in vitro. J Physiol (Lond) 402:731–739

Hounsgaard J, Nicholson C (1990) The isolated turtle brain and the physiology of neuronal circuits. In: Jahnsen H (ed) Preparations of vertebrate central nervous system in vitro. Wiley, Chichester, pp 279–294

Hsu SM, Raine L, Fanger H (1981) The use of avidin-biotin-peroxidase complex (ABC) in immunoperoxidase techniques. A comparison between ABC and unlabelled antibody (PAP) procedures. J Histochem Cytochem 29: 577–580

Hubbard JI, Llinás R, Quastel DMJ (1969) Electrophysiological analysis of synaptic transmission. Williams and Wilkins, Baltimore

Inoué S (1986) Video microscopy. Plenum, New York

Izzo PN (1991) A note on the use of biocytin in anterograde tracing studies in the central nervous system: applications at both light and electron microscopic level. J Neurosci Methods 36:155–166

Jahnsen H (1990) Preparations of vertebrate central nervous system in vitro. Wiley, Chichester

Jenkins R, McMahon SB, Bond AB, Hunt SP (1993) Expression of c-Jun as a response to dorsal root and peripheral nerve section in damaged and adjacent intact primary sensory neurons in the rat. Eur J Neurosci 5:751–759

John H, Birnstiel M, Jones K (1969) RNA-DNA hybrids at the cytological level. Nature 223:582–587

Joosten EAJ, van der Ven PFM, Hooiveld MHW, ten Donkelaar HJ (1991) Induction of corticospinal target finding by release of a diffusible, chemotropic factor in cervical spinal grey matter. Neurosci Lett 128:25–28

Karamian OA, Kozhanov VM, Chmykhova NM (1991) Relation between structural and release parameters at the young rat sensorimotor connection. Neuroscience 43:577–584

Karrer R, Cohen J, Tueting P (1984) Brain and information. Event-related potentials. Ann NY Acad Sci, Vol 425

Kater SB, Nicholson C (eds) (1973) Intracellular staining in neurobiology. Springer, Berlin Heidelberg New York

Katz B (1962) The transmission of impulses from nerve to muscle, and the subcellular unit of synaptic action. Proc R Soc Lond B 155:455–477

Kawagoe KT, Zimmerman JB, Wightman RM (1993) Principles of voltammetry and microelectrode surface states. J Neurosci Methods 48:225–240

Keifer J, Houk JC (1989) An in vitro preparation for studying motor pattern generation in the cerebellorubrospinal circuit of the turtle. Neurosci Lett 97:123–128

Keifer J, Vyas D, Houk JC (1992) Sulforhodamine labeling of neural circuits engaged in motor pattern generation in the in vitro turtle brainstem-cerebellum. J Neurosci 12:3187–3199

Kerkut GA, Bagust J (1995) The isolated mammalian spinal cord. Prog Neurobiol 46:1–48

Ketchum KL, Haberly LB (1993) Membrane currents evoked by afferent fiber stimulation in rat piriform cortex. 1. Current source-density analysis. J Neurophysiol 69:248–260

Kim JH, Dunn MB, Hua Y, Rydberg J, Yae H, Elias SA, Ebner TJ (1989) Imaging of cerebellar surface activation in vivo using voltage sensitive dyes. Neuroscience 31:613–623

King MA, Louis PM, Hunter BE, Walker DW (1989) Biocytin: a versatile anterograde neuroanatomical tract-tracing alternative. Brain Res 497:361–367

Kita H, Armstrong W (1991) A biocytin-containing compound N-(2-aminoethyl) biotinamide for intracellular labeling and neuronal tracing studies: comparison with biocytin. J Neurosci Methods 37:141–150

Kitai ST, Bishop GA (1981) Horseradish peroxidase: intracellular staining of neurons. In: Heimer L, RoBards MJ (eds) Neuroanatomical tract-tracing methods. Plenum, New York, pp 263–277

Kitai ST, Penny GR, Chang HT (1989) Intracellular labeling and immunocytochemistry. In: Heimer L, Záborsky L (eds) Neuroanatomical tract-tracing methods 2. Plenum, New York, pp 173–179

Klüver H, Barrera E (1953) A method for the combined staining of cells and fibers in the nervous system. J Neuropathol Exp Neurol 12:400–403

Kriegstein AR, Connors BW (1986) Cellular physiology of the turtle visual cortex: synaptic properties and in intrinsic circuitry. J Neurosci 7:2488–2492

Kristensson K, Olsson Y (1971) Retrograde axonal transport of protein. Brain Res 29:363–365

Kuypers HGJM, Huisman AM (1984) Fluorescent neuronal tracers. In: Fedoroff S (ed) Labeling methods applicable to the study of neuronal pathways. Academic, New York, pp 307–340 (Advances in cell neurobiology, vol 5)

Kuypers HGJM, Ugolini G (1990) Viruses as transneuronal tracers. Trends Neurosci 13:71–75

Kuypers HGJM, Bentivoglio M, Catsman-Berrevoets CE, Bharos TB (1980) Double retrograde neuronal labeling through divergent axon collaterals, using two fluorescent tracers with the same excitation wavelength which label different features of the cell. Exp Brain Res 40:383–392

Kyuhou SI, Okada YC (1993) Detection of magnetic evoked fields associated with synchronous population activities in the transverse CA1 slice of the guinea pig. J Neurophysiol 70:2665–2668

Larson-Prior LJ, Ulinski PS, Slater NT (1991) Excitatory amino acid receptor-mediated transmission in geniculocortical and intracortical pathways within visual cortex. J Neurophysiol 66:293–306

Lasek RJ, Katz MJ (1987) Mechanisms at the axon tip regulate metabolic processes critical to axonal elongation. Prog Brain Res 71:49–60

Lasek RJ, Joseph BS, Whitlock DG (1968) Evaluation of a radioautographic neuroanatomical tracing method. Brain Res 8:319–336

LaVail JH, LaVail MM (1972) Retrograde axonal transport in the central nervous system. Science 176:1415–1417

Lavallée M, Schanne M, Hébert NC (eds) (1969) Glass microelectrodes. Wiley, New York

Leonard CM (1979) Degeneration methods in neurobiology. Trends Neurosci 2:156–159

Li C-L, McIlwain H (1957) Maintenance of resting membrane potentials in slices of mammalian cerebral cortex and other tissue in vitro. J Physiol (Lond) 139:179–190

Lichtman JW, Sutherland WJ, Wilkinson RS (1989) High-resolution imaging of synaptic structure with a simple confocal microscope. New Biol 1:75–82

Lieberman AR (1971) The axon reaction: a review of the principal features of perikaryal responses to axon injury. Int Rev Neurobiol 14:49–124

Lieke EE, Frostig RD, Arieli A, Ts'o DYHR, Grinvald A (1989) Optical imaging of cortical activity: real-time imaging using extrinsic dye-signals and high resolution imaging based on slow intrinsic-signals. Annu Rev Physiol 51:543–559

Lindvall O, Björklund A (1974) The glyoxylic acid fluorescence histochemical method: a detailed account of the methodology for the visualization of central catecholamine neurons. Histochemistry 39:97–127

Lindvall O, Björklund A (1978) Organization of catecholamine neurons in the rat central nervous system. In: Iversen LL, Iversen SD, Snyder SH (eds) Chemical pathways in the brain. Plenum, New York, pp 139–231 (Handbook of psychopharmacology, vol 9)

Ling G, Gerard RW (1949) The normal membrane potential of frog sartorius fibers. J Cell Comp Physiol 34:383–396

Ljungdahl Å, Hökfelt T, Goldstein M, Park D (1975) Retrograde peroxidase tracing of neurons combined with transmitter histochemistry. Brain Res 84:313–319

Llewellyn-Smith IJ, Pilowsky P, Minson JB (1992) Retrograde tracers for light and electron microscopy. In: Bolam JP (ed) Experimental neuroanatomy. A practical approach. IRL Press, Oxford, pp 31–59

Llinás R, Nicholson C (1974) Analysis of field potentials in the central nervous system. In: Stevens CF (ed) Handbook of electroencephalography and clinical neurophysiology, vol 2, part B. Elsevier, Amsterdam, pp 61–83

Llinás R, Yarom Y, Sugimori M (1981) Isolated mammalian brain in vitro: new technique for analysis of electrical activity of neuronal circuit function. Fed Proc 40:2240–2245

Lorente de Nó R (1947a) Action potential of the motoneurones of the hypoglossus nucleus. J Cell Comp Physiol 29:207–287

Lorente de Nó R (1947b) A study of nerve physiology. Studies from the Rockefeller Institute, vol 132

Lowe J, Cox G (1990) Neuropathological techniques. In: Bancroft JD, Stevens A, Turner DR (eds) Theory and practice of histological techniques. Churchill Livingstone, Edinburgh, pp 343–378

Loyez M (1920) Coloration des fibres nerveuses par la méthode à l'hematoxyline au fèr après inclusion à la celloidine. C R Séanc Soc Biol Fil 62:511

Lubínska L (1964) Axoplasmic streaming in regenerating and in normal nerve fibres. Prog Brain Res 13:1–66

Luksch H, Walkowiak W, Muñoz A, ten Donkelaar HJ (1996) The use of in vitro preparations of the isolated amphibian CNS in neuroanatomy and electrophysiology. J Neurosci Methods 70:91–102

Lumsden AGS, Davies AM (1983) Earliest sensory nerve fibres are guided to peripheral targets by attractants other than nerve growth factor. Nature 306:786–788

Lutz PL, Rosenthal M, Sick TJ (1985) Living without oxygen: turtle brain as a model of anaerobic metabolism. Mol Physiol 8:411–525

Magrassi L, Purves D, Lichtman JW (1987) Fluorescent probes that stain living nerve terminals. J Neurosci 7:1207–1214

Manns M, Fritzsch B (1991) The eye in the brain: retinoic acid effects morphogenesis of the eye and pathway selection of axons but not the differentiation of the retina in Xenopus laevis. Neurosci Lett 127:150–154

Marchi V, Algeri G (1885) Sulle degenerazioni discendenti consecutive a lesioni sperimentale in diverse zone della corteccia cerebrale. Riv Sper Freniatria Med Leg 11:492–494

Marinesco G (1898) Veränderungen der Nervencentren nach Ausreissung der Nerven mit einigen Erwägungen betreffs ihrer Natur. Neurol Zentralbl 17:882–890

McIlwain H, Buchel L, Cheshire JD (1951) The inorganic phosphate and phosphocreatine of brain especially during metabolism in vitro. Biochem J 48:12–20

Meister M, Wong RO, Baylor DA, Shatz CJ (1991) Synchronous bursts of action potentials in ganglion cells of the developing mammalian retina. Science 252:939–943

Meredith GE, Arbuthnott GW (eds) (1993) Morphological investigations of single neurons in vitro. Wiley, Chichester (IBRO handbook series: methods in the neurosciences, vol 16)

Merrill EG, Ainsworth A (1972) Glass-coated platinum-plated tungsten microelectrodes. Med Biol Eng 10:662–672

Mesulam M-M (1978) Tetramethylbenzidine for horseradish peroxidase neurohistochemistry. A non-carcinogenic blue reaction-product with superior sensitivity for visualizing neural afferents and efferents. J Histochem Cytochem 26:106–117

Mesulam M-M (ed) (1982) Tracing neural connections with horseradish peroxidase. Wiley, Chichester (IBRO handbook series: methods in the neurosciences, vol 1)

Midtgaard J (1994) Processing of information from different sources: spatial synaptic integration in the dendrites of vertebrate CNS neurons. Trends Neurosci 17:166–173

Millhouse OE (1981) The Golgi methods. In: Heimer L, RoBards MJ (eds) Neuroanatomical tract-tracing methods. Plenum, New York, pp 311–344

Misgeld U (1992) Hippocampal slices. In: Kettenmann H, Grantyn R (eds) Practical electrophysiological methods. Wiley-Liss, New York, pp 41–53

Mitzdorf U (1985) Current source-density method and application in cat cerebral cortex: investigation of evoked potentials and EEG-phenomena. Physiol Rev 65:37–100

Miyakawa H, Ross WN, Jaffe D, Callaway JC, Lasser-Ross N, Lisman JE, Johnston D (1992) Synaptically activated increases in Ca^{2+} concentration in hippocampal CA1 pyramidal cells are primarily due to voltage-gated Ca^{2+} channels. Neuron 9:1163–1173

Møllgård K, Balslev Y, Stagaard-Janas M, Treherne JM, Saunders NR, Nicholls JG (1994) Development of spinal cord in the isolated CNS of a neonatal mammal (the opossum Monodelphis domestica) maintained in longterm culture. J Neurocytol 23:151–165

Moore RY (1981) Fluorescence histochemical methods. In: Heimer L, RoBards MJ (eds) Neuroanatomical tract-tracing methods. Plenum, New York, pp 441–482

Moore RY, Bloom FE (1978) Central catecholamine neurons systems: anatomy and physiology of the dopamine systems. Annu Rev Neurosci 1:129–169

Morgan JI, Curran T (1991) Stimulus-transcription coupling in the nervous system: Involvement of the inducible proto-oncogenes fos and jun. Annu Rev Neurosci 14:421–451

Morgan JI, Curran T (1995) Proto-oncogenes. Beyond second messengers. In: Bloom FE, Kupfer DJ (eds) Psychopharmacology: the fourth generation of progress. Raven, New York, pp 631–642

Mori K, Shepherd GM (1979) Synaptic excitation and long-lasting inhibition of mitral cells in the in vitro turtle olfactory bulb. Brain Res 172:155–159

Mühlethaler M, de Curtis M, Walton K, Llinás R (1993) The isolated and perfused brain of the guinea pig. Eur J Neurosci 5:915–926

Näätänen R, Ilmoniemi RJ, Alho K (1994) Magnetoencephalography in studies of human cognitive brain function. Trends Neurosci 17:389–395

Nance DM, Burns J (1990) Fluorescent dextrans as sensitive anterograde neuroanatomical tracers: applications and pitfalls. Brain Res Bull 25:139–145

Nauta WJH (1950) Ueber die sogenannte terminale Degeneration in Zentralnervensystem und ihre Darstellung durch Silberimprägnation. Schweiz Arch Neurol Psychiatr 66:353–376

Nauta WJH, Gygax PA (1951) Silver impregnation of degenerating axon terminals in the central nervous system. 1. Technic. 2. Chemical notes. Stain Technol 26:5–11

Nauta WJH, Gygax PA (1954) Silver impregnation of degenerating axons in the central nervous system: a modified technique. Stain Technol 29:91–93

Neher E, Sakmann B (1976) Single channel currents recorded from membrane of denervated frog muscle fibres. Nature 260:799–801

Nicholls JG, Stewart RR, Erulkar SD, Saunders NR (1990) Reflexes, fictive respiration and cell division in the brain and spinal cord of the newborn opossum, Monodelphis domestica. J Exp Biol 152:1–15

Nicholson C (1973) Theoretical analysis of field potentials in anisotropic ensembles of neuronal elements. IEEE Trans Biomed Eng 20:278–288

Nicholson C (1993) Ion-selective microelectrodes and diffusion measurements as tools to explore the brain cell microenvironment. J Neurosci Methods 48:199–213.

Nicholson C, Freeman JA (1975) Theory of current source-density analysis and determination of conductivity tensor for anuran cerebellum. J Neurophysiol 38:356–368

Nicholson C, Kater SB (1973) The development of intracellular staining. In: Kater SB, Nicholson C (eds) Intracellular staining in neurobiology. Springer, Berlin Heidelberg New York, pp 1–19

Nicholson C, Llinás R (1971) Field potentials in the alligator cerebellum and theory of their relationship to Purkinje cell dendritic spikes. J Neurophysiol 34:509–531

Nicholson C, Phillips JM (1981) Ion diffusion modified by tortuosity and volume fraction in the extracellular microenvironment of the rat cerebellum. J Physiol (Lond) 321:225–257

Nicholson C, Rice ME (1988) Use of ion-selective microelectrodes and voltammetric microsensors to study brain cell microenvironment. In: Boulton AA, Baker GB, Walz W (eds) Neuromethods: the neuronal microenvironment. Humana, Clifton, pp 247–361

Nicholson C, Rice ME (1991) Diffusion of ions and transmitters in the brain cell microenvironment. In: Fuxe K, Agnati LF (eds) Volume transmission in the brain: novel mechanisms for neural transmission. Raven, New York, pp 279––294 (Advances in neuroscience, vol 1)

Nishikawa KC (1987) Staining amphibian peripheral nerves with Sudan black B: progressive vs. regressive methods. Copeia 2:489–491

Nissl F (1885) Über die Untersuchungsmethoden der Grosshirnrinde. Neurol Zentralbl 4:500–501

Nissl F (1892) Über die Veränderungen der Ganglienzellen am Facialiskern des Kaninchens nach Ausreissung der Nerven. Allg Z Psychiat 48:197–198

Nissl F (1894) Ueber die sogenannten Granula der Nervenzellen. Neurol Zentralbl 13:676–688

Nonidez JF (1939) Studies on the innervation of the heart. I. Distribution of the cardiac nerves with special reference to the identification of the sympathetic and parasympathetic postganglionics. Am J Anat 65:361–413

Nuñez PL (1981) Electric fields of the brain. The neurophysics of EEG. Oxford University Press, New York

Ochs S (1982) Axoplasmic transport and its relation to other nerve functions. Wiley, New York

Ochs S, Burger E (1958) Movement of substance proximo-distally in nerve axons as studied with spinal cord injections of radioactive phosphorus. Am J Physiol 194:499–506

O'Donovan MJ (1989) Motor activity in the isolated spinal cord of the chick embryo: synaptic drive and firing pattern of single motoneurons. J Neurosci 9:943–958

O'Donovan MJ, Landmesser LT (1987) The development of hindlimb motor activity studied in an isolated preparation of the chick spinal cord. J Neurosci 7:3256–3264

O'Donovan MJ, Ho S, Sholomenko G, Yee W (1993) Real-time imaging of neurons retrogradely and anterogradely labelled with calcium-sensitive dyes. J Neurosci Methods 46:91–106

O'Donovan MJ, Ho S, Yee W (1994) Calcium imaging of rhythmic network activity in the developing spinal cord of the chick embryo. J Neurosci 14:6354–6369

Okada YC, Nicholson C (1988) Magnetic field associated with transcortical currents in turtle cerebellum. Biophys J 53:723–731

O'Leary DDM, Koester SE (1993) Development of projection neuron type, axon pathways, and patterned connections of the mammalian cortex. Neuron 10:991–1006

Pal J (1887) Ein Beitrag zur Nervenfarbetechnik. Z Wiss Mikrosk 4:92–96

Pardue ML, Gall JG (1969) Molecular hybridization of radioactive DNA to the DNA of cytological preparations. Proc Natl Acad Sci U S A 64:600–604

Parent A, Poitras D, Dubé L (1984) Comparative anatomy of central monoaminergic systems. In: Björklund A, Hökfelt T (eds) Classical transmitters in the CNS, part I. Elsevier, Amsterdam, pp 409–439 (Handbook of chemical neuroanatomy, vol 2)

Perkel DH (1970) Spike trains as carriers of information. In: Schmitt FO, Quarton GC, Melnechuk T, Adelman G (eds) The neurosciences. Second Study Program. Rockefeller University Press, New York, pp 587–596

Pitman RM, Tweedle CD, Cohen MJ (1972) Branching of central neurons: intracellular cobalt injection for light and electron microscopy. Science 176:412–414

Plenz D, Aertsen A (1993) Current source density profiles of optical recording maps – a new approach to the analysis of spatio-temporal neural activity patterns. Eur J Neurosci 5:437–448

Purkyně JE (1838) Bericht über die Versammlung deutscher Naturforscher und Ärzte in Prag im September, 1837. Prague Pt3, Sec 5, A. Anat Physiol Verh: 177–180

Rager G, Laussmann S, Gallyas F (1979) An improved silver stain for developing nervous tissue. Stain Technol 4:193–200

Ramón y Cajal S (1904) Quelques méthodes de coloration des cylindres axes, des neurofibrilles et des nids nerveux. Trav Lab Rech Biol 3:1–7

Ramón y Cajal S (1928) Degeneration and regeneration of the nervous system (translated and edited by RM May). Oxford University Press, London (extended reprint: J De Felipe and EG Jones (eds) 1991, Oxford University Press, New York)

Ramón-Moliner E (1970) The Golgi-Cox technique. In: Nauta WJH, Ebbesson SOE (eds) Contemporary research methods in neuroanatomy. Springer, Berlin Heidelberg New York, pp 32–55

Reil JC (1809) Untersuchungen über den Bau des grossen Gehirns im Menschen. Vierte Fortsetzung VIII. Arch Psychol (Halle) 9:136–146

Remak R (1836) Vorläufige Mittheilung mikroscopischer Beobachtungen über den innern Bau der Cerebrospinalnerven und über die Entwicklung ihrer Formelemente. Arch Anat Physiol: 145–161

Rich M, Lichtman JW (1989a) In vivo visualization of pre- and postsynaptic changes during synapse elimination in reinnervated mouse muscle. J Neurosci 9:1781–1805

Rich M, Lichtman JW (1989b) Motor nerve terminal loss from degenerating muscle fibers. Neuron 3:677–688

Robinson LR (1969) Bulbospinal fibers and their nuclei of origin in Lacerta viridis demonstrated by axonal degeneration and chromatolysis respectively. J Anat (Lond) 105:59–88

Romijn HJ, de Jong BM, Ruijter JM (1988) A procedure for culturing rat neocortex explants in a serum-free nutrient medium. J Neurosci Methods 23:75–83

Ross WN, Lasser-Ross N (1992) High time resolution imaging of calcium transients with a CCD camera. In: Kettenmann H, Grantyn R (eds) Practical electrophysiological methods. Wiley-Liss, New York, pp 378–382

Rovainen CM (1967a) Physiological and anatomical studies on large neurons of central nervous system of the sea lamprey (Petromyzon marinus). I. Müller and Mauthner cells. J Neurophysiol 30:1000–1023

Rovainen CM (1967b) Physiological and anatomical studies on large neurons of central nervous system of the sea lamprey (Petromyzon marinus). II. Dorsal cells and giant interneurons J Neurophysiol 30:1024–1042

Ruda M, Coulter JB (1982) Axonal and transneuronal transport of wheat germ agglutinin demonstrated by immunocytochemistry. Brain Res 249:237–246

Sagar SM, Sharp FR (1993) Early response genes as markers of neural activity and growth factor action. In: Seil F (ed)

Neural injury and regeneration. Raven, New York, pp 273–284 (Advances in neurology, vol 59)

Sagar SM, Sharp FR, Curran T (1988) Expression of c-fos protein in brain: metabolic mapping at the cellular level. Science 240:1328–1331

Sakmann B, Neher E (1983) Single-channel recording. Plenum, New York

Salzberg BM (1989) Optical recording of voltage changes in nerve terminals and in fine neuronal processes. Annu Rev Physiol 51:507–526

Sarrafizadeh R, Houk JC (1994) Anatomical organization of the limb premotor network in the turtle (Chrysemys picta) revealed by in vitro transport of biocytin and neurobiotin. J Comp Neurol 344:137–159

Saunders NR, Balkwill P, Knott G, Habgood MD, Møllgård K, Treherne JM, Nicholls JG (1992) Growth of axons through a lesion in the intact CNS of fetal rat maintained in long-term culture. Proc R Soc Lond B 250:171–180

Sawchenko PE, Gerfen CR (1985) Plant lectins and bacterial toxins as tools for tracing neuronal connections. Trends Neurosci 8:378–384

Scheibel ME, Scheibel AB (1970) The rapid Golgi method. Indian summer or renaissance. In: Nauta WJH, Ebbesson SOE (eds) Contemporary research methods in neuroanatomy. Springer, Berlin Heidelberg New York, pp 1–11

Schmued L, Kyriakidis K, Heimer L (1990) In vivo anterograde and retrograde axonal transport of the fluorescent rhodamine-dextran-amine, fluoro-ruby, within the CNS. Brain Res 526:127–134

Schwab ME, Thoenen H (1976) Electron microscopic evidence for a transsynaptic migration of tetanus toxin in spinal cord motoneurons: an autoradiographic and morphometric study. Brain Res 105:213–227

Segundo JP (1970) Communication and coding by nerve cells. In: Schmitt FO, Quarton GC, Melnechuk T, Adelman G (eds) The neurosciences. Second study program. Rockefeller University Press, New York, pp 569–586

Sharp FR, Sagar SM, Hicks K, Lowenstein D, Hisanaga KC (1991) c-fos mRNA, Fos, and Fos-related antigen induction by hypertonic saline and stress. J Neurosci 11:2321–2331

Sheng M, Greenberg ME (1990) The regulation and function of c-fos and other immediate early genes in the nervous system. Neuron 4:477–485

Shepherd GM (ed) (1990) The synaptic organization of the brain, 3rd edn. Oxford University Press, New York

Shepherd GM (1991) Foundations of the neuron doctrine. Oxford University Press, New York (History of neuroscience, vol 6)

Sherrington CS (1906) The integrative action of the nervous system. Cambridge University Press, Cambridge

Skirboll L, Hökfelt T, Norell G, Phillipson O, Kuypers HGJM, Bentivoglio M, Catsman-Berrevoets CE, Visser TJ, Steinbusch H, Verhofstad A, Cuello AC, Goldstein M, Brownstein M (1984) A method for specific transmitter identification of retrogradely labelled neurons: immunofluorescence combined with fluorescence tracing. Brain Res Rev 8:99–127

Skirboll L, Thor K, Helke C, Hökfelt T, Robertson B, Long R (1989) Use of retrograde fluorescent tracers in combination with immunohistochemical methods. In: Heimer L, Záborsky L (eds) Neuroanatomical tract-tracing methods 2. Plenum, New York, pp 5–18

Smeets WJAJ, Reiner A (eds) (1994) Phylogeny and development of catecholamine systems in the CNS of vertebrates. Cambridge University Press, Cambridge

Smith JC, Feldman JL (1987) In vitro brainstem-spinal cord preparations for study of motor systems for mammalian respiration and locomotion. J Neurosci Methods 21:321–333

Smith Y, Bolam JP (1992) Combined approaches to experimental neuroanatomy: combined tracing and immunocytochemical techniques for the study of neuronal microcircuits. In: Bolam JP (ed) Experimental neuroanatomy. A practical approach. IRL Press, Oxford, pp 239–266

Sokoloff L, Reivich M, Kennedy CH, DesRosiers MH, Patlak CS, Pettigrew KD, Sakurada O, Shinohara M (1977) The (^{14}C) deoxyglucose method for the measurement of local cerebral glucose utilization: theory, procedure, and normal values in the conscious and anesthetized albino rat. J Neurochem 28:897–916

Somogyi P (1978) The study of Golgi stained cells and of experimental degeneration under the electron microscopy: a direct method for the identification in the visual cortex of three successive links in a neuron chain. Neuroscience 3:167–180

Somogyi P, Hodgson AJ, Smith AD (1979) An approach to tracing neuron networks in the cerebral cortex and basal ganglia. Combination of Golgi staining, retrograde transport of horseradish peroxidase and anterograde degeneration of synaptic boutons in the same material. Neuroscience 4:1805–1852

Spillane JD (1981) The doctrine of the nerves. Oxford University Press, Oxford

Spenger C, Braschler UF, Streit J, Lüscher H-R (1991) An organotypic spinal cord – dorsal root ganglion – skeletal muscle coculture of embryonic rat. I. The morphological correlates of the spinal reflex arc. Eur J Neurosci 3:1037–1053

Steinbusch HWM (ed) (1987) Monoaminergic neurons: light microscopy and ultrastructure. Wiley, Chichester (IBRO handbook series: methods in the neuroscience, vol 10)

Steinbusch HWM, Tilders FJH (1987) Immunohistochemical techniques for light-microscopical localization of dopamine, noradrenaline, adrenaline, serotonin and histamine in the central nervous system. In: Steinbusch HWM (ed) Monoaminergic neurons: light microscopy and ultrastructure. Wiley, Chichester, pp 125–166

Steinbusch HWM, Verhofstad AAJ, Joosten HWJ (1978) Localization of serotonin in the central nervous system by immunohistochemistry: description of a specific and sensitive technique and some applications. Neuroscience 3:811–819

Sternberger LA (1979) Immunocytochemistry, 2nd edn. Wiley, New York

Sternberger LA, Hardy PH, Cuculis JJ, Meyer HG (1970) The unlabeled antibody-enzyme method of immunohistochemistry. Preparation and properties of soluble antigen-antibody complex (horseradish peroxidase-antihorseradish peroxidase) and its use in identification of spirochetes. J Histochem Cytochem 18:315–333

Stewart WW (1978) Functional connections between cells as revealed by dye-coupling with a highly fluorescent naphthalimide tracer. Cell 14:741–759

Stoppini L, Buchs P-A, Muller D (1991) A simple method for organotypic cultures of nervous tissue. J Neurosci Methods 37:173–182

Strack AM, Loewy AD (1990) Pseudorabies virus: a highly specific transneuronal cell body marker in the sympathetic nervous system. J Neurosci 10:2139–2147

Straka H, Dieringer N (1993) Electrophysiological and pharmacological characterization of vestibular inputs to identified frog abducens motoneurons and internuclear neurons in vitro. Eur J Neurosci 5:251–260

Streit J, Spenger C, Lüscher H-R (1991) An organotypic spinal cord – dorsal root ganglion – skeletal muscle coculture of embryonic rat. II. Functional evidence for the formation of spinal reflex arcs in vitro. Eur J Neurosci 3:1054–1068

Stretton AO, Kravitz EA (1968) Neuronal geometry: determination with a technique of intracellular dye injection. Science 162:132–134

Sugihara I, Lang EJ, Llinás R (1993) Uniform olivocerebellar conduction time underlies Purkinje cell complex spike synchronicity in the rat cerebellum. J Physiol (Lond) 470:243–271

Sugimori M, Llinás RR (1990) Real-time imaging of calcium entry in mammalian cerebellar Purkinje cells in vitro. Proc Natl Acad Sci U S A 87:5084–5088

Syková E (1992) Ionic and volume changes in the microenvironment of nerve and receptor cells. Springer, Berlin Heidelberg New York (Progress in sensory physiology, vol 13)

Syková E, Svoboda J, Polák J, Chvátal A (1994) Extracellular volume fraction and diffusion characteristics during progressive ischemia and terminal anoxia in the spinal cord of the rat. J Cerebral Blood Flow Metab 14:301–311

Tank DW, Sugimori M, Connor JA, Llinás RR (1988) Spatially resolved calcium dynamics of mammalian Purkinje cells in cerebellar slice. Science 242:773–777

ten Donkelaar HJ (1976) Descending pathways from the brain stem to the spinal cord in some reptiles. I. Origin. J Comp Neurol 167:421–442

Tenke CE, Schroeder CE, Arezzo JC, Vaughan HG jr (1993) Interpretation of high-resolution current source density profiles: a simulation of sublaminar contributions to the visual evoked potential. Exp Brain Res 94:183–192

Trojanowski JQ, Gonatas JO, Gonatas NK (1982) Horseradish peroxidase (HRP) conjugates of cholera toxin an lectins are more sensitive retrogradely transported markers than free HRP. Brain Res 231:33–50

Türck L (1849) Über sekundäre Erkrankung einzelner Rückenmarksstränge und ihrer Fortsetzungen zum Gehirne. Z Kais Kön Ges Ärzte Wien 1:173–176

Ugolini G, Kuypers HGJM, Strick PL (1989) Transneuronal transfer of herpes virus from peripheral nerves to cortex and brain stem. Science 243:89–91

Valentino KL, Eberwine JH, Barchas DJ (1992) In situ hybridization: application to neurobiology, 2nd edn. Oxford University Press, New York

Vallee RB, Bloom GS (1991) Mechanisms of fast and slow axonal transport. Annu Rev Neurosci 14:59–92

Valverde F (1970) The Golgi method. A tool for comparative structural analysis. In: Nauta WJH, Ebbesson SOE (eds) Contemporary research methods in neuroanatomy. Springer, Berlin Heidelberg New York, pp 12–31

van Mier P, Lichtman JW (1994) Regenerating muscle fibers induce directional sprouting from nearby nerve terminals: studies in living mice. J Neurosci 14:5672–5686

Veenman CL, Reiner A, Honig MC (1992) Biotinylated dextran amine as an anterograde tracer for single- and double-labeling studies. J Neurosci Methods 41:239–244

von Gerlach J (1858) Mikroskopische Studien aus dem Gebiet der menschlichen Morphologie. Enke, Erlangen

von Gudden B (1870) Experimentaluntersuchungen über das peripherische und centrale Nervensystem. Arch Psychiatr 2:693–724

Wahle P (1994) Combining non-radioactive in situ hybridization with immunohistological and anatomical techniques. In: Wisden W, Morris BJ (eds) In situ hybridization protocols for the brain. Academic, London, pp 98–120

Wallén P (1993) Analysing the three-dimensional structure of functionally identified neurons using confocal microscopy. In: Meredith GE, Arbuthnott GW (eds) Morphological investigations of single neurons in vitro. Wiley, Chichester, pp 140–154 (IBRO handbook series: methods in the neurosciences, vol 16)

Wallén P, Carlsson K, Liljeborg A, Grillner S (1988) Three-dimensional reconstruction of neurons in the lamprey spinal cord in a whole-mount, using a confocal laser scanning microscope. J Neurosci Methods 24:91–100

Waller A (1850) Experiments on the section of the glossopharyngeal and hypoglossal nerve of the frog, and observations of the alterations produced thereby in the structure of their primitive fibers. Philos Trans 140:423–469

Wan XST, Trojanowski JQ, Gonatas JO (1982) Cholera toxin and wheat germ agglutinin conjugates as neuroanatomical probes: their uptake and clearance, transganglionic and retrograde transport and sensitivity. Brain Res 243:215–224

Wan XST, Liang F, Moret V, Wiesendanger M, Rouiller EM (1992) Mapping of the motor pathways in rats: C-fos induction by intracortical microstimulation of the motor cortex correlated with efferent connectivity of the site of cortical stimulation. Neuroscience 49:749–761

Weigert K (1884) Ausführliche Beschreibung der in No. 2 dieser Zeitschrift erwähnten neuen Färbungsmethode für das Centralnervensystem. Fortsch Med 2:190–191

Weil AA (1928) A rapid method for staining myelin sheaths. Arch Neurol Psychiatr 20:392–393

Weiss P, Hiscoe HB (1948) Experiments on the mechanism of nerve growth. J Exp Zool 107:315–395

Wilkinson DG (ed) (1992) In situ hybridization. A practical approach. IRL Press, Oxford

Williams PL, Warwick R (1975) Functional neuroanatomy of man. Churchill Livingstone, Edingburgh

Williamson SJ, Romani G, Kaufman L, Modena I (eds) (1983) Biomagnetism. Plenum, New York

Williamson SJ, Hoke M, Stroink G, Kotani M (eds) (1989) Advances in biomagnetism. Plenum, New York

Woelcke M (1942) Eine neue Methode der Markscheidenfärbung. J Physiol Neurol 51:199–202

Woodward SKA, Treherne JM, Knott GW, Fernandez J, Varga ZM, Nicholls JG (1993) Development of connections by axons growing through injured spinal cord of neonatal opossum in culture. J Exp Biol 176:77–88

Wouterlood FG (1988) Anterograde neuroanatomical tracing with Phaseolus vulgaris-leucoagglutinin combined with immunocytochemistry of gamma-amino butyric acid, choline acetyltransferase or serotonin. Histochemistry 89:421–428

Wouterlood FG, Jorritsma-Byham B (1993) The anterograde neuroanatomical tracer biotinylated dextran-amine: comparison with the tracer PHA-L in preparations for electron microscopy. J Neurosci Methods 48:75–87

Wouterlood FG, Jorritsma-Byham B, Goede PH (1990) Combination of anterograde tracing with Phaseolus vulgaris-leucoagglutinin, retrograde fluorescent tracing and fixed-slice intracellular injection of lucifer yellow. J Neurosci Methods 33:207–217

Wouterlood FG, Pattiselanno A, Jorritsma-Byham B, Arts MPM, Meredith GE (1993) Connectional, immunocytochemical and ultrastructural characterization of neurons injected intracellularly in fixed brain tissue. In: Meredith GE, Arbuthnott GW (eds) Morphological investigations of single neurons in vitro. Wiley, Chichester, pp 47–74 (IBRO handbook series: methods in the neurosciences, vol 16)

Yamamoto C, McIlwain H (1966) Electrical activities in thin sections from the mammalian brain maintained in chemically defined media in vitro. J Neurochem 13:1333–1343

Young WS (1990) In situ hybridization histochemistry. In: Björklund A, Hökfelt T, Wouterlood F, van den Pol AN (eds) Analysis of neuronal microcircuits and synaptic interactions. Elsevier, Amsterdam, pp 481–512 (Handbook of chemical neuroanatomy, vol 8)

Záborsky L, Heimer L (1989) Combinations of tracer techniques, especially HRP and PHA-L, with transmitter identification for correlated light and electron microscopic studies. In: Heimer L, Záborsky L (eds) Neuroanatomical tract-tracing methods 2. Plenum, New York, pp 49–96

Záborsky L, Alheid GF, Heimer L (1985) Mapping of transmitter-specific connections: Simultaneous demonstration of anterograde degeneration and changes in the immunostaining pattern induced by lesions. J Neurosci Methods 14:255–266

Appendix – (Stereotaxic) Atlases. A bibliography of (stereotaxic) brain atlases arranged by chapter

Lampreys

Schober W (1964) Vergleichend-anatomische Untersuchungen am Gehirn der Larven und adulten Tiere von Lampetra fluviatilis (Linné 1758) und Lampetra planeri (Bloch 1784). J Hirnforsch 7:107–209

Cartilaginous fishes

Northcutt RG (1978) Brain organization in the cartilaginous fishes. In: Hodgson ES, Mathewson RF (eds) Sensory biology of sharks, skates and rays. US Government Printing Office, Washington DC, pp 117–193

Smeets WJAJ, Nieuwenhuys R, Roberts BL (1983) The central nervous system of cartilaginous fishes. Structure and functional correlation. Springer, Berlin Heidelberg New York

Ray-finned fishes

Braford MR jr, Northcutt RG (1983) Organization of the diencephalon and pretectum in ray-finned fishes. In: Davis RE, Northcutt RG (eds) Fish neurobiology, vol 2: higher brain areas and functions. University of Michigan Press, Ann Arbor, pp 117–164

Nieuwenhuys R, Pouwels E (1983) The brain stem of actinopterygian fishes. In: Northcutt RG, Davis RE (eds) Fish neurobiology, vol 1. University of Michigan Press, Ann Arbor, pp 25–87

Teleosts

Anken RH, Rahmann H (1994) Brain atlas of the adult swordtail fish Xiphophorus helleri and of certain developmental stages. Fischer, Stuttgart

Maler L, Sas E, Johnston S, Ellis W (1991) An atlas of the brain of the electric fish Apteronotus leptorhynchus. J Chem Neuroanat 4:1–38

Meek J, Joosten HWJ, Steinbusch HWM (1989) The distribution of dopamine-immunoreactivity in the brain of the mormyrid teleost Gnathonemus petersii. J Comp Neurol 281:362–383

Peter RE, Gill VE (1975) A stereotaxic atlas and technique for forebrain nuclei of the goldfish, Carassius auratus. J Comp Neurol 159:69–102

Wullimann MF, Rupp B, Reichert H (1996) Neuroanatomy of the zebrafish brain. A topological atlas. Birkhäuser, Basel

Urodeles

Herrick CJ (1930) The medulla oblongata of Necturus. J Comp Neurol 50:1–96

Herrick CJ (1933) The amphibian forebrain. VI. Necturus. J Comp Neurol 58:1–288

Herrick CJ (1948) The brain of the tiger salamander. University of Chicago Press, Chicago

Anurans

Hoffmann A (1973) Stereotaxic atlas of the toad's brain. Acta Anat 84:416–451

Kemali M, Braitenberg V (1969) Atlas of the frog's brain. Springer, Berlin Heidelberg New York

Wada M, Urano A, Gorbman A (1980) A stereotaxic atlas for diencephalic nuclei of the frog, Rana pipiens. Arch Histol Jpn 43:157–193

Reptiles

Distel H (1976) Behavior and electrical brain stimulation in the green iguana, Iguana iguana L. I. Schematic brain atlas and stimulation device. Brain Behav Evol 13:421–450

Greenberg N (1982) A forebrain atlas and stereotaxic technique for the lizard, Anolis carolinensis. J Morphol 174:217–236

Powers AS, Reiner (1980) A stereotaxic atlas of the forebrain and midbrain of the eastern painted turtle (Chrysemys picta picta). J Hirnforsch 21:125–159

Smeets WJAJ, Hoogland PV, Lohman AHM (1986) A forebrain atlas of the lizard Gekko gecko. J Comp Neurol 254:1–19

Birds

Karten HJ, Hodos W (1967) A stereotaxic atlas of the pigeon (Columba livia). Johns Hopkins Press, Baltimore

Stokes TM, Leonard CM, Nottebohm F (1974) The telencephalon, diencephalon, and mesencephalon of the canary, Serinus canaria, in stereotaxic coordinates. J Comp Neurol 156:337–374

van Tienhoven A, Juhásh LP (1962) The chicken telencephalon, diencephalon and mesencephalon in stereotaxic coordinates. J Comp Neurol 118:185–197

Zweers GA (1971) A stereotaxic atlas of the brainstem of the mallard (Anas platyrhynchos L). A stereotaxic apparatus for birds and an investigation of the individual variability of some head structures. Van Gorcum, Assen

Mammals

Marsupials

Oswaldo-Cruz E, Rocha-Miranda CE (1968) The brain of the opossum (Didelphis marsupialis) in stereotaxic coordinates. Inst Biofisica, Univ Fed Rio de Janeiro, Rio de Janeiro

Rodents (selected)

de Groot J (1959) The rat forebrain in stereotaxic coordinates. Noord-Hollandse Uitgevers Maatschappij, Amsterdam

Franklin KBJ, Paxinos G (1996) The mouse brain in stereotactic coordinates. Academic, San Diego

Knigge KM, Joseph SA (1968) A stereotaxic atlas of the brain of the golden hamster. In: Hoffmann RA, Robinson PF, Magalhaes H (eds) The golden hamster. Its biology and use in medical research. Iowa State University Press, Ames

König JFR, Klippel RA (1963) The rat brain. A stereotaxic atlas of the forebrain and lower parts of the brainstem. Williams and Wilkins, Baltimore

Luparello TJ (1967) Stereotaxic atlas of the forebrain of the guinea pig. Karger, Basel

Paxinos G, Watson C (1986) The rat brain in stereotaxic coordinates, 2nd edn. Academic, Sydney

Pellegrino LJ, Pellegrino AS, Cushman AJ (1979) A stereotaxic atlas of the rat brain, 2nd edn. Plenum, New York

Rössner W (1965) Stereotaktischer Hirnatlas vom Meerschweinchen. Pallas, Lochham

Sidman RL, Angevine JB jr, Taber Pierce E (1971) Atlas of the mouse brain and spinal cord. Harvard University Press, Cambridge

Slotnick BM, Leonard CM (1975) A stereotaxic atlas of the albino mouse forebrain. Publication (ADM) 75–100, US Dept Health Educ Welfare, Rockville MD

Swanson LW (1992) Brain maps. Structure of the rat brain. Elsevier, Amsterdam

Wünscher W, Schober W, Werner L (1965) Architektonischer Atlas vom Hirnstamm der Ratte. Hirzel, Leipzig

Lagomorphs

Fifková E, Maršala J (1960) Stereotaxic atlas for the rabbit. In: Bureš J, Petrán M, Zachar J (eds) Electrophysiological methods in biological research. Publication House Czechoslovak Acad Sci, Prague

McBride RL, Klemm WR (1968) Stereotaxic atlas of rabbit brain, based on the rapid method of photography of frozen unstained sections. Comm Behav Biol Part A 2:179–215

Carnivores

Berman AL (1968) The brain stem of the cat. A cytoarchitectonic atlas with stereotaxic coordinates. University of Wisconsin Press, Madison

Berman AL, Jones EG (1982) The thalamus and basal telencephalon of the cat. A cytoarchitectonic atlas with stereotaxic coordinates. University of Wisconsin Press, Madison

Dua-Sharma S, Sharma KN, Jacobs HL (1970) The canine brain in stereotaxic coordinates. MIT Press, Cambridge

Lim RKS, Liu C, Moffit RL (1960) A stereotaxic atlas of the dog's brain. Thomas, Springfield

Singer M (1962) The brain of the dog in section. Saunders, Philadelphia

Snider RS, Niemer WT (1961) A stereotaxic atlas of the cat brain. University of Chicago Press, Chicago

Primates

Emmers R, Akert K (1963) A stereotaxic atlas of the brain of the squirrel monkey (Saimiri sciureus). University of Wisconsin Press, Madison

Kusama T, Mabuchi M (1970) Stereotaxic atlas of the brain of Macaca fuscata. University Park Press, Baltimore

Manocha SL, Shantha TR, Bourne GH (1968) A stereotaxic atlas of the brain of the cebus monkey (Cebus apella). Clarendon, Oxford

Saavedra JP, Mazzuchellu AL (1969) A stereotaxic atlas of the brain of the marmoset (Hapale jacchus). J Hirnforsch 11:105–122

Shantha TR, Manocha SL, Bourne GH (1968) A stereotaxic atlas of the Java monkey brain (Macaca irus). Karger, Basel

Snider RS, Lee JC (1961) A stereotaxic atlas of the monkey brain (Macaca mulatta). University of Chicago Press, Chicago

Tigges J, Shantha TR (1969) A stereotaxic atlas of the tree shrew (Tupaia glis). Williams and Wilkins, Baltimore

II. SPECIALISED PART

CHAPTER 8

Introduction

R. NIEUWENHUYS

8.1
Choice of Material

This section surveys the structure and function of the central nervous system in the various groups of vertebrates. Since it would be impossible to review all the available data on every species examined, we had to be selective. Our choices were based on the following considerations. (a) The species to be dealt with should represent all major groups. (b) Preference should be given to species in which the CNS has been extensively studied, both morphologically and physiologically. (c) The book should be useful as an introductory guide to experimental studies; thus, easily obtainable species should be selected whenever possible.

Taking these three points into consideration, we arrived at the 17 core species indicated with an asterisk in Table 8.1. I will now briefly comment on these species and on their taxonomic and phylogenetic relationships. The latter are indicated in the cladogram shown in Fig. 8.1.

The vertebrates, defined as the members of the large subphylum Craniata, are commonly grouped with the members of a much smaller subphylum, the Acrania, to form the phylum Chordata. Although the theme of this book is the CNS of the vertebrates, we consider it important to include a single species of the other chordate subphylum, namely the lancelet, *Branchiostoma lanceolatum*. It is now thought that this remarkable acraniate has retained many features common to primitive chordates and may well represent one stage in the evolution of the craniates (Young 1962). We shall see that the tiny brain of the lancelet differs considerably from that of all vertebrates, although its well-developed spinal cord is quite similar.

The subphylum Craniata or Vertebrata can be divided into two superclasses, the Agnatha (jawless vertebrates) and the Gnathostomata (vertebrates with jaws). Among the jawless vertebrates are the oldest fossil vertebrates known, the ostracoderms. These small, heavily armoured creatures are found in Silurian and Devonian deposits. It is thought that certain of the ostracoderms, upon losing their dermal armour, gave rise to the cyclostomes (Romer 1969). The two recent representatives of the class are the lampreys (Petromyzontoidea) and the hagfishes (Myxinoidea), which are superficially similar but differ considerably in their detailed morphology. Although these two groups are known to have had long separate histories (Jarvik 1964; Bardack 1991), according to some authors (Jarvik 1968b; Schaeffer and Thomson 1980; Yalden 1985), the cyclostomes are nevertheless a monophyletic group which represents the sister-group of gnathostomes. Others (Janvier 1981; Forey 1984; Jefferies 1986) believe that hagfishes are the sister-group of all other craniates, which are collectively termed the myopterygians and include both lampreys and gnathostomes. In our series, *Lampetra fluviatilis* represents the Petromyzontoidea, whereas the pacific hagfish, *Eptatretus stouti*, represents the Myxinoidea. It will be seen that the brains of these two species show several striking differences.

The development of the gnathostomes was a major event in vertebrate evolution, but, unfortunately, their origin is not well documented by the fossil record (Schaeffer 1969). It is known, however, that various groups of fishes possessing rudimentary jaws were present at the end of the Silurian. These are now considered to form a special class of vertebrates, the Placodermi, but it is not certain that they form a truly natural group (Romer 1962). Although these fishes were exceedingly prominent in the Devonian, they became extinct before the end of the Palaeozoic. It is generally accepted that a group of placoderms gave rise to the cartilaginous fishes or Chondrichthyes. Fossils of these fishes appear in the Devonian. They flourished through the Carboniferous to the present day, where they

Table 8.1. Classification of the species selected

Phylum chordata

SUBPHYLUM	ACRANIA	
Superclass	Cephalochordata	The lancelet, *Branchiostoma lanceolatum**
SUBPHYLUM	CRANIATA	
Superclass	Agnatha	
Class	Cyclostomata	
Subclass	Petromyzontoidea	The lamprey, *Lampetra fluviatilis**
Subclass	Myxinoidea	The pacific hagfish, *Eptatretus stouti**, the atlantic hagfish, *Myxine glutinosa**
Superclass	Gnathostomata	
Class	Chondrichthyes	
Subclass	Elasmobranchii	The spiny dogfish, *Squalus acanthias**, the lesser spotted dogfish, *Scyliorhinus canicula*, the thornback ray, *Raja clavata*
Subclass	Holocephali	The ratfish, *Hydrolagus collei*
Class	Osteichthyes	
Subclass	Brachioperygii	The reedfish, *Erpetoichthys calabaricus**
Subclass	Actinopterygii	
Superorder	Chondrostei	The shovelnose sturgeon, *Scaphirhynchus platorynchus**
Superorder	Holostei	The bowfin, *Amia calva*
Superorder	Teleostei	The rainbow trout, *Salmo gairdneri**, the catfish, *Ictalurus nebulosus*, the mormyrid, *Gnathonemus petersii* and several other species
Subclass	Dipnoi	The South American lungfish, *Lepidosiren paradoxa**, the Australian lungfish, *Neoceratodus forsteri*
Subclass	Crossopterygii	The coelacanth, *Latimeria chalumnae**
Class	Amphibia	
Subclass	Lissamphibia	
Superorder	Caudata	The tiger salamander, *Ambystoma tigrinum*
Superorder	Anura	The green fog, *Rana esculenta**, the bull frog, *Rana catesbeiana*, the clawed toad, *Xenopus laevis*
Class	Reptilia	
Subclass	Anapsida	
Order	Chelonia	The tortoise, *Testudo hermanni**
Subclass	Lepidosauria	
Order	Rhynchocephalia	The tuatara, *Sphenodon punctatus*
Order	Squamata	The common iguana, *Iguana iguana*, the tegu, *Tupinambis tequixin**, the chameleon, *Chamaeleo lateralis*, the amphisbaenian, *Monopeltis guentheri*, the python, *Python reticulatus*
Subclass	Archosauria	
Order	Crocodilia	The spectacled caiman, *Caiman crocodilus*, the alligator, *Alligator mississippiensis*
Class	Aves	
Superorder	Neognathae	The pigeon, *Columba livia**, and several other species
Class	Mammalia	
Superorder	Metatheria	
Order	Marsupialia	The opossum, *Didelphis virginiana**
Superorder	Eutheria	
Order	Carnivora	The cat, *Felis catus**
Order	Primates	The rhesus monkey, *Macaca mulatta**

* Seventeen core species discussed primarily.

are represented by the modern sharks and rays, united in the subclass Elasmobranchii, and by the ratfishes or chimaeras, which constitute the subclass Holocephali. We have included the spiny dogfish, *Squalus acanthias*, a primitive chondrichthyan, in our series.

Not only the cartilaginous fishes or Chondrichthyes, but also the bony fishes or Osteichthyes attained prominence in the Devonian. The latter class can be subdivided into four groups, the Brachiopterygii (arm-finned fishes), the Actinopterygii (ray-finned fishes), the Dipnoi (lungfishes) and the Crossopterygii (tassel-finned fishes), which are ranked here as subclasses. Probably all four of these

groups are descended from a common ancestral stock of which we have no direct knowledge.

One group of the Osteichthyes that presents us with taxonomic difficulties is the Polypteriformes. This is a small group of African fresh-water fish, having only some ten extant species allocated in two genera, *Polypterus* and *Erpetoichthys*. Formerly, it was believed that the polypteriform fishes were crossopterygians, but at present they are either assigned to the Actinopterygii or considered as an isolated group. The adherents of the Actinopterygii theory (Romer 1962; Moy-Thomas and Miles 1971; Patterson 1982; Lauder and Liem 1983) believe that the Polypteriformes, like the chondrostean Acti-

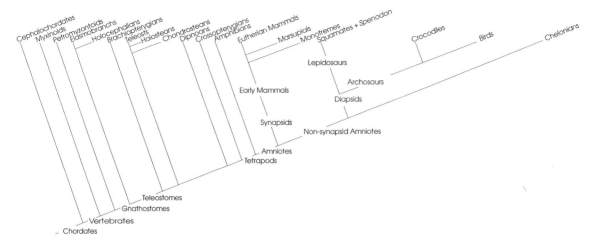

Fig. 8.1. Cladogram of extant groups of chordates (based mainly on Carroll 1988 and Butler 1994)

nopterygii (see below), are descendents of the primitive Palaeozoic Palaeoniscoidei. Those who claim a separate status for this group of fishes (Stensiö 1921; Jarvik 1942, 1968a,b; Daget 1958; Nelson 1969; Bjerring 1985) emphasise that their structure differs profoundly from that of all other living and fossil fishes. The structure of the CNS of the Polypteriformes bears little relation to that of the Actinopterygii. Consequently, we have placed the reedfish, *Erpetoichthys calabaricus*, the species representing Polypteriformes, in a separate subclass, the Brachiopterygii, in accordance with Stensiö and the other authors mentioned above.

The Actinopterygii constitute by far the largest group of bony fishes and are usually subdivided into three superorders, the Chondrostei, Holostei and Teleostei. Interestingly, these three superorders may be considered as representing three subsequent stages or gradations of actinopterygian evolution (Romer 1962, 1969).

It is generally agreed that the chondrosteans and teleosts form monophyletic groups, but during the past several decades there has been some dispute on the phylogenetic status of the holosteans. This group is represented in the recent fauna by two genera: *Amia*, with a single species: the bowfin, *Amia calva*, and *Lepisosteus*, with eight species (gars). Patterson (1973) extensively reviewed the structure of living holosteans and teleosts and arrived at the conclusion that *Amia* is more closely related to teleosts than either is to gars and that, hence, *Amia* is the recent sister-group of teleosts. He suggested that *Amia*, gars and teleosts be reclassified as a new infraclass, the Neopterygii, containing two divisions: the Ginglymodi (*Lepisosteus*) and Halecostomi (*Amia* plus teleosts). In contrast, on the basis

of fossil evidence, Olsen (1984) held that gars are more closely related to teleosts than is *Amia*. Jollie (1984) claimed that the morphological evidence available from the recent forms favours the monophyly of the holosteans, and molecular sequence data (Le et al. 1993; Lecointre et al. 1993) also lend some support to the pairing of *Amia* and *Lepisosteus*. Patterson (1994) recently considered the 'holostean problem' unresolved. In the present work, *Amia* and *Lepisosteus* will be regarded as members of the Holostei, because their brains exhibit a number of unique, derived characters not shared by chondrosteans or teleosts.

As mentioned earlier, the Chondrostei are descendents of the primitive Palaeozoic Palaeoniscoidei. During early and middle Mesozoic times the Chondrostei were supplanted by the Holostei, which in turn were largely replaced in the last phase of the Mesozoic era and the Cenozoic era by the expanding and now abundant Teleostei. In the core group we have included a chondrostean, the shovelnose sturgeon, *Scaphirhynchus platorynchus*, and a 'generalised' teleostean, the trout *Salmo gairdneri*. Special attention will also be paid to the holostean *Amia calva*.

The lungfish or Dipnoi form an extremely ancient group of fish that appeared in the lower Devonian and reached the zenith of its evolution in Late Devonian and Carboniferous times (Moy-Thomas and Miles 1971). Like the brachiopterygians, the dipnoans have a very limited representation in the recent fauna. The living lungfish constitute three genera: the African genus, *Protopterus*, with some four species; the South American genus, *Lepidosiren*, with a single species, and the Australian genus, *Neoceratodus*, also with a single species.

The extant species comprise a monophyletic group (Nelson 1989). *Lepidosiren paradoxa* is included in the core list of the present work.

The fourth ostheichthyan subclass, the Crossopterygii, survives only as a single species, the coelacanth *Latimeria chalumnae*. This rare fish inhabits the waters around the Comoros, a small group of islands lying midway between the northern tip of Madagascar and the African coast. We have included this interesting fish, which can be aptly designated as a mesozoic relic, in our series.

The crossopterygians and dipnoans are often classified together as Sarcopterygii or fleshy-finned fishes (Romer 1962, 1969; Lauder and Liem 1983; Carroll 1988). There can be no doubt that the terrestrial vertebrates developed from some Palaeozoic sarcopterygian in which the fleshy fins were gradually transformed into tetrapod limbs. Most palaeontologists believe that the Palaeozoic crossopterygians were the direct ancestors of terrestrial vertebrates. It is noteworthy, however, that the coelacanths, to which *Latimeria* belongs, represent only a side branch of the crossopterygians and are not in the main line of evolution to the tetrapods. The other sarcopterygians, the dipnoans, adhere closely in many embryonic features to the patterns found in living amphibians. This resemblance, most evident in the CNS, has been considered by Bertmar (1968) and others to have arisen from parallel evolution. Rosen et al. (1981) claimed that the dipnoans and the tetrapods share numerous derived characters, thus reviving the nineteenth-century idea that dipnoans, and not crossopterygians, are close to the ancestry of tetrapods. Their arguments have been countered by numerous authors, however, among them Jarvik (1981), Holmes (1985), Schultze (1986) and Panchen and Smithson (1987).

The rise of the amphibians, the first group of terrestrial vertebrates, from some unknown sarcopterygian stock must have occurred in the Palaeozoic era, as unmistakable amphibians, with fully developed tetrapod limbs, are known from the last part of the Devonian. These early amphibians were members of a large group, the extinct labyrinthodonts. Contemporary amphibians are represented by three superorders, the Caudata (salamanders and newts), the Anura (frogs and toads), and the wormlike Gymnophiona. The fossil record of these groups is highly incomplete, and fossil forms intermediate between them and the Palaeozoic amphibians have not been found. However, because the three superorders of recent amphibians share a considerable number of derived features, they are regarded as forming a monophyletic group (Rage 1985; Milner 1988; Bolt 1991; Trueb and Cloutier 1991a,b). In the present work the amphibians are represented by two species, the tiger salamander, *Ambystoma tigrinum*, and the green frog, *Rana esculenta*.

Reptiles, birds and mammals constitute a monophyletic group, the Amniota, that is distinguished from amphibians by having achieved a fully terrestrial way of life. Amniotes can be broadly classified on the basis of the structure of the temporal region of their skull. Three groups, the Anapsida, the Synapsida and the Diapsida are distinguished. In the Anapsida the skull roof behind the orbits is completely covered with bone. The Synapsida have a single pair of openings in the lower part of the skull, whereas the Diapsida are characterised by the presence of two pairs of temporal openings. According to Carroll (1988), the ancestral group that gave rise to all modern amniotes were a single family in the Captorhinomorpha order. The skull of these primitive amniotes, like that of their Palaeozoic amphibian contemporaries, was of the anapsid type.

The first group to diverge from the captorhinomorph stock were the Synapsida. These are grouped in two orders, the Pelycosauria and the Therapsida (forming together the mammal-like reptiles). The mammals evolved from the Therapsida.

Later, a second group, the Diapsida, diverged form the basal anapsid stock. The Diapsida can be divided into two subgroups, the Lepidosauria and the Archosauria. The lepidosaurs are represented in the modern fauna by the lizards and snakes (united in the Squamata order) and by *Sphenodon* of New Zealand, a mesozoic relic. The archosaurs include the extinct dinosaurs and pterosaurs, and the living crocodiles. The birds arose from a group of small, bipedal, insectivorous dinosaurs.

Turtles (order: Chelonia) form the third major group of modern amniotes. The history of this group can be traced back to the Upper Triassic, in which they already showed a typical chelonian appearance. How they evolved to that stage from their Carboniferous captorhinomorph ancestors is unknown. However, because they did not develop the typical temporal openings occurring in other reptilian groups, turtles are usually included in the subclass Anapsida. According to Carroll (1988), however, the assignment of the turtles to the Anapsida is somewhat arbitrary, since their overall anatomy and way of life differ markedly from those of the early anapsids.

The earliest known mammalian fossils date from the late Triassic period, some 160 million years ago. This group did not become abundant, however, before the end of the Cretaceous period (about 75 million years ago). All mammals originated from the therapsids, an advanced group of mammal-like reptiles. The living mammals are divided into three groups, the monotremes or Prototheria, the marsupials or Metatheria, and the placental mammals or Eutheria. The monotremes, to which the duckbill (*Platypus*) and the spiny anteater (*Tachyglossus*) belong, have scarcely any fossil record. Because they differ in so many respects from all other mammals, it is believed that they diverged very early from the mammalian stock. The phylogenetic relationship of the living marsupials is in question, but they are probably more closely related to the placental mammals than to the monotremes. Probably, the marsupials and the placental mammals diverged shortly before the mid-Cretaceous essentially simultaneously from a common stem-group of poorly known, primitive mammals.

From this brief survey of the evolution of amniotes, which is based largely on Carroll (1988), it appears that the birds arose from archosaurs and the mammals form synapsids. This means that in a strictly phylogenetic classification the reptiles form a paraphyletic group that can be defined only as amniotes that are neither birds nor mammals (Fig. 8.1).

Two reptiles have been included in the series of core species, the spur-trailed Mediterranean tortoise, *Testudo hermanni* (order: Chelonia) and the tegu lizard, *Tupinambis teguixin* (order: Squamata).

It is not known which of the many orders of living birds, if any, can be considered as direct, relatively unchanged descendants of the earliest birds (Bock 1969; Carroll 1988). Consequently, we have included only a single bird, the pigeon, *Columba livia* in the core list. This species has been selected mainly because its CNS has been thoroughly analysed with modern experimental neuroanatomical techniques.

As regards the selection of mammalian species, lack of material prevented us from including a monotreme. The marsupials will be represented by the North American opossum, *Didelphis virginiana*, a species that has retained a large number of primitive mammalian brain characters. Therefore, a study of this 'living fossil' should provide us with an illustration of the organisation of the CNS of the earlier marsupials.

Finally, two common and well-known species, the cat *Felis domestica* and the rhesus monkey, *Macaca mulatta*, will represent the placental mammals in our series. Because the insectivores are among the most primitive living placentals, it would have been appropriate to include a member of this group. Two reasons prevented us from doing so: (a) the CNS of insectivores is much like that of the marsupials, and (b) the CNS of the insectivores

has not been as extensively and systematically explored with experimental hodological techniques as that of the marsupial selected. The three mammals selected – opossum, cat and rhesus monkey – will confront us with the organisation of the mammalian brain at three grades: primitive, intermediate, and advanced. It should be emphasised, however, that this hierarchy does not imply any phylogenetic sequence.

8.2 Illustrations

The structure of the CNS of the 17 species introduced above will be discussed in the ensuing chapters. These surveys rely largely on data derived from the literature. In contrast, most of the morphological illustrations are originals, especially designed for this work. Coherence was sought by preparing sets of standard illustrations for all chapters. These sets include:

1. A drawing of the entire animal. Some of these pictures were adapted from the literature, but most are originals.
2. Simple sketches showing the brain in position.
3. Half-tone drawings showing the dorsal, ventral, lateral and medial aspects of the brain. A few of these drawings are based on figures in the literature, a few others on macroscopic preparations; most were based on carefully prepared graphical and three-dimensional reconstructions derived from microscopic sections.
4. Sequences of transversely oriented microscopical sections through the spinal cord and brain. These sections were drawn from original preparations. They were selected from continuous histological series of Nissl-stained sections alternating with either Klüver-Barrera or Bodian preparations or both. Illustrations of these sections involved the following sequence of steps: (a) Photographs were made of selected Nissl preparations and of the adjacent Klüver-Barrera or Bodian preparations. (b) Under microscopic control, the various areas of grey and white matter were delineated on the photographs. In this way outline drawings for the final figures were obtained. (c) The fibre composition and pattern of the various tracts and more diffuse areas of white matter were analysed in the Klüver-Barrera or Bodian material. Additional series, stained according to Häggquist's methylene blue-eosin technique, were employed in the analysis of the white matter of reptiles and mammals. The fibres were generally graded into two or three groups: coarse, (medium) and thin. By using contours or

dots, and lines of corresponding widths, the results were represented in the right half on the drawings. (d) From adjacent Nissl preparations, samples of the various nuclei and other cellular areas were drawn and transferred to their appropriate positions in the left halves of the drawings. It is to be emphasised that in these drawings both the fibre system and the cellular pattern are depicted *semi-diagrammatically*; the individual fibres and cells are generally drawn at a magnification exceeding that of the sections as a whole.

5. A topological chart showing the dispositions of cell masses in the brain stem. These charts were prepared according to a special procedure that facilitates the study of the zonal pattern of the brain stem. This procedure involves two fundamental steps: first, the cell masses and large individual cells are projected upon the ventricular surface; next, the ventricular surface, with its sulci and the projections of the cell masses marked upon it, is flattened out by subjecting it to a one-to-one continuous topological transformation. The projection of the cell masses upon the ventricular surface is based on radially arranged projection curves, which form part of the natural coordinate system of the brain (cf. Chaps. 5 and 6). The topological reconstruction procedure is detailed in Nieuwenhuys (1974).

6. One or more topographical reconstructions of the cell masses and principal fibre paths in the diencephalon.

7. Some diagrammatic representations of the most important fibre systems in the brain.

8.3
Scope

In the preceding paragraphs we have discussed the choice of species and the types of illustrations that together constitute the morphological core of the second section of this book. With this approach we seek to present a view of the structural diversity of the CNS of vertebrates, a diversity that, in light of our knowledge of the phylogeny, can be cautiously placed in an evolutionary perspective. We hardly need to point out that within any comparative approach, however condensed, attention should not be confined to representative species of different groups, but should also be focussed on the spectrum of variability within the various taxa. It is at this level of comparison that structural differences can often be correlated with functional specialisations, which in turn can be placed in the perspective of their adaptive significance. In order to elucidate this second major aspect of comparative neuroanatomy, in many chapters we have gone beyond

the 'core species' to discuss and depict pertinent structural features of the CNS of other animals. The names of these additional species have been included in Table 8.1. The distinction of these two categories of species is entirely artificial, of course, and serves only to keep this book within reasonable limits.

As in the general introduction, functional and structural aspects will be dealt with together. It is important to note that the comparative study of function in the nervous system of vertebrates has not been pursued as comprehensively as has structural research. Most comparative neurophysiological information is a by-product of the exploitation of some particularly advantageous feature in a particular species. Because of this fragmentary comparative neurophysiological knowledge, some chapters lack any reference to functional aspects, whereas in others only a brief functional commentary can be presented.

In concluding this introduction we point out that, although structure and function will be main topics in the ensuing chapters, attention will also be paid to the ontogenesis of the CNS of the species or groups dealt with. The discussion devoted to this subject, however, is limited to brief illustrations of the principles of the morphogenesis and histogenesis of the brain set forth in Chaps. 4 and 5 in the first section of this book.

References

Bardack D (1991) First fossil hagfish (Myxinoidea): a record from the Pennsylvanian of Illinois. Science 254:701–703

Bertmar G (1968) Lungfish phylogeny. In: Ørvig T (ed) Current problems of lower vertebrate phylogeny. Fourth Nobel symposium. Almqvist and Wiksell, Stockholm, pp 259–283

Bjerring HC (1985) Facts and thoughts on piscine phylogeny. In: Forman RE, Gorbman A, Dodd JM, Rolsson R (eds) Evolutionary biology of primitive fishes. Plenum, New York, pp 31–57

Bock WJ (1969) The origin and radiation of birds. In: Petras JM, Noback CR (eds) Comparative and evolutionary aspects of the vertebrate central nervous system. Ann N Y Acad Sci 167:147–155

Bolt JR (1991) Lissamphibian origins. In: Schultze HP, Trueb L (eds) Origins of the higher groups of tetrapods. Controversy and consensus. Comstock, Ithaca/London, pp 194–222

Butler AB (1994) The evolution of the dorsal thalamus of jawed vertebrates, including mammals: cladistic analysis and a new hypothesis. Brain Res Rev 19:29–65

Carroll RL (1988) Vertebrate paleontology and evolution. Freeman, New York

Daget J (1958) Sous-classe des brachioptérygiens. Traité de Zool 13:2500–2521

Forey PL (1984) Yet more reflections on agnathan-gnathosome relationships. J Vertebr Paleontol 4:330–343

Holmes EB (1985) Are lungfishes the sister group of tetrapods? Biol J Linn Soc 25:379–397

Janvier P (1981) The phylogeny of the Craniata, with particular reference to the significance of fossil 'agnathans'. J Vertebr Paleontol 1:121–159

Jarvik E (1942) On the structure of the snout of crossopterygians and lower gnathostomes in general. Zool Bidr Uppsala 21:237–675

Jarvik E (1964) Specializations in early vertebrates. Ann Soc R Zool Belg 94:11–95

Jarvik E (1968a) The systematic position of the Dipnoi. In: Ørvig T (ed) Current problems of lower vertebrate phylogeny. Fourth Nobel Symposium. Almqvist and Wiksell, Stockholm, pp 223–245

Jarvik E (1968b) Aspects of vertebrate phylogeny. In: Ørvig T (ed) Current problems of lower vertebrate phylogeny. Fourth Nobel Symposium. Almqvist and Wiksell, Stockholm, pp 497–527

Jarvik E (1981) (Review of) Lungfishes, tetrapods, paleontology, and plesiomorphy. Syst Zool 30:378–384

Jefferies RPS (1986) The ancestry of vertebrates. British Museum of Natural History and Press Syndicate Cambridge University, Cambridge

Jollie M (1984) Development of cranial and pectoral girdle bones of Lepisosteus with a note on scales. Copeia 1984:497–502

Lauder GV, Liem KF (1983) The evolution and interrelationships of the actinopterygian fishes. Bull Mus Comp Zool 150:95–197

Le HLV, Lecointre G, Perasso R (1993) A 28S rRNA-based phylogeny of the Gnathostomes: fist steps in the analysis of conflict and congruence with morphologically based cladograms. Mol Phylogen Evol 2:31–51

Lecointre G, Philippe H, Le HLV, Le Guyader H (1993) Species sampling has a major impact on phylogenetic inference. Mol Phylogen Evol 2:205–224

Milner AR (1988) The relationships and origin of living amphibians. In: Benton MJ (ed) The phylogeny and classification of tetrapods, vol 1, amphibians, reptiles, birds systematics association, special vol 35A. Clarendon, Oxford, pp 59–102

Moy-Thomas JA, Miles RS (1971) Palaeozoic fishes. Chapman and Hall, London

Nelson GJ (1969) Origin and diversification of teleostean fishes. In: Petras JM, Noback CR (eds) Comparative and evolutionary aspects of the vertebrate central nervous system. Ann NY Acad Sci 167:18–30

Nelson G (1989) Phylogeny of major fish groups. In: Fernholm B, Bremer K, Jörnvall H (eds) The hierarchy of life. Elsevier, Amsterdam, pp 325–336

Nieuwenhuys R (1974) Topological analysis of the brain stem: a general introduction. J Comp Neurol 156:255–276

Olsen PE (1984) The skull and pectoral girdle of the parasemionotid fish Watsonulus eugnathoides from the Early Triassic Sakamena Group of Madagascar, with comments on the relationships of the holostean fishes. J Vertebr Paleontol 4:481–499

Panchen AL, Smithson TR (1987) Character diagnosis, fossils and the origin of tetrapods. Biol Rev Camb Philos Soc 62:341–438

Patterson C (1973) Interrelationships of holosteans. In: Greenwood PH, Miles RS, Patterson C (eds) Interrelationships of fishes. Academic, London, pp 233–305

Patterson C (1982) Morphology and interrelationships of primitive actinopterygian fishes. Am Zool 22:241–259

Patterson C (1994) Bony fishes In: Prothero DR, Schoch RM (eds) Major features of vertebrate evolution. Short courses in paleontology, no 7. Paleontological Society, Knoxville, Tennessee, pp 57–84

Rage JC (1985) Origine et phylogénie des Amphibiens. Bull Soc Herp Fr 34:1–19

Romer AS (1962) The vertebrate body. Saunders, Philadelphia

Romer AS (1969) Vertebrate history with special reference to factors related to cerebellar evolution. In: Llínas R (ed) Neurobiology of cerebellar evolution and development. American Medical Association, Chicago, pp 1–18

Rosen DE, Forey PL, Gardiner BG, Patterson C (1981) Lungfishes, tetrapods, paleontology, and plesiomorphy. Bull Am Mus Nat Hist 167:163–275

Schaeffer B (1969) Adaptive radiation of the fishes and the fish-amphibian transition. In: Petras JM, Noback CR (eds) Comparative and evolutionary aspects of the vertebrate central nervous system. Ann N Y Acad Sci 167:5–17

Schaeffer B, Thomson KS (1980) Reflections on agnathan-gnathostome relationships. In: Jacobs LL (ed) Aspects of vertebrate history. Museum of Northern Arizona Press, Flagstaff, pp 19–33

Schultze HP (1986) Dipnoans as sarcopterygians. In: Bemis WE, Burggren WW, Kemps NE (eds) The biology and evolution of lungfishes. Liss, New York, pp 39–74 (Journal of morphology, supplement 1)

Stensiö EA (1921) Triassic fishes from Spitzbergen, part I. Holzhausen, Vienna

Trueb L, Cloutier R (1991a) Toward an understanding of the amphibians: Two centuries of systematic history. In: Schultze HP, Trueb L (eds) Origins of the higher groups of tetrapods. Controversy and consensus. Comstock, Ithaca/London, pp 175–193

Trueb L, Cloutier R (1991b) A phylogenetic investigation of the inter- and intrarelationships of the Lissamphibia (Amphibia: Temnospondyli). In: Schultze HP, Trueb L (eds) Origins of the higher groups of tetrapods. Controversy and consensus. Comstock, Ithaca/London, pp 223–313

Yalden DW (1985) Feeding mechanisms as evidence for cyclostome monophyly. Zool J Linn Soc 84:291–300

Young JZ (1962) The life of vertebrates. Clarendon, Oxford

Amphioxus

R. Nieuwenhuys

9.1
Introduction

The lancelet *Branchiostoma lanceolatum*, or amphioxus as it is commonly called, is a small, translucent animal some 4–6 cm in length. Although amphioxus resembles a fish, it has a much simpler organisation (Fig. 9.1). There is no true head, and paired special sense organs are entirely lacking. Its body axis is formed by a well-developed notochord, extending from the very tip of the rostrum (Fig. 9.2) to the end of the tail. As its body shape suggests, the animal is able to swim effectively. The propulsive force is provided by the patterned contractions of the body musculature, which is segmented into a series of some 60 myotomes on either side of the body. The myotomes of the right side alternate with those of the left. There are two modes of undulatory movements during forward swimming, fast and slow, that appear to depend on a differentiation of the muscle fibres, as in cyclostomes and in certain groups of fish (Guthrie 1975). In addition to forward swimming, amphioxus displays backward swimming and is also able to burrow rapidly in the sand with either the rostrum or the tail leading.

Amphioxus lives in shallow waters in various regions of the world. It spends most of its time buried in the sand, with only its rostral end protruding above the surface. In this position it feeds by extracting small particles from a stream of water drawn in by ciliary action. The mouth is surrounded by a circle of stiffened cirri, provided with sensory cells. The buccal cavity is short and leads into the large, elongated pharynx, which extends about half the total length of the body.

Bilaterally, the wall of the pharynx is perforated by a large number of pharyngeal slits (100 or more pairs) and supported by an equal number of pharyngeal bars located between adjacent pharyngeal slits. These slits are lined with cilia, which generate the water current supplying the animal with food. Externally, the pharyngeal slits are covered by folds of the body wall, which enclose a large chamber termed the atrium. The latter surrounds the pharynx and vents to the outside by the atriopore, a small opening situated at some distance rostral to the anus. Thus, the water current driven by the cilia enters the mouth opening, passes through the pharynx, enters the atrium through the pharyngeal slits, and leaves via the atriopore. The atrium is strongly muscled, which allows its volume to be reduced to produce an exhalant water current.

A ciliated groove (endostyle or hypobranchial groove) located in the floor of the pharynx produces mucoid sheets which are propelled rostrally and dorsally along the medial surfaces of the pharyngeal bars. The individual sheets merge into a dorsally located pharyngeal groove (the epibranchial groove) and are then transported caudally into the intestine. Thus food particles swept into the pharynx by the water movement created by the beat of

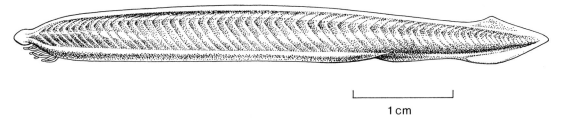

1 cm

Fig. 9.1. The lancelet *Branchiostoma lanceolatum*

the pharyngeal cilia are trapped by the moving mucoid sheets and ingested.

In summary, the lancelet is able to move actively through the water but, unlike aquatic vertebrates, does not employ this faculty to actively seek food; rather, it spends most of its life largely burrowed in the sand, filtering food from the water. The enormously developed pharyngeal basket with its surrounding atrium is obviously a specialisation related to this mode of life.

The genus *Branchiostoma* encompasses, apart from the European *B. lanceolatum*, seven other species, among them *B. floridae* and *B. belcheri*.

Together with the genus *Epigonichthys* (six species), the genus *Branchiostoma* constitutes a separate subphylum, the Cephalochordata or Acrania. This subphylum forms the sister-group of the subphylum Vertebrata or Craniota (Wada and Satoh 1994). Together with two other subphyla, the Hemichordata and the Tunicata, the Cephalochordata and the Vertebrata constitute the phylum Chordata. All living cephalochordates resemble each other closely. The main difference between the genera *Branchiostoma* and *Epigonichthys* is that the latter have gonads only on the right side.

There is a vast literature on the morphology of amphioxus, and this also holds true for the CNS. Physiological studies of amphioxus have been largely limited to analyses of locomotion and electrophysiological investigations of the muscles. The latter include both the propulsive body musculature and the unusual contractile notochord, as well as the atrial system. The only physiological studies of the CNS are the preliminary investigations of Guthrie and collaborators, which have been summarised by Guthrie (1975).

9.2
Gross Features of the CNS

The lancelet possesses a hollow nerve cord situated directly dorsal to the notochord throughout the animal's extent. This cord attains its largest size in the middle of the body, from where it gradually tapers toward its rostral and caudal ends. In the adult animal the cord approximates a triangle in cross-section, with a dorsally directed apex and a slightly concave base (Figs. 9.16, 9.17), the latter resting directly on the notochord. The ventricular cavity consists of a slitlike central canal, which extends from the thin roof plate, constituting the apex, into the centre of the cord. The most dorsal and ventral parts of this canal are somewhat dilated.

The cord is connected with the periphery through a bilateral series of some 60 segmentally arranged spinal nerves consisting exclusively of dorsal roots. These nerves issue opposite the

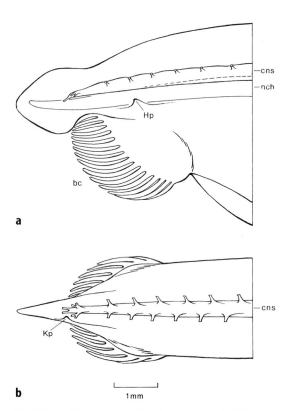

Fig. 9.2a,b. Diagrams showing the rostral end of the lancelet *Branchiostoma lanceolatum* with the central nervous system (*cns*) and the notochord (*nch*) in position. **a** Lateral view, **b** dorsal view. *bc*, Buccal cirri; *Hp*, Hatschek's pit; *Kp*, Kölliker's pit

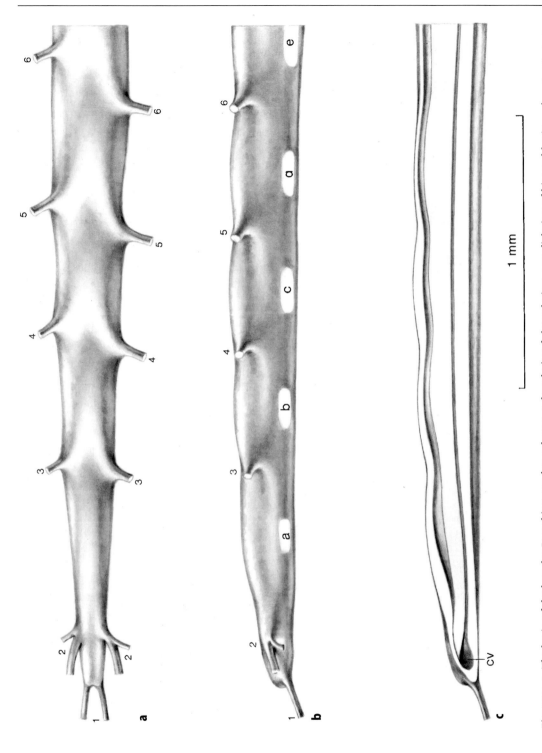

Fig. 9.3a–c. The brain of the lancelet *Branchiostoma lanceolatum*. a dorsal view, b lateral view, c medial view of bisected brain. *a, b, c* etc. Contact zones of 'ventral roots'; *cv,* cerebral vesicle; *1, 2, 3* etc. nerve roots

intermuscular septa, in which they pass peripherally. They alternate because the left half of the body is shifted somewhat rostrally with respect to the right half (Rohde 1888b). Ventral nerves of the cord are unmistakably present but have an unusual origin. Contrary to the descriptions of many authors, among them Retzius (1891), Ariëns Kappers et al. (1936), and Bone (1960b), they do not consist of axons, but rather of long processes of the myotomal muscle cells which approach the cord (Rolph 1876; Schneider 1879; Rohde 1888a; Flood 1966). The 'ventral nerves' lie opposite the myotomes from which they receive fibres; hence, they alternate with the dorsal nerves. Because of the asymmetry indicated above, it frequently happens that a dorsal nerve on one side lies at the same level as a 'ventral nerve' on the other side (Fig. 9.17). The dorsal and ventral nerves are entirely separate in their peripheral distribution.

Although the rostral part of the nerve cord does not show any enlargements, it differs sufficiently from the remainder to warrant its designation as brain (Fig. 9.3). However, the exact boundary between brain and spinal cord has so far not been determined (cf. Sect. 9.5). In the present account the brain will be defined provisionally as the part of the nerve cord situated in front of a transverse plane passing between the fifth and the sixth nerves (Fig. 9.23a).

In the caudal part of the brain the roof plate is thickened to accommodate a bilaterally expanding sheet of large elements known as Joseph cells (Figs. 9.13, 9.18, 9.23). Directly ventral to this cellular sheet the ventricular cavity shows a number of irregular lateral expansions (Franz 1923; von Ubisch 1937).

The rostral part of the brain is commonly designated as the cerebral vesicle because its thin walls surround a small but distinct ventricular widening (Figs. 9.19, 9.23). The first pair of nerves arises from the ventral side of the most rostral part of the brain. The much thicker and often doubled second pair enters the brain dorsally at the transitional area between its rostral and caudal regions (Figs. 9.3, 9.12). These two pairs of nerves are purely sensory; they supply the end of the fin surrounding the most rostral part of the notochord. The remaining nerves emanating from the brain (i.e. pairs 3–6) supply the buccal cavity and the perioral region. They carry afferent impulses from tactile receptors and chemoreceptor organs. Unlike the first and second nerves, the third and subsequent nerves have visceral rami containing both afferent and efferent fibres. These fibres connect the neuraxis with the so-called atrial nervous system (see Sect. 9.8). As a consequence of the large size and extent of the pharyngeal basket and its surrounding atrium in amphioxus, no less than 37 nerves (the third to the 39th) are related to the branchial region (Ariëns Kappers 1934). The right and left nerves one and two arise symmetrically; however, in the next few pairs, the nerves on the right-hand side gradually shift backward until, at the level of the sixth nerve, the alternating pattern characteristic of the entire spinal cord is attained (Fig. 9.3). 'Ventral nerves' extend medially from the four most rostral myotomes and contact the caudal part of the brain.

9.3
Development of the CNS

In early larval stages, the anlage of the CNS of amphioxus has the shape of a simple tube formed of a single layer of cuboid epithelial cells. Rostrally, this tube has an anterior neuropore; at its caudal end it curves ventrally and continues through a neurenteric canal into the gut (Fig. 9.4). The mode in which this neural tube develops differs from that seen in vertebrates. In the latter, the lateral edges of the neural plate fold upward and meet dorsally, where the somatic and neural ectoderm of both

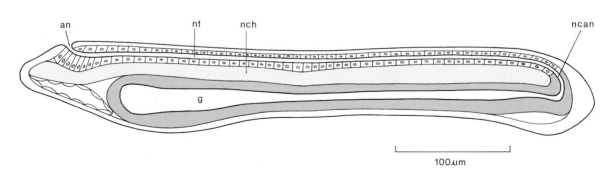

Fig. 9.4. Median section through a young *Branchiostoma* larva. *an*, Anterior neuropore; *g*, gut; *ncan*, neurenteric canal; *nch*, notochord; *nt*, neural tube (modified from Hatschek 1882, Fig. 60)

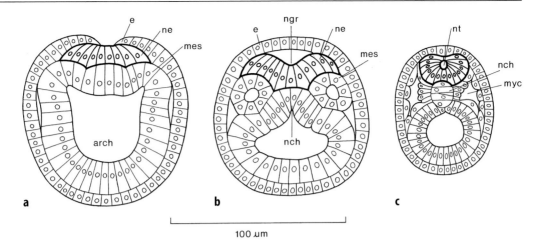

Fig. 9.5a–c. Stages in the development of *Branchiostoma* as seen in transverse sections. **a** Stage of one somite; **b** five somites; **c** nine somites. *arch*, Archenteron; *e*, ectoderm; *mes*, mesoderm; *myc*, myocoele; *nch*, notochord; *ne*, neurectoderm; *ngr*, neural groove; *nt*, neural tube (redrawn from Hatschek 1882, Figs. 77, 93, 115)

sides unite (see Chap. 4). In amphioxus, an upward folding of the tissue at the junction of the somatic and neural ectoderm also occurs (Fig. 9.5a). Then, however, the lateral borders of the neural groove separate from the somatic ectoderm, and the free margins of the latter overgrow the anlage of the CNS and meet dorsally (Fig. 9.5b). Beneath this now-continuous ectodermal layer, the neuroectodermal groove transforms itself secondarily into a tube (Kowalewski 1876; Hatschek 1882; Fig. 9.5c).

During further development, the rostral part of the neural tube widens to form the anlage of the cerebral vesicle. The ventricular cavity of the latter extends ventrally, and, in its caudal wall, a highly characteristic structure, consisting of tall cylindrical cells, forms a slight ventricular elevation termed the tuberculum posterius (Fig. 9.13) by von Kupffer (1893, 1906). Rostrally, the anterior neuropore closes, but its position remains marked on both the ectodermal and the neural side by short, funnel-shaped structures. During later development these structures are displaced to the left side; hence, they are not present in the median section shown in Fig. 9.13. The epidermal remainder of the anterior neuropore is represented in the adult by a ciliated groove, known as Kölliker's pit (Fig. 9.2). Several authors (Kölliker 1843; Rohon 1882; von Kupffer 1893, 1906) considered this pit to be an *olfactory organ*, homologous to the olfactory groove of vertebrates. The conical neural process extending from the cerebral vesicle toward this pit was interpreted by Rohon (1882) and von Kupffer (1893, 1906) as being an unpaired olfactory lobe.

Lacalli et al. (1994) presented a detailed description of the anterior end of the nerve cord of 12.5-day-old larvae, with a total length of 1.75 mm, based on serial transmission electron microscopy and three-dimensional reconstruction (Figs. 9.6, 9.7). In the stage studied the cell bodies are arranged in a single ventricular layer around the central canal and the main fibre tracts occupy a ventrolateral position. All the cells of the neural tube (except for those in the 'tectum'; see below) are in direct contact with the canal and their cilia project into it. All the cells are uniciliate. The cells situated in front of the level of the infundibular organ have cilia that point anteriorly, whereas the cilia of those situated behind that level point posteriorly (Fig. 9.7). Neurons are scattered along the cord singly or in small clusters. Some large, well-differentiated neurons lie ventrally, immediately adjacent to the floorplate. The latter is formed by small, angular cells that form a row, one or two cells wide, along the base of the cord. A row of flattened dorsal cells forms the dorsal closure of the ventricular cavity. A large commissure, designated as the principal commissure, is situated in the floorplate, immediately behind the infundibular organ. Several smaller commissural connections are found more rostrally in the ventral wall of the brain vesicle (Fig. 9.7).

In a subsequent study of the same larval material, Lacalli (1996) noted that the cells surrounding the cerebral ventricle fall broadly into two categories depending on the types of processes they form. Particularly common are cells that look like neurosensory cells, resembling those found in invertebrates generally and in the olfactory epithelium of vertebrates. Each has an axon that forms terminals and, in some cases, synapses. Otherwise,

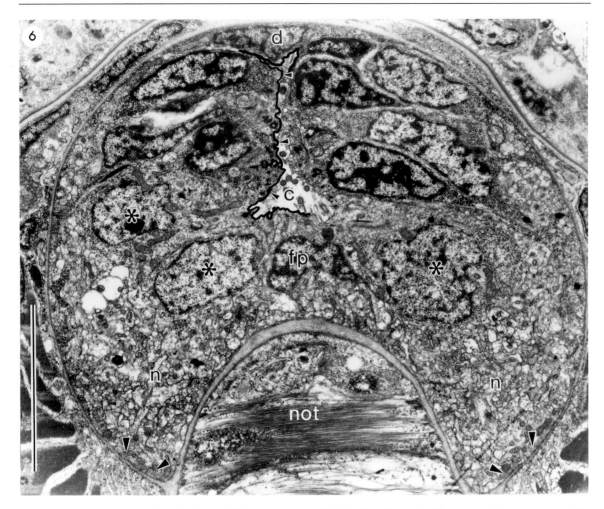

Fig. 9.6. Transverse section through the caudal brain region of a 14-day larva of *Branchiostoma floridae*. The surface of the central canal is traced on one side for emphasis (*solid line, small arrows*). It is capped at the *top* by a single dorsal cell (*d*). The ventral midline is occupied by a single floorplate cell (*fp*). The cells marked by an *asterisk* are neurons. The ventrolateral nerve tracts (*n*) are shown, each with a superficial zone in which neuromuscular junctions are made (between *arrowheads*). *Scale bar*=5 μm (reproduced from Lacalli et al. 1994, Fig. 3)

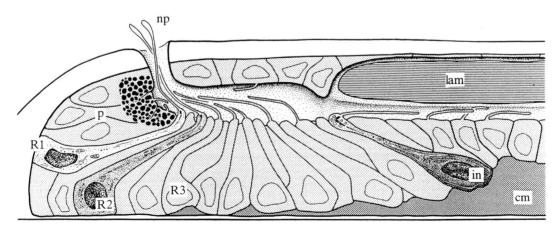

Fig. 9.7. Median section through the rostral part of the brain of a 12.5-day-old amphioxus larva. *cm*, Principal commissure; *in*, infundibular organ; *lam*, lamellar body; *np*, neuropore; *p*, pigment cell cluster; *R1-R3*, transversely oriented cell rows (reproduced from Lacalli et al. 1994, Fig. 6)

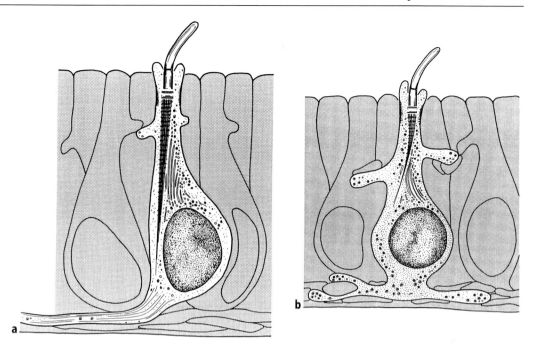

Fig. 9.8a,b. The two main types of nerve cells encountered in the rostral part of the cerebral ventricle. **a** A cell like those in row 4, with a basal axon and a few lateral processes. **b** A cell like those in row 3, with no obvious axon but with extensive subapical, lateral and basal processes, some expanded as terminals (reproduced from Lacalli 1996, Fig. 32)

these cells have few other processes except, in some instances, in the subapical zone (Fig. 9.8a). Less common are cells without axons but with far more extensively developed short, undifferentiated basal, lateral and subpial processes (Fig. 9.8b). Vesicles may be present, usually as dense-core granules, but there are no obvious junctional specialisations. The cells in the caudal part of the brain showed in general a more distinct neuronal differentiation. Synapses, often with large concentrations of clear vesicles, are very common in this region. Lacalli (1996) concluded from these findings that the rostral part of the brain is mainly subserves slow, integrative activities involving extensive reciprocal interactions between cells, whereas in the caudal part more specific and faster modes of transmission are employed.

Lacalli et al. (1994) observed four groups of (putative) *visual elements* in the CNS of larval amphioxus:

1. The dorsal ocelli or organs of Hesse, each of which consists of a pigment cup enclosing a receptor cell. These organs are confined to the spinal cord (Fig. 9.23).
2. Clusters of giant, microvillate Joseph cells, which together form a dorsally situated complex in the caudal part of the brain (Fig. 9.13).
3. The dorsal lamellar body, made up of cells with

cilia that expand into flattened lamellae. This lamellar body is situated directly in front of the complex of Joseph cells (Figs. 9.7, 9.9).
4. A complex consisting of pigment-containing cells and some adjacent transversely oriented rows of non-pigmented elements, which constitute together the most rostral part of the larval neuraxis. The apical portions of the pigment cells together form the rostral wall of the anterior neuropore (Figs. 9.7, 9.12).

All of these groups of (putative) visual elements will be further discussed in later sections. Suffice it to mention here that the dorsal lamellar body is much more conspicuous in early larvae than in the adult state, and that Lacalli et al. (1994) designated the rostral complex, including the pigmented cells, as the frontal eye.

In the subsequent study already mentioned, Lacalli (1996) provided further details on the *frontal eye* and described the structure and connections of two hitherto unknown centres in the brain, which he designated as the tectum and the primary motor centre. Moreover, he subjected the ventrolateral nerve tract, which represents the principal communication channel of the larval brain, to a detailed analysis. The frontal eye appeared to be composed, from rostral to caudal, of a cluster of pigment cells, two rows of receptor cells and a clus-

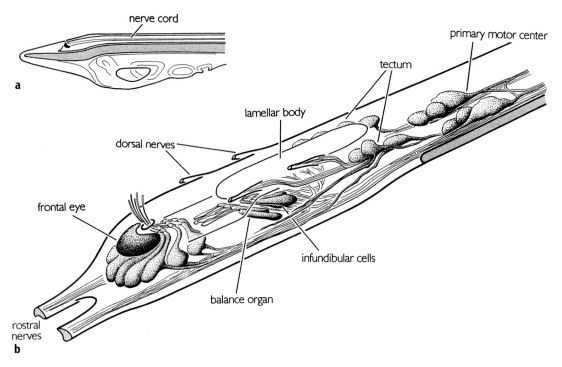

nerve cord

primary motor center

tectum

lamellar body

dorsal nerves

frontal eye

infundibular cells

balance organ

rostral
nerves

a

b

Fig. 9.9a,b. Stereogram of the anterior region of the CNS of a 12.5-day amphioxus larva (reproduced from Lacalli 1996, Fig. 2)

ter of neurons. The row 1 and row 2 receptor cells both send their axons into the ventrolateral tract (Fig. 9.9). Large terminals containing both small clear vesicles and dense-core granules arise at intervals from the row 1 fibres. Contacts are formed with processes of several other cell types, but specialised synaptic junctions were not observed. The axons of the row 2 cells are thinner than those of the row 1 cells and contain exclusively dense-core vesicles. Elements situated in the neuron cluster of the frontal eye send short processes into the domain of the ventrolateral tract. The processes emanating from more rostrally situated (row 3) neurons contain scattered dense-core granules, but those from more caudal (row 4) neurons are filled with small clear vesicles.

Bilateral rows of three to five small neurons, situated directly behind the lamellar organ, were interpreted by Lacalli (1996) as being the homologue of the tectum mesencephali of craniotes (Figs. 9.9, 9.10, 9.12). Fibres from the, often paired, dorsally situated second nerve pass through the tectum, where they participate in the formation of a series of glomerular synaptic zones, and then proceed to the primary motor centre.

The *tectal cells* have two principal processes, a long rostral one and a short caudal one. The rostral processes pass along the lateral wall of the cerebral vesicle and sooner or later joint the ventrolateral

bundle. The proximal parts of these fibres are elaborated into series of club-shaped synaptic terminals, which are the main input into the tectal synaptic zone. The caudal processes enter the primary motor centre (Fig. 9.9). Some synapse *en route* with the ascending dendrite of the left giant cell (Fig. 9.12).

The *primary motor centre* is a distinct zone of large ventral neurons, arranged mostly in pairs, that begins just behind the level of the tectum (Figs. 9.9, 9.10, 9.12). It comprises typical motor neurons that innervate the most rostral somites, and three consecutive pairs of large projection neurons, the axons of which descend to the spinal cord. The elements forming the third pair fall, because of their extraordinarily large size, into the category of giant cells. Both giant cells are provided with a large, forward-projecting dendrite that enters the ventrolateral bundle. The dendritic trunk of the left cell extends much further rostrally than that of the right one and attains the level of the rostral eye (Figs. 9.9, 9.10, 9.12). The dendrites from both cells receive synaptic input all along their length from different sources, including a series of synapses from sensory fibres which enter the ventrolateral bundle via the first nerve. Each of the giant cells has a coarse axon that branches immediately after its emergence from the soma; one branch descends in the ipsilateral ventrolateral bundle while the other

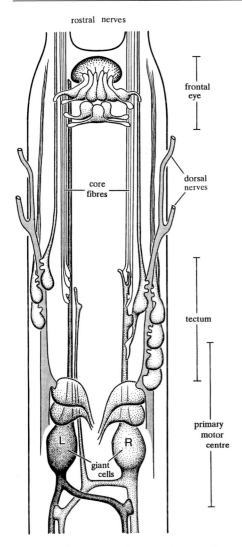

Fig. 9.10. The patterns of contact between the rostral nerves, the frontal eye, the dorsal nerves, the tectum and the primary motor centre in dorsal view (reproduced from Lacalli 1996, Fig. 31)

minate as the bundle passes through the rostral part of the brain, but a core of three to four fibres remains that can be traced to caudal parts of the brain, where they terminate in synapses with dendrites form the two giant cells (Figs. 9.10, 9.11).

Judging from their morphology, it appears that the large projection neurons in the primary motor centre are involved in coordinating the response to sensory input by generating signals that attain the spinal cord along the caudal part of the ventrolateral bundle. Figure 12 summarises the direct and indirect input pathways to the primary motor centre so far discovered by Lacalli (1996). These include: (a) direct contacts from core rostral fibres to the dendrite of the giant cell on both sides, (b) indirect contacts with receptor and nerve cells of the frontal eye via the core rostral fibres, (c) direct contacts from R1 receptor cells to the dendrite of the left giant cell, (d) indirect contacts from the same cells via the tectal cells, (e) an indirect pathway from the second nerve via the tectum, (f) a direct pathway from the second nerve to the three projection neurons in the primary motor centre, and (g) contacts of tectal efferents with the ascending dendritic trunk of the left giant cell.

Larvae of the stage examined by Lacalli (1996) can generate two quite different *locomotory responses*: bouts of swimming and the startle response. Lacalli considered it likely that the giant cells trigger the startle response. These cells innervate both sides of the cord massively, their axons and terminals being the largest in the cord.

The further development and fate of the three large coordinating brain cells described by Lacalli (1996) remains an intriguing problem. Possibly, they correspond to some ventrally situated larger cells observed by Bone (1960b, p. 46), "whose processes cross the floor of the central canal, in a manner reminiscent of the axons of the Rohde cells."

Finally, it may be mentioned that Lacalli (1996) discovered in the early larval stage studied by him a row of cells with unusually large, swollen cilia, situated just in front of the infundibular organ (Fig. 9.9). He suggested that these cells may be involved in either *motion* or *gravity* detection, and pointed out that they may well represent an analogue of the balance organ of tunicate larvae.

During later development the parts of the CNS surrounding the cerebral vesicle remain epithelial, but the wall of the remainder of the nerve cord differentiates into a central zone of grey and a wide, peripheral zone of fibres. Details of this development are not known; however, the studies of von Kupffer (1893, 1906) suggest that the histogenetic processes observed in vertebrates (cf. Chap. 5) are also operative in amphioxus.

decussates and descends in the contralateral bundle. Each contralateral branch forms immediately after crossing a large synapse with the ipsilateral fibre (Fig. 9.10). The axons of the other two pairs of large projection neurons all cross and descend in the contralateral ventrolateral bundle.

The *ventrolateral bundle* is a typical open-fibre system (cf. Chap. 3, Sect. 3.5.4), in the most rostral part of which rostrally directed long dendrites of tectal and giant projection cells are contacted by axons originating from the frontal eye and by general somatosensory fibres. The latter arise from clusters of neurosensory cells situated at the rostral tip of the larva, which enter the bundle via the rostral nerves (Figs. 9.9–9.11). Most of these fibres ter-

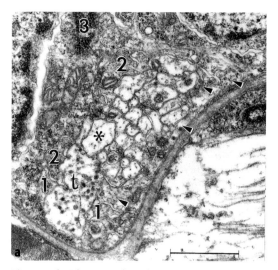

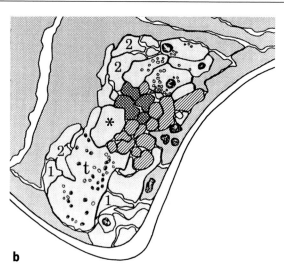

Fig. 9.11a,b. The ventrolateral tract in a 12.5-day larva of *Branchiostoma floridae*. **a** Transverse section at the level of the row 3 cells of the frontal eye; for orientation cf. Fig. 9.7. **b** Diagram of the same section. In both figures the axial tectal fibre is indicated by an *asterisk*; row 1 and row 2 fibres are indicated by *numbers*, and *t* indicates a large row 1 terminal, contacting the tectal fibre. The *arrows* in **a** indicate basal pro-cesses from row 3 cells. In **b** surrounding cells and processes are *shaded* in a darker tone than the tract; rostral fibres are *hatched*; core rostral fibres are *cross-hatched*. Some of the vesicle-containing profiles directly dorsal to the bundle of core rostral fibres are probably row 1 terminals, just like t. *Scale bar* = 1 µm (reproduced from Lacalli 1996, Figs. 11, 14)

9.4
Overall Histological Pattern

In the adult CNS the neurons vary widely in size and shape; many of them have one or more broad processes which reach the surface of the ventricular cavity, and frequently neuronal processes or even the cell bodies themselves are found in the lumen of the ventricular space (Figs. 9.16–9.18). According to Bone (1960b), these peculiar relations reflect the fact that the fluid of the ventricular cavities is probably the only site where the neurons can exchange metabolites; blood vessels or specialised glial elements for neuron nutrition are entirely lacking in the CNS of amphioxus. A rudimentary circulatory system does exist for the rest of the body, but it is maintained by several contractile organs instead of

a heart, and no respiratory pigment is present in the blood. The nerve fibres, none of which are enveloped by glial cells, show a marked heterogeneity in size (Fig. 9.20a). Guthrie (1975) reported that synaptic differentiations are observed only occasionally in either the fibrous or cellular regions of the spinal cord. Most of these structures occur in the ventral fibrous regions of the cord and appear as axo-axonic junctions. Many of the large axons show zones of very close apposition, presumably representing sites for electrotonic coupling. These observations do not tally with those of Ruiz and Anadón (1989), who stated (without providing further details) that in the spinal cord chemical synapses are very common at many places. In the brain, synapses are found more frequently than in the spinal cord (Guthrie 1975), but most of them are at a

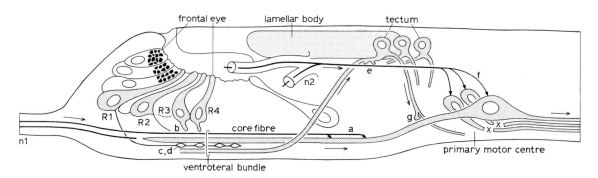

Fig. 9.12. Summary diagram showing the patterns of connectivity in the brain of a 12.5-day larva of *Branchiostoma floridae*, described by Lacalli (1996)

relatively low level of differentiation. Quite often, pre- and postsynaptic membranes cannot be clearly differentiated because the cytoplasm contains granules on both sides. Often, the number of granules is larger on one side than on the other. In the presumably presynaptic endings thus characterised, mitochondria are lacking (Meves 1973).

Most of the glial cells of amphioxus are ependyma, formed by the cells that line the ventricular cavity. The processes of these elements diverge to the surface, where their endfeet constitute the external limiting membrane. These processes run in bundles or fasciae, which are arranged in a more or less regular manner so that they incompletely divide the neuraxis into different areas (Bone 1960b).

9.5
Interpretation of the Rostral Part of the Neuraxis

Numerous authors have attempted to subdivide the nerve cord of amphioxus into a rostral brain section and a caudal spinal section, and to make more detailed comparisons between the putative cephalochordate and the vertebrate brain. The ensuing historical survey will show that these attempts have led to widely diverging results.

Rohde (1888b) contended that the disappearance of the dorsal cells of Joseph marks the caudal end of the brain and the beginning of the spinal cord, which latter structure is indicated further by the appearance of the largest and the most rostral giant cell. (As will be discussed in Sect. 9.6.6, these conspicuous elements were later named after Rohde).

Von Kupffer (1906) thought that the presence and location of the neuroporic recess, the olfactory organ, and the tuberculum posterius warranted the conclusion that the cerebral vesicle of amphioxus represents the archencephalon and the remainder of the brain the deuterencephalon of vertebrates (cf. Fig. 9.13 with Chap. 4, Fig. 4.5a). This interpretation has been adopted by several authors, including Ariëns Kappers (1929). Boeke (1902, 1908) came to a similar interpretation based on an analysis of the structure that von Kupffer regarded as the tuberculum posterius. He found that the cells in this structure show a striking resemblance to the sensory epithelium in the saccus vasculosus of fishes. Since the latter organ is an extension of the infundibular region, Boeke termed it the *infundibular organ*, a name that has found general acceptance.

A salient feature of the infundibulum and of the entire hypothalamo-hypophysial complex of vertebrates is that it occupies a prechordal position. Figure 13 shows clearly that in amphioxus the "tuberculum posterius" of von Kupffer (containing the infundibular organ of Boeke), and in fact the entire brain vesicle, is situated dorsal to, rather than in front of, the notochord. On the basis of these positional relations, Delsman (1913) rejected von Kupffer's (1906) interpretation of the parts of the brain of amphioxus. Because of its suprachordal position he regarded the entire brain area of amphioxus as representing the deuterencephalon of vertebrates. A homologue of the archencephalon of vertebrates is, according to Delsman, entirely lacking in amphioxus.

As we have seen, von Kupffer's interpretation of the rostral part of the brain of amphioxus as representing the archencephalon of vertebrates was based not only on the position of the "tuberculum posterius", but also on the presence of a putative

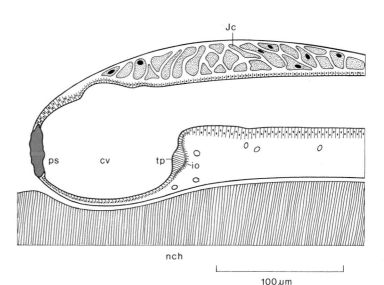

Fig. 9.13. Median section through the rostral part of the CNS of a *Branchiostoma* specimen of 25 mm total length. *cv*, Cerebral vesicle; *io*, infundibular organ; *Jc*, Joseph cells; *nch*, notochord; *ps*, pigment spot; *tp*, tuberculum posterius, according to the interpretation of von Kupffer (modified from von Kupffer 1893, Fig. 21)

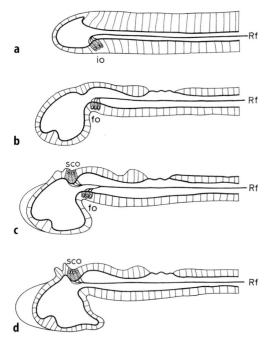

Fig. 9.14a–d. Schematic median sections showing the organs producing Reissner's fibre (*Rf*) in various chordates. **a** Adult amphioxus with fibre-forming infundibular organ (*io*). **b** Embryonic craniate (*Salmo*) with juvenile flexural organ (*fo*). **c** Slightly older stage than the previous one; flexural and subcommissural organs (*sco*) both contribute to the fibre. **d** Adult craniate; the fibre is produced solely by the subcommissural organ (modified from Olsson 1993, Fig. 2)

olfactory organ, Kölliker's pit, directly in front of the cerebral vesicle. It is important from both a morphological and a functional point of view that Tjoa and Welsh (1974) established that the epithelial cells lining Kölliker's pit lack the ultrastructural features characteristic of an olfactory function.

The interpretation of Boeke's infundibular organ entered a new phase with the investigations of Olsson and Wingstrand (1954). Using Gomori's chrome alum hematoxylin-phloxine technique, these authors demonstrated that: (a) the cells of the infundibular organ contain secretory material, (b) the secretion is released into the cerebrospinal fluid, and (c) Reissner's fibre originates from the surface of the organ and has the same staining properties as the secretion of the organ (Fig. 9.14a).

Reissner's fibre is a thread of non-cellular material passing backward from the brain into the central canal of the spinal cord. Reissner's fibre also occurs in vertebrates; however, in this group it arises from the secretory cells of the subcommissural organ, a specialised ependymal region in the roof of the brain below the posterior commissure (for review see Oksche 1969; Sterba 1969; Leonhardt 1980; Fig. 9.14d).

Evaluating the combination of features mentioned above, Olsson and Wingstrand (1954) suggested that the *infundibular organ* of amphioxus represents the functional and morphological counterpart to the hypothalamo-hypophysial neurosecretory system of vertebrates. In their opinion, Reissner's fibre in amphioxus is unique in starting from the ventrally situated infundibular organ, rather than from a dorsally situated structure as in vertebrates. Hofer (1959) confirmed the observations of Olsson and Wingstrand (1954); however, he pointed out that the cells of the infundibular organ of amphioxus are specialised ependymal elements from which a neurosecretory neuronal system cannot be derived. In Hofer's opinion, the infundibular organ of amphioxus has nothing to do with the infundibulum of vertebrates. Further studies revealed that the subcommissural organ of gnathostomes and the infundibular organ of amphioxus closely resemble each other, not only structurally, but also ultrastructurally (Olsson 1962; Obermüller-Wilén 1976) and immunohistochemically (Sterba et al. 1983). Hence, a classical problem of homology: two organs showing a striking structural similarity and delivering an identical product, i.e. Reissner's fibre, occupy obviously different topological positions (Fig. 9.14a,d). The solution to the enigmatic situation came from Olsson's (1956, 1993) studies on *Salmo*, which showed that in early larval stages of this teleost Reissner's fibre originates from a group of secretory cells, situated ventrally at the level of the tuberculum posterius (Fig. 9.1b). During a brief, subsequent developmental phase this juvenile flexural organ and the subcommissural organ both contributed to the production of one and the same Reissner's fibre (Fig. 9.14c). In later stages the flexural organ was not observed to contribute any longer to the fibre. Olsson (1993) concluded that, phylogenetically, the subcommissural organ is exclusively a vertebrate structure, and that cephalochordates never pass the infundibular organ/flexural organ stage, which is purely embryonic in vertebrates.

Olsson and Wingstrand's (1954) presumption that the infundibular organ of amphioxus may correspond to the vertebrate hypothalamo-hypophysial neurosecretory system (HHNS) was not corroborated by the ultrastructural observations of Obermüller-Wilén (1976). She found that the cells of the infundibular organ are not in receipt of any synaptic contacts and do not show any sign of basal release. However, a few year later, Obermüller-Wilén (1979) reported the presence of another possible homologue of the vertebrate HHNS in the brain of amphioxus, comprising bilateral dorsal and ventral groups of neurosecretory cells. The ventral groups appeared to be situated in the vicinity of the infundibular organ. The axons of both groups were observed to contain secretory granules and to establish contacts with the ventral brain surface. The axons of the dorsal groups were innervated by numerous, catecholaminergic, *boutons en passant* (Fig. 9.25).

Strong evidence in support of the concept that a long rostral portion of the nerve cord of amphioxus is homologous to the brain of vertebrates has come from recent studies on the expression patterns of *homeotic genes*. The genome of amphioxus has a single *Hox* gene cluster, which is organised similar to the four mammalian *Hox* clusters and contains homologues of at least the first ten paralogous groups of vertebrate *Hox* genes in a colinear array (Garcia-Fernández and Holland 1994).

Holland et al. (1992, 1994) identified an amphioxus *Hox* gene, *AmphiHox 3*, which is homologous to mouse *Hox-2.7 (Hox B3)*. In situ hybridisation revealed region-specific expression in the nerve cord of amphioxus neurulae, later embryos and larvae (Fig. 9.15). The anterior limit to expression in the nerve cord appeared to be at the level of the boundary between the fourth and the fifth somite at the neurula stage (13–17 h postfertilisation). Between 20 and 24 h postfertilisation a pigment spot (most probably corresponding to an accumulation of ocelli, R.N.) appears in the cord at the level of somite 5. The anterior limit of *AmphiHox 3* expression is situated just in front of this spot and remains there during further development. Because the anterior expression boundary of mouse *Hox-2.7 (Hox B3)* and related genes in the vertebrate CNS have an expression boundary at the anterior limit of rhombomere 5 in the hindbrain, the conclusion that the brain of amphioxus extends much further caudally than the cerebral vesicle seems to be warranted.

Finally, it may be mentioned that Lacalli (1996) and Lacalli et al. (1994), who made a detailed study of the rostral part of the nerve cord of some amphioxus larvae, suggested the following homologies:

1. The frontal eye represents, on account of its position, structure and connections, the counterpart of the paired eyes, the optic stalks and the chiasmatic ridge of craniotes.

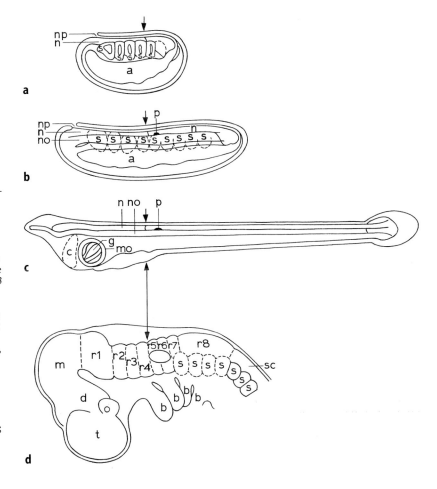

Fig. 9.15a–d. Homologous *Hox* gene expression limits in amphioxus and craniotes. **a** Neurula-stage amphioxus embryo (13 h postfertilisation); **b** amphioxus embryo 24 h postfertilisation; **c** 3.5-day amphioxus larva. The three embryos are drawn to the same scale. *Arrowheads* indicate the anterior limit to *AmphiHox3* neural expression. **d** Chick embryo for comparison; the *double-headed arrow* relates the anterior neural expression limit of *AmphiHox3* with that shown by its vertebrate homologues. *a*, Archenteron; *b*, branchial arch; *c*, club-shaped gland; *d*, diencephalon; *g*, first gill slit; *m*, mesencephalon; *mo*, mouth; *n*, nerve cord; *no*, notochord; *np*, neuropore; *p*, pigment spot; *r1, r2* etc., rhombomeres; *s*, somite; *sc*, spinal cord; *t*, telencephalon (redrawn from Holland et al. 1992, Fig. 4)

2. The lamellar body may be homologised on positional and structural grounds to the pineal organ of anamniotes.

3. The file of small neurons situated dorsally, directly caudal to the lamellar body, is designated as tectum because, with regard to its position, afferent and efferent connections, it corresponds to the centre of the same name in craniotes.

4. The motoneurons forming part of the primary motor centre, which innervate the first two somites, may well be homologous to the cells innervating the extrinsic eye muscles in vertebrates, whereas the somewhat more caudally situated large projection neurons within that centre presumably correspond to the giant reticulospinal elements in the lamprey.

If we survey the data and opinions reviewed above, we may draw the following conclusions:

1. The brain of amphioxus is not confined to the area surrounding the brain vesicle.

2. The boundary between brain and spinal cord may be provisionally positioned between the fifth and the sixth nerve pairs. This boundary marks the end of the Joseph cells and the beginning of the giant cells of Rohde, and most probably corresponds to the anterior expression limit of the amphioxus *Hox* gene *AmphiHox3*. This boundary is to be regarded as provisional, because (a) neither the Joseph nor the Rohde cells have distinct craniote homologues, and (b) knowledge concerning the expression patterns of additional amphioxus *Hox* genes, particularly of homologues of vertebrate *Hox 2.1 (B5)* and *Hox 2.6 (B4)* genes (cf. Chap. 4, Fig. 4.34), will be needed for the definitive determination of the 'encephalo'-spinal boundary.

3. The entire brain occupies a suprachordal position but is, nevertheless, subdivisible into a rostral archencephalon and a caudal deuterencephalon. This subdivision rests on the following (possible) homologies of cephalochordate and craniote structures: (a) infundibular organ – juvenile flexural organ; (b) lamellar organ – pineal organ; (c) dorsal and ventral clusters of neurosecretory cells – hypothalamo-hypophysial system; (d) rostral eye – anlage of paired eyes.

4. The presence of putative homologues of the craniote tectum and the craniote rhombencephalic reticular formation in amphioxus suggests that in the future, if further evidence becomes available (e.g. on the expression patterns of homeobox genes), a subdivision of the deuterencephalon of amphioxus into a rostral prosencephalon and a caudal mesencephalon will appear to be justified.

5. Homologues of craniote telencephalic structures have so far not been detected in amphioxus.

Abbreviations

1, 2 etc.	nerve roots
aRa	anterior Rohde axon(s)
aRc	anterior Rohde cell
aRc1	first anterior Rohde cell
c	collaterals
cs	commissural cell
dcc	dorsal commissural cell
epc	epithelial cells
fnrl	fibres of the first nerve root
gglc	ganglion cells
io	infundibular organ
Jc	Joseph's cells
mRa	median Rohde axon
nr	nerve root
pc	pigment cells
pRa	posterior Rohde axons
ps	pigment spot
smb	somatomotor bundle
smc 1, 2 etc.	somatomotor cell types 1, 2 etc.
ssb	somatosensory bundle
ssc 1, 2 etc.	somatosensory cell types 1, 2 etc.
v	ventricular cavity
vc	vertical cell
vmc	visceromotor cell
vr	"ventral root"
vsf	viscerosensory fibres

9.6
Spinal Cord

9.6.1
Introductory Notes

The microstructure of the spinal cord of the lancelet has been the subject of numerous investigations. Most notable are those of Rohde (1888a,b), Retzius (1891), Johnston (1905) and, in particular, Bone (1959, 1960a,b). The studies of the last-mentioned author on both larval and adult material have greatly extended our knowledge of the neurohistology of amphioxus, and the following synopsis is based largely on his work.

Figures 9.16 and 9.17 show that in the spinal cord the cellular elements are almost all situated in the immediate vicinity of the cleftlike central canal. The regions of the cord situated lateral to this narrow cellular area are occupied largely by longitudinally running fibres of many sizes. Most conspicuous among these fibres are the coarse axons of Rohde (1888a,b), which constitute five distinct bundles in

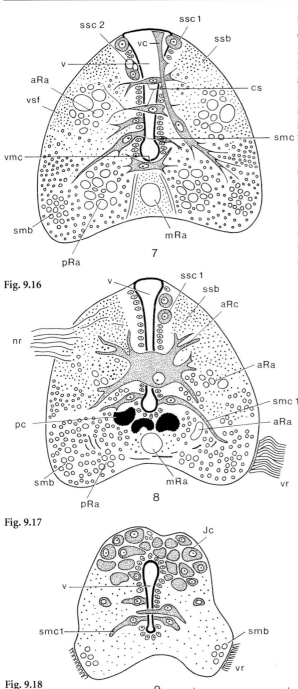

Fig. 9.16

Fig. 9.17

Fig. 9.18

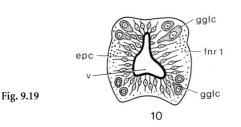

Fig. 9.19

the cord (Fig. 9.16): two on each side and the fifth, consisting of one huge axon, in the median plane immediately below the central canal. These Rohde fibres are often compared to the giant fibres of annelids. Another feature reminiscent of the worms is the presence in the cord of simply organised eye spots, each consisting of a photoreceptor cell and a surrounding, cuplike pigment cell (Fig. 9.16). The fine structure of these photoreceptor cells was studied by Eakin and Westfall (1962), who showed that an irregular array of tubules situated superficially in the zone adjacent to the pigment cells probably constitutes the photoreceptive membrane. The physiological role of these cells was further substantiated by Guthrie (1975), who recorded light-evoked slow potentials from the immediate vicinity of the pigmented cells. Guthrie also traced the axons of some of these light-perceiving elements to the ventrolateral part of the cord, where they descend to a position close to the collateral fibres of somatic motor neurons before breaking up into longitudinal processes.

9.6.2
Spinal Nerves

The nerve roots of the cord of amphioxus contain fibres of several types. Bone (1960a) distinguished a general cutaneous, a visceral sensory and a visceral motor component, a composition that agrees with the earlier and less substantiated view of Johnston (1905). The somatosensory skin fibres end generally as free terminals in relation to the elements of the single-layered epidermis, although encapsulated nerve endings occur in certain regions. The visceral afferent fibres are the axons of peripheral

◀ **Figs. 9.16–9.19.** A series of transverse sections through the CNS of *Branchiostoma*. **Fig. 9.16:** section through the cord in the mid-region of the body; **Fig. 9.17:** section through the rostral part of the cord, at the level of one of the anterior Rohde cells; **Fig. 9.18:** section through the caudal region of the brain; **Fig. 9.19:** section through the middle of the brain vesicles. **Figs. 9.18** and **9.19** are drawn from semithin sections stained with toluidine blue. For abbreviations see list on p. 378 (**Fig. 9.16** is based on Fig. 2 of Bone 1960b; **Fig. 9.17** is based on some sections stained according to Palmgren, kindly made available by Dr. Q. Bone)

neurons of the atrial nervous system and the visce-romotor fibres are efferent to that system. Several authors (Rohde 1888b; Hatschek 1892; Johnston 1905) described scattered or agglomerated cells along the nerve roots or their branches, corresponding to the elements that constitute the spinal ganglia of vertebrates. However, the work of Retzius (1891) and Bone (1960b) has adequately shown that any traces of spinal ganglia are entirely absent in amphioxus. The cells that constitute the somatosensory and the visceromotor systems are all situated within the spinal cord.

9.6.3
Somatosensory System

The somatosensory system was well characterised by Retzius (1891), who described the types of cells that contribute most of the peripheral fibres of this system. His observations were confirmed and extended by Bone (1960a,b).

The most common somatosensory elements are bipolar in shape and constitute a continuous column alongside the central canal near the dorsal surface of the cord. The cells of these column are oriented longitudinally, and their rostral and caudal poles send out offshoots which form a distinct somatosensory bundle (Figs. 9.16, 9.17, 9.21). One fibre from each cell dichotomises in the vicinity of a nerve root; then, one branch passes out of the cord and the other continues in its longitudinal course. Almost certainly it is the peripheral processes from these cells that form the majority of the nerve fibres terminating in the skin (Bone 1959).

In each segment there are a few cells that conform to the pattern just sketched but must be considered a separate type (Bone 1960b). These elements, indicated in Figs. 9.16 and 9.21 as somatosensory type 2 cells, are larger than the common cells of the somatosensory column, and there is a conspicuous vacuole in the perikaryon. Bone mentions, in addition, two other kinds of large cells that contribute fibres to the nerve roots. These elements occur, often associated, singly in each segment. Their central connections are imperfectly known.

Besides the elements termed here somatosensory type 1 cells (Figs. 9.16 and 9.21), there is another common type of cell that sends a fibre out of the nerve root (Fig. 9.21: SSC3). The bipolar or tripolar cell bodies of these elements lie roughly transversely across the central canal; their central processes go, contrary to those of the somatosensory type 1 cells, to the side opposite that of the nerve root through which their peripheral fibre issues.

This brief survey illustrates the complexity of the primary somatosensory system of amphioxus.

Although the peripheral distribution of the fibres of this system is not exactly known, the subepithelial plexus almost certainly contains fibres derived from several different types of centrally located primary afferent neurons (Bone 1960a).

9.6.4
Visceromotor System

Unlike the somatosensory cells, the cells that give rise to the visceromotor fibres lie far removed from the level of the nerve roots. These elements constitute the ventralmost part of the narrow grey zone and are located close to the bottom of the somewhat widened ventral part of the ventricle (Bone 1960b). The somatomotor cells line the lateral aspect of this widened part of the ventricular cleft; thus, in the cord of amphioxus, there exists the peculiar and exceptional condition that the visceromotor elements lie ventral to the somatomotor system (Fig. 9.16).

The visceromotor column consists of large and small multipolar cells, the latter being the more numerous. The axons of these cells ascend to the nerve roots and reach, by way of their ventral rami, the atrial region. Most of them innervate the pterygial muscle, which forms the floor of the atrium. According to Bone, it is likely that some of the dendrites of the visceromotor cells synapse directly with the visceral afferent fibres that enter the cord. These fibres, which form the third component of the nerve roots, originate from the peripherally located neurons of the atrial nervous system. The central course of these visceral afferents is unclear; Bone presumes, however, that these fibres assemble lateral to the dorsal bundle of Rohde axons (Fig. 9.16). In larval acraniates, he (Bone 1959) found a bundle of visceral fibres occupying a similar position.

9.6.5
Internuncial Cells

Ventral to the somatosensory and dorsal to the somatomotor system, cells are found that vary widely in form and size, but probably all are internuncial in function. The axons of several of these elements have been traced to the ventrolateral area of the cord, where they come close to the more peripheral parts of somatomotor neurons (Guthrie 1975). A large proportion of these cells lie across the central canal, thus forming a protoplasmic commissure; others lie to one side of the ventricle but have broad processes that terminate in the wall of the central canal (Fig. 9.16). The more dorsally situated elements of this intermediate cell group are clearly related to the somatosensory system. Among them there are conspicuous dorsal commissural cells that extend their dendrites among the somatosensory columns on either side of the cord (Fig. 9.21). A more detailed description of the other cell types that occur in the intermediate grey of the cord of amphioxus is provided in Bone's (1960b) paper. Here we mention only the remarkable vertical cells that occur segmentally in the cord between the dorsal roots (Fig. 9.16). Dorsally, each cell consists of a long process that terminates upon the roof of the cord; the ventral part is provided with dendrites which extend among the axons of the somatomotor bundle, to be discussed later. The spread of their processes suggests that the vertical cells form part of a somatic sensorimotor arc.

9.6.6
Somatomotor System

The somatomotor system is located lateral to the ventral part of the ventricle and lies immediately dorsal to the visceromotor column (Fig. 9.16). The somatomotor neurons are slender, pyramidal cells which are arranged in peculiar triangular patterns. The bases of the cells themselves are broadened and form part of the wall of the central canal; their apices are directed laterally and taper gradually into a nerve fibre. The base of the triangular or fan-shaped pattern in which these cells are arranged is formed by the bases of the individual cells; the apex

of the triangle consists of their somewhat converging, laterally directed processes. The fans of somatomotor cells are arranged segmentally; i.e. at the level of each 'ventral nerve' there is a triangular pool which directs its apex towards that root. Because the anterior roots alternate (Fig. 9.22a), these fans alternate as well. This is not the whole story, however. There also seem to be fans of somatomotor cells which lie opposite to the one just described and whose apices are midway between adjacent 'ventral nerves'. Passing downward on one side of the cord we thus meet alternating 'segmentally' and 'intersegmentally' placed fans of motor cells. These two types of somatomotor cell pools do not differ in position alone; they also contain cells peculiar to each. The relations just sketched were partly described by Retzius (1891). Bone (1960b) supplemented these observations and unravelled the remarkable way in which the somatomotor cells distribute their central processes. He found that the laterally directed offshoots of these cells reach the ventrolateral angle of the cord, where they bend longitudinally, forming a distinct bundle of coarse fibres (Figs. 9.16, 9.17, 9.21). Guthrie (1975) observed that these fibres extend posteriorly for two to three neural segments, and that in each segmental motor contact zone only a few motor neurons are synaptically linked with the muscle fibres constituting the 'ventral nerves'. The fibres of the somatomotor bundle have both short branches to the ventral contact zones and numerous long collat-

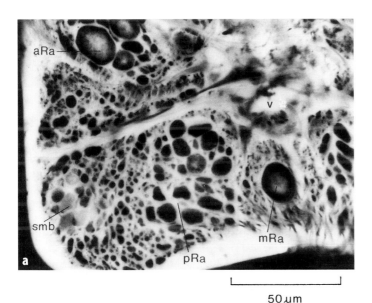

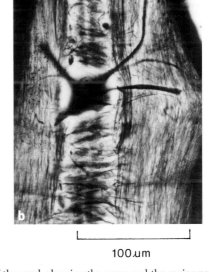

50 μm   100 μm

Fig. 9.20a,b. Photomicrographs showing some features of the structure of the spinal cord of *Branchiostoma*. **a** Transverse section showing the left ventral quadrant of the cord; note the considerable differences in diameter of the axons. **b** Horizontal section of the cord, showing the soma and the major part of the dorsal dendrites of the most anterior Rohde cell. For abbreviations see list on p. 378 (Palmgren preparations made available by Dr. Q. Bone)

erals that cross the cord floor to reach the contralateral somatomotor bundle (Fig. 9.21). I shall not describe the various cell types that constitute a somatomotor pool (cf. Bone 1960b), except to mention that in each fan there is a single large cell which, like some somatosensory cells, has a large vacuole in its cytoplasm (Fig. 9.21). It is also worth noting that the peripheral offshoot of a small proportion of the somatic motor cells branches in the vicinity of a 'ventral nerve' but does not give rise to a longitudinal fibre for the somatomotor bundle.

A few interesting preliminary electrophysiological results regarding the spinal cord of amphioxus have been described by Guthrie (1975). He recorded fast ($2-4\,\mathrm{m\,s^{-1}}$) and slow ($0.8-1.5\,\mathrm{m\,s^{-1}}$) waves of conduction using microelectrodes and direct stimulation. The fast waves were probably generated in the large axons of the Rohde cells.

The *muscle segments* of amphioxus contain two types of flattened, striated muscle cells: (a) elements with abundant sarcoplasm and glycogen granules and (b) elements with little sarcoplasm or glycogen. The former are found in the lateral lamellae of the myotomes, whereas the latter predominate in the medial lamellae. It has already been mentioned that in amphioxus the ventral nerves actually consist of medial projections from the myotomal muscle cells. Within each 'ventral nerve' two kinds of fibres can be identified. Each of these corresponds peripherally with muscle fibres of one of the two types discussed above and is opposed centrally to a particular compartment of the zone where these processes contact the terminals of the motor neurons. The muscle fibres constituting the lateral myotomal lamellae are involved in fast movements. They send thin fibres to the surface of the cord, where they are apposed to short, dorsal compartments of large boutons filled with synaptic vesicles about 600 nm in diameter. The muscle fibres making up the medial myotomal lamellae subserve slow movements. Their coarse, medially extending processes are apposed to a long ventral compartment of small boutons tightly packed with vesicles about 100 nm in diameter (Flood 1966, 1968). It remains to be established which neuron types or neural groups contribute terminals to these dorsal and ventral bouton compartments.

The muscle fibres themselves have been studied with microelectrode techniques by several groups, most notably Hagiwara and associates (Hagiwara and Kidokoro 1971; Hagiwara et al. 1971). These authors concluded that the Ca^{2+} necessary for muscular twitch enters across the membrane of the cell, in parallel with the Na^+ associated with the action potential. This mechanism resembles that found in crustacea rather than that in vertebrates; in the lat-

ter, release of Ca^{2+} from internal stores plays a crucial role in muscle contraction. In vertebrates, the release of internal Ca^{2+} is probably triggered via the T-tubule system, which is absent in amphioxus.

Before we leave the spinal motor system it should be mentioned that, curiously, the *notochord* of *Branchiostoma* consists of flat, disk-shaped muscle cells. Like the myotomal muscle cells, these notochordal elements send processes to the surface of the spinal cord. These processes terminate in groups on the ventral surface of the cord, where they are apposed to large terminals originating as short collaterals of axons running longitudinally in the cord (Flood 1970). The latter are assembled in a well-defined tract, but their cells of origin are unknown. The transmitter in the myotomal, as well as in the chordal, motor systems is presumably cholinergic, because in both systems the neuromuscular junctions stain positively for cholinesterase (Flood 1970, 1974). The muscles of the notochord are apparently composed of paramyosin and consequently resemble molluscan structures rather than the usual vertebrate pattern (Guthrie and Banks 1970). Contractions of the notochord increase its stiffness. The functional significance of this adjustability of notochordal turgor could be related to forward and backward swimming or to burrowing movements (Flood 1975).

9.6.7
Rohde Cells and Their Axons

During the course of this survey of the structure of the acraniate cord we have encountered several fibre bundles: (a) a dorsally located somatosensory tract, (b) a ventrolaterally situated somatomotor bundle, and (c) a viscerosensory tract lying intermediate between the other two. All of these tracts are primary pathways; i.e., their fibres form part of neurons which are either directly receptive or directly effective in character. The only long-axon internuncial system so far known is derived from the Rohde cells. These are larger multipolar elements lying across the midline in the dorsal region of the cord and extending their dendrites to both sides. The Rohde cells, which generally are located in the vicinity of a dorsal nerve, form a rostral and a caudal group. The first cell of the rostral group is the largest neuron in the nervous system of amphioxus (Figs. 9.20b, 9.22b) and lies, according to Rohde (1888b), at the level of the left sixth dorsal nerve; the second giant cell is found immediately behind the right sixth dorsal nerve; the third is located at the level of the seventh dorsal nerve, and so on (Fig. 9.22b). The total number of elements in the rostral group is 12, the last of which lies in the vicinity of the right 11th dorsal nerve. No giant cells occur in the 12th to the 39th segments, but between the latter and the 60th segment they are present again. In this caudal region, 14 Rohde cells are present (for details see Rohde 1888a,b and Franz 1923). The course of the axons of the cells is highly peculiar. The neuraxis of the foremost cell arches to the right and ventrally, then bends longitudinally and passes down the middle of the cord (Figs. 9.16, 9.17, 9.20a). The more caudally situated Rohde cells send their axons alternately to the left and right; these cross in the floor of the spinal cord and form two bundles of descending fibres (Figs. 9.16, 9.17, 9.20a, 9.22). The neuraxes of the posterior groups of Rohde axons show similar relations but ascend in the cord. They also constitute distinct bundles which lie, as Rohde himself demonstrated, ventral to the descending ones (Figs. 9.16, 9.17). According to Rohde's descriptions and illustrations, the caudally and rostrally directed groups of axons both extend throughout most of the cord. However, the light- and electron-microscopic studies of Guthrie (1975) have shown that the axons of the rostral group of cells do not extend beyond myotome 37 and that many of the caudal cells have axons that do not extend further than five neural segments rostral of their perikarya.

Several authors have observed collaterals of the giant fibres; for instance, Rohde (1888b) mentions that all of the colossal neuraxes have fine collaterals. These are, according to him, particularly numerous from the fibre of the foremost cell but only scanty from the other giant axons. Retzius' (1891) description and figures suggest that collaterals appear only on the initial part of the Rohde axons, and Bone (1960b) came to a similar conclusion. Figure 9.22a has been redrawn from Retzius. It shows that the laterally directed arches of Rohde fibres emit two collaterals which run longitudinally in opposite directions. The remaining collaterals are smaller and come from the segment between the arch and the perikaryon.

Ruiz and Anadón (1989) studied the fine structure of Rohde cells and their processes. Chemical synapses were not found over the perikarya and dendrites of these elements, but contacts resembling gap junctions appeared to be common. The median Rohde axon was observed to establish minute efferent *en-passant* synapses with thin axonal processes. A few reciprocal synapses between Rohde axons and other elements of unknown nature were also observed.

What is the function of the Rohde cells? Some authors (Tagliani 1897; Heymans and van der Stricht 1898) believed that they are motor elements. In contrast, Ariëns Kappers et al. (1936) thought

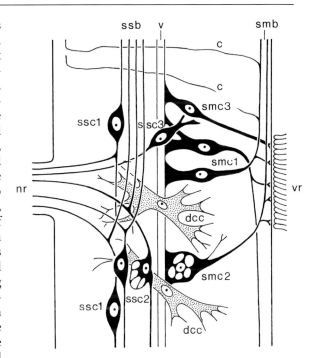

Fig. 9.21. Horizontal projection of a number of neurons in the spinal ord of *Branchiostoma*; for abbreviations see list on p. 378 (based on Bone 1960b, Figs. 3 and 5)

that these elements form a secondary sensory system, comparable to the arcuate fibres of vertebrates. Bone (1959, 1960b) pointed out that the dendrites of the Rohde cells ramify among the fibres of the somatosensory tracts, and that their axons give off many collaterals in the region of the somatomotor tract. This led Bone to conclude that the Rohde cells are both internuncial elements, intervening in the somatic sensorimotor arc, and coordinating elements, controlling the normal swimming pattern of the animal. Bone also drew attention to the close proximity of the dendrites of the visceromotor cells and the median giant fibre. He suggested that this fibre may play a role in coordinating the contractions of the pterygial muscle. Guthrie (1975) doubted this, pointing out that the median giant fibre provides the most rapid conduction pathway in the neuraxis of amphioxus, and that the rate at which the waves of contraction pass through the atrial musculature is much slower than might be expected from a direct involvement of the giant fibre.

9.6.8
Caudal Part of the Spinal Cord

Neurons are lacking in the most caudal part of the spinal cord. In this region the cord forms an ependymal tube, the filum terminale, which ends in an

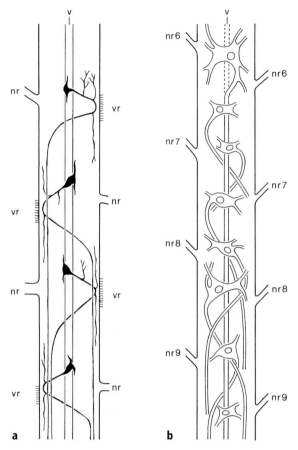

Fig. 9.22a,b. Rohde cells in the spinal cord of *Branchiostoma* in dorsal view. **a** Some Rohde cells and their processes, as observed in a methylene blue preparation (redrawn from Retzius 1891, Fig. 1). **b** The most rostrally situated elements of the anterior group of Rohde cells (based on Franz 1923, Fig. 24). For abbreviation see list on p. 378

ampullar enlargement. The latter often bends upward above the notochord (Ariëns Kappers et al. 1936). Obermüller-Wilén and Olsson (1974) found that Reissner's fibre forms in the ampulla a condensed caudal mass. Ependymal cells surrounding the ampulla appeared to be specialised to engulf and digest the disintegrated fibre material. Ruiz and Anadón (1991b) reported that nerve fibres, coursing through the filum terminale and caudal ampulla, end on the basal lamina of the ependymal cells, forming neuro-connective structures. The terminal boutons observed appeared to be divisible into several classes according to their vesicle content. They concluded that in amphioxus a primitive neurosecretory system similar to the fish urophysis may well be present.

9.7
Brain

9.7.1
Introductory Notes

Although certain structures in the brain, such as the infundibular organ, have attracted considerable attention, in general the rostral part of the neuraxis of amphioxus has been studied much less than the spinal cord. The brain is subdivided here into a rostral, archencephalic and a caudal, deuterencephalic region. The rostral region encompasses the epithelial structures surrounding the cerebral ventricle, plus some adjacent structures which probably have a vertebrate archenecephalic counterpart, as for instance the lamellar complex and the dorsal and ventral groups of neurosecretory cells. The caudal region is taken as extending to a plane situated between the fifth and the sixth nerves and is coextensive with the area occupied by the large cells of Joseph (Fig. 9.23).

9.7.2
Caudal Brain Region

As in the spinal cord, the lateral and ventral parts of the caudal brain region are occupied by fibres and neuropil. Most fibres contain many granular and vesicular inclusions. Meves (1973) reported six different types of granules and vesicles occurring in different combinations in the various fibres. She also reported that some fibre profiles show local membrane thickening suggestive of synaptic differentiations. Meves' findings differ somewhat from those of Guthrie (1975), who reported that synaptic structures are frequent in the brain. Apart from elements also occurring at the spinal level (e.g. motor and commissural neurons, Fig. 9.9), the caudal region of the brain contains various cell types that are not found further caudalward. Prominent among these are the cells of Joseph, the B cells and large ventral cells, first described by Bone (1960b).

The *Joseph cells* constitute a large, elongated, dorsally situated group extending throughout the caudal brain region (Figs. 9.13, 9.18, 9.23, 9.24). They have been described under different names: for instance, *Ganglienzellen* (von Kupffer 1893, 1906), *dorsale Riesenzellen* (Edinger 1906), *Gehirnzellen* (Franz 1923), and large dorsal cells (Guthrie 1975). Ariëns Kappers et al. (1936) introduced the term Joseph cells, after the author who made the first detailed studies of these elements (Joseph 1904). The bipolar or polygonal perikarya of Joseph cells are 15–25 µm in diameter. Most of them have a long rostral and a shorter ventral process (Edin-

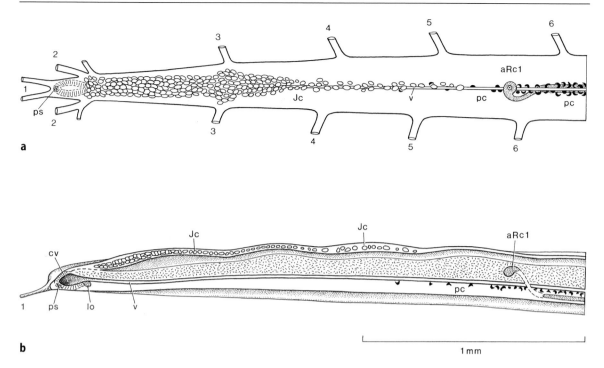

Fig. 9.23a,b. Dorsal (**a**) and medial (**b**) views of the rostral part of the *Branchiostoma* CNS, showing several structural features in projection. For abbreviations see list on p. 378

ger 1906), but it is difficult to distinguish the axon from the dendrites. Guthrie (1975) observed that some of the ventral processes connect with a median zone of small cells in the caudal part of the brain. Joseph (1904) reported that the large dorsal cells structurally resemble the spinal light-sensitive elements, and the electron-microscopic studies of Welsh (1968) revealed that they contain a typical photoreceptor organelle consisting of a brushborder-like array of microvilli. Watanabe and Yoshida (1986) demonstrated that this organelle contains a rhodopsin-like substance and that its ultrastructure may be influenced by photic treatments.

The *B cells* of Bone (1959, 1960b) are bipolar elements that form a group situated under the dorsal mass of Joseph cells. They send out one process through the first or second nerve. These peripheral processes can be followed to the edge of the rostrum, where they contact the sense cells of the so-called corpuscles of de Quatrefages (1845). The other process of the B cells runs caudally and joins the dorsal somatosensory tract. In larval specimens, Bone (1959) traced some of these fibres caudally beyond the level of the anus, but in the adult he remained unable to trace them for more than a few segments backward along the cord (Bone 1960b). It seems likely that the B cells form part of a system conveying somatosensory data from the

rostrum to the spinal cord. (In his paper, Bone termed the common spinal bipolar elements contributing fibres to the somatosensory system A cells, hence the designation B cell for their encephalic equivalents.) Lacalli (1996) considers it likely that the B cells are equivalent to the tectal cells observed by him in early larvae (Figs. 9.9, 9.10).

The *large ventral cells* of Bone (1960b) are, according to him, few in number. The dendrites ramify in the dorsal somatosensory columns. Their axonal processes cross the floor of the central canal in a manner reminiscent of the axons of the Rohde cells and pass backward along the ventral columns of the cord. Bone considered it likely that these elements, like the B cells, coordinate the activity of the segmental systems of cells in the spinal cord.

9.7.3
Rostral Brain Region

The walls of the cerebral ventricle are formed mostly of a single layer of ciliated epithelial cells. Meves (1973) distinguished five groups of cells passing from dorsocaudal to ventrocaudal with different structural and ultrastructural characteristics (Fig. 9.24). Prominent among these is the rostrally situated group three, the elements of which contain numerous pigment granules. Together, these elements constitute the 'pigment spot' described by

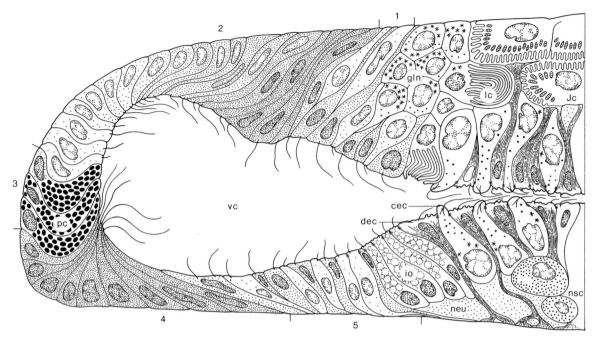

Fig. 9.24. Semidiagrammatic representation of a paramedian sagittal section through the most rostral part of the *Branchiostoma* CNS. *cec,* Clear epithelial cells; *dec,* dark epithelial cells; *gln,* glycogen containing neurons; *io,* infundibular organ; *Jc,* Joseph cells; *lc,* lamellar cells; *neu,* neuropil; *nsc,* neurosecretory cells; *pc,* pigment cells; *vc,* ventricular cavity; 1, 2, 3, 4, 5, zones of different types of epithelial cells (redrawn from Meves 1973, Fig. 9)

the early students of the amphioxus brain. Lacalli et al. (1994) suggested that the pigmented cells constitute, together with adjacent non-pigmented cells in the basal wall of the brain vesicle, a photoreceptive complex, which they designated as the frontal eye. In larval amphioxus the non-pigmented cells form three transverse rows, R1–3 (Fig. 9.7). According to Lacalli et al. (1994), the cells forming the first two rows presumably represent receptor cells, whereas the cells behind these might constitute the beginning of an interneuronal relay system. In a subsequent study of the same material, Lacalli (1996) distinguished a fourth row of elements belonging to the frontal eye (Fig. 9.12) and presented data on the interneuronal relations of the second row of receptor cells and the neuronal elements of row 3 and row 4, suggesting a retina-like microcircuitry. As has already been discussed in Sect. 9.4 and depicted in Fig. 9.12, the output of the frontal eye is conducted, according to Lacalli's (1996) analysis, either directly, or indirectly via a tectum-like structure, to a deuterencephalic motor coordinating centre.

The area surrounding the caudal part of the cerebral ventricle contains three cell complexes, which have more or less distinct counterparts in the craniote brain: the group of lamellar cells, the infundibular organ, and some groups of neurosecretory cells. These complexes have already been

dealt with in Sect. 9.5; hence, I confine myself here to a few additional remarks.

The group of *lamellar cells* is situated dorsally, in front of the complex of Joseph cells (Figs. 9.7, 9.24). It is composed of fairly large elements, each of which projects a highly modified cilium into the ventricle, where it expands into densely packed, flattened lamellae (Meves 1973; Ruiz and Anadón 1991a). Similar cells have been found in ascidian tadpoles (Barnes 1974) and in the lateral and pineal eyes of vertebrates. Lacalli et al. (1994) discovered that the group of lamellar cells is much larger in early larvae than in the adult stage. Given the fact that amphioxus larvae can regulate their position in the water column on a diurnal basis (Wickstead and Bone 1959; Webb 1969), Lacalli et al. (1994) conjectured that the lamellar body, with its huge area of membrane surface, functions as a high-sensitivity, non-directional photoreceptor.

The secretory, highly cylindrical cells of the *infundibular organ* produce Reissner's fibre, a thread of non-cellular material present in all vertebrates, extending backward throughout the ventricular system of the neuraxis. In amphioxus (cf. Sect. 9.6.8), and in most vertebrates, it is collected and absorbed in a saclike widening of the central canal at the caudal end of the spinal cord. The function of Reissner's fibre – first described by Ernst

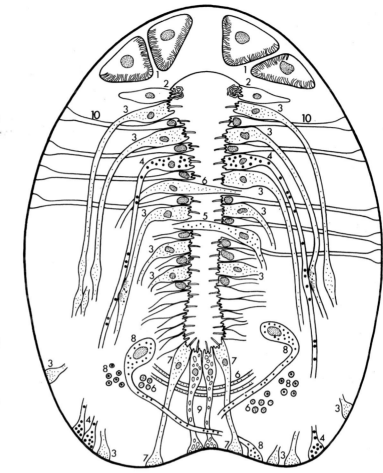

Fig. 9.25. Semidiagrammatic transverse section through the caudal part of the cerebral vesicle of amphioxus, showing the location and interrelationship of the various types of neurons and receptor cells present at that level. *1*, Joseph cells; *2*, lamellar photoreceptor cells; *3*, catecholaminergic cells, fibres and terminals; *4*, neuroendocrine FSH-immunoreactive cells, fibres and terminals; *5*, prolactin-immunoreactive cell; *6*, calcitonin-immunoreactive cells; *7*, catecholaminergic neuroendocrine cells; *8*, serotoninergic cells, fibres and terminals; *9*, infundibular cells; *10*, tanycytes (reproduced from Obermüller-Wilén 1984, Fig. 1)

Reissner (1860) in the lamprey spinal cord – is unknown. It has been suggested that the fibre acts as a detoxicant for the cerebrospinal fluid; harmful molecules in the liquor would bind to it and be degraded with the fibre material in the spinal terminal ampulla (Olsson 1956, 1958, 1993; Sterba 1969; Ermisch 1973).

Meves (1973) described some small groups of large neurons with coarse processes situated caudal to the infundibular organ, close to the central canal (Fig. 9.24: Ns). Because the perikarya of these elements contain numerous electron-dense granules (average diameter 255 nm) which are surrounded by a unit membrane, she considered them *neurosecretory cells*.

Obermüller-Wilén (1979) observed two groups of neurosecretory cells in the brain of amphioxus, a larger, dorsal one and a smaller, ventral one. The dorsal group is situated dorsolaterally on both sides of the ventricle in the middle part of the rostral brain region (Fig. 9.25: 4). It consists of large cells which are characteristically surrounded by up to five or six layers of thin lamellae formed by pro-

cesses of adjacent tanycytes. Their axons run ventrocaudally and contact the basal lamina of the brain surface. The perikarya as well as the axons of these elements contain numerous granules, the perikaryal ones having a diameter in the range of 85–160 nm. The axons of these dorsally situated cells are contacted by numerous *boutons en passant* and *boutons terminaux* of varying size, containing clear synaptic vesicles (50–55 nm) and dense-cored vesicles (90–100 nm) (Fig. 9.25: 3). Histochemical evidence suggested that the terminals just discussed are catecholaminergic. Their parent cell bodies appeared to lie scattered among the large neurons of the dorsal group (Fig. 25: 3). In a later study, Obermüller-Wilén (1984) demonstrated that the cells of the dorsal group display considerable ultrastructural changes before, during, and after the spawning period, implicating them in spawning and gametogenesis. An antiserum raised against human follicle-stimulating hormone (FSH) reacted positively with these cells.

The *ventral group*, which is also paired, consists of catecholaminergic cells. It is situated in the vicinity of the infundibular organ (Fig. 9.25: 7). The axons of the neurons forming this group pass ventrocaudally and contact, like those of the dorsal cells, the basal brain lamina. The perikarya and axons of these neurons contain globular, electron-dense granules with a diameter of 75–165 nm. Synaptic endings contacting these cells were not observed. The relationship between the dorsal and ventral cell groups just discussed and the neurosecretory and other cell groups described by Meves (1973; Fig. 9.24) is unknown. Obermüller-Wilén (1979, 1984) observed that the basal lamina on the ventral brain surface is contacted not only by processes of the dorsal, peptidergic (Fig. 9.25: 4) and ventral catecholaminergic cell groups (Fig. 9.25: 7), but also by terminal profiles of other catecholaminergic (Fig. 9.25: 3) and of serotoninergic cells (Fig. 9.25: 8). She considered it likely that all of these terminals release their products through the basal lamina, from where they diffuse to the blood space which borders the ventrolateral part of the brain.

Two types of epithelial cells designated by Meves (1973) as dark cells and clear cells (*helle Zellen*) deserve mentioning. According to Meves, these cells form large groups in the wall of the central canal. She depicted these cells as being situated directly caudal to the level of the infundibular organ but did not indicate their caudal extent. The *dark cells* (Fig. 9.24: dec), which contain many filaments and a dark cytoplasm, have distinct peripheral processes, some of which extend to the surface of the brain. The *clear cells* (Fig. 9.24: cec) are provided with plump cytoplasmatic processes protruding into the ventricular cavity. Finally, it should be mentioned that, according to Meves (1973), the transitional area between the rostral and caudal brain regions contains numerous cells with large amounts of glycogen (Fig. 9.24: gln).

Guthrie (1975) recorded a variety of electrical responses from amphioxus following stimulation of the brain. He also noted that when the brain was ablated, lancelets were more easily aroused to swim. These observations hint at the possibility of supraspinal control in this primitive animal.

9.8
Atrial Nervous System

The lancelet possesses a complex peripheral nervous system lying just under the thin sheet of epithelium lining the atrium and covering the various organs within it. This so-called atrial nervous system contains both efferent and afferent components that are connected to the CNS by way of the spinal nerves. As already mentioned, the efferent fibres originate from a column of cells that constitutes the most ventral part of the spinal grey matter. They pass without synaptic interruption to the striated fibres of the pterygial muscle that forms the atrial floor, and there is also a large ciliary-motor component which controls the action of the cilia on the pharyngeal bars (Bone 1961). The visceral afferent fibres are the axons of peripheral neurons lying directly under the atrial epithelium. These neurons do not occur before the formation of the atrium. Interestingly, they have been shown to migrate out along the spinal nerves in the course of metamorphosis (Bone 1958). The peripheral afferent elements include unipolar neurons, most of which lie on the surface of the pterygial muscle, and multipolar neurons, which are abundant on the foregut (Bone 1961). Guthrie (1975) has presented evidence suggesting that the unipolar elements are mechanoreceptors, responsive to rapid increases in tension of the atrial muscle. We will not discuss the functions of the atrial system beyond remarking that it is concerned mainly with the regulation of the various stages of the feeding process (Bone 1961).

The atrial nervous system of amphioxus has often been considered as the homologue of the autonomic system of vertebrates. Boeke (1935), one of the prominent advocates of this homology, even went so far as to recognise specific homologues of the plexuses of Auerbach and Meissner in the gut of amphioxus. However, as indicated above, the neurons of the atrial system are probably sensory elements and not, as Boeke supposed, motor cells similar to those of the craniate enteric plexus. Consequently, Bone (1961) considered it unsafe to make a direct comparison between the acraniate atrial nervous system and the craniate autonomic system. Rather, he believed that this system is a specialisation that has evolved in relation to the peculiar feeding habits of the Acrania.

9.9
Localisation of Neurotransmitters and Neuropeptides

In the literature some scattered data are found on the localisation of neurotransmitters in the CNS of amphioxus, and Uemura et al. (1994) recently devoted an immunohistochemical study to the distribution of some neuropeptides in this form. They reported the presence of six separate peptidergic neuronal populations in *Branchiostoma belcheri*, containing arginin-vasopressin, oxytocin, angiotensin II, cholecystokinin, urotensin I and FMRF amide. No colocalisation of these peptides was observed in any neurons of the nerve cord. The most salient data concerning the localisation of neuromediators in the CNS of amphioxus are

Table 9.1. Localisation of neuromediators in the CNS of amphioxus

Neuromediator		Source
Acetylcholine	– Motoneurons	Flood (1974)
Catecholamines	– Elongated, csf-contacting cells in the dorsal part of the brain vesicle, giving rise to ventrocaudally running, varicose axons (Fig. 9.26: cp,f; Fig. 9.25: 3)	Obermüller-Wilén (1979) Obermüller-Wilén and van Veen (1981)
	– Elongated, csf-contacting cells in the ventral and ventrocaudal part of the brain vesicle, contacting the floor of the brain (Fig. 9.25: 7; Fig. 9.26: vp)	
	– Cells in the dorsal part of the caudal brain region, ventral to Joseph cells (Fig. 9.26: pp)	
Serotonin	– Cells in the most rostral part of the brain; these cells reportedly represent the row 2 cells of the frontal eye (Fig. 9.7)	Holland and Holland (1993) Lacalli (1996) Obermüller-Wilén (1984)
	– Subependymal cells in the lateral and ventral parts of the brain, the axons of which form two caudally directed bundles on either side of the central canal (Fig. 9.25: 8)	
Follicle-stimulating hormone (FSH)	– A group of large neurosecretory cells situated dorsolaterally in the wall of the cerebral vesicle (Fig. 9.25: 4)	Obermüller-Wilén (1984)
Calcitonin	– Large perikarya located around the posterior part of the cerebral ventricle, some of which cross the ventricle (Fig. 9.25: 6)	Obermüller-Wilén (1984)
	– A bilateral bundle of axons in the ventral part of the caudal brain region (Fig. 9.25: 6)	
Prolactin	– Perikarya located intra-ependymally around the caudal part of the cerebral vesicle, some crossing the ventricle (Fig. 9.25: 5)	Obermüller-Wilén (1984)
Arginine vasopressin (AVP)	– Small cells in the spinal white matter throughout the length of the cord (Figs. 9.27–9.29)	Uemura et al. (1994)
	– Coarse fibres coursing longitudinally through the rostral spinal cord, often contacting Rohde axons (Figs. 9.30, 9.31)	
Oxytocin (OT)	– A few cells along the central canal	Vallet and Ody (1985) Uemura et al. (1994)
	– Small cells at the lateral margin of the grey matter throughout the length of the spinal cord (Figs. 9.27, 9.28, 9.32)	
	– Short axons in the ventral part of the brain	
Angiotensin II (ANGII)	– Small cells in the lateral grey matter of the caudal spinal cord (Figs. 9.27, 9.28, 9.33)	
Cholecystokinin (CCK)	– A bundle of fibres running beneath the grey matter of the rostral spinal cord (Figs. 9.34–9.36)	
Carp urotensin I (UI)	– Medium-sized commissural cells situated in the caudal part of the brain (Figs. 9.28, 9.37, 9.38), giving rise to axons passing caudally along the ventrolateral surface of the spinal cord (Fig. 9.39)	
	– Small cells in the caudal (Fig. 9.40), particularly the caudalmost, portion of the cord (Fig. 9.41)	
FMFRamide	– A small group of large commissural cells in the caudal part of the brain (Figs. 9.28, 9.42, 9.43)	
	– Fibres running along the ventrolateral surface of the brain (Fig. 9.43)	
	– Coarse and thin fibres in the ventral part of the brain (Fig. 9.44)	
	– Numerous cells, just lateral to the grey matter, throughout almost the entire spinal cord (Figs. 9.45, 9.46)	

assembled in Table 9.1. For details, the publications quoted should be consulted.

9.10
Concluding Comments: Amphioxus and Its Relationship to the Craniates

It is natural to seek comparisons between the CNS of amphioxus and that of anamniote craniates. The organisation of the spinal *somatic system* suggests two reasons why this system in the adult amphioxus might be expected to resemble the corresponding system of craniate larvae, rather than the adult forms. The first is that the spinal somatic systems of anamniote craniate larvae are built on a strikingly

uniform plan, which is most probably an ancient one in the craniates (Whiting 1955; Bone 1960b). The second reason is that amphioxus shows the same kind of somatomotor activity that is seen in craniate larvae. The animal always responds to external stimuli with sinuous movements of the whole body; localised reflexes are entirely lacking (ten Cate 1938a,b). Although the somatic system in the cord of the adult amphioxus exhibits a much more complicated pattern than that of the early vertebrate embryo, there do appear to be some striking points of resemblance. In both forms the primary sensory cells are situated within the spinal cord, and a large proportion of the intramedullary sensory cells of amphioxus, i.e. the somatosensory type

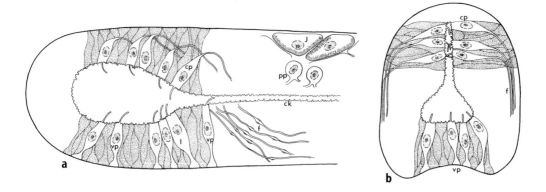

Fig. 9.26a,b. Schematic representation of the location of cate-cholaminergic neurons in the brain of amphioxus. **a** Sagittal section through the brain; **b** transverse section through the caudal part of the brain vesicle. *ck*, Central canal; *cp*, dorsola-teral group of catecholaminergic perikarya; *f*, catecholamin-ergic fibres in the neuropil; *I*, infundibular organ; *J*, Joseph cell; *pp*, catecholaminergic cells in caudal portion of the brain; *vp*, ventral group of catecholaminergic perikarya (reproduced from Obermüller-Wilén and van Veen 1981, Fig. 1)

1 elements (Fig. 9.21), resemble the Rohon-Beard cells of the craniate embryo, sending out one branch through the spinal nerves and contributing another to a longitudinally directed somatosensory pathway. The somatomotor elements of amphioxus and those of early larvae of anamniote vertebrates differ in shape but resemble each other closely with regard to the direction of their principal axonal process. In both groups these cells send out longitudinally running axons which come together to form a ventrolaterally situated somatomotor bundle, and in both groups the fibres contacting the muscles are collaterals from these longitudinal stem fibres. A third point of resemblance is that in both groups, vertical internuncial elements link the somatosensory columns with the somatomotor elements.

Nothing is known of the organisation of the *visceral system* in the early larvae of craniates; hence, comparisons like those made above for the somatic system cannot be made for the visceral system. All we know is that the visceral system of amphioxus does not show any particular resemblance to the corresponding system of adult craniates. The marked differences between the atrial system of amphioxus and the peripheral autonomic system of craniates have already been pointed out, and it should be emphasised that the visceral motor cells of amphioxus differ in two notable respects from the autonomic preganglionic cells of craniates: (a) they constitute a single median column which is situated ventral rather than dorsal to the somatomotor elements, and (b) their axons impinge directly on the effector organs, not indirectly via postganglionic elements.

The large *Rohde cells* obviously belong to the somatic mechanism of the spinal cord of amphioxus, but there are various suggestions regarding to which specific neuronal elements in the craniate central nervous system they correspond. According to Ariëns Kappers et al. (1936), the axons of the Rohde cells constitute fibre systems comparable to the secondary afferent pathways for vital or protopathic sensibility in vertebrates. Bone (1959) observed that in larval lancelets the Rohde cells

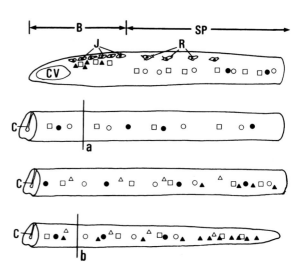

Fig. 9.27. Schematic representation of the CNS of *Branchiostoma belcheri*. The distribution of perikarya immunoreactive for AVP (*open circles*), OT (*closed circles*), ANGII (*open triangles*), UI (*closed triangles*) and FMRF amide (*open squares*) is indicated. *a,b*, Levels of cross-sections shown in Fig. 9.28. *B*, Brain; *C*, central canal; *CV*, cerebral ventricle; *J*, Joseph cells; *R*, Rohde cells; *SP*, spinal cord (reproduced from Uemura et al. 1994, Fig. 1)

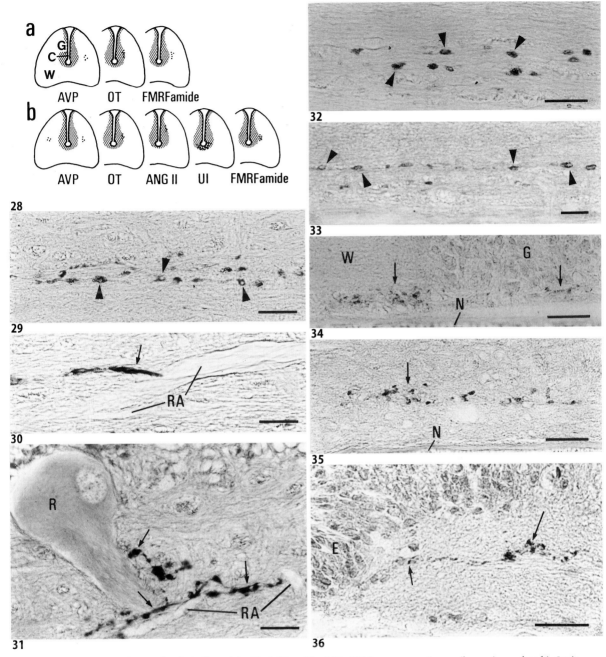

Fig. 9.28. Transverse sections at levels indicated in Fig. 9.27, showing the location of the cells immunoreactive for the neuropeptides indicated. *C*, Central canal; *G*, grey matter; *W*, white matter; *shaded area*, area of grey matter (reproduced from Uemura et al. 1994, Fig. 2)

Fig. 9.29. AVP-immunoreactive fibres and perikarya. Sagittal section of the lateral part of the anterior spinal cord. *Bar* = 20 μm (Figs. 9.29–9.46 are all from Uemura et al. 1994, Figs. 3–20)

Fig. 9.30. A coarse AVP-immunoreactive fibre (*arrow*) in contact with an axon of a Rohde cell (*RA*). Sagittal section of the lateral part of the anterior spinal cord. *Bar* = 20 μm

Fig. 9.31. Coarse AVP-immunoreactive fibres (*arrows*) in contact with the proximal portion of the axon of a Rohde cell (*R*) and with other Rohde axons (*RA*). Sagittal section of the anterior portion of the spinal cord. *Bar* = 20 μm

Fig. 9.32. OT-immunoreactive perikarya (*arrowheads*). Sagittal section of the caudal portion of the spinal cord. *Bar* = 20 μm

Fig. 9.33. AngII-immunoreactive perikarya (*arrowheads*) in the posterior portion of the spinal cord. Parasagittal section. *Bar* = 20 μm

Fig. 9.34. A bundle of CCK-immunoreactive fibres (*arrows*). Parasagittal section of the rostral portion of the spinal cord. *Bar* = 20 μm. *G*, Grey matter; *N*, notochord; *W*, white matter

Fig. 9.35. CCK-immunoreactive fibres (*arrows*) running in the white matter. Sagittal section. *N*, Notochord. *Bar* = 20 μm

Fig. 9.36. CCK-immunoreactive fibres (*arrows*) that extend as far rostrally as the ependymal layer (*E*) of the cerebral ventricle. Median section. *Bar* = 20 μm

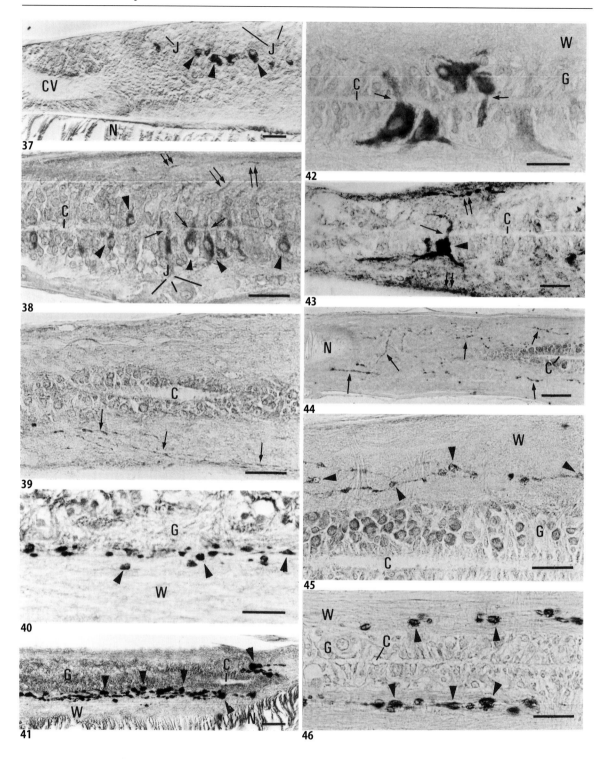

◀ **Fig. 9.37.** UI-immunoreactive perikarya (*arrowheads*). Paramedian section through the brain. *CV*, Cerebral ventricle; *J*, Joseph cells; *N*, notochord. *Bar* = 20 μm

Fig. 9.38. UI-immunoreactive perikarya (*arrowheads*) and fibres (*double arrows*) in the brain. Processes of the cells (*arrows*) cross the slitlike central canal (*C*). Horizontal section under a mass of Joseph cells (*J*). *Bar* = 20 μm

Fig. 9.39. UI-immunoreactive fibres (*arrows*) in the ventrocaudal region of the brain. Horizontal section. *C*, Central canal. *Bar* = 20 μm

Fig. 9.40. UI-immunoreactive perikarya (*arrowheads*). Parasagittal section of the caudal region of the spinal cord. *G*, Grey matter; *W*, white mater. *Bar* = 20 μm

Fig. 9.41. UI-immunoreactive perikarya (*arrowheads*) and fibres. Parasagittal section of the most caudal portion of the spinal cord. *C*, Central canal; *G*, grey matter; *N*, notochord; *W*, white matter. *Bar* = 20 μm

Fig. 9.42. FMRF amide-immunoreactive commissural cells. One of their processes (*arrows*) crosses the central canal (*C*). Horizontal section of the brain. *G*, Grey matter; *W*, white matter. *Bar* = 20 μm

Fig. 9.43. An FMRF amide-immunoreactive perikarya (*arrowhead*), one of the processes of which crosses the central canal (*C*). Immunoreactive fibres (*double arrows*) run near the lateral surface of the brain. *Bar* = 20 μm

Fig. 9.44. FMRF amide-immunoreactive fibres (*arrows*). Horizontal section through the ventral area of the brain. *C*, Central canal; *N*, notochord. *Bar* = 20 μm

Fig. 9.45. FMRF amide-immunoreactive perikarya (*arrowheads*) and fibres in the rostral portion of the spinal cord. Horizontal section. *C*, Central canal; *G*, grey matter; *W*, white matter. *Bar* = 20 μm

Fig. 9.46. FMRF amide-immunoreactive perikarya (*arrowheads*) in the caudal part of the spinal cord. Horizontal section. *C*, Central canal; *G*, grey matter; *W*, white matter. *Bar* = 20 μm

have a peripheral process passing out through the spinal root opposite the perikaryon. On that account, Bone considered these elements equivalent to the Rohon-Beard cells of vertebrate embryos. Guthrie (1975) and Nieuwenhuys (1977) compared the anterior and posterior Rohde cells, respectively, with the Müller and the giant interneurons of the lamprey. I am of the opinion that the Rohde cells combine functions which are accomplished in more complex neuron systems by at least two different sets of neurons (i.e. secondary sensory and motor coordinating elements). Therefore, I believe that these cells are not directly comparable to any particular group of neurons present in the craniate CNS. Indeed, their possible multifunctional role suggests some similarity with the neurons of invertebrates.

The *brain* of amphioxus is extremely small, and the fact that the three major subdivisions of the craniate brain, i.e. prosencephalon, mesencephalon and rhombencephalon, cannot be delineated is doubtless related to the absence of the three large paired special sense organs (for olfaction, vision

and equilibrium) which have shaped both the head and the brain of the latter subclass. Yet, a subdivision into a rostral archencephalon and a caudal deuterencephalon, as first proposed by von Kupffer (1906), seems to be justified. As discussed, this subdivision is based on the presence in the amphioxus nerve cord of some probable homologues of craniote diencephalic structures. Thus, the cephalochordate infundibular organ is homologous to the juvenile flexural organ of craniotes (Olsson 1993; Fig. 9.14), the lamellar complex corresponds positionally and structurally to the anamniote pineal eye (Lacalli et al. 1994), and the neurosecretory cell groups of amphioxus may well represent at least part of the vertebrate hypothalamic-hypophysial system (Obermüller-Wilén 1979). Recent data regarding the expression of the homeotic gene *AmphiHox3* imply that (a) a transverse plane passing through the nerve cord between the fifth and sixth nerve pairs (Figs. 9.3, 9.23) corresponds to a lower rhombencephalic level in craniotes, and that (b) in amphioxus the cord region situated between this plane and the caudal ends of the lamellar organ dorsally and the infundibular organ ventrally roughly corresponds to the brain stem, i.e. the rhombencephalon plus the mesencephalon of craniotes. The presence of a small primordial tectum and of some large cells, presumably representing part of the rhombencephalic reticular formation, tallies with this interpretation. However, a structure comparable to the cerebellum is entirely lacking, and the large, dorsally situated Joseph cells (Figs. 9.13, 9.18, 9.23) are remarkably similar to photoreceptor cells of annelids and molluscs but have no vertebrate equivalent (Welsch 1968).

Olsson (1986, 1992) noted that the first two *nerve pairs* of amphioxus, which convey impulses from mechanoreceptors in the snout to the brain, may correspond to general cutaneous branches of the trigeminal, and possibly also the facial nerve.

The physiological experiments of Guthrie (1975) suggested that the brain may be capable of modulating the operation of the cord. The primary motor centre described by Lacalli (1996) most probably represents a supraspinal coordinating centre upon which somatosensory and visual projections converge (Fig. 9.12). However, whether sensory projections of spinal origin attain this coordinating centre or any other part of the brain is presently unknown.

Catecholaminergic and peptidergic *neurosecretory cells* form a functional complex in the caudal part of the brain vesicle that may correspond to the hypothalamic regulating system in vertebrates. Processes of many cells of this complex participate in the formation of a neurohaemal contact zone which, according to Obermüller-Wilén (1984) is comparable to the median eminence and the neurohypophysis in vertebrates.

A heavily ciliated, left-sided, dorsal extension of the oral cavity, known as *Hatschek's pit* (Hatschek 1884; Fig. 9.2), has been considered to be homologous to the adenohypophysis of craniotes. Some of its cells contain granules which may be released internally as endocrine substances (Tjoa and Welsch 1974), and the epithelium lining this pit has been found to react with antibodies against gonadotropins (Nozaki and Gorbman 1992), LHRH, TRH (Chang et al. 1985) and CCK (Sahlin 1988; Nozaki and Gorbman 1992). However, Uemura et al. (1994) found no immunoreactivity in it for the neuropeptides examined in their study (see Table 9.1). Whatever the morphological significance of Hatschek's pit may be, functionally it is important that this structure is widely separated from the neurosecretory area of the brain by the notochord and that a neural or vascular link between the two has not been reported so far.

Finally, it should be noted that the close apposition to the ventricular cavity of all epithelial and most neuronal elements in the CNS of amphioxus, although necessitated by the absence of a vascular system, might allow the cavity to mediate neurohumoral communication between cells.

Acknowledgement. I thank Dr. Quentin Bone for critically reading the manuscript of this chapter.

References

Ariëns Kappers CU (1929) The evolution of the nervous system. Bohn, Haarlem

Ariëns Kappers CU (1934) Feinerer Bau und Bahnverbindungen des Zentralnervensystems. In: Bolk L, Göppert E, Kallius E, Lubosch W (eds) Handbuch der vergleichenden Anatomie. Zerebrospinales Nervensystem. Urban and Schwarzenberg, Berlin, Vol II, 1 pp 319–486

Ariëns Kappers CU, Huber GC, Crosby EC (1936) The comparative anatomy of the nervous system of vertebrates, including man, vol 1. MacMillan, New York

Barnes SN (1974) Fine structure of the photoreceptor of the ascidian tadpole during development. Cell Tissue Res 155:27–45

Boeke J (1902) Über das Homologon des Infundibularorgans bei Amphioxus lanceolatus. Anat Anz 15:411–414

Boeke J (1908) Das Infundibularorgan im Gehirn des Amphioxus. Anat Anz 32:473–488

Boeke J (1935) The autonomic (enteric) nervous system of Amphioxus lanceolatus. J Microsc Sci 77:623–658

Bone Q (1958) Synaptic relations in the atrial nervous system of amphioxus. J Microsc Sci 99:243–261

Bone Q (1959) The central nervous system in larval acraniates. J Microsc Sci 100:509–527

Bone Q (1960a) A note on the innervation of the integument in amphioxus, and its bearing on the mechanism of cutaneous sensibility. J Microsc Sci 101:371–379

Bone Q (1960b) The central nervous system in amphioxus. J Comp Neurol 115:27–64

Bone Q (1961) The organization of the atrial nervous system of amphioxus (Branchiostoma lanceolatum (Pallas). Philos Trans R Soc Lond B 243:241–269

Chang CY, Liu YX, Zhu YT, Zhu HH (1985) The reproductive endocrinology of amphioxus. In: Carlick DG, Kroner PI (eds) Frontiers in physiological research. Australian Academy of Science, Canberra, pp 79–86

Delsman HC (1913) Ist das Hirnbläschen des Amphioxus dem Gehirn der Kranioten homolog? Anat Anz 44:481–497

de Quatrefages MA (1845) Mémoire sur le système nerveux et sur l'histologie du Branchiostoma ou Amphioxus. Ann Sci Nat 3:197–248

Eakin RM, Westfall JA (1962) Fine structure of photoreceptors in amphioxus. J Ultrastruct Res 6:531–539

Edinger L (1906) Einiges vom 'Gehirn' des Amphioxus. Anat Anz 28:417–428

Ermisch A (1973) Zur Charakterisierung des Komplexes Subcommissuralorgan-Reissnerscher Faden und seiner Beziehungen zum Liquor unter besonderer Berücksichtigung autoradiographischer Untersuchungen sowie funktioneller Aspekte. Wiss Z Karl Marx Univ, Leipzig, Math Naturwiss R 22:297–336

Flood PR (1966) A peculiar mode of muscular innervation in amphioxus. Light and electron microscopic studies on the so-called ventral roots. J Comp Neurol 126:181–218

Flood PR (1968) Structure on the segmental trunk muscle in amphioxus. With notes on the course and 'endings' of the so-called ventral root fibres. Z Zellforsch 84:389–416

Flood PR (1970) The connection between spinal cord and notochord in amphioxus (Branchiostoma lanceolatum). Z Zellforsch 103:115–128

Flood PR (1974) Histochemistry of cholinesterase in amphioxus (Branchiostoma lanceolatum (Pallas)). J Comp Neurol 157:407–438

Flood PR (1975) Fine structure of the notochord of amphioxus. In: Barrington EJW, Jefferies RPS (eds) Protochordates. Symposia of the Zoological Society of London, vol 36. Academic, London, pp 81–104

Franz V (1923) Haut, Sinnesorgane und Nervensystem der Akranier. Jen Z Naturwiss 59:401–526

Garcia-Fernández, Holland PWH (1994) Archetypical organization of the amphioxus Hox gene cluster. Nature 370:563–566

Guthrie DM (1975) The physiology and structure of the nervous system of amphioxus (the lancelet), Branchiostoma lanceolatum pallas. In: Barrington EJW, Jefferies RPS (eds) Protochordates. Symposia of the Zoological Society of London, vol 36. Academic, London, pp 43–80

Guthrie DM, Banks JR (1970) Function and physiological properties of a fast paramyosin muscle – the notochord of amphioxus (Branchiostoma lanceolatum). J Exp Biol 52:125–138

Hagiwara S, Kidokoro Y (1971) Na and Ca components of action potential in amphioxus muscle cells. J Physiol (Lond) 219:217–232

Hagiwara S, Henkart MP, Kidokoro Y (1971) Excitation-contraction coupling in amphioxus muscle cells. J Physiol (Lond) 219:233–251

Hatschek B (1882) Studien über die Entwicklung von Amphioxus. Arb Zool Inst Univ Wien 4:1–88

Hatschek B (1884) Mitteilungen über Amphioxus. Zool Anz 7:86–127

Hatschek B (1892) Die Metamerie des Amphioxus und des Ammocoetes. Anat Anz 7:89–81

Heymans JF, van der Stricht O (1898) Le sytème nerveux de l'amphioxus et en particulier sur la constitution et la genèse des racines sensibles. Mémoires couronnés et Mémoires des savants étrangers de l'Akad Roy Bruxelles, vol 56, pp 1–74

Hofer H (1959) Über das Infundibularorgan und den Reissnerschen Faden von Branchiostoma lanceolatum. Zool Jahrb 77:465–490

Holland ND, Holland LZ (1993) Serotonin-containing cells in the nervous system and other tissues during ontogeny of a lancelet, Branchiostoma floridae. Acta Zool (Stockh) 74:195–204

Holland PWH, Holland LZ, Williams NA, Holland ND (1992) An amphioxus homeobox gene: sequence conservation, spatial expression during development and insights into vertebrate evolution. Development 116:653–661

Holland PWH, Garcia-Fernández J, Holland LZ, Williams NA, Holland ND (1994) The molecular control of spatial patterning in amphioxus. J Mar Biol Ass (UK) 74:49–60

Johnston JB (1905) The cranial and spinal ganglia, and the viscero-motor roots in amphioxus. Biol Bull 9:113–127

Joseph H (1904) Über eigentümliche Zellstrukturen im Zentralnervensystem von Amphioxus. Verh Anat Ges 18:16–26

Kölliker A (1843) Über des Geruchsorgan von Amphioxus. Arch Anat Physiol Berl 32–35

Kowalewski A (1876) Studien über den Amphioxus lanceolatus. Mém Acad Imp St Pétersbourg 19:1–29

Lacalli T (1996) Frontal eye circuitry, rostral sensory pathways, and brain organization in amphioxus larvae: evidence from 3D reconstructions. Philos Trans R Soc Lond B 351:243–263

Lacalli TC, Holland ND, West JE (1994) Landmarks in the anterior central nervous system of amphioxus larvae. Philos Trans R Soc Lond B 344:165–185

Leonhardt H (1980) Ependym und circumventriculäre Organe. In: Oksche A, Vollrath L (eds) Neuroglia I. Springer, Berlin Heidelberg New York, pp 177–665 (Handbuch der mikroskopischen Anatomie des Menschen, vol IV, part 10)

Meves A (1973) Elektronenmikroskopische Untersuchungen über die Zytoarchitektur des Gehirns von Branchiostoma lanceolatum. Z Zellforsch 139:511–532

Nieuwenhuys R (1977) The brain of the lamprey in a comparative perspective. Ann N Y Acad Sci 299:97–145

Nozaki M, Gorbman A (1992) The question of functional homology of Hatschek's pit of amphioxus (Branchiostoma belcheri) and the vertebrate adenohypophysis. Zool Sci 9:387–395

Obermüller-Wilén H (1976) The infundibular organ of Branchiostoma lanceolatum. Acta Zool (Stockh) 57:211–216

Obermüller-Wilén H (1979) A neurosecretory system in the brain of the lancelet, Branchiostoma lanceolatum. Acta Zool (Stockh) 60:187–196

Obermüller-Wilén H (1984) Neuroendocrine systems in the brain of the lancelet, Branchiostoma lanceolatum (Cephalochordata). Thesis, Stockholm

Obermüller-Wilén H, Olsson R (1974) The Reissner's fiber termination in some lower chordates. Acta Zool (Stockh) 55:71–79

Obermüller-Wilén H, van Veen T (1981) Monoamines in the brain of the lancelet, Branchiostoma lanceolatum: a fluorescence-histochemical and electron-microscopical investigation. Cell Tissue Res 221:245–256

Oksche A (1969) The subcommissural organ. J Neuro Visc Relat [Suppl] 9:111–139

Olsson R (1956) The development of Reissner's fibre in the brain of the salmon. Acta Zool (Stockh) 37:235–250

Olsson R (1958) Studies on the subcommissural organ. Acta Zool (Stockh) 39:71–102

Olsson R (1962) The infundibular cells of amphioxus and the question of fibre-forming secretions. Arkiv Zool 15:347–356

Olsson R (1986) Basic design of the chordate brain. In: Uyeno T, Arai R, Taniuchi T, Matsuura K (eds) Indo-Pacific fish biology: proceedings of the 2nd International Conference on Indo-Pacific Fishes. Ichthyological Society of Japan, Tokyo, pp 86–93

Olsson R (1992) Reconstructing the 'Eochordate' brain. J Gen Biol (Moscow) 53:350–361

Olsson R (1993) Reissner's fiber mechanisms: some common denominators. In: Oksche A, Rodríguez EM, Fernández-Llebrez P (eds) The subcommissural organ. Springer, Berlin Heidelberg New York, pp 33–39

Olsson R, Wingstrand KG (1954) Reissner's fiber and the infundibular organ in amphioxus. Publ Biol Station Bergen 14:3–14

Reissner E (1860) Beiträge zur Kenntnis vom Bau des Rückenmarkes von Petromyzon fluviatilis. Arch Anat Physiol Wiss Med, Leipzig, pp 545–588

Retzius G (1891) Zur Kenntniss des Centralnervensystems von Amphioxus lanceolatus. Biol Untersuch NF 2:29–46

Rohde E (1888a) Histologische Untersuchungen über das Nervensystem von Amphioxus. Zool Anz 11:190–196

Rohde E (1888b) Histologische Untersuchungen über das Nervensystem von Amphioxus lanceolatus. Schneiders Zool Beitr 2:169–211

Rohon JV (1882) Untersuchungen über Amphioxus lanceolatus. Denkschr Kaiserl Akad Wiss Wien 45:1–64

Rolph W (1876) Untersuchungen über den Bau des Amphioxus lanceolatus. Morphol Jahrb 2:87–164

Ruiz MS, Anadón R (1989) Some observations on the fine structure of the Rohde cells of the spinal cord of the amphioxus, Branchiostoma lanceolatum (Cephalochordata). J Hirnforsch 30:671–677

Ruiz S, Anadón R (1991a) The fine structure of lamellate cells in the brain of amphioxus (Branchiostoma lanceolatum, Cephalochordata). Cell Tissue Res 263:597–600

Ruiz MS, Anadón R (1991b) Ultrastructural study of the filum terminale and caudal ampulla of the spinal cord of amphioxus (Branchiostoma lanceolatum Pallas). Acta Zool (Stockh) 72:63–71

Sahlin K (1988) Gastrin/CCK-like immunoreactivity in Hatschek's groove of Branchiostoma lanceolatum (Cephalochordata). Gen Comp Endocrinol 70:436–441

Schneider A (1879) Beiträge zur vergleichenden Anatomie und Entwicklungsgeschichte der Wirbeltiere. Reimer, Berlin

Sterba G (1969) Morphologie und Funktion des Subcommissuralorgans. In: Sterba G (ed) Zirkumventriculäre Organe und Liquor. Fischer, Jena, pp 17–32

Sterba G, Fredriksson G, Olsson R (1983) Immunocytochemical investigations of the infundibular organ in amphioxus (Branchiostoma lanceolatum; Cephalochordata). Acta Zool (Stockh) 64:149–153

Tagliani G (1897) Considerazioni morfologiche intorno alle cellule nervose colossali dell' Amphioxus lanceolatus e alle cellule nervose giganti del midollo spinale di alcuni Teleostei. Monit Zool Ital 8:264–275

Ten Cate J (1938a) Zur Physiologie des Zentralnervensystems des Amphioxus (Branchiostoma lanceolatus). I. Die reflektorische Tätigkeit des Amphioxus. Arch Neerl Physiol 23:409–415

Ten Cate J (1938b) Contribution à la physiologie du système nerveux central du Amphioxus (Branchiostoma lanceolatus). II. Les mouvements ondulatoires et leurs innervations. Arch Neerl Physiol 23:416–423

Tjoa LT, Welsch U (1974) Electron microscopical observations on Kölliker's and Hatschek's pit and on the wheel organ in the head region of amphioxus (Branchiostoma lanceolatum). Cell Tissue Res 153:175–188

Uemura H, Tezuka Y, Hasegawa C, Kobayashi H (1994) Immunohistochemical investigation of neuropeptides in the central nervous system of the amphioxus, Branchiostoma belcheri. Cell Tissue Res 277:279–287

von Kupffer K (1893) Studien zur vergleichenden Entwicklungsgeschichte des Kopfes der Cranioten, Heft 1. Lehman, Munich, pp 1–95

von Kupffer K (1906) Die Morphogenie des Centralnervensystems. In: Hertwig O (ed) Handbuch der vergleichenden und experimentellen Entwicklungsgeschichte der Wirbeltiere, vol II, part 3. Fischer, Jena, pp 1–272

Vallet PG, Ody MG (1985) Oxytocinergic-like cells and fibers in the nervous system of amphioxus (Branchiostoma lanceolatum Pallas). Experientia 41:776–777

Von Ubisch LV (1937) Ist das 'Gehirn' von Branchiostoma primitiv oder rudimentar? Z Wiss Zool 150:155–187

Wada H, Satoh N (1994) Details of the evolutionary history from invertebrates to vertebrates, as deduced from the sequences of 18S rDNA. Proc Natl Acad Sci U S A 91:1801–1804

Watanabe T, Yoshida M (1986) Morphological and histochemical studies on Joseph cells of amphioxus, Branchiostoma belcheri Gray. Exp Biol 46:67–73

Webb JE (1969) On the feeding and behavior of the larva of Branchiostoma lanceolatum. Mar Biol 3:58–72

Welsch U (1968) Die Feinstruktur der Josephschen Zellen im Gehirn von Amphioxus. Z Zellforsch 86:252–261

Whiting HP (1955) Functional development in the nervous system. In: Waelsch H (ed) Biochemistry in the developing nervous system. Academic Press, New York, pp 85–103

Wickstead JH, Bone Q (1959) Ecology of acraniate larvae. Nature 184:1849–1851

CHAPTER 10

Lampreys, Petromyzontoidea

R. NIEUWENHUYS AND C. NICHOLSON

10.1
Introduction

The lampreys represent the most primitive group of presently living vertebrates. They are water inhabitants with elongated, eel-like bodies which lack paired fins (Fig. 10.1). In contrast to amphioxus, the head of the lamprey bears a number of special sense organs (nose, eyes, ears). The information gathered by these organs is relayed over the cranial nerves to centres in the enlarged rostral part of the CNS. There is a single nasal orifice high on top of the head and slightly behind this opening; a patch of pigment-free skin marks the position of the well-developed third or pineal eye. The animals lack jaws, having instead a large disc-shaped sucking mouth with many horny teeth. Many, but not all adult lampreys are predacious. The predacious varieties attach themselves to fish using their sucking mouths; then they produce a wound by rasping movements of a tongue-like structure which bears numerous sharp denticles. Finally, the lamprey ingests the blood and tissue fragments of its prey.

The body musculature of lampreys consists of a bilateral series of segmentally arranged myotomes, within which the muscle fibres are arranged in subunits, each comprising several internal sheets of central fibres and an external sheet of parietal fibres (Hardisty and Rovainen 1982). Physiologically, the

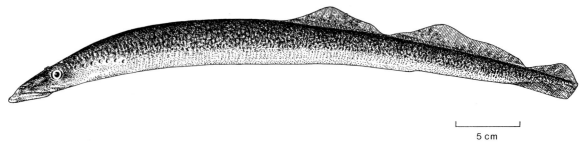

5 cm

Fig. 10.1. Lake lamprey, *Petromyzon marinus* (modified from Gage 1928)

parietal fibres are classified as of the slow type, whereas the central fibres are of the twitch type (Terävainen 1971). Lampreys swim well; like amphioxus and many fish, they swim by producing myotomal contractions that alternate in opposite segments and sweep caudally along the body (Rovainen 1979b; Bowtell and Williams 1991).

The biology of the lampreys has been described at length by Hardisty (1979) and in the multi-volume series entitled *The Biology of Lampreys* edited by M.W. Hardisty and I.C. Potter and published by Academic Press, London (Hardisty and Potter 1982). The distribution of lampreys is limited to the temperate zones of both hemispheres, where they inhabit both salt and fresh water. During the spring, lampreys migrate to shallow-water streams to spawn. There, they locate a rocky suitable bottom and excavate a nest by removing stones with their sucking mouths. Lampreys spawn but once and then die. In about 2 or 3 weeks the eggs hatch, and several days later the larvae move out of the nest. The larvae differ so much in structure and mode of life from the adults that they were once considered a separate species: *Ammocoetes branchialis*; hence, lamprey larvae are still designated for short as ammocoetes. After an initial 'proammo-coete' stage, i.e. the time during which the yolk is absorbed, the wormlike, blind and toothless ammo-coetes move to silted areas of brooks and streams, where they live in tubes with only their mouth protruding. They are filter feeders, using this mechanism to strain small organisms, much as does amphioxus. After several years of larval life (Rubinson 1990 noted, however, that in *Petromyzon marinus* metamorphosis does not usually occur before the fifth year and, remarkably, may be delayed until as late as the 18th year), the ammocoetes enter metamorphosis and a number of dramatic changes occur. The mouth, which is initially horseshoe-shaped with an upper and lower lip, transforms into a disc-shaped sucker mouth, while in the throat a rasping, tongue-like structure appears and, most essential for the location of potential prey, the pre-

viously rudimentary olfactory organ and lateral eyes differentiate and become functional. The transition from larva to adult occurs in about 2 months, during which the animals remain burrowed under the sand and gravel for protection. Following metamorphosis, the young lampreys migrate downstream to commence the active, predacious phase of their life history. This predacious phase continues for 1–4 years; then the animals return to their birthplace to lay their eggs for a new generation. However, certain species of lampreys never prey upon fish, but reproduce and die shortly after metamorphosis; these are the so-called brook lampreys (e.g. *Lampetra planeri*) that remain in fresh water.

From the beginning of comparative neurobiology, the CNS of the lamprey has attracted the attention of numerous researchers. Owing to its allegedly primitive characteristics and simplicity, it was expected that analysis of the brain and spinal cord of these forms would yield the 'prototype' of the vertebrate neuraxis (Ariëns Kappers 1929; Heier 1948). Hence, many so-called evolutionary trends such as progressive migration, differentiation, segregation and specialisation of neurons and neuron groups were believed to originate in lamprey-like brains. It is known, however, that the lampreys have a long, independent phylogenetic history, and no recent group of vertebrates, let alone the entire vertebrate kingdom, can be said to have arisen from them. Nevertheless, the lampreys do retain more features of the presumed ancestral craniates than do any other members of the group and therefore deserve special attention. Their CNS may not embody the 'prototype' of the craniote brain; however, it does offer an excellent starting point for the study of the development, structure and function of the vertebrate neuraxis.

Our ensuing survey is based on a review by one of us (Nieuwenhuys 1977) and on an analysis of Nissl-stained and silver-impregnated material of the river lamprey, *Lampetra fluviatilis*. These data are supplemented by the results of numerous recent experimental neuroanatomical and physiological studies from a number of laboratories. Due attention has been paid to the pioneering investigations of Carl Rovainen at Washington University, St Louis, and the subsequent extraordinarily extensive functional studies carried out by Sten Grillner and his associates at the Karolinska Institute in Stockholm, together with histological analyses by Tomas Hökfelt and colleagues at the same institution.

It should be noted that our present-day knowledge of the lamprey CNS is based on the study of many separate species. It stands to reason that these species differ sufficiently to be distinguished by the taxonomist. However, neuroanatomically there are no appreciable differences among them; consequently, we will use the term 'lamprey' generally without any further qualification.

10.2
Gross Features of the CNS

In the lamprey the spinal cord is by far the largest part of the CNS. It is a flattened, ribbon-like structure which extends throughout the length of the vertebral canal. The brain is slender and extremely small (Figs. 10.2, 10.3). In adult specimens of *Lampetra fluviatilis* its total length is approximately 8 mm whereas its greatest width and height are about 2.5 mm. Notwithstanding its small dimensions, the principal parts of the vertebrate neuraxis, namely, the rhombencephalon, mesencephalon, diencephalon and telencephalon, can be readily identified in the lamprey (Fig. 10.3). The cerebellum is represented by a small, transversely oriented plate of tissue (Fig. 10.3a,d: 40). The telencephalon is tilted upward, a displacement which is usually ascribed to the enhanced development of the olfactory organ. The dorsal surface of the brain is largely covered by extensive choroid plexuses, which are attached to the neural walls of the rhombencephalon, mesencephalon, diencephalon and telencephalon. The rhombencephalic and mesencephalic plexuses are richly folded (Fig. 10.3d). The mesencephalic plexus is a unique feature of lampreys; it does not occur in any other group of vertebrates. The membranous diencephalic roof forms a dorsal expansion, which is known as the saccus dorsalis (Fig. 10.3d). The well-developed pineal and parapineal organs rest upon the dorsal surface of this structure. The telencephalon is partly evaginated into a pair of hollow lobes. Small interventricular foramina (Fig. 10.3d) connect the large, unpaired telencephalic ventricle (i.e. the direct rostral continuation of the third ventricle) with the lateral ventricles. Remarkably, the olfactory bulbs exceed the cerebral hemispheres in size.

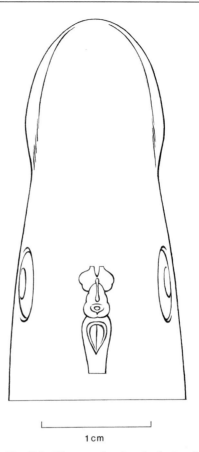

Fig. 10.2. Diagram showing the brain of the lamprey *Lampetra fluviatilis* in position

Distinct dorsal and ventral roots emerge from the spinal cord, but these do not join to form spinal nerves. The cranial nerves show the typical vertebrate pattern, except for the abducens nerve, which exits from the rhombencephalon ventrolaterally, in close association with the motor trigeminal root, rather than ventrally as in gnathostomes (Fig. 10.3b,c). The pineal and parapineal organs are connected with the habenular region of the diencephalon by a pair of stalks, which are known as the pineal and parapineal nerves. Like the optic nerves these structures are not true nerves but rather tubular outgrowths of the brain wall.

10.3
Development and Overall Histological Pattern

10.3.1
Early Development

In the lamprey the CNS originates as a median ingrowth of the dorsal ectoderm which gradually detaches itself from the surface to form a cord of

1　nervus olfactorius
2　telencephalon
3　bulbus olfactorius
4　fissura circularis
5　cerebral hemisphere
6　primordium hippocampi
7　foramen interventriculare
8　recessus preopticus
9　lamina terminalis
10　commissura interbulbaris
11　commissura anterior
12　diencephalon
13　ventriculus tertius
14　pineal organ
15　parapineal organ
16　nervus pinealis
17　nervus parapinealis
18　tela chorioidea diencephali
19　saccus dorsalis
20　ganglion habenulae
21　commissura habenulae
22　thalamus
23　hypothalamus
24　nervus opticus
25　chiasma opticum
26　commissura postoptica
27　recessus postopticus

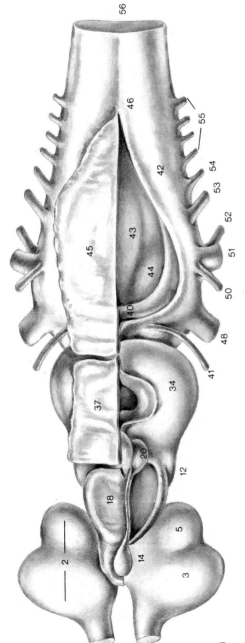

a

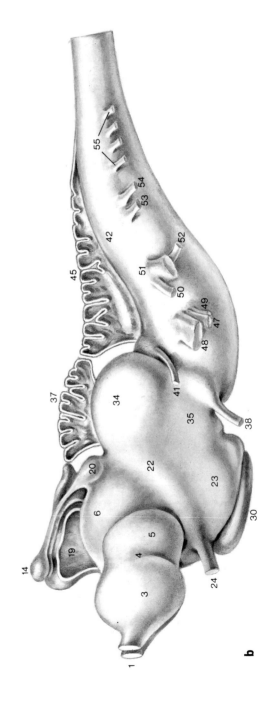

b

28 infundibulum
29 neurohypophysis
30 adenohypophysis
31 commissura posterior
32 tuberculum posterius
33 mesencephalon
34 tectum mesencephali
35 tegmentum mesencephali
36 ventriculus mesencephalicus
37 tela chorioidea mesencephalica
38 nervus oculomotorius
39 isthmus rhombencephali
40 cerebellum
41 nervus trochlearis
42 rhombencephalon
43 ventriculus quartus
44 eminentia trigemini
45 tela chorioidea rhombencephali
46 obex
47 nervus trigeminus, radix motoria
48 nervus trigeminus, radix sensoria
49 nervus abducens
50 nervus lineae lateralis anterior
51 nervus octavus
52 nervus facialis
53 nervus lineae lateralis posterior
54 nervus glossopharyngeus
55 nervus vagus
56 medulla spinalis

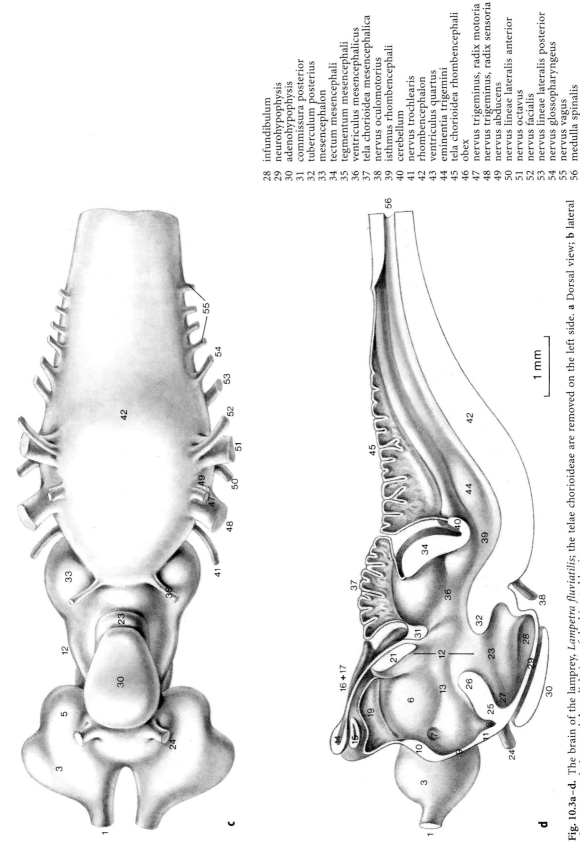

Fig. 10.3a–d. The brain of the lamprey, *Lampetra fluviatilis*; the telae chorioideae are removed on the left side. **a** Dorsal view; **b** lateral view; **c** ventral view; and **d** medial view of the bisected brain.

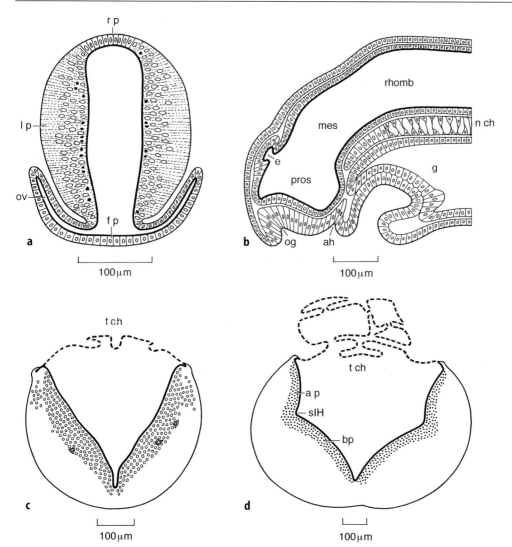

Fig. 10.4a–d. The development of the brain of the lamprey, *Lampetra planeri.* **a** Transverse section through the prosencephalon of a 7-day embryo. **b** Median section through the brain of a 3-mm larva. **c** Transverse section through the rhombencephalon of a 20-mm larva. **d** Transverse section through the rhombencephalon of a 14-cm larva. *ah,* Ade- nohypophysial anlage; *ap,* alar plate; *bp,* basal plate; *e,* epi- physial anlage; *fp,* floor plate; *g,* gut; *lp,* lateral plate; *mes,* mesencephalon; *nch,* notochord; *og,* olfactory groove; *ov,* optic vesicle; *pros,* prosencephalon; *rhomb,* rhombencepha- lon; *rp,* roof plate; *slH,* sulcus limitans of His, *tch,* tela chorioi- dea. (Modified from von Kupffer 1906, Figs. 42, 43, 55, 64)

solid cells. After the separation from the ectoderm has been completed, a central ventricular cavity is secondarily formed by the confluence of a series of clefts which have arisen independently from each other (von Kupffer 1906). As in all vertebrates the immature neural tube consists of a pair of lateral plates that are connected dorsally by a roof plate and ventrally by a floor plate (Fig. 10.4a). The roof and floor plates remain thin, but the lateral plates thicken considerably in most places. Initially, they consist of a pseudostratified epithelium, but after a few days a wide ventricular matrix layer can be del- imited from a narrow marginal zone. Mitoses are found exclusively in the proximity of the ventricular surface (von Kupffer 1906; Pfister 1971a). Before hatching, the first neuroblasts appear at the exter- nal border of the matrix layer and the first sets of functioning neurons are already present in the CNS of young proammocoetes. Whiting (1948) analysed these precociously differentiating elements at the spinal level and found: (a) dorsal cells, (b) large internuncial cells and (c) primitive motoneurons (Fig. 10.5). The dorsal cells constitute two rows, one on each side of the midline. Each of these neurons has three large processes; one fibre proceeds ros- trally, another caudally, whereas the third process

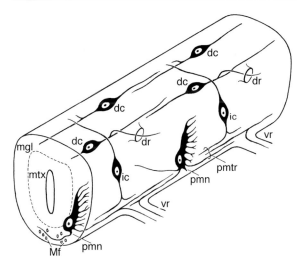

Fig. 10.5. Stereogram of the spinal cord of a proammocoete, showing the relations between the neurons described in the text. *dc*, dorsal cell; *dr*, dorsal root; *ic*, internuncial cell; *Mf*, Müller fibres; *mgl*, marginal layer; *mtx*, matrix layer; *pmn*, primitive motor neurons; *pmtr*, primitive motor tract; *vr*, ventral root. (Based on Whiting 1948, Fig. 17 and Bone 1960, Fig. 7)

leaves the cord at its dorsolateral edge and goes to a muscle segment, or to the skin or both. These elements most probably represent the Rohon-Beard cells, which persist in the adult lamprey. The large internuncial cells occupy a lateral position. Their

dendrites are in a clear relationship with the axons of the dorsal cells of both sides of the cord. Their axons pass ventrally and join the longitudinal motor tract. These large internuncial elements usually occur singly, one per segment. The ventrolaterally situated primitive motoneurons have many laterally extending dendrites, while their main axons participate in the formation of a ventrolaterally situated motor tract. The peripheral fibres to the motor roots are given off as collaterals of these longitudinally passing fibres. Not only the spinal elements just discussed, but also the large Müller and Mauthner cells of the brain stem and their descending axons differentiate very early in the lamprey (Whiting 1948, 1957) as do the so-called giant interneurons of the cord (Whiting 1948; Rovainen 1979b). It may be assumed that the three cell types depicted in Fig. 10.5 together constitute a primitive somatic sensorimotor reflex arc and that the large Müller and Mauthner cells coordinate the earliest swimming movements under the control of the brain. Nakao and Ishikawa (1987a–d) studied the development of the caudal spinal cord in larvae of *Lampetra japonica* of 13-mm and 21-mm total length at the ultrastructural level (Fig. 10.6a). The caudal cord of 13-mm larvae contained three types of differentiated neurons: Rohon-Beard cells, internuncial cells, and primitive motor neurons. Two axonal bundles, the large dorsolateral bundle (dlb)

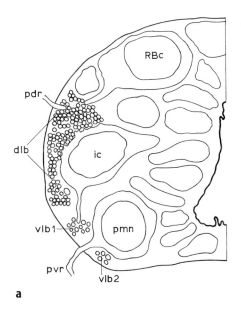

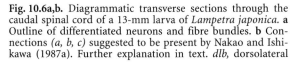

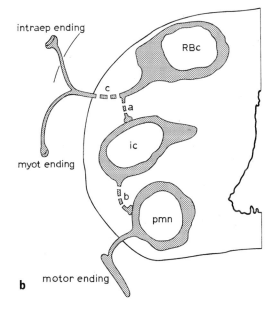

Fig. 10.6a,b. Diagrammatic transverse sections through the caudal spinal cord of a 13-mm larva of *Lampetra japonica*. a Outline of differentiated neurons and fibre bundles. b Connections *(a, b, c)* suggested to be present by Nakao and Ishikawa (1987a). Further explanation in text. *dlb*, dorsolateral bundle; *ic*, internuncial cells; *intraep*, intraepidermal; *myot*, myotomal; *pdr*, primitive dorsal root; *pmn*, primitive motor neuron; *pvr*, primitive ventral root; *RBc*, Rohon-Beard cell; *vlb*, ventrolateral bundles. (Based on electron micrographs and diagrams in Nakao and Ishikawa 1987a)

and the much smaller ventrolateral bundle (vlb) could be distinguished. In the caudal cord the vlb divided into two subsystems, vlb1 and 2. Primitive dorsal (pdr) and ventral roots (pvr) were present.

The Rohon-Beard cells contribute a process to the dlb, which bifurcates into two processes that extend rostrally and caudally within that bundle. The Rohon-Beard cells appeared to be the only neurons that contribute longitudinal processes to the dlb. The dlb gives rise to fibres which emerge segmentally at the levels of the myosepta from the cord, thus forming the primitive dorsal roots already mentioned. The pdr fibres bifurcate and terminate in two different ways; those of one group pierce the dermis to end as intraepidermal free endings, whereas those of the other group form varicosities that lie within depressions on the surface of myotomes.

The internuncial cells are closely associated with the dlb. Their dorsolateral surface has numerous small dendritic projections, which extend into the dlb. Nerve endings contained within the dlb make synaptic junctions with these dendritic projections as well as on the intervening surface of the soma. The internuncial cells send their axonal process ventrally to the vlb.

The primitive motor neurons are directly associated with the vlb and receive numerous synaptic

nerve endings from it. Their axons leave the cord, contributing to the primitive ventral roots. They form synaptic junctions or nonsynaptic contacts with myotomal muscle cells.

Nakao and Ishikawa (1987a) proposed a possible neural chain between Rohon-Beard cells, intercalated cells and primitive motoneurons and they also conjectured that the fibres forming the primitive dorsal roots are extramedullary processes of the Rohon-Beard cells (Fig. 10.6b).

Some neuroepithelial cells in the dorsal wall of the caudal neural tube migrate into the extramedullary space as free cells and probably correspond to neural crest cells and participate in the formation of the spinal ganglia (Nakao and Ishikawa 1987b). The central processes of the developing spinal ganglion cells enter the spinal cord along the fibres of the primitive dorsal roots, thus forming the definitive dorsal roots (Nakao and Ishikawa 1987c,d). The primitive dorsal root fibres are temporary structures which, according to Nakao and Ishikawa (1987a,b), have completely disappeared from the spinal cord of 21-mm larvae. This observation raises the question as to how the early embryonic Rohon-Beard cells relate to the dorsal cells observed by previous authors in the spinal cord of ammocoetes and adult lampreys (Figs. 10.5, 10.11–10.13). Nakao and Ishikawa (1987a) sug-

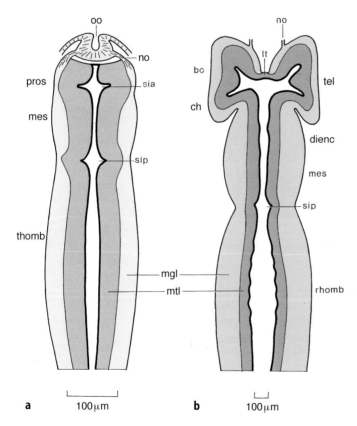

Fig. 10.7a,b. Horizontal sections through the brains of larval lampreys, *Lampetra planeri.* **a** Specimen of 6 mm. **b** Specimen of 55 mm. *bo,* bulbus olfactorius; *ch,* cerebral hemispheres; *dienc,* diencephalon; *lt,* lamina terminalis; *mes,* mesencephalon; *mgl,* marginal layer; *mtl,* mantle layer; *no,* nervus olfactorius; *oo,* olfactory organ (unpaired!); *pros,* prosencephalon; *rhomb,* rhombencephalon; *sia,* sulcus intraencephalicus anterior; *sip,* sulcus intraencephalicus posterior; *tel,* telencephalon. (Modified from von Kupffer 1906, Figs. 48, 56)

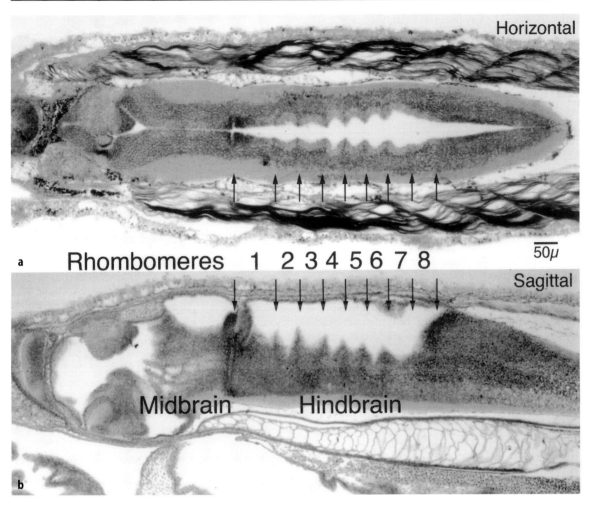

Horizontal

a   Rhombomeres   1 2 3 4 5 6 7 8 $\overline{50\mu}$

Sagittal

Midbrain   Hindbrain

b

Fig. 10.8a,b. Rhombomeres in brain of 9- to 12-mm larva of *Petromyzon marinus.* **a** Horizontal view (reduced silver stain). **b** Sagittal view (phenol red stain). Rhombomeres are labelled *1–8.* (Courtesy of Dr. R. G. Baker, New York University Medical Center)

gested two possible scenarios for the fate of the Rohon-Beard cells: (a) By the 21-mm stage the Rohon-Beard cells have been completely replaced by newly formed dorsal cells. (b) The Rohon-Beard cells have become transformed into dorsal cells. The latter hypothesis would imply that the Rohon-Beard cells lose their peripheral offshoots during early ontogenesis and regrow these processes in a later developmental phase. The present authors consider it likely that the Rohon-Beard cells retain their peripheral processes and persist as dorsal cells in the adult stage.

10.3.2
Morphogenesis

Figure 10.5 shows that the spinal cord is initially more or less cylindrical in shape. During further development the cord is gradually transformed into a flattened, ribbon-like structure, a process which is completed before the end of the larval period. In the brain area the primordial prosencephalon widens and a slight flexura cranialis develops (Fig. 10.4b). However, this flexure soon disappears again; this is why during most of the larval period all parts of the brain can be surveyed in single horizontal sections (Fig. 10.7). During the time when the early brain anlage remains in the form of a tube, two deep, transverse ventricular sulci appear; these are known as the sulcus intraencephalicus anterior and posterior. These sulci mark the boundaries between, respectively, the telencephalon and the diencephalon, and the mesencephalon and the rhombencephalon (von Kupffer 1906). The presence of neuromeres in the brain of larval lampreys has been reported by several authors. According to von Kupffer (1906), these structures are confined to the rhombencephalon (cf. Fig. 10.7b), but Bergquist and Källén (1953a,b) observed them throughout the brain (cf. Chap. 4, Fig. 4.21). Gilland and Baker (1995) recently studied the development of the hindbrain in larvae of *Petromyzon marinus*, using serial sections, scanning electron microscopy and micro injection of fluorescent dyes and biocytin for neuronal tracing. They observed 8 rhombomeres (numbered Rh1-Rh8) in 9- to 12-mm larvae and indicated how the various cranial nerve nuclei and large reticular cells relate to the rhombomeres (Fig. 10.8). They noticed that some of the embryonic features of segmental organisation persist throughout larval and adult periods, and claimed that "rhombomeres provide not merely a blueprint for connectional development, but also a stable neuroepithelial framework for functional circuitry" (Gilland and Baker 1995, p. 779).

The evagination of the telencephalon (Fig. 10.7b) and the divergence of the lateral walls in the rhombencephalic area (Fig. 10.4c,d) are salient features of the later morphogenesis of the brain. A longitudinal sulcus limitans, dividing the lateral walls into a dorsal alar plate and a ventral basal plate, appears late in development (Fig. 10.4d). It is noteworthy that neither in larvae (von Kupffer 1906) nor in the adult stage (Nieuwenhuys 1972; Fig. 10.30) does this groove extend beyond the rhombencephalon.

10.3.3
Histogenesis

The histogenesis of the CNS of the lamprey follows the general vertebrate pattern outlined in Chap. 5. The matrix layer is very wide initially (Fig. 10.4a), but gradually decreases in thickness and ultimately disappears completely. It seems generally true that the proliferative activity ceases prior to metamorphosis (Rovainen 1979b), but the time at which this complete exhaustion of the matrix is attained differs from place to place. The studies of Pfister (1971b,c) revealed that in two centres, the telencephalic primordium hippocampi and the mesencephalic tectum, the matrix continues to proliferate for a particularly long time. The prolonged mitotic activity in the hippocampal primordium is not well understood, but in the tectum this phenomenon is doubtless related to the late outgrowth and differentiation of this primarily visual centre (Kennedy and Rubinson 1977, 1978; Rubinson 1990; Fig. 10.39). As stated in the introduction, it is only after metamorphosis that the visual system of the lamprey becomes functional.

10.3.4
Overall Histological Pattern

In the CNS of the lamprey most of the neuronal perikarya are located in a continuous zone of central grey (Figs. 10.11–10.23). The embryonic mantle layer retains its structural appearance in the adult animal. Most of the somata in this central grey zone are small and granular, but the layer as a whole is not homogeneous. A considerable number of cell masses can be delimited within it. Most appear as local condensations of small cells, but some are composed of larger elements. A characteristic feature of the lamprey CNS is the occurrence of some very prominent giant neurons which are confined to the brain stem (Figs. 10.15, 10.16, 10.18, 10.19, 10.30, 10.34) and spinal cord. The positional relations, structure and functions of these giant elements were thoroughly analysed by Rovainen and his collaborators (Rovainen 1967a,b, 1974a,b, 1978, 1979b; Rovainen et al. 1973). Golgi studies (Johnston 1902, 1912; Heier 1948; Larsell 1967; Figs. 10.29, 10.42) have revealed that some of the small neurons are confined to the grey zone, but that most elements extend their dendrites into the fibre zone located at the periphery of the neuraxis. Fibres of very variable diameters occur in this zone, all of which are devoid of myelin sheaths. Many previous researchers, among them Johnston (1902), Tretjakoff (1909a,b), Ariëns Kappers (1920), Stefanelli (1934), Pearson (1936), Heier (1948) and Larsell (1967), have attempted to trace fibre systems through the 'white' matter of the brain and cord of the lamprey using normal silver-impregnated or methylene blue-stained material. Numerous tracts and diffuse pathways were described by these authors, but during the past decade, many early studies have been superseded by experimental neuroanatomical studies. We will give preference to experimentally established hodological data, but the most salient findings of previous investigators will also be mentioned.

Two general comments should be made about the 'white' matter. The first is that, although the great majority of the neuronal somata in the lamprey's neuraxis are located in a zone of central grey, a number of neurons and neuron groups have migrated away from the ventricular surface and are embedded in the stratum 'album'. The second is that, since most neurons extend their dendrites into the peripheral fibre zone, where all fibres are unmyelinated, potentialities for synaptic relationships exist throughout that zone, including both dendrites and perikarya. Thus, Owsiannikow (1903) observed that large, ventrally situated cells in the cord are surrounded by a dense meshwork of fine fibres, and the electron-microscopic studies of

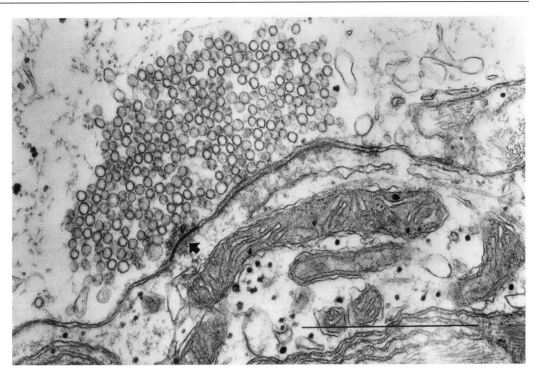

Fig. 10.9. Electron micrograph of a synapse from a Müller axon *(above)* to an unidentified dendrite *(below)*. The thickened region with close apposition of pre-and postsynaptic membranes *(arrow)* is a possible gap junction. (Courtesy of Dr. C. M. Rovainen, Washington University School of Medicine)

Schultz et al. (1956) revealed that the surface of these, as well as the large spinal dorsal cells, exhibit many irregularities and indentations that are apparently sites of synaptic connections. Moreover, Stefanelli and Caravita (1970) reported that the perikarya of the large neurons present in the vestibular nuclei envelop expansions of the incoming, primary afferent vestibular fibres.

10.3.5
Synapses

Work on the ultrastructure of synapses in the lamprey CNS has been focussed mainly on the contacts made by Müller fibres and other large axons. In the spinal cord these fibres make typical chemical synaptic junctions, characterised by the presence of clusters of vesicles and local membrane thickenings (Bertolini 1964; Smith et al. 1970; Christensen 1976; Rovainen 1974a,b; Wickelgren 1977a; Wickelgren et al. 1985). The synaptic vesicles associated with the Müller fibres are round, whereas in some other coarse and medium-sized fibres flattened vesicles occur (Smith 1971; Pfenninger and Rovainen 1974). Many synapses in the CNS of the lamprey are morphologically mixed, i.e. they contain both the specialisations for chemical transmission and small areas where the pre- and postsynaptic membranes are closely apposed (Fig. 10.9). Such gap junctions have been observed between Müller fibres and postsynaptic dendrites (Rovainen 1974b), between giant interneurons (Rovainen 1974a; Christensen 1976) and between primary afferent vestibular fibres and cells in the vestibular nuclei (Stefanelli and Caravita 1970). These mixed synapses constitute the morphological substrate of the dual electrical-chemical excitatory postsynaptic potentials (EPSPs) that have been recorded from various elements in the brain and cord of the lamprey (Batueva and Shapovalov 1977a,b; Christensen 1983; Brodin et al. 1994).

The transmitters at many chemical synapses have been identified. They are detailed in Table 10.3 (see also Brodin and Grillner 1990). Glutamate synapses predominate as the excitatory transmitter and GABA and glycine as inhibitory transmitters, with acetylcholine being the mediator at the neuromuscular junction. A variety of monoamines and neuropeptides have also been identified using immunohistochemistry (Table 10.3). The complex properties of the different NMDA and AMPA receptors that constitute glutamate responsiveness, as well as other transmitters, play important roles in the control and maintenance of the network activity underlying swimming (Grillner et al. 1991; Grillner and Matsushima 1991; Travén et al. 1993). In gen-

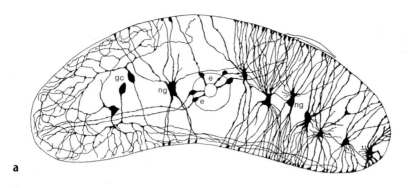

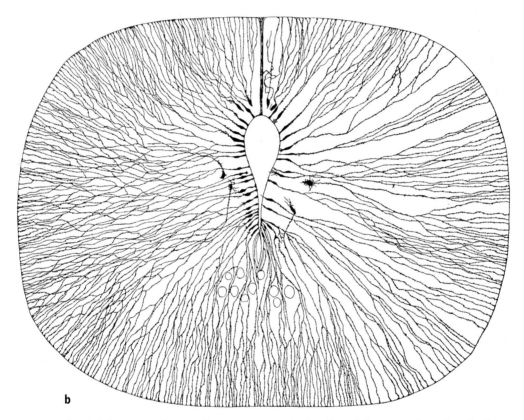

Fig. 10.10a,b. Glial elements in the central nervous system of the lamprey as seen in Golgi preparation. **a** Cross section through the caudal part of the spinal cord. **b** Cross section through the most caudal part of the rhombencephalon, showing the radial pattern of ependymal cells. *e,* ependymal cells; gc, ganglion cells; *ng,* neuroglial cells. (From Retzius 1893a, Table V, Fig. 1 and Table VI, Fig. 2)

eral, the transmitters of the lamprey follow a pattern similar to that established in other vertebrates.

There is evidence that, apart from classical synaptic neurotransmission, non-synaptic interneuronal communication also occurs in the CNS of the lamprey. Thus, Christenson et al. (1990) reported that serotonin-immunoreactive varicosities in the spinal cord were not seen to make synaptic contact with surrounding neuronal elements, even when followed through serial sections. This mode of action is substantiated by the finding that serotonin has complex effects on spinal circuitry, acting on calcium-dependent potassium channels at postsynaptic sites (Wikström et al. 1995) and by modulation of presynaptic glutaminergic transmission (Shupliakov et al. 1995). Further evidence for nonsynaptic interactions have been provided by Brodin et al. (1990b), who found that most neurotensin-immunoreactive terminal structures participating in the formation of the lateral marginal plexus of the spinal cord are not associated with synaptic specialisations. Electrophysiological studies (Barthe and Grillner 1995) revealed that neurotensin can modulate the activity of spinal neurons and alter the course of fictive locomotion.

10.3.6
Glia, Vascularisation and Extracellular Space

The glial cells in the lamprey's CNS were studied extensively by Retzius (1893a) with the Golgi technique. In the spinal cord he found two types of cells which he considered to be neuroglial and ependymal cells (Fig. 10.10a). The somata of the neuroglial cells are located in the narrow strip of spinal grey matter, from where they extend tufts of ramifying processes towards both the dorsal and the ventral surface of the cord. The spinal ependymal cells have a short central process and a long peripheral process. The former is in contact with the central canal, the latter branches only rarely and extends towards the surface. The glial system of the brain consists almost exclusively of ependymal cells. It is noteworthy that the peripheral processes of brain ependyma, contrary to those in the cord, branch profusely and have numerous small excrescences (Fig. 10.10b). At the external surface of the brain and cord the processes of the glial elements widen into end-feet, which, as an ensemble, constitute the membrana limitans gliae externae. At the spinal level, these end-feet are rich in mitochondria and glycogen granules (Bertolini 1964).

Ultrastructurally, lamprey glial cells contain dense accumulations of intermediate filaments (IFs; Bertolini 1964; Lurie et al. 1994). Merrick et al. (1995) presented immunohistochemical evidence indicating that the predominant IF proteins in glial cells throughout the nervous system of larval and adult lampreys are cytokeratins or related proteins. The presence of glial fibrillary acid protein (GFA)-immunoreactive material was not demonstrated (see, however, Wasowicz et al. 1994). The authors considered the widespread presence of glial cells containing keratin-like IFs in the lamprey a primitive feature, which may have significance for the extraordinary ability of lamprey spinal axons to regenerate. For data concerning the morphological and functional regeneration of the lamprey spinal cord after injury, the reader is referred to the reviews by Rovainen (1979b, 1982) and to the recent studies of McClellan and collaborators (McClellan 1988a,b, 1992, 1994; Davis and McClellan 1993, 1994a,b; Davis et al. 1993).

Blood vessels do not enter the spinal cord, and several authors, among them Tretjakoff (1909a,b) and Hibbard (1963a,b), believed that the peculiar ribbon-like appearance of the cord is associated with this absence of intramedullary blood vessels; the flattened shape would facilitate the access of oxygen and nutrients from the external vascular network. Against this view, Ariëns Kappers et al. (1936) argued that the spinal cord of myxinoids has both intramedullary blood vessels and a flattened shape. The brain of the lamprey is infiltrated by numerous looplike capillaries with an approximately radial pattern. The lamprey has a blood-brain barrier, located at the capillary endothelium and choroid plexus, that is very similar to the blood-brain barrier in other vertebrates (Bundgaard 1982; Bundgaard and Van Deurs 1982).

The extracellular space of the lamprey spinal cord assumes unusual significance because of the absence of vascularisation, so all metabolites and possibly some chemical signals must diffuse in from the external vascular network. Quantitative measurements of diffusion have not been made, but Wald and Selzer (1981) noted the rapid effect of bath-applied tetrodotoxin on neuronal activity in the isolated cord. Wald and Selzer (1981) did make careful measurements of the volume fraction of the extracellular space in the isolated cord, using radio-labelled inulin. They found that the extracellular volume fraction was 32%-33% in larval lampreys and 18%-19% in adults, and they hypothesised that the change in relative volume occurred with larval transformation to adulthood. These results are very similar to those in earlier studies of the developing and adult mammalian brain and to recent data obtained in the cerebral cortex of the rat using the tetramethylammonium method (Lehmenkühler et al. 1993). Thus the extracellular space of the lamprey spinal cord does not seem to exhibit any unusual characteristics that would compensate for a lack of vascularisation.

Abbreviations

adhypoph	adenohypophysis
anh	anterior part of neurohypophysis
aoctl	area octavolateralis
bol	bulbus olfactorius
ca	commissura anterior
cap	pl capillary plexus
cb	cerebellum
cc	canalis centralis
cd	commissura dorsalis
cgl	corpus geniculatum laterale
chab	commissura habenularum
chopt	chiasma opticum
cmsp	columna motoria spinalis
cp	commissura posterior
cpo	commissura postoptica
cpri	commissura preinfundibularis

cpoi	commissura postinfundibularis	mra	mesencephalic reticular area
crcb	crista cerebellaris	msah	meso-adenohypophysis
ctp	commissura tuberculi posterioris	mtah	meta-adenohypophysis
dc	dorsal cell(s)	Mth	cell of Mauthner
'dh'	'dorsal horn'	Mü,Mü1,etc.	cells of Müller
dhy	nucleus dorsalis hypothalami	neurhypoph	neurohypophysis
disthgr	dorsal isthmic grey	n pin	nervus pinealis
dmtn	dorsomedial telencephalic neuropil	nucb	nucleus cerebelli
dnVI	decussating fibres of nervus abducens	nucp	nucleus commissurae posterioris
doml	decussation of tractus octavomesencephalicus lateralis	nucpo	nucleus commissurae postopticae
		nudorshy	nucleus dorsalis hypothalami
domm	decussation of tractus octavomesencephalicus medialis	nufd	nucleus funiculi dorsalis
		nuflm	nucleus of the fasciculus longitudinalis medialis
dors	nucleus dorsalis areae octavolateralis		
dosm	decussation of tractus octavospinalis medialis	nuoma	nucleus octavomotorius anterior
		nuomi	nucleus octavomotorius intermedius
ec	edge cell	nuomp	nucleus octavomotorius posterior
effviii	efferent octavus cells	nupret	nucleus pretectalis
epend	ependyma	nurdV	nucleus of the radix descendens nervi trigemini
epith	epithalamus		
fai	fibrae arcuatae internae	nutp	nucleus tuberculi posterioris
fbt	fasciculus basalis telencephali (tr. olfacto-thalamicus et hypothalamicus + tr. striothalamicus)	nuventrhy	nucleus ventralis hypothalami
		nterm	nervus terminalis
		nI	nervus olfactorius
fcirc	fissura circularis	nII	nervus opticus
fd	funiculus dorsalis	nIII	nervus oculomotorius
fi	foramen interventriculare	nVm	nervus trigeminus, radix motoria
fl	funiculus lateralis	nVI	nervus abducens
flm	fasciculus longitudinalis medialis	nVIII	nervus octavus
fMth	fibre of Mauthner	octcb	tractus octavocerebellaris
fMü	fibres of Müller	olhyd	tractus olfacto-hypothalamicus dorsalis
fr	fasciculus retroflexus		
fv	funiculus ventralis	olhyv	tractus olfacto-hypothalamicus ventralis
gcrh	griseum centrale rhombencephali		
glol	glomeruli olfactorii	oll	tractus olfactorius lateralis
grc	granular cell(s)	olm	tractus olfactorius medialis
hab	ganglion habenulae	olnf	olfactory nerve fibres
hpmb	hippocampo-pretecto-midbrain bundle	olthd	tractus olfacto-thalamicus dorsalis
		olthv	tractus olfacto-thalamicus ventralis
hypoth	hypothalamus	oml	tractus octavomesencephalicus lateralis
int	nucleus intermedius areae octavolateralis	opt	tractus opticus
		optax	tractus opticus axialis
ip	nucleus interpeduncularis	optl	tractus opticus lateralis
isth	isthmus rhombencephali	osc	organon subcommissurale
lamt	lamina terminalis	pah	pro-adenohypophysis
lincb	tractus lineocerebellaris	parapin org	parapineal organ
lint	lateral interneuron	pdist	pars distalis of adenohypophysis
ll	lemniscus lateralis	phip	primordium hippocampi
lm	lemniscus medialis	pinorg	pineal organ
lobship	lobus subhippocampalis	pinf	nucleus postinfundibularis
lobX	lobus vagi	pnerv	pars nervosa of hypophysis
lsh	lobus subhippocampalis	pnh	posterior neurohypophysis
lsp	lemniscus spinalis	po	tractus postopticus (direct and crossed tectothalamic fibres)
mc	mitral cells		
mlr	mesencephalic locomotor region	ppald	primordium pallii dorsalis
mn	motoneuron	ppir	primordium piriforme

pr	nucleus preopticus
pretect	area pretectalis
ranllad	radix ascendens nervi lineae lateralis anterioris, pars dorsalis
ranllav	radix ascendens nervi lineae lateralis anterioris, pars ventralis
ranllp	radix ascendens nervi lineae lateralis posterioris
raVIII	radix ascendens nervi octavi
rdnll	radix descendens nervi lineae lateralis
rdnllp	radix descendens nervi lineae latera-lis posterioris
rdV	radix descendens nervi trigemini
rdVIII	radix descendens nervi octavi
ri	nucleus reticularis inferior
rinf	recessus infundibuli
rlt	recessus lateralis tecti
rm	nucleus reticularis medius
rmes	nucleus reticularis mesencephali
rmV	radix motoria nervi trigemini
rpo	recessus postopticus ventriculi tertii
rpost	recessus posterior ventriculi tertii
rs	nucleus reticularis superior
rsV	radix sensoria nervi trigemini
rv(1)	radix ventralis nervi spinalis (1)
rX	radix nervi vagi
s"a"c	stratum "album" centrale
saccd	saccus dorsalis
s"a"p	stratum "album" periventriculare
sep	septum
sfgs	stratum fibrosum et griseum superfi-ciale
sgc	stratum griseum centrale
sgp	stratum griseum periventriculare
siv	sulcus intermedius ventralis
slH	sulcus limitans of His
sm	stria medullaris
smi	sulcus medianus inferior
sms	sulcus medianus superior
sop	stratum opticum
ssh	sulcus subhabenularis = s. dience-phalicus dorsalis
str	corpus striatum
strth	tractus striothalamicus
t	taenia
tb	tractus tectobulbaris
tchmes	tela chorioidea mesencephali
tchrh	tela chorioidea rhombencephali
tchtel	tela chorioidea telencephali
tect	tectum mesencephali
teg, tegmes	tegmentum mesencephali
tegisth	tegmentum isthmi
tegl	nucleus tegmentalis lateralis
tegmot	tegmentum motoricum
thd	thalamus dorsalis = pars dorsalis thalami

thv	thalamus ventralis = pars ventralis thalami
tp	tuberculum posterius
tsc	torus semicircularis
vbol	ventriculus bulbi olfactori
velma	velum medullare anterius
ventr	nucleus ventralis areae octavolateralis
vhy	nucleus ventralis hypothalami
vimp	ventriculus impar telencephali
vl	ventriculus lateralis
vmes	ventriculus mesencephali
vq	ventriculus quartus
vt	ventriculus tertius
IIIc	nucleus nervi oculomotorii, pars caudalis
IIIi	nucleus nervi oculomotorii, pars intermedia
IIIp	nucleus nervi oculomotorii, pars periventricularis
IIIs	nucleus nervi oculomotorii, pars superficialis
IV	nucleus nervi trochlearis
Vm	nucleus motorius nervi trigemini
VI	nucleus nervi abducentis
VIdm	nucleus nervi abducentis, pars dorsomedialis
VIvl	nucleus nervi abducentis, pars ventrolateralis
VIIm	nucleus motorius nervi facialis
IX	nucleus motorius glossopharyngei
Xmc	nucleus motorius nervi vagi caudalis
Xmr	nucleus motorius nervi vagi rostralis
1, 2, etc.	Müller cells

10.4
Spinal Cord

10.4.1
Introductory Note

Just behind the obex the spinal cord is ellipsoidal in cross-section (Fig. 10.13), but throughout the rest of its extent this part of the neuraxis appears flattened and ribbon-like (Figs. 10.11, 10.12). Dorsal and ventral roots are present, but these do not emerge at the same transverse levels and do not unite to form mixed spinal nerves.

Electron microscopy of dorsal and ventral roots in the lamprey has shown that axons are unmyelin-ated and ensheathed individually by Schwann cells (Peters 1960) and measure about 2–20 µm in diam-eter. Small bundles of thin axons (less than 0.5 µm) are also present in the dorsal and ventral roots (Nakao and Suzuki 1978, 1980). These small axons, which bear varicosities with large dense-core vesi-cles, may be part of the loosely organised auto-

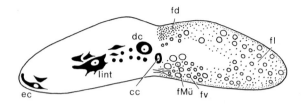

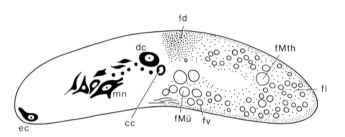

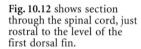

Figs. 10.11–10.14. A series of transverse sections through the spinal cord and brain of the lamprey, *Lampetra fluviatilis.* The *left half* of each figure shows the cell picture, based on Nissl-stained sections; the *right half* shows the fibre systems, based on adjacent Bodian-stained sections.

Fig. 10.11 shows section through the spinal cord at the level of the vent.

Fig. 10.12 shows section through the spinal cord, just rostral to the level of the first dorsal fin.

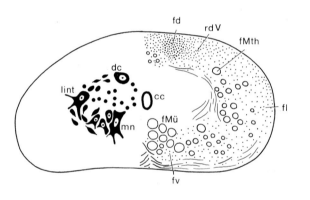

500 μm

Fig. 10.13 shows section through the spinal cord just caudal to the obex.

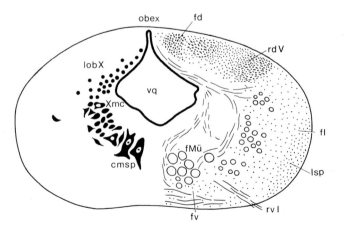

Fig. 10.14 shows section through the obex. For a list of the abbreviations used in figures, see p. 409

Figs. 10.15–17. A series of transverse sections through the spinal cord and brain of the lamprey *Lampetra fluviatilis.*

Fig. 10.15 shows a section through the caudal part of the rhombencephalon, at the level of entrance of the most rostral root of the vagus nerve.

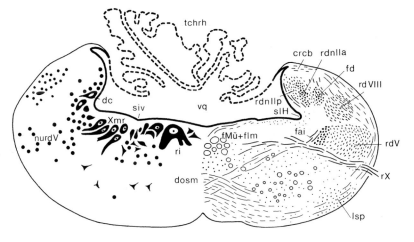

Fig. 10.16 shows a section through the intermediate part of the rhombencephalon, at the level of entrance of the eighth nerve.

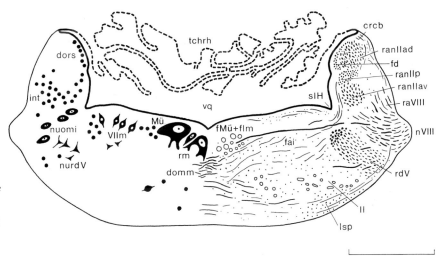

500 μm

Fig. 10.17 shows a section through the rostral part of the rhombencephalon, at the level of the motor nucleus and root of the fifth nerve. For a list of the abbreviations used in figures, see p. 409

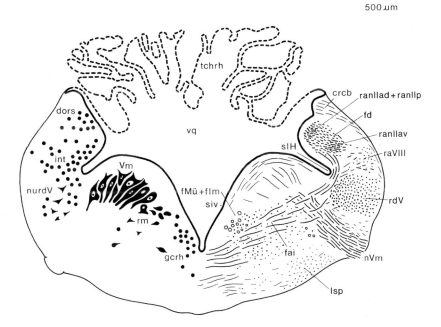

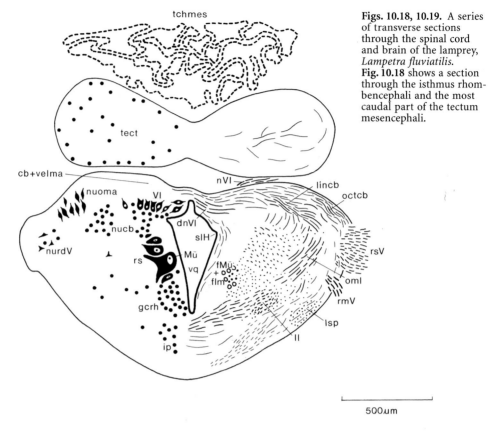

Figs. 10.18, 10.19. A series of transverse sections through the spinal cord and brain of the lamprey, *Lampetra fluviatilis.* **Fig. 10.18** shows a section through the isthmus rhombencephali and the most caudal part of the tectum mesencephali.

500 μm

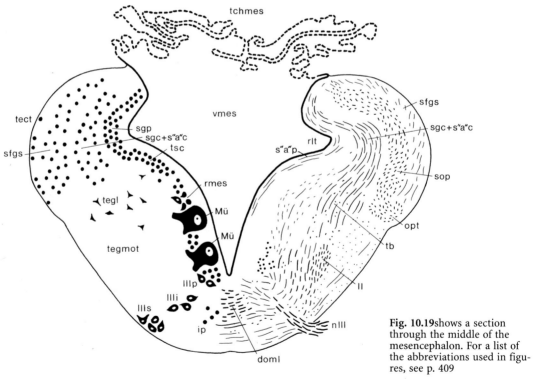

Fig. 10.19 shows a section through the middle of the mesencephalon. For a list of the abbreviations used in figures, see p. 409

Figs. 10.20, 10.21. A series of transverse sections through the spinal cord and brain of the **Fig. 10.20** shows a section through the commissura posterior.

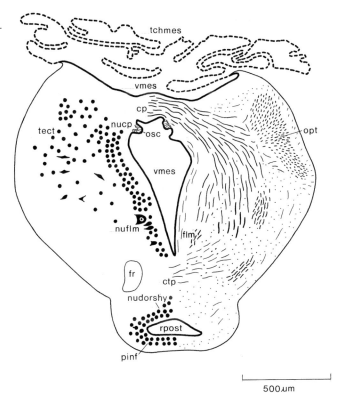

500 µm

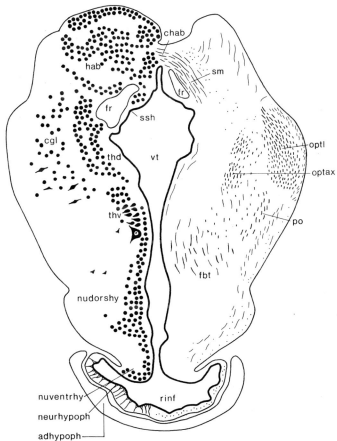

Fig. 10.21 shows a section through the diencephalon. For a list of the abbreviations used in figures, see p. 409

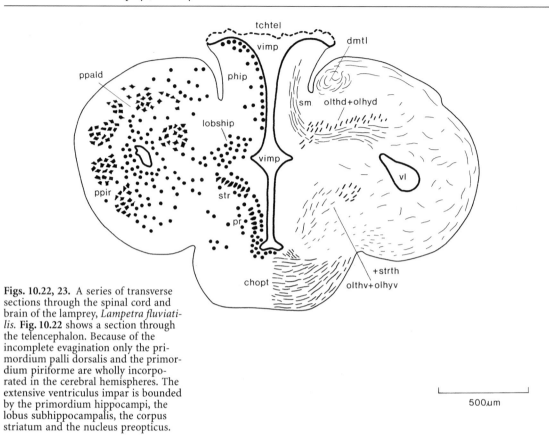

Figs. 10.22, 23. A series of transverse sections through the spinal cord and brain of the lamprey, *Lampetra fluviatilis*. **Fig. 10.22** shows a section through the telencephalon. Because of the incomplete evagination only the primordium palli dorsalis and the primordium piriforme are wholly incorporated in the cerebral hemispheres. The extensive ventriculus impar is bounded by the primordium hippocampi, the lobus subhippocampalis, the corpus striatum and the nucleus preopticus.

500μm

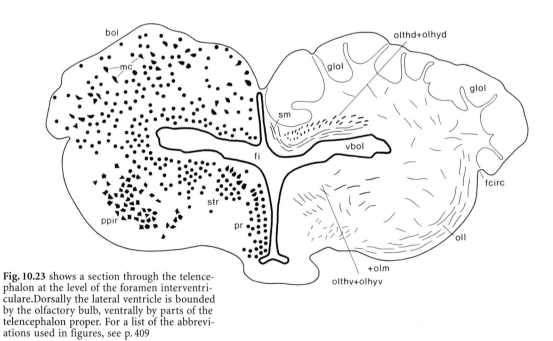

Fig. 10.23 shows a section through the telencephalon at the level of the foramen interventriculare. Dorsally the lateral ventricle is bounded by the olfactory bulb, ventrally by parts of the telencephalon proper. For a list of the abbreviations used in figures, see p. 409

nomic nervous system in the lamprey (see Nakao and Ishikawa 1982), but their origin is unknown. Rovainen and Dill (1984) counted the axons in electron-microscopic sections of ventral roots at branchial, midtrunk, precloacal and postcloacal levels in three different lamprey species. The number of large axonal profiles per ventral root in both *Ichthyomyzon unicuspis* and *I. castaneus* appeared to be about 70 and that in *Petromyzon marinus* about 100. The number of small axonal profiles per ventral root in *Ichthyomyzon* and *Petromyzon marinus* amounted to roughly 5. Knowing that motoneurons have one axon each in an ipsilateral ventral root, and assuming that all of the large, individual ensheathed axons in ventral roots of the lamprey are from motoneurons to myotomal or fin muscles, Rovainen and Dill (1984) used the data summarised above to estimate the number of motoneurons in the lamprey spinal cord. The total estimated numbers of myotomal and fin motoneurons were as high as 13 000 in the spinal cord of adult *Ichthyomyzon* and 24 000 in adult *Petromyzon*. The spinal cord comprises approximately 100 segments (Brodin and Grillner 1990).

Transverse sections through the spinal cord (Figs. 10.11–10.13) show that the grey substance forms bilateral winglike expansions, which are surrounded by a zone of 'white' matter comprising a mantle of longitudinally running fibres. The spinal grey substance contains cells differing greatly in size and shape. These include somatomotor neurons, dorsal cells and several types of intrinsic elements. Rovainen (1979b) estimated the total number of nerve cells in the lamprey spinal cord to be about 10^5, with approximately 500 per hemisegment.

10.4.2
Somatomotor Neurons

The somatomotor neurons together make up a distinct, laterally situated column on either side of the cord. Some of these elements are large, but most are of medium size (Terävänen and Rovainen 1971a). Separate motoneurons in the spinal cord innervate the slow, parietal fibres and the central twitch fibres (Terävänen and Rovainen 1971a). A third distinguishable class of motoneurons innervate the dorsal fins (Rovainen and Birnberger 1971). The parietal muscle fibres are true slow fibres since they do not generate action potentials, only excitatory junctional potentials (Terävänen 1971; Terävänen and Rovainen 1971b). Twitch fibres do have overshooting action potentials, with the interesting feature that only the outer fibres are innervated while the inner ones rely on electrical coupling for excitation

(Terävänen 1971). The fin muscles probably contain both fast and slow fibres (Rovainen and Birnberger 1971).

The dendritic trees of the myotomal motoneurons are large and often spread over the entire ipsilateral 'white' matter, except for the dorsal column (Terävänen and Rovainen 1971a; Wallén et al. 1985). Motoneurons innervating different parts of the myotome have a different morphology. Thus motoneurons supplying the ventral third of the myotome have a dense fanlike dendritic tree, whereas motoneurons innervating the dorsal third of the myotome have a less dense and more widespread dendritic tree. Motoneurons supplying muscle fibres near the ventral or dorsal midline may have dendrites crossing the midline (Wallén et al. 1985; Fig. 10.24: MMN). Tretjakoff (1909a) observed that some of the dendrites of the somatomotor neurons branch around the very coarse reticulospinal Müller axons that are concentrated in the ventromedial part of the cord (Figs. 10.11–10.14). This agrees with the findings of Rovainen (1974b), who recorded monosynaptic EPSPs in myotomal motoneurons following the stimulation of specific reticulospinal axons. Rovainen found that the first component of the EPSP was mediated by electrical coupling and that the later components were generated by a chemical synapse. These results were confirmed and extended by Batueva and Shapovalov (1977a,b), who also showed that several giant reticulospinal fibres can make electrochemical synapses with a single motoneuron. Further observations revealed that the chemical component was glutaminergic and showed use-dependent facilitation, but the electrical component did not alter with repeated activation (Brodin et al. 1994). Electrochemical synapses on motoneurons are not restricted to contacts from the descending giant fibres but may also be formed by slower descending fibres, primary afferents and recurrent collaterals (Shapovalov 1977). It is noteworthy that the Mauthner axon also makes excitatory synapses on myotomal motoneurons (Rovainen 1974b). Motoneurons situated in the most rostral part of the cord have been shown to be excited directly by both ipsilateral and crossed vestibulospinal fibres (Rovainen 1979a). Slow motoneurons are probably more numerous than fast ones. Both types show spontaneous inhibitory postsynaptic potentials (IPSPs), in addition to EPSPs, but the former do not seem to represent recurrent inhibition (Terävänen and Rovainen 1971b).

Lampreys have either one or two continuous dorsal fins along the midline of the caudal half of their body (Fig. 10.1). The muscles at the base of the fin are composed of short fibres parallel to the fin rays and act to bend the fins to the same side (Rovainen 1983a). Motoneurons innervating the fin muscles are located, just like the myotomal motoneurons, in the ipsilateral lateral cell column. However, these two cell types differ from one another in several respects (Shupliakov et al. 1992; cf. also Rovainen and Birnberger 1971 and Rovainen and Dill 1984): (a) The fin motoneurons are clearly smaller than the myotomal motoneurons. (b) Unlike myotomal motoneurons, which are closely spaced in the lateral cell column, fin motoneurons are distributed along the spinal cord separately or in pairs. (c) Unlike myotomal motoneurons, many fin motoneurons extend dendritic branches into the dorsal funiculi, where they make synaptic contact with primary afferent fibres. (d) Whereas myotomal motoneurons send numerous dendrites to the most lateral part of the lateral funiculus, fin motoneurons have few dendrites in this region. (e) Contrary to myotomal motoneurons, fin motoneurons make very few synaptic contacts with the descending reticulospinal axons located in the ventral funiculus.

Shupliakov et al. (1992) further divided the motoneurons innervating fin muscles into two types, based on their dendritic morphology. Type I cells have a widespread dendritic tree in the rostrocaudal direction, which are, with few exceptions, completely restricted to the ipsilateral side. Type II fin motoneurons have long dendrites crossing the median plane above and below the central canal (Fig. 10.24: FMN I, II).

During fictive locomotion (see Sect. 10.4.8 for definition), dorsal fin muscles are active in antiphase to the burst activity of the ipsilateral myotomes (Buchanan and Cohen 1982). Moreover, there is physiological evidence suggesting that both types of fin motoneurons are driven by the contralateral part of the segmental motor network and that they may receive similar synaptic input, despite their differences in dendritic morphology (Shupliakov et al. 1992).

The large reticulospinal Müller neurons M3 and I1 have been shown to provide a polysynaptic excitation of contralateral fin motoneurons in contrast to the monosynaptic excitation of ipsilateral myotomal motoneurons (Rovainen 1978).

10.4.3
Dorsal Cells and Dorsal Roots

In the dorsomedial part of the spinal grey, the large *dorsal cells* (up to 60 μm in diameter) are found (Figs. 10.11–10.13). As in the larva, they form two distinct longitudinal rows on either side of the midline. Selzer (1979), who studied the large elements in whole mounted spinal cords of large larval sea

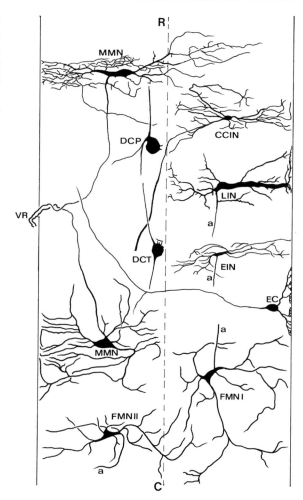

Fig. 10.24. Neurons in the spinal cord of the lamprey viewed from the dorsal side, reconstructed after intracellular injection with lucifer yellow. *CCIN,* crossed caudal interneuron (Rovainen 1983); *DCP,* dorsal cell, 'pressure' type (Christenson et al. 1988b); *DCT,* dorsal cell, 'touch' type (Christenson et al. 1988b); *EC,* edge cell (Grillner et al. 1984); *EIN,* excitatory interneuron (Buchanan et al. 1989); *FMNI,II,* fin motoneurons, types I and II (Shupliakov et al. 1992); *LIN,* lateral interneuron (van Dongen et al. 1985b); *MMN,* myotomal motoneurons (Brodin and Grillner 1990); *VR,* ventral root; *R,* rostral; *C,* caudal; *a,* axon. (Redrawn from the sources indicated)

lampreys (*Petromyzon marinus*), found that the dorsal cells numbered between 1645 and 1900 per spinal cord. Their density in various spinal segments varied considerably, the average being eight per hemisegment. Dorsal cells were not paired, and the number of dorsal cells on either side of the midline in any spinal segment fluctuated greatly. These dorsal cells appear in longitudinal sections as fusiform elements having two coarse processes, which originate from their rostral and caudal apices and run in the dorsal funiculus of the spinal cord (Fig. 10.24: DCP, DCT). In addition to these longitudinal processes, the dorsal cells have a process that leaves the spinal cord via a dorsal root (Kutschin 1863; Freud 1877, 1878; Rovainen 1967a; Christenson et al. 1988b). Some dorsal cells send out processes three or more dorsal roots both rostral and caudal to the cell body (Tang and Selzer 1979). These peripherally directed processes end as mechanoreceptors in the skin (Martin and Wickelgren 1971); consequently, the processes that remain in the cord are the efferents of these cells.

The experiments of Martin and Wickelgren (1971) have confirmed the suspicion of earlier researchers(Ariëns Kappers et al. 1936; Whiting 1948; Rovain 1967b) that the dorsal cells represent first-order sensory neurons. On the basis of the responses of dorsal cells to mechanical stimulation of the skin, they were subdivided into three different functional types: 'touch' cells, 'pressure' cells, and 'nociceptive' cells. The physiological properties of the three cell types differ. The touch (T) cells have a small receptive field and adapt rapidly so that only one or two action potentials are generated by a stimulus. The T cells comprise 10 % of the dorsal cell population. Pressure (P) cells and nociceptive (N) cells have large receptive fields, adapt slowly and fire at a frequency related to the applied stimulus pressure. N cells require much more pressure to activate them than do P cells. P cells comprise 50 %, N cells 40 %, of the dorsal cell population. It should be mentioned, however, that Johnels (1958) has provided morphological evidence that some of the dorsal cells are visceral efferents. Christenson et al. (1988a,b) re-studied the morphological and physiological properties of the dorsal cells in the lamprey spinal cord. They confirmed the presence of T and P cells but remained unable to distinguish, within the group of slowly adapting cells, a separate category of N cells (Christenson et al. 1988b; Fig. 10.24: DCT, DCP). The authors injected physiologically identified dorsal cells with lucifer yellow for light microscopy and with histidine-rich protein (HRP) for combined light and electron microscopy. T cells appeared to be contacted by numerous synaptic boutons containing spherical vesicles, but P cells were found to be completely devoid of such contacts. No other morphological differences between T and P cells were detected ultrastructurally or under the light microscope (Christenson et al. 1988a). While both T and P cells have high-threshold calcium channels, the membrane properties of T and P cells differ primarily through the presence of low-voltage calcium channels in only the T cells (Christenson et al. 1993).

By means of dye injection and electrical stimulation, Rovainen (1967b) showed that the length of the ascending and descending processes of dorsal cells varies considerably. Stimulation of these elements appeared to produce both EPSPs and IPSPs in the large internuncial elements, known as giant interneurons (see below). Some of the EPSPs were monosynaptic and had both electrical and chemical components (Rovainen 1974a). These EPSPs are depressed by excitatory amino acid antagonists, which suggests that dorsal cells use glutamate as a neurotransmitter (Brodin et al. 1987). Although dorsal cells are not devoid of synaptic innervation (see above), they have appeared to be insensitive to a variety of putative transmitters (Martin et al. 1970; Homma and Rovainen 1978). Thompson (1990) presented evidence suggesting that stimulation of dorsal cells may induce a pattern of respiratory motor activity known as 'arousal respiration'. Recent results from El Manira et al. (1996) indicate that dorsal cells make monosynaptic connections with fin motoneurons and confirm that they have a role similar to that of stretch receptors in the lamprey or muscle spindles in other vertebrates.

The fibres of the extramedullary and intramedullary primary sensory cells together constitute a distinct fibre bundle, indicated in Figs. 10.11–10.13 as the dorsal funiculus. Many of these fibres terminate in the caudal rhombencephalon, but some continue further rostrally to reach the octavolateral area and the cerebellum (Ronan and Northcutt 1990; Dubuc et al. 1993b; see below). Martin and Bowsher (1977) provided neurophysiological evidence that in the ammocoete the axons of many dorsal cells extend as far rostral as the isthmus region.

The peripherally directed fibres of the dorsal cells are large and prominent but represent only a minority of the fibres in the dorsal roots. Most of these processes originate from extramedullary elements, which are aggregated in true spinal ganglia. Freud (1878) estimated the total number of fibres in a dorsal root of the lamprey as 50 and believed that about one fifth of them originate from intramedullary cells. According to Rovainen (1983a), the total number of axons in sections of dorsal roots is within the range of 24–60. The spinal ganglion cells are of two main types, large cells with a diameter of about 50 µm and small elements with a diameter of 10–20 µm (Freud 1878; Van Dongen et al. 1985a). Many of the small neurons contain serotonin (Van Dongen et al. 1985a), and it has been shown that these serotoninergic elements also contain calcitonin gene-related peptide (CGRP) and bombesin (Brodin et al. 1988a). The function of these small cells has not been studied; conceivably, they mediate nociception and temperature sensibility, by analogy with the condition in jawed vertebrates (Brodin and Grillner 1990).

The central fibres of the spinal ganglion cells

bifurcate in the cord into ascending and descending branches and take up a position just lateral to the longitudinal fibres of the dorsal cells. Their exact rostral and caudal extent is unknown.

The dorsal funiculus has a high density of longitudinally oriented axons, which vary in diameter between 1 and 4 μm. Dendrites of fin motoneurons (Shupliakov et al. 1992) and giant interneurons (Rovainen 1974a) are known to extend into the dorsal funiculi. Christenson et al. (1988b) studied the ultrastructure of the dorsal funiculus and found that many of its fibres have output synapses with spherical vesicles. However, neither the nature of the presynaptic elements (dorsal cells or spinal ganglion cells or both), nor that of the postsynaptic elements could be established. According to Tretjakoff (1909a), the central processes of the spinal ganglion cells send branches to the periphery of the cord, where they synapse with the dendrites of motor and internuncial cells.

10.4.4
Intrinsic Neurons

Intrinsic neurons, that is, elements of which neither the dendrites nor the axons leave the CNS, are abundant in the spinal cord of the lamprey. First of all, there are numerous small cells that make up most of the spinal grey. These elements appear in methylene blue or Golgi preparations as spindle-shaped cells that send out two or three dendrites to ramify in the 'white' matter. The dendritic branches of these small cells spread in a plane perpendicular to the longitudinal axis of the cord, and one of the dendrites often passes, either dorsal or ventral to the central canal, in the direction of the other side of the cord. Tretjakoff (1909a) observed that such dendrites frequently connect circumscribed areas of the 'white' matter. Thus he found many such cells where one dendrite ramified in the dorsolateral funiculus while the other branched in the ventral funiculus. He also saw numerous elements whose dendrites connected the dorsolateral funiculus with the ventrally located reticulospinal fibres. Tretjakoff thought that these small cells had no axons, and he named them amacrine cells, after similar cells in the retina. As early as 1893, Retzius observed small elements (*Ganglienzellen*) in the cord of the lamprey, the axons of which he traced to the contralateral lateral funiculus (Retzius 1893a; Fig. 10.10a). His observations agree with those of Tang and Selzer (1979), who reported the presence of many small rostrally projecting interneurons with decussating axons.

In addition to the small elements just discussed, the lamprey's cord contains several types of larger intercalated neurons. The structural and functional relations of these elements were clarified by Rovainen (1967b, 1974a,b, 1982, 1983), Rovainen et al. (1973), Selzer (1979) and Tang and Selzer (1979), Buchanan (1982), Buchanan and Cohen (1982), Buchanan and Grillner (1987, 1988), Buchanan et al. (1989), Grillner et al. (1982, 1984) and Viana Di Prisco et al. (1990). These authors distinguished, among others, giant interneurons, lateral interneurons, edge cells, crossed caudal interneurons, excitatory interneurons and small ipsilateral inhibitory interneurons. Some salient features of these elements will now be briefly reviewed.

The *giant interneurons* are located in the caudal half to two thirds of the cord. In the analysis of Selzer (1979), carried out on wholemounts of spinal cords from larval sea lampreys, the number of giant interneurons varied from 16 to 22 per spinal cord and did not show a segmental distribution. According to Rovainen (1967b), they number about 12–35 per animal. The dendrites of the giant interneurons extend in all directions from the cell body; one or two of them cross the median plane dorsal to the central canal. The axons of the giant interneurons cross ventral to the central canal and pass rostralward in the dorsal part of the lateral funiculus. Rovainen et al. (1973) traced one of these axons uninterruptedly from the spinal cord into the lateral part of the rhombencephalon, but they considered it likely that the axons of the other giant interneurons also extend into the brain.

Because of their large size, giant interneurons have been popular for physiological studies. Rovainen and co-workers (Rovainen et al. 1973; Rovainen 1974a,b) identified four sources of input to these cells. These inputs comprise: dorsal cells, unidentified interneurons excited by dorsal cells, other giant cells, and one pair of Müller cell axons, designated I2 (see later). The latter input is mediated by an electrochemical synapse (Rovainen 1967a), and this relationship was subjected to detailed theoretical and experimental analysis using cable theory, intracellular recording and HRP injection (Christensen and Teubl 1979a,b). Another example of a combined electrochemical synapse onto a giant cell has also been examined at both the electrophysiological and ultrastructural level (Christensen 1976). This junction was originally also ascribed to the Müller fibre but was later considered to be an example of a connection between giant interneurons (Rovainen 1979b). Giant interneurons excite more rostral giant interneurons monosynaptically by a presumed excitatory amino acid synapse (Homma 1981) and the Müller cell B3 in the brain (Rovainen 1974a,b). They respond to a variety of putative transmitters including glutamate, glycine, GABA, alanine and taurine (Martin et al. 1970; Homma and Rovainen 1978; Homma 1979).

Rovainen (1979b, 1982, 1983) regarded the giant interneurons as secondary mechanosensory neurons which relay activity rapidly from the tail region toward the head. Mechanical stimulation of the skin elicits barrages of excitatory synaptic potentials (EPSPs) in giant interneurons (Terävainen and Rovainen 1971b), which are produced monosynaptically by dorsal cells or disynaptically from such cells via unidentified interneurons (Rovainen 1967b, 1974a). Giant interneurons are not directly involved in swimming (Rovainen 1982, 1983). They do not interact synaptically with motoneurons or with interneurons that are phasically active during locomotion (Rovainen 1974a; Buchanan 1982), and they do not elicit movements of the body when stimulated intracellularly (Terävainen and Rovainen 1971b). During swimming they lack any significant phasic potential changes or periodic activity (Buchanan and Cohen 1982).

The *lateral interneurons* (LIN: Fig. 10.24) are characterised by their lateral position in the spinal grey (Fig. 10.11, 10.13) and by their large lateral dendrite extending to the edge of the spinal cord. Selzer (1979) found that the lateral cells numbered between 64 and 69 per spinal cord. There was no apparent segmental organisation or alteration in their positions. They were located only in the rostral two thirds of the spinal cord, their distribution overlapping slightly with that of the giant interneurons. The dendritic trees of the LINs are usually confined to the ipsilateral half of the spinal cord and their axons extend caudally in the ipsilateral lateral funiculus as far as the tail region (Rovainen 1974a; Tang and Selzer 1979). Stimulation of the axons of cells of the rhombencephalic reticular formation produces monosynaptic EPSPs in the LINs via Müller cell axons B2, B3 and B4 (Rovainen 1974b), which will be discussed later. Injection of LINs with a fluorescent dye revealed that these elements extend several dendritic processes into the region of the large reticulospinal (Müller) fibres and spread longitudinally along these fibres over about 150 μm (Ringham 1975). The dendrites and axons are also coupled both electrically and chemically. The electrical synapse in this case has been studied and found to be highly rectifying, so that orthodromic conduction is favoured over antidromic (Ringham 1975). The favourable geometry of this synapse has also permitted a detailed evaluation of the synaptic transfer characteristics of the chemical component (Martin and Ringham 1975). During swimming activity, lateral interneurons show phasic changes in potential which are nearly identical to those in myotomal motoneurons (Buchanan and Cohen 1982). The LINs are excited and inhibited polysynaptically by dorsal cells (Rovainen 1974a) and exert an inhibitory action on myotomal motoneurons and crossed caudal interneurons (Buchanan 1982). They are thought to promote undulatory postures, by producing relaxation of ipsilateral caudal myotomes via inhibition of their motoneurons and contralateral caudal excitation by disinhibition of motoneurons through crossed caudal interneurons (Buchanan 1982; Rovainen 1982, 1983).

The *edge cells* are situated outside the grey matter in the lateral funiculi (Figs. 10.11, 10.12). Large elements of this type are found in the gill region, but smaller ones are distributed all along the cord. These elements are rather numerous, about 20 per hemisegment (Rovainen 1979b). They have a characteristic, horizontally oriented dendritic tree, the terminal branches of which ramify in the lateral marginal zone of the spinal cord (Grillner et al. 1984; Viana Di Prisco et al. 1990; Fig. 10.24: EC). Although most edge cells project rostrally, as many as 20% may have a caudal projection or both rostral and caudal projections. They project equally to the ipsilateral and contralateral spinal hemicord, but their processes do not extend more than a few segments (Tang and Selzer 1979). Edge cells are also designated as stretch receptor neurons because they sense the lateral bending of the cord during locomotion (Grillner et al. 1982, 1984). (Lampreys lack muscle spindles or other muscle receptors in the trunk muscles.) In order to elucidate the synaptic effects of edge cells, Viana Di Prisco et al. (1990) performed paired intracellular recordings and staining with lucifer yellow. Edge cells with an ipsilaterial axon produced EPSPs in ipsilateral motoneurons and interneurons of the locomotor network, whereas edge cells with a contralateral axon elicited monosynaptic IPSPs in contralateral neurons, including contralateral edge cells. Some edge cells are excited by stimulation of the axon of the Müller I2 cell and by giant interneurons (Rovainen 1974a,b, 1979b). Wallén et al. (1995) investigated the modulation of edge cells during NMDA-evoked fictive locomotion (cf. Sect. 10.4.8). Most of the edge cells appeared to receive periods of alternating excitation and inhibition, occurring during the ipsilateral and the contralateral ventral root bursts, respectively. The locomotion-related inhibition in edge cells appeared to have a dual origin: glycinergic neurons providing the phasic inhibition and GABAergic neurons exerting a tonic inhibition via GABA$_A$ receptors. Recent evidence (Anadón et al. 1995) indicates that the edge cells of the lamprey are similar to the marginal cells found in the spinal cords of four species of elasmobranchs as well as marginal nucleus cells of urodeles and snakes.

Crossed caudal interneurons (CCIN) resemble myotomal motoneurons, with regard to soma size and pattern of dendritic ramification (Fig. 10.24). They receive monosynaptic excitation from the ipsilateral Müller cell B1 and polysynaptic excitation and inhibition from dorsal cells (Buchanan 1982). Their axons project contralaterally and caudally in the spinal cord, extending over 19 segments or more. Most CCINs are inhibitory (glycinergic) to myotomal motoneurons, lateral interneurons and other CCINs (Buchanan 1982, 1986). Others excite contralateral motoneurons and are inhibited, perhaps disynaptically, by the ipsilateral Müller cell I1 (Buchanan, personal communication, cited from Rovainen 1982).

Excitatory interneurons (EIN; Fig. 10.24) have small cell bodies, a transversely oriented ipsilateral dendritic tree and thin, ipsilateral descending axons extending up to six to seven segments. EINs receive monosynaptic reticulospinal input and polysynaptic input from skin afferents. They employ an excitatory amino acid as neurotransmitter and activate both motoneurons and inhibitory premotor interneurons (i.e. LINS and CCINS). During locomotion they are phasically active and clearly form part of the network underlying locomotion (Buchanan and Grillner 1987; Buchanan et al. 1989; see Section 4.8).

Small ipsilateral inhibitory interneurons (sIIN) mediating glycinergic IPSPs in motoneurons were described by Buchanan and Grillner (1988). Morphologically, these elements closely resemble the excitatory interneurons (EINs) described above. Like the latter they are phasically active during locomotion.

Three types of *GABA-immunoreactive interneurons* have been identified in the spinal cord of the lamprey, i.e. (a) small bipolar cells in the entrance zone of the dorsal root, (b) multipolar neurons in the lateral column, and (c) small, CSF-contacting neurons surrounding the central canal (Brodin et al. 1990c; Batueva et al. 1990; Fig. 10.54). These cells give rise to a dense plexus of GABA-immunoreactive fibres in the ventral and lateral funiculi. A particularly dense GABA-immunoreactive plexus is situated in the lateral spinal margin, i.e. the domain of the stretch receptor neurons known as edge cells. Small neurons surrounding the central canal, which co-localise GABA and somatostatin, were found to project heavily to the lateral marginal plexus (Christenson et al. 1991). Both GABA and somatostatin hyperpolarise edge cells, but by effects on different types of ionic channels (Christenson et al. 1991).

The GABAergic neurons intrinsic to the lamprey spinal cord apparently exert complex modulatory effects (Leonard and Wickelgren 1986) at both pre- and postsynaptic sites on interneurons active during locomotion and on mechanosensory dorsal cells, mediated both by ionotropic GABA$_A$ and metabotropic GABA$_B$ receptors (Alford et al. 1991; Alford and Grillner 1991; Tegnér et al. 1993; Matsushima et al. 1993; Christenson et al. 1993).

A population of small, *dopaminergic, CSF-contacting neurons* is located directly ventral to the central canal. A sparse plexus of dopaminergic fibres is situated ventral to these cells (McPherson and Kemnitz 1994).

Serotoninergic neurons form a ventromedian column throughout the spinal cord (Fig. 10.54). These cells give rise to a dense plexus in the medial part of the ventral funiculus (Harris-Warwick and Cohen 1985; Van Dongen et al. 1985a,b). Some of these neurons co-contain a tachykinin-like peptide (Van Dongen et al. 1985a). Motoneurons and lateral interneurons extend some of their dendrites into the dense ventromedial serotoninergic plexus, where they come in close apposition to serotoninergic varicosities (Van Dongen et al. 1985b). Ultrastructural analysis has shown that serotoninergic varicosities do not make synaptic contact with surrounding elements in either the ventromedial plexus or other parts of the spinal cord (Christenson et al. 1990). Schotland et al. (1995) demonstrated that the serotoninergic neurons just described also contain dopamine. Thus, it appears that these serotoninergic elements and the dopaminergic cells described by McPherson and Kemnitz (1994) in fact form a single population.

Peptidergic neurons. Small neuropeptide Y (Van Dongen et al. 1985a; Brodin et al. 1989b) and neurotensin-immunoreactive neurons (Brodin et al. 1990b) occur in the dorsomedial part of the spinal grey matter. Their function is unknown, but neurotensin has been shown to modulate the activity of spinal neurons and fictive locomotion (Barthe and Grillner 1995).

Other spinal neurons. In addition to the elements discussed above, Selzer (1979) observed a particular type of intrinsic spinal neuron which he termed an oblique bipolar cell. These elongated, spindle-shaped neurons are located in the most medial part of the grey matter at all levels of the spinal cord. Experiments with the retrograde tracer technique (Tang and Selzer 1979) indicated that these elements project rostrally for long distances in the contralateral hemicord.

Finally, it is a possible reflection of the work that remains to be done that Ohta et al. (1991), using retrograde labelling techniques, detected a large population (180–300 cells per hemisegment) of small cells with contralaterally projecting axons.

These neurons differ from CCINs because their axon bifurcates into a rostrally directed main axon and a caudally projecting collateral, and both of these axonal branches extend over fewer than five segments. Because Rovainen (1979b) estimated that each hemisegment contains approximately 500 neurons, this new population is quantitatively very important. The functional significance of these neurons remains to be determined.

10.4.5
'White Matter'

The 'white matter' of the lamprey's spinal cord contains fibres of very diverse diameters, ranging from 100 to 0.1 μm (Bertolini 1964; Chap. 3, Fig. 3.3). There is no trace of a myelin sheath around any of them; all are bounded only by a plasma membrane. As in all vertebrates, the spinal matter can be subdivided into ventral, lateral, and dorsal funiculi, separated by the entry zones of the ventral and dorsal roots. In the ventral funiculus a conspicuous concentration of very coarse fibres is found. The lateral funiculus contains a continuum of fibre diameters ranging from very coarse to extremely thin. In the dorsal funiculus conspicuous coarse fibres are lacking (Figs. 10.11–10.13, 10.34). Buchanan and Grillner (1988) studied the organisation of neural elements immunoreactive to ten different peptides in the lamprey spinal cord. It appeared that the various sets of peptidergic fibres are mostly diffusely distributed over the white matter. GABA- and serotonin-immunoreactive fibres show a similar characteristic (Van Dongen et al. 1985a; Harris-Warwick and Cohen 1985; Brodin et al. 1990b; Batueva et al. 1990; Fig. 10.54). Terminal axonal branches of spinal GABA-, somatostatin-, and neurotensin-immunoreactive fibres participate in the formation of a dense marginal plexus in the most lateral part of the lateral funiculus (Fig. 10.54).

10.4.6
Descending Fibre Systems

It has been known since the classic studies of Johnston (1902) and Tretjakoff (1909a,b) that the reticular formation of the brain stem sends a number of axons into the spinal cord. Collectively, these axons undoubtedly constitute the largest and the most important system descending from the brain to the spinal level. Among the neurons giving origin to this reticulospinal system there are a few prominent, very large elements, the so-called Müller cells. Stefanelli (1934) counted ten, Rovainen (1967a) eight (M1–3 I1, B1–4; cf. Fig. 10.34), and Nieuwenhuys (1972) seven (M1–7; cf. Fig. 10.34) of these

cells on either side of the brain (these elements will be discussed more extensively in Sect. 10.5.8.1). These elements give rise to very coarse descending axons. In the brain stem these fibres are concentrated in the fasciculus longitudinalis medialis, with which they descend into the ventral funiculus of the spinal cord (Tretjakoff 1909b; Martin 1979). Some of them maintain their ventromedial position throughout the cord, but others shift gradually from the ventral to the lateral funiculus (Rovainen et al. 1973). The axons of most of the smaller reticular elements also descend via the medial longitudinal fascicle into the ventral funiculus, but those originating from the most caudal part of the reticular formation (vagal group of Rovainen 1967a; nucleus reticularis inferior of the present account; Fig. 10.30) pass directly to the dorsal part of the lateral funiculus (Rovainen et al. 1973). It has been demonstrated (Rovainen 1974b) that following stimulation of the axons of certain Müller cells monosynaptic EPSPs can be recorded in myotomal motoneurons and, as already mentioned, in lateral interneurons. Dual electrical and chemical synapses appeared to be the most common mode of transmission between Müller axons and their target cells (Rovainen 1974b). All of the larger reticulospinal axons, with conduction velocities of 1 m s^{-1} or more, activate excitatory amino acid receptors and are, hence, probably glutamatergic (Buchanan et al. 1987b; Ohta and Grillner 1989). Although stimulation of individual Müller axons may lead to distinctive movements, such as flexion of the body and tail, movements of the fins and propagated undulation (Rovainen 1967a), the fact that most unitary EPSPs are subthreshold makes it likely that simultaneous activity of many neurons is required to produce normal swimming (Rovainen 1974b; see also Cohen and Wallén 1980).

Apart from the reticulospinal fibres already mentioned, the lateral funiculi contain the large axons of the Mauthner cells. This is a pair of very large neurons, situated lateral to the reticular formation at the level of entrance of the eighth nerve (Figs. 10.30, 10.34a). Whereas the axons of the Müller and other reticular elements remain ipsilateral, the axons of the Mauthner cells decussate immediately behind the level of their somata. Stimulation of Mauthner fibres elicits distinct body contractions and tail flexures (Rovainen 1967a).

A third system, descending from supraspinal levels towards the cord, is formed by a number of medium-sized axons that originate from two vestibular nuclei, known as the nucleus octavomotorius intermedius and posterior (Figs. 10.16, 10.30). Stefanelli (1934), who studied non-experimental silver-impregnated material, observed that the for-

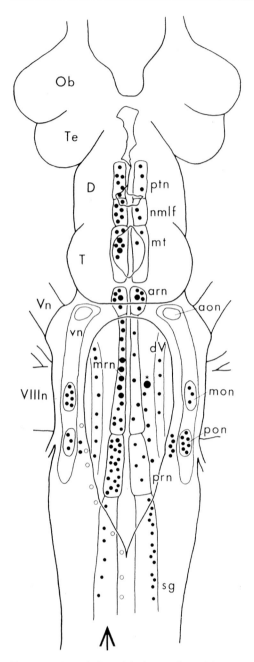

Fig. 10.25. Dorsal view of the brain of an adult silver lamprey, *Ichthyomyzon unicuspis*, showing the location and laterality of the cells of origin for the descending spinal projections. *Filled circles* represent neurons retrogradely labelled by HRP implanted in the left half of the rostral spinal cord *(arrow)*. *Large circles* indicate Müller and Mauthner (*) cells. *Open circles* represent labelled dorsal cells in the spinal cord and rhombencephalic alar plate. *aon,* anterior octavomotor nucleus; *arn,* anterior reticular nucleus; *D,* diencephalon; *dV,* descending trigeminal tract; *mon,* intermediate octavomotor nucleus; *mrn,* middle reticular nucleus; *mt,* mesencephalic tegmentum; *nmlf,* nucleus of the medial longitudinal fasciculus; *Ob,* olfactory bulb; *pon,* posterior octavomotor nucleus; *prn,* posterior reticular nucleus; *ptn,* nucleus of the posterior tubercle; *sg,* spinal grey; *T,* mesencephalic tegmentum; *Te,* telencephalic hemisphere; *Vn,* trigeminal nerve; *VIIIn,* octaval nerve (Reproduced from Ronan 1989)

mer cell mass gives rise to an ipsilateral tractus octavospinalis lateralis, the latter to a crossed tractus octavospinalis medialis. According to Rovainen (1979a) and Rovainen et al. (1973), neither of these tracts extends beyond the most rostral part of the spinal cord.

The findings just discussed have been confirmed and extended by the experimental studies of Ronan (1989) and Swain et al. (1993), in which HRP was used as a retrograde tracer. Ronan (1989) studied the origins of the descending spinal pathways in sea lampreys (*Petromyzon marinus*) and silver lampreys (*Ichthyomyzon unicuspis*), following unilateral implants of HRP in the spinal cord between the 5th and 11th spinal segments (Fig. 10.25). Numerous labelled cells were found in all parts of the rhombencephalic and mesencephalic reticular formation, mainly ipsilaterally. The posterior or inferior reticular nucleus appeared to be the most heavily labelled cell group in the brain. Labelled reticular elements included the Müller and Mauthner cells. HRP-filled cells were also present in the lateral margin of the contralateral spinal grey, the vagal lobe (mainly contralaterally), the posterior octavomotor nucleus (mainly contralaterally), the intermediate octavomotor nucleus (mainly ipsilaterally), the descending trigeminal tract, and a diencephalic cell group, the nucleus of the posterior tubercle. Spinal and rhombencephalic dorsal cells were also labelled. These data confirm the presence of large, mainly ipsilateral, reticulospinal projections and of smaller direct and crossed vestibulospinal projections; moreover, they reveal the presence of small trigeminospinal and vago- or solitariospinal fibre contingents.

Swain et al. (1993) also studied the brain neurons with spinal projections in a lamprey (larval sea lampreys, approximately 5 years old). However, in order to establish the distribution of the terminations along the spinal cord of the encephalospinal projections, they injected HRP at six different levels in the spinal cord, ranging from 10 % to 75 % of body length from the anterior end. They divided the spinal projection system into ten bilateral groups, one corresponding to the nucleus of the posterior tubercle, a second corresponding to the inferior octavomotor nucleus, and the remaining eight falling within the confines of the reticular formation of the brain stem. The authors estimated that there are approximately 2000 neurons in the larval lamprey brain that project to the spinal cord. Müller and Mauthner axons were found to extend throughout the length of the cord. None of the ten cell groups projected exclusively to one level. In most of the cell groups the number of cells labelled by injections at progressively more caudal spinal levels declined steadily, the fall-off in axonal projections being steepest between the last gill (~15% body length) and the rostral end of the abdomen (~25% body length). The inferior octavomotor nucleus was found to project primarily to the gill region. Wannier (1994) obtained comparable results in a study of the reticular formation. The caudal extent of single reticulospinal axons was established by stimulating them antidromically while recording from their soma. It was found that most mesencephalic reticulospinal neurons project to the end of the spinal cord, while considerable proportions of the neurons in the superior, middle and inferior rhombencephalic reticular nuclei (18%, 17% and 36%, respectively) terminate in the anterior or middle region of the spinal cord.

In order to determine the developmental changes in the reticulospinal system, Swain et al. (1995) ex-

amined the brains of larval, transforming and adult sea lampreys, following injections of HRP into the spinal cord at 25% of body length. It appeared that in the lamprey, transformation from the larval to the adult form is not accompanied by a substantial increase in the number of reticulospinal neurons, but that the distance of projection of many of these neurons increases progressively with maturation.

Several neuroactive compounds have been shown to be present in the reticulospinal system. Following experiments in which tritiated D-aspartate was injected in the rostral spinal cord, labelled cells were observed in the mesencephalic as well as in the superior, middle and inferior rhombencephalic reticular nuclei (Brodin et al. 1988b; Fig. 10.53d). Combination of retrograde tracers with immunohistochemistry revealed the presence of serotoninergic reticulospinal neurons

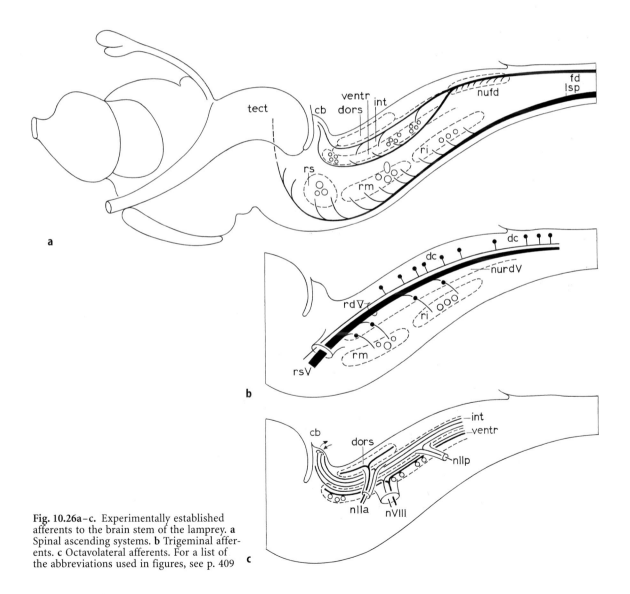

Fig. 10.26a–c. Experimentally established afferents to the brain stem of the lamprey. **a** Spinal ascending systems. **b** Trigeminal afferents. **c** Octavolateral afferents. For a list of the abbreviations used in figures, see p. 409

in the superior and inferior rhombencephalic nuclei (Brodin et al. 1986), peptide YY-immunoreactive descending neurons in the superior rhombencephalic reticular nucleus (Brodin et al. 1989b), and cholecystokinin-immunoreactive reticulospinal neurons in the inferior rhombencephalic nucleus (Brodin et al. 1988a; Fig. 10.53d). The latter probably contact motoneurons and giant interneurons (Ohta et al. 1988). Wannier et al. (1995) recently showed that there was an inhibitory component of the reticulospinal system that acted on spinal neurons via glycinergic receptors.

Bussières and Dubuc (1995) examined the vestibulospinal projections in *Ichthyomyzon unicuspis*, using cobalt-lysine as a retrograde tracer, and by injecting lucifer yellow in antidromically identified vestibulospinal neurons. They found that the intermediate and inferior octavomotor nuclei contain approximately 150 and 70 vestibulospinal neurons, respectively, and that the descending axons of these neurons emit ascending collateral branches which project towards the reticular nuclei.

10.4.7
Ascending Fibre Systems

Two large fibre systems, the dorsal column pathway and the spinal lemniscus, ascend from the spinal cord to the brain stem (Fig. 10.26a). The experimental studies of Ronan and Northcutt (1990) and Dubuc et al. (1993b) have shown that the fibres of the *dorsal column pathway* ascend in the dorsal funiculus, reach the level of the obex and continue into the ipsilateral rhombencephalic alar plate. Gradually diminishing in size, the dorsal column pathway extends throughout the length of the latter structure. At mid-rhombencephalic levels it splits up into separate upper and lower bundles. The fibres of the lower bundle pass along the ventral border of the ventral octavolateral nucleus, whereas those of the upper bundle course rostrally between the dorsal and intermediate octavolateral nuclei. In the most rostral part of the rhombencephalon the fibres of both bundles arch dorsomedially and enter the cerebellum, where some cross the midline. Dubuc et al. (1993b) observed that dorsal column fibres ramify extensively among cells situated in the alar plate, immediately in front of the obex. According to the authors mentioned, these cells represent the dorsal column nucleus, observed in elasmobranchs (Ebbesson and Hodde 1981) and amphibians (Urbán and Székely 1982). Further rostrally, dorsal column fibres intermingle with cells of the nucleus octavolateralis ventralis and octavomotorius intermedius and posterior (Dubuc et al. 1993b).

Although the fibre composition of the dorsal column pathway is imperfectly known, there is anatomical (Ronan and Northcutt 1990) and physiological evidence (Dubuc et al. 1993b) that a population of primary spinal afferents attains the rhombencephalon by way of this pathway, and that axons of spinal dorsal cells, carrying touch and pressure information, form part of this population (Martin and Bowsher 1977; Christenson et al. 1988a,b).

The recent findings of Dubuc et al. (1993b) that cells situated in the caudal rhombencephalon form the principal target of the dorsal column projection are remarkable. Formerly, it was generally held that the dorsal column and its rhombencephalic relay represent a phylogenetically young system, occurring only in amniotes. Ebbesson (1969) believed that the dorsal column-medial lemniscus-thalamocortical system has evolved principally in relation to the evolution of limbs. Because the cells discovered by Dubuc et al. (1993b) occupy a topological position corresponding to that of the dorsal column nuclei of elasmobranchs, amphibians and amniotes, their tentative interpretation appears justified. Following large HRP injections in the dorsal thalamus in lampreys, Ronan and Northcutt (1990) observed labelled cells in the contralateral obex region. This observation suggests the presence of a lemniscus medialis system. More information is available on another output channel of this newly discovered cell group. Dubuc et al. (1993a,b) reported that cells belonging to this group project to the inferior reticular nuclei, exerting an inhibitory influence on neurons in that centre, with glycine as a neurotransmitter.

Several authors (Ariëns Kappers 1920; Ariëns Kappers et al. 1936; Pearson 1936; Larsell 1967) have described fibres that ascend from the lateral funiculus to the brain stem. These fibres, which collectively form the spinal lemniscus, are believed to reach the octavolateral area (Pearson 1936), the cerebellum (Heier 1948; Pearson 1936; Larsell 1967), the roof of the midbrain (Heier 1948; Pearson 1936; Ariëns Kappers 1947) and even the dorsal thalamus (Heier 1948). According to Ariëns Kappers (1920), the bulk of these fibres constitute a decussating secondary sensory system, conveying general tactile, pain and temperature stimuli to the brain. Possible sources of this crossed, secondary sensory system include the giant interneurons (Rovainen 1974a), the small *Ganglienzellen* of Retzius (1893a; Fig. 10.10a) and the oblique bipolar cells described by Selzer (1979) and Tang and Selzer (1979).

The experimental neuroanatomical study of Ronan and Northcutt (1990) has shown that the spinal lemniscus in the lamprey forms a large system of rather loosely arranged fibres, which ascends

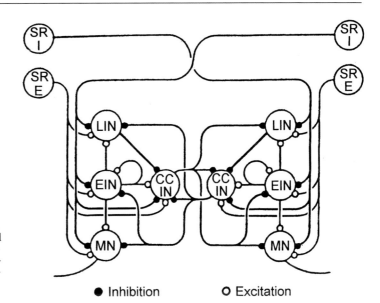

Fig. 10.27. Segmental network of spinal cord. See text for details. Abbreviations: *LIN*, lateral inhibitory interneurons; *EIN*, excitatory interneuron; *MN*, motoneuron; *CCIN*, crossed caudal inhibitory interneuron; *SRI*, stretch receptor, inhibitory; *SRE*: stretch receptor, excitatory. (From Travén et al. 1993, Fig. 10).

from the lateral funiculus of the spinal cord to the brain, proceeding rostrally in the lateral part of the rhombencephalic basal plate. Throughout the rhombencephalon axons, or their branches, extend dorsomedially from the bundle, without attaining the periventricular cell plate, however. Most likely these fibres enter into synaptic contact with the entities of branchiomotor and reticular elements (Fig. 10.26a). The most rostrally extending spinal lemniscus fibres fan out in the isthmic tegmentum. A few, originating from high cervical segments, may project to the tectum. The presence of a direct spinothalamic projection could not be confirmed.

Physiological experiments have shown that spinal neurons, the axons of which ascend in the spinal lemniscus, convey information concerning the progress of the different phases of the locomotor cycle to the brain stem reticular nuclei (Vinay and Grillner 1992).

Finally it should be emphasised that most ascending and descending fibres of the cord are probably propriospinal fibres of varying length.

10.4.8
Neural Network Underlying Locomotion and Supraspinal Control in the Cord

The lamprey has become a model system for understanding locomotion. What commenced as a study of relatively simple spinal reflexes by Rovainen and colleagues has been transformed by Grillner and co-workers into detailed models of complex intrinsic oscillators and their supraspinal control.

During the past 15 years or so, the studies of spinal circuitry have been greatly aided by preparations that demonstrate 'fictive locomotion'. In the context of work on the lamprey this has come to signify an in vitro preparation that exhibits all the rhythmic activity patterns of the intact animal while remaining fixed in location and so accessible to detailed electrical recording. The first such preparations were described by Poon (1980) and Cohen and Wallén (1980) and consisted of an isolated spinal cord bathed in a physiological saline to which was added an excitatory amino acid to provide the necessary excitation to the pattern generators. Later work (Wallén and Williams 1984) demonstrated that the activity in the isolated preparation was indeed similar to that in the intact, swimming animal. This type of preparation has been extended to include brain stem control centres (e.g. Hagevik and McClellan 1994a).

Early electrophysiological observations by Teräväinen and Rovainen (1971b) distinguished a local and a long reflex in the cord. The local reflex was elicited by stimulating the skin of the body or tail and produced a contraction of nearby myotomes and an inhibition of motoneurons to contralateral muscles. The long reflex was brought into play by stimulating the tail and resulted in bilateral activation of myotomes several centimeters rostral to the site of stimulation. Birnberger and Rovainen (1971) described another spinal reflex that moved the dorsal fin to the ipsilateral side after stimulation of the ventrolateral skin. An interesting feature of the reflex was its ability to habituate upon repeated activation. As might be expected, sensory dorsal cells seemed to be active during all these reflexes but little else could be determined. Rovainen (1983a) summarised progress on interpreting the

Transmitters

● — ＜ Glutamate - excitatory
● — • Glycine - inhibitory
● — ► Ach - Motoneuron
►— ┣ Stretch receptor

Fig. 10.28. Coupling of segmental networks and brain stem centres. Each *box* contains the four neurons illustrated in Fig. 10.27 and represents a spinal segment. Segments are coupled by excitatory interconnections and receive both excitatory and inhibitory input from stretch receptors. Reticulopinal control is exerted by centres in the brain stem. Note that inhibitory interneurons and 5-HT neurons within each segment have been omitted for clarity. Transmitters indicated by *symbols*. See Fig. 10.27, for abbreviations. See text for further details. (Modified from Grillner and Matsushima 1991, Fig. 2)

circuitry underlying locomotion and published a final contribution to the problem in 1985. In this (Rovainen 1985a) he hypothesised that two groups of

proprriospinal interneurons were involved in coordination of swimmimg movements: crossed ascending interneurons excited in phase with nearby motoneurons which might entrain contralateral pattern generators and short, commissural interneurons that inhibited contralateral generators.

The basic circuits and synaptic mechanisms underlying locomotion in the lamprey are known now in considerable detail and are surveyed in a number of reviews (Grillner et al. 1991, 1995; Grillner and Matsushima 1991; Wallén 1994). The lamprey swims by propagating a mechanical wave down the body by means of alternate contractions of the myotomes of the left and right sides at frequencies of between 0.2 and 10 Hz. This is brought about by coupling a sequence of intrinsic segmental oscillators or central pattern generators, with appropriate phase lag. The movement is initiated and further modulated by reticulospinal projections and by sensory input at the segmental level. A pictorial synthesis of the spinal and supraspinal systems will be given in Sect. 10.11, based on summary Figs. 10.56–10.58. Here we shall focus on the intrinsic spinal circuitry.

An idealised segmental circuit (Grillner et al. 1995) is shown in Fig. 10.27. Three classes of bilateral neurons form an intrinsic oscillator: the excitatory interneurons (EIN), the lateral inhibitory interneurons (LIN) and the crossed caudal inhibitory interneurons (CCIN). In common with all excitatory synapses involved in the cord, the EIN cells are excitatory, release glutamate and activate both kainate/AMPA receptors and NMDA receptors on the ipsilateral LIN and CCIN and on the final common path of the cord, the myotomal motoneuron (MMN). Motoneurons, however, do not form part of the oscillatory mechanism.

Both the LIN and CCIN mediate their inhibition through glycinergic synapses. The LIN inhibit ipsilateral CCIN but the CCIN inhibit all three classes of contralateral interneuron. In addition the CCIN mediate part of the mechanism that ensures that the two sides of the cord exhibit opposite behaviour at any instant by inhibiting the contralateral CCIN.

Postural information is needed to refine the oscillatory cycle as the animal moves. Two classes of edge cells accomplish this. The stretch-receptor neurons of the excitatory (SRE) type excite all ipsilateral interneurons while the inhibitory (SRI) type inhibit all contralateral neurons. The basic electrophysiological properties of the interneurons, motoneurons and edge cells (as well as the giant interneurons) are quite similar and are mainly distinguished by size-related properties (Buchanen 1993).

As noted above, it is known from experiments of Poon (1980) and Cohen and Wallén (1980) where isolated cords were bathed in excitatory amino acids that all that is necessary to initiate the swimming activity is to depolarize many of the neurons in the circuit, and that such artificial depolarisation mimics that produced in the intact animal and by electrically evoked fictive locomotion (Buchanen and Kasicki 1995). In the intact animal such depolarisation is achieved under natural conditions by descending reticulospinal projections that excite the LIN, EIN, CCIN and MMN. Both NMDA and non-NMDA receptors were found to be involved in the descending initiation of locomotor activity (Traven et al. 1993; Hagevik and McClellan 1994a,b). Two functional classes of reticulospinal projection can be identified (Fig. 10.28; Grillner et al. 1995), a tonic (R_t) and a phasic (R_{ph}) pathway. The phasic pathway is under the further influence of ascending, ipsilateral excitation and contralateral inhibition from the spinal interneurons.

Further modulation of the spinal central pattern generator is provided by serotonin and intrinsic GABAergic interneurons. Such modulatory functions, coupled with the intrinsic properties of the neurons, play an important role in bringing about reliable termination of the oscillatory burst activity in each segment. Three types of propriospinal GABA-immunoreactive neurons have been identified (Brodin et al. 1990c). These apparently exert complex modulatory effects (Leonard and Wickelgren 1986) at both pre- and postsynaptic sites on both interneurons and motoneurons. The effects are mediated both by ionotropic $GABA_A$ and metabotropic $GABA_B$ receptors (Alford et al. 1991; Alford and Grillner 1991; Tegnér et al. 1993; Matsushima et al. 1993; Christenson et al. 1993). Serotoninergic fibres of spinal (Fig. 10.54) and supraspinal origin (Fig. 10.53d) end as varicosities without evident postsynaptic sites (Christenson et al. 1990) and modulate glutaminergically activated central pattern generation (Harris-Warwick and Cohen 1985) through a mechanism involving calcium-dependent K^+ channels (Wallén et al. 1989; El Manira et al. 1994). There is also evidence for the involvement of dopamine (Cohen and Wallén 1980), possibly mediated by small cells slightly ventral to the central canal (McPherson and Kemnitz 1994). As noted earlier, some of the dopamine is apparently co-localised with the serotonin in midline neurons, but acts to reduce Ca^{2+} entry during action potentials and consequently also brings about a reduction in the Ca^{2+}-mediated K^+ influx (Schotland et al. 1995).

While a single segment contains the central pattern generator responsible for alternating bursts of motoneuron activity, swimming requires that a wave of such activity propagates along the animal.

This is brought about by intersegmental coupling of the local oscillators with controlled phase lag. It appears that this is accomplished in a very simple way by means of collaterals from the excitatory interneurons in each segment extending to the adjacent rostral and caudal segments (Fig. 10.28). This excitatory coupling might be expected to lead to all segments being active at once. In fact, though, it has been shown (Matsushima and Grillner 1990, 1992) that when the rostral-most oscillator is the first to be excited, subsequent segments follow with a constant lag; this has been termed the 'trailing lag hypothesis' (Matsushima and Grillner 1990; Grillner and Matsushima 1991). A great merit of this scheme is that backward swimming can be initiated by simply causing the most caudal segment to be active first.

By means of computer simulations with realistic membrane properties and synaptic activation (Cohen et al. 1982; Grillner and Matsushima 1991; McClellan and Jang 1993; Moore and Buchanen 1993; Travén et al. 1993; Hagevik and McClellan 1994b; Jung et al. 1996) it has been shown that the mechanisms outlined above are capable of generating rhythmic activity. More complex long-range intersegmental coupling may also be involved (Mellen et al. 1995).

10.4.9
Reissner's Fibre and the Caudal End of the Spinal Cord

The bilateral subcommissural organs of the lamprey consist of highly cylindrical cells and are, as their name indicates, situated directly ventral to the posterior commissure in the synencephalic region of the brain (see Sect. 10.7.4). The secretory material of the subcommissural organs is released into the cerebrospinal fluid and becomes condensed into a thread-like structure known as Reissner's fibre (RF; Reissner 1860; Oksche 1969; Sterba 1969; Leonhardt 1980). This fibre extends from its site of origin through the ventricular cavity of the brain stem and the central canal of the spinal cord into the ampulla caudalis, i.e. a terminal dilatation of the central canal (Olsson 1955). Newly released secretory material is continuously added to the rostral portion of RF, effecting a constant rostro-caudal growth of this non-cellular structure. In the ampulla caudalis, RF material accumulates as an amorphic and irregular terminal mass, the so-called massa caudalis (Olsson 1955, 1958; Hofer 1963).

The structure of the ampulla caudalis and adjacent structures has been studied in larval lampreys at the light- and electron-microscopical level by

Hofer et al. (1984; *Lampetra planeri*), Rodríguez et al. (1987; *Geotria australis*) and Peruzzo et al. (1987; *G. australis*). The principal results of these authors may be summarised as follows. The ampulla caudalis is spherical in shape, having an internal diameter of 20–35 μm, which was approximately 2.5 times wider than the lumen of the central canal. The ventral and lateral walls of the ampulla caudalis are formed by a compact layer of tall ependymal cells, endowed with junctional complexes, numerous cilia and a few microcilli. Dorsally, the ampulla is lined either by a loose layer of cells devoid of junctional complexes or by a thin but compact layer of ependymal cells. Dorsal to the ampulla a number of large, irregularly shaped cavities, bordered by slender cells of unknown nature are found. These cavities or lacunae communicate directly with the lumen of the ampulla caudalis via openings in the dorsal wall of the latter. The lacunae communicate with structures resembling blood capillaries or lymphatic vessels, and these vessels, in turn, are in direct communication with typical blood capillaries. Using an antiserum raised against bovine RF, Peruzzo et al. (1987) demonstrated that in larval lampreys, RF and its massa caudalis are strongly immunoreactive, and that the ampulla caudalis, the lacunae, the modified vessels as well as the typical blood capillaries are all filled with a flocculent, immunoreactive material. No immunoreactive material was found outside these structures. These observations corroborate the suggestion of several previous authors, among them Olsson (1955), Wislocki et al. (1956), Hofer (1963), Oksche (1969) and Hofer et al. (1984), that RF material arriving in the ampulla caudalis is ultimately discharged into the blood stream.

The function of Reissner's fibre and its associated structures in the lamprey, as in other species, remains an enigma.

10.5
Rhombencephalon

10.5.1
Introductory Note

The rhombencephalon is well developed and makes up one half the length of the entire brain (Fig. 10.3). It includes bilaterally a horizontal basal plate and a vertically oriented alar plate. Throughout the rhombencephalon the boundary between the basal and alar plates is marked by a distinct ventricular groove, the sulcus limitans of His (Figs. 10.15–10.18). Passing rostrally from the obex the alar plates gradually diverge, but in the rostral part of the rhombencephalon these structures approach

each other again. They finally fuse, forming a small cap over the rostral part of the fourth ventricle (Figs. 10.3a, 10.18). The remaining part of the fourth ventricle is covered by a richly folded chooid roof, which is attached to the dorsal aspect of the alar plates (Fig. 10.3a,b,d). The somewhat tapered, most rostral part of the rhombencephalon is known as the isthmus rhombencephali. Fiure 10.3 shows that the fifth to the tenth cranial nerves issue from the rhombencephalon. The fibres of the abducens nerve exit the brain immediately caudoventral to the motor trigeminal root (Stefanelli 1934; Heier 1948; Finger and Rovainen 1978; Fritzsch et al. 1990).

10.5.2
Alar Plate

According to the classic studies of the American school of comparative neuroanatomists (see Chap. 4), this structure typically comprises three longitudinal sensory zones or columns in all vertebrates. These are, from ventral to dorsal: (a) the viscerosensory zone which, by way of the fasciculus communis, receives fibres from the afferent divisions of VII, IX, and X; (b) the general somatosensory zone, which is chiefly related to the sensory fibres of V; (c) the special somatosensory zone, which is made up of the nuclei of termination of VIII and (in aquatic vertebrates) of the related lateral line nerves. In the lamprey these zones have been identified by Johnston (1902) and by several subsequent researchers (Pearson 1936; Barnard 1936; Larsell 1967). Their position, however, diverges somewhat from the dorso-ventral pattern indicated above.

10.5.3
Viscerosensory Zone

Johnston (1902) found that in the caudal part of the alar plate a large area of periventricular grey is related to the entering fibres of VII, IX and X. He termed this area the lobus vagi (Figs. 10.14, 10.30). According to Saito (1930) the viscerosensory zone can be recognized throughout the rhombencephalon as a strip of periventricular grey embraced between the sulcus limitans and a more dorsally situated groove, which he designated as the sulcus intermedius dorsalis. Barnard (1936) agreed with Saito, although the former did not mention the presence of a dorsal limiting groove. Nieuwenhuys (1972) remained unable to delimit the viscerosensory zone as a morphological entity. In the caudal part of the rhombencephalon the periventricular grey was found to be very diffusely organised, without showing any morphological landmarks. More rostrally, the area described as viscerosensory by Saito (1930) and Barnard (1936) could not be delimited from the area octavo-lateralis. It may be safely assumed that the vagal lobe of the lamprey receives general visceral sensory fibres via the VIIth, XIth and Xth nerves, as well as by special visceral sensory (i.e. taste) fibres, but it is not possible to separate centrally these two fibre contingents.

Baatrup presented morphological (Baatrup 1983a,b) and physiological evidence (Baatrup 1985) for the existence of a sense of taste in lampreys. Using scanning- and transmission-electron microscopy he identified papilla-like structures corresponding to previous light-microscopic descriptions (Retzius 1893b; Fahrenholz 1936) on the gill arches of larvae and in the branchial tube of adult lampreys (*P. Marinus*, *L. fluviatilis*, *L. planeri*). These structures are, just like the taste buds of gnathostomes, composed of sensory and supporting cells extending through the entire height of the epithelium; however, they differ from the latter in that the sensory cells are ciliated rather than bearing microvilli. The connection between the terminal buds and the brain is formed by fibres travelling with branches of the IXth and Xth cranial nerves. Recordings from the base of isolated terminal buds revealed that these structures are sensitive to chemical stimulation (Baatrup 1985). On the basis of these results, Baatrup (1985) suggested a functional relationship between the lamprey terminal buds and the taste buds of gnathostomes.

The oral disc in adult lampreys is bordered by numerous conical papillae, in the epidermis of which occur two types of solitary receptor cells, i.e. Merkel cells and elements bearing apical microvilli (Whitear and Lane 1981, 1983). The papillae are innervated by the Vth cranial nerve (Matthews and Wickelgren 1978). Baatrup and Døving (1985) presented electrophysiological evidence suggesting that Merkel cells respond to mechanical stimulation, whereas the villous cells respond to chemical stimuli. The functional spectrum of the latter differed, however, from that of the pharyngeal terminal buds.

The primary afferent fibres innervating the terminal buds reach the vagal lobe via the visceral rami of the IXth and Xth nerves. The central relations of the special viscerosensory trigeminal fibres, which carry impulses from the villous cells in the oral disc papillae, are unknown.

The efferent connections of the vagal lobe have not been studied with experimental techniques so far. However, Johnston (1902) used evidence from comparative studies to surmise that a polysynaptic gustatory pathway ascends from this lobe to the hypothalamus and, thence, to the so-called primordium hippocampi.

10.5.4
General Somatosensory System

The peripheral part of the general somatosensory system (Fig. 10.26b) is formed by the fibres of the trigeminal nerve. The cell bodies of these fibres are bilaterally located in a pair of large ganglia. Matthews and Wickelgren (1978) studied the physiology of neurons in these ganglia with intracellular electrodes. They distinguished four classes of sensory cells – touch (T), pressure (P), nociceptive (N) and pit organ (PO or neuromast) – on the basis of sensitivity to mechanical stimulation, rate of adaptation and response to electric and thermal stimuli. The T, P and N cells had several similarities with the dorsal cells of the cord. The PO cells, surprisingly, did not respond well to typical water-jet stimulation and it is likely that the adequate stimulus remains to be discovered (Matthews and Wickelgren 1978).

The central processes of the somata in the trigeminal ganglia enter the brain where they constitute a compact, caudally passing bundle, the radix descendens of V. This bundle is situated ventrolateral to the sulcus limitans and occupies a submeningeal position throughout its extent (Northcutt 1979a; Koyama et al. 1987; Anadón et al. 1989; Figs. 10.13–10.18). Small neurons are present within the confines of the tractus descendens of V. Although these elements never form a distinct aggregation, they are indicated in Figs. 10.13–10.18 as the 'nucleus' descendens of V. It has been suggested on the basis of non-experimental material (Ariëns Kappers et al. 1936; Heier 1948) that the efferent fibres of these cells establish reflex connections with the motor nuclei of V and VII and with reticulospinal neurons, and that they also participate in the formation of a diffuse ascending system, the general lemniscus spinalis + bulbaris, which will be further considered below. Some cells in the nucleus descendens have been shown to project to the spinal cord (Ronan 1989). Electrical stimulation of the trigeminal nerve elicits disynaptic and polysynaptic excitation and inhibition of reticulospinal neurons on both sides of the brain (Rovainen 1967a; Wickelgren 1977a). Viana Di Prisco et al. (1995), who also stimulated the trigeminal nerve electrically, recorded such mixed responses from reticulospinal neurons situated in the middle and inferior rhombencephalic reticular nuclei. Their results suggest that this trigemino-reticular input is conveyed via interneurons that use excitatory and inhibitory amino acid neurotransmitters. These interneurons may well be located in the 'nucleus' descendens (Fig. 10.26b).

A principal sensory trigeminal nucleus has not been identified so far in lampreys. However, Northcutt (1979a) surmised that the cell mass generally considered as the cerebellar nucleus (Figs. 10.18, 10.30) and some more caudally situated cells which cap the descending trigeminal fibres possibly represent a principal sensory nucleus. Huard et al. (1995) recently reported that, following injection of cobalt lysine as a retrograde tracer, numerous cells in the principal sensory nucleus on both sides are labelled, but details on the localisation of that nucleus are lacking.

It should be noted that the ventral position of the cellular elements constituting the general somatic afferent zone does not accord with the concept that, in the alar plate, the somatic cells lie dorsal to the visceral ones. It is possible, however, that the matrix of the general somatic afferent cells is situated in the dorsal part of the alar plate, and that these elements migrate downward during ontogenesis. Hugosson (1957) has presented evidence for such a migration in the chick.

The caudal and ventral parts of the alar plates contain a number of large cells. These elements show, in both their size and shape, a striking resemblance to the dorsal cells in the spinal cord (Figs. 10.15, 10.30). Finger and Rovainen (1982: *Lampetra lamottenii, Ichthyomyzon unicuspis, Petromyzon marinus*) and Koyama et al. (1987: *Lampetra japonica*) observed that rhombencephalic and rostral spinal dorsal cells are labelled after HRP application to the ophthalmic branch of the trigeminal nerve. Anadón et al. (1989) counted the rhombencephalic dorsal cells in larval sea lampreys and found that their number varied among individuals, ranging between 21 and 34 elements on each side. Application of HRP in the orbit labelled most ipsilateral rhombencephalic dorsal cells and up to 31 ipsilateral spinal dorsal neurons. The labelled rhombencephalic elements had a coarse, rostrally directed process that at some distance from the cell body gave rise to a thinner caudally directed process. Both processes were observed to join the descending trigeminal root. The coarse rostrally directed processes of rhombencephalic and spinal dorsal cells formed a distinct group located in the most medial part of the trigeminal descending root. The somata of most rhombencephalic dorsal cells had short processes that could be traced to neighbouring dorsal cells with which they apparently made contacts. Anadón et al. (1989) concluded from their experiments that in larval lampreys the rhombencephalic dorsal cells and the most rostral spinal dorsal cells together form a trigeminal com-

ponent, for which they proposed the name 'primary medullary and spinal nucleus of the trigeminus'. Finger and Rovainen (1982) demonstrated that rhombencephalic dorsal cells can be activated by electrical stimulation of the ophthalmic nerve, while Rovainen and Yan (1985) evoked sensory responses from such cells by mechanical stimulation of the skin of the sucker and head. Most of the responsive cells tested could be classified as pressure-sensitive, and a few as nociceptive. According to Rovainen and Yan (1985) these experiments indicate that some of the rhombencephalic dorsal cells of the lamprey, which project into the branches of the trigeminal nerve, have sensory functions similar to those of dorsal cells in the spinal cord (see Sect. 10.4.3).

It is interesting to note that the mesencephalic nucleus of V, a group of primary afferent neurons which is present in all gnathostomes, is entirely lacking in the lamprey. Finally it may be mentioned that the recently discovered dorsal column nucleus (Dubuc et al. 1993a,b), which has already been discussed in Sect. 10.4.7, also belongs to the rhombencephalic general somatosensory system.

10.5.5
Special Somatosensory Zone

The special somatosensory zone, or area octavolateralis, occupies almost the entire rostral part of the rhombencephalic alar plate (Fig. 10.29). This

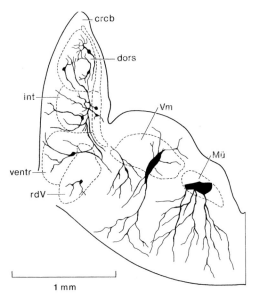

Fig. 10.29. Composite drawing of neurons observed in transversely sectioned Golgi preparations of the brain of the lamprey, *Lampetra fluviatilis* at the level of the entrance of the eighth nerve. For a list of the abbreviations used in figures, see p. 409 (Based on Heier 1948, Fig. 21b)

area contains three longitudinally arranged cell masses, which have been described under somewhat different names in the literature (Pearson 1936; Larsell 1947a, 1967; Heier 1948). In accordance with Nieuwenhuys (1967b, 1972, 1977) they are designated here as nucleus dorsalis, nucleus intermedius and nucleus ventralis of the area octavo-lateralis. The first two nuclei represent the end stations of the lateral line nerve fibres. These fibres enter the brain via two nerves, the posterior and anterior lateral line nerves (Fig. 10.3) and the latter can be further subdivided into separate dorsal and ventral roots. Centrally, these fibres bifurcate into ascending and descending branches; peripherally they are in contact with the lateral line receptors or neuromast organs. These are little patches of sensory cells arranged in rows in the skin of the head and body. The dorsal and intermediate nuclei border the ventricle and comprise, in addition to a thin, compact lamina of central grey, a zone of more laterally situated neuropil, in which scattered cells are embedded (Figs. 10.16, 10.17). Together these two nuclei constitute the area linea lateralis. Dorsolaterally this area is covered by a thin zone of neuropil, which is known as the crista cerebellaris. This zone is composed of (a) terminal branchings of root fibres of the lateral line nerves, (b) axons of small cells of the dorsal and intermediate nuclei, and (c) terminal dendritic processes of larger neurons contained within the same nuclei. The name of this zone refers to its striking resemblance with the molecular layer of the cerebellum, with which it is in continuity. To date the function of the lateral line system has received relatively little attention. Intracellular recording from lateral line ganglion cells (the PO cells mentioned earlier) showed that these elements are very sensitive to touch on the inside of the pit organ, and also usually respond to pressure on the skin around the organ but responded poorly to jets of water from a syringe onto the neuromast or to water-borne vibrations (Matthews and Wickelgren 1978). Electrical stimulation of the posterior lateral line nerve elicits EPSPs and IPSPs in various reticulospinal cells on both sides of the rhombencephalon. The Müller cells of the bulbar group, the cell I2 and the Mauthner cells are excited the most strongly (Rovainen 1979b; Fig. 10.34). Such activity is carried by the efferents of the lateral line nerve centres to the reticular formation. These efferents form part of the system of internal arcuate fibres (Figs. 10.15–10.17), which will be further discussed below.

The lateral line system also appears to mediate electroreception (Bodznick and Northcutt 1981; Bodznick and Preston 1983; Fritzsch et al. 1984). In the experiments of Bodznick and Northcutt, lampreys responded to cathodal (outside negative) low-frequency electric fields with an intensity as low as $0.1 \mu V$ cm^{-1} and, with higher fields, electrical responses could be recorded in the dorsal rhombencephalon, the torus semicircularis and the tectum. Later studies (Ronan and Bodznick 1986) showed that the electroreceptors were the so-called end buds, goblet-shaped cells found in the epidermis of mature lampreys. Ronan (1988) showed that the primary electrosensory afferents are components of the anterior lateral line nerve which project, via the dorsal root of that nerve, to the nucleus dorsalis. Some of the endings may form 'giant terminals' some 10–30 μm in diameter (Kishida et al. 1988; Koyama et al. 1993). Ronan (1988) also showed physiological evidence for electroreception in ammocoetes, but the receptors were not identified. Electroreception is found in most groups of jawed fishes and in urodeles as well. In all of these groups the lateral line area is divided into dorsal and intermediate nuclei, and in all of them the electroreceptive afferents project by way of a separate dorsal root of the anterior lateral line nerve to the dorsal nucleus, whereas the mechanoreceptive lateral line fibres converge upon the intermediate nucleus (for references and details see Ronan and Northcutt 1987). From these data it may be inferred that the projections to the intermediate nucleus of lampreys are also mechanosensory. The wide distribution of electroreception over jawed anamniotes and its existence in the lamprey shows that it may be one of the more fundamental sensory modalities of aquatic forms (Bullock et al. 1983).

The central projections of the lateral line nerves and their peripheral branches have been experimentally examined in adult (Ronan and Northcutt 1987: *Ichthyomyzon unicuspis, Petromyzon marinus*; Koyama et al. 1990: *Lampetra japonica*) and larval lampreys (González and Anadón 1992: *P. marinus*), utilizing silver impregnation of degenerating fibres and axonal transport of HRP. These studies revealed that, as already mentioned, the electroreceptive dorsal nucleus is exclusively innervated by fibres which enter the brain by way of the dorsal root of the anterior lateral line nerve (ALLN). These fibres terminate ipsilaterally throughout the longitudinal extent of the neuropil zone of the nucleus. Most fibres entering the brain via the dorsal root of the ALLN originate from ganglion cells, the peripheral axonal processes of which innervate postcephalic electroreceptors, which are known to be widely distributed over the trunk and tails of lampreys (Bodznick and Preston 1983). The peripheral processes reach the postcephalic electroreceptors by way of a recurrent branch which connects the ALLN ganglion with the large trunk lateral line nerve. The latter nerve, which lies alongside the vertebral column throughout the length of the trunk and tail, also carries fibres innervating the mechanoreceptive neuromast organs of the body. These fibres do not enter the recurrent branch of the ALLN, but rather join the posterior lateral line nerve (PLLN). Separate HRP labelling of afferents to the dorsal nucleus in three peripheral ALLN branches (buccal, superficial ophthalmic, recurrent) gave no indication of a somatotopic arrangement of these projections.

The mechanoreceptive intermediate nucleus receives three contingents of lateral line nerve fibres, which all terminate throughout the longitudinal extent of the neuropil lateral to the periventricular cell plate of that nucleus: (a) afferents from the ipsilateral PLLN, terminating in the medial neuropil adjacent to the cell plate, (b) fibres from the ipsilateral ventral root of the ALLN, projecting to the most lateral part of the neuropil, and (c) fibres from the contralateral PLLN, terminating in the central zone of the neuropil. The fibres of the lattermost contingent, which has been observed only in adult lampreys, continue beyond the rostral limit of the intermediate nucleus and decussate in the dorsal part of the cerebellar fibre layer, after which they descend to the contralateral intermediate nucleus. Some primary ALLN afferents also enter the cerebellar fibre layer but, unlike the PLLN projection, remain on the ipsilateral side.

The mechanoreceptive neuromasts are, contrary to those of gnathostomes, not innervated by central efferent neurons (Yamada 1973; Fritzsch et al. 1984; Ronan and Northcutt 1987; Koyama et al. 1990; González and Anadón 1992). Figure 10.26c summarises the lateral line nerve projections.

The fibres of the octavus nerve are peripherally connected with the labyrinth, which may be considered as a specialised part of the lateral line system. It contains cristae with hair cells that respond to rotatory movements of the head and maculae that respond to vibrations and tilt (Rovainen 1979b). The central projections of the octavus nerve have been experimentally examined in *Ichthyomyzon unicuspis* (Northcutt 1979b; axon degeneration technique) and in *Lampetra japonica* (Koyama et al. 1989; HRP technique). It appeared that the primary octavus afferents form ascending and descending roots, both of which terminate in the nucleus ventralis. The latter nucleus occupies a position between the nucleus intermedius and the radix descendens of V. Apart from rather diffusely arranged small cells, this nucleus contains three local aggregations of larger elements, which are known as the nuclei octavomotorii (Tretjakoff 1909b; Stefanelli 1934; Larsell 1947a, 1967; Figs. 10.17, 10.19,

10.30) or as the vestibular nuclei (Stefanelli and Caravita 1970). Since the octavus fibres that terminate in the nucleus ventralis and in the nuclei octavomotorii are all vestibular in nature, these nuclei may be designated collectively as the area vestibularis. It is noteworthy that the eighth nerve is composed of coarse and thin fibres, both of which bifurcate upon entering the brain. Within the ascending and descending octavus roots the thin fibres occupy a dorsal position while the coarse fibres are located ventrally. The coarse fibres establish peculiar synaptic contacts with the cells of the octavomotor nuclei. Large fusiform expansions of these primary afferent fibres are actually enveloped by the neuronal somata of these nuclei (Stefanelli and Caravita 1970). The labyrinth contains two types of ganglion cells, large cells which innervate the cristae of the ampullae, and small cells which innervate the maculae (Lowenstein et al. 1968, *L. fluviatilis*). It seems likely that the large cells correspond to the coarse octavus fibres and the small cells to the thin fibres. If this is correct, cristae afferents terminate ventrally and maculae afferents dorsally within the ventral nucleus (Koyama et al. 1989).

Some octavus fibres ascend to the cerebellum, where they terminate in the granular as well as in the molecular layers. A few fibres cross the median plane to terminate within the molecular layers of the contralateral cerebellum (Koyama et al. 1989). Figure 10.26c summarises the central projections of the octavus nerve.

Experimental studies employing retrograde tracers (Koyama et al. 1989, *L. japonica*; Fritzsch et al. 1989, *L. fluviatilis*) have shown that the labyrinth receives an efferent projection, which originates from a group of about 20 cells, located between the motor nuclei of V and VII, directly lateral to the soma of the Mauthner cell (Fig. 10.30). The nucleus ventralis of the area octavolateralis has been shown to receive, in addition to octavus fibres, a projection from the dorsal column pathway (Dubuc et al. 1993b; Fig. 10.26a).

Our knowledge of the efferent connections of the area linea lateralis and the area vestibularis is mainly based on the studies of Stefanelli (1934, 1937), Pearson (1936), Heier (1948) and Larsell (1967). These studies, which were all based on non-experimental material, can be summarised as follows:

1. Fibres originating from both areas pass rostrally to the cerebellum, where some of them decussate in a distinct commissure, the commissura vestibulolateralis (Fig. 10.18).

2. Fibres likewise originating from the vestibular as well as the lateral line areas arch around the

laterral angle of the fourth ventricle and pass medially as internal arcuate fibres (Figs. 10.15–10.17). Some of these fibres enter the ipsilateral or contralateral medial longitudinal fascicle in which they may ascend, descend or bifurcate and do both. These fibres form part of reflex paths to reticular and motor centres in the brain stem. However, most of the internal arcuate fibres that arise from the lateral line and vestibular centres cross over and proceed to the ventrolateral part of the basal plate, where they enter the lemniscus bulbaris. This is a diffuse ascending pathway which, as already mentioned, also contains fibres originating from the contralateral sensory nucleus of V and from the spinal cord. According to Heier (1948) the lemniscus lateralis, i.e. the vestibulolateral component of the bulbar lemniscus, terminates mainly in the torus semicircularis and, to a lesser extent, in the tectum mesencephali and in the caudal part of the dorsal thalamus.

3. The efferent fibres from the octavomotor nuclei follow different courses (Stefanelli 1937; Heier 1948; Larsell 1967). It has already been mentioned (cf. Sect. 10.4.6) that axons arising from the posterior nucleus constitute a crossed octavospinal system, whereas the intermediate nucleus gives rise to ipsilateral octavospinal fibres. Other axons or axonal branches from the intermediate nucleus pass rostrally in the medial longitudinal fascicle, attaining mesencephalic levels. We designate these fibres collectively as the tractus octavomesencephalicus medialis. The axons arising from the anterior octavomotor nucleus, finally, follow a rostromedial course and decussate in the base of the midbrain at the level of exit of the oculomotor nerve. This projection is termed here tractus octavomesencephalicus lateralis (Figs. 10.18, 10.19). The fibres of the octavo-mesencephalic tracts terminate in the nuclei innervating the external eye muscles and in several reticular centres. Direct synaptic contact between octavomesencephalic fibres and some large Müller cells has been demonstrated. Before turning to a discussion of the physiology of the important reflex systems of which the octavomotor and octavoreticular projections form part, we note that, according to Heier (1948), the fibres of the lateral octavo-mesencephalic tract issue terminals to the oculomotor nuclei and ascend to both the torus semicircularis and to the dorsal thalamus.

The central physiology of the vestibular reflexes, via ocular, reticulospinal and direct spinal systems, has been established in some detail. An interesting feature of the vestibular apparatus in lampreys is the absence of a horizontal vestibular canal (Lowenstein et al. 1968). Apparently, rotations in this plane are detected by the vertical cristae in each ampulla (Lowenstein 1970); nevertheless,

lampreys are able to maintain accurate control of pitch and roll under a variety of conditions (Ullén et al. 1995a).

Vestibulospinal reflexes can be elicited by mechanical stimulation of the labyrinth or electrical stimulation of the vestibular nerve. Such reflexes are probably mediated by octavomotor cells. The behavioral sequelae of these reflexes are conjugate eye movements; however there is neither inhibition of opposing reflexes nor nystagmus. Consequently, the eye reflexes of the lamprey are probably comparable to the simple excitatory pathways from single ampullae to individual eye muscles of mammals (Rovainen 1976). Visual input to the reticulospinal complex also plays a significant part in controlling body motion (Deliagina et al. 1993), and lampreys show complex behavioural responses to light (Ullén et al. 1995b).

Stimulation of the labyrinths also elicits more far-reaching reflexes that descend to the spinal level. Both early and late EPSPs are seen in myotomal motoneurons and interneurons in the rostral cord. The early EPSPs are produced by monosynaptic electrical excitation from the vestibulospinal tracts, while late ones are correlated with the discharges of reticulospinal neurons (Rovainen 1979a).

Among the reticulospinal cells, those of the 'V group' are strongly excited by electrical and rotatory stimulation of the labyrinths. Some of the excitation is mediated via electrical transmission from vestibular axons. Mauthner cells fire in response to vibration of the otic capsules or stimulation of the anterior vestibular nerves of either side; on the ipsilateral side this is again mediated electrically. Müller cells M3 and M4 (I1 in Rovainen's nomenclature) are excited by movement of the contralateral labyrinth, but inhibited by ipsilateral vestibular stimulation. Bulbar Müller cells respond to more general vibration and rotation (Rovainen 1979a).

Bussières and Dubuc (1992) studied the activity of vestibulospinal neurons in *Ichthyomyzon unicuspis* by intracellular recording during fictive locomotion. Most cells projecting to the ipsilateral side of the cord were maximally depolarised during discharges from ipsilateral rostral spinal ventral roots and showed a minimum during contralateral activity, whereas contralaterally projecting cells generally showed an opposite pattern, i.e. their peak of depolarisation occurred during contralateral activity. The source of this modulation is not known; however, it is known that the chemical component of the vestibular input to the posterior rhombencephalic nucleus is mediated by an NMDA metabotropic receptor that leads to suppression of other inputs (Alford and Dubuc 1993).

Orlovski et al. (1992) and Deliagina et al. (1992a,b) analysed the influence of the labyrinth on different groups of reticulospinal neurons by recording their responses to natural stimulation of vestibular receptors, evoked by rotation around the longitudinal axis (roll) and the transverse axis (pitch). The neurons tested exhibited both static and dynamic responses, i.e. they were both position and motion sensitive. The most characteristic response in neurons belonging to all four reticular nuclei (mesencephalic and superior, middle and inferior rhombencephalic) was an excitation with contralateral roll. It seemed that pitch and roll could trigger the spinal locomotor central pattern generator, which, by sending 'efference copy' signals back to the brain stem, produced modulation of reticulospinal neurons in relation to the locomotor rhythm (Orlovsky et al. 1992). Further details of the underlying vestibular circuitry were provided by Deliagina (1995), who studied the effects of labyrinthectomy.

In summary, several, apparently redundant, vestibulospinal pathways exist in the lamprey. These systems are involved in neck reflexes, undulatory movements and righting reflexes. Some of the pathways are relatively direct but others are transmitted through the reticulospinal system and indicate that at least one of the functions of the latter complex is related to elaboration of vestibular reflexes.

10.5.6
Basal Plate

The rhombencephalic basal plate is clearly divisible into a medial, somatomotor zone and a lateral, visceromotor zone. The somatomotor zones of both sides are separated from each other in the midline by the sulcus medianus inferior; in the middle part of the rhombencephalon their boundary with the visceromotor zones is marked by the sulcus intermedius ventralis (Fig. 10.30). Although the somatomotor zone does contain some primary efferent centres, by far the largest part is occupied by three motor centres of higher order that together constitute the rhombencephalic reticular formation. Related to the fact that the visceromotor zone terminates at the caudal level of the isthmus rhombencephali, we find that the somatomotor zone widens there, touching the anterior end of the sulcus limitans. Rostrally the somatomotor zone is broadly continuous with the tegmentum of the midbrain. However, the ventromedial part of this transitional area contains a large nucleus, which cannot be subsumed under the heading somatomotor. This cell mass, the nucleus interpeduncularis (Figs. 10.18, 10.19), receives fibres from the habenula via the fas-

ciculus retroflexus. Its efferents pass caudally in the diffuse tractus interpedunculobulbaris, the site of termination of which is unknown.

10.5.7
Somatomotor Nuclei

The rhombencephalic somatomotor centres include the rostral part of the spinal motor column, the abducens nucleus, the caudal oculomotor nucleus and the trochlear nucleus. The rostral part of the spinal motor column occupies the caudo-medial part of the basal plate (Figs. 10.18, 10.19). According to Johnston (1902) the large neurons in this area constitute the motor nucleus of XII, but it is known (cf. e.g. Black 1917, 1920) that this nerve occurs only in anurans and amniotes. Investigations with the retrograde tracer technique (Finger and Rovainen 1978) have shown that the motoneurons of the abducens nerve are situated medial and ventromedial to the facial nucleus. These investigations also suggested that the most rostral part of the rhombencephalic basal plate contains a separate caudal oculomotor nucleus which contributes fibres to the contralateral oculomotor nerve (Fig. 10.30).

The trochlear nucleus of the lamprey was discussed as early as 1883, when Ahlborn remarked upon the exceptionally far dorsal position of this somatomotor nucleus in this animal (Fig. 10.14). Along with Addens (1933), Ahlborn regarded the nucleus as being intracerebral. Larsell (1947b, 1967), however, thought that most of the trochlear nucleus is not located in the cerebellum proper, but rather in the transitional area between the cerebellum and the tectum mesencephali, i.e. the velum

Fig. 10.30. Chart of the brain stem (rhombencephalon + mesencephalon) of the lamprey, *Lampetra fluviatilis*. By means of a one-to-one topologic transformation the ventricular surface with its sulci is flattened out into a plane. The *thick vertical line* that constitutes the axis of the figure represents the sulcus medianus inferior. The *curved lines* that constitute the lateral limits of the figure represent the taenia mesencephali and the taenia rhombencephali (*continuous lines*) and the sulcus medianus superior (*dashed lines*). The remaining sulci are indicated by *thick curved* lines. Cell masses and large individual cells have been projected upon the ventricular surface and their outlines have been included in the transformation. The *thin continuous curved lines* indicate the boundaries of periventricular cell masses; the outlines of reticular nuclei are indicated by *thin, broken dashed curved lines*. The positions of migrated nuclei are indicated by *dotted curved lines*. The somata of large neuronal elements are shown in solid black. The numbers of cells present in some nuclei are indicated by *numerals* enclosed within the outline of the structure. For a list of the abbreviations used in figures, see p. 409. The approximate position of the nucleus nervi oculomotorii caudalis and of the nucleus nervi abducentis is indicated on the basis of the work of Finger and Rovainen 1978; modified from Nieuwenhuys 1972)

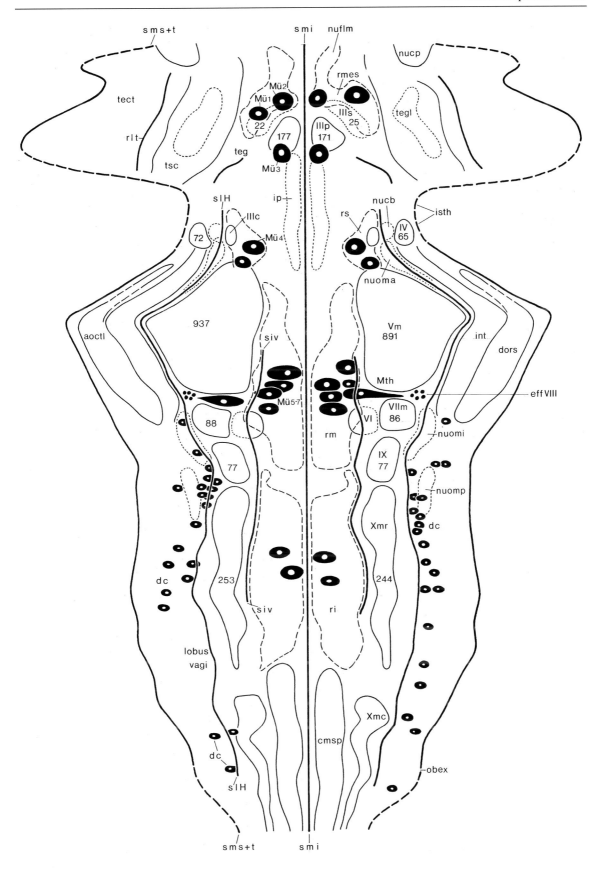

medullare anterius. Larsell thought that the cells of this nucleus are derived from the basal plate, from where they migrate dorsalward. Hugosson (1957), on the contrary, believed that the nucleus in question arises from an alar plate region, which, curiously enough, corresponds to the viscerosensory zone. The trochlear nucleus has been shown to contribute fibres to both the ipsilateral and the contralateral trochlear nerve (Finger and Rovainen 1978).

The innervation of the ocular muscles in the lamprey, *Lampetra fluviatilis,* has been subjected to a detailed analysis with retrograde tracers by Fritzsch and Sonntag (1988) and Fritzsch et al. (1990). The main results of these studies have been summarised in Table 10.1 and Fig. 10.31. Lampreys have six ocular muscles, most of which can be homologized to gnathostome eye muscles. However, the medial rectus muscle of jawed vertebrates appears to have no counterpart in lampreys; hence only three eye muscles are innervated by the oculomotor nerve in lampreys. A separate ventrolateral portion of the abducens nucleus, not described before, appears to innervate the caudal rectus muscle of lampreys. According to Fritzsch et al. (1990) this cell mass and muscle correspond to the accessory abducens nucleus and retractor bulbi muscle, respectively, of tetrapods. The existence of a separate caudal oculomotor nucleus, as claimed by Finger and Rovainen (1978), could not be confirmed. Neurons in all eye muscle nuclei, including the trochlear, extend dendrites to the medial longitudinal fasciculus (Fig. 10.32). Dorsally directed dendrites of trochlear neurons reach the area of lateral-line and retinofugal fibres (Fritzsch and Sonntag 1988), whereas the dendrites of the neurons in the oculomotor nuclei extend mainly laterally in the mesencephalic tegmentum (Fritzsch et al. 1990). Within the latter domain they may enter

into contact with retinotegmental fibres, which form a diffuse basal optic root (Fig. 10.38).

The development of the oculomotor system has been studied with retrograde tracers by Rodicio et al. (1992), Fritzsch and Northcutt (1993) and Pombal et al. (1994). Rodicio et al. (1992) found that immature trochlear motoneurons have ventricular attachments. Using this feature as an indicator of the site of origin of these neurons, they established that most trochlear neurons originate from the region between the sulcus limitans and the sulcus intermedius dorsalis, i.e. the rostral continuation of the viscerosensory zone. Fritzsch and Northcutt (1993), on the other hand, presented evidence suggesting that the matrix zone from which the trochlear neurons originate is situated in the tegmentum of the midbrain, close to the site of origin of the oculomotor nuclei, and that these cells migrate from there to the velum medullare anterius. So, curiously enough, the contradictory views of Hugosson (1957) and Larsell (1967) on the origin of the trochlear nucleus have both received some experimental support. Fritzsch and Northcutt (1993) also presented evidence suggesting that the motor neurons innervating the contralateral superior rectus muscle reach their contralateral position by perikaryal translocation within the dendrite that crosses the midline.

The main results of the study of Pombal et al. (1994) may be summarised as follows (Fig. 10.33). The ocular motor nuclei, like the retinopetal system, differentiate in early larvae before the development of the lateral optic tract. In the smallest larvae investigated (10–19 mm in length), the oculomotor and abducens neurons were ipsilateral to the site of tracer application, whereas trochlear neurons were contralateral. These motoneurons did not have any dendritic processes as yet. In larvae more than

Table 10.1 Ocular muscles and their innervation in the lamprey[a]

Ocular muscle	Abbreviation	Innervated by
Dorsal recuts m. (superior rectus m., s, h)[b]	d.r.	Nucleus III periventricularis, contralateral
Rostral rectus m. (inferior rectus m., h)	r.r	Nucleus III superficialis, ipsilateral
Rostral oblique m. (inferior oblique m., s, h)	r.o.	Nucleus III intermedia, ipsilateral
— (medial rectus m., h)		
Caudal oblique m. (superior oblique m., s, h)	c.o.	Nucleus IV, mainly contralateral
Ventral rectus m. (posterius rectus m., h)	v.r.	Nucleus VI dorsomedialis, ipsilateral
Caudal rectus m. (retractor bulbi m., h)	c.r.	Nucleus VI ventrolateralis, ipsilateral

[a] Based on Fritzsch et al. (1990) to which the reader is referred for references and details.
[b] Synonym(s) (s) and/or homologues (h) of the ocular muscles in the lamprey are added in brackets.

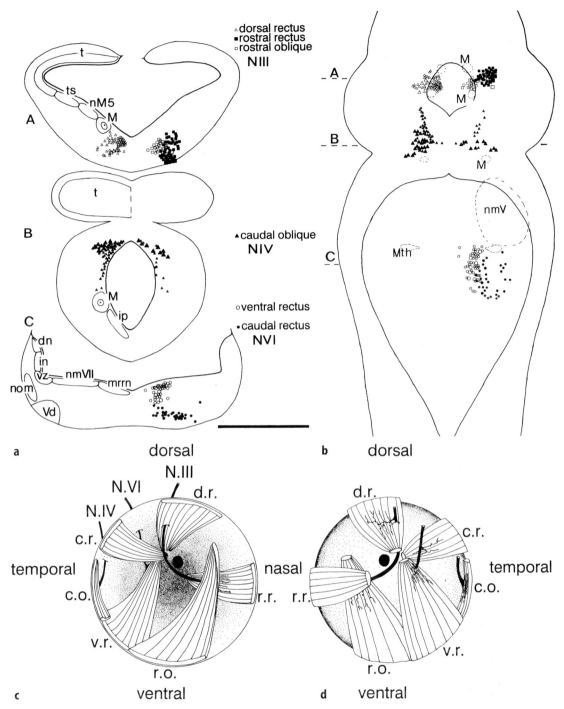

Fig. 10.31a-d. Survey of the location of ocular motoneurons and ocular muscles in the lamprey, *Lampetra fluviatilis*. The distribution of motoneurons which innervate ocular muscles is shown in transverse (**a**) and horizontal (**b**) camera lucida reconstructions. Ipsilaterally projecting neurons are plotted in the right halves and contralaterally projecting neurons in the left halves of these figures. The arrangement and topography of the ocular muscles and nerves are show in a right orbit viewed with the eye removed (**c**) and on a right eye viewed from the orbit (**d**). *Filled circle* indicates the optic nerve. *Bar* equals 1 mm. *cr*, caudal oblique muscle; *c.o.*, caudal oblique muscle; *c.r.*, caudal rectus muscle; *dn*, dorsal nucleus; *d.r.*, dorsal rectus muscle; *ip*, interpeduncular nucleus; *M*, Müller cells; *mrrn*, medial rhombencephalic reticular nucleus; *Mth*, Mauthner cell; *nM5*, nucleus of M5 of Schober (tegmentum mesencephali); *nmV*, *nmVII*, trigeminal and facial motor nuclei, respectively; *nom*, nucleus octavomotorius medius; *N III, IV, VI*, oculomotor, trochlear and abducens nerve, respectively; *rma*, reticular mesencephalic area; *r.o.*, rostral oblique muscle; *r.r.*, rostral rectus muscle; *t*, tectum; *ts*, torus semicircularis; *Vd*, descending trigeminal tract; *v.r.*, ventral rectus muscle. (Reproduced from Fritzsch et al. 1990)

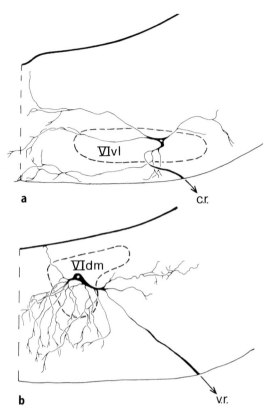

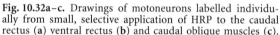

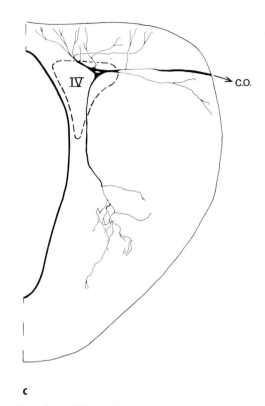

Fig. 10.32a–c. Drawings of motoneurons labelled individually from small, selective application of HRP to the caudal rectus (**a**) ventral rectus (**b**) and caudal oblique muscles (**c**).

For a list of the abbreviations used in figures, see p. 409 (Modified from Fritzsch et al. 1990)

19 mm in length, ipsilateral as well as contralateral components were found in the oculomotor and trochlear nuclei; dendrites were present, and their length and branching increased with larval age. An adult-like pattern of cellular arrangement and dendritic arborization was observed in larvae of about 45–60 mm in length. The question as to whether or not the contralateral oculomotor and ipsilateral trochlear neurons cross the midline during ontogenesis could not be answered.

10.5.8
Reticular Centres

10.5.8.1
Subdivision and Classification of Large Neurons

The positional relations of the rhombencephalic reticular centres, i.e. the nucleus reticularis inferior, medius and superior, are shown in Fig. 10.30. Together with the nucleus reticularis mesencephali these cell masses constitute the reticular formation of the brain stem. The cells in this formation vary considerably in size (Figs. 10.15–10.18). Many of

them are larger than the primary somatomotor neurons, and some are so conspicuous that they deserve to be named 'giant cells'. Such cells – the so-called Müller cells – are constant in position and generally occur bilaterally in pairs (Saito 1928; Stefanelli 1933a,b, 1934). Since there is no unanimity in the literature concerning the exact number of these Müller cells, one of us (Nieuwenhuys 1972) subjected the somata of all large reticular elements in the mesencephalon and rhombencephalon of five specimens of *Lampetra fluviatilis* to a quantitative and statistical analysis. Seven pairs of cells appeared to be significantly larger than the remaining reticular elements. Hence, the suggestion was made that the term Müller cells be reserved for these seven elements. Figure 10.30 shows that two of these elements, M1 and M2, are situated within the nucleus reticularis mesencephali. The third one, M3, occupies a slightly more caudal position and lies just behind the periventricular part of the oculomotor nucleus. The fourth Müller cell (M4) forms part of the nucleus reticularis superior, and the remaining three elements (M5-M7) are situated in the nucleus reticularis medius. The dendrites of the

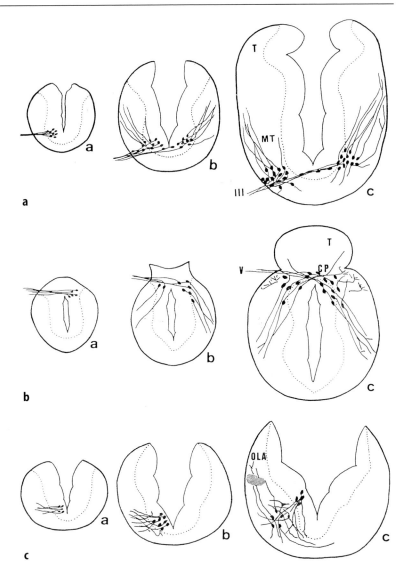

Fig. 10.33a–c. Schematic drawings of transverse sections showing the location and morphology of labelled oculomotor (**a**), trochlear (**b**) and abducens (**c**) motoneurons after applications of HRP into the left orbit. *a,* larvae of 10–19 mm; *b,* larvae of 20–45 mm; *c,* larvae of more than 45 mm in length. *Shaded area* indicates descending trigeminal tract. *CP,* cerebellar plate; *III,* oculomotor nerve; *IV,* trochlear nerve; *MT,* mesencephalic tegmentum; *OLA,* octavolateral area; *T,* mesencephalic tectum. (Reproduced from Pombal et al. 1994)

Müller and other reticular elements extend widely into the 'white' matter of the basal plate (Tretjakoff 1909b; Heier 1948; Martin 1979; Fig. 10.27). The axons of the Müller cells enter the ipsilateral medial longitudinal fasciculus, with which they descend to the ventral funiculus of the spinal cord (Tretjakoff 1909b; Stefanelli 1933b, 1934; Figs. 10.11–10.19). As far as we know, the axons of the remaining reticular elements also descend to the ipsilateral half of the spinal cord (cf. Martin 1979), hence the Müller and other reticular cells are collectively termed reticulospinal neurons. It is noteworthy that the initial parts of the axons of the reticulospinal cells are narrow; only at the spinal level do a number of these processes attain the dimensions of the giant fibres (Bertolini 1964; Rovainen et al. 1973; Martin 1979).

The area directly lateral to the nucleus reticularis medius contains a conspicuous neuron that has a large, lateral dendrite extending toward the entering fibres of the octavus nerve and is characteristically situated between the motor nuclei of V and VII. Johnston (1902) and Pearson (1936) suggested that this cell is homologous with the Mauthner cell of fishes and amphibians. This homology was also considered possible by Stefanelli (1933a), but he emphasised that the axon of this cell passes ipsilaterally to the spinal cord, whereas the Mauthner axon in other species decussates and descends contralaterally. However, Whiting (1957) and Rovainen (1967a) have conclusively shown that the axon of the cell in question does decussate, hence they labelled it a Mauthner cell. Figures 10.12, 10.13 and 10.22 show that the axon passes caudally in the lateral funiculus of the spinal cord.

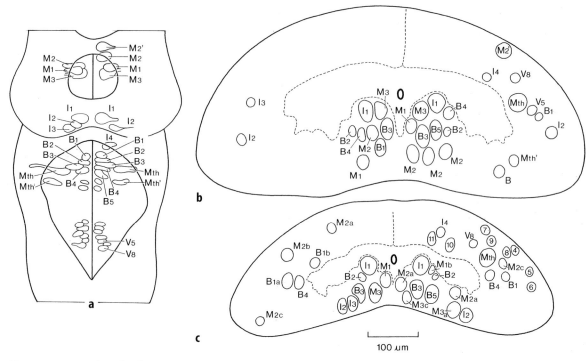

Fig. 10.34a–c. Tracing of individual axons in the spinal cord of the sea lamprey, *Petromyzon marinus*. **a** Brain stem of a small adult specimen reconstructed from serial transverse sections. Large reticulospinal neurons of the mesencephalic *(M)*, isthmic *(I)*, bulbar *(B)*, Mauthner *(Mth)*, and vagal *(V)* groups are labelled. Numbers with labelled cells are consecutive and are arbitrary for the B and V cells since several were omitted. The right side of the figure represents the position of individually traced axons in the spinal cord: just behind the obex (**b**) and at the level of the first dorsal fin (**c**). Lettered axons were traced from neurons in **a**, and numbered axons from giant interneurons situated on the left side and in the caudal half of the cord. Branches of axons are indicated by *a*, *b* and *c*. (Based on Rovainen et al. 1973, Figs. 1, 3)

Caudal to the Mauthner neuron, Rovainen (1967a) found a second, less conspicuous, large cell that he labelled Mth'. This element, which had been previously described by Stefanelli (1933a,b, 1934), is located between the motor nuclei of VII and IX (Fig. 10.34a). The axon of the caudal cell is not as large as the one from the rostral element but has a similar, crossed projection through the caudal rhombencephalon into the lateral funiculus of the spinal cord (Rovainen 1967a; Rovainen et al. 1973). Although the Mth and Mth' cells are both situated within the lateral, visceromotor rather than within the medial, somatomotor territory, these elements are discussed here because they are generally considered as forming part of the reticular formation.

Ultrastructural studies (Wickelgren 1977a) have shown that there are two types of synaptic ending on Müller cells, one containing round vesicles and the other containing ellipsoidal vesicles. These terminals were intermixed over the surface of the cell bodies and dendrites with no obvious segregation.

Rovainen and his collaborators (Rovainen 1967a, 1974b, 1979a,b, 1982, 1983; Rovainen et al. 1973) extensively studied the structure and function of the reticulospinal neurons in the lamprey, particularly the large elements. With regard to the latter they developed a nomenclature which differs in some respects from the one presented above (cf. Fig. 10.34).

In the rhombencephalon these authors observed three condensations of large reticular elements which they called the isthmus group I, the bulbar group B, and the vagal group V. These groups fall clearly within the confines of the nucleus reticularis superior, medius and inferior of the present account. A fourth group of large cells was identified by Rovainen and co-workers, which they termed the midbrain group, M; these lie in the mesencephalic reticular nucleus and are described in the next section.

Defining Müller neurons as the parent cells of large, uncrossed reticulospinal axons that lie in the ventral funiculus for at least part of their projection in the spinal cord Rovainen (1978, 1979b) counted nine pairs of Müller cells. Three of these, M1, M2 and M3, are situated in the tegmentum of the midbrain and clearly correspond to the elements indicated by us with the same labels. Element M4 of our numbering scheme corresponds to cell I1 of Rovainen. According to that author the isthmus region contains another very large reticular cell, which he

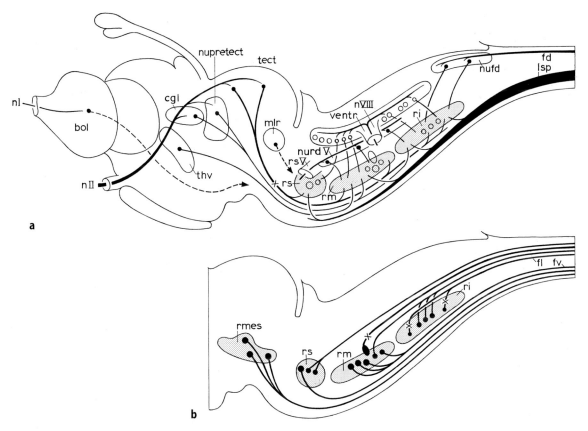

Fig. 10.35a,b. The principal connections of the reticular nuclei in the brain stem. **a** Afferents. **b** Descending efferents. For a list of the abbreviations used in figures, see p. 409

designated as I2. According to our analysis, the nucleus reticularis medius contains three true Müller cells, M5-M7, but Rovainen counted four of these elements in the corresponding area, which he labelled B1-B4.

Because frequent reference will be made hereafter to the four reticular centres of the lamprey, their names will be abbreviated in the text as follows: nucleus reticularis mesencephali, NRmes; nucleus reticularis rhombencephali superior, medius and inferior, NRRS, NRRM and NRRI, respectively.

10.5.8.2
Input to Reticulospinal Neurons

Olfactory nerve stimulation produces long latency excitation in Müller cells (Wickelgren 1977a). The pathways by which the impulses reach these cells are unknown (Fig. 10.35a). Regarding input from the ventral thalamus, Pombal et al. (1995) reported that injection of a retrograde tracer in the rhombencephalic reticular nuclei labelled cells in the ventral thalamus, and that electrical stimulation of this thalamic centre elicited both monosynaptic and

long-latency amino acid-mediated EPSPs in reticulospinal neurons.

Concerning input from the 'brain stem locomotor region', this region was identified by McClellan and Grillner (1984), who described it as a zone extending "from the mesencephalon, near the torus semicircularis, caudally through the rhombencephalon in parallel strips about 200–300 μm, lateral to the midline" (McClellan and Grillner 1984, p. 357). Electrical microstimulation of this strip in an in vitro preparation of the central nervous system of the lamprey elicited 'swimming motor activity' recorded from spinal ventral roots, but the nature of the structures stimulated remained obscure. Sirota et al. (1995) observed that electrical stimulation of a periventricular area in the caudal mesencephalon, directly in front of the isthmus, elicited well-coordinated swimming in semi-intact larval specimens of *Petromyzon marinus*. The activity began within a few seconds after stimulation. The authors concluded that the tegmentum of the midbrain, just like that of higher vertebrates, contains an area that is involved in the control of locomotor activity. The efferents of this area are unknown in the lamprey. Possibly it exerts its influence via the nearby NRRS, another nucleus which has been suggested to be involved in the command system for locomotion (Swain et al. 1993).

Stimulation of an optic nerve or natural stimulation of visual receptors in one eye excites Müller cells bilaterally (Rovainen 1967a; Wickelgren 1977a). These impulses are believed to reach the Müller cells via the tectobulbar tract, a large fibre system which originates from the deeper layers of the tectum and descends through the area directly ventral and lateral to the medial longitudinal fasciculus. The most effective visual stimulus is a shadow falling across the eye (Wickelgren 1977b). Recent physiological (Deliagina et al. 1993) and combined physiological and experimental neuroanatomical studies (Ullén et al. 1995b; Zompa and Dubuc 1995) have shown that visual input may reach rhombencephalic reticulospinal neurons via at least three different routes: (a) decussating retinotectal fibres, contacting tectal neurons and giving rise to decussating tectoreticular fibres; (b) a retino-thalamo-reticular projection, which is synaptically interrupted in the lateral geniculate nucleus (cf. Sects. 10.7.2 and 10.8.3); and (c) a retino-pretecto-reticular projection.

Electrical stimulation of the posterior lateral line nerve elicits EPSPs in the Müller cell I2 and in the anterior Mauthner cell and bulbar Müller cells on each side of the brain in both larval and adult lampreys (Rovainen 1983).

Reticulospinal cells of all types are excited or inhibited by stimulation of the octavus nerves (Rovainen 1967a; Wickelgren 1977a). The Mauthner cells are known to receive direct excitatory synapses from vestibular afferents (Rovainen 1979a). The vestibular input to rostral Müller cells is mediated by cells situated in the vestibular area, probably by the large contralateral octavomotor neurons (Rovainen 1978). Stimulation of the ipsilateral labyrinth produces IPSPs in M1-M4, while most bulbar Müller cells (M5-M7) also receive excitation from the ipsilateral anterior vestibular nerve, probably through a disynaptic relay in the octavomotor region (Rovainen 1978). Cells of the vagal group (inferior reticular nucleus) are excited via the ipsilateral anterior or contralateral posterior nerve via electrical synapses. Reticulospinal neurons in all four reticular nuclei (NRMes, NRRS, NRRS, NRRI) have been shown to respond to natural stimulation of vestibular receptors evoked by rotation along the longitudinal and transverse axes (Orlovski et al. 1992; Deliagina et al. 1992a,b). The NRRI receives direct projections from the ipsilateral intermediate octavomotor nucleus and from the ipsilateral and contralateral ventral octavolateral and inferior octa-

vomotor nucleus (Dubuc et al. 1993b). Stimulation of the latter nucleus elicits large EPSPs in reticulospinal neurons situated in the NRRI (Dubuc et al. 1993b).

Mechanical stimulation of the sucker mouth, head, gills or activation of the trigeminal nerve produces both EPSPs and IPSPs in Müller and other reticular cells (Rovainen 1967a; Wickelgren 1977a,b). The synaptic latencies of these potentials suggest a disynaptic pathway in which neurons of the nucleus of the tractus descendens nervi trigemini are intercalated (Rovainen 1978). These responses have both early and late components; the latter seem associated with arousal of the animal (Rovainen 1978). Mechanical or electrical stimulation of the snout, or direct stimulation of the sensory division of the trigeminal nerve, elicits a withdrawal response followed by escape swimming (McClellan 1984). Viana Di Prisco and Dubuc (1995) and Viana Di Prisco et al. (1995) presented evidence suggesting that the trigeminal inputs triggering these responses are conveyed to reticulospinal neurons in the NRRM and NRRI via interneurons using excitatory and inhibitory amino acid transmitters.

Two major ascending spinal systems project to the lamprey brain stem, i.e. the dorsal column pathway and the spinal lemniscus (cf. Sect. 10.4.7), and the reticulospinal neurons receive a prominent input from both of these systems. Impulses travelling along the dorsal column are relayed in the dorsal column nucleus, and neurons in the latter have been shown to project to the NRRI, exerting an inhibitory influence, mediated by glycine, on its constituent neurons (Dubuc et al. 1993a,b). The spinal lemniscus induces oligo- or polysynaptic responses in reticulospinal neurons (Dubuc and Grillner 1987). Physiological experiments have shown that spinobulbar axons passing rostrally in the spinal lemniscus signal different aspects of the locomotor pattern to the reticular formation. In in vitro brain stem/spinal cord preparations, some of these fibres were observed to discharge in phase with either the ipsilateral or the contralateral ventral root bursts, while others discharged with either of the transition phases between these two bursts (Vinay and Grillner 1992).

10.5.8.3
Efferents of the Reticular Formation

Although it seems likely that some reticular neurons project to higher centres of the brain (cf. Sect. 10.8.3), there can be no doubt that most of its smaller neurons, and all of its large, individually identifiable cells project to the spinal cord (cf. Sect. 10.4.6). On the basis of the histological tracing

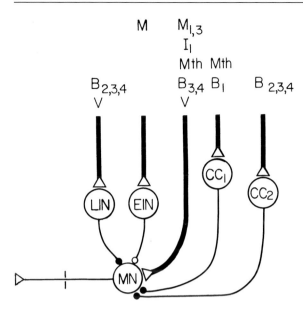

Fig. 10.36. Synaptic connections between large reticulospinal and different types of spinal neurons. *B1-B4,* bulbar Müller (Mü) cells; *CC1, 2,* inhibitory premotor interneurons with crossed, caudal axons; *EIN,* excitatory premotor interneuron; *I1,* isthmic Mü cell; *LIN,* lateral inhibitory interneurons; *M, M1, 3,* mesencephalic Mü cells; *MN,* motoneurons; *Mth,* Mauthner cell; *V,* 'vagal' cells in the inferior rhombencephalic reticular nucleus. (Modified from Brodin et al. 1988b)

study of Rovainen et al. (1973; Fig. 10.34), the reticulospinal projection may be subdivided into a medial and a lateral system (Fig. 10.35b). The medial reticulospinal system, which descends in the ventral funiculus, consists of the coarse axons of most Müller cells, including M1–3, I1 and B2, B3, and B5. The lateral reticulospinal system, which is less compact, is mainly situated in the dorsolateral part of the lateral funiculus. It contains, apart from the very coarse Mauthner fibre, at least 20 coarse and medium-sized fibres originating from the ipsilateral rhombencephalic reticular formation. The analysis of Rovainen et al. (1973) revealed that at least seven to ten large and medium-sized axons in each lateral funiculus originate from cells situated in the vagal (V) group (i.e. the nucleus reticularis inferior). A smaller number of these lateral axons originate from cells in the bulbar group (i.e. the nucleus reticularis medius). Finally, about four medium-sized lateral axons could be traced from cells situated in the isthmus region, within the confines of the nucleus reticularis superior.

All large and numerous smaller neurons located in all four reticular nuclei use an excitatory amino acid as a neurotransmitter (Ohta and Grillner 1989; Brodin et al. 1989a). Monosynaptic reticulospinal EPSPs have been recorded in motoneurons as well as in the various types of premotor neurons form-

ing part of the spinal locomotor network (Rovainen 1974b; Buchanan 1982; Ohta and Grillner 1989; Brodin et al. 1988b; Fig. 10.28). These synapses frequently utilize both electrical and chemical transmission (Rovainen 1974b). Brodin et al. (1988b) have pointed out that the different classes of spinal neurons receive a specific reticular input. Thus, motoneurons are excited by the cells M1, M3, I1, Mth and some large neurons of the bulbar and vagal groups, whereas inhibitory interneurons receive input from B1–4 and large neurons of the vagal group (Fig. 10.36).

Apart from numerous elements using excitatory amino acids as neurotransmitters, cells containing serotonin, cholecystokinin (CCK) and a pancreatic polypeptide/peptide YY-like compound (PYY) also contribute to the reticulospinal projection. It is remarkable that the serotoninergic descending projection arises from two bilateral cell groups situated within the NRRS and NRRI, respectively (Fig. 10.53d). Serotoninergic cell-containing raphe nuclei, as found in jawed vertebrates as well as in myxinoids, are lacking in the lamprey (Steinbusch and Nieuwenhuys 1979; Pierre et al. 1992). CCK- and PYY-containing neurons form subpopulations within the NRRI and NRRS, respectively. The serotoninergic, CCK- and PYY-containing fibres are all thin and diffusely distributed over the spinal 'white' matter. The serotonin- and CCK-containing fibres occur in the ventral and lateral funiculi, but those containing PYY are confined to the lateral funiculus (Brodin et al. 1988a, 1989a,b; Figs. 10.53, 10.54).

Finally, there is some evidence for potentiation and habituation of Müller cells following cranial nerve stimulation (Wickelgren 1977b). Use-dependent modulation of the glutaminergic EPSPs from reticulospinal neurons have also been reported (Brodin et al. 1994). The Müller and Mauthner cells have also provided the first intracellular studies of neuronal pH regulation in a vertebrate (Chesler and Nicholson 1985; Chesler 1986).

10.5.9
Visceromotor Zone

The visceromotor zone extends from the isthmus region caudally to the level of the obex, where it passes gradually over into the spinal grey matter. Its intermediate part is embraced between the sulcus intermedius ventralis and the sulcus limitans. Contrary to the somatomotor zone, the visceromotor zone consists largely of primary motor elements. The latter constitute a series of five distinct cell masses, the efferent nuclei of V, VII and IX, and the rostral and caudal nucleus of X (Fig. 10.30). The motor nucleus of V is extraordinarily large and makes a ventricular elevation, known as the eminentia trigemini (Figs. 10.3a, 10.17). This nucleus innervates the muscles of the organs used in feeding, i.e. the sucker mouth, the rasping organ and the pharynx.

Motoneurons of the lateral part of the motor trigeminal nucleus also innervate the velum, a structure primarily responsible for the unidirectional respiration in the larval lamprey (Homma 1975; Rovainen 1977). In the adult, respiration becomes tidal and is mediated by respiratory motoneurons that innervate the muscles of the branchial basket and are located in the nuclei of VII, IX and X. During breathing these neurons exhibit periodic and powerful bursts of EPSPs, which are elicited by unidentified generator neurons in the rhombencephalon (Rovainen 1974c, 1977, 1982). Adult lampreys breathe by a tidal motion of water, into and out of seven gill pores and pouches on each side of the head. The gill pouches consist of cartilaginous bags surrounded by muscle fibres. During exhalation the muscle fibres contract synchronously to empty the pouches. Inhalation, which is passive, is produced by the elastic recoil of the cartilaginous baskets surrounding the gill pouches (Rovainen 1982). The neurons innervating the muscles of the branchial baskets are located in the motor nuclei of VII, IX and X. During breathing these neurons exhibit periodic and powerful bursts of EPSPs, which are elicited by rhombencephalic generator neurons, the exact location of which is unknown (Rovainen 1974c, 1977, 1982). In later studies Rovainen concluded that conventional inhibition is not essential to the neural oscillator underlying respiration (Rovainen 1983b) and that the motor pattern for respiration is partly generated and coordinated in the rostral half of the medulla and is transmitted to respiratory motoneurons via descending pathways (Rovainen 1985b).

The physiological experiments of Russell (1986) suggested that the central pattern generator for the respiratory rhythm in adult lampreys comprises at least six different types of interneurons and that certain components of this pattern generator are located in the area surrounding the motor nucleus of V. Brodin and Grillner (1990) considered it likely that the respiratory motor pattern is produced by interaction between groups of excitatory interneurons at different levels of the rhombencephalon, each of which in isolation can produce a respiratory burst pattern. It is noteworthy that respiration in the hagfish is accomplished in the same mode as that of the larval lamprey, i.e. by an unidirectional velar mechanism, rather than by the reciprocal branchial motion found in adult lampreys.

The motor nuclei of VII, IX and X presumably contain, in addition to branchial motoneurons, pre-ganglionic, general visceral efferent elements (Johnston 1905); however, the location of these cells is not known. Addens (1933) believed that the caudal nucleus of X represents a splanchnic centre and thus consists entirely of preganglionic cells.

10.6
Cerebellum

10.6.1
Structure of the Cerebellum

The cerebellum is very small and consists of a thin, transversely oriented lamina of tissue that bridges the most rostral part of the fourth ventricle (Figs. 10.3a, 10.18). The cerebellum of the lamprey, as that of all other vertebrates, is doubtless a derivative of the rhombencephalic alar plate, but as regards the components of the latter that contribute to it, two different opinions have been expressed. Johnston (1902) believed that the cerebellum of cyclostomes, and that of other vertebrates as well, develops only from the special somatosensory zone; but Larsell (1947a, 1967) conjectured that the general somatosensory zone is also involved in the formation of the cerebellum. A contribution of the viscerosensory zone to the cerebellum has not been described so far, although Barnard (1936) observed that the visceral afferent grey passes "insensibly into the cerebellar grey". Using our material from adult specimens of *Lampetra fluviatilis*, we got the impression that the cerebellum of the lamprey is made up of a complete segment of the alar plate. However, concerning its continuity with special cellular areas, we were able to confirm the observation of Pearson (1936) that the nucleus intermedius of the area octavo-lateralis passes directly over into the cerebellar periventricular grey (Fig. 10.30). The cellular elements making up the cerebellar grey do not differ materially from those in the lateral line centres. Small, granule-like cells are present, which send fine axons into the external, molecular layer. The latter also receives dendrites of larger elements, the axons of which enter the anterior part of the internal arcuate system (Larsell 1967). Some authors (Johnston 1902; Ariëns Kappers et al. 1936; Larsell 1967) considered these larger elements as the 'forerunners' of the Purkinje cells of gnathostomes. However, these elements are neither arranged in a distinct layer, nor are their dendritic trees confined to a single plane. It is likely that the large 'Purkinje cells' observed in the lamprey by Schaper (1899) and Saito (1928) were actually elements of the trochlear nucleus.

In addition to a layer of periventricular grey the cerebellum of the lamprey contains a small group of migrated cells that Rüdeberg (1961) and Larsell (1967) considered to be 'a precursor' of the cerebellar nuclei of gnathostomes. This nucleus is situated ventromedial to the nucleus octavomotorius anterior (Fig. 10.18). Larsell thought it likely that axons of primitive Purkinje cells terminate in it and that its efferents contribute to a fibre bundle that ascends from the cerebellum to the tegmentum mesencephali.

10.6.2
Cerebellar Connections

The fibre connections of the cerebellum have been studied in normal material by Clark (1906), Pearson (1936), Larsell (1947a), Heier (1948) and others. Since these studies have been ably reviewed by Larsell (1967), we confine ourselves here to a brief survey. For a diagrammatic representation of the principal pathways the reader is referred to Nieuwenhuys (1967b). The following afferent systems have been described:

1. Primary fibres from the lateral line and octavus (=vestibular) nerves. The presence of primary lateral line fibres and octavus fibres passing to the cerebellum has been experimentally confirmed (Northcutt 1979b; Ronan and Northcutt 1987; Koyama et al. 1989, 1990; cf. Sect. 10.5.5 and Fig. 10.26c). Moreover, it has been shown that some fibres ascend in the spinal dorsal column and reach the cerebellum (Ronan and Northcutt 1990; Dubuc et al. 1993b; cf. Sect. 104.7 and Fig. 10.26a).
2. Secondary fibres from the area octavo-lateralis. These fibres terminate both ipsilaterally and contralaterally. The decussating fibres form a large commissure, the commissura vestibulo-lateralis, in the most dorsocaudal part of the cerebellum.
3. Primary and secondary trigeminocerebellar fibres. The existence of a direct projection of trigeminal afferents could not be experimentally confirmed, however (Northcutt 1979a).
4. Spinocerebellar fibres ascend from the lateral funiculus of the spinal cord (Pearson 1936; Heier 1948). These fibres and the secondary octavolateral and trigeminal fibres mentioned above form part of an assemblage of loosely arranged axons, situated ventromedial to the tractus descendens of the trigeminal nerve. This bundle is known under various names. It is designated here as the general spinal and bulbar lemniscus. Most of its fibres terminate in the tectum rather than in the cerebellum. Larsell (1947a) has pointed out that some of the spinocerebellar

fibres decussate, constituting a separate commissura cerebelli, situated rostroventral to the large commissura vestibulo-lateralis.
5. Tectocerebellar fibres pass caudally through the velum medullare anterius (Pearson 1936).
6. A lobo-cerebellar tract, connecting the inferior parts of the hpothalamus with the cerebellum, was first described by Johnston (1902). Several authors have confirmed the existence of this tract in adult (Clark 1906; Pearson 1936) or larval lampreys (Tretjakoff 1909b; Larsell 1967), but Heier (1948) was unable to find it.

The efferent fibres of the cerebellum form a bundle of internal arcuate fibres, which is a direct continuation of the system of arcuate fibres that originate in the area octavo-lateralis (Larsell 1947a, 1967). A rostroventrally directed component of this system fans out in the tegmentum of the midbrain, thus representing the beginning of the brachium conjunctivum of higher vertebrates (Larsell 1947a, 1967). The more caudally directed cerebellar arcuate fibres constitute, according to Larsell, crossed and uncrossed pathways that descend into the rhombencephalon. Pearson (1936) thought that these fibres form a tractus cerebellomotorius, distributing the nuclei of III and the motor nuclei of the rhombencephalon to the medial longitudinal fasciculus. Pearson also said that the cerebellar efferent system contains a cerebellotectal component which enters the superficial layers of the roof of the midbrain. Heier (1948) thought it likely that fibres of cerebellar origin pass to the tectum as well as to the torus semicircularis and also described a direct projection from the cerebellum to the dorsal thalamus.

10.7
Mesencephalon

10.7.1
Introductory Note

The mesencephalon of the lamprey is divisible into two principal regions, a dorsal region, comprising the tectum mesencephali and the torus semicircularis, and a ventral region, which is termed the tegmentum mesencephali (Fig. 10.19). These dorsal and ventral regions may be considered as rostral continuations of the rhombencephalic alar and basal plates, respectively. Several authors (Johnston 1912; Saito 1930; Heier 1948) have traced the sulcus limitans (the ventricular landmark indicating the boundary of the alar and basal plates) throughout the midbrain. Schober (1964), on the other hand, found that this groove extends only in the caudal

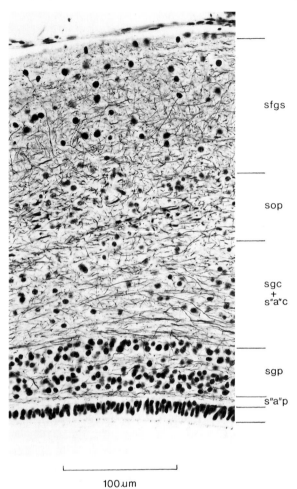

sfgs

sop

sgc
+
s"a"c

sgp

s"a"p

100 µm

Fig. 10.37. Transverse section through the tectum mesencephali of the lamprey, *Lampetra fluviatilis.* Silver impregnation according to Bodian. For a list of the abbreviations used in figures, see p. 409

part of the midbrain, and according to the observations of one of us (Nieuwenhuys 1972; Fig. 30) the sulcus limitans terminates in the isthmus region. The lateral recess of the mesencephalic ventricular cavity, which was considered by Addens (1933) to represent the mesencephalic continuation of the sulcus limitans, is definitely a separate groove. A unique feature of the mesencephalon of the lamprey is that the central part of its roof is formed by a choroid plexus (Figs. 3d, 10).

10.7.2
Torus Semicircularis, the Tectum Mesencephali and the Visual System

The torus semicircularis consists of a zone of grey that is a direct ventral continuation of the stratum griseum periventriculare tecti (Fig. 10.15). Accord-

ing to Heier (1948) the fibre connections of the torus closely resemble those of the tectum, the main difference being that the tectum receives a large number of optic and general somatosensory fibres and fewer fibres from the vestibular and lateral line centres, whereas for the torus the reverse condition prevails.

The tectum mesencephali is a large, distinctly laminated structure that, as its name implies, occupies the dorsal part of the midbrain. Contrary to Heier's (1948) description, its ventral boundary does not coincide with the recessus lateralis, but is situated somewhat further ventrally (Figs. 10.19, 10.30). Since optic tract fibres are quantitatively its most important afferent system, this centre is generally referred to as the tectum opticum. However, it should be appreciated that the tectum receives inputs from most, if not all, other somatic sensory modalities and therefore represents a general centre for somatic sensory correlation, rather than an optic relay station (Ariëns Kappers et al. 1936; Heier 1948).

The structure of the tectum mesencephali has been extensively described by Johnston (1902), Ariëns Kappers et al. (1936) and Heier (1948). Its layers are, from external to internal (Figs. 10.19, 10.37):

1. Stratum fibrosum et griseum superficiale, containing fine optic tract fibres and terminals, and some neurons
2. Stratum opticum, through which coarse and medium-sized optic fibres pass, which either terminate in contact with the intrinsic neurons scattered among them, or pass to the underlying layer
3. Stratum 'album' et griseum centrale, a layer which also receives afferents from other sources and whose constituent neurons contribute to the efferent tracts of the tectum
4. Stratum griseum periventriculare, consisting of densely packed cells, which are arranged in several sublayers and the elements of which send their dendrites towards the surface
5. Stratum 'album' periventriculare, made up by afferent fibres from diencephalic areas and by afferents from and efferents to the torus semicircularis

The retinal projections were studied experimentally by Northcutt and Przybylski (1973) and Kennedy and Rubinson (1977). Most of the optic nerve fibres reaching the midbrain appeared to terminate contralaterally, constituting a continuous layer throughout the superficial two thirds of the tectum. Ipsilaterally, only a restricted zone located at the ventrolateral margin of the tectum was found to

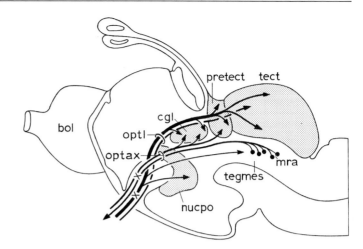

Fig. 10.38. Diagram showing experimentally established retinofugal and retinopetal projections. Based on various sources mentioned in the text. For a list of the Abbreviations Used in Figures, see p. 409

receive a retinal input. Other afferents to the tectum originate, according to Heier (1948), from the spinal cord, from the lateral line and vestibular centres, and from the sensory nucleus of V. These afferents decussate as arcuate fibres and ascend in the lemniscus spinalis and bulbaris. A small cerebellotectal projection has been described (Pearson 1936), and Heier reported, in addition, afferents from the ganglion habenulae, the dorsal thalamus, and the pineal and parapineal organs. The latter afferents originate from accumulations of ganglion cells in these organs and pass centrally in the so-called parietal nerve.

The most prominent efferent system of the tectum is the tractus tectobulbaris. Its fibres assemble in the stratum fibrosum profundum, then they course ventralward, undergo a partial decussation in the base of the midbrain and finally turn caudally to form a bundle situated just ventral to the fasciculus longitudinalis medialis. It has already been mentioned that the fibres of the tractus tectobulbaris synapse with the cells of the reticular formation (Fig. 10.35a). Other tectal efferents include, according to Heier (1948), a tract that descends to the tegmentum mesencephali and a number of fibres that distribute to the diencephalon. Most of these ascending efferents form a fibre system that Heier termed the tractus tecto-thalamicus et hypothalamicus rectus et cruciatus. Reportedly, this tract originates in the tectum and in the torus semicircularis. It passes rostrally along the medial aspect of the optic tract and terminates in the dorsal thalamus, the ventral thalamus and the hypothalamus. Some of its fibres decussate in the commissura postoptica, a large commissure situated caudal to the chiasma opticum (Fig. 10.45). Commissural fibres, passing rostral and caudal to the membranous part of the mesencephalic roof, interconnect the right and left halves of the tectum.

Because the tectum is by far the largest and the most differentiated visual centre, it is appropriate to summarise here the morphology and some aspects of the physiology of the central visual system. The lamprey possesses lateral eyes and two additional photosensitive systems: a dermal light sense, associated with the lateral line nerve, and the pineal eye. Of these systems, the lateral eyes are the most important.

The optic nerve enters the brain ventrorostrally. It contains about 35 000 thin fibres (Öhman 1977), some 750 of which are retinopetal (Vesselkin et al. 1980). Most of the optic nerve fibres decussate in the rostral part of the optic chiasma, after which they form the large lateral optic tract. This tract passes superficially, via the thalamic and pretectal regions, to the tectum mesencephali. A smaller contingent of optic nerve fibres decussates more dorsally and caudally in the chiasma and forms the axial optic tract. This tract passes dorsocaudalward through the thalamic complex of migrated cell masses, formed by the nucleus of Bellonci and the anterior and posterior geniculate bodies, and attains the pretectal region (Heier 1948). A limited number of optic nerve fibres join the ipsilateral lateral olfactory tract. The retinopetal optic nerve fibres are largely crossed.

The retinofugal and retinopetal projections have been studied experimentally by a number of investigators using different techniques (Northcutt and Przybylski 1973; Kennedy and Rubinson 1977; Repérant et al. 1980; Vesselkin et al. 1980, 1984, 1989; Rio et al. 1993; Fig. 10.38). These studies have shown that the lateral optic tract distributes mainly, but not exclusively, crossed fibres to the thalamic complex of migrated cells, the pretectum and the tectum mesencephali. The lateral optic tract also projects bilaterally to the nucleus of the postoptic

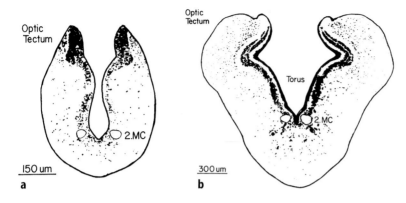

Fig. 10.39a,b. Transverse sections through the midbrain of a larval (**a**) and a metamorphosing lamprey (**b**). Note the differences in size and differentiation of the tectum. (Reproduced from Kennedy and Rubinson 1977)

commissure and sends some solitary, diffuse fibres to the tegmentum of the midbrain. The fibres of the axial optic tract terminate mainly in the dorsal thalamus and in the pretectum. The retinopetal fibres originate from two groups of mesencephalic cells, one situated in the periventricular grey, the other more laterally in what has been called the mesencephalic reticular area (Repérant et al. 1980). The axons of these neurons join the axial optic tract and project bilaterally to the retina, although with a predominant contralateral input. A considerable number of the mesencephalic retinopetal neurons has been shown to be GABAergic (Rio et al. 1993).

The visual system as a whole develops slowly over several years and only becomes functional at metamorphosis (for a review, see Rubinson 1990). In early larvae the eyes are separated from the surface by a dense dermal covering and only a small central zone of retina, immediately surrounding the optic nerve head, is differentiated (Studnicka 1912; Dickson and Collard 1979). During metamorphosis the eyes enlarge greatly and emerge to the surface. Differentiation of cells in the peripheral zone of the

retina occurs over a long period of time. Amacrine cells, horizontal cells and ganglion cells develop gradually during the larval period. Retinal development is arrested during the premetamorphic period, to be resumed and completed during metamorphosis with the differentiation of photoreceptor and bipolar cells (Rubinson and Cain 1989).

The development of the tectum has been studied by many authors (Schober et al. 1964; Pfister 1971c; Kennedy and Rubinson 1977, 1984; De Miguel and Anadón 1987; Rubinson 1990). In young larvae the tectum is small and exhibits a dense periventricular cellular layer, flanked by a zone of loosely scattered cells (Fig. 10.39a). During metamorphosis the tectum grows both by mitoses and by radial migration of cells (Pfister 1971c), and the typical tectal lamination becomes identifiable (Fig. 10.39b).

Axon degeneration studies (Kennedy and Rubinson 1977) suggested that during larval life only a contralateral retinofugal projection is present, but a more recent tracer study (De Miguel et al. 1990) revealed the presence of both contralateral and ipsilateral retinofugal projections. This study also

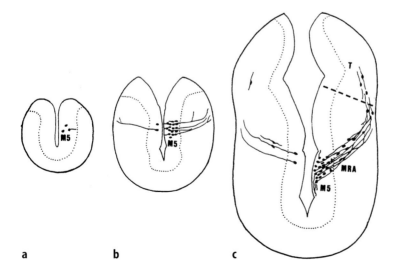

Fig. 10.40a–c. Schematic drawings of transverse sections of the midbrain of larval lampreys, showing the location of labelled retinopetal cells after injection of HRP into the orbit. **a** Larvae of 11–20 mm. **b** Larvae of 21–40 mm. **c** Larvae of more than 18 mm in length. The dotted line in **c** indicates the tectotegmental boundary. Note the labelled cells in the tectum in **c**. *M5*, mesencephalic tegmental nucleus; *MRA*, mesencephalic reticular area; *T*, tectum. (Reproduced from Rodicio et al. 1992)

showed that the development of the axial optic tract precedes that of the lateral tract.

The development of the retinopetal system has been studied by De Miguel et al. (1990) and Rodicio et al. (1995) by retrograde transport of HRP injected into the eye. These investigations have shown that the retinopetal neurons arise from a common anlage in the matrix zone of the mesencephalic tegmentum and that late in development a transient population of retinopetal neurons is present in the ventral zone of the tectum (Fig. 10.40).

Only simple electrophysiology has been carried out in the lamprey. Flashes to the eye produce activity in the tectum (Rovainen 1979b) and some later effects are seen on reticulospinal neurons (Wickelgren 1977a). Long latency responses can also be seen in the forebrain. It has also been noted that illumination of one eye modulates postural control by pathways that do not involve the tectum opticum (Deliagina et al. 1993).

In addition to the lateral eyes, the lamprey has two other photosensitive organs, the pineal complex and the dermal photoreceptors. The pineal complex is involved in the generation of circadian rhythms, increasing locomotor activity at night and depressing it during daytime. The structure and connection of the pineal complex are discussed in Sect. 10.8.2.

Dermal photoreceptors are located along the base of the caudal fin and, curiously enough, their signals are transmitted to the brain stem via the trunk and the posterior lateral line nerves (Kleerekoper 1972). Little is known about the receptors and the site of termination of their central projections is unknown. Activation of dermal photoreceptors evokes locomotion without any preferential orientation relative to the source of the light (Ullén et al. 1993). The responses and pathways mediating the reaction to illumination of dermal photoreceptors in the tail have begun to yield to electrophysiological studies (Deliagina et al. 1995).

10.7.3
Tegmentum Mesencephali

The tegmentum mesencephali contains the following discrete cell masses: (a) the nucleus tegmentalis lateralis, (b) the oculomotor nuclei, (c) the nucleus reticularis mesencephali, and (d) the rostral part of the nucleus interpeduncularis (Fig. 10.30). The last of these nuclei has already been discussed; the remaining ones will now be briefly considered.

The nucleus tegmentalis lateralis is a group of rather diffusely arranged cells, situated ventrolateral to the torus semicircularis (Fig. 10.19). It has not been described before.

The oculomotor neurons together constitute the most rostral somatic efferent centre of the brain. They are arranged in three cell masses that form a continuous chain, extending from the ventricular surface to the meningeal surface. These cell masses are known as the nucleus periventricularis, the nucleus intermedius and the nucleus superficialis (Addens 1933; Heier 1948; Fig. 10.19). The cells of the former give rise to crossed fibres, which decussate in the so-called chiasma oculomotorii; the axons of the latter two nuclei leave the brain ipsilaterally. Remarkably, some cells of the superficial nucleus extend outside of the brain among the root fibres of the oculomotor nerve. The fibres of the tractus octavomesencephali pass through the oculomotor nuclei and form large calyciform endings on the surface of their neurons (Heier 1948). A visceral efferent oculomotor centre ('nucleus of Edinger-Westphal') has not been observed in the lamprey. For a discussion of the innervation of the ocular muscles, we refer to Sect. 10.5.7.

The area situated ventral to the torus semicircularis has been designated by Heier as the tegmentum motoricum mesencephali. Caudally this area passes over into the undifferentiated central grey of the isthmus region. The tegmentum motoricum consists largely of medium-sized cells; however, within its confines a distinct group of larger elements, the nucleus reticularis mesencephali, can be delimited (Figs. 10.19, 10.30). It has already been mentioned that this cell mass contains two giant neurons, the Müller cells M1 and M2.

The tegmentum motoricum is, as its name implies, primarily concerned with motor coordination. According to Heier (1948) its multifarious afferents include:

1. Fibres descending from the ventral thalamus and, possibly, from the telencephalon. The latter would form a tractus olfacto-tegmentalis.
2. Fibres descending from the hypothalamus. They have been designated here as the tractus hypothalamo-tegmentalis. Ariëns Kappers et al. (1936) mention the presence of two different descending hypothalamic pathways, the tractus lobo-bulbaris and the tractus mamillo-peduncularis. They traced the fibres of both of these systems as far back as the level of the nuclei of X. Heier described a tractus mamillo-tegmentalis, whose extremely thin fibres make contact with the dendrites of the cells of the motor tegmentum and some of which also reach the nucleus interpeduncularis.
3. Fibres originating from the nucleus commissurae posterioris. The latter is situated dorsally in the synencephalon (Figs. 10.20, 10.30).
4. A tractus tecto- et torotegmentalis, descending along the external surface of the mesencephalic grey.

5. The ascending fibres of the lemniscus spinalis et bulbaris ascend through the tegmentum motoricum, and the same holds true for the axons of the tractus octavomesencephali.

6. A crossed cerebellotegmental pathway has been described by Larsell (1947a, 1967).

7. Ascending visceral fibres. Concerning these fibres Heier (1948 p. 77) states: "Fibres from the visceral centres in the medulla [=rhombencephalon] undoubtedly pass rostral through the tegmentum mesencephali. They could not be traced as a definite tract, but are probably present among the fibres of the other ascending tracts".

As regards the tegmental efferents, the cells of the nucleus reticularis mesencephali send their axons caudally in the fasciculus longitudinalis medialis. According to Heier the majority of the smaller cells of the tegmentum motoricum mesencephali also

contribute to this descending pathway, but some project rostrally, to the hypothalamus.

It has been experimentally established that the tegmentum of the midbrain receives diffuse projections from the retina (Kosareva 1980; De Miguel et al. 1990) and from the pineal complex (Puzdrowski and Northcutt 1989). The pinealotegmental projection is strongly developed in larval lampreys, in which it specifically approaches the mesencephalic periventricular tegmental grey (Yáñez et al. 1993). As already discussed (Sect. 10.7.2), this zone of grey, also known as area M5 of Schober (1964; Fig. 10.41), is the principal source of a mainly serotoninergic retinopetal projection (Repérant et al. 1980; De Miguel et al. 1990; Rio et al. 1993). Numerous cells in the nucleus of the FLM and in the mesencephalic tegmental periventricular grey have been demonstrated to project, mainly ipsilaterally, to the spinal cord (Ronan 1989).

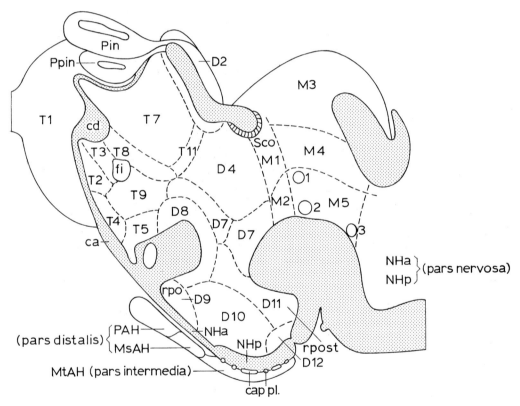

Fig. 10.41. The ventricular surface of the rostral part of the brain of the brook lamprey, *Lampetra planeri*, showing the position of cell masses of as delineated and labelled by Schober (1964). *T1*, bulbus olfactorius; *T2*, nucleus septi; *T3*, nucleus olfactorius anterior; *T4*, nucleus commissurae anterioris; *T5*, nucleus preopticus; *T7*, primordium hippocampi; *T8*, lobus subhippocampalis; *T9*, corpus striatum; *T11*, eminentia thalami; *D2*, ganglion habenulae; *D4*, nucleus dorsalis thalami; *D6*, nucleus ventralis thalami; *D7*, nucleus tuberculi posterioris; *D8*, nucleus commissurae postopticae; *D9*, nucleus commissurae preinfundibularis; *D10*, nucleus ventra-

lis hypothalami; *D11*, nucleus dorsali hypothalami; *D12*, nucleus commissurae postinfundibularis; *M1*, nucleus commissurae posterioris; *M2*, nucleus fasciculi longitudinalis medialis; *M3*, tectum opticum; *M4*, torus semicircularis; *M5*, tegmentum motoricum; *1, 2, 3*, Müller cells; *MsAH*, meso-adenohypophysis; *MtAH*, meta-adenohypophysis; *NHa*, neurohypophysis, pars anterior; *NHp*, neurohypophysis, pars posterior; *PAH*, pro-adenohypophysis; *Pin*, pineal organ; *Ppin*, parapineal organ; *Sco*, subcommissural organ; *ca*, commissura anterior; *cd*, commissura dorsalis; *fi*, foramen interventriculare. (Modified from Schober 1964)

10.7.4
Synencephalon

Although the synencephalon is the product of the most caudal prosomere, it is common practice to treat this brain segment as forming part of the midbrain (cf. Chap. 4, Sects. 4.5 and 4.7). It comprises two parts, the dorsal area pretectalis and the ventral nucleus of the FLM. Its roof plate is considerably thickened by the large posterior commissure (Fig. 10.20). The paramedian parts of this commissure are ventrally covered by the subcommissural organ.

The area pretectalis contains, according to Heier (1948), a single cell mass, the nucleus commissurae posterioris, in which three layers, the periventricular grey, the stratum centrale and the stratum superficiale can be distinguished. Similar observations were made by Kuhlenbeck (1956). According to Schober (1964; Fig. 10.41) the periventricular and migrated cells in the pretectal area constitute together a cytoarchitectonic unit, which he designated as nucleus commissurae posterioris, nucleus pretectalis or M1. Kosareva (1980) considered only the migrated pretectal cells as nucleus of the posterior commissure, whereas Puzdrowski and Northcutt (1989) distinguished a periventricular and a superficial pretectum. In gnathostomes, three pretectal nuclei – periventricular, central and superficial – can generally be distinguished (Fite 1985). According to the present authors, these cell masses are also present in the lamprey, but in this form the central and superficial nuclei cannot be sharply delineated from each other.

Heier (1948) studied the fibre connections of the pretectal area in non-experimental silver material. He reported that this area receives impulses from the lateral as well as from the axial optic tracts and that its efferents pass to the tectum, the torus semicircularis, the dorsal and ventral thalamus, and the hypothalamus. Some fibres were observed to join the FLM. The existence of primary optic projections to the pretectum passing via the lateral and axial optic tracts has been experimentally confirmed by several authors (see Sect. 10.7.2). According to Kosareva (1980), the pretectum is the principal target of the axial optic tract.

The nucleus of the FLM represents the rostral part of the nucleus reticularis mesencephali (Figs. 10.21, 10.30). The afferent and efferent connections of this cell mass have already been discussed in Sect. 10.7.3.

The posterior commissure is situated between the pineal recess and the beginning of the lamina chorioidea of the midbrain. This commissure comprises, according to Heier (1948), the commissure fibres from the pretectal area, the pars dorsalis thalami, the tectum and the torus semicircularis, decussating habenulo-tectal, tecto-habenular, thalamo-tectal and tecto-thalamic fibres, as well as decussating efferents from the parietal organs. Only the existence of the component last mentioned has been experimentally confirmed (cf. Sect. 10.8.2).

The subcommissural organ is paired in the lamprey. It protrudes, crest-like, from the angle between the posterior commissure and the lateral walls of the synencephalon. It is composed of slender cylindrical cells, arranged in a four to five nuclei wide pseudostratified epithelium (Adam 1956; Oksche 1969; Sterba 1969). The subcommissural organ produces in the lamprey (and in all other craniates) a thread of solidified mucus, which passes backward from the brain into the central canal of the spinal cord. This so-called fibre of Reissner was first observed in the spinal cord of the lamprey (Reissner 1860). Its formation and fate have been discussed in Sect. 10.4.9.

10.8
Diencephalon

10.8.1
Introductory Note

The diencephalon is composed of four regions, which are, from dorsal to ventral, the epithalamus, the thalamus dorsalis, the thalamus ventralis and the hypothalamus (Figs. 10.21, 10.42). In the lamprey Heier (1948) considered the ventricular groove, which separates the dorsal thalamus from the ventral thalamus, as a part of the sulcus limitans of His. Consequently, Heier regarded the dorsal part of the diencephalon as belonging to the alar plate and the ventral part as a basal-plate derivative. Heier arrived at the following conclusions about the individual diencephalic regions: (a) the epithalamus does not correspond to any zones of the brain stem; (b) the dorsal part of the thalamus dorsalis belongs to the same longitudinal zone as the tectum mesencephali, whereas the ventral part of the thalamus dorsalis and the torus semicircularis are parts of another zone; (c) rostrally the thalamus dorsalis borders on the pallial part of the telencephalon; (d) the thalamus ventralis forms part of a zone to which the striatal part of the telencephalon and the tegmentum motoricum mesencephali also belong; and (e) the nucleus preopticus, the hypothalamus and the nucleus interpeduncularis are components of the most ventral zone of the brain.

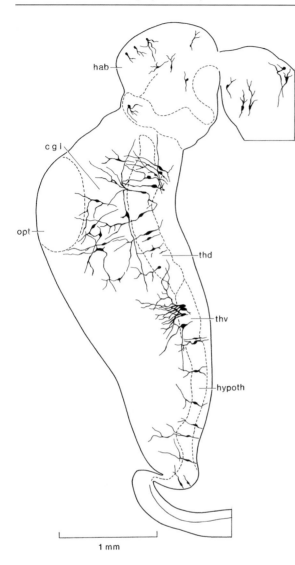

Fig. 10.42. Composite drawing of neurons observed in transversely sectioned Golgi preparations of the brain of the lamprey, *Lampetra fluviatilis* through the middle of the diencephalon. For a list of the Abbreviations Used in Figures, see p. 409 (Based on Heier 1948, Fig. 10b)

10.8.2
Epithalamus

The epithalamus comprises the pineal complex and the ganglia habenulae. The pineal complex in lampreys is a photosensory organ (Pu and Dowling 1981) located at the dorsal midline of the head at the same level as the lateral eyes, beneath a patch of translucent skin. It lies in a depression on the dorsal surface of the saccus dorsalis, i.e. the bulging ependymal roof to the third ventricle (Fig. 10.3c,d). In lampreys, the pineal complex comprises a dorsally situated pineal organ and, in all genera except *Mor-*

dacia, a more ventrally situated parapineal organ (Hardisty 1979; Fig. 10.3c,d). The pineal organs develop ontogenetically as bilateral evaginations of the diencephalic roof, the pineal organ from the right side, the parapineal from the left. Both organs consist of a widened vesicular rostral part and a narrow, tube-like caudal part. The lumina in the rostral parts persist throughout development, but the caudal parts soon become solid. The latter merge into a single pineal stalk which is attached to the roof of the brain between the habenular and posterior commissures.

In the pineal organ the lumen separates a thin, nonpigmented dorsal wall from a thicker, pigmented ventral wall. These dorsal and ventral walls are known as the pellucida and the pineal retina. The pineal retina comprises an apical layer made up of photoreceptor cells of different types and supporting cells, and a basal layer composed of pineal ganglion cells. The parapineal organ shows a similar overall structure, but a region comparable to the pellucida of the pineal organ is lacking.

Although much remains to be learned concerning the functions of the pineal complex in lampreys, it is well established that these organs register environmental photic information and that this information is transformed into neural, and presumably also into neuroendocrine signals, which are relayed to the brain. Because the lateral eyes and retinae are poorly developed during the long larval period, the pineal and parapineal organs represent the principal photoreceptive structures during that period.

The cells in the pineal retina include type I, type II and neurosensory cells with long axons (Meiniel 1980, 1981; Meiniel and Hartwig 1980; Cole and Youson 1982; Samejima et al. 1989). Type I cells are typical photoreceptor cells, which closely resemble the core photoreceptors in the lateral eyes of non-mammalian vertebrates (Pu and Dowling 1981) in that they are provided with large, well developed outer segments, composed of disks continuous with the plasma membrane. Their ramifying basal processes make ribbon synapses onto the dendrites of the ganglion cells. These elements are capable of transducing a photic stimulus into an electrical message. They respond to photic stimulation with graded hyperpolarizations, modulating the firing frequency of the action potentials of the ganglion cells (Pu and Dowling 1981; Morita et al. 1985; Tamotsu and Morita 1986; Uchido et al. 1992).

Type II cells differ from type I cells in that their photoreceptor apparatus is poorly developed, and their basal processes lack the extensive ramifications and synaptic ribbons noted in type I cells. Moreover, and importantly, type II cells, contrary to type I cells, synthesise serotonin (Meiniel 1980;

Meiniel and Hartwig 1980; Cole and Youson 1982; Tamotsu et al. 1990). Apart from being an important neurotransmitter, serotonin is the precursor of the pineal hormone melatonin. In many vertebrates this hormone is involved in the mechanisms that cause circadian and seasonal rhythms. Melatonin is synthesised according to a circadian rhythm in the lamprey pineal organ (Joss 1973); accordingly, hydroxyindole-O-methyltransferase (HIOMT), an enzyme involved in the synthesis of melatonin, shows a rhythmic change in level under alternate light:dark conditions. Bolliet et al. (1993) found that both light and temperature influenced the secretion of melatonin from cultured pineal complexes of the lamprey, *Petromyzon marinus*.

In type II cells serotonin may play a role as a neurotransmitter in the synaptic transfer of information from these cells to pineal ganglion cells. This conjecture gains support from the observation of Tamotsu et al. (1990) that the basal processes of some serotonin-immunoreactive photoreceptors are closely apposed to serotonin-immunoreactive cells clearly identified as intrapineal second-order neurons (see below). However, in the type II cells serotonin may also be used as a precursor for the production of the melatonin which is released as a neurohormone into the cerebrospinal fluid or into general circulation. In this case the indoleaminergic photoreceptive cells in the lamprey pineal complex would represent photoneuroendocrine cells capable of transducing light stimuli directly into a neuroendocrine response (Meiniel 1980, 1981; Tamotsu et

al. 1990). Type II cells possess some ultrastructural features commensurate with endocrine activity, such as the presence of a well-developed Golgi apparatus giving rise to dense-core vesicles and an amorphous electron-dense material. Furthermore their basal processes contain an abundance of both clear and dense-core vesicles and frequently display a vascular polarity by being often closely aligned to regions of the basement membrane that are near capillaries (Cole and Youson 1982).

Retrograde tracer experiments have shown that some (less than 1%) of the total number of pineal photoreceptor cells have long axonal processes that enter the pineal stalk. The ganglion cells are concentrated in the basal layer of the pineal retina. They have one or a few ramifying dendrites which contact the basal processes of the photoreceptor cells and their axons leave the pineal organ via the pineal stalk. Tamotsu et al. (1990) demonstrated the presence of a population of serotonin-immunoreactive neurons which may receive input from pineal photoreceptors. It is not clear whether these cells establish pinealofugal projections to the brain or represent interneurons comparable to the serotonin-immunoreactive amacrine cells of the retina.

The structure of the parapineal retina is similar to that of the pineal retina. However, in the parapineal retina, type I cells and ganglion cells are sparse, and the photoreceptive part of the type II cells is much reduced, at least in *Petromyzon marinus*. On this account, Cole and Youson (1982) con-

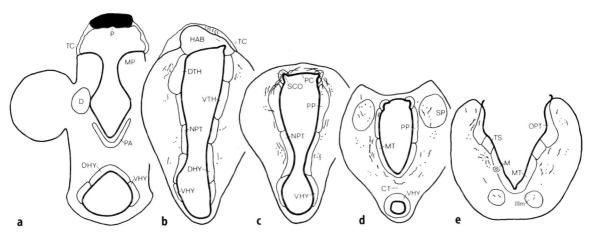

Fig. 10.43. Charting of the central projections of the pineal complex in the silver lamprey, *Ichthyomyzon unicuspis* following HRP inoculation of the pineal complex, *P* (*solid black region in a*), from rostral (*a*) to caudal (*e*). *Short wavy lines* represent anterogradely labelled fibres. *Small dots* represent labelled terminals and synapses en passant. *CT*, tegmental commissure; *D*, dorsal pallium; *DHY*, dorsal hypothalamus; *DTH*, dorsal thalamus; *HAB*, habenula; *IIIm*, oculomotor nucleus; *LP*, lateral pallium; *M*, Müller cell; *MP*, medial pallium; *MT*, midbrain tegmentum; *NPT*, nucleus of the posterior tuberculum; *OPT*, optic tectum; *PA*, preoptic area; *PC*, posterior commissure; *PP*, periventricular pretectum; *SCO*, subcommissural organ; *SP*, superficial pretectum; *TC*, tela choroidea; *TS*, torus semicircularis; *VHY*, ventral hypothalamus; *VTH*, ventral thalamus. (Reproduced from Puzdrowski and Northcutt 1989)

sidered the parapineal retina in this species as a regressed or rudimentary structure.

The pineal complex in lampreys is involved in the control of nocturnal pealing of the skin (Eddy 1972) and in the regulation of circadian locomotor activity (Morita et al. 1992) Pinealectomy six months prior to normal metamorphosis completely prevents this transition (Eddy 1969; Cole and Youson 1981) and a similar operation in the adult form can delay sexual maturity in both males and females (Eddy 1972).

Puzdrowski and Northcutt (1989) have studied the central projections of the pineal complex of the silver lamprey, *Ichthyomyzon unicuspis*, by injection of HRP. This study revealed that the efferent fibres of the pineal and parapineal organs form a single tract, the pineal tract, which enters the brain at the level of the posterior commissure (Fig. 10.43). The entering pineofugal fibres appeared to pass laterally through the posterior commissure and into the subcommissural organs (Fig. 10.43c). After having passed through these organs the pineofugal fibres were observed to divide into rostro-ventrally directed fibres that appeared to project to the dorsal thalamus, the ventral thalamus, the nucleus of the tuberculum posterius and the dorsal hypothalamus (Fig. 10.43b,c). On the other hand, the caudoventrally directed fibres could be traced to the periventricular and superficial pretectum, the tectum, the midbrain tegmentum and the oculomotor nucleus (Fig 10.43d,e).

Yáñez et al. (1993) studied the neural projections of the pineal organ in larvae of the sea lamprey, *Petromyzon marinus*, by means of anterograde and retrograde tracing with the fluorescent lipophilic dye DiI. It appeared that the pineofugal projections are well developed in these larvae and extended, just as in adult lampreys, from the posterior commissure rostrally in the diencephalon and caudally into the midbrain. The rostrally directed fibres, which were found to be much smaller in number than the caudal ones, could be traced to the pretectal area and the dorsal thalamus. The majority of the pineofugal fibres projected ventrocaudally to the tectum and to the mesencephalic tegmentum, particularly to the M5 area of Schober and to the mesencephalic reticular area. Some small, retrogradely labelled neurons were observed in the dorsal tegmental area. The authors noted in these larvae an extensive overlap between pineofugal projections and primary visual centres.

The ganglion habenulae is delimited from the dorsal thalamus by a distinct sulcus subhabenularis. It contains a periventricular and a superficial cell layer; the latter has greatly hypertrophied on the right side (equivalent to the left side in Figs. 10.17, 10.20b). This unilateral hypertrophy accounts for the marked asymmetry of the habenular ganglia in the lamprey. Some authors (Ariëns Kappers 1947; Heier 1948) thought that this asymmetry is due to the asymmetrical development of the parietal organs. However, this is unlikely because the habenular nuclei hardly receive any fibres from these organs.

The connections of the habenula were studied in non-experimental material by Johnston (1912) and Heier (1948). They reported that the ganglion or nucleus habenulae receives fibres from almost all parts of the telencephalon by way of the stria medullaris (Figs. 10.21, 10.22). This tract decussates partly in the commissura habenulae. It is accompanied by fibres that arise in telencephalic areas of one side and decussate and terminate in the corresponding telencephalic areas of the other side (Johnston 1902). The anterograde tracer studies of Northcutt and Puzdrowski (1988) and Polenova and Vesselkin (1993) have shown that fibres originating from the olfactory bulb decussate in the habenular commissure and subsequently pass to the primordium piriforme, the striatum and the septum as well as to the ventral hypothalamus and the nucleus of the tuberculum posterius. Fibres arising from the evaginated parts of the telencephalon, i.e. the primordium piriforme and the primordium pallii dorsalis, were found to pass via the habenular commissure to the contralateral hippocampal primordium.

Efferent fibres leave the habenula by way of the fasciculus retroflexus or tractus habenulo-interpeduncularis (Figs. 10.20, 10.21). Like the ganglia this bundle shows a marked asymmetry. It descends through the pars dorsali thalami and the tuberculum posterius to the ventral surface of the brain. After having decussated in the rostral part of the tegmentum of the midbrain the fibres of the fasciculus retroflexus terminate in the nucleus interpeduncularis. Heier (1948) observed in non-experimental silver material that the fibres take a spiral course through the interpeduncular nucleus, decussate again and pass caudally into the base of the rhombencephalon, some of them reaching the caudal border of this region.

The afferent and efferent connections of the habenula have been studied experimentally in the larval sea lamprey, *Petromyzon marinus*, by Yáñez and Anadón (1994), using HRP and DiI labelling. The main results of this study may be summarised as follows: (a) Afferents to the habenular ganglia arise almost exclusively from two neighbouring areas, the lobus subhippocampalis and the rostral part of the dorsal thalamus. (b) Fibres originating from the evaginated parts of the telencephalon decussate in the habenular commissure. They use this commissure as a way to reach the contralateral hemisphere and probably do not form synapses within the habenular ganglion. (c) The efferents assemble in the fasciculus retroflexus and course to the neuropil of the interpeduncular nucleus. This neuropil occupies a superficial position throughout its extent. It comprises a commissural region in the rostral mesencephalon, two long bilateral areas extending from the basal midbrain into the rostral rhombencephalon. At the level of entrance of the trigeminal nerve these two formations converge in the ventral midline to form a dense unpaired neuropil. These findings largely confirm the observations of Heier (1948), though not his suggestion that habenular efferents may reach the caudal rhombencephalon.

10.8.3
Thalamus Dorsalis

The dorsal thalamus contains a wide, compact plate of central grey and, external to that, a zone of more loosely arranged cells (Fig. 10.21). Schober (1964) subdivided the periventricular grey or nucleus dorsalis thalami on topographical grounds into a pars subhabenularis, a pars medius and a pars caudalis. Many authors considered the external zone to represent a primordial corpus geniculatum laterale (Herrick and Obenchain 1913; Kuhlenbeck 1929; Saito 1930; Schober 1964). According to Heier (1948) this external zone can be divided into three rostrocaudally arranged nuclei, which he termed nucleus of Bellonci, corpus geniculatum anterius, and corpus geniculatum posterius. The boundary between the last two is marked by the fasciculus retroflexus. Caudally the corpus geniculatum posterius cannot be sharply delimited from the superficial pretectum. Figure 10.42 shows that the neurons in the dorsal thalamus are provided with long, sparsely branching dendrites.

From a comparative neuroanatomic point of view a thorough knowledge of the fibre connections of the dorsal thalamus of the lamprey is of great interest. Before the appearance of Heier's extensive monograph in 1948, next to nothing was known of these connections. According to Heier, the dorsal thalamus has a particularly multifarious input and output; however, since his work is entirely based on non-experimental silver material it should be considered with caution. In what follows the various thalamic afferents and efferents described by Heier (1948) will be reviewed. Next, these findings will be considered in light of the results of experimental studies on thalamic connections in the lamprey. Heier described the following afferents:

1. Spinothalamic and general bulbothalamic fibres, the latter originating from the nucleus sensorius of V. Reportedly these two fibres contingents ascend in the spinal and general bulbar lemniscus and terminate in all three of the nuclei, constituting the external thalamic cell zone.

2. Fibres originating from the cerebellum and from the area octavolateralis. The latter were observed to terminate in the corpus geniculatum posterius.

3. The tectum mesencephali gives rise to two different thalamopetal projections, the tractus tecto-thalamicus rectus et cruciatus and the tractus tecto-thalamicus et -hypothalamicus rectus et cruciatus. The former takes a dorsal course, passing through the region of the posterior commissure, where some of its fibres decussate. The latter passes rostroventrally along the medial surface of the tractus opticus and decussates partly in the commissura postoptica. This tract is believed to be in contact with all parts of the dorsal thalamus.

4. Retinothalamic fibres. The fibres of the optic nerve decussate almost completely in the chiasma opticum, after which they proceed as the optic tract dorsocaudally along the external surface of the diencephalon, to reach the tectum mesencephali (Figs. 10.19–10.22, 10.42). During their course through the diencephalon, the fibres of the optic tract establish synaptic connections with the nucleus of Bellonci, the corpus geniculatum anterius and the corpus geniculatum posterius.

5. A hypothalamo-thalamic connection is made up by fine axons that ascend from the dorsal part of the hypothalamus to the posterior part of the central grey of the dorsal thalamus. Heier regarded this pathway as being comparable with the fasciculus mamillothalamicus (tract of Vicq d'Azyr) of mammals.

6. A tractus habenulothalamicus accompanies the fasciculus retroflexus. Its fibres are in contact with dendrites of cells situated in the central grey.

7. Fibres from the parietal organs reach, either directly or via a relay in the habenula, the dorsal thalamus.

8. A tractus olfactothalamicus connects the telencephalon proper, and possibly the olfactory bulb as well, with the nucleus of Bellonci.

The efferents of the dorsal thalamus project, according to Heier (1948), to all parts of the diencephalon, and also to the adjoining parts of the mesencephalon and the telencephalon. Thus he describes: (a) ipsilateral connections with the haben-

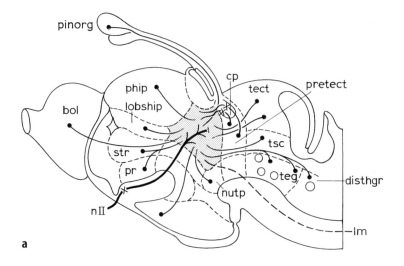

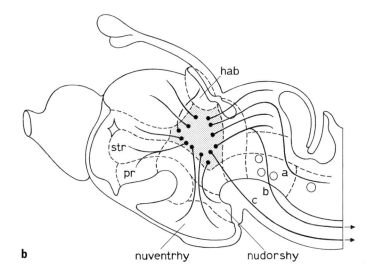

Fig. 10.44a,b. Experimentally established connections of the lamprey dorsal thalamus. **a** Afferents. **b** Efferents. *a, b,* and *c* represent the thalamotegmental projections described in the text. For a list of the abbreviations used in figures, see p. 409 (Based on Polenova and Vesselkin 1993; Yáñez and Anadon 1994; Yáñez et al. 1993; Northcutt and Wicht 1996; and Puzdrowski and Northcutt 1989)

ula and with the ventral thalamus; (b) crossed and uncrossed connections with the tectum mesencephali, the tegmentum mesencephali and the hypothalamus; (c) commissural fibres, interconnecting the dorsal thalami of both sides (these fibres probably pass through both the habenular and the posterior commissures); and (d) some fibres, which pass ventrally, decussate in the postoptic commissure and fan out in the caudal part of the telencephalon. They constitute together the tractus thalamofrontalis.

In Heier's (1948, p. 138) opinion, the dorsal thalamus represents a 'high sensory correlation centre', receiving afferent impulses from the olfactory epithelium, the eyes, the parietal organs and from all sensory nuclei of the mesencephalon and the rhombencephalon. Furthermore, Heier points out

that associative fibre systems connect the dorsal thalamus with sensory correlation centres in the telencephalon and the mesencephalon and that the information processed in the dorsal thalamus is relayed to various coordinating centres situated in the ventral part of the brain.

Experimental data indicating that the dorsal thalamus receives a strong projection from the retina and also receives efferents from the parietal organs has been reviewed in Sects. 10.7.2 and 10.8.2. The connections of the dorsal thalamus of the river lamprey, *Lampetra fluviatilis,* have been experimentally studied by Polenova and Vesselkin (1993), using HRP as a tracer. Their principal findings may be summarized as follows: The dorsal thalamus receives afferents from the ipsilateral olfactory bulb, the subhippocampal lobe, the hippocampal primordium, the contralateral thalamus, the contralateral pretectal area, the tectum, the torus semicircularis and the tegmentum motoricum mesencephali (Fig. 10.44a). Following large HRP injections in the diencephalon, Ronan and North-cutt (1990) found labelled cells in the contralateral obex region. It seems likely that these cells represent the dorsal column nucleus, which has recently been described by Dubuc et al. (1993a,b). Taken together, these findings suggest the presence of a lemniscus medialis in the lamprey. The thalamic efferents include fibres passing to the habenular ganglion (Yáñez and Anadón 1994), the hippocampal primordium, the subhippocampal lobe, the hypothalamus and the tectum (Fig. 10.44b). All of these projections are bilateral, but mainly ipsilateral. Polenova and Vesselkin described, in addition, the following three descending thalamic projections (Fig. 10.44b: a,b,c): (a) fibres descending in the caudal part of the tectum, which pass to the tegmentum motoricum mesencephali, where they turn caudally to attain the rostral rhombencephalon; (b) fine fibres descending from the tectum to the neuropil of the tegmentum motoricum mesencephali, where some of them contact the large Müller cells; (c) Diffusely arranged fibres which descend to the tuberculum posterius, from where they pass to the rostral spinal cord, via the tegmentum of the brain stem. The connections of the dorsal thalamus in the silver lamprey, *Ichthyomyzon unicuspis,* have recently also been experimentally analyzed by Northcutt and Wicht (1996), who used the carbocyanine dye DiI as a tracer. They found that this centre makes reciprocal connections with all divisions of the pallium, the striatum, the preoptic area and the hypothalamus, projects to the septum and receives afferents from the dorsal isthmal grey, the midbrain tegmentum and the tectum. In Table 10.2 the findings of Heier (1948) are compared with those obtained with experimental techniques. It is noteworthy that the existence of general somatic and visceral projections from the spinal cord and the medulla oblongata to the dorsal thalamus has not been confirmed by experimental studies and that the projections from the thalamus to the telencephalon are more substantial than was anticipated by Heier (1948). The experimental confirmation of the existence of sizable thalamohypothalamic and thalamotegmental projections also deserves mentioning.

The dorsal thalamus is relatively well developed in the lamprey and its periventricular grey (Schober 1964), as well as its zone of migrated cells (Heier 1948), have been subdivided into several moieties. Heier (1948) noticed that most afferents terminate in specific parts of the dorsal thalamus. Thus, he

Table 10.2. Connections of the dorsal thalamus in the lamprey

Projections/areas of origin/termination	Heier 1948	Experimental	Source
Afferents			
Spinothalamic	+	−	
Dorsal column nucleus	−	+	Ronan and Northcutt (1990)
General bulbothalamic	+	−	
Octavolateral area	+	−	
Cerebellum	+	−	
Dorsal isthmic grey	−	+	Northcutt and Wicht (1996)
Tectum mesencephali	+	+	Polenova and Vesselkin (1993)
Torus semicircularis	−	+	Polenova and Vesselkin (19939
Tegmentum mesencephali	−	+	Polenova and Vesselkin (1993)
Pretectal area	−	+	Polenova and Vesselkin (1993)
Parietal organs	+	+	Puzdrowski and Northcutt (1989)
Habenula	+	−	
Hypothalamus	+	+	Northcutt and Wicht (1996)
Retina	+	+	Vesselin et al. (1989)
			Rio et al. (1993)
Pallium	−	+	Polenova and Vesselkin (1993)
Striatum	−	+	Northcutt and Wicht (1996)
Preoptic area	−	+	Northcutt and Wicht (1996)
Olfactory bulb	?	+	Polenova and Vesselkin (1993)
Efferents			
Pallium	+	+	Polenova and Vesselkin (1993)
Striatum	−	+	Northcutt and Wicht (1996)
Preoptic area	−	+	Northcutt and Wicht (1996)
Habenula	+	+	Yáñez and Anadón (1994)
Ventral thalamus	+	−	
Hypothalamus	+	+	Polenova and Vesselkin (1993)
Tegmentum mesencephali	+	+	Polenova and Vesselkin (1993)
Rhombencephalon	−	+	Polenova and Vesselkin (1993)
Spinal cord	−	+	

mentioned that fibres originating from the octavo-lateral area terminate in the corpus geniculatum posterius and that efferents from the hypothalamus project to the caudal part of the thalamic periventricular grey. In the experimental studies published so far, no mention is made of efferent systems projecting to specific parts of the dorsal thalamus, save for the notion that the retinothalamic fibres mainly contact the migrated cells. Butler (1994) stated, on the basis of an extensive cladistic analysis, that the dorsal thalamus of jawed vertebrates comprises two basic divisions – the collothalamus and the lemnothalamus – that receive their predominant input from the tectum and from lemniscal pathways, including the optic tract, respectively. It has been established experimentally that in the lamprey the dorsal thalamus receives substantial projections from the tectum and from the optic tract. However, it is not known whether these two projections terminate in different thalamic regions, and therefore a subdivision into a collothalamus and a lemnothalamus cannot be made. The fibres originating from cells in the rostral and caudal parts of the tegmentum of the midbrain may well represent a reticulothalamic projection. From a comparative neuroanatomical point of view it is of particular interest that the dorsal thalamus projects to the telencephalon, and that those telencephalic areas that are most strongly innervated by the thalamus, i.e. the hippocampal primordium and the subhippocampal lobe, receive only weak projections from the olfactory bulb (Polenova and Vesselkin 1993; Fig. 10.50). Thus, there can be no doubt that non-olfactory impulses arrive at the telencephalon. Visual impulses may either travel by the direct retino-geniculo-telencephalic pathway or by the indirect retino-tecto-thalamo-telencephalic pathway. There is physiological evidence suggesting that in the tectum the visual impulses are correlated with other somatosensory as well as with viscerosensory impulses (Karamian et al. 1984). Finally, impulses from the vestibular and lateral line organs may attain the telencephalon via the torus semicircularis and the dorsal thalamus.

10.8.4
Thalamus Ventralis

The thalamus ventralis is a rather narrow strip of central grey situated between the thalamus dorsalis and the hypothalamus. Its widened caudal portion, which surrounds the rostrodorsal part of the tuberculum posterius, has been designated by Heier (1948) and Schober (1964) as the nucleus tuberculi posterioris (Fig. 10.41). Most of the neurons in the ventral thalamus are somewhat larger than those in the other diencephalic regions (Figs. 10.21, 10.42).

Stefanelli (1934) considered this region to be the most rostral part of the reticular formation. As regards its fibre connections, Heier (1948) found that the ventral thalamus receives fibres from all parts of the telencephalon, from the ganglion habenulae, the dorsal thalamus and the hypothalamus, and from the tectum mesencephali. Most of its efferent fibres pass caudally and join the fasciculus longitudinalis medialis, but some descend toward the hypothalamus. On the basis of these findings Heier (1948, p. 144) regarded the ventral thalamus as a "higher motor coordinating centre which receives impulses from the whole telencephalon and from all sensory regions of the brain, and conveys the impulses to lower coordinating centres, to motor centres and to the hypothalamus."

Baumgarten (1972), who analysed the central nervous system of the lamprey with the histofluorescence technique of Falck and Hillarp, demonstrated the presence of a catecholaminergic cell group in the region of the tuberculum posterius, which he designated as nucleus paratubercularis posterior. This cell group was found to send efferent fibres to many different centres, among them the striatum, the preoptic nucleus and the tectum. Baumgarten (1972) pointed out that in gnathostomes the region corresponding to the tuberculum posterius in the lamprey is incorporated into the midbrain. On that account he regarded the nucleus paratubercularis posterior as a precursor of the mesencephalic catecholaminergic system. The observations of Baumgarten (1972) have been confirmed by Pierre et al. (1994), who used an antibody against tyrosine hydroxylase, an enzyme involved in the biosynthesis of catecholamines. To date, the connections of the ventral thalamus and the nucleus of the tuberculum posterius have not been studied with axon degeneration or tracer techniques.

10.8.5
Hypothalamus

The hypothalamus forms the largest part of the diencephalon. It surrounds the ventral part of the third ventricle which widens ventrally into the infundibular recess; rostrally it extends as a postoptic recess underneath the optic chiasm, and caudally it forms the posterior recess, which is situated ventral to the tuberculum posterius (Figs. 10.20, 10.21, 10.41). Throughout the hypothalamus the neuronal perikarya are concentrated in a compact layer of periventricular grey. Figure 10.42 shows that the neurons in the hypothalamus are small; they send one or a few dendritic branches into the external neuropil and fibre zone. Many of these neurons have a central process that extends between the ependymal cells toward the ventricular surface. According to Heier (1948) the hypothalamic grey can be subdivided into a dorsal zone, comprising the nucleus commissurae postopticae and the nucleus dorsalis hypothalami, and a ventral zone, formed by the nuclei preinfundibularis, ventralis hypothalami and postinfundibularis. Schober (1964) adopted this subdivision, although he admits that the boundaries between most of these cell masses are indistinct (Fig. 10.41). The nucleus preopticus, which morphologically forms part of the telencephalon, will be treated here together with the diencephalon. It consists of some compact laminae of cells, which are situated at some distance from the ventricular surface (Figs. 10.22, 10.23). Heier (1948) did not observe any true neurons in the wall surrounding the recessus infundibularis. In fact, the basal part of this wall represents the neurohypophysis (see below).

Immunohistochemical studies have revealed the presence of numerous neuromediator-specific neuron groups in the preoptico-hypothalamic continuum. For references and details on this subject, the reader is referred to Table 10.3. Suffice it to mention here that, in the preoptic nucleus, neurons containing the following neuromediators have been found: catecholamines (mainly dopamine), neurotensin, gonadotropin-releasing hormone, growth hormone, prolactin, met-enkephalin and arginine vasotocin. The nucleus of the postoptic commissure contains neurotensin- and β-endorphin-immunoreactive cells, whereas in the dorsal and ventral hypothalamic nuclei the presence of cells, containing catecholamines (mainly dopamine), serotonin, histamine, cholecystokinin, neurotensin, somatostatin, and β-endorphin has been demonstrated (Fig. 10.52).

The neurons in the preoptic nucleus, containing gonadotropin-releasing hormone, growth hormone, protactin, met-enkephalin and arginine vasotocin, form part of the brain-pituitary neurosecretory system. Their axons project, via the preopticohypophysial tract to the neurohypophysis.

There is evidence suggesting that the activity of certain neuronal populations in the preoptico-hypothalamic continuum is controlled by steroid hormones. In larval and adult lampreys, Kim et al. (1980, 1981) reported that cells in the preoptic nucleus and in the ventral hypothalamic nucleus concentrate estradiol. Many of the preoptic neurons concentrating this hormone clearly formed part of the preoptico-hypophysial neurosecretory system. Smaller groups of estrogen target cells appeared to be present in the striatum and in the dorsal thalamus.

There is physiological evidence suggesting that extraocular and extrapineal photoreceptors are present in all groups of non-mammalian vertebrates, and that these photoreceptors reside within the basal part of the brain (for reviews, see Groos 1982; Foster et al. 1993). Using antibodies raised against different visual pigment opsins and an antibody against a certain part of the α-subunit of retinal G protein (α-transducin), García-Fernández and Foster (1994) demonstrated in larval lampreys that cells endowed with photoreceptor markers are not only present in the retina and in the pineal and parapineal organs, but also in the hypothalamus. These 'deep encephalic photoreceptors' appeared to be CSF-contacting cells located within the nucleus of the postoptic commissure and in the ventral hypothalamic nucleus.

According to Heier's (1948) analysis, the hypothalamus receives afferents from the dorsal thalamus, the tectum, the midbrain tegmentum and from all parts of the telencephalon. Most of the fibres arising from the two sources first mentioned assemble in two bundles, which Heier (1948) termed tractus thalamo-hypothalamicus rectus et cruciatus and tractus tecto-thalamicus et hypothalamicus rectus et cruciatus. These tracts, which pass rostroventrally along the optic tract, are believed to terminate in all parts of the hypothalamus. They decussate partly in the postoptic commissure. Other fibres originating from the tectum and the dorsal thalamus run directly external to the central grey to reach the hypothalamus, and the same holds true for the tegmentohypothalamic projection.

The fibres descending from the telencephalon originate from all areas of that brain part, including the olfactory bulb, and constitute a large but diffuse bundle, known as the tractus olfactohypothalamicus (Figs. 10.22, 10.23). According to Heier (1948) the fibres of this pathway fan out over the entire hypothalamus.

The most caudoventral part of the telencephalon, the nucleus preopticus (Figs. 10.22, 10.23, 10.41), contributes fibres to the tractus olfactohypothalamicus, but this nucleus also gives rise to a well-developed neurosecretory pathway that passes through the hypothalamus to the neurohypophysis. The axons of this tractus preopticohypophyseus terminate between the processes of the specialised ependymal cells that make up the part of the pituitary gland under discussion (Sterba 1972; Polenov et al. 1974; Tsuneki and Gorbman 1975a,b).

Experimental tracer studies have demonstrated that the hypothalamus receives afferents from the following sources: (a) the parietal organs (Puzdrowski and Northcutt 1989; Fig. 10.43), (b) the dorsal thalamus (Polenova and Vesselkin 1993; Fig. 10.44), (c) the nervus terminalis (Northcutt and

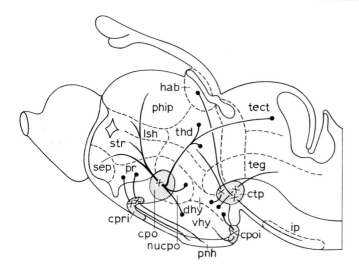

Fig. 10.45. Commissures in the ventral part of the diencephalon and some of the projections decussating through these commissures. For a list of the abbreviations used in figures, see p. 409 (Based on Heier 1948).

Puzdrowski 1988; Fig. 10.48), (d) the olfactory bulb (Northcutt and Puzdrowski 1988; Polenova and Vesselkin 1993; Fig. 10.49), and (e) the hippocampal, dorsal pallial and piriform primordia (Polenova and Vesselkin 1993; Fig. 10.51). Finally, the nucleus of the postoptic commissure has been shown to receive a retinal projection via the lateral optic tract (Vesselkin et al. 1980; Fig. 10.38).

Heier (1948) described the following hypothalamic efferent systems: (a) the tractus hypothalamo-tegmentalis, projecting to the tegmentum motoricum of the midbrain and to the rhombencephalic basal plate (Fig. 10.28c); (b) diffusely arranged fibres which ascend to the dorsal and ventral parts of the thalamus, and (c) the tractus pallii and the tractus hypothalamo-olfactorius. The latter two systems deserve some comment. The tractus pallii was described by Johnston (1902, 1912) as a bundle of fibres passing from the hypothalamus to the primordium hippocampi, an area constituting the most dorsal part of the telencephalon. In Johnston's opinion this tract forms the final link in a polysynaptic gustatory pathway that originates in the lobus vagi and ascends via the hypothalamus to terminate in the telencephalon. According to Heier (1948, p. 151), the tractus pallii forms part of a diffuse tractus hypothalamo-olfactorius, the fibres of which "seem to reach all telencephalic nuclei apart from the bulb." Reportedly, most of the fibres of this tract decussate in the commissura postoptica.

The efferent connections of the lamprey hypothalamus have not been studied so far with axon degeneration or tracer techniques. It is noteworthy that Polenova and Vesselkin (1993) did not observe any labelled cells in the hypothalamus following HRP injections in the hippocampal primordium. This negative finding casts doubt on the existence of

Johnston's (1902, 1912) tractus pallii in the lamprey.

Studies with immunohistochemical techniques have revealed the presence of several neuromediator-specific hypothalamic efferent systems. Steinbusch and Nieuwenhuys (1979) reported the presence of a serotoninergic system projecting from cells in the ventral thalamus and hypothalamus to all regions of the telencephalon; Brodin et al. (1990a; Fig. 10.52d) found that histaminergic cells situated in the caudal hypothalamus project to the striatum and the septal nucleus, and the existence of a neurotensinergic projection with similar areas of origin and termination has been demonstrated by Brodin et al. (1990b; Fig. 10.52e).

The ventral diencephalon contains four commissures, two large ones, the commissura postoptica and the commissura tuberculi posterioris, and two much smaller ones, the commissurae pre- and post-infundibulares (Fig. 10.45). The postoptic commissure occupies the dorsocaudal part of the commissure bed, which also contains the optic chiasm. The preinfundibular commissure surrounds the rostral part of the postoptic recess. The postinfundibular commissure is situated in the caudal wall of the infundibular recess, and the posterior tubercular commissure is, as its name indicates, situated in the tuberculum posterius. Heier (1948) presents extensive descriptions of the various fibre contingents participating in the formation of these commissures. No attempt will be made to summarise this part of Heier's work here, only to mention that, according to him: (a) numerous fibres originating from the tectum, the migrated thalamic cell groups, and the hypothalamus decussate in the postoptic commissure and thence ascend to the telencephalon; (b) a certain proportion of the fibres of the pre-opticohypophysial tract decussates in the preinfundibular commissure; (c) the postinfundibular commissure consists mainly of fibres interconnecting centres situated in its immediate vicinity, and (d) the commissure of the tuberculum posterius contains, apart from some decussating axons of the fasciculus retroflexus, fibres connecting the dorsal hypothalamic nucleus with the contralateral dorsal thalamus and midbrain tegmentum (Fig. 10.45) (designated as tractus mammillo-tegmentalis cruciatus). It is remarkable that, apart from Heier (1948), several other previous authors (Johnston 1912; Herrick and Obenchain 1913) interpreted the caudal part of the dorsal hypothalamic nucleus as a primordial corpus mamillare.

Experimental data on the fibre systems decussating through the ventral diencephalic commissures are scant. However, the recent HRP study by Polenova and Vesselkin (1993) has shown that secondary olfactory fibres pass via the postoptic commissure to the contralateral hypothalamus and that fibres with the same destination, which originate in the various sectors of the pallium, decussate in the postoptic as well as in the posterior tubercular commissures. The same authors also established that the dorsal thalamus projects via the supraoptic commissure to the contralateral subhippocampal lobe.

Before closing this survey of the hypothalamus, it should be mentioned that Ariëns Kappers et al. (1936) and Heier (1948) characterised this part of the brain as a centre for the correlation of olfactory impulses with other sensory stimuli, which, via its descending pathways, discharges mainly to the visceral centres of the rhombencephalon. Because experimental data on the caudal projections of the hypothalamus are lacking, this view remains entirely hypothetical.

10.8.6
Neurohypophysis

Although the structure and functions of the pituitary fall outside the scope of the present work, some notes on the neurohypophysis are appropriate. It has already been mentioned that in the lamprey the neurohypophysis constitutes the basal wall of the third ventricle. It is essentially made up of a single layer of specialised ependymal cells (tanycytes), consisting of a perikaryon and a slender process. The perikarya line the ventricle, whereas their processes are directed peripherally, forming a palisade-like pattern (Sterba 1972). The space between these ependymal processes is filled with peptidergic and monoaminergic axons and axonal endings. The peptidergic fibres are of two types, A1 and A2, which are characterised by the

presence of neurosecretory granules of 140–220 nm and 100–150 nm diameter, respectively (Polenov et al. 1974; Tsuneki and Gorbman 1975a,b). These fibres form the preoptic nucleus and constitute the preoptico-hypophysial neurosecretory system. Neurons containing arginine vasotocin (Rurak and Perks 1976, 1977; Goossens et al. 1977), met-enkephalin (Nozaki and Gorbman 1984), growth hormone (Wright 1986), prolactin (Wright 1986) and gonadotropin-releasing hormone (King et al. 1988) have been demonstrated to contribute to this system. The relation between these various hormones and the types of neurosecretory granules found in the neurohypophysis are not known.

The monoaminergic fibres in the neurohypophysis originate from the cells situated in the dorsal and ventral nuclei of the hypothalamus. These fibres, which are also referred to as B-type fibres, contain granules 80–100 nm in diameter.

The nerve terminals in the neurohypophysis contain, in addition to the granules indicated above, small, clear vesicles with a diameter of 30–60 nm. Clusters of these vesicles near the plasma membrane mark sites of possible synaptoid contact between nerve terminals and tanycytes (Polenov et al. 1974; Tsuneki and Gorbman 1975a,b; Belenky et al. 1979b).

The neurohypophysis is differentiated into anterior and posterior regions. The anterior region is co-extensive with the pars distalis of the adenohypophysis, from which it is separated by a layer of connective tissue (Fig. 10.41). The posterior region contains a much larger number of neurosecretory fibres than the anterior region and is closely apposed to the meta-adenohypophysis. The thin connective tissue layer between these two structures contains a plexus of widened sinusoid capillaries. The meta-adenohypophysis is penetrated by nerve fibres. The anterior region of the neurohypophysis is considered as the primordium of the median eminence of tetrapods, although any trace of a vascular link with the pro- and meso-adenohypophysis is lacking (Tsuneki and Gorbman 1975a; Belenky et al. 1979b). The posterior part is considered homologous to the neural lobe or pars nervosa of tetrapods (Sterba 1972; Tsuneki and Gorbman 1975a).

Many episodes in the life cycle of the lamprey, such as the onset of metamorphosis, upstream migration and spawning are seasonal, suggesting that these events may be triggered by environmental factors and mediated through the hypothalamo-hypophysial system, but there is as yet no experimental evidence for such a hypothalamic control of hypophysial function (Hardisty 1979). It is possible that humoral factors released from the neurohypophysis may reach the glandular elements of the adenohypophysis. Ultrastructural studies have shown that on the ventral surface of the neurohypophysis many neurosecretory terminals are in direct contact with the basement membrane of the connective tissue layer that covers the pituitary, and it has been suggested that both peptide neurohormones and monoamines may diffuse through this connective tissue layer and thus affect the function of the glandular cells (Tsuneki and Gorbman 1975a; Belenky et al. 1979a). At the surface of the posterior part of the neurohypophysis, peptides and monoamines may be released into the sinusoid capillaries present there and thus reach the general circulation (Polenov et al. 1974).

10.9
Telencephalon

10.9.1
Introductory Note

The telencephalon of the lamprey comprises three parts: the olfactory bulbs, the cerebral hemispheres, and the telencephalon medium. The olfactory bulbs

are even larger than the cerebral hemispheres, from which they are marked off by a shallow groove, the fissura circularis (Figs. 10.3, 10.23, 10.46). During ontogenesis the bulbs and the hemispheres have bulged out laterally; because of this evagination process they have become hollow structures surrounding the lateral ventricle (Figs. 10.22, 10.23, 10.46). However, a considerable part of the walls of the anlage of the telencephalon has not participated in this evagination. This unevaginated portion represents the telencephalon medium. Its walls border the narrow, but dorsoventrally very extensive, ventriculus impar, which is a direct rostral continuation of the diencephalic ventricular cavity (Figs. 10.22, 10.23, 10.41, 10.46). By means of the foramen interventriculare, the ventriculus impar communicates bilaterally with the lateral ventricle (Figs. 10.23, 10.46). Rostrally, the walls of the telencephalon impar are connected by the lamina terminalis. The ventral and dorsal parts of this lamella are widened to form the beds of the anterior and dorsal commissures (Fig. 10.41). Dorsally, the ventriculus impar is closed by a membranous roof, which expands to form the so-called saccus dorsalis. The parietal organs lie in a slight depression on the dorsal surface of this membranous structure (Fig. 10.3c).

The structure of the telencephalon of the lamprey has been studied by numerous authors (Johnston 1902, 1912; Schilling 1907; Tretjakoff 1909b; Herrick and Obenchain 1913; Heier 1948; Schober 1964; Nieuwenhuys 1967a). The general conclusion is that the neuronal perikarya form a zone of central grey, although in several telencephalic areas this zone is wider than elsewhere in the brain. The fibre systems of the telencephalon consist of very diffusely arranged axons, most of which are of fine calibre.

10.9.2
Olfactory Bulb

The organisation of the olfactory bulb resembles that of the corresponding structure of other anamniotes (Heier 1948; Iwahori et al. 1987). The fibres of the olfactory nerve spread over the entire surface of the bulb, forming a plexiform stratum nervosum. The end branches of the olfactory nerve fibres arch inward from this zone and contribute to the glomeruli, which form the second layer of the bulb. Large mitral cells lie between and directly central to the glomeruli (Figs. 10.23, 10.46, 10.47a). These elements have two or more primary dendrites which form bushy branches within a single glomerulus. The axon of the mitral cells, which arise either from the soma or from the proximal portion of a den-

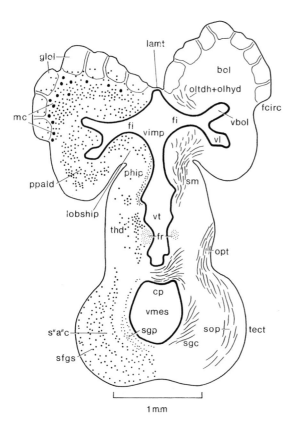

Fig. 10.46. Horizontal section through the rostral part of the brain of the lamprey, *Lampetra fluviatilis.* For a list of the abbreviations used in figures, see p. 409

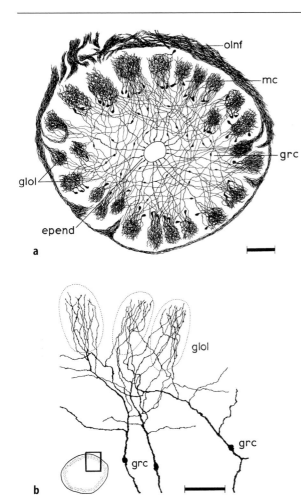

olnf

mc

grc

glol

epend

a

glol

grc

grc

b

Fig. 10.47a,b. The olfactory bulb of the lamprey, *Lampetra japonica*. **a** Transverse sections through the caudal part of the bulb. **b** Distribution of granule cell dendrites over some olfactory glomeruli. For a list of the abbreviations used in figures, see p. 409 [Drawings of rapid Golgi preparations reproduced from Iwahori et al. (1987, Figs. 2, 9)]

drite, pass to the deeper zone of the olfactory bulb (Fig. 10.47b, open triangle) and eventually leave that structure. A number of mitral cell axons pass by way of the dorsal commissure (Fig. 10.41) to the contralateral side of the telencephalon. The deeper zone of the olfactory bulb is occupied by granule cells. These elements have two to four, often very long, dendrites that diverge and may break up into widely separated parts of the glomerular layer.

According to Heier (1948), the granular layer of the olfactory bulb is separated from the centres in the cerebral hemisphere by a transitional area, which he termed the nucleus olfactorius anterior. The position of this nucleus corresponds roughly to the fissura circularis.

The extensive development of the olfactory system in the lamprey indicates that olfaction is a pri-

mary sensory modality. Relatively little is known about its physiology, however. It seems that the olfactory system does not become functional until metamorphosis (Kleerekoper 1972; Rovainen 1979b). The olfactory receptors are depolarized by several amino acids (Rovainen 1979a) and the animal shows distinct behavioural patterns in response to various odors in the water (Kleerekoper 1972). After stimulation of the olfactory nerve, extracellular field potentials are seen in the olfactory bulb, the cerebral hemisphere and the hippocampal region (Bruckmoser 1971). The bulbar and hippocampal responses are quite comparable to those seen in other vertebrates, reflecting the high level of development in these regions of the lamprey. In contrast, the responses of the cerebral hemisphere are less characterised and probably reflect the rather diffuse organisation of this region (Bruckmoser 1971). Excitation of the olfactory nerve also excites Müller cells, but after a substantial delay (Wickelgren 1977a).

10.9.3
Pallium

The telencephalon proper can be divided into a dorsal, pallial region and a ventral, subpallial region. Both of these regions lie partly within the cerebral hemispheres and partly within the telencephalon medium (Figs. 10.22, 10.23, 10.41). The pallium comprises four areas: the primordium hippocampi, the lobus subhippocampalis, the primordium pallii dorsalis, and the primordium piriforme. In the subpallium three areas, the corpus striatum, the septum and the nucleus preopticus, can be distinguished (Heier 1948; Schober 1964).

The primordium hippocampi, which forms the most dorsal part of the pallium, lies entirely within the telencephalon medium. The majority of its neuronal perikarya are situated within a narrow, compact layer of periventricular grey, but some have migrated outward into the stratum 'album' (Fig. 10.22). The morphological identity of this structure has been much debated in the literature. The term primordium hippocampi, indicating that the structure belongs to the telencephalon and forms part of the pallium, was introduced by Johnston (1912). His interpretation has been followed by many later investigators (Herrick and Obenchain 1913; Heier 1948; Schober 1964). Other authors, however, (Holmgren 1922; Bergquist 1932; Schwab 1973) considered that the primordium hippocampi of Johnston belongs to the diencephalon.

The lobus subhippocampalis is a strip-like zone that separates the primordium hippocampi from the evaginated part of the pallium. Ventrally, over a considerable distance, this lobe borders on the corpus striatum (Figs. 10.22, 10.41: T8). Its neurons are arranged in a wide and rather diffuse zone of central grey.

The primordium pallii dorsalis and the primordium piriforme together constitute the fully evaginated portion of the pallium. Heier (1948) and Schober (1964) described these two areas as structurally different entities, but according to our observations, they form a cytoarchitectonic unit. In these areas the layer of central grey is particularly wide and can be subdivided into two zones, an inner zone consisting of diffusely arranged cells and an outer zone made up by smaller and larger cell clusters (Figs. 10.22, 10.23). Some previous authors (Studnicka 1895; Mayer 1897; Tretjakoff 1909b) interpreted the outer zone as a true cortex.

In anamniote gnathostomes the pallium comprises three principal parts, the lateral or piriform pallium, the dorsal or general pallium and the medial or hippocampal pallium. As indicated by their nomenclature, Heier (1948) and Schober (1964) were of the opinion that homologues of these three pallial formations are present in the lamprey. Northcutt and Puzdrowski (1988) and Northcutt and Wicht (1996) concurred with this view; however, they considered it likely that the lateral pallium in the lamprey is subdivisible into a ventral and a dorsal part (Figs. 10.48b, 10.49b–d), the former corresponding to the primordium piriforme and the latter to the primordium pallii dorsalis of Heier (1948) and Schober (1964). The subhippocampal lobe of the authors just mentioned is, according to Northcutt and Puzdrowski (1988) and Northcutt and Wicht (1996), probably homologous to the dorsal pallium of other vertebrates. We will return to the interpretation of the parts of the lamprey pallium at the end of the present section.

A distinct zone of neuropil, ventrally bounded by a thin cell plate is situated in the dorsomedial part of the hemisphere, immediately lateral to the primordium hippocampi. This formation has been interpreted by Edinger (1908, Fig. 262) as 'Episphärium-Anfang' (literally translated, the beginning of the 'epispherium'), and as a caudally situated glomerulus of the olfactory bulb by Heier (1948), Schober (1964) and Nieuwenhuys (1977). However, HRP labelling of the olfactory epithelium has shown that the olfactory nerve does not terminate in this dorsomedial neuropil (Fig. 10.48a), and injections of the same tracer into the olfactory bulb resulted in the labelling of this formation throughout its entire rostrocaudal extent (Northcutt and Puzdrowski 1988; Fig. 10.49a–d).

10.9.4
Subpallium

The corpus striatum lies, like the lobus subhippocampalis, partly in the cerebral hemisphere and partly in the telencephalon medium. In the former it is bounded by the primordium piriforme, and in the latter it occupies a position between the lobus subhippocampalis and the nucleus preopticus. The ventricular groove that marks the boundary between the corpus striatum and the lobus subhippocampalis represents, according to Heier (1948) the most rostral part of the sulcus limitans. The striatal grey is arranged in some layers of closely packed cells, the innermost of which is separated from the ventricular surface by a wide stratum 'album' periventriculare (Figs. 10.22, 1023). Pombal et al. (1995) reported that the striatum receives numerous terminal arborizations with immunoreactivity to dopamine, serotonin, neurotensin or histamine in the periventricular neuropil zone. According to these authors, the dopaminergic projection to the striatum originates from neurons located in the caudal-medio-basal diencephalon, within the confines of the ventral thalamus and hypothalamus. These elements presumably correspond to the tyrosine hydroxylase-positive neurons observed by Pierre et al. (1994) in the nucleus of the tuberculum posterius and in the dorsal hypothalamus (Fig. 10.52b).

The septum is represented by a narrow zone of closely packed cells, which borders on the most rostral part of the ventriculus impar (Fig. 10.41: T2). The nucleus preopticus, finally, forms the most ventrally located telencephalic centre. It consists of some compact laminae of cells, which are situated at some distance from the ventricular surface (Figs. 10.22, 10.23).

It has already been mentioned that the subpallium contains a large group of neurosecretory cells, which produce the peptide neurohormone arginine vasotocin (Goossens et al. 1977). This group is not confined to the preoptic region but extends over the corpus striatum and even over the rostral part of the ventral thalamus. Many of the elements in this group are characterised as liquor contacting nerve cells, i.e. they have a dendrite which penetrates the ependymal layer ending with a small globule in the ventricular cavity (Sterba 1972). Their axons pass to the infundibular region and constitute the large preoptico-hypophysial neurosecretory system. In addition to this pathway, the peptidergic cell group also gives rise to scattered neurosecretory fibres which enter different regions of the telencephalon, diencephalon and brain stem (Sterba 1972; Goossens et al. 1977).

The Golgi studies of Johnston (1902, 1912) and Heier (1948) have revealed that the neurons of the various telencephalic areas usually have a few sparsely branching dendrites that extend into the superficial stratum 'album' or remain within the confines of the central grey. Many dendrites transgress the cytoarchitectonic boundaries of the areas in which their parent cells lie. Johnston (1912) noted that the dendrites of the neurons in the primordium hippocampi are densely covered with knobbed spines. According to him, such spiny cells are found nowhere else in the brain of the lamprey, but the same type of neuron is characteristic of the corresponding telencephalic areas of *Acipenser*, *Amia*, and *Rana*.

10.9.5
Telencephalic Fibre Connections

10.9.5.1
Non-experimental Studies

The fibre connections of the telencephalon have been thoroughly analysed by Heier (1948). That author described a tractus olfactorius, a tractus olfacto-habenularis, a tractus olfacto-thalamicus et hypothalamicus, and a tractus strio-thalamicus et hypothalamicus (Figs. 10.22, 10.23).

The tractus olfactorius consists of diffusely arranged fibres that stream backward and distribute over all parts of the telencephalon proper. Part of its fibres terminate in the anterior olfactory nucleus, and, conversely, fibres from this nucleus join the tractus olfactorius. A certain proportion of these secondary and tertiary olfactory fibres decussate in the telencephalic commissures, i.e. the commissura dorsalis and the commissura anterior (Fig. 10.41).

The tractus olfacto-habenularis or stria medullaris is composed of fine fibres, which originate from the various parts of the pallium and probably from the septum and the nucleus preopticus as well. The tract passes dorsocaudally through the primordium hippocampi to the nucleus habenulae. Reportedly, some of its fibres terminate in the rostral part of the dorsal thalamus.

The tractus olfacto-thalamicus et hypothalamicus consists partly of coarse axons and partly of fibres of a finer calibre. The coarse axons originate from the large mitral cells of the olfactory bulb and assemble in the vicinity of the commissura dorsalis telencephali, through which many of them decussate. From there, these axons descend in two separate bundles, which pass dorsal and ventral to the foramen interventriculare through the wall of the telencephalon medium (Figs. 10.22, 10.23). In the

most caudal region of that brain area the two bundles reunite and, arching ventrally, enter the ventral thalamus and, thence, the hypothalamus. During their trajectory through the telencephalon, the coarse axons of the mitral cells are joined by numerous finer fibres, originating from all telencephalic centres. The pathway under discussion receives some fibres from the contralateral olfactory bulb and from the contralateral septum, which pass through the anterior commissure. Its fibres fan out over the entire hypothalamus, but terminate partly in the ventral thalamus.

The tractus strio-thalamicus et hypothalamicus consists of fibres that join the medial part of the tractus olfacto-thalamicus et hypothalamicus during its course through the corpus striatum. These fibres are reported to terminate in the ventral thalamus, in the hypothalamus and, possibly, in the tegmentum motoricum mesencephali (Heier 1948). Johnston (1902, 1912) emphasised that the efferents of the corpus striatum pass to the thalamus and not to the hypothalamus.

10.9.5.2
Experimental Studies

Until recently, the existence of a nervus terminalis was not established in the lamprey. In gnathostomes the olfactory mucosa is innervated by peripheral branches of the terminal nerve. Applying HRP onto the olfactory epithelium of larval (von Bartheld et al. 1987; von Bartheld and Meyer 1988) and adult lampreys, Northcutt and Puzdrowski (1988) labelled fibres, the course of which is comparable to that of the central projections of the nervus terminalis in other anamniotes, such as the cladistian *Polypterus* (von Bartheld and Meyer 1986, 1988) orthe dipnoan *Protopterus* (von Bartheld and Meyer 1988), and in urodeles (McKibben 1911). These fibres pass through the ventromedial wall of the olfactory bulb without terminating there, and course via the medial subpallium and the preoptic region to the hypothalamus (Fig. 10.48). According to von Bartheld and Meyer (1988) most fibres terminate there, but Northcutt and Puzdrowski (1988) observed that fibres of the nervus terminalis arborize extensively among cells of the striatum and the preoptic region, and project only diffusely to the hypothalamus and the tuberculum posterius. The diffusely arranged fibres of the terminal nerve clearly follow the trajectory of the olfactohypothalamic tract, observed in non-experimental material (Figs. 10.21–10.23).

Anterogradely labelled fibres were not observed to enter the optic nerve following application of HRP into the olfactory epithelium, nor were retro-

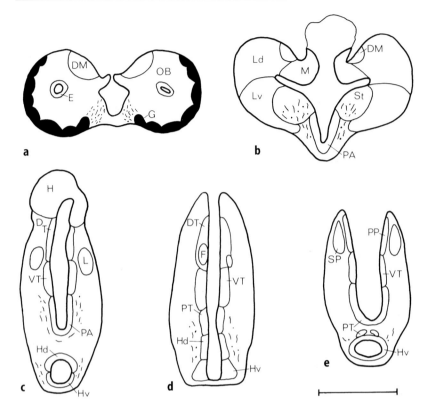

Fig. 10.48a–e. Charting of the projections of the left and right nervus terminalis following application of HRP onto the olfactory epithelium. *Dashed lines* and *dots* indicate fibres of passage and terminals, respectively. *Solid black regions* in **a** denote labelled olfactory glomeruli. Bar scale equals 500 µm. Note that according to Eisthen and Northcutt (1996) the fibres depicted do not represent a terminal nerve, but rather an extrabulbar olfactory pathway. *DM,* dorsomedial neuropil; *DT,* dorsal thalamus; *E,* ependyma; *F,* fasciculus retroflexus; *G,* glomerular layer of olfactory bulb; *H,* habenular nuclei; *Hd,* dorsal hypothalamic nucleus; *Hv,* ventral hypothalamic nucleus; *L,* lateral thalamic neuropil; *Ld,* pars lateralis of lateral pallium; *Lv,* pars ventralis of lateral pallium; *M,* medial pallium; *OB,* olfactory bulb; *PP,* periventricular pretectal nucleus; *PT,* tuberculum posterius; *SP,* superficial pretectal nucleus; *St,* striatum; *VT,* ventral thalamic nucleus. (Reproduced from Northcutt and Puzdrowski 1988, Fig. 4)

gradely labelled cell bodies or any labelled fibres observed by Northcutt and Puzdrowski (1988) following application of HRP into the optic nerve. Hence there is no evidence for the presence of retinopetal projections of the nervus terminalis in the lamprey, although they have been observed in other vertebrates (as e.g. teleosts: Münz et al. 1982; Stell et al. 1984). Also in contradistinction to other vertebrates, there is no immunohistochemical evidence in lampreys for luteinizing hormone-releasing hormone (LHRH)-positive perikarya and fibres in the nervus terminalis in the lamprey (Nozaki 1985).

Most recently, Eisthen and Northcutt (1996) questioned whether or not a true nervus terminalis is present in lampreys. In silver lampreys (*Ichthyomyzon unicuspis*) they did not observe either gonadotropin-releasing hormone (GnRH) or FMRFamide-like immunoreactivity in the telencephalic regions that represent the typical path of terminal nerve fibres, and they also failed to locate a terminal nerve ganglion. In the light of these findings, Eisthen and Northcutt (1996) concluded that lampreys lack a terminal nerve and that the fine bundle previously described by von Bartheld et al. (1987), von Bartheld and Meyer (1988) and Northcutt and Puzdrowsky (1988) represents an extrabulbar olfactory pathway.

The secondary olfactory connections in the lamprey have been studied experimentally by Northcutt and Puzdrowski (1988) and by Polenova and Vesselkin (1993) following injections of HRP in the olfactory bulb. These authors found that the lateral, dorsomedial and ventral streams of labelled fibres left the olfactory bulb.

The lateral fibres course caudally and project mainly to the primordium piriforme and the primordium pallii dorsalis, i.e. the pars ventralis and pars dorsalis of the lateral pallium of Northcutt and Puzdrowski (1988; Fig. 10.49b–d). Sizable numbers of fibres turn medially, pass through the subhippocampal lobe (the dorsal pallium of Northcutt and Puzdrowski 1988), and enter the primordium hippocampi or medial pallium (Fig. 10.49c,d). Synapses en passant are probably formed in both the subhippocampal lobe and the primordium hippocampi, but distinct terminal fields, similar to those in the evaginated part of the hemisphere, were not seen in these areas. The fibres passing through the primordium hippocampi ascend to the epithalamic region, decussate in the habenular commissure, and subsequently course to, and terminate in, the primordium piriforme, striatum and septal nucleus, as well as in the ventral hypothalamus and the tuberculum posterius (Fig. 10.49). Other fibres, belong-

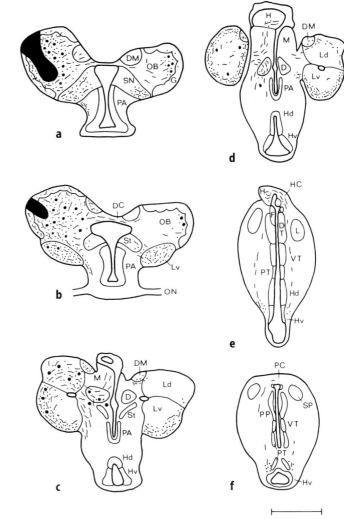

Fig. 10.49. Charting of the secondary olfactory projections following a unilateral HRP injection *(solid black region)* into the olfactory bulb of the silver lamprey, from rostral (*a*) to caudal (*f*). *Dashed lines* and *large dots* indicate fibres of passage and retrogradely labelled cell bodies, respectively. *Small dots* indicate anterogradely labelled terminals. *Bar scale* equals 500 µm. *D,* dorsal pallium; *DM,* dorsomedial neuropil; *DT,* dorsal thalamus; *F,* fasciculus retroflexus; *G,* glomerular layer of olfactory bulb; *H,* habenular nuclei; *HC,* habenular commissure; *Hd,* dorsal hypothalamic nucleus; *Hv,* ventral hypothalamic nucleus; *L,* lateral thalamic neuropil; *Ld,* pars lateralis of lateral pallium; *Lv,* pars ventralis of lateral pallium; *M,* medial pallium; *OB,* olfactory bulb; *ON,* optic nerve; *PA,* preoptic area; *PP,* periventricular pretectal nucleus; *PT,* tuberculum posterius; *SN,* septal nucleus; *SP,* superficial pretectal nucleus; *St,* striatum; *VT,* ventral thalamic nucleus. (Reproduced from Northcutt and Puzdrowski 1988, Fig. 2)

ing to the lateral contingent, continue ventrally and caudally to terminate in the ipsilateral tuberculum posterius and ventral hypothalamus (Fig. 10.49d–f).

Many dorsomedial secondary olfactory fibres enter the ipsilateral dorsomedial neuropil where they terminate. Other dorsomedial fibres enter the contralateral olfactory bulb via the dorsal commissure and terminate among the cells immediately beneath the glomeruli (Fig. 10.49a,b). Secondary olfactory fibres that exit the bulb ventrally enter and terminate extensively in the ipsilateral septal nucleus and along the lateral edge of the preoptic region (Fig. 10.49a,b).

Following unilateral injections of HRP into the olfactory bulb, Northcutt and Puzdrowski (1988) and Polenova and Vesselkin (1993) observed retrogradely labelled perikarya in the ipsilateral primordium piriforme, primordium pallii dorsalis and

subhippocampal lobe, as well as in the contralateral olfactory bulb (Fig. 10.49b–d). Within the contralateral bulb, Polenova and Vesselkin (1993) observed only medium-sized and bipolar labelled elements, but no labelled mitral cells.

The projections of the primordial piriform, dorsal and hippocampal pallia, and of the dorsal thalamus have been studied by Polenova and Vesselkin (1993) and Northcutt and Wicht (1996) using HRP and the carbocyanine dye DiI as tracers, respectively (Fig. 10.51). The connections of the piriform and dorsal pallia primordia appeared to be practically identical. Afferents of this pallial complex include bilateral fibres from the olfactory bulb, and fibres from the habenular nuclei, the septum, the striatum, the hippocampal primordium and the dorsal and ventral thalami. The efferents of the piriform and dorsal pallial primordia form rostrally, dorsomedially and ventrocaudally directed fibre

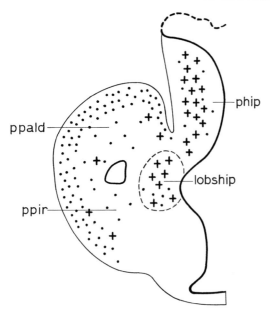

Fig. 10.50. Secondary olfactory projections *(black dots)* and ascending thalamic projections *(crosses)* in the telencephalon of the lamprey. For a list of the abbreviations used in figures, see p. 409. (Redrawn from Polenova and Vesselkin 1993, Fig. 11)

contingents. Most rostrally directed fibres terminate in the caudal two thirds of the olfactory bulb, but some terminate in the septum. The dorsomedially directed fibres followed the stria medullaris and ramify in the hippocampal primordium and in the pretectum or decussate via the habenular commissure and terminate in the contralateral habenular nucleus and hippocampal primordium. The fibres forming the ventrocaudal fibre contingent terminate in the ipsilateral striatum, preoptic nucleus and hypothalamus, as well as in the contralateral hypothalamus after having decussated in the postoptic commissure.

Following injection of tracer in the hippocampal primordium, Northcutt and Wicht (1996) observed labelled cells in the olfactory bulb, the piriform and dorsal pallial primordia, the septum, the striatum, the preoptic nucleus, the habenular nucleus, the dorsal and ventral thalami, the hypothalamus and the dorsal isthmic grey. In all of these centres labelled cells were observed on both sides, with an ipsilateral preponderance. Moreover, labelled cells were observed in the ipsilateral caudal midbrain tegmentum. The efferents of the hippocampal primordium could be traced rostrally, dorsocaudally and ventrally. The rostrally directed fibres could be followed to the piriform and dorsal pallial primordia and to the ipsilateral olfactory bulb. The caudally directed fibres form a distinct bundle, which passes through the dorsal thalamus, to which it

contributes, to reach the pretectal area, where most of its fibres terminate. The remaining fibres project to the optic tectum, as well as to the torus semicularis (Polenova and Vesselkin 1993) and the mesencephalic reticular area (Northcutt and Wicht 1996). Polenova and Vesselkin (1993) designated this fibre system, which has not been described before, as the hippocampal-pretecto-midbrain bundle. The ventrally directed hippocampal efferents pass to the preoptic area where approximately half of them decussate in the postoptic commissure. A certain proportion of both the decussating and the non-decussating fibres arch rostrally to terminate in the preoptic nucleus and in the septum. Other contraleral and ipsilateral hippocampal efferents pass ventrocaudally, and ramify in the hypothalamus.

According to Polenova and Vesselkin (1993), all parts of the dorsal thalamus, viz. the anterior and posterior parts of the periventricular dorsal thalamic nucleus, the nucleus of Bellonci and the primordial rostral and caudal geniculate bodies were found to project to the telencephalon. These ascending thalamic fibres terminated heavily in the hippocampal primordium and in the subhippocampal lobe and sparsely in the piriform and dorsal pallial primordia. All four of these areas also receive secondary olfactory fibres, but the areas with heavy thalamic projections receive only sparse olfactory projections and vice versa (Fig. 10.50). A single subzone, i.e. the dorsal part of the subhippocampal lobe, appeared to receive a strong thalamic projections, but no direct olfactory input. Northcutt and Wicht (1996) confirmed that the hippocampal primordium receives a strong projection from the dorsal thalamus and that the piriform and dorsal pallial primordia, on the contrary, are only sparsely innervated by ascending thalamic afferents. These authors also reported direct projections from the dorsal thalamus to the striatal neuropil and preoptic region.

If we survey the fibre connections just discussed (Fig. 10.51) it appears that the telencephalon of the lamprey is dominated by the strongly developed olfactory system. However, given the facts that first, fibres originating from the dorsal thalamus ascend to the pallium, particularly to the hippocampal primordium and the subhippocampal lobe, and second, the dorsal thalamus receives (a) visual afferents from the retina and the optic tectum, (b) octavolateral afferents from the torus semicircularis, and (c) somatosensory afferents from the dorsal column nucleus (cf. Sect. 10.8.3 and Fig. 10.44a), it seems likely that in the pallium of the lamprey olfactory stimuli are correlated with sensory impulses from other sources. The dorsal thalamus has also been found to project to the striatum and the

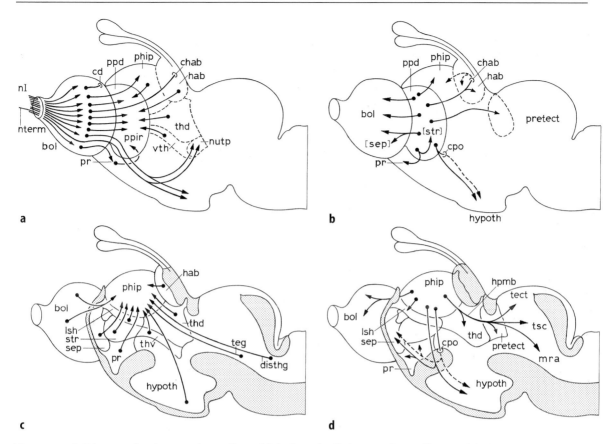

Fig. 10.51a–d. Diagrams showing experimentally established telencephalic connections. **a** Extra-bulbar olfactory pathway, until recently considered as nervus terminalis, secondary olfactory connections and the afferents of the piriform and dorsal pallial primordia. **b** Efferents of the piriform and dorsal pallial primordia. **c** Afferents of the hippocampal primordium. **d** Efferents of the hippocampal primordium. For a list of the abbreviations used in figures, see p. 409 (Based on Northcutt and Puzdrowski 1988; Polenova and Vesselkin 1993; and Northcutt and Wicht 1996)

preoptic region but these subpallial fields receive hardly any secondary olfactory fibres.

In this chapter the pallium of the lamprey has been subdivided, in accordance with Heier (1948), Schober (1964) and Nieuwenhuys (1977), into four areas or fields: (a) the piriform primordium, (b) the dorsal pallial primordium, (c) the primordium hippocampi (considered to be homologous to the lateral or piriform pallium, the dorsal or general pallium and the medial or hippocampal pallium, respectively, of anamniote gnathostomes), and (d) the subhippocampal lobe, which is intercalated between the hippocampal and the dorsal pallial primordia. As already indicated in Sect. 10.9.3, Northcutt and Puzdrowski (1988) and Northcutt and Wicht (1996) wish to interpret the piriform plus the primordial dorsal pallium as the lateral pallium and the subhippocampal lobe as the dorsal pallium. Several features argue in favour of this new interpretation: (a) Structurally, the piriform and the dorsal pallial primordia resemble each other closely. (b) The same holds true for the fibre connections of these parts: both receive a strong projection from the olfactory bulb, and both project heavily to the hippocampal primordium or medial pallium. (c) If we accept the interpretation of the dorsal pallial primordium as the dorsal part of the lateral pallium, the topological position of the subhippocampal lobe corresponds to that of the dorsal pallium of gnathostomes. (d) Though the fibre connections of the subhippocampal lobe are still imperfectly known, the findings of Polenova and Vesselkin (1993) that the dorsal part of this structure receives a strong thalamic projection, but no direct olfactory input are also in line with this new interpretation.

10.10
Localisation of Neuromediators

During the past several decades, numerous studies have appeared on the localisation of neurotransmit-

Table 10.3. Localisation of neuromediators and related substances in the CNS of the lamprey

Neuromediator		Source
Acetylcholine (ACH)	A high concentration of AChE in the spinal motor column, the nuclei of III and IV, the branchiomotor nuclei, the torus semicircularis and the striatum	Wächtler (1974, 1983)
NADPH-diaphorase (NO-synthase)	Glomerular layer of olfactory bulb, pineal and parapineal organs, habenular region, Müller and Mauthner cells	Schober et al. (1994)
Aspartate	Inferior, middle and mesencephalic reticular nuclei, including Müller and Mauthner cells, as well as medium-sized and small neurons within these cell masses (Fig. 53a)	Brodin et al. (1989a)
Glycine	Small neurons 5–10 μm of unknown nature, evenly distributed throughet the spinal grey matter (44–86 per segment), being more numerous in caudal segments	Sheridan et al. (1984)
GABA	(1) scattered small ellipsoid cells (10–20 μm) in the dorsal part of the spinal grey matter, (2) multipolar cells (15–25 μm) in the spinal lateral grey column, (3) small, spherical CSF-contacting neurons (10–15 μm) surrounding the spinal cental canal, (4) fibres in all parts of the spinal cord, and (5) dense plexuses in the lateral spinal margin and the most dorsomedial part of the lateral funiculus (Fig. 54).	Brodin et al. (1990c) Batueva et al. (1990)
	Retinopetal cells in the periventricular zone of the mesencephalic tegmentum (65 %) and in the neuropil and fibre zone situated ventrolaterally to that zone (15 %)	Rio et al. (1993)
GABA + Somatostatin	Columnar cells surrounding the central canal, sending their processes to the lateral spinal margin to form a dense plexus around the edge cell dendrites	Christenson et al. (1991)
Catecholamines (Tyrosine hydroxlase)	Numerous small, bipolar cells in the olfctory bulb and in the anterior olfactory nucleus; large and small cells in the preoptic nucleus, some in contact with CSF; cells in the dorsal and ventral hypothalamic nucleus, many in contact with CSF; cells in the nucleus of the tuberculum posterius; some scattered cells in the medial and lateral nuclei of the posterior commissure; cells situated in the ventral and dorsal part of the isthmic tegmentum; CSF-contacting cells in the caudal part of the rhombencephalic basal plate, situated ventrally to the motor nucleus of X (Fig. 52b); small CSF-contacting cells situated ventrally to the spinal central canal	Pierre et al. (1994)
	dense fibre plexuses in the olfactory bulb, the striatum, the septum, the preoptico-hypothalamic continuum, the medial mesencephalic tegmentum, the rhombencephalic basal plate; numerous fibres in the spinal ventral funiculus	
Catecholamines + Neuropeptides	Tyrosine hydroxylase-immunoreactive (TH-ir) cells located ventrally to the spinal central canal also contain neurotensin, tachykinin, somatostatin and porcine peptide YY (PYY)	Van Dongen et al. (1986) Buchanan et al. (1987b) Pierre et al. (1994)
Dopamine (DA)	Chemical concentration of dopamine in the brain is much higher than that of noradrenaline, suggesting that most TH-ir cells are dopaminergic	Baumgarten (1972)
	The pattern of DA and TH-immunireactivity is almost identical Small CSF-contacting celle (8–10 μm) form a colateral column just ventral to the spinal central canal; a sparse plexus of fibres is situated ventral to these cells	Pierre et al. (1994) McPherson and Kemnitz (1994)
Noradrenaline (NA)	Comparison of TH-ir and DA-ir neuronal populations reveals that the TH-ir cells in the isthmic region and in the rhombencephalon proper are most probably dopaminergic	Pierre et al. (19949)
Serotonin or 5-hydroxytryptamine (5-HT)	Numerous CSF-contacting cells in the dorsal and ventral hypothalamic nuclei (10–12 μm), particularly in the caudal, postinfundibular part of the latter (Fig. 55); some cells in the ventral thalamic nucleus; scattered cells in the dorsal thalamic nucleus, in the lateral geniculate body and in medial and lateral nuclei of the posterior commissure (13–18 μm); cells in the pineal organ; some cells in the periventricular layer of the tectum and the mesencephalic tegmentum; numerous cells in the tegmentum isthmi and in the dorsal or subcerebellar isthmic grey; small cells in the medial part of the rhombencephalic central grey, at the levels of the motor nuclei of V and X	Pierre et al. (1994)
	Cells in the superior reticular nucleus and in the caudomedial part of the inferior reticular nucleus (Figs. 52c, 53b) Fibres and terminals distributed widely throughout the brain; a relatively dense innervation of the primordium hippocampi and the striatum; dense fibre plexuses in the hypothalamus, the striatum of the tectum, the medial part of the mesencephalic tegmentum, and the medial part of the rhombencephalic basal plate	Brodin et al. (1986) Brodin et al. (1988b) Pierre et al. (1992)

Table 10.3.

Neuromediator		Source
	Dense plexuses surrounding the somata of the neurons in the superior and middle reticular nuclei, including Müller and Mauthner cells; fibres in cranial nerves III–X	Viana Di Prisco et al. (1994)
	Cells forming a ventromedian column over the entire length of the spinal cord; these cells send a diffuse projection of processes throughout the spinal cord; ventral processes of these cells form a very dense ventromedial plexus of varicosities, contacting dendrites of montoneurons and lateral cells; 5-HT-ir varicosities in the spinal cord have no synaptic specialisations; the elements of the ventromedian column send fibres into the spinal dorsal and ventral roots; the spinal dorsal root ganglia contain 5-HT-positive cells (Fig. 54).	Harris-Warwick et al. (1985) Van Dongen et al. (1985 a, b) Christensen et al. (1990)
	Cells in the caudal part of the inferior reticular nucleus and in the superior reticular nucleus send small diameter axons to the spinal cord, which descend in the lateral and ventromedial columns	Brodin et al. (1986, 1988a)
Serotonin + dopamine	Cells in the spinal ventromedian column giving rise to the dense ventromedial plexus co-contain and co-release serotonin and dopamine	
Serotonin + neuropeptides	Some 5-HT-ir cells in the spinal ventromedian column also contain tachykinin	Van Dongen et al. (1985a)
	Small cell bodies in spinal ganglia projecting to dorsal column and dorsal contain, in addition to 5-HT, calcitonin gene-related peptide (CGRP) and bombesin	Brodin et al. (1988a)
Histamine	Numerous bipolar CSF-contacting neurons in the dorsal and ventral hypothalamic nuclei and in the postinfundibular commissural nucleus (10–15 μm); some cells in the ventral part of the isthmic tegmentum (10–15 μm); some cells in the ventral part of the isthmic tegmentum (10–15 μm); fibres ascending from the hypothalamus to the telencephalon where they terminate in the striatum and in the septal nucleus (Fig. 52d)	Brodin et al. (1990a)
Tachykinin (TK)	Some cells in the ventromedian column of the spinal cord contain TK as well as 5-HT; longitudinally running fibres in the spinal cord, some of which enter a dorsal root; a few of these fibres contain both TK and 5-HT	Van Dongen et al. (1985a) Van Dongen et al. (1986)
	Some cells in the superior and middle rhombencephalic reticular nuclei	Nozaki and Gorbman (1986)
Neuropeptide Y (NPY) and related peptides	Small bipolar neurons (10–15 μm) forming a column in the dorsomedial part of the spinal grey matter; these cells give rise to fibres which descend mainly in the area of the dorsal horn	Van Dongen et al. (1985a) Brodin et al. (1989b)
	Small cells (10–20 μm) scattered near the medial border of the alar plate, at the level of the motor nucleus of X; cells in the rostral part of the rhombencephalic alar plate extending medially in the superior reticular nucleus; small cells in the lateral part of the mesencephalic tegmentum; numerous fibres throughout the rhombencephalon and in the dorsal and lateral parts of the mesencephalon (Fig. 54c)	
Porcine peptide YY (PYY) and related peptides	Small, rounded cells (10–20 μm) in the middle reticular nucleus; larger cells (20–40 μm) in the rostral portion of the motor nucleus of V, and in and around the superior reticular nucleus; most of these larger elements project to the spinal cord, their axons descending in the lateral funiculus; small to intermediate cells (15–30 μm) are clustered in the mesencephalic reticular nucleus (Fig. 53c)	Brodin et al. (1989b)
Cholecystokinin (CCK)	A dorsal root–dorsal column system of fibres originating from cell bodies in the spinal dorsal root ganglia; fibres descending in the lateral and ventromedial columns of the spinal cord, which originate from large cell bodies (20–40 μm) in the ventromedial part of the inferior reticular nucleus; smaller cells in the rostral part of the middle reticular nucleus; small to medium-sized cells in the mesencephalic reticular nucleus (Fig. 53b); numerous cells in the ventral hypothalamic nucleus and in the pre- and postcommissural infundibular nuclei dense fibre plexuses in the hypothalamus and in the striatum	Brodin et al. (1988a, b)
	The fibres descending from the inferior reticular nucleus to the spinal cord probably contact spinal motoneurons and giant relay interneurons	Ohta et al. (1988)
Neurotensin (NT)	A cluster of CSF-contacting neurons in the preoptic nucleus; a large number of small (8–15 μm), CSF-contacting neurons in the periventricular area of the hypothalamus, including the dorsal and ventral hypothalamic nuclei and the postinfundibular commissural nucleus; some small cells in the postoptic commissural nucleus and in the ventral thalamic nucleus; a group of small neurons (10–15 μm) in the rostromedial part of the mesencephalic tegmentum, along the ventral aspect of the nucleus of the fasciculus longitudinalis medialis (Fig. 52e)	Brodin et al. (1990b)

Table 10.3.

Neuromediator		Source
Somatostatin (SRIF)	Small CFS-contacting neurons (10–15 µm) along the dorsal and lateral aspects of the spinal canal; some larger cells (15–15 µm) in the ventral aspect of the dorsal horn Dense fibre plexuses in the corpus striatum, the caudal hypothalamus, the ventral mesencephalon and the dorsal horn and lateral marginal zone of the spinal cord (Fig. 54) Numerous CSF-contacting cells throughout the ventral hypothalamic nucleus; a group of cells in the dorsal thalamic nucleus, adjacent to the fasciculus retroflexus; cells along the dorsal edge of the pellucida within the pineal organ; cells in the isthmic tegmentum and in the rhombencephalic basal plate (Fig. 52f) Fibres were found throughout most of the brain, predominating within the preoptic nucleus, ventral hypothalamic area and infundibular nucleus	Wright (1986) Cheung et al. (1990)
Gonadotropin-releasing hormone (GnRH)	Numerous CSF-contacting neurons (13–18 µm) in the preoptic nucleus forming two groups, one in the most ventral part of the nucleus, the other more dorsally situated Fibres project with the preoptico-hypophysial tract to the neurohypophysis	King et al. (1988) Crim et al. (1979)
Growth hormone (GH)	Numerous neurons in the rostral and middle regions of the preoptic nucleus Fibres pass with the preoptico-hypophysial tract to the neurohypophysis (Fig. 52c)	Wright (1986)
Prolactin (PRL)	A few cells in the preoptic nucleus, projecting via the preoptico-hypophysial tract to the neurohypophysis (Fig. 52c)	Wright (1986)
FMRFamide	Two groups of cells in the preoptic nucleus, one immediately rostral, the other dorsal to the optic chiasma; cell groups in the rostral and caudal part of the ventral hypothalamic nucleus, and a large third group in the most dorsal part of that nucleus, extending into the nucleus of the postoptic commissure Fibres distributed widely in the brain, but particularly in the hypothalamus; fibres pass caudally from the hypothalamus and reach the spinal cord (Figs. 52g, 54)	Ohtomi et al. (1989)
β-Endorphin (END)	Numerous cells in the dorsal and ventral hypothalamic nuclei and in the nucleus of the postoptic commissure; a cluster of cells in the lateral part of the dorsal thalamic nucleus; fibres in almost all parts of the brain, particularly in the ventral hypothalamus, the preoptic region and the geniculate body (Fig. 52h)	Nozaki and Gorbman (1984)
Met-enkephalin	Some neurons in the preoptic nucleus; fibres in the hypothalamus and in the neurohypophysis	Nozaki and Gorbman (1984)
Arginine vasotocin	A large group of CSF-contacting neurosecretory cells extending over the preoptic nucleus, the striatum and the ventral thalamus Fibres contribute the preoptico-hypophysial tract; scattered fibres, spreading over telencephalon, diencephalon and brain stem A dense plexus is formed in the lateral part of the isthmic tegmentum	Sterba (1972) Goossens et al. (1977)
Urotensin I (UI)/ Cortocotropin-releasing factor (CRF)	A few primarily longitudinally oriented fibres in the 'white' matter of the caudal spinal cord	Hoheisel et al. (1978) Onstott and Elde (1986)

ters and related substances in the brain of the lamprey. In the preceding sections of this chapter reference has been made to many of these studies. A summary of our present-day knowledge of the chemoarchitecture of the central nervous system of the lamprey is presented in Table 10.3 and the accompanying Figs. 10.52–10.55.

10.11
Overview of Functional Circuits

The preceding sections have shown that, due mainly to the studies of Rovainen (summarised in 1979b, 1982, 1983) and of Grillner and his associates (reviewed in Brodin and Grillner 1990; Grillner and Matsushima 1991; and Grillner et al. 1991, 1995), considerable insight has been gained into the mechanisms of locomotion and its control. In this section we give a pictorial overview of neuronal circuitry.

Most of the information has been detailed earlier in the chapter (see Sect. 10.4.8 and Fig. 10.28) and the figure legends here briefly review the salient facts. Figure 10.56 depicts the segmental network responsible for producing alternating burst activity, forming the basis of locomotor activity. Only three

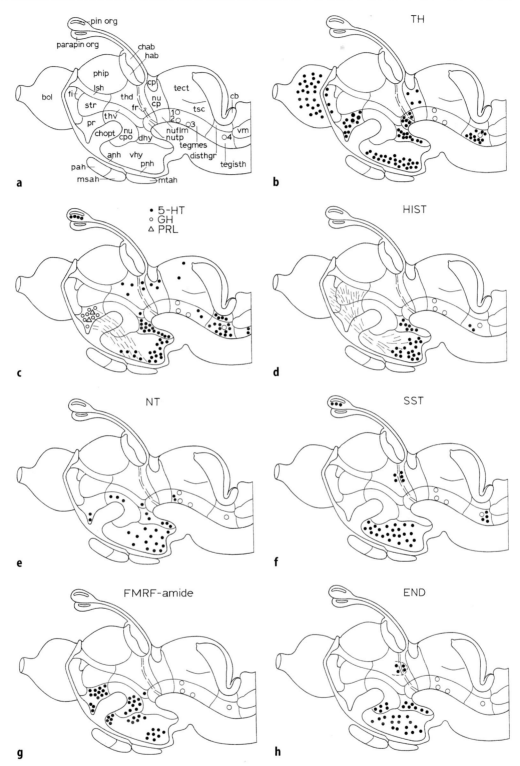

Fig. 10.52a–h. Schematic representation of the distribution of cells containing particular neuromediators or neuromediator-related substances in the rostral part of the brain of the lamprey, *Lampetra fluviatilis*, as they project on a medial view of the bisected brain. **a** Location of principal cell masses. **b** Tyrosine-hydroxylase *(TH)*. **c** Serotonin *(5-HT)*. **d** Histamine *(HIST)*. **e** Neurotensin *(NT)*. **f** Somatostatin *(SST)*. **g** FMRFamide. **h** β-endorphin *(END)*. For a list of the abbreviations used in figures, see p. 409 (Based on sources specified in Table 10.3)

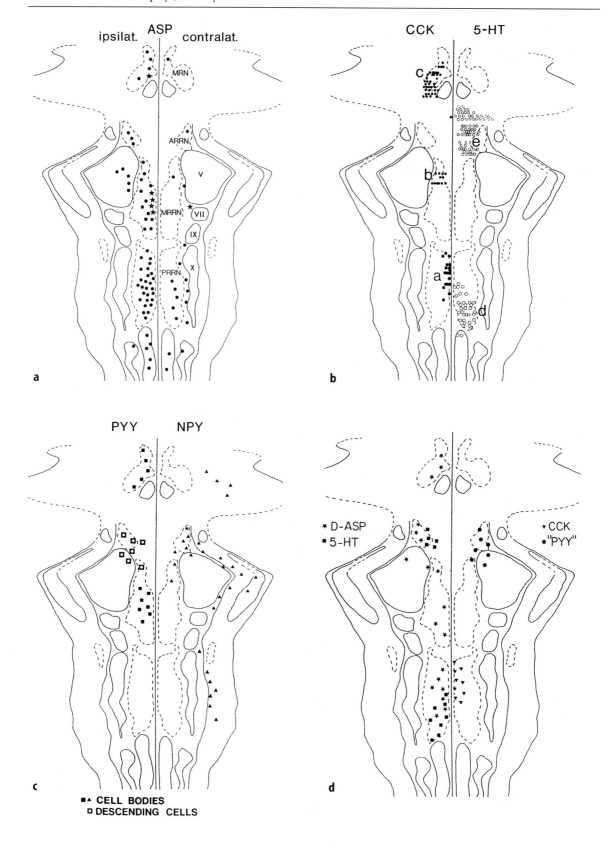

CELL BODIES
DESCENDING CELLS

◀ **Fig. 10.53a–d.** Schematic representation of the distribution of cells containing particular neuromediators in the brain stem of the lamprey, n *Lampetra fluviatilis* as they project onto the topological map shown in Fig. 30. **a** Radio-labelled cells following an injection of *3H*-D-aspartate *(ASP)* in the left side of the rostral spinal cord; *stars* represent large identifiable Müller and Mauthner cells; other neuron types are represented by *dots;* **b** CCK- and 5-HT-immunoreactive cells; groups of CCK-ir cells are present in the inferior *(a)* and middle *(b)* reticular nuclei of the rhombencephalon and in the mesencephalic reticular nucleus *(c);* groups of 5-HT-containing cells are located in the caudal part of the inferior reticular nucleus *(d),* and in the isthmic region *(e),* within and medial to the superior reticular nucleus of the rhombencephalon; **c** PYY- and NPY-immunoreactive cells; open squares indicate large to medium-sized somata projecting to the spinal cord; **d** Reticulospinal neurons labelled by transmitter-related markers. *ARRN,* anterior rhombencephalic reticular nucleus; *MRRN,* middle rhombencephalic reticular nucleus; *PRRN,* posterior rhombencephalic reticular nucleus. (Reproduced from the following sources: a, Brodin et al. 1989a; b, Brodin et al. (1988a); c, Brodin et al. 1989b; d, Brodin et al. 1988b.)

classes of interneuron are required (LIN, CCIN and EIN) but they must reciprocally innervate their counterparts on the contralateral part of the cord. Figure 10.57 adds stretch receptor neurons (SRE and SRI) to the segmental circuit, which provide feedback on the posture of the animal and influence the locomotor cycle. Figure 10.58 introduces a crucial element, a descending excitatory influence onto all the interneurons, without which oscillation would not occur, since it is this excitation that provides the substrate for the action of the crossed, reciprocal inhibition. Note that, as discussed in Sect. 10.4.8, and shown in Fig. 10.29, for actual swimming to occur, adjacent segments must be coupled via collaterals of excitatory interneurons. Figure 10.59 introduces spinal and supraspinal somatosensory systems; the structure and function of these systems is still not well known. Finally, Fig. 10.60 brings the vestibulospinal and vestibuloreticulospinal systems into focus. These systems are crucial to any prolonged swimming activity because they provide essential information and reflex pathways to maintain the orientation and direction the body (Sect. 10.5.5). Such orientation and direction is also aided by cues from the illumination of the animal which are transduced by one or more of the various light-sensing organs (eyes, pineal gland, dermal photoreceptors and diencephalic cells) of the lamprey.

10.12
Concluding Comments: The Lamprey and Its Relationship to the Gnathostomes

In this final section of the chapter we place some aspects of the structural and functional organisation of the CNS of the lamprey in a comparative perspective.

10.12.1
Size of the Brain and Gross Morphology

The brain of the lamprey is very small, indeed much smaller than that of any gnathostome. Ebbesson and Northcutt (1976) have shown that the brain to body ratio of *Petromyzon marinus* is at least 500 % less than that of any jawed vertebrates.

In their gross morphology the brains of the various groups of anamniotes show profound differences. In all anamniotes a considerable part of the brain is in direct connection with the periphery through the cranial nerves, and many of the gross neuromorphological differences are directly related to the development and differentiation of particular sense organs. Thus, the size of the olfactory bulb reflects the size of the olfactory organ and the size of the eyes is correlated with the size of the tectum mesencephali. A similar relationship exists between the development and number of taste organs and the size of the visceral sensory zone of the rhombencephalon and between the differentiation of the lateral line system and the rhombencephalic somatosensory zone. In many groups the development of the lateral line system is also reflected in the size of the cerebellum or parts thereof; this is particularly evident where an electrosensory modality is present. In the lamprey, electrosensitivity has been established but much remains unknown about the system. The olfactory bulbs of the lamprey are large, but they share this feature with almost all other groups of anamniotes. The only sense organs that show a remarkable development in the lamprey are the pineal and parapineal organs. These organs may have influenced the size of the habenular nuclei somewhat (Heier 1948), but in general it seems that the brain of the lamprey does not show any marked group-specific, sense organ-bound 'hypertrophies'. In this respect the brain of the lamprey may be designated as 'generalised'. Among the anamniotes a similar paucity of sense organ-bound 'hypertrophies' is observed in dipnoans and in urodele amphibians. It is with the brain of these groups that the brain of the lamprey shows the greatest gross morphological similarity.

Apart from differences in proportional size, the various parts of the brains of anamniotes also show considerable differences in shape. This is particularly true of the cerebellum and the telencephalon. In so far as shape is concerned the cerebellum is doubtless the most variable part of the central nervous system of anamniotes and indeed, of the vertebrates in general (Nieuwenhuys 1967b). In dipnoans and amphibians the cerebellum is small and shows a plate-like configuration. In elasmobranchs, holosteans, teleosts, and in the crossopterygian

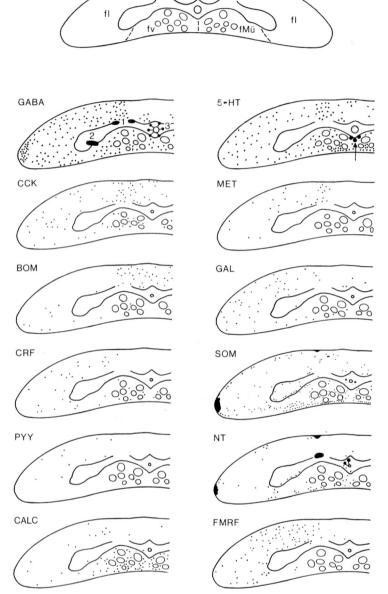

Fig. 10.54. Distribution of neurome-diator-specified cells and fibres in the spinal cord of the lamprey. *Top* figure: subdivision of white matter (drez, dorsal root entry zone). The cells and fibres were visualised with antibodies raised against *GABA:* serotonin *(5-HT)*, metor-phamide *(MT)*, bombesin *(BOM)*, gala-nin *(GAL)*, corticotropin-releasing factor *(CRF)*, somatostatin *(SOM)*, peptide YY *(PYY)*, neurotensin *(NT)*, calcitonin *(CALC)* and Phe-Met-Arg-Phe-amide *(FMRF)*. 1,2,3 types of GABAergic cells. *Arrow* indicates ventromedian column of serotoninergic cells. For a list of the abbreviations used in figures, see p. 409 (GABAergic cells and fibres drawn from photomicrographs in Brodin et al. 1990c; serotoninergic cells and fibres drawn from photomicrographs in Van Dongen et al. 1985a; the remaining figure parts are redrawn from Buchanan et al. (1987b).

Latimeria the cerebellum has evaginated, whereas in brachiopterygians and chondrosteans the structure under discussion has invaginated into the ventricular system. As we have seen, in the lamprey the cerebellum is represented only by a small lamella that bridges the most rostral part of the fourth ventricle. It clearly resembles the cerebellum of urodeles, although auriculae, that is, rostrolateral cerebellar extensions that link urodeles with most other anamniotes, are lacking in the lamprey.

The telencephalon of most vertebrates is evag-

inated; that is to say, its walls surround bilateral extensions of the ventricular cavity of the neuraxis. In all vertebrates, however, part of the telencephalon remains unevaginated. In this unevaginated part the telencephalic walls may maintain the early embryonic tube-shaped condition of the neuraxis: a pair of vertically oriented lateral walls or side plates bound a median ventricular cavity. The portion of the telencephalon showing this condition is called the telencephalon medium, and its ventricular cavity the ventriculus impar.

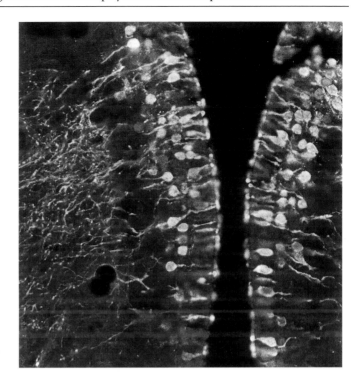

Fig. 10.55. Serotonin-containing neurons and fibres in the hypothalamus of the lamprey, *Lampetra fluviatilis*, as visualised by immunofluorescence using a direct antibody against serotonin. Note that the dendritic processes of these elements are provided with bulb-like endings which protrude into the ventricular lumen. (Courtesy of Dr. H. W. M. Steinbusch, University of Maastricht) 400x

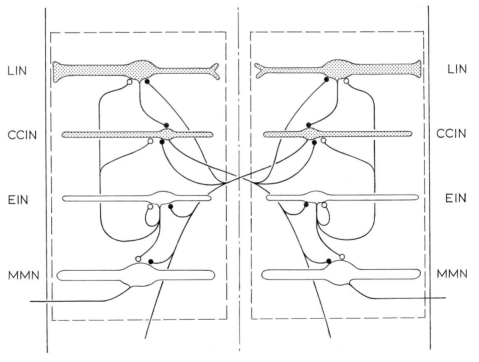

Fig. 10.56. The segmental network responsible for producing alternating burst activity forming the basis of locomotion. Three classes of interacting interneuron, the premotor excitatory interneurons *(EIN)*, the crossed caudal interneurons *(CCIN)* and the lateral interneurons *(LIN)*, constitute the two parts of the network on either side of the spinal midline. The rhythmic output of the network is via the EINs which activate myotomal motoneurons *(MMNs)*, resulting in ipsilateral burst discharges. The CCINs and the LINs are also activated by the EINs. The CCINs inhibit all types of network neurons on the contralateral side. The LINs, which discharge only at the peak of their depolarization, inhibit the CCINs, leading to disinhibition of the contralateral side. Provided that the background excitation is sufficient, the neurons on the contralateral side will start discharging, thereby inhibiting neurons on the initially active side. The burst termination is not only achieved by the activity of the LINs, but also by summation of postspike afterhyperpolarization in the locomotor network interneurons. The EINs use glutamate and the CCINs and LINs use glycine as their neurotransmitters.

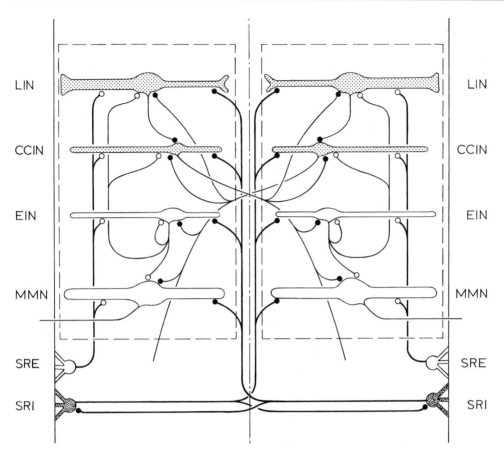

Fig. 10.57. The stretch receptor neurons and their influence on the locomotor cycle. During normal swimming the undulatory movements of the body are sensed by stretch receptor neurons (edge cells) located along the lateral margin of the spinal cord (Grillner et al. 1984). These neurons are of two types, an excitatory type *(SRE)*, which excites myotomal motoneurons and interneurons of the locomotor network on the ipsilateral side, and an inhibitory type *(SRI)*, which inhibits neurons on the contralateral side (Viana Di Prisco et al. 1990). The stretch receptor neurons inhibit network neurons responsible for their stretch in the active, contralateral half of the segments and provide excitation to the neurons in the ipsilateral half which, at the moment of maximal stretch, already are on the verge of becoming active (Travén et al. 1993; Grillner et al. 1995). In addition, local GABAergic and monoaminergic neurons (containing both serotonin and dopamine) exert a phasic modulation on the neurons of the spinal locomotor networks, and the monoaminergic neurons also modulate synaptic efficacy between reticulospinal and motoneurons (Grillner et al. 1995). *LIN,* lateral interneurons; *CCIN,* crossed caudal interneurons; *EIN,* excitatory interneurons; *MMN,* myotomal motoneurons; *SRE,* stretch receptors, excitatory; *SRI,* stretch receptors, inhibitory

In the lamprey the olfactory bulbs are clearly evaginated, whereas the telencephalon proper is partly evaginated and partly represents a telencephalon medium. The same can be said of the telencephalon of elasmobranchs, dipnoans, amphibians, and anamniotes. However, it should be emphasised that in the lamprey the telencephalon impar is vastly more extensive than in any of these groups. In the lamprey, the septum, the primordium hippocampi, and a part of the corpus striatum are situated in the telencephalon medium, whereas, in all the groups of gnathostomes mentioned, the homologues of these structures are entirely incorporated in the evaginated cerebral hemispheres.

10.12.2
Neurons

Although the neurons in the CNS of the lamprey vary considerably in size, the great majority of them share the following two characteristics: their somata are located in a continuous zone of central grey, and they extend most of their dendrites into the peripheral fibre zone of the neuraxis. In our opinion both of these characteristics may be considered truly primitive (primitive in the sense of reflecting the ancestral condition); they are also prevalent features in the brains of dipnoans and urodele amphibians.

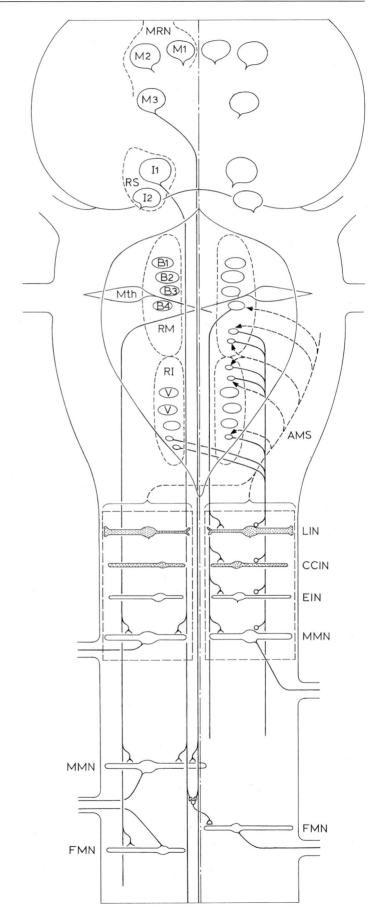

Fig. 10.58. Reticulospinal neurons involved in activating and controlling the spinal loco-motor networks. These neurons are aggre-gated in four centres, the mesencephalic reticular nucleus *(MRN)* and the superior, middle and inferior rhombencephalic reticu-lar nuclei *(RS, RM and RI, respectively).* All of these centres contain, in addition to some very large Müller cells *(M, B, V),* numerous smaller, reticulospinal neurons. Neurons in the *RM* and *RI,* among which some large Müller cells *(B3, B4),* projecting to the spinal cord, activate myotomal motoneurons *(MMNs)* and all types of interneurons of the spinal locomotor networks *(EIN, CCIN, LIN)* (McClellan and Grillner 1984; Ohta and Grillner 1989). These reticulospinal synaptic connections are glutamatergic, and there is both an NMDA and a non-NMDA (AMPA-kainate) component of the excitatory post-synaptic potential (EPSP) and, in addition, an electrical component (Wallén 1994). The reticulospinal neurons in RM and RI are rhythmically modulated during fictive swimming (Kasicki and Grillner 1986; Kasicki et al. 1989). This modulation origi-nates from the spinal cord locomotor net-works and not from the brain stem itself (Dubuc and Grillner 1987). Spinobulbar neurons, the axons of which form part of the spinal lemniscus, have been shown to convey information about the progress of the different phases of the locomotor cycle (Vinay and Grillner 1992), and hence most probably form part of this ascending modu-latory system *(AMS).* The synaptic relations of these neurons have not yet been deter-mined, however. The axons of the Müller cells make mixed chemical and electrical synapses throughout the length of the spinal cord. The axons of Müller cells *M3* and *I1* excite ipsilateral myotomal motoneurons and contralateral fin motoneurons *(FMNs),* the latter through unidentified interneurons. The Mauthner axon crosses the midline and descends to synapse directly with both myo-tomal and fin motoneurons (Rovainen 1982, 1983).

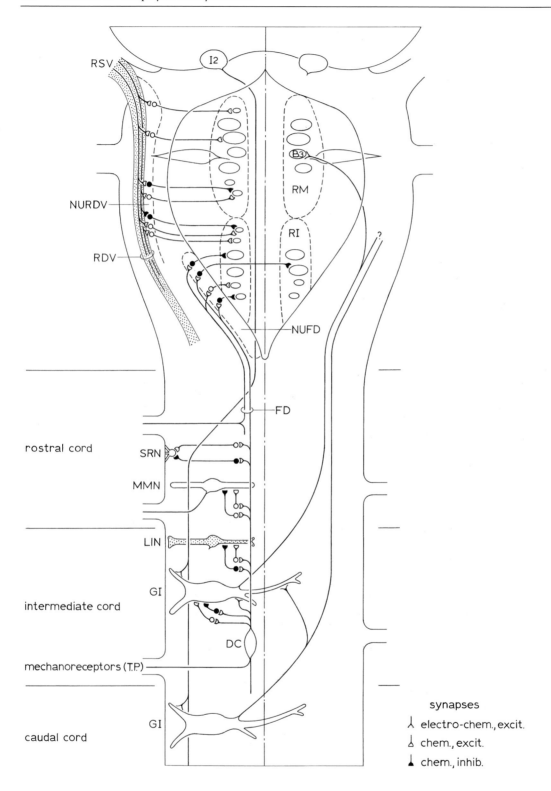

synapses

λ electro-chem., excit.

λ chem., excit.

$\blacktriangle$ chem., inhib.

◄ **Fig. 10.59.** Spinal and supraspinal mechanosensory systems. The dorsal cells *(DC)* are glutamatergic primary sensory neurons, the soma of which is located in the spinal cord. They convey information from touch and pressure *(T.P.)* skin receptors. Their main axons ascend in the spinal dorsal funiculus *(FD)*. Dorsal cells produce monosynaptic EPSPs in giant interneurons *(GI)* and generate EPSPs and IPSPs that are most likely polysynaptic and mediated by unknown interneurons, with *LINs* and *SRNs* (Rovainen 1967b). GIs are located in the caudal cord and receive inputs both from DCs and more caudal GIs. They are secondary mechanosensory neurons which relay activity rapidly from the tail towards the head (Rovainen 1982). The axons of GIs ascend on the contralateral side and reach the brain, but the only identified target is Müller cell *B3* (Rovainen 1974a,b). GIs are not directly involved in swimming (Rovainen 1982). Reticulospinal neurons situated in RI receive excitatory and inhibitory synaptic inputs from spinal dorsal roots and dorsal funiculi via neurons situated in the nucleus of the dorsal funiculus *(NUFD)*. The neurons involved in this ascending system convey somesthetic information and rely on excitatory and inhibitory amino acids (Dubuc et al. 1993a,b). Trigeminal afferents convey comparable mechanosensory inputs from the head region to reticulospinal neurons in *RM* and *RI*, via excitatory and inhibitory interneurons. The location of these interneurons is unknown; in the figure they have been postulated to be situated in the nucleus of the radix descendens of the trigeminal nerve *(NURDV)*. Excitatory and inhibitory amino acids are also involved in mediating synaptic transmission of these trigeminal inputs to the reticular formation (Viana Di Prisco et al. 1995). *LIN*, lateral interneurons; *MMN*, myotomal motoneurons; *SRN*, stretch receptor neuron

The small size and the simple histological organisation of the lamprey brain and also the basic similarity between this brain and that of the gnathostomes led investigators like Ariëns Kappers (1929) and Heier (1948) to conclude that the brain of the lamprey embodies the prototype of the craniote brain. However, in the introduction to the present chapter it has been suggested that during the long, independent phylogenetic history of the lampreys the structure of the central nervous system may well have become altered. Conversely, it is possible that features which the lampreys had in common with the (extinct) central craniate ancestors may well have been lost at a pre-gnathostome level.

The lamprey possesses three types of giant cells: the Müller and Mauthner cells of the rhombencephalon and the giant interneurons of the cord. Giant cells have been found in many other animals, most notably invertebrates, where they frequently have the status of an 'identified neuron'. Such neurons can be demonstrated to be both anatomically and functionally similar from one animal to the next within, and sometimes across, species. The Müller and Mauthner cells appear to have this property in the lamprey (although the studies of Rovainen differ slightly from those of Nieuwenhuys in regard to the number of Müller cells). The large reticulospinal neurons can be distinguished by the characteristics of their somata and dendrites (Martin 1979). A study by Selzer (1979) revealed, however, that the giant interneurons of the lamprey cord do not have fixed positions and consequently cannot be identified as individuals from one animal to the next.

Looking at the question of homologies of the cells of the lamprey with groups other than the petromyozonts, we note that although amphioxus possesses giant Rohde cells in its cord, these cannot be considered as homologous to the giant interneurons of the lamprey (Chap. 9). Homologues of the Müller and Mauthner cells are not present in amphioxus either, probably because of the rudimentary state of development of the rostral neuraxis in that species.

Rather surprisingly, Müller and Mauthner cells are absent in myxinoids. Müller cells, at least as defined here, cannot be identified in Gnathostomes either. In contrast, Mauthner cells can be found in all ichthyopsid gnathostomes, in fact from the lamprey to the anuran tadpole. Perhaps not surprisingly there are significant differences in both anatomy and physiology amongst the Mauthner cells encountered in this wide range of animals (Faber and Korn 1978).

The evolutionary significance of identifiable cells is hard to judge. Investigators are attracted to these elements because of their experimental utility, but they seem only to be prevalent in relatively simple nervous systems, particularly invertebrates. In the case of an element such as the Mauthner cell which mediates the escape reflex in many aquatic species, reliance on but one pair of neurons must introduce a substantial degree of vulnerability; this may be the reason for the apparent duplication of systems with evolution and consequent loss of the identifiable element.

10.12.3
Axons

All of the axons found in the nervous system of the lamprey are unmyelinated. A similar condition is found in the myxinoids, the other group of extant cyclostomes, as well as in the cephalochordates. While this is characteristic of cells with short axons and certain thin-fibred pathways in all gnathostomes, it is unusual to find medium-sized and coarse axons devoid of myelin. This is because, in terms of velocity of conduction and energy requirement, myelinated axons are superior to naked ones. We agree with Bone (1963) and Bullock et al. (1984) that the complete absence of myelin sheaths in the nervous system of cyclostomes is a primitive characteristic.

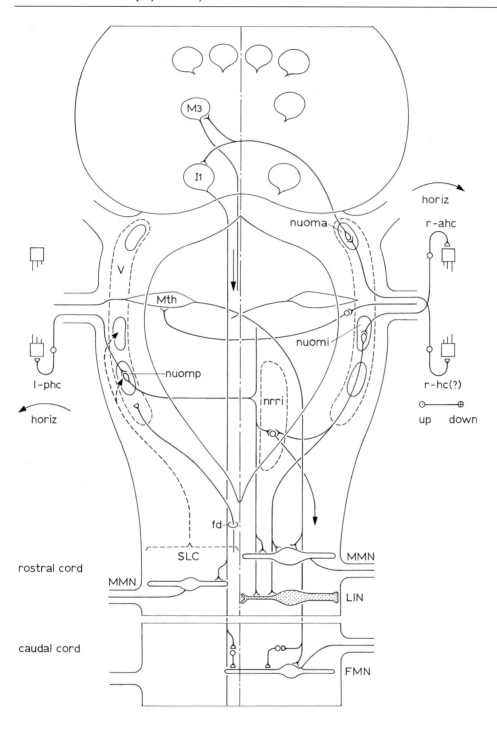

◀ **Fig. 10.60.** Vestibulospinal and vestibuloreticulospinal systems (Rovainen 1979a,b, 1982). Vestibular fibres terminate mainly in the nucleus ventralis *(v)* of the area octavolateralis. Within this nucleus three clusters of large cells, known as the nucleus octavomotorius anterior, intermedius and posterior *(nuoma, nuomi* and *nuomp,* respectively) can be distinguished.
(1) *Reflex pathways associated with horizontal rotation.* Horizontal rotation *(horiz)* in a clockwise direction activates the right hand anterior hair cells *(r-ahc)* of the vertical crista which leads to excitation of nuomi cells that send ipsilateral efferents to ipsilateral *MMNs* and *LINs*. Reticulospinal neurons in the *nrri* are powerfully co-activated by the descending pathways from the nuomi. Horizontal rotation in a counter-clockwise direction activates the left hand posterior hair cells (l-phc) that relay through nuomp cells of that side; the efferents from this nucleus cross the midline and make connections with MMNs, LINs and nrri cells in a similar manner to the connections described above for the r-ahc reflex.
(2) *Reflex pathways associated with vertical rotation.* Downward rotation of the otic capsule, so that the right hand side of the animal descends with respect to the longitudinal axis, excites hair cells of the ventral cristae. Efferents from these cells excite *nuoma* cells that, in turn, excite contralateral Müller cells *M3* and *I1*. The axons of these Müller cells excite ipsilateral *MMNs* and contralateral *FMNs,* the latter via polysynaptic pathways. Dorsal column fibres reach the octavomotor nuclei (Dubuc et al. 1993b), and it has also been established that vestibulospinal neurons in the *nuomi* and *nuomp* are phasically modulated during fictive locomotion (Bussières and Dubuc 1992). This modulation presumably comes from the spinal locomotor circuitry *(SLC),* but its exact morphological substrate is unknown.
(3) The *Mauthner cell (Mth)* is the only reticulospinal neuron to receive a direct input from the vestibular receptors. This ipsilateral input is carried by the anterior vestibular nerve. A contralateral input appears to relay through contralateral vestibular interneurons. In contrast to the situation in teleosts, both these inputs are excitatory. The *Mth* axon crosses the midline and descends to synapse directly with *MMNs* and *FMNs.*

10.12.4
Fibre Tracts

As regards the evolutionary development of fibre systems, Herrick (1948, p. 33) remarked upon the following trend: "In phylogeny the long, well-organized tracts seem to have been formed by a concentration of the fibres of the neuropil." Evaluating the situation in the tiger salamander, an animal considered by him as a very 'generalized' vertebrate, Herrick stated: ".... in Amblystoma, even in the adult stage, there are few tracts which are compactly fasciculated and free from functional connection with the surrounding neuropil. Most of the long, well-fasciculated tracts have some myelinated fibres that seem to have functional connections only at their ends; but these are accompanied by others, which are without myelin and are provided with numberless collaterals tied into the enveloping neuropil." According to Heier (1948), in the lamprey a considerably more primitive condition prevails. In this form a number of long fibre systems is present, but their constituent axons are very diffusely arranged and it may be assumed that they transmit impulses to the peripherally extending dendrites of all the grisea that they pass. Interestingly, Heier observed that, in many regions, long dendrites constitute an essential part of the fibre tracts. Jansen (1930) observed similar relations in the brain of *Myxine.*

10.12.5
Synapses

Our knowledge of the synapses in the CNS of the lamprey is fragmentary and, since the same holds true for all groups of non-mammalian gnathostomes, no general comparative statements on these structures can be made at present. One feature is striking, however, namely, the presence of mixed or dual synapses having both electrical and chemical modes of transmission. Thus in one survey, Rovainen (see Bennett and Goodenough 1978, p. 432) notes that dual synapses occur between (Rovainen's nomenclature): Müller axons I1, M1,M3, B3, B4 and myotomal motoneurons; Müller axons B2, B3, B4, V axons and lateral interneurons; Müller axon I2 and both giant interneurons and the edge cells; giant interneurons and giant interneurons; and sensory dorsal cells and giant interneurons.

The above conclusions have been largely substantiated, both physiologically and anatomically. Prior to these studies one of the few authenticated examples of a dual synapse was that found in the chick ciliary ganglion and described by Martin and Pilar (1963). This and other possible examples, known at the time, are discussed by Rovainen (1974b). Subsequently the dual synapses of the lamprey cord have been subjected to detailed electrophysiological analysis (Shapovalov 1977, 1980; Batueva and Shapovalov 1977a, b; Christensen 1983; Brodin et al. 1994).

The question of dual synapses, both in the lamprey and elsewhere, has been reviewed by Shapovalov (1980). The author noted that in general the prevalence of electrical junctions seemed to diminish both with phylogenetic and ontogenetic progression. Moreover, Shapovalov drew attention to the fact that pure electrical coupling seemed to occur mainly between homotypic cells, while dual action synapses were found between heterotypic elements. The author also conjectured that dual synapses represented a key microstructure in the evolution of the synapse.

10.12.6
Functional Significance of the Brain

Notwithstanding its small size and its relatively simple structural organisation, there can be no doubt that the brain of the lamprey, like that of gnathostomes, functions as a superimposed integrative centre, in which information gathered by the various special sense organs situated in the head is processed, correlated at two or three different levels, and relayed to a set of coordinating neurons which, in turn, control somatomotor activity at both spinal and supraspinal levels. Two centres involved in the processing of sensory information, namely, the bulbus olfactorius and the tectum 'opticum' share a number of characteristic structural features with their homologues in gnathostomes and, hence, may be expected to display functional similarities as well. The correlation of sensory impulses (originating not only from the special sense organs in the head but also from the spinal cord) occurs (a) at the level of the reticular formation, and (b) in the tectum mesencephali. The latter exerts this correlative function in addition to that of a visual centre (hence, we prefer the term tectum mesencephali over tectum opticum). We have seen that the dorsal thalamus of the lamprey represents a third and still higher centre where sensory information is correlated. The well-developed reticular formation, with both large and small neurons, obviously functions as a premotor final common path for most of the information processed in the brain and it does so at 'different levels'. That is to say it receives relatively direct information from, for instance the lateral line and vestibular systems, but also further processed information from the olfactory and visual systems, for example information which on its arrival in the reticular centres may already have been correlated with stimuli from other sources.

The functional organisation just surveyed is clearly reflected in the macrocircuitry of the lamprey brain, i.e. in the disposition of the fibre connections which link its various centres. From a comparative neurobiological point of view it is now highly important that many of the fibre systems present in the lamprey are also found in all gnathostomes and may thus be considered as representing the 'basic wiring' of the vertebrate brain. It should be emphasised, however, there are still considerable gaps in our knowledge of the functional organisation of the CNS of the lamprey. Despite the presence of a well-developed preoptico-hypothalamic neurosecretory system and of other structural features suggesting a functional relationship between brain and pituitary, the actual existence of this relationship remains to be demonstrated. Another issue is that, although we know much about the organisation of the larger spinal neurons in functional systems, the nature of many smaller interneurons is still unknown. To put it in Rovainen's (1979b, pp. 1022, 1065) words: "Further progress on the organization of the vertebrate spinal cord requires the identification of additional types of interneurons and the determination of their synaptic connections," and even: "In my opinion, the most important problem in lamprey neurobiology is the identification of spinal interneurons." As noted elsewhere in this chapter, since Rovainen wrote this, there has been an enormous effort by Grillner and others to understand the mechanisms of locomotion in the lamprey. It appears increasingly likely that a considerable depth of understanding of the mechanism of control of locomotion can be achieved by assigning many apparently discrete neurons to a few broad functional classes. In fact if it proves possible to sustain the present pace of progress, the central nervous system of the lamprey may soon be one of the best understood of all the vertebrates.

References

Adam H (1956) Der III. Ventrikel und die mikroskopische Struktur seiner Wände bei Lampetra (Petromyzon) fluviatilis L und Myxine glutinosa L, nebst einigen Bemerkungen über das Infundibularorgan von Branchiostoma (Amphioxus) lanceolatum pall. In: Ariëns Kappers J (ed) Progress in neurobiology. Proceedings of the first international meeting of neurobiologists. Elsevier, Amsterdam, pp 146–158

Addens JL (1933) The motor nuclei and roots of the cranial and first spinal nerves of vertebrates. I. Introduction and cyclostomes. Z Anat Entw Gesch 101:307–410

Ahlborn F (1883) Untersuchungen über das Gehirn der Petromyzonten. Z Wiss Zool 39:191–294

Alford S, Dubuc R (1993) Glutamate metabotropic receptor-mediated depression of synaptic inputs to lamprey reticulospinal neurones. Brain Res 605:175–179

Alford S, Grillner S (1991) The involvement of GABA$_B$ receptors and coupled G-proteins in spinal GABAergic presynaptic inhibition. J Neurosci 11:3718–3726

Alford S, Christenson J, Grillner S (1991) Presynaptic GABA$_A$ and GABA$_B$ receptor-mediated phasic modulation in axons of spinal motor interneurons. Eur J Neurosci 3:107–117

Anadón R, De Miguel E, Gonzalez-Fuentes MJ, Rodicio C (1989) HRP study of the central components of the trigeminal nerve in the larval sea lamprey: organization and homology of the primary medullary and spinal nucleus of the trigeminus. J Comp Neurol 283:602–610

Anadón R, Molist P, Pombal MA, Rodriguez-Moldes I, Rodico MC (1995) Marginal cells in the spinal cord of four elasmobranchs (Torpedo marmorata, T. torpedo, Raja undulata and Scyliorhinus canicula): evidence for homology with lamprey intraspinal stretch receptor neurons. Eur J Neurosci 7:934–943

Ariëns Kappers CU (1920) Die vergleichende Anatomie des Nervensystems der Wirbeltiere und des Menschen, vol 1. Bohn, Haarlem

Ariëns Kappers CU (1929) The evolution of the nervous system. Bohn, Haarlem

Ariëns Kappers CU (1947) Anatomie comparée du système nerveux. Bohn, Haarlem

Ariëns Kappers CU, Huber GC, Crosby EC (1936) The comparative anatomy of the nervous system of vertebrates, including man, vol 1. MacMillan, New York

Baatrup E (1983a) Ciliated receptors in the pharyngeal terminal buds of larval Lampetra planeri (Bloch) (Cyclostomata). Acta Zool (Stockh) 64:67–75

Baatrup E (1983b) Terminal buds in the branchial tube of the brook lamprey Lampetra planeri (Bloch) – putative respiratory monitors. Acta Zool (Stockh) 64:139–147

Baatrup E (1985) Physiological studies on the pharyngeal terminal buds in the larval brook lamprey, Lampetra planeri (Bloch). Chem Senses 10:549–558

Baatrup E, Døving KB (1985) Physiological studies on solitary receptors of the oral disc papillae in the adult brook lamprey, Lampetra planeri (Bloch). Chem Senses 10:559–566

Barnard JW (1936) A phylogenetic study of the visceral afferent areas associated with the facial, glossopharyngeal, and vagus nerves, and their fiber connections. The efferent facial nucleus. J Comp Neurol 65:503–603

Barthe JY, Grillner S (1995) Neurotensin-induced modulation of spinal neurons and fictive locomotion in the lamprey. J Neurophysiol 73:1308–1312

Batueva IV, Shapovalov AI (1977a) Electrotonic and chemical EPSPs evoked in lamprey motoneurons by descending tract and dorsal root afferent stimulation. Neirofiziologiya 9:512–517 (English translation: Plenum, New York, 1978)

Batueva IV, Shapovalov AI (1977b) Synaptic effects evoked in motoneurons by direct stimulation of single presynaptic fibers in the lamprey. Neirofiziologiya 9:390–396 (English translation: Plenum, New York, 1978)

Batueva IV, Suderevskaya EI, Vesselkin NP, Pierre J, Repérant J (1990) Localisation of GABA-immunopositive cells in the river lamprey spinal cord. J Hirnforsch 31:739–745

Baumgarten HG (1972) Biogenic monoamines in the cyclostome and lower vertebrate brain. Prog Histochem Cytochem 4:1–90

Belenky MA, Konstantinova MS, Polenov AL (1979a) The hypothalamo-hypophysial system of the lamprey, Lampetra fluviatilis L. II. The proximal neurosecretory contact region. Cell Tissue Res 204:319–331

Belenky MA, Chetverukhin VK, Polenov AL (1979b) The hypothalamo-hypophysial system of the lamprey, Lampetra fluviatilis L. III. High resolution radioautography of monoaminergic structures in neurohemal regions. Cell Tissue Res 204:333–342

Bennett MVL, Goodenough DA (1978) Gap junctions, electrotonic coupling, and intercellular communication. Neurosci Res Prog Bull 16:373–486

Bergqvist H (1932) Zur Morphologie des Zwischenhirns bei niederen Wirbeltieren. Acta Zool 13:57–304

Bergqvist H, Källén B (1953a) Studies on the topography of the migration areas in the vertebrate brain. Acta Anat 17:353–369

Bergqvist H, Källén B (1953b) On the development of neuromeres to migration areas in the vertebrate cerebral tube. Acta Anat 18:66–73

Bertolini B (1964) Ultrastructure of the spinal cord of the lamprey. J Ultrastruct Res 11:1–24

Birnberger KL, Rovainen CM (1971) Behavioral and intracellular studies of a habituating fin reflex in the sea lamprey. J Neurophysiol 34:983–989

Black D (1917) The motor nuclei of the cerebral nerves in phylogeny: a study of the phenomena of neurobiotaxis. I. Cyclostomi and pisces. J Comp Neurol 27:467–564

Black D (1920) The motor nuclei of the cerebral nerves in phylogeny. III. Reptilia. J Comp Neurol 32:61–98

Bodznick D, Northcutt RG (1981) Electroreception in lampreys: evidence that the earliest vertebrates were electroreceptive. Science 212:465–467

Bodznick D, Preston DG (1983) Physiological characterization of electroreceptors in the lampreys Ichthyomyzon unicuspis and Petromyzon marinus. J Comp Physiol 152:209–217

Bolliet V, Ali MA, Anctil M, Zachmann A (1993) Melatonin secretion in vitro from the pineal complex of the lamprey Petromyzon marinus. Gen Comp Endocrinol 89:101–106

Bone Q (1960) The central nervous system in amphioxus. J Comp Neurol 115:27–64

Bone Q (1963) The central nervous system. In: Brodal A, Fänge R (eds) The biology of myxine. Universitetsforlaget, Oslo, pp 50–91

Bowtell G, Williams TL (1991) Anguilliform body dynamics – modelling the interaction between muscle activation and body curvature. Philos Trans R Soc Lond Ser B Biol Sci 334:385–390

Brodin L, Grillner S (1990) The lamprey CNS in vitro, an experimentally amenable model for synaptic transmission and integrative functions. In: Jahnsen H (ed) Preparations of vertebrate central nervous system in vitro. Wiley, Chichester, UK, pp 103–153

Brodin L, Buchanan JT, Hökfelt T, Grillner S, Verhofstad AAJ (1986) A spinal projection of 5-hydroxytryptamine neurons in the lamprey brainstem; evidence from combined retrograde tracing and immunohistochemistry. Neurosci Lett 67:53–57

Brodin L, Christenson J, Grillner S (1987) Single sensory neurones activate excitatory amino acid receptors in the lamprey spinal cord. Neurosci Lett 75:75–79

Brodin L, Buchanan JT, Hökfelt T, Grillner S, Rehfeld JF, Frey P, Verhofstad AAJ, Dockray GJ, Walsh JH (1988a) Immunohistochemical studies of cholecystokinin-like peptides and their relation to 5-HT CGRP, and bombesin immunoreactivities in the brainstem and spinal cord of lampreys. J Comp Neurol 271:1–18

Brodin L, Grillner S, Dubuc R, Ohta Y, Kasicki S, Hökfelt T (1988b) Reticulospinal neurons in lamprey: transmitters, synaptic interactions and their role during locomotion. Arch Ital Biol 126:317–345

Brodin L, Ohta Y, Hökfelt T, Grillner S (1989a) Further evidence for excitatory amino acid transmission in lamprey reticulospinal neurons: selective retrograde labeling with (^{3}H)D-aspartate. J Comp Neurol 281:225–233

Brodin L, Rawitch A, Taylor T, Ohta Y, Ring H, Hökfelt T, Grillner S, Terenius L (1989b) Multiple forms of pancreatic polypeptide-related compounds in the lamprey CNS: partial characterization and immunohistochemical localization in the brain stem and spinal cord. J Neurosci 9:3428–3442

Brodin L, Hökfelt T, Grillner S, Panula P (1990a) Distribution of histaminergic neurons in the brain of the lamprey Lampetra fluviatilis as revealed by histamine-immunohistochemistry. J Comp Neurol 292:435–442

Brodin L, Theordorsson E, Christenson J, Cullheim S, Hökfelt T, Brown JC, Buchan A, Panula P, Verhofstad AAJ, Goldstein M (1990b) Neurotensin-like peptides in the CNS of lampreys. Chromatographic characterization and immunohistochemical localization with reference to aminergic markers. Eur J Neurosci 2:1095–1109

Brodin L, Dale N, Christenson J, Storm-Mathisen J, Hökfelt T, Grillner S (1990c) Three types of GABA-immunoreactive cells in the lamprey spinal cord. Brain Res 508:172–175

Brodin L, Shupliakov O, Pieribone VA, Hellgren J, Hill RH (1994) The reticulospinal glutamate synapse in lamprey: plasticity and presynaptic variability. J Neurophysiol 72:592–604

Bruckmoser P (1971) Elektrische Antworten im Vorderhirn von Lampetra fluviatilis L. bei Reizung des Nervus olfactorius. Z Vergl Physiol 75:69–85

Buchanan JT (1982) Identification of interneurons with contralateral, caudal axons in the lamprey spinal cord: synaptic interactions and morphology. J Neurophysiol 47:961–975

Buchanan JT (1986) Premotor interneurons in the lamprey spinal cord: morphology, synaptic interactions and activities during fictive swimming. In: Grillner S, Stein PSG, Stuard DG, Forssberg H, Herman RM (eds) Neurobiology of vertebrate locomotion. MacMillan, London, pp 321–334

Buchanan JT (1993) Electrophysiological properties of identified classes of lamprey spinal neurons. J Neurophysiol 70:2313–2325

Buchanan JT, Cohen AH (1982) Activities of identified interneurons, motoneurons, and muscle fibers during fictive swimming in the lamprey and effects of reticulospinal and dorsal cell stimulation. J Neurosphysiol 47:948–960

Buchanan JT, Grillner S (1987) Newly identified 'glutamate interneurons' and their role in locomotion in the lamprey spinal cord. Science 236:312–314

Buchanan JT, Grillner S (1988) A new class of small inhibitory interneurones in the lamprey spinal cord. Brain Res 438:404–407

Buchanan JT, Grillner S (1991) 5-Hydroxytryptamine depresses reticulospinal excitatory postsynaptic potentials in motoneurons of the lamprey. Neurosci Lett 112:71–74

Buchanan JT, Kasicki S (1995) Activities of spinal neurons during brain stem-dependent fictive swimming in Lamprey. J Neurophysiol 73:80–87

Buchanan JT, Brodin L, Dale N, Grillner S (1987a) Reticulospinal neurons activate excitatory amino acid receptors. Brain Res 408:321–325

Buchanan JT, Brodin L, Hökfelt T, Van Dongen PAM, Grillner S (1987b) Survey of neuropeptide-like immunoreactivity in the lamprey spinal cord. Brain Res 408:299–302

Buchanan JT, Grillner S, Cullheim S, Risling M (1989) Identification of excitatory interneurons contributing to generation of locomotion in lamprey: structure, pharmacology, and function. J Neurophysiol 62:59–69

Bullock TH, Bodznick DA, Northcutt RG (1983) The phylogenetic distribution of electroreception: evidence for convergent evolution of a primitive vertebrate sense modality. Brain Res Rev 6:25–46

Bullock TH, Moore JK, Fields RD (1984) Evolution of myelin sheaths: both lamprey and hagfish lack myelin. Neurosci Lett 48:145–148

Bundgaard M (1982) Brain barrier systems in the lamprey. I. Ultrastructure and permeability of cerebral blood vessels. Brain Res 240:55–64

Bundgaard M, Van Deurs B (1982) Brain barrier systems in the lamprey. II. Ultrastructure and permeability of the choroid plexus. Brain Res 240:65–75

Bussières N, Dubuc R (1992) Phasic modulation of vestibulospinal neuron activity during fictive locomotion in lampreys. Brain Res 575:174–179

Bussières N, Dubuc R (1995) Morphology and axonal trajectories of vestibulospinal neurones in lampreys. Soc Neurosci Abstr 21:142

Butler AB (1994) The evolution of the dorsal thalamus of jawed vertebrates, including mammals: cladistic analysis and a new hypothesis. Brain Res Rev 19:29–65

Cheung R, Plisetskaya EM, Youson JH (1990) Distribution of two forms of somatostatin in the brain, anterior intestine, and pancreas of adult lampreys (Petromyzon marinus). Cell Tissue Res 262:283–292

Chesler M (1986) Regulation of intracellular pH in reticulospinal neurones of the lamprey, Petromyzon marinus. J Physiol (Lond) 381:241–261

Chesler M, Nicholson C (1985) Regulation of intracellular pH in vertebrate central neurons. Brain Res 325:313–316

Christensen BN (1976) Morphological correlates of synaptic transmission in lamprey spinal cord. J Neurophysiol 39:197–212

Christensen BN (1983) Distribution of electrotonic synapses on identified lamprey neurons: a comparison of a model prediction with an electron microscopic analysis. J Neurophysiol 49:705–716

Christensen BN, Teubl WP (1979a) Estimates of cable parameters in lamprey spinal cord neurones. J Physiol (Lond) 297:299–318

Christensen BN, Teubl WP (1979b) Localization of synaptic input on dendrites of a lamprey spinal cord neurone from physiological measurements of membrane properties. J Physiol (Lond) 297:319–333

Christenson J, Boman A, Lagerbäck PA, Grillner S (1988a) The dorsal cell, one class of primary sensory neuron in the lamprey spinal cord. I. Touch, pressure but no nociception – a physiological study. Brain Res 440:1–8

Christenson J, Lagerbäck PA, Grillner S (1988b) The dorsal cell, one class of primary sensory neuron in the lamprey spinal cord. II. A light- and electron-microscopical study. Brain Res 440:9–17

Christenson J, Cullheim S, Grillner S, Hökfelt T (1990) 5-Hydroxytryptamine immunoreactive varicosities in the lamprey spinal cord have no synaptic specializations – an ultrastructural study. Brain Res 512:201–209

Christenson J, Alford S, Grillner S, Hokfelt T (1991) Co-localized GABA and somatostatin use different ionic mechanisms to hyperpolarize target neurons in the lamprey spinal cord. Neurosci Lett 134:93–97

Christenson J, Hill RH, Bongianni F, Grillner S (1993) Presence of low voltage activated calcium channels distinguishes touch from pressure sensory neurons in the lamprey spinal cord. Brain Res 608:58–66

Clark WB (1906) The cerebellum of Petromyzon fluviatilis. J Anat 40:318–325

Cohen AH, Wallén P (1980) The neuronal correlate of locomotion in fish. 'Fictive swimming' induced in an in vitro preparation of the lamprey spinal cord. Exp Brain Res 41:11–18

Cohen AH, Holmes PJ, Rand RH (1982) The nature of the coupling between segmental oscillators of the lamprey spinal generator for locomotion: a mathematical model. J Math Biol 13:345–369

Cole WC, Youson JH (1981) The effect of pinealectomy, continuous light, and continuous darkness on metamorphosis of anadromous sea lampreys, Petromyzon marinus L. J Exp Zool 218:397–404

Cole WC, Youson JH (1982) Morphology of the pineal complex of the anadromous sea lamprey, Petromyzon marinus L. Am J Anat 165:131–63

Crim JW, Urano A, Gorbman A (1979) Immunocytochemical studies of luteinizing hormone-releasing hormone in brains of agnathan fishes. I. Comparisons of adult pacific lamprey (Entosphenus tridentata) and the pacific hagfish (Eptatretus stouti). Gen Comp Endocrinol 37:294–305

Davis GR Jr, McClellan AD (1993) Time course of anatomical regeneration of descending brainstem neurons and behavioral recovery in spinal-transected lamprey. Brain Res 602:131–137

Davis GR Jr, McClellan AD (1994a) Long distance axonal regeneration of identified lamprey reticulospinal neurons. Exp Neurol 127:94–105

Davis GR Jr, McClellan AD (1994b) Extent and time course of restoration of descending brainstem projections in spinal cord-transected lamprey. J Comp Neurol 344:65–82

Davis GR Jr, Troxel MT, Kohler VJ, Grossmann EM, McClellan AD (1993) Time course of locomotor recovery and functional regeneration in spinal-transected lamprey: kinematics and electromyography. Exp Brain Res 97:83–95

Deliagina TG (1995) Vestibular compensation in the lamprey. Neuroreport 6:2599–2603

Deliagina TG, Orlovsky GN, Grillner S, Wallén P (1992a) Vestibular control of swimming in lamprey. II. Characteristics of spatial sensitivity of reticulospinal neurons. Exp Brain Res 90:489–498

Deliagina TG, Orlovsky GN, Grillner S, Wallén P (1992b) Vestibular control of swimming in lamprey. III. Activity of vestibular afferents: convergence of vestibular inputs on reticulospinal neurons. Exp Brain Res 90:499–507

Deliagina TG, Grillner S, Orlovsky GN, Ullén F (1993) Visual input affects the response to roll in reticulospinal neurons of the lamprey. Exp Brain Res 95:421–428

Deliagina TG, Ullén F, González MJ, Ehrsson H, Orlovsky GN, Grillner S (1995) Initiation of locomotion by lateral line photoreceptors in lamprey: behavioural and neurophysiological studies. J Exp Biol 198:2581–2591

De Miguel E, Anadón R (1987) The development of the retina and the optic tectum of Petromyzon marinus L. J Hirnforsch 28:445–456

De Miguel E, Rodicio MC, Anadón R (1990) Organization of the visual system in larval lampreys: an HRP study. J Comp Neurol 302:529–542

Dickson DH, Collard TR (1979) Retinal development in the lamprey (Petromyzon marinus L): premetamorphic ammocoete eye. J Anat 154:321–336

Dubuc R, Grillner S (1987) Spinal cord input to reticulospinal neurones in the lamprey. Acta Physiol Scand 129:28A

Dubuc R, Bongianni F, Ohta Y, Grillner S (1993a) Dorsal root and dorsal column mediated synaptic inputs to reticulospinal neurons in lampreys: involvement of glutamatergic, glycinergic, and GABAergic transmission. J Comp Neurol 327:251–259

Dubuc R, Bongianni F, Ohta Y, Grillner S (1993b) Anatomical and physiological study of brainstem nuclei relaying dorsal column inputs in lampreys. J Comp Neurol 327:260–270

Ebbesson SOE (1969) Brain stem afferents from the spinal cord in a sample of reptilian and amphibian species. Ann N Y Acad Sci 167:80–101

Ebbesson SOE, Hodde KC (1981) Ascending spinal systems in the nurse shark, Ginglymostoma cirratum. Cell Tissue Res 216:313–331

Ebbesson SOE, Northcutt RG (1976) Neurology of anamniotic vertebrates. In: Masterton RB, Bitterman ME, Campbell CBG, Hotton N (eds) Evolution of brain and behavior in vertebrates. Erlbaum, Hilsdale, pp 115–146

Eddy JMP (1969) Metamorphosis and the pineal complex in the brook lamprey, Lampetra planeri. J Endocrinol 44:451–452

Eddy JMP (1972) The pineal complex. In: Hardisty MW, Potter IC (eds) The biology of lampreys, vol 2. Academic, London, pp 91–103

Edinger L (1908) Vorlesungen über den Bau der Nervösen Zentralorgane. Vogel, Leipzig

Eisthen H, Northcutt RG (1996) Silver lampreys (Ichthyomyzon unicuspis) lack a gonadotrophin-releasing hormone- and FMRFamide-immunoreactive terminal nerve. J Comp Neurol 370:159–172

El Manira A, Tegnér J, Grillner S (1994) Calcium-dependent potassium channels play a critical role for burst termination in the locomotor network in lamprey. J Neurophysiol 72:1852–1861

El Manira A, Shupliakov O, Fagerstadt P, Grillner S (1996) Monosynaptic input from cutaneous sensory afferents to fin motoneurons in lamprey. J Comp Neurol 369:533–542

Faber DS, Korn H (1978) Neurobiology of the Mauthner cell. Raven, New York

Fahrenholz C (1936) Die sensiblen Einrichtungen der Neunaugenhaut. Z Mikrosk Anat Forsch 40:323–380

Finger TE, Rovainen CM (1978) Retrograde HRP labeling of the oculomotoneurons in adult lampreys. Brain Res 154:123–127

Finger TE, Rovainen CM (1982) Spinal and medullary dorsal cell axons in the trigeminal nerve in lampreys. Brain Res 240:331–333

Fite KV (1985) Pretectal and accessory-optic visual nuclei of fish, amphibia and reptiles: theme and variations. Brain Behav Evol 26:71–90

Foster RG, García-Fernández JM, Provencio I, DeGrip WJ (1993) Opsin localization and chromophore retinoids identified within the basis brain of the lizard Anolis carolinensis. J Comp Physiol 172:33–45

Freud S (1877) Über den Ursprung der hinteren Nervenwurzeln im Rückenmark von ammocoetes (Petromyzon planeri). Sitzungsber Akad Wiss Wien 75:15–30

Freud S (1878) Über Spinalganglien und Rückenmark des Petromyzon. Sitzungsber Akad Wiss Wien 78:81–167

Fritzsch B, Northcutt RG (1993) Origin and migration of trochlear, oculomotor and abducent motor neurons in Petromyzon marinus L. Dev Brain Res 74:122–126

Fritzsch B, Sonntag R (1988) The trochlear motoneurons of lampreys (Lampetra fluviatilis): location, morphology and numbers as revealed with horseradish peroxidase. Cell Tissue Res 252:223–229

Fritzsch B, Crapon de Caprona MD, Wachtler K, Kortje KH (1984) Neuroanatomical evidence for electroreception in lampreys. Z Naturforsch, Section C: Biosci 39:856–858

Fritzsch B, Dubuc R, Ohta Y, Grillner S (1989) Efferents to the labyrinth of the river lamprey (Lampetra fluviatilis) as revealed with retrograde tracing techniques. Neurosci Lett 96:241–246

Fritzsch B, Sonntag R, Dubuc R, Ohta Y, Grillner S (1990) Organization of the six motor nuclei innervating the ocular muscles in lamprey. J Comp Neurol 294:491–506

Gage SH (1928) The lampreys of New York State. Life history and economics. Biological survey of the Oswego river system, supplement to the 17th annual report, New York State Conservation Dept, 1927. Lyon, Albany, pp 158–191

García-Fernández JM, Foster RG (1994) Immunocytochemical identification of photoreceptor proteins in hypothalamic cerebrospinal fluid-contacting neurons of the larval lamprey (Petromyzon marinus). Cell Tissue Res 275:319–326

Gilland E, Baker R (1995) Organization of rhombomeres and brainstem efferent neuronal populations in larval sea lamprey, Petromyzon marinus. Soc Neurosci Abstr 21:779

González MA, Anadón R (1992) Primary projections of the lateral line nerves in larval sea lamprey, Petromyzon marinus L: an HRP study. J Hirnforsch 33:185–194

Goossens N, Dierickx K, Vandesande F (1977) Immunocytochemical demonstration of the hypothalamo-hypophysial vasotocinergic system of Lampetra fluviatilis. Cell Tissue Res 177:317–323

Grillner S, Matsushima T (1991) The neural network underlying locomotion in lamprey – synaptic and cellular mechanisms. Neuron 7:1–15

Grillner S, McClellan A, Sigvardt K (1982) Mechanosensitive neurons in the spinal cord of the lamprey. Brain Res 235:169–173

Grillner S, Williams T, Lagerback PA (1984) The edge cell, a possible intraspinal mechanoreceptor. Science 233:500–503

Grillner S, Wallén P, Brodin L, Lansner A (1991) Neuronal network generating locomotor behavior in lamprey: circuitry, transmitters, membrane properties, and simulation. Annu Rev Neurosci 14:169–199

Grillner S, Deliagina T, Ekeberg Ö, El Manira A, Hill RH, Lansner A, Orlovsky GN, Wallén P (1995) Neural networks that co-ordinate locomotion and body orientation in lamprey. Trends Neurosci 18:270–279

Groos G (1982) The comparative physiology of extraocular photoreception. Experientia 38:989–1128

Hagevik A, McClellan AD (1994a) Role of excitatory amino acids in brainstem activation of spinal locomotor networks in larval lamprey. Brain Res 636:147–152

Hagevik A, McClellan AD (1994b) Coupling of spinal locomotor networks in larval lamprey revealed by receptor blockers for inhibitory amino acids: Neurophysiology and computer modeling. J Neurophysiol 72:1810–1829

Hardisty MW (1979) Biology of the cyclostomes. Chapman and Hall, London

Hardisty MW, Potter IC (eds) The biology of lampreys, vol 4A. Academic, London

Hardisty MW, Rovainen CM (1982) Morphological and functional aspects of the muscular system. In: Hardisty MW, Potter IC (eds) The biology of lampreys, vol 4A. Academic, London, pp 137–231

Harris-Warrick RM, Cohen AH (1985) Serotonin modulates the central pattern generator for locomotion in the isolated lamprey spinal cord. J Exp Biol 116:27–46

Heier P (1948) Fundamental principles in the structure of the brain. A study of the brain of Petromyzon fluviatilis. Acta Anat [Suppl] VI:1–213

Herrick CJ (1948) The brain of the tiger salamander. University of Chicago Press, Chicago

Herrick CJ, Obenchain JB (1913) Notes on the anatomy of a cyclostome brain: Ichthyomyzon concolor. J Comp Neurol 23:635–675

Hibbard E (1963a) The vascular supply to the central nervous system of the larval lamprey. Am J Anat 113:93–99

Hibbard E (1963b) Regeneration in the severed spinal cord of chordate larvae of Petromyzon marinus. Exp Neurol 7:175–185

Hofer H (1963) Neuere Ergebnisse zur Kenntnis des Subkommissuralorganes, des Reissnerschen Fadens und der Massa caudalis. Verh Zool Ges 431–440

Hofer H, Meinel W, Erhardt H, Wolter A (1984) Preliminary electron-microscopical observations on the ampulla caudalis and the discharge of the material of Reissner's fibre into the capillary system of the terminal part of the tail of ammocoetes (Agnathi). Gegenbaurs Morphol Jahrb (Leipz) 130 (1):77–110

Hoheisel G, Rühle HJ, Sterba G (1978) The reticular formation of lampreys (Petromyzonidae) – a target area for exohypothalamic vasotocinergic fibres. Cell Tissue Res 189:331–345

Holmgren N (1922) Points of view concerning forebrain morphology in lower vertebrates. J Comp Neurol 34:391–440

Homma S (1975) Velar motoneurons of lamprey larvae. J Comp Physiol 104:175–183

Homma S (1979) Conductance changes during bath application of β-alanine and taurine in giant interneurons of the isolated lampreys spinal cord. Brain Res 173:287–293

Homma S (1981) Effects of DL-aminoadipate on synaptic transmission in spinal interneurones of the lamprey. J Comp Physiol 143:423–426

Homma S, Rovainen CM (1978) Conductance increases produced by glycine and γ-aminobytyric acid in lamprey interneurons. J Physiol (Lond) 279:231–252

Huard H, Lund JP, Dubuc R (1995) A study of trigeminal premotor neurones in lampreys. Soc Neurosci Abstr 21:142

Hugosson R (1957) Morphologic and experimental studies on the development and significance of the rhombencephalic longitudinal cell columns. Thesis, Lund

Iwahori N, Kiyota E, Nakamura K (1987) A Golgi study on the olfactory bulb in the lamprey, Lampetra japonica. Neurosci Res 5:126–139

Jansen J (1930) The brain of Myxine glutinosa. J Comp Neurol 49:359–507

Johnels AG (1958) On the dorsal ganglion cells of the spinal cord in lampreys. Acta Zool 39:201–216

Johnston JB (1902) The brain of Petromyzon. J Comp Neurol 12:2–86

Johnston JB (1905) The cranial nerve components of Petromyzon. Morphol Jahrb 34:149–203

Johnston JB (1912) The telencephalon in cyclostomes. J Comp Neurol 22:341–404

Joss JMP (1973) The pineal complex, melatonin, and color change in the lamprey Lampetra. Gen Comp Endocrinol 21:188–195

Jung R, Kiemel T, Cohen AH (1996) Dynamic behavior of a neural network model of locomotor control in the lamprey. J Neurophysiol 75:1074–1086

Karamian AI, Vessekin NP, Agayan AL (1984) Electrophysiological and behavioral studies of the optic tectum in cyclostomes, chap 2. In: Vanegas (ed) Comparative neurology of the optic tectum. Plenum, New York

Kasicki S, Grillner S (1986) Müller cells and other reticulospinal neurones are phasically active during fictive locomotion in the isolated nervous system of the lamprey. Neurosci Lett 69:239–243

Kasicki S, Grillner S, Ohta Y, Dubuc R, Brodin L (1989) Phasic modulation of reticulospinal neurones during fictive locomotion and other types of spinal motor activity in lamprey. Brain Res 484:203–216

Kennedy MC, Rubinson K (1977) Retinal projections in larval, transforming and adult sea lamprey, Petromyzon marinus. J Comp Neurol 171:465–480

Kennedy MC, Rubinson K (1978) The structure of the optic tectum in the sea lamprey, Petromyzon marinus. Anat Rec 190:441–442

Kennedy M, Rubinson K (1984) Development and structure of the lamprey optic tectum. In: Vanegas H (ed) Comparative neurology of the optic tectum. Plenum, New York, pp 1–13

Kim YS, Stumpf WE, Reid FA, Sar M, Selzer ME (1980) Estrogen target cells in the forebrain of river lamprey Ichthyomyzon unicuspis. J Comp Neurol 191:607–613

Kim YS, Stumpf WE, Sar M, Reid FA, Selzer ME, Epple AW (1981) Autoradiographic studies or estrogen target cells in the forebrain of larval lamprey, Petromyzon marinus. Brain Res 210:53–60

King JC, Sower SA, Anthony ELP (1988) Neuronal systems immunoreactive with antiserum to lamprey gonadotropin-releasing hormone in the brain of Petromyzon marinus. Cell Tissue Res 253:1–8

Kishida R, Koyama H, Goris RC (1988) Giant lateral-line afferent terminals in the electroreceptive dorsal nucleus of lampreys. Neurosci Res 6:83–87

Kleerekoper H (1972) The sense organs. In: Hardisty MW, Potter IC (eds) The biology of lampreys, vol 2. Academic, London, pp 373–404

Kosareva AA (1980) Retinal projections in lamprey (Lampetra fluviatilis). J Hirnforsch 21:243–256

Koyama H, Kishida R, Goris RC, Kusunoki T (1987) Organization of sensory and motor nuclei of the trigeminal nerve in lampreys. J Comp Neurol 264:437–448

Koyama H, Kishida R, Goris RC, Kusunoki T (1989) Afferent and efferent projections of the VIIIth cranial nerve in the lamprey Lampetra japonica. J Comp Neurol 280:663–671

Koyama H, Kishida R, Goris RC, Kusunoki T (1990) Organization of the primary projections of the lateral line nerves in the lamprey Lampetra japonica. J Comp Neurol 295:277–289

Koyama H, Kishida R, Goris R, Kusunoki T (1993) Giant terminals in the dorsal octavolateralis nucleus of lampreys. J Comp Neurol 335:245–251

Kuhlenbeck H (1929) Über die Grundbestandteile des Zwischenhirnbauplans der Anamnier. Morphol Jahrb 63:50–95

Kuhlenbeck H (1956) Die Formbestandteile der Regio praetectalis des Anamnier-Gehirns und ihre Beziehungen zum Hirnbauplan. Fol Anat Jpn 28:23–44

Kutschin K (1863) Über den Bau des Rückenmarkes der Neunaugen (in Russian). Thesis, Kasan. (Reviewed by L Stieda.) Arch Mikr Anat 2:525–530

Larsell O (1947a) The cerebellum of myxinoids and petromyzonts, including developmental stages in the lampreys. J Comp Neurol 86:395–445

Larsell O (1947b) The nucleus of the IVth nerve in petromyzonts. J Comp Neurol 86:447–466

Larsell O (1967) The comparative anatomy and histology of the cerebellum from myxinoids through birds. University of Minnesota Press, Minneapolis

Lehmenkühler A, Syková E, Svoboda J, Zilles K, Nicholson C (1993) Extracellular space parameters in the rat neocortex and subcortical white matter during postnatal development determined by diffusion analysis. Neuroscience 55:339–351

Leonhardt H (1980) Organum subcommissurale. In: Oksche A (ed) Handbuch der Mikroskopischen Anatomie des Menschen, vol 4: Nervensystem, part 10: Neuroglia I. Springer, Berlin Heidelberg New York, pp 472–504

Leonard JP, Wickelgren WO (1986) Prolongation of calcium action potentials by γ-aminobutyric acid in primary sensory neurones of lamprey. J Physiol (Lond) 375:481–497

Lowenstein O (1970) The electrophysiological study of the responses of the isolated labyrinth of the lamprey (Lampetra fluviatilis L) to angular acceleration, tilting and mechanical vibration. Proc R Soc Lond [Biol] 174:419–434

Lowenstein O, Osborne MP, Thornhill RA (1968) The anatomy and ultrastructure of the labyrinth of the lamprey (Lampetra fluviatilis L). Proc R Soc Lond [Biol] 170:113–134

Lurie DI, Pijak DS, Selzer ME (1994) The structure of reticulospinal axon growth cones and their cellular environment during regeneration in the lamprey spinal cord. J Comp Neurol 344:559–580

Martin RJ (1979) A study of the morphology of the large reti- culospinal neurons of the lamprey ammocoete by intracel- lular injection of procion yellow. Brain Behav Evol 16:1–18

Martin RJ, Bowsher D (1977) An electrophysiological investiga- tion of the projection of the intramedullary primary afferent cells of the lamprey ammocoete. Neurosci Lett 5:39–43

Martin AR, Pilar G (1963) Dual mode of synaptic trans- mission in the avian ciliary ganglion. J Physiol (Lond) 168:443–463

Martin AR, Ringham GL (1975) Synaptic transfer at a verte- brate central nervous system synapse. J Physiol (Lond) 25:409–426

Martin AR, Wickelgren WO (1971) Sensory cells in the spinal cord of the sea lamprey. J Physiol (Lond) 212:65–83

Martin AR, Wickelgren WO, Beránek R (1970) Effects of ion- tophoretically applied drugs on spinal interneurons of the lamprey. J Physiol (Lond) 207:653–665

Matsushima T, Grillner S (1990) Intersegmental co- ordination of undulatory movements – a "trailing oscill- ator" hypothesis. Neuroreport 1:97–100

Matsushima T, Grillner S (1992) Neural mechanisms of inter- segmental coordination in lamprey: local excitability changes modify the phase coupling along the spinal cord. J Neurophysiol 67:373–388

Matsushima T, Tegnér J, Hill RH, Grillner S (1993) GABA$_B$ receptor activation causes a depression of low-voltage and high-voltage activated Ca^{2+} currents, postinhibitory rebound, and postspike afterhyperpolarization in lamprey neurons. J Neurophysiol 70:2606–2619

Matthews G, Wickelgren WO (1978) Trigeminal sensory neu- rons of the sea lamprey. J Comp Physiol 123:329–333

Mayer F (1897) Das Centralnervensystem von Ammocoetes. I. Vorder-, Zwischen- und Mittelhirn. Anat Anz 13:649–657

McClellan AD (1984) Descending control and sensory gating of 'fictive' swimming and turning responses elicited in an in vitro preparation of the lamprey brainstem/spinal cord. Brain Res 302:151–162

McClellan AD (1988a) Functional regeneration of descending brainstem command pathway for locomotion demon- strated in the in vitro lamprey CNS. Brain Res 448:339–345

McClellan AD (1988b) Brainstem command system for locomotion in the lamprey: localization of descending pathways in the spinal cord. Brain Res 457:338–349

McClellan AD (1992) Functional regeneration and recovery of locomotor activity in spinally transected lamprey. J Exp Zool 261:274–287

McClellan AD (1994) Time course of locomotor recovery and functional regeneration in spinal cord-transected lamprey: in vitro preparations. J Neurophysiol 72:847–860

McClellan AD, Grillner S (1984) Activation of 'fictive swim- ming' by electrical microstimulation of brainstem locomo- tor regions in an in vitro preparation of the lamprey cen- tral nervous system. Brain Res 300:357–361

McClellan AD, Jang WC (1993) Mechanosensory inputs to the central pattern generators for locomotion in the lamprey spinal cord: resetting, entrainment, and computer model- ing. J Neurophysiol 70:2442–2454

McKibben PS (1911) The nervus terminalis in urodele Amphibia. J Comp Neurol 21:261–310

McPherson DR, Kemnitz CP (1994) Modulation of lamprey fictive swimming and motoneuron physiology by dop- amine, and its immunocytochemical localization in the spinal cord. Neurosci Lett 166:23–26

Meiniel A (1980) Ultrastructure of serotonin-containing cells in the pineal organ of Lampetra planeri (Petromyzonti- dae). Cell Tissue Res 207:407–427

Meiniel A (1981) New aspects of the phylogenetic evolution of sensory cell lines in the vertebrate pineal complex. In: Oksche A, Pévet P (eds) The pineal organ: photobiology – biochrono- metry – endocrinology. Elsevier, Amsterdam, pp 27–48

Meiniel A, Hartwig HG (1980) Indoleamines in the pineal complex of Lampetra planeri (Petromyzontidae): a fluores- cence microscopic and microspectrofuorimetric study. J Neural Transm 48:65–83

Mellen N, Kiemel T, Cohen AH (1995) Correlational analy- sis of fictive swimming in the lamprey reveals strong functional intersegmental coupling. J Neurophysiol 73:1020–1030

Merrick SA, Pleasure SJ, Lurie DI, Pijak DS, Selzer ME, Lee VMY (1995) Glial cells of the lamprey nervous system con- tain keratin-like proteins. J Comp Neurol 355:199–210

Moore LE, Buchanan JT (1993) The effects of neurotransmit- ters on the integrative properties of spinal neurons in the lamprey. J Exp Biol 175:89–114

Morita Y, Tabata M, Tamotsu S (1985) Intracellular response and input resistance change of pineal photoreceptors and ganglion cells. Neurosci Res [Suppl] 2:79–88

Morita Y, Tabata M, Uchida K, Samejima M (1992) Pineal- dependent locomotor activity of lamprey, Lampetra japo- nica, measured in relation to LD cycle and circadian rhythmicity. J Comp Physiol A 171:555–562

Münz H, Claas B, Stumpf WE, Jennes L (1982) Centrifugal innervation of the retina by luteinizing hormone-releasing hormone (LHRH)-immunoreactive telencephalic neurons in teleostean fishes. Cell Tissue Res 222:313–323

Nakao T, Ishizawa A (1982) An electron microscopic study of autonomic nerve cells in the cloacal region of the lamprey, Lampetra japonica. J Neurocytol 11:517–532

Nakao T, Ishizawa A (1987a) Development of the spinal ner- ves in the lamprey: I. Rohon-Beard cells and interneurons. J Comp Neurol 256:342–355

Nakao T, Ishizawa A (1987b) Development of the spinal ner- ves in the lamprey: II. Outflows from the spinal cord. J Comp Neurol 256:356–368

Nakao T, Ishizawa A (1987c) Development of the spinal ner- ves in the lamprey: III. Spinal ganglia and dorsal roots in 26-day (13 mm) larvae. J Comp Neurol 256:369–385

Nakao T, Ishizawa A (1987d) Development of the spinal ner- ves of the larval lamprey: IV. Spinal nerve roots of 21-mm larval and adult lampreys, with special reference to the relation of meninges to the root sheath and the perineu- rium. J Comp Neurol 256:386–399

Nakao T, Suzuki S (1978) Sympathetic nerves in the branchial region of lamprey, Lampetra japonica. Acta Anat Nippon 53:51–52

Nakao T, Suzuki S (1980) The structure and innervation of the cloacal region of lamprey. Acta Anat Nippon 55:438

Nieuwenhuys R (1967a) Comparative anatomy of olfactory centres and tracts. Prog Brain Res 23:1–64

Nieuwenhuys R (1967b) Comparative anatomy of the cerebel- lum. Prog Brain Res 25:1–93

Nieuwenhuys R (1972) Topological analysis of the brain stem of the lamprey Lampetra fluviatilis. J Comp Neurol 145:165–178

Nieuwenhuys R (1977) The brain of the lamprey in a compar- ative perspective. Ann N Y Acad Sci 299:97–145

Northcutt R (1979a) Experimental determination of the pri- mary trigeminal projections in lampreys. Brain Res 163:323–327

Northcutt RG (1979b) Central projections of the eighth cra- nial nerve in lampreys. Brain Res 167:163–167

Northcutt RG, Przybylski RJ (1973) Retinal projections in the lamprey Petromyzon marinus L. Anat Rec 175:400

Northcutt RG, Puzdrowski RL (1988) Projections of the olfactory bulb and nervus terminalis in the silver lamprey. Brain Behav Evol 32:96–107

Nortcutt RG, Wicht H (1996) Afferent and efferent connec- tions of the lateral and medial palloia of the silver lamprey. Brain Behav Evol (in press)

Nozaki M (1985) Tissue distribution of hormonal peptides in primitive fishes. In: Foreman D, Gorbman A, Dodd JM, Olsson R (eds) Evolutionary biology of primitive fishes. Plenum, New York, pp 433–454

Nozaki M, Gorbman A (1984) Distribution of immunoreac- tive sites for several components of pro-opiocortin in the pituitary and brain of adult lampreys, Petromyzon marinus and Entosphenus tridentatus. Gen Comp Endocrinol 53:335–352

Nozaki M, Gorbman A (1986) Occurrence and distribution of substance P-related immunoreactivity in the brain of adult lampreys. Gen Comp Endocrinol 62:217–229

Öhman P (1977) Fine structure of the optic nerve of Lampetra fluviatilis (Cyclostomi). Vision Res 17:719–722

Ohta Y, Grillner S (1989) Monosynaptic excitatory amino acid transmission from the posterior rhombencephalic reticular nucleus to spinal neurons involved in the control of locomotion in lamprey. J Neurophysiol 62:1079–1089

Ohta Y, Brodin L, Grillner S, Hökfelt T, Walsh JH (1988) Possible target neurons of the reticulospinal cholecystokinin (CCK) projection to the lamprey spinal cord: immunohistochemistry combined with intracellular staining with lucifer yellow. Brain Res 445:400–403

Ohta Y, Dubuc R, Grillner S (1991) A new population of neurons with crossed axons in the lamprey spinal cord. Brain Res 564:143–148

Ohtomi M, Fujii K, Kobayashi H (1989) Distribution of FMRFamide-like immunoreactivity in the brain and neurohypophysis of the lamprey, Lampetra japonica. Cell Tissue Res 256:581–584

Oksche A (1969) The subcommissural organ. J Neurol Visc Relat [Suppl] 9:111–139

Olsson R (1955) Structure and development of Reissner's fibre in the caudal end of amphioxus and some lower vertebrates. Acta Zool (Stockh) 36:167–198

Olsson R (1958) Studies on the subcommissural organ. Acta Zool (Stockh) 39:71–102

Onstott D, Elde R (1986) Immunohistochemical localization of urotensin I/corticotropin-releasing factor, urotensin II, and serotonin immunoreactivities in the caudal spinal cord of nonteleost fishes. J Comp Neurol 249:205–225

Orlovsky GN, Deliagina TG, Wallén P (1992) Vestibular control of swimming in lamprey. I. Responses of reticulospinal neurons to roll and pitch. Exp Brain Res 90:479–488

Owsiannikow P (1903) Das Rückenmark und das Verlängerte Mark des Neunauges. Mem Acad Imp Sci St Petersbourg 14:1–32

Pearson AA (1936) The acustico-lateral centers and the cerebellum, with fiber connections, of fishes. J Comp Neurol 65:201–294

Peruzzo B, Rodríguez S, Delannoy L, Hein S, Rodríguez EM, Oksche A (1987) Ultrastructural immunocytochemical study of the massa caudalis of the subcommissural organ-Reissner's fiber complex in lamprey larvae (Geotria australis): evidence for a terminal vascular route of secretory material. Cell Tissue Res 247:367–376

Peters A (1960) The structure of peripheral nerves of the lamprey (Lampetra fluviatilis). J Ultrastruct Res 4:349–359

Pfenninger KH, Rovainen CM (1974) Stimulation- and calcium-dependence of vesicle attachment sites in the presynaptic membrane; a freeze-cleave study on the lamprey spinal cord. Brain Res 72:1–23

Pfister C (1971a) Die Matrix im Gehirn von Neunaugenembryonen (Lampetra planeri) (Bloch 1874). Z Mikrosk Anat Forsch 4:485–492

Pfister C (1971b) Die Matrixentwicklung in Tel- und Diencephalon von Lampetra planeri (Bloch) (Cyclostomata) im Verlaufe des Individualzyklus. J Hirnforsch 13:363–375

Pfister C (1971c) Die Matrixentwicklung in Mes- und Rhombencephalon von Lampetra planeri (Bloch) (Cyclostomata) im Verlaufe des Individualzyklus. J Hirnforsch 13:377–383

Pierre J, Réperant J, Ward R, Vesselkin NP, Rio JP, Miceli D, Kratskin I (1992) The serotoninergic system of the brain of the lamprey, Lampetra fluviatilis: an evolutionary perspective. J Chem Neuroanat 5:195–219

Pierre J, Rio JP, Mahouche M, Repérant J (1994) Catecholamine systems in the brain of cyclostomes, the lamprey, Lampetra fluviatilis. In: Smeets WJAJ, Reiner A (eds) Phylogeny and development of catecholamine systems in the CNS of vertebrates. Cambridge University Press, Cambridge, pp 7–19

Polenov AL, Belenky MA, Konstantinova MS (1974) The hypothalamo-hypophysial system of the lamprey, Lam-

petra fluviatilis L. Cell Tissue Res 150:505–519

Polenova OA, Vesselkin NP (1993) Olfactory and nonolfactory projections in the river lamprey (Lampetra fluviatilis) telencephalon. J Hirnforsch 34:261–279

Pombal MA, Rodicio MC, Anadón R (1994) Development and organization of the ocular motor nuclei in the larval sea lamprey, Petromyzon marinus L: an HRP study. J Comp Neurol 341:393–406

Pombal MA, El Manira A, Orlovsky G, Grillner S (1995) Identification of the striatum and its inputs, and the role of the ventral thalamus in the control of reticulospinal neurons and locomotion in lamprey. Soc Neurosci Abstr 21:142

Poon MLT (1980) Induction of swimming in lamprey by L-dopa and amino acids. J Comp Physiol 136:337–344

Pu GA, Dowling JE (1981) Anatomical and physiological characristics of pineal photoreceptor cells in the larval lamprey Petromyzon marinus. J Neurophysiol 46:1018–1038

Puzdrowski RL, Northcutt RG (1989) Central projections of the pineal complex in the silver lamprey Ichthyomyzon unicuspis. Cell Tissue Res 225:269–274

Reissner E (1860) Beiträge zur Kenntnis vom Bau des Rückenmarkes von Petromyzon fluviatilis. Arch Anat Physiol Wiss Med, Leipzig, pp 545–588

Repérant J, Vesselkin NP, Ermakova TV, Kenigfest NB, Kosareva AA (1980) Radioautographic evidence for both orthograde and retrograde axonal transport of labeled compounds after intraocular injection of [³H]proline in the lamprey (Lampetra fluviatilis). Brain Res 200:179–183

Retzius G (1893a) Ependym und Neuroglia bei den Cyclostomen. Biol Untersuch (Stockh) 5:15–18

Retzius G (1893b) Über Geschmacksknospen bei Petromyzon. Biol Untersuch (Stockh) 5:69–70

Ringham GL (1975) Localization and electrical characteristics of a giant synapse in the spinal cord of the lamprey. J Physiol (Lond) 251:395–407

Rio JP, Vesselkin NP, Kirpitchnikova E, Kenigfest NB, Versaux-Botteri C, Repérant J (1993) Presumptive GABAergic centrifugal input to the lamprey retina: a double-labeling study with axonal tracing and GABA immunocytochemistry. Brain Res 600:9–19

Rodicio MC, De Miguel E, Pombal MA, Anadón R (1992) The origin of trochlear motoneurons in the larval sea lamprey, Petromyzon marinus L. An HRP study. Neurosci Lett 138:19–22

Rodicio MC, Pombal MA, Anadón A (1995) Early development and organization of the retinopetal system in the larval sea lamprey, Petromyzon marinus L: an HRP study. Anat Embryol (Berl) 192:517–526

Rodríguez S, Rodríguez PA, Banse C, Rodríguez EM, Oksche A (1987) Reissner's fiber, massa caudali and ampulla caudalis in the spinal cord of lamprey larvae (Geotria australis). Light-microscopic immunocytochemical and lectinhistochemical studies. Cell Tissue Res 247:359–366

Ronan M (1988) Anatomical and physiological evidence for electroreception in larval lampreys. Brain Res 448:173–177

Ronan M (1989) Origins of the descending spinal projections in petromyzontid and myxinoid agnathans. J Comp Neurol 281:54–68

Ronan MC, Bodznick D (1986) End buds: non-ampullary electroreceptors in adult lampreys. J Comp Physiol [A] 158:9–15

Ronan M, Northcutt RG (1987) Primary projections of the lateral line nerves in adult lampreys. Brain Behav Evol 30:62–81

Ronan M, Northcutt G (1990) Projections ascending from the spinal cord to the brain in petromyzonid and myxinoid agnathans. J Comp Neurol 291:491–508

Rovainen CM (1967a) Physiological and anatomical studies on large neurons of central nervous system of the sea lamprey (Petromyzon marinus). I. Müller and Mauthner cells. J Neurophysiol 30:1000–1023

Rovainen CM (1967b) Physiological and anatomical studies on large neurons of central nervous system of the sea lamprey (Petromyzon marinus). II. Dorsal cells and giant interneurons. J Neurophysiol 30:1024–1042

Rovainen CM (1974a) Synaptic interactions of identified cells in the spinal cord of the sea lamprey. J Comp Neurol 154:189–206

Rovainen CM (1974b) Synaptic interactions of reticulospinal neurons and nerve cells in the spinal cord of the sea lamprey. J Comp Neurol 154:207–224

Rovainen CM (1974c) Respiratory motoneurons in lampreys. J Comp Physiol 94:57–68

Rovainen CM (1976) Vestibulo-ocular reflexes in the adult sea lamprey. J Comp Physiol 112:159–164

Rovainen CM (1977) Neural control of ventilation in the lamprey. Fed Proc 36:2386–2389

Rovainen CM (1978) Müller cells, 'Mauthner' cells, and other identified reticulospinal neurons in the lamprey. In: Faber DS, Korn H (eds) Neurobiology of the Mauthner cell. Raven, New York, pp 245–269

Rovainen CM (1979a) Electrophysiology of vestibulospinal and vestibuloreticulospinal systems in lampreys. J Neurophysiol 42:745–766

Rovainen CM (1979b) Neurobiology of lampreys. Physiol Rev 59:1007–1077

Rovainen CM (1982) Neurophysiology. In: Hardisty MW, Potter IC (eds) The biology of lampreys, vol 4A. Academic, London, pp 1–136

Rovainen CM (1983a) Identified neurons in the lamprey spinal cord and their roles in fictive swimming. In: Roberts A, Roberts B (eds) Neural origin of rhythmic movements. The Society for Experimental Biology Symposium 37, pp 305–330

Rovainen CM (1983b) Generation of respiratory activity by the lamprey brain exposed to picrotoxin and strychnine, and weak synaptic inhibition in motoneurons. Neuroscience 10:875–882

Rovainen CM (1985a) Effects of groups of propriospinal interneurons on fictive swimming in the isolated spinal cord of the lamprey. J Neurophysiol 54:959–977

Rovainen CM (1985b) Respiratory bursts at the midline of the rostral medulla of the lamprey. J Comp Physiol [A] 157:303–309

Rovainen CM, Birnberger KL (1971) Identification and properties of motoneurons to fin muscle of the sea lamprey. J Neurophysiol 34:974–982

Rovainen CM, Dill DA (1984) Counts of axons in electron microscopic sections of ventral roots in lampreys. J Comp Neurol 225:433–440

Rovainen CM, Yan Q (1985) Sensory responses of dorsal cells in the lamprey brain. J Comp Physiol [A] 156:181–183

Rovainen CM, Johnson PA, Roach EA, Mankovsky JA (1973) Projections of individual axons in lamprey spinal cord determined by tracings through serial sections. J Comp Neurol 149:193–202

Rubinson K (1990) The developing visual system and metamorphosis in the lamprey. J Neurobiol 21:1123–1135

Rubinson K, Cain H (1989) Neural differentiation in the retina of the larval sea lamprey (Petromyzon marinus). Vis Neurosci 3:241–248

Rubinson K, Kennedy MC (1979) The organization of the optic tectum in larval, transforming and adult sea lamprey, Petromyzon marinus. In: Freeman RD (ed) Developmental neurobiology of vision. Plenum, New York, pp 359–369

Rüdeberg SI (1961) Morphogenetic studies on the cerebellar nuclei and their homologization in different vertebrates including man. Thesis, Lund

Rurak DW, Perks AM (1976) The neurohypophysial principle of the Western brook lamprey, Lampetra richardsonii. Studies in the adult. Gen Comp Endocrinol 29:301–312

Rurak DW, Perks AM (1977) The neurohypophysial principle of the Western brook lamprey, Lampetra richardsonii. Studies in the ammocoete larva. Gen Comp Endocrinol 31:91–100

Russell DF (1986) Respiratory pattern generation in adult lampreys (Lampetra fluviatilis): interneurons and burst resetting. J Comp Physiol [A] 158:91–102

Saito T (1928) Über die Müllerschen Zellen im Gehirn des japanischen Flussneunauges (Entosphenus japonicus Martens). Folia Anat Jpn 6:457–475

Saito T (1930) Über das Gehirn des japanischen Flussneunauges (Entosphenus japonicus Martens). Folia Anat Jpn 8:189–263

Samejima M, Tamotsu S, Watanabe K, Morita Y (1989) Photoreceptor cells and neural elements with long axonal processes in the pineal organ of the lamprey, Lampetra japonica, identified by use of the horseradish peroxidase method. Cell Tissue Res 258:219–224

Schaper A (1899) Zur Histologie des Kleinhirns der Petromyzonten. Anat Anz 16:439–446

Schilling K (1907) Über das Gehirn von Petromyzon fluviatilis. Abh Senckenb Naturforsch Ges [Frankf A M] 30:423–446

Schober W (1964) Vergleichend-anatomische Untersuchungen am Gehirn der Larven und adulten Tiere von Lampetra fluviatilis und Lampetra planeri. J Hirnforsch 7:107–209

Schober A, Malz CR, Schober W, Meyer DL (1994) NADPH-diaphorase in the central nervous system of the larval lamprey (Lampetra planeri). J Comp Neurol 345:94–104

Schotland J, Shupliakov O, Wikström M, Brodin L, Srinivasan M, You ZB, Herrera-Marschitz M, Zhang W, Hökfelt T, Grillner S (1995) Control of lamprey locomotor neurons by colocalized monoamine transmitters. Nature 374:266–268

Schultz RE, Berkowitz EC, Pease C (1956) The electron microscopy of the lamprey spinal cord. J Morphol 98:251–273

Schwab ME (1973) Some new aspects about the prosencephalon of Lampetra fluviatilis L. Acta Anat 86:353–375

Selzer ME (1979) Variability in maps of identified neurons in the sea lamprey spinal cord examined by a wholemount technique. Brain Res 163:181–193

Shapovalov AI (1977) Interneuronal synapses with electrical and chemical mechanisms of transmission and evolution of the central nervous system. Zhur Evolyut Biokhim Fiziol 13:621–632

Shapovalov AI (1980) Interneuronal synapses with electrical dual and chemical mode of transmission in vertebrates. Neuroscience 5:1113–1124

Sheridan PH, Youngs LJ, Krieger NR, Selzer ME (1984) Glycine uptake by lamprey spinal neurons demonstrated by light microscopic autoradiography. J Comp Neurol 223:252–258

Shupliakov O, Wallén P, Grillner S (1992) Two types of motoneurons supplying dorsal fin muscles in lamprey and their activity during fictive locomotion. J Comp Neurol 321:112–123

Shupliakov O, Pierbone VA, Gad H, Brodin L (1995) Synaptic vesicle depletion in reticulospinal axons is reduced by 5-hydroxytryptamine: direct evidence for presynaptic modulation of glutaminergic transmission. Eur J Neurosci 7:1111–1116

Sirota M, Viana Di Prisco G, Dubuc R (1995) Electrical microstimulation of mesencephalic locomotor region elicits controlled swimming in semi-intact lampreys. Soc Neurosci Abstr 21:142

Smith DS (1971) On the significance of cross-bridges between microtubules and synaptic vesicles. Philos Trans R Soc Lond [Biol] 261:395–405

Smith DS, Järlfors U, Beránek R (1970) The organization of synaptic axoplasm in the lamprey (Petromyzon marinus) central nervous system. J Cell Biol 46:199–219

Stefanelli A (1933a) Numero, grandezza e forma di alcuni peculiari elementi nervosi dei Petromizonti. Z Zellforsch 18:146–165

Stefanelli A (1933b) Le cellule e le fibre di Müller dei Petromizonti. Arch Ital Anat Embryol 31:519–548

Stefanelli A (1934) I centri tegmentali dell'encefalo dei Petromizonti. Arch Zool Ital 20:117–202

Stefanelli A (1937) Il sistema statico dei Petromizonti (sistema laterale, sistema vestibolare, cervelletto). I. Centri nervosi e vie centrali. Arch Zool Ital 24:209–273

Stefanelli A, Caravita S (1970) Ultrastructural features of the synaptic complex of the vestibular nuclei of Lampetra planeri (Bloch). Z Zellforsch 108:282–296

Steinbusch HWM, Nieuwenhuys R (1979) Serotonergic neuron systems in the brain of the lamprey, Lampetra fluviatilis. Anat Rec 193:693–694

Stell WK, Walker SE, Chohan KS, Ball AK (1984) The goldfish nervus terminalis: an LHRH- and FMRF-amide-immunoreactive olfactoretinal pathway. Proc Natl Acad Sci U S A 81:940–944

Sterba G (1969) Morphologie und Funktion des Subcommissuralorgans. In: Sterba G (ed) Zirkumventrikuläre Organe und Liquor. Int Symp Schloss Reinhardsbrunn 1968. Fischer, Jena, pp 17–32

Sterba G (1972) Neuro- and gliasecretion. In: Hardisty MW, Potter IC (eds) The biology of lampreys, vol 2. Academic, London, pp 69–89

Studnicka FK (1895) Beiträge zur Anatomie und Entwicklungsgeschichte des Vorderhirns der Cranioten. Sitzungsber K Böhm Gesell Wiss Math Naturwiss Kl Abt 1:1–41

Studnicka FK (1912) Über die Entwicklung und die Bedeutung der Seitenaugen von Ammocoetes. Anat Anz 41:561–578

Swain GP, Snedeker JA, Ayers J, Selzer ME (1993) Cytoarchitecture of spinal-projecting neurons in the brain of the larval sea lamprey. J Comp Neurol 336:194–210

Swain GP, Ayers J, Selzer ME (1995) Metamorphosis of spinal-projecting neurons in the brain of the sea lamprey during transformation of the larva to adult: normal anatomy and response to axotomy. J Comp Neurol 362:453–467

Tamotsu S, Morita Y (1986) Photoreception in pineal organs of larval and adult lampreys, Lampetra japonica. J Comp Physiol [A] 159:1–5

Tamotsu S, Korf HW, Morita Y, Oksche A (1990) Immunocytochemical localization of serotonin and photoreceptor-specific proteins (rod-opsin, S-antigen) in the pineal complex of the river lamprey, Lampetra japonica, with special reference to photoneuroendocrine cells. Cell Tissue Res 262:205–216

Tang D, Selzer E (1979) Projections of lamprey spinal neurons determined by the retrograde axonal transport of horseradish peroxidase. J Comp Neurol 188:629–646

Tégner J, Matsushima T, El Manira A, Grillner S (1993) The spinal GABA system modulates burst frequency and intersegmental coordination in the lamprey: differential effects of GABA$_A$ and GABA$_B$ receptors. J Neurophysiol 69:647–657

Teräväinen H (1971) Anatomical and physiological studies on muscles of lamprey. J Neurophysiol 34:954–973

Teräväinen H, Rovainen CM (1971a) Fast and slow motoneurons to body muscle of the sea lamprey. J Neurophysiol 34:990–999

Teräväinen H, Rovainen CM (1971b) Electrical activity of myotomal and sensory dorsal cells during spinal reflexes in lampreys. J Neurophysiol 34:999–1009

Thompson KJ (1990) Control of respiratory motor pattern by sensory neurons in spinal cord of lamprey. J Comp Physiol [A] 166:675–684

Travén HGC, Brodin L, Lansner A, Ekeberg O, Wallén P, Grillner S (1993) Computer simulations of NMDA and non-NMDA receptor-mediated synaptic drive: sensory and supraspinal modulation of neurons and small networks. J Neurophysiol 70:695–709

Tretjakoff D (1909a) Das Nervensystem von Ammocoetes. I. Das Rückenmark. Arch Mikrosk Anat 73:607–680

Tretjakoff D (1909b) Das Nervensystem von Ammocoetes. II. Gehirn. Arch Mikrosk Anat 74:636–779

Tsuneki K, Gorbman A (1975a) Ultrastructure of the anterior neurohypophysis and the pars distalis of the lamprey, Lampetra tridentata. Gen Comp Endocrinol 25:487–508

Tsuneki K, Gorbman A (1975b) Ultrastructure of pars nervosa and pars intermedia of the lamprey, Lampetra tridentata. Cell Tissue Res 157:165–184

Uchida K, Nakamura T, Morita Y (1992) Signal transmission from pineal photoreceptors to luminosity-type ganglion cells in the lamprey, Lampetra japonica. Neuroscience 47:241–247

Ullén F, Orlovsky GN, Deliagina TG, Grillner S (1993) Role of dermal photoreceptors and lateral eyes in initiation and orientation of locomotion in lamprey. Behav Brain Res 54:107–110

Ullén F, Deliagina TG, Orlovsky GN Grillner S (1995a) Spatial orientation in the lamprey. 1. Control of pitch and roll. J Exp Biol 198:665–673

Ullén F, Deliagina TG, Orlovsky GN, Grillner S (1995b) Spatial orientation in the lamprey. 2. Visual influence on orientation during locomotion and in the attached state. J Exp Biol 198:675–681

Urbán L, Székely G (1982) The dorsal column nuclei of the frog. Neuroscience 7:1187–1196

Van Dongen PAM, Hökfelt T, Grillner S, Verhofstad AAJ, Steinbusch HWM, Cuello AC, Terenius L (1985a) Immunohistochemical demonstration of some putative neurotransmitters in the lamprey spinal cord and spinal ganglia: 5-hydroxytryptamine-, tachykinin-, and neuropeptide-Y-immunoreactive neurons and fibers. J Comp Neurol 234:501–522

Van Dongen PAM, Hökfelt T, Grillner S, Verhofstad AAJ, Steinbusch HWM (1985b) Possible target neurons of 5-hydroxytryptamine fibers in the lamprey spinal cord: immunohistochemistry combined with intracellular staining with lucifer yellow. J Comp Neurol 234:523–535

Van Dongen PAM, Theodorsson-Norheim E, Brodin L, Hökfelt T, Grillner S, Peters A, Cuello AC, Forssmann WG, Reinecke M, Singer EA, Lazarus LH (1986) Immunohistochemical and chromatographic studies of peptides with tachykinin-like immunoreactivity in the central nervous system of the lamprey. Peptides 7:297–313

Vesselkin NP, Ermakova TV, Repérant J, Kosareva AA, Kenigfest NB (1980) The retinofugal and retinopetal systems in Lampetra fluviatilis. An experimental study using radioautographic and HRP methods. Brain Res 195:453–460

Vesselkin NP, Repérant J, Kenigfest NB, Miceli D, Ermakova TV, Rio JP (1984) An anatomical and electrophysiological study of the centrifugal visual system in the lamprey (Lampetra fluviatilis). Brain Res 292:41–56

Vesselkin NP, Repérant J, Kenigfest NB, Rio JP, Miceli D, Shupliakov OV (1989) Centrifugal innervation of the lamprey retina. Light- and electron-microscopic and electrophysiological investigation. Brain Res 493:51–65

Viana Di Prisco G, Dubuc R (1995) A study of synaptic responses in lamprey reticulospinal neurones elicited by cutaneous stimulation. Soc Neurosci Abstr 21:142

Viana Di Prisco G, Wallén P, Grillner S (1990) Synaptic effects of intraspinal stretch receptor neurons mediating movement-related feedback during locomotion. Brain Res 530:161–166

Viana Di Prisco G, Ohta Y, Bongianni F, Grillner S, Dubuc R (1995) Trigeminal inputs to reticulospinal neurones in lampreys are mediated by excitatory and inhibitory amino acids. Brain Res 695:76–80

Vinay L, Grillner S (1992) Spino-bulbar neurons convey information to the brainstem about different phases of the locomotor cycle in the lamprey. Brain Res 582:134–138

Von Bartheld CS, Meyer DL (1986) Central projections of the nervus terminalis in the bichir, Polypterus palmas. Cell Tissue Res 244:181–186

Von Bartheld CS, Meyer DL (1988) Central projections of the nervus terminalis in lampreys, lungfishes, and bichirs. Brain Behav Evol 32:151–159

Von Bartheld CS, Lindörfer HW, Meyer DL (1987) The nervus terminalis also exists in cyclostomes and birds. Cell Tissue Res 250:431–434

Von Kupffer K (1906) Die Morphogenie des Centralnerven-systems. In: Hertwig O (ed) Handbuch der vergleichenden und experimentellen Entwicklungslehre der Wirbeltiere, vol II, part 3. Fischer, Jena, pp 1–272

Wächtler K (1974) The distribution of acetylcholinesterase in the cyclostome brain. I. Lampetra planeri (L). Cell Tissue Res 152:259–270

Wächtler K (1983) The acetylcholine-system in the brain of cyclostomes with special references to the telencephalon. J Hirnforsch 24:63–70

Wald U, Selzer M (1981) The inulin space of the lamprey spinal cord. Brain Res 208:113–122

Wallén P (1994) Sensorimotor integration in the lamprey locomotor system. Eur J Morphol 32:168–175

Wallén P, Williams TL (1984) Fictive locomotion in the lamprey spinal cord in vitro compared with swimming in the intact and spinal animal. J Physiol (Lond) 347:225–239

Wallén P, Grillner S, Feldman J, Bergelt S (1985) Dorsal and ventral myotome motoneurons and their input during fictive locomotion in lamprey. J Neurosci 5:654–651

Wallén P, Buchanan JT, Grillner S, Hill RH, Christenson J, Hökfelt T (1989) Effects of 5-hydroxytryptamine on the afterhyperpolarization, spike frequency regulation, and oscillatory membrane properties in lamprey spinal cord neurons. J Neurophysiol 61:759–68

Wallén P, Vinay L, Barthe JY, Grillner S (1995) Locomotor-related modulation of stretch receptor neurons in the lamprey. Soc Neurosci Abstr 21:152

Wannier T (1994) Rostro-caudal distribution of reticulospinal projections from different brainstem nuclei in the lamprey. Brain Res 666:275–278

Wannier T, Orlovsky G, Grillner S (1995) Reticulospinal neurones provide monosynaptic glycinergic inhibition of spinal neurones in lamprey. Neuroreport 6:1597–1600

Wasowicz M, Pierre J, Repérant J, Ward R., Vesselkin NP, Versaux-Botteri C (1994) Immunoreactivity to glial fibrillary acid protein (GFAP) in the brain and spinal cord of the lamprey (Lampetra fluviatilis). J Brain Res 35:71–78

Whitear M, Lane EB (1981) Fine structure of Merkel cells in lampreys. Cell Tissue Res 220:139–151

Whitear M, Lane EB (1983) Oligovillous cells of the epidermis: sensory elements of lamprey skin. J Zool (Lond) 199:359–384

Whiting HP (1948) Nervous structure of the spinal cord of the young larval brook-lamprey. Q J Microsc Sci 89:359–385

Whiting HP (1957) Mauthner neurones in young larval lampreys (Lampetra spp). Q J Microsc Sci 98:163–178

Wickelgren WO (1977a) Physiological and anatomical characteristics of reticulospinal neurones in lamprey. J Physiol (Lond) 270:89–114

Wickelgren WO (1977b) Post-tetanic potentation, habituation and facilitation of synaptic potentials in reticulospinal neurones of lamprey. J Physiol (Lond) 270:115–131

Wickelgren WO, Leonard JP, Grimes MJ, Clard RD (1985) Ultrstuctural correlates of transmitter release in presynaptic areas of lamprey reticulospinal axons. J Neurosci 5:1188–1201

Wikström M, Hill R, Hellgren J, Grillner S (1995) The action of 5-HT on calcium-dependent potassium channels and on the spinal locomotor network in lamprey is mediated by 5-HT$_{1A}$-like receptors. Brain Res 678:191–199

Wislocki GB, Leduc EH, Mitchele AJ (1956) On the ending of Reissner's fiber in the filum terminale of the spinal cord. J Comp Neurol 104:493–517

Wright GM (1986) Immunocytochemical demonstration of growth hormone, prolactin and somatostatin-like immunoreactivities in the brain of larval, young adult and upstream migrant adult sea lamprey, Petromyzon marinus. Cell Tissue Res 246:23–31

Yamada Y (1973) Fine structure of the ordinary lateral line organ. I. The neuromast of lamprey, Entosphenus japonicus. J Ultrastruct Res 43:1–17

Yáñez J, Anadón R (1994) Afferent and efferent connections of the habenula in the larval sea lamprey (Petromyzon marinus L): an experimental study. J Comp Neurol 345:148–160

Yáñez J, Anadón R, Holmqvist BI, Ekström P (1993) Neural projections of the pineal organ in the larval sea lamprey (Petromyzon marinus L) revealed by indocarbocyanine dye tracing. Neurosci Lett 164:213–216

Zompa IC, Dubuc R (1995) Optic nerve and tectal inputs to reticulospinal neurones in lampreys. Soc Neurosci Abstr 21:142

CHAPTER 11

Hagfishes (Myxinoidea)

H. WICHT AND R. NIEUWENHUYS

11.1
General Anatomy, Ecology, Behaviour, Taxonomy

Hagfishes (Myxinoidea, myxinoids), sometimes also referred to as slime eels, are benthic marine organisms that occur in temperate or cold water at depths varying from approximately 10 to 1000 m. The following features are characteristic of myxinoids (Fig. 11.1): the body is cylindric and elongated; the adults of most hagfish species achieve a body length of 40–50 cm. Paired fins are lacking, but the tail displays an unpaired sagittal fin fold that is stabilised by cartilaginous rays dorsally. However, in contrast to lampreys (Petromyzontoidea) and jawed vertebrates (Gnathostomata), there are no muscles associated with these fin rays (Janvier 1981). The skeleton is entirely cartilaginous, the notochord persists throughout life, and neural and hemal arches as well as vertebrae are lacking. In contrast to lampreys, hagfishes are direct developers; the eggs are polylecithalic and large. Fertilisation and embryonic development occur externally (Dean 1899). The nasal opening is unpaired and located at the tip of the snout. It is surrounded by two pairs of barbels which contain mechano- and chemoreceptors (Georgieva et al. 1979; Andres and von Düring 1993a; Fig. 11.2a,b). An unpaired nasal duct connects the nasal opening with the nasal cavity, and, unlike in lampreys, there is an unpaired naso-pharyngeal duct which connects the nasal cavity with the pharynx (Fig. 11.2c). In some species of hagfishes (genus *Eptatretus*) the number of gill pairs is higher than in any other craniate (up to 15) and variable, even within one species (Worthington 1905a). The gills are contained in pouches, and each of these pouches connects with the pharynx by way of a short afferent duct. In hagfishes of the family Eptatretidae (formerly also called Bdellostomatidae or Polistotrematidae) the efferent ducts from the gills open separately to the exterior (Fig. 11.1), whereas in the second familiy of hagfishes, the Myxinidae, these ducts have a common

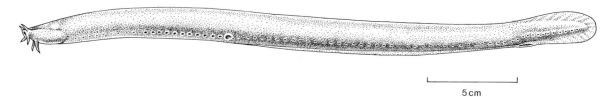

5 cm

Fig. 11.1. The Pacific hagfish, *Eptatretus stouti* (redrawn from Jensen 1966)

external aperture. Although true hinged jaws are lacking, hagfishes possess a very effective biting apparatus. It consists of a left and a right 'tongue pad' equipped with rows of pointed horny teeth. This apparatus is contained in the floor of the buccal cavity (Fig. 11.2c); during feeding it can be projected through the mouth opening (Fig. 11.2a,b). Segmentally arranged slime glands, each equipped with its own efferent duct, occur on each side of the body. On irritation, mediated by the contraction of striated muscles which surround each gland, these glands release mucin (a glycoprotein) and a keratin-like intermediate filament thread into the surrounding water, resulting in the rapid formation of a highly viscous slime. The whole process of slime formation is a matter of seconds and the total volume of slime produced by one animal can amount to several litres (Blackstad 1963; Koch et al. 1991). Thus, hagfishes can enclose themselves in a slimy cocoon which forms a very effective protective device to deter predators.

Very little is known about the ecology and the behaviour of hagfishes (see Strahan 1963, and Jensen 1966, for reviews). The Atlantic hagfish, *Myxine glutinosa*, seems to prefer soft mud or clay bottoms, where it spends most of its time buried in the mud with only the tip of the head protruding. On the other hand, the Pacific hagfish, *Eptatretus stouti*, is also taken from rocky bottoms and seems to prefer them when kept in aquaria (Worthington 1905a). In captivity, hagfishes have been observed to attack and kill living fish by entering their body cavity and eating them 'inside out' (Doflein 1898; this behaviour has led to the erroneous notion that hagfishes are intestinal parasites). They are also known for attacking dead and moribund fishes and fishes caught in traps or on lines; however, their normal diet seems to consist mainly of small invertebrates (Strahan 1963). Judging from the anatomical organisation of their sensory systems (see below), chemical and tactile stimuli seem to be the most important for the localisation of prey. Hagfishes spend most of their time on the ocean floor; nevertheless, they are elegant swimmers, moving by lateral undulations like true eels. Most curiously, the animals are able to loop their body into a knot

and to roll this knot from the tail to the head. This manoeuvre allows the hagfish to escape from the cocoon of its own slime and also to apply leverage when tearing at the surface of larger pieces of prey (Jensen 1966). It is not commonly known that hagfishes also play a certain commercial role: particularly in Korea and Japan, there is a sizeable industry which manufactures small leather items such as wallets and purses from the skin of hagfishes (Kato 1989; Gorbman et al. 1990).

The sensory systems of hagfishes are known mainly from anatomical studies. The tactile system is highly developed and displays a number of specialised receptors, some of which resemble Vater-Pacinian corpuscles of other craniates (Andres and von Düring 1993b). The striated muscles have a rich supply of encapsulated nerve endings which are similar to Ruffini-type stretch receptors (von Düring and Andres 1994). Both the peripheral and the central olfactory apparatus are large and very complex (see Døving and Holmberg 1974 for a preliminary electrophysiological study). Recent research suggests that hagfishes possess an equally elaborate gustatory system with receptors distributed over the entire skin of the head and body (Braun and Northcutt 1997b). The visual apparatus, on the other hand, is poorly developed (Fernholm and Holmberg 1975): relatively small paired lateral eyes are present in all hagfishes, but lens, iris, internal and external eye muscles and the cranial nerves which are associated with these muscles (i.e. the oculomotor, the trochlear, and the abducent nerve) are lacking. In Eptatretidae, the eyes are located under an opaque translucent patch of skin; in Myxinidae, they are buried under layers of muscle. Nevertheless, Myxinidae are able to discriminate light from darkness by skin photoreceptors which are concentrated at the anterior end of the head and around the cloacal region; the posterior group of photoreceptors is supplied by rami of the spinal nerves (Newth and Ross 1955; Steven 1955). Eptatretid hagfishes also possess a well-developed photoreceptive system in the skin (Patzner 1978). Parietal organs are lacking in all hagfishes. The labyrinth is simple and consists of a single torus-shaped structure. Its ventromedial wall contains a single

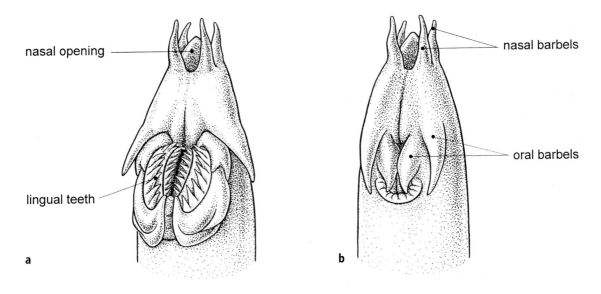

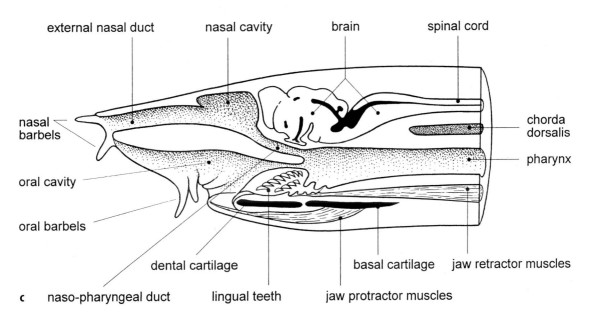

Fig. 11.2a–c. Ventral view of the head of the Atlantic hagfish, *Myxine glutinosa*, with the jaws in protracted (**a**) and retracted (**b**) position (redrawn from Dawson 1963). **c** Mid-sagittal section through the head of a myxinoid (based on our own observations and on Dawson 1963)

macula communis. An anterior and a posterior dilatation of the torus represent the ampullae, each provided with a ring-shaped crista, which, however, is devoid of a cupula (Lowenstein and Thornhill 1970). As compared with that of other craniates, the sensitivity of the inner ear of hagfishes is rather low (McVean 1991). A unique, presumably sensory system occurs in the head of eptatretid hagfishes. A number of short, densely innervated grooves is found in the skin of the head, rostral and caudal to the eye. Their sensory capacities are entirely unknown. The innervation of the grooves derives from two cranial nerves which resemble lateral line nerves (Kishida et al. 1987; Braun et al. 1993; Braun and Northcutt 1997a), but typical mechanosensitive lateral line organs (i.e., neuromasts) do not occur in those grooves (Fernholm 1985) and there is no anatomical and no electrophysiological evidence (Bullock et al. 1983) for electroreception. Nevertheless, lateral line placodes and developing neuromasts have been observed in embryos of Pacific hagfish (von Kupffer 1900; Wicht and Northcutt 1994a, 1995), and it is possible that the embryonic anlagen of the lateral line system transform into the enigmatic system of nerves and grooves in the adults.

The morphology of the CNS of myxinoids has been the subject of a considerable number of papers (for an annotated bibliography of studies published up to 1963, see Peters and Bone 1963). Prominent among these are von Kupffer's (1900, 1906) and Conel's (1929, 1931) studies on the development of the CNS and Worthington's (1905b) and Lindström's (1949) studies of the cranial nerves, the classic accounts of Edinger (1906), Holmgren (1919), and Jansen (1930) on the brain of *Myxine glutinosa*, and Nansen's (1886)[1] and Bone's (1963a,b) analyses of the structure of the spinal cord of the same species. Naturally, the present summary rests on the foundations provided by the work of these authors.

In the past two decades several researchers have developed a new interest in the anatomy of the nervous system of hagfishes. In part, this interest may be attributed to the new anatomical techniques which have become available, but we also have witnessed a change of perspective with regard to the interpretation of anatomical features of hagfishes. The classical authors expected hagfishes (and lampreys, which were believed to be closely related to hagfishes) to display a very general, primitive, or, in other words, plesiomorphic brain organisation which could be used to illustrate the starting point of craniate brain evolution. However, recent research on the taxonomic position of hagfishes and lampreys suggests that lampreys are more closely related to gnathostomes than to hagfishes, and that hagfishes are a very early offspring from the main craniate lineage (see Sect. 11.10). The most recent taxonomic schemes (see Fig. 11.26, bottom) suggest that myxinoids are the survivors of a group which split off from the remaining craniates before the 'invention' of a number of typical craniate characters, such as external and internal eye muscles, cellular bone, vertebral column, and mineralised exoskeleton, characters which are

shared by most recent (and many fossil) craniate taxa. Thus, the myxinoids have had a very long and independent evolutionary history, and it cannot be assumed a priori that their CNS represents the plesiomorphic craniate pattern. From this point of view, it is not very astonishing that hagfishes display a number of CNS features which are highly 'unusual', and the classical authors have had some difficulties in their attempts to interpret these characters within their scheme of a 'generalised brain'.

Thus, throughout the following summary, we will accept and use the descriptive framework laid down by the classical authors; particularly Jansen's (1930) treatise 'The Brain of *Myxine glutinosa*' deserves to be mentioned as a valuable source of information. Citations of more recent publications will be included to accommodate the data from experimental anatomical studies and, where necessary, disparities and disagreements between the classical and the modern interpretation of hagfish CNS characters will be mentioned and discussed. A list of abbreviations is provided on page 508.

11.2
Gross Features of the CNS

The *spinal cord* of hagfishes, like that of lampreys is a flattened, ribbon-like structure (Fig. 11.8a). Dorsal and ventral roots can be clearly recognised. In lampreys as well as in cephalochordates, these roots stay separate, and the same is true for the dorsal and ventral roots in the tail region of hagfishes. In the more anterior body segments, however, branches of the dorsal and ventral roots fuse to form mixed spinal nerves (Goodrich 1937; Peters 1963). Superficially, this resembles the typical gnathostome pattern, but the details of the spatial arrangement of the fused components are very different from those seen in gnathostomes, in particular with respect to the position of the intersegmental arteries (Fig. 11.4a,b). Goodrich (1937) thus concluded that hagfishes and gnathostomes have acquired mixed spinal nerves independently.

The *brain* of myxinoids is surprisingly large; the degree of encephalisation is well within the range of teleost fishes and amphibians and exceeds that of lampreys by far (Platel and Delfini 1981). The brain is unusually compact and appears to be compressed along the rostro-caudal axis (Figs. 11.2c, 11.3a–d). More or less transversely oriented grooves mark the external boundaries between its four principal parts: rhombencephalon, mesencephalon, diencephalon and telencephalon; a cerebellum is lacking. Due to the compression of the entire neuraxis and the peculiar organisation of the ventricular system, the internal boundaries of the respective brain

---

[1] Fridtjof Nansen (1861–1930) is better known as a polar researcher ('Fram'-expedition of 1895/96) and as the recipient of the Nobel peace prize in 1922. His paper of 1886, which contains numerous observations on the spinal cord of *Myxine glutinosa*, is a scientific landmark and a 'tour de force' comparable to his famous polar expeditions. In this paper, he not only established the role of the neuropil for signal processing in the CNS; he also rejected the then prevailing theory of reticularism and showed, for the first time and in *Myxine*, that the centripetal processes of the dorsal root ganglion cells bifurcate into ascending and descending branches after entering the dorsal funiculus of the spinal cord.

parts do not always coincide with the external sulci. Internally, the most striking feature is the condition of the ventricular system. Throughout most of the hindbrain it takes the form of a rather narrow canal and in the forebrain it is restricted to isolated cavities (Fig. 11.3d). Choroid plexuses are entirely lacking; thus, the production of the cerebrospinal fluid must occur elsewhere, most likely in the brain itself (Murray et al. 1975).

The *rhombencephalon* is the largest division of the brain. Caudally it merges into the spinal cord, while its rostral limits are marked dorsally and ventrally by the isthmic fissure and the plica encephali ventralis, respectively. Rostrolaterally, the rhombencephalon expands into two massive horns which embrace the mesencephalon. At the tip of these horns, the trigeminal nerve enters the brain.

The *mesencephalon* is small and wedge-shaped. Dorsally, the mesencephalic tectum is visible externally (Fig. 11.3a); on the ventral side this brain part is confined to a small medial bulge (corresponding to the interpeduncular nucleus) that forms part of the caudal wall of the plica encephali (Fig. 11.3d).

The *diencephalon* is demarcated externally from the mesencephalon by the di-mesencephalic sulcus and from the telencephalon by the telo-diencephalic sulcus. Internally, however, the diencephalon extends far more rostrally than is indicated by the sulcus telo-diencephalicus. The fused right and left habenular ganglia are intercalated between the two thalamic lobes and the cerebral hemispheres (Fig. 11.3a). The tiny optic nerves enter the base of the diencephalon near the midline (arrow in Fig. 11.3c); notably, the decussation of the optic nerves occurs *inside* the brain.

The *telencephalon* is well developed. It comprises two large cerebral hemispheres and a pair of olfactory bulbs. The latter closely appose the nasal cavity (also see Fig. 11.21) and are externally demarcated from the hemispheres by circular fissures. Internally, the olfactory bulbs are wedged into the solid cerebral hemispheres; hence, they are larger than suggested by their external appearance.

Due to the progressive obliteration of the ventricular system during ontogeny (see Sect. 11.3), the organisation of that system in the tel- and diencephalon is somewhat variable between individuals of different age and between Myxinidae and Eptatretidae. In Myxinidae, as shown in Fig. 11.3d, the ventricular cavities in the tel- and diencephalon consist of the blind-ending rostral continuation of the mesencephalic ventricle dorsally and two isolated cavities ventrally. The ventral cavities are the preoptic recess, anteriorly (representing a remnant of the formerly large ventriculus impar telencephali, see Fig. 11.6a), and the hypothalamic ven-

tricle, caudally. The neurohypophysis is connected to the brain via a short, thin and hollow stalk. Through that stalk, the hypothalamic ventricle communicates with the lumen of the neurohypophysis, which has the form of a dorso-ventrally flattened sac at the base of the diencephalon. The adenohypophysis is situated ventral to the neurohypophysis, a thin sheet of connective tissue separating the two (Fig. 11.23). In Eptatretidae, the same ventricular cavities can be observed in the forebrain, but the system is slightly less reduced and displays an additional cavity immediately ventral to the habenular ganglion: the subhabenular ventricle (see Figs. 11.16, 11.22).

11.3
Development and Overall Histological Pattern

11.3.1
Development

Only three embryos of *Myxine glutinosa* have ever been found and inspected (Holmgren 1946; Fernholm 1969), but the ontogenesis of the CNS of *Eptatretus stouti* is known in some detail from the studies of von Kupffer (1900, 1906) and Conel (1929, 1931)[2]. According to the latter author, the early morphogenesis of the neuraxis in this species follows the general vertebrate pattern of primary neurulation; i.e. there is initially a neural plate, followed by a neural groove, the folds of which eventually meet and fuse dorsally to form a closed but hollow neural tube. It is not known whether secondary neurulation (see Sect. 11.10) occurs in the tail region of hagfishes. From a comparative point of view these observations are interesting, because in lampreys the formation of the neural tube takes place in a fundamentally different way (von Kupffer 1906; see Chap. 10 and Sect. 11.10).

In early stages the neural tube is rather wide and has a pointed apex which is still in direct contact with the ectoderm and represents the processus neuroporicus. Soon a small plica encephali becomes discernable on the ventral side and marks

---

[2] Adult specimens of hagfishes can be obtained quite easily by trapping; however, the collection of fertilised eggs and embryos is a tortuous enterprise which requires special techniques (Dean et al. 1897). Practically all the known embryos of *Eptatretus stouti* were collected in the Monterey Bay in 1896 (Dean et al. 1897), in 1898 (Doflein 1898), and in 1930 (Conel 1942). There have been no successful attempts to collect embryos since, and, as hagfishes do not reproduce in captivity, the aged and weathered histological material produced from those embryos is the only source of information on the development of myxinoids. See Wicht and Tusch (1997) for a more general review of the ontogeny of the head of myxinoids.

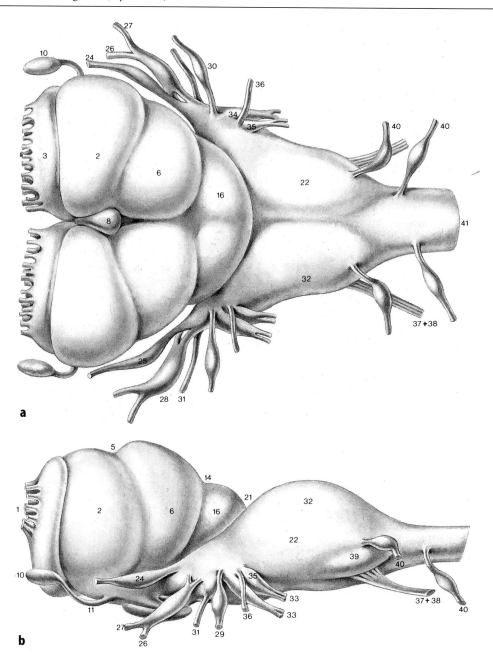

Fig. 11.3a–d. The brain of the Atlantic hagfish, *Myxine glutinosa*, in dorsal (**a**), lateral (**b**), and ventral view (**c**). **d** Medial view of the bisected brain. (Drawings are based on Retzius 1893b and Jansen 1930; terminology for the cranial nerves is from Lindström 1949.) The *arrow* in **c** marks the entrance of the optic nerve which enters the brain prior to its decussation. *Numbers*:

1 nervus olfactorius
2 telencephalon
3 bulbus olfactorius
4 ventriculus praeopticus
5 sulcus telo-diencephalicus
6 diencephalon
7 ventriculus diencephalicus
8 ganglion habenulae
9 ventriculus hypothalamicus
10 eye
11 nervus opticus

12 neurohypophysis
13 adenohypophysis
14 sulcus di-mesencephalicus
15 plica encephalica ventralis
16 mesencephalon
17 ventriculus mesencephalicus
18 recessus dorsalis ventriculi mesencephalici
19 recessus ventralis ventriculi mesencephalici
20 nucleus interpeduncularis
21 fissura isthmi
22 rhombencephalon

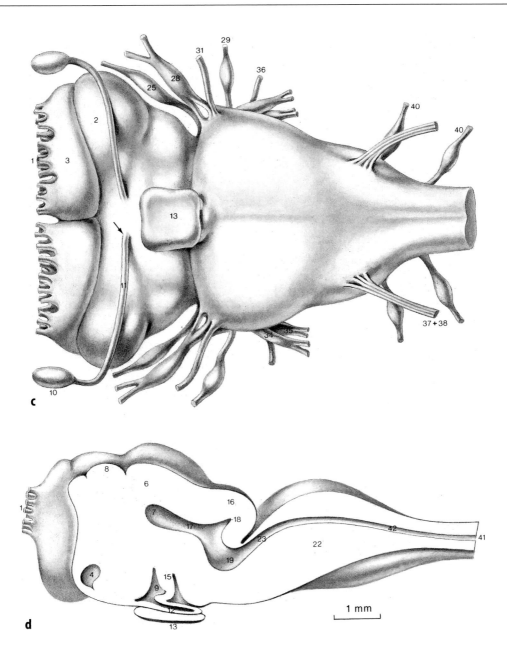

23 ventriculus quartus
24 nervus ophthalmicus (trigemini)
25 ganglion nervi ophthalmici (trigemini)
26 nervus velobuccalis et nervus dentalis
   (trigemini)
27 nervus externus (trigemini)
28 ganglion nervi velobuccalis, nervi dentalis et nervi
   externi (trigemini)
29 nervus buccalis
30 ganglion nervi buccalis
31 nervus trigeminus, radix motoria

32 nucleus radicis descendentis nervi trigemini
33 nervus octavus
34 ganglion utriculi
35 ganglion sacculi
36 nervus facialis
37 nervus glossopharyngeus
38 nervus vagus
39 lobus nervi vagi
40 nervi spino-occipitales
41 medulla spinalis
42 canalis centralis

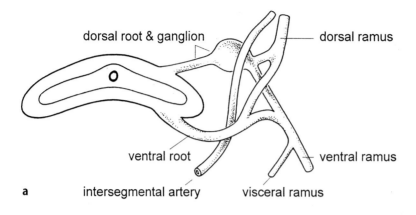

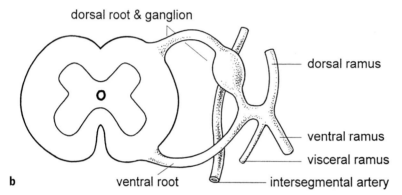

Fig. 11.4a,b. The mixed spinal nerves of myxinoids (a) in comparison to the spinal nerves of gnathostomes (b). Note the different position of the intersegmental arteries in relation to the spinal nerve roots, indicating that dorsal and ventral roots from different segmental levels fuse to form mixed spinal nerves in hagfishes and gnathostomes (based on Goodrich 1937)

the boundary between the anlagen of the archencephalon and the deuterencephalon. In slightly older embryos (Fig. 11.5a), the tripartitioning of the brain into prosencephalon, mesencephalon and rhombencephalon can be readily recognised. The anlage of the rhombencephalon is large and distinctly segmented. The prosencephalon is represented by a wide, flattened structure which, due to the strong flexure of the rostral end of the neural axis, is situated ventral to the rather small mesencephalon. It is important to note that this pronounced flexure will be reduced later in ontogeny; i.e. the prosencephalic vesicle will perform a rotation of more than 90°. Thus, prosencephalic regions which occupy a ventral position at this stage will shift to an anterior position after the rotation, while dorsal regions will be shifted into posterior positions.

Conel (1929, 1931) believed that external forces play an important role in the shaping of the myxinoid CNS. He considered that during early development the brain is dorsoventrally compressed between the tough eggshell on one side and the resistant yolk mass on the other side. Through these external forces the parts of the brain wall undergoing particularly active linear growth are

unable to expand laterally and are forced into a folded pattern. The morphogenesis of the prosencephalic vesicle is heavily affected by this process of folding, and it results in a forebrain which – prior to the above-mentioned rotation – displays paired dorsal and paired ventral evaginations (Fig. 11.6a). Based on this observation (initially made by von Kupffer 1906), one may conclude that hagfishes possess *four* hemispheric lobes and *four* lateral ventricles, two on each side (Tusch et al. 1995). The dorsal pair of lobes will develop into the thalamic part of the diencephalon; the ventral pair of lobes will give rise to the telencephalic hemispheres and the olfactory bulbs. Furthermore, the ventral wall of the diencephalic lobe and the dorsal wall of the telencephalic lobe eventually fuse, and extensive neuronal proliferation and migration occur in this zone of fusion. During this process, both diencephalic and telencephalic matrix zones (Conel 1931) contribute to the formation of a large nuclear mass that will occupy the centre of the forebrain in adults, i.e. the central prosencephalic complex (see Figs. 11.6a, 11.16, 11.17, 11.21). As mentioned above, the brain axis is straightened out in later developmental stages, so that the telencephalic and the dience-

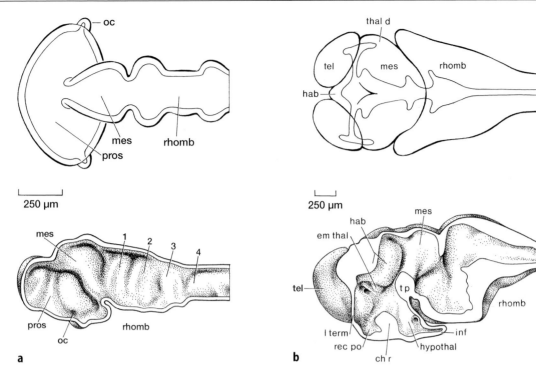

Fig. 11.5a,b. An early (**a**) and a late (**b**) stage in the development of the brain of the Pacific hagfish, *Eptatretus stouti*. A dorsal view of the developing brain is shown in the *upper row*; the extension of the ventricular system is indicated by the *thin inner line*. The *lower row* depicts medial views of bisected brains. The brain shown in **a** is in the process of formation of the ventral encephalic flexure and rotation of the prosencephalon, the one in **b** shows the conditions after the rotation reversal. *1–4*, Rhombomeres (based on wax-plate reconstructions by Conel 1929, 1931)

phalic hemispheres come to lie in front of the mesencephalon (Fig. 11.5b).

The ontogeny of the rhombencephalon is still poorly understood. In early embryonic stages, the lateral walls of the rhombencephalon show a differentiation into rostrocaudally arranged neuromeres, as in other craniates (Fig. 11.5a); however, the exact number of neuromeres and their relation to structures in the adult hindbrain are unknown. In transverse sections, the early embryonic rhombencephalon shows a pattern of longitudinal ridges or zones separated by ventricular grooves (von Kupffer 1900, 1906, Fig. 11.6b). This pattern is strongly reminiscent of that found in the rhombencephalon of other anamniote groups, and it should be recalled that these zones and the intervening grooves form the basis for morphological comparisons (cf. Chap. 4 and Sect. 11.5).

However, the hindbrain of adult hagfishes shows marked topographical differences as compared with that of other craniates (see Sect. 11.5), and it has been suggested (Nishizawa et al. 1988; Matsuda et al. 1991) that these differences result from a series of unusual morphogenetic events. In most craniates, once the rhombencephalic neural tube and its longitudinal zones have formed, the rhombencephalon undergoes a process of *eversion* or 'folding out' of its lateral walls; during that process the roof plate is stretched out and transformed into the chorioid plexus of the fourth ventricle. In hagfishes, however, the rhombencephalon may undergo a process of *inversion* or 'folding in' of its lateral walls instead (Nishizawa et al. 1988; Matsuda et al. 1991), and this inversion may be responsible for the atypical topographical relations which are encountered in the adult hindbrain (see Sect. 11.5), as well as for the absence of a chorioid plexus.

Once this unique developmental 'blueprint' has been laid down, there is a massive increase in the thickness of the brain walls which is brought about by extensive proliferation and migration of neuroblasts. During this process, the ventricular cavities of the hindbrain are reduced to narrow canals; in the forebrain most of the ventricular cavities, in particular the lateral telencephalic and diencephalic ventricles, lose their lumen. However, the ventricular cavities do not disappear without traces: rows of transformed ependymal cells indicate their former site (Edinger 1906; Wicht and Northcutt 1992a; Fig. 11.21), and these vestiges of the ventricular system can be used to clarify topological relationships in the adult forebrain (see Sect. 11.8.1).

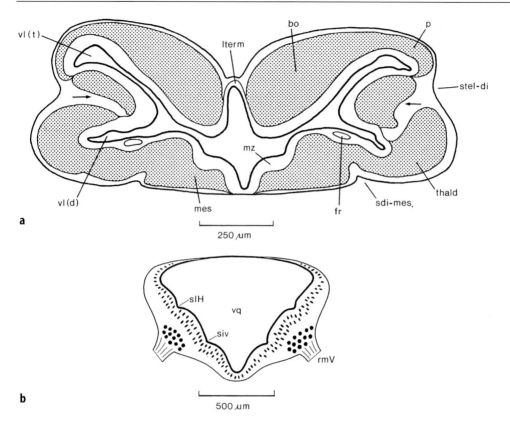

Fig. 11.6a A slightly oblique transverse section through the prosencephalic vesicle of an embryo (*Eptatretus stouti*) somewhat older than the one depicted in Fig. 11.5a. Note that the orientation of the figure is unusual, in that the ventral surface of the embryonic brain, occupied by the olfactory bulb (*bo*), is turned upward, while the dorsal surface, occupied by the thalamus (*thald*) and the mesencephalon (*mes*), is turned downward. This orientation of a transverse section through the (rotated) prosencephalic hemisphere of the embryo allows a direct comparison with a horizontal section through the adult forebrain, after the rotation reversal, as shown in Fig. 11.21. Migrated cell masses are *shaded*, the periventricular matrix zone (*mz*) and the marginal layer are *unshaded*. Note that both the thalamic (*thald*) and the telencephalic (*p, bo*) walls of the prosencephalic vesicle are evaginated at this stage; thus, both contain lateral ventricles, *vl(d)* and *vl(t)*. The *small arrows* point at the primordial central prosencephalic complex. On the *right hand side* (which is slightly more posterior than the *left*) it is evident that the central prosencephalic complex receives contributions from both diencephalic and telencephalic matrix zones. For additional abbreviations, see list (drawn from an actual section of embryo no. 2343, Dean-Conel collection, Museum of Comparative Zoology, Boston, Mass., USA).
b Transverse section through the rhombencephalon of an embryo slightly older than the one depicted in Fig. 11.5a. Note the distinct longitudinal ridges and the limiting sulcus of His (*slH*). The branchiomotor nuclei have already started to migrate towards the meningeal surface. For additional abbreviations, see list (based on von Kupffer 1906)

11.3.2
Overall Histological Pattern, Glia, Ependyma, and Blood-Brain Barrier

The grey matter of the spinal cord is surrounded by a zone of nerve fibres (Fig. 11.8a), but in the brain this simple pattern, which prevails throughout the CNS of many anamniotes, is not maintained. Rather, in most parts of the brain of myxinoids the neurons have spread over almost the entire width of the walls (Figs. 11.10–11.18, 11.21, 11.23, 11.24). The Golgi studies of Holmgren (1919) and Jansen (1930) have shown that most neurons have relatively long, sparsely branching dendrites (Fig. 11.7). Giant neurons such as Mauthner and typical Müller cells, which are prominent in the brain of lampreys, do not occur in the brain of myxinoids; their spinal cord, however, contains conspicous giant interneurons. A striking similarity between lampreys and myxinoids, however, is that in both groups all axons are devoid of a myelin sheath (Bullock et al. 1984; Waehneldt et al. 1986). Consequently, typical oligodendral glia is absent.

The *glia* in the spinal cord of *Myxine glutinosa* was studied by Nansen (1886) and Retzius (1891, 1893a), who found the same two types of glial cells that are present in lampreys, i.e., ependymal cells and tufted neuroglial cells. In a later study Retzius (1921) mentioned the presence of yet a third type of glia in the spinal cord of *Myxine*, which he designated as astroglial cells. The somata of the astroglial cells are confined to the grey matter and give rise to

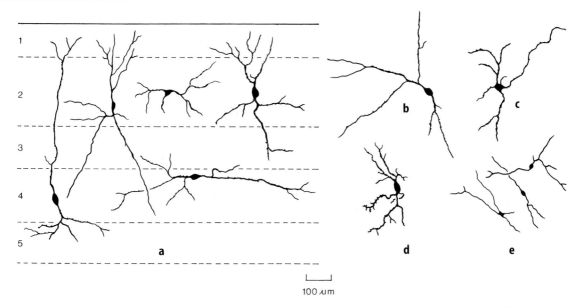

Fig. 11.7a–e. Neurons in various parts of the brain of the Atlantic hagfish, *Myxine glutinosa*, as seen in Golgi preparations. **a** Pallial cortex, **b** central prosencephalic complex, **c** preoptic area, **d** ventral thalamic area, **e** infundibular hypothalamus; *1–5*: pallial layers (based on Jansen 1930)

several rather short, irregularly arranged processes. Interestingly, Retzius observed that these astrocytes are much more numerous in the rhombencephalon (the only part of the brain studied by him) than in the spinal cord. More recently, Wicht et al. (1994) performed an immunocytochemical investigation of the glia in the brain and anterior spinal cord of the Pacific hagfish, *Eptatretus stouti*. Using two glia-specific proteins (glial fibrillary acidic protein and glutamine synthetase) as markers for glial cells, they confirmed Retzius' early observations of astrocyte-like glial cells. They also noted that in the brain astroglia-like cells were far more numerous than ependymal glia – another marked difference to lampreys, where ependymal glial cells prevail (Wasowicz et al. 1994). There are no reports on microglia in the brain of myxinoids. However, Chiba and Honma (1986) observed cells in the hypothalamic ventricle and in the neurohypophysis that apparently migrate through the neuronal tissues and ventricles and show signs of phagocytotic activity. They resemble granulocytes of the peripheral blood and may thus play the role of the microglia in the brain of myxinoids.

The *ependymal lining* of the ventricular system varies in appearance in different regions of the brain (Jansen 1930; Adam 1956, 1963a; Wicht and Northcutt 1992a). The most rostral, tube-like part of the common ventricle of the brain stem (Fig. 11.3d) is surrounded by high cylindrical ependymal cells, which show distinct signs of an intense secretory activity. Together, these cells constitute

the subcommissural organ, which gives rise to Reissner's fibre. Even though the exact chemical nature of Reissner's fibre in hagfishes has not yet been determined, it probably consists of glycoproteins as in other craniates, since it cross-reacts with antibodies directed against material extracted from Reissner's fibre of mammals (Pérez et al. 1993). The fibre passes backwards, through the rhombencephalic ventricle and the central canal of the cord, to end in a small terminal widening of the latter (Vigh and Vigh-Teichmann 1992). Elsewhere in the brain, the ependymal lining is represented by a cuboidal or flat epithelium. In areas where the ventricular cavity is lost during ontogeny, strings and rows of cells that display dense nuclei with irregular outlines represent a transformed ependyma (Edinger 1906; Wicht and Northcutt 1992a; Fig. 11.21).

The brain and the spinal cord of hagfishes are both penetrated by *blood vessels* and, at least in the brain, these vessels give rise to an anastomosing network of capillaries; this is in contrast to the situation in lampreys, where the capillaries form simple, non-anastomosing loops (Scharrer 1962). Ultrastructural studies (Mugnaini and Walberg 1965; Murray et al. 1975) have shown that the capillary endothelial cells differ markedly from those of other craniates, as they contain numerous vesicles and tubules which open both to the luminal and to the abluminal surfaces of the endothelial cells. Tight junctions between adjoining endothelial membranes are rare, but the adjacent membranes are interdigitated and the capillaries are generally not fenestrated. An initial study of the rate and extent of penetration of several isotopic test substances (Murray et al. 1975) seemed to indicate that in hagfishes a blood-brain barrier was poorly developed or absent. Therefore, Murray et al. (1975) suggested that the unusual vesicles and tubules of the endothelial cells play a role in the passage of materials between plasma and brain. In a more recent study (Cserr and Bundgard 1984), however, the blood-brain barrier of hagfishes was found to be intact, and it was also shown that the vesicles and tubules of the endothelial cells do not participate in trans-endothelial transport processes. Their role thus remains enigmatic.

Abbreviations

Pallial layers (numerical)

1–5	strata pallii
1	stratum album superficiale
2co	stratum griseum superficiale, pars compacta
2l	stratum griseum superficiale, pars lateralis
2mc	stratum griseum superficiale, pars magnocellularis
2pc	stratum griseum superficiale, pars parvocellularis
3	stratum album intermedium
4l	stratum griseum intermedium, pars lateralis
4m	stratum griseum intermedium, pars medialis
5	stratum album et griseum profundum

Other structures (alphabetical)

a	axon
adhypoph	adenohypophysis
aoctl	area octavolateralis
aoctla	area octavolateralis, pars anterior
aoctlp	area octavolateralis, pars posterior
ap	area praetectalis
bo	bulbus olfactorius
boci	stratum cellulare internum bulbi olfactorii
bogl	stratum glomerulosum bulbi olfactorii
bom	stratum mitrale bulbi olfactorii
bopg	stratum periglomerulosum bulbi olfactorii
caa1	cell with ascending axon, type 1
caa2	cell with ascending axon, type 2
cc	canalis centralis
cd	commissura dorsalis
ch r	chiasmatic ridge
chab	commissura habenularum
cib	commissura interbulbaris

cmsp	columna motoria medullae spinalis
cnud	central prosencephalic complex, nucleus dorsalis
cnum	central prosencephalic complex, nucleus medialis
cnuvl	central prosencephalic complex, nucleus ventrolateralis
cpo	commissura posterior
cpoo	commissura postoptica
crsa	coarse reticulospinal axons
ctm	commissura tecti mesencephali
ctp	commissura tuberculi posterioris
cv	commissura ventralis
dfr	decussatio fasciculi retroflexi
di	diencephalon
drf	dorsal root fibre
ec	edge cell
em thal	eminentia thalami (=nucleus subhabenularis of Wicht and Northcutt 1992a)
fai	fibrae arcuatae internae
fbt	fasciculus basalis telencephali
fd	funiculus dorsalis
fis	fissura isthmi
fl	funiculus lateralis
flm	fasciculus longitudinalis medialis
fr	fasciculus retroflexus (=tractus habenulo-interpeduncularis)
fv	funiculus ventralis
gnbucc	ganglion nervi buccalis
gc	giant cell
gnlla	ganglion nervi lineae lateralis anterioris
gnllp	ganglion nervi lineae lateralis posterioris
gnsp	ganglion nervi spinalis
grc	griseum centrale
gsac	ganglion sacculi sive ganglion posterius nervi octavi
gutr	ganglion utriculi sive ganglion anterius nervi octavi
gV	ganglia nervi trigemini
gVophth	ganglion nervi ophthalmici (trigemini)
gVvb/d/ext	ganglion nervi velobuccalis, nervi dentalis et nervi externi (trigemini)
hab	ganglion habenulae
hem	hemisphaera
hypothal	hypothalamus
inf	infundibulum
ir	nucleus raphes inferior
IX, Xm	nucleus motorius nervi glossopharyngei et nervi vagi
lcoer	locus coeruleus
ldic	large dorsal internuncial cell
lemb	lemniscus bulbaris

lemsp	lemniscus spinalis	rec po	recessus praeopticus
lterm	lamina terminalis	ret	retina
mes	mesencephalon	rhomb	rhombencephalon
myel	myelencephalon	ri	formatio reticularis, pars inferior
mz	matrix zone	rlvhy	recessus lateralis ventriculi hypothalamici
nbucc	nervus buccalis (Myxinidae), corresponds to nervus lineae lateralis anterioris (Eptatretidae)	rm	formatio reticularis, pars medialis
		rmes	formatio reticularis, pars mesencephalica
neurhypoph	neurohypophysis	rmV	radix motoria nervi trigemini
nIX,X	nervus glossopharyngeus, vagus	rnll	radix nervorum lineae lateralis
nlla	nervus lineae lateralis anterioris	rs	formatio reticularis, pars superior
nolf	nervus olfactorius	rs IX,X	radix sensoria nervi glossopharyngei et nervi vagi sive fasciculus communis
nop	nervus opticus		
nu 'a'	nucleus 'a' of Kusunoki et al. (1982)		
nua	nucleus anterior	rv	radix ventralis nervi spinalis
nucpo	nucleus commissurae posterioris	rvvmes	recessus ventralis ventriculi mesencephalici
nudc	nuclei columnae dorsalis		
nudi	nucleus diffusus	sdi-mes	sulcus di-mesencephalicus
nue	nucleus externus	sdic	small dorsal internuncial cell
nuflm	nucleus fasciculi longitudinalis medialis	se	septum
		siv	sulcus intermedioventralis
nui	nucleus internus	slH	sulcus limitans His
nuinf	nucleus infundibularis	smc1	somatomotor cell, type 1
nuip	nucleus interpeduncularis	smc2	somatomotor cell, type 2
nupc	nucleus paracommissuralis	spocc	nervi spino-occipitales
nuperiX	nucleus perivagalis	sr	nucleus raphes superior
nupo	nuclei praeoptici	st	striatum
nurdV	nucleus radicis descendentis nervi trigemini	stel-di	sulcus telo-diencephalicus
		tbspc	tractus bulbo-spinalis cruciatus
nus IX,X	nucleus sensorius nervi glossopharyngei et nervi vagi sive lobus vagi	tbspr	tractus bulbo-spinalis rectus
		tect	tectum mesencephali
nush	nucleus subhabenularis	tegmv	tegmentum ventrale
nut	nucleus triangularis	tel	telencephalon
nutpl	nucleus lateralis tuberculi posterioris	thald	thalamus dorsalis
nutpm	nucleus medialis tuberculi posterioris	toll	tractus olfactorius lateralis
nVext	nervus externus (trigemini)	tollp	tractus olfactorius lateralis, pars profunda
nVII	nervus facialis		
nVIII	nervus octavus	tolls	tractus olfactorius lateralis, pars superficialis
nVm	radix motoria nervi trigemini		
nVophth	nervus ophthalamicus (trigemini)	tolm	tractus olfactorius medialis
nVs	radix sensoria nervi trigemini	tolv	tractus olfactorius ventralis
nVvb/d	nervus velobuccalis et nervus dentalis (trigemini)	top	tractus opticus
		tp	tuberculum posterius
oc	optic cup	tpht	tractus pallio-habenularis et -thalamicus
oe	epithelium olfactorium		
p	pallium	ttb	tractus tecto-bulbaris
pev	plica encephalica ventralis	vdi	ventriculus diencephalicus
proh	tractus praeoptico-hypophyseos	vhy	ventriculus hypothalamicus
pros	prosencephalon	VIIm	nucleus motorius nervi facialis
ra	nuclei raphes	vinf	ventriculus infundibularis
rd	radix dorsalis nervi spinalis	vl(d)	ventriculus lateralis, pars diencephalica
rdV	radix descendens nervi trigemini		
raVIII	radix ascendens nervi octavi	vl(t)	ventriculus lateralis, pars telencephalica
rdVIII	radix descendens nervi octavi		
rdvmes	recessus dorsalis ventriculi mesencephalici	vmc	visceromotor cell

vmes	ventriculus mesencephalicus
Vmm	nucleus motorius magnocellularis nervi trigemini
Vmpa	nucleus motorius parvocellularis nervi trigemini, pars anterior
Vmpp	nucleus motorius parvocellularis nervi trigemini, pars posterior
vpo	ventriculus praeopticus
vq	ventriculus quartus
vsh	ventriculus subhabenularis

11.4
Spinal Cord

11.4.1
Introductory Notes

It has already been mentioned that the spinal cord of myxinoids, like that of lampreys, is a flattened structure (Fig. 11.8a). Only in its most rostral portion, where it approaches the rhombencephalon, does the flattening become less pronounced (Fig. 11.9). The flattening occurs relatively late in development, since von Kupffer (1906) observed that the cord is cylindrical during the embryonic stages.

The central band of grey matter of the adult spinal cord contains motoneurons and interneurons of different types, but they all seem to occur at random along the cord without any hint of a segmental arrangement (Bone 1963b). As in other anamniotes, most of the cells in the cord of hagfishes extend their dendrites into the peripheral fibre zone. On topographical grounds, the latter may be subdivided into dorsal, lateral and ventral funiculi (Fig. 11.8a). The dorsal roots enter and ventral roots leave the cord about halfway between its median plane and its lateral borders. The following account of the fine structure is based almost entirely on Bone's (1963b) description of the spinal cord of *Myxine glutinosa*.

11.4.2
Ventral Roots and Motoneurons

The ventral roots contain fibres of different diameters. Bone (1963b) found that in the mid-trunk region each ventral root contains about 140 fibres. Before dealing with the cells giving rise to these fibres we will discuss their target. It is known that the myotomal muscles are composed of two types of muscle fibres, i.e., fast-twitch ('white') and slow non-twitch ('red') fibres. Bone (1963a) observed that these two types of muscle fibres have two different types of nerve endings that are invariably

supplied by two different sorts of axons. His observations received support from the ultrastructural and fluorescence-histochemical investigations of Korneliussen and Nicolaysen (1973), who also showed that the terminals on 'white' muscle fibres differ from those on 'red' muscle fibres. Both types of endings contain clear vesicles with a diameter of 45–50 nm, but in those on the 'white' muscle fibres only a few (1–2 %) large, dense-core vesicles (diameter 80–110 nm) were found, whereas such vesicles appeared to be very numerous (up to 40 %) in the terminals on the 'red' muscle fibres. Examination with the formaldehyde-induced fluorescence (FIF) technique showed a yellow fluorescence of the terminals on the 'red' muscle fibres, suggesting that the dense-core vesicles in these terminals may contain serotonin.

The skin of myxinoids contains a plexus of subcutaneous neurons, and a similar, particularly rich, neuronal network surrounds the capsules of the large slime glands. Structurally, these plexuses are reminiscent of the peripheral autonomic system (Peters 1963). The uni- and bipolar neurons that constitute both the subcutaneous and the slime-gland plexuses receive pericellular terminations from the axons of central cells (Bone 1963a). The plexus around the slime glands probably innervates the muscle fibres in the capsule of these organs. The physiology of the other parts of the subcutaneous plexus of myxinoids is entirely unknown. Bone (1963a) suggested that it may play some role in the regulation of the flow of venous blood through the extensive system of subcutaneous sinuses which is present in these animals.

In the spinal cord, Bone (1963a) found three distinct types of cells which have axons that leave the cord. The first and by far the most abundant type of cell is situated in the lateral part of the cord (Fig. 11.8, smc1). The main dendritic branches of these elements extend medially and laterally along the margin of grey and white matter, sending numerous branches into the latter. The axons arise from the medially directed dendritic trunk. Bone (1963b) suggested that these elements supply the fast, 'white' muscle fibres of the myotomes.

The second type of primary efferent neurons is smaller and less abundant than the first type. These elements lie medial to the exit of the ventral root from the spinal cord (Fig. 11.8, smc2). According to Bone (1963b), these neurons may innervate the slow 'red' fibres of the myotomes. It has been mentioned above that the nerve terminals on these muscle fibres contain numerous dense-core vesicles that show a yellow fluorescence with the FIF technique. Interestingly, with the same technique, Ochi et al. (1979) found numerous small neurons emitting a

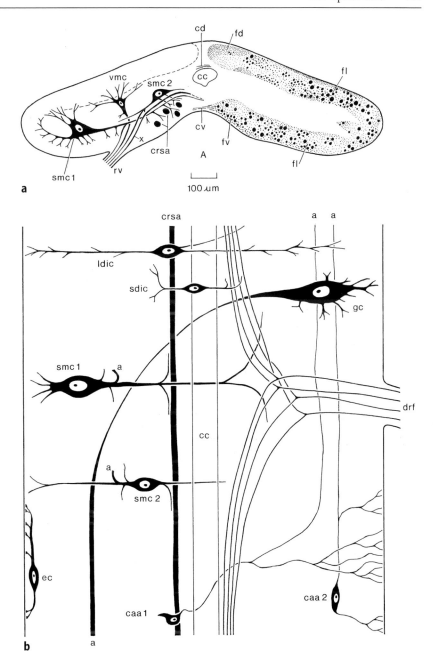

Fig. 11.8a,b. Organisation of the spinal cord of the Atlantic hagfish, *Myxine glutinosa.* **a** Cross-section showing several cell types on the *left* and the fibre spectrum on the *right. x,* Single large fibre of unknown origin in each ventral root; other abbreviations, see text and page 508. **b** Horizontal projection of some cells and fibres (based on Retzius 1891 and Bone 1963b)

yellow fluorescence in the medial part of the cord. Taken together, these observations make it likely that Bone's smc2 cells do indeed supply the slow muscle fibres and that serotonin may play a role in the transmission of impulses to these fibres.

Multipolar neurons in the dorsal part of the grey above the site of exit of the ventral roots constitute the third type of motoneuron (Fig. 11.8a, vmc). These elements are less numerous than the other types and, according to Bone (1963b), may be visceromotor cells, with peripheral connections to the neurons forming the subcutaneous and slime-gland

plexuses. It is worth mentioning the observation made by Bone (1963b) that each ventral root contains a solitary coarse fibre (Fig. 11.8a, x) which takes a longitudinal course within the ventral funiculus of the cord. The peripheral and central relations of these fibres remain to be determined.

11.4.3
Dorsal Roots

The dorsal roots are formed mainly, and perhaps exclusively, by the central axonal processes of extra-

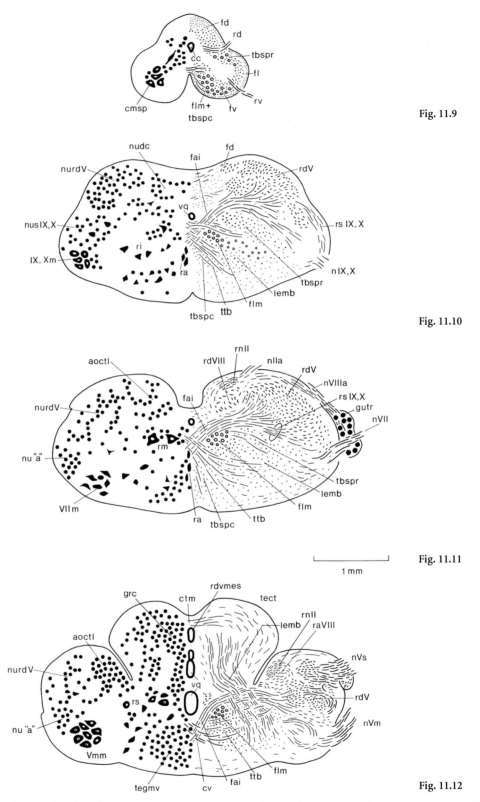

Fig. 11.9

Fig. 11.10

1 mm

Fig. 11.11

Fig. 11.12

Figs. 11.9–11.18. Caudo-rostral series of transverse sections through the CNS of the Pacific hagfish, *Eptatretus stouti*. Major nuclear centres are shown on the *left*, major fibre systems on the *right*. The *interrupted line* in Figs. 11.15–11.17 represents the position of transformed ependymal cells. These cells indicate the site of embryonic ventricular cavities that are lost in the adults

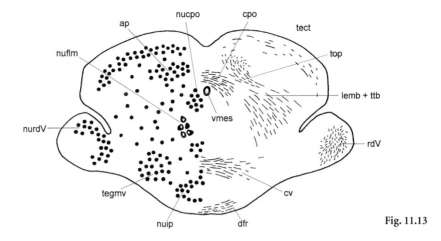

Fig. 11.13

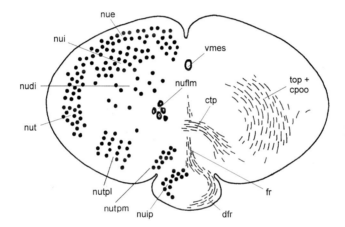

Fig. 11.14

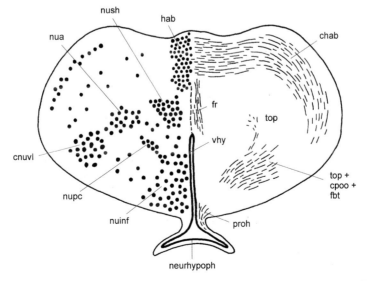

Fig. 11.15

1 mm

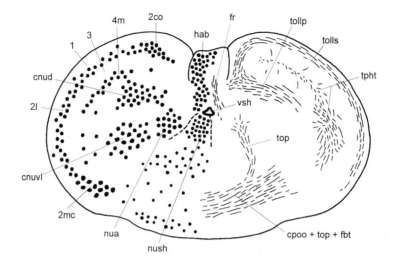

Fig. 11.16

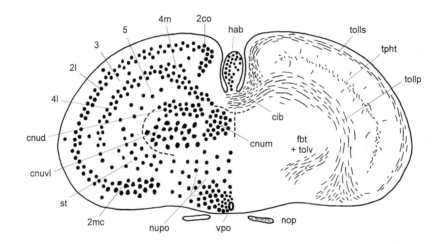

Fig. 11.17

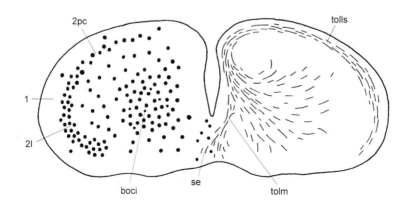

Fig. 11.18

1 mm

medullary cells situated in spinal ganglia. Peripherally, these cells supply fibres to the skin and muscles. Those that end on muscles have branching terminations considered by Bone (1963a) to be probably proprioceptive. Upon entering the cord, the central processes of the spinal ganglion cells bifurcate into ascending and descending branches that course in the dorsal funiculus (Fig. 11.8b, drf). This bifurcation of dorsal root fibres, which is now known to be characteristic of all craniates, was actually observed for the first time by Nansen (1886) in *Myxine* and confirmed by Retzius (1891) a few years later. In a more recent experimental study, Ronan and Northcutt (1990) were able to show that ascending branches of dorsal root ganglion cells in anterior body segments ascend to the medulla oblongata (see Sect. 11.4.5). Remarkably, intraspinal sensory cells (large dorsal cells) are lacking both in the embryonic (von Kupffer 1906) and in the adult (Bone 1963b) spinal cord; this is one of the most striking differences between hagfishes and lampreys.

11.4.4
Internuncial Cells

The intrinsic neurons in the cord were studied first by Retzius (1891) and later by Bone (1963b). Retzius used the Golgi and methylene blue techniques, Bone additionally employed silver impregnations, and their studies revealed the following types of internuncial elements:

1. Numerous small, bipolar cells with transversely oriented dendritic trees, described by Retzius (1891), which may be partially identical to Bone's (1963b) small dorsal internuncial cells (Fig. 11.8b, sdic). These cells are commonly found in the dorsomedial part of the spinal grey, with their dendrites linking together the dorsal funiculi of either side. The axons of these elements were not identified.
2. Unipolar and
3. Longitudinally oriented bipolar cells, both having an ascending axon (Fig. 11.8b, caa1, caa2).
4. Large dorsal internuncial cells (Fig. 11.8b, ldic) which lie under the dorsal funiculus and have long, transversely oriented dendritic stems were described by Bone (1963b). Their axon enters the commissura ventralis, where it is lost to view. We consider it likely that these elements correspond to the medium-sized tri- or quadropolar cells of Retzius (1891).
5. The cell bodies of the giant cells (Fig. 11.8b, gc) lie in the lateral or intermediate parts of the grey. Their dendrites radiate into the dorsolateral and ventral funiculi. The axons of the giant cells pass through the ventral commissure and then turn to take a longitudinal course. The axons of the more rostral giant cells turn caudally to descend in the spinal cord, whereas those of the more caudal cells appear to ascend. Several pairs of these elements approximate an alternating arrangement in each segment. Bone (1963b) suggested that the giant cells are involved in coordinating successive groups of motoneurons along the spinal cord. Corresponding elements are present in the spinal cord of lampreys, but in the latter animals the giant interneurons are confined to the caudal half to two thirds of the spinal cord.
6. Edge cells are found outside the grey in the most lateral part of the spinal white matter. Their dendrites are directed laterally and contact the outer limiting membrane via thick, club-like processes. This arrangement suggests that the edge cells act as proprioceptive stretch receptors which record the varying tension of the limiting membrane during the undulating swimming movements. Bone (1963b) distinguished two types of such edge cells. The cell body of the first type (as shown in Fig. 11.8b, ec) is located very close to the lateral surface of the cord and the dendrites show a longitudinal orientation. The axons of this cell type were not identified. Bone (1963b) compared these cells to Cajal's marginal cells in the substantia gelatinosa of gnathostomes. The cell bodies of the second type of edge cell (not shown in Fig. 11.8) are located slightly more medially and display a medially directed axon. Bone compared these cells to the lateral arcuate or 'edge cells' of lampreys. He suggested that, similar to the edge cells of lampreys (cf. Chap. 10), these cells may play a role in the reciprocal excitation and inhibition of the motoneurons on either side of the spinal cord that occur during the swimming movements.

11.4.5
Ascending and Descending Fibre Systems

The spinal white matter is made up of longitudinal axons of varying diameter interlaced with the externally directed dendrites of the spinal neurons. A fair number of large-diameter axons can be found dispersed in the lateral and ventral funiculi; this is in marked contrast to the situation in lampreys and gnathostomes, where such large axons are located predominantly in the ventral funiculus and in the medial longitudinal fascicle.

The *descending supraspinal projections* of *Eptatretus stouti* have been studied experimentally by Ronan (1989) and the results of this study are summarised in Fig. 11.19c. The *mesencephalic* and the *rhombencephalic reticular formations* are the main source of such projections and they project bilaterally to the spinal cord. At the di-mesencephalic border there is a condensation of relatively large reticular cells which form the nucleus of the medial longitudinal fascicle. These cells give rise to an uncrossed reticulospinal projection. Other large and very large neurons are dispersed throughout the reticular formation. Some of them probably give rise to the coarse axons which can be found in the lateral and ventral funiculi. The majority of these large cells display an uncrossed projection to the spinal cord. According to Jansen (1930), the reticulospinal fibres may be grouped in two bundles, the fasciculus longitudinalis medialis and the tractus bulbospinalis rectus. The former continues into the funiculus ventralis of the spinal cord, whereas the latter descends into the funiculus lateralis.

In addition to the reticular formation, the *octavolateral area* gives rise to fibres that descend to the spinal cord. Ipsilaterally, these fibres arise from the anterior magnocellular octavomotor nucleus, contralaterally, from the posterior magnocellullar octavomotor nucleus (Ronan 1989). These crossed fibres pass ventrally as internal arcuate fibres, most of which bifurcate after crossing. The ascending branches are directed to the mesencephalic tectum (see Sect. 11.5.2); the descending branches collect into the tractus bulbospinalis cruciatus, a bundle that passes spinalward near the median plane and enters the funiculus ventralis of the spinal cord (Jansen 1930). Ronan (1989) also identified other supraspinal projections which arise from the superior and inferior nuclei of the raphe, and from a group of cells situated medial to the vagal motor nucleus (the perivagal nucleus). Tecto-, rubro-, diencephalo-, and telencephalo-spinal projections were not demonstrated in hagfishes. Likewise, the Mauthner cells typical of lampreys and gnathostomes do not occur in hagfishes.

The system of *ascending spinal projections* has been investigated experimentally in *Eptatretus stouti* by Ronan and Northcutt (1990) and the results of this study are summarised in Fig. 11.19b. There is a well-developed dorsal column pathway which occupies the dorsal funiculus and terminates bilaterally in the dorsal column nuclei and the octavolateral area of the medulla oblongata. At least some of these afferents to the rhombencephalon derive from dorsal root ganglion cells of the anterior spinal cord, but second-order intraspinal projection neurons participate in the formation of the dorsal column system as well. Other ascending spinal projections which derive from intraspinal cells travel in the lateral and ventral funiculi and consti-

tute the spinal lemniscus. These projections terminate in the reticular nuclei and in the mesencephalic tectum. Notably, this projection is predominantly uncrossed and does not ascend to thalamic levels, in contrast to the situation in other craniates.

11.5
Rhombencephalon

11.5.1
Introductory Notes

Descriptions of the rhombencephalon usually rest on the assumption that the walls of the rhombencephalon of craniates consist of a dorsoventral series of longitudinal zones or columns which are formed early in ontogeny and which are connected to specific sensory sources or muscular targets in the periphery of the head and body. These zones are, in a dorsoventral sequence:

1. The special somatosensory zone or octavolateral area, connected to the sensory systems deriving from the octaval and dorsolateral placodes, i.e. the inner ear and the lateral line system.
2. The general somatosensory zone, connected to the extero- and proprioceptors in skin and muscles of the head and the branchial region.
3. The viscerosensory zone, which is supposed to receive information from proprioceptors in the digestive tract and which also receives information from the 'special' viscerosensory apparatus (i.e. the gustatory system, even though the latter may not be confined to the digestive tract, but may be dispersed over the entire body surface as in some actinopterygians).
4. The special and general visceromotor zones, containing efferents to muscles which are supposed to derive from lateral plate (i.e. hypomeric) mesoderm; the general zone contains efferent (autonomic, parasympathetic, preganglionary) neurons concerned with the control of the smooth musculature of the digestive system, while the special zone (also called the branchiomotor zone) is connected to the 'specialised visceral' (striated) muscles of the branchial region.
5. The somatomotor zone, containing the motor neurons for the innervation of the striated muscles which derive from paraxial (epimeric, somitic) mesoderm, i.e. the external eye muscles and a certain group of muscles in the floor of the pharyngeal cavity (hypobranchial muscles of anamniotes, hypoglossal muscles of amniotes).

In the following account the rhombencephalon of hagfishes will be described according to the classical scheme above. However, the following caveats need to be mentioned.

The above-mentioned scheme (also known as the 'doctrine of functional components'), which was developed around the turn of the century by W.H. Gaskell, O.S. Strong, C.J. Herrick, and J.B. Johnston (see Chap. 4, Sect. 4.6.2), is in need of some revision. Recent embryological work on the development of the mesoderm has shown that the branchial ('special visceral') muscles do *not* derive from lateral plate mesoderm, but from the paraxial mesoderm of the head (from the so-called somitomeres), as do the somatomotor muscles of the eye and tongue (e.g. Noden 1983, 1991). Thus, the distinction between a 'special visceromotor' and 'somatomotor zone' cannot be upheld on embryological grounds.

In hagfishes, as well as in other craniates, the initial dorsoventral arrangement of the longitudinal zones undergoes some spatial reorganisation during ontogeny. In most craniates, this reorganisation is well documented and involves (a) a transformation of a dorsoventral sequence into a lateromedial one, due to the eversion ('folding out') of the embryonic rhombencephalic neural tube, and (b) shifts of relative positions of individual zones due to the migration of the neurons which constitute an individual zone. The results of this process are most evident from the adult positions of the general somatosensory and the branchiomotor columns, which move lateral and ventral from their points of origin, while the other zones more or less maintain their initial periventricular positions. In hagfishes, however, these transformations are poorly documented and poorly understood. As mentioned above (see Sect. 11.3.1), the rhombencephalic neural tube of hagfishes may undergo a process of inversion rather than an eversion; furthermore, practically all neurons in all columns migrate from their site of origin and the routes of migration are unknown. Further complications arise from the fact that a true somatomotor zone cannot be recognised in hagfishes, since the muscular targets of that zone (extrinsic eye muscles and hypobranchial musculature) are not present. In addition, the viscerosensory and the gustatory systems (see below) show a number of unique features which make it questionable whether these systems are homologues of their namesakes in other craniates.

Nevertheless, short of an alternative morphological scheme and of detailed data on the ontogeny and connectivity, and for the sake of the consistency of the present work, the rhombencephalon of hagfishes will be described according to the classical scheme. However, the inherent problems and deviations from that scheme which have been mentioned above will be presented in some more detail in the respective sections.

11.5.2
Octavolateral Area

The octavolateral area occupies a dorsomedial position in the horns of the rhombencephalon, extending along the isthmic fissure (Figs. 11.11, 11.12, 11.19b, 11.21, aoctl). This area consists mostly of rather small neurons, but there are some larger cells scattered along its whole length. These elements probably represent homologues of the intermediate and posterior octavomotor nuclei of lampreys, since both give rise to a projection to the spinal cord (see Sect. 11.4.5 and below). Based on its cytoarchitecture and connections (see below), the octavolateral area can be divided roughly into a medial, a ventral, and a dorsal nucleus (Kishida et al. 1987).

The *ventral nucleus* receives its afferents from the inner ear. The eighth nerve consists of two separate roots and ganglia, the utricular nerve and ganglion anteriorly, and the saccular nerve and ganglion posteriorly (Figs. 11.3a,c, 11.21). The utricular nerve and ganglion supply the anterior ampulla of the labyrinth, the posterior ampulla and the macula communis are supplied by the saccular nerve and ganglion. Centrally, the fibres of both divisions of the eighth nerve traverse the large descending nucleus of the trigeminal nerve and enter the ventral nucleus of the octavolateral area, where they bifurcate into ascending and descending branches (Amemiya et al. 1985, Fig. 11.19b). Efferent projections from the brain to the labyrinth have not been observed (Amemiya et al. 1985).

The *medial nucleus* of the octavolateral area is the target of the lateral line afferents. Two (possibly three; see below) lateral line nerves are present in eptatretid hagfishes (Fig. 11.19b). The anterior nerve of the lateral line enters the rhombencephalon dorsally at the level of the utricular ganglion; its ganglion cells are found in a ganglion dorsal and posterior to the trigeminal ganglion. Peripherally, this nerve supplies the preocular set of skin grooves (see Sect. 11.1).[3] It is noteworthy that this nerve also seems to carry a general cutaneous component, since it also contributes fibres to the sensory nucleus of the trigeminus (Kishida et al. 1987; Nishizawa et al. 1988; Fig. 11.19a). The posterior nerve of the lateral line is much smaller and enters the brain dorsal to the saccular ganglion (Fig. 11.19b). It supplies the postocular set of skin

---

[3] It should be mentioned here that the lateral line 'canals' observed by Ayers and Worthington (1907) are histological artefacts. A description of the anatomy of the peripheral lateral line system can be found in Kishida et al. (1987), Wicht and Northcutt (1995), and Braun and Northcutt (1997a).

grooves, and its fibres also terminate in the medial nucleus of the octavolateral area (Kishida et al. 1987). Notably, the posterior nerve of the lateral line carries two ganglia, one immediately dorsal and caudal to the otic capsule (shown in Fig. 11.19b), and another more peripheral one (Wicht and Northcutt 1995; Braun and Northcutt 1997a) not shown in Fig. 11.19. Since two separate lateral line placodes have been observed in the postocular region of embryonic hagfishes (Wicht and Northcutt 1995), it is possible that the posterior lateral line nerve and its two ganglia do not represent a natural unit, but correspond to two lateral line nerves of other craniates.

The data on the lateral line system summarised above refer exclusively to hagfishes of the familiy Eptatretidae. The lateral line system of the second family, the Myxinidae, is poorly understood. As mentioned in the Sect. 11.1, they do not possess the pre- and postocular skin grooves typical of Eptatretidae. They do, however, possess a nerve very similar to the anterior nerve of the lateral line of Eptatretidae, the buccal nerve (Fig. 11.3a), which innervates the skin in the preocular region (Lindström 1949). The central connections of that nerve have not been determined experimentally.

In addition to the lateral line afferents, the octavolateral area also receives ascending spinal projections via the dorsal column pathway (Fig. 11.19b). These fibres terminate in the *dorsal nucleus* of the octavolateral area (Ronan and Northcutt 1990).

According to Jansen (1930), the *efferents* from the octavolateral area enter the ventral commissure of the brain stem and dichotomise after crossing into ascending and descending branches. The *ascending branches* contribute to the bulbar lemniscus, a large fibre system that contains numerous fibres from the nucleus of the descending tract of the trigeminus. Throughout the rhombencephalon this fibre system lies ventrolateral to the medial longitudinal fascicle (Figs. 11.10, 11.11). However, in the most rostral part of the rhombencephalon it makes an abrupt dorsal turn to enter the mesencephalic tectum, where most of its fibres terminate (Figs. 11.12, 11.20). Again according to Jansen (1930), some branches of the bulbar lemniscus pass even further rostrally to the thalamus. The *descending branches* of the efferents from the octavolateral area collect into the bulbospinal tract which passes to the ventral funiculus of the spinal cord. Recent experimental studies have confirmed most of Jansen's observations on the efferents of the octavolateral area. As mentioned above, descending spinal projections arise from the magnocellular elements dispersed throughout the octavolateral area. Based on the laterality of these projections, an anterior magnocellular octavomotor nucleus with ipsilateral projections may be distinguished from a posterior magnocellular octavomotor nucleus with contralaterally descending projections (Ronan 1989, Fig. 11.19c). The presence of a crossed octavo-tectal projection was confirmed by Amemiya (1983). However, the existence of the octavo-thalamic projection observed by Jansen still awaits experimental proof.

The organisation of the octavolateral area of hagfishes is somewhat difficult to compare with that of other craniates. The ventral nucleus is probably homologous to the same-named nucleus of other craniates, as evidenced from its connections with the inner ear. The medial (sometimes also called 'intermediate') nucleus of the octavolateral area of other craniates receives (mechanoreceptive) lateral line afferents, as does the medial nucleus of hagfishes; thus, the two are probably homologues. The dorsal nucleus of the octavolateral area of hagfishes is enigmatic. In craniates which possess an electroreceptive lateral line (e.g. lampreys, see Chap. 10) this nucleus receives the electroreceptive afferents. Hagfishes are not electroreceptive (Bullock et al. 1983), and the only known input to the dorsal nucleus is an ascending spinal projection (see above). The homology of this nucleus is therefore unclear.

11.5.3
General Somatosensory Zone

One of the most characteristic features of the rhombencephalon of myxinoids is the magnitude of the sensory component of the trigeminal nerve and its centre of termination (Figs. 11.3, 11.10–11.13, 11.21, nurd). The experimental studies of Nishizawa et al. (1988) and Ronan (1988) form the basis of the following description. The complex of the trigeminal nerve consists of four major subdivisions, namely, the ophthalmic, external, and velobuccal/dental nerves; in addition, there is a massive nervus palatinus which innervates the roof of the mouth cavity (Lindström 1949). With respect to the regions they innervate, the first three nerves are comparable to the ophthalmic, maxillar, and mandibular nerves of the trigeminus of other craniates, respectively. The nervus palatinus which also innervates chemoreceptors in the mouth cavity (Braun and Northcutt 1997b) has no obvious homologue in the trigeminal complex of other craniates.

The vast majority of fibres in these nerves stem from sensory neurons in the large ganglia of the trigeminal nerve; however, some neurons in the utricular ganglion also send their peripheral processes into the branches of the trigeminus (Fig. 11.19a).

Furthermore, some intramedullary sensory cells which are located in the most ventral subdivision of the sensory nucleus of the trigeminus were observed by Ronan (1988) after application of tracers to the peripheral branches of the trigeminus.[4] In the rhombencephalon the central processes of these sensory neurons form a large, dorsolaterally situated descending root of the trigeminus (Figs. 11.10–11.13), the longest fibres of which extend into the dorsal part of the spinal lateral funiculus (Fig. 11.19a). There is a well-developed somatotopical arrangement: fibres of the nervus velobuccalis/dentalis (comparable to the mandibular nerve of other craniates) occupy the ventrolateral part of the radix descendens; the intermediate part stems from fibres of the nervus externus (comparable to the maxillary nerve), and the most dorsal and medial fibres belong to the nervus ophthalmicus. Notably, this somatotopy is just the reverse of that seen in other craniates, since the ophthalmic fibres normally occupy the most ventral and the mandibular fibres the most dorsal position in the descending root of the trigeminus. It should also be noted that the descending root of the trigeminus is not entirely a system of 'general somatosensory' afferents: hagfishes possess numerous taste-bud-like structures (the so-called 'Schreiner-organs', Schreiner 1919; Braun and Northcutt 1997b) in the skin of the barbels surrounding the nasal aperture (Blackstad 1963; Georgieva et al. 1979; von Düring and Andres 1997), in the skin of the head and in the anterior (buccal and nasal) regions of the pharyngeal cavity (Braun and Northcutt 1997b). These organs are supplied by rami of the trigeminal nerve (Nishizawa et al. 1988; Braun and Northcutt 1997b). Thus, parts of the sensory nucleus of the trigeminus must be rated as a combined somato- and viscerosensory

centre. In addition, the most ventral parts of the sensory nucleus of the trigeminus receive collateral fibres from the sensory root of the glossopharyngeal and vagal nerves, which is situated immediately ventral to that nucleus (see below).

In marked contrast to other craniates, there are no ascending primary trigeminal fibres; hence, a principal and a mesencephalic sensory nucleus of the trigeminus are entirely lacking. The descending sensory nucleus of the trigeminus, however, is well developed. It consists of rather small cells, which are arranged in strands along and among the fibre bundles (Figs. 11.10–11.13, 11.21, nurd V). The efferents of this nucleus pass ventromedially as internal arcuate fibres and join the bulbar lemniscus after having decussated in the ventral commissure (Figs. 11.12, 11.13, 11.20). Their target is the contralateral mesencephalic tectum (Amemiya 1983; Ronan 1989). Notably, a distinct trigemino-spinal projection, which occurs in lampreys and many gnathostomes, is lacking in hagfishes (Ronan 1988).

More posteriorly, the general somatosensory zone is represented by the dorsal column nuclei (Figs. 11.10, 11.19b). They receive ascending bilateral projections from the spinal cord and project to the contralateral mesencephalic tectum via the bulbar lemniscus (Amemiya 1983; Fig. 11.20). It should be mentioned here that these nuclei were interpreted by Ayers and Worthington (1911) and by Jansen (1930) as being the nuclei of the solitary tract. Recent experimental studies (see following section) have shown that this interpretation is probably not correct.

11.5.4
The Viscerosensory Zone

As mentioned above, a large number of viscerosensory (gustatory) afferents seem to reach the brain via the trigeminal nerve. Other viscerosensory fibres which probably innervate posterior pharyngeal (branchial) and intestinal surfaces travel in the glossopharyngeal and vagal nerves; it is not known whether the facial nerve carries a (viscero-) sensory component. The topographical organisation of the sensory root of the vagal and glossopharyngeal nerves in the rhombencephalon is unique in hagfishes, and a recent experimental study (Matsuda et al. 1991) has shown that Jansen's (1930) description of that system is in need of a thorough revision. Matsuda et al. (1991) demonstrated that the afferents which travel in the vagal and glossopharyngeal nerves enter an area which is wedged between the branchiomotor nuclei ventrally and the sensory nucleus of the trigeminus dorsally at the lateral surface of the rhombencephalon (Figs. 11.10,

---

[4] There are a number of problems connected with this observation, and they are important enough to justify a footnote. Prima facie – as justly suggested by Ronan (1988) – these cells observed in *Eptatretus stouti* might be regarded as homologues of the dorsal cells (=*Hinterzellen*) of lampreys, particularly since Ronan observed that they are very similar in shape and size (up to 60 μm in diameter) to the sensory ganglion cells in the nearby utricular ganglion. Furthermore, Ronan was right when he noted that the dorsal cells of lampreys, similar to the cells he observed in hagfishes, send peripheral processes into the trigeminal nerve. However, in a study published simultaneously on the trigeminal system of *Eptatretus burgeri* (Nishizawa et al. 1988), no such cells were found even though the tracing technique was almost identical in both studies. Even in nonexperimental material, the dorsal cells of lampreys are easily recognised by their size. However, in our normal material of juveniles and adults of *Eptatretus stouti*, *Eptatretus sinus*, and *Myxine glutinosa*, we did not find any cells as large as the ones seen by Ronan (1988), and they were also not mentioned by Jansen (1930). We are thus hesitant to accept the presence of intramedullary sensory cells as a general feature of the brain stem of hagfishes, even though we can presently not offer a satisfying explanation for Ronan's observation.

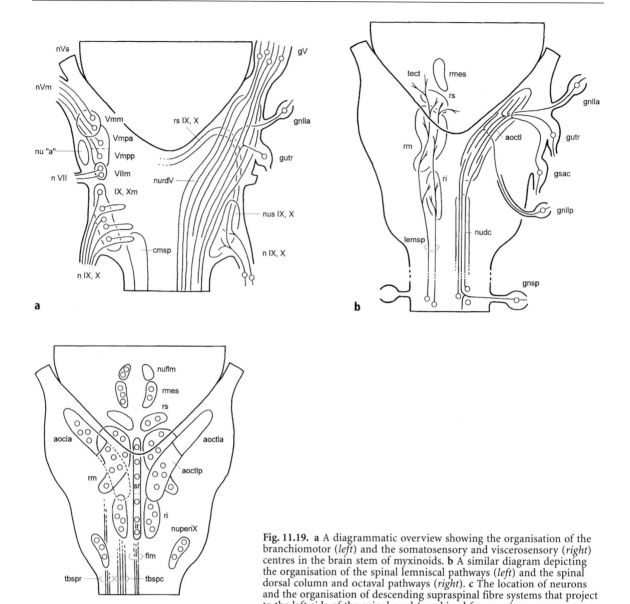

Fig. 11.19. a A diagrammatic overview showing the organisation of the branchiomotor (*left*) and the somatosensory and viscerosensory (*right*) centres in the brain stem of myxinoids. **b** A similar diagram depicting the organisation of the spinal lemniscal pathways (*left*) and the spinal dorsal column and octaval pathways (*right*). **c** The location of neurons and the organisation of descending supraspinal fibre systems that project to the left side of the spinal cord (combined from various sources, see text)

11.19a, nus IX,X). This area, called the lobus vagi or sensory nucleus of the vagus, contains numerous small neurons. Notably, this area occupies a sub-meningeal position, in contrast to other craniates, where the viscerosensory zone is located close to the ventricular surface of the rhombencephalon. The afferent fibres (termed radix sensoria nervi glossopharyngei et nervi vagi, rs IX,X) continue rostrally in a position *ventral* – instead of dorsal, as claimed by Jansen (1930) – to the sensory nucleus of the trigeminus towards the medullary horns (Figs. 11.10, 11.11, 11.19a). As mentioned above, the

ventral parts of the sensory nucleus of the trigeminus receive a number of collaterals from the sensory root of the vagus and glossopharyngeus. At the level of the motor nucleus of the trigeminus the sensory fibres sweep medially and caudally (Fig. 11.19a). Some fibres decussate in a commissure located beneath the caudal pole of the optic tectum (Matsuda et al. 1991). There are no experimental data on the efferent connections of the viscerosensory zone.

The viscerosensory system of hagfishes is very difficult to compare with that of other craniates.

There, the viscerosensory zone (represented by the solitary tract and nucleus) is a more or less U-shaped structure which embraces the calamus scriptorius of the fourth ventricle. The arms of the U point rostrally, decussating primary afferent fibres together with the nucleus commissuralis of Cajal connect both arms and thus form the commissura infima (Halleri) of the obex. The entire system is located close to the ventricular surface. In hagfishes, the viscerosensory nuclei have migrated from their periventricular site of origin towards the periphery of the rhombencephalon, where they can be found wedged between the general somatosensory nuclei dorsally and the branchiomotor nuclei ventrally. From a topological and embryological point of view (see Sect. 11.5.1), this is exactly the position in which one would expect to find a migrated viscerosensory column. However, the viscerosensory zone of hagfishes also shows a number of features that cannot be explained as the result of a simple radial migration of the viscerosensory neuroblasts. It should be recalled that the presence of gustatory afferents in the trigeminal nerve (see above) is an unusual feature; it should also be noted that the overall topography of the sensory root of the vagus and glossopharyngeus of hagfishes is different from that found in any other craniate. In contrast to the typical U-shaped appearance described above, the viscerosensory zone is M-shaped in hagfishes (Fig. 11.19a). The arms of the M point caudally, and the commissural connection between the arms does not occur in the caudal (obex) region of the rhombencephalon, but in an anterior region that would typically be occupied by the cerebellum in other craniates. Hagfishes possess no cerebellum (see Sect. 11.6), and it has been suggested (Matsuda et al. 1991) that the peculiarities in the organisation of the viscerosensory zone in hagfishes represent a transformation of the general craniate *Bauplan*, which is due to the absence of a cerebellum. However, it is also possible that the organisation of the viscerosensory system of hagfishes results from an evolutionary history that is independent of the evolution of viscerosensation in other craniates.

11.5.5
Motor nuclei

The branchiomotor nuclei are situated in the ventrolateral part of the rhombencephalon, where they constitute a distinct column of cells that is almost continuous from a cytoarchitectural point of view (Figs. 11.10–11.12, 11.19a). The axons of many of the cells within this column make a long dorsomedial loop before entering the peripheral nerves. This diversion indicates that during ontogenesis

these cells issued their axon before starting their ventrolateral migration. Röthig and Ariëns Kappers (1914) and Black (1917) believed that the ventrolateral shift of the visceromotor centres is a neurobiotactic phenomenon, brought about by the dominant influence exerted on these cells by the large descending root of the trigeminus. Jansen (1930), however, argued against this explanation, noting that the dendrites of the branchiomotor cells do not extend specifically toward the fibres of this tract. The same observation has been made in more recent experimental studies (see next paragraph) which have shown that the dendrites of the branchiomotor neurons are directed mainly ventromedially into the reticular formation.

Figure 11.19a shows the disposition of the branchiomotor nuclei in *Eptatretus* as determined experimentally by Kishida et al. (1986) and Matsuda et al. (1991). These authors distinguished, from rostral to caudal, a magnocellular trigeminal nucleus, an anterior and a posterior parvocellular trigeminal nucleus, a facial motor nucleus (which cannot be distinguished from the trigeminal nuclei in standard cytoarchitectural preparations), and a combined motor nucleus of the glossopharyngeal and vagal nerves. In contrast to earlier claims based on non-experimental studies (Addens 1933), the more recent studies did not allow the distinction between autonomic (parasympathetic) and branchiomotor elements within the branchiomotor column. If preganglionic parasympathetic motoneurons exist in the brain stem of hagfishes at all, they do not form nuclei of their own but are dispersed among the branchiomotor cells.

Because eye muscles and their nerves are absent in myxinoids, a true somatomotor zone cannot be observed in the anterior parts of the rhombencephalon. More posteriorly (Fig. 11.19a), however, the rostral part of the spinal somatomotor column extends into the caudal part of the rhombencephalon. It there forms the motor nuclei of the so-called spino-occipital nerves (Worthington 1905b; Matsuda et al. 1991), two nerves that innervate longitudinal muscles on the lateroventral side of the head.

Another peculiarity of hagfishes is the presence of an enigmatic nucleus located at the lateral surface of the brain stem, wedged between the parvocellular component of the trigeminal motor nucleus ventrally and the sensory nucleus and tract of the trigeminus dorsally (Figs. 11.11, 11.12, 11.19a). This nucleus was first described by Kusunoki et al. (1982), based on its high content of AChE, and was subsequently termed nucleus 'a' of Kusunoki (Kishida et al. 1986). This nucleus is also rich in neuropeptides (see Sect. 11.9.3). So far, there is no information on the connections of that nucleus;

however, the tracing studies carried out in the past decade have shown that it is neither a primary motor nor a primary sensory centre and it also does not seem to have projections that leave the rhombencephalon. Judging from this information and from its position between the sensory and the motor zone of the medulla, it might be interpreted as an integrative centre coordinating the activity in the sensory centres dorsally with the motor zone ventrally, or, in other words, it may represent a laterally displaced part of the rhombencephalic reticular formation. This interpretation is highly speculative, however.

11.5.6
Reticular Formation

The ventromedial region of the rhombencephalon is occupied largely by the reticular formation, i.e. the nucleus motorius tegmenti of Jansen (1930). Although the cells in this region are diffusely arranged, three moieties, termed here nucleus reticularis superior, medius, and inferior, can be delimited (Figs. 11.10–11.12, 11.19b,c). The first two of these nuclei contain numerous large neurons. Several authors, among them Black (1917) and Jansen (1930), have designated these elements as Müller cells. However, because these elements do not meet some of our criteria for Müller cells (constant position, occurrence in pairs, cf. Chap. 10), we prefer to term them simply large reticular neurons. It should be noted once more that Mauthner cells are also absent in myxinoids; this again contrasts with the situation in petromyzonts. The orientation of the dendrites of the large reticular cells strongly suggests that they receive their inputs from the spinal lemniscus, from the tectobulbar tract, and from the efferents of the sensory trigeminal nucleus and the octavolateral area (internal arcuate fibres, bulbar lemniscus). As mentioned in the preceding section, experimental studies have shown that all of the reticular nuclei send efferent projections to the spinal cord and that they also receive ascending spinal fibres via the spinal lemniscus (Ronan 1989; Ronan and Northcutt 1990, Figs. 11.19b,c).

As a final remark on the reticular formation, it should be noted that the rhombencephalic raphe of myxinoids, contrary to that of petromyzonts, contains distinct superior and inferior nuclei raphes (Figs. 11.10, 11.11). As mentioned earlier, the nuclei of the raphe project to the spinal cord (Ronan 1989, Fig. 11.19c).

11.5.7
Rhombencephalic Fibre Systems

The fibre systems of the rhombencephalon may be divided into:

1. The primary afferent bundles, which travel in the cranial nerves
2. Ascending afferents from the spinal cord which travel in the dorsal column pathway and in the spinal lemniscus
3. Descending afferents from the mesencephalon, which are supposed to arise from the tectum and the interpeduncular nucleus
4. Descending efferents to the spinal cord, which travel in the crossed and uncrossed bulbospinal tracts
5. Ascending efferents to the mesencephalon, which form the bulbar lemniscus

All of these fibre systems, except for the ones mentioned under 3, have been dealt with in the previous sections; consequently, we will limit our comments to the tectobulbar tract and the interpeduncular nucleus and its efferents.

The tectobulbar tract is composed of rather coarse fibres that descend from the tectum and pass caudally through the tegmentum and the rhombencephalon where the bundle runs ventrolateral to the medial longitudinal fasciculus (Figs. 11.10–11.12). The interpeduncular nucleus occupies a submeningeal position in the most ventral part of the mesencephalon (Figs. 11.13, 11.14). This nucleus receives afferents from the habenular region via the fasciculus retroflexus. Its very fine efferents fan out into the tegmentum of the midbrain as a tractus interpedunculo-tegmentalis; others pass caudally into the rhombencephalon as a tractus interpedunculo-bulbaris (Jansen 1930). Again, it should be noted that this information is based on nonexperimental material; neither the tecto- nor the interpedunculo-bulbar projection have been confirmed experimentally.

11.6
Cerebellum

The question of whether or not the myxinoids possess a cerebellum has received widely different answers in the literature. Holm (1901) and Edinger (1906) believed that a cerebellum was entirely lacking, but Holmgren (1919) claimed that the structure which we have identified as the mesencephalic tectum (Fig. 11.3a) actually represented a relatively large cerebellum. According to Holmgren, this 'cerebellum' contains a distinct commissure and closely resembles, with regard to both its afferent and efferent connections, its homologue in petro-myzonts. Like Holmgren (1919), Jansen (1930) stud-ied *Myxine* material, but the latter arrived at a widely different interpretation regarding the cere-bellum. Jansen pointed out that in myxinoids the lateral line organs are poorly developed and that the internal ear of this group is the simplest of any vertebrate. By way of comparison, Jansen noted that lampreys have a much better developed octavolater-al system than do myxinoids, but nevertheless a small cerebellum; consequently, it would be very surprising to find a large cerebellum in myxinoids. Accordingly, Jansen believed that the region denoted by Holmgren as 'cerebellum' represents the roof of the midbrain, thus supporting the interpre-tation given earlier by Holm and Edinger. More recent experimental data have confirmed the inter-pretation of Holm, Edinger, and Jansen. The 'cere-bellum' of Holmgren does not receive an octaval input typical for a vestibulocerebellum (see Sect. 11.5.2.), but it does receive a bilateral input from the retina (Kusunoki and Amemiya 1983; Wicht and Northcutt 1990), which identifies it as the homologue of the optic tectum of other crani-ates.

Larsell (1947), who studied the brain of *Eptatre-tus*, also concluded that an externally visible cere-bellum was lacking. He claimed, however, that a tiny homologue of a cerebellum was present at the meso-rhombencephalic border in the isthmic region, ventral to the caudal pole of the mesence-phalic tectum. In this area, he observed a small commissure and strands of cells accompanying that commisssure. These cells were regarded by Larsell as medially directed extensions of the octavolateral area, and he concluded that the commissure and accompanying cells he had found constituted the precursor of the octavolateral part of the cerebel-lum of petromyzonts. Bone (1963b) looked for Lar-sell's commissure in *Myxine*, but he remained unable to discern such a cerebellar rudiment in the latter species. In addition, the tracing studies on the connections of the octaval and lateral line nerves in *Eptatretus* (see Sect. 11.5.2) have shown that neither of these nerves projects to Larsell's cerebellum, and we must thus conclude that a cerebellum is actually lacking in hagfishes.

11.7
Mesencephalon, Pretectum, Posterior Tuberculum

It has already been mentioned that the mesen-cephalon of myxinoids is only relatively small. As in other parts of the brain, the cells are scattered over almost the entire span of the walls. Within this wide zone of grey the chief fibre systems occupy a posi-tion intermediate between the small ventricular cavity and the meningeal surface (Fig. 11.12). The diffusely arranged cells situated in the dorsal part of the midbrain represent the mesencephalic tec-tum. The tectum displays three indistinct layers, namely, an inner periventricular layer consisting of small cells, a thick cellular and fibrous layer which contains clusters of cells of various sizes and shapes (which is here called the central grey), and a cell-poor outer marginal layer (Iwahori et al. 1996). The outer layer receives a bilateral, but predominantly contralateral projection from the retinae (Kusunoki and Amemiya 1983; Wicht and Northcutt 1990; Fig. 11.20). The tectum also receives crossed projec-tions from the octavolateral area, from the sensory nucleus of the trigeminus, from the dorsal column nuclei via the bulbar lemniscus (Fig. 11.20), and from the contralateral tectum. Uncrossed afferents arise from the spinal cord via the spinal lemniscus (Figs. 11.19b, 11.20), from the infundibular hypo-thalamus and from the paracommissural nucleus of the thalamus (Amemiya 1983; Ronan 1988; Ronan and Northcutt 1990). Efferent projections from the mesencephalic tectum may reach the rhomben-cephalon via the tectobulbar tract (Jansen 1930); a tecto-spinal projection which occurs in many other craniates, is notably lacking in hagfishes and in lampreys (Ronan 1989).

The region ventral to the level of the mesence-phalic ventricle represents the mesencephalic teg-mentum. Apart from scattered small cells, this region contains distinct large cells which form the reticular mesencephalic nucleus. At the rostral bor-der of the tegmentum these large cells are con-densed and form the nucleus of the medial longitu-dinal fascicle (Figs. 11.13, 11.14, 11.21). As men-tioned above, these reticular elements are the source of a descending spinal projection (Ronan 1989). Ventrally, the tegmenta of both sides are con-nected by the ventral tegmental commissure (Figs. 11.12, 11.13). More anteriorly, at the meso-diencephalic border, dorsal to the plica encephalica ventralis, the commissural fibres condense to form a distinct commissure of the posterior tuberculum (Fig. 11.14). As in lampreys, there is no cytoarchi-tectural and no connectional (Ronan 1989) evi-dence for the presence of a nucleus ruber. Accord-ing to the analysis of Jansen (1930), the tegmentum of the midbrain receives afferents from a variety of sources, i.e. the rhombencephalic reticular forma-tion, the interpeduncular nucleus, the hypothala-mus, the dorsal thalamus, the cerebral hemispheres, and even from the olfactory bulbs. A few labelled fibres arising from the olfactory bulb have indeed been found in the most anterior parts of the teg-mentum after injections of tracers into the olfactory

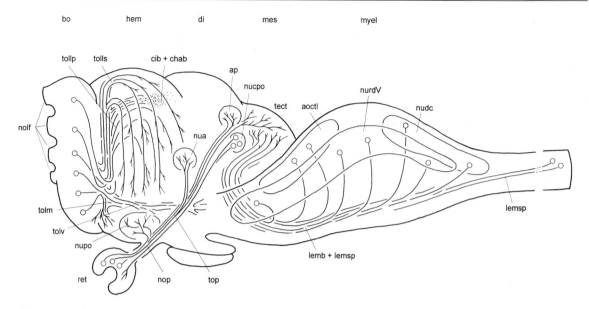

Fig. 11.20. A diagrammatic overview showing the organisation of secondary olfactory, visual, bulbar, and spinal lemniscal pathways in myxinoids (combined from various sources, see text)

bulb (Wicht and Northcutt 1993); however, it must be noted again that Jansen's other observations on the connections of the tegmentum still await experimental proof.

The most ventral part of the mesencephalon is occupied by the interpeduncular nucleus (Figs. 11.13, 11.14). It should be noted that Jansen (1930) identified a posterior ('rhombencephalic') and an anterior ('mesencephalic') part of that nucleus. Jansen's mesencephalic part of the interpeduncular nucleus corresponds to the medial nucleus of the posterior tuberculum in our interpretation (Fig. 11.14), since it displays ascending connections to the cerebral hemispheres (Wicht and Northcutt 1992a; Wicht et al. 1996). Jansen's posterior interpeduncular nucleus corresponds to our interpeduncular nucleus, as a whole. There are no experimental data on the connections of that nucleus; however, since a massive fasciculus retroflexus which enters this nucleus may be observed in standard histological material, it is very likely that the habenula is the main afferent source of the interpeduncular nucleus, as is the case in other craniates.

The posterior tubercular area (Fig. 11.14) consists of two nuclei, a well-defined medial nucleus [corresponding to the 'mesencephalic part of the interpeduncular nucleus' of Jansen (1930)] and an ill-defined lateral nucleus, connected across the midline by the well-developed commissure of the posterior tuberculum. As in other craniates, these nuclei display an ascending connection to the tel-

encephalon (Wicht and Northcutt 1992a; Wicht et al. 1996).

The pretectum (Fig. 11.13) is characterised by a conspicuous small-celled nucleus which is found laterally and dorsally adjacent to the posterior commissure, the area praetectalis of Jansen (1930). This nucleus was shown by Kusunoki and Amemiya (1983) and by Wicht and Northcutt (1990) to receive a dense bilateral retinal input. More ventrally, the nucleus of the posterior commissure of Jansen (1930) can be found. This nucleus gives rise to a prominent, predominantly contralateral retinopetal projection. In addition, retinopetal projections arise from an ill-defined cell group laterally adjacent to the nucleus of the posterior commissure (Wicht and Northcutt 1990). A very similar situation is encountered in lampreys, where two rostral mesencephalic cell groups give rise to a retinopetal projection: the so-called cell group M5 of Schober (1964) and the laterally adjacent mesencephalic reticular formation (Vesselkin et al. 1980, 1984). Thus, the retinopetal systems of hagfishes and lampreys probably are homologues, in spite of the fact that the two species differ considerably with respect to other features of their visual systems.

11.8
Prosencephalon: Diencephalon, Central Prosencephalic Complex, Telencephalon

11.8.1
Introductory Notes –
General Morphological Interpretation

Due to the extremely compact structure of the brain and the obliteration of the greater part of the ventricular system, most of the usual landmarks for the determination of the main diencephalic and telencephalic subdivisions are hard to recognise in myxinoids. However, as noted already by Edinger (1906), the ventricular cavities which are so pronounced in the embryonic forebrain (Figs. 11.5, 11.6a) do not disappear without traces: chords and rows of transformed ependymal cells indicate their former site (also see Fig. 11.21). Since information on the position of cell groups relative to ventricular landmarks is critical in any attempt to homologise prosencephalic cell groups, we provide a reconstruction of the ventricular system of the adult brain, including the vestigial parts of the ventricular system that have lost their lumen during ontogeny. This reconstruction is shown in Fig. 11.22. Together with the sections of the embryonic brain as shown in Figs. 11.5 and 11.6a, the reconstruction suggests the following general morphological interpretation of the myxinoid forebrain. The dorsal parts of both the telencephalon and the diencephalon undergo an evagination during ontogeny. This evagination results in the formation of telencephalic *and* diencephalic lateral ventricles and hemispheres; in the adult these ventricles loose their lumen but their former sites are still identifiable. The region between the diencephalic (posterior) and telencephalic (anterior) lateral ventricles (cf. Fig. 11.21) arises from a fusion of the anterior wall of the diencephalic vesicle with the posterior wall of the telencephalic vesicle (see Sect. 11.3.1). It contains a number of nuclei that are here termed the central prosencephalic complex. Since it is currently not possible to decide whether individual subnuclei of this complex are of diencephalic or telencephalic origin, the central prosencephalic complex will be dealt with in a separate section. The 'telencephalon proper' (i.e. those parts of the telencephalic vesicles that do not participate in that fusion) comprises the regions located dorsal, lateral, and anterior to the remnants of the telencephalic lateral ventricles; the 'diencephalon proper' consists of the regions located dorsal, lateral, and caudal to the remnants of the diencephalic lateral ventricles. Dorsally, the habenular commissure may serve as a landmark for the separation of diencephalic and telencephalic regions. The infundibular hypothalamus is the region which surrounds the hypothalamic ventricle and its recesses; the preoptic region is found lateral and anterior to the preoptic ventricle. The latter two regions are separated by the large postoptic commissure.

11.8.2
Epithalamus

There are no parietal organs in hagfishes; i.e., the paraphysis, the saccus dorsalis, the photoreceptive pineal (=epiphysis cerebri) and parapineal organs, which play a key role in the maintenance of circadian rhythms in other craniates, are lacking entirely in hagfishes. This absence has also been confirmed in immunohistochemical experiments using antibodies against photoreceptor-specific proteins: while the antibodies labelled the retinal photoreceptors, they failed to produce any staining in the epithalamic area (Wicht, unpublished observation). In accordance with these anatomical observations, the destruction of the epithalamic area does not affect the circadian rhythms of hagfishes (Ooka-Souda et al. 1993).

The habenular complex is large and its macroscopic appearance is that of a single body, wedged between the cerebral hemispheres and the thalamic lobes (Figs. 11.3a, 11.15–11.17). However, it is known that this structure is of a bilateral origin (Conel 1931). Microscopically, a habenular corpus and lateral and ventral habenular nuclei can be distinguished on each side. While the latter two nuclei are more or less symmetrical, the habenular corpus displays a striking asymmetry. The left corpus is small and displaced anteriorly, the right corpus is large and displaced posteriorly. All the habenular nuclei consist of small, densely packed cells; the habenular corpus displays distinct cell-free patches of neuropil. The epithalamus is traversed by two commissures, namely, the habenular commissure (Fig. 11.15) anteriorly and the commissura tecti diencephali posteriorly. The habenular commissure pierces the posterior part of the habenular complex; the commissura tecti diencephali is an ill-defined region of white matter between the the thalamic lobes.

According to Jansen (1930), the habenular complex receives afferent fibres from the following sources: (a) the olfactory bulb (via the superficial and deep portions of the lateral olfactory tract that decussate in the interbulbar and habenular commissures), (b) the pallial cortex (via the olfacto-habenular tract), (c) the central prosencephalic complex, (d) the preoptic region, and (e) the external and internal thalamic nuclei. It must be noted

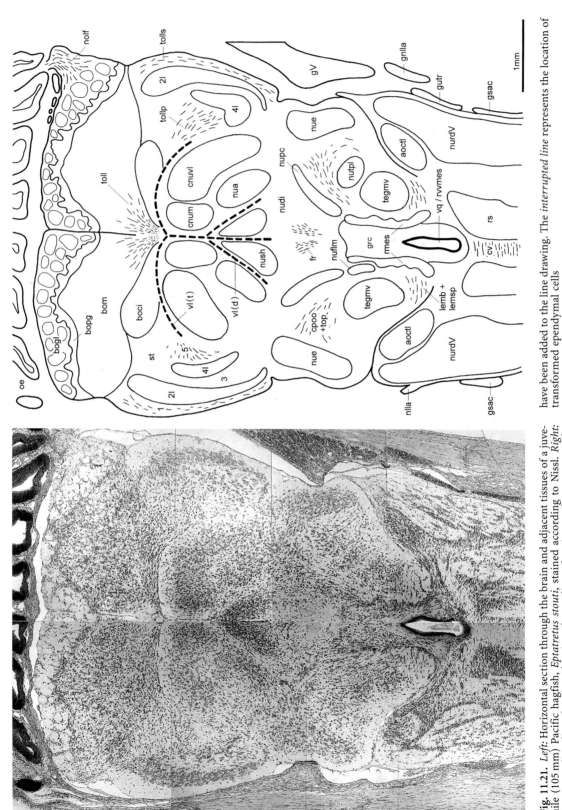

Fig. 11.21. *Left:* Horizontal section through the brain and adjacent tissues of a juvenile (105 mm) Pacific hagfish, *Eptatretus stouti*, stained according to Nissl. *Right:* Corresponding line drawing showing the outline of major nuclear centres. The position of some major fibre systems which are not visible in the section on the left have been added to the line drawing. The *interrupted line* represents the location of transformed ependymal cells

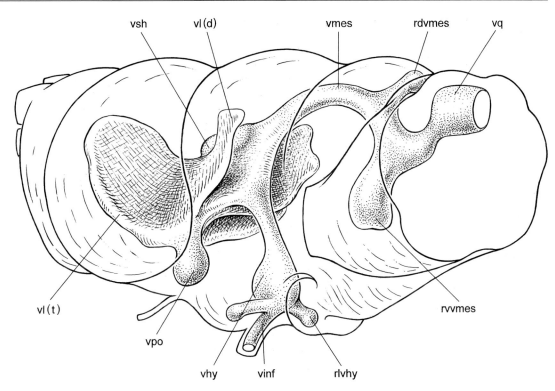

Fig. 11.22. A reconstruction of the ventricular system (*solid body*) in the brain (*outlines*) of a juvenile (105 mm) Pacific hagfish (*Eptatretus stouti*), based on serial horizontal sections and viewed from a position lateral, posterior, and slightly ventral to the brain. Anterior is to the *left*. Those parts of the ventricular system which retain a lumen and a typical ependyma are represented by *stippling*; the parts which loose their lumen during development and which are characterised by a transformed ependyma (compare with Fig. 11.21) are represented by the *cross-hatched areas*

that, so far, only the projections from the olfactory bulb and from the central prosencephalic complex have been confirmed experimentally (Wicht and Northcutt 1993; Amemiya and Northcutt 1996).

The largest efferent system of the habenular nuclei is the fasciculus retroflexus, a mostly thin-fibred bundle that passes caudoventrally to end in the interpeduncular nucleus (Figs. 11.13–11.16). As in gnathostomes, the terminal parts of its fibres follow a peculiar course; they cross and recross the median plane and coil around the interpeduncular nucleus before terminating in it. Other efferents from the habenular nuclei pass to the cell masses making up the ventral series of thalamic nuclei. Again, these observations of Jansen (1930) are based on non-experimental material of *Myxine glutinosa*.

11.8.3
Thalamus

Due to the obliteration of the ventricular system and the scarcity of connectional data, it is currently impossible to delineate with certainty a dorsal from a ventral thalamus in myxinoids. Thus, the entire thalamic area will be treated as a unit in the subsequent description, which is based mainly on Wicht and Northcutt (1992a). From a topographical point of view, the thalamus may be defined as comprising those nuclear masses that are located posterior, dorsal, and lateral to the obliterated diencephalic lateral ventricles (see Figs. 11.6, 11.21, 11.22). In addition, the large fibre system of the postoptic commissure, which also contains the fibres of the small optic tract, may be used as a landmark. Throughout its course from the chiasmatic region to the pretectum and tectum (Figs. 11.14–11.16) this commissure defines an oblique plane through the diencephalon. Thus the diencephalic nuclei may be grouped into a series that is dorsal to that plane (the dorsally situated thalamic nuclei), a series that is in the plane of the commissure (the ventrally situated thalamic nuclei), and the cell masses that are located ventral to that plane (the infundibular area).

The dorsally situated series of thalamic nuclei comprises the external, the internal, and the subhabenular thalamic nuclei (Figs. 11.14–11.16). The external thalamic nucleus occupies the dorsal aspect of the diencephalic lobes; it is bordered by the habenular commissure anteriorly and the pretectal area caudally. Ventral and medial to the external thalamic nucleus the internal thalamic nucleus is found. Rostrally, this nucleus is continuous with a conspicuous condensation of cells immediately ventral to the habenular complex, which is termed the subhabenular nucleus (eminentia thalami of Jansen 1930). Anteriorly, the subhabenular nucleus borders the remnants of the diencephalic lateral ventricles. In the median plane, the left and right subhabenular nuclei are separated by the remnants of the ventriculus impar (Fig. 11.21).

In the plane defined by the commissura postoptica and the optic tract, the following nuclei are found: lateral to the commissure there is a dense-celled nucleus with a distinct triangular outline, termed the triangular thalamic nucleus (Fig. 11.14). This nucleus was identified by Jansen (1930) as the nucleus of the posterior tuberculum; however, the available connectional data (see below) do not support this hypothesis. The postoptic commissure itself is populated by numerous neurons which intermingle with the commissural fibres and this group of cells has been termed the intracommissural nucleus of the thalamus. Directly medial to the commissure, a dense sheet of cells forms the paracommissural thalamic nucleus (Fig. 11.15). Still more medially, an ill-defined aggregate of neurons forms the diffuse nucleus of the thalamus. In the median plane the left and right diffuse nuclei of the thalamus are separated by remnants of the ventricular system and the fasciculi retroflexi (Fig. 11.21).

Jansen (1930) gave a detailed account of the fibre connections of the various diencephalic grisea based on observations he made in *Myxine glutinosa*; however, many of his observations have not been confirmed in experimental studies. We will therefore limit the following account to a brief summary of data from experimental studies. It is noteworthy that – in contrast to other craniates – neither the spinal, nor the bulbar, nor the trigeminal lemniscus ascend to diencephalic levels (Ronan 1988; Ronan and Northcutt 1990), nor do any of the diencephalic nuclei project to the spinal cord (Ronan 1989). The retinal input to the thalamic nuclei listed above is also quite restricted: the optic tract travels together with the fibres of the postoptic commissure; thus, the intracommissural nucleus

probably receives a retinal input. A few retinofugal fibres have been observed in the triangular thalamic nucleus (Kusunoki and Amemiya 1983; Wicht and Northcutt 1990). A small number of secondary olfactory fibres reaches the diffuse, the intra- and paracommissural, and the triangular nuclei of the thalamus (Wicht and Northcutt 1993).

Wicht et al. (1996) recently investigated the thalamo-telencephalic connections in *Eptatretus stouti*. There is an ascending projection from the internal and external thalamic nuclei, and the main target of these projections is the laminated pallium. Thalamo-striatal projections are sparse. However, the thalamic nuclei, particularly the subhabenular and the external thalamic nuclei, do receive a strong bilateral input from the pallium.

11.8.4
Hypothalamus, Preoptic Area, Optic Nerve and Tract

The hypothalamus of myxinoids is but slightly differentiated. It surrounds the hypothalamic ventricle and its recesses (Figs. 11.15, 11.23). The ventricle is lined by a dense layer of cells that represent intermingled ependymal and neuronal elements. Some of the neurons in this layer display cerebrospinal fluid (CSF) contacts (Jansen 1930), and many of these CSF-contacting cells contain monoamines (see below). However, these CSF-contacting cells do not form circumscribed circumventricular organs, but are dispersed throughout the entire periventricular layer. More laterally, a relatively ill-defined mass of migrated neurons represents the infundibular hypothalamic nucleus. There are few experimental data on the connections of the infundibular hypothalamus in myxinoids, but it has recently been shown to possess reciprocal connections with the central prosencephalic complex (Amemiya and Northcutt 1996) and to project to the mesencephalic tectum (Amemiya 1983). Hypothalamo-spinal projections are lacking (Ronan 1989).

The most rostral part of the hypothalamus contains a conspicuous commissural system, the postoptic commissure (Fig. 11.16). The crossing of fibres occurs near the ventral surface of the brain, immediately caudal to the entrance of the optic nerves. From there the bundle courses in a dorsal, lateral, and caudal direction through the diencephalon, passing laterally to the basal forebrain bundle. The origins and targets of the majority of the fibres in this commissure are unknown; however, it is known that the optic tract accompanies this commissural system on its way to the mesencephalic tectum (Kusunoki and Amemiya 1983; Wicht

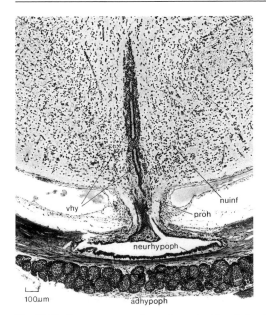

Fig. 11.23. Transverse section through the infundibular hypothalamus and the neuro-/adenohypophysial complex of a Pacific hagfish, *Eptatretus stouti*. Bodian preparation

and Northcutt 1990). The tiny optic nerves enter the ventral surface of the brain slightly rostral to the postoptic commissure and immediately caudal to the preoptic recess of the third ventricle (Fig. 11.17). Unlike in any other craniate, the optic nerves enter the brain *before* their decussation (Fig. 11.3c), which occurs in a chiasma located *inside* the brain. After entering the brain, the optic nerves give off numerous fibres to the cell masses which flank the preoptic recess laterally and anteriorly, i.e. the preoptic nuclei (Figs. 11.17, 11.20). Similar to the situation in gnathostomes, the preoptic region of eptatretid hagfishes seems to contain a circadian pacemaker (Ooka-Souda and Kabasawa 1988; Ooka-Souda et al. 1993) which may be triggered by those retinal inputs. After a partial decussation in the optic chiasm the fibres of the optic tract join the postoptic commissure, with which they ascend to their diencephalic, pretectal, and tectal targets (see above and Fig. 11.20). Other connections of the preoptic area include reciprocal connections with the central prosencephalic complex (Amemiya and Northcutt 1996) and projections arising from the dorsal series of thalamic nuclei, the pallium, and the olfactory bulb (Wicht and Northcutt 1993; Wicht et al. 1996; Wicht, unpublished observation). It should be noted that Jansen (1930) interpreted major parts of our preoptic region as the 'area basalis', i.e. the striatum. The connectional and physiological data which are summarised above (i.e. the presence of retinofugal

fibres and a circadian pacemaker) clearly speak against this interpretation.

11.8.5
Neuro- and Adenohypophysis, Neuroendocrine System

In myxinoids the neurohypophysis is represented by a dorsoventrally flattened sac attached to the ventral surface of the hypothalamus by a hollow stalk (Figs. 11.3d, 11.15, 11.23). Through this stalk the cavity of the neurohypophysis communicates with the hypothalamic ventricle. The walls of the neurohypophysis are built up of a single layer of tanycytes, ependyma-like glial cells, each of which consists of a perikaryon and a slender, peripherally directed process. Together, the perikarya form the inner zone of the neurohypophysis. The outer zone consists of the ependymal processes interspersed with a dense plexus of axons and terminals. The adenohypophysis is located ventral to the neurohypophysis. It consists of small nests of cells which are separated from each other and from the neurohypophysis above by layers of connective tissue (Fig. 11.23). The embryonic origin of the adenohypophysis is not entirely clear. According to von Kupffer (1900), the adenoypophysis (as well as the olfactory placode) derives from the *ectodermal* lining of a stomodeal recess that is found underneath the developing head, while Gorbman (1983) brought forward some evidence for a (unique) *endodermal* origin of the stomodeum, the olfactory placode, and the adenohypophysis in myxinoids. Our own observations in hagfish embryos (Wicht, unpublished observation) are in accord with the more 'conventional' view of von Kupffer (1900).

It is difficult to relate the existing data on the anatomy of the neuroendocrine system of myxinoids to the classical ideas on the organisation of such systems. These ideas and their anatomical substrates shall be recalled briefly here. Historically, a 'preoptico-neurohypophysial endocrine axis' is distinguished from a 'hypothalamo-adenohypophysial endocrine axis'. The former consists of cells in the preoptic region which produce the nonapeptide neurohormones of the oxytocin and vasopressin family (vasopressin = adiuretin = antidiuretic hormone = ADH). These hormones are transported via the tractus preoptico-hypophyseos to the neural lobe of the neurohypophysis, where they are released into the general circulation.

The hypothalamo-adenohypophysial endocrine axis is thought to regulate the endocrine activity of the adenohypophysis, which, in turn, controls the activity of other endocrine glands via its own hormones. The central nervous control of adenohypophysial activity occurs via a number of peptides (releasing hormones or -factors) which are proriety of preoptic and infundibular (tuberal) nuclei. The 'link' between the releasing-factor-producing neurons and the adenohypophysis may be established in different ways. In teleost fishes, the preoptic/infundibular endocrine neurons send their axons directly into the adenohypophysis. In amniotes, they project to a specialised region of the neurohypophysis (the so-called eminentia mediana). There, the releasing factors are given off into a portal system of vessels which carry them to the adenohypophysis. Finally, as is the case in lampreys, simple diffusion between the neuro- and the adenohypophysis may establish another route of hormonal communication between the two structures (Nozaki et al. 1994).

A number of data suggest that the hypothalamo-adenohypophysial endocrine axis may be absent in hagfishes. Many (but not all; see below) species of hagfishes lack a clear-cut anatomical connection between the neuro- and the adenohypophysis, the two being separated by a barrier of connective tissue (Adam 1963b; Holmes and Ball 1974). The number of neuroactive peptides (i.e. potential releasing hormones) which can be detected in the pituitary of hagfishes is surprisingly small and the levels of these peptides are quite low (Nozaki 1985). In addition, Matty et al. (1976) showed that removal of the hypophysis in *Eptatretus burgeri* had no noticeable effects on the gonads. Nozaki (1985) found no evidence for the presence of adrenocorticotrophic hormones in the adenohypohysis of hagfishes. Thus, the pituitary of hagfishes may also not play a role in the regulation of the hormone production of the adrenocorticoid glands. One might therefore argue that the hypothalamo-adenohypophysial-gonadal axis is lacking in hagfishes (Dodd and Dodd 1985; Tsuneki 1988; Sower 1990).

Some other data, however, speak in favour of the presence of a hypothalamo-adenohypophysial axis. From a purely morphological point of view, the connective tissue barrier between the neuro- and the adenohypophysis seems to prevent a hormonal communication between the two structures. Nevertheless, it is possible that neurohormones released from the neurohypophysis reach the adenohypophysis simply by diffusion through the connective tissue. In fact, both Nozaki et al. (1975) and Tsukahara et al. (1986) have shown that the connective tissue between the neuro- and adenohypophysis forms little if any barrier for the diffusion of even high-molecular-weight proteins from the neuro- to the adenohypophysis. Furthermore, in

Eptatretus burgeri, Kobayashi and Uemura (1972) and Tsuneki et al. (1976) observed some 20 small vessels passing from the ventral wall of the neurohypophysis to the adenohypophysis. The ventral part of the neurohypophysis, which lies dorsal to the adenohypophysis, has been claimed to have an ultrastructure comparable to that of the median eminence. In this region of *Eptatretus burgeri* Kobayashi and Uemura (1972) and Tsuneki et al. (1976) observed presumably neurosecretory axons containing relatively large (60–160 nm) secretory granules. These axons appeared to have terminals, often synaptoid in character, on the processes of the tanycytes, and the end-feet of the latter lie against the connective tissue that separates the neurohypophysis from the adenohypophysis, thus establishing a possible route for the release of hypothalamic factors into the adenohyophysis. Evidence for the presence of releasing hormones stems from immunocytochemistry. At least two substances which share epitopes with typical releasing hormones (gonadotropin-releasing hormone, somatostatin) have been detected in preoptic/hypothalamic perikarya and in the neurohypophysis (Blähser et al. 1989; Nozaki 1985; Braun et al. 1995; Sower et al. 1995). The adenohypophysial hormones of hagfishes – if they exist – are almost entirely unknown. Jirikowski et al. (1984) found some adenohypophysial cells that showed a positive immunoreaction for FMRFamide, but is is not known whether this substance plays a hormonal role in myxinoids. It is possible that the adenohypophysial hormones of hagfishes differ chemically from those of other craniates, and the (possible) endodermal origin (see above) of the myxinoid adenohypophysis may explain this difference.

Early evidence for the presence of a preoptico-neurohypophysial endocrine axis in hagfishes stems from data obtained with the paraldehyde-fuchsin staining method (Adam 1956, 1963b; Ollson 1959). This method allows the demonstration of neuropeptides containing disulphide-bridges, i.e., oxytocin and vasopressin, and their respective neurophysins which are the typical neurohormones of the preoptico-neurohypophysial system. In hagfishes, as well as in other craniates, the cells of origin of this system are situated laterally and caudally adjacent to the preoptic recess. In contrast to the situation in other craniates, where perikarya of the neurosecretory cells are often very large, they are about the same size as ordinary neurons located in this region in hagfishes (Tsuneki et al. 1974). Most of the neurosecretory fibres end in the dorsal wall of the neurohypophysis, which on this account may be regarded as the homologue of the neural lobe of gnathostomes. In this region, a rich accumulation of fuchsinophilic neurosecretory material lies in a dense bed of capillary loops draining into the general circulation (Holmes and Ball 1974). These observations suggest that neurosecretory material the preoptic region enters the general circulation via a neurohemal contact zone in the dorsal part of the neurohypophysis. The neurohormone in this preoptico-neurohypophysial system is arginine-vasotocin (Rurak and Perks 1974; Nozaki and Gorbman 1983; Nozaki 1985), and, similar to the situation in other craniates, this substance has an antidiuretic effect in hagfishes (Adam 1963b). Neurohormones of the oxytocin/mesotocin group seem to be entirely absent (Nozaki 1985).

11.8.6
Eyes and Retina

As mentioned in Sect. 11.1, the eyes show varying degrees of development in different genera of hagfishes (Fernholm and Holmberg 1975). Generally, species of the genus *Eptatretus* have relatively well-developed paired lateral eyes. Like in all hagfishes, however, the eyes lack lens, iris, external, and internal eye muscles. The eye muscles and the iris are missing throughout all stages of embryonic development; however, a lens placode occurs in the ectoderm adjacent to the optic cup in early embryos of *Eptatretus stouti* (von Kupffer 1900; Stockard 1907). This placode disappears without giving rise to a lens, and it has been speculated (Wicht and Northcutt 1995) that the placode may play a role in the prevention of skin pigmentation above the eye cup, thus producing the opaque eye spot typical of eptatretid hagfishes (Fig. 11.1). The eptatretid retina is inverted and layered, but the layering differs from that of a typical gnathostomian retina (see below). The neural retina is surrounded by an outer 'pigment' epithelium which consists of cuboidal cells, but, unlike in other craniates, this epithelium is *unpigmented* (Allen 1905; Dücker 1924; Holmberg 1971; Fernholm and Holmberg 1975). The receptor cells display well-developed outer segments containing stacks of membrane discs; these outer segments are attached to the receptor cell bodies by way of a connecting cilium (Holmberg 1971). The apical tips of the outer segments are often in close contact with the cells of the 'pigment' epithelium, and there are some signs of disc shedding and phagocytotic activity of the 'pigment'-epithelial cells (Holmberg 1971). The nuclei of the receptor cells are contained in an outer nuclear layer, which is separated from the inner nuclear layer by a thin outer plexiform layer (Fernholm and Holmberg 1975). Synaptic contacts between the receptor cells and the postsynapic

elements occur in the outer plexiform layer by means of spherical 'synaptic bodies' (Holmberg 1971); however, typical synaptic ribbons are absent. The (dendritic) morphology of the postsynaptic elements in the inner nuclear layer is unknown; thus, it is not clear whether bipolar, horizontal, and amacrine cells occur at all. About 25 % of the perikarya in this layer are those of glial Müller cells (Dücker 1924). Unlike in other craniates, there is no clear-cut inner plexiform layer, no ganglionic layer and no layer of optic nerve fibres. Instead, the area between the inner nuclear layer and the well-developed vitreal body is occupied by a thick fibrous layer with relatively large cell bodies randomly interspersed; these large elements probably represent ganglion cells, but their dendritic morphology is also entirely unknown. Notably, the retina of *Eptatretus stouti* is vascularised by numerous capillaries (Fernholm and Holmberg 1975).

In myxinid hagfishes, as exemplified by *Myxine glutinosa* (Retzius 1893b; Dücker 1924; Holmberg 1970), the eyes are hidden under layers of muscle and there is no unpigmented eye spot. The layering of the retina is much less pronounced than in Eptatretidae, and the membrane stacks in the outer segments of the photoreceptors show highly irregular and disorganised 'whorls' (Holmberg 1970). The Golgi study done by Retzius (1893b) revealed that the cellular morphology of the other retinal elements is barely comparable to the 'classical' appearance of retinal neurons. A vitreal body is absent.

The central connections of the retina of Eptatretidae have been dealt with above. In *Myxine glutinosa*, the retina displays essentially the same connections, but the number of retinofugal and retinopetal fibres is much lower than in *Eptatretus stouti* (Wicht, unpublished observation). From an anatomical point of view, the eyes of hagfishes do not seem to be capable of perceiving forms and shapes, and due to the absence of a pigmented epithelium, even the directional sensitivity may be rather limited. However, hagfishes in general and Eptatretidae in particular are certainly not 'blind' (Greene 1925): apart from their skin light sense (see Sect. 11.1), their eyes may play an important role in the maintenance of circadian (Ooka-Souda et al. 1993) and circannual (Fernholm 1974; Wicht and Northcutt 1990) rhythms.

11.8.7
Central Prosencephalic Complex

The central prosencephalic complex occupies a position between the vestiges of the telencephalic lateral ventricles anteriorly and those of the diencephalic lateral ventricles posteriorly. It consists of a number of nuclei, namely, the dorsal, the large-celled ventrolateral, and the medial nucleus (Figs. 11.15–11.17). In addition, a cell group that was formerly assigned to the thalamus (the anterior thalamic nucleus, Wicht and Northcutt 1992a) should be included in this complex, since this nucleus is clearly located rostral to the remnants of the diencephalic lateral ventricles (Fig. 11.21).

Historically – and up to the present day – the homology of the central prosencephalic complex has been the subject of many debates and a source of confusion and misunderstandings (reviewed extensively in Wicht and Northcutt 1992a, cf. also Chap. 6, Sect. 11.2.6.4). While all researchers have more or less agreed on the boundaries of this conspicuous cell group, there have been widely different proposals concerning its homology. It has been identified with a plethora of di- and telencephalic structures of other craniates, i.e., the nucleus rotundus (Holm 1901), the thalamus (Edinger 1906; Crosby and Schnitzlein 1974), the medial pallium (Holmgren 1919; Conel 1929, 1931; Jansen 1930), the eminentia thalami (Holmgren 1946; Kuhlenbeck 1977), and the striatum (Wächtler 1975).

In spite of a number of new topological, connectional and histochemical data it is still not possible to propose detailed homologies of the individual nuclei of this complex to single diencephalic or telencephalic nuclei of other craniates. It is clear, however, that Jansen's (1930) 'classical' proposal is in need of critical revision. He suggested a homology between the nuclei of the central prosencephalic complex and the medial pallium of other craniates. However, the topological and embryological data presented above (see Sects. 11.3.1 and 11.8.1) clearly speak against this homology. The complex as a whole is not located medial and dorsal to the lateral ventricles, but posterior to them. Unlike the medial pallium, it does not arise from the telencephalic vesicle alone. Due to the presence of paired rostral and caudal evaginations of the prosencephalic vesicle, and to the fusion of the anterior diencephalic with the posterior telencephalic walls, the complex probably receives contributions from both telencephalic and diencephalic matrix zones (see Fig. 11.6a). When it is first discernible in the embryo, the anlage of the central prosencephalic complex is actually located almost entirely in the anterior diencephalic wall (Conel 1931).

The few available connectional data also shed little light on the homology of the central prosencephalic complex. The dorsal nucleus which borders the laminated pallium receives a bilateral secondary olfactory input via the deep part of the lateral olfactory tract (Wicht and Northcutt 1993; Fig. 11.17). The anterior nucleus receives a bilateral retinal input (Wicht and Northcutt 1990; Figs. 11.16, 11.20). The connections of the medial and ventrolateral nuclei have been investigated by Amemiya and Northcutt (1996). They observed weak reciprocal connections with the septum, the habenula, and the most anterior part of the subhabenular nucleus of the thalamus. Stronger reciprocal connections were found with the preoptic nuclei and the infundibular hypothalamus. Injections of tracers into the thalamus (Wicht et al. 1996) did not reveal any ascending thalamic projections to the central prosencephalic complex.

Thus, the central prosencephalic complex shows some connectional characters which are similar to those of a typical medial pallium (reciprocal connections with the septum, the preoptic region, and the hypothalamus, moderately strong secondary olfactory input), but it also displays other characters which are dissimilar (absence of an input from the main olfactorecipient lateral pallium, absence of ascending thalamic projections). Finally, the histochemical data (see Sect. 11.9) all speak against a homology of the central prosencephalic complex with the medial pallium.

Given these data, we feel confident in rejecting Jansen's (1930) hypothesis, and we suggest the following interpretation of the complex: As a whole, and as is evident from its unique ontogeny, the complex must be rated as a myxinoid autapomorphy, i.e., it has no homologue in other craniates. Nevertheless, individual nuclei of the complex may be homologous to other nuclei in other craniates. Such a homology may be recognisable if a particular nucleus of the complex can be traced back to a particular (diencephalic or telencephalic) matrix zone, and if it displays certain connectional and/or histochemical characters which match with a nucleus of other craniates. In that respect, the anterior nucleus with its bilateral retinal input (Wicht and Northcutt 1990) might actually be a homologue of the same-named nucleus of other craniates; currently, however, the other subdivisions of the central prosencephalic complex cannot be homologised with any certainty.

11.8.8
Olfactory Bulb, Secondary Olfactory Tracts, Terminal Nerve

The *olfactory bulbs* consist of five layers which are arranged in a rostro-caudal sequence and which may be imagined as a series of bowls stacked with their convexities pointing rostrally (Fig. 11.21). The rostral surface of the olfactory bulbs is in close contact with the large nasal sac, from which they receive numerous (approximately ten per side) olfactory nerve bundles. These bundles enter the olfactory bulb at its dorsal and lateral margins and form a meshwork of fibres which covers the rostral surface of the bulb, thus forming the nervous layer. A typical and well-developed glomerular layer occurs caudally adjacent to the nervous layer; the subsequent thin, but densely packed layer of cells has been termed the periglomerular layer. The mitral layer is the thickest of all the layers of the olfactory bulb; still more caudally, the internal cellular layer is found. This last layer of the olfactory bulb is bordered caudally by the remnants of the telencephalic lateral ventricles (Fig. 11.21). A distinct secondary olfactory fibre layer, typical for the bulbs of many gnathostomes, is missing in hagfishes.

Holmgren (1919) and Jansen (1930) described the cell types of the olfactory bulb in *Myxine glutinosa*. Mitral cells which extend their dendrites into the glomeruli occur in all layers of the olfactory bulb. Two types of stellate cell can be distinguished, one that contacts the glomeruli with its dendrites and another that does not. The axons of the mitral cells and the two types of stellate cells together form the *secondary olfactory tracts*. In addition, the olfactory bulb contains other, multipolar elements which may play a role in interglomerular association (Jansen 1930); these cells may represent a homologue of the periglomerular cells of other craniates.

The observations of Jansen (1930) on the course and extent of the secondary olfactory projections have been largely confirmed by tract-tracing studies (Wicht and Northcutt 1993, Fig. 11.20), but some additional connections have also become evident. In the mitral layer, the secondary olfactory fibres initially converge towards the medial side of the bulb and a small medial olfactory tract splits off medially and ventrally to innervate the septal region (Fig. 11.18). The majority of the fibres, however, continue dorsally and laterally and form the lateral olfactory tract. This tract soon divides into a deep part, which runs in the fifth pallial layer, and a superficial part in the first pallial layer (Fig. 11.17). Thus, the pallium is 'sandwiched' between the deep and the superficial olfactory tract

and all of its subdivisions (see below) receive secondary olfactory inputs. The projection from the bulb onto the pallium is bilateral: a massive interbulbar commissure is found underneath the rostral pole of the habenula (Fig. 11.17). A large number of fibres of the lateral olfactory tract decussate in this commissure; some fibres also decussate in the habenular commissure, which is located a short distance caudal to the interbulbar commissure. The fibres thus reach the contralateral pallium, where they also divide into a deep and a superficial tract, innervating the same targets as the fibres on the ipsilateral side.

In addition to the medial and the lateral olfactory tracts, there is a ventral olfactory tract which accompanies the basal forebrain bundle and gives off fibres to the preoptic and the hypothalamic area; a few fibres even reach the anterior tegmentum of the midbrain (Fig. 11.20). A number of cells which project to the olfactory bulb are found in the vicinity of the ventral olfactory tract. There is no evidence for the presence of a *terminal nerve* and an *olfactoretinal system* in myxinoids.

The organisation of the secondary olfactory system of hagfishes shows a number of peculiarities. The presence of a medial and a ventral olfactory tract (innervating the septum and the basal forebrain, respectively) is a feature common to many craniates. However, the organisation of the lateral olfactory tract of hagfishes is unique. In all craniates, the lateral olfactory tract occupies a submeningeal position (comparable to the position of the superficial part of the lateral olfactory tract of hagfishes), but the lateral olfactory tract also displays a deep component in myxinoids that is located close to the (former) ventricular surface of the lateral ventricles (e.g. Fig. 11.17). Such a deep (periventricular) olfactory tract does not occur in any other craniate.

11.8.9
Pallium

The pallium covers the dorsolateral aspect of the hemispheres in a position mainly dorsal and lateral to the remnants of the lateral ventricles (Figs. 11.16–11.18). The cytoarchitectural organisation of the pallium is remarkable since it displays a cortical and laminar organisation. The neurons which constitute the pallium have migrated towards the surface of the brain, where they form distinct grey layers which alternate with layers of white matter (Figs. 11.21, 11.24). All in all, five layers, numbered 1 through 5, may be distinguished (Jansen 1930; Wicht and Northcutt 1992a).

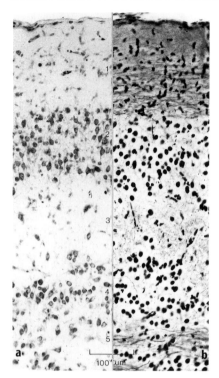

Fig. 11.24a,b. Transverse sections of the pallial cortex of a Pacific hagfish, *Eptatretus stouti*, stained according to Nissl (a) and Bodian (b). *1–5*, Pallial layers

The superficial layer of white matter (1) contains only very few neurons and consists mainly of the fibres of the superficial part of the lateral olfactory tract (Fig. 11.24). The next deeper layer, the superficial layer of grey matter (2, Fig. 11.24), can be subdivided into several subregions based on differences in the packing density and orientation of the cell bodies. These subdivisions are the parvocellular part (2pc, Fig. 11.18) anteriorly, the lateral part (2l, Figs. 11.16–11.18), the compact part (2co, Figs. 11.16, 11.17) medially, and the magnocellular part (2mc, Figs. 11.16, 11.17) ventrally. To a certain degree, these multiple subdivisions of layer 2 are also evident from its chemoarchitecture. Thus, a relatively large number of substance P-positive cells are found in the magnocellular, the parvocellular and the anterior lateral part of this layer; in addition, enkephalin-positive cells are found predominantly in the parvocellular and anterior lateral part of pallial layer 2 (Wicht and Northcutt 1994b). The following layer, the intermediate white layer (3, Fig. 11.24), appears to consist mainly of palliofugal fibres (pallio-habenular and -thalamic tracts, Figs. 11.16, 11.17) and of commissural fibres that connect the pallia of both hemispheres via the interbulbar commissure. The intermediate layer of grey matter (4, Fig. 11.24) consists of a laminar

lateral (4l, Fig. 11.17) and a nuclear medial (4m, Figs. 11.16, 11.17) part. Finally, there is a deep layer of mixed grey and white matter, the chief fibrous component of which is formed by the axons of the deep part of the lateral olfactory tract (5, Fig. 11.24). Medially, this layer borders the remnants of the telencepahlic lateral ventricles (Fig. 11.17).

The cytological structure of the various layers of the pallium was described in great detail by Holmgren (1919) and Jansen (1930). From their studies it appears that the cells of the second and fourth layer are similar. Most of them are bipolar or tripolar, but multipolar elements also occur. Many of the cells are arranged perpendicular to the surface and some extend their dendrites through all the pallial layers. Other pallial cells do not show any preference in the spread of their dendrites, and still others have their dendrites oriented parallel to the surface (Fig. 11.7a).

The secondary olfactory projections to the pallium have been dealt with above; thus, only the non-olfactory connections need be considered here (Wicht et al. 1996; Wicht, unpublished observation). Throughout its entire extent, the laminated pallium receives relatively weak ascending afferents from the external and the internal thalamic nuclei. Notably, these projections are not focussed onto a particular pallial field. In addition, the medial and the lateral nuclei of the posterior tubercular area project to the pallium; other afferents arise from the preoptic nuclei. The main efferents of the pallium are directed to the preoptic nuclei, to the habenula, the dorsal thalamus, and the contralateral pallium (via the interbulbar commissure).

The comparison of pallial organisation in myxinoids and other craniates is problematic. In all gnathostomes (except teleosts) the pallium can be divided on topographical and cytoarchitectural grounds into medial, dorsal, and lateral fields with typical connectional properties. In hagfishes, however, this tripartite subdivision is not evident from the cytoarchitecture or from the connections of the laminated pallium. Thus, two hypotheses need to be considered:

1. The pallium of hagfishes is a field homologue of two or all three pallial fields of gnathostomes. In this case, one would expect that the pallium of hagfishes, as a whole, has properties which appear as separate characters of individual pallial fields in other craniates. This is indeed the case with regard to some features of the pallium of hagfishes: it combines characters of the gnathostomian lateral pallium (e.g. secondary olfactory afferents) with those of the medial pal-

lium (e.g. ascending thalamic afferents, interhemispherical connections, fornical connections to the preoptic area, long descending connections). One might thus regard the pallium of hagfishes as an undivided field homologue of the medial and lateral pallia of other craniates and lampreys, in which these pallial fields exist as separate cytoarchitectural entities with separate connectional properties (Polenova and Vesselkin 1993; Northcutt and Wicht 1997).

2. It is equally likely that the pallial organisation of hagfishes is entirely autapomorphic. In this case the pallium should display characters which are not found in any other craniate. In some respects, this is so. The laminar organisation of the pallium itself and the presence of multiple subfields within each lamina (particularly evident in layer 2) are unique features. From a connectional point of view, the presence of a deep lateral olfactory tract (see above) is another unique character which does not occur in any other craniate.

The solution of the problem will require additional connectional and embryological studies. In particular, it will have to be clarified whether and how the multiple subdivisions of the pallium relate to connectional or chemoarchitectural features. In addition, it will have to be investigated whether the adult pallium arises from one or from many anlagen in the embryo.

11.8.10
Subpallium

In contrast to the pallium, which is highly developed in hagfishes, the subpallium is but slightly differentiated and consists of a septum and a striatum. The unpaired septum is wedged between the olfactory bulbs on the ventral side of the brain, immediately rostral to the preoptic area (Fig. 11.18). It receives a secondary olfactory projection via the medial olfactory tract (Wicht and Northcutt 1993, Figs. 11.18, 11.20) and has weak reciprocal connections with the medial and ventrolateral nuclei of the central prosencephalic complex (Amemiya and Northcutt 1996). In spite of the positional and connectional similarities to the septum of other craniates, the septum of myxinoids shows none of the typical chemoarchitectural features; i.e., it has very low contents of AChE, tyrosine hydroxylase, and neuroactive peptides (Wicht and Northcutt 1994b). Thus, the proposed homology to the gnathostome septum (Wicht and Northcutt 1992a) is somewhat doubtful.

The striatum occupies a position ventral, lateral,

and anterior to the remnants of the lateral telencephalic ventricles (Figs. 11.17, 11.21). Caudally, it merges with the preoptic area without a sharp boundary. In histological preparations stained for fibre tracts, the basal forebrain bundle can be traced to and from this area into the diencephalon. To date, there are no experimental data on its connections. It is noteworthy, however, that the striatum, as well as the other parts of the telencephalon, does not have long descending connections to the rhombencephalon and spinal cord, as is the case in many other craniates.

11.9
Chemoarchitecture of the Brain

11.9.1
Cholinergic Neuronal Populations

The cholinergic neuronal populations are known from histochemical studies utilising AChE activity as a marker for cholinergic systems (Wächtler 1975, 1983; Kusunoki et al. 1981, 1982; Wicht and Northcutt 1994b; see Fig. 11.25a). However, this method does not allow us to differentiate with certainty between neuronal populations that are cholinergic and those that are postsynaptic to a cholinergic input. Data on the distribution of the synthetic enzyme of ACh, i.e. cholineacetyltransferase, are currently not available. Thus, the following description of the cholinergic system needs to be read under the caveat that some 'AChE-positive' neurons may actually be postsynaptic to other cholinergic cells.

Throughout the rhombencephalon, the cell bodies of the branchiomotor nuclei display strong AChE activity, as do the large reticular cells wich are dispersed throughout the mesencephalon and rhombencephalon. A conspicuous condensation of AChE-positive cells is found in the rostral rhombencephalon, slightly dorsal to the motor nuclei of the trigeminal and facial nerves. It is known as nucleus 'a' of Kusunoki (see also Figs. 11.11, 11.12); in fact, this nucleus was first described on account of its high content of AChE. Even though nucleus 'a' shares this character with the motor and premotor centres (branchiomotor neurons and large reticular cells) of the rhombencephalon, the connectional data indicate that it is neither a motor nor a premotor centre (see Sect. 11.5.5).

In the mes- and prosencephalon, dispersed AChE-positive cells and fibres are found in most areas, including the various layers of the pallium. The habenula and the fasciculus retroflexus display strong activity. The highest activity of AChE, both in the neuropil and in cell bodies, however, is found

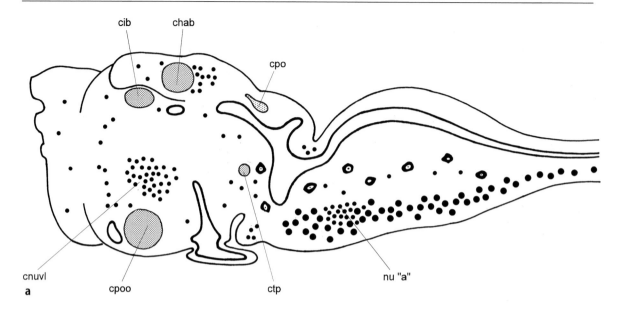

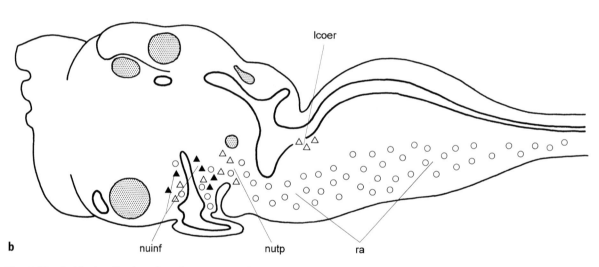

Fig. 11.25a–d. The localisation of perikarya positive for certain neurotransmitters/-modulators and neurohormones, projected onto the mid-sagittal plane. *Shaded areas* represent major commissures which have been added for the sake of orientation. **a** Cholinergic perikarya. **b** Monoaminergic perikarya: *circles*, serotoninergic cells; *open triangles*, cells positive for tyrosine hydroxylase; *solid triangles*, dopaminergic cells. **c** 'hypothalamo-hypophysial' neuropeptides: *triangles*, gonadotropin-releasing hormone-positive cells; *circles*, vasotocin-positive cells; *dots*, somatostatin-positive cells. **d** 'Other' neuropeptides: *open triangles*, FMRF-positive cells; *solid triangles*, enkephalin-positive cells; *circles*, substance P-positive cells; *dots*, neuropeptide Y-positive cells; *stars*, natriuretic peptide-positive cells (combined from various sources, see text)

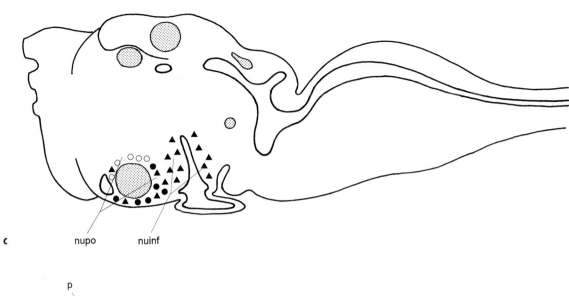

c   nupo   nuinf

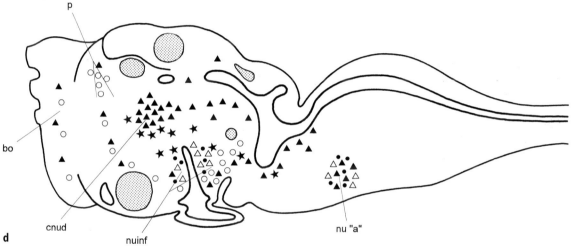

d

bo

p

cnud

nuinf

nu "a"

Fig. 11.25c, d

in the ventrolateral subnucleus of the central prosencephalic complex. Together with the data on the
peptidergic system (see Sect. 11.9.3) this speaks
against the pallial nature of that complex, since the
highest content of AChE in the telencephalon is typically found in the subpallium, i.e. the septum and
striatum. Nevertheless, the nucleus cannot be interpreted as being the homologue of either the septum
or the striatum, since the available topological and
connectional data speak against this interpretation
(see Sect. 11.8.7).

11.9.2
Monoaminergic Neuronal Populations

The distribution of monoamine-containing systems
in the brain was initially studied using monoamine-
oxidase (MAO) activity as a marker for those sys-

tems (Kusunoki et al. 1981, 1982); more recently,
data stemming from immunocytochemical investigations have become available (Kadota 1991;
Kadota et al. 1993; Wicht and Northcutt 1994b; see
Fig. 11.25b).

Serotoninergic cells occur in the diencephalon
and surround the various recesses of the hypothalamic ventricle; many of those cells contact the CSF
via typical, club-like processes. More caudally, the
majority of the serotoninergic cells is found in the
median plane ventral to the ventricular system, i.e.
in the superior and inferior nuclei of the raphe, corresponding to the mammalian cell groups B1-B7.
These cells do not display CSF contacts. Notably, the
presence of a serotoninergic midline raphe is a
character shared by hagfishes and all other craniates, but not by lampreys. In the latter group, the
raphe does not contain any serotoninergic cells

(Steinbusch and Nieuwenhuys 1979; Pierre et al. 1992; cf. Chap. 10, Sect. 10.10).

The catecholaminergic (Kadota et al. 1993; Wicht and Northcutt 1994b) system is much less wide-spread; actually, it is more restricted than in most other craniates. The synthetic enzyme for DOPA, tyrosine hydroxylase (TH), is found in cells which surround the ventral parts and recesses of the hypothalamic ventricle and – like the seroton-inergic cells – these cells often display CSF contacts. More caudally, TH-positive perikarya occur in the nuclei of the posterior tuberculum. As mentioned above, these nuclei give rise to an ascending projection to the forebrain (Wicht et al. 1996). They might therefore be homologues of cell groups with similar connectional and neurochemical properties that are found at the di-/mesencephalic junction in other craniates, such as the catecholaminergic cell groups of the posterior tubercular area in many anamni-otes and/or the well-known substantia nigra and ventral tegmental area (A8-A10) of amniotes. In the brain stem, another group of weakly TH-positive cells is found in an ill-defined region medial to the octavolateral area and ventral to the caudal pole of the mesencephalic tectum (Wicht, unpublished observation). These cells might constitute a homo-logue of the catecholaminergic locus coeruleus (A6) of other craniates; however, there are no connec-tional data to support this claim.

Perikarya containing dopamine seem to be restricted to the infundibular hypothalamus, where they occur in positions similar to those of the TH-positive hypothalamic cells (Kadota et al. 1993).

As mentioned above, the distribution of cate-cholamines in the brain of myxinoids is quite restricted. In most other craniates investigated to date, the olfactory bulb contains a relatively large number of catecholaminergic cells and the subpal-lium receives a heavy catecholaminergic innerva-tion. There are no such cells in the olfactory bulb of myxinoids, and the catecholaminergic innervation of the subpallium is present, but weak (Wicht and Northcutt 1994b).

11.9.3
Peptidergic Neuronal Populations

In the past two decades, an amazing array of neu-roactive peptides has been discovered in the brains of craniates, and it appears that most neurons in the CNS utilise – in addition to the 'classical' neuro-transmitters – at least one peptide for neuro-transmission or neuromodulation. The 'systemat-ics' of neuropeptides, their grouping according to anatomical or biochemical criteria, is still a field of very active research. The situation is complicated

by the fact that many of these peptides undergo complicated processes of post-transcriptional pro-cessing and cleavage. In addition, a single biologi-cally active neuropeptide, such as gonadotropin-releasing hormone (GnRH), may occur in isoforms that show varying amino acid sequences, and these isoforms may or may not co-exist in the brain of a given species. Therefore, the immunohistochemical studies of the anatomy of peptidergic systems of hagfishes in particular and craniates in general suf-fer from the problem of antibody specifity: a posi-tive result may be due to a reaction of the antibody with the 'target' peptide of that antibody, however, it may also result from cross-reactions with precur-sors and isoforms. Similarly, a negative result may be due not to the absence of the neuropeptide, but to the presence of an isoform which is not recog-nised by the antibody. Since most antibodies are developed in mammals, it may not be surprising that some 'common' neuropeptides, like most pro-opiomelanocortin-derived peptides, have not been detected in the brain of myxinoids.

For the purpose of the present survey, we have grouped the available data on the peptidergic system quite roughly and arbitrarily into 'hypothalamo-hypophysial' and 'other' neuropep-tides (see Fig. 11.25c and d). Among the classical hypophysial neuropeptides and -hormones, only arginine-vasotocin, somatostatin, and GnRH have been discovered in the brain of hagfishes. However, it is not known whether they play the same role as neurohormones and releasing factors, respectively, that they do in other craniates (see above, Sect. 11.8.5). Vasotocin and somatostatin are found in perikarya of the preoptic nuclei (Nozaki and Gorbman 1983; Nozaki 1985); GnRH-positive peri-karya also occur in the preoptic nuclei and in the infundibular hypothalamic nucleus (Braun et al. 1995; Sower et al. 1995). The neurohypophysis receives a dense innervation by vasotocin and GnRH-positive fibres and terminals, but there are no somatostatin-positive fibres in the neurohy-pophysis. There is a widespread system of GnRH-positive fibres and terminals throughout the entire brain, indicating a possible neuromodulatory role of this substance in myxinoids. Notably, there are no GnRH-positive cell bodies in the vicinity of the olfactory nerve, at the base of the olfactory bulb, or in the basal telencephalon, and there are no GnRH-positive fibres in the optic nerves (Braun et al. 1995). These GnRH-positive structures are com-monly regarded as indicators for the presence of a terminal nerve; therefore, it may be concluded that this nerve is lacking in myxinoids, particularly since another 'marker' for the terminal nerve, FMRFamide, is also not found in the above-mentioned locations (Jirikowski et al. 1984; Wicht and Northcutt 1992b).

The 'other' neuropeptides (Fig. 11.25d, data from Jirikowski et al. 1984; Dores and Gorbman 1990; Chiba and Honma 1992; Donald et al. 1992; Wicht and Northcutt 1992b; Chiba et al. 1993; Wicht and Northcutt 1994b) are distributed widely throughout the entire brain, with conspicuous condensations of positive cell bodies in three regions. (a) The infundibular hypothalamic nucleus contains the greatest variety of neuropeptides. However, (b) perikarya of nucleus 'a' of Kusunoki also contain at least three neuropeptides (FMRFamide, NPY, and both met- and leu-enkephalin); in addition, the neuropil of that nucleus receives a dense innervation by substance P-positive fibres. (c) Finally, there is a dense aggregation of neuropeptide-positive structures in the subnuclei of the central prosencephalic complex. A very large number of met- and leu-enkephalin-positive cells and terminals is found in the dorsal and ventrolateral subnuclei of that complex. In addition, there are cells positive for natriuretic peptide in the ventrolateral subnucleus. The entire complex also receives a dense innervation by substance P-positive fibres and terminals. The occurrence of such a large number of neuroactive peptides and the high density of neuropeptide-positive structures in the central prosencephalic complex adds further evidence to support the claim that this complex is not homologous to pallial structures of other craniates (see Sect. 11.8.7), since those typically contain only few dispersed neuropeptide-positive structures, as does the laminated pallium of hagfishes.

11.9.4
Other Neurotransmitters, Amino Acids

Other neurotransmitters, in particular the inhibitory γ-aminobutyric acid (GABA) and other amino acids such as taurine, aspartate, glycine and glutamate, have not been investigated in the brain of hagfishes. However, since the degrading enzyme of glutamate, glutamine synthetase, has been located in glial cells and processes in practically all parts of the brain of hagfishes (Wicht et al. 1994), it can be assumed that glutamate plays an important role as a widespread neurotransmitter.

11.10
Phylogenetic Interpretation and Concluding Remarks

The foregoing analysis of the myxinoid CNS should have made clear that they do *not* "...have...a simpler pattern, a more primitive arrangement of brain parts than...the higher forms previously studied..." (Worthington 1905b) and that they do *not* "...reveal great concordance with the conditions in generalized vertebrate brains, like those of petromyzonts and Amphibia" (Jansen 1930). It rather appears as if the concluding comment of an almost forgotten study on the brain of myxinoids (Sanders 1894) was correct: "At any rate, the conclusion to which I come, is that, judging from the nervous system, *Myxine* is an offshoot and is not in the main phylon of the vetebrata [sic]." Sanders misspelled vertebrata, but he was right after all.

Hagfishes and lampreys, formerly believed to be closely allied (hence the old taxon 'Cyclostomata/ Agnatha'), are the sole survivors of a once flourishing radiation of ante-Devonian jawless craniates. The common ancestor of all craniates doubtlessly lived among this Silurian and Devonian stock of heterostracans, osteostracans, conodonts, galeaspids, anaspids, pituriospids, arandaspids and thelodonts. However, recent paleontological, comparative anatomical and physiological research (Løvtrup 1977; Janvier 1974, 1981, 1995; Jefferies 1986; Bardack 1991; Briggs 1992; Sansom et al. 1992; Aldridge et al. 1993; Forey and Janvier 1993; Gabbott et al. 1995) has shown that hagfishes split off from that lineage before the 'invention' of a number of typical craniate characters such as cellular bone (possessed, for example, by conodonts and gnathostomes and believed to be secondarily lost in lampreys), external eye muscles (also present in conodonts, lampreys and gnathostomes), neural and hemal vertebral arches (shared by lampreys and gnathostomes), and active osmoregulation (shared by lampreys and gnathostomes). Therefore, more recent taxonomical schemes (Fig. 11.26, bottom) suggest that lampreys are much more closely related to gnathostomes than hagfishes are to either group, and a new taxon 'Myopterygii', comprising lampreys and gnathostomes, has been established.[5] The name of the taxon is derived from another character shared by lampreys and gnathostomes, i.e., the presence of striated muscle at the base of the fin rays. Myopterygians and hagfishes together form the Craniota; the sister-group of the latter taxon are the Cephalochordata. Cephalochordata and Craniota together form the Chordata.

With regard to the phylogenetic interpretation of hagfish CNS characters, it must therefore be borne in mind that myxinoids have had an evolutionary history that has been independent from that of myopterygians for at least 500 million years –

---

[5] Alternative views of craniate interrelationships are discussed in Yalden (1985), Caroll (1988), and Stock and Whitt (1992).

	Distribution of 3 hypothetical characters a, b, c				Cladistic analysis: polarity of characters *at the level of the taxon Craniota*
	Cephalochordata	Myxinoidea	Petromyzontoidea	Gnathostomata	
(i)	any character or character absent	a	a	a	a is plesiomorphic
(ii)	character absent	a a	a b	b a	a is probably plesiomorphic
	a a	a a	a b	b a	a is probably plesiomorphic
	b b	a a	a b	b a	polarity cannot be resolved
(iii)	character absent	a	b	b	polarity cannot be resolved
	a	a	b	b	a is probably plesiomorphic
	b	a	b	b	b is probably plesiomorphic
(iv)	character absent	a	b	c	polarity cannot be resolved
	one of the characters (e.g. a) is present	a	b	c	a is probably plesiomorphic
(v)	character absent	a	b	a and b	polarity cannot be resolved
	a	a	b	a and b	a is probably plesiomorphic
	b	a	b	a and b	b is probably plesiomorphic

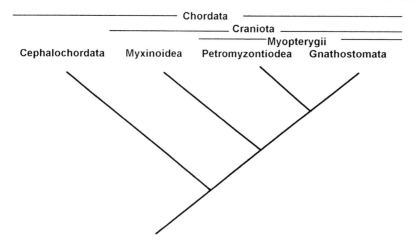

Fig. 11.26. *Bottom*: A cladogram showing the genealogical and taxonomical interrelationships of the major Recent chordate taxa. (from various sources, see text). *Top*: Table shows a 'distribution matrix' of three hypothetical characters *a*, *b*, and *c* among the major chordate taxa. The column on the *right side* contains the results of a cladistic analysis of the respective character distribution in each corresponding row. The analysis is carried out at the level of the taxon Craniota. See text for a detailed description

plenty of time for the acquisition of unique and derived characters. Hence, they cannot be taken as a model for a primitive craniate. They can, however, be compared with other craniates, and a hypothetical model of an ancestral craniate brain, a 'morphotype' (Northcutt 1985; Wicht and Northcutt 1992a), can be constructed from such a comparison. The morphotype of the craniate brain consists

of all brain characters which are primitive for that group; thus, it is necessary to determine the polarity (primitive or plesiomorphic vs. derived or apomorphic) of differences in character states between the individual taxa. This is done by examining and comparing the distribution of the character differences utilising a cladogram. The 'rules' of the comparison are relatively simple and are determined by cladistic (probabilistic) reasoning (see Fig. 26, also see Wicht and Northcutt 1992a and Chap. 6, Sect. 6.3.3); i.e., one assumes that a certain character out of a set of characters is plesiomorphic if this assumption minimises the number of evolutionary transformations required to explain the distribution of the characters among the Recent taxa. If one tries to reconstruct the morphotype of Craniota, it is also necessary to analyse the characters of the sister-group of Craniota, the Cephalochordata ('out-group comparison'). Quite frequently and not surprisingly, however, such an out-group comparison fails, because – as compared with craniates – the head and brain of cephalochordates are relatively undifferentiated; they thus lack many of the characters or even entire character complexes that are typical of craniates. Theoretically, the following distribution of characters will lead to the following conclusions:

I.   Any character 'a' which is present in all three craniate taxa is probably primitive for craniates, irrespective of the presence or absence or state of the character in cephalochordates ([i] in Fig. 11.26).

II. Any character 'a' which is present in hagfishes and either lampreys or gnathostomes is probably primitive for craniates, if cephalochordates do not display the character complex. If cephalochordates do display a similar character, the polarity can be established only if they possess character 'a', which is then plesiomorphic for craniates. If they display 'b' or another character, the polarity cannot be resolved ([ii] in Fig. 11.26).

III. The polarity of characters shared by lampreys and gnathostomes but not by hagfishes can be established only if cephalochordates display one of the characters. If they do, the character they possess is probably plesiomorphic for craniates. Irrespective of the condition in cephalochordates, characters shared by lampreys and gnathostomes are plesiomorphic for the taxon Myopterygii ([iii] in Fig. 11.26).

IV. Hagfishes, lampreys, and gnathostomes may display entirely different characters. In this case, a cladistic analysis cannot reveal the polarity of the characters at the level of craniates unless one of the characters is present in cephalochordates ([iv] in Fig. 11.26).

V.   Quite frequently, one encounters characters which are shared between hagfishes and some gnathostomes, and lampreys and some other gnathostomes, but not between hagfishes and lampreys ([v] in Fig. 11.26). Under these circumstances, it is very difficult to determine the polarity of the characters, as events of conver-

gence and parallelism must be taken into account. Since lampreys and gnathostomes are more closely related to each other than to hagfishes, one might argue that parallelism is more likely to occur among the closely related taxa (the Myopterygii), thus favouring the assumption that the 'hagfish' character is primitive and the 'lamprey' character derived (alternatively, a parallelism between hagfishes and gnathostomes would have to be considered). However, a strict cladistic solution to the problem requires the analysis of the next out-group, the cephalochordates. As mentioned above, this analysis frequently fails because cephalochordates lack many of the character complexes typical of craniates.

The results of the cladistic reconstruction of the morphotypes of the craniate and the myopterygian brains are summarised below and in Table 11.1.

I. The polarity of a large number of CNS characters can be determined because they are present in all three craniate taxa, and they must therefore be regarded as plesiomorphic for craniates. The division of the CNS into brain and spinal cord, the vascularisation of the brain, and the histological differentiation into glial and neuronal cells are such characters. Plesiomorphic spinal characters are the possession of 'true' ventral roots consisting of axons arising from motoneurons in the spinal grey, sensory ganglia in the dorsal roots of the spinal nerves, reticulo-, octavo-, and raphe-spinal projections, ascending spino-bulbar and spino-reticular projections. Plesiomorphic brain characters are the division of the brain into rhombencephalon, mesencephalon, diencephalon and telencephalon, the presence of cranial nerves I, II, V, VII, VIII, IX, and X and of at least two lateral line nerves. The morphotype includes a longitudinal arrangement of sensory (dorsal) and branchiomotor (ventral) zones in the rhombencephalon and ascending secondary sensory fibres from the octavolateral and the trigeminal sensory columns to the mesencephalic tectum and tegmentum. In the forebrain, plesiomorphic characters are the presence of a habenula, fasciculus retroflexus, and interpeduncular nucleus, and bilateral retinofugal projections to preoptic, thalamic, pretectal and tectal areas. A specialised neuroendocrine area at the base of the forebrain, consisting of preoptic and infundibular nuclei which show a high content of neuropeptides, and a neurohypophysis are also part of the morphotype. The presence of at least one circumventricular organ, i.e. the subcommissural organ, and Reissner's fibre is plesiomorphic for craniates. Furthermore, an olfactory bulb with typical glomerular

Table 11.1. Craniate and myopterygian morphotypes

	Craniate morphotype	Myopterygian morphotype
General	CNS divided into brain and spinal cord, neurons and glia present, axons unmyelinated, vascularised brain	
Ventricular system	?	Well developed in the adult, plexus chorioidei present
Spinal cord	*Present:* dorsal and ventral roots, dorsal root ganglia, intraspinal sensory cells, giant interneurons	
Ascending and descending spinal fibre systems	*Present:* dorsal column pathway to dorsal (alar) rhombencephalic areas, spinal lemniscal pathway to reticular nuclei and mesencephalic tectum; reticulo-spinal, raphe-spinal, octavo-spinal projections *Absent:* spinal lemniscal pathway to diencephalon, tecto-spinal, and telencephalo-spinal projections	
	?	*Present:* (limited) diencephalo-spinal projections, trigemino-spinal projections
Rhombencephalon	*Present:* cranial nerves V, VII, VIII, IX, X, two lateral line nerves, dorsal sensory areas (sensory nucleus of the trigeminus, ventral and medial nuclei of the octavolateral area, dorsal column nuclei), ventral branchiomotor column, intrarhombencephalic sensory cells	
	?	*Present:* cranial nerves IV, VI, Mauthner cells, Müller cells, 'typical' viscerosensory system, 'typical' somatotopy in sensory nucleus of V, dorsal nucleus of the octavolateral area
Cerebellum	?	Possibly present as a small vestibulocerebellum
Mesencephalon	*Present:* division into tectum and tegmentum, ventrally situated interpeduncular nucleus, posterior tubercular cell groups with ascending projections to telencephalon, distinct posterior commissure at rostral border of tectum, distinct commissure of the tuberculum posterius at the rostral border of the tegmentum, subcommissural organ and Reissner's fibre *Absent:* distinct nucleus ruber and rubro-spinal projections	
	?	*Present:* cranial nerve III
Diencephalon	*Present:* photosensitive parietal organ(s), habenula and fasciculus retroflexus, dorsal thalamic areas with ascending projections to telencephalon, neuroendocrine areas in hypothalamus, adeno- and neurohypophysis, habenular and postoptic commissures *Absent:* spinal lemniscal projections to diencephalon	
Eye and visual system	*Present:* bilateral retinal projections to preoptic, thalamic, pretectal and tectal areas, retinofugal projection arising from anterior tegmentum	
	?	*Present:* lens and external eye muscles, external optic chiasm
Telencephalon and olfactory bulb	*Present:* division into pallium and subpallium, sessile olfactory bulbs with typical glomerular organisation *Absent:* terminal nerve, olfacto-retinal projections	
	?	*Present:* medial and lateral pallial fields *Absent:* central prosencephalic nuclear complex
Telencephalic fibre systems	*Present:* extensive bilateral secondary olfactory projections (decussation in an interbulbar commissure), descending pallial projections to diencephalon, pretectum and tectum, ascending thalamic projections to pallial areas *Absent:* telencephalo-spinal projections	
	?	*Present:* medial pallium main target of ascending diencephalic projections, lateral pallium main target of secondary olfactory projections *Absent:* deep part of lateral olfactory tract

organisation that gave rise to bilateral secondary olfactory projections to the pallium was also present in the last common ancestor of craniates, and an optic chiasm, postoptic, posterior, and habenular commissures and the commissure of the posterior tuberculum were present as well.

II. There is a relatively small number of characters which are shared by myxinoids and lampreys that are either absent or similar in cephalochordates. These characters are probably plesiomorphic for craniates and include the absence of myelin sheaths in both the central and peripheral nervous system, the presence of giant interneurons in the spinal cord, the presence of an interbulbar commissure and of extensive secondary olfactory projections to the pallium, as well as the presence of a retinopetal projection arising from cells in the anterior mesencephalic tegmentum. A distinct nucleus ruber and a rubro-spinal projection are lacking in hagfishes and lampreys; thus, the lack of this system is also part of the craniate morphotype. Hagfishes and lampreys also share a flattened spinal cord. Since the spinal hord is more or less tubelike in both cephalochordates and gnathostomes, the polarity of that character cannot be resolved. Hagfishes and gnathostomes both display an ascending spino-tectal projection; thus, this charcter is probably plesiomorphic for craniates.

III. There are two characters which are present in cephalochordates, lampreys, and gnathostomes but absent in hagfishes. These characters must therefore be regarded as plesiomorphic for craniates, and their absence in hagfishes is probably due to a secondary loss. From the presence of intramedullary sensory cells (cells of Bone) in the nervous system of cephalochordates and the presence of very similar cells (dorsal cells, *Hinterzellen*) in the hindbrain and spinal cord of adult lampreys and in the brains of gnathostomes (cells of Rohon-Beard, transient sensory cells) it can be concluded that hagfishes have lost this character secondarily. Similarly, the absence of parietal photosensitive organs (i.e. pineal and parapineal organs) may be an apomorphic feature of hagfishes. Myopterygians, and lampreys in particular, possess well-developed parietal organs. The mid-dorsal region of the neural tube of cephalochordates is occupied by an unpaired photoreceptive organ (the so-called lamellar body) that displays a number of topological and ultrastructural similarities to the pineal organ of myopterygians (cf. Chap. 9), and it has been suggested (Lacalli et al. 1994; Lacalli 1996) that this organ may be a homologue of the pineal organ in myopterygians.

Plesiomorphic *myopterygian* characters (i.e. characters present in lampreys and gnathostomes)

are the absence of a central prosencephalic nuclear complex and a deep (periventricular) lateral olfactory tract. The last common ancestor of myopterygians possessed a rhomb- and mesencephalic somatomotor column and cranial nerves III, IV, and VI, and a cerebellum, even though it may have been very small and restricted to the presence of a vestibulocerebellum as in lampreys (see Chap. 10). It also possessed a trigemino-spinal projection, Müller and Mauthner cells projecting to the spinal cord, an ectodermal adenohypophysis, chorioid plexuses and a well-developed ventricular system, an external optic chiasm, an electroreceptive dorsal nucleus in the octavolateral area, and at least two pallial fields (medial and lateral pallia), which may be distinguished based on their cytoarchitecture and connectivity. Again, it should be noted that all these 'familiar' characters constitute a part of the *myopterygian* morphotype; however, they may or may not be plesiomorphic for *craniates*.

IV. Many characters are found in this category, among them, by definition, all the micro- and macroscopic features which allow one to recognise and distinguish the brains of each group. Since cephalochordates lack many of the character complexes, it is frequently impossible to determine their polarity.

One of the most interesting differences among all three taxa concerns the developmental fate of the prosencephalic vesicle. All craniates initially form an undivided prosencephalic vesicle. In lampreys, the (telencephalic) hemispheres form by a restricted lateral evagination from that vesicle. Only the lateral pallium and the olfactory bulb participate in this evagination; the medial and dorsal pallia, as well as the septum and large parts of the striatum, remain in the unevaginated telencephalon impar, flanking the third ventricle (cf. Chap. 10). In gnathostomes (with the exception of actinopterygians) the hemispheres form in a similar manner, but the medial and dorsal pallia, as well as the septum and the major part of the striatum, participate in the evagination. The pattern in myxinoids is entirely different (see above, Sect. 11.3.1): The prosencephalic vesicle gives rise to paired rostral and caudal evaginations, leading to the formation of telencephalic and diencephalic hemispheres and probably also causing the formation of the central prosencephalic nuclear complex. Together with the mode of formation of the hemispheres in actinopterygians, which is again entirely different (cf. Chap. 15), this comparison indicates that even early 'gross-morphogenetical' events are less stable and uniform than generally believed.

The mode of formation of spinal nerves is different among all three taxa. In myxinoids and gnathostomes dorsal and ventral roots fuse to form mixed spinal nerves, but the pattern of fusion is different (see Sect. 11.2 and Fig. 11.4). In lampreys, the roots do not fuse at all. The ventral and dorsal 'roots' of the cord of cephalochordates also do not fuse, thus indicating that the condition seen in lampreys is primitive for craniates. However, it should be borne in mind that the ventral and dorsal 'roots' of cephalochordates might not be strictly homologous to the spinal nerve roots of craniates, since they consist mainly of the processes of intramedullary sensory cells (dorsal 'roots'), and of processes of muscle cells that synapse on motoneurons at the ventral surface of the cord (ventral 'roots').

V. Finally, a large number of characters, among them some of the most interesting ones with regard to the phylogeny of the craniate brain, fall into this most problematic category.

In myxinoids and most gnathostomes, neurulation and formation of the neural tube occur by an invagination of the neural plate and folds. The ventricular cavity can thus be regarded as an enclosed, formerly external space, as evidenced, for example, by the presence of anterior and posterior neuropores. In lampreys and in teleost gnathostomes the neural tube develops from a solid neural keel and the ventricular cavities develop secondarily. At first view, this seems to be a most striking difference; however, it is frequently overlooked that both patterns actually co-exist in many gnathostomes: the posterior neural tube in the lumbar and tail region of gnathostomes develops by a process called secondary neurulation (Griffith et al. 1992). This form of neurulation involves not the formation of a neural plate, but the 'cavitation' of an initially solid cylinder of cells which derive from the mesenchymal tail bud, thus leading to the formation of a neural tube in a manner similar to the one seen in lampreys and teleosts. Neurulation in cephalochordates follows yet another process: a neural plate is formed, but it is overgrown by general ectoderm and displaced ventrally as a plate. The formation of the neural tube occurs only after the plate has separated from the ectoderm above (Hertwig 1890). Thus, it is not possible to determine the plesiomorphic mode of neurulation for craniates with certainty. However, the 'instability' of this developmental character among craniates, together with the fact that it does not seem to have a profound impact on the general morphological appearance, suggests that the mode of neurulation may not be one of the main shaping factors of the CNS.

Myxinoids and many gnathostomes share a pattern of extensive neuronal migration and hypertrophy of the brain walls, which results in a quite complicated cytoarchitectural appearance. These characters are coupled with a relatively high brain weight, the presence of astroglia, and the occurrence of capillary networks. Lampreys and some other gnathostomes, in particular urodele amphibians, lungfishes, and polypteriform bony fishes, share thin-walled brains with little or no neuronal migration in which radial glia and capillary loops prevail. Obviously, it would be highly desirable to determine the polarity of these sets of characters, since they are intimately connected with the general question for the evolution of complexity. In other words: is the history of the craniate brain to be written as a series of inventions which led to ever-increasing complexity, or have there been complications and simplifications? Did the last common ancestor of craniates have a brain that was more lamprey- or hagfish-like?

As mentioned above, strict cladistic reasoning (comparison with the out-group, the cephalochordates) is of little help with regard to this question. If we had more information on the selective pressures that cause the formation of complexity, we might be able to distinguish parallelism from homology. However, we are almost totally ignorant of these pressures. Are they external, does complication arise as response to a complex environment? If so, why do the bottom-dwelling inhabitants of a presumably dull, dark, and stable environment like hagfishes have a brain so much more complex than that of lampreys, which are active predators with a complicated life history? Are the selective pressures internal, the result of an interaction of the brain with itself or with the brains of members of the same species? We know nothing about the intellectual, social, and cultural life of hagfishes. Or, to phrase it in the vernacular of biology: we know almost nothing about the ethology and ecology of hagfishes.

Returning to the realm of neuroanatomy, a closer inspection of the characters under consideration may provide some tentative answers. A close look at some of the complex structures in the brains of hagfishes, e.g. the laminated pallium and the somatotopic organisation of the nucleus of the trigeminus, reveals that the 'special qualities' (Remane 1956; cf. Chap. 6, Sect. 6.2.6.4) of their organisation are different from complex structures in gnathostomes. In hagfishes, the presence of a large and laminated pallium is clearly coupled with the presence of massive deep and superficial olfactory tracts, with the presence of heavy pallio-pallial projections, and with the formation of a pallial 'output layer' (layer 3). In gnathostomes, hypertrophied pallia (dorsal areas in teleosts, dorsal ventricular ridge in reptiles and birds, neocortex in mammals) seem to occur in conjunction with massive ascending inputs that are lacking in hagfishes. Similarly, the large, migrated sensory nucleus of the trigeminus is, at first sight, similar to the migrated sensory trigeminal nuclei of many gnathostomes; however, the topography of the incoming sensory fibres is just reversed (see Sect. 11.5.3). Thus, it seems likely that complexity has evolved independently in these cases.

This is not to say, however, that the simple state of characters seen in lampreys and a minority of gnathostomes is primitive. If there is parallelism in complexity, there might also be parallelism in simplicity, caused by paedomorphosis and neoteny. Due to the nature of simplicity, differences in 'special quality' are harder to recognise, since simple characters have fewer attributes – nevertheless, they are present. Thus, both lampreys and urodele amphibians display a relatively similar periventricular arrangement of cell groups in their diencephala. Still, a connectional analysis reveals that urodeles possess a number of well-circumscribed 'hidden nuclei' in their periventricular grey that are not evident on cytoarchitectural analysis, but that may be defined based on their connectional properties (Wicht and Himstedt 1986, 1988). Similar hidden nuclei do not seem to exist in lampreys (Northcutt and Wicht 1997).

In summary, we can be fairly sure that complexity in hagfishes and in many gnathostomes has arisen independently. However, we cannot be sure whether this complexity arose from the modification of an already complex common ancestor, or whether it results from a modification of a simple morphotype. Even if we take refuge in the principle that complex structures – of necessity – must derive from simple ones, this does not imply that the particular type of simplicity seen in lampreys is actually plesiomorphic for craniates as a whole. Again, we should not forget that lampreys and hagfishes are only the surviving tips of the iceberg of an extinct Devonian and Permian biodiversity which we will never be able to fully appreciate. In a way, hagfishes are a living caveat to any attempt to underestimate the potential diversity, heterogeneity, and complexity of that ensemble.

References

Adam H (1956) Der III. Ventrikel und die mikroskopische Struktur seiner Wände bei Lampetra (Petromyzon) fluviatilis L. und Myxine glutinosa L., nebst einigen Bemerkungen über das Infundibularorgan von Branchiostoma (Amphioxus) lanceolatum Pall. Elsevier, Amsterdam, pp 146–158 (Progress in Neurobiology, Vol 1)

Adam H (1963a) Brain ventricles, ependyma, and related structures. In: Brodal A, Fänge R (eds) The biology of myxine. Universitetsforlaget, Oslo, pp 137–149

Adam H (1963b) The pituitary gland. In: Brodal A, Fänge R (eds) The biology of Myxine. Universitetsforlaget, Oslo, pp 457–476

Addens JL (1933) The motor nuclei and roots of the cranial and first spinal nerves of vertebrates. I. Introduction and cyclostomes. Z Anat Entw Gesch 101:307–410

Aldridge RJ, Briggs DGE, Smith MP, Clarkson ENK, Clark NDL (1993) The anatomy of conodonts. Philos Trans R Soc Lond B 340:405–421

Allen BM (1905) The eye of Bdellostoma stouti. Anat Anz 26:208–211

Amemiya F (1983) Afferent connections of the tectum mesencephali in the hagfish Eptatretus burgeri: an HRP study. J Hirnforsch 24:225–236

Amemiya F, Northcutt RG (1996) Afferent and efferent connections of the central prosencephalic nucleus in the Pacific hagfish. Brain Behav Evol 47:149–155

Amemiya F, Kishida R, Goris RC, Onishi H, Kusunoki T (1985) Primary vestibular projections in the hagfish, Eptatretus burgeri. Brain Res 337:73–79

Andres KH, von Düring M (1993a) Cutaneous and subcutaneous sensory receptors of the hagfish Myxine glutinosa with special respect to the trigeminal system. Cell Tissue Res 274:353–366

Andres KH, von Düring M (1993b) Lamellated receptors in the skin of the hagfish, Myxine glutinosa. Neurosci Lett 151:74–76

Ayers H, Worthington J (1907) The skin end-organs of the trigeminus and lateralis nerves of Bdellostoma dombeyi. Am J Anat 7:327–336

Ayers H, Worthington J (1911) The finer anatomy of the brain of Bdellostoma dombeyi. II. The fasciculus communis system. J Comp Neurol 21:593–616

Bardack D (1991) First fossil hagfish (Myxinoidea): a record from the Pennsylvanian of Illinois. Science 254:701–703

Black D (1917) The motor nuclei of the cerebral nerves in phylogeny: a study of the phenomena of neurobiotaxis. I. Cyclostomi and pisces. J Comp Neurol 27:467–564

Blackstad T (1963) The skin and its slime glands In: Brodal A, Fänge R (eds) The biology of Myxine. Universitetsforlaget, Oslo, pp 195–230

Blähser S, King JA, Kuenzel WJ (1989) Testing of arg-8-gonadotropin releasing hormone-directed antisera by immunological and immunocytochemical methods for use in comparative studies. Histochemistry 93:39–48

Bone Q (1963a) Some observations on the peripheral nervous system of the hagfish, Myxine glutinosa. J Marine Biol Ass UK 43:31–47

Bone Q (1963b) The central nervous system. In: Brodal A, Fänge R (eds) The biology of Myxine. Universitetsforlaget, Oslo, pp 50–91

Braun CB, Northcutt RG (1997a) The lateral line system of hagfishes (Craniata: Myxinoidea). Acta Zool (in press)

Braun CB, Northcutt RG (1997b) Cutaneous exteroreceptors and their innervation in the Pacific hagfish, Eptatretus stouti. In: Jørgensen J, Weber R, Lomholt J, Malte H (eds) The biology of hagfishes. Chapman and Hall, London (in press)

Braun CB, Wicht H, Northcutt RG (1993) Evidence that hagfish possess a degenerate lateral line system. Soc Neurosci Abstr 19:159

Braun CB, Wicht H, Northcutt RG (1995) Distribution of gonadotropin-releasing hormone immunoreactivity in the brain of the Pacific hagfish, Eptatretus stouti (Craniata: Myxinoidea). J Comp Neurol 353:464–476

Briggs DEG (1992) Conodonts: a major extinct group added to the vertebrates. Science 256:1285–1286

Bullock TH, Bodznick D, Northcutt RG (1983) The phylogenetic distribution of electroreception: evidence for convergent evolution of a primitive vertebrate sense modality. Brain Res Rev 6:25–46

Bullock TH, Moore JK, Fields RD (1984) Evolution of myelin sheaths: both lamprey and hagfish lack myelin. Neurosci Lett 48:145–148

Caroll RL (1988) Vertebrate paleontology and evolution. Freeman, New York (German translation: 1993, Thieme, Stuttgart)

Chiba A, Honma Y (1986) Fine structure of the granulocytes occurring in the hypothalamic-hypophyseal ventricle and neurohypophysis of the hagfish, Paramyxine atami. Jpn J Ichthyol 33:174–179

Chiba A, Honma Y (1992) FMRFamide-immunoreactive structures in the brain of the brown hagfish, Paramyxine atami: relationship with neuropeptide Y-immunoreactive structures. Histochemistry 98:33–38

Chiba A, Honma Y, Oka S (1993) Immunohistochemical localization of neuropeptide Y-like substance in the brain and hypophysis of the brown hagfish Paramyxine atami. Cell Tissue Res 271:289–295

Conel JL (1929) The development of the brain of Bdellostoma stouti. I. External growth changes. J Comp Neurol 47:343–403

Conel JL (1931) The development of the brain of Bdellostoma stouti. II. Internal growth changes. J Comp Neurol 52:365–499

Conel JL (1942) The origin of the neural crest. J Comp Neurol 76:191–215

Crosby EC, Schnitzlein HN (1974) The comparative anatomy of the telencephalon of the hagfish, Myxine glutinosa. J Hirnforsch 15:211–236

Cserr HF, Bundgaard M (1984) Blood-brain interfaces in vertebrates: a comparative approach. Am J Physiol 246:R277-R288

Dawson JA (1963) The oral cavity, the 'jaws' and the horny teeth of Myxine glutinosa. In: Brodal A, Fänge R (eds) The biology of Myxine. Universitetsforlaget, Oslo, pp 231–255

Dean B (1899) On the embryology of Bdellostoma stouti. A general account of myxinoid development from the egg and segmentation to hatching. In: Festschrift zum siebenzigsten Geburtstag von Carl von Kupffer, Gustav Fischer, Jena, pp 221–276

Dean B, Harrington NR, Calkins GN, Griffin BB (1897) The Columbia University zoölogical expedition of 1896. With a brief account of the work of collecting in Puget Sound and on the Pacific coast. Trans N Y Acad Sci 16:33–42

Dodd JM, Dodd MHI (1985) Evolutionary aspects of reproduction in cyclostome fishes. In: Foreman RE, Gorbman A, Dodd JM, Olsson R (eds) Evolutionary biology of primitive fishes. Plenum, New York, pp 295–319

Doflein F (1898) Bericht über eine wissenschaftliche Reise nach Californien (Mittheilungen über die Erlangung von Eiern und Embryonen von Bdellostoma). Sitzungsber Ges Morphol Physiol München 14:105–118

Donald JA, Vomachka AJ, Evans DH (1992) Immunohistochemical localization of natriuretic peptides in the brains and hearts of the spiny dogfish Squalus acanthias and the Atlantic hagfish Myxine glutinosa. Cell Tissue Res 270:535–545

Dores RM, Gorbman A (1990) Detection of met-enkephalin and leu-enkephalin in the brain of the hagfish, Eptatretus stouti, and the lamprey, Petromyzon marinus. Gen Comp Endocrinol 77:489–499

Døving KB, Holmberg K (1974) A note on the function of the olfactory organ of the hagfish Myxine glutinosa. Acta Physiol Scand 91:430–432

Dücker M (1924) Über die Augen der Zyklostomen. Jenaische Z Med Naturwiss 60:471–530

Edinger L (1906) Über das Gehirn von Myxine glutinosa. Abhandlungen der königlich preußischen Akademie der Wissenschaften aus dem Jahre 1906. Kön Akad Wissensch, Berlin, pp 1–36

Fernholm B (1969) A third embryo of myxine: considerations on hypophyseal ontogeny and phylogeny. Acta Zool 50:169–177

Fernholm B (1974) Diurnal variations in the behaviour of the hagfish Eptatretus burgeri. Marine Biol 27:351–356

Fernholm B (1985) The lateral line system of cyclostomes. In: Foreman RE, Gorbman A, Dodd JM, Olsson R (eds) Evolutionary biology of primitive fishes. Plenum, New York, pp 113–122

Fernholm B, Holmberg K (1975) The eyes in three genera of

hagfish (Eptatretus, Paramyxine and Myxine): a case of degenerative evolution. Vision Res 15:253–259

Forey P, Janvier P (1993) Agnathans and the origin of jawed vertebrates. Nature 361:129–134

Gabbott SE, Aldridge RJ, Theron JN (1995) A giant conodont with preserved muscle tissue from the upper Ordovician of South Africa. Nature 374:800–803

Georgieva V, Patzner RA, Adam H (1979) Transmissions- und rasterelektronenmikroskopische Untersuchung an den Sinnesknospen und Tentakeln von Myxine glutinosa L. (Cyclostomata). Zool Scripta 8:61–67

Goodrich ES (1937) On the spinal nerves of the Myxinoidea. Q J Micr Sci 80:153–158

Gorbman A (1983) Early development of the hagfish pituitary gland: evidence for the endodermal origin of the adenohypophysis. Am Zool 23:639–654

Gorbman A, Kobayashi H, Honma Y, Matsuyama M (1990) The hagfishery of Japan. Fisheries 15:12–18

Greene CW (1925) Notes on the olfactory and other physiological reactions of the California hagfish. Science 61:68–70

Griffith CM, Wiley MJ, Sanders EJ (1992) The vertebrate tail bud: three germ layers from one tissue. Anat Embryol (Berl) 185:101–114

Hertwig O (1890) Lehrbuch der Entwicklungsgeschichte des Menschen und der Wirbelthiere. Fischer, Jena

Holm JF (1901) The finer anatomy of the nervous system of Myxine glutinosa. Morphol Jahrb 29:365–401

Holmberg K (1970) The hagfish retina: fine structure of retinal cells in Myxine glutinosa, L., with special reference to receptor and epithelial cells. Z Zellforsch 111:519–538

Holmberg K (1971) The hagfish retina: electron microscopic study comparing receptor and epithelial cells in the Pacific hagfish, Polistotrema stouti, with those in the Atlantic hagfish, Myxine glutinosa. Z Zellforsch 121:249–269

Holmes RL, Ball JN (1974) The pituitary gland: a comparative account. Cambridge University Press, Cambridge

Holmgren N (1919) Zur Anatomie des Gehirns von Myxine. Kungliga Svenska Vetenskapsakademiens Handlingar 60/7:1–96

Holmgren N (1946) On two embryos of Myxine glutinosa. Acta Zool 27:1–90

Iwahori N, Nakamura K, Tsuda A (1996) Neuronal organization of the optic tectum in the hagfish, Eptatretus burgeri: a Golgi-study. Anat Embryol 193:271–279

Jansen J (1930) The brain of Myxine glutinosa. J Comp Neurol 49:359–507

Janvier P (1974) The structure of the naso-hypophyseal complex and the mouth in fossil and extant cyclostomes, with remarks on Amphiaspiforms. Zool Scripta 3:193–200

Janvier P (1981) The phylogeny of the craniata, with particular reference to fossil 'agnathans'. J Vertebr Paleontol 1:121–159

Janvier P (1995) Conodonts join the club. Nature 374:761–762

Jefferies RPS (1986) The ancestry of vertebrates. British Museum of Natural History and Press Syndicate University Cambridge, Cambridge

Jensen D (1966) The hagfish. Sci Am 214:82–91

Jirikowski G, Erhardt G, Grimmelikhuijzen CJP, Triepel J, Patzner RA (1984) FMRF-amide-like immunoreactivity in brain and pituitary gland of the hagfish Eptatretus burgeri (Cyclostomata). Cell Tissue Res 237:363–366

Kadota T (1991) Distribution of 5-HT (serotonin) immunoreactivity in the central nervous system of the inshore hagfish, Eptatretus burgeri. Cell Tissue Res 266:107–116

Kadota T, Goris RC, Kusunoki T (1993) Dopamine neurons in the hagfish brain. Anat Rec [Suppl] 1:69

Kato S (1989) Report on a trip to Japan and South Korea, June 1989. Unpublished memorandum for the United States Department of Commerce, National Oceanic and Atmospheric Administration, National Marine Fisheries Service, Southwest Region

Kishida R, Onishi H, Nishizawa H, Kadota T, Goris RC, Kusunoki T (1986) Organization of the trigeminal and facial motor nuclei in the hagfish, Eptatretus burgeri: a retrograde HRP study. Brain Res 385:263–272

Kishida R, Goris RC, Nishizawa H, Koyama H, Kadota T, Amemiya F (1987) Primary neurons of the lateral line nerves and their central projections in hagfishes. J Comp Neurol 264:303–310

Kobayashi H, Uemura H (1972) The neurohypophysis of the hagfish, Eptatretus burgeri (Girard). Gen Comp Endocrinol 3 [Suppl]:114–124

Koch EA, Spitzer RH, Pithawalla RB, Downing SW (1991) Keratin-like components of gland thread cells modulate the properties of mucus from hagfish (Eptatretus stouti). Cell Tissue Res 264:79–86

Korneliussen H, Nicolaysen K (1973) Ultrastructure of four types of striated muscle fibers in the Atlantic hagfish (Myxine glutinosa L). Z Zellforsch 143:273–290

Kuhlenbeck H (1977) The central nervous system of vertebrates, Vol 5/I: Derivatives of the prosencephalon: diencephalon and telencephalon. Karger, Basel

Kusunoki T, Amemiya F (1983) Retinal projections in the hagfish, Eptatretus burgeri. Brain Res 262:295–298

Kusunoki T, Kadota T, Kishida R (1981) Chemoarchitectonics of the forebrain of the hagfish, Eptatretus burgeri. J Hirnforsch 22:285–298

Kusunoki T, Kadota T, Kishida R (1982) Chemoarchitectonics of the brain stem of the hagfish, Eptatretus burgeri, with special reference to the primordial cerebellum. J Hirnforsch 23:109–119

Lacalli TC (1996) Frontal eye circuitry, rostral sensory pathways and brain organization in amphioxus larvae: evidence from 3D reconstructions. Philos Trans R Soc Lond B 351:243–263

Lacalli TC, Holland ND, West JE (1994) Landmarks in the anterior nervous system of amphioxus larvae. Philos Trans R Soc Lond B 344:165–185

Larsell O (1947) The cerebellum of myxinoids and petromyzontids including developmental stages in the lampreys. J Comp Neurol 86:395–445

Lindström T (1949) On the cranial nerves of the cyclostomes with special reference to n. trigeminus. Acta Zool 30:315–458

Løvtrup S (1977) The phylogeny of vertebrata. Wiley, London

Lowenstein O, Thornhill RA (1970) The labyrinth of Myxine: anatomy, ultrastructure, and electrophysiology. Proc R Soc Lond B 176:21–42

Matsuda H, Goris RC, Kishida R (1991) Afferent and efferent projections of the glossopharyngeal-vagal nerve in the hagfish. J Comp Neurol 311:520–530

Matty AJ, Tsuneki K, Dickhoff WW, Gorbman A (1976) Thyroid and gonadal function in hypophysectomized hagfish, Eptatretus stouti. Gen Comp Endocrinol 30:500–516

McVean A (1991) The semicircular canals of the hagfish Myxine glutinosa. Proc R Soc Lond B 224:213–222

Mugnaini E, Walberg F (1965) The fine structure of the capillaries and their surroundings in the cerebral hemispheres of Myxine glutinosa (L). Z Zellforsch 66:333–351

Murray M, Jones H, Cserr HF, Rall DP (1975) The blood-brain barrier and ventricular system of Myxine glutinosa. Brain Res 99:17–33

Nansen F (1886) The structure and combination of the histological elements of the central nervous system. Bergens Museums Aarsberetning, pp 29–215

Newth DR, Ross DM (1955) On the reaction to light of Myxine glutinosa L. J Exp Biol 32:4–21

Nishizawa H, Kishida R, Kadota T, Goris RC (1988) Somatotopic organization of the primary sensory trigeminal neurons in the hagfish, Eptatretus burgeri. J Comp Neurol 267:281–295

Noden DM (1983) The embryonic origin of avian cephalic and cervical muscles and associated connective tissues. Am J Anat 168:257–276

Noden DM (1991) Vertebrate craniofacial development: the relation between ontogenetic process and morphological outcome. Brain Behav Evol 38:190–225

Northcutt RG (1985) The brain and the sense organs of the earliest craniates: reconstruction of a morphotype. In: Foreman RE, Gorbman A, Dodd JM, Olsson R (eds) Evolutionary biology of primitive fishes. Plenum, New York, pp 81–112

Northcutt RG, Wicht H (1997) Afferent and efferent connections of the lateral and medial pallia of the silver lamprey. Brain Behav Evol 49:1–19

Nozaki M (1985) Tissue distribution of hormonal peptides in primitive fishes. In: Foreman RE, Gorbman A, Dodd JM, Olsson R (eds) Evolutionary biology of primitive fishes. Plenum, New York, pp 433–454

Nozaki M, Gorbman A (1983) Immunocytochemical localization of somatostatin and vasotocin in the brain of the Pacific hagfish, Eptatretus stouti. Cell Tissue Res 229:541–550

Nozaki M, Fernholm B, Kobayashi H (1975) Ependymal absorption of peroxidase into the third ventricle of the hagfish, Eptatretus burgeri (Girard). Acta Zool 56:265–269

Nozaki M, Gorbman A, Sower SA (1994) Diffusion between the neurohypophysis and the adenohypophysis of lampreys, Petromyzon marinus. Gen Comp Endocrinol 96:385–391

Ochi J, Yamamoto T, Hosoya Y (1979) Comparative study of the monoamine system in the spinal cord of the lamprey and hagfish. Arch Histol Jpn 42:327–336

Ollson R (1959) The neurosecretory hypothalamus system and the adenohypophysis of Myxine. Z Zellforsch 51:97–107

Ooka-Souda S, Kabasawa H (1988) Circadian rhythms in locomotor activity of the hagfish, Eptatretus burgeri. III. Hypothalamus: a locus of the circadian pacemaker? Zool Sci [Japan] 5:437–442

Ooka-Souda S, Kadota T, Kabasawa H (1993) The preoptic nucleus: the probable location of the circadian pacemaker of the hagfish, Eptatretus burgeri. Neurosci Lett 164:33–36

Patzner RA (1978) Experimental studies on the light sense in the hagfish, Eptatretus burgeri and Paramyxine atami. Helgoländer Wiss Meeresunters 31:180–190

Pérez J, Grondona JM, Cifuentes M, Nualart F, Fernández-Llebrez P, Rodríguez EM (1993) Immunochemical analysis of the dogfish subcommissural organ. In: Oksche A, Rodríguez EM, Fernández-Llebrez P (eds) The subcommissural organ – an ependymal brain gland. Springer, Berlin Heidelberg New York, pp 99–107

Peters A (1963) The peripheral nervous system. In: Brodal A, Fänge R (eds) The biology of Myxine. Universitetsforlaget, Oslo, pp 92–123

Peters A, Bone Q (1963) The nervous system and the sense organs – general introduction. In: Brodal A, Fänge R (eds) The biology of Myxine. Universitetsforlaget, Oslo, pp 47–149

Pierre J, Repérant J, Ward R, Vesselkin NP, Rio J-P, Miceli D, Kratskin I (1992) The serotoninergic system of the brain of the lamprey, Lampetra fluviatilis: an evolutionary perspective. J Chem Neuroanat 5:195–219

Platel R, Delfini C (1981) L'encéphalization chez la Myxine (Myxine glutinosa L). Analyse quantifiée des principales subdivisiones encéphaliques. Cah Biol Mar 22:407–430

Polenova OA, Vesselkin NP (1993) Olfactory and non-olfactory projections in the river lamprey (Lampetra fluviatilis) telencephalon. J Hirnforsch 34:261–279

Remane A (1956) Die Grundlagen des natürlichen Systems, der vergleichenden Anatomie und der Phylogenetik. Akademische Verlagsgesellschaft Geest und Portig, Leipzig

Retzius G (1891) Zur Kenntniss des centralen Nervensystems von Myxine glutinosa. Biol Unters NF 2:47–53

Retzius G (1893a) Studien über Ependym und Neuroglia. Biol Unters NF 5:9–26

Retzius G (1893b) Das Gehirn und das Auge von Myxine. Biol Unters NF 5:55–68

Retzius G (1921) Weitere Beiträge zur Kenntnis von dem Bau und der Anordnung des Ependyms und der sämtlichen

Neuroglia, besonders bei den niederen Vertebraten. Biol Unters NF 19:1–26

Ronan M (1988) The sensory trigeminal tract of Pacific hagfish. Primary afferent projections and neurons of the tract nucleus. Brain Behav Evol 32:169–180

Ronan M (1989) Origin of descending spinal projections in petromyzontid and myxinoid agnathans. J Comp Neurol 281:54–68

Ronan M, Northcutt RG (1990) Projections ascending from the spinal cord to the brain in petromyzontid and myxinoid agnathans. J Comp Neurol 291:491–508

Röthig P, Ariëns Kappers CPU (1914) Further contributions to our knowledge of the brain of Myxine glutinosa. Proc Kon Ned Akad Wet 17:2–12

Rurak DW, Perks AM (1974) The pharmacological characterisation of arginine vasotocin in the pituitary of the Pacific hagfish (Polistotrema stouti). Gen Comp Endocrinol 22:480–488

Sanders A (1894) Researches in the nervous system of Myxine glutinosa. Williams and Norgate, London

Sansom IJ, Smith MP, Armstrong HA, Smith MM (1992) Presence of the earliest vertebrate hard tissues in conodonts. Science 256:1308–1310

Scharrer E (1962) Brain function and the evolution of cerebral vascularization. The James Arthur Lecture on the Evolution of the Human Brain of 1960. American Museum of Natural History, New York, pp 1–32

Schober W (1964) Vergleichend anatomische Untersuchungen am Gehirn der Larven und adulten Tiere von Lampetra fluviatilis (Linné 1758) und Lampetra planeri (Bloch 1748). J Hirnforsch 7:9–209

Schreiner KE (1919) Zur Kenntnis der Zellgranula. Untersuchungen über den feineren Bau der Haut von Myxine glutinosa. Arch Mikrosk Anat 92:1–63

Sower SA (1990) Gonadotropin-releasing hormone in primitive fishes. Prog Clin Biol Res 324:393–398

Sower SA, Nozaki M, Knox CJ, Gorbman A (1995) The occurrence and distribution of GnRH in the brain of Atlantic hagfish, an Agnatha, determined by chromatography and immunocytochemistry. Gen Comp Endocrinol 97:300–307

Steinbusch HWM, Nieuwenhuys R (1979) Serotonergic neuron systems in the brain of the lamprey, Lampetra fluviatilis. Anat Rec 193:693–694

Steven DM (1955) Experiments on the light sense of the hag, Myxine glutinosa. Linn J Exp Biol 32:22–38

Stock DW, Whitt GS (1992) Evidence from 18 s ribosomal RNA sequences that lampreys and hagfishes form a natural group. Science 257:787–789

Stockard C (1907) The embryonic history of the lens in Bdellostoma stouti in relation to recent experiments. Am J Anat 6:511–515

Strahan R (1963) The behaviour of Myxine and other myxinoids. In: Brodal A, Fänge R (eds) The biology of Myxine. Universitetsforlaget, Oslo, pp 22–32

Tsukahara T, Gorbman A, Kobayashi H (1986) Median eminence equivalence of the neurohypophysis of the hagfish, Eptatretus burgeri. Gen Comp Endocrinol 61:348–354

Tsuneki K (1988) The neurohypophysis of cyclostomes as a primitive hypothalamic center of vertebrates. Zool Sci (Jpn) 5:21–32

Tsuneki K, Urano A, Kobayashi H (1974) Monoamine-oxidase and acetylcholinesterase in the neurohypophysis of the hagfish, Eptatretus burgeri. Gen Comp Endocrinol 22:480–488

Tsuneki K, Adachi T, Ishii S, Oota Y (1976) Morphometric classification of neurosecretory granules in the neurohypophysis of the hagfish, Eptatretus burgeri. Cell Tissue Res 166:145–157

Tusch U, Wicht H, Korf H-W (1995) Observations on the topology of the forebrain of the Pacific hagfish, Eptatretus stouti. J Anat 187:227–228

Vesselkin NP, Ermakova TV, Repérant J, Kosareva AA, Kenigfest NB (1980) The retinofugal and retinopetal systems in

Lampetra fluviatilis. An experimental study using radioautographic and HRP methods. Brain Res 195:453–460

Vesselkin NP, Repérant J, Kenigfest NB, Miceli D, Ermakova TV, Rio JP (1984) An anatomical and electrophysiological study of the centrifugal visual system in the lamprey (Lampetra fluviatilis). Brain Res 292:41–56

Vigh B, Vigh-Teichmann I (1992) Cytochemistry of CSF-contacting neurons and pinealocytes. Prog Brain Res 91:299–306

von Düring M, Andres KH (1994) Topography and fine structure of proprioreceptors in the hagfish, Myxine glutinosa. Eur J Morphol 32:248–256

von Düring M, Andres KH (1997) Skin sensory organs in the Atlantic hagfish, Myxine glutinosa. In: Jørgensen J, Weber R, Lomholt J, Malte H (eds) The biology of hagfishes. Chapman and Hall, London (in press)

von Kupffer C (1900) Studien zur vergleichenden Entwicklungsgeschichte des Kopfes der Kranioten. Heft 4: Zur Kopfentwicklung von Bdellostoma. Lehmann, München

von Kupffer C (1906) Ontogenie des Zentralnervensystems. In: Hertwig O (ed) Handbuch der vergleichenden und experimentellen Entwickelungslehre der Wirbeltiere, vol II/3. Fischer, Jena

Wächtler K (1975) The distribution of acetylcholinesterase in the cyclostome brain. II. Myxine glutinosa. Cell Tissue Res 159:109–120

Wächtler K (1983) The acetylcholine-system in the brain of cyclostomes with special reference to the telencephalon. J Hirnforsch 24:63–70

Waehneldt TV, Matthieu J-M, Jeserich G (1986) Appearance of myelin during vertebrate evolution. Neurochem Int 9:463–474

Wasowicz M, Pierre J, Repérant J, Ward R, Vesselkin NP, Versaux-Bottieri C (1994) Immunoreactivity to glial fibrillary acidic protein (GFAP) in the brain and spinal cord of the lamprey (Lampetra fluviatilis). J Brain Res 35:71–78

Wicht H, Himstedt W (1986) Two thalamo-telencephalic pathways in a urodele, Triturus alpestris. Neurosci Lett 68:90–94

Wicht H, Himstedt W (1988) Topologic and connectional analysis of the dorsal thalamus of Triturus alpestris (Amphibia, Urodela, Salamandridae). J Comp Neurol 267:545–561

Wicht H, Northcutt RG (1990) Retinofugal and retinopetal projections in the Pacific hagfish, Eptatretus stouti. Brain Behav Evol 36:315–328

Wicht H, Northcutt RG (1992a) The forebrain of the Pacific hagfish: a cladistic reconstruction of the ancestral craniate forebrain. Brain Behav Evol 40:25–64

Wicht H, Northcutt RG (1992b) FMRF-like immunoreactivity in the brain of the Pacific hagfish, Eptatretus stouti. Cell Tissue Res 270:443–449

Wicht H, Northcutt RG (1993) Secondary olfactory projections and pallial topography in the Pacific hagfish, Eptatretus stouti. J Comp Neurol 337:529–542

Wicht H, Northcutt RG (1994a) Observations on the development of the lateral line system in the Pacific hagfish (Eptatretus stouti, Myxinoidea). Eur J Morphol 32:257–261

Wicht H, Northcutt RG (1994b) An immunohistochemical study of the telencephalon and the diencephalon in a myxinoid jawless fish, the Pacific hagfish, Eptatretus stouti. Brain Behav Evol 43:140–161

Wicht H, Northcutt RG (1995) Ontogeny of the head of the Pacific hagfish (Eptatretus stouti, Myxinoidea): development of the lateral line system. Philos Trans R Soc Lond B 349:119–134

Wicht H, Tusch U (1997) Ontogeny of the head and nervous system of myxinoids – a brief review. In: Jørgensen J, Weber R, Lomholt J, Malte H (eds) The biology of hagfishes. Chapman and Hall, London (in press)

Wicht H, Derouiche A, Korf H-W (1994) An immunocytochemical investigation of glial morphology in the Pacific hagfish: radial and astrocyte-like glia have the same phylogenetic age. J Neurocytol 23:565–576

Wicht H, Northcutt RG, Korf H-W (1996) The evolution of the forebrain: new data on forebrain connectivity in jawless craniates. Verh Anat Ges 178 [Suppl]:47

Worthington J (1905a) Contribution to our knowledge of the myxinoids. Am Naturalist 39:625–663

Worthington J (1905b) The descriptive anatomy of the brain and cranial nerves of Bdellostoma dombeyi. Q J Micr Sci 49:137–181

Yalden DW (1985) Feeding mechanisms as evidence for cyclostome monophyly. Zool J Linn Soc 84:291–300

CHAPTER 12

Cartilaginous Fishes

W.J.A.J. SMEETS

12.1
Introduction

The class Chondrichthyes or cartilaginous fishes comprises about 800 species belonging to two major radiations that have diverged over 350 million years ago, i.e. the elasmobranchs and the holocephalians (Compagno 1977; Carroll 1988). Fossil evidence suggests that living holocephalians are a much older form than modern elasmobranchs, which arose approximately 200 million years ago (Carroll 1988). The living elasmobranchs (sharks, skates and rays) comprise four major superorders: Squalomorphii, Galeomorphii, Squatinomorphii and Batoidea (Table 12.1). The elasmobranchs are widely distributed, as they are marine in habitat, except for one species, *Carcharinus leucas*, which lives in Lake Nicaragua and in estuaries of large rivers, such as the Ganges, the Mississippi and the Zambesi. There are about 350 described living species of sharks, ranging in size from the giant whale shark (up to 15 m long) and basking shark (up to 10 m long) to tiny ones such as *Squaliolus laticaudus*, which in adulthood measures 15 cm in length. Paradoxically, the largest sharks mentioned above are plankton feeders and quite harmless. Of the 350 species of living sharks, no more than 35 have been implicated in attacks on humans (Gilbert 1984).

Cartilaginous fishes can be distinguished from the Osteichthyes or bony fishes by the following characteristics (Schaeffer and Williams 1977):

1. A cartilaginous endoskeleton that is calcified in a polygonal pattern
2. A dermal skeleton represented by denticles with a particular pattern of enameloid, dentine and basal tissue

Table 12.1 Classification of the cartilaginous fishes (largely based on Compagno 1977)

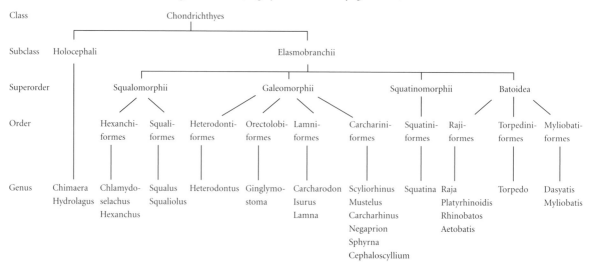

3. Pelvic claspers of unique design

4. Labial cartilages

5. A triangular basal cartilage behind the single dorsal fin spine

6. Ceratotrichia in paired and unpaired fins

7. A rectal salt gland

Early in this century, the cartilaginous fishes were considered to be generalised primitive vertebrates and were thus important to morphologists for typological comparisons with other vertebrates. However, it appears that they are less simple in organisation than had been supposed and that they possess relatively large brains, elegant sensory systems, sophisticated reproductive mechanisms, specialized osmoregulatory adaptations and, in some species, even thermoregulation (Gilbert 1984).

Investigations by Ebbesson and Northcutt (1976), Bauchot et al. (1976) and Northcutt (1977, 1978) have shown that the brain-body weight ratio in some elasmobranchs is comparable to that of birds and mammals and exceeds that of most other non-mammalian species. Traditionally, large brains have been thought to reflect complex behaviour, but in the elasmobranchs it is difficult to make any statements about behaviour as so little work has been done (Gilbert 1984; Klimley 1994). Northcutt (1977, 1978) divided the sharks into two groups: squalomorph sharks, which have low brain-body weight ratios, and galeomorph sharks, which have brain-body weight ratios that are two to six times larger. A similar variation can be seen within the superorder Batoidea, as the Rajiformes have low brain-body weight ratios, whereas the more advanced Myliobatiformes show the highest brain-body weight ratios known for elasmobranchs.

A comparative study by Kreps (1981) of elasmobranch and bony fish brain lipids (phospholipids, glycolipids) has shown that, biochemically, the brain of sharks and rays is much closer to that of amniotes than to the anamniotes and has also confirmed the existence of markedly different primitive and more advanced groups of sharks and rays.

There is abundant anecdotal and experimental evidence of the importance of chemical sensing to elasmobranchs, particularly of olfaction (Tester 1963; Hodgson et al. 1967; Hodgson and Mathewson 1971; Kleerekoper 1978). The sense of smell is mediated through the receptor cells of the olfactory epithelium, while taste originates in the oral and buccal cavities. Taste is restricted primarily to aspects of food finding and feeding, but smell, while certainly utilised in food detection and location, is also important in reproductive and social behaviour.

The eyes of elasmobranchs are prominent, although not particularly large, and are placed laterally on the head (except in Batoidea) to provide some degree of binocular overlap in the visual field. They are thought to be important for predatory behaviour, particularly at close quarters. The eyes can be rotated by the extrinsic eye muscles to maintain a constant visual field during locomotion and turning. The rod-rich retina provides the eye with high sensitivity, which can be enhanced in dim light by tapetal mirrors that reflect light from the choroid back to the rods (for reviews see Gruber and Cohen 1978; Heath 1990; Hueter 1990; Ripps and Dowling 1990; Sivak 1990).

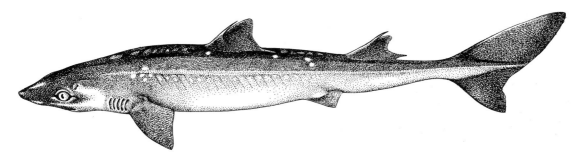

Fig. 12.1. Habitus of the spiny dogfish, *Squalus acanthias*

It is now clearly recognised that sound plays a significant role in the lives of sharks and other elasmobranchs and is used by them to locate food sources and other objects. Acoustic playback techniques have revealed that sounds below 1000 Hz, particularly low frequencies (around 40 Hz), are attractive to these fishes (Myrberg 1978; Corwin 1981b, 1983).

The lateral line sensory system is well developed in cartilaginous fishes and was recognised by Lorenzini long ago as comprising two components, canals and gelatinous tubules. In most species, the canals are enclosed, opening at intervals to the water. The lateral line canal system is thought to be involved in the detection of water displacements set up close to the surface of the fish. It is now appreciated that the other component of the lateral line system, the ampullae of Lorenzini, provides an extremely sensitive sensory system that is capable of responding to very small electric currents. This possibility was first demonstrated by Dijkgraaf and Kalmijn (1962) and was then firmly established by Kalmijn's subsequent experiments, which revealed the extreme electrosensitivity of elasmobranchs. These animals appear to be able to respond to gradients as low as 0.005 µV/cm. Cartilaginous fishes seem to use this sensory system to detect living food and to determine the direction and polarity of electric fields in the ocean (for a review, see Kalmijn 1978).

Electrosensitivity in teleosts is usually associated with the production of discharges from electric organs, both systems being used for navigation. No shark is known to have an electric organ, but electric organs are found in the batoids. In Torpedoids, the electric organ situated in the pectoral discs is substantial in size, and powerful discharges can be produced; these are apparently used to paralyse prey during feeding (Belbenoit and Bauer 1972). A much weaker discharge is emitted from the tail electric organ of rajoids, which may play a role in electrolocation (Baron et al. 1985; Bratton and Ayers 1987).

Compared to other anamniotes, the cartilaginous fishes show two striking reproductive features:

1. Fertilisation is always internal and is achieved by a pair of characteristic male copulatory organs (the claspers) developed from the pelvic fins.
2. Embryonic development is oviparous in some (e.g. *Scyliorhinus* and *Raja*), ovoviviparous in most (e.g. *Squalus*) and truly viviparous in a few others (e.g. *Mustelus*).

Horny egg capsules are produced by oviparous species, whereas in ovoviviparous species the embryo escapes from the egg capsule early, continues development in the mother's oviduct and is not born until it is fully formed. A special yolk-sac placenta is formed in viviparous species (for a review, see Wourms 1977). A peculiar feature of some mackerel (Lamnidae family) and tresher sharks (Alopiidae family) is the occurrence of intra-uterine cannibalism (Gilbert 1984).

Osmoregulation in cartilaginous fishes is characterised by urea retention, a feature which is shared by coelacanths and lungfishes. However, the way in which the urea is retained is different (Pang et al. 1977).

Most cartilaginous fishes are poikilothermic and have body temperatures approximately the same that of the water in which they swim. A few, however, such as the thresher and several lamnid sharks, possess counter-current heat-exchange mechanisms for conserving metabolic heat and raising their body temperatures above that of the surrounding water. Since warm muscles produce more power, these fast-swimming sharks possess a decisive advantage over the slower-moving bony fishes on which they frequently prey. Carey and Teal (1969) found that two sharks, *Isurus oxyrynchus* and *Lamna nassus*, were able to maintain their body temperatures 7–10 °C above the ambient temperature. Behavioural thermoregulation in sharks has been reported for the horn shark *Heterodontus francisci* (Crawshaw and Hammel 1973) and for the smooth dogfish *Mustelus canis* (Casterlin and Reynolds 1979).

Sharks and sawfishes, guitarfishes and torpedoids all propel themselves by undulation of the posterior part of the trunk. Skates and stingrays send waves of undulation along the enlarged pectoral fins, from front to rear, but some rays and the holocephalians swim by bird-like flapping of their pectoral fins. Some elasmobranchs swim ceaselessly, while other show diurnal patterns of locomotor activity (Nelson and Johnson 1970; Finstad and Nelson 1975). However, in all elasmobranchs, the sustained movements are produced by contractions of the lateral "red" body musculature, whereas the larger mass of "white" myotomal muscle is reserved for brief, rapid movements such as attack or escape (Bone 1978).

The core species of this chapter, *Squalus acanthias* or spiny dogfish, is world-wide in distribution; it lives in open water, feeds voraciously on a great variety of other fish, such as herring and cod, and migrates extensively both in European and North American waters. It is characterised by a spine in front of each dorsal fin and shows the classical heterocercal tail (Fig. 12.1). *Squalus* has been the subject of many neuro-anatomical studies and, although it has been considered a primitive shark (Northcutt 1977, 1978), in general terms it shows all the features of the brain of cartilaginous fishes. A more comprehensive study of the brains of chondrichthyans, including the sharks *Squalus acanthias* and *Scyliorhinus canicula*, the ray *Raja clavata* and the holocephalian *Hydrolagus collei*, has been published previously (Smeets et al. 1983).

12.2
Gross Morphology and Overall Histological Pattern

12.2.1
Gross Morphology

As in most sharks, the brain of *Squalus* is tightly enclosed by the neurocranium (Fig. 12.2). The same is true of the brain of holocephalians, but in batoids, it occupies only part of the neurocranial cavity (Bundgaard and Cserr 1991). The brain of *Squalus* is large and well developed (Fig. 12.3). The spinal cord passes imperceptibly into the rhombencephalon, a structure which, as its name implies, surrounds the rhomboid-shaped fourth ventricle. The floor of this ventricular cavity is predominantly made up of the rhombencephalic basal plates, whereas the alar plates form the lateral walls. The boundary between the basal and alar plates throughout most of the rhombencephalon is marked by a ventricular groove, the sulcus limitans of His. Rostrolaterally, the fourth ventricle widens on either side to form a recessus lateralis, enclosed by the auriculae cerebelli. The ventricle gradually

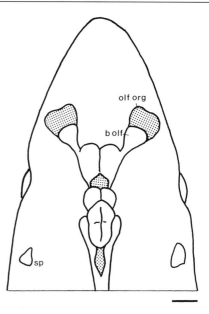

Fig. 12.2. Brain of *Squalus acanthias* in position, dorsal view. *b olf*, bulbus olfactorius; *olf org*, olfactory organ; *sp*, spiraculum. *Bar*, 10 mm

diminishes in the most rostral part of the rhombencephalon, and the region which surrounds this tapering part of the ventricle is known as the isthmus rhombencephali. Dorsally, the fourth ventricle is covered by a folded, highly vascular tela choroidea (Figs. 12.3).

The cerebellum of *Squalus* consists of a central unpaired corpus cerebelli and the paired auriculae. The evaginated corpus cerebelli, enclosing a large ventricular cavity, expands rostrally over the tectum mesencephali and caudally over the fourth ventricle (Fig. 12.3d). The auriculae can be subdivided into a rostromedial upper leaf and a caudolateral lower leaf (Fig. 12.3a,c). The upper leaves of both auricles are connected over the fourth ventricle by a band of nervous tissue, known as the lower lip. Caudolaterally, the auriculae merge into the dorsal parts of the lateral walls of the rhombencephalon, which constitute the prominent areae octavolaterales.

The most rostral part of the brain stem, i.e. the mesencephalon, comprises the tegmentum mesencephali and the tectum mesencephali, externally demarcated from each other by a groove, the sulcus tectotegmentalis (Fig. 12.3c). The tegmentum is caudally continuous with the rhombencephalic basal plate. The tectum, which forms the roof of the midbrain, is large and differentiated into bilateral lobes, which surround expansions of the mesencephalic ventricle. Caudally, the tectum passes into the corpus cerebelli via the relatively thin velum medullare anterius (Fig. 12.3d). This velum constitutes the most dorsal part of the isthmus rhomb-encephali. Rostrally, the dorsal border of the diencephalic-mesencephalic boundary is formed by the commissura posterior.

The diencephalon of the adult *Squalus* retains the tube shape of the early embryo. The thickened lateral walls enclose a slit-like median ventricle, which opens ventrocaudally into lateral expansions within the hypothalamus. The diencephalon, as in other anamniotes, is wedge-shaped in the lateral view with the apex directed dorsally (Fig. 12.3d). The short dorsal border displays the distinct habenular ganglia, which are connected dorso-caudally by the commissura habenulae. Rostral to this commissure, the diencephalon is roofed by a membranous structure, the tela diencephali, which passes rostrally into a deep, transverse fold, the velum transversum (Fig. 12.3d). The diencephalic roof again thins out caudal to the habenular commissure to form a tube-like evagination, the epiphysis.

The rostral boundary of the diencephalon extends from the lateral attachment of the velum transversum through to the chiasma opticum. Rostral to the chiasm, the diencephalic base thins out and forms the preoptic recess and thickens again further rostrally to hold the anterior commissure. The caudal boundary of the diencephalon passes from the posterior commissure to the deep, transverse external fold of the brain floor known as the plica encephali ventralis. Rostral to this fold, there is a marked intraventricular protrusion of the brain wall, the tuberculum posterius. This structure divides off a separate recess, the infundibulum, from the main ventricular system, which tapers caudally into the cavity of the saccus vasculosus (Fig. 12.3d). The recesses of the hypothalamic ventricle are the recessus mamillaris, a dorsal ventricular outpouching appearing caudally in the tuberculum posterius, and the bilateral recesses, which are surrounded by a massive nervous wall and form the lobi inferiores hypothalami. The floor plate of the diencephalon is thin except for its rostral part, which contains the optic chiasm and the post-optic commissure. Part of the diencephalic floor contributes to the neural part of the hypophysis. The epithelial part of the hypophysis, the adenohypophysis, lies against the ventrocaudal surface of the diencephalon.

Two external sulci may be useful for orientation. The first is the sulcus tectodiencephalicus, which marks the boundary between the tectum and the diencephalon and which is caudally continuous with the sulcus tectotegmentalis. The second is the sulcus thalamohypothalamicus, which separates externally the thalamus from the hypothalamus (Fig. 12.3c).

The telencephalon of *Squalus* is well developed and can be readily subdivided into three parts:

1. The bulbus olfactorius
2. The lobus hemisphericus
3. The telencephalon impar

The bulbus olfactorius, which has an ovoid shape, lies with its rostral or rostroventral surface near to the olfactory organs, and thus the primary olfactory nerve fibres are only short. The bulb is connected to the lobus hemisphericus by a distinct pedunculus olfactorius, which consists mainly of secondary olfactory fibres. The lateral ventricle in the lobus hemisphericus of *Squalus* is wide, and the ventricular surface of the neural wall contains a number of distinct grooves. On the basis of these grooves, the lobus hemisphericus has been roughly subdivided into a dorsal pallium and a ventral subpallium.

A remarkable feature of the chondrichthyan forebrain is the fusion of the medial walls of both hemispheres. Both pallial and subpallial zones are involved in this fusion. Johnston (1911) supposed that the connection of both hemispheres is the result of a secondary fusion of the meningeal surfaces of the medial walls of the hemispheres. However, Bäckström (1924) concluded from his embryological studies that the fusion of the hemispheres results largely from a migration of nervous elements into the initially membranous structures that interconnect the two telencephalic halves.

The most caudal portion of the telencephalon is formed by the telencephalon impar. This structure is dorsally covered by a membranous roof, the tela telencephali, and by the caudalmost part of the pallium. Ventrally, the two halves of the telencephalon are connected by a well-developed anterior commissure.

If the gross morphology of the brain of *Squalus* is compared with that of other cartilaginous fishes (Fig. 12.4), it appears that two brain divisions differ considerably: the cerebellum and the telencephalon. The brain stem shows great similarities in all chondrichthyans studied, except for the extraordinarily developed electric lobe in *Torpedo* and the fusion of the dorsal alar plates in *Hydrolagus*. In the cerebellum, it is particularly the corpus cerebelli that varies in external form and size between species. In general, it can be stated that holocephalians and squalomorph sharks have a non-convoluted corpus cerebelli, whereas in galeomorph sharks the cerebellar corpus is convoluted with hypertrophy, which results in asymmetry (Northcutt 1978; Sato et al. 1983). Within the subdivision of the batoids, the more primitive Rajiformes and Torpediniformes have a slightly convoluted cerebellum, while the more advanced batoids, such as *Myliobatis*,

1   Bulbus olfactorius
2   pedunculus olfactorius
3   nervus terminalis
4   lobus hemisphericus telencephali
5   telencephalon impar
6   commissura anterior
7   diencephalon
8   paraphysis cerebri
9   pineal stalk
10 pineal end vesicle
11 commissura habenulae
12 recessus preopticus

13 nervus opticus
14 chiasma opticum
15 commissura posterior
16 tuberculum posterius
17 mesencephalic ventricle
18 foramen interventriculare
19 recessus infundibuli
20 lobus inferior
21 lobus medianus
23 lobus rostralis
24 lobus neuro-intermedius
25 tectum mesencephali

26 nervus oculomotorius
27 nervus trochlearis
28 corpus cerebelli
29 upper leaf of auricle
30 lower leaf of auricle
31 tela choroidea of rhombencephalic ventricle
32 rhombencephalon
33 rhombencephalic ventricle
34 recessus lateralis of rhombencephalic ventricle
35 medulla spinalis
36 canalis centralis

37 nervus trigeminus
38 anterior lateral line nerve
39 dorsal root of anterior lateral line nerve
40 ventral root of anterior lateral line nerve
41 nervus abducens
42 nervus facialis
43 nervus octavus
44 nervus glossopharyngeus
45 posteror lateral line nerve
46 nervus vagus
47 radix ventralis 1

0.5 cm

b

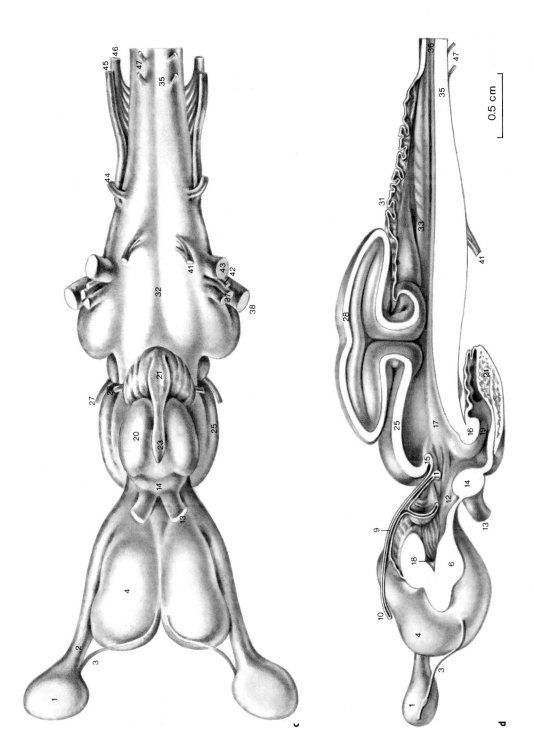

Fig. 12.3. a Dorsal, **b** ventral and **d** midsagittal views of the brain of *squalus acanthias.*

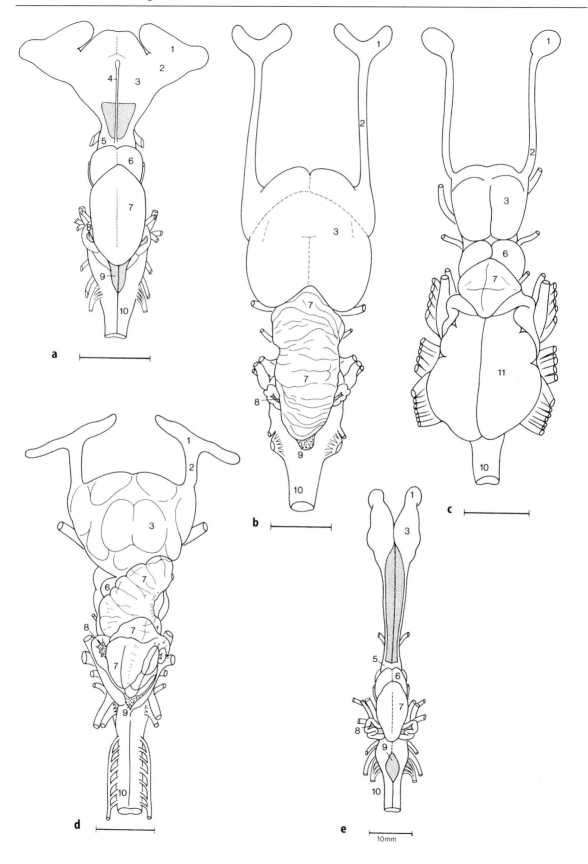

◀ **Fig. 12.4a–e.** Dorsal view of the brains of several species of cartilaginous fishes showing the variation in external brain morphology. **a** *Scyliorhinus canicula.* **b** *Ginglymostoma cirratum.* **c** *Torpedo nobiliana.* **d** *Myliobatis californica.* **e** *Hydrolagus collei. 1,* Bulbus olfactorius; *2,* pedunculus olfactorius; *3,* lobus hemisphericus telencephali; *4,* pineal stalk; *5,* diencephalon; *6,* tectum mesencephali; *7,* corpus cerebelli; *8,* auricula cerebelli; *9,* ventriculus quartus; *10,* medulla spinalis; *11,* lobus electricus

possess a complexely convoluted, asymmetrical cerebellar corpus.

There are considerable structural differences between the telencephala of chondrichthyan species with respect to olfactory bulb insertion, length of olfactory peduncles, degree of fusion between the hemispheric lobes, degree of reduction of the telencephalic ventricle and the relative weight of the telencephalon. These differences will only be mentioned briefly here; for a more detailed description, the reader is referred to the study by Smeets et al. (1983).

The insertion of the olfactory peduncles varies from a rostral position in *Hydrolagus* to a lateral one in *Scyliorhinus* (Figs. 12.4, 12.5). The length of the olfactory peduncle is long in batoids and some sharks, such as *Squalus* and *Ginglymostoma*, but short in *Scyliorhinus*. In the Holocephali, the bulbs

pass imperceptibly into the telencephalic hemispheres. The degree of fusion of the hemispheric walls varies from species to species. In primitive sharks, such as *Clamydoselachus* and *Hexanchus*, it is very restricted, but in sharks such as *Mustelus*, *Scyliorhinus* and *Ginglymostoma* and in batoids, the medial hemisphere walls are completely united. *Squalus* occupies an intermediate position in this respect, whereas fusion is completely lacking in the Holo cephali.

The early development of the telencephalon of batoids closely resembles that of the sharks. Later in development, however, the two groups diverge with regard to the extent of the ventricular cavities. Whereas in most sharks the lateral ventricles remain wide, in batoids they become gradually reduced to narrow, slit-like spaces. In fully grown rays, the lateral ventricles are further obliterated and are merely represented by two short, horn-shaped rudiments (Fig. 12.5). Solid hemispheres are also found in certain shark species, such as the nurse shark *Ginglymostoma*.

The relative development of the telencephalon in chondrichthyans varies considerably. In *Squalus*, the telencephalon constitutes 24% of the total brain weight, in *Scyliorhinus* 30%, and in *Carcharinus* and *Sphyrna* more than 50% (Northcutt 1978).

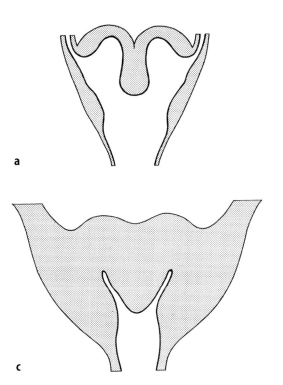
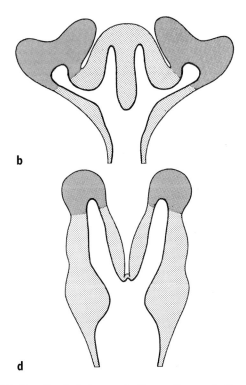

a

b

c

d

Fig. 12.5a–d. Horizontal sections through the forebrain of the sharks **a** *Squalus acanthias* and **b** *Scyliorhinus canicula,* of **c** the skate *Raja clavata* and of **d** the holocephalian *Hydrolagus collei. Heavily shaded parts* indicate olfactory bulbs; *lightly shaded parts* indicate cerebral hemispheres. Note the variation in the extent of the ventricles

12.2.2
Overall Histological Pattern

In contrast to the cyclostomes, the white matter of the chondrichthyan spinal cord comprises numerous myelinated fibres (Waechneldt 1990; Spivack et al. 1993). The spinal grey matter contains cells of a wide size range, although in adults the larger elements are mainly confined to the ventral horns.

Whereas the neuronal perikarya in the brain stem of many groups of anamniotes (e.g. cyclostomes, lungfishes, amphibians) are chiefly located in a continuous zone of central grey, in cartilaginous fishes these elements are spread through the entire width of the wall, a condition similar to that found in amniotes (i.e. reptiles, birds and mammals). However, in amniotes the grey matter of the brain stem is segregated to a considerable extent into discrete nuclei, whereas in cartilaginous fishes a more diffuse architecture prevails. The nerve cells in the brain stem of *Squalus* show considerable variation in size, ranging from 8 to 80 µm. Smeets et al. (1983) subdivided the neurons into four size categories: small (8–20 µm), medium-sized (20–30 µm), large (30–43 µm) and very large (43–80 µm).

Golgi studies on the chondrichthyan brain stem (Houser 1901; Ramón-Moliner and Nauta 1966) and our own silver-impregnated material indicate that cells of the medium-sized, large and very large categories have scanty, long, rectilinear dendrites with relatively few branches. This means that the great majority of the elements in the brain stem of these fishes are "leptodendritic" and "isodendritic", following the general categories used by Ramón-Moliner (1967, 1969) and Ramón-Moliner and Nauta (1966). The long dendrites of these cells form bundles which invade and subdivide the white matter. This highly characteristic reticular organisation is found throughout almost the entire extent of the rhombencephalon and the tegmentum mesencephali. Thus the dendrites of neurons in these parts of the chondrichthyan brain extend far beyond the cytoarchitectonic boundaries of the various nuclei, a fact which is of great importance for the interpretation of the results of experimental hodological studies.

In the diencephalon, most cells maintain a periventricular position, possibly indicating a more primitive condition than that observed in the brain stem. The variation in cell size is also less than in the brain stem, as almost all cell masses are composed of cells that fall into the class of small neurons. In the telencephalon, the cells are spread in most places through the entire width of the wall. The variation in cell size, as in the diencephalon, is much less than in the brain stem. A recent Golgi study by Manso and Anadón (1993) has revealed a much greater variety of cell types in the telencephalon of a shark than that shown in older studies (e.g. Houser 1901).

The studies by Horstmann (1954) and Klatzo (1967) revealed that the elasmobranch brain comprises at least two types of glial elements, i.e. tanycytes and astrocytes. The tanycytes are characterised by long processes than span the entire width of the neural wall and ramify only slightly. Astrocytes do not possess such elongated processes, but have rather numerous, short ramifications. Although some astrocytes, like most tanycytes, lie within the ependymal layer, the majority of astrocytes are found in the proximity of neurons, blood vessels and the meningeal surface. In addition to the two glia types mentioned above, Klatzo (1967) identified oligodendrocytes and microgliocytes. A striking feature observed by Klatzo in the central and rostrodorsal region of the telencephalon of *Negaprion* is a clustering of glial cells, including astrocytes, oligodendrocytes and microgliocytes, around nerve cells. A similar satellitosis has been recognised in some telencephalic areas of *Raja clavata* (Smeets et al. 1983).

12.3
Development

12.3.1
Introductory Notes

The embryonic development of cartilaginous fishes varies considerably in length (for a review, see Wourms 1977), as the rate of development is temperature dependent and also species specific. A short gestation period of 2–4 months has been reported for viviparous rays such as *Myliobatis* and *Dasyatis*. An intermediate duration ranging from 4 to 8 months has been generally found in the genus *Raja*, while a much longer incubation time has been reported for the holocephalian *Hydrolagus*, i.e. 9–12 months. In the common sharks, an incubation time of 6–8 months and 8–12 months has been found in *Scyliorhinus* and *Mustelus*, respectively The 22- to 24-month gestation period in *Squalus* is the longest documented so far.

Since an exact determination of the age of the embryos is impossible, the total body length has mostly been used to indicate the level of development (e.g. Balfour 1878; von Kupffer 1906; Bergquist 1932). Recently, a serious attempt has been undertaken by Ballard et al. (1993) to provide a table of normal stages for *Scyliorhinus*; with slight modifications, this can be adapted to other chondrichthyan fishes as well.

The basis of our knowledge of the ontogenesis of the elasmobranchs was laid down in the comprehensive monography by Balfour (1878). Subsequently, the classical study by von Kupffer (1906) has to be considered as providing a major contribution to our understanding of the morphogenesis of the central nervous system. In a detailed study, von Kupffer described the development of the brain and spinal cord of cartilaginous fishes using embryological material mainly derived from *Squalus*. Later, specific studies appeared on the development of the cerebellum (Palmgren 1921, *Squalus*), mesencephalon (Palmgren 1921, *Squalus*; Farner 1978b,c, *Scyliorhinus*), diencephalon (Bergquist 1932, *Squalus*; Farner 1978a, *Scyliorhinus*) and telencephalon (Bäckström 1924, several species; Holmgren 1922, *Squalus, Chimaera*).

12.3.2
Morphogenesis

The following description of the morphogenesis of the brain of *Squalus* is largely based on the work by von Kupffer (1906). After the neural plate is closed, the neural tube is still attached to the epidermal ectoderm on its cephalic part (Fig. 12.6a). At this stage, the brain can be subdivided into an archencephalon and a deutero-encephalon. The archencephalon is characterised by a widened ventricle which is bended ventrally. On the ventral side, the two divisions are separated from each other by a distinct brain fold, the plica encephali ventralis. The dorsal surface of the brain shows a more gradual flexure. Even at this stage, in embryos with a total length of 4 mm, two lateral evaginations of the archencephalon, i.e. the optic vesicles, can be recognised. In the region which later in development becomes the rhombencephalon, six neuromeres can be distinguished. The first, rostralmost one of these neuromeres provides the cerebellum; the second is primarily connected with the maxillo-mandibular ganglion of the trigeminal nerve, the fourth with the octavo-facial nerve root and the fifth with the glossopharyngeal nerve. The third neuromere does not display a connection with the periphery. In the rostral part of the brain, the neuromeres are less distinct at this stage. This is due to the following factors:

1. The increase in length of this brain part by the broadening of the plica encephali ventralis
2. The increase in the cephalic flexure, which results in inconstant, mechanically raised flexures and folds on the ventral side of the brain
3. The development of the optic vesicles

In *Squalus* embryos of 7–8 mm, a distinct constriction of the neural tube, i.e. the fissura rhombomesencephalica, indicates the boundary between the rhombencephalon and mesencephalon. At the same time, the brain becomes rostrally detached from the epidermal ectoderm. However, only when the embryos are 10 mm in length can five neuromeres separated by external grooves be recognised in the region of the prosencephalon and mesencephalon. This subdivision can best be seen in sagittal sections (Fig. 12.7), since the transverse grooves and ridges lie in the lateral wall of the neural tube and do not continue to the dorsal median plane. From rostral to caudal, these five neuromeres represent the telencephalon, the diencephalon encompassing parencephalon and synencephalon, and the mesencephalon consisting of two neuromeres. In the midsagittal section, the subdivision into telencephalon and diencephalon can also be seen (Fig. 12.6b). A caudally directed evagination indicates the initial development of the epiphysis. On the ventral side of the prosencephalon, a recessus preopticus can already be recognised. The caudal boundary of the mesencephalon is only ventrally marked by a groove, the fissura rhombomesencephalica. Dorsally, the mesencephalic roof plate merges into the lamina ependymalis, which covers the rhombencephalon. On basis of the commissura posterior, von Kupffer (1906) considered the synencephalon as belonging to the diencephalon.

In embryos of 16–17 mm in length (Fig. 12.6c), the cephalic flexure is at its greatest. The telencephalon evaginates no further in a rostral direction, but it evaginates dorsalward and now partly covers the diencephalon. At the transition zone between the telencephalon and diencephalon, the membranous roof of the prosencephalon forms a fold, the velum transversum (Fig. 12.6c). The rostral end of the brain, i.e. the lamina terminalis, increases considerably in thickness, and the processus neuroporicus, i.e. the site at which the neural tube is attached to the epithelial ectoderm, has disappeared. Caudal to the developing epiphysis, the commissura posterior can be clearly recognised, forming the dorsal diencephalic-mesencephalic boundary. At this stage, a dorsal boundary between the mesencephalon and rhombencephalon manifests itself as the plica rhombomesencephalica. Caudal to this fold lies the lamina cerebelli, which later gives rise to the cerebellum. Ventrally, a new groove, the sulcus intraencephalicus posterior, appears.

At the time at which the embryos reach a body length of about 23 mm, the picture changes rapidly, with the appearance of white matter and the formation of commissures. Slightly later in development, in embryos of 27 mm in length (Fig. 12.6d), both the roof plate (pallium) and the lamina terminalis

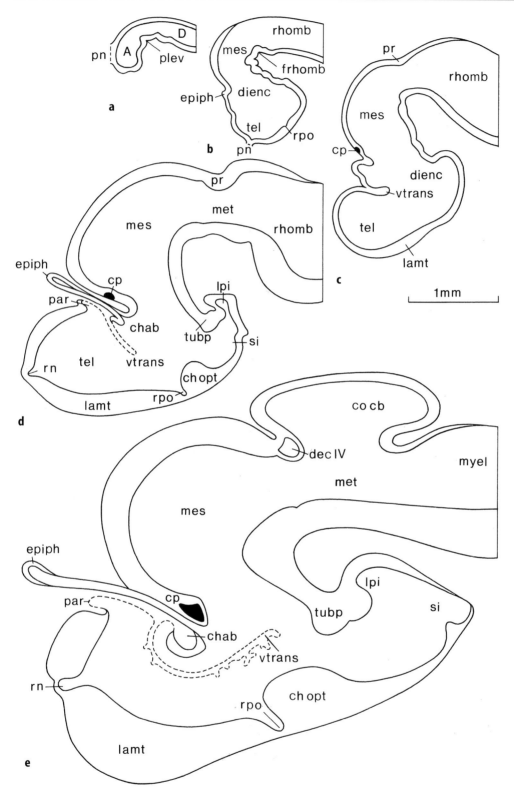

Fig. 12.6a–e. Developmental stages of the brain of *Squalus acanthias* embryos with a total length of **a** 4 mm, **b** 10 mm, **c** 16–17 mm, **d** 27 mm and **e** 70 mm. *A*, archencephalon; *D*, deuterencephalon; *dienc*, diencephalon; *frhomb*, fissura rhombomesencephalica; *lamt*, lamina terminalis; *lpi*, lobus posterior infundibuli; *mes*, mesencephalon; *met*, metencephalon; *myel*, myelencephalon; *par*, paraphysis; *pn*, processus neuroporicus; *pr*, plica rhombomesencephalica; *rn*, recessus neuroporicus; *si*, saccus infundibuli. (Redrawn from von Kupffer 1906)

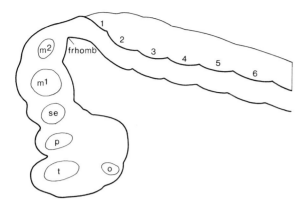

Fig. 12.7. Sagittal section through the brain of a *Squalus* embryo 10 mm in length showing the rostral neuromeres. *frhomb*, fissura rhombomesencephalic; m^1, m^2, mesencephalic neuromeres; *o*, optic vesicle; *p*, parencephalic neuromere; *se*, synencephalic neuromere; *t*, telencephalic neuromere; *1–6*, rhombencephalic neuromeres. (Redrawn from von Kupffer 1906)

(subpallium) have increased in thickness and in the mid-sagittal plane are separated from each other by the recessus neuroporicus. In terms of its position, this recess resembles the processus neuroporicus, which is no longer present, and, according to von Kupffer (1906), its appearance indicates that this part of the neural wall differs from that of other parts of the telencephalon. Because this region does not take part in the formation of cell layers, a conical introversion is formed. Caudally, the pallium does not reach the basis of the velum transversum. The intervening part of the membranous roof of the forebrain evaginates dorsally to form the paraphysis. In the diencephalon, the following structures can be recognised on the dorsal side passing from rostral to caudal: a short, dorsally directed fold, i.e. the parencephalon; commissura habenulae; the epiphysis; and the roof of the synencephalon. Ventrally, a distinct chiasma ridge has developed. The infundibular region can be divided into two parts: the lobus posterior, which later forms the recessus mamillaris, and the saccus infundibuli, which becomes the saccus vasculosus in adults. The tectum mesencephali has risen considerably and its neural wall is thickened. In the rhombencephalon, two subdivisions can be recognised: a narrow, rostral, canal-like metencephalon and a caudal myelencephalon containing a much wider ventricular cavity. The metencephalon, which develops from the first rhombencephalic neuromere, gives rise to the cerebellum. The two neuromeres of the mesencephalon, M_1 and M_2, can no longer be recognised as separate entities. According to Palmgren (1921), almost the whole mesencephalon is made up of the M_1 neuromere, whereas the rudiment of the M_2 seg-

ment forms a boundary plane between the mesencephalon and metencephalon and its reduced ventricle coincides with the sulcus intraencephalicus posterior of von Kupffer (1906).

Further development of the brain in *Squalus* involves thickening of the neural walls of the telencephalon and mesencephalon and, in particular, the development of the cerebellum (Fig. 12.6e). The cerebellar anlage lengthens and the caudodorsal portions protrude to form the auricles, while the corpus cerebelli develops from a dorsal evagination of the rostromedial parts.

Studies by Sterzi (1912) have shown that in sharks the telencephalic side plates bulge out laterally and that the olfactory bulbs are formed later through an outgrowth of the lateral wall of the expanded portions. During the formation of these bulbs, another outpouching appears, more rostral and lateral, thus forming the lateral lobes or hemispheres. Late in development, the olfactory bulbs separate more and more from the hemispheres, and the interconnections, the crura olfactoria or olfactory peduncles, become stalks of considerable length.

The development of the brain of *Scyliorhinus*, as described by Farner (1978a), essentially resembles that of *Squalus*. However, from normal, adult material we know that, although evagination is the essential morphogenetic process in the telencephalon of cartilaginous fishes, particularly in this brain part considerable structural differences exist compared with *Squalus*. These differences concern insertion of olfactory bulbs, size of the ventricles and degree of fusion of the medial hemispheric walls.

12.3.3
Histogenesis

The following description of the histogenesis of the brain of *Squalus* is based on the work of von Kupffer (1906), Palmgren (1921), Holmgren (1922), Bäckström (1924) and Bergquist (1932). In embryos of 10 mm in length, only an almost undifferentiated, periventricular matrix layer can be recognised in the neural tube, although a few differentiating cells occur in the outer zone of this layer. Numerous mitoses can be observed in the matrix layer, but only a few fibres begin to develop.

In embryos of 22 mm in length, the matrix layer can be subdivided into an inner stratum cellulare ependymale and an outer stratum cellulare internum. At some places, a stratum cellulare externum can already be recognised. From this developmental stage onwards, white matter appears, which results in a considerable thickening of the neural wall.

Great mitotic activity is found in the pallial ventricular layer of *Squalus* embryos of 30 mm in length, but differentiation of the primordial cortex does not begin until the embryos reach a body length of 33 mm. The subpallial structures generally develop somewhat earlier than the pallial ones.

Embryos of about 40 mm in length show a subdivision of the tegmentum of the rhombencephalon into longitudinal columns. In the tegmentum mesencephali, Palmgren (1921) distinguished longitudinal columns which he called ventral, medial, lateral and dorsal zones. In somewhat older embryos, 50 mm in length, the development of nuclei is almost complete. A similar subdivision of the rhombencephalon was described by Hugosson (1957) in *Torpedo*. According to the latter author, this subdivision proves, in general terms, that the somatomotor nuclei arise from the ventral column, the visceromotor nuclei from the ventrolateral column, the viscerosensory nuclei from the dorsolateral column and the somatosensory nuclei from the dorsal column. There are, however, several exceptions to this general rule, e.g. the Edinger-Westphal nucleus and the octaval nuclei.

More recently, Farner (1978a) carried out a well-documented study on the development of the brain of *Scyliorhinus* with an emphasis on diencephalic and mesencephalic structures. The histogenesis found in *Scyliorhinus* embryos is essentially the same as in *Squalus*. Farner mentioned that the development of the telencephalon and spinal cord shows some delay when compared with that of the tegmentum mesencephali and the tegmentum rhombencephali. The development of the dorsal horn of the spinal cord occurs after that of the ventral horn.

In the transition zone between the rhombencephalon and medulla spinalis of *Scyliorhinus* embryos of about 6 mm in length, large, dorsal sensory neurons appear. They have peripheral processes which provide sensory endings to the skin. These neurons are usually named after Rohon (1884) and Beard (1896), who first described them for a number of species, including elasmobranchs. In *Scyliorhinus*, the Rohon-Beard cells possess well-myelinated axons and lie in two rows in the dorsal median plane of the spinal cord. The number of these cells is highest when the embryos reach a body length of 39 mm, but it decreases in the subsequent developmental stages. It is noteworthy that axial body movements have already started in 6-mm *Scyliorhinus* embryos; thus the synchronous appearance of both Rohon-Beard cells and axial body movements suggests that these transitory large cells play a role in embryonic movement.

Abbreviations

A	nucleus A
a	sulcus a
apdl	area periventricularis dorsolateralis
app	area periventricularis pallialis
apvl	area periventricularis ventrolateralis
asb	area superficialis basalis
aur	auricula cerebelli
aur ll	auricula cerebelli, lower leaf
aur up	auricula cerebelli, upper leaf
B	nucleus B
b	sulcus b
brc	brachium conjunctivum
C1	nucleus C1
C2	nucleus C2
c	sulcus c
ca	commissura anterior
cans	commissura ansulata
cbb	tractus cerebellovestibularis et cerebellobulbaris rectus
cc	canalis centralis medullae spinalis
ccp	corpus commune posterius
Cer	nucleus cerebelli
ch opt	chiasma opticum
chab	commissura habenulae
cmsp	columna motoria spinalis
co cb	corpus cerebelli
colfi	commissura olfactoria inferior
colfs	commissura olfactoria superior
cp	commissura posterior
cpo	commissura postoptica
cr cb	crista cerebellaris
cst	commissura superior telencephali
ctect	commissura tecti mesencephali
ctrans	commissura transversa
cv	cornu ventrale
cvt	commissura ventralis telencephali
D	nucleus D
d	sulcus d
ddp	decussatio dorsalis pallii
dec tpal	decussatio tractus pallii
dec IV	decussatio nervi trochlearis
E	nucleus E
e	sulcus e
emth	eminentia thalami
emvcb	eminentia ventralis cerebelli
Ent	nucleus entopeduncularis
epiph	epiphysis
EW	nucleus of Edinger-Westphal
F	nucleus F
f	sulcus f
farc	fibrae arcuatae
fbt	fasciculus basalis telencephali
Fd	nucleus funiculi dorsalis
fd	funiculus dorsalis

fi	foramen interventriculare (Monroi)	Nob	nucleus opticus basalis
Fl	nucleus funiculi lateralis	Nph	nucleus periventricularis hypothalami
fl	funiculus lateralis		
flm	fasciculus longitudinalis medialis	Nsc	nucleus suprachiasmaticus
fpd	fasciculus predorsalis	Nscd	nucleus septi caudodorsalis
fretr	fasciculus retroflexus	Nscv	nucleus septi caudoventralis
fsol	fasciculus solitarius	Nsl	nucleus septi lateralis
fSt	fasciculus medianus of Stieda	Nsm	nucleus septi medialis
fstr	fibrae striatae	Nsma	nucleus septi medialis, pars anterior
fv	funiculus ventralis	Nsmd	nucleus septi medialis, pars dorsalis
fvm	fasciculus ventromedialis of Schroeder and Ebbesson	Nsv	nucleus sacci vasculosi
		nterm	nervus terminalis
G	nucleus G	nII	nervus opticus
Gc	griseum centrale rhombencephali	nIII	nervus oculomotorius
H	nucleus H	nIV	nervus trochlearis
Hab	ganglion habenulae	nV	nervus trigeminus
hypoph	hypophysis	nVI	nervus abducens
Ic	nucleus intercollicularis	nVII	nervus facialis
Ica	nucleus interstitialis commissurae anterioris	nVIII	nervus octavus
		nIX	nervus glossopharyngeus
Ifbt	nucleus interstitialis fasciculi basalis telencephali	nX	nervus vagus
		olcb	tractus olivocerebellaris
inf	infundibulum	Oli	oliva inferior
Ipd	nucleus interpeduncularis, pars dorsalis	oph	organon periventriculare hypothalami
Ipv	nucleus interpeduncularis, pars ventralis	osc	organon subcommissurale
		P	nucleus P
Is	nucleus isthmi	P1, P2 etc.	area pallialis 1, 2 etc.
lb	tractus lobobulbaris	P9m	area pallialis 9, pars medialis
leml	lemniscus lateralis	P9l	area pallialis 9, pars lateralis
lemV	lemniscus trigeminalis	pal	pallium
lih	lobus inferior hypothalami	Pdc	pallium dorsale, pars centralis
lob vlm	lobus vestibulolateralis, pars medialis	Pds	pallium dorsale, pars superficialis
lobX	lobus vagi	Pdsp	pallium dorsale, pars superficialis posterior
lowl	lower leaf of auricula cerebelli		
lsp	lemniscus spinalis	pg	prominentia granularis
M	nucleus M	Pl	pallium laterale
msp	medulla spinalis	plev	plica encephali ventrale
N	nucleus N	Pm	pallium mediale
N dors	nucleus dorsalis areae octavolateralis	Po	nucleus preopticus
		pret	area pretectalis
N interm	nucleus intermedius areae octavolateralis	Pretc	nucleus pretectalis centralis
		Prets	nucleus pretectalis superficialis
Ncp	nucleus commissurae posterioris	Rai	nucleus raphes superior
Nflm	nucleus fasciculi longitudinalis medialis	ranllav	radix ascendens nervi lineae lateralis anterioris, pars ventralis
nllad	nervus lineae lateralis anterior, pars dorsalis	ranllp	radix ascendens nervi lineae lateralis posterioris
		Ras	nucleus raphes superior
nllav	nervus lineae lateralis anterior, pars ventralis	raVIII	radix ascendens nervi octavi
		rb	area retrobulbaris
nllp	nervus lineae lateralis posterior	rd	radix dorsalis nervi spinalis
Nlobl	nucleus lobi lateralis	rdnllav	radix descendens nervi lineae lateralis anterioris, pars ventralis
Nlobld	nucleus lobi lateralis, pars dorsalis		
Nloblm	nucleus lobi lateralis, pars medialis	rdV	radix descendens nervi trigemini
Nloblv	nucleus lobi lateralis, pars ventralis	rdVIII	radix descendens nervi octavi
Nlt	nucleus lateralis tuberis		
Nmh	nucleus medius hypothalami		

rhomb	rhombencephalon
Ri	nucleus reticularis inferior
Ris	nucleus reticularis isthmi
Rm	nucleus reticularis medius
rmam	recessus mamillaris
rmeV	radix mesencephalica nervi trigemini
rmVII	radix motoria nervi facialis
rmX	radix motoria nervi vagi
rne	recessus neuroporicus externus
rni	recessus neuroporicus internus
rpo	recessus preopticus
Rs	nucleus reticularis superior
rsp	tractus rubrospinalis
Rub	nucleus ruber
rv	radix ventralis nervi spinalis
sdd1	sulcus diencephalicus dorsalis 1
sdd2	sulcus diencephalicus dorsalis 2
sdm	sulcus diencephalicus medius
sdv	sulcus diencephalicus ventralis
sg	substantia gelatinosa
sgr	stratum granulare cerebelli
shyp	sulcus hypothalamicus
sid	sulcus intermedius dorsalis
sih	sulcus intrahabenularis
siv	sulcus intermedius ventralis
slH	sulcus limitans of His
sll	sulcus limitans lateralis
slm	sulcus limitans medialis
sm	stria medullaris
smi	sulcus medianus inferior
smol	stratum moleculare cerebelli
sms	sulcus medianus superior
sos	sulcus organi subcommissuralis
sP	stratum Purkinje
SP1, SP2 etc.	area subpallialis 1, 2 etc.
spal	subpallium
spcb	tractus spinocerebellaris
spo	sulcus preopticus
spth	tractus spinothalamicus
ssh	sulcus subhabenularis
std	sulcus tectodiencephalicus
stdl	sulcus telencephalicus dorsolateralis
stdm	sulcus telencephalicus dorsomedialis
sth	sulcus thalamohypothalamicus
str	striatum
stt	sulcus tectotegmentalis
stv	sulcus telencephalicus ventralis
stvm	sulcus telencephalicus ventromedialis
sv	saccus vasculosus
t	taenia
tchdi	tela chorioidea diencephali
tchrh	tela chorioidea rhombencephali
tb	tractus tectobulbaris
tbv	tractus tectobulbaris ventralis
tect	tectum mesencephali
tecthab	tractus tectohabenularis
Tegl	nucleus tegmentalis lateralis
tegm	tegmentum mesencephali
telenc	telencephalon
th	thalamus
thd	thalamus dorsalis
thdl	thalamus dorsalis, pars lateralis
thdm	thalamus dorsalis, pars medialis
thdmd	thalamus dorsalis, pars dorsomedialis
thdmv	thalamus dorsalis, pars ventromedialis
thtect	tractus thalamotectalis
thtegm	tractus thalamotegmentalis
thtel	tractus thalamotelencephalicus
thv	thalamus ventralis
thvl	thalamus ventralis, pars lateralis
thvm	thalamus ventralis, pars medialis
tolf	tractus olfactorius
tolfl	tractus olfactorius lateralis
tolfm	tractus olfactorius medialis
tomsc	tractus olfactorius medialis septi cruciatus
topt	tractus opticus
toptb	tractus opticus basalis
toptl	tractus opticus lateralis
toptm	tractus opticus medialis
Torsc	nucleus tori semicircularis
torsc	torus semicircularis
tpal	tractus pallii
tpall	tractus pallii, pars lateralis
tpalm	tractus pallii, pars medialis
tpalp	tractus pallii, pars periventricularis
tpals	tractus pallii, pars superficialis
tpohyp	tractus preopticohypophyseus
tps	tractus pallialis septi
tpsl	tractus pallialis septi, pars lateralis
tpsm	tractus pallialis septi, pars medialis
tsv	tractus sacci vasculosi
tt	fasciculus tectotegmentalis
ttd	fasciculus tectotegmentalis dorsalis
tti	fasciculus tectotegmentalis intermedius
ttv	fasciculus tectotegmentalis ventralis
Tubp	nucleus tuberculi posterioris
tubp	tuberculum posterius
upl	upper leaf of auriculae cerebelli
vbo	ventriculus bulbi olfactorii
vimp	ventriculus impar telencephali
vl	ventriculus lateralis telencephali
vma	velum medullare anterius
vtrans	velum transversum
zll	zona limitans lateralis
zlm	zona limitans medialis
III	nucleus nervi oculomotorii
IV	nucleus nervi trochlearis
Vd	nucleus descendens nervi trigemini
Vm	nucleus motorius nervi trigemini

Vme	nucleus mesencephalicus nervi trige-mini
Vs	nucleus princeps nervi trigemini
VI	nucleus nervi abducentis
VIIm	nucleus motorius nervi facialis
VIIIa	nucleus octavus ascendens
VIIId	nucleus octavus descendens
VIIIm	nucleus octavus magnocellularis
IXm	nucleus motorius nervi glossopha-ryngei
Xm	nucleus motorius nervi vagi
Xml	nucleus motorius nervi vagi, pars lateralis
Xmm	nucleus motorius nervi vagi, pars medialis

12.4
Spinal Cord

12.4.1
Introductory Notes

The spinal cord of chondrichthyan fishes merges rostrally with the rhombencephalon and ends caudally in the tail fin, surrounded by lymph tissue. It retains the tube shape of the early embryo, being slightly flattened (*Squalus*) or cylindrical (*Raja*) (Fig. 12.8). The spinal cord is widest at its junction with the rhombencephalon and gradually tapers in the caudal region. There are no regional enlargements of the cord. It consists of a series of segments, essentially identical in organisation, that match the segmental divisions of the body musculature.

As in other vertebrates, each segment of the spinal cord gives rise to a pair of dorsal and a pair of ventral roots. The ventral roots of each segment attach to the spinal cord slightly more rostrally than do the dorsal roots. After passing through the vertebral cartilage, the dorsal and ventral roots become enclosed in a common connective sheath and form the mixed spinal nerve, from which dorsal, intermediate and ventral rami fork off. The mixed nerve also gives rise to a small ramus communicans, carrying sympathetic fibres to the chain of sympathetic ganglia; there is no grey ramus communicans (Young 1933a).

Although the dorsal and ventral root fibres intermix in the peripheral nerves of most chordates, in many elasmobranchs they remain separate, although they are contained within a common nerve (Roberts 1969; Coggeshall et al. 1978); only in some rays (*Raja*) do sensory and motor fibres appear to intermix within the nerve (Roberts 1969).

Studies by Coggeshall et al. (1978) and Leonard et al. (1978a) revealed several interesting features of spinal cord organisation in the stingray *Dasyatis*. Firstly, the numbers of dorsal root ganglion cells, dorsal root axons, ventral root axons and "motor" cells in the ventral horn increase steadily as the animals increase in size. This increase in axonal and neuronal numbers persists much further into adult life than is the case for other vertebrates that have been studied from this point of view. A similar increase in number and size of spinal motoneurons and axons has been reported for *Scyliorhinus* (Mos and Williamson 1986). Secondly, unmyelinated fibres are virtually absent from these nerves. This observation is in sharp contrast with the condition found in mammals, in which two morphologically distinct populations of sensory neurons have been identified: those with myelinated axons and large-diameter cell bodies (A fibres) and those with unmyelinated axons and small diameter somata (C fibres) (Willis and Coggeshall 1991). The virtual absence of unmyelinated fibres in the dorsal root has recently been confirmed by Snow et al. (1993) in a quantitative study in which the proportion of myelinated versus unmyelinated sensory fibres in the dorsal roots of three species of elasmobranch fishes was ascertained. In support of the virtual lack of unmyelinated fibres, the latter authors found no evidence of a bimodal distribution in the diameters of the dorsal root ganglia cells. At the present time, it is not possible to say whether the information usually carried by the unmyelinated fibres in mammals (C-fibre system) is lacking in cartilaginous fishes. The afferent fibres respond to low- and high-threshold mechanical input (Leonard et al. 1978c), but do not respond to thermal or chemical stimuli. These data suggest that the absence of unmyelinated afferents may reflect an absence of certain categories of sensory input to the spinal cord. Finally, physiological examination shows that a compound action potential with a single peak is obtained from the ventral root and motor part of the peripheral nerve and a potential with two peaks from the dorsal root and sensory part of the peripheral nerve. According to Coggeshall et al. (1978), the unimodal distribution of the ventral root presumably implies either that there is only one class of motor axons in these roots or that, if these are several categories, their size distribution forms a continuum. According to the same author, on basis of their conduction velocities the two peaks of the dorsal root seem to correspond to the A-α, β and A-δ sensory fibres in mammals.

Though the great majority of the sensory nerves have cell bodies located in the dorsal root ganglia close to the vertebral column, at least some sensory fibres appear to enter the spinal cord via a few ventral roots (Grillner et al. 1976; *Squalus*). Using elec-

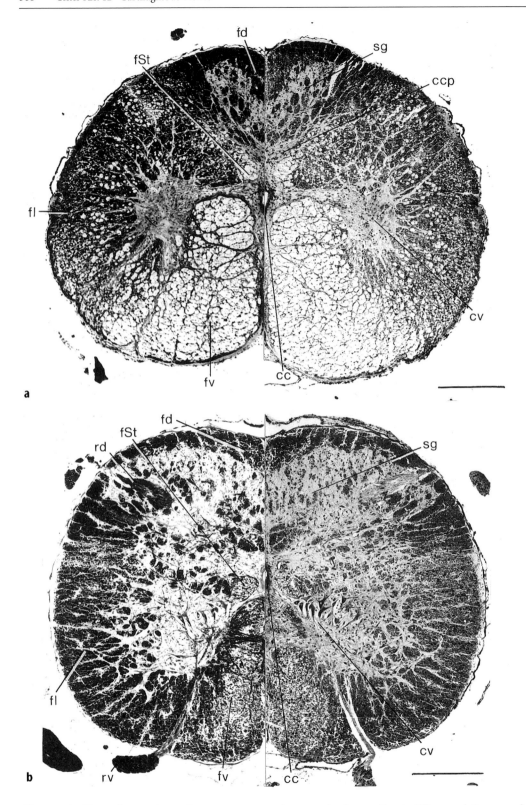

Fig. 12.8a,b. Photomicrographs of transverse sections through the 'cervical' cord of **a** the shark *Squalus acanthias* and **b** the skate, *Raja clavata*. The *left half* of each section is stained according to Klüver and Barrera, whereas the *right half* is stained with cresyl violet. *Bar*, 500 μm

trophysiological methods, Roberts (1969) found that, in *Scyliorhinus*, the area of the body which is innervated by any one spinal nerve (the dermatome) is trapezoidal in shape and extends over at least six body segments, four of which are caudal to the root from which recordings were made.

General sensory information is provided by the "free-nerve" terminations of nerve fibres, which form a network close to the skin surface. Specialized nerve terminals are found deeper in the skin, in connective myocommata, between vertebrae and among fin muscle fibres. The corpuscular endings of Wunderer (1908) provide general mechanoreceptive information and may serve a proprioceptive function; the intermuscular endings of Poloumordwinoff (1898) found in the pectoral fins of rays are stretch receptors. Muscle spindles and Golgi tendon organs are apparently absent from chondrichthyan musculature.

12.4.2
General Organisation

As in all gnathostomes, the grey matter of the chondrichthyan spinal cord is arranged in dorsal and ventral horns. The dorsal horns, which are separated by small dorsal funiculi, fuse central as a mass, i.e. the corpus commune posterius of Keenan (1928), just dorsal to the central canal (Figs. 12.9–12.12). Keenan thought that almost the entire dorsal horn of these fishes constitutes a substantia gelatinosa. Comparison of the spinal cord of various chondrichthyans shows that there are large differences between species in the extent and shape of the dorsal and ventral horns (Fig. 12.8; Smeets et al. 1983; Cameron et al. 1990).

The grey matter contains cells of widely differing sizes, but the larger elements are, in adult forms, mainly confined to the ventral horns. In early developmental stages, the dorsal region of the chondrichthyan cord contains numerous Rohon-Beard cells, but these elements disappear far before the adult stage has been reached. On the basis of stimulation experiments and embryological observations, the neurons of the ventral horn in *Squalus* have been subdivided by Glees (1940) into three groups:

1. A ventrolateral group of somatomotor cells
2. An intermediate group of commissural neurons
3. A dorsomedial group of preganglionic sympathetic elements

An experimental horseradish peroxidase (HRP) study by Droge and Leonard (1983a) also revealed the existence of three distinct cell groups in the ventral horn of the stingray *Dasyatis*:

1. A lateral group consisting of motoneurons which innervate the dorsal (elevator) pectoral fin muscles
2. A dorsomedial group of neurons innervating the ventral (depressor) pectoral fin muscles
3. A ventral group of neurons which innervate the epaxial muscles

In the tail region of rays (Rajidae), the innervation of the small electric organs is provided by motoneurons which are much larger than adjacent motoneurons that innervate conventional musculature (Ewart 1893).

In all chondrichthyans, the tail region of the spinal cord contains cells that form a caudal neurosecretory unit, the urophysis. One element of this, the Dahlgren cell, is a truly enormous neuron (in *Raja* up to 300 μm diameter), extending several short processes to the vascular meninx (Dahlgren 1914; Speidel 1922; Onstott and Elde 1986). The urophysis is thought to serve a role in osmoregulation (Bern 1969).

All fibres contained within the dorsal root of chondrichthyans originate in adults from extramedullary spinal ganglion cells. After their entrance into the spinal cord, the dorsal root fibres bifurcate and run rostrally and caudally. These ascending and descending axons constitute the small dorsal funiculi and various so-called intercornual bundles. The lengths of axons constituting the dorsal funiculi have never been determined, but it is known that the dorsal funiculi show no clear frontal accumulation and distinct funicular nuclei are therefore lacking. Ariëns Kappers et al. (1936) found that in the ray the dorsal root enters the dorsal horn through its lateral aspect and then divides into a dorsally projecting bundle of fine afferents that terminate in the substantia gelatinosa and into a ventrally projecting bundle of coarser afferents that end deeper in the dorsal horn. This finding has received electrophysiological support from a study by Leonard et al. (1978b). The latter study also revealed evidence that, in *Dasyatis*, monosynaptic activation of motoneurons from dorsal root stimulation occurs, although most of the reflex discharge is polysynaptic.

As already mentioned, the dorsal funiculus is small and contains only thin fibres (Figs. 12.10–12.12). The lateral funiculus also consists mainly of thin fibres, but coarse fibres are found in the fasciculus medianus of Stieda (1873) lying in the deeper zone, dorsolateral to the central canal, and more diffusely arranged in the outer zone. The majority of coarse axons in the chondrichthyan cord, however, are confined within the area of the ventral funiculus.

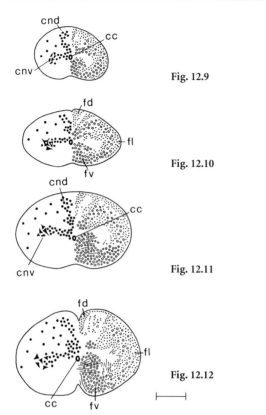

Figs. 12.9–12.12. A series of transverse sections through the spinal cord of the spiny dogfish. **Fig. 12.9.** Lumbal level. **Fig. 12.10.** The mid-thoracic level. **Fig. 12.11.** The pectoral level. **Fig. 12.12.** The cervical level. *Left,* cell picture, based on a Nissl-stained series; *right,* fibre systems, based on Bodian and Klüver-Barrera-stained series. *Bar,* 1 mm

12.4.3
Ascending Fibre Systems

So far, the spinal ascending pathways to the brain of chondrichthyan fishes have been studied experimentally in the dogfish *Scyliorhinus* (Hayle 1973a,b) and the nurse shark *Ginglymostoma* (Ebbesson and Hodde 1981) and are shown in Fig. 12.13a. Hayle (1973a) could not identify a distinct dorsal column nucleus, although he could trace some degenerating fibres in the dorsal funiculus. Most of the ascending fibres, however, were found in the lateral funiculus, and these were collectively described by Hayle as the lemniscus spinalis. He found ascending spinal pathways reaching to the predominantly ipsilateral rhombencephalon and mesencephalon and ending as mossy fibres in the cerebellar corpus. Ebbesson and Hodde (1981), in a degeneration study in *Ginglymostoma*, found that the spinal lemniscus comprises three components in the brain stem: one that reaches the cerebellum (also receiving a small input from the dorsal

funiculus), a second that projects to the mesencephalon and to the thalamus and a third that ends in the rhombencephalon, the latter being far the largest. Most of the spinorhombencephalic fibres terminate in the reticular formation, but some project to the motor nucleus of the vagus. The projection to the cerebellum terminates as mossy fibres among the granule cells of the corpus cerebelli and also distributes to the cerebellar nuclei. A retrogradely tracing study by Fiebig (1988) revealed that, in *Platyrhinoidis*, small and large neurons at the base of the dorsal and ventral horns projected to the cerebellar corpus. The spinal projection to the mesencephalon passes to the deep layers of the tectum, the nucleus intercollicularis, the periaqueductal grey and the nucleus tegmentalis lateralis. The neurons which project to the tectum are diffusely arranged in the dorsal part of the cervical cord (Smeets 1982, *Scyliorhinus, Raja*). In *Ginglymostoma*, the projection to the thalamus terminates within the central thalamic nucleus; in *Scyliorhinus*, Hayle (1973a) was unable to find a spinal projection to the thalamus, but R.L. Boord (unpublished observation) found a thalamic projection in the clearnose skate after spinal cord hemisections.

12.4.4
Descending Fibre Systems

Long-axonal pathways descending from the brain have been identified in *Scyliorhinus* and *Raja* in material labelled retrogradely with HRP (Smeets and Timerick 1981; Timerick et al. 1992). These pathways are shown in Fig. 12.13b. Following HRP injections in the spinal cord, labelled cells were found in the rhombencephalon, mesencephalon and diencephalon. The rostralmost neurons that send axons to the spinal cord lie in the thalamus ventralis pars medialis, the nucleus periventricularis hypothalami and the nucleus commissurae posterioris (the latter two only in *Scyliorhinus*). Projections from the mesencephalon arise from the nucleus of the fasciculus longitudinalis medialis, the tectum mesencephali, the nucleus intercollicularis and the tectotegmental junction zone and from diffusely arranged tegmental neurons. A contralateral rubrospinal pathway was identified in *Raja*, but not in *Scyliorhinus*.

The major descending pathways from the rhombencephalon originate from reticular and octaval neurons. The nuclei B, F and G, the nucleus tractus descendens nervi trigemini, the nucleus funiculi lateralis and the nucleus tractus solitarii also send fibres into the cord.

Selective injections of HRP at four different levels in the spinal cord of the dogfish *Scyliorhinus*,

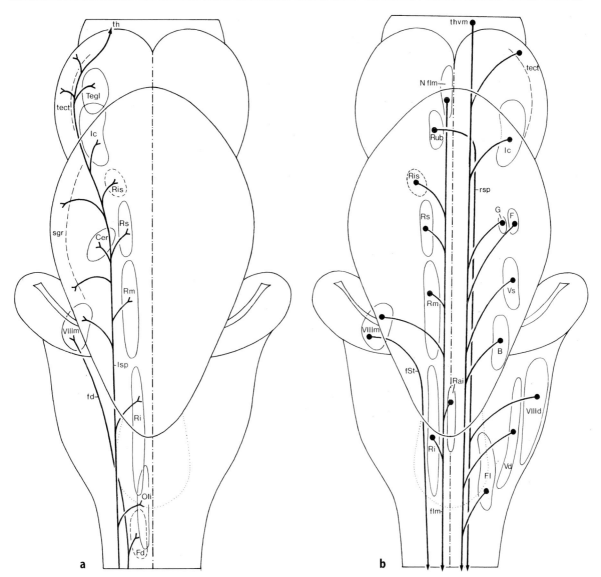

Fig. 12.13a,b. Experimentally verified connections of the spinal cord in cartilaginous fish. **a** Ascending spinal pathways. **b** Pathways descending to the spinal cord. Only the main ipsilateral pathways are illustrated, although most projections are, in fact, bilateral

i.e. cervical (segments 3–6), pectoral (segments 16–18), pelvic (segments 34–36) and caudal (segments 60–80) cord, have provided more detailed information about supraspinal control of the spinal cord (Timerick et al. 1992). Of the three diencephalic nuclei, only the nucleus thalamus ventralis pars medialis projects further than cervical regions reaching pelvic levels. From the midbrain, the tectospinal projection, which arises from tectal and tectotegmental regions, reaches only as far as the cervical cord. Supraspinal inputs which do not project beyond the pectoral cord originate from cell bodies in the nucleus octavus descendens, the nucleus descendens nervi trigemini and the nuclei

B, F and G (see Fig. 12.13b). Only reticular cells in the midbrain and rhombencephalon as well as cells in the magnocellular octaval nucleus reach caudal spinal cord levels. Timerick and colleagues concluded that only a few brain stem nuclei could directly affect the whole spinal cord, but many nuclei might have an impact on its rostralmost regions. It should be noted that the latter authors found labelled spinal interneurons with long descending axons that reach the caudalmost cord levels.

For the most part, the route followed by descending fibres in the spinal cord and the sites of termination are unknown. However, it seems probable that the reticular nuclei and some of the octaval

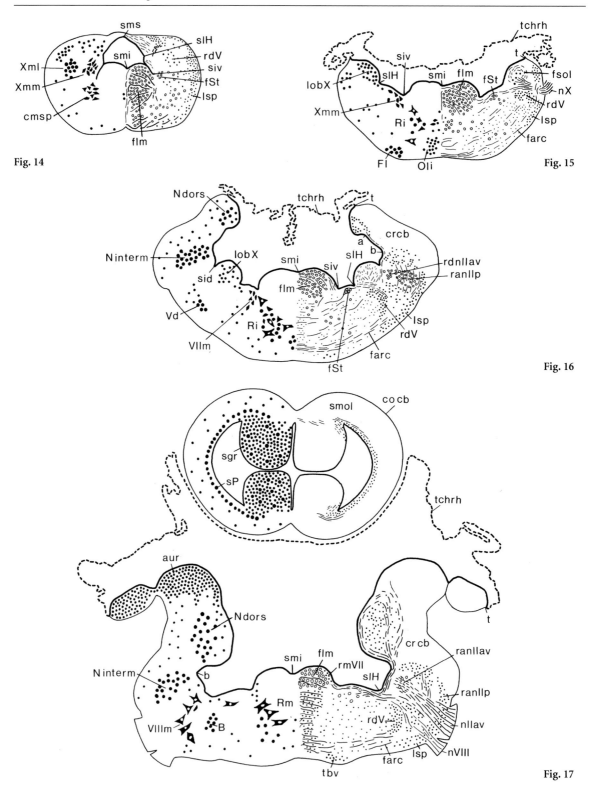

Fig. 14

Fig. 15

Fig. 16

Fig. 17

Figs. 12.14–12.21. A series of transverse sections through the brain stem of *Squalus acanthias.* The levels of the sections are indicated in Fig. 12.22. For further details, see the legend to Figs. 12.9–12.12

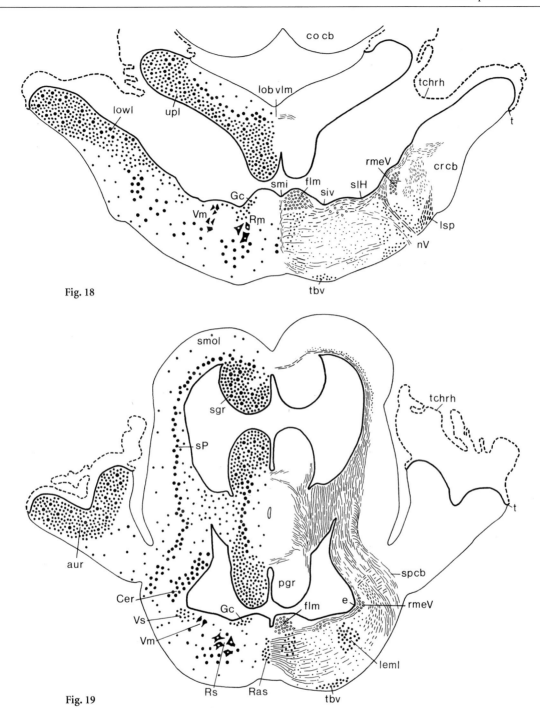

Fig. 18

Fig. 19

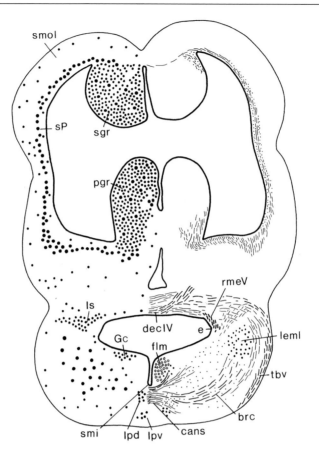

Fig. 20

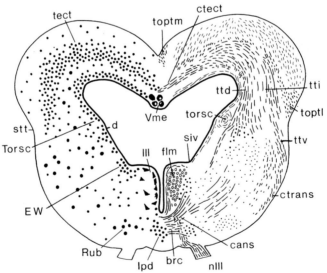

Fig. 21

nuclei project via the fasciculus longitudinalis medialis and that some of the efferents of the nucleus octavus magnocellularis descend by way of the fasciculus medianus of Stieda. The fasciculus longitudinalis medialis is evidently the principal encephalospinal pathway in cartilaginous fishes. This bundle is very large in *Squalus*, *Scyliorhinus* and *Hydrolagus*, but less well developed in *Raja* (Smeets et al. 1983). It extends from the rostral part of the mesencephalon throughout the brain stem and continues caudally in the ventral funiculus of the spinal cord (Figs. 12.9–12.12, 12.14–12.21). Most of its fibres are coarse and well myelinated. It mainly receives contributions from the midbrain reticular formation, i.e. the nucleus of the fasciculus longitudinalis medialis, from the rhombencephalic median and medial reticular formation and from the nucleus octavus magnocellularis (Fig. 12.13b). In Holocephali, the Mauthner cells also contribute their axons to the fasciculus longitudinalis medialis.

The fasciculus medianus of Stieda is formed by large axons of the magnocellular octaval nucleus (Smeets and Timerick 1981; Smeets et al. 1983) and extends a long way caudally into the spinal cord (Figs. 12.9–12.12, 12.14–12.16). This bundle contains about 50 fibres in *Raja* and *Hydrolagus* and about 25 fibres in *Squalus* and *Scyliorhinus* (Smeets et al. 1983).

In a preliminary report, Livingston et al. (1983), using the HRP technique combined with lesions rostral to the injection site, have demonstrated the trajectories of several descending pathways in the spinal cord of the stingray *Dasyatis*. According to these authors, the rhombencephalic reticular formation projects via the dorsolateral, lateral and ventrolateral funiculus, whereas the octavospinal projections course mainly through the more ventral funiculi. A contralateral rubrospinal pathway was demonstrated as projecting by way of the dorsolateral funiculus. A study by Williams et al. (1984) revealed that stimulation of intermediate portions of the lateral funiculus in the rostral spinal cord of *Dasyatis* produced swimming-like rhythmic activity, whereas lesions of this same area produced severe swimming deficits. Together, these findings suggest that the descending supraspinal pathways that play a role in initiating swimming movements project through the intermediate portions of the lateral funiculus.

12.4.5
Chemoarchitecture

As in other vertebrates, increasing interest is being shown in the occurrence and distribution of the various candidate neurotransmitters in the chon-drichthyan central nervous system. In the last decade, data about neurotransmitters in the spinal cord of cartilaginous fishes are steadily accumulating by means of immunohistochemical studies (Ritchie and Leonard 1982, 1983a; Ritchie et al. 1984; Onstott and Elde 1986; Roberts and Meredith 1987; Cameron et al. 1990).

The distribution and origin of four peptide neurotransmitter candidates of primary afferents to the spinal cord, i.e. substance P (SP), somatostatin, cholecystokinin (CCK) and vasoactive intestinal polypeptide (VIP), were studied by Ritchie and Leonard (1983a) in *Dasyatis* using the peroxidase-antiperoxidase (PAP) technique. SP-like immunoreactivity was densely distributed in the superficial aspect of the substantia gelatinosa, particularly laterally, and was scattered in the deeper part of the substantia gelatinosa, in the intermediate zone and in the ventral horn. The distribution of somatostatin-, CCK- and VIP-like immunoreactivity was similar, but different from that of SP. Stained fibres appeared to issue from a prominent tract in the dorsolateral funiculus to form a plexus at the lateral margin of the deeper part of the substantia gelatinosa. From there, the fibres spread dorsally and medially through the substantia gelatinosa to terminate in a thin band at the superficial margin of this structure. Somatostatin- and CCK-like immunoreactivity was more dense in the lateral third of this band, whereas VIP immunoreactivity was more diffusely distributed within this structure. Somatostatin-positive cell bodies were observed in the ventral horn, in the deep dorsal horn and in the ependymal layer and are considerable in number. CCK-positive cells, which were few in number, were observed in the medial part of the ventral horn, while VIP-positive cells were located subjacent to the superficial aspect of the substantia gelatinosa and within the dorsolateral funiculus. Ritchie and Leonard (1983a) also demonstrated that, of the four peptides studied, only SP appears to be a candidate primary afferent neurotransmitter.

A relatively well studied neurotransmitter in the spinal cord of cartilaginous fishes is serotonin. Serotoninergic cells, terminals and axons have been immunocytochemically demonstrated in the spinal cord of *Dasyatis* (Ritchie and Leonard 1982). Serotonin-like immunoreactivity, as demonstrated by the PAP technique, was found in cell bodies lying along the ventral median fissure, within the ventral funiculus, lateral or ventral to the central canal, at the ventromedial margin of the ventral horn and within the meninges covering the ventral surface of the spinal cord. Serotonin-immunopositive cells were found at all segmental levels of the spinal cord, and no segmental variation was observed in the pattern of immunostaining. Serotoninergic terminals were densely distributed throughout the superficial portion of the substantia gelatinosa, whereas the immediately subjacent region contained stained fibres but few terminals. The remaining regions of the dorsal horn and the entire ventral horn contained a dense plexus of fine fibres and terminals. Serotoninergic fibres were also observed in the funicular regions, except for the ventromedial portion of the ventral funiculus and the dorsal funiculus.

By transectioning the spinal cord and comparing the distribution of serotonin-immunoreactive elements caudal and rostral to the transection, Ritchie et al. (1984) demonstrated that the descending pathways provide virtually the entire serotoninergic innervation of the dorsal horn, the intermediate zone and the dorsal and lateral portions of the ventral horn. The intrinsic spinal serotoninergic system consists of cells distributed in the ventromedial spinal cord that have processes extending longitudinally in a ventral submeningeal fibre network.

Using an antibody against dopamine, Roberts and Meredith (1987) found immunoreactive somata ventral to, and in close association with, the central canal in the spinal cord of the sandy ray *Raja radiata*. The majority of the cell bodies lie ventral to the central canal, but in the rostral part of the cord, some cell bodies lie dorsal to the canal. Typically, two to three dopamine-immunoreactive cell bodies are seen in each 40-μm-thick section. These cells form an almost continuous column that extends throughout the rostrocaudal length of the cord. A typical feature of each cell is that it sends a single, relatively thick process through the ependymal layer to the lumen of the central canal, where it probably contacts the cerebrospinal fluid. These cells also give rise to one or more fine processes which run in a ventral or lateral direction, away from the central canal.

More recently, the cytoarchitecture and chemoarchitecture of the dorsal horn in four species of elasmobranch fishes have been studied in detail (Cameron et al. 1990). The distribution of immunoreactivity to serotonin, SP, somatostatin, calcitonin gene-related peptide, neuropeptide Y and bombesin was determined in three species of rays (*Dasyatis fluviorum*, *Aetobatis narinari*, and *Rhinobatis battilum*) and one species of shark (*Carcharinus melanopterus*). In addition, some spinal cord sections of the shovelnose ray (*Rhinobatis*) were treated with antibodies against met-enkephalin. In all species, many fibers and varicosities immunoreactive to SP, calcitonin gene-related peptide and bombesin were found in the outer part of the substantia gelatinosa, whereas smaller numbers of

fibres were observed in the nucleus proprius. Immunoreactivity to somatostatin consisted of coarse fibre bundles which entered the dorsal horn at the nucleus proprius and radiated dorsally to the substantia gelatinosa. Serotonin- and neuropeptide Y-immunoreactive fibres were found in all regions of the dorsal horn, but were more concentrated in the outer part of the substantia gelatinosa. Met-enkephalin immunoreactivity in the dorsal horn of the spinal cord of the shovelnose ray was concentrated in the lateral third of the substantia gelatinosa and to a lesser extent in the nucleus proprius. In mammals, with the exception of calcitonin gene-related peptide, these peptides have been largely associated with unmyelinated primary sensory neurons. Considering that in elasmobranchs unmyelinated primary sensory fibres are virtually lacking, and yet the same peptides are still present, it is suggested that, at least in elasmobranchs, peptides are associated with myelinated primary sensory fibres and are involved in functions other than nociception.

The presence of the caudal neurosecretory system has been well established morphologically, and its histological organisation has been determined in several species of elasmobranchs (Fridberg 1962; Bern 1969). In addition, moderate to strong urotensin I (bladder-contracting activity)-immunoreactive neuronal structures have been found in the caudal spinal cords of several elasmobranch fishes (Chan and Ho 1969; Bern et al. 1973). More recently, Onstott and Elde (1986) have extended these observations to the sister group, i.e. the Holencephali. Of the two major families of biologically active peptides (urotensins I and II), only urotensin II immunoreactive cells were observed in all cartilaginous fishes studied. Urotensin I immunoreactivity was detected in *Raja*, but not in *Squalus*, *Dasyatis* or *Hydrolagus* (Onstott and Elde 1986). The latter study also suggested the possibility of a serotoninergic modulation of the neurosecretory cells in the urophysis.

12.4.6
Functional Correlations

The spinal cord of chondrichthyans plays an important role in body movement maintenance and control, in coordinating the viscera through the sympathetic system, in osmoregulation via the urophysis (Bern 1969) and in the firing of the electric organ in the Rajidae (Bennett 1971).

Locomotor body movements are the basis of most of the behaviour of elasmobranch fishes, and in most species (sharks and *Torpedo*) movement is brought about by lateral body undulation produced by alternating activity in the axial muscles on both sides. In contrast, in most batoids and in the holocephalian fish, locomotion involves an elevation-depression movement of the enlarged, paired pectoral fins, produced by alternating activity in the dorsal and ventral muscles (Droge and Leonard 1983a,b). The importance of the spinal cord in determining the fundamental locomotor pattern is evident from the fact that a dogfish in which the spinal cord has been totally severed close to the brain can perform continuous locomotor movements that are essentially of normal form (for a review, see Wallén 1982; Roberts and Williamson 1983). Such movements have proper intersegmental coordination, and the capacity to produce the rhythmic alternating activity appears not to be restricted to any specific part of the cord; in dogfish, motor rhythms can be produced by as few as eight segments of isolated spinal cord (Grillner 1974). It is not yet known whether there is one generator in each segment or whether a more complex situation exists, e.g. generators in nearby segments overlapping somewhat. Some evidence has been presented for the existence of one generator within each spinal segment in the stingray *Dasyatis* by Droge and Leonard (1983c). These authors have also provided some evidence that separate oscillators for elevators and depressors may be present within one spinal segment.

Although spinal preparations of rays do not show spontaneous locomotor movements, motor rhythms can be elicited by stimulation of the mesencephalic locomotor region (MLR; Leonard et al. 1979). According to Grillner and Kashin (1976), the difference between fishes with and without spontaneous spinal swimming is presumably due not to any fundamental difference in organisation of the nervous system, but rather to a difference in the degree of excitability in the interneuronal network responsible for generating the movements.

As shown by Timerick et al. (1992), several brain stem nuclei project directly to the segments from which the pectoral fin muscles receive their innervation (segments 4–12) and may, therefore, be expected to participate in the control of the movement of these fins. For example, fin movements evoked by vestibular stimulation (Timerick et al. 1990; Roberts et al. 1991) may depend on the strong projections from the magnocellular and descending octaval nuclei. The more caudal regions of the cord receive strong projections only from the nucleus of the fasciculus longitudinalis medialis and the medius and inferior subdivisions of the medial reticular formation. Only these brain stem nuclei can exert direct control over the whole cord and may,

therefore, have an important role in the coordination of locomotion (see also Bernau et al. 1991).

12.5
Brain Stem

12.5.1
Introductory Notes

The brain stem comprises the rhombencephalon and the mesencephalon; caudally, it grades into the spinal cord, while rostrally it borders on the diencephalon. As in all vertebrates, the brain stem of cartilaginous fishes harbours the centres of origin and termination of all cranial nerves except for the olfactory nerve. In cartilaginous fish, the brain stem contains a fairly well developed reticular formation and a number of relay centres and their associated ascending and descending connections. Although ontogenetically and phylogenetically a derivative of the rostral part of the rhombencephalon, the cerebellum is not included in the brain stem and will be considered separately (see Sect. 6). However, the cerebellum and the brain stem are strongly interconnected, and many of the cerebellar afferent systems originate, and almost all cerebellar efferent fibres terminate, within the brain stem. Although it belongs to the brain stem, the tectum mesencephali will also be dealt with separately (see Sect. 7).

12.5.2
Zonal Arrangement of Brain Stem Nuclei

As in other vertebrates, the afferent centres in the brain stem of cartilaginous fishes are located in the dorsally and laterally situated alar plate, wheras the efferent centres lie in the medially and ventrally located basal plate. The boundary between these two plates in cartilaginous fishes is marked by a distinct ventricular groove, the sulcus limitans of His (Fig. 12.22). The "motor" basal and "sensory" alar plates containing the cranial nerve nuclei can be subdivided into separate longitudinal zones which, as pointed out by Herrick, Johnston and others, are related specifically to one of the four main categories (somatomotor, visceromotor, viscerosensory and somatosensory). The ventricular surface of the rhombencephalon of cartilaginous fishes is marked by elongated bulges or ridges, separated by distinct ventricular sulci (Figs. 12.14–12.19, 12.22). It is not surprising, therefore, that the sulcal pattern of the chondrichthyan brain stem has been used many times to illustrate the zonal arrangements of brain stem cell masses (e.g. Kuhlenbeck 1927, 1973).

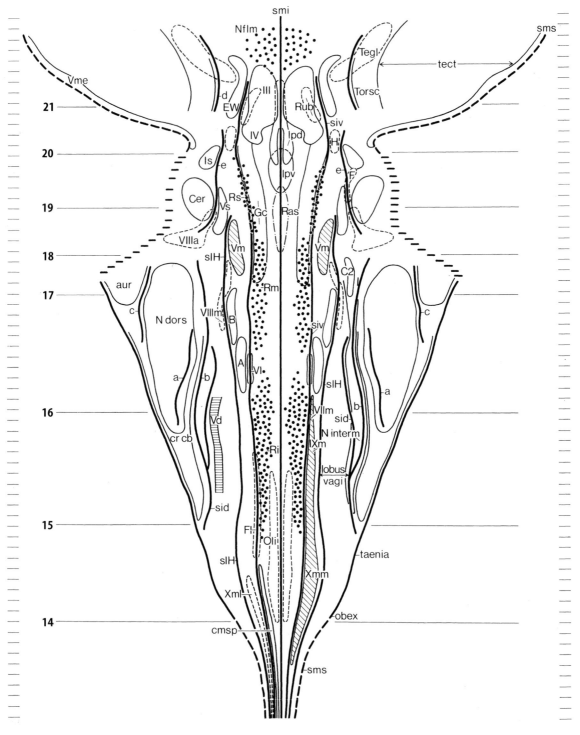

Fig. 12.22. Topological reconstruction of the brain stem of the shark *Squalus acanthias*. The *heavy line* which constitutes the axis of the figure represents the sulcus medianus inferior. The *curves* which constitute the lateral limits of the figure represent the taenia rhombencephali (*continuous parts*) and the sulcus medianus superior (*dashed parts*). The *interruptions* in these curves, marked by *short horizontal lines*, indicate the areas where parts of the auriculae and of the corpus cerebelli have been omitted from the reconstruction. The *remaining heavy curves* indicate the position of ventricular sulci. The *thin, continuous curves* indicate the boundaries of periventricular cell masses; the outlines of migrated cell masses are indicated by *interrupted curves*. The *black dots* give an impression of the distribution and density of the very large reticular cells. For the sake of clarity, the cell masses are not always drawn bilaterally. *Numbers on the scale* at the side of the drawing correspond to atlas figure of the same number. The distance between sections, indicated by *small horizontal bars*, equals 450 μm

The sulcus limitans of His, which separates the basal from the alar plate, extends from the spinal level to the pretrigeminal region (Figs. 12.15–12.18, 12.22). The basal plate can be subdivided into a medial area ventralis and a lateral area intermedioventralis separated from each other by a distinct groove, the sulcus intermedius ventralis, which extends rostrally to the level of the oculomotor nucleus (Figs. 12.15–12.22). Along most of the rhombencephalon, the alar plate contains likewise two longitudinally arranged cell zones, the area intermediodorsalis and the area dorsalis. The boundary between these areas is marked by the sulcus intermedius dorsalis (Figs. 12.16, 12.17, 12.22). In addition to these long, principal grooves, a number of shorter, accessory grooves appear to be present (Smeets et al. 1983). In the isthmus region and in the mesencephalon, the four zones are less readily recognised, but at least one, the area ventralis, extends throughout the midbrain (Figs. 12.20, 12.21). The various zones, their relations to particular nerves or nerve components and their constituent centres will now be briefly reviewed (Fig. 12.22).

12.5.2.1
Area Ventralis/Somatomotor Zone

The area ventralis contains a large fibre bundle, i.e. the fasciculus longitudinalis medialis, together with a number of cell masses. Its functional designation as a somatomotor zone is appropriate in so far as it contains the rostral end of the spinal motor column, the nucleus nervi abducentis and almost the entire rhombencephalic median and medial reticular formation. The latter is believed to project to somatomotor centres in the spinal cord via the fasciculus longitudinalis medialis. Moreover, the area ventralis may be considered to be continuous with the medial parts of the tegmenti isthmi and the tegmentum mesencephali, and these regions contain the somatomotor nuclei of the oculomotor and trochlear nerves as well as the nucleus fasciculi longitudinalis medialis, which probably belongs to the somatomotor coordinating apparatus. The nucleus ruber, which probably also forms part of the somatomotor mechanism, is likewise situated within the medial part of the tegmentum mesencephali. However, there are several centres within the confines of the rhombencephalic area ventralis and its rostral continuation which cannot be considered to be somatomotor in function. For example, the inferior olive and the griseum centrale are known to project to the cerebellum (Fiebig 1988), while the efferent connections of the interpeduncular nucleus are still entirely unknown. Finally, it cannot be excluded

that elements of the medial reticular formation serve as links in ascending projections.

12.5.2.2
Area Intermedioventralis/Visceromotor Zone

The area intermedioventralis contains the motor nuclei of the trigeminal, facial, glossopharyngeal and vagal nerve. Together, these centres constitute a visceromotor column, of which the mesencephalic Edinger-Westphal nucleus may be considered to be the rostral extreme. However, the nucleus funiculi lateralis, the nucleus nervi abducentis and particularly the nucleus reticularis superior encroach upon the area intermedioventralis from the medial side. This area also contains several nuclei (nuclei A and B and some nuclei in the isthmic region), the connections of which are not known. Consequently, the term visceromotor zone is not applicable to the whole of the area intermedioventralis.

12.5.2.3
Area Intermediodorsalis/Viscerosensory Zone

The area intermediodorsalis is predominantly the common centre of termination for general visceral afferent and special visceral afferent fibres that enter the brain by way of the facial, glossopharyngeal and vagal nerves. However, the general somatosensory nucleus of the trigeminal nerve is found in the superficial part of the intermediodorsal area, and its rostralmost part contains the nucleus octavus magnocellularis. The latter nucleus receives direct octavus nerve afferent fibres and thus represents a special somatosensory centre. On the other hand, since the magnocellular octaval nucleus projects directly to somatic efferent centres in the spinal cord, it may equally be considered to be part of the brain stem motor apparatus. Nevertheless, it is obvious that the term viscerosensory zone is not applicable to the whole of the area intermediodorsalis.

12.5.2.4
Area Dorsalis/Somatosensory Zone

The area dorsalis is entirely occupied by two large, lateral line nerve centres, the nucleus dorsalis and nucleus intermedius areae octavolateralis (Figs. 12.16, 12.17, 12.22). It is therefore appropriate to refer to this area as a special somatosensory zone.

12.5.2.5
Zonal Arrangement in the Isthmus and Mesencephalon

The zonal pattern is less clear in the subcerebellar and isthmus regions than in the caudal parts of the rhombencephalon of the chondrichthyans that have been studied. In *Squalus*, sulcus e (Fig. 12.22) may be considered to constitute the boundary between the intermediodorsal and the dorsal area. Thus the medial part of the nucleus octavus ascendens and nucleus V would fall within the intermediodorsal area, whereas the nucleus cerebelli and the nucleus isthmi would be contained within the area dorsalis. However, in other elasmobranchs the sulcal pattern in the rostral part of the rhombencephalon is even more complex (see Smeets et al. 1983).

The general arrangement of the cell masses in the mesencephalon of cartilaginous fishes does not exhibit a clear morphological zonal pattern. However, this part of the brain may be subdivided into primarily motor and sensory zones. The motor zone, which includes the nuclei of nerves III and IV, the Edinger-Westphal nucleus, the nucleus of the fasciculus longitudinalis medialis and the nucleus ruber, occupies the medial part of the tegmentum mesencephali. The sensory zone comprises the lateral part of the tegmentum and the tectum. In *Squalus*, as in other cartilaginous fishes studied (Smeets et al. 1983), the boundary between these two zones is marked on the ventricular side by sulcus d, which has, however, been demonstrated not to be a rostral continuation of the sulcus limitans of His found in the rhombencephalon (Fig. 12.22).

12.5.3
Organisation and Connections of the Sensory Trigeminal Nuclei

In *Squalus*, three general somatosensory nuclei related to the trigeminal nerve can be distinguished. These are the nucleus mesencephalicus, the nucleus princeps and the nucleus of the tractus descendens nervi trigemini.

The *nucleus mesencephalicus nervi trigemini* consists of a continuous dorsomedian strip of very conspicuous cells extending throughout the length of the tectum (Figs. 12.21, 12.22). These cells are spherical or ovoid in shape and have fine granules of Nissl substance distributed throughout the cytoplasm, with large particles located at the periphery. Many of the cells make contact with the ventricular cerebrospinal fluid and also have a close connection to the tectal blood supply (Witkovsky and Roberts 1975; MacDonnell 1980a,b, 1984, 1989). Since the mesencephalic trigeminal cells cluster around the

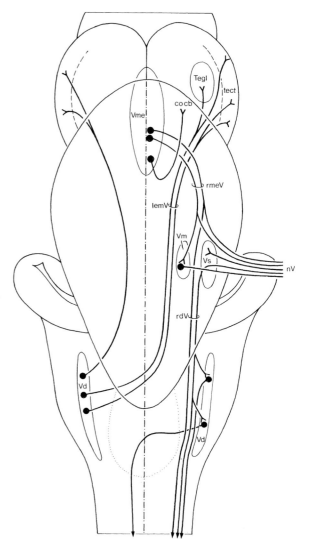

Fig. 12.23. Experimentally verified primary and secondary connections of the trigeminal nerve, as viewed from the dorsal surface of the brain

midline, this was designated the midline ridge formation (MRF) by MacDonnell (1980a,b). Significant variation in the structure of the MRF has been noted among elasmobrach species. At least three major types of MRF were distinguished in 15 species of sharks, while such a structure has not yet been observed in skates or rays (for details, see MacDonnell 1984). However, the presence of the MRF in holocephalian fishes suggests that this proposed circumventricular organ is a primitive feature of cartilaginous fish.

The *tractus mesencephalicus nervi trigemini*, formed by the major processes of the mesencephalic trigeminal neurons, has been studied by Johnston (1905), Weinberg (1928) and, more recently, Witkovsky and Roberts (1975) and Roberts and Witkovsky (1975). The axons of the mesencephalic trigeminal neurons collect at caudal mesencephalic levels in a distinct bundle, passing via the posterior wall of the tectum and the velum medullare anterius to the tegmentum rhombencephali (Figs. 12.18–12.21, 12.23). According to Roberts and Witkovsky (1975), a number of neurons in the posterior part of the mesencephalic trigeminal nucleus projects to the anterior lobe of the corpus cerebelli. The majority of the mesencephalic trigeminal axons, however, pass alongside the motor trigeminal nucleus, issuing fibres to the motor nucleus and to the trigeminal nerve (Fig. 12.23). Some fibres are found to enter the descending trigeminal tract and to descend to cervical spinal cord levels (Smeets and Timerick 1981; Smeets et al. 1983; Timerick et al. 1992).

The *nucleus princeps nervi trigemini* is lateral to the motor trigeminal nucleus and consists of small cells. In *Squalus*, this cell mass occupies a periventricular position (Fig. 12.19). Although experimental evidence is lacking, it is thought that an appreciable number of trigeminal fibres terminate at the level of entrance of nerve V in the trigeminal princeps nucleus.

Most fibres entering nerve V constitute a superficially situated radix descendens which can be traced as far caudal as the obex region (Figs. 12.14–12.18, 12.23). Throughout its extent, this bundle contains medium-sized neuronal perikarya. The cells are diffusely scattered in the rostralmost and caudalmost parts of the radix descendens, but the neurons in the intermediate part are sufficiently compact to be considered a nucleus (Figs. 12.16, 12.22).

The efferent connections of the *nucleus tractus descendens nervi trigemini* comprise "reflex" fibres and fibres ascending to higher levels of the brain. According to Wallenberg (1907) and Ariëns Kappers et al. (1936), the reflex fibres pass from this nucleus to the ventral horns of the spinal cord and to the efferent nuclei of the rhombencephalon. Smeets and Timerick (1981) confirmed experimentally the existence of a spinal projection of the nucleus of the descending trigeminal tract. Fibres ascending to higher levels of the brain have been described in *Scyliorhinus stellaris* by Wallenberg (1907) to reach the tectum mesencephali, the nucleus tegmentalis lateralis and, for a small part, the caudal part of the thalamus region. A tectal projection has been confirmed experimentally (Smeets 1982, *Scyliorhinus*, *Raja*). A cerebellar projection originating from both the descending and principal sensory trigeminal nucleus has been experimentally revealed by Fiebig (1988) in the thornback guitarfish *Platyrhinoidis triseriata*.

12.5.4
Solitary Tract and Related Nuclei

The viscerosensory fibres of nerves X, IX and VII terminate in a longitudinal protrusion of the lateral wall of the fourth ventricle, which may be termed the viscerosensory column or the vagal lobe (Figs. 12.15, 12.16, 12.22). The vagal lobe consists almost entirely of small cells that vary in density throughout the nucleus. As in other anamniotes, the viscerosensory area in chondrichthyans cannot be subdivided into separate nuclei that are specific for taste or general visceral sensibility.

Many of the primary viscerosensory fibres end at their level of entrance, but a certain portion descends in the lobe and constitutes the fasciculus solitarius (Fig. 12.16). At caudal rhombencephalic levels, the number of axons in the solitary tract decreases as they terminate on cells of the surrounding nucleus fasciculi solitarii. The nucleus could not be distinguished as a separate cell mass in our material, although in some sections an accumulation of cells around the solitary tract was recognised. Only a relatively small solitary tract remains at the level of the obex. Slightly more caudally, at the commissura infima (Halleri), the fascicles of both sides fuse and exchange fibres, which then ascend in the contralateral fascicle (Ariëns Kappers et al. 1936).

By means of anterograde transport of HRP, Barry (1987a) largely confirmed the central connections of the nerves IX and X. In the clearnose skate *Raja eglanteria*, glossopharyngeal and vagal afferents were found to enter the brain stem in all except the last vagal rootlet and to course medially to the ipsilateral viscerosensory column. In register with the terminal fields of the corresponding cranial nerves, a facial, a glossopharyngeal and a vagal lobe were recognised in the viscerosensory column. Most glossopharyngeal afferents terminate in the glossopharyngeal lobe at the level of entrance of nerve IX rootlets (Barry 1987a). Only a few glossopharyngeal nerve fibres contribute to the solitary tract. Occasionally, fibres ascend to terminate in more rostral parts of the viscerosensory column. Vagal afferents also show a small ascending component and a rather dense terminal field at the level of entrance of nerve X rootlets. In contrast to the glossopharyngeal nerve, vagal afferent fibres contribute extensively to the solitary tract. Vagal afferents could be traced to the commissura infima, where they terminated in the nucleus of the commissura. Although a number of fibres were found to decussate within the commissura infima, the existence of ascending fibers in the contralateral solitary tract could not be confirmed.

Both glossopharyngeal and vagal afferents were found to terminate in the area of the dendrites of the dorsalmost cells of the ipsilateral motor nucleus of nerve IX and dorsal motor nucleus of nerve X, suggesting that a monosynaptic input of nerve IX and X afferents to motoneurons exists. No glossopharyngeal or vagal afferent fibres enter the trigeminal tract or the dorsolateral funiculus of the spinal cord; thus there appear to be no somatic sensory components in nerve IX or X.

The efferent connections of the viscerosensory area are still obscure. It is known in teleosts (Finger 1978; Morita et al. 1980) and amphibians (Herrick 1948) that the efferent fibres of the viscerosensory area ascend to a centre situated in the rostralmost part of the rhombencephalon, which is termed the nucleus visceralis secundarius because of its specific input. However, such an ascending viscerosensory pathway could not be observed in our *Squalus* material, although Kusunoki et al. (1973) identified such a nucleus, on the basis of its high monoamine oxidase activity, ventral to the nucleus cerebelli in *Mustelus manazo*. Descending efferent projections of the solitary tract nucleus have been traced to cervical spinal cord levels in *Scyliorhinus* (Smeets and Timerick 1981; Timerick et al. 1992).

A study of the chemoarchitecture of the viscerosensory area of cartilaginous fishes by Stuesse et al. (1992) revealed that the number of lobes apparently corresponds to the number of gill arches. The latter authors found four bulges on the medullary surface in the sagittally sectioned brain stem of *Hydrolagus,* whereas five bulges were identified in a corresponding position in *Squalus*. Further studies in elasmobranchs with six or seven gill arches are needed to draw final conclusions about the relationship between the number of viscerosensory lobes and gill arches.

12.5.5
Octavolateral Systems

The octavolateral system is a special sensory system found in most anamniotic vertebrates. The system comprises the inner ear, the mechanoreceptors of the lateral line and the electroreceptors. Behavioural and anatomical studies (e.g. Roberts 1978; Bullock et al. 1982; Bodznick and Schmidt 1984; Dunn and Koester 1984; Bleckmann et al. 1989; Puzdrowski and Leonard 1993) have favoured a subdivision of the octavolateral system into three separate sensory systems: the electrosensory, the lateral line and the octaval systems. The octaval system itself may be further divided into two functional systems, the vestibular and auditory systems.

There are two lateral line nerves in cartilaginous fishes, the anterior and the posterior nerve. Central to the ganglion, the anterior lateral line nerve divides and enters the rhombencephalon as separate dorsal and ventral roots (Fig. 12.3c). The dorsal root projects to the dorsal nucleus of the octavolateral area, whereas the ventral root sends its fibres to the nucleus intermedius areae octavolateralis. Studies by McCready and Boord (1976), Boord and Campbell (1977), Bodznick and Northcutt (1980), Bullock et al. (1983), Koester (1983) and Puzdrowski and Leonard (1993) have presented circumstantial evidence that those fibres carried by the dorsal root innervate the electroreceptive ampullae of Lorenzini, while those that innervate mechanoreceptive canal and pit organs of the head are carried by the ventral root. The posterior lateral line nerve which innervates canal neuromasts and pit organs of the trunk and tail, and therefore is only mechanoreceptive, enters the rhombencephalon as a single root just dorsal to the glossopharyngeal nerve (Fig. 12.3c).

The nucleus dorsalis (Figs. 12.16, 12.17) is situated at the level of entrance of the anterior lateral line nerve and contains medium-sized cells which are multipolar and resemble cerebellar Purkinje cells in form (Paul and Roberts 1977). Rostrally, this nucleus is replaced by the densely packed granule cells which form the inner zone of the lower leaf of the cerebellar auricle (Fig. 12.18). The nucleus intermedius is a very elongated cell mass which, throughout its extent, lies closely to the sulcus intermedius dorsalis (Figs. 12.16, 12.17, 12.22). Its constituent elements are similar to those of the dorsal nucleus. Rostrally, the nucleus intermedius is continuous with the grey matter of the auriculae cerebelli, whereas caudally it extends to the caudalmost part of the crista cerebellaris. The region occupied by the nucleus intermedius in *Squalus* contains at rostral levels a concentration of cells, i.e. nucleus C2, which will be discussed later in connection with the octaval nuclei.

Both lateral line nuclei are covered dorsally by a highly characteristic layer, the crista cerebellaris or cerebellar crest, which closely resembles the molecular layer of the cerebellum, with which it is contiguous rostrally. The cerebellar crest contains numerous parallel-running unmyelinated axons which arise from granule cells of the auricles (Boord 1977; Paul and Roberts 1977; Montgomery 1981).

12.5.5.1
Sense Organs and Peripheral Nerves of the Lateral Line System

In accordance with the terminology of Dijkgraaf (1963), the component sense organs can be divided into ordinary and ampullary categories. The former includes the canal and pit organs of the head and body, while the ampullary group, restricted in cartilaginous fishes to the head, is represented by the ampullae of Lorenzini.

The ordinary lateral line component comprises the neuromasts that are found in a series of superficially located canals (Figs. 12.24, 12.25) and the pit organs (or "free neuromasts") that are spread over the body surface between pairs of modified scales. Lateral line neuromasts are sensitive to external water displacements, whereas the pit organs are

activated by both mechano- and chemostimulation. The biological significance of this sensitivity still remains unclear (Roberts 1978, 1981). A specialized receptor, the spiracular organ, is associated with the spiracular gill slit in all cartilaginous fishes studied (Barry and Boord 1984). In elasmobranchs, the spiracular organ is a tube or pouch, blind at one end, that opens directly onto the medial wall of the spiracular cleft in skates, rays and spiny dogfishes or into an anterior diverticulum of the spiracular cleft in carcharhinid sharks (Norris and Hughes 1920; Barry and Boord 1984; Barry et al. 1988a). The presence of hair cells and patterns of the central projections suggest that the elasmobranch spiracular sense organs are mechanoreceptors that are rela-

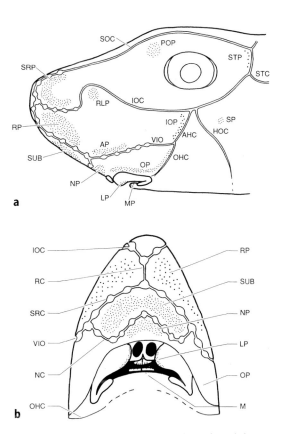

Fig. 12.25a,b. Distribution of the head canals and the system of ampullae of Lorenzini of the ratfish *Hydrolagus collei*. **a** Lateral view of head. **b** Ventral view. The major groups of pores on the skin of the head are shown as *dots* and *small circles*. Lateral line canals are shown as *paired lines* with enlargements. *AHE*, angular segment of the hyomandibular canal; *AP*, angular pore field; *HOC*, hyomandibular canal; *IOP*, infra-orbital pore field; *IOC*, infra-orbital canal; *LP*, labial pores; *M*, mouth; *MP*, mandibular pores; *NC*, nasal canal; *NP*, nasal pores; *OHC*, oral segment of hyomandibular canal; *OP*, oral pores; *POP*, preorbital pores; *RC*, rostral canal; *RP*, rostral pore field; *RLP*, rostrolateral pores; *SOC*, supraorbital canal; *SP*, spiracular pores; *SRP*, suprarostral pore field; *STC*, supratemporal canal; *STP*, supratemporal pores; *SUB*, subrostral pores; *SRC*, subrostral canal; *VIO*, ventral segment of infra-orbital canal. (After Fields et al. 1993)

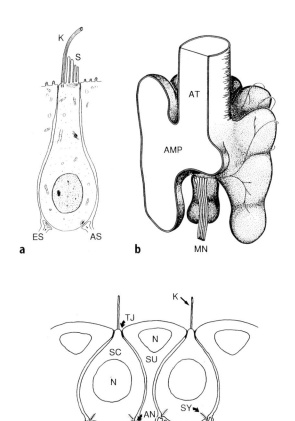

Fig. 12.24a–c. Ratfish *Hydrolagus collei*. **a** Mechanosensory lateral line neuromast sense cell. **b,c** Electroreceptive ampulla of Lorenzini and its sense cell. Note the differences between the two types of cells in shape, kinocilium, efferent innervation and presence of a presynaptic ribbon. *AMP*, ampulla; *AN*, afferent nerve ending; *AS*, afferent synapse with sensory nerve fibre terminal; *AT*, ampullary tube; *ES*, efferent synapse with centrifugal fibre; *K*, kinocilium; *MN*, myelinated nerve; *N*, nucleus; *S*, stereocilia; *SC*, sense cell; *SU*, supporting cell; *SY*, synapse; *TJ*, tight junction. (After Fields et al. 1993)

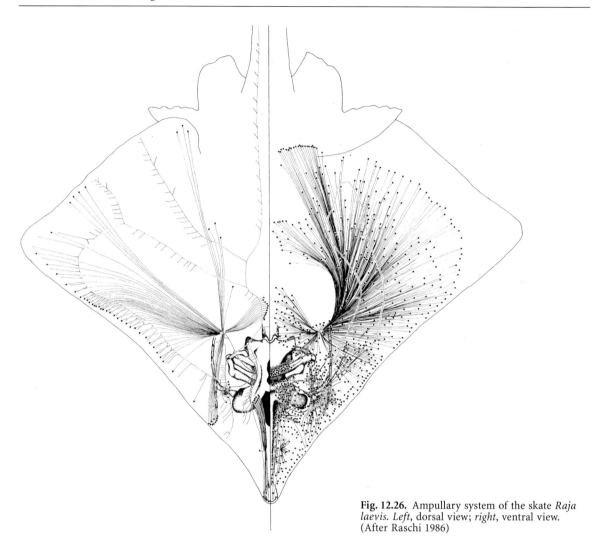

Fig. 12.26. Ampullary system of the skate *Raja laevis. Left,* dorsal view; *right,* ventral view. (After Raschi 1986)

tively insensitive to electrical stimuli, vibration or water movement, but that respond to movements of the hyomandibula-cranial joint (Barry et al. 1988b). The spiracular organ is, therefore, considered as a sensitive joint receptor.

The ampullae of Lorenzini contain ciliated sense cells in an alveolate-shaped epithelium, which communicates to the surface through a jelly-filled tube (Fig. 12.24). There are several characteristic differences between the neuromast sense cell and the electroreceptive sense cell, including cell shape, kinocilium and efferent innervation (Fig. 12.24). Ampullary sense organs are distributed in groups over the head of cartilaginous fishes with their pores in clusters and innervated by the buccal, hyomandibular and superficial ophthalmic branches of the anterior lateral line nerve (Figs. 12.25–12.27). A comparison of the distribution patterns of the ampullae of Lorenzini among 40 species of

skates suggested a close relationship between inferred electroreceptive capabilities and feeding mechanisms (Raschi 1986). The general distribution of the ampullary pores on deep dwelling rajoids appears to compensate for reduced visual input, whereas their relative densities are a measure of the system's resolution and reflect major differences in feeding strategies.

12.5.5.2
Central Projections of the Lateral Line Nerves

Several experimental studies have revealed the central projections of the anterior and posterior lateral line nerves (Boord and Campbell 1977; Koester 1983; Bodznick and Schmidt 1984; Puzdrowski and Leonard 1993). The dorsal root of the anterior lateral line nerve, which carries electrosensory information, enters the nucleus dorsalis areae octa-

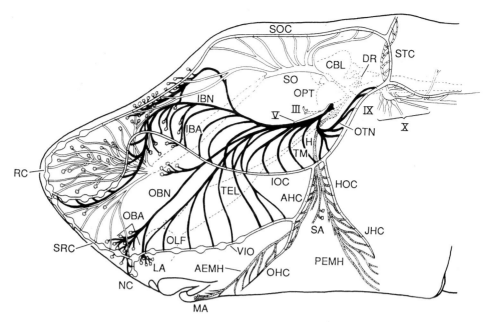

Fig. 12.27. Innervation of the lateral line canal system and the ampullary groups on the head of the ratfish *Hydrolagus collei. Unfilled nerves,* branches of the superficial ophthalmic nerve; *black nerves,* branches of the buccal nerve; *stippled nerves,* branches of the hyomandibular nerve; *striped nerves,* branches of the supratemporal nerve. *III,* oculomotor nerve; *V,* trigeminal nerve; *IX,* glossopharyngeal nerve; *X,* vagus nerve; *AEMH,* anterior external mandibular branch of hyomandibular nerve; *CBL,* cerebellum; *DR,* dorsal root of anterior lateral line nerve; *H,* hyomandibular trunk; *IBA,* inner buccal ampullae; *IBN,* inner buccal nerve; *JHC* jugal segment of HOC; *LA* labial group of outer buccal ampullae; *MA,* mandibular ampullae; *OBA,* outer buccal ampullae; *OBN,* outer buccal nerve; *OLF,* olfactory lobe; *OPT,* optic lobe; *OTN,* otic branch of buccal nerve; *PEMH,* posterior external mandibular branch of hyomandibular nerve; *SA,* spiracular ampullae; *SO,* superficial ophthalmic nerve; *TEL,* telencephalon; *TM,* telencephalon median. For the other abbreviations, see Fig. 12.25. (After Fields et al. 1993)

volateralis and bifurcates to form ascending and descending bundles (Fig. 12.28, 12.29). The fibres of the ventral anterior lateral line nerve were shown to bifurcate into descending and ascending branches within the nucleus intermedius. The fibres of the posterior lateral line nerve ascend within the territory of the nucleus intermedius, but run lateral to the fibres from the anterior lateral line nerve (Fig. 12.16, 12.29). Somatotopic organisation of the lateral line terminals has also been reported by Bodznick and Schmidt (1984) in *Raja* and by Puzdrowski and Leonard (1993) in *Dasyatis*. The fibres from both lateral line nerves terminate around the cells that lie among the ascending and descending tracts; there is no evidence of a projection to the crista cerebellaris. Koester (1983) reports that some mechanoreceptive lateral line fibres terminate on the nucleus octavus magnocellularis and that some others pass through the lateral line nuclei and project to the granule cells of the auricles. Mechanoreceptive fibres from the anterior and posterior lateral line nerves project to the more lateral portions of the lower leaf of the auricles, whereas the electroreceptive fibres take a more medial course (Schmidt and Bodznick 1987). Selective labelling of the pos-

terior lateral line, the external mandibular ramus and the superficial ophthalmic-buccal rami revealed a topographical representation of the electro- and mechanoreceptors in the dorsal and, to a lesser extent, intermediate nuclei of the areae octavolateralis (Fig. 12.30; Bodznick and Schmidt 1984; Puzdrowski and Leonard 1993).

In our material, numerous fibres can be recognised which collect along the medial surface of the lateral line area and subsequently pass ventrally and medially as arcuate fibres (Figs. 12.17, 12.18). In the rostral rhombencephalon, these arcuate fibres form a distinct lemniscus lateralis, which can be traced to the mesencephalic tegmentum (Figs. 12.19–12.21, 12.28). Secondary connections of the lateral line centres have been studied experimentally by Boord and Northcutt (1982) in *Raja* using autoradiography and degeneration techniques. They found that efferent axons from each nucleus give rise to commissural components and contribute to an ipsilateral and contralateral lemniscus lateralis (Fig. 12.28). The commissural fibres project to the equivalent nucleus on the contralateral side. The lemniscus lateralis ascends to the mesencephalon and terminates within the nucleus

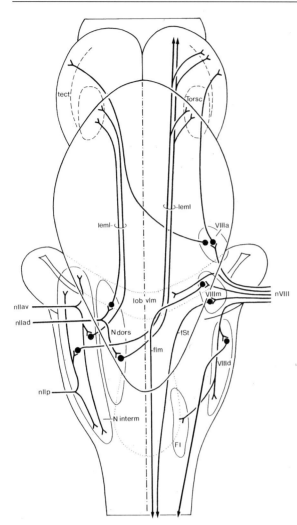

Fig. 12.28. Projections of primary and secondary connections of the nervus lineae lateralis and nervus octavus, as viewed from the dorsal surface of the brain

important aspect of sensory processing in the brain of cartilaginous fish.

Boord and Northcutt (1982) found that a substantial number of fibres from the dorsal and intermediate nucleus of the octavolateral area terminated bilaterally within the central zone of the tectum (Fig. 12.28), a finding which has been corroborated by Smeets (1982) for *Scyliorhinus* and *Raja*.

12.5.5.3
Sense Organs and Peripheral Nerves of the Nervus Octavus

The nervus octavus conveys information from the labyrinth to the brain. The labyrinth of cartilaginous fishes has the same basic form as is found in other vertebrates; it consists of two large sacs – the sacculus (with lagena) and utriculus – and three semicircular canals (Fig. 12.31). Within the labyrinth are located a number of mechanoreceptive sense organs (the maculae and cristae). Because of the accessibility of the nerve and end-organs, the isolated elasmobranch labyrinth has been a classic preparation in vestibular physiology (see Lowenstein 1974; O'Leary et al. 1976; Lowenstein and Compton 1978; Montgomery 1980). It was found that the three semicircular canals form a transducer system that detects angular acceleration of the head, while the detection of gravity and linear acceleration is performed by the sacculus, utriculus and lagena.

Recordings of microphonic potentials from the labyrinth and of compound potentials and single-unit activity from nerve VIII all suggest that the maculae sacculi and neglecta function as vibration detectors and play an important role in sound detection (Lowenstein and Roberts 1951; Fay et al. 1974; Corwin 1981a). Quantitative studies of the

of the lateral line lemniscus and a dorsolateral tegmental area. It is noteworthy that the segregation of electro- and mechanoreceptive inputs, as revealed in the rhombencephalon, is maintained, as the fibre contingent from the nucleus dorsalis lies more lateral than that from the intermediate nucleus. Electrophysiological studies have confirmed the segregation of the two lateral line modalities within the mesencephalon (Schweitzer 1983, 1986; Bleckmann et al. 1987). Moreover, the latter studies also revealed the maintenance of the topography, as found in the termination sites of the lateral line nerve rami, throughout their projections to the midbrain. The highly structured organisation of both the first-order (rhombencephalic) and second-order (midbrain) centres of the electro- and mechanoreceptive lateral line inputs suggests that preservation and calculation of spatial information is an

Fig. 12.29. Selected transverse sections in a rostrocaudal ▶ sequence through the brain stem and cerebellum of the skate *Raja eglanteria. Left panel,* degenerating axons (*dashed lines, large dots*) and terminal degeneration (*small dots*) observed after lesions of the dorsal and ventral branches of the anterior lateral line nerve; *middle panel,* the same after lesioning the posterior lateral line nerve; *right panel,* degeneration pattern observed after lesions of the octaval nerve. *AON,* anterior octaval nucleus; *dALLN,* dorsal root of anterior lateral line nerve; *DGR,* dorsal granular ridge; *DN,* dorsal nucleus areae octavolateralis; *DON,* descending octaval nucleus; *DV,* descending trigeminal tract; *EN,* efferent octavolateralis nucleus; *IN,* intermediate nucleus areae octavolateralis; *IRF,* inferior reticular formation; *LG,* lateral granular layer of the vestibulolateral cerebellar lobe; *LL,* lower lip; *MN,* magnocellular octaval nucleus; *MRF,* medial reticular formation; *PL,* Purkinje-like cell zone; *PON,* posterior octaval nucleus; *PV,* principal trigeminal nucleus; *vALLN,* ventral root of the anterior lateral line nerve; *VIII,* octaval nerve; *X,* vagal nerve. (Modified from Koester 1983)

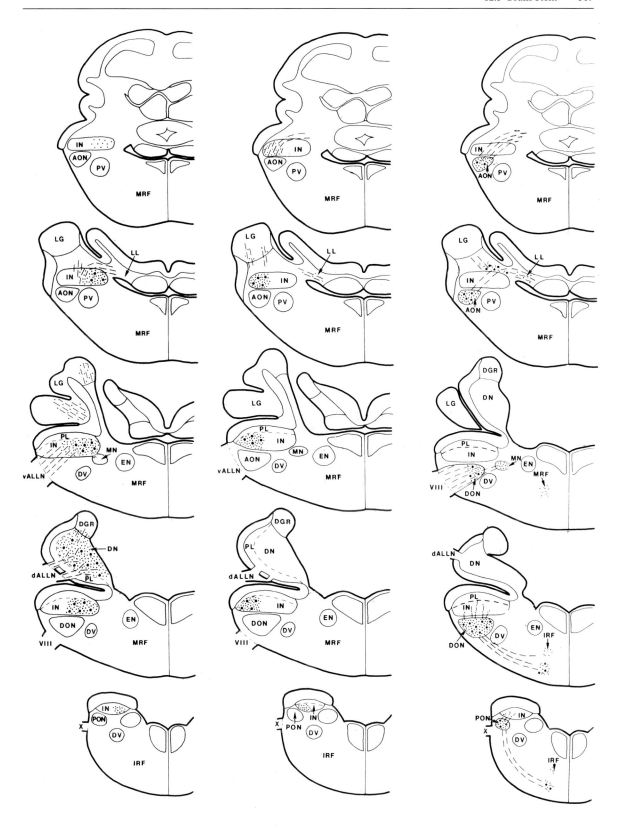

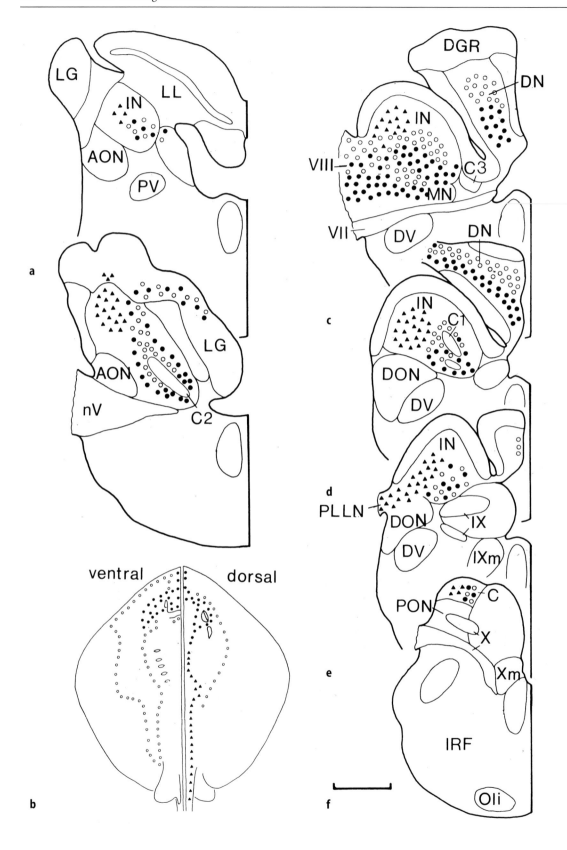

◀ **Fig. 12.30.** Selected transverse sections in a rostrocaudal sequence through the hindbrain of the stingray *Dasyatis sabina*, illustrating the afferent projections of the posterior lateral line nerve (*triangles*), external mandibular ramus (*open circles*) and superficial ophthalmic-buccal rami (*filled circles*). The drawing at the *lower left* shows the topographic organisation of these projections. The topographic arrangement of the nasal group (*filled circles*) and hyoid-mandibular groups (*open circles*) of electroreceptors on the dorsal nucleus are also illustrated. *C*, caudal nucleus; *C1-C3*, nucleus C1-C3; *IX*, glossopharyngeal nerve; *IXm*, glossopharyngeal nucleus; *PLLN*, posterior lateral line nerve; *Xm*, vagal motor nucleus. For the other abbreviations, see Fig. 12.29. *Bar*, 1 mm. (Modified from Puzdrowski and Leonard 1993)

hair cells in the macula neglecta revealed that in skates (*Raja clavata*) their number increases from approximately 500 cells at birth to 6000 at 7 years of age (Corwin 1983, 1985). For further details about this postembryonic growth, the mechanisms involved in the shaping of the macula during development and sex differences in the macula neglecta and the associated nerve, the ramus neglectus, the reader is referred to Corwin (1985) and Barber et al. (1985).

The various end-organs of the labyrinth are innervated by two branches of the octavus nerve: the *ramus posterior*, which innervates the crista of the posterior vertical canal and the maculae of the sacculus, lagena and neglecta, and the *ramus anterior*, which consists of fibres supplying the cristae of the anterior vertical and horizontal canals and the macula utriculi (Fig. 12.31). In adult chondrichthyans, these two rami intermingle on entering the rhombencephalon to such an extent that they can-

not be separated (Boord and Roberts 1980; Koester 1983; Barry 1987b).

12.5.5.4
Central Projections of the Octaval Nerve

Experimental studies of the central projections of the octaval nerve in *Raja* (Koester 1983; Barry 1987b) and *Rhinobatos* (Dunn and Koester 1987) have revealed that first-order octaval afferents, upon entering the medulla, divide into ascending, descending and medially directed fascicles (Fig. 12.28). First-order octaval afferents project ipsilaterally to five primary octaval nuclei (Koester 1983; Barry 1987b): the nucleus octavus magnocellularis, the nucleus octavus descendens, the nucleus octavus posterior, the nucleus octavus ascendens and the nucleus octavus periventricularis. Additional projections were found to the reticular formation, the nucleus intermedius areae octavolateralis and the cerebellar vestibulolateral lobe.

The *nucleus octavus magnocellularis* is situated at the level of entrance of the nerve VIII close to the descending trigeminal tract (Figs. 12.17, 12.22). It consists of very large bi- and tripolar cells, although the neurons are smaller in the rostral and caudal parts. The *nucleus octavus ascendens* is situated dorsolateral to the rostralmost part of the sulcus limitans. Its posterior margin is adjacent to the entrance of nerve VIII, and it extends rostrally as far as the caudal part of the nucleus cerebelli (Fig. 12.22). The neurons of the ascending octaval nucleus vary in form, although their irregularly

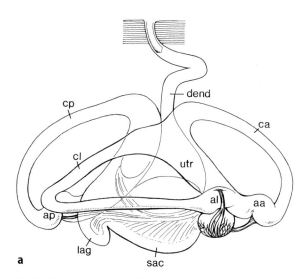

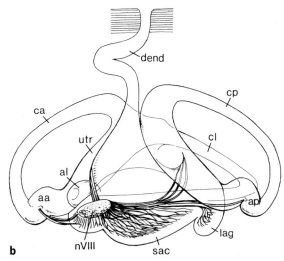

Fig. 12.31a,b. Labyrinth of *Squalus acanthias* seen from **a** the lateral and **b** the medial view. *aa*, ampulla canalis anterioris; *al*, ampulla canalis lateralis; *ap*, ampulla canalis posterioris; *ca*, canalis anterior; *cl*, canalis lateralis (or horizontalis); *cp*, canalis posterior; *dend*, ductus endolymphaticus; *lag*, lagena; *nVIII*, nervus octavus; *sac*, sacculus; *utr*, utriculus. (Modified from Retzius 1881)

shaped cell bodies are predominantly multipolar (Montgomery and Roberts 1979) and extend along the ascending branches of nerve VIII fibres. The *nucleus octavus descendens* is not readily distinguished in our material, although its limits are clearly evident from experimental studies (Montgomery and Roberts 1979; Boord and Roberts 1980; Northcutt 1980; Barry 1987b; Housley and Montgomery 1983; Koester 1983). This nucleus, which begins at the level of entry of the octavus nerve and descends in the rhombencephalic wall to about the level of the entry of the vagus nerve, lies ventral to the nucleus intermedius and lateral to the descending trigeminal tract (Fig. 12.16, 12.29). Because of its vague boundaries, this nucleus is not drawn in the topological reconstruction. In Osteichthyes, the caudalmost part of the descending projection has been separated as a distinct cell mass, the nucleus caudalis (McCormick 1981). Northcutt (1980) and Koester (1983) recognised a similar division in elasmobranchs (*Platyrhinoidis*, *Raja*), but we remained unable to detect such a posterior octaval nucleus in *Squalus*. The same holds true for the periventricular octaval nucleus, which was recognised on the basis of the termination of a few primary octaval afferents by Barry (1987b) and Puzdrowski and Leonard (1993) (Fig. 12.30), but not by Koester (1983).

As already mentioned, the region occupied by the nucleus intermedius areae octavolateralis contains a small accumulation of cells in its rostral part (Fig. 12.22). A similar cell mass has been found in *Scyliorhinus*, whereas in *Raja* and *Hydrolagus* an additional accumulation of cells has been recognised at more caudal levels (Smeets et al. 1983). The nucleus intermedius areae octavolateralis lies just rostral to the entrance of the central branch of the anterior lateral line nerve (nucleus C2), whereas the other cluster of cells (nucleus C1) is situated at the level of entry of the posterior lateral line nerve. It is thought that these two nuclei are equivalent to cell plate "X" of Northcutt (1978). Recently, Puzdrowski and Leonard (1993) studied the octavolateralis systems in the stingray *Dasyatis* and found a third aggregate of cells within the octavolateral area, which they have labelled C3. This cell aggregate lies at the ventromedial edge of the octavolateral region extending rostrocaudally from the entrance of the facial nerve to the level of the rostral pole of the visceral sensory lobe (Fig. 12.30c).

The central projections of the nervus octavus have been investigated in several elasmobranch species, including *Scyliorhinus* (Montgomery and Roberts 1979; Boord and Roberts 1980), *Platyrhinoidis* (Northcutt 1980; Plassmann 1982), *Rhinobatos* (Dunn and Koester 1984, 1987), *Raja* (Koester

1983; Barry 1987b) and *Dasyatis* (Puzdrowski and Leonard 1993). The results of these studies revealed a basically identical organisation of octaval afferents in all elasmobranch species studied. On entering the rhombencephalon, the fibre bundles divide into medial, ascending and descending branches. Fibres of the medial pathway terminate about the soma and lateral dendrites of the neurons of the magnocellular nucleus. Fibres in the descending branch terminate in the descending octavus nucleus, while fibres of the ascending tract project to the ascending nucleus and to the granule cells of the medial part of the lobus vestibulolateralis (lower lip) of the cerebellum. Boord and Roberts (1980), Northcutt (1980), Housley and Montgomery (1983), Koester (1983) and Dunn and Koester (1984) reported that some primary fibres leave the descending tract and distribute to the ventromedial part of the rhombencephalon. The latter fibres probably terminate in a group of cells adjacent to the spinal lemniscus, the nucleus funiculi lateralis and the inferior reticular formation (Dunn and Koester 1984).

Single octaval nerve branch labelling with HRP has revealed that each primary nucleus receives afferent input from each labyrinthine end-organ (Barry 1987b). An exception may be the absence of a macula neglecta input to the magnocellular nucleus. However, within the ascending, descending and, to a lesser extent, posterior and magnocellular nucleus, the input of the labyrinthine end-organs is largely nonoverlapping (Fig. 12.32). Semicircular canal cristae terminate ventrally, saccular and lagenar afferents dorsally and utricular afferents laterally; macula neglecta afferents course ventrally, but terminate largely dorsally within these nuclei. Primary afferent input to the reticular formation and the nucleus funiculi lateralis is predominantly from the horizontal canal crista, whereas the densest projections to the intermediate nucleus originate from the utriculus and the sacculus (Barry 1987b).

An attempt to answer the important question of whether there is a separation between auditory and equilibrium modalities in the brain stem has been made by Corwin and Northcutt (1982). Based on metabolic markers (deoxyglucose autoradiography) and electrophysiological evidence, these authors suggested that the cell aggregates C1 and C2 in the brain stem of *Platyrhinoidis* (their cell plate "X") were part of an auditory lemniscal pathway projecting to the midbrain homologous to the superior olive in mammals. HRP injections into the medial mesencephalic nucleus revealed that cell groups C1 and C2 did indeed contain secondary ascending cells. However, since these cell groups lie within the

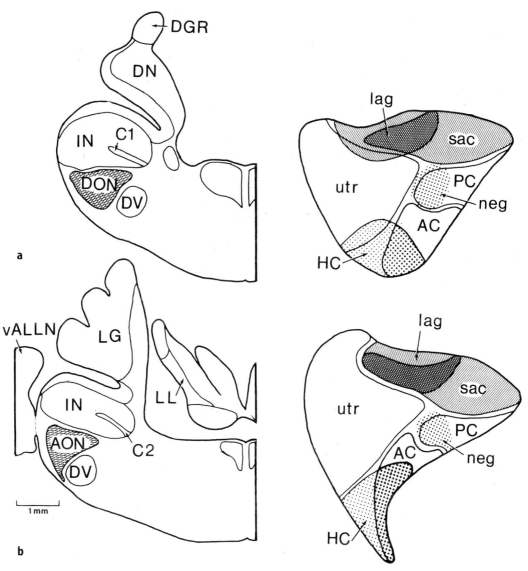

Fig. 12.32a,b. Transverse section through the rhombencephalon at the level of **a** the descending octaval nucleus and **b** that of the anterior or ascending octaval nucleus. *Left*, a hemisection illustrating the location of the octaval nuclei; *right*, an enlargement of the nuclei showing the relative distribution of the primary afferent fibres and terminals arising from the various end-organs of the labyrinth. For the sake of clarity, fibres from the macula neglecta, posterior crista and utriculus which terminate in other parts of the nuclei are omitted. *AC*, anterior crista; *HC*, horizontal crista; *neg*, macula neglecta; *PC*, posterior crista. For the other abbreviations, see Figs. 12.29 and 12.31. (Modified from Barry 1987b)

treminal fields of the anterior lateral line nerve, it seems more likely that they are related to the mechanosensory lateral line system rather than the auditory system (Bodznick and Schmidt 1984; Puzdrowski and Leonard 1993). The latter hypothesis is supported by preliminary single-unit electrophysiological recording experiments which showed that cells in C1 and C2 can be driven by stimulation of the anterior lateral line nerve, but not by stimulation of nerve VIII (Puzdrowski and Leonard 1993).

12.5.5.5
Secondary Connections
of the Primary Octaval Nuclei

Data about secondary projections of the primary octaval nuclei are scanty (Fig. 12.28). From HRP injections into the spinal cord of *Scyliorhinus* and *Raja* (Smeets and Timerick 1981; Barry 1987b; Timerick et al. 1992), it was established that cells in the magnocellular octaval nucleus project bilaterally to the spinal cord by way of the fasciculus medianus of Stieda and the fasciculus longitudinalis medialis. The descending octaval nucleus, in particular cells in its rostral portion, also project to the spinal cord. Cerebellar projections have been reported for the ascending and descending octaval nuclei (Fiebig 1988). An electrophysiological study by Plassmann (1982) revealed that the descending magnocellular and ascending octaval nuclei of *Platyrhinoidis* project to their contralateral homologues via second-order commissural fibres and that both the ipsilateral and contralateral ascending and magnocellular nuclei pass on information via secondary neurons to the oculomotor area and the torus semicircularis in the midbrain. Ascending projections to the midbrain arise from the ascending and descending octaval nuclei, but particularly from nucleus C1 and C2 located in the intermediate nucleus of the octavolateral area (Smeets 1982; Barry 1987b).

12.5.5.6
Efferent Supply to the Lateral Line and Labyrinthine Sense Organs

An important aspect of the innervation of many labyrinthine and lateral line receptors is the efferent supply which originates in the brain and terminates on the peripheral sense organs. An octavolateralis efferent nucleus has been identified by Meredith and Roberts (1986) in *Scyliorhinus* by means of application of HRP to the exposed octaval and lateral line nerves. The efferent nucleus, approximately 2.3 mm long, is loosely arranged with no distinct boundaries. Caudally it is contiguous with the facial motor nucleus, whereas rostrally it reaches the level of entrance of nerve VIII. The efferent octavolateral nucleus consists of cells which have dendrites that extend ventrally to the medial reticular formation, laterally to nerve VIII nuclei and medially to the medial longitudinal fascicle. Other features of the octavolateral efferent nuclei include the following (Meredith and Roberts 1986):

1. They project bilaterally.
2. They are contacted by afferent fibres of nerve VIII, but not by lateral line fibres.
3. There is no clear segregation of neurons projecting to the lateral line or to the inner ear, and there is evidence that some neurons innervate both the ear and the lateral line.
4. The number of efferent cells is small. For example, the posterior lateral line nerve innervates about 8000 hair cells and contains about 1000 axons, of which 20 are efferent in nature. Similarly, nerve VIII of *Scyliorhinus* contains approximately 10 000 axons, of which only approximately 20 are efferent.

It is suggested, therefore, that the effect of the efferent fibres on the sense organs is probably widespread and non-specific (Meredith and Roberts 1986). Before closing this section on the octavolateral systems, it should be mentioned that, in adult *Squalus* material, Mauthner cells are lacking. These very large neurons, which are evidently associated with nerve VIII and project to the spinal cord, are found in many groups of fishes. However, a pair of large neurons was observed by Larsell (1967) in *Squalus* embryo, and these were shown by Bone (1977) to have some features typical of Mauthner cells; in particular, they have large dendrites and crossing axons. Later in development, these Mauthner cells disappear in *Squalus*. Smeets et al. (1983) were able to distinguish a pair of Mauthner cells in the holocephalian fish *Hydrolagus*. The axons of these cells cross immediately caudal to the cell body and course within the fasciculus longitudinalis medialis to the spinal cord. However, contrary to what has been found in bony fishes, their axon diameters are no larger than adjacent axons in the ventral funiculus.

12.5.6
Cranial Nerve Motor Nuclei

The cranial motor nuclei are confined to the basal plate of the brain stem. The area ventralis harbors the somatomotor nuclei, while the intermedioventral area contains the visceromotor and branchiomotor nuclei.

12.5.6.1
Somatomotor Nuclei

From caudal to rostral, the somatic efferent cell masses include the rostral extension of the spinal motor column and the cells of origin of the abducens, trochlear and oculomotor nerves. All of these lie close to the median plane and, although of variable topographical position, show a distinct relationship to the fasciculus longitudinalis medialis. The spinal motor column and the abducens nucleus lie lateral to the fasciculus, whereas the trochlear and oculomotor nuclei are situated mainly dorsal and dorsomedial to the bundle. The spinal motor column continues for some distance rostral to the obex. It consists of large to very large tripolar and quadrupolar cells (Fig. 12.14).

As in other vertebrates, the six eye muscles are innervated by motoneurons located in the oculomotor nucleus (superior rectus, inferior rectus, inferior oblique and medial rectus), the trochlear nucleus (superior oblique) and the abducens nucleus (lateral rectus). A nucleus nervi abducentis is hard to identify in Nissl material of *Squalus*, but in Klüver-Barrera and Bodian sections the roots of the nervus abducens can be traced towards the lateral aspect of the medial longitudinal fasciculus, where an ill-defined group of medium-sized to large neurons may be identified as the nucleus of that nerve. Experimental studies have revealed a similar position of abducens motoneurons in the brain of the carpet shark *Cephaloscyllium* (Montgomery and Housley 1983); however, in the smooth dogfish *Mustelus*, Graf and Brunken (1984) identified a tightly clustered cell group in a more ventrolateral position as the abducens nucleus. In batoids, the abducens motoneurons are scattered over a larger area extending from a ventrolateral position near the base of the rhombencephalon to a dorsomedial position close to the medial longitudinal fascicle (Rosiles and Leonard 1980; Graf and Brunken 1984).

The nucleus nervi trochlearis and the nucleus nervi oculomotorii are well developed and together constitute a continuous aggregation of rather loosely arranged, fusiform and triangular cells. In Nissl-stained material, both nuclei have been identified in a periventricular position in *Squalus*, but in *Raja* the trochlear nucleus was found to lie ventrolateral to the fasciculus longitudinalis medialis; in the latter species, a sharp separation between oculomotor and trochlear nucleus could therefore be made (Smeets et al. 1983). However, our current knowledge of the extraocular motor nuclei of elasmobranchs, largely based on the experimental studies by Rosiles and Leonard (1980) and Graf and Brunken (1984), has become more complicated. Graf and Brunken, studying oculomotor organisation in a shark (*Mustelus*) and a skate (*Raja*), found that in both species the inferior oblique and inferior rectus muscles received an ipsilateral innervation, whereas the superior rectus, the medial rectus and the superior oblique muscles were innervated contralaterally. In the skate, the latter authors divided the medial rectus motoneuron group into a dorsomedial and a ventrolateral subdivision. In the shark, the ventral subdivision is lacking, suggesting that it is not a feature common to all elasmobranchs. Rosiles and Leonard (1980), studying the stingray *Dasyatis*, found a pattern that largely resembled that of *Raja*, although they did not find a distinct dorsomedial subdivision of the medial rectus group. Although some species differences may exist, a notable feature of oculomotor organisation in elasmobranch fishes is the contralateral position of the medial rectus motoneurons, which is in sharp contrast to what is found in all other vertebrates studied. This may have consequences for the circuitry responsible for the pro-duction of conjugate compensatory eye movements in the horizontal plane (see Graf and Brunken 1984; Puzdrowski and Leonard 1994). It cannot be excluded that motoneurons of the ventral subdivision intermingle with cells of the nucleus ruber in *Raja* (Smeets et al. 1983).

12.5.6.2
Branchiomotor Nuclei

The motor nuclei of the nervus vagus (X), nervus glossopharyngeus (IX), nervus facialis (VII) and nervus trigeminus (V) constitute a column which shows a wide gap at the level of entrance of the octavus nerve (Figs. 12.13, 12.14, 12.16, 12.18, 12.19). The motor nucleus of V lies rostral to this gap, while the other nuclei lie caudal to it. The motor nuclei of X and IX occupy a periventricular position, but the motor nuclei of VII and part of that of V have migrated away from the ventricular surface. The nucleus motorius nervi vagi, the nucleus motorius nervi glossopharyngei and the nucleus motorius nervi facialis consist of large bipolar, tripolar or (less frequently) quadrupolar cells. The area lateral to the caudal part of the visceromotor column in *Squalus* contains a distinct aggregation of large, bipolar and triangular cells which was considered to represent a motor nucleus of X and was therefore named the nucleus motorius nervi vagi, pars lateralis (Smeets and Nieuwenhuys 1976). The vagal part of the visceromotor column of *Squalus* has been designated the nucleus motorius nervi vagi pars medialis.

In the shark *Cetorhinus* (Addens 1933), the vagal part of the visceromotor column has been divided into separate rostral and caudal parts, and it is suggested that the rostral portion is special visceromotor or branchiomotor in nature, whereas the caudal part represents a general visceromotor or splanchnic centre. Smeets et al. (1983) observed that the vagal nucleus of *Hydrolagus* can be readily divided into a caudodorsal and a rostroventral nucleus, the latter being continuous with the nuclei of IX and VII. A similar subdivision into a dorsal and a ventral nucleus was recognised at some levels in *Raja*, the ventral portion containing somewhat larger cells than the the dorsal portion (Smeets et al. 1983).

During the last decade, experimental evidence of the organisation of the visceromotor centres in the brain of cartilaginous fishes has become available (Barrett and Taylor 1985a-c; Withington-Wray et al. 1986; Barry 1987a). Withington-Wray et al. (1986), studying the brain of the dogfish *Scyliorhinus*, found that the vagus leaves the brain as a series of rootlets that fuse together as they pass through the

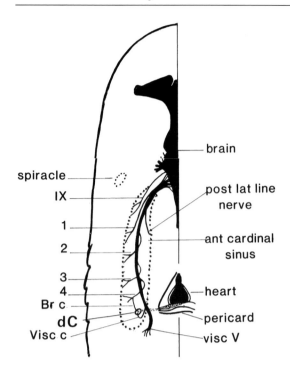

Fig. 12.33. Dorsal view of the left side of the head of the lesser spotted dogfish *Scyliorhinus canicula* showing the peripheral path of the glossopharyngeal (*IX*) and vagus (*X*) cranial nerves in relation to the anterior cardinal sinus. The vagus divides to form branchial branches 1–4, which innervate the intrinsic respiratory muscles of either side of gill clefts 2–5, respectively. The first gill cleft is innervated by the glossopharyngeal nerve. The vagus also sends two branches to the heart, i.e. the branchial cardiac (*Br c*) and visceral cardiac (*Visc c*) branches, which enter the ductus cuvieri (*dC*) and run toward the heart. The remaining part of the vagus nerve constitutes the visceral branch (*visc V*), which innervates the viscera. (Modified from Withington-Wray et al. 1986)

chondrocranium. It then divides to form branchial branches 1–4, which contain motor fibers that innervate the intrinsic respiratory muscles of gill arches 2–5, respectively (Fig. 12.33). The first gill arch is innervated by the glossopharyngeal nerve. Each gill arch receives a pre- and post-trematic ramus. On each side of the fish, the vagus sends two branches to the heart: the branchial cardiac branch, which arises from the fourth branchial ramus, and the visceral cardiac branch, which arises from the visceral ramus of the vagus. The vagal fibres that continue into the viscera form the visceral branch, which innervates the gut as far caudal as the pylorus and the rostral part of the spiral intestine (Young 1933a, 1980). Selective retrograde filling of the various branches of the vagal nerve (Withington-Wray et al. 1986) revealed that the motoneurons of the vagal motor column are arranged as four distinct groups: a dorsomedial and a ventromedial division caudal to the obex, and a rostromedial and a lateral division rostral to the

obex. Neurons in the dorsomedial division supply the heart and the viscera, whereas the ventromedial subdivision supplies the viscera. The lateral subdivision contains neurons that contribute to the innervation of the heart. The rostromedial subdivision is the exclusive supplier of the innervation of the gill arches, but also sends some axons into the visceral branch.

The study by Withington-Wray et al. also confirmed the existence of a topographical representation of the vagus nerve in the vagal motor column. Neurons supplying the gastrointestinal tract are located caudally, those supplying the cardiac nerves lie in the mid-portion of the column and those innervating the gill arches lie in the rostral part of the column. However, some overlap occurs between the pools of neurons supplying adjacent branches of the vagus. As Withington-Wray et al. (1986) pointed out, it is still premature to compare the subdivisions found by experimental back-filling of branches with those observed in Nissl-stained material. In support of this notion are the results of a study dealing with vagal efferents in a skate (*Raja*) by Barry (1987a). Whereas Withington-Wray et al. (1986) identified four subdivisons of the vagal motoneuron column in *Scyliorhinus*, Barry (1987a) found only two divisions of the motor vagal nucleus in *Raja*, i.e. a dorsal and a ventrolateral group. Barry did not observe dorsal and ventral portions in the dorsal motor nucleus, although he noted that cells in the dorsal portion of the dorsal motor nucleus caudal to the obex were smaller than those in the ventral portion, thus resembling the substantially larger size of cells in the ventromedial subdivision of *Scyliorhinus*. The ventrolateral group of *Raja* is probably partly homologous to the lateral subdivision of the vagal motor nucleus in the shark, but selective backfills of the branches of the vagal nerve in the skate are needed before conclusions can be drawn. Barry (1987a) noted that the most caudal vagal rootlet contained only motor components and labeled it provisionally as the accessory nerve root. The latter author also predicted that the caudal part of the ventrolateral vagal group may contain the motoneurons which supply the accessory nerve. More recently, Boord and Sperry (1991) detected in *Raja* a homologue of the cucullaris (trapezius) of sharks as described previously (Tanaka 1988). In a subsequent paper, Sperry and Boord (1992) have experimentally confirmed the location of the accessory nerve root in the caudal part of the ventrolateral vagal nucleus.

Barry (1987a) also studied the organisation of nerve IX. Glossopharyngeal motoneurons are largely confined to the visceromotor column wrapped around the medial border of Stieda's fasciculus. Only a few cell bodies are scattered ventro-laterally. The dendrites of nerve IX motoneurons course ventrolaterally, reaching almost as far as the meningeal surface. The axons of the glossopharyn-geal and vagal nerve often course medially to make a hairpin loop over the medial longitudinal fascicle and return laterally to join the other efferent fibres (Barry 1987a).

The rostral component of the branchiomotor column, i.e. the motor nucleus of V, lies directly medial to the most rostral part of the sulcus limi-tans and consists of rather loosely arranged, large to very large neurons of similar appearance to those in the more caudally situated visceromotor centres.

Rostral to the motor nucleus of VII, a conspicu-ous cell mass labelled nucleus A is found. It consists of very large, elongated, spindle-shaped neurons that are oriented in the transverse plane and lie parallel to the ventricular surface (Smeets and Nieuwenhuys 1976). This nucleus could only be identified in *Squalus* (Smeets et al. 1983). A corre-sponding cell mass has been observed in the cros-sopterygian fish *Latimeria* (Kremers and Nieuwen-huys 1979). The connections of the nucleus A are entirely unknown.

Finally, it should be mentioned that the rhomb-encephalon of the electric rays (torpedoids) in this region has been extensively modified because of the large electric lobes, which account for nearly 60 % of the total weight of the brain (Roberts and Ryan 1975). The homology of these lobes is much de-bated, but it is probable that in *Torpedo* its constitu-ent electromotor neurons are derived from the motor nuclei of nerves VII, IX and X (Ewart 1890; Hugosson 1957).

12.5.6.3
Visceromotor Nuclei

An Edinger-Westphal nucleus consisting of small, spherical and ellipsoidal cells, situated dorsolateral to the somatomotor nucleus of III, was found by Smeets and Nieuwenhuys (1976) in *Squalus* and *Scyliorhinus*. In mammals, this nucleus supplies vis-ceral efferent fibres in the oculomotor nerve that bring about the constriction of the pupil on illu-mination of the eye (the light reflex). The elasmo-branch eye also constricts the pupil when it is illu-minated. The iris sphincter is directly light respon-sive (Young 1933b) and is innervated by oculomo-tor neurons. The iris constricts fastest in diurnal sharks, more slowly in nocturnal sharks and slowest of all in Rajiformes (Kuchnow 1971). These differences may reflect the relative importance of neurally mediated compared with local responses of the iris. It is noteworthy that an Edinger-

Westphal nucleus can be delineated in *Squalus* and *Scyliorhinus*, but not in *Raja* and *Hydrolagus* (Smeets et al. 1983).

12.5.7
Reticular Formation

The reticular formation of cartilaginous fishes is usually subdivided into three longitudinal zones: median, medial and lateral. The median zone is confined to the rhombencephalon and consists of cells situated in or near the raphe. The medial zone constitutes by far the largest part of the chondrich-thyan reticular formation and extends throughout the brain stem. The lateral zone cannot be sharply delimited and lies mostly in the rhombencephalon in the vicinity of the sulcus limitans.

The *median reticular zone* is composed of two nuclei: the nucleus raphes superior and the nucleus raphes inferior. Of these, the former contains small, ellipsoidal elements situated in or immediately adjacent to the rhombencephalic raphe at the level of the motor trigeminal nucleus (Figs. 12.13, 12.19). A nucleus raphes inferior could not be seen in *Squalus*, but in *Raja*, *Hydrolagus* and *Scyliorhinus* such a nucleus was easily identified on the basis of their large and very large constituent elements, which lay in or immediately adjacent to the rhomb-encephalic raphe at mid- and caudal rhombence-phalic levels (Smeets et al. 1983). The inferior raphe nucleus of *Hydrolagus* can be divided into a rostral part of very large neurons and a caudal part charac-terised by large neurons (Smeets et al. 1983). Sub-stantial species differences in the appearance of the inferior raphe nucleus are also noted in reptiles (Newman and Cruce 1982; Newman et al. 1983).

The *medial reticular zone* extends almost the entire length of the brain stem and consists of rather loosely arranged, very large cells that occupy a position ventrolateral to the fasciculus longitudi-nalis medialis. In *Squalus*, the rhombencephalic part of the medial reticular zone comprises three distinct accumulations of cells: nucleus reticularis inferior, medius and superior (Figs. 12.15–12.19, 12.22). Considerable variation in the number of cells within each subdivision has been noted between species, but also between individuals of the same species (Smeets et al. 1983).

The nucleus of the fasciculus longitudinalis medialis constitutes the most rostral part of the medial reticular zone and consists of very large, tri-angular and polygonal cells which are rather dif-fusely arranged. In *Squalus*, this nucleus is situated for the most part anterior to the oculomotor nucleus (Fig. 12.22), but in *Raja* these neurons lie at the same level as the motor nucleus of nerve III.

The isthmic basal plate contains a cell mass labelled nucleus H. The nucleus consists of medium-sized cells that lie ventral to the lateral lemniscus. According to its position, this nucleus may represent a nucleus of the lemniscus lateralis or a more condensed part of the isthmic reticular formation, i.e. the nucleus reticularis isthmi. A study by Smeets and Timerick (1981) showed that descending pathways to the spinal cord originate from this area.

The *lateral reticular zone* is probably represented in cartilaginous fishes by the small and medium-sized cells that lie ventral and ventrolateral to the sulcus limitans. In *Squalus*, this area contains a distinct cell mass termed nucleus B (Smeets and Nieuwenhuys 1976) at the level of entrance of nerve VIII (Figs. 12.17, 12.22). This nucleus may well represent a more condensed part of the lateral reticular zone. An experimental study by Smeets and Timerick (1981) has shown that this nucleus projects to the spinal cord. Newman et al. (1983) noted that the elasmobranch nucleus B topographically resembled the lateral part of the nucleus reticularis medius.

Afferent Connections of the Reticular Formation. As shown in Fig. 12.34, currently no experimental data on the connections of the nucleus raphes superior of cartilaginous fishes are available. The inferior raphe nucleus, in contrast, has been shown to receive an input from the tectum mesencephali (Smeets 1981b, *Scyliorhinus, Raja*) and to project extensively to the spinal cord (Smeets and Timerick 1981; Ritchie et al. 1984; Timerick et al. 1992).

The afferent connections of the rhombencephalic medial reticular formation arise from neurons in the spinal cord, cerebellum, tectum and telencephalon (Fig. 12.34). Spinoreticular projections have been revealed by experimental studies carried out by Ebbesson (1972) and Hayle (1973a), who found that the long, peripherally extending dendrites of the large and very large neurons of the medial reticular formation made contact with fibres that ascend from the spinal cord. Electrophysiological evidence for such connections has been provided by Paul and Roberts (1978) in *Scyliorhinus*. The cells of the medial reticular formation also contact the direct and crossed fibres that originate from the tectum mesencephali (Smeets 1981b) and nucleus cerebelli (Ebbesson and Campbell 1973), a cerebelloreticular pathway that has been confirmed electrophysiologically by Paul and Roberts (1978). As mentioned previously, the inferior reticular nucleus receives primary afferent fibres of the octaval nerve (Dunn and Koester 1984). Finally, evidence for a direct telencephaloreticular projection has been obtained by Ebbesson (1972) for *Ginglymostoma* and by

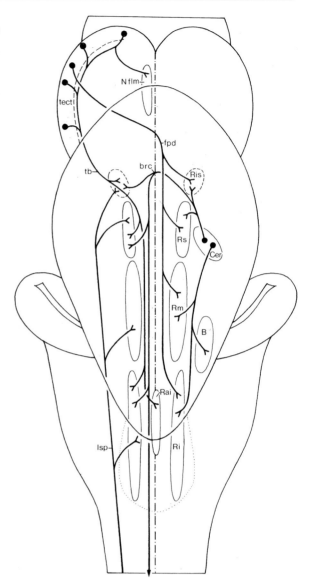

Fig. 12.34. Afferent connections of the reticular formation, as viewed from the dorsal surface of the brain

Smeets (1983) for *Scyliorhinus*. However, the exact site of origin of this pathway is not known, because the lesions in these experimental cases involved a large part of the telencephalon.

Efferent Connections of the Reticular Formation. Apart from the spinal projections that have already been dealt with in Sect. 4, the reticular formation projects to the cerebellum and the tectum mesencephali. A reticulocerebellar projection was suggested by Paul and Roberts (1975, *Scyliorhinus*) based on electrophysiological recordings. More recently, Fiebig (1988) found retrogradely labelled cells in the reticular formation of the midbrain and the isthmic region as well as in the inferior reticular nucleus after HRP had been injected in the corpus cerebelli of the thornback guitarfish *Platyrhinoidis*. A reticulocerebellar connection was demonstrated in *Scyliorhinus* and *Raja* by Smeets (1982). After tectal HRP injections, labelling was found in some medium-sized reticular cells.

12.5.8
Some Remaining Brain Stem Nuclei

In addition to the nuclei described in previous sections, the brain stem of cartilaginous fishes contains several other cell masses. These are the oliva inferior and the nucleus funiculi lateralis in the caudal rhombencephalon, the griseum centrale rhombencephali, the nucleus interpeduncularis, the nucleus cerebelli, the nucleus isthmi and nucleus F in the rostral part of the rhombencephalon and the nucleus tori semicircularis, the nucleus tegmentalis lateralis and the nucleus ruber in the tegmentum mesencephali.

The *oliva inferior* is an elongated nucleus, which occupies a paramedian position in the caudal part of the rhombencephalon. It comprises small, round, and ellipsoid cells (Figs. 12.15. 12.22). An experimental study of afferent connections in the guitarfish *Platyrhinoidis* (Fiebig 1988) revealed an exclusively contralateral projection from the inferior olive to the cerebellar corpus. Furthermore, the same study provided evidence that two populations of inferior olive cells project in medial and lateral sagittal zones onto the molecular layer of the cerebellar corpus. This topographical arrangement in the olivocerebellar projection in principle resembles the longitudinal zones described for the corresponding projection in the mammalian cerebellum. However, whether olivocerebellar fibres terminate as climbing fibres on Purkinje cell dendrites in cartilaginous fishes remains doubtful (Paul 1982).

The *nucleus funiculi lateralis* (Fig. 12.16, 12.22) is a rather compact cell mass which occupies a superficial position, ventrolateral to the inferior olive. Its neuronal elements are large (27 µm) and round or ellipsoidal in shape. Smeets and Nieuwenhuys (1976) termed this cell mass the nucleus funiculus lateralis because of its striking positional correspondence to the nucleus of the same name in the mammalian brain stem. Another term used by Fiebig (1988) for this cell mass is lateral reticular nucleus. Experimental studies have shown that the lateral funicular nucleus of cartilaginous fishes probably receives inputs from the spinal cord (Ebbesson 1972; Hayle 1973a) and from the primary octaval (horizontal canal) nerve (Dunn and Koester 1984; Barry 1987b). In turn, the nucleus projects to the spinal cord (Smeets and Timerick 1981) and to the cerebellar corpus (Fiebig 1988).

The *griseum centrale rhombencephali* and *nucleus interpeduncularis* are both located in the rostralmost part of the hindbrain, and both consist of granular cells (Figs. 12.18–12.22). The former occupies a periventricular position, while the latter is clearly detached from the central grey. The nucleus interpeduncularis can be subdivided into a superficially situated pars ventralis and a slightly more dorsal pars dorsalis. The topological map shows that the pars ventralis is located in the isthmus region, whereas the pars dorsalis extends from the rostralmost part of the isthmus into the tegmentum mesencephali (Fig. 12.22).

The *nucleus cerebelli* in *Squalus* lies in a protrusion formed by the basal part of the cerebellar peduncle called the eminentia ventralis cerebelli. The nucleus consists of a large cluster of diffusely arranged, medium-sized cells (Fig. 12.19). Edinger (1901) and several later workers considered this nucleus to be a subcerebellar structure, but studies on its ontogenetic development by Rüdeberg (1961) have shown that it originates from the cerebellar anlage and lies within the cerebellum. Experimental studies have revealed that the cerebellar nucleus is the target of the Purkinje cell axons which constitute the sole output of the cerebellar corpus (Ebbesson and Campbell 1973; Paul and Roberts 1984).

The *nucleus isthmi* is an accumulation of small, spherical cells situated directly rostral to the nucleus cerebelli (Figs. 12.20, 12.22). Johnston (1905) suggested that this nucleus receives afferent fibres from the vagal lobe and sends an efferent projection to the hypothalamus. He regarded this nucleus as being homologous to the nucleus visceralis secundarius of other vertebrates. Ariëns Kappers (1921, 1947) believed this cell mass to be a correlation centre for octavolateral and optic impulses. However, contrary to what has been found in other anamniotes, experimental studies by Smeets (1981b, 1982) in *Scyliorhinus* and *Raja* did not provide any evidence of a reciprocal connection between the tectum mesencephali and the nucleus isthmi.

The nucleus F (according to the terminology of Smeets et al. 1983) consists of relatively densely packed cells and lies ventrolateral to the nucleus cerebelli. This nucleus is a rather constant feature of the chondrichthyan brain and has a strong projection to the spinal cord (Smeets and Timerick 1981; Timerick et al. 1992) and a much weaker projection to the cerebellum (Fiebig 1988). Experimental studies also revealed a strong projection of nucleus F to the lobus inferior of the hypothalamus in the clearnose skate *Raja eglanteria* (Smeets and Boord

1985). Because of this connection and its position, the nucleus F resembles the nucleus visceralis secundarius of other vertebrates, but an input from the viscerosensory area has yet to be demonstrated. However, most likely the nucleus F consists of several cell populations. Cells in the caudal part of the nucleus project primarily to the spinal cord and cerebellum, while cells in the rostral part innervate predominantly the hypothalamic inferior lobe.

In *Squalus*, the ventricular surface of the lateral part of the tegmentum shows a slight eminence, which is known as the torus semicircularis (Smeets and Nieuwenhuys 1976). This eminence contains a *nucleus tori semicircularis*, which consists of a subependymal layer of small, granular cells and an ill-defined deeper zone of scattered slightly larger elements (Figs. 12.20, 12.22). This nucleus is thought to receive afferent inputs from the octavolateral area. Experimental studies (Boord and Northcutt 1982; Barry 1987b), however, revealed that the octavolateral nuclei project predominantly contralaterally via the lemniscus lateralis to a subtectal region which does not coincide with the nucleus tori semicircularis, indicated as such by Smeets et al. (1983) in *Raja*.

The *nucleus tegmentalis lateralis* is a large aggregation of small, ellipsoidal cells, situated in the lateral part of the tegmentum. This nucleus lies in the rostral half of the mesencephalic tegmentum. Comparing the location of the nucleus tegmentalis lateralis with that of the terminal field of the secondary lateral line projections, it seems that the former lies ventral to the latter. Little is known about the connections of the lateral tegmental nucleus except for a strong, ipsilateral projection to the lobus inferior hypothalami (Smeets and Boord 1985).

A *nucleus ruber* was demonstrated in *Squalus* and *Scyliorhinus* by Smeets and Nieuwenhuys (1976) lying lateral to nerve III (Figs. 12.21, 12.22). In *Hydrolagus*, and particularly in *Raja*, this nucleus is better developed than in the sharks studied (Smeets et al. 1983). In line with these observations are the results of spinal cord injections of HRP (Smeets and Timerick 1981), which revealed labelled neurons in the contralateral nucleus ruber of *Raja*, but not in *Scyliorhinus*. Paul and Roberts (1978) were also unable to demonstrate the presence of a rubrospinal tract in *Scyliorhinus* using electrophysiological techniques, although all neurons of the nucleus ruber could be activated from the contralateral cerebellar nucleus. It therefore seems that, although a nucleus ruber is present in *Scyliorhinus*, it does not provide a rubrospinal projection. However, such a pathway evidently exists in *Raja*. This result might be explained by the consi-

derable difference in the locomotion of the two species.

12.5.9
Chemoarchitecture

The number of studies dealing with the chemoarchitecture of the brain stem in cartilaginous fishes is steadily increasing. The monoaminergic systems, in particular, have received much attention. Hitherto, the distribution of serotonin-immunoreactive neuronal elements in the brain stem has been studied in *Dasyatis* (Ritchie et al. 1983), *Platyrhinoidis* (Stuesse et al. 1990), *Scyliorhinus* (Yamanaka et al. 1990), *Heterodontus* (Stuesse et al. 1991a), *Raja* and *Myliobatis* (Stuesse et al. 1991b), *Hydrolagus* (Stuesse and Cruce 1991) and *Squalus* (Stuesse and Cruce 1992). Immunocytochemical studies of the distribution of catecholamines in the brain stem of cartilaginous fishes have been carried out (Meredith and Smeets 1987, *Raja*; Stuesse et al. 1990, *Platyrhinoidis*; Stuesse et al. 1991a, *Heterodontus*; Stuesse and Cruce 1991, *Hydrolagus*; Stuesse and Cruce 1992, *Squalus*; Stuesse et al. 1994, several species).

Moreover, the distribution of several peptides has been studied in more or less detail, including SP (Stuesse et al. 1992, several species; Rodríguez-Moldes et al. 1993, *Scyliorhinus*), enkephalin (Stuesse and Cruce 1991, *Hydrolagus*; Stuesse et al. 1991a, *Heterodontus*; Stuesse et al. 1991b, *Raja*, *Myliobatis*; Stuesse and Cruce 1992, *Squalus*), neuropeptide Y (Vallarino et al. 1988a, *Scyliorhinus*; Chiba and Honma 1992, *Scyliorhinus*), gonadotropin-releasing hormone (Wright and Demski 1991, several species; Lovejoy et al. 1992, *Squalus, Bathyraja*) and somatostatin (Chiba et al. 1989, *Mustelus*). Furthermore, the distribution of calbindin-D$_{28K}$ (Rodríguez-Moldes et al. 1990) has been studied in the brain of the dogfish *Scyliorhinus*.

Despite the quantity of immunohistochemical studies, there are not enough data for drawing hard conclusions. This is because (a) only a limited number of species has been studied, (b) most studies have focused on distribution of cell bodies and provide no details about fibre distribution and (c) some reported differences are certainly due to technical procedures, e.g. different antibodies. Nevertheless, some general comments on the chemoarchitecture of the brain stem of cartilaginous fishes can be made.

Serotonin. Serotoninergic cell bodies are mainly confined to the raphe nuclei and the reticular formation in the brain stem of cartilaginous fishes.

In the sharks *Squalus* and *Heterodontus*, a putative dorsal raphe nucleus was identified on the basis of the position of serotoninergic cells in the central grey of the metencephalon dorsal and medial to the medial longitudinal fasciculus (Stuesse et al. 1991a; Stuesse and Cruce 1992). In skates and rays, however, as well as in the holocephalian *Hydrolagus*, such a cell group could not be found. Stuesse and colleagues have used the distribution of serotoninergic and leu-enkephalin-immunoreactive cell bodies for a further subdivision of the raphe nuclei following the nomenclature used for mammalian brains, but connectional data are needed to substantiate these putative homologies.

Catecholamines. Catecholamine cell bodies occur at caudal rhombencephalic levels in the viscerosensory area, the vagal motor nucleus and the ventrolateral tegmentum (Stuesse et al. 1994). The cell bodies have been roughly divided into two groups, a ventrolateral and a dorsomedial group, which are referred to as the A1 and A2 cell group, respectively, following the terminology proposed by Hökfelt et al. (1984). As in other vertebrates (for a review, see Smeets and Reiner 1994), these cell groups may contain dopamine, noradrenaline or adrenaline, but the exact nature of the catecholamine involved has yet to be identified.

A rather unexpected finding was the demonstration of dopamine-immunoreactive cell bodies in the midbrain of *Raja radiata* by Meredith and Smeets (1987), since it was previously thought that catecholaminergic midbrain cell groups were not present in anamniotes (Parent et al. 1984). Two distinct populations of dopamine-immunoreactive cell bodies are, however, present in the rostral mesencephalic tegmentum of cartilaginous fishes, one located along the midline, the other lying mainly lateral to the nucleus ruber. On the basis of their positions, these cell groups have been labelled as the ventral tegmental area and substantia nigra, respectively, following the terminology used for mammalian catecholamine cell groups. The ventral tegmental area cell group contains a high density of neurons whose cell processes are predominantly oriented dorsoventrally. The number of dopamine cells in the presumed homologue of the mammalian substantia nigra is considerably smaller than that in the midline group and the cell processes are mainly oriented mediolaterally. Similar cell groups have been demonstrated in all elasmobranchs studied so far, but not in holocephalians (Stuesse et al. 1994). We will discuss these midbrain dopamine cell groups and their bearing on the evolution of basal ganglia organisation in more detail in relation to the telencephalic basal ganglia structures (see Sect. 12.9).

Substance P. As shown in the study by Rodríguez-Moldes et al. (1993), only a few SP-immunoreactive (SP-IR) cell bodies are found in the brain stem of *Scyliorhinus*, in the presumed Edinger-Westphal nucleus. In contrast, distinct plexuses of SP-IR fibres were observed in the ventrolateral mesencephalic tegmentum, the isthmic region, the descending trigeminal root, the visceromotor column, the viscerosensory area and the raphe region throughout the brain stem. The immunoreactive fibres in the ventrolateral midbrain tegmentum resemble the descending SP-IR basal ganglia projection fibres of amniotes and will be discussed in more detail in Sect. 9.

Enkephalin. Although studies by Stuesse and colleagues have focused on the distribution of leu-enkephalin in the brain in a variety of cartilaginous fishes, only some knowledge on the distribution of cell bodies has been gained. In general, enkephalinergic perikarya are found in the same brain stem nuclei that serotoninergic cell bodies are present in, i.e. in the reticular formation.

12.5.10
Functional Correlations

At the present time, we have little idea whether the brain stem regions of cartilaginous fishes have essentially the same roles as those in mammals, as very few experimental studies have been carried out. For example, apart from anecdotal observations, there is little known about sleep-wakefulness states in elasmobranchs or, indeed, whether these terms have any meaning for these animals. It is known that some species show rhythmical patterns of locomotor activity, e.g. *Heterodontus, Cephaloscyllium* (Nelson and Johnson 1970; Finstadt and Nelson 1975), but it is uncertain how these patterns relate to sleep, and there are no indications of how brain stem actions might affect these patterns. We know, however, that the brain stem plays a significant role in determining the state of the spinal cord and is important for the control of locomotion.

12.5.10.1
Brain Stem and Locomotion

When the brain stem in batoids is separated from the spinal cord by a transection made at the obex, the resulting spinal preparation does not show the spinal "shock" typical of mammals, but instantly displays a wide range of reflex movements. Spontaneous locomotor movements, however, are absent. In contrast, in the shark-like elasmobranchs, spontaneous locomotor movements of essentially normal form appear immediately after spinal section, a result which suggests that the motor programme for undulatory motion is produced by spinal cord circuits. In species such as *Scyliorhinus*, in which locomotor activity varies diurnally, these centres are presumably restrained by pathways that descend from the brain. The sites of origin of these pathways are unknown, but transections made at various levels through the brain stem of *Scyliorhinus* indicate that they are located in the rhombencephalon close to the root of nerve VIII. The study by Paul and Roberts (1979) on cerebellar function in *Scyliorhinus* suggests that the inhibitory impact on spinal cord centres is modulated by the cerebellum.

Inhibitory and excitatory regions related to locomotion have been described in the mammalian brain stem (Mori et al. 1978, 1980), in which it has been shown that activation of the lateral reticular formation results in controlled locomotion. This region extends as a continuous zone, rostral to the cuneiform nucleus, in the mesencephalon. Stimulation of this zone, the MLR, induces locomotion in the decerebrated cat. Kashin et al. (1974) reported that, in teleost fish, electrical stimulation of a limited region in the caudal part of the tegmentum evokes coordinated fin and body movements. They suggest that this region is equivalent to the mammalian MLR.

Recently, the MLR in the Atlantic stingray *Dasyatis sabina* was identified and characterised by Bernau et al. (1991). Electrical stimulation (50–100 μA, 60 Hz) of the midbrain in decerebrated, paralysed animals was used to elicit locomotion monitored as alternating activity in nerves innervating an antagonist pair of elevator and depressor muscles. Effective sites for evoking locomotion in the midbrain included the caudal portion of the interstitial nucleus of the medial longitudinal fasciculus and an area immediately lateral to it (cuneiform and subcuneiform nuclei according to Bernau et al. 1991). The MLR of *Dasyatis* does not include the nucleus ruber, the tectum mesencephali or the medial or lateral mesencephalic nuclei. Electrical stimulation in the MLR evokes locomotion in both the ipsilateral and contralateral pectoral fin, while stimulation of the medullary reticular formation evokes locomotion only in the contralateral fin. Moreover, Bernau et al. (1991) found that the medial reticular formation in the rostral medulla had to be lesioned bilaterally to abolish locomotion evoked by electrical stimulation in the MLR. Such locomotion could also be abolished by lesioning the ipsilateral medullary reticular formation plus interruption of fibres that project from the MLR to the contralateral midbrain. The MLR of *Dasyatis*

appears, therefore, to be similar to the lateral component of the mammalian MLR and to the MLR in other nonmammalian vertebrates (Bernau et al. 1991).

12.5.10.2
Brain Stem and Electric Organ Discharge

A modified motor control function for the brain stem is seen in those elasmobranchs that possess electric organs. Weak electric discharges are produced in Rajiformes from the small electric organs that are located in the tail. In torpedoids, the pectoral discs contain large electric organs that produce substantial shocks. Both types of electric organ are activated only intermittently. In torpedoids, this discharge is important for prey capture (Belbenoit and Bauer 1972), whereas the weak electric discharge in Rajidae plays a role in electrolocation (Baron et al. 1985; Bratton and Ayers 1987).

As in other electric fishes (teleosts), the coordination of electric organ discharge from the electric organs is achieved in most cases by a series of cells that are activated in sequence. The cells which initiate the discharge are called "control" cells or "pacemakers", and the following neurons are "relay" cells (Bennett 1968). In *Torpedo*, the electromotoneurons are contained in the rhombencephalic electric lobes; in *Raja*, they lie in the spinal motor column. In both species, the electro-motoneurons are activated by a nucleus located in the inferior raphe nucleus. Little experimental work has been done on the properties of this nucleus (Szabo 1955), and it remains uncertain whether the nucleus is the "command" centre for electric discharge in these fishes. One somewhat surprising feature of the organisation of this centre, particularly in *Torpedo*, is the apparent lack of gap junctions on these cells (Fox 1977; Nakajima 1970) and presumably, therefore, the absence of electronic coupling between the neurons, which is such a striking feature of the command centres in other electric fishes (Bennett 1968).

12.5.10.3
Brain Stem and the Common Mode Rejection Within the Electrosensory System

As already noted, the electrosensory system of cartilaginous fishes is the best studied system with respect to anatomical, electrophysiological and behavioural aspects. The sense organs, i.e. the ampullae of Lorenzini, have been shown to operate as electroreceptors. Elasmobranchs encounter electric fields of a uniform nature in the natural environment due to the motion of sea water through the earth's geomagnetic field (von Arx 1962) and due to the fields the fishes induce in themselves while swimming through the earth's magnetic field (Kalmijn 1978). Kalmijn (1982) has also shown that the stingray *Urolophus halleri* can detect uniform electric fields and determine the field's polarity and direction with a detection threshold of 0.005 µV/cm. Neuropharmacological analysis of the synaptic transmission in the ampullae of Lorenzini of the skate *Raja clavata* has suggested that the most likely transmitter at the afferent synapse is L-glutamate, L-aspartate or a substance of similar nature (Akoev et al. 1991).

One of the major functions of the electrosensory system is its use in prey detection. The bioelectric fields which extend into the water around aquatic animals are detected by the ampullary system, and the appropriate behaviour is initiated. Such fields attenuate rapidly with distance, and the range of the system is limited to approximately 30 cm for relatively small prey. One problem with this system is that the animal itself produces fields with the same characteristics as those that need to be detected. It seems likely that the electrosensory system would require some mechanisms to cancel out afferent input caused by its own fields to preserve its sensitivity to external stimuli. Kalmijn (1974) suggested that the grouping of the ampullae into clusters may enable the electroreceptor system to operate differentially and to suppress interference from respiratory movements by common mode rection. The hypothesis of a common mode rejection mechanism was tested in later electrophysiological studies by Montgomery (1984), Bodznick and Montgomery (1992), Bodznick et al. (1992) and Montgomery and Bodznick (1993, 1994) and is now generally accepted. A common mode mechanism is possible because all electroreceptors, regardless of their orientation or location on the body surface, are stimulated at the same phase and amplitude by ventilatory movement, whereas they are affected differentially by extrinsic fields. Suppression of this noise is achieved by sensory processing within the the dorsal nucleus of the lateral line area. The most simple model of the circuitry underlining common mode rejection, as proposed by Montgomery and Bodznick (1993, 1994), is that direct afferent input impinges on the basal dendrites of the ascending efferent neurons located in the peripheral zone of the dorsal nucleus. The spiny apical dendrites of the ascending efferent neurons extending into the cerebellar crest are contacted by the parallel fibers of granule cells from distinct populations in the vestibulocerebellum. For further details of the model of such an adaptive filter, the reader is referred to the paper by Montgomery and Bodznick (1994).

12.6
Cerebellum

12.6.1
Introductory Notes

The cerebellum of cartilaginous fishes is considerably larger and further developed than that of cyclostomes. It is difficult to delineate the extent of the cerebellum in fishes, because the easily recognised corpus cerebelli lies adjacent to other regions which are essentially cerebellar-like in organisation. These regions, the auricles and lateral line areas, relate to the octavolateral sensory systems and were considered by Larsell (1967) to form the octavolateral lobe. Nicholson et al. (1969), following Johnston (1902), considered this part of the brain to be the phylogenetic base of the vertebrate cerebellum.

In early development, the cerebellum of cartilaginous fishes consists of a simple plate-like structure, similar to the cerebellum of the adult lamprey, but its form changes as development proceeds. Two processes take place which give the chondrichthyan cerebellum its characteristic appearance:

1. Bilaterally, a rostrolaterally directed lengthening and outpouching of the caudolateral parts of the cerebellar anlage, a process in which not only the nervous tissue, but also the adjacent choroid tissue is involved
2. A dorsalward evagination of the rostromedial parts of the cerebellar plate

The former event leads to the formation of the auriculae, while the latter gives rise to the corpus cerebelli.

The auricles are entirely covered on their dorsal surface by extensions of the tela choroidea of the fourth ventricle and can be subdivided on each side into a rostromedial upper leaf and a caudolateral lower leaf (Figs. 12.3, 12.17–12.19). The lower leaf is contiguous caudally with the areae octavolaterales, while the upper leaves of both auricles unite medially as the "lower lip", which bridges the fourth ventricle.

The unpaired corpus cerebelli encloses a large ventricular cavity in *Squalus* and extends rostrally over the tectum mesencephali and caudally over the lower lip. The basal part of the corpus forms a distinct stalk, the peduncle, which joins rostrally with the tectum via a very short velum medullare anterius and caudally with the lower lip and upper leaves of the auricles.

Considerable variation in external form and size of the cerebellar corpus between chondrichthyan species has been reported (Northcutt 1978; Sato et al. 1983), ranging from an almost smooth cerebellar

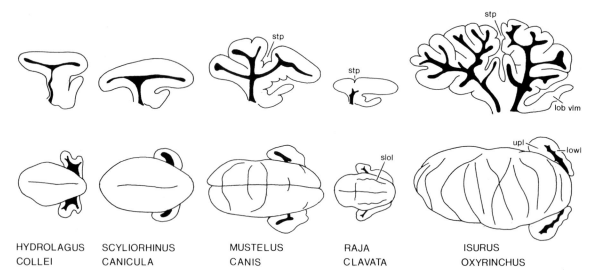

Fig. 12.35. Dorsal and mid-sagittal views of the cerebellum of selected cartilaginous fishes, showing increasing extent of foliation of the cerebellar wall. *slol*, sulcus longitudinalis lateralis; *stp*, sulcus transversus primus

surface in such species as *Scyliorhinus* to a highly convoluted cerebellar surface in species such as *Dasyatis* and *Isurus* (Fig. 12.35). In these latter species, transverse grooves divide the corpus into numerous sublobuli. Voorhoeve (1917) thought that the variation in the size of the corpus was in some way related to the size of the body of the fish, but recent investigations by Northcutt (1978) revealed considerable evidence against such a simple correlation. Considerable intraspecific variation of cerebellar morphology also occurs (Puzdrowski and Leonard 1992). This variation, as noted in *Dasyatis*, could not be related to size or sex of the animal.

12.6.2
Structure

In its internal organisation, the chondrichthyan cerebellar corpus is very similar to that of other gnathostomes. Its wall is differentiated into four layers: the innermost granular layer, the fibre zone, the layer of Purkinje cells and the outer molecular layer. The granule cells are concentrated in two longitudinal ridges, the prominentiae granulares (Nieuwenhuys 1967; Alvarez and Anadón 1987), which are situated on either side of the median plane. They begin immediately behind the trochlear decussation, curve along the inside of the wall of the corpus cerebelli and terminate as two slight ventral protrusions of the lower lip (Fig. 12.18). The granule cells of these ventral protrusions are laterally contiguous with the granule cells of the auricular upper leaves (Fig. 12.18). The granule cells (Fig. 12.36) are small and densely packed and have

three to six dendrites that emerge randomly from the somata, sometimes bifurcate once or twice and terminate as large, convoluted claws (Schaper 1898; Houser 1901; Nicholson et al. 1969). The axon of the granule cell arises either from the soma or from the proximal region of a dendrite and then passes to the molecular layer, where it divides as a T to give parallel fibres, lying in the transverse plane of the cerebellum. In addition to the granular cells, the prominentiae granulares also contain larger neurons, assumed to be Golgi cells (Nicholson et al. 1969; Eccles et al. 1970; Alvarez-Otero and Anadón 1992). Golgi cells probably receive input from mossy fibres, and the more superficial elements of this type extend into the molecular layer, where they are in contact with parallel fibres.

In the lateral parts of the corpus cerebelli, where granule cells are absent, a zone of fibres can be distinguished beneath the Purkinje cells. This zone includes incoming afferent fibres and outgoing Purkinje cell axons, the latter of which passing ventrally to the cerebellar nucleus.

The Purkinje cells form a regular layer, one or two cells thick, which extends throughout the entire wall of the corpus cerebelli, with the exception of the paramedian regions (Figs. 12.17, 12.19, 12.20). In holocephalian fish, however, the arrangement of Purkinje cells is less precise (Braak 1967; Smeets et al. 1983), as they do not form a layer but are scattered throughout the molecular layer. Studies by Schaper (1898), Houser (1901) and Nicholson et al. (1969) revealed that the Purkinje cell has an elongated body from which several thick, spineless processes emerge (Fig. 12.36). Each process gives rise

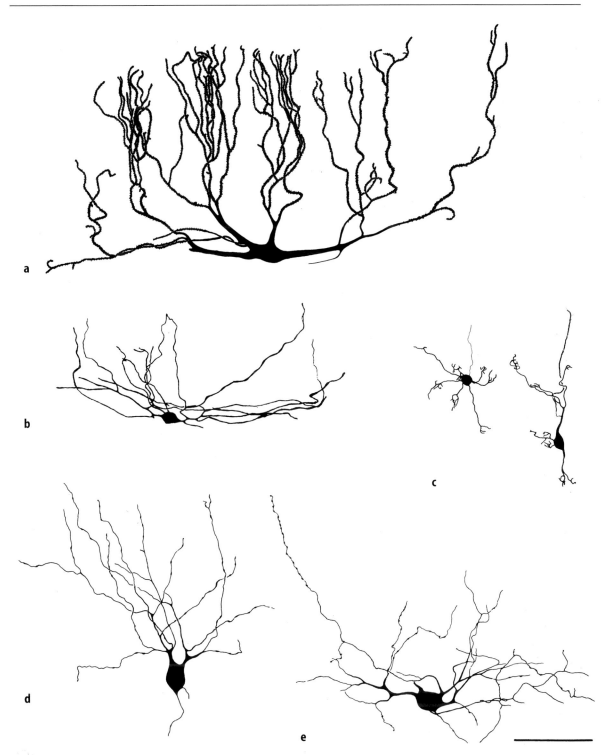

Fig. 12.36a-e. Camera lucida drawings of examples of cerebellar neurons of *Platyrhinoidis*. Golgi impregnation. **a** Purkinje cell. **b** Stellate cell. **c** Granule cells. **d,e** Golgi cells. *Bar*, 100 μm. (Combined figures from Nicholson et al. 1969)

to a number of ascending spiny dendrites, which together form a dendritic tree that is quite markedly flattened in the parasagittal plane.

The molecular layer contains several elements:

1. Axons ascending from granule cells, and their branches, i.e. parallel fibres
2. Purkinje cell dendrites
3. Possibly climbing fibres
4. Stellate-type interneurons

A thorough discussion of whether climbing fibres are present in the cerebellum of fishes was provided by Paul (1982). The oliva inferior, the chief source of climbing fibres in the mammal, has long been recognised in many elasmobranchs. Recent studies in *Platyrhinoidis* have shown this nucleus to label retrogradely with HRP injected into the cerebellum (Fiebig 1988), suggesting that axons of olivary neurons pass to the cerebellum. However, according to Paul (1982), it remains uncertain how they terminate there. Recently, Alvarez-Otero et al. (1993), using the Cajal reduced silver technique and Golgi methods, revealed axons with thick swellings ("pincushions") close to the base of Purkinje cells. By means of the electron microscope, the same authors demonstrated that flattened fibres formed extensive synaptic contacts with these Purkinje cell pincushions. On the basis of the ultrastructural features, Alvarez-Otero et al. (1993) proposed that these fibres are a primitive type of climbing fibres resembling the condition found in the early developing climbing fibres of mammals.

Stellate cells in chondrichthyan fishes can be divided into two categories (Schaper 1898; Houser 1901; Nicholson et al. 1969). The first category comprises superficially located cells with dendrites passing predominantly parallel to the sagittal plane (Fig. 12.36). Their axons characteristically bear small swellings. The second category is represented by more deeply located, fusiform cells with dendrites flattened horizontally. Unlike mammalian basket cells, they do not form pericellular baskets.

Electrical stimulation of the surface of the corpus has been carried out for several elasmobranch species and evokes a series of potentials that can be recorded with microelectrodes inserted into the molecular layer. These compound potentials are thought to result from a volley of parallel fibre activity, conducting at around 0.2 m/s (Paul 1969), and the subsequent excitation of the Purkinje cell and stellate cell dendrites. The precise interpretation of these potentials has caused some controversy (see Paul 1982).

Studies by Montgomery (1978, 1981, 1982) have extended our knowledge of the organisation of the auricles. The three component regions, the upper and lower leaves and the lower lip (pars medialis of Larsell 1967), differ in their internal organisation and connections. The lower lip shows a typical cerebellar construction with a molecular layer, conspicuous Purkinje cells and a granular layer. The granule cells do not form a layer beneath the Purkinje cells, but also cluster as median accumulations, the granular cords or prominentiae granulares, which join with the prominences of the cerebellar corpus (Fig. 12.18). The lateral portion of the upper leaf is composed exclusively of granule cells and perhaps Golgi cells and is not covered by a molecular layer. Only a few Purkinje cells are found in the lower leaf, but in the lower lip they form a well-defined layer (except for in *Hydrolagus*).

12.6.3
Cerebellar Afferent Connections

Data on afferent fibre connections to the chondrichthyan cerebellum have been presented by a great number of authors using normal material (e.g. Voorhoeve 1917; Ariëns Kappers et al. 1936; Larsell 1967; Nieuwenhuys 1967). More recently, our knowledge of cerebellar afferents has been extended by experimental studies carried out by Hayle (1973b), Boord and Campbell (1977), Montgomery (1978), Boord and Roberts (1980), Paul and Roberts (1981), Ebbesson and Hodde (1981) and Fiebig (1988). These findings are summarised in Fig. 12.37a. The following afferents to the chondrichthyan cerebellar regions have been described: primary fibres of the lateral line nerves, primary fibres of the octavus nerve to the lower lip, fibres originating in the area octavolateralis, trigeminocerebellar fibres, tractus spinocerebellaris, tractus olivocerebellaris, tractus mesencephalocerebellaris and tractus lobocerebellaris. The first six projections have already been described elsewhere, whereas the existence of the last two projections, i.e. tractus mesencephalocerebellaris and tractus lobocerebellaris, is suggested on basis of normal material (see Smeets et al. 1983). The latter pathway was demonstrated by Smeets and Boord (1985) in *Raja* as originating from the most dorsomedial part of the inferior lobe and terminating in the ipsilateral granular ridge of the rostral pole of the cerebellar corpus.

An electrophysiological study by Tong and Bullock (1982) has revealed that the cerebellum of the thornback ray *Platyrhinoidis* receives input from a variety of sensory systems, including electrosensory, mechanosensory and tactile-proprioceptive information. Discrete areas of the cerebellum are responsive to these modalities, and the areas show limited overlap. The visual and tactile-pro-

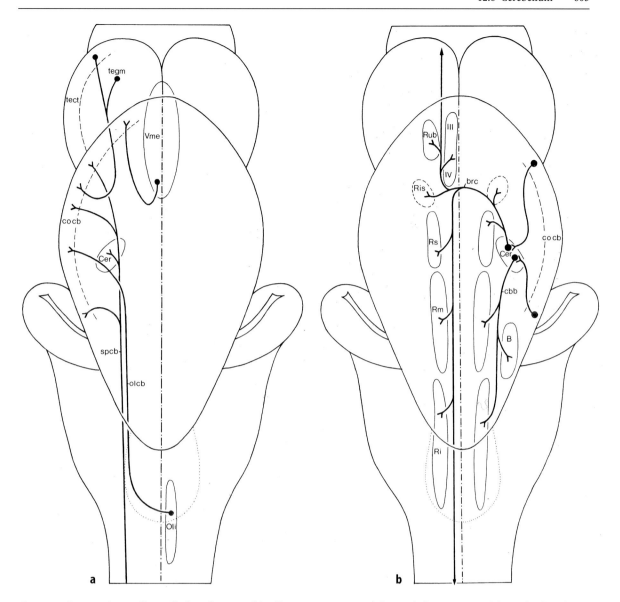

Fig. 12.37a,b. Experimentally verified **a** afferent and **b** efferent connections of the cerebellum, as viewed from the dorsal surface of the brain

prioceptive areas are large, while the electrosensory area is small. The tail is represented only in the posterior lobe, whereas the trigeminal input reaches both the posterior and anterior lobe, suggesting some topographical representation (Tong and Bullock 1982). Recently, Conley and Bodznick (1994) demonstrated electrophysiologically the presence of proprioceptive and electroreceptive representations in the cerebellar dorsal granular ridge of *Raja*.

12.6.4
Cerebellar Efferent Connections

Experimental studies by Ebbesson and Campbell (1973) and Paul and Roberts (1984) have revealed that the Purkinje cell axons in elasmobranchs do not project directly to the brain stem, but terminate in the cerebellar nuclei located in the cerebellar peduncle. The nucleus cerebelli in turn gives rise to two distinct fibre systems (Fig. 12.37b):

1. The tractus cerebellobulbaris rectus, which terminates mainly in the ipsilateral rhombencephalic reticular formation

2. The brachium conjunctivum

According to Ebbesson and Campbell (1973), the latter pathway appears to issue a few pre-terminal fibres to the ipsilateral reticular formation, just before it decussates. Immediately after crossing, the tract splits into a large ascending and a small descending pathway. The latter descends to the caudal rhombencephalon and gives off fibres to the medial reticular formation. The ascending crossing component of the brachium conjunctivum provides a massive input to the nucleus ruber and also distributes fibres to the trochlear and oculomotor nuclei and the adjacent central grey. A very small number of fibres of the brachium conjunctivum passes into the dorsomedial parts of the diencephalon.

12.6.5
Functional Correlations

Only a few neurophysiological investigations of the elasmobranch cerebellum have been performed, and the majority of these were designed to elucidate interneuronal connections. Nevertheless, these physiological studies, when taken together with anatomical work, show that the elasmobranch cerebellum contains the same kind of elements as the mammalian cerebellum and has similar interactions (Nicholson et al. 1969).

Although it is widely held that the cerebellum plays an important role in motor control, it still remains unclear how this is achieved. The profound locomotor disturbances seen in mammals following cerebellectomy are primarily the result of loss of postural control. This problem is, of course, considerably simplified for aquatic animals because of the support provided by the dense medium. This raises the intriguing question of why the cerebellum is so well developed in cartilaginous fishes, which perform relatively stereotyped movements, uncomplicated by postural problems.

Studies on the consequence of cerebellar ablation for a reflex pectoral fin movement in *Scyliorhinus* have provided some insight into the significance of the cerebellum (Paul and Roberts 1979). It was shown that the reflex was considerably attenuated in cerebellectomised fishes, suggesting that the cerebellum plays a modulating rather than an initiating role in locomotor control. The motor loops that generate movements of the fins and body are located in the spinal cord, and the "gain" of these circuits is regulated by the rhombencephalic reticular formation. The latter in turn is controlled by cerebellar action (Paul and Roberts 1983, 1984). This might mean that different regions of the body are regulated by selected areas of the cerebellum, which effectively holds a map of the body's motor activities. Consequently, complex movements might require elaborated cerebella. This may help us to understand why elasmobranch cerebella are relatively so much larger than that of most teleosts, as the latter, in possessing neutral buoyancy, do not have the problem of matching thrust against lift which confronts the swimming elasmobranch.

12.7
Tectum Mesencephali

12.7.1
Introductory Notes

In the majority of cartilaginous fishes, the tectum mesencephali is strongly developed and differentiated into two bilateral lobes which surround expansions of the ventricular cavity. The tectum mesencephali is the chief centre of termination for retinofugal fibres and, as in other fishes, generally appears to be well developed in those species with large eyes. In addition to visual information, the tectum mesencephali receives octavolateralis, vestibular, trigeminal and spinal cord inputs, as mentioned previously. Thus the widely used term optic tectum is inappropriate and should be avoided. The efferent tectal pathways pass to all main regions of the brain, except for the telencephalon.

12.7.2
Structure

The mesencephalon of chondrichthyan fishes is well developed and can be divided into a dorsal tectum mesencephali and a ventral tegmentum mesencephali. The boundary between these two regions is externally marked by a groove, the sulcus tectotegmentalis (Fig. 12.21). On the ventricular surface, the boundary between these two regions is not indicated by any gross morphological landmark, although the ventricular surface contains two sulci, i.e. sulcus intermedius ventralis and sulcus d. The former is a rostral extension of the rhombencephalic sulcus with the same name, whereas the latter lies more laterally and rostrally and forms the ventral boundary of the torus semicircularis (Fig. 12.22).

Although tectal lamination in chondrichthyans is not as distinctive as reported for teleosts, it is generally possible to distinguish six tectal layers in these fishes (Witkovsky et al. 1980; Smeets et al. 1983; Manso and Anadón 1991). The most primitive condition occurs in the holocephalian *Hydrolagus*, in which most of the tectal neurons are confined within a wide periventricular layer. Among elas-

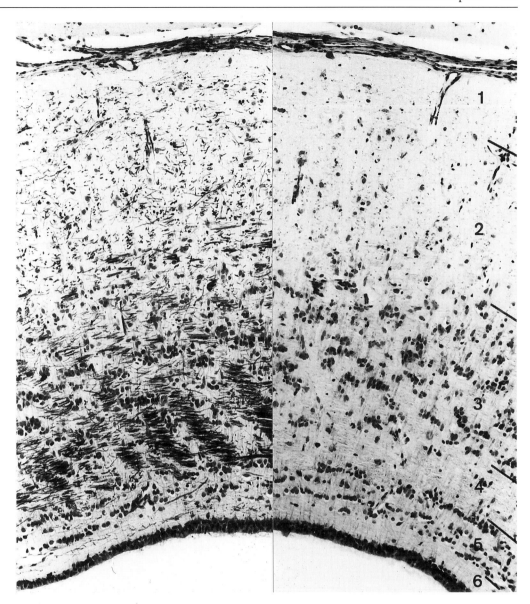

Fig. 12.38. Photomicrographs of transverse sections showing the tectal lamination in the dogfish *Squalus acanthias*. *Left*, Klüver and Barrera staining; *right*, Nissl staining. *Bar*, 100 μm

mobranchs, the squalomorph shark *Squalus* probably exhibits the most primitive tectal pattern, although extensive cellular migration away from the ventricular surface has taken place (Fig. 12.38). In the galeomorph shark *Scyliorhinus*, and especially in the batoid *Raja*, tectal cells have migrated still further away from the ventricular surface (Smeets et al. 1983). These findings support Northcutt's suggestion (1979) that the evolution of the tectal visual circuitry in cartilaginous fishes may have occurred by migration and increase of neuronal number rather than by development of new visual pathways.

As reviewed by Smeets et al. (1983) and Réperant

et al. (1986), the nomenclature used for tectal layers varies considerably not only between species, but also between authors (Table 12.2). For the present review, the tectal layers of *Squalus* and *Scyliorhinus* are indicated as follows passing from the meningeal to ventricular surfaces (Figs. 12.38, 12.39):

1. A *stratum medullare externum*, containing a few neurons, scattered between a network of unmyelinated fibres. Numerous retinofugal fibres terminate in this layer (Northcutt 1979).
2. A *zona externa of the stratum cellulare externum*, consisting of numerous pyramidal and

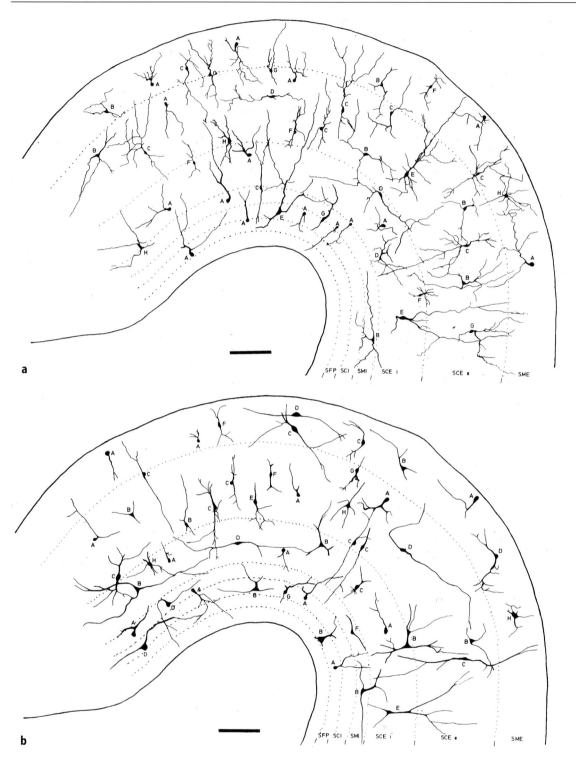

Fig. 12.39a,b. Camera lucida drawings of cell types in the tectum mesencephali of the shark *Scyliorhinus canicula,* as shown by means of **a** the Golgi-aldehyde or **b** the Golgi-Cox procedures. *A*, monopolar cells; *B*, triangular cells; *C*, radial bipolar cells; *D*, horizontal fusiform cells; *E*, large tectal cells; *F*, small tectal cells; *G*, pyriform cells; *H*, stellate cells; *SCEe,* stratum cellulare externum zona externa; *SCEi,* stratum cellulare externum zona interna; *SCI,* stratum cellulare internum; *SFP,* stratum fibrosum periventriculare; *SME,* stratum medullare externum; *SMI,* stratum medullare internum. *Bar,* 200 μm. (Combined figures from Manso and Anadón 1991)

Table. 12.2. Comparison of nomenclatures for tectal lamination of elasmobranchs (modified from Réperant et al. 1986)

Ginglymostoma	Scyliorhinus	Squalus	Raja	Scyliorhinus, Raja	Scyliorhinus
(Schroeder and Ebbeson 1975; Luiten 1981a)	(Farner 1978b)	(Northcutt 1979)	(Witkovsky et al. 1980)	(Smeets 1981a; Smeets et al. 1983)	(Repérant et al. 1986)
Stratum medullare externum Layer A	Layer 13	Marginal layer *(Superficial tectal zone)*	Marginal layer *(Superficial zone)*	Stratum medullare externum Layer 1	Stratum marginale Layer 1
Stratum cellulare externum zona externa Layer B	Layers 12–10 *(Stratum superficiale)*	Layer 6	Optic fiber layer	Zona externa Layer 2 *(Stratum cellulare externum)*	Stratum fibrosum et griseum superficiale Layer 2 — pars externa / pars interna
		Layer 5		Zona interna Layer 3	Stratum griseum superficiale Layer 3
Stratum cellulare externum zona interna Layer C	Layer 9	Layer 4 *(Central tectal zone)*			Stratum fibrosum et griseum centrale Layer 4
Stratum medullare internum Layer D	Stratum centrale Layers 8–6	Layer 3	Central zone	Stratum medullare internum Layer 4	Stratum album centrale Layer 5
Stratum cellulare internum	Layers 5–3 *(Stratum periventriculare)*	Layer 2 *(Periventricular tectal zone)*	Periventricular zone	Stratum cellulare internum Layer 5	Stratum griseum periventriculare Layer 6
Layer E	Layer 2	Layer 1		Stratum fibrosum periventriculare Layer 6	Stratum fibrosum periventriculare Layer 7

bipolar cells. Both types of cell have superficial dendrites, which branch repeatedly into layer 1 and receive retinal inputs, and deeper dendrites, which extend into layers 3 and 4, where they receive other sensory input. Some of the cells of this layer give rise to retinopetal fibres (Luiten 1981a, *Ginglymostoma*).

3. A *zona interna of the stratum cellulare externum*, containing neurons which are generally somewhat larger and more densely packed than those in layer 2. Their dendrites extend in all directions, but stay mainly in layer 3. It is generally assumed that the axons of the cells in layer 3 contribute to the main tectal pathways.

4. A *stratum medullare internum*, containing only a few cells. The main efferent tectal fibres course through this layer.

5. A *stratum cellulare internum*, consisting of various cell types. Tectal neurons in this layer have piriform perikarya and processes that often arborise horizontally. In addition, this layer con-

tains the very large cells of the mesencephalic trigeminal nucleus.

6. A *stratum fibrosum periventriculare* has not been distinguished as a separate entity in the chondrichthyan brain by most authors. The origin and termination of the unmyelinated fibres in this layer are unknown.

A Golgi study of the tectum mesencephali of *Scyliorhinus* by Manso and Anadón (1991) revealed eight types of cells (Fig. 12.39), of which at least six show dendritic specializations such as spines in the form of "drumsticks" and thin varicose appendages. The latter authors concluded, therefore, that the organisation of the tectum mesencephali of elasmobranchs is more varied and complex than was thought previously. Moreover, Fig. 12.39 underscores that, for a more complete picture of tectal organisation, application of different Golgi techniques is necessary.

12.7.3
Afferent Connections

The majority of tectal afferent fibres is constituted by retinofugal fibres, which in chondrichthyan fishes have been studied experimentally by numerous authors (Ebbesson and Ramsey 1968; Graeber and Ebbesson 1972; Northcutt 1979, 1990; Ebbesson and Meyer 1980; Northcutt and Wathey 1980; Witkovsky et al. 1980; Luiten 1981a; Smeets 1981a; Jen et al. 1983; Repérant et al. 1986; Northcutt 1990). It was found that tectal layers 1, 2 and, to a lesser extent, 3 receive input from the retinal ganglion cells. Variations in the laminar distribution of the

retinal projection fibres have been reported; these are mainly a consequence of a different tectal terminology, technique or species (for reviews, see Smeets et al. 1983; Réperant et al. 1986). In addition to the visual input, the tectum mesencephali receives fibres from all main parts of the brain and spinal cord (Hayle 1973a; Ebbesson and Hodde 1981; Boord and Northcutt 1982; Smeets 1982); these are shown in Fig. 12.40a. Spinal neurons which project to the tectum appear to be confined to the dorsal half of the "cervical" cord (Smeets 1982). Rhombencephalic cell masses that are known to project to the tectum are the reticular formation, the nucleus tractus descendens nervi

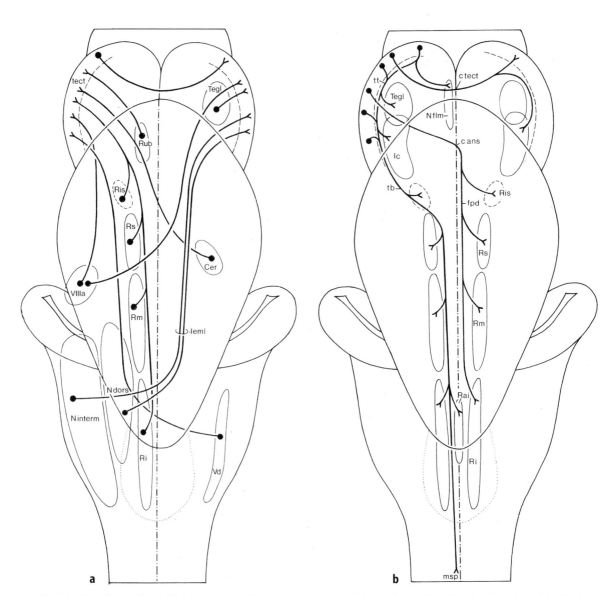

Fig. 12.40a,b. Experimentally verified connections of the tectum mesencephali, as viewed from the dorsal surface of the brainstem. **a** Afferent projections originating from the brain stem. **b** Tectal efferent projections to the brain stem and spinal cord

trigemini, the nucleus dorsalis and intermedius areae octavolateralis, the nucleus octavus ascendens, the nucleus cerebelli and the nucleus funiculi lateralis. Midbrain structures which project to the tectum are the contralateral tectal half, the nucleus tegmentalis lateralis, the ventrolateral tegmental area and the nucleus ruber. In the diencephalon, the pretectal area, the thalamus dorsalis pars medialis, the thalamus ventralis pars lateralis and the nucleus medius hypothalami are sources of pathways projecting to the chondrichthyan tectum. Telencephalic neurons which project to the tectum are found in the central dorsal pallium. Of all these tectal afferent projections, those from the nucleus medius hypothalami, pallium dorsale pars centralis and nucleus cerebelli (only in *Scyliorhinus*) are strong.

12.7.4
Efferent Connections

The tractus tectobulbaris is the most prominent efferent pathway of the tectum (Figs. 12.17–12.21, 12.41). This descending pathway can be subdivided into an ipsilateral and a contralateral component (Smeets 1981b). The ipsilateral descending pathway can be divided into fasciculi tectotegmentales ventrales and intermedii. A number of these fibres terminate in the tegmentum, but the majority continue further caudalward as the tractus tectobulbaris ventralis and intermedius. The intermediate portion of the fasciculi tectotegmentales in *Scyliorhinus* does not reach rhombencephalic levels, whereas in *Raja* the same portion forms the main descending tectal pathway (Smeets 1981b). The tractus tectobulbaris ventralis and tectobulbaris intermedius distribute their fibres to the reticular formation of the isthmus region, the nucleus reticularis superior, medius and inferior and to the nucleus raphes inferior. The contralateral descending tectobulbar component is formed by the fasciculi tectotegmentales dorsales, which, after decussating in the commissura ansulata, continue caudalward as the fasciculus predorsalis. This fasciculus is situated ventrolateral to the fasciculus longitudinalis medialis and distributes its fibres to the medial reticular formation (Figs. 12.18–12.21, 12.41).

Apart from descending projections, the following three tectal efferent projections have been experimentally confirmed (Ebbesson 1972; Smeets 1981b; Fig. 12.40b):

1. An ascending projection both ipsi- and contralateral to the pretectal area, the dorsomedial region of the thalamus and the lateral geniculate body (nucleus pretectalis superficialis of the present study)

2. A commissural projection to the contralateral tectal half, some fibres of which can be traced to the contralateral tegmentum mesencephali
3. A rather diffuse, ipsilateral tectotegmental system

A direct tectospinal pathway originating from neurons in the stratum cellulare internum and reaching cervical spinal cord levels has also been recognised (Smeets and Timerick 1981; Timerick et al. 1992).

Finally, two striking differences between the tectum mesencephali of *Hydrolagus* and that of the elasmobranchs should be mentioned:

1. Most tectal neurons are situated in a periventricular plate.
2. Some optic nerve fibres do not enter the tectum via the deep tectal zone (zona interna of the stratum cellulare externum), but via the superficial stratum medullare externum.

12.7.5
Pretectum and Accessory Optic System

The *pretectum* constitutes the dorsal part of the diencephalic-mesencephalic transitional region (or synencephalon) and lies, as its name implies, rostral to the tectum. The region has distinct boundaries in the median plane, but laterally it merges with other areas of the brain. The habenular commissure forms the rostral boundary, the sulcus organi subcommissuralis marks the ventral boundary and the dorsal border is provided by the surface of the brain. On the lateral border, the pretectal region merges ventrally with the thalamus and dorsally with the tectum. The caudal extent is limited ventrally by the rostral wall of the wide ventricle of the tectum, and dorsally it is continuous with the tectal grey.

The pretectal region can be subdivided into three longitudinal areas that lie on top of one another:

1. The area pretectalis sensu strictiori
2. The commissura posterior
3. The organon subcommissurale

The *area pretectalis* sensu strictiori, which constitutes the dorsalmost zone of the pretectal region, extends throughout the length of the latter. Northcutt (1979) divided the pretectal area in *Squalus* into three cellular groups – the periventricular pretectal nucleus, the central pretectal nucleus and the superficial pretectal nucleus – all of which receive retinofugal fibres. He believed that these three nuclei correspond topologically to the rostral continuation of the periventricular, central and superficial tectal layers. In our normal material, the *periventricular pretectal* and the *central pretectal* nuclei

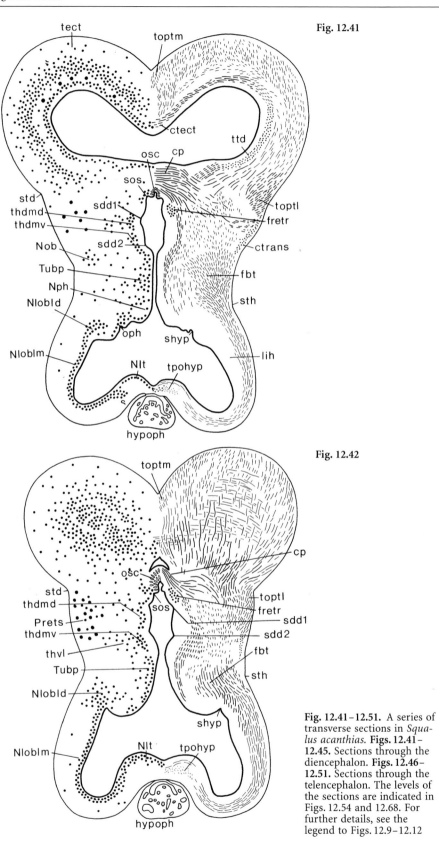

Fig. 12.41

Fig. 12.42

Fig. 12.41–12.51. A series of transverse sections in *Squalus acanthias*. **Figs. 12.41–12.45.** Sections through the diencephalon. **Figs. 12.46–12.51.** Sections through the telencephalon. The levels of the sections are indicated in Figs. 12.54 and 12.68. For further details, see the legend to Figs. 12.9–12.12

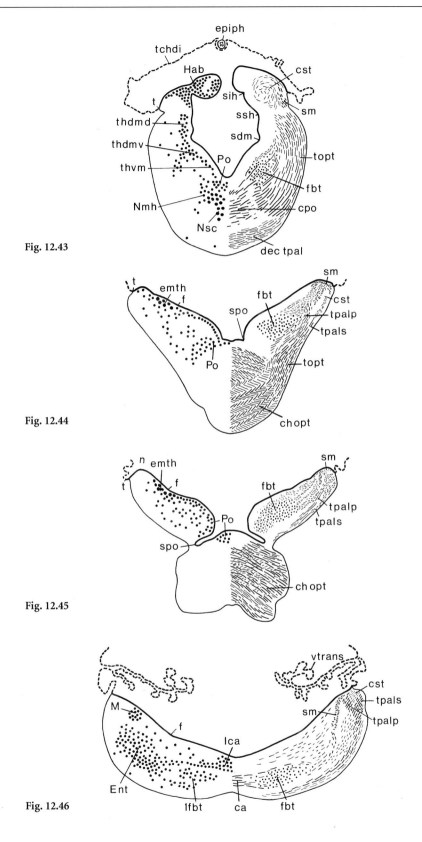

Fig. 12.43

Fig. 12.44

Fig. 12.45

Fig. 12.46

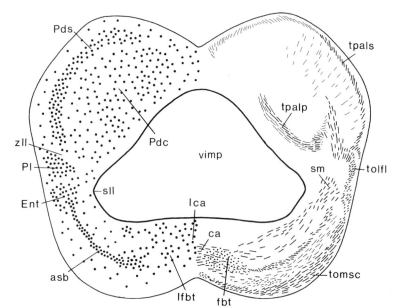

Fig. 12.47

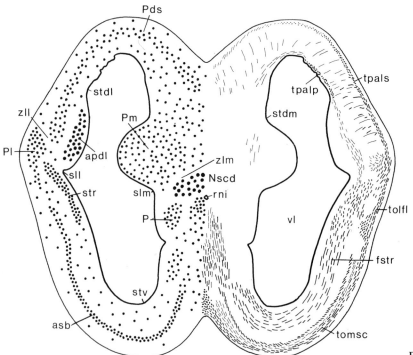

Fig. 12.48

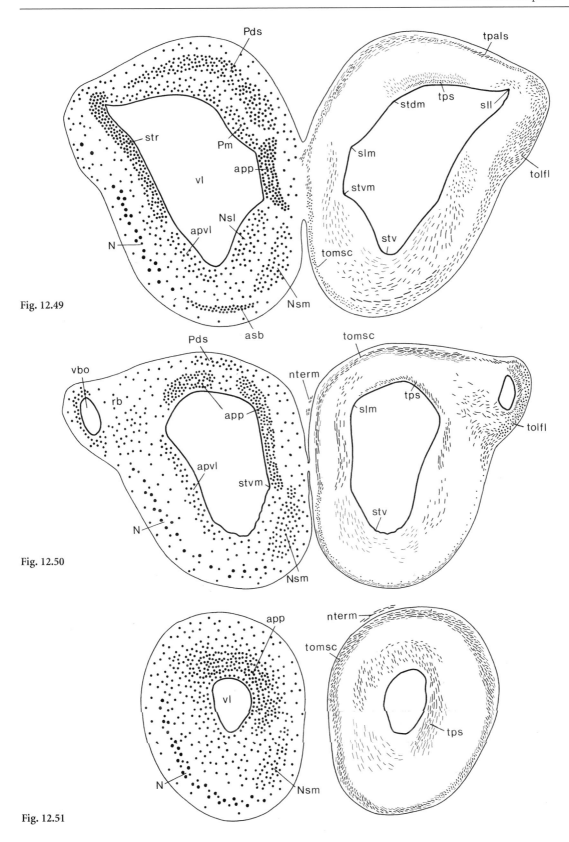

Fig. 12.49

Fig. 12.50

Fig. 12.51

could not be delineated as distinct entities, although at some levels their positions can be roughly indicated (Figs. 12.41, 12.42). The *nucleus pretectalis superficialis* is the only nucleus in the chondrichthyan diencephalon which comprises medium-sized cells. This nucleus, which was previously labelled as corpus geniculatum laterale by Smeets et al. (1983), has migrated from the periventricular zone and is embedded in the stream of optic nerve fibres (Fig. 12.42). Although this nucleus has now been termed the nucleus pretectalis superficialis, following the terminology of Northcutt (1979), it is still possible that at least a part of this cell mass may be equivalent to the lateral geniculate body of amniotes. Luiten (1981a,b) divided a cell mass that has about the same position in the diencephalon of the nurse shark *Ginglymostoma* into two, calling only the dorsal part the lateral geniculate body. This portion does not receive direct retinal projections, but does have tectal input, whereas the ventral part, i.e. the ventrolateral optic nucleus, receives a strong retinal input. Both (sub)nuclei project to the caudal part of the telencephalic pallium. Although Luiten (1981a) proposed that the ventrolateral optic nucleus should be more appropriately termed the lateral geniculate nucleus because of its direct retinal input and its projection to the telencephalon, he did not do so in his subsequent paper (Luiten 1981b). Degeneration studies by Smeets (1981a,b) in *Scyliorhinus* and *Raja* revealed both retinal and tectal inputs to the nucleus, labelled here as the superficial pretectal nucleus. These differences may be explained by considering the brain of *Ginglymostoma* to be more advanced (Northcutt 1977, 1978) than that of *Scyliorhinus* and *Raja*. According to the parcellation theory of Ebbesson (1980a, p. 183), "neural systems evolve not by the mixing of systems, but by differentiation and parcellation, which involves competition of inputs, the redistribution of inputs, and the loss of connections". Experimental studies, particularly HRP injections in the caudal pallium of *Squalus*, *Scyliorhinus* and *Raja*, will be needed to test the validity of this theory with regard to the lateral geniculate body in elasmobranchs.

The middle layer of the pretectal region, the bed of the commissura posterior, consists chiefly of decussating fibres, but also contains the *nucleus commissurae posterioris*. This nucleus is formed of diffusely spread cells that lie among the ventrolaterally directed fibres of the commissure. However, because of the diffuse arrangement of the neurons, we have been unable to delineate this nucleus.

The third and ventralmost layer of the pretectal region is made up of the *organon subcommissurale* (Figs. 12.40a, 12.41, 12.42). This organ is formed of tall, thin columnar cells which lie within the ependymal layer. Histochemical, autoradiographic and ultrastructural studies have revealed that the subcommissural organ cells synthesise and secrete glycoproteins (Rodríguez et al. 1992). The secretory products are released into the cerebrospinal fluid (CSF) of the diencephalic ventricle and condense to form a filament, i.e. Reissner's fibre, which passes caudalward through the ventricles of the brain and the central canal of the cord and attaches to a mass of mesenchymatous tissue at the end of this canal (for reviews, see Gabriel 1970; Ziegels 1976, 1979; Oksche 1993; Grondona et al. 1994). The function of the subcommissural organ is still unknown, although recent studies suggest that the secretion by subcommissural organ cells plays a role in the CSF circulation. A maldevelopment of the subcommissural organ during fetal life or alteration of the organ by X-irradiation was found to lead to aqueductal stenosis and hydrocephalus in rats and mice (Overholser et al. 1954; Takeuchi and Takeuchi 1986; Jones and Bucknall 1988). Recently, Cifuentes et al. (1994) studying CSF flow through the central canal of normal rats and rats which were immunologically deprived of Reissner's fibre, found a significant decrease in the flow of intraventricularly injected tracers to the distal part of the central canal of the spinal cord. The latter authors hypothesised, therefore, that Reissner's fibre plays a major role in the main bulk flow of CSF along the canal.

Ventral Synencephalic Region. The cell masses which will be discussed in this section are the nucleus opticus basalis, the nucleus tuberculi posterioris and the nucleus sacci vasculosi. An *accessory optic system* was demonstrated in an experimental study of the retinal efferents in *Squalus* (Northcutt 1979). It consists of two subdivisions, a rostrodorsal and a caudoventral accessory nucleus. An accessory optic system or basal optic system appears to be a common feature of elasmobranch fishes. A comparison of results obtained by means of degeneration techniques with those of autoradiographic methods reveals that the latter demonstrate better the existence of the accessory optic system (see Smeets 1981a; Réperant et al. 1986). Recently, a basal optic system has been recognised in the holocephalian fish *Hydrolagus collei* by means of the autoradiographic method (R.G. Northcutt and W.J.A.J. Smeets, unpublished observations, Fig. 12.52). Although ventral and dorsal divisions of the accessory nuclei are easily recognised in experimental material of all cartilaginous fishes studied so far, these nuclei do not form distinct cytoarchitectonic entities. At the moment, some experimental evidence exists that these presumed basal optic nuclei project to the cerebellum (Fiebig 1988), as has been demonstrated for some teleosts (Finger and Karten 1978).

12.7.6
Functional Correlations

Its central position, its multimodal input and its extensive connections with other centres suggest that the tectum mesencephali plays an important coordinating role in the behaviour of cartilaginous fishes. However, the exact function of this region and its behavioural significance are largely unknown. Tectal ablation studies in elasmobranchs have given confusing results, some workers finding that tectal ablations produce motor deficits, while others report little change (for a review, see Bullock 1984). Graeber et al. (1973) reported that removal of parts of the tectum in the nurse shark *Ginglymostoma* had only an initial impact on locomotor motor performance and that even after extensive tectal ablation the fish could still learn visual discriminations. Indeed, there was little evidence to suggest any substantial postoperative visual discrimination deficit (Graeber 1978).

The anatomical studies have shown that the chondrichthyan tectum receives a major visual projection, predominantly in its superficial laminae, while other non-visual information, such as auditory, mechanoreceptive, electroreceptive, trigeminal and somatosensory information is relayed to deep tectal layers. Some electrophysiology has been carried out and has shown some properties of tectal neurons responsive to light flashes, cutaneous mechanical stimuli, electric current in the water and acoustic stimuli (Bullock 1984). Multimodal

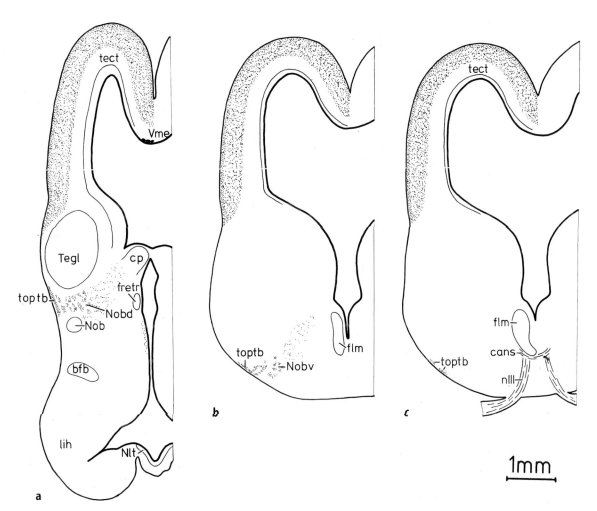

Fig. 12.52a–c. Transverse sections through **a** the caudal diencephalon and **b,c** midbrain of the ratfish *Hydrolagus collei*, illustrating the presence of a distinct basal optic tract as demonstrated by means of autoradiographic methods. *Nob*, nucleus opticus basalis; *Nobd*, nucleus opticus basalis pars dorsalis; *Nobv*, nucleus opticus basalis pars ventralis. (From R.G. Northcutt and W.J.A.J. Smeets, unpublished)

units were not found in the tectum, but more recent work in skates by Bodznick (1990) suggests that both electrosensory and visual stimuli reach the same deep tectal neurons. In addition, the study by Bodznick showed not only a spatiotopical organisation of retinotectal fibres, but also the presence of a spatiotopic tectal map of the electrosense. In both maps, the region of space near the horizon is greatly over-represented, which may indicate the importance of this region of space in the skate's natural orientation. A similar retinotectal projection pattern has been reported for the lemon shark *Negaprion brevirostris* (Hueter 1990). In the latter species, a prominent visual streak – a horizontal band of higher cell density – in both cone and ganglion cell layers is observed. Cone densities range from approximately 6500 cones per mm^2 along the horizontal meridian to less than 500 cones per mm^2 in the dorsal and ventral periphery (Hueter 1990). The presence of a visual streak in the lemon shark retina indicates that spatial vision is not simply a uniform, non-specialized feature of the visual system in this species or in cartilaginous fishes in general (Collin 1988; Hueter 1990). As has been noted before (e.g. Hueter 1990; McFarland 1990), assessment of the quality of vision in chondrichthyan fishes should consider the sensory niches of these animals in their specific marine environments as well as the phylogenetic and ontogenetic diversity within this class.

12.8
Diencephalon

12.8.1
Introductory Notes

Unlike other regions of the brain, the diencephalon in cartilaginous fishes has essentially retained the embryonic tube-like shape. Its thickened wall surrounds a narrow ventricle that expands ventrocaudally within the hypothalamus to form the lobi inferiores hypothalami. Most authors (e.g. Kuhlenbeck 1929a; Ariëns Kappers et al. 1936; Gerlach 1947) have subdivided the chondrichthyan diencephalon into four principal parts: epithalamus, thalamus dorsalis, thalamus ventralis and hypothalamus. Such a quadripartitioning can also be made in the rostral half of the diencephalon of *Squalus* (Smeets et al. 1983).

The diencephalic ventricular surface contains a rather distinct sulcal pattern (Figs. 12.41–12.45, 12.53). Eleven sulci can be recognised in *Squalus*, of which the sulcus subhabenularis, the sulcus organi subcommissuralis and its accessory component mark the boundary between epithalamus and dorsal thalamus. The sulcus diencephalicus medius, which was thought to divide the thalamus into a pars dorsalis and a pars ventralis (Kuhlenbeck 1929a, 1973; Gerlach 1947), does not coincide with the boundary between these two regions in *Squalus*, as can be seen in Fig. 12.53a. In addition, the sulcus diencephalicus ventralis, which is a short, shallow groove, does not separate the ventral thalamus from the hypothalamus as was previously reported (Kuhlenbeck 1929a; Gerlach 1947).

In addition to a subdivision of the diencephalon into epithalamus, dorsal thalamus, ventral thalamus and hypothalamus, the cell masses of this brain region can also be subdivided according to their degree of migration:

1. A zone situated immediately lateral to the ventricular ependyma
2. A zone lateral to the inner ventricular zone but still a part of the ventricular grey, called the outer ventricular zone
3. Migrated nuclei which are either completely separate from, or retain only a slight connection with, the ventricular zones

12.8.2
Cell Masses and Their Connections

12.8.2.1
Epithalamus

The epithalamus comprises the epiphysis and the ganglia habenulae. The epiphysis or pineal organ is a long, thin, tubular structure, the closed distal end of which rests against the roof of the brain case (Rüdeberg 1969). The proximal end opens into the third ventricle between the habenular and posterior commissures (Fig. 12.53). Using light and electron microscopy techniques, Rüdeberg concluded that the pineal parenchyma consists of receptor cells, supporting cells and ganglion cells. The receptor cells have well-developed outer segments which show opsin immunoreactivity (Vigh-Teichmann et al. 1983), suggesting a photoreceptive capacity of the organ. Electrophysiological evidence that the pineal organ in *Scyliorhinus* is a very sensitive photoreceptor organ was provided by Hamasaki and Streck (1971), whereas Wilson and Dodd (1973a) found that, in the same species, the pineal organ evokes a paling response in darkness. It has been shown that the axons of the pineal ganglion cells give rise to a diffuse *tractus pinealis*, which passes to the posterior commissure and, to a lesser extent, to the habenular commissure (Rüdeberg 1969). However, their exact termination site has not been identified.

The *ganglia habenulae* (Fig. 12.43), which are well developed in chondrichthyans, constitute the rostral part of the epithalamus. In all species studied, the left habenular ganglion was found to be larger than the right one (see Smeets et al. 1983). However, differences in histological structure between the two ganglia and their most important efferent connection, i.e. the fasciculus retroflexus, are more considerable. In the left ganglion, cells lying within a dense network of myelinated fibres are somewhat larger and more loosely arranged than the other cells of the ganglion. The latter cells are of the characteristically small habenular type and lie tightly packed together, just like all the cells of the right ganglion, which also lacks a network of myelinated fibres. The chief efferent projection of the ganglion habenulae is formed by the *fasciculus retroflexus of Meynert*. The right and left bundles are asymmetrically developed, with the left bundle containing far more myelinated fibres. The fasciculus retroflexus passes ventrocaudalward through the diencephalon, occupying during this course a position close to the ventricular wall. During its course to the interpeduncular nucleus, the diencephalic part of the fibre bundle protrudes into the ventricle as a slight eminence, the ventral border of which is formed by the sulcus diencephalicus dorsalis 1. The fasciculus retroflexus terminates after decussation in the nucleus interpeduncularis (Fig. 12.54c).

The quantitatively most important afferent connection of the ganglion habenulae is formed by the *stria medullaris*, which is largely of telencephalic origin (see Sect. 9.3). Additional afferent connections originating from tectal and tegmental midbrain regions have been reported by Addens (1945) and were termed *tractus tectohabenularis* and *tractus tegmentohabenularis*, respectively. Although in sagittal sections of *Squalus* fibres connecting the habenular ganglia with the mesencephalon can be recognised, the direction of the fibres is still not certain.

12.8.2.2
Thalamus Dorsalis

The dorsal thalamus can be subdivided into two parts, a pars medialis, which consists of small densely packed periventricular cells, and a pars lateralis, which contains slightly larger and more diffusely arranged cells. The thalamus dorsalis pars medialis extends from the level of the optic chiasm to the diencephalic-mesencephalic boundary. In *Squalus*, this cell mass can be further subdivided into a pars dorsalis and a pars ventralis because of the greater density and slightly more intensive

staining of the ventral part (Figs. 12.41–12.43, 12.53a). The boundary between the two parts is, at caudal levels, marked by the sulcus diencephalicus dorsalis 2. The thalamus dorsalis pars lateralis lies lateral to the caudal half of the thalamus dorsalis pars medialis.

In *Scyliorhinus*, it has been demonstrated that the lateral part of the thalamus dorsalis receives a contralateral retinal projection (Smeets 1981a), whereas the medial part projects to the tectum (Smeets 1982). Spinothalamic (Ebbesson and Hodde 1981, *Ginglymostoma*), cerebellothalamic (Ebbesson and Campbell 1973, *Ginglymostoma*) and tectothalamic (Smeets 1981b, *Scyliorhinus, Raja*) projections to the dorsal thalamic region have been reported, indicating that the dorsal thalamus may be considered as a relay centre for several sensory modalities to the telencephalon.

12.8.2.3
Thalamus Ventralis

The ventral thalamus is composed of two parts, a periventricular pars medialis and a migrated pars lateralis (Figs. 12.42, 12.43, 12.53). The former consists of loosely arranged laminae of small cells, the latter of diffusely arranged small cells. The thalamus ventralis pars lateralis caps the fasciculus basalis telencephali.

Rostral to the thalamus ventralis pars medialis, a distinct nucleus can be delineated which, following Bergquist (1932) and Kuhlenbeck (1977), has been termed *eminentia thalami* (Figs. 12.44, 12.45, 12.53a). This cell mass is distinguished on account of the larger size and the more intensive staining of its constituent neuronal elements. A comparable nucleus is found in *Scyliorhinus* and *Raja*, but not in *Hydrolagus* (Smeets et al. 1983).

The thalamus ventralis pars lateralis receives a retinal input (Northcutt 1979; Smeets 1981a) and projects in turn to the tectum (Smeets 1982). The connections of the medial part of the ventral thalamus are largely unknown; only a descending efferent pathway to the spinal cord has been experimentally established (Smeets and Timerick 1981).

12.8.2.4
Hypothalamus and Preoptic Region

In *Squalus*, the following hypothalamic cell masses can be delineated: nucleus preopticus, nucleus suprachiasmaticus, nucleus medius hypothalami, nucleus lateralis tuberis and nucleus lobi lateralis hypothalami. Although it belongs morphologically to the telencephalon, the *nucleus preopticus* is included here because of its functional relationship with the hypophysis (see below). Rostrally, it merges into the interstitial nucleus of the commissura anterior; caudally, its boundary is formed ventrally by the optic chiasm and dorsally by differences in the cytoarchitecture of the preoptic area and thalamus (Figs. 12.44, 12.45, 12.53a). A subdivision into a pars magnocellularis and a pars parvocellularis, as found by Charlton (1932), cannot be made.

The *nucleus suprachiasmaticus* consists of cells which are rostrally scattered among the fibres of the decussating optic tract and the commissura postoptica (Figs. 12.43, 12.53a). Caudally, this nucleus is replaced by the *nucleus medius hypothalami.* The two nuclei can be easily distinguished from each other by the orientation of the bipolar cells of the nucleus suprachiasmaticus, converging toward the ventralmost part of the ventricular surface. The nucleus suprachiasmaticus receives both ipsilateral and contralateral retinal projections (Ebbesson and Ramsey 1968; Northcutt 1979, 1990; Smeets 1981a).

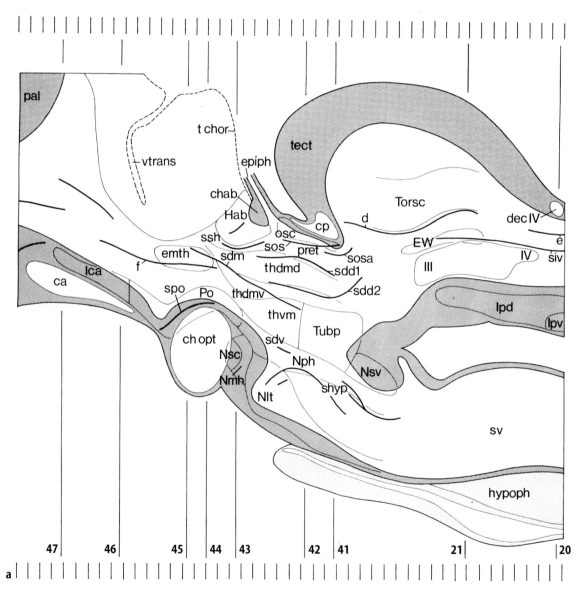

Fig. 12.53. Topographical reconstructions showing the cell masses in the diencephalon of *Squalus acanthias*, projected upon a mid-sagittal plane. **a** Cell masses that lie periventricularly or in the median plane. **b** Migrated cell masses. The *heavy continuous lines* represent the ventricular sulci, whereas the cell masses are represented by *thin, continuous curves.* The *numbers at the base* correspond to the figures of the same number showing the selected transverse sections. The distance between sections, indicated by *small, vertical bars,* equals 300 μm

The *nucleus medius hypothalami* varies considerably between chondrichthyan fishes (Smeets et al. 1983). In *Squalus* (Fig. 12.53a) and *Hydrolagus*, it is only small and does not extend far caudalward. In *Scyliorhinus*, it is better developed, whereas in *Raja* the nucleus medius hypothalami shows the greatest degree of development, in cell number, extent and cell type. Afferent connections of this cell mass are unknown, but efferent projections to the tectum (Smeets 1982) and hypophysis (Wilson and Dodd 1973b) have been reported.

Apart from the nucleus suprachiasmaticus and the nucleus medius hypothalami, the postchiasmatic region also contains the *nucleus periventricularis hypothalami* and the *nucleus lateralis tuberis* (Figs. 12.41, 12.42, 12.53a). Both nuclei surround the recessus infundibuli and lie within the periventricular zone. It is known that, in *Scyliorhinus*, the periventricular hypothalamic nucleus projects to the spinal cord (Smeets and Timerick 1981), whereas the nucleus lateralis tuberis is thought to be involved in hypophysial control (Knowles 1965).

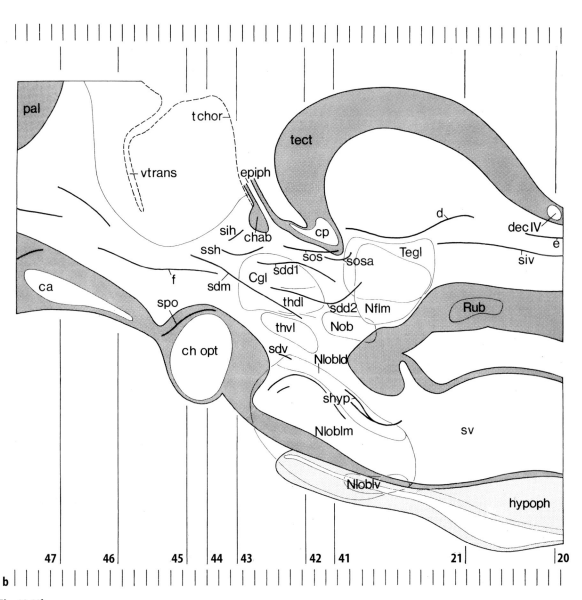

Fig. 12.53b

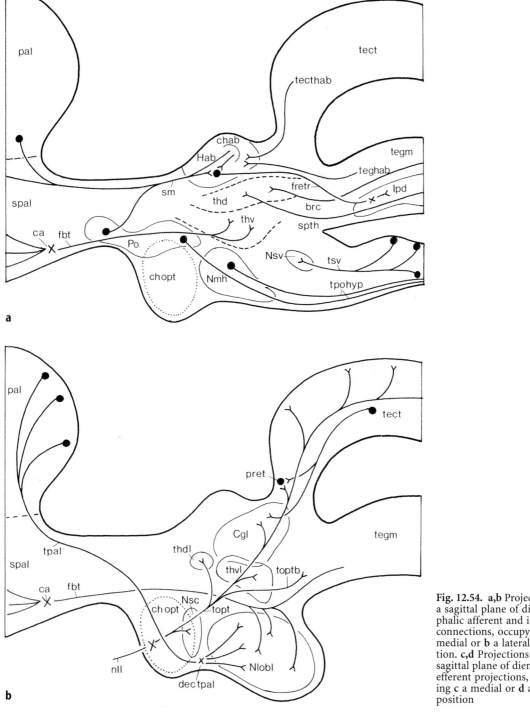

Fig. 12.54. a,b Projections on a sagittal plane of diencephalic afferent and intrinsic connections, occupying **a** a medial or **b** a lateral position. **c,d** Projections on a sagittal plane of diencephalic efferent projections, occupying **c** a medial or **d** a lateral position

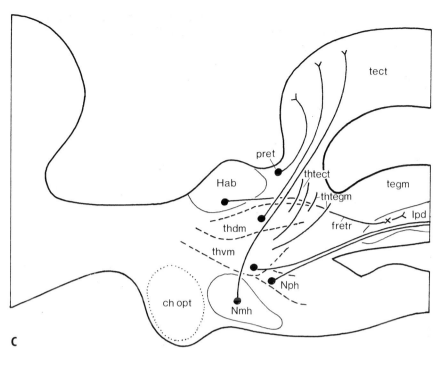

Fig. 12.54c

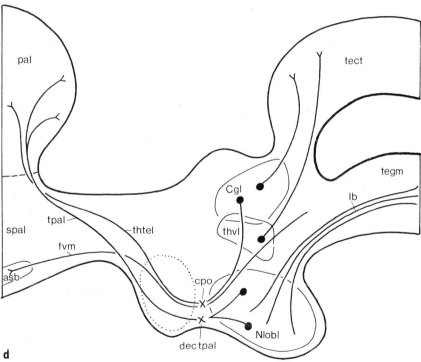

Fig. 12.54d

One of the most distinct cell masses of the hypothalamic area in *Squalus* is the *nucleus lobi lateralis hypothalami*. This nucleus consists of densely packed cells which cap the lateral expansion of the hypothalamic ventricle. It can be subdivided into a pars dorsalis, a pars intermedius and a pars ventralis. The cells of the dorsal and ventral parts are notably less densely packed than those of the intermediate part (Figs. 12.41, 12.42, 12.53b). On the basis of normal material, at least two afferent connections of the nucleus lobi lateralis inferior of *Squalus* can be recognised: the tractus pallii and the basal forebrain bundle. The *tractus pallii* has its origin in the roof of the telencephalon and was first described by Edinger (1892) as the *Mantelbündel*. This tract, which has been noted by all later students of the chondrichthyan brain (e.g. Johnston 1911; Bäckström 1924; Ariëns Kappers et al. 1936; Smeets et al. 1983) is a well-defined tract of predominantly myelinated fibres. The tract passes through the telencephalon impar, keeping a position lateral to the basal forebrain bundle, and then runs superficial to, or intermingles with, lateral fibres of the optic tract. Finally, the fibres of the tractus pallii decussate largely in the decussatio tractus pallii, caudoventral to the optic chiasm, and terminate in the rostral half of the nucleus lobi lateralis.

The *fasciculus basalis telencephali* or basal forebrain bundle originates predominantly from subpallial telencephalic regions, in particular the striatum and the area superficialis basalis. After partial crossing in the commissura anterior, the bundle enters the diencephalon to distribute its fibres mainly to the more superficial and caudal parts of the lobi inferiores hypothalami. In our normal material of *Squalus*, especially in sagittal sections, numerous fibres – both myelinated and unmyelinated – run from the caudal part of the lobus inferior hypothalami to the midbrain tegmentum.

The connections of the inferior lobe of the clearnose skate *Raja eglanteria* have been studied using the HRP technique (Smeets and Boord 1985) and shown to be both ascending and descending (Fig. 12.55). The ascending efferent connections consist of fibres that course within the basal forebrain bundle and pallial tract distributing to subpallial and pallial areas, whereas the descending efferent fibres course within the tractus lobobulbaris and the tractus lobocerebellaris. The lobobulbar tract can be traced to mid-rhombencephalic levels and appears to project to the lateral portion of the reticular formation. The afferent connections of the inferior lobe arise mainly from the midbrain tegmentum and the telencephalon. The major midbrain input is from cells which lie in the ipsilateral ventrolateral tegmentum, including the nucleus tegmentalis lateralis of Smeets et al. (1983) and nucleus F. Telencephalic inputs to the inferior lobe have their origin in cells of the preoptic area, area superficialis basalis and several areas in the dorsal and medial pallium. The pallial tract, the basal forebrain bundle and the lobobulbar tract seem to consist of both ascending and descending fibres.

Because of the almost complete obliteration of the lateral recesses of the hypothalamic ventricle and the more diffuse arrangement of the cells, the inferior lobe of the skate cannot readily be compared with that of *Squalus*, but similar connections may well be expected.

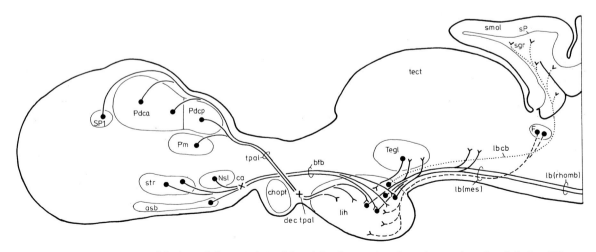

Fig. 12.55. Main connections of the hypothalamic inferior lobe of the skate *Raja eglanteria*. For the sake of clarity, different kinds of lines have been used. (From Smeets and Boord 1985)

12.8.2.5
Diencephalic Commissures

The *commissura habenulae* consists of myelinated and unmyelinated fibres of the stria medullaris. The unmyelinated fibres are probably true commissural fibres that form a commissura superior telencephali (Ariëns Kappers et al. 1936), whereas the unmyelinated fibres arise from the stria medullaris, which decussates in the commissura habenulae and terminates in the contralateral ganglion habenulae.

The *commissura posterior* consists of a large mass of fibres forming a dorsal convex arch just ventral to the pretectal region. It is compact in the midplane but sprays out laterally, with most of its fibres going in a ventrocaudal direction (Figs. 12.41, 12.42). No definite connections of the posterior commissure could be distinguished, but the fibres possibly relate to all of the nearby cellular regions.

The *commissura postoptica* lies ventrocaudal and dorsocaudal to the optic chiasm (Fig. 12.43). Its ventral portion is formed by crossing fibres of the tractus thalamotelencephalicus, whereas its dorsal portion is made up by commissural fibres between the rostral hypothalamic regions and may also contain crossing fibres of the commissura transversa.

The *commissura transversa*, sometimes infelicitously termed commissura postoptica (Ariëns Kappers et al. 1936), crosses immediately behind the optic chiasm in the dorsal part of the commissura postoptica. Although in our material the commissura transversa could easily be traced over some distance (Figs. 12.21, 12.41), its origin and termination could not be determined.

The *commissura preinfundibularis* lies caudal to the decussatio tractus pallii in the ventral part of the hypothalamus and probably consists of crossing fibres of the tractus preopticohypophyseus.

The *commissura postinfundibularis*, which can be subdivided into a pars superior and a pars inferior, lies in the tuberculum posterius. The fibres in the superior part form a connection between the caudal cell masses of the hypothalamus, whereas the pars inferior is made up by decussating fibres of the tractus sacci vasculosi.

12.8.3
Hypothalamo-hypophyseal Relationships

The *hypothalamus* influences behaviour by way of direct neural action on other neural centres and because of its powerful endocrine function. It collates information from the external and internal environment and causes the release of hormones from the hypophysis that bring about appropriate responses to environmental change. In this way, processes such as skin colour (Sumpter et al. 1984) and reproduction may be synchronised with environmental events.

The *hypophysis* of elasmobranchs is always made up of two structurally distinct parts: the adenohypophysis and the neurohypophysis. The adenohypophysis consists of a pars distalis and a pars intermedia. The pars distalis can be further subdivided into a dorsal lobe, which includes a narrow rostral lobe (zones 1–4; Fig. 12.56) and a wider caudal part or median lobe (zones 5–8; Fig. 12.56), and a ventral lobe. The ventral lobe is attached to the dorsal lobe by an epithelial stalk. The arrangement is somewhat different in Holocephali. In this group, a follicular glandular structure, the *Rachendachhypophyse*, lying far ahead of the main gland outside the cranium in the roof of the mouth, replaces the ventral lobe. A study by Dodd et al. (1980) in *Hydrolagus* has revealed that the *Rachendachhypophyse* is the main source of the gonadotropic hormone just like the ventral lobe in elasmobranchs. The pars intermedia of the adenohypophysis is intimately connected with the pars nervosa of the neurohypophysis; together, they constitute the neuro-intermediate lobe. The degree of intermingling of the two components varies in different cartilaginous fishes, but the pars intermedia is always penetrated to some extent by neural tissue.

The neurohypophysis can be divided into the following:

1. The neural lobe, which is derived from the ventral wall of the saccus infundibuli and is in close contact with the pars intermedia, forming the neuro-intermediate lobe
2. The hypophysial stem, which carries axons from the hypothalamus
3. The median eminence

For a more detailed description of the different parts of the hypophysis and the various adenohypophysial hormones and their place of secretion, the reader is referred to reviews by Perks (1969), Holmes and Ball (1974) and Smeets et al. (1983).

Hypothalamic control of the hypophysis is achieved in two ways:

1. Via the tractus preopticohypophyseus
2. Via the portal system

The *tractus preopticohypophyseus* was first described by Scharrer (1952) and Bargmann (1953) in *Scyliorhinus*. According to those authors, the fibres of this tract arise from the well-developed, neurosecretory preoptic nucleus. Caudal to the optic chiasm, the fibres of this tract converge and form a discrete tract close to the mid-line (Figs. 12.41, 12.42), which passes through the thin

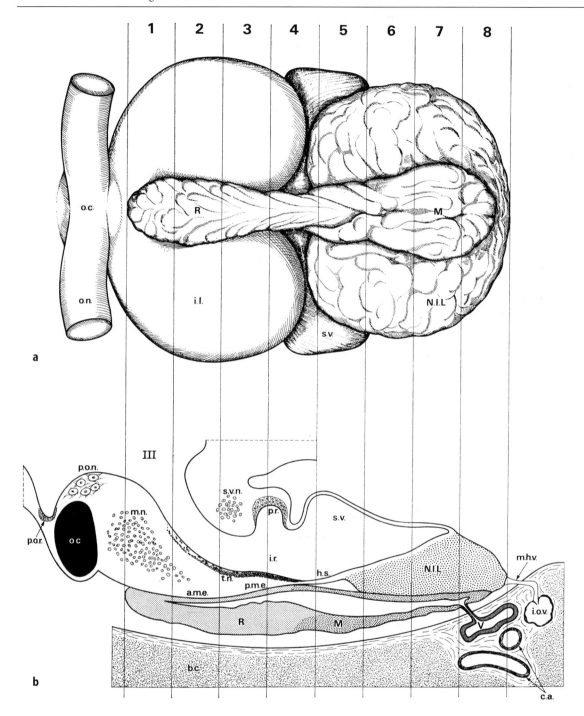

Fig. 12.56a,b. The hypothalamus and hypophysis of *Scylio-rhinus canicula.* **a** Ventral view. The regions mentioned in the text are indicated *1–8.* **b** Sagittal section. *a.m.e.,* eminentia mediana anterior; *b.c.,* base of cranium; *c.a.,* carotid arteries; *h.s.,* hypophysial stem; *i.l.,* lobus inferior; *i.o.v.,* interorbital vein; *i.r.,* recessus infundibularis; *M,* median lobe; *m.h.v.,* median hypophyseal vein; *m.n.,* nucleus medius hypothalami; *NIL,* neuro-intermediate lobe; *o.c.,* chiasma opticum; *o.n.,* nervus opticus; *p.m.e.,* eminentia mediana posterior; *p.o.n.,* nucleus preopticus; *p.o.r.,* recessus preopticus; *p.r.,* recessus posterior (mamillaris); *R,* rostral lobe; *s.v.,* saccus vasculosus; *s.v.n.,* nucleus sacci vasculosi; *t.n.,* nucleus lateralis tuberis; *III,* third ventricle; *V,* ventral lobe. (After Knowles et al. 1975)

hypophysial stem toward the neuro-intermediate lobe. There is evidence that some axons leave the main tract and form terminals on portal capillaries in the median eminence (for a review, see Perks 1969), but the majority of the axons of the tractus preopticohypophyseus terminate in the neuro-intermediate lobe. Knowles (1965) distinguished two types of fibres that innervate the cells of the neuro-intermediate lobe, the main source of melanocyte-stimulating hormone (MSH) activity. The neurosecretory type A fibres originate from the preoptic nucleus and terminate on the synthetic pole of these cells, whereas the aminergic type B fibres possibly arise from the nucleus lateralis tuberis and terminate on the release pole of the same cells. Knowles interpreted these observations as indicating that synthesis and release are under independent control. More recently, however, evidence has been obtained that the synthetic pole is innervated by both type A and B fibres (Meurling and Björklund 1970; Knowles et al. 1970). Many studies have shown that lesioning of the tractus preopticohypophyseus leads to ultrastructural signs of hyperactivity in the intermedia cells and to continuous release of MSH, a result that suggests that there is an inhibitory hypothalamic control of the secretion of MSH (Mellinger 1963, 1964; Chevins and Dodd 1970; Meurling et al. 1969; Meurling 1972; Wilson et al. 1974). Selective sectioning of the type A neurosecretory fibres in *Raja* has little effect on MSH secretion, whereas sectioning of type B tracts or treatment with reserpine results in sustained MSH secretion (Meurling et al. 1969). Therefore, type A neurosecretory fibres do not seem to be involved in the control of the pars intermedia, but type B aminergic fibres inhibit both the synthesis and release of MSH in *Raja*. Further support for the view that amines are involved in the inhibitory control of MSH release in elasmobranchs has been provided by Wilson and Dodd (1973b), who have also shown that the aminergic type B fibres originate from the nucleus medius hypothalami.

Since there is no direct innervation of the pars distalis, it seems that hypothalamic control is achieved through a *portal system*. Evidence for the existence of such a system has been provided by studies carried out by Meurling (1960, 1967), Mellinger (1964), Sathyanesan (1965), Jasinsky and Gorbman (1966) and Knowles et al. (1975). According to Meurling (1967), the median eminence of the chondrichthyan hypothalamus is that part of the ventral hypothalamic wall situated between the lobi inferiores that contains a specialized external zone formed by neurosecretory and glial processes. It also contains a capillary plexus which drains toward the adenohypophysis and is considered to be the primary plexus of the pituitary portal system. The portal supply to the pars distalis is similar to that seen in primitive actinopterygians and dipnoans and, as in those groups, is to be correlated with the absence of direct innervation of the pars distalis (Ball and Baker 1969). However, the reported existence of a portal supply to the neurointermediate lobe of *Raja* (Chevins 1968) is puzzling in view of the rich innervation of this region. The functional significance of this part of the portal system, therefore, needs further investigation. As yet there is still no experimental evidence for hypothalamic control of the pituitary via the median eminence in elasmobranch fishes.

The *nucleus tuberculi posterioris* is a rather ill-defined cell mass situated in the tuberculum posterius and in the area immediately rostral to this conspicuous encephalic landmark (Figs. 12.41, 12.42, 12.53a). Rostrally, it is continuous with the thalamus ventralis pars medialis, and caudally it merges into the tegmental grey. No experimental data are available with respect to its connections.

The *saccus vasculosus* is a highly specialized hypothalamic caudal diverticulum of the third ventricle of fishes. It consists of a neural epithelium with three types of cells, i.e. coronet cells, CSF-contacting cells and supporting cells. There are several hypotheses about the function of the saccus vasculosus, varying from an organ for the perception of liquor pressure and chemoreception (Dammerman 1910; Watanabe 1966; von Harrach 1970) to an organ involved in a transcellular ion transport process between the CSF and the blood (van de Kamer et al. 1973).

The *nucleus sacci vasculosi* lies in the ventral part of the tuberculum posterius (Fig. 12.53a). In *Squalus*, the nuclei of both sides are fused to form an unpaired nucleus sacci vasculosi, a feature also reported by Bergquist (1932), whereas the nucleus is paired in other cartilaginous fishes (Smeets et al. 1983). An afferent connection of this cell mass, which can be clearly seen in normal material, is the *tractus sacci vasculosi*. This tract, which was first reported by Dammerman (1910), is a well-developed unmedullated bundle. Previously, it was thought that the tractus sacci vasculosi was formed by the processes of the coronet cells of the saccus vasculosus (Smeets et al. 1983). These coronet or crown cells have a cubic or cylindric shape and sense hairs with knob-shaped terminations turned toward the lumen of the sac. More recently, however, an ultrastructural study has provided evidence that the CSF-contacting cells, and not the coronet cells, are likely to be the origin of the tract (Rodríguez-Moldes and Anadón 1988). Moreover, the fibres of the tractus sacci vasculosi were found to form conventional synapses with dendrites and perikarya of the nucleus tractus sacci vasculosi (Molist et al. 1992).

12.8.4
Retina and Diencephalic Visual Centres

As in other vertebrates, the retina of cartilaginous fishes consists of two synaptic layers (the outer and inner plexiform layers) interspersed between three cellular layers (the outer and inner nuclear and ganglion cell layers). Several studies have dealt with putative neurotransmitters and revealed the involvement of glutamic acid decarboxylase (GAD), dopamine, serotonin and neuropeptides (Ritchie and Leonard 1983b; Bruun et al. 1984, 1985; Brunken et al. 1986). Figure 12.57 shows the distribution of cells that are immunoreactive for antisera to GAD, tyrosine hydroxylase, serotonin and leu-enkephalin in several species of cartilaginous fishes. Most cell bodies that are immunoreactive to any of these antisera lie in the amacrine layer of the inner nuclear layer, and only a few serotonin-immunoreactive cell bodies are found in the ganglion cell layer (*displaced amacrine cells*). It is obvious from this figure that, although the same transmitter systems occur in the retina of each species, there is considerable interspecies variation in the cell types involved. For example, the retina of *Mustelus* contains both leu-enkephalin-immunoreactive interplexiform and amacrine cells, while those of *Squalus* and *Raja* possess only leu-enkephalin-immunoreactive amacrine cells. Moreover, differences in the distribution of immunoreactive fibres with respect to the sublaminae in the inner plexiform layer are easily recognised. Although serotonin-immunoreactive fibres have been occasionally found in the optic nerve fibre layer (Ritchie and Leonard 1983b; Brunken et al. 1986), they were not observed to arise from perikarya in the ganglion cell layer. Thus the neurotransmitter involved in the primary visual projection fibres is still unknown.

The retinal input to diencephalic visual centres and the midbrain has already been mentioned in the description of the respective cell groups and brain regions. However, it seems appropriate to make a few additional comments on this system at this point. For excellent reviews, the reader is referred to papers by Repérant et al. (1986) and Northcutt (1990). During the last 15 years, it has become clear that the primary retinal efferents and higher-order visual pathways of cartilaginous fishes are more elaborate than previously thought. Studies using tracing methods that are more sensitive than the previously used degeneration techniques have revealed that most cartilaginous fishes possess ten primary visual centres in addition to the optic tectum (Table 12.3). Moreover, many species appear to have not only contralateral, but also ipsilateral retinal projections to these nuclei. Currently, there is also good evidence that many, if not all, cartilaginous fishes possess two separate visual pathways: a retino-thalamo-telencephalic and a retino-tecto-thalamo-telencephalic pathway. Even though the exact sites of termination in the telencephalon are unclear, these findings do not support Ebbesson's claim that cartilaginous fishes have a thalamic organisation that differs from that of most other vertebrates.

12.8.5
Chemoarchitecture

The presence of a variety of chemical substances has been demonstrated immunohistochemically in the diencephalon of elasmobranch fishes. These substances include catecholamines (Meredith and Smeets 1987; Northcutt et al. 1988; Molist et al. 1993; Stuesse et al. 1994), serotonin (Ritchie et al. 1983), vasotocin (Vallarino et al. 1990b), neuropeptide Y (Vallarino et al. 1988a; Chiba and Honma 1992), enkephalin (Vallarino et al. 1994), somatostatin (Chiba et al. 1989), adrenocorticotropin (Vallarino and Ottonello 1987), corticotropin-releasing factor hormone (Vallarino et al. 1989), bombesin-like peptide (Vallarino et al. 1990a), FMRFamide (Phe-Met-Arg-Phe-NH_2; Chiba et al. 1991), sauvagine/urotensin I (Vallarino et al. 1988b) and calbindin-D_{28K} (Rodríguez-Moldes et al. 1990). Apart from the monoamines and neuropeptide Y, only a single study has been devoted to the mapping of each of the other chemical substances, making it hazardous to draw any general conclusions.

Dopamine. Several dopamine cell groups are found within the diencephalon (Meredith and Smeets 1987). The rostralmost dopamine cell populations are present in the preoptic and suprachiasmatic nuclei. Caudal to these groups, dopamine neurons lie in the lateral confines of the nucleus medius hypothalami and, more dorsally, at the mid-diencephalic level, within the paraventricular hypothalamic organ. The latter group consists of an extremely dense population of intensely stained dopamine cells. In the lateral part of the organ, round to fusiform somata, with processes oriented mediolaterally, are arranged in layers. In the central part, dopamine neurons line the floor of the ventricle and are of the liquor-contacting type. The latter cells have not been reported to be immunoreactive in studies using tyrosine hydroxylase antisera

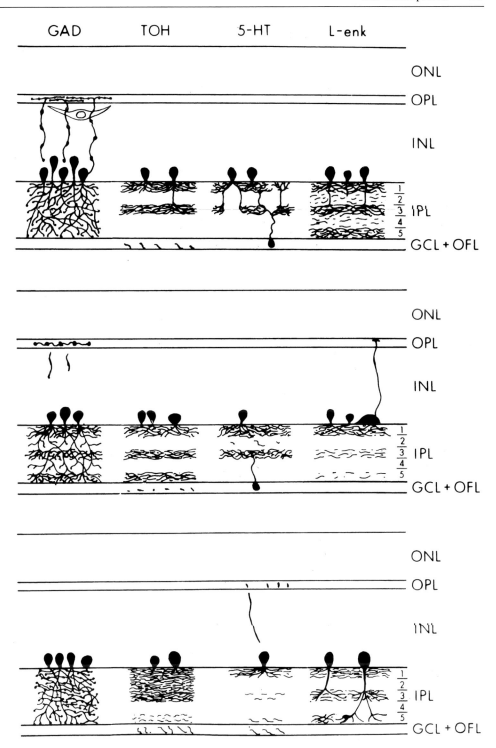

Fig. 12.57. Cellular localisation of four transmitter candidates in the retina of the skate *Raja erinacea* (*top*) and the sharks *Mustelus canis* (*middle*) and *Squalus acanthias* (*bottom*). Note the variation in the morphology of amacrine cells using the same transmitter and the presence of other cell types in some species. *GAD*, glutamic acid decarboxylase; *GCL*, ganglion cell layer; *INL*, inner nuclear layer; *L-enk*, leucine-enkephalin; *IPL*, inner plexiform layer; *OFL*, optic fbre layer; *ONL*, outer nuclear layer; *OPL*, outer plexiform layer; *TOH*, tyrosine hydroxylase; *5-HT*, serotonin. (Modified from Brunken et al. 1986)

Table 12.3. Comparison of nomenclature for retino-recipient nuclei of elasmobranchs (modified after Northcutt 1990)

Platyrhinoidus (Northcutt and Wathey 1980)	*Ginglymostoma* (Luiten 1981a)	*Scyliorhinus, Reja* (Smeets 1981a)	*Sycliorhinus* (Repérant et al. 1986)	*Heterodontus* (Fiebig and Northcutt 1995)
Anterior thalamic	Rostral part of pretectal area and dorsal part of ventrolateral optic	Dorsolateral thalamic	n. opticus dorsalis posterior thalami	Anterior thalamic
			n. opticus dorsolateralis anterior thalami	Intermediate thalamic
Ventrolateral thalamic	Ventrolateral optic	Ventrolateral thalamic	n. opticus ventralis thalami	Ventrolateral thalamic
Ventromedial thalamic	Periventricular gray		Thalamus ventralis pars medialis	Ventromedial thalamic
Superficial pretectal	Rostrolateral part of ventrolateral optic	Lateral geniculate body	n. thalamicus tractus optici marginalis	Superficial pretectal
Central pretecal	Posterior optic	Central pretectal	n. opticus pretectalis centralis	Central pretectal
Periventricular pretectal	Pretectal area	Pretectal	n. opticus commissurae posterioris	Periventricular pretectal
Dorsal accessory	Charted, but not named	Basal optic	n. opticus tegmenti mesencephali dorsalis	Dorsal accessory
Ventral accessory	Basal optic		n. opticus tegmenti mesencephali ventralis	Ventral accessory
Suprachiasmatic		Preoptic	Suprachiasmaticus	Preoptic

(Stuesse et al. 1994). Similar results, i.e. CSF-contacting cells that are dopamine immunopositive, but tyrosine hydroxylase immunonegative, have been reported in a corresponding position in the diencephalon of other vertebrates (Smeets and González 1990; González and Smeets 1994; Meek 1994; Smeets 1994). It has been proposed that these cells accumulate their dopamine from the ventricle (Nakai et al. 1977; Smeets et al. 1991).

Other well-elaborated dopamine cell groups are found in the lateral wall of the infundibular recess, along the mamillary recess and in the posterior tubercle nucleus. In addition, a few dopamine cells are present in the nucleus of hypothalamic inferior lobe. An extra-hypothalamic diencephalic DA cell group is found in the habenular region and consists of weakly immunoreactive cells which rostrally lie adjacent to the habenular ganglion, but more caudally are included within the habenular commissure.

As shown in Table 12.4, cell bodies which are immunoreactive for the various other antisera studied are almost exclusively confined to preoptic and hypothalamic nuclei. An exception to this are the somatostatin-immunoreactive cells in the ventral thalamus (Chiba et al. 1989). Most of the chemical substances studied, such as neuropeptide Y, dopamine, somatostatin, corticotropin-releasing factor, bombesin, adenocorticotropic hormone (ACTH), FMRFamide and arginine-vasotocin (AVT) have been found to project to the hypophysis, suggesting a role in neuroendocrine functions.

12.8.6
Functional Correlations

In mammals, the regions contained within the diencephalon are associated with complex functions and are involved in various behavioural activities. The thalamus is a major subcortical centre that relays sensory information from all over the body to the telencephalon. The hypothalamus, in association with the limbic centres of the telencephalon, maintains homeostatic control of body functions and mediates "motivated" behaviours such as feeding, escape, attack, aggression and sex.

Our knowledge of the functions of these parts of the brain in cartilaginous fishes is very limited at present. However, the hypothalamus is well developed, and it is conceivable that, in association with the hypophysis, it is involved in the regulation of skin colour, growth, salt and water balance and sexual development. Stimulation of the inferior lobe of the hypothalamus via implanted electrodes in a free-swimming nurse shark *Ginglymostoma* elicited feeding responses, a result suggesting that this zone may regulate feeding behaviour (Demski 1977).

For cartilaginous fishes, the traditional view has been that the telencephalon and diencephalon are dominated by olfactory input and that relationships are established in subsequent evolution with other sensory systems. However, during the last decade, several studies have revealed that the thalamus of

Table 12.4. Location of the various monoamines and peptides in the hypothalamus of elasmobranchs

	NPY	AVT	ACTH	Sauvagine/urotensin I	FMRF-amide	Bombesin	CRF	Somato-statin	DA	5-HT
N. preopticus	+	+	–	–	+	+	+	+	+	+
N. lobi lateralis hypothalami	+	–	+	–	+	–	–	–	–	–
N. lateralis tuberis	+	–	+	–	+	–	+	+	–	–
N. tuberculi posteriori	–	–	–	+	–	–	–	+	+	–
N. medius hypothalami	–	–	–	–	+	–	–	–	+	–
N. sacci vasculosi	–	–	–	+	–	–	–	–	–	+
Organon periventriculare hypothalami	–	–	–	–	–	–	–	–	+	+
Recessus mamillaris	–	–	–	–	–	–	–	+	+	+

NPY, neuropeptide Y; AVT, arginine-vasotocin; ACTH, adenocorticotrope hormone; FMRF-amide, oligopeptide (Phe-Met-Arg-Phe-NH$_2$); CRF, corticotropin-releasing hormone; DA, dopamine; 5-HT, 5-hydroxytryptamine.

cartilaginous fishes plays an important role as a relay centre of multimodal sensory input to the telencephalon. In the preceding sections, it has been shown that information from the spinal cord, cerebellum, octavolateralis systems, tectum and retina reach the thalamus of elasmobranch fishes (Fig. 12.58). With the exception of monoaminergic fibres (Ritchie et al. 1983; Meredith and Smeets 1987), all ascending fibres appear to reach the telencephalon through a thalamic relay. In turn, the majority of the telencephalic descending fibres project only to the diencephalon, suggesting that the elasmobranch diencephalon functions as a relay centre in both sensory and motor pathways.

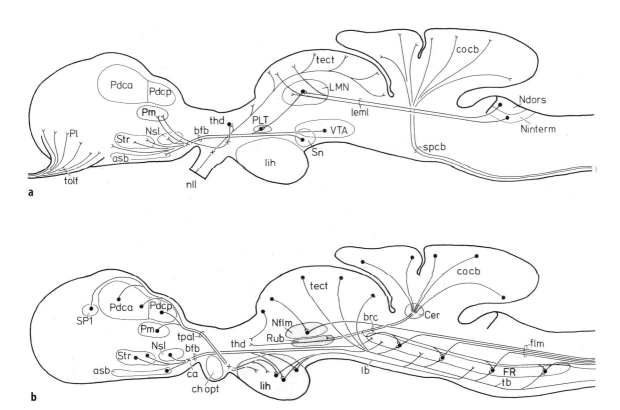

Fig. 12.58. a Main afferent and **b** main efferent fibre systems of the forebrain and midbrain of cartilaginous fishes, as revealed by experimental studies. (After Smeets 1990)

12.9
Telencephalon

12.9.1
Introductory Notes

Studies on the structure of the telencephalon of cartilaginous fishes, based on normal material, have been carried out by Johnston (1911), Ariëns Kappers and Carpenter (1911), Holmgren (1922), Bäckström (1924), Kuhlenbeck (1929b, 1973, 1977), Gerlach (1947), Faucette (1969a,b), Kuhlenbeck and Niimi (1969), Williams (1973), Northcutt (1978, 1981), Ebbesson (1980b), Smeets et al. (1983) and Smeets (1990). However, experimental work on the telencephalon is more limited (Ebbesson and Heimer 1970; Ebbesson and Schroeder 1971; Schroeder and Ebbesson 1974; Graeber et al. 1978; Luiten 1981b; Smeets 1983; Smeets and Boord 1985).

The telencephalon of *Squalus* can be readily divided into three parts (Fig. 12.3). From caudal to rostral, these are as follows:

1. The telencephalon impar
2. The lobus hemisphericus telencephali
3. The bulbus olfactorius

The *telencephalon impar* is that portion of the telencephalon which, as its name implies, encloses an unpaired ventricle. Its caudal boundary can be defined as a plane passing from the attachment of the velum transversum to the caudal pole of the optic chiasm. Rostrally, at the level of the foramen of Monro, it merges into the basal part of the lobus hemisphericus. The *lobus hemisphericus* can be subdivided into a dorsal pallium and a ventral subpallium. The boundary between these two regions is marked more or less by sulci limitantes and cell-free zones, i.e. zonae limitantes. The third telencephalic subdivision, the *bulbus olfactorius*, can be clearly delimited from the telencephalic hemisphere in *Squalus* on account of its long peduncle.

The *telencephalic sulcal pattern* has received little attention in the older literature, and apart from Kuhlenbeck (1929b, 1973, 1977) and his students (Gerlach 1947; Kuhlenbeck and Niimi 1969), workers on the chondrichthyan telencephalon have only incidentically labelled the ventricular sulci. The following distinct grooves could be recognised in our *Squalus* material (for more details, see Smeets et al. 1983).

A *sulcus limitans lateralis* is very distinct and can be recognised throughout the whole extent of the lobus hemisphericus. This sulcus marks the position of a more or less distinct zona limitans lateralis on the ventricular surface. The sulcus limitans lateralis can, therefore, be considered as the ventricular

landmark of the pallio-subpallial boundary in the lateral neural wall. A *sulcus limitans medialis* is represented by a distinct groove in the medial hemispheric wall (Figs. 12.48–12.50). At caudal telencephalic levels, just like the sulcus limitans lateralis, this sulcus marks the pallio-subpallial boundary and the ventricular starting-point of a more or less distinct zona limitans medialis.

A *sulcus telencephalicus dorsomedialis*, forming the dorsal boundary of the medial pallium (Figs. 12.48, 12.49), and a *sulcus telencephalicus ventromedialis*, marking the ventral boundary of the area periventricularis pallialis (Figs. 12.49, 12.50), was also distinguished. Moreover, a *sulcus telencephalicus dorsolateralis* is found in the rostral part of the telencephalon impar and in the caudal part of the lateral hemispheric wall and forms the dorsal boundary of the area periventricularis dorsolateralis (Fig. 12.48). A *sulcus telencephalicus ventralis* extends throughout the ventral hemispheric wall (Figs. 12.48–12.50).

The sulcal patterns of *Squalus* and *Scyliorhinus* are essentially the same, whereas the sulcal pattern of *Hydrolagus* is only to some extent comparable with that of sharks studied (Figs. 12.62–12.64). In *Raja*, a sulcal pattern is absent due to the extreme reduction of the ventricular cavity (Figs. 12.59–12.61).

12.9.2
Olfactory System

Histologically, four layers can be distinguished in the olfactory bulb:

1. A stratum nervosum, made up – as in all vertebrates – of intertwined primary olfactory fibres
2. A stratum glomerulosum, consisting of several layers of well-defined glomeruli
3. A stratum mitrale, built up of large polymorph elements
4. A wide stratum granulare

The axons of the mitral cells traverse the granular layer obliquely and assemble in the periventricular zone of the bulb, forming the secondary olfactory tracts. It is thought that the granular cells have an inhibitory action on the mitral cells (for details, see Andres 1970, 1975).

The olfactory bulb of many elasmobranch fishes is morphologically subdivided into distinct units or sub-bulbs immediately adjacent to the olfactory epithelium (Dryer and Graziadei 1993). This morphological compartmentalisation varies from a succession of swellings along the olfactory rosette in *Sphyrna* to two independent subunits in *Rhizoprionodon*. Dryer and Graziadei (1993) found a secto-

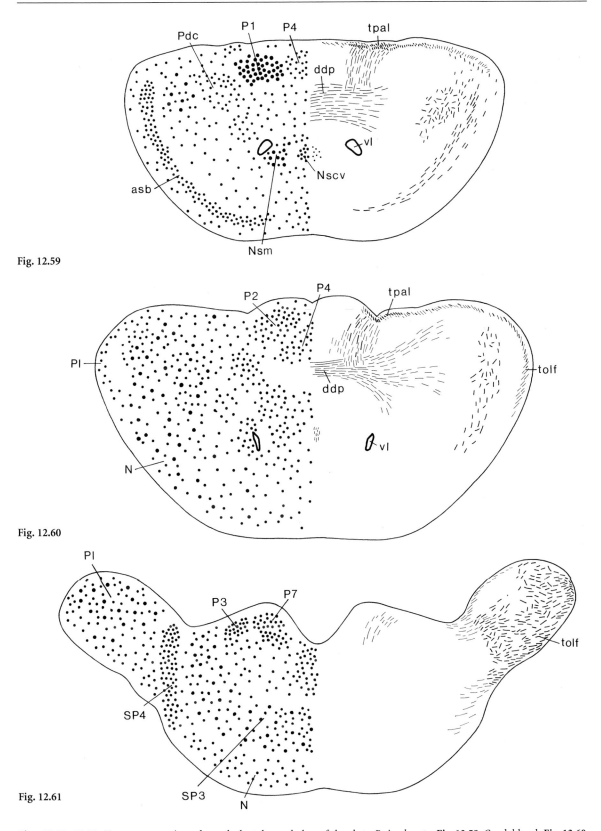

Fig. 12.59

Fig. 12.60

Fig. 12.61

Figs. 12.59–12.61. Transverse sections through the telencephalon of the skate *Raja clavata*. **Fig. 12.59.** Caudal level. **Fig. 12.60.** Intermediate level. **Fig. 12.61.** Rostral level. For further details, see Figs. 12.9–12.12

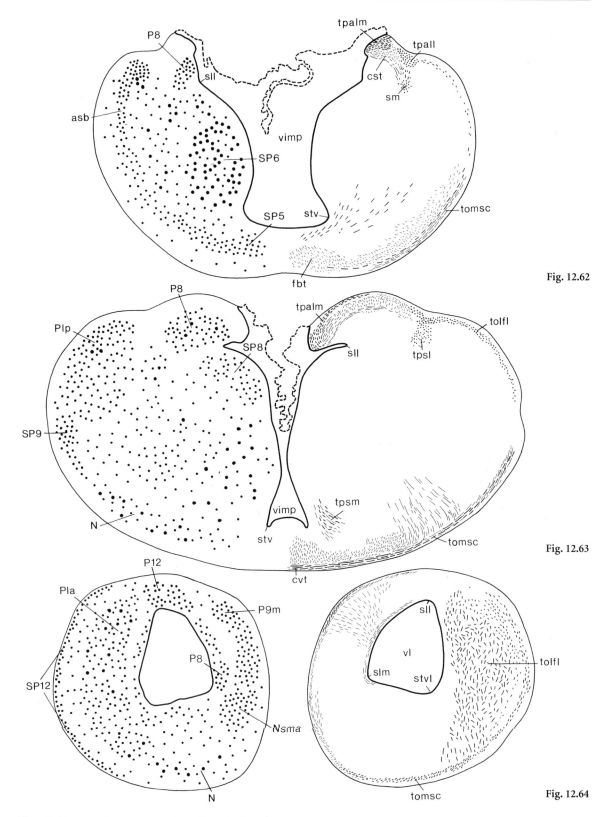

Figs. 12.62–12.64. Transverse sections through the telencephalon of the ratfish *Hydrolagus collei*. **Fig. 12.62.** Caudal level. **Fig. 12.63.** Intermediate level. **Fig. 12.64.** Rostral level. For further details, see Figs. 12.9–12.12

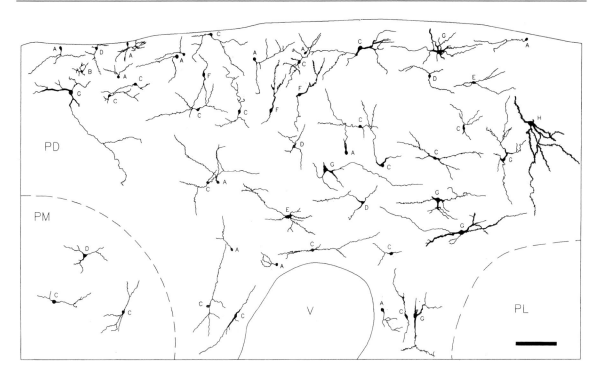

Fig. 12.65. Different cell types in the dorsal (*PD*) and medial (*PM*) pallium of the telencephalon of *Scyliorhinus canicula*, as revealed by the Golgi-aldehyde method. *A*, monopolar cells; *B*, granule cells; *C*, bipolar cells; *D*, small triangular cells; *E*, small stellate cells; *F*, piriform cells; *G*, large stellate neurons; *H*, primitive pyramidal neuron. *Scale*, 200 µm. (From Manso and Anadón 1993)

rial arrangement of the primary projections: anterograde tracers in the medial part of the olfactory epithelium resulted in labelled fibres reaching the medial part of the olfactory bulb, while lateral olfactory epithelial injections labelled fibres in the lateral part of the olfactory bulb. The same authors also recognised two types of mitral cells based on their dendritic morphology. Type L cells are characterised by a loose dendritic organisation, whereas type T cells exhibit a dense, bush-like dendritic arborisation. Basal dendrites are lacking in both types of mitral cells.

In the classical studies on the chondrichthyan forebrain (Edinger 1908; Johnston 1911; Bäckström 1924; Ariëns Kappers et al. 1936), but also in some more recent studies (Williams 1973; Kuhlenbeck 1977), almost the entire telencephalon has been thought to be dominated by the olfactory system and hence considered to be a rhinencephalon. However, experimental studies (Ebbesson and Heimer 1970; Bruckmoser 1973; Bruckmoser and Dieringer 1973; Vesselkin and Kovacevic 1973; Northcutt 1978; Ebbesson 1980b; Smeets 1983) have revealed that, in fact, only a restricted area of the forebrain receives secondary olfactory fibres.

In *Squalus*, the compact mass of secondary olfactory fibres divides on reaching the lateral hemispheric wall into two parts, the tractus olfactorius medialis and lateralis. The fibres of the *tractus olfactorius medialis* (Figs. 12.50, 12.51) intermingle with the myelinated fibres in the superficial zone of the pallium and become indistinguishable from these. The myelinated fibres in the superficial pallial zone have been thought to represent secondary olfactory tracts (Johnston 1911; Bäckström 1924), but an experimental study in *Scyliorhinus* (Smeets 1983) revealed that the secondary olfactory projections are restricted and that the fibres in the superficial pallial zone should be considered as intrinsic telencephalic connections (Fig. 12.67). The *tractus olfactorius lateralis*, which is much larger than the medial tract, courses caudally and maintains a superficial position in the ventrolateral hemispheric wall in *Squalus* (Figs. 12.47–12.50). The lateral olfactory tract sends some fibres to the area retrobulbaris and striatum, but the majority terminate in the lateral pallium and in a zone superficial to the lateral part of the area superficialis basalis. So far, *Scyliorhinus* is the only species in which the existence of a contralateral secondary olfactory projection has been experimentally confirmed (Fig. 12.67). This contralateral projection could be

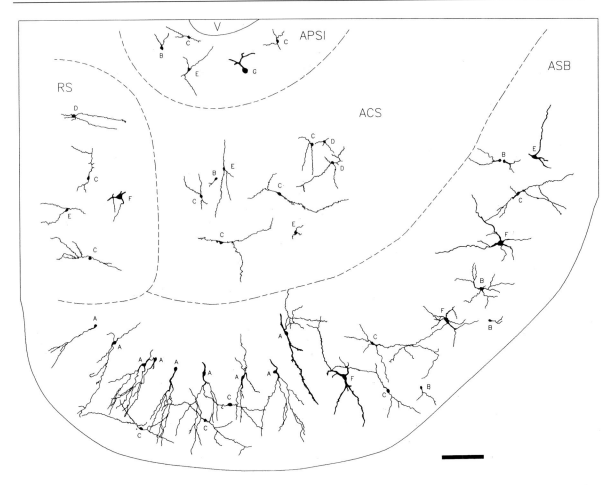

Fig. 12.66. Different cell types in the area superficialis basalis (*ASB*), area centralis subpallialis (*ACS*), area periventricularis subpallialis inferior (*APSI*) and regio septalis (*RS*) of *Scyliorhinus canicula*, as revealed by the Golgi-aldehyde method. *A*, large radial cells; *B*, monopolar cells; *C*, bipolar cells; *D*, small stellate cells; *E*, small triangular cells; *F*, large triangular and stellate cells; *G*, large neurons of the APSI. For the sake of clarity, the most abundant cell types of the ASB (types *A* and *C*) are represented on the *left* and the less abundant on the *right*. Scale, 200 μm. (From Manso and Anadón 1993)

traced to the superficial pallial zone via the *commissura olfactoria inferior* and to the contralateral olfactory bulb via the *commissura olfactoria superior*. In *Raja*, the secondary olfactory projections are more restricted than those in *Scyliorhinus*. The lack of evidence for a projection to the lateral part of the area superficialis basalis, as reported in all sharks studied experimentally, is especially notable. Although no experimental data are available, it does seem unlikely that the lateral olfactory tract of *Hydrolagus* (Figs. 12.60, 12.61) reaches the area superficialis basalis, and in this respect this tract resembles the equivalent structure in *Raja*.

Another connection between the olfactory bulb and the hemisphere is formed by the *nervus terminalis*. Except for birds, a terminal nerve and ganglion have been identified in representatives of each vertebrate class; however, only in elasmobranchs,

not in all species studied, is the terminal nerve completely isolated from the adjacent olfactory tract (Demski and Fields 1988). In *Squalus*, this nerve, which consists of unmyelinated fibres, passes from the olfactory bulb over the dorsal surface of the lobus hemisphericus to the cleft between the two hemispheres (Figs. 12.50, 12.51). It then enters the medial hemispheric wall and becomes unrecognisable as it intermingles with other unmyelinated fibres. Extensive studies on the nervus terminalis made by Locy (1905), Johnston (1911) and Bäckström (1924) have revealed that considerable variability exists between cartilaginous fishes in the course and entry of this nerve. It is notable that variablity is evident even within the same species, as in the specimen of *Squalus* which was used to prepare the figures on gross morphology, the nervus terminalis enters the brain ventrally between

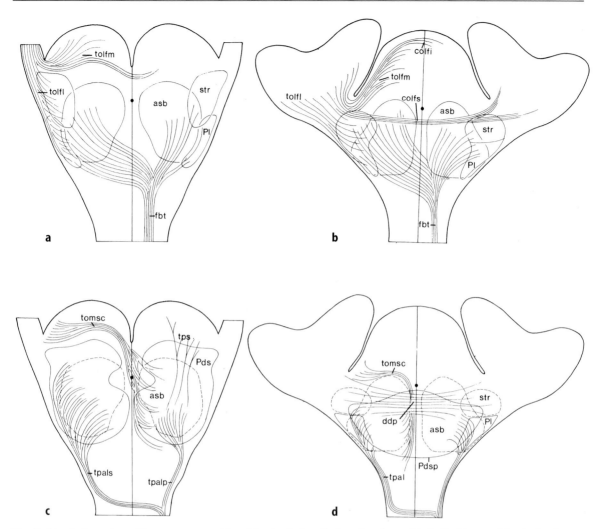

Fig. 12.67. a,b Secondary olfactory tracts and basal forebrain bundle in **a** *Squalus* and **b** *Scyliorhinus*. **c,d** Intrinsic telencephalic connections and tractus pallii in **c** *Squalus* and **d** *Scyliorhinus*. Dorsal view of the brain

the subpallial parts of the hemisphere, whereas in the specimen used for the transverse sections a dorsal approach can be recognised.

The nervus terminalis has been found to contain neuropeptide luteinising hormone-releasing hormone (LHRH) as well as one or more peptides that cross-react with antisera against molluscan cardioactive peptide FMRFamide (Nozaki et al. 1984; Stell 1984; Stell et al. 1984). Dense-cored vesicle-containing axons, cell bodies and endings of the terminal nerve have been studied using the electron microscope by Demski and Fields (1988). The vesicles were observed in two different cell types: one with a polymorphic nucleus, and the other with an oval nucleus. It is not yet known whether these different types of cells are related to the two neuropeptides (LHRH, FMRFamide) that have been localised in the nerve. Demski and Fields also noted that the terminal nerve fibres run near to blood vessels and appear to end adjacent to endothelial cells, suggesting the possibility of neurosecretion from this nerve into the cerebral circulation. The functional significance of the terminal nerve is still obscure, although Demski and Northcutt (1983) have provided some evidence that one of its functions may include chemoreception of pheromones. The involvement of LHRH and FMRFamide peptides suggests that the terminal nerve plays a role in seasonal reproductive development and sex-steroid release (Dodd et al. 1983).

12.9.3
Pallium and Its Connections

Although Johnston (1911) considered the entire roof of the chondrichthyan telencephalon to be a

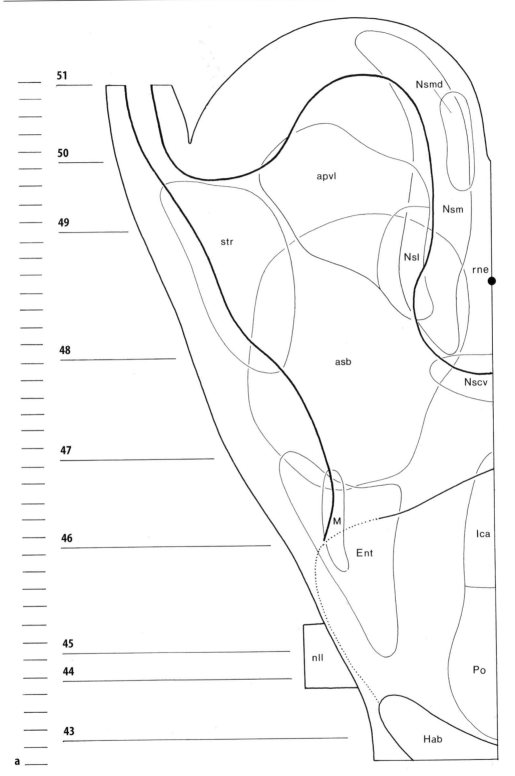

Fig. 12.68a,b. Topographical reconstructions showing the cell masses in the telencephalon of *Squalus acanthias*, as projected upon a horizontal plane. **a** Cell masses of the telencephalon impar and of the subpallium. **b** Cell masses of the pallium. *Heavy continuous lines* represent the outlines of the ventricle, *medium continuous curves* represent the contours of the external telencephalic features and the *dotted line* indicates the taeniae. The *remaining thin lines* represent the cell masses. *Numbers at the side* of the drawing correspond to the figures of the same number showing selected transverse sections. The distance between sections, indicated by *small, horizontal bars*, equals 300 μm

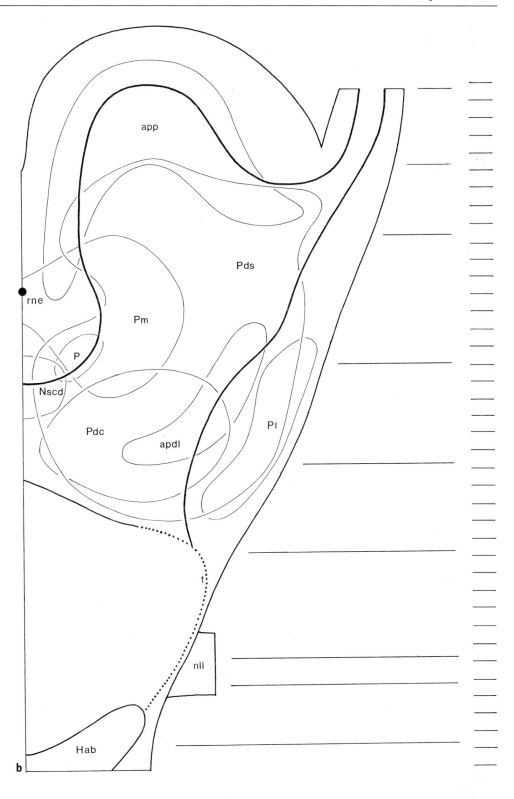

primordium hippocampi, embryological studies by Holmgren (1922) and Bäckström (1924) revealed that, during early ontogenesis, three pallial formations can be clearly recognised. They termed these three formations the medial or hippocampal pallium, the dorsal or general pallium and the lateral or piriform pallium. According to Holmgren and Bäckström, these subdivisions become less distinct with increasing age. Studies on adult material have confirmed the subdivision of the pallium into three components; however, there is much disagreement about the extent and the boundaries of these parts (see Kuhlenbeck 1977; Northcutt 1978, 1981; Smeets et al. 1983; Northcutt et al. 1988).

The caudal part of the telencephalic roof in *Squalus* is almost completely occupied by the *pallium dorsale* (Figs. 12.47–12.49). This cell mass can be subdivided into inner and outer laminae separated from each other by a zone that contains few cells. The former is termed the pallium dorsale, pars centralis, whereas the latter has been termed the pallium dorsale, pars superficialis. The pars centralis occupies a restricted area in the caudal part of the telencephalic roof, while the pars superficialis extends rostrally beyond the level of the neuroporus externus (Fig. 12.68b). A remarkable feature of the pallium dorsale pars superficialis is the presence of fibres which traverse the lateral portion of this cell mass to join the tractus pallii.

Apart from the superficial and central part, the dorsal pallium contains two other cell masses. The first of these, the *area periventricularis pallialis*, lies in the rostral part of the pallium and has a laminar character. It covers the ventricle on the medial and dorsal side and contains small granular cells that form clusters like those of the striatum. Caudally, this area divides into two parts, one medial and the other dorsal to the ventricle (Figs. 12.49–12.51, 12.68b). The medial component, which extends further caudalwards than the dorsal one, was interpreted by Johnston (1911) and Northcutt (1981) to be a septal nucleus and, more recently, to be the medial subdivision of the lateral pallium (Northcutt et al. 1988).

The second, remaining cell mass in the pallium dorsale is formed by a periventricular cell zone in the caudal half of the pallium, termed *area periventricularis dorsolateralis*. This cell mass is well developed in *Squalus* and causes a slight protuberance into the ventricle (Fig. 12.48).

The *pallium mediale* or *hippocampale* in *Squalus* consists of rather small, ellipsoidal cells and fuses with its counterpart across the midline (Figs. 12.48, 12.49, 12.68b). It is relatively small when compared with the pallium dorsale and causes a ventricular protrusion of the dorsomedial wall into the telencephalic ventricle.

In *Squalus*, a *pallium laterale* with a triangular form can be recognised lying at the same level as, but dorsolateral to, the area superficialis basalis (Fig. 12.47). Although this cell mass lies ventral to the zona limitans lateralis, we consider it to be the pallium laterale, as did Northcutt (1977). The principal reason for doing so is that this cell mass receives the main olfactory input, just like the lateral pallium of tetrapods. In older studies (Johnston 1911; Holmgren 1922), this cell mass was labelled as the nucleus olfactorius lateralis. The region labelled as the pallium laterale by Johnston and Holmgren is represented in our study by the lateral part of the pallium dorsale.

Finally, one cell mass needs to be mentioned which lies at about the same level as the nucleus septi, pars caudodorsalis and which has been termed *nucleus P* (Fig. 12.48). This cell mass is very distinct, consists of small, rather densely packed cells and lies lateral to the recessus neuroporicus.

A recent Golgi study by Manso and Anadón (1993) has revealed a wide variety of cell types in the pallium of the dogfish *Scyliorhinus canicula*. At least eight different types were found in the dorsal pallium (Fig. 12.65), and six in the lateral pallium. Remarkably, both the dorsal and lateral pallium contain a type of primitive pyramidal cell characterised by the dense appearance of its thorny dendrites. As shown in Fig. 12.65, the different cell types are not homogeneously distributed throughout the dorsal pallium. For example, monopolar cells, granule cells and primitive pyramidal cells are primarily found superficially in the dorsal pallium, whereas large stellate cells occur primarily in its intermediate portion and in the area periventricularis pallialis. On the other hand, bipolar cells are widely distributed throughout the dorsal pallium. In contrast to the dorsal and lateral pallium, the medial pallium contains only two cell types, i.e. bipolar cells and small triangular cells.

If we compare the pallial cell masses of *Squalus* with those found in the other cartilaginous fishes studied, it becomes obvious that, although some minor differences exist, the pallium of *Scyliorhinus* closely resembles that of *Squalus*. In contrast, the pallia of *Raja* and *Hydrolagus* contain several cell masses which have no obvious homologue in the pallium of the two sharks studied. A pallium dorsale pars centralis was recognised in *Raja* (Fig. 12.59), whereas the pallial areas labelled P1, P2 and, to a lesser extend, P3 (Figs. 12.59–12.61) may be considered to be equivalent to the caudal half of the pallium dorsale pars superficialis of sharks. The latter assumption is based on the characteristic course of the fibres, which traverse these areas before joining the pallial tract (Figs. 12.59, 12.60).

Because of the same reason, the areas labelled P8 and P9 m (Figs. 12.62–12.64) in *Hydrolagus* are considered to represent a pallium dorsale pars superficialis. The pallial area P7 in *Raja* (Fig. 12.61) may, on account of its position and lack of traversing fibres, be compared with the rostral part of the pallium dorsale, pars superficialis of sharks, whereas area P4 (Fig. 12.59) can be considered to be part of the pallium mediale.

The lateral pallium of *Raja* is more extensive than in *Squalus* and *Scyliorhinus*. It is noteworthy that it consists of numerous large neurons surrounded by satellite glial cells. The lateral pallium of *Hydrolagus* is formed by a large cell mass which lies in the lateral hemispheric wall and extends caudally into the telencephalon impar. On the basis of the size of its constituent neurons, it can be subdivided into a rostral and a caudal portion.

Connections. The most obvious fibre system in the pallium of elasmobranch fishes is the *tractus pallii*. The pallial tract contains both myelinated and unmyelinated fibres and, as already mentioned, is characterised by the peculiar way in which its fibres join or leave the tract in the telencephalic roof. Although the diencephalic portion of this tract does now show great differences, its telencephalic part appears to vary considerably between species (Smeets et al. 1983). In *Scyliorhinus* and *Raja*, only a single tractus pallii has been observed, whereas in *Squalus* and *Hydrolagus* this tract consists of two components at levels rostral to the optic chiasm. Because of their position in *Squalus*, these components have been termed pars superficialis and pars periventricularis (Figs. 12.46, 12.47). According to Bäckström (1924), the tractus pallii, pars periventricularis represents the tractus pallii rectus, whereas the superficial component constitutes the pars cruciatus (Fig. 12.67). At the level of the chiasma opticum, both components intermingle and become indistinguishable.

According to our observations, another distinct fibre system, the *stria medullaris*, collects fibres from the caudal parts of both the pallium and the subpallium. The exact site of origin of these fibres could not be determined. The stria medullaris, in passing to the diencephalon, lies in close apposition to the tractus pallii (Figs. 12.43–12.46), and the possibility cannot be excluded that these two systems interchange fibres, as reported by Kuhlenbeck (1977). A detailed study of the stria medullaris complex was performed by Johnston (1911), who recognised at least seven components in *Squalus* and *Scyliorhinus*: a tractus septohabenularis, a tractus olfactohabenularis lateralis, a tractus olfactohabenularis posterior, a tractus corticohabenularis, a

tractus taeniae, a tractus habenulothalamicus and a commissura pallii posterior. With the exception of the latter, which is equivalent to the commissura superior telencephali in the present study, none of these components could be traced with certainty to their site of origin.

12.9.4
Subpallium and Telencephalon Impar
with Their Connections

Because some subpallial cell masses extend beyond the caudal boundary of the hemispheres, i.e. the foramen of Monro, we will include in this section the cell masses of the telencephalon impar as well as those of the subpallial division. These cell masses are also represented together in the topographical reconstruction (Fig. 12.68a).

The most caudal telencephalic cell mass, the nucleus preopticus, has already been described in connection with the diencephalon. The preoptic nucleus merges rostrally into the *nucleus interstitialis commissurae anterioris*, a cell mass which has been identified as a cluster of small, rather densely packed, round to ellipsoidal cells lying dorsal to the anterior commissure (Figs. 12.46, 12.47).

The fasciculus basalis telencephali contains throughout its extent numerous cells which accumulate in the rostral part of the telencephalon impar. We have labelled this accumulation in the transverse sections as the *nucleus interstitialis fasciculi basalis telencephali* (Figs. 12.46, 12.47).

In the caudal part of the telencephalon impar, a cell mass can be recognised which was previously termed the area somatica by Johnston (1911), but now is labelled as the *nucleus entopeduncularis* (Figs. 12.46, 12.47). The latter term is used here in the strictly descriptive sense of "lying within the peduncle" and, at least for the moment, is not intended to imply any homology with the nucleus of the same name in tetrapods. Rostrally, the nucleus entopeduncularis is replaced by the *area superficialis basalis*, which shows a distinctly laminar organisation (Figs. 12.47–12.49). In Nissl-stained material, the latter nucleus consists of small cells arranged in a U-shaped layer and occupies the whole ventral outer zone of the subpallium. At rostral telencephalic levels, the area superficialis basalis is replaced by another superficial cell layer. Several authors (e.g. Williams 1973; Northcutt 1978) considered this layer to be the rostral continuation of the area superficialis basalis, but we consider this designation inappropriate, because the cells are larger and more diffusely arranged than those of the area superficialis basalis and, moreover, the two cell masses differ with respect to their fibre connections. We have designated this layer, which could not be exactly delineated, as *nucleus N* in the transverse sections (Figs. 12.49–12.51).

Dorsal to both the rostral pole of the nucleus entopeduncularis and the caudal part of the area superficialis, a small cell mass can be recognised in *Squalus*, which has been labelled *nucleus M* and which consists of rather small cells (Fig. 12.47). This nucleus has a periventricular position adjacent to the stria medullaris and the taenia choroidea.

In *Squalus*, a conspicuous cell mass is found in the ventrolateral wall of the hemisphere, where it causes a slight protrusion (Figs. 12.48, 12.49). This cell mass, which lies rostral to the foramen of Monro, is labelled as *striatum*. It consists of numerous clusters of small granular cells and exhibits a laminar organisation. Also in the ventrolateral wall, but medial to the striatum, a cell mass was delineated which has been designated descriptively as the *area periventricularis ventrolateralis* (Figs. 12.49, 12.50).

In the telencephalic hemisphere, the *septal region* comprises both the medial and the ventromedial portion of the subpallium. In *Squalus*, five subdivisions were recognised; from caudal to rostral, these are the nucleus septi caudodorsalis, nucleus septi caudoventralis, nucleus septi medialis, nucleus septi lateralis and nucleus septi medialis, pars dorsalis. Their positions are shown in the transverse sections (Figs. 12.48–12.51) and in the topographical reconstruction (Fig. 12.68a). It is noteworthy that, in *Squalus*, part of the septal region, i.e. the nucleus septi caudodorsalis, arches around the foramen of Monro and apparently occupies a pallial position.

Golgi methods have revealed a variety of cell types in the basal forebrain (Fig. 12.66). At least six cell types were distinguished by Manso and Anadón (1993) in the area superficialis basalis. The most abundant cell types are large radial cells and bipolar cells. Furthermore, nucleus N was found to lack the large radial cells that are characteristic of the area superficialis basalis. This finding supports the hypothesis (Smeets et al. 1983) that nucleus N and the area superficialis basalis are two different regions. The two cell types revealed by means of the Golgi method in the interstitial nucleus of the basal forebrain bundle and in the entopeduncular nucleus are large radial cells and bipolar cells, making these nuclei resemble the area superficialis basalis. However, the two nuclei differ from the area superficialis basalis not only in that they lack the other cell types found in the latter area, but also in their chemoarchitecture (see Sect. 9.5).

If we compare the telencephalon impar and subpallial region of *Squalus* with those of *Scyliorhinus*, it appears that there is no difficulty in homologising the various cell masses. However, the cell masses found in the corresponding regions of *Raja* and *Hydrolagus* are more problematic. The cell masses in the telencephalon impar of the four species studied can be easily compared, except for the elongated preoptic area in *Hydrolagus*. In all four species, this part of the telencephalon contains a nucleus preopticus, a nucleus interstitialis fasciculi basalis telencephali, a nucleus entopeduncularis, a nucleus M, an area superficialis basalis and, more rostrally, a nucleus N. A striatum could not be recognised with certainty in *Raja* and *Hydrolagus*. In *Raja*, Northcutt (1978) labelled an area as striatum which is equivalent to the U-shaped area subpallialis 3 (SP3; Fig. 12.61) of Smeets et al. (1983). However, this cell mass differs from the striatum in sharks in that it consists of large cells accompanied by glial elements (satellitosis) and occupies a more rostral topological position. Moreover, no fibres originating from this cell mass could be traced toward the basal forebrain bundle. The large intraventricular protrusion of the lateral telencephalic wall in holocephalian fish (Figs. 12.62, 12.63) has been considered as striatal swellings (Ariëns Kappers and Carpenter 1911; Holmgren 1922; Faucette 1969a,b). It was found that, in *Hydrolagus*, this prominent bulge contains at least three distinct cell masses, which were provisionally labelled as area subpallialis 6 (SP6), 7 (SP7) and 8 (SP8) by Smeets et al. (1983). The most conspicuous of these, SP6, is characterised by medium-sized, fusiform and triangular cells and lies in the caudal part of the protrusion (Fig. 12.62). Rostralward, the lateral part of this area is replaced by a cell mass indicated as SP7, which consists of small, densely packed cells. In a dorsolateral position, the rostral part of the bulge contains a cell mass that has a periventricular position lying just below the sulcus limitans lateralis (Fig. 12.63). It is as yet still uncertain whether any of these three structures can be compared with the shark's striatum.

In addition to the structures mentioned above, the subpallial area of *Raja* and *Hydrolagus* contains a number of cell masses which can be hardly homologised with any structure in the telencephalon of sharks. For further details, the reader is referred to the study by Smeets et al. (1983).

Connections. The *fasciculus basalis telencephali* constitutes the major connection of the basal forebrain with more caudally situated brain regions. This fibre system is well developed in *Squalus* and consists of both myelinated and unmyelinated fibres. In normal material, it is obvious that the fibres form a well-defined tract at the level of the caudal pole of the area superficialis basalis (Fig. 12.47). Before passing caudalward in the basal forebrain bundle, the majority of these fibres decussate in the anterior commissure. Bäckström (1924) described 11 components within the basal forebrain bundle originating from both pallial and subpallial regions. However, our normal material suggests that, in *Squalus*, only the striatum and the area superficialis basalis contribute a substantial amount of fibres to the basal forebrain bundle (Fig. 12.67). A experimental study by Smeets (1983) has shown that in *Scyliorhinus*, after telencephalic lesions that include a part of the striatum, numerous degenerating fibres can be traced to the basal forebrain bundle and that the majority of these fibres decussate in the anterior commissure.

According to Johnston (1911), in addition to descending components, the basal forebrain bundle contains fibres which ascend from several diencephalic regions to the pallium to form a *tractus thalamocorticalis* in the lateral part of the basal forebrain bundle. Experimental studies on *Ginglymostoma* (Ebbesson and Schroeder 1971; Schroeder and Ebbesson 1974) and on *Raja* (Smeets and Boord 1985) have confirmed the existence of ascending fibres within the confines of the basal forebrain bundle.

Here, the following *intrinsic telencephalic connections* will be discussed. In a previous section, we have seen that the secondary olfactory projections in cartilaginous fishes are limited in their telencephalic distribution. The same holds true for the projections to and from more caudal brain regions. There are numerous telencephalic fibres that can neither be classified as direct afferent nor as efferent fibres. The majority of these are unmyelinated and form a dense meshwork within the telencephalon. However, some collect as rather distinct fibre bundles which can be traced for some distance within the telencephalon, although their exact sites of origin and termination cannot be determined in our material. In the following account, we will use the terminology of Bäckström (1924) wherever possible. This does not imply that we fully agree with him; we merely wish to avoid a further accumulation of novel names. The following fibre systems will be discussed: the tractus olfactorius medialis septi cruciatus, the tractus pallialis septi, the fornix and the decussatio dorsalis pallii.

The *tractus olfactorius medialis septi cruciatus* is described by Bäckström (1924) as consisting of secondary olfactory fibres which separate from the tractus olfactorius medialis and curve medially along the anterior surface of the brain until they reach the midplane, where they arch caudoventrally through the septum and cross, ventral to the anterior commissure, to the opposite side. The bundle then turns dorsolateralward and rostralward and finally distributes its fibres to the striatum, the area superficialis basalis and the lateral pallium. In *Squalus* and *Scyliorhinus*, a fibre system is found that, with regard to its course, resembles the tractus olfactorius medialis septi cruciatus of Bäckström (Figs. 12.47–12.51). This tract is particularly distinct and well-myelinated in *Squalus*, but in *Scyliorhinus* it consists almost exclusively of unmyelinated fibres which, according to our observations, all remain ipsilateral. Our results differ from those of Bäckström in two respects:

1. Experimental studies on *Scyliorhinus* (Smeets 1983) have shown that this tract does not degenerate after olfactory bulb lesions.
2. Careful analysis of the tractus olfactorius medialis septi cruciatus in normal material of *Squalus* reveals a decrease in the number of fibres at midhemispheric levels which does not tally with the description given by Bäckström (1924).

The latter finding might explain why Johnston (1911) considered the rostral part of this tract in *Squalus* to be a tractus olfacto-corticalis mediodorsalis and the caudal part an extension of the tractus olfactorius lateralis.

The *tractus pallialis septi* in *Squalus* consists of myelinated fibres and can be traced from caudal hemispheric levels, where its fibres intermingle with those of the tractus pallii, pars periventricularis, to the rostral part of the telencephalon, where it distributes to the ventrolateral hemispheric wall (Figs. 12.49, 12.50).

A presumed *fornix* was described by Johnston (1911) and Bäckström (1924) in several cartilaginous fishes. According to these authors, fibres collect from the medial pallium of either side and pass medioventrally to form two compact bundles lateral to the recessus neuroporicus (Fig. 12.48). As they descend caudalward, they gradually take a position ventral to this recess and finally join the fasciculus basalis telencephali. They pass with this bundle to a hypothalamic centre, which Johnston and Bäckström considered to be a primordal corpus mammillare. In *Squalus*, fibres were traced from the pallium to caudal telencephalic levels which resemble the fornix as described by Johnston and Bäckström, but we were unable to trace these fibres to the hypothalamus.

A small pallial commissure was described by Johnston (1911) in *Squalus*, but in our normal material the existence of such a commissure could not be confirmed. In contrast, in both *Scyliorhinus* and *Raja* (Figs. 12.59, 12.60), a strongly developed *decussatio dorsalis pallii* was recognised (Smeets et al. 1983). Most of the fibres constituting this decussation are myelinated in *Scyliorhinus*, but unmyelinated in *Raja*. Although the various components of the decussatio dorsalis pallii are hard to trace to their sites of origin or termination, it may well be that this complex consists of both decussating and true commissural fibres which connect the lateral pallium, striatum, area superficialis basalis and dorsal pallium.

12.9.5
Chemoarchitecture

From the previous sections, it has become clear that our current knowledge of the telencephalon of cartilaginous fishes is still limited. The main reason for this is that degeneration techniques are not suitable for unravelling telencephalic connections in chondrichthyans (Smeets 1983). Moreover, the HRP tracing technique is not appropriate either, since HRP seems to diffuse widely throughout the telencephalon even after small injections, making delimitation of the injection site impossible. This diffusion is particularly strong in animals with reduced ventricles. It is, therefore, not surprising that the application of immunohistochemical techniques has been considered as an alternative way of understanding the structure and organisation of the telencephalon of cartilaginous fishes.

Hitherto, studies have appeared on the telencephalic distribution of catecholamines (Meredith and Smeets 1987; Northcutt et al. 1988; Stuesse et al. 1994), serotonin (Ritchie et al. 1983; Northcutt et al. 1988; Yamanaka et al. 1990), neuropeptide Y (Vallarino et al. 1988a; Chiba and Honma 1992), SP (Northcutt et al. 1988; Rodríguez-Moldes et al. 1993), enkephalins (Northcutt et al. 1988; Vallarino et al. 1994), somatostatin (Chiba et al. 1989), FMRFamide (Chiba et al. 1991) and gonadotropin-releasing hormone (Lovejoy et al. 1992). The present review does not intend to enumerate all nuclei and fibre systems that are immunoreactive to the antisera developed against the monoamines and peptides mentioned above, because the data are too limited for a meaningful discussion. However, some comments on the chemoarchitecture of certain cell masses, possibly homologous to basal ganglia structures in amniotes, would seem to be warranted.

In reptiles, birds and mammals, a descending forebrain projection containing SP terminates among or is contiguous with the dopaminergic cell bodies in the tegmentum of the midbrain. Similarly, in all amniotes studied, the dopaminergic cell bodies in the midbrain project back to the basal forebrain, supplying the striatum and the nucleus accumbens with a dense dopaminergic innervation (Parent 1986; Smeets 1992). Thus the intriguing question is raised as to whether such an organisation also exists in cartilaginous fishes. The rather unexpected finding of distinct midbrain dopaminergic cell groups and a strong dopaminergic innervation of basal forebrain regions, in particular the area superficialis basalis (Meredith and Smeets 1987), points to a basal ganglia organisation that might be similar to that observed in amniotes, although other data urge us to be cautious in making such straightforward comparisons. For example, in contrast to the condition found in amniotes, the midbrain dopaminergic cell groups of elasmobranchs do not extend caudally beyond the level of the exit of the oculomotor nerve. Furthermore, the area superficialis basalis also contains dense plexuses of serotonin-immunoreactive (Ritchie et al. 1983; Yamanaka et al. 1990) and SP-immunoreactive fibres (Northcutt et al. 1988; Rodríguez-Moldes et al. 1993). Northcutt et al. (1988) suggested that the area superficialis basalis is homologous to the globus pallidus of amniotes on the basis of the SP fibre plexus and the presence of neurons positive for LANT6, a neurotensin-like hexapeptide which is also found in the globus pallidus of amniotes (Reiner and Carraway 1987). However, the presence of a strong dopaminergic and serotoninergic plexus is not in agreement with such a hypothesis. The area superficialis basalis of elasmobranchs is most likely to consist of striatal and pallidal elements which have not yet segregated. More detailed information on the chemoarchitecture and connections of cell bodies in the basal forebrain of cartilaginous fishes is obviously needed before final conclusions can be drawn. From the foregoing, it is also clear that, although immunohistochemistry may be very helpful in comparative neuroanatomy, it is certainly not the best way to define corresponding areas in different species, as it is well known that closely related species show considerable variation in neurotransmitter systems (e.g. Smeets and Reiner 1994).

12.9.6
Functional Correlations and Concluding Remarks

From the previous account, it is clear that cartilaginous fishes possess several very sophisticated sensory systems, such as olfaction, vision, hearing and the lateral line system. It is known that the olfactory receptors of sharks are highly efficient detectors of waterborn chemicals. Neuroanatomically, the importance of olfaction is underscored by the existence of well-developed olfactory bulbs, which seem to give rise to higher-order olfactory fibres reaching almost every region of the telencephalon. Thus it is used to be thought that cartilaginous fishes were "smell animals". However, experimental neuroanatomical and electrophysiological studies have demonstrated that only a restricted area in the telencephalon receives secondary olfactory input and that even tertiary projections seemed to be limited in character. Thus the label 'rhinencephalon', so often used to indicate the chondrichthyan telencephalon, is inapplicable.

Although olfaction plays a very important role in the detection of food sources, evidence is accumulating that hearing is also involved in locating food sources and possibly even other objects, such as competitors and predators. It has been shown that the maculae sacculi and neglecta of the inner ear may serve as auditory receptors. Fibres originating from these auditory receptors, together with vestibular fibres, form the octaval nerve, which enters the brain as a single structure. Both auditory and vestibular fibres project to all brain stem octaval nuclei, but some topical organisation appears to exist in both the descending and the ascending octaval nucleus. Almost nothing is known about the higher-order projection of the octaval nuclei, although evoked potential recordings and an experimental study using 2-deoxyglucose autoradiography have indicated that auditory information reaches the telencephalon (Corwin and Northcutt 1982).

Whereas olfaction and hearing play important roles in reception at long range, the lateral line system and vision are effective over short distances. Both mechanoreceptive and electroreceptive lateral line information is available in cartilaginous fishes. Mechanoreceptive information from the canal and pit organs of the head and body reach the brain by way of the posterior and ventral anterior lateral line nerve, whereas electroreceptive information is carried from the ampullae of Lorenzini in the head to the brain stem by the dorsal anterior lateral line nerve. There is a distinct separation of mechanoreceptive from electroreceptive inputs of lateral line receptors in the intermediate and dorsal nuclei of the octavolateral area. It has been experimentally confirmed that the lateral line nerves have not only separated mechanoreceptive and electroreceptive projections at the first-order level, but also that the secondary lateral line projections to higher brain levels, via the lateral lemniscus, retain this separate character. Evoked potential recordings and experimental neuroanatomical studies have provided evidence that both lateral line sensory modalities reach the telencephalon (Bodznick and Northcutt 1984; Bleckmann et al. 1987, 1989; Smeets and Northcutt 1987). Although we know that the lateral line neuromasts are sensitive to external water dis-

placements and that the pit organs are activated by both mechano- and chemostimulation, the biological significance of this sensitivity still remains unclear. With regard to electroreception, it is now firmly established that cartilaginous fishes seem to use the ampullae of Lorenzini to detect living food and to determine the direction and polarity of electric fields in the ocean.

Cartilaginous fishes swim in nearly all major habitats in the marine environment, and some of them are restricted to fresh water. Moreover, they show a wide variation in behaviour, and it seems that some species are apparently highly dependent on vision, while others are more dependent on non-visual cues. The variations in habitat and behaviour are reflected in the variation of the eyes of the various chondrichthyan species. However, in general the following statements can be made:

1. Cartilaginous fishes possess unique ocular features which reflect a long and independent phylogenetic history.
2. The chondrichthyan eye cannot be considered as primitive, but shows many similarities with the eye of amniotes, especially mammals.

However, in contrast to what has been found in mammals, the optic nerve decussates almost completely in the optic chiasm and projects to several diencephalic nuclei and, in particular, to the tectum mesencephali. Evidence has been presented that, at least in some chondrichthyans, two visual pathways exist: a retino-thalamo-telencephalic pathway and a retino-tecto-thalamo-telencephalic pathway. Anatomical (Ebbesson and Schroeder 1971; Schroeder and Ebbesson 1974; Luiten 1981b) and electrophysiological (Cohen et al. 1973; Vesselkin and Kovacevic 1973) evidence has been provided in support of the telencephalic visual function in cartilaginous fish. Moreover, behavioural experiments by Graeber et al. (1973, 1978) and Graeber (1978, 1980) have underscored a role of the telencephalon in visual information processing. It was found that free-swimming nurse sharks (*Ginglymostoma*) with complete tectal ablations could still learn to perform simple visual discriminations, such as the distinction between black and white or horizontal and vertical stripes. This result implies that the tectum mesencephali is not the only brain centre to influence the visually guided behaviour of elasmobranchs. Partial or complete lesioning of the pallial central nucleus results in an inability to perform discriminations between black and white or between horizontal and vertical stripes better than by chance. These results imply that the central pallial nucleus plays a role in the control of visually guided behaviour in nurse sharks.

From the preceding account, it it clear that, because of its multimodal input, the chondrichthyan telencephalon should be considered as a higher-order coordinating centre rather than a rhinencephalon. Besides the telencephalon, there are two other important coordinating centres, the tectum mesencephali and the cerebellum. In addition to visual input, the midbrain tectum receives input from, among others, the lateral line system, the trigeminal nerve and the spinal cord. The cerebellum also receives visual, electrosensory, mechanosensory and tactile-proprioceptive input.

We do not know as yet exactly how the coordinating centres mentioned above may influence behaviour in cartilaginous fishes. From a neuroanatomical point of view, it seems quite obvious that midbrain tectum and cerebellum may have a direct influence on motor behaviour by way of their strong efferent connections with the brain stem medial reticular formation, which in turn projects heavily onto the spinal motoneurons. Another possibility might be that both centres project via thalamic nuclei to the telencephalon.

As far as we know, there is no direct connection of the telencephalon to the spinal cord, as in mammals. Efferent telencephalic pathways in the nurse shark can be traced as far as the obex, issuing fibres to the reticular formation, but in *Raja* it seems that the descending pathways do not pass beyond the isthmic level. However, the pallial areas which receive the various sensory modalities may project to the hypothalamic inferior lobe by way of the pallial tract and via the basal forebrain bundle. The lobe gives rise to a strong lobobulbar tract, the fibres of which terminate in the lateral reticular formation of the brain stem. It is not known whether the ventral thalamus also forms a relay centre for telencephalic efferent fibres.

Our knowledge of the central nervous system of cartilaginous fishes has been increasing rapidly in recent years, thanks to new experimental neuroanatomical tracing techniques, but our understanding of the organisation of the brain and behaviour of these species remains limited. We have gained some insight into the way in which cartilaginous fishes detect their food sources, but we know almost nothing about other aspects of their behavior. The requisite studies are not easy to perform, but they are essential for a better understanding of the biological significance of the notable interspecies variation evident within brain structures.

References

Addens JL (1933) The motor nuclei and roots of the cranial and first spinal nerves of vertebrates. I. Introduction and cylcostomes. Z Anat Entw Gesch 101:307–410

Addens JL (1945) The nucleus of Bellonci and adjacent cell groups in Selachians. I. The ganglion habenulae and its connections. Proc Kon Ned Akad Wet Amsterdam 48:345–359

Akoev G, Andrianov GN, Szabo T, Bromm B (1991) Neuropharmacological analysis of synaptic transmission in the Lorenzinian ampulla of the skate Raja clavata. J Comp Physiol [A] 168:639–646

Alvarez R, Anadón R (1987) The cerebellum of the dogfish, Scyliorhinus canicula: a quantitative study. J Hirnforsch 28:133–137

Alvarez-Otero R, Anadón R (1992) Golgi cells of the cerebellum of the dogfish, Scyliorhinus canicula (elasmobranchs): a Golgi and ultrastructural study. J Hirnforsch 33:321–327

Alvarez-Otero R, Regueira SD, Anadón R (1993) New structural aspects of the synaptic contacts on Purkinje cells in an elasmobranch cerebellum. J Anat 182:13–21

Andres KH (1970) Anatomy and ultrastructure of the olfactory bulb in fish, amphibia, reptiles, birds and mammals. In: Wolstenholm GEW, Knight J (eds) Taste and smell in vertebrates. Churchill, London, pp 177–196 (Ciba Foundation symposia)

Andres KH (1975) Neue morphologische Grundlagen zur Physiologie des Riechens und Schmeckens. Arch Otorhinolaryngol 210:1–41

Ariëns Kappers CU (1921) Die vergleichende Anatomie des Nervensystems der Wirbeltiere und des Menschen, part 2. Bohn, Haarlem

Ariëns Kappers CU (1947) Anatomie comparée du système nerveux. Bohn, Haarlem

Ariëns Kappers CU, Carpenter FW (1911) Das Gehirn von Chimaera monstrosa. Folia Neurobiol 5:127–160

Ariëns Kappers CU, Huber GC, Crosby EC (1936) The comparative anatomy of the nervous system of vertebrates, including man, vol 1. MacMillan, New York

Bäckström K (1924) Contributions to the forebrain morphology in selachians. Acta Zool 5:123–240

Balfour FM (1878) A monograph on the development of elasmobranch fishes. MacMillan, London

Ball JN, Baker BI (1969) The pituitary gland: anatomy and histophysiology. In: Hoar WS, Randall DJ (eds) Fish physiology, vol 2. Academic, New York, pp 1–111

Ballard WW, Mellinger J, Léchenault H (1993) A series of normal stages for development of Scyliorhinus canicula, the lesser spotted dogfish (Chondrichthyes: Scyliorhinidae). J Exp Zool 267:318–336

Barber VC, Yake KI, Clark VF, Pungur J (1985) Quantitative analyses of sex and size differences in the macula neglecta and ramus neglectus in the inner ear of the skate, Raja ocellata. Cell Tissue Res 241:597–605

Bargmann W (1953) Über das Zwischenhirn-Hypophysensystem von Fischen. Z Zellforsch 38:275–298

Baron VD, Brown GR, Mikhailenko NA, Orlov AA (1985) About the ordinary skates electrolocation function (in Russian). Dokl Acad Sci USSR 280:240–243

Barrett DJ, Taylor EW (1985a) Spontaneous efferent activity in branches of the vagus nerve controlling ventilation and heart rate in the dogfish. J Exp Biol 117:433–448

Barrett DJ, Taylor EW (1985b) The location of cardiac vagal preganglionic neurones in the brain stem of the dogfish Scyliorhinus canicula. J Exp Biol 117:449–458

Barrett DJ, Taylor EW (1985c) The characteristics of cardial vagal preganglionic motoneurones in the dogfish. J Exp Biol 117:459–470

Barry MA (1987a) Central connections of the IXth and Xth cranial nerves in the clearnose skate, Raja eglanteria. Brain Res 425:159–166

Barry MA (1987b) Afferent and efferent connections of the primary octaval nuclei in the clearnose skate, Raja eglanteria. J Comp Neurol 266:457–477

Barry MA, Boord RL (1984) The spiracular organ of sharks and skates: Anatomical evidence indicating a mechanoreceptive role. Science 226:990–992

Barry MA, Hall DH, Bennett MVL (1988a) The elasmobranch spiracular organ. I. Morphological studies. J Comp Physiol [A] 163:85–92

Barry MA, White RL, Bennett MVL (1988b) The elasmobranch spiracular organ. II. Physiological studies. J Comp Physiol [A] 163:93–98

Bauchot R, Platel R, Ridet J-M (1976) Brain-body weight relationships in Selachii. Copeia 1976:305–309

Beard J (1896) The history of a transient nervous apparatus in certain Ichthyopsida. An account of the development and degeneration of ganglion cells and nerve-fibres. I. Raja batis. Zool Jahrb 9:319–426

Belbenoit P, Bauer R (1972) Video recordings of prey capture behaviour and associated electric organ discharge of Torpedo Marmorata. Marine Biol 17:93–99

Bennett MVL (1968) Neural control of electric organs. In: Ingle D (ed) The central nervous system and fish behaviour. University Press, Chicago, pp 147–169

Bennett MVL (1971) Electric organs. In: Hoar WS, Randall DJ (eds) Fish physiology, vol 5. Academic, New York, pp 347–491

Bergquist H (1932) Zur Morphologie des Zwischenhirns bei niederen Wirbeltieren. Acta Zool 13:57–304

Bern HA (1969) Urophysis and caudal neurosecretory system. In: Hoar WS, Randall DJ (eds) Fish physiology, vol 2. Academic, New York, pp 399–416

Bern HA, Gunther R, Johnson DW, Nishioka RS (1973) Occurrence of urotensin II (bladder-contracting activity) in the caudal spinal cord of anamniote vertebrates. Acta Zool 15:15–19

Bernau NA, Puzdrowski RL, Leonard RB (1991) Identification of the midbrain locomotor region and its relation to descending locomotor pathways in the Atlantic stingray, Dasyatis sabina. Brain Res 557:83–94

Bleckmann H, Bullock TH, Jorgensen JM (1987) The lateral line mechanoreceptive mesencephalic, diencephalic, and telencephalic regions in the thornback ray, Platyrhinoidis triseriata (Elasmobranchii). J Comp Physiol [A] 161:67–84

Bleckmann H, Weiss O, Bullock TH (1989) Physiology of lateral line mechanoreceptive regions in the elasmobranch brain. J Comp Physiol [A] 164:459–474

Bodznick D (1990) Elasmobranch vision: multimodal integration in the brain. J Exp Zool [Suppl] 5:108–116

Bodznick D, Montgomery JC (1992) Suppression of ventilatory reafference in the elasmobranch electrosensory system: medullary neuron receptive fields support a common mode rejection mechanism. J Exp Biol 171:127–137

Bodznick D, Northcutt RG (1980) Segregation of electro- and mechanoreceptive inputs to the elasmobranch medulla. Brain Res 195:313–321

Bodznick D, Northcutt RG (1984) An electrosensory area in the telencephalon of the little skate, Raja erinacea. Brain Res 298:117–124

Bodznick D, Schmidt AW (1984) Somatotopy within the medullary electrosensory nucleus of the little skate, Raja erinacea. J Comp Neurol 225:581–590

Bodznick D, Montgomery JC, Bradley DJ (1992) Suppression of common mode signals within the electrosensory system of the little skate, Raja erinacea. J Exp Biol 171:107–125

Bone Q (1977) Mauthner neurons in elasmobranchs. J Mar Biol Ass UK 57:253–259

Bone Q (1978) Locomotor muscle. In: Hoar WS, Randall DK (eds) Fish physiology, vol 7. Academic, New York, pp 361–424

Boord RL (1977) Auricular projections in the clearnose skate, Raja eglanteria. Am Zool 17:887

Boord RL, Campbell CBG (1977) Structural and functional organization of the lateral line system of sharks. Am Zool 17:431–441

Boord RL, Northcutt RG (1982) Ascending lateral line pathways to the midbrain of the clearnose skate, Raja eglanteria. J Comp Neurol 207:274–282

Boord RL, Roberts BL (1980) Medullary and cerebellar projections of the statoacoustic nerve of the dogfish Scyliorhinus canicula. J Comp Neurol 193:57–68

Boord RL, Sperry DG (1991) Topography and nerve supply of the cucullaris (trapezius) of skates. J Morphol 207:165–172

Braak H (1967) Purkinje cell bodies in the stratum moleculare of Chimaera monstrosa. Anat Anz 120:357–359

Bratton B, Ayers J (1987) Observations on the organ discharge of the skate species (Chondrichthyes, Rajidae) and its relationship to behaviour. Environ Biol Fish 4:241–254

Bruckmoser P (1973) Beziehungen zwischen Struktur und Funktion in der Evolution des Telencephalon. Verh Dtsch Zool Ges 66:219–229

Bruckmoser P, Dieringer N (1973) Evoked potentials in the primary and secondary olfactory projection areas of the forebrain in Elasmobranchia. J Comp Physiol [A] 87:65–74

Brunken WJ, Witkovsky P, Karten HJ (1986) Retinal neurochemistry of three elasmobranch species: an immunohistochemical approach. J Comp Neurol 243:1–12

Bruun A, Ehinger B, Systsma VM (1984) Neurotransmitter localization in the skate retina. Brain Res 295:233–248

Bruun A, Ehinger B, Sytsma V, Tornqvist K (1985) Retinal neuropeptides in the skates, Raja clavata, R radiata, R oscellata (Elasmobranchii). Cell Tissue Res 241:17–24

Bullock TH (1984) Physiology of the tectum mesencephali in elasmobranchs. In: Vanegas H (ed) Comparative neurology of the optic tectum. Plenum, New York, pp 47–68

Bullock TH, Northcutt RG, Bodznick DA (1982) Evolution of electroreception. Trends Neurosci 5:50–53

Bullock TH, Bodznick DA, Northcutt RG (1983) The phylogenetic distribution of electroreception: evidence for convergent evolution of a primitive vertebrate sense modality. Brain Res Rev 6:25–46

Bundgaard M, Cserr HF (1991) Barrier membranes at the outer surface of the brain of an elasmobranch, Raja erinacea. Cell Tissue Res 265:113–120

Cameron AA, Plenderleith MB, Snow PJ (1990) Organization of the spinal cord in four species of elasmobranch fish: cytoarchitecture and distribution of serotonin and selected neuropeptides. J Comp Neurol 297:201–218

Carey FG, Teal JM (1969) Mako and porbeagle: warm-bodied sharks. Comp Biochem Physiol [A] 28:199–204

Carroll RL (1988) Vertebrate paleontology and evolution. Freeman, New York

Casterlin ME, Reynolds WW (1979) Shark (Mustelus canis) thermoregulation. Comp Biochem Physiol [A] 64:451–453

Chan DKO, Ho MW (1969) Pressor substances from the caudal neurosecretory system of teleost and elasmobranch fish. Gen Comp Endocrinol 13:498

Charlton HH (1932) Comparative studies on the nucleus preopticus pars magnocellularis and the nucleus lateralis tuberis in fishes. J Comp Neurol 54:237–276

Chevins PFD (1968) The anatomy and physiology of the pituitary complex in the genus Raja, Elasmobranchii. PhD thesis, Leeds University

Chevins PFD, Dodd JM (1970) Pituitary innervation and control of colour change in skates, Raja naevus, R clavata, R montagui, and R radiata. Gen Comp Endocrinol 15:232–241

Chiba A, Honma Y (1992) Distribution of neuropeptide Y-like immunoreactivity in the brain and hypophysis of the cloudy dogfish, Scyliorhinus torazame. Cell Tissue Res 268:453–461

Chiba A, Honma Y, Ito S, Honma S (1989) Somatostatin-immunoreactivity in the brain of the gummy shark, Mustelus manazo Bleeker, with special regard to the hypothalamo-hypophyseal system. Biomed Res 10 [Suppl 3]:1–12

Chiba A, Oka S, Honma Y (1991) Immunocytochemical distribution of FMRFamide-like substance in the brain of the cloudy dogfish, Scyliorhinus torazame. Cell Tissue Res 265:243–250

Cifuentes M, Rodríguez S, Pérez J, Grondona JM, Rodríguez EM, Fernández-Llebrez P (1994) Decreased cerebrospinal fluid flow through the central canal of the spinal cord of rats immunologically deprived of Reissner's fibre. Exp Brain Res 98:431–440

Coggeshall RE, Leonard RB, Applebaum ML, Willis WD (1978) Organization of peripheral nerves and spinal roots of the Atlantic stingray, Dasyatis sabina. J Neurophysiol 41:97–107

Cohen DH, Duff TA, Ebbesson SOE (1973) Electrophysiological identification of a visual area in shark telencephalon. Science 182:492–494

Collin SP (1988) The retina of the shovel-nosed ray, Rhinobatos batillum (Rhinobatidae): morphology and quantitative analysis of the ganglion, amacrine and bipolar cell populations. Exp Biol 47:195–207

Compagno LJV (1977) Phyletic relationships of living sharks and rays. Am Zool 17:303–322

Conley RA, Bodznick D (1994) The cerebellar dorsal granular ridge in elasmobranch has proprioceptive and electroreceptive representations and projects homotopically to the medullary electrosensory nucleus. J Comp Physiol [A] 174:707–721

Corwin JT (1981a) Peripheral auditory physiology in the lemon shark; evidence of parallel otolithic and nonotolithic sound detection. J Comp Physiol [A] 142:379–390

Corwin JT (1981b) Audition in elasmobranchs. In: Tavolga WN, Popper AN, Fay RR (eds) Hearing and sound communication in fishes. Springer, Berlin Heidelberg New York, pp 81–105

Corwin JT (1983) Postembryonic growth of the macula neglecta auditory detector in the ray, Raja clavata: continual increases in haircell number, neural convergence, and physiological sensitivity. J Comp Neurol 217:345–356

Corwin JT (1985) Auditory neurons expand their terminal arbors throughout life and orient toward site of postembryonic hair cell production in the macula neglecta in elasmobranchs. J Comp Neurol 239:445–452

Corwin JT, Northcutt RG (1982) Auditory centers in the elasmobranch brain stem: deoxyglucose autoradiography and evoked potential recording. Brain Res 236:261–273

Crawshaw LI, Hammel HT (1973) Behavioral temperature regulation in the California hornshark, Heterodontus francisci. Brain Behav Evol 7:447–452

Dahlgren U (1914) On the electric motor nerve centers in the skates (Rajidae). Science 40:862–863

Dammerman KW (1910) Der Saccus vasculosus der Fische, ein Tieforgan. Z Wiss Zool 96:654–726

Demski LS (1977) Electrical stimulation of the shark brain. Am Zool 17:487–500

Demski LS, Fields RD (1988) Dense-cored vesicle-containing components of the terminal nerve of sharks and rays. J Comp Neurol 278:604–614

Demski LS, Northcutt RG (1983) The terminal nerve: a new chemosensory system in vertebrates? Science 220:435–437

Dijkgraaf S (1963) The functioning and significance of the lateral line organs. Biol Rev 38:51–105

Dijkgraaf S, Kalmijn AJ (1962) Verhaltensversuche zur Funktion des Lorenzinischen Ampullen. Naturwissenschaften 49:400

Dodd JM, Dodd MHJ, Jenkins N (1980) Presence of a gonadotropin in the Rachendachhypophyse of the pituitary gland of the rabbitfish Hydrolagus collei (Chondrichthyes: Holocephali). Gen Comp Endocrinol 40:342–343

Dodd JM, Dodd MHI, Duggan RT (1983) Control of reproduction in elasmobranch fishes. In: Rankin JC, Pitcher TC, Duggan RT (eds) Control processes in fish physiology. Wiley, New York, pp 221–285

Droge MH, Leonard RB (1983a) Organization of spinal motor nuclei in the stingray, Dasyatis sabina. Brain Res 276:201–211

Droge MH, Leonard RB (1983b) Swimming pattern in intact and decerebrated stingrays. J Neurophysiol 50:162–177

Droge MH, Leonard RB (1983c) Swimming rhythm in decerebrated, paralyzed stingrays: normal and abnormal coupling. J Neurophysiol 50:178–191

Dryer L, Graziadei PPC (1993) A pilot study on morphological compartmentalization and heterogeneity in the elasmobranch olfactory bulb. Anat Embryol (Berl) 188:41–51

Dunn RF, Koester DM (1984) Primary afferent projections of the octavus nerve to the inferior reticular formation and adjacent nuclei in the elasmobranch, Rhinobatos sp. Brain Res 323:354–359

Dunn RF, Koester DM (1987) Primary afferent projections to the central octavus nuclei in the elasmobranch, Rhinobatos sp, as demonstrated by nerve degeneration. J Comp Neurol 260:564–572

Ebbesson SOE (1972) New insights into the organization of the shark brain. Comp Biochem Physiol [A] 42:121–129

Ebbesson SOE (1980a) The parcellation theory and its relation to interspecific variability in brain organization evolutionary and ontogenetic development, and neuronal plasticity. Cell Tissue Res 213:179–212

Ebbesson SOE (1980b) On the organization of the telencephalon in elasmobranchs. In: Ebbesson SOE (ed) Comparative neurology of the telencephalon. Plenum, New York, pp 1–16

Ebbesson SOE, Campbell CBG (1973) On the organization of cerebellar efferent pathways in the nurse shark (Ginglymostoma cirratum). J Comp Neurol 152:233–254

Ebbesson SOE, Heimer L (1970) Projections of the olfactory tract fibers in the nurse shark (Ginglymostoma cirratum). Brain Res 17:47–55

Ebbesson SOE, Hodde KC (1981) Ascending spinal systems in the nurse shark, Ginglymostoma cirratum. Cell Tissue Res 216:313–331

Ebbesson SOE, Meyer DL (1980) The visual system of the guitarfish (Rhinobatos productus). Cell Tissue Res 206:243–250

Ebbesson SOE, Northcutt RG (1976) Neurology of anamniotic vertebrates. In: Masterton RB, Bitterman ME, Campbell CBG, Hotton N (eds) Evolution of brain and behavior in vertebrates. Erlbaum, Hilsdale, pp 115–146

Ebbesson SOE, Ramsey JS (1968) The optic tracts of two species of sharks (Galeocerdo cuvieri and Ginglymostoma cirratum). Brain Res 8:36–53

Ebbesson SOE, Schroeder D (1971) Connections of the nurse shark's telencephalon. Science 173:254–256

Eccles JC, Táboriková H, Tsukahara N (1970) Responses of the granule cells of the selachian cerebellum (Mustelus canis). Brain Res 17:87–102

Edinger L (1892) Untersuchungen über die vergleichende Anatomie des Gehirns. 2. Das Zwischenhirn. I. Das Zwischenhirn der Selachier und der Amphibien. Abh Senckenberg Naturforsch Gesellsch 18:1–55

Edinger L (1901) Das Cerebellum von Scyllium canicula. Arch Mikr Anat 58:661–678

Edinger L (1908) Vorlesungen über den Bau der Nervösen Zentralorgane des Menschen und der Tiere. Vogel, Leipzig

Ewart JC (1890) The cranial nerves of the torpedo. Proc R Soc Lond B 47:290–291

Ewart JC (1983) The electric organ of the skate: note on an electric centre in the spinal cord. Proc R Soc Lond B 53:388–391

Farner H-P (1978a) Untersuchungen zur Embryonalentwicklung des Gehirns von Scyliorhinus canicula (L). I. Bildung der Hirngestalt, Migrationsmodi und -phasen, Bau des Zwischenhirns. J Hirnforsch 19:313–332

Farner H-P (1978b) Untersuchungen zur Embryonalentwicklung des Gehirns von Scyliorhinus canicula (L). II. Das Tectum opticum und dessen Stratifikation. J Hirnforsch 19:333–344

Farner H-P (1978c) Untersuchungen zur Embryonalentwicklung des Gehirns von Scyliorhinus canicula (L). III. Das optische System und angrenzende Nuclei im mesencephalen Tegmentum. J Hirnforsch 19:405–414

Faucette JR (1969a) The olfactory bulb and medial hemisphere wall of the rat-fish, Chimaera. J Comp Neurol 137:377–406

Faucette JR (1969b) The accessory olfactory bulbs and the lateral telencephalic wall of the rat-fish, Chimaera. J Comp Neurol 137:407–432

Fay RR, Kendall JI, Popper AN, Tester AL (1974) Vibration detection by the macula neglecta of sharks. Comp Biochem Physiol [A] 47:1235–1240

Fiebig E (1988) Connections of the corpus cerebelli in the thornback guitarfish, Platyrhinoidis triseriata (elasmobranchii): a study with WGA-HRP and extracellular granule cell recording. J Comp Neurol 268:567–583

Fields RD, Bullock TH, Lange GD (1993) Ampullary sense organs, periheral, central and behavioral electroreception in chimaeras (Hydrolages, Holocephali, Chondrichthyes). Brain Behav Evol 41:269–289

Finger TE (1978) Gustatory pathways in the bullhead catfish. II. Facial lobe connections. J Comp Neurol 180:691–706

Finger TE, Karten HJ (1978) The accessory optic system in teleosts. Brain Res 153:144–149

Finstad WD, Nelson DR (1975) Circadian activity rhythm in the hornshark, Heterodontus francisci: effect of light intensity. Bull South Calif Acad Sci 74:20–26

Fox GQ (1977) The morphology of the oval nuclei of neonatal Torpedo marmorata. Cell Tissue Res 178:155–167

Fridberg G (1962) The caudal neurosecretory system in some elasmobranchs. Gen Comp Endocrinol 2:249–265

Gabriel KH (1970) Vergleichend-histologische Studien am Subcommissuralorgan. Anat Anz 127:129–170

Gerlach J (1947) Beiträge zur vergleichenden Morphologie des Selachierhirnes. Anat Anz 96:79–165

Gilbert PW (1984) Biology and behavior of sharks. Endeavour 8:179–187

Glees P (1940) Der periphere und zentrale Anteil des sympatischen Nervensystems der Selachier. Acta Neerl Morphol 3:209–248

González A, Smeets WJAJ (1994) Catecholamine systems in the CNS of amphibians. In: Smeets WJAJ, Reiner A (eds) Phylogeny and development of catecholamine systems in the CNS of vertebrates. Cambridge University Press, Cambridge, pp 77–102

Graeber RC (1978) Behavioral studies correlated with central nervous system integration of vision in sharks. In: Hodgson ES, Mathewson RF (eds) Sensory biology of sharks, skates, and rays. US Government Printing, Washington DC, pp 195–225

Graeber RC (1980) Telencephalic function in elasmobranchs. A behavioral perspective. In: Ebbesson SOE (ed) Comparative neurology of the telencephalon. Plenum, Washington DC, pp 17–39

Graeber RC, Ebbesson SOE (1972) Retinal projections in the lemon shark (Negaprion brevirostris). Brain Behav Evol 5:461–477

Graeber RC, Ebbesson SOE, Jane JA (1973) Visual discrimination in sharks without optic tectum. Science 180:413–415

Graeber RC, Schroeder DM, Jane JA, Ebbesson SOE (1978) Visual discrimination following partial telencephalic ablations in nurse sharks (Ginglymostoma cirratum). J Comp Neurol 180:325–344

Graf W, Brunken WJ (1984) Elasmobranch oculomotor organization: anatomical and theoretical aspects of the phylogenetic development of vestibulo-oculomotor connectivity. J Comp Neurol 227:569–581

Grillner S (1974) On the generation of locomotion in the spinal dogfish. Exp Brain Res 20:459–470

Grillner S, Kashin S (1976) On the generation and performance of swimming in fish. In: Herman R, Grillner S, Stein P, Stuart D (eds) Neural control of locomotion, vol 18. Plenum, New York, pp 181–202

Grillner S, Perret C, Zangger P (1976) Central generation of locomotion in the spinal dogfish. Brain Res 109:255–269

Grondona JM, Fernández-Llebrez P, Pérez J, Cifuentes M, Pérez-Fígares Rodríguez EM (1994) Class-specific epitopes detected by polyclonal antibodies against the secretory products of the subcommissural organ of the dogfish Scyliorhinus canicula. Cell Tissue Res 276:515–522

Gruber SH, Cohen JL (1978) Visual system of the elasmobranchs: state of the art 1960–1975. In: Hodgson ES, Mathewson RF (eds) Sensory biology of sharks, skates and rays. US Government Printing Office, Washington DC, pp 11–105

Hamasaki DI, Streck P (1971) Properties of the epiphysis cerebri of the small-spotted dogfish shark, Scyliorhinus caniculus L. Vision Res 11:189–198

Hayle TH (1973a) A comparative study of spinal projections to the brain (except cerebellum) in three classes of poikilothermic vertebrates. J Comp Neurol 149:463–476

Hayle TH (1973b) A comparative study of spinocerebellar systems in three classes of poikilothermic vertebrates. J Comp Neurol 149:477–495

Heath AR (1990) The ocular tapetum lucidum: a model system for interdisciplinary studies in elasmobranch biology. J Exp Zool [Suppl] 5:41–45

Herrick CJ (1948) The brain of the tiger salamander. University of Chicago Press, Chicago

Hodgson ES, Mathewson RF (1971) Chemosensory orientation in sharks. Ann NY Acad Sc 188:175–182

Hodgson ES, Mathewson RF, Gilbert PW (1967) Electroencephalographic studies of chemoreception in sharks. In: Gilbert PW, Mathewson RF, Rall DP (eds) Sharks, skates and rays. Johns Hopkins Press, Baltimore, pp 491–502

Hökfelt T, Martensson R, Björklund A, Kleinau S, Goldstein M (1984) Distributional maps of tyrosine-hydroxylase-immunoreactive neurons in the rat brain. In: Handbook of chemical neuroanatomy, vol 2. Classical neurotransmitters in the CNS, part I. Elsevier, Amsterdam, pp 277–379

Holmes RL, Ball JN (1974) The pituitary gland. Cambridge University Press, London

Holmgren N (1922) Points of view concerning forebrain morphology in lower vertebrates. J Comp Neurol 34:391–440

Horstmann E (1954) Die Faserglia des Selachiergehirns. Z Zellforsch 30:588–617

Houser GL (1901) The neurones and supporting elements of the brain of a selachian. J Comp Neurol 11:65–175

Housley GD, Montgomery JC (1983) Central projections of vestibular afferents from the horizontal semicircular canal in the carpet shark Cephaloscyllium isabella. J Comp Neurol 221:154–162

Hueter RE (1990) Adaptations for spatial vision in sharks. J Exp Zool [Suppl] 5:130–141

Hugosson R (1957) Morphologic and experimental studies on the development and significance of the rhombencephalic longitudinal cell columns. PhD thesis, Lund University

Jasinkski A, Gorbman A (1966) Hypothalamo-hypophyseal vascular and neurosecretory links in the ratfish, Hydrolagus collei (Lay and Bennett). Gen Comp Endocrinol 6:476–490

Jen LS, So K-F, Yew DT, Lee M (1983) An autoradiographic study of the retinofugal projections in the shark, Hemiscyllium plagiosum. Brain Res 274:135–139

Johnston JB (1902) The brain of Petromyzon. J Comp Neurol 7:2–82

Johnston JB (1905) The radix mesencephalica trigemini. The ganglion isthmi. Anat Anz 87:364–379

Johnston JB (1911) The telencephalon of selachians. J Comp Neurol 21:1–113

Jones HC, Bucknall RM (1988) Inherited prenatal hydrocephalus in the X-Tx rat: a morphological study. Neuropathol Appl Neurobiol 14:263–274

Kalmijn AJ (1974) The detection of electric fields from inanimate and animate sources other than electric organs. In: Fessard A (ed) Handbook of sensory physiology, vol III/3. Springer, Berlin Heidelberg New York, pp 147–200

Kalmijn AJ (1978) Electric and magnetic sensory world of sharks, skates and rays. In: Hodgson ES, Mathewson RF (eds) Sensory biology of sharks, skates and rays. US Government Printing Office, Washington DC, pp 507–528

Kalmijn AJ (1982) Electric and magnetic field detection in elasmobranch fishes. Science 218:916–917

Kashin SM, Feldman AG, Orlovsky GN (1974) Locomotion of fish evoked by electrical stimulation of the brain. Brain Res 82:41–47

Keenan E (1928) The phylogenetic development of the substantia gelatinosa Rolandi, part I: fishes. Proc Kon Ned Akad Wet Amsterdam 31:837–854

Klatzo I (1967) Cellular morphology of the lemon shark brain. In: Gilbert PW, Mathewson RF, Rall DP (eds) Sharks, skates and rays. Johns Hopkins Press, Maryland, pp 341–359

Kleerekoper H (1978) Chemoreception and its interaction with flow and light perception in the locomotion and orientation of some elasmobranchs. In: Hodgson ES, Mathewson RF (eds) Sensory biology of sharks, skates and rays. US Government Printing Office, Washington DC, pp 269–329

Klimley AP (1994) The predatory behavior of the white shark. Am Sci 82:122–133

Knowles F (1965) Evidence for a dual control, by neurosecretion of hormone synthesis and hormone release in the pituitary of the dogfish, Scyliorhinus stellaris. Philos Trans R Soc Lond B 249:435–456

Knowles F, Weatherhead B, Martin R (1970) The ultrastructure of neurosecretory fibre terminals after zinc-iodine-osmium impregnation. In: Bargmann W, Scharrer B (eds) Aspects of neurosecretion. Springer, Berlin Heidelberg New York, pp 159–165

Knowles F, Vollrath L, Meurling P (1975) Cytology and neuroendocrine relations of the pituitary of the dogfish, Scyliorhinus canicula. Proc R Soc Lond B 191:507–525

Koester DM (1983) Central projections of the octavolateralis nerves of the clearnose skate, Raja eglanteria. J Comp Neurol 221:199–215

Kremers JWPM, Nieuwenhuys R (1979) Topological analsyis of the brain stem of the crossopterygian fish Latimeria chalumnae. J Comp Neurol 187:613–638

Kreps EM (1981) Brain lipids of elasmobranchs (an essay on comparative neurobiology). Comp Biochem Physiol 688:363–367

Kuchnow KP (1971) The elasmobranch pupillary response. Vision Res 11:1395–1406

Kuhlenbeck H (1927) Vorlesungen über das Zentralnervensystem der Wirbeltiere. Fischer, Jena

Kuhlenbeck H (1929a) Über die Grundbestandteile des Zwischenhirnbauplans der Anamnier. Morphol Jahrb 63:50–95

Kuhlenbeck H (1929b) Die Grundbestandteile des Endhirns im Lichte der Bauplanlehre. Anat Anz 67:1–51

Kuhlenbeck H (1973) Overall morphological pattern. In: Kuhlenbeck H (ed) The central nervous system of vertebrates, vol 3/II. Karger, Basel

Kuhlenbeck H (1977) Derivatives of the prosencephalon: diencephalon and telencephalon. In: Kuhlenbeck H (ed) The central nervous system of vertebrates, vol 5/I Karger, Basel

Kuhlenbeck H, Niimi K (1969) Further observations on the morphology of the brain in the holocephalian elasmobranchs Chimaera and Callorhynchus. J Hirnforsch 11:267–314

Kusunoki T, Tsuda Y, Takashima F (1973) The chemoarchitectonics of the shark brain. J Hirnforsch 14:13–36

Larsell O (1967) The comparative anatomy and histology of the cerebellum from myxinoids through birds. University of Minnesota Press, Minneapolis

Leonard RB, Coggeshall RE, Willis WD (1978a) A documentation of an age related increase in neuronal and axonal members in the stingray, Dasyatis sabina Lesener. J Comp Neurol 179:13–22

Leonard RB, Rudomín P, Willis WD (1978b) Central effects of volleys in sensory and motor components of peripheral nerve in the stingray, Dasyatis sabina. J Neurophysiol 41:108–125

Leonard RB, Kenshalo DR Jr, Willis WD (1978c) Receptive field properties of primary afferents in the atlantic stingray, Dasyatis sabina. Soc Neurosci Abstr 4:46

Leonard RB, Rudomín P, Droge MH, Grossman AE, Willis WD (1979) Locomotion in the decerebrate stingray. Neurosci Lett 14:315–319

Livingston CA, Williams BJ, Ritchie TC, Leonard RB (1983) The origin and trajectories of descending pathways in the spinal cord of the stingray, Dasyatis sabina. Soc Neurosci Abstr 9:285

Locy WA (1905) On a newly recognized nerve connected with the forebrain of selachians. Anat Anz 26:33–36

Lovejoy DA, Ashmead BJ, Coe IR, Sherwood NM (1992) Presence of gonadotropin-releasing hormone immunoreactivity in dogfish and skate brains. J Exp Zool 263:272–283

Lowenstein O (1974) Comparative morphology and physiology. In: Kornhuber MH (ed) Handbook of sensory physiology, vol 6. Springer, Berlin Heidelberg New York, pp 75–120

Lowenstein O, Compton GJ (1978) A comparative study of the responses of isolated first-order semicircular canal afferents to angular and linear acceleration, analysed in the time and frequency domains. Proc R Soc Lond B 202:313–338

Lowenstein O, Roberts TDM (1951) The localization and analysis of the responses to vibration from the isolated elasmobranch labyrinth. A contribution to the problem of the evolution of hearing in vertebrates. J Physiol (Lond) 114:471–489

Luiten PGM (1981a) Two visual pathways to the telencephalon in the nurse shark (Ginglymostoma cirratum). I. Retinal projections. J Comp Neurol 196:531–538

Luiten PGM (1981b) Two visual pathways to the telencephalon in the nurse shark (Ginglymostoma cirratum). II. Ascending thalamo-telencephalic connections. J Comp Neurol 196:539–548

MacDonnell MF (1980a) Mesencephalic trigeminal nucleus in sharks. A light microscopic study. Brain Behav Evol 17:152–163

MacDonnell MF (1980b) Cerebrospinal fluid contacting and supra-ependymal mesencephalic trigeminal cells in the blue and mako sharks. A scanning electron microscopic study. Brain Behav Evol 17:164–177

MacDonnell MF (1984) Circumventricular mesencephalic trigeminal midline ridge formation in cartilaginous fishes: species variations. Brain Behav Evol 24:124–134

MacDonnell MF (1989) Sub/supraependymal axonal net in the brains of sharks and probably targets in parasynaptic relationship. Brain Behav Evol 34:201–211

Manso MJ, Anadón R (1991) The optic tectum of the dogfish Scyliorhinus canicula L: a Golgi study. J Comp Neurol 307:335–349

Manso MJ, Anadón R (1993) Golgi study of the telencephalon of the small-spotted dogfish Scyliorhinus canicula L. J Comp Neurol 333:485–502

McCormick CA (1981) Central projections of the lateral line and eighth nerves in the bowfin, Amia calva. J Comp Neurol 197:1–15

McCready PJ, Boord RL (1976) The topography of the superficial roots and ganglia of the anterior lateral line nerve of the smooth dogfish, Mustelus canis. J Morphol 150:527–538

McFarland WN (1990) Light in the sea: The optical world of elasmobranchs. J Exp Zool [Suppl] 5:3–12

Meek J (1994) Catecholamines in the brains of Osteichthyes (bony fishes). In: Smeets WJAJ, Reiner A (eds) Phylogeny and development of catecholamine systems in the CNS of vertebrates. Cambridge University Press, Cambridge, pp 49–76

Mellinger JCA (1963) Étude histophysiologique du système hypothalamo-hypophysaire de Scyliorhinus caniculus (L) en état de melanodispersion permanente. Gen Comp Endocrinol 3:26–45

Mellinger JCA (1964) Les relations neuro-vasculo-glandulaires dans l'appareil hypophysaire de la rousette, Scyliorhinus caniculus (L) (poissons elasmobranches). Arch Anat Histol Embryol 47:1–201

Meredith GM, Roberts BL (1986) Central organization of the efferent supply to the labyrinthine and lateral line receptors of the dogfish. Neuroscience 17:225–233

Meredith GE, Smeets WJAJ (1987) Immunocytochemical analysis of the dopamine system in the forebrain and midbrain of Raja radiata: evidence for a substantia nigra and ventral tegmental area in cartilaginous fish. J Comp Neurol 265:530–548

Meurling P (1960) Presence of a pituitary portal system in elasmobranches. Nature 187:336–337

Meurling P (1967) The vascularization of the pituitary in elasmobranchs. Sarsia 28:1–104

Meurling P (1972) Control of pars intermedia in large embryos of the spiny dogfish, Squalus acanthias. Gen Comp Endocrinol 18:609

Meurling P, Björklund A (1970) The arrangement of neurosecretory and catecholamine fibres in relation to the pituitary intermedia cells of the skate, Raja radiata. Z Zellforsch Mikr Anat 108:81–93

Meurling P, Fremberg M, Björklund A (1969) Control of MSH release in the intermediate lobe of Raja radiata (Elasmobranchii). Gen Comp Endocrinol 13:520

Molist P, Rodríguez-Moldes I, Anadón R (1992) Immunocytochemical and electron-microscopic study of the elasmobranch nucleus sacci vasculosi. Cell Tissue Res 270:395–404

Molist P, Rodríguez-Moldes I, Anadón R (1993) Organization of catecholaminergic systems in the hypothalamus of two elasmobranch species, Raja undulata and Scyliorhinus canicula. A histofluorescence and immunohistochemical study. Brain Behav Evol 41:290–302

Montgomery JC (1978) Dogfish vestibular system. PhD thesis, Bristol University

Montgomery JC (1980) Dogfish horizontal canal system: responses of primary afferent, vestibular and cerebellar neurons to rotational stimulation. Neuroscience 5:1761–1769

Montgomery JC (1981) Origin of the parallel fibers in the cerebellar crest overlying the intermediate nucleus of the elasmobranch hindbrain. J Comp Neurol 202:185–191

Montgomery JC (1982) Functional organization of the dogfish vestibulocerebellum. Brain Behav Evol 20:118–128

Montgomery JC (1984) Noise cancellation in the electrosensory system of the thornback ray; common mode rejection of input produced by the animal's own ventilatory movement. J Comp Physiol [A] 155:103–111

Montgomery JC, Bodznick D (1993) Hindbrain circuitry mediating common mode suppression of ventilatory reafference in the electrosensory system of the little skate Raja erinacea. J Exp Biol 183:203–215

Montgomery JC, Bodznick D (1994) An adaptive filter that cancels self-induced noise in the electrosensory and lateral line mechanosensory systems of fish. Neurosci Lett 174:145–148

Montgomery JC, Housley GD (1983) The abducens nucleus in the carpet shark Cephaloscyllium isabella. J Comp Neurol 221:163–168

Montgomery JC, Roberts BL (1979) Organization of vestibular afferents to the vestibular nuclei of the dogfish. Brain Behav Evol 16:81–98

Mori S, Nishimura H, Kurakami C, Yamamura T, Aoki M (1978) Controlled locomotion in the mesencephalic cat: distribution of facilitatory and inhibitory regions within pontine tegmentum. J Neurophysiol 41:1580–1591

Mori S, Nishimura H, Aoki M (1980) Brain stem activation of the spinal stepping generator. In: Hobson JA, Brazier MAB (eds) The reticular formation revisited: specifying function for a nonspecific system. Raven, New York, pp 241–260

Morita Y, Ito H, Masai H (1980) Central gustatory paths in the crucian carp Carassius carassius. J Comp Neurol 191:119–132

Mos W, Williamson R (1986) A quantitative analysis of the spinal motor pool and its target muscle during growth in the dogfish, Scyliorhinus canicula. J Comp Neurol 248:431–440

Myrberg AA Jr (1978) Underwater sound – its effect on the behavior of sharks. In: Hodgson ES, Mathewson RF (eds) Sensory biology of sharks, skates, and rays. US Government Printing Office, Washington DC, pp 391–417

Nakai Y, Ochiai H, Shioda S (1977) Cytological evidence for different types of cerebrospinal fluid-contacting subependymal cells in the preoptic and infundibular recesses of the frog. Cell Tissue Res 176:317–334

Nakajima Y (1970) Fine structure of the medullary command nucleus of the electric organ of the skate. Tissue Cell 2:47–59

Nelson DR, Johnson RH (1970) Diel activity rhythms in the nocturnal botton-dwelling sharks, Heterodontus francisci and Cephaloscyllium ventriosum. Copeia 1970:732–739

Newman DB, Cruce WLR (1982) The organization of the reptilian brainstem reticular formation: a comparative study using Nissl and Golgi techniques. J Morphol 173:325–349

Newman DB, Cruce WLR, Bruce LL (1983) The sources of supraspinal afferents to the spinal cord in a variety of limbed reptiles. I. Reticulospinal systems. J Comp Neurol 215:17–32

Nicholson C, Llinás R, Precht W (1969) Neural elements of the cerebellum in elasmobranch fishes; structural and functional characteristics. In: Llinás R (ed) Neurobiology of cerebellar evolution and development. American Medical Association, Chicago, pp 215–243

Nieuwenhuys R (1967) Comparative anatomy of the cerebellum. In: Fox CA, Snider RS (eds) The cerebellum. Elsevier, Amsterdam, pp 1–93 (Progress in brain research, vol 25)

Norris HW, Hughes SP (1920) The cranial, occipital and anterior spinal nerves of the dogfish, Squalus acanthias. J Comp Neurol 31:293–402

Northcutt RG (1977) Elasmobranch central nervous system organization and its possible evolutionary significance. Am Zool 17:411–429

Northcutt RG (1978) Brain organization in the cartilaginous fishes. In: Hodgson ES, Mathewson RF (eds) Sensory biology of sharks, skates, and rays. US Government Printing Office, Washington DC, pp 117–193

Northcutt RG (1979) Retinofugal pathways in fetal and adult spiny dogfish, Squalus acanthias. Brain Res 162:219–230

Northcutt RG (1980) Central auditory pathways in anamniotic vertebrates. In: Popper AN, Fay RR (eds) Proceedings in life science: comparative studies of hearing in vertebrates. Springer, Berlin Heidelberg New York, pp 79–118

Northcutt RG (1981) Evolution of the telencephalon in nonmammals. Annu Rev Neurosci 4:301–350

Northcutt RG (1984) Evolution of the vertebrate central nervous system: patterns and processes. Am Zool 24:701–716

Northcutt RG (1990) Visual pathways in elasmobranchs: Organization and phylogenetic implications. J Exp Zool [Suppl] 5:97–107

Northcutt RG, Wathey JC (1980) Guitarfish possess ipsilateral as well as contralateral retinofugal projections. Neurosci Lett 20:237–242

Northcutt RG, Reiner A, Karten HJ (1988) Immunohistochemical study of the telencephalon of the spiny dogfish, Squalus acanthias. J Comp Neurol 277:250–267

Nozaki M, Tsukahara T, Kobayashi H (1984) An immunocytological study of the distribution of neuropeptides in the brain of fish. Biomed Res [Suppl] 4:135–145

Oksche A (1993) Phylogenetic and conceptual aspects of the subcommissural organ. In: Oksche A, Rodríguez EM, Fernández-Llebrez P (eds) The subcommissural organ. An ependymal brain gland. Springer, Berlin Heidelberg New York, pp 23–32

O'Leary DP, Dunn RF, Honrubia V (1976) Analysis of afferent responses from isolated semicircular canal of the guitarfish using rotational acceleration white-noise inputs. I. Correlation of response dynamics with receptor innervation. J Neurophysiol 39:631–644

Onstott D, Elde R (1986) Immunohistochemical localization of urotensin I/corticotropin-releasing factor, urotensin II, and serotonin immunoreactivities in the caudal spinal cord of nonteleost fishes. J Comp Neurol 249:205–225

Overholser MD, Whitley JR, O'Dell BL, Hogan AG (1954) The ventricular system in hydrocephalic rat brains produced by a deficiency of vitamin B12 and folic acid in the maternal diet. Anat Rec 120:917–934

Palmgren A (1921) Embryological and morphological studies on the midbrain and cerebellum of vertebrates. Acta Zool 2:1–94

Pang PKT, Griffith RW, Atz JW (1977) Osmoregulation in elasmobranchs. Am Zool 17:365–377

Parent A (1986) Comparative neurobiology of the basal ganglia. Wiley, New York, pp xiv, 335

Parent A, Poitras D, Dubé L (1984) Comparative anatomy of central monoaminergic systems. In: Björklund A, Hökfelt T (eds) Classical transmitters in the CNS, part I. Elsevier, Amsterdam, pp 409–439 (Handbook of chemical neuroanatomy, vol 2)

Paul DH (1969) Parallel fibre response in the elasmobranch cerebellum. J Physiol (Lond) 202:110–112

Paul DH (1982) The cerebellum of fishes: a comparative neurophysiological and neuroanatomical review. Adv Comp Physiol Biochem 8:111–177

Paul DH, Roberts BL (1975) Connections between the cerebellum and the reticular formation in the dogfish Scyliorhinus caniculata. J Physiol (Lond) 249:62–63P

Paul DH, Roberts BL (1977) Studies on a primitive cerebellar cortex. I. The anatomy of the lateral-line lobes of the dogfish, Scyliorhinus canicula. Proc R Soc Lond B 195:453–466

Paul DH, Roberts BL (1978) Organization of the reticular formation in the dogfish Scyliorhinus canicula. J Physiol (Lond) 280:71P-72P

Paul DH, Roberts BL (1979) The significance of cerebellar function for a reflex movement of the dogfish. J Comp Physiol 134:69–74

Paul DH, Roberts BL (1981) The activity of cerebellar neurones of an elasmobranch fish (Scyliorhinus canicula) during a reflex movement of a fin. J Physiol (Lond) 321:369–383

Paul DH, Roberts BL (1983) The activity of cerebellar nuclear neurones in relation to stimuli which evoke a pectoral fin reflex in dogfish. J Physiol (Lond) 342:465–481

Paul DH, Roberts BL (1984) Projections of cerebellar Purkinje cells in the dogfish, Scyliorhinus. Neurosci Lett 44:43–46

Perks AM (1969) The neurohypophysis. In: Hoar WS, Randall DJ (eds) Fish physiology, vol II: the endocrine system. Academic, New York, pp 112–205

Plassmann W (1982) Central projections of the octaval system in the thornback ray Platyrhinoidis tiseriata. Neurosci Lett 32:229–233

Poloumordwinoff D (1898) Recherches sur les terminaisons nerveuses sensitives dans les muscles striés volontaires. Trav Soc Sci Arcachon 3:73–79

Puzdrowski RL, Leonard RB (1992) Variations in cerebellar morphology of the Atlantic stingray, Dasyatis sabina. Neurosci Lett 135:196–200

Puzdrowski RL, Leonard RB (1993) The octavolateral systems in the stingray, Dasyatis sabina. I. Primary projections of the octaval and lateral line nerves. J Comp Neurol 332:21–37

Puzdrowski RL, Leonard RB (1994) Vestibulo-oculomotor connections in an elasmobranch fish, the Atlantic stingray, Dasyatis sabina. J Comp Neurol 339:587–597

Ramón-Moliner E (1967) La différentiation morphologique des neurones. Arch Ital Biol 105:149–188

Ramón-Moliner E (1969) The leptodendritic neuron: its distribution and significance. Ann NY Acad Sci 167:65–70

Ramón-Moliner E, Nauta WJH (1966) The isodendritic core of the brain stem. J Comp Neurol 126:311–335

Raschi W (1986) A morphological analysis of the ampullae of Lorenzini in selected skates (Pisces, Rajoidei). J Morphol 189:225–247

Reiner A, Carraway RE (1987) Immunohistochemical and biochemical studies on Lys[8]-Asn[9]-neurotensin[8–13] (LANT6)-related peptides in the basal ganglia of pigeons, turtles, and hamsters. J Comp Neurol 257:453–476

Repérant J, Miceli D, Rio JP, Peyrichoux J, Pierre J, Kirpitchnikova E (1986) The anatomical organization of retinal projections in the shark Scyliorhinus caniculata with special reference to the evolution of the selachian primary visual system. Brain Res Rev 11:227–248

Retzius G (1881) Das Gehörorgan der Wirbeltiere. Samson and Wallin, Stockholm

Ripps H, Dowling JE (1990) Structural features and adaptive properties of photoreceptors in the skate retina. J Exp Zool [Suppl] 5:46–54

Ritchie TC, Leonard RB (1982) Immunocytochemical demonstration of serotonergic cells, terminals and axons in the spinal cord of the stingray, Dasyatis sabina. Brain Res 240:334–337

Ritchie TC, Leonard RB (1983a) Immunohistochemical studies on the distribution and origin of candidate peptidergic primary afferent neurotransmitters in the spinal cord of an elasmobranch fish, the atlantic stingray (Dasyatis sabina). J Comp Neurol 213:414–425

Ritchie TC, Leonard RB (1983b) Immunocytochemical demonstration of serotonergic neurons and processes in the retina and optic nerve of the stingray, Dasyatis sabina. Brain Res 267:352–356

Ritchie TC, Livingston CA, Hughes MG, McAdoo DJ, Leonard RB (1983) The distribution of serotonin in the CNS of an elasmobranch fish: Immunocytochemical and biochemical studies in the atlantic stingray, Dasyatis sabina. J Comp Neurol 213:414–425

Ritchie TC, Roos LJ, Wiliams BJ, Leonard RB (1984) The descending and intrinsic serotoninergic innervation of an elasmobranch spinal cord. J Comp Neurol 224:395–406

Roberts BL (1969) The spinal nerves of the dogfish, Scyliorhinus. J Mar Biol Ass UK 49:51–75

Roberts BL (1978) Mechanoreceptors and the behaviour of elasmobranch fishes with special reference to the acoustico-lateralis system. In: Hodgson ES, Mathewson RF (eds) Sensory biology of sharks, skates and rays. US Government Printing Office, Washington DC, pp 331–390

Roberts BL (1981) Central processing of acousticolateralis signals in elasmobranchs. In: Tavolga WN, Popper AN, Fay RR (eds) Hearing and sound communication in fishes. Springer, Berlin Heidelberg New York, pp 357–373

Roberts BL, Meredith GM (1987) immunohistochemical study of a dopaminergic system in the spinal cord of the ray, Raja radiata. Brain Res 437:171–175

Roberts BL, Ryan KP (1975) Cytological features of the giant neurons controlling electric discharge in the ray Torpedo. J Mar Biol Ass UK 55:123–131

Roberts BL, Timerick SJ, Paul DH (1991) Circuits for vestibular control of pectoral fin muscles in dogfish. In: Bush B, Armstrong DM (eds) Locomotor neural mechanisms in arthropods and vertebrates. Manchester University Press, Manchester, pp 285–291

Roberts BL, Williamson RM (1983) Motor pattern formation in the dogfish spinal cord. In: Roberts A, Roberts B (eds) Neural origin of rhythmic movements. Society of Experimental Biology, Great Britain, pp 331–350

Roberts BL, Witkovsky P (1975) A functional analysis of the mesencephalic nucleus of the fifth nerve in the selachian brain. Proc R Soc Lond B 190:473–495

Rodríguez EM, Oksche A, Hein S, Yulis CR (1992) Cell biology of the subcommissural organ. Int Rev Cytol 135:39–121

Rodríguez-Moldes MI, Anadón R (1988) Ultrastructural study of the evolution of globules in coronet cells of the saccus vasculosus of an elasmobranch (Scyliorhinus canicula L), with some observations on cerebrospinal fluid-contacting neurons. Acta Zool (Stockh) 69:217–224

Rodríguez-Moldes I, Timmermans JP, Adriaensen D, De Groodt-Lasseel MHA, Scheuermann DW, Anadón R (1990) Immunohistochemical localization of calbindin-D_{28K} in the brain of a cartilaginous fish, the dogfish (Scyliorhinus canicula L). Acta Anat 137:293–302

Rodríguez-Moldes I, Manso MJ, Becerra M, Molist P, Anadón R (1993) Distribution of substance P-like immunoreactivity in the brain of the elasmobranch Scyliorhinus canicula. J Comp Neurol 335:228–244

Rohon JV (1884) Zur Histiogenese des Rückenmarkes der Forelle. Sitz Ber Math Phys Kl Konigl Bayr Akad Wiss 14:39–56

Rosiles JR, Leonard RB (1980) The organization of the extraocular motor nuclei in the atlantic stingray, Dasyatis sabina. J Comp Neurol 193:677–687

Rüdeberg C (1969) Light and electron microscopic studies on the pineal organ of the dogfish, Scyliorhinus canicula (L.). Z Zellforsch 96:548–581

Rüdeberg S-I (1961) Morphogenetic studies on the cerebellar nuclei and their homologization in different vertebrates including man. PhD thesis, Lund University

Sathyanesan AG (1965) The hypophysis and hypothalamo-hypophyseal system in the chimaeroid fish Hydrolagus collei (Lay and Bennett) with a note on their vascularization. J Morphol 166:413–449

Sato Y, Takatsuji K, Masai H (1983) Brain organization of sharks, with special reference to archaic species. J Hirnforsch 24:289–295

Schaeffer B, Williams M (1977) Relationships of fossil and living elasmobranchs. Am Zool 17:293–302

Schaper A (1898) The fine structure of the selachian cerebellum (Mustelus vulgaris) as shown by chrome-silver impregnation. J Comp Neurol 8:1–120

Scharrer E (1952) Das Hypophysen-Zwischenhirnsystem von Scyllium stellare. Z Zellforsch 37:196–204

Schmidt AW, Bodznick D (1987) Afferent and efferent connections of the vestibulolateral cerebellum of the little skate, Raja erinacea. Brain Behav Evol 30:282–302

Schroeder DM, Ebbesson SOE (1974) Nonolfactory telencephalic afferents in the nurse shark (Ginglymostoma cirratum). Brain Behav Evol 9:121–155

Schweitzer J (1983) The physiological and anatomical localization of two electroreceptive diencephalic nuclei in the thornbak ray, Platyrhinoidis triseriata. J Comp Physiol A 153:331–341

Schweitzer J (1986) Functional organization of the electroreceptive midbrain in an elasmobranch (Platyrhinoidis triseriata). J Comp Physiol [A] 158:43–58

Sivak JG (1990) Elasmobranch visual optics. J Exp Zool [Suppl] 5:13–21

Smeets WJAJ (1981a) Retinofugal pathways in two chondrichthyans, the shark Scyliorhinus canicula and the ray Raja clavata. J Comp Neurol 195:1–11

Smeets WJAJ (1981b) Efferent tectal pathways in two chondrichthyans, the shark Scyliorhinus canicula and the ray Raja clavata. J Comp Neurol 195:13–23

Smeets WJAJ (1982) The afferent connections of the tectum mesencephali in two chondrichthyans, the shark Scyliorhinus canicula and the ray Raja clavata. J Comp Neurol 205:139–152

Smeets WJAJ 1983 The secondary olfactory connections in two chondrichthyans, the shark Scyliorhinus canicula and the ray Raja clavata. J Comp Neurol 218:334–344

Smeets WJAJ (1990) The telencephalon of cartilaginous fishes. In: Jones EG, Peters A (eds) Cerebral cortex, vol 8A: comparative structure and evolution of cerebral cortex, part I. Plenum, New York, pp 3–30

Smeets WJAJ (1992) Comparative aspects of basal forebrain organization in vertebrates. Eur J Morphol 30:23–36

Smeets WJAJ (1994) Catecholamine systems in the CNS of reptiles: structure and functional correlations. In: Smeets WJAJ, Reiner A (eds) Phylogeny and development of catecholamine systems in the CNS of vertebrates. Cambridge University Press, Cambridge, pp 103–133

Smeets WJAJ, Boord RL (1985) Connections of the lobus inferior hypothalami of the clearnose skate, Raja eglanteria. J Comp Neurol 234:380–392

Smeets WJAJ, González A (1990) Are putative dopamine-accumulating cell bodies in the hypothalamic periventricular organ a primitive brain character of nonmammalian vertebrates? Neurosci Lett 114:248–252

Smeets WJAJ, Nieuwenhuys R (1976) Topological analysis of the brain stem of the sharks Squalus acanthias and Scyliorhinus canicula. J Comp Neurol 165:333–368

Smeets WJAJ, Northcutt RG (1987) At least one thalamotelencephalic pathway in cartilaginous fishes projects to the medial pallium. Neurosci Lett 78:277–282

Smeets WJAJ, Reiner A (1994) Catecholamines in the CNS of vertebrates: current concepts of evolution and functional significance. In: Smeets WJAJ, Reiner A (eds) Phylogeny and development of catecholamine systems in the CNS of vertebrates. Cambridge University Press, Cambridge, pp 463–481

Smeets WJAJ, Timerick SJB (1981) Cells of origin of pathways descending to the spinal cord in two chondrichthyans, the shark Scyliorhinus canicula and the ray Raja clavata. J Comp Neurol 202:473–491

Smeets WJAJ, Nieuwenhuys R, Roberts BL (1983) The central nervous system of cartilaginous fishes. Structure and functional correlations. Springer, Berlin Heidelberg New York

Smeets WJAJ, Kidjan M, Jonker AJ (1991) α-MPT does not affect dopamine levels in the periventricular organ of lizards. Neuroreport 2:369–372

Snow PJ, Plenderleith MB, Wright LL (1993) Quantitative study of primary sensory neurone populations of three species of elasmobranch fish. J Comp Neurol 334:97–103

Speidel CC (1922) Further comparative studies in other fishes of cells that are homologous to the large irregular glandular cells in the spinal cord of the skates. J Comp Neurol 34:303–317

Sperry DG, Boord RL (1992) Central location of the motoneurons that supply the cucullaris (trapezius) of the clearnose skate, Raja eglanteria. Brain Res 582:312–319

Spivack WD, Zhong N, Salerno S, Saavedra RA, Gould RM (1993) Molecular cloning of the myelin basic proteins in the shark, Squalus acanthias, and the ray, Raja erinacea. J Neurosci Res 35:577–584

Stell WK (1984) Luteinizing hormone-releasing hormone (LHRH)- and pancreatic polypeptide (PP)-immunoreactive neurons in the terminal nerve of spiny dogfish, Squalus acanthias. Anat Rec 208:173A–174A

Stell WK, Walker SE, Chohan KS, Ball AK (1984) The goldfish nervus terminalis: an LHRH- and FMRF-amide-immunoreactive olfactory pathway. Proc Natl Acad Sci USA 81:940–944

Sterzi G (1912) Il sistema nervosos centrale dei Vertebrati. Pesci, vol II. Draghi, Padova

Stieda L (1873) Über den Bau des Rückenmarkes der Rochen und der Haie. Z Wiss Zool 23:435–442

Stuesse SL, Cruce WLR (1991) Immunohistochemical localization of serotoninergic, enkephalinergic, and catecholaminergic cells in the brainstem of a cartilaginous fish, Hydrolagus collei. J Comp Neurol 309:535–548

Stuesse SL, Cruce WLR (1992) Distribution of tyrosine hydroxylase, serotonin, and leu-enkephalin immunoreactive cells in the brainstem of a shark, Squalus acanthias. Brain Behav Evol 39:77–92

Stuesse SL, Cruce WLR, Northcutt RG (1990) Distribution of tyrosine hydroxylase- and serotonin-immunoreactive cells in the central nervous system of the thornback guitarfish, Platyrhinoidis triseriata. J Chem Neuroanat 3:45–58

Stuesse SL, Cruce WLR, Northcutt RG (1991a) Localization of serotonin, tyrosine hydroxylase, and leu-enkephalin immunoreactive cells in the brainstem of the horn shark, Heterodontus francisci. J Comp Neurol 308:277–292

Stuesse SL, Cruce WLR, Northcutt RG (1991b) Serotoninergic and enkephalinergic cell groups in the reticular formation of the bat ray and two skates. Brain Behav Evol 38:39–52

Stuesse SL, Stuesse DC, Cruce WLR (1992) Immunohistochemical localization of serotonin, leu-enkephalin, tyrosine hydroxylase, and substance P within the visceral sensory area of cartilaginous fish. Cell Tissue Res 268:305–316

Stuesse SL, Cruce WLR, Northcutt RG (1994) Localization of catecholamines in the brains of Chondrichthyes (cartilaginous fishes). In: Smeets WJAJ, Reiner A (eds) Phylogeny and development of catecholamine systems in the CNS of vertebrates. Cambridge University Press, Cambridge, pp 21–47

Sumpter JP, Denning-Kendall PA, Lowry PJ (1984) The involvement of melanotrophins in physiological colour change in the dogfish Scyliorhinus canicula. Gen Comp Endocrinol 56:360–367

Szabo T (1955) Quelques precisions sur le noyau de commande centrale de la décharge electrique chez la Raie (Raja clavata). J Physiol (Paris) 47:283–285

Tanaka S (1988) A macroscopical study of the trapezius muscle of sharks with reference to the topographically related nerves and vein. Anat Anz 165:1–21

Takeuchi IK, Takeuchi YK (1986) Congenital hydrocephalus following X-irradiation of pregnant rats on an early gestational day. Neurobehav Toxicol Teratol 8:143–150

Tester AL (1963) Olfaction, gustation, and the common chemical sense in sharks. In: Gilbert PW (ed) Sharks and survival. Heath, Lexington, pp 255–282

Timerick SJB, Paul DH, Roberts BL (1990) Dynamic characteristics of vestibular-driven compensatory fin movements of the dogfish. Brain Res 516:318–321

Timerick SJB, Roberts BL, Paul DH (1992) Brainstem neurons projecting to different levels of the spinal cord of the dogfish Scyliorhinus canicula. Brain Behav Evol 39:93–100

Tong SL, Bullock TH (1982) The sensory functions of the cerebellum of the thornback ray, Platyrhinoidis triseriata. J Comp Physiol [A] 148:399–410

Vallarino M, Ottonello I (1987) Neuronal localization of immunoreactive adrenocorticotropin-like substance in the hypothalamus of elasmobranch fishes. Neurosci Lett 80:1–6

Vallarino M, Danger JM, Fasolo A, Pelletier G, Saint-Pierre S, Vaudry H (1988a) Distribution and characterization of neuropeptide Y in the brain of an elasmobranch fish. Brain Res 448:67–76

Vallarino M, Ottonello I, D'Este L, Renda T (1988b) Sauvagine/urotensin I-like immunoreactivity in the brain of the dogfish, Scyliorhinus canicula. Neurosci Lett 95:119–124

Vallarino M, Fasolo A, Ottonello I, Perroteau I, Tonon MC, Vandesande F, Vaudry H (1989) Localization of corticotropin-releasing hormone (CRF)-like immunoreactivity in the central nervous system of the elasmobranch fish, Scyliorhinus canicula. Cell Tissue Res 258:541–546

Vallarino M, D'Este L, Negri L, Ottonello I, Renda T (1990a) Occurrence of bombesin-like immunoreactivity in the brain of the cartilaginous fish, Scyliorhinus canicula. Cell Tissue Res 259:177–181

Vallarino M, Viglietti-Panzica C, Panzica GC (1990b) Immunocytochemical localization of vasotocin-like immunoreactivity in the brain of the cartilaginous fish, Scyliorhinus canicula. Cell Tissue Res 262:507–513

Vallarino M, Bucharles C, Facchinetti F, Vaudry H (1994) Immunocytochemical evidence for the presence of met-enkephalin and leu-enkephalin in distinct neurons in the brain of the elasmobranch fish Scyliorhinus canicula. J Comp Neurol 347:585–597

van de Kamer JC, Wilschut IJC, Heussen AMA (1973) On the presence and localization of glycogen accumulations inside coronet cells of the saccus vasculosus of the dogfish (Scyliorhinus caniculus). Z Zellforsch 140:277–290

Vesselkin NP, Kovacevic N (1973) Nonolfactory telencephalic afferent projections in elasmobranch fishes. Zh Evol Ion Biokhim Fiziol (Tome) 9:585–592

Vigh-Teichmann I, Vigh B, Manzano e Silva MJ, Aros B (1983) The pineal organ of Raia clavata: opsin immunoreactivity and ultrastructure. Cell Tiss Res 228:139–148

Von Arx WS (1962) An introduction to physical oceanography. Addison-Wesley, Reading-London

Von Harrach M (1970) Elektronenmikroskopische Beobachtungen am Saccus vasculosus einiger Knorpelfische. Z Zellforsch 105:188–209

Von Kupffer C (1906) Die Morphogenie der Centralnervensystems. In: Hertwig O (ed) Handbuch der vergleichenden und experimentellen Entwickelungslehre der Wirbeltiere, vol II, part 3. Fischer, Jena, pp 1–272

Voorhoeve JJ (1917) Over den Bouw van de kleine Hersenen der Plagiostomen. Thesis, Amsterdam University

Waehneldt TV (1990) Phylogeny of myelin proteins. Ann NY Acad Sci 605:15–28

Wallén P (1982) Spinal mechanisms controlling locomotion in dogfish and lamprey. Acta Physiol Scand [Suppl] 503:3–45

Wallenberg A (1907) Beiträge zur Kenntnis des Gehirns der Teleostier und Selachier. Anat Anz 31:369–399

Watanabe A (1966) The saccus vasculosus of the ray (Dasyatis akajei). Arch Hist Japon 27:427–449

Weinberg E (1928) The mesencephalic root of the fifth nerve. A comparative study. J Comp Neurol 46:249–405

Williams BJ, Livingston CA, Leonard RB (1984) Spinal cord pathways involved in initiation of swimming in the stingray, Dasyatis sabina: spinal cord stimulation and lesions. J Neurophysiol 51:578–591

Williams HT (1973) The telencephalon of the newborn dogfish shark, Squalus acanthias. J Hirnforsch 14:261–285

Willis WD, Coggeshall RE (1991) Sensory mechanisms of the spinal cord. Plenum, New York

Wilson JF, Dodd JM (1973a) The role of the pineal complex and lateral eyes in the colour change response of the dogfish, Scyliorhinus canicula L. J Endocrinol 58:591–598

Wilson JF, Dodd JM (1973b) Effects of pharmacological agents on the in vivo release of melanophore-stimulating hormone in the dogfish, Scyliorhinus canicula. Gen Comp Endocrinol 20:556–566

Wilson JF, Goos HJT, Dodd JM (1974) An investigation of the neural mechanisms controlling the colour change responses of the dogfish Scyliorhinus canicula L by mesencephalic and diencephalic lesions. Proc R Soc Lond B 187:171–190

Withington-Wray DJ, Roberts BL, Taylor EW (1986) The topographical organization of the vagal motor column in the elasmobranch fish, Scyliorhinus canicula L. J Comp Neurol 248:95–104

Witkovsky P, Roberts BL (1975) The light microscopical structure of the mesencephalic nucleus of the fifth nerve in the selachian brain. Proc R Soc Lond B 190:457–471

Witkovsky P, Powell CC, Brunken WJ (1980) Some aspects of the organization of the optic tectum of the skate Raja. Neuroscience 5:1989–2002

Wourms JP (1977) Reproduction and development in chondrichthyan fishes. Am Zool 17:379–410

Wright DE, Demski LS (1991) Gonadotropin hormone-releasing hormone (GnRH) immunoreactivity in the mesencephalon of sharks and rays. J Comp Neurol 307:49–56

Wunderer H (1908) Über Terminal Körperchen der Anamnien. Arch Mikr Anat 71:504–569

Yamanaka S, Honma Y, Ueda S, Sano Y (1990) Immunohistochemical demonstration of serotonin neuron system in the central nervous system of the Japanese dogfish, Scyliorhinus torazame (Chondrichthyes). J Hirnforsch 31:385–397

Young JZ (1933a) The autonomic nervous system of Selachians. Q J Microsc Sc 75:571–624

Young JZ (1933b) Comparative studies on the physiology of the iris. I. Selachians. Proc R Soc Lond B 112:228–241

Young JZ (1980) Nervous control of stomach movements in dogfishes and rays. J Mar Biol Assoc UK 60:1–17

Ziegels J (1976) The vertebrate subcommissural organ: a structural and functional review. Arch Biol 87:429–476

Ziegels J (1979) The subcommissural organ of submammalian vertebrates: a histochemical study. J Hirnforsch 20:11–18

CHAPTER 13

Brachiopterygian Fishes

R. NIEUWENHUYS

13.1
Introduction

The bony fishes, or Osteichthys, constitute by far the largest class of extant vertebrates, encompassing more than 25 000 species. The great majority of these species belong in the large subclass Actinopterygii, or ray-finned fishes. However, the Osteichthyes encompass three other subclasses: the Brachiopterygii, or arm-finned fishes (also designated as Cladistia); the Dipnoi, or lungfishes; and the Crossopterygii, or tassel-finned fishes. Although each of these three subclasses includes only a limited number of species, these are of great zoological and evolutionary interest. In this chapter, we focus on the Brachiopterygii, which contain the single family Polypteridae, with two genera: *Polypterus* and *Erpetoichthys* (previously indicated as *Calamoichthys*). The genus *Polypterus*, the bichirs, comprises ten species, all with an elongated body and a similar appearance. The other genus consists of a single species, the reedfish *Erpetoichthys calabaricus*. The body of the reedfish is eel-shaped (Fig. 13.1), but it is otherwise similar to the bichirs morphologically and ecologically.

The body of a brachiopterygian has a covering of very hard rhomboid scales. As the family name, Polypteridae, implies, the dorsal fin comprises a row of five to 18 separate finlets, each with a spine supporting a small, membranous web (Fig. 13.1). The paired fins consist of a proximal fleshy lobe and a fan-shaped distal array of fin rays. The pectoral fins are used to propel the body during slow swimming and to support the forebody when the fish rests on the bottom. Thus Poll (1965) characterized the brachiopterygians as 'bipodal fish'.

Brachiopterygians are able to breathe air and have a pair of well-vascularised lungs which arise from a diverticulum in the floor of the pharynx. The lungs enable the fish to live in muddy, oxygen-poor waters and are also a survival aid during seasonal droughts.

The brachiopterygians inhabit the fresh waters of tropical Africa. They are nocturnal, bottom-dwelling predators that feed primarily on other fishes. Their eyes are moderately developed, but their olfactory organs are unusually large and consist of a pair of longitudinally folded sacs whose external openings are located at the end of peculiar olfactory tentacles (Figs. 13.1, 13.2). Taste buds are found only in the mouth cavity. In addition to the usual hydrodynamic pressoreceptor organs, the lateral line system has a number of ampullary electroreceptors (Roth 1973). These receptors, known

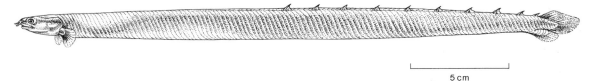

Fig. 13.1. The reedfish *Erpetoichthys calabaricus*

as the organs of Fahrenholz, are confined to the head and concentrated on the tip of the snout. Roth and Tscharntke (1976) considered it likely that the mouth is surrounded by an electric field, generated by current flow between it and the gill slits, as in a number of other fishes. This electrical field would be distorted each time the fish comes to the surface to breathe, and a possible function of the electroreceptors could be to sense this distortion and thus to indicate the proximity of the surface. The electroreceptive system in brachiopterygians might also aid in prey detection, as has been demonstrated in sharks (Kalmijn 1978).

The literature on the central nervous system in the brachiopterygians is rather scanty. Waldschmidt (1887) briefly described the gross morphology of the brain. The structure of the brain stem was analysed by Kenemans (1980) and by Nieuwenhuys and Oey (1983). Observations on the cerebellum were made by van der Horst (1919, 1925) and Nieuwenhuys (1967b). The diencephalon was described by Jeener (1930), Bergquist (1932), Nieuwenhuys and Bodenheimer (1966) and Braford and Northcutt (1983). Holmgren (1922), Nieuwenhuys (1962, 1963, 1969) and Northcutt and Braford (1980) studied the telencephalon, and Senn (1976a,b) presented a brief overview of the structure of the brain as a whole. A group of Italian investigators (Mazzi et al. 1977, 1978; Fasolo et al. 1978) analysed the structure of the tectum mesencephali, the preoptic area and the hypothalamus using the Golgi technique. Experimental neuroanatomical studies are few. So far, the following fibre connections have been analysed with axon degeneration or tracer techniques: retinofugal pathways (Repérant et al. 1979, 1981; Braford and Northcutt 1983), telencephalic afferents and efferents (Northcutt 1981; Braford and Northcutt 1978; Holmes and Northcutt 1995), secondary olfactory connections (Braford and Northcutt 1974; von Bartheld and Meyer 1986b) and the central projections of the terminal nerve (von Bartheld and Meyer 1986a; von Bartheld et al. 1988). Reiner and Northcutt (1992) studied the distribution of some neuroactive substances in the prosencephalon of *Polypterus senegalus* using immunohistochemical techniques. The literature on the ontogenesis of the

brachiopterygian brain is confined to the study by Nieuwenhuys et al. (1969) on telencephalic development.

The following survey is based on a previous review (Nieuwenhuys 1983), on the above references and on an extensive collection of histological material from the reedfish *Erpetoichthys calabaricus*. A number of serially sectioned larvae of *Polypterus senegalus* were also available for developmental study. To our knowledge, no physiological studies on the central nervous system of any brachiopterygian have been reported thus far.

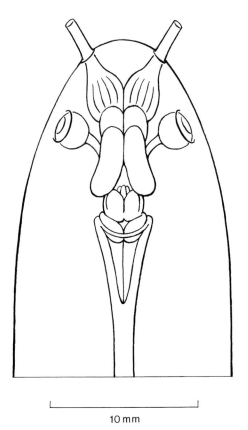

Fig. 13.2. Brain, eyes and olfactory organs of the reedfish *Erpetoichthys calabaricus* (juvenile specimen) in position

13.2
Gross Morphology

In its overall appearance, the slender, elongated brain in brachiopterygians somewhat resembles the brain in dipnoans (Figs. 13.2, 13.3). Platel et al. (1977) and Northcutt and Braford (1980) reported that the relative brain weight of *Polypterus* is higher than that of the primitive actinopterygians such as *Acipenser*, *Scaphirhynchus* and *Amia*, but not as high as that in many teleosts.

The *rhombencephalon* is well developed and includes a bilateral horizontal basal plate and an alar plate that is vertically oriented throughout most of its extent. The ventricular surface of the alar plate has two horizontally oriented, elongated bulges, the lobus vagi and the lobus lineae lateralis (Fig. 13.3d). Rostrally, the alar plates curve around lateral recesses of the fourth ventricle, thus constituting the homologues of the auriculae in many other groups of fishes. Dorsomedially, these auriculae continue into a transversely oriented band of tissue which constitutes the caudalmost part of the cerebellum. Notwithstanding its superficial position, this band represents the so-called lower lip of the cerebellum in cartilaginous fishes. Caudally, the rhombencephalon grades into the cylindrically shaped spinal cord. The rhombencephalic ventricular cavity can be divided into a rather narrow rostral part and a much wider caudal part, which is covered dorsally by a slightly folded tela choroidea (Fig. 13.3a,d).

The *cerebellum* invaginates into the ventricular cavity. Due to this unusual configuration, only its lateral and caudal parts are externally visible. Its central part protrudes into the fourth ventricle, and its rostral part thrusts forward under the tectum mesencephali. The brachiopterygian cerebellum, as a whole, encloses an extension of the extraencephalic space, thus displaying a remarkable mirror image of the cerebellum of cartilaginous fishes (Fig. 13.3d).

In most vertebrates, the two halves of the cerebellum are more or less extensively fused in the median plane, but the cerebellum in brachiopterygians is a paired structure, the two halves of which are connected only by a thin, curved lamella (Fig. 13.3d).

The medium-sized *mesencephalon* comprises a ventral tegmentum and a dorsal tectum. The medial part of the tegmentum is continuous caudally with the rhombencephalic basal plate. The lateral part of the tegmentum has two protrusions, one ventricular and one external. The ventricular protrusion is the torus semicircularis, which is rather small in brachiopterygians, and the external protrusion is

the strongly developed torus lateralis (Fig. 13.3a–c). The tectum is evaginated into bilateral halves. Dorsocaudally, the tectum grades into the cerebellum via a short velum medullare anterius. Caudally, the midbrain is demarcated from the rhombencephalon by the fissura rhombomesencephalica. The dorsal part of this groove is situated between the tectum and the cerebellum; its ventral part traverses the caudal margin of the torus lateralis (Fig. 13.3b). The frontal pole of the tectum extends to a level rostral to the commissura posterior, which is clearly visible in the median plane (Fig. 13.3d). The rostral boundary of the midbrain can be defined as a plane passing from the commissura posterior to a deep, ventral fold, the plica encephali ventralis (Fig. 13.3d).

As in most anamniotes, the *diencephalon* is wedge-shaped, with the apex directed dorsally. The short dorsal border is constituted by the ganglia habenulae, which are connected dorsocaudally by the habenular commissure (Fig. 13.3d). The habenular ganglia display a marked asymmetry, as the right ganglion is much larger than the left (Fig. 13.3a). Dorsal and rostral to the habenular ganglia, the diencephalon is covered by a membranous structure, the tela diencephali, which passes rostrally into a deep, transverse fold called the velum transversum. The velum is attached to the lateral brain wall in front of the habenulae, and this point defines the dorsal border of the diencephalic-telencephalic boundary. Caudal to the habenular commissure, the diencephalic roof becomes membranous again to form a finger-shaped evagination, the epiphysis. The thickened lateral walls of the diencephalon bound the slit-like third ventricle; ventrocaudally, this ventricle expands laterally within the hypothalamus. The hypothalamus extends caudally beneath the rostral part of the brain stem. Its thin-walled caudal part, the infundibulum, continues into the saccus vasculosus. The pituitary lies against the ventral surface of the infundibulum. Remarkably, a small but distinct dorsal evagination of the pharyngeal wall, i.e. a persistent pouch of Rathke, penetrates the base of the skull and indents the pituitary tissue (Figs. 13.3d; see also Fig. 13.25). The hypothalamic ventricle is largely separated from the main part of the diencephalic ventricular cavity by a large mediobasal protrusion, the tuberculum posterius. The diencephalic-telencephalic boundary can be considered to pass from the lateral attachment of the velum transversum through the chiasma opticum, which constitutes a high, transverse ridge in the floor of the diencephalon.

The *telencephalon* is relatively large and elongated. Its rostral part consists of two well-developed, hollow olfactory bulbs. The ventral parts

Fig. 13.3a–d. The brain of the reedfish *Erpetoichthys calabaricus*; the telae chorioideae are removed on the left side. **a** Dorsal view. **b** Lateral view. **c** Ventral view. **d** Medial view of the bisected brain.

1	nervus olfactorius
2	telencephalon
3	bulbus olfactorius
4	pallium
5	subpallium
6	tractus pallii
7	tela telencephali
8	taenia
9	septum ependymale
10	lamina terminalis
11	commissura anterior
12	recessus preopticus
13	diencephalon
14	ganglion habenulae
15	commissura habenulae
16	epiphysis
17	tela diencephali
18	saccus dorsalis
19	velum transversum
20	thalamus dorsalis
21	thalamus ventralis
22	hypothalamus
23	sulcus hypothalamicus
24	nervus opticus
25	chiasma opticum
26	tractus opticus marginalis
27	tractus opticus medialis
28	tractus opticus lateralis
29	torus lateralis
30	commissura postoptica

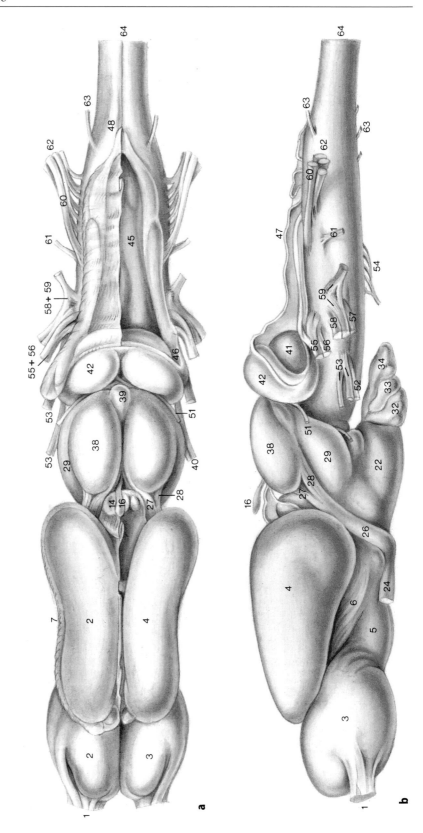

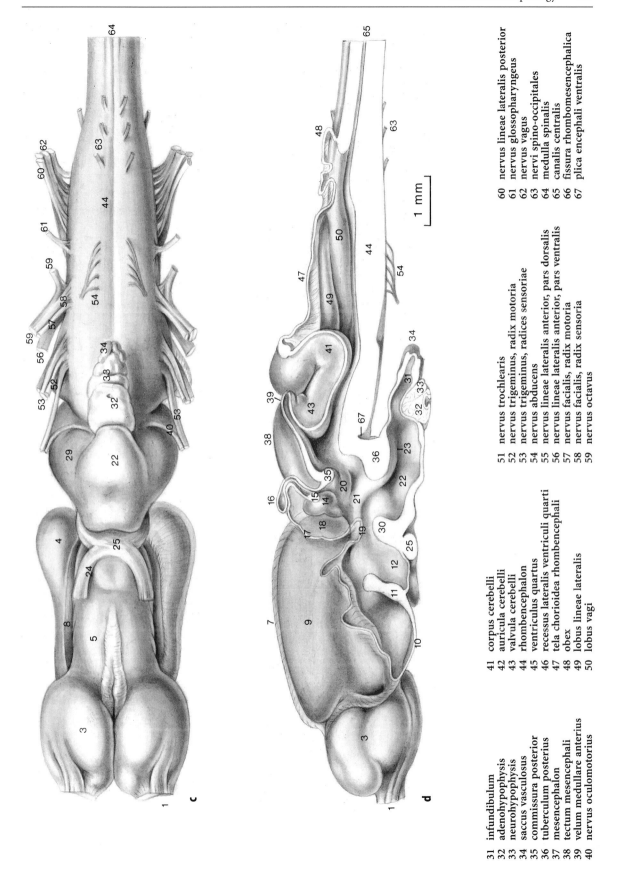

31 infundibulum
32 adenohypophysis
33 neurohypophysis
34 saccus vasculosus
35 commissura posterior
36 tuberculum posterius
37 mesencephalon
38 tectum mesencephali
39 velum medullare anterius
40 nervus oculomotorius

41 corpus cerebelli
42 auricula cerebelli
43 valvula cerebelli
44 rhombencephalon
45 ventriculus quartus
46 recessus lateralis ventriculi quarti
47 tela chorioidea rhombencephali
48 obex
49 lobus lineae lateralis
50 lobus vagi

51 nervus trochlearis
52 nervus trigeminus, radix motoria
53 nervus trigeminus, radices sensoriae
54 nervus abducens
55 nervus lineae lateralis anterior, pars dorsalis
56 nervus lineae lateralis anterior, pars ventralis
57 nervus facialis, radix motoria
58 nervus facialis, radix sensoria
59 nervus octavus

60 nervus lineae lateralis posterior
61 nervus glossopharyngeus
62 nervus vagus
63 nervi spino-occipitales
64 medulla spinalis
65 canalis centralis
66 fissura rhombomesencephalica
67 plica encephali ventralis

of the telencephalon proper form the lateral walls of a median telencephalic ventricle, which passes caudally into the diencephalic third ventricle. However, the dorsal parts of the telencephalic walls are everted, i.e. folded laterally, and these everted parts gradually increase in length rostrocaudally. It is interesting to note that the dorsocaudal parts of the telencephalic walls are recurved and surround part of the extraencephalic space. (This 'pseudo-ventricle' is clearly visible in Fig. 13.18.) Due to the eversion, the telencephalic roof plate becomes a wide, membranous structure which covers the dorsal parts of the telencephalon and attaches to the lateral margins of the walls (see Figs. 13.19, 13.20). This tela telencephali forms a long, median fold, which separates the telencephalic ventricles into two parts (Fig. 13.3d).

Caudoventrally, the two telencephalic hemispheres are connected by the *commissura anterior*. The portion of the telencephalic ventricle that lies between the anterior commissure and the optic chiasm is called the recessus preopticus (Fig. 13.3d).

13.3
Development and Overall Histological Pattern

13.3.1
Development

Analysis of serially sectioned larvae of *Polypterus senegalus* reveals that the development of this brain generally follows a pattern typical of most vertebrates. However, there are a few points which deserve mention. Firstly, in early larval stages the brain stem is extraordinarily curved (Fig. 13.4b,c). During later development, this curvature gradually decreases (Fig. 13.4d), and in the adult it has disappeared entirely (Fig. 13.3d).

A second feature is the unusual morphogenesis of the cerebellum. As in all vertebrates, this brain area develops bilaterally from the rostralmost portions of the rhombencephalic alar plates. Due to the strong curvature of the brain stem, these bilateral cerebellar anlagen initially occupy a primarily subtectal position (Fig. 13.4a–c). At an early stage of development, a small, retrotectal indentation is apparent in that part of the rhombencephalic roof plate which later forms part of the cerebellum (Fig. 13.4b,c). Later in development, the bilateral halves of the cerebellar plate curve inward to fill the space created by the decreasing curvature of the brain stem. The rostral part of the rhombencephalic roof plate is involved in this invagination (Fig. 13.4d), thus forming the peculiar extraencephalic, but intracerebellar space (Fig. 13.3d).

Finally, the remarkable development of the tel-

encephalon deserves some comment (for details, see Nieuwenhuys et al. 1969). In the earliest stage at our disposal, the forebrain is tube-shaped. The side-walls or side-plates of the tube are thickened and contain a multilayered epithelium, from which the nervous tissue of the telencephalon will appear.

Dorsally, the side-plates are connected by a narrow membrane, the lamina supraneuroporica. The ventral connection of the side-plates is somewhat thicker than the lamina supraneuroporica (Fig. 13.4e). The morphogenesis of the telencephalon appears to pass through three subsequent phases: the first parallels the early development seen in most other vertebrates; in the second, a pattern also typical of actinopterygian fishes becomes manifest; the third is unique to brachiopterygians. During the first phase, a slight folding-inward of the lateral walls occurs (Fig. 13.4f); however, during the second phase, this inversion, which marks the beginning of an evagination of the cerebral hemispheres in other vertebrates, disappears completely. Instead, in *Erpetoichthys*, the lateral walls begin to protrude into the ventricular cavity, and their dorsal borders separate gradually, with consequent broadening of the lamina supraneuroporica (Fig. 13.4 g,h). During further development, the appearance of longitudinal grooves in the meningeal surface of the lateral walls (the sulci externi) marks the beginning of a recurvature or eversion of the dorsal parts of these walls (Fig. 13.4i,j). The taeniae, i.e. the lines of attachment of the lamina supraneuroporica, which shifted laterally during an earlier phase, are now displaced ventrally. Consequently, the lamina supraneuroporica gradually transforms into a membrane which envelops the dorsal parts of the telencephalon.

In both brachiopterygians and actinopterygians, eversion is the principal morphogenetic event in the development of the forebrain. However, in actinopterygians, the eversion is accompanied by considerable thickening of the dorsal parts of the tel-

Fig. 13.4a–j. Development of the brain of the Senegal bichir ▶ *Polypterus senegalus*. **a** Transverse section through the mesencephalon, rhombencephalon and cerebellar anlage of a larva of 8.3 mm. **b** Paramedian section through the brain of a larva of 7.5 mm. **c** Sagittal section through the brain of a larva of 9.8 mm. **d** Paramedian section through the brain of a larva of 28 mm. **e** Transverse section through the prosencephalon of a just hatched larva of 3.8 mm. **f** Transverse section through the telencephalon of a larva of 5.7 mm. **g–j** Transverse sections through the telencephalon at the level of the anterior commissure of larvae of 8.3 mm, 10 mm, 17 mm and 24 mm, respectively. *inv*, inversion; *lp*, lateral plate; *lsn*, lamina supraneuroporica; *lt*, lamina terminalis; *olf*, olfactory organ; *P*, pallium; *S*, subpallium; *sext*, sulcus externus. The *arrow* indicates the retrotectal indentation of the roof plate which will be incorporated in the cerebellum

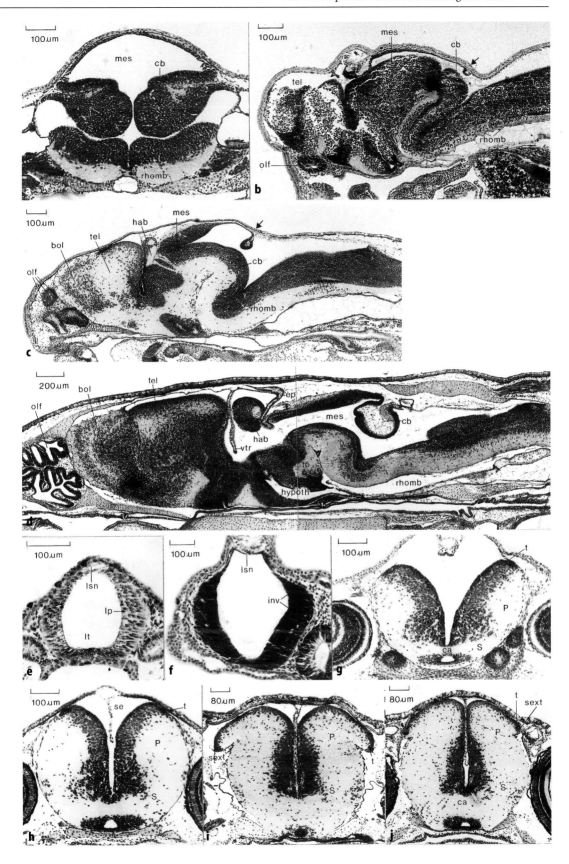

encephalic walls, whereas in brachiopterygians, these walls grow in a ventrolateral direction rather than showing an increase in thickness. Hence, the dorsal parts of the telencephalic lateral plates in adult brachiopterygians are relatively thin, recurved lamellae (see Figs. 13.19, 13.20), which differ morphologically from their solid actinopterygian homologues.

The bulbi olfactorii initially develop as thickenings of the rostral parts of the telencephalic walls and later assume the configuration of separate evaginations with ventricular cavities of their own (Fig. 13.3d; see also Fig. 13.21).

Finally, it should be noted that the considerable lengthening and stretching of the brain during development (see Figs. 13.3d, 13.4b–d) is part of a process involving the head as a whole. In this context, it is remarkable that the olfactory organ, whose early embryonic anlage is situated ventral to the forebrain (Fig. 13.4b), comes to lie entirely rostral to the forebrain in the adult animal (Fig. 13.2).

13.3.2
Overall Histological Pattern

In brachiopterygians, most of the neuronal perikarya retain their embryonic periventricular position; consequently, a deep stratum griseum and a superficial stratum album can be distinguished throughout the entire central nervous system. Differences in size, density and arrangement of the cells make it possible to subdivide the grey zone into separate neuronal populations. Although most of the cell masses form part of the central grey, a number have migrated away from the ventricular surface and are embedded in the stratum album or occupy a superficial position. Prominent among these are the large nucleus tori lateralis in the midbrain (see Figs. 13.14, 13.15) and the stratum cellulare externum in the olfactory bulb (see Fig. 13.21).

Golgi preparations show that the cells situated in the central grey have long dendrites that arborise in the peripherally located stratum album (Fig. 13.5), which comprises fibres of varying diameter. The rhombencephalic basal plate and the spinal cord contain numerous heavily myelinated, coarse fibres, but in the remaining parts of the central nervous system thinner fibres prevail. At more rostral levels of the brain, a number of fibre systems consist mainly or exclusively of unmyelinated axons.

As only a few fibre pathways in brachiopterygian brains have been analysed experimentally, our hodological knowledge of this group is very limited. It should be kept in mind that the descriptions of fibre systems in the following sections derive from analysis of non-experimental material unless stated otherwise.

The somata of the ependymal gliocytes lining the ventricle vary in shape from flattened to cylindrical. Their peripherally directed processes are more or less rough, mossy or thorny. Some of these processes traverse the entire thickness of the brain wall, terminating with conical end-feet at the meningeal surface (Fig. 13.5d). We have not seen elements that could be identified as free neuroglial cells. Throughout most of the brain stem, a narrow, cell-free zone separates the ependymal somata from the neuronal grey.

13.4
Spinal Cord

There are no reports on the brachiopterygian spinal cord in the literature. Our own *Erpetoichthys* material (Figs. 13.6, 13.7) reveals that in the high cervical cord the grey matter has a characteristic butterfly shape (Fig. 13.7), but at lower levels the ventral and dorsal horns are represented by isolated islands of grey without a zona intermedia (Fig. 13.6).

The lateral parts of the ventral horn contain large neurons, including motoneurons, as well as funicular cells. Both of these cell types extend their dendrites into the white matter, mainly into the lateral funiculi. The main dendritic trunks of many of these large elements are convex, parallel to the border of the grey matter. A ventromedially directed process of the motoneurons characteristically arches around the myelin sheath of the ipsilateral Mauthner fibre (Fig. 13.5a).

The white matter contains scattered neurons, one of which – a large funicular or commissural element – is shown in Fig. 13.5a. Our Golgi material was not adequate to determine whether the long, laterally extending dendrites of the spinal neurons constitute a marginal dendritic plexus.

The medial parts of the ventral and dorsal horns consist mainly of small neurons. In the dorsal horn, occasional large, round and ellipsoid elements may well represent persisting Rohon-Beard cells. The central canal is surrounded by cylindrical ependymal cells.

Fig. 13.5a–e. Cells observed in transversely sectioned Golgi preparations of the central nervous system of the reedfish *Erpetoichthys calabaricus.* **a** Section through the cervical spinal cord. **b** Section through the caudalmost part of the rhombencephalon. **c** Section through the caudal part of the rhombencephalon. **d** Section through the ventral part of the hypothalamus. **e** Section through the medial part of the pallium. *ax*, axon; *cmsp*, neuron of columna motoria spinalis; *ep*, ependymal gliocytes; *fcn*, funicular or commissural neurons; *int*, neurons of nucleus intermedius; *lcn*, liquor-contacting neuron; *mn*, motoneuron; *P1, P2*, neurons in pallial fields P1 and P2, respectively; *ri*, neuron of nucleus reticularis inferior; *Xm*, neuron of nucleus motorius nervi vagi. (From Nieuwenhuys 1983)

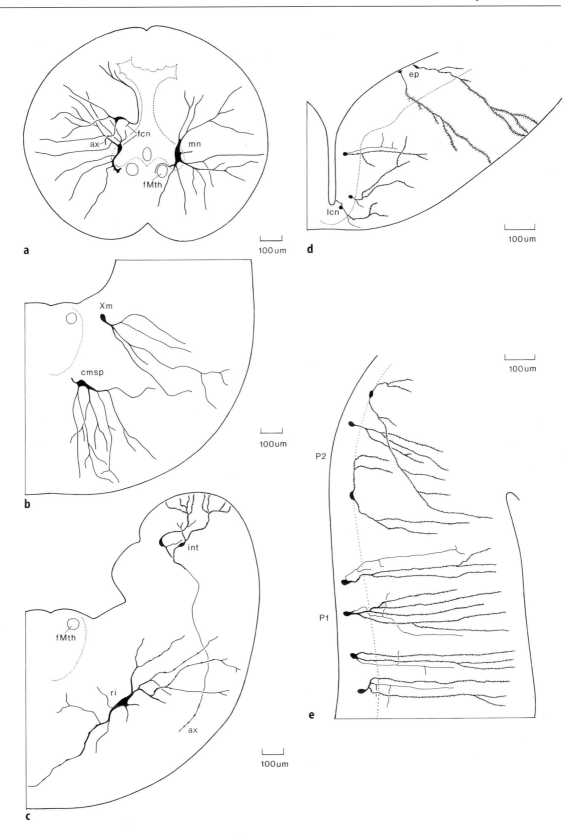

a

b

c

d

e

Abbreviations

a,b,c	subzones of sgp
ath	nucleus anterior thalami
aur	auricula cerebelli
bol	bulbus olfactorius
ca	commissura anterior
cans	commissura ansulata
cb	cerebellum
cbll	cerebellar 'lower lip'
cc	canalis centralis
chab	commissura habenulae
chopt	chiasma opticum
cmsp	columna motoria spinalis
cnd	cornu dorsale
cnv	cornu ventrale
cocb	corpus cerebelli
cP	cells of Purkinje
cp	commissura posterior
cpal	commissura pallii
cpth	nucleus centralis posterior thalami
crcb	crista cerebellaris
ctm	commissura tecti mesencephali
ctp	commissura tuberculi posterioris
dfmt	decussation of fasciculus medialis telencephali
dienc	diencephalon
dors	nucleus dorsalis areae octavolateralis
dpth	nucleus dorsalis posterior thalami
ees	extraencephalic space
emmed	eminentia mediana
ent	nuclei entopedunculares
entc	nucleus entopeduncularis caudalis
entd	nucleus entopeduncularis dorsalis
entv	nucleus entopeduncularis ventralis
ep	epiphysis
fae	fibrae arcuatae externae
fai	fibrae arcuatae internae
fd	funiculus dorsalis
fhyt	fibrae hypothalamotegmentales
fl	funiculus lateralis
flm	fasciculus longitudinalis medialis
flt	fasciculus lateralis telencephali
fmt	fasciculus medialis telencephali
fMth	fibre of Mauthner
fr	fasciculus retroflexus
fs	fasciculus solitarius
ft	fasciculi tegmentales
fttd	fibrae tectotegmentales dorsales
ftth	fibrae tectothalamicae
fttv	fibrae tectotegmentales ventrales
fv	funiculus ventralis
gcm	griseum centrale mesencephali
gll	glomerular layer
hab	ganglion habenulae
hypoth	hypothalamus
inf	infundibulum
int	nucleus intermedius areae octavolateralis
ip	nucleus interpeduncularis
ith	nucleus intermedius thalami
lcoe	locus coeruleus
ll	lemniscus lateralis
lobll	lobus lineae lateralis
lobX	lobus vagi
lv	nucleus lateralis valvulae
medsp	medulla spinalis
mes	mesencephalon
meV	nucleus mesencephalicus nervi trigemini
Mth	cell of Mauthner
nil	neurointermediate lobe of pituitary
nllad	nervus lineae lateralis anterior, pars dorsalis
nllav	nervus lineae lateralis anterior, pars ventralis
nllp	nervus lineae lateralis posterior
nspoc	nervus spino-occipitalis
nufl	nucleus funiculi lateralis
nuflm	nucleus of the fasciculus longitudinalis medialis
nufs	nucleus fasciculi solitarii
numm	nucleus medianus magnocellularis
numtp	nucleus medianus tuberculi posterioris
nuptp	nucleus periventricularis tuberculi posterioris
nurdV	nucleus of the radix descendens nervi trigemini
nutbcp	nucleus tuberculi posterioris
nutsc	nucleus tori semicircularis
nI	nervus olfactorius
nII	nervus opticus
nIII	nervus oculomotorius
nIV	nervus trochlearis
nVm	nervus trigeminus, radix motoria
nVs	nervus trigeminus, radix sensibilis
nVI	nervus abducens
nVII	nervus facialis
nVIIm	nervus facialis, radix motoria
nVIIs	nervus facialis, radix sensibilis
nVIII	nervus octavus
nIX	nervus glossopharyngeus
nX	nervus vagus
ohd	orohypophysial duct
oli	oliva inferior
oll	tractus olfactorius lateralis
olm	tractus olfactorius medialis
opt	tractus opticus
optl	tractus opticus lateralis
optm	tractus opticus medialis
optma	tractus opticus marginalis
opv	organon paraventriculare

osc	organon subcommissurale	sgc	stratum griseum centrale
P1, P2 etc.	pallial fields	sgp	stratum griseum periventriculare
pal	pallium	sgr	stratum granulare
pd	pars distalis of pituitary	shy	sulcus hypothalamicus
pdhy	pars dorsalis hypothalami	sid	sulcus intermedius dorsalis
plencv	plica encephali ventralis	sih	sulcus intrahabenularis
plxm	plexus marginalis	sis	sulcus isthmi
pm	nucleus preopticus magnocellularis	siv	sulcus intermedius ventralis
pol	layer of primary olfactory fibres	slH	sulcus limitans of His
pp	nucleus preopticus parvocellularis	slmes	sulcus lateralis mesencephali
prom	nucleus profundus mesencephali	slt	sulcus limitans telencephali
prt	regio pretectalis	sm	stria medullaris
prtp	nucleus pretectalis periventricularis	smarg	stratum marginale
prts	nucleus pretectalis superficialis	smi	sulcus medianus inferior
prtsc	nucleus pretectalis supracommissuralis	smol	stratum moleculare
psv	pseudoventricle	sms	sulcus medianus superior
pv	portal vessels	sol	layer of secondary olfactory fibres
pvhy	pars ventralis hypothalami	sop	stratum opticum
rai	nucleus raphes inferior	ssh	sulcus subhabenularis = s. diencephali-
ranllav	radix ascendens nervi lineae lateralis anterioris, pars ventralis		cus dorsalis
		t	taenia
ranllp	radix ascendens nervi lineae lateralis posterioris	tbc	tractus tectobulbaris cruciatus
		tbr	tractus tectobulbaris rectus
ras	nucleus raphes superior	tchrh	tela chorioidea rhombencephali
rb	area retrobulbaris	td	tela diencephali
rdnllav	radix descendens nervi lineae lateralis anterioris, pars ventralis	tect	tectum mesencephali
		teg	tegmentum mesencephali
rdnllp	radix descendens nervi lineae lateralis posterioris	tel	telencephalon
		thd	thalamus dorsalis = pars dorsalis tha-
rdV	radix descendens nervi trigemini		lami
rhomb	rhombencephalon	thv	thalamus ventralis = pars ventralis tha-
ri	nucleus reticularis inferior		lami
rism	nucleus reticularis isthmi et mesence- phali	tl	nucleus tori lateralis
		tp	tuberculum posterius
rlvq	recessus lateralis ventriculi quarti	trpal	tractus pallii
rm	nucleus reticularis medius	tt	tela telencephali
rmeV	radix mesencephalicus nervi trigemini	V	area ventralis telencephali
rmtp	recessus medianus tuberculi posterioris	valvcb	valvula cerebelli
rmV	radix motoria nervi trigemini	vbol	ventriculus bulbi olfactorii
rmVII	radix motoria nervi facialis	Vd	dorsal part of V
rnllad	radix nervi lineae lateralis anterioris, pars dorsalis	veltr	velum transversum
		vem	nucleus vestibularis magnocellularis
Rp	Rathke's pouch	vep	nucleus vestibularis parvocellularis
Rpr	remnant of Rathke's pouch	visc	nucleus visceralis secundarius
rpr	recessus preopticus	Vl	lateral part of V
rs	nucleus reticularis superior	vlth	nucleus ventrolateralis thalami
rv(1)	radix ventralis nervi spinalis (1)	vmth	nucleus ventromedialis thalami
S	subpallium	Vn	nucleus situated in the dorsalmost part of V
sac	stratum album centrale		
sce	stratum cellulare externum	Vp	postcommissural part of V
sch	nucleus suprachiasmatis	vq	ventriculus quartus
sci	stratum cellulare internum	Vs	supracommissural part of V
sdm	sulcus diencephalicus medius	vt	ventriculus tertius
sdv	sulcus diencephalicus ventralis	Vv	ventral part of V
se	septum ependymale	zld	zona limitans diencephali
sfgs	stratum fibrosum et griseum superficiale	III	nucleus nervi oculomotorii

IV	nucleus nervi trochlearis
Vm	nucleus motorius nervi trigemini
VI	nucleus nervi abducentis
VIIm	nucleus motorius nervi facialis
VIIIa	nucleus anterior nervi octavi
IXm	nucleus motorius nervi glossopharyngei
Xm	nucleus motorius nervi vagi

Fibres of the spinal white matter are extremely variable in size. The ventral funiculi contain a number of coarse fibres, among which the bilateral Mauthner fibres are most conspicuous, although several ventral funicular fibres have nearly the same diameter (Fig. 13.6).

The deepest parts of the lateral funiculi also contain numerous coarse fibres. In the intermediate part of the lateral funiculi, fibres of varying diameter are found, but in the superficial part of the lateral funiculi and throughout the dorsal funiculi, thin fibres prevail (Fig. 13.6).

Little is known about the origin and projection of the funicular fibres. However, we consider it likely that the coarse fibres in the ventral funiculi originate from the large rhombencephalic reticular neurons and that the coarse fibres in the deepest part of the lateral funiculi represent, at least in part, a vestibulo-spinal system. Whether the dorsal funiculi contain any long ascending axons is open to question.

Neurosecretory cells are present in the caudal part of the brachiopterygian spinal cord (Bern 1969). Onstott and Elde (1986) observed some urotensin II-immunoreactive cells in the caudalmost five to six segments of the cord of the reedfish *Erpetoichthys calabaricus*. The relatively small (20–28 μm), round perikarya of these cells were situated lateral and ventrolateral to the central canal. The lateral- and ventralmost parts of the caudal cord contained numerous longitudinally oriented urotensin II-immunoreactive fibres, and along the ventral surface of the caudal cord intermittent areas of bright urotensin II immunofluorescence were observed. These structural features resemble those found in the urophysis of teleosts (see Sect. 15.2.7).

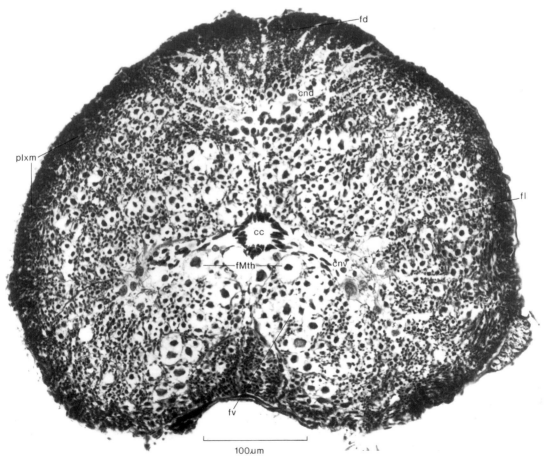

Fig. 13.6. Transverse section through the spinal cord of the reedfish *Erpetoichthys calabaricus*, slightly in front of the level of the first dorsal finlet (see also Fig. 13.1). Bodian preparation

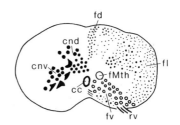

Fig. 13.7

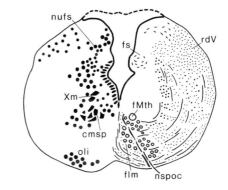

Fig. 13.8

Fig. 13.9

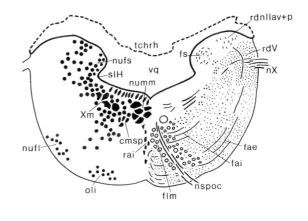

Figs. 13.7–13.21. Series of transverse sections through the spinal cord and brain of the reedfish *Erpetoichthys calabaricus.* The *left half* of each figure shows the cell picture, based on Nissl-stained sections, the *right half* shows the fibre systems, based on adjacent Klüver-Barrera- and Bodian-stained sections. **Fig. 13.7.** Section through the cervical spinal cord. **Fig. 13.8.** Section through the rhombencephalon, just rostral to the obex. **Fig. 13.9.** Section through the caudal part of the rhombencephalon, at the level of the most caudal vagal root. **Fig. 13.10.** Section through the intermediate part of the rhombencephalon, at the level of entrance of the glossopharyngeal nerve. **Fig. 13.11.** Section through the intermediate part of the rhombencephalon, at the level of entrance of the octavus and anterior lateral line nerves. **Fig. 13.12.** Section through the rostral rhombencephalon, at the level of the motor trigeminal nucleus. **Fig. 13.13.** Section through the isthmus rhombencephali. **Fig. 13.14.** Section through the caudal mesencephalon. **Fig. 13.15.** Section through the rostral mesencephalon. **Fig. 13.16.** Section through the commissura posterior. **Fig. 13.17.** Section through the diencephalon. **Fig. 13.18.** Section through the caudal telencephalon. **Fig. 13.19.** Section through the telencephalon at the level of the commissura anterior. **Fig. 13.20.** Section through the telencephalon halfway between the commissura anterior and the bulbus olfactorius. **Fig. 13.21.** Section through the caudal parts of the olfactory bulbs and the most rostral part of the telencephalon proper

1mm

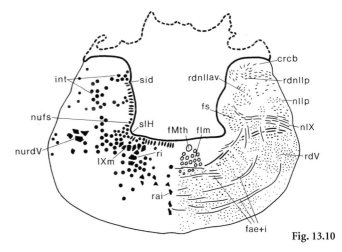

Fig. 13.10

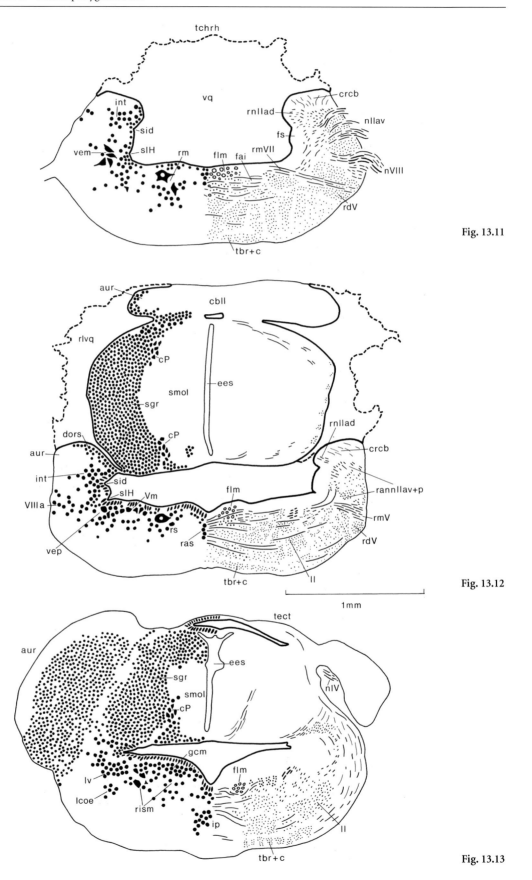

Fig. 13.11

Fig. 13.12

1mm

Fig. 13.13

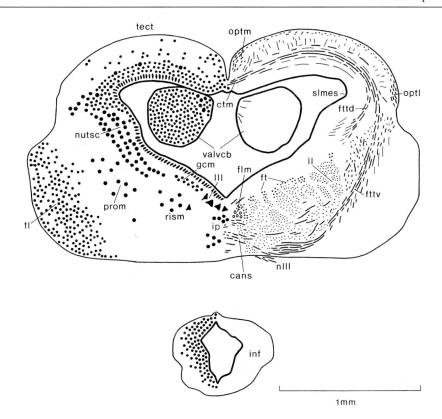

Fig. 13.14

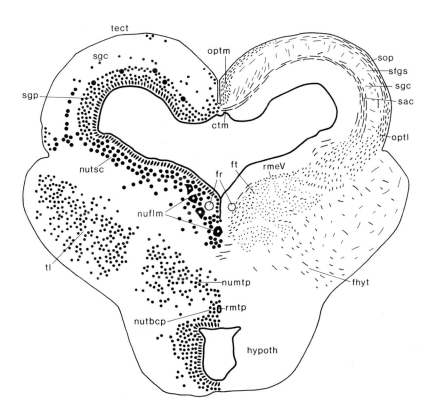

Fig. 13.15

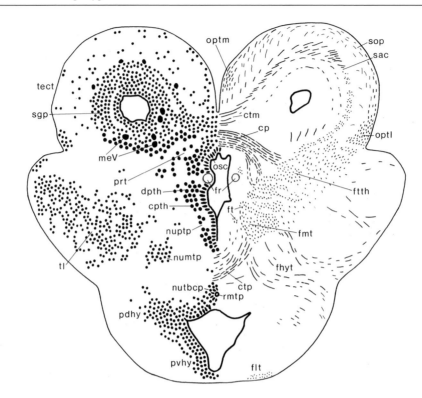

Fig. 13.16

1mm

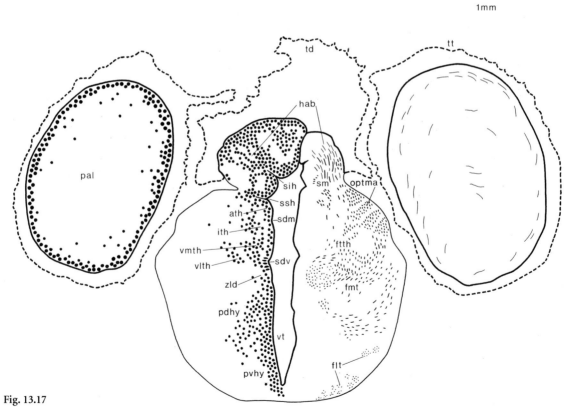

Fig. 13.17

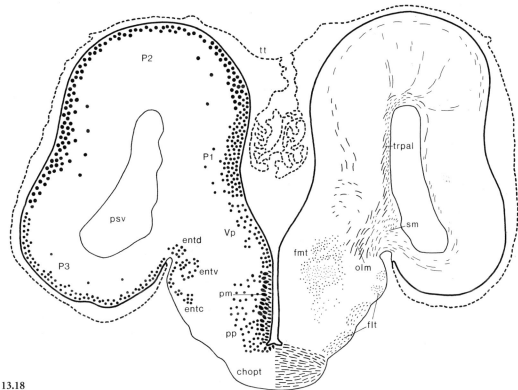

Fig. 13.18

1 mm

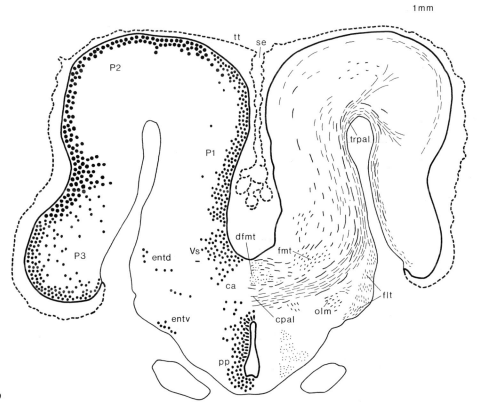

Fig. 13.19

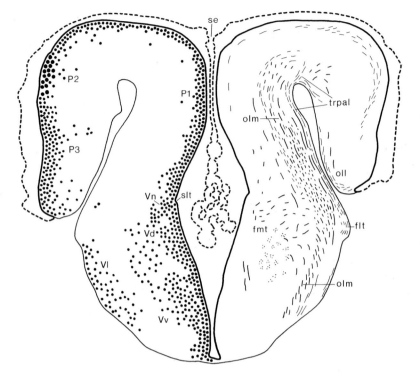

1mm

Fig. 13.20

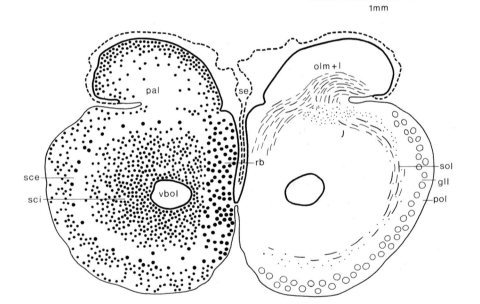

Fig. 13.21

However, whether this area is functionally equivalent to the urophysis remains to be determined. In the reedfish, Onstott and Elde (1986) did not detect any urotensin I immunoreactivity in the 14 caudalmost spinal cord segments.

13.5
Rhombencephalon

13.5.1
Introductory Notes

Throughout most of the ventricular surface of the rhombencephalon, a distinct sulcus limitans demarcates the primarily motor basal plate from the alar plate, which is primarily sensory (Fig. 13.22). Within the basal plate, a medial area ventralis and a lateral area intermedioventralis can be discerned. In *Polypterus*, the boundary between these two areas is indicated by a long sulcus intermedius ventralis (Kenemans 1980). A corresponding sulcus is present in *Erpetoichthys*; however, in this species the groove is short and does not coincide exactly with the boundary between the ventral and intermedioventral zones (Figs. 13.10, 13.22). The alar plate also contains two longitudinally arranged cell zones, the area intermediodorsalis and the area dorsalis. In *Polypterus*, a long, distinct sulcus intermedius dorsalis throughout the rostral rhombencephalon marks the boundary between these two alar zones. In *Erpetoichthys*, this groove is clearly present, but much shorter (Figs. 13.12–13.19, 13.22). The four morphological zones described for the basal and alar plates roughly correspond to the well-known functional columns of Herrick and Johnston (see Sect. 4.6.2). However, the following survey will reveal several important exceptions.

13.5.2
Area Dorsalis

The area dorsalis is aptly designated as somatosensory, as its constituent grisea represent the termination of the special somatosensory lateral line nerves. These comprise a rather small nervus lineae lateralis anterior, pars dorsalis, the much larger nervus lineae lateralis anterior, pars ventralis and the nervus lineae lateralis posterior (Fig. 13.3a–c). Fibres of the nervus lineae lateralis anterior, pars dorsalis bifurcate and terminate in the nucleus dorsalis areae octavolateralis, a cell mass confined to the rostral part of the rhombencephalon (Fig. 13.12). On comparative grounds, it may be expected that this nucleus and its afferent nerve root is connected with the organs of Fahrenholz, i.e. the electroreceptors located on the rostral part of

the head (see McCormick 1978, 1981). Fibres of the ventral part of anterior lateral line and those of the posterior lateral line nerves bifurcate into ascending and descending branches on entering the brain and terminate in the very elongated nucleus intermedius areae octavolateralis (Figs. 13.9, 13.12, 13.22).

The most dorsal part of the area dorsalis is occupied by the crista cerebellaris, a zone of neuropil that closely resembles the cerebellar molecular layer with which it is rostrally continuous (Figs. 13.10–13.12). Medium-sized, Purkinje-like neurons situated in the nucleus intermedius send long, branching dendrites into the cerebellar crest (Fig. 13.5c). The main axonal component of the cerebellar crest are presumably the axons of small, granular cells present in the cerebellar auricle.

Large numbers of fibres descend as fibrae arcuatae from the lateral line centres toward the rhombencephalic basal plates (Figs. 13.5c, 13.8, 13.11). Although experimental evidence is lacking, it is likely that many of these fibres cross the median plane and ascend in the contralateral lemniscus lateralis (Figs. 13.12, 13.14) to a mesencephalic centre, the torus semicircularis (Figs. 13.14, 13.15).

13.5.3
Area Intermediodorsalis

According to the classical doctrine of functional columns, the intermediodorsal column is viscerosensory in nature. Although its mediocaudal and rostralmost portions are indeed viscerosensory, its caudolateral and intermediate parts appear to be somatosensory.

After entering the brain, the afferent components of cranial nerves VII, IX and X constitute a distinct descending bundle, the fasciculus solitarius. This bundle most probably contains both general viscerosensory and special viscerosensory (i.e. taste) fibres. Its centre of termination is the nucleus fasciculi solitarii, a large elongated cell mass which occupies a periventricular position in the caudal half of the intermediodorsal zone (Figs. 13.8, 13.11, 13.22).

In other fishes, particularly teleosts, the nucleus of the solitary tract sends its efferent fibres primarily to a centre in the rostralmost part of the rhombencephalon; because of its specific input, this centre is called the nucleus visceralis secundarius. We were unable to trace the efferents emanating from the nucleus of the solitary tract in *Erpetoichthys*. However, in this species and in *Polypterus* (Kenemans 1980), a distinct cell group situated in the ventrolateral part of the isthmus region probably represents the nucleus visceralis secundarius in

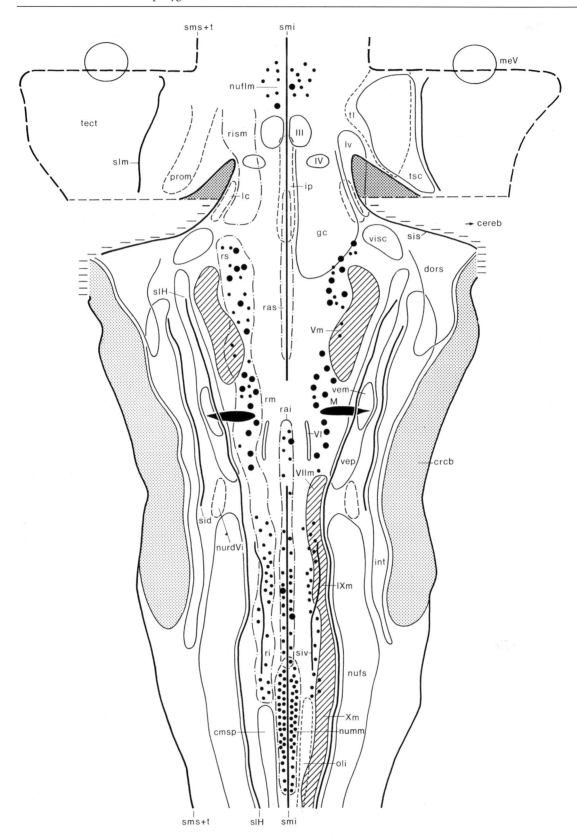

Fig. 13.22. Topological chart of the brain stem of the reedfish *Erpetoichthys calabaricus*. (Modified from Nieuwenhuys and Oey 1983)

other groups of fishes. This interpretation is based only on the topological position of the putative secondary visceral nucleus in brachiopterygians, but suggests that the viscerosensory zone comprises two centres: one in the caudal portion and the other in the rostralmost portion of the rhombencephalon (Fig. 13.22).

The trigeminal nerve in brachiopterygians consists of three separate parts: a ventral motor root and two sensory roots, the dorsorostral nervus ophthalmicus profundus and the dorsocaudal nervus maxillomandibularis (Fig. 13.3a–c). After entering the brain, the two sensory roots constitute the large, distinct radix descendens nervi trigemini, situated in the superficial part of the area intermediodorsalis and extending to the caudalmost parts of the rhombencephalon (Figs. 13.8–13.12). Throughout its extent, this bundle contains small, scattered neurons that represent the nucleus descendens of the trigeminal nerve, but are too diffusely arranged to be designated as such. However, at mid-rhombencephalic levels, the descending trigeminal root contains a small group of large neurons that we have labelled the nucleus descendens nervi trigemini (Fig. 13.10). This nucleus gives rise to a small, but distinct bundle of arcuate fibres. It should be noted that the small and large cells located in the descending root of the trigeminal nerve are the only general somatosensory elements observed in the rhombencephalon of *Erpetoichthys*. Neither a nucleus funiculi dorsalis nor a nucleus princeps nervi trigemini could be distinguished in this species.

The special somatosensory area octavolateralis comprises three cell masses: the nucleus dorsalis, nucleus intermedius and nucleus ventralis. As noted above, the dorsal and intermediate nuclei are situated in the area dorsalis and receive their afferents from the lateral line system. The nucleus ventralis, however, is located in the area intermediodorsalis and represents the centre of termination for the nervus octavus. Upon entering the brain, most of the octavus fibres bifurcate into ascending and descending branches which synapse with cells of the nucleus ventralis (areae octavolateralis). This nucleus comprises two subunits which are referred to here as the nucleus vestibularis magnocellularis and nucleus vestibularis parvocellularis (Figs. 13.11, 13.12, 13.22). It should be noted, however, that the octavus nerve in fishes is composed of both auditory and vestibular fibres, but that, with the exception of teleosts (see Sect. 15.3.4.1), it has not been determined whether these two types of fibres project to separate parts of the octavolateral area. Efferents from the nucleus ventralis pass as fibrae arcuatae internae to the ipsilateral and con-

tralateral fasciculus longitudinalis medialis (flm) in which they ascend, descend or bifurcate (Fig. 13.11). Other efferent fibres from the nucleus ventralis descend laterally to the fasciculus longitudinalis medialis and to the ipsilateral half of the spinal cord; still others probably join the contralateral lemniscus lateralis, with which they ascend to the torus semicircularis.

The description of the octavolateral nuclei presented above agrees with that of Kenemans (1980), but differs somewhat from those of Senn (1976a) and McCormick (1982). The nucleus of Deiters, described by Senn, clearly corresponds to our nucleus vestibularis magnocellularis, but the cell masses he designated as the nucleus vestibularis and nucleus vestibularis anterior represent lateral line centres and correspond largely to our nucleus intermedius. McCormick (1982) observed seven octavolateral nuclei. Three of these, the dorsal, medial and magnocellular vestibular nuclei, correspond to the dorsal, intermediate and magnocellular nuclei, respectively, in our description; however, the remaining four, i.e. the nucleus caudalis and the posterior, descending and anterior octavus nuclei, have not been delineated by us. We were unable to find equivalents of the nucleus caudalis and the posterior octavus nucleus. The descending octavus nucleus is represented by scattered cells lateral to the fasciculus solitarius and is visible in Figs. 13.9 and 13.10. The anterior octavus nucleus most probably corresponds to a group of small, superficially situated cells labelled VIIIa in Fig. 13.12.

In summary, the rhombencephalic area intermediodorsalis appears to be functionally heterogeneous. Its caudomedial and rostralmost parts are occupied by viscerosensory centres; its superficial zone is dominated by the general somatosensory descending root of the fifth nerve, and the intervening zone contains some special somatosensory cell masses.

13.5.4
Area Intermedioventralis

The lateral part of the rhombencephalic basal plate is occupied by a longitudinal zone of large cells extending from the obex to the isthmic region. This zone, which consists of large, special visceromotor or branchiomotor elements, is interrupted at the level of entrance of the eighth nerve (Fig. 13.22). Its rostral part represents the nucleus motorius nervi trigemini (Fig. 13.12), whereas its caudal part harbours the motor nuclei of VII, IX and X (Figs. 13.8–13.11). Golgi material reveals that the large branchiomotor neurons extend their dendrites peripherally into a rather narrow zone of the stratum album (Fig. 13.5b). In addition, the area intermedioventralis contains numerous small, scattered cells. Some of these elements probably represent the undifferentiated lateral part of the reticular formation; others, particularly those in the caudal rhombencephalon, may be splanchnic, preganglionic elements.

Mauthner cells are located largely within the intermedioventral zone, in the gap between the motor nucleus of the trigeminal nerve and the caudal branchiomotor column (Fig. 13.22). As in other groups of fishes, the somato-dendritic complex of these neurons is oriented mainly in the transverse plane. There are some ventral dendrites, and the complex grades medially and laterally into a huge dendritic trunk. The lateral dendritic trunk contacts entering fibres of the eighth nerve via typical club-like endings. The medial dendritic trunk branches profusely in the stratum album of the medial part of the basal plate (Kenemans 1980). The axon arises from the dorsomedial aspect of the soma. It acquires a myelin sheath, decussates and passes caudally in the dorsal parts of the fasciculus longitudinalis medialis and the ventral funiculus to the caudal end of the spinal cord (Figs. 13.6–13.10). The axon hillock and proximal segment are invested by a dense plexus of unmyelinated fibres constituting the so-called axon cap (Kenemans 1980).

13.5.5
Area Ventralis

The area ventralis contains the following (Fig. 13.22):

1. Two somatomotor centres, the rostral part of the spinal motor column and the abducens nucleus
2. The median and medial parts of the reticular formation
3. A cell mass that most probably represents the oliva inferior

That part of the spinal motor column extending into the brain gives rise to three spino-occipital nerve roots (Figs. 13.3c, 13.8). These are considered to be spinal roots that have become 'encephalised' due to the incorporation of trunk sclerotomal tissue within the skull. A typical neuron of one of the spino-occipital nerve roots is shown in Fig. 13.5b.

The small nucleus nervi abducentis consists of a number of diffusely arranged neurons situated ventrolateral to the fasciculus longitudinalis medialis. The axons of these cells constitute some five extremely fine rootlets, which unite to form a single nerve (Fig. 13.3c).

The median reticular formation comprises three centres: the nucleus medianus magnocellularis, the

nucleus raphes inferior and the nucleus raphes superior. The *nucleus medianus magnocellularis* is a large, elongated cell mass situated in the most caudal part of the rhombencephalon. It extends bilaterally over the medial part of the fasciculus longitudinalis medialis (Figs. 13.9, 13.22). This remarkable nucleus, which does not correspond to any known cell mass in other vertebrates, was first described by Kenemans (1980), who observed that its neurons send their axons caudalward in the fasciculus longitudinalis medialis.

The *nuclei raphes superior* and *inferior* are situated in the rostral and caudal rhombencephalic raphe area, respectively (Figs. 13.10, 13.13). The inferior raphe nucleus contains a number of large, scattered neurons. Horseradish peroxidase (HRP) tracing techniques (Northcutt 1981) have established that the superior raphe nucleus projects bilaterally to the telencephalon.

The medial reticular formation is a continuous cell column extending nearly the entire length of the rhombencephalon. The neurons of this column vary widely in size. Plotting of the large and very large reticular cells reveals that these elements are not equally dispersed. Thus they have been divided into three moieties: the nucleus reticularis inferior, nucleus reticularis medius and nucleus reticularis superior (Nieuwenhuys and Oey 1983; Figs. 13.10–13.12, 13.22). Many of the large and very large reticular elements spread their dendrites over a remarkably wide area (Fig. 13.5c). Axons of the reticular elements frequently enter the fasciculus longitudinalis medialis, with which they descend to the spinal cord. The medial reticular zone probably subserves a somatomotor coordinating function; however, it is also possible that some elements in this zone serve as links in ascending projections.

The only centre located in the area ventralis that is definitely not somatomotor in nature (in accordance with the classical doctrine of the functional columns) is the inferior olive. This elongated, small-celled nucleus occupies a superficial position in the caudal part of the rhombencephalon. Although we were unable to determine a crossed projection to the molecular layer of the cerebellum, its characteristic topological position justifies identification of this nucleus as the inferior olive.

13.5.6
Isthmus Rhombencephali

The isthmus region, i.e. the tapering, rostralmost part of the rhombencephalon, contains neither cranial nerve nuclei nor a distinct zonal pattern. In addition to the nucleus visceralis secundarius and the nucleus raphes superior described above, this region harbours the following centres: the nucleus interpeduncularis, the griseum centrale, the nucleus reticularis isthmi, the locus coeruleus and the nucleus lateralis valvulae (Fig. 13.22).

The *nucleus interpeduncularis* is an elongated, unpaired cell mass, situated in the raphe area about halfway between the ependymal and the meningeal surfaces. This nucleus, which rostrally extends into the mesencephalon, receives efferents from the fasciculus retroflexus (Figs. 13.13, 13.14).

The *griseum centrale* is represented by a compact cell plate occupying a periventricular position in the tegmentum isthmi and the adjacent tegmentum mesencephali (Figs. 13.13, 13.14). The fibre connections of this conspicuous structure are entirely unknown.

The *nucleus reticularis isthmi* constitutes a rostral continuation of the rhombencephalic medial reticular zone which, in turn, grades into the mesencephalic reticular nucleus. The isthmal reticular nucleus consists of scattered, rather small cells located subjacent to the griseum centrale (Figs. 13.13, 13.14).

The *locus coeruleus* (Figs. 13.13, 13.22) is a small, but distinct cell group located ventrolateral to the lateral edge of the rostralmost part of the fourth ventricle. HRP tracings demonstrate that this cell group projects bilaterally to the telencephalon (Northcutt 1981).

Finally, the *nucleus lateralis valvulae* consists of small, compactly arranged cells located in the lateral part of the tegmentum isthmi, adjacent to the area in which the tegmentum lies contiguous to the rostroventral part of the cerebellum (Fig. 13.14). The fibre connections of this cell mass have not been established; however, its actinopterygian homologue has been proven to project to the cerebellum.

13.5.7
Summary of Rhombencephalic Fibre Systems

As noted, experimental studies of the rhombencephalic fibre connections are lacking. Observations based on non-experimental material discussed above are summarized here and supplemented with additional data:

1. The rhombencephalic alar plate contains a number of conspicuous, longitudinally arranged fibre systems, all of which consist of primary afferent axons. These systems are the viscerosensory fasciculus solitarius, the general somatosensory radix descendens nervi trigemini a number of special somatosensory bundles composed of bifurcating axons of the lateral line and octavus nerves. Longitudinal pathways, consisting of fibres of higher order (e.g. spinocerebellar, bulbocerebellar), are presumably present, but have not been identified.

2. Both general and special somatosensory cell masses in the alar plate give rise to large numbers of arcuate fibres that pass to the basal parts of the rhombencephalon. It is likely that the clearly distinguishable lateral lemniscus, which terminates in the subtectal torus semicircularis, is composed of both crossed arcuate and uncrossed fibres arising from the lateral line and octavus centres.

3. Two cell masses located in the rostralmost part of the rhombencephalon, i.e. the (possibly serotoninergic) nucleus raphes superior and the (possibly noradrenergic) locus coeruleus, project directly to the telencephalon (Northcutt 1981; Holmes and Northcutt 1995).

4. The fasciculus longitudinalis medialis is the principal fibre system descending from the brain stem to the spinal cord. It is composed primarily of reticulospinal and vestibulospinal fibres.

5. The vestibular centres also give rise to fibres that ascend in the fasciculus longitudinalis medialis and fibres that descend laterally to the fasciculus longitudinalis medialis to the spinal cord.

6. Axons originating from the deeper layers of the tectum pass as tectobulbar fibres to the rhombencephalon, where they constitute a crossed and uncrossed pathway descending in the medial part of the basal plate (Figs. 13.11–13.13).

7. As in all vertebrates, a thin-fibred fasciculus retroflexus descends from the diencephalic ganglion habenulae to the nucleus interpeduncularis (Figs. 13.16, 13.17).

13.6
Cerebellum

As already noted, the brachiopterygian cerebellum invaginates into the fourth ventricle and, due to this remarkable configuration, only its caudal portion is externally visible. This caudal portion surrounds the lateral recesses of the fourth ventricle and is probably homologous to the auriculae in cartilaginous fishes and most other anamniotes. It is important to note that, in addition to strands of tissue immediately adjacent to the ventricular cavity, the brachiopterygian auriculae comprise a pair of massive bilateral lobes that approach each other in the area directly caudal to the tectum (Fig. 13.3a). These lobes contain a large mass of densely packed, granular cells, covered externally by a molecular layer (Fig. 13.13). Dorsocaudally, the auriculae are interconnected by a transversely oriented structure; its superficial position notwithstanding, this structure represents the chondrichthyan 'lower lip' (Fig. 13.12). The large masses of granular cells in the rostral parts of the auriculae are probably homologous to the eminentiae granulares found in most actinopterygians (van der Horst 1925; Nieuwenhuys 1967b).

The invaginated central body of the cerebellum consists of bilaterally symmetrical halves, connected by a thin, curved lamella (Fig. 13.3d). The central body contains, bilaterally, a molecular layer, a zone of Purkinje cells, and a zone of granular cells, although these structures are not arranged in the usual laminated pattern. The granular cells are concentrated in the lateral parts of the central body; they surround slit-like lateral extensions of the extraencephalic space (see Nieuwenhuys 1967b) and bound the ventricular surface laterally, dorsally and ventrally (Fig. 13.12). Medially, these zones of granular cells are covered by a molecular layer. The perikarya of the Purkinje cells form a curved, triangular band which partly separates the molecular and granular zones (Figs. 13.12, 13.13). Throughout the central body of the cerebellum, these bilateral bands parallel each other; however, dorsocaudally, they diverge and arch laterally over the dorsal and lateral surfaces of the eminentiae granulares.

Several authors (van der Horst 1919, 1925; Larsell 1967; Kenemans 1980) have suggested that the caudal and rostral parts of the central body represent the corpus and valvula, respectively, of the actinopterygian cerebellum. Although in brachiopterygians the rostral and caudal parts of the central body cannot be demarcated from each other on the basis of any external or internal feature, this interpretation is supported by the existence of a homologue of the actinopterygian nucleus lateralis valvulae, located in the transitional area between the rostroventral cerebellum and the isthmal tegmentum (Fig. 13.13).

In the ventral part of the cerebellum in most groups of fishes, a nucleus cerebelli can be distinguished. We were unable to find this cell mass in *Erpetoichthys*, although Kenemans (1980) mentions its presence in *Polypterus*. In our non-experimental material, we observed large numbers of fibres connecting the brain stem with the cerebellum. Because the exact sites of origin and termination of these fibres could not be determined, they are not labelled in the illustrations.

13.7
Mesencephalon

The mesencephalon can be divided into a dorsal tectum and a ventral tegmentum. The boundary between these two regions is marked on the ventricular side by a deep groove, the sulcus lateralis mesencephali (Figs. 13.14, 13.15).

The caudomedial part of the *tectum* consists of a thin, ependymal lamella (Fig. 13.13), but throughout its other parts, six layers can be distinguished. From external to internal, these layers are as follows (Figs. 13.15, 13.23):

1. An extremely thin *stratum marginale* (sm), which contains few cells. In our Klüver-Barrera and Bodian preparations, no fibres were observed in this layer.
2. The *stratum opticum* (sop), consisting of both myelinated and unmyelinated fibres.
3. The *stratum fibrosum et griseum superficiale* (sfgs), which contains scattered cells but is primarily composed of neuropil.

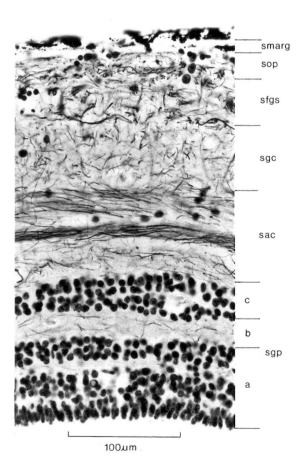

Fig. 13.23. Transverse section through the tectum mesencephali of the reedfish *Erpetoichthys calabaricus*. Bodian preparation

4. The *stratum griseum centrale* (sgc), composed of cells of varying size and density and intervening neuropil.

5. The *stratum album centrale* (sac), which mainly comprises coarse, myelinated fibres. Medially, fibres of this layer constitute the commissura tecti; laterally, they form the tectal efferent systems. There is frequent and considerable overlap between layers 4 and 5.

6. The *stratum griseum periventriculare* (sgp) is particularly wide and can be divided into three layers: (a) an inner layer of two to three small-celled laminae, (b) a narrow fibre zone composed mainly of unmyelinated fibres and (c) an external layer containing small cells and numerous larger elements of varying shape. The largest tectal neurons are found in this layer; in the rostral tectum, a group of these large neurons constitutes the nucleus mesencephalicus of the trigeminal nerve (Fig. 13.16).

The laminar pattern described above corresponds closely to those reported by Senn (1976b), Repérant et al. (1981) and Northcutt (1983), but it differs somewhat from that of Mazzi et al. (1977). These authors included our stratum album centrale in the stratum fibrosum et griseum superficiale, and they interpreted the thin fibre zone within the stratum griseum periventriculare (sublayer 6) as stratum album centrale.

Retinofugal fibres constitute the most prominent tectal *afferent system*. After crossing in the chiasma opticum, most of these fibres pass into the tractus opticus along the lateral surface of the diencephalon. On approaching the rostral pole of the tectum, the tractus opticus splits into distinct medial and lateral optic tracts (Figs. 13.3a,b, 13.13, 13.14). Repérant et al. (1979, 1981) and Braford and Northcutt (1983) analysed the distribution and termination of retinofugal fibres in the tectum with amino acid autoradiography. Their findings can be summarised as follows: The fibres of the tractus opticus lateralis project mainly to the stratum marginale in which they constitute a thin, continuous layer of terminals. Below this layer, a terminal-free zone is present. Heavy labelling was found in a wide band comprising the stratum opticum and the stratum fibrosum et griseum superficiale. Some optic terminals were observed in the stratum griseum centrale, the stratum album centrale and the most superficial part of the stratum griseum periventriculare. In addition to these contralateral projections, a small ipsilateral projection is mainly confined to the rostromedial portions of the stratum marginale and the stratum album centrale.

As noted above, the stratum griseum periventriculare is extraordinarily wide. As Repérant et al.

(1981) pointed out, this feature is not found in other groups of fishes, but is typical of anuran amphibians. The nature of the stratum marginale is uncertain. In the tectum of holosteans and teleosts, a distinct superficial zone, closely resembling the cerebellar molecular layer, is called the stratum marginale. It contains abundant fine, unmyelinated axons originating from the torus longitudinalis, a bilateral extension of the medial tectum into the mesencephalic ventricle. Brachiopterygians do not possess a torus longitudinalis (Mazzi et al. 1977; Northcutt 1983); thus it is unlikely that the stratum marginale in this group is comparable to the layer with the same name in holosteans and teleosts. As the brachiopterygian stratum marginale contains a band of optic terminals, Repérant et al. (1981) suggested that this layer is comparable to the stratum zonale in anurans.

In most vertebrate groups, the tectum receives afferents from several sources other than the retina. Although it is likely that such non-retinal afferents are also present in brachiopterygians, we were unable to demonstrate them clearly in our non-experimental material.

The *efferent connections* of the tectum include groups of ascending, decussating and descending fibres which form part of the stratum album centrale within the tectum. The ascending fibres can be traced from the rostral tectum to the pretectal area and the dorsal and ventral thalamus. However, their precise terminations have not been determined. The decussating fibres constitute the well-developed commissura tecti (Figs. 13.14, 13.15). The descending tectal efferents comprise two fibre streams: the fibrae tectotegmentales dorsales and the fibrae tectotegmentales ventrales (Fig. 13.14). The fibrae tectotegmentales dorsales pass medially below the subtectal periventricular grey and can be traced to the nucleus tori semicircularis and the nucleus profundus mesencephali. The fibrae tectotegmentales ventrales form a conspicuous bundle which swings ventromedially through the tegmentum mesencephali (Fig. 13.14). Some of these fibres remain ipsilateral to constitute a tractus tectobulbaris rectus, while others cross in the commissura ansulata and descend into the rhombencephalon as the tractus tectobulbaris cruciatus.

The *tegmentum mesencephali* can be divided into lateral and medial zones. The lateral zone contains three cell masses: the nucleus tori semicircularis, the nucleus profundus mesencephali and the nucleus tori lateralis.

The *nucleus tori semicircularis* consists of several laminae of compact cells and is continuous laterally with the stratum griseum periventriculare of the tectum.

The *nucleus profundus mesencephali* is an aggregation of rather diffusely arranged cells, situated ventrolateral to the nucleus tori semicircularis and confined to the caudal and intermediate parts of the tegmentum mesencephali.

The *nucleus tori lateralis* is by far the largest tegmental cell mass. As its name implies, it is situated in the torus lateralis, a conspicuous external protrusion of the mesencephalic wall (Figs. 13.3b,c, 13.14, 13.15). Rostrally, it extends into the lateral wall of the diencephalon (Fig. 13.16). Reiner and Northcutt (1992) pointed out that the lateral toral nucleus, though occupying a position reminiscent of that of the substantia nigra, is devoid of tyrosine hydroxylase-positive perikarya, poor in substance P-positive fibres and relatively poor in leu-enkephalin-positive fibres (see Sect. 13.10 and Fig. 13.29b) and is thus immunohistochemically clearly distinct from the substantia nigra.

The *medial tegmental zone* is a rostral continuation of the rhombencephalic area ventralis. It contains three distinct nuclei: the nucleus nervi oculomotorii, nucleus nervi trochlearis and the large-celled nucleus of the fasciculus longitudinalis medialis (Fig. 13.15). Three other cell masses, discussed in a previous section, extend from the rhombencephalon into the medial tegmental zone of the midbrain: the nucleus interpeduncularis, the griseum centrale and the nucleus reticularis isthmi et mesencephali (Fig. 13.22). Scattered cells located in the paramedian tegmental zone may well represent a mesencephalic raphe nucleus.

Little is known about the fibre connections of the mesencephalic tegmentum. The compact lemniscus lateralis, which originates from the rhombencephalic octavolateral area, can be traced to nucleus tori semicircularis which, as noted, also receives fibres from the tectum. On the basis of non-experimental materal, Senn (1976a) suggested that the large nucleus tori lateralis is connected with the hypothalamus, nucleus profundus mesencephali, the caudal part of the tectum and the cerebellum. Experiments using the HRP technique have shown that the nucleus profundus mesencephali, the mesencephalic raphe cells and the nucleus tori lateralis project directly to the telencephalon, and it has been recently reported by Holmes and Northcutt (1995) that the latter nucleus receives a bilateral projection from the pallial fields P2 and P3. Silver-impregnated sagittal sections from the brains of *Polypterus* and *Erpetoichthys* suggest that numerous diffusely arranged fibres connect the hypothalamus with the mesencephalic tegmentum, although the polarity of these fibres is not known. Axons from the nucleus of the fasciculus longitudinalis medialis join the bundle of that name. In the medial part of the mesencephalic tegmentum, the zone adjacent to the periventricular grey contains numerous distinct bundles composed of fibres of varying diameter. The exact origin of these fasciculi tegmentales (Figs. 13.14, 13.15) could not be determined in our material; however, some of their constituent fibres appear to originate from the telencephalon and ventral thalamus. The thin-fibred fasciculus retroflexus can be clearly traced through the midbrain. Due to the marked asymmetry of the habenular ganglia, the right bundle is much larger than the left (Figs. 13.15, 13.16). Finally, it should be noted that Repérant et al. (1981) and Braford and Northcutt (1983) presented experimental evidence suggesting that an accessory optic tract in *Polypterus* terminates diffusely in the tegmentum mesencephali.

13.8
Diencephalon

13.8.1
Introductory Notes

The diencephalon can be divided into four principal zones or regions: the epithalamus, dorsal thalamus, ventral thalamus and hypothalamus. Two other regions, the regio pretectalis and the regio tuberculi posterioris, also lie within the diencephalon. Ventricular sulci, so prominent in many other groups of anamniotes, are only moderately developed in the diencephalon of *Polypterus* and are even less distinct in that of *Erpetoichthys*. Most of the diencephalic neurons are clustered in a periventricular grey zone, although a number of migrated cell masses occur in the thalamic, pretectal and tubercular regions.

It is noteworthy that, according to current views, extensively discussed in Chap. 4, Sect. 4.5, most of the diencephalic zones, though topographically oriented more or less horizontally, are to be considered as derivatives of initially transversely oriented neuromeres. In accordance with these views, Northcutt (1995) considered it likely that, in *Polypterus* (see Fig. 7A in Northcutt 1995) and in gnathostomes in general, an anterior parencephalic segment has given rise to the ventral thalamus, a posterior parencephalic segment has given rise to the habenular nuclei, the dorsal thalamus and the posterior tubercular nuclei, and a synencephalic segment has given rise to the pretectal nuclei and the nucleus of the fasciculus longitudinalis medialis. The present author concurs with this interpretation, with the reservation that the posterior tubercular nuclear complex might well be a derivative of both the anterior and posterior parencephala (see Bergquist and Källén 1954; see also Figs. 4.19, 4.20).

The diencephalic fibre systems in brachiopterygians are not well known, but a number of 'fibre streams' are apparent in non-experimental material, and some experimental hodological data are available. The centres and connections of the following complexes will be discussed below: the epithalamus, the dorsal thalamus and pretectum, the ventral thalamus and posterior tuberculum and the hypothalamus. These descriptions will be followed by comments on hypothalamo-hypophysial relationships and a general survey of the main fibre systems. The nucleus of the fasciculus longitudinalis medialis, which, as indicated above, belongs to the diencephalon, has already been discussed in Sect. 13.7.

13.8.2
Epithalamus

The epithalamus comprises the epiphysis and the ganglia habenulae. The epiphysis is rather long and tube-shaped with a slightly widened distal end (Fig. 13.3d). Its walls contain undifferentiated ependymal cells and more specialized elements which presumably represent sensory cells. The epiphysial stalk contains a number of nerve fibres, which probably connect this organ with the habenular ganglion and with the subcommissural organ (Ariëns Kappers 1965). The latter consists of a patch of columnar ependymal cells extending from the caudal end of the habenular commissure to the caudal end of the posterior commissure (Figs. 13.16, 13.24a).

The ganglia habenulae constitute two distinct protuberances into the ventricular cavity, bounded ventrally by a distinct sulcus subhabenularis or sulcus dorsalis diencephali. The medial surface of the habenulae is usually bordered by a sulcus intrahabenularis (Fig. 13.24a). The small neurons of the habenulae are tightly packed in a wide layer bounding the extensive ependymal surface of the ganglia (Fig. 13.17). Caudally, the two ganglia are connected by a large commissura habenulae. As in all vertebrates, afferent fibres reach the habenular ganglion by way of the stria medullaris, whereas habenular efferents assemble in the fasciculus retroflexus.

The stria medullaris forms in the dorsocaudal telencephalon and ascends along the rostrocaudal aspect of the ventral thalamus to the habenular commissure (Figs. 13.17, 13.18). It is unclear how many fibres synapse with habenular cells and how many constitute a true telencephalic commissure, but both systems may exist in the stria. The stria medullaris contains four fibre contingents:

1. A large number of rather diffusely arranged fibres from the caudomedial and caudolateral parts of the pallium
2. Some fibres from the preoptic region
3. An assemblage of compactly arranged fibres originating from the lateral subpallial nucleus and the nuclei entopedunculares (Fig. 13.24c)
4. Secondary olfactory fibres

The existence of this olfactory component has been experimentally confirmed by von Bartheld and Meyer (1986b), who found that fibres of the medial olfactory tract merge in the caudal telencephalon with fibres of the lateral tract and that these secondary olfactory fibres decussate in the habenular commissure. After decussation, these fibres turn rostrally on the other side, to terminate in the entopeduncular nuclei and in the lateral part of the area ventralis.

The fasciculus retroflexus gathers its fibres from all parts of the habenular ganglion and passes out of this structure caudomedially. Like the ganglia, the fascicular bundle is asymmetrical, the right side being much larger than the left. After passing caudally along the ventricular surface, the bundles unite in the rostroventral part of the tegmentum and then terminate in the nucleus interpeduncularis (Figs. 13.16, 13.24c).

13.8.3
Thalamus Dorsalis and Pretectum

The periventricular grey of the thalamus is differentiated into cellular laminae in a pattern that strongly resembles that found in chondrostean and holostean fishes and anuran amphibians. Jeener (1930) noted a break in the outer cellular zones of the thalamus, dividing this area into separate dorsal and ventral portions. This subdivision will be followed in the description below (the entire periventricular thalamic zone was considered as a single cytoarchitectonic entity in a study by Nieuwenhuys and Bodenheimer 1966). The dorsal thalamus has been further subdivided, distinguishing a rostrodorsal nucleus anterior thalami, a rostroventral nucleus intermedius thalami, a caudodorsal nucleus dorsalis posterior and a caudoventral nucleus centralis posterior thalami (Northcutt 1981; Braford and Northcutt 1983; Figs. 13.16, 13.17, 13.24a). Experimental neuroanatomical studies have revealed that the nucleus anterior thalami receives a strong direct retinal input (Repérant et al. 1981; Braford and Northcutt 1983) and projects bilaterally to the telencephalon (Northcutt 1981), thus forming part of a retino-thalamo-telencephalic circuit. In this respect, the brachiopterygian ante-

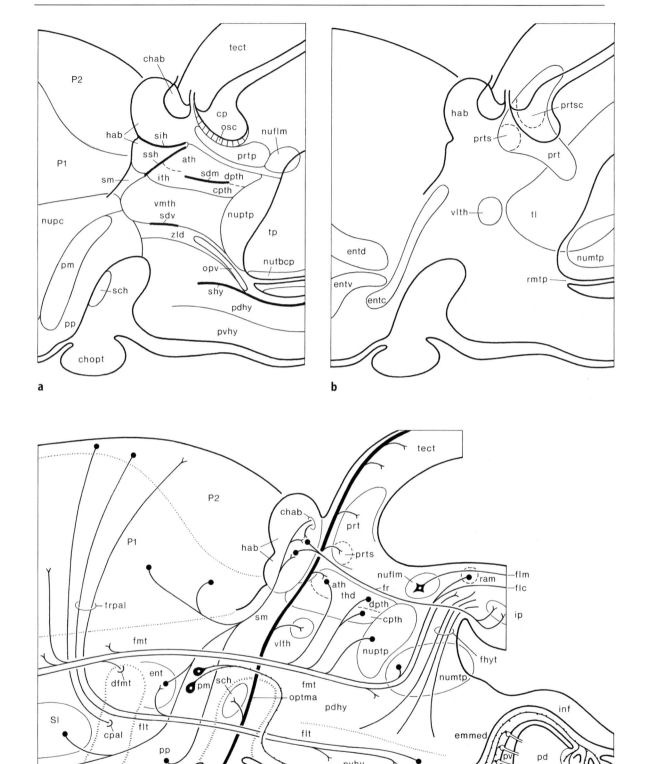

◀ **Fig. 13.24a–c.** Graphical reconstructions of the diencephalon and the adjacent mesencephalic and telencephalic regions of the reedfish *Erpetoichthys calabaricus*. **a** Periventricular cell masses. **b** Migrated cell masses. **c** Fibre connections. based on normal materal and on the experimental neuroanatomical studies carried out by Northcutt (1981), Braford and Northcutt (1978, 1983) and Repérant et al. (1981). See Fig. 13.30 for the results of more recent experimental neuroanatomical studies. *A*, type A fibres forming tractus preopticohypophyseus; *B*, type B fibres terminating in eminentia mediana; *flc*, fibres ascending from locus coeruleus; *ram*, mesencephalic raphe cells. (Reproduced from Nieuwenhuys 1983)

rior thalamic nucleus closely resembles the dorsal lateral geniculate nucleus in terrestrial vertebrates (Northcutt 1981; see also Northcutt and Butler 1980). The dorsal posterior and central posterior thalamic nuclei also project to the telencephalon (Northcutt 1981). The afferents of these nuclei have not been experimentally determined. In holosteans, however, the dorsal posterior nucleus receives a bilateral tectal input (Northcutt and Butler 1980), and the central posterior nucleus receives a bilateral input from the torus semicircularis (Braford and McCormick 1979). These data strongly suggest that the dorsal posterior and central posterior nuclei in brachiopterygians are homologous to the sauropsid nucleus rotundus and nucleus medialis thalami, respectively (Northcutt 1981). According to recent, as yet unpublished observations by P.H. Holmes and R.G. Northcutt (see Northcutt 1995), in *Polypterus* the anterior, dorsal posterior and central posterior thalamic nuclei project to the dorsal part of the area ventralis.

The following four nuclei have been distinguished in the brachiopterygian pretectum:

1. The *nucleus pretectalis periventricularis*. Directly dorsal to the fasciculus retroflexus and lateral to the subcommissural organ, a strip of periventricular grey extends caudally into the tegmentum mesencephali. This cell mass, which has been considered part of the dorsal thalamus (Nieuwenhuys and Bodenheimer 1966), has also been termed the nucleus pretectalis periventricularis (Braford and Northcutt 1983).
2. The *nucleus pretectalis centralis*. This nucleus consists of somewhat scattered cells, which lie lateral to the periventricular pretectum. It receives a retinal input; however, this is less dense that that to the periventricular or superficial pretectum (Braford and Northcutt 1983).
3. The *nucleus pretectalis superficialis*. At the site of bifurcation of the optic tract, a small, superficially situated cell mass has been designated the corpus geniculatum laterale (Nieuwenhuys and Bodenheimer 1966; Repérant et al. 1981). This nucleus receives a strong contralateral projection

from the retina (Repérant et al. 1981; Braford and Northcutt 1983); however, it does not project to the telencephalon (Northcutt 1981). Because of its positional relations and the absence of a telencephalic projection, it is likely that this so-called lateral geniculate body actually represents a pretectal cell group. Accordingly, it is termed the nucleus pretectalis superficialis, following Braford and Northcutt (1983) (see Fig. 13.24a,c).

4. The *nucleus pretectalis supracommissuralis*. The most dorsal part of the pretectal region is occupied by a large mass of small, densely clustered cells. This cell mass has been referred to as the nucleus pretectalis (Nieuwenhuys and Bodenheimer 1966) and the nucleus commissurae posterioris (Repérant et al. 1981), but is referred to here as the nucleus pretectalis supracommissuralis (Braford and Northcutt 1983). Like the nucleus pretectalis superficialis, the supracommissural nucleus receives a strong projection from the retina. Unlike the other three pretectal nuclei, the brachiopterygian supracommissural nucleus has no distinct counterpart in actinopterygians.

13.8.4
Thalamus Ventralis and Regio Tuberculi Posterioris

The nucleus ventromedialis thalami occupies the periventricular zone of the ventral thalamus and is separated ventrally from the hypothalamus by a cell-poor zona limitans. For a short distance, its ventral boundary is also marked by a ventricular groove, the sulcus diencephalicus ventralis. The more dorsally situated sulcus diencephalicus medius does not coincide with the boundary between the dorsal and ventral thalamic areas (Figs. 13.17, 13.24a). According to Jeener (1930), the rostralmost part of the ventromedial thalamic nucleus represents the eminentia thalami in urodeles. No experimental data concerning the fibre connections of this centre are available.

In the region of the ventral thalamus, a small, well-defined group of cells is situated directly lateral to the periventricular grey zone. This cell mass, the nucleus ventrolateralis thalami (Braford and Northcutt 1983), receives a strong contralateral projection from the retina (Repérant et al. 1981; Braford and Northcutt 1983).

The regio tuberculi posterioris contains three cell masses: the nucleus periventricularis tuberculi posterioris, the nucleus tuberculi posterioris and the nucleus medianus tuberculi posterioris. The huge nucleus tori lateralis, which according to Bergquist (1932) is of mesencephalic origin, also encroaches upon this region (Figs. 13.15, 13.16).

The nucleus periventricularis tuberculi posterioris consists of a thin central grey layer situated between the periventricular thalamic centres rostrally and the tuberculum posterius caudally. Rostroventrally, it borders the so-called paraventricular organ, an elongated outpouching of the ventricular wall (Figs. 13.16, 13.24a). Most of the cells in the nucleus periventricularis tuberculi posterioris are rather small, but large neurons do occasionally occur.

The immunohistochemical investigations carried out by Reiner and Northcutt (1992) and Piñuela and Northcutt (1995) revealed that, in *Polypterus*, the periventricular nucleus of the posterior tubercle (PPN) contains numerous dopamine- and serotonin-positive perikarya and that this nucleus and the region lateral to it are rich in tyrosine hydroxylase-, serotonin-, substance P- and leu-enkephalin-positive fibres (see Fig. 13.29b). Comparable observations have been made in the holostean *Lepisosteus* and in several teleosts (for references, see Reiner and Northcutt 1992; see also Chap. 15). With regard to these immunohistochemical features, the brachiopterygian and actinopterygian posterior tubercle resembles the ventral tegmental-substantia nigra (VT/SN) complex of amniotes. Although tyrosine hydroxylase- and serotonin-positive neurons are absent from the rostral tegmentum in *Polypterus*, and although the substance P- and leu-enkephalin-positive fibre plexuses in the tubercular region of that species are probably of striatal origin (see below), as they are known to be in amniotes, Reiner and Northcutt (1992) were reluctant to consider the posterior tubercle homologous to the amniote ventral tegmental-substantia nigra complex, particularly because the former is situated in the diencephalon and the latter in the midbrain. In their opinion, it would be of particular interest to know the ontogenetic source of the dopaminergic/tyrosine hydroxylase-positive neurons of the posterior tubercle and those that populate the ventral tegmental-substantia nigra complex in amniotes.

The small, elongated nucleus tuberculi posterioris flanks the recessus medianus, a long, caudally directed offshoot of the third ventricle which penetrates the tuberculum posterius in the median plane (Figs. 13.16, 13.24a).

The nucleus medianus tuberculi posterioris is a large cell mass situated ventromedial to the nucleus tori lateralis. The two sides of this nucleus fuse in the median plane of the tuberculum posterius (Figs. 13.16, 13.17). The nucleus tori lateralis has already been discussed in Sect. 13.7, but it is noteworthy that both this nucleus and the nucleus medianus consist of clusters of rather small cells that, according to Northcutt (1981), project bilaterally to the telencephalon. Holmes and Northcutt (1995) recently reported that the nucleus medianus of the posterior tuberculum is a major source of afferents to the pallial fields P2 and P3.

13.8.5
Hypothalamus

The hypothalamus is a relatively simple structure containing a zone of small, densely packed neuronal somata in no obvious arrangement. Throughout most of its extent, the hypothalamus can be divided into a dorsal part and a ventral part. The dorsal part is characterised by the presence of a layer of cells situated directly lateral to the periventricular grey (Figs. 13.16, 13.17). Hypothalamic neurons are small, mostly piriform elements with poorly branched, peripherally directed dendrites (Fig. 13.5d). Many of these cells have a central, liquor-contacting process (Fasolo et al. 1978).

Little is known about the fibre connections of the hypothalamus. Analysis of non-experimental histological material suggests that fibres of the medial forebrain bundle connect the dorsal part of the hypothalamus with the mediobasal part of the telencephalon, whereas the ventral part of the hypothalamus is strongly interconnected with the dorsal, pallial part of the telencephalon by the lateral forebrain bundle. Both forebrain bundles are discussed in more detail below. A large fibre stream can be traced from the brain stem into the hypothalamus. The caudal extent of this pathway could not be determined, but the fibres pass dorsal to the commissura ansulata and bend ventralward into the hypothalamus, passing between the nucleus tori lateralis and the nucleus medianus tuberculi posterioris and rostral to the latter. The fibres fan out in the dorsal parts of the hypothalamus (Figs. 13.16, 13.24c). The polarity of these fibres is unknown; however, they may well include ascending tertiary gustatory fibres and descending projections connecting the hypothalamus with the rostral rhombencephalon, as such connections have been described in most groups of fishes.

13.8.6
Hypothalamo-hypophysial Relationships

The pituitary gland lies against the caudoventral surface of the hypothalamus (Fig. 13.3b,d). It consists of a rostral pars distalis, a pars intermedia and a caudal pars nervosa. Long, canal-like processes of the infundibular recess extend into the pars ner-

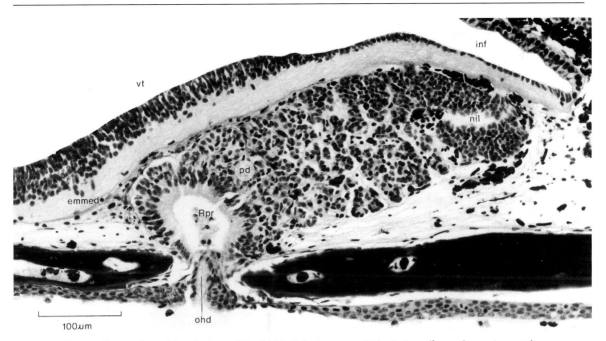

Fig. 13.25. Paramedian section of the pituitary of the bichir *Polypterus ornatipinnis.* Juvenile specimen, Azan stain

vosa. As already noted, the rostral portion of the pars distalis contains a remarkable remnant of Rathke's pouch, which opens into the oral cavity (Fig. 13.25). Just rostral to the pituitary gland, the hypothalamic floor forms a median eminence. The space between this eminence and the pars distalis of the pituitary gland contains blood vessels forming part of a typical hypothalamo-hypophysial portal system (Kerr 1968; Lagios 1968). Two types of neurosecretory fibres pass rostrocaudally through the hypothalamic floor (Lagios 1968). The first type, designated as *type A fibres*, pass through the internal layer of the median eminence and enter the pars nervosa of the pituitary gland, in which they form specialized pericapillary contacts. These contacts may well be the release site of neurohypophysial hormones into the bloodstream (Holmes and Ball 1974). Two octapeptides, arginine-vasotocin and isotocin, are known to occur in *Polypterus* (Sawyer 1969). The origin of these type A fibres has not been determined; however, it is reasonable to assume that they originate from a distinct magnocellular, neurosecretory nucleus in the preoptic region (Figs. 13.18, 13.24a). Type A fibres in the neural lobe contain electron-lucent secretion granules ranging from 91 to 106 nm in diameter and many clear vesicles averaging 45 nm in diameter. Other type A fibres with larger secretion granules (122–129 nm in diameter) form close neurosecretory contacts with the pars intermedia (Lagios 1968).

Type B fibres constitute the second type of axon traversing the floor of the hypothalamus. In addition to small 'synaptic' vesicles, these fibres contain electron-dense secretion granules ranging from 91 to 106 nm in diameter (Lagios 1968). Type B fibres enter the external palisade layer of the median eminence, where they form a plexus. Portal blood vessels form a capillary bed within the palisade layer, and the type B fibres make typical synapse-like contacts with the basement membrane of the perivascular space surrounding these portal capillaries (Lagios 1968). The type B neurosecretory material is carried by the portal blood to the pars distalis of the pituitary gland; a direct innervation of the pars distalis is entirely lacking. The origin of type B fibres is unknown. Brachiopterygians lack a nucleus lateralis tuberis, a major source of type B fibres in other fishes, and Holmes and Ball (1974) suggested that the preoptic nucleus might well be the source of such fibres in this group. Fasolo et al. (1978) observed that liquor-contacting neurons located in the caudal portion of the hypothalamus send their axons toward the vascular apparatus in the median eminence.

13.8.7
Diencephalic Fibre Systems

A number of diencephalic fibre systems have already been discussed in previous sections: the stria medullaris, the fasciculus retroflexus, the hypothalamotegmental fibres and the hypothalamohypophysial connections. In this section, we will describe the remaining diencephalic pathways: the tractus opticus, the medial and lateral forebrain bundles and the commissural systems.

The *optic nerve* decussates largely at the chiasm, and the majority of its fibres form a massive tractus opticus marginalis. Upon reaching the dorsocaudal part of the thalamus, this bundle splits into medial and lateral optic tracts, both of which terminate in the tectum (Fig. 13.3b). Some of the optic nerve fibres that decussate in the dorsalmost part of the chiasm form a separate axillary or axial optic tract. After crossing, most of the axial fibres run in fascicles through the medial forebrain bundle and then rejoin the main part of the optic tract. Experimental anatomical studies (Repérant et al. 1981; Braford and Northcutt 1983) demonstrated that the following diencephalic centres receive a retinal projection via the optic tract (Fig. 13.24c):

1. The *nucleus suprachiasmaticus*, a periventricular cell mass that surrounds the caudalmost part of the preoptic recess.
2. The *nucleus ventrolateralis thalami.*
3. The *nucleus anterior thalami* and the neuropil directly adjacent to that cell mass, i.e. the area optica dorsolateralis thalami of Repérant et al. (1981) and the nucleus a + b of Braford and Northcutt (1983).
4. The *nucleus pretectalis superficialis*, i.e. the corpus geniculatum laterale of Nieuwenhuys and Bodenheimer (1966) and Repérant et al. (1981).
5. The *nucleus pretectalis supracommissuralis*, i.e. the nucleus pretectalis of Nieuwenhuys and Bodenheimer (1966) and the nucleus commissurae posterioris of Repérant et al. (1981). However, Braford and Northcutt did not report optic terminals to this nucleus, but did report terminals to the adjacent central pretectal nucleus.

All of the above-mentioned centres receive a contralateral retinal projection; the first, third and fifth also receive an ipsilateral projection.

The fasciculus medialis telencephali or *medial forebrain bundle* is a large fibre system that extends from the subpallial part of the telencephalon rostrally to the regio tuberculi posterioris caudally. In the telencephalon, it runs lateral to the dorsal part of the area ventralis and its supracommissural and postcommissural caudal extensions. In the diencephalon, it passes along the zona limitans and the adjacent zones of thalamic and hypothalamic periventricular grey (Figs. 13.16, 13.20). Its dorsal fibres, which can be traced rostrally into the medial pallial field P1 (Figs. 13.19, 13.20), are coarse and well myelinated; however, most of its fibres are thin

and poorly myelinated. In the precommissural part of the subpallium, the bundle is not easily delineated from the medial olfactory tract, and caudally it contributes fibres to the fasciculi tegmentales (Figs. 13.15, 13.16). During its course, the bundle passes and contributes fibres to three commissures: the commissura anterior, the commissura postoptica and the commissura tuberculi posterioris, the contribution to the anterior commissure being by far the largest (Fig. 13.19). On the basis of non-experimental silver material, it has been suggested that the medial forebrain bundle receives and discharges fibres throughout its course and contacts the laterally directed dendrites of many periventricular grisea (Nieuwenhuys and Bodenheimer 1966). HRP tracing techniques have established that several diencephalic and mesencephalic centres project to the telencephalon (Northcutt 1981). The combined data indicate that telencephalic afferents from at least two of these centres – nucleus tori lateralis and nucleus medianus – pass through the medial forebrain bundle.

The fasciculus lateralis telencephali or *lateral forebrain bundle* is formed by pallial and subpallial components which join in the caudal telencephalon (Figs. 13.18, 13.20). The bundle proceeds caudalward, piercing the optic tract, to the ventral part of the hypothalamus (Figs. 13.16, 13.18). The fine fibres of the larger pallial component are mainly related to the dorsal pallial field P2 and form a large, submeningeal bundle designated as the tractus pallii (Figs. 13.18, 13.20). The fibres of the subpallial component are related to the nuclei entopedunculares, three submeningeal centres located in the preoptic region. Axon degeneration studies have shown that both pallial and subpallial components of the lateral forebrain bundle contain ascending fibres (Braford and Northcutt 1978). Using HRP and DiI as tracers, Holmes and Northcutt recently analysed the afferent and efferent connections of the pallial fields P2 and P3 in *Polypterus senegalus*. According to this analysis, which so far has only been published in abstract form (Holmes and Northcutt 1995), the nucleus medianus of the posterior tuberculum projects strongly to pallial fields P2 and P3, while sparser projections to the same fields arise from the ipsilateral caudal dorsal hypothalamic nucleus and the contralateral nucleus medianus. Efferents from P2 and P3, which could be traced to the diencephalon, appeared to terminate in the dorsal hypothalamic nucleus. These findings suggest that the pallial component of the lateral forebrain bundle is related to the tubercular and dorsal hypothalamic regions, and, as far as the pallial fields P2 and P3 are concerned, *not* to the ventral hypothalamus, as indicated in Fig. 13.24c.

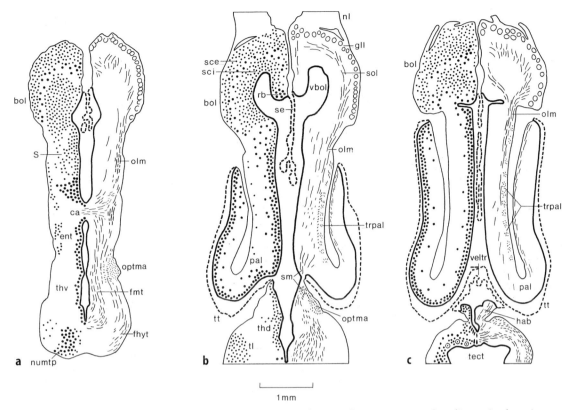

1mm

Fig. 13.26a–c. Horizontal sections through the rostral parts of the brain of the reedfish *Erpetoichthys calabaricus*. The *left half* of each figure shows the cell picture, based on Nissl-stained sections, the *right half* shows the fibre systems, based on adjacent Klüver-Barrera- and Bodian-stained sections. **a** Section through the ventral part. **b** Section through the intermediate part. **c** Section through the dorsal part

In addition to the commissura habenulae and the chiasma opticum discussed above, the diencephalon contains three other commissures: the commissura posterior, the commissura postoptica and the commissura tuberculi posterioris (Figs. 13.16, 13.24c). Our non-experimental material reveals no definite connections of these commissures, except for the contribution of the medial forebrain bundle to the commissura postoptica and the commissura tuberculi posterioris already noted.

13.9
Telencephalon

13.9.1
Introductory Notes

The morphogenesis and gross morphology of the telencephalon have been discussed in Sects. 13.2 and 13.3. Again, it should be noted that its rostral-most area is evaginated and its dorsal portions everted, as is evident in Figs. 13.18, 13.21 and, particularly, 13.26. The telencephalon has four principal parts:

1. The olfactory bulbs, which occupy almost the entire evaginated portion
2. The ventromedial subpallium or area ventralis telencephali
3. The caudoventral regio preoptica
4. The dorsally situated, everted pallium

These structures and their fibre connections will be discussed below. It should also be noted, however, that quantitative analysis indicates that the olfactory bulbs and the telencephalon proper in brachiopterygians are relatively much larger than their respective homologues in actinopterygian fishes (Northcutt and Braford 1980).

13.9.2
Olfactory Bulb

The olfactory bulbs attach to the rostroventral part of the telencephalon proper and are laminated concentrically. In centripetal order, the layers are as follows (Nieuwenhuys 1963, 1967a; Figs. 13.21, 13.26):

1. Layer of primary olfactory fibres
2. Glomerular layer

3. Stratum cellulare externum
4. Layer of secondary olfactory fibres
5. Stratum cellulare internum
6. Ependymal layer

The external cell layer is particularly well developed and contains numerous small elements, as well as large mitral cells. The mediocaudal portion of the evaginated walls lacks glomeruli and is thus considered part of the telencephalon proper. Nieuwenhuys (1963) held that the cell mass in this area represents part of the septum in other vertebrates, while Braford designated it as the nucleus retrobulbaris (Fig. 13.21). The immunohistochemical study carried out by Reiner and Northcutt (1992) has shown that substance P- and tyrosine hydroxylase-positive fibres are abundant in the olfactory bulb and have somewhat complementary distributions. Moreover, the external cellular layer contains numerous tyrosine hydroxylase-positive, presumably dopaminergic perikarya (see Fig. 13.28a).

Secondary olfactory fibres stream backward from the fibre layer of the bulb and gradually assemble in two fibre systems: the large tractus olfactorius medialis and the much smaller tractus olfactorius lateralis (Figs. 13.18–13.21, 13.26). For *Polypterus*, a degeneration study (Braford and Northcutt 1974; see also Northcutt and Braford 1980) and a tracer study with HRP and cobalt (von Bartheld and Meyer 1986b) of the secondary olfactory connections are available. Both studies agree in describing the medial olfactory tract as occupying a lateral position in the area ventralis and extending dorsally throughout the superficial zone of region P1. During its course, this tract does not give rise to projections to the various cell masses within the area ventralis, but only to the areas lateral to the ventral, dorsal and supracommissural parts of the area ventralis and to the nucleus situated in its dorsalmost part and medial to its lateral part. Most fibres of the medial olfactory tract terminate in Pl, i.e. the homologue of the medial zone of the area dorsalis (Dm) in actinopterygians.

The lateral olfactory tract, not detected in degeneration experiments (Braford and Northcutt 1974), appears to course through and to project to the ventral portion of region P3, i.e. the zone situated close to the line of attachment of the membranous telencephalic roof (von Bartheld and Meyer 1986b).

Fibres of the medial olfactory tract enter the diencephalon, where they divide into a major portion ascending to the epithalamus and a minor portion continuing its caudal course. The major portion crosses to the contralateral side in the habenular commissure, after which its fibres pass back to the telencephalon, terminating in the entopeduncu-lar nuclei and in the lateral part of the area ventralis. The rest of the diencephalic secondary olfactory fibres form a periventricular terminal field in the border zone between the ventral thalamus and the hypothalamus (von Bartheld and Meyer 1986b).

Apart from their primary olfactory input, the olfactory bulbs receive afferents from their ipsilateral subpallium. These ascending bulbar afferents arise mainly from the lateral part of the dorsal part of the area ventralis, but following injections of tracer in the olfactory bulb, the ventral, supracommissural and postcommissural parts of the area ventralis also contain a few labelled cells (von Bartheld and Meyer 1986b).

The central projections of the nervus terminalis in *Polypterus palmas* were experimentally examined by von Bartheld and Meyer (1986a, 1988). Following unilateral injections of HRP in the olfactory mucosa, these authors observed several small bundles of labeled fibres coursing caudally through the area ventralis. These fibres were found to terminate in the ventral and supracommissural parts of the area ventralis as well as in the preoptic region. Some fibres which decussated in the anterior commissure appeared to project to the corresponding telencephalic areas on the contralateral side. A few fibres could be traced to the diencephalon, where they terminated in the border zone between the ventral thalamus and hypothalamus. Injections of HRP into the olfactory bulb and into the eye of control specimens yielded no evidence for the presence of an olfactoretinal projection. The location of the ganglion cells of the terminal nerve could not be determined.

13.9.3
Subpallium

The subpallium or area ventralis telencephali contains a continuous layer of periventricular grey and a submeningeal cellular zone. Cytoarchitecturally, the periventricular grey can be differentiated into a pars dorsalis and a pars ventralis (Fig. 13.20). Both cell masses extend from the retrobulbar area to the ridge of the anterior commissure. Dorsal to the commissure, the pars dorsalis extends into nucleus supracommissuralis (Fig. 13.19), which in turn is caudally continuous with the nucleus postcommissuralis (Fig. 13.18). In the precommissural area, the dorsalmost part of the subpallial central grey is formed by a dense cluster of cells, designated in some studies as Vn (Nieuwenhuys 1963; Northcutt and Braford 1980; Fig. 13.20). Like the periventricular nuclei, the superficial subpallial cell zone, or lateral part of the area ventralis (Nieuwenhuys 1966; Fig. 13.20), extends from the olfactory bulb to

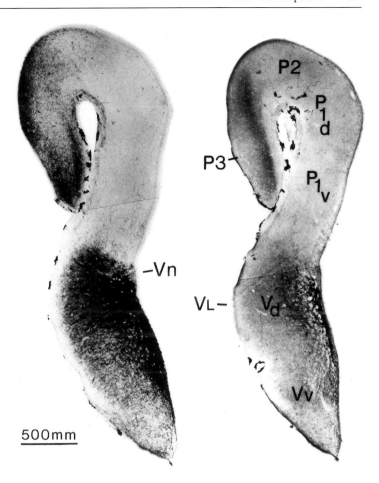

Fig. 13.27. Photomicrographs of sections through the telencephalon of the Senegal bichir *Polypterus senegalus*, labelled immunohistochemically for substance P (*left*) and leu-enkephalin (*right*). (Reproduced from Reiner and Northcutt 1992)

500mm

the level of the commissural ridge. Experimental studies carried out by Braford and Northcutt (1974) and von Bartheld and Meyer (1986b) have shown that secondary olfactory fibres form a sheet situated directly medial to the lateral part of the area ventralis. Older (Braford and Northcutt 1978; Northcutt 1981) and more recent experimental studies (Holmes and Northcutt 1995; P.H. Holmes and R.G. Northcutt, unpublished, see Northcutt 1995) have shown that the brachiopterygian subpallium is extensively and reciprocally connected with both the pallium and the diencephalon. Some of the fibre systems entertaining these relations will be discussed below.

Due to the unusually unevaginated, everted condition of the brachiopterygian cerebral hemispheres, comparison of its various grisea with those in vertebrates with evaginated hemispheres is not possible on a topographical basis, but only on a topological basis (Nieuwenhuys 1966). Following this principle, we arrive at the following, tentative interpretation of the cell masses in the brachiopterygian area ventralis, as compared to those in the subpallium of gnathostomes with evaginated forebrains:

The ventral part is probably homologous to the lateral septal nucleus, whereas the dorsal part may be homologous to the corpus striatum; the submeningeal, lateral part of the area ventralis corresponds topologically to the tuberculum olfactorium and may also comprise the homologue of the medial septal nucleus in evaginated forebrains. This interpretation is compatible with the results of recent immunohistological (Reiner and Northcutt 1992; Piñuela and Northcutt 1995) and experimental hodological studies (Holmes and Northcutt 1995; P.H. Holmes and R.G. Northcutt, unpublished, see Northcutt 1995). As regards immunohistochemistry, the neuropil of the dorsal part, like that of the striatum of chondrichthyans, dipnoans and amniotes, is rich in substance P-, leu-enkephalin-, dopamine- and serotonin-positive fibres (Figs. 13.27, 13.28). The majority of the dopamine-positive fibres most probably originate from cells situated in the periventricular nucleus of the posterior tuberculum. Another immunohistochemical feature which the pars dorsalis shares with the striatum of gnathostomes with evaginated forebrain is the presence of substance P-positive

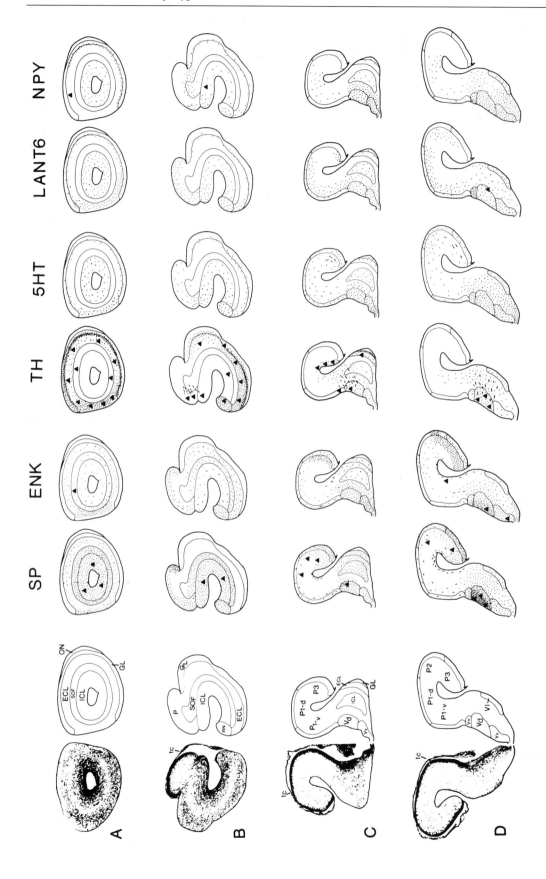

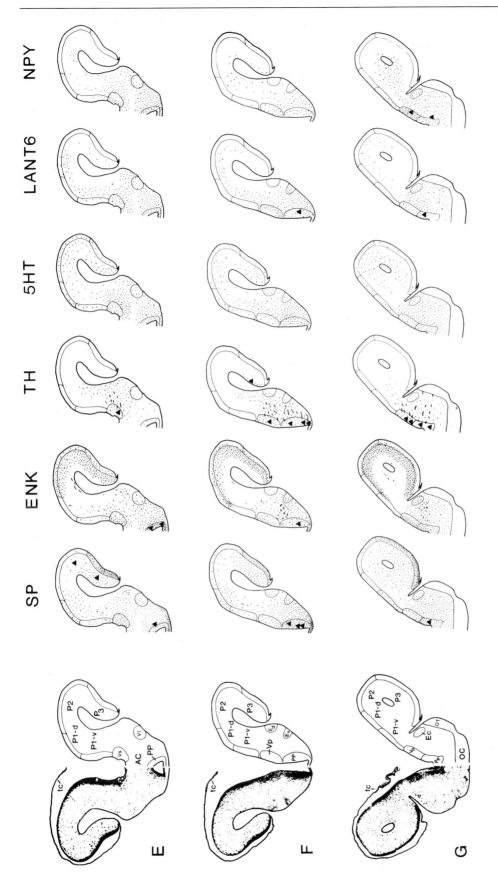

Fig. 13.28A–G. Semi-diagrammatic representation of the distribution of perikarya (*triangles*) and fibres (*dots*) containing substance P (*SP*), leu-enkephalin (*ENK*), tyrosine hydroxylase (*TH*), serotonin (*5HT*) and the neurotensin-related hexapeptides *LANT6* and neuropeptide Y (*NPY*) in a series of transverse sections through **a,b** the olfactory bulb and **c–g** the telencephalic hemispheres of the Senegal bichir *Polypterus senegalus*. *AC*, anterior commissure; *ECL*, external cell layers of the olfactory bulb; *Ec, Ed, Ev*, caudal, dorsal and ventral entopeduncular nuclei; *GL*, glomerular layer of the olfactory bulb; *ICL*, internal cell layer of the olfactory bulb; *OC*, optic chiasm; *ON*, olfactory nerve; *OT*, optic tract; *P*, pallium; *P1–d*, dorsal part of the first pallial zone; *PP*, periventricular preoptic nucleus; *RN*, retrobulbar nucleus; *SOF*, secondary olfactory fibre layer of the olfactory bulb; *tc*, tela choroidea. (From Reiner and Northcutt 1992)

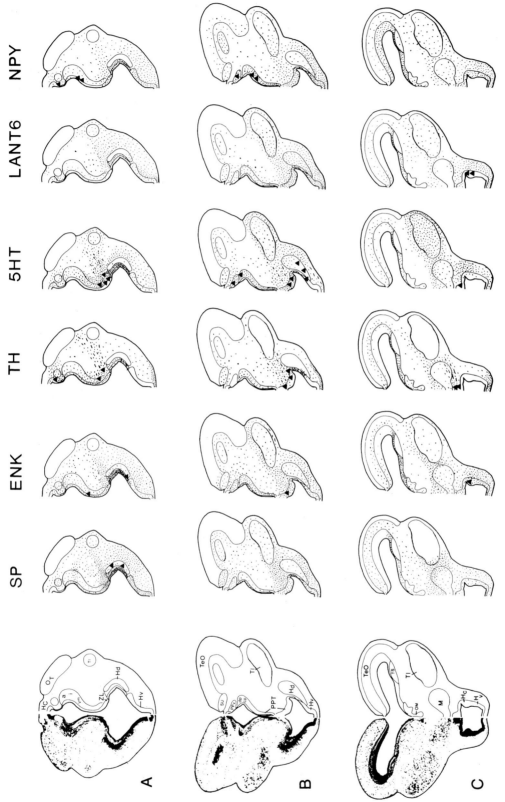

Fig. 13.29A–C. Semi-diagrammatic representation of the distribution of perikarya (*triangles*) and fibres (*dots*) containing substance P (*SP*), leu-enkephalin (*ENK*), tyrosine hydroxylase (*TH*), serotonin (*5HT*) and the neuropeptides *LANT6* and neuropeptide Y (*NPY*) in a series of transverse sections through **a**,**b** the posterior tubercle and **c** the rostral midbrain of the Senegal bichir *Polypterus senegalus*. *a*, anterior thalamic nuclei; *cp*, central posterior thalamic nucleus; *dp*, dorsal posterior thalamic nucleus; *HC*, habenular commissure; *Hd*, dorsal hypothalamus; *Hv*, ventral hypothalamus; *i*, intermediate thalamic nucleus; *M*, median nucleus of the posterior tubercle; *OM*, oculomotor nuclear complex; *OT*, optic tract; *PPT*, periventricular nucleus of the posterior tubercle; *PPv*, ventral periventricular pretectal nucleus; *SU*, supracommissural pretectal nucleus; *TeO*, tectum opticum; *Tl*, lateral toral nucleus; *Ts*, torus semicircularis; *Vm*, ventromedial thalamic nucleus; *ZL*, zona limitans diencephali. (Reproduced from Reiner and Northcutt 1992)

cells, giving rise to a bundle which descends to the posterior tubercle/tegmentum (Reiner and North-cutt 1992). As regards hodology, the ventral and lateral parts of the area ventralis are reciprocally connected with the pallial complex P2-P3 (Holmes and Northcutt 1995). These connections probably represent the septohippocampal and hippocampo-septal projections, found in evaginated telen-cephala. The pars dorsalis has been shown to receive afferents from the anterior, dorsal posterior and central posterior thalamic nuclei, is reciprocally connected with the olfactory bulb, the pallial complex P2-P3, the posterior tubercular region and the hypothalamus and projects to the midbrain teg-mentum (Holmes and Northcutt 1995, and unpub-lished results quoted in Northcutt 1995). Except for the connections with the pallial complex P2-P3, this pattern of connectivity corresponds to that of the striatum in urodele and anuran amphibians (see Chaps. 18 and 19). Experimental data on the con-nections of the striatum in other anamniotes with evaginated telencephala are not available. The inter-pretation of the parts of the brachiopterygian area ventralis, which – it should be re-emphasized – is tentative, differs from that of Northcutt and Braford (1980), but corresponds largely to that of Northcutt (1995).

13.9.4
Regio Preoptica

The periventricular grey of the preoptic region can be divided into three parts which occupy more or less rostrodorsal, intermediate and caudoventral positions: the nucleus postcommissuralis, nucleus preopticus magnocellularis and nucleus preopticus parvocellularis (Figs. 13.18, 13.19, 13.24a). Golgi studies (Mazzi et al. 1978) have shown that the magnocellular nucleus consists of multipolar neu-rons of the liquor-contacting type. Contact with the ventricular lumen is established directly by the soma or by means of a central process. The den-dritic tree consists of two or more infrequently branching dendrites which pass laterally. Numerous cerebrospinal fluid (CSF)-contacting neurons also occur in the parvocellular nucleus.

The submeningeal zone of the preoptic region contains three cell masses: the nucleus entopedunc-ularis dorsalis, ventralis and caudalis (Figs. 13.18, 13.19, 13.24b). The caudal entopeduncular nucleus extends into the thalamus and connects with the ventricular grey just rostral to the nucleus ventro-medialis thalami.

The immunohistochemical studies carried out by Reiner and Northcutt (1992) and Piñuela and Northcutt (1995) have shown that the periventricu-lar preoptic region contains substance P-, leu-enkephalin-, dopamine-, LANT6- and neuropeptide Y-positive perikarya and that fibres containing sub-stance P, leu-enkephalin and serotonin are abun-dant in this region (Fig. 13.28).

Little is known about the fibre connections of the preoptic periventricular grey. The neurons in this region extend their dendrites into the medial fore-brain bundle and presumably also discharge their axons into this bundle. Other efferents of the pre-optic periventricular grey probably pass to the habenular ganglia via the stria medullaris (Fig. 13.24c). The magnocellular preoptic elements may project to the pituitary gland, although it has not been established whether their axons are actu-ally continuous with the neurosecretory fibres pass-ing via the hypothalamic floor to the neuro-intermediate lobe.

It has been demonstrated experimentally that the entopeduncular nuclei receive projections from the olfactory bulb (Braford and Northcutt 1974) and, via the lateral forebrain bundle, from the hypothal-amus (Braford and Northcutt 1978). Examination of non-experimental material suggests that efferents from these nuclei join the subpallial component of the stria medullaris.

The apparent homologue of the entopeduncular nuclear complex in actinopterygian fishes has been described under various names: area somatica (Johnston 1911), nucleus entopeduncularis (Shel-don 1912; Schnitzlein 1962) and nucleus entope-duncularis anterior (Hocke Hoogenboom 1929). Northcutt (1995) suggested that both the postcom-missural and supracommissural parts of the area ventralis may be homologous to the basal portions of the amygdala in other gnathostomes.

13.9.5
Pallium

The large, thin-walled pallium has a simple histo-logical organisation. The majority of its perikarya are concentrated in a narrow periventricular layer, which, due to eversion, occupies a superficial posi-tion (Figs. 13.17–13.21). Deep to this zone, only sparsely distributed cells are found. Previous pub-lications have suggested that the periventricular pallial grey layer is uniform throughout (Nieuwen-huys 1962, 1963, 1966, 1969). However, re-examination of our *Polypterus* and *Erpetoichthys* material indicates that this layer comprises three zones, as suggested by Holmgren (1922) and Bra-ford and Northcutt (1974, 1978). These zones, plus their adjacent neuropil and fibre fields, are desig-nated here as P1, P2 and P3 (Braford and Northcutt 1974, 1978). The medial (P1) and lateral (P3) pallial zones contain small, darkly staining cells, whereas the intervening cell layer (P2) is composed of slightly larger and lighter-staining elements. In *Erpetoichthys*, P2 cannot be recognised in the rostralmost part of the pallium (Fig. 13.21); halfway between the olfactory bulb and the anterior commissure, it occupies only a small part of the lateral pallium (Fig. 13.20); more caudally, it progressively enlarges to extend over the dorsal and medial parts of the pallium (Figs. 13.18, 13.19). Golgi preparations revealed that the perikarya bordering the ventricle radiate a limited number of sparsely ramifying dendrites toward the meningeal surface. In area P2, elements with tangentially or obliquely coursing dendrites were observed (Fig. 13.5e). The axons of the periventricular elements usually issue from the cell bodies and have numerous collateral branches, which, at least in P1, constitute a dense plexus in a wide zone adjacent to the cell layer.

The following positional, immunohistochemical and hodological data are relevant to the morphological interpretation of the brachiopterygian pallial fields:

1. With regard to their topological position – and not their topographical position, as Northcutt (1995) claims – P1, P2 and P3 correspond to the lateral or piriform, dorsal or general and medial or hippocampal pallial fields, respectively, of gnathostomes with evaginated forebrains.
2. Reiner and Northcutt (1992) applied immunohistochemical techniques to the study of the telencephalon of *Polypterus*, using antibodies against six different neuroactive principles (see Sect. 13.10 and Figs. 13.27, 13.28). They found that P1 is distinguishable from the other pallial fields in that it generally possesses lower levels of all six substances examined. In the remaining part of the pallium, local differences in the labelling pattern of several of the substances examined were observed, but these differences did not clearly match the cytoarchitectonic boundary between P2 and P3.
3. P1 receives a prominent input from the olfactory bulb (Braford and Northcutt 1974; von Bartheld and Meyer 1986b), a property which it shares with the lateral pallium of gnathostomes with evaginated forebrains. Other afferents of P1 are presently unknown. The tracer study carried out by Holmes and Northcutt (1995) has shown that P1 projects to P2 and/or P3.
4. Holmes and Northcutt (1995) studied the connections of the pallial fields P2 and P3 (collectively designated by them as the 'nonolfactory pallium'), using HRP and DiI as tracers. It appeared that the pallial field P1, the subpallial,

ventral part of the area ventralis and the diencephalic nucleus medianus of the posterior tubercle are the major sources of afferents to this complex; in addition, it appeared that sparser projections to this complex arise ipsilaterally from the dorsal and lateral parts of the area ventralis and from the dorsal hypothalamic nucleus, and contralaterally from the ventral and dorsal parts of the area ventralis and from the nucleus medianus, as well as from the superior raphe and the locus coeruleus. Efferent projections from P2 and P3 appeared to terminate ipsilaterally in the dorsal, ventral and lateral parts of the area ventralis and the preoptic area, and contralaterally in the ventral and lateral parts of the area ventralis and the preoptic area after decussating in the anterior commissure. Bilaterally descending fibres continue into the diencephalon, where they divide into a ventral component projecting to the dorsal hypothalamic nucleus and a dorsal component terminating in the lateral toral complex. The commissural fibres, passing via the anterior commissure, were found to interconnect the P2-P3 complexes of both sides.

On surveying these data, it may be concluded that the brachiopterygian pallial field P1 is homologous to the lateral pallium of gnathostomes with evaginated forebrains. This interpretation, made on positional and hodological grounds, is also compatible with immunohistochemical data, because studies performed by Reiner and Northcutt (1987) and Northcutt et al. (1988) have revealed that the lateral pallial fields of lungfishes and sharks are poor in most of the neuroactive principles which also show a low concentration in P1. As regards P2 and P3, it may be concluded that these fields are probably homologous to the dorsal and medial pallial areas of gnathostomes with evaginated forebrains, respectively, but that more selective hodological data are required to fully establish their identity. The functional implications of the hodological data reviewed above will be discussed in the final section of the present chapter.

13.10
Immunohistochemical Data

The reports by Reiner and Northcutt (1992) and Piñuela and Northcutt (1995) are, to my knowledge, the only immunohistochemical studies on the brachiopterygian brain published so far. Reiner and Northcutt (1992) studied the distribution of the following six neuroactive principles in the brain of *Polypterus senegalus*: substance P, leu-enkephalin, tyrosine hydroxylase, serotonin, the neurotensin-related hexapeptide LANT6 and neuropeptide Y. In the preceding sections, frequent references have been made to this study. Figs. 13.28, 13.29 present a pictorial overview of its results.

In a study by Piñuela and Northcutt (1995), an antibody against dopamine was used to determine the distribution of this neurotransmitter in the telencephalon and diencephalon of *Polypterus senegalus*. Not surprisingly, the results showed a considerable overlap with those obtained with an antityrosine hydroxylase serum in the previous study. A high density of both perikarya and fibres labelled for dopamine was found in the periventricular pretectal nucleus, the periventricular nucleus of the posterior tuberculum and the basal hypothalamus. In the telencephalon, dopamine immunoreactivity appeared to be weaker and restricted to the area ventralis.

13.11
Concluding Remarks

In the above sections, we have attempted to summarise what is known about the neurobiology of brachiopterygian fishes, information that is entirely confined to morphological data. In this section, we will attempt to place these data in a functional perspective and will briefly discuss the implications of their neuromorphology on the currently still enigmatic taxonomy of these fishes.

Polypterids, as noted, are nocturnal, bottom-dwelling predators that feed primarily on other fish. The relative development of their various special sense organs suggests that olfaction is the primary sensory modality used in feeding. The olfactory organs are exceptionally large (Fig. 13.2), as are the olfactory bulbs (Fig. 13.3), and both are remarkably well differentiated histologically (Fig. 13.21). Experimental anatomical investigations (Braford and Northcutt 1974; von Bartheld and Meyer 1986b) have shown that the olfactory bulbs project to some subpallial centres, but that their main target is the medial part of the pallium, P1. Of the remaining pallial fields, P2 and P3, only the latter receives a limited secondary olfactory projection. It is, however, important to note that P2 and P3 receive strong projections from pallial field P1 as well as from the subpallial, ventral part of the area ventralis, both of which are targets of bulbar efferents (Fig. 13.30).

The size and differentiation of the tectum and the nuclei innervating the eye muscles indicate a moderately developed visual system. Visual impulses enter the tectum by way of the optic tracts and are probably correlated with stimuli from other sensory systems before information is relayed to the premotor reticular centres in the brain stem via the tectobulbar tracts.

Both the olfactory and visual systems undoubtedly influence reproductive behavior via the pituitary gland. The preoptic and suprachiasmatic nuclei may well serve as links in routes to this organ.

The size of the primary targets of the gustatory system (the lobus vagi), the mechanoreceptive lateral line system (the nucleus intermedius) and the electroreceptive lateral line system (the nucleus dorsalis) suggests that all of these special sense systems are at best moderately developed.

The results of experimental neuroanatomical studies indicate that the telencephalon of polypterids is not only involved in the processing of olfactory information (Fig. 13.30). Fibres originating from three dorsal thalamic centres, the nucleus anterior, the nucleus dorsalis posterior and the nucleus centralis posterior reach the dorsal part of the area ventralis, which represents the striatum; the nucleus medianus of the posterior tuberculum projects strongly to the pallial fields P2 and/or P3, and striatum and pallium are both in receipt of fibres originating from the hypothalamus. One of the dorsal thalamic nuclei, the nucleus anterior thalami, receives a direct input from the retina via the optic tract; hence it may be concluded that secondary visual information must reach the striatum, but the exact nature of the other projections ascending to the telencephalon is uncertain. It has been experimentally established that the nucleus dorsalis posterior thalami in chondrichthyans (Smeets 1981), holosteans (Northcutt and Butler 1980) and anuran amphibians (Masino and Grobstein 1990) receives afferents from the tectum mesencephali, and that in chondrichthyans (Boord and Northcutt 1982, 1988; Barry 1987), holosteans (Braford and McCormick 1979), teleosts (Echteler 1984; Striedter 1991) and anurans (Neary 1974) the torus semicircularis projects to the nucleus centralis posterior thalami. In adult ranid anurans, the torus semicircularis is the target of auditory projections, but in other anurans (such as *Xenopus*) and in most fishes, this centre receives not only an auditory, but also a strong lateral line input. If the projections discussed above are also present in polypterids, this would mean that, in this group, the dorsal part of the area ventralis not only receives a secondary visual input, but is also the target of an indirect visual circuit passing via the tectum and of ascending auditory and lateral line projections.

The nucleus medianus tuberculi posterioris forms the principal source of ascending afferents to the pallium in polypterids. At present, all that can be said about the possible nature of this projection is that, in teleosts, the various nuclei of the large

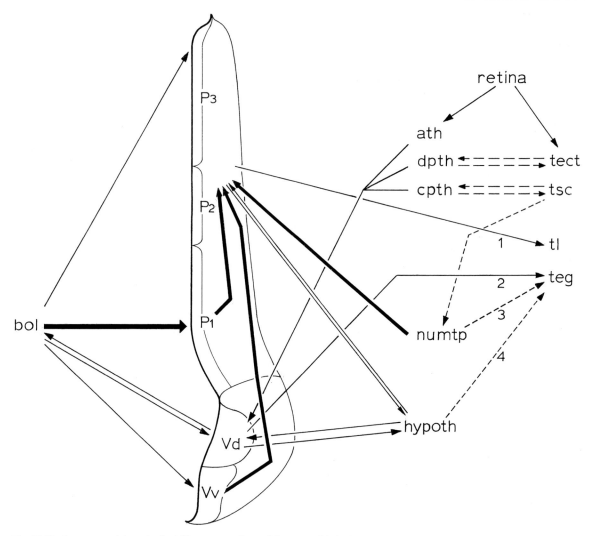

Fig. 13.30. Summary of the principal fibre connections of the cell masses in the brachiopterygian telencephalon. The pallium is unfolded. *Continuous lines* indicate experimentally established connections. (Based on data from von Bartheld and Meyer 1986b; Northcutt 1995; Holmes and Northcutt 1995)

preglomerulosus complex, which are ontogenetically also derived from the posterior tubercular region and which also project to the pallium, serve as relays in ascending auditory, lateral line and tectal projections (see Sect. 15.5.4.1 for details and references).

Experimental neuroanatomical findings in chondrichthyans (Smeets and Boord 1985) and teleosts (Morita et al. 1980, 1983; Finger 1983; Wulliman 1988) suggest that the hypothalamo-telencephalic projections in polypterids possibly convey gustatory information to the telencephalon.

In summary, the data reviewed indicate that, in polypterids, the subpallial, dorsal part of the area ventralis receives a secondary optic input and that, both in this centre and in the pallial fields P2 and

P3, olfactory information is correlated with information from other sensory modalities.

The (partly hypothetical) paths along which information processed in the telencephalon and diencephalon may reach the pre-motor centres in the brain stem are numbered 1–4 in Fig. 13.30.

The organisation of the motor system does not require much comment. As noted, the somatomotor centres innervating the external eye muscles are rather small. With regard to the branchiomotor nuclei, the motor nucleus of the trigeminal nerve, which innervates the jaw musculature, is strongly developed, as in most predatory fishes. The medial reticular formation, which presumably serves to relay information from the cerebellum, tectum, diencephalon and telencephalon to the spinal motor apparatus, does not differ significantly from that observed in other gnathostome fishes. However, it should be added that the large, compact nucleus medianus magnocellularis, situated in the caudal rhombencephalon, probably belongs to the pre-motor apparatus of the brain stem. This intriguing centre is unique to polypterids, and its specific function is so far unknown.

The taxonomical position of the polypterids is controversial. According to Lauder and Liem (1983b, p. 108), "probably no other group of living fishes has been placed, at one time or another, in so many widely differing taxonomic groups". In the nineteenth century, the polypterids were considered to be related to the crossopterygians (e.g. Huxley 1861). In the Devonian period, the crossopterygians were by far the commonest bony fishes. Until 1938, this group was generally thought to have long been extinct; in that year, however, a strange fish caught off the coast of southeast Africa proved to be a surviving crossopterygian, *Latimeria chalumnae* (see Chap. 17). In the early twentieth century, Goodrich (1907, 1928) suggested that the polypterids exhibit numerous paleaoniscoid features and therefore must be related to the actinopterygians. The palaeoniscoids form another group of once abundant paleozoic fishes. They have long been considered to represent the stem group of the actinopterygians or ray-finned fishes, to which, not only the abundant teleosts, but also the holosteans and chondrosteans belong (see Chaps. 14, 15). Although the extinct palaeoniscoids probably form a polyphyletic group (see Patterson 1994), Goodrich's suggestion that the polypterids should be classified as actinopterygians was accepted by numerous authors (Berg 1958; Gardiner 1967; Moy-Thomas and Miles 1971; Patterson 1982, 1994; Lauder and Liem 1983a,b). Most of these authors held that the polypterids are the most primitive of the living actinopterygians. Other authors, including Jarvik (1947, 1968), Daget (1958), Poll (1965), Nelson (1969) and Bjerring (1985), however, maintained that the polypterids lack specific affinities with the actinopterygians and placed them in a separate division of the Osteichthyes, which, in accordance with Stensiö (1921), is commonly referred to as the Brachiopterygii. The latter author placed the Brachiopterygii as a separate subclass intermediate between actinopterygians and crossopterygians. Cytological and molecular data have not yielded concordant results with regard to the taxonomic position of polypterids either. Without going into details, the following should be mentioned:

1. The study of *Polypterus* chromosomes and DNA/cell values showed striking similarities with lungfishes and revealed marked differences with actinopterygians (Bachmann 1972; Denton and Howell 1973).
2. According to Matthieu et al. (1986), the chemical composition of the myelin in the central nervous system of *Polypterus* not only reveals its close relationship to chondrosteans, but also demonstrates the existence of a relationship with tetrapods and their allies, the lungfishes.
3. 'Molecular phylogenetic trees' based on 28S rRNA sequences (Le et al. 1993) support the position of polypterids as actinopterygians, but suggest that in such trees they may fall above, rather than below the chondrosteans (Patterson 1994).

In view of the uncertainty sketched above, an inquiry into the bearing of neural characters on the various hypotheses concerning the phylogenetic affinity of the polypterids could be of great interest.

References

Ariëns Kappers J (1965) Survey of the innervation of the epiphysis cerebri and the accessory pineal organs of vertebrates. Prog Brain Res 10:87–153

Bachmann K (1972) The nuclear DNA of Polypterus palmas. Copeia 2:363–365

Barry MA (1987) Afferent and efferent connections of the primary octaval nuclei in the clearnose skate, Raja eglanteria. J Comp Neurol 266:457–477

Berg LS (1958) System der Rezenten und Fossilen Fischartigen und Fische. Deutscher Verlag der Wissenschaften, Berlin

Bergquist H (1932) Zur Morphologie des Zwischenhirns bei niederen Wirbeltieren. Acta Zool 13:57–304

Bergquist H, Källén B (1954) Notes on the early histogenesis and morphogenesis of the central nervous system in vertebrates. J Comp Neurol 100:627–659

Bern HA (1969) Urophysis and caudal neurosecretory system. In: Hoar WS, Randall DJ (eds) Fish physiology, vol II. Academic, New York, pp 399–418

Bjerring HC (1985) Facts and thoughts on piscine phylogeny. In: Foreman RE, Gorbman A, Dodd JM, Olsson R (eds) Evolutionary biology of primitive fishes. Plenum, New York, pp 31–57

Boord RL, Northcutt RG (1982) Ascending lateral line pathways to the midbrain of the clearnose skate, Raja eglanteria. J Comp Neurol 207:274–282

Boord RL, Northcutt RG (1988) Medullary and mesencephalic pathways and connections of lateral line neurons of the spiny dogfish Squalus acanthias. Brain Behav Evol 32:76–88

Braford MR Jr, McCormick CA (1979) Some connections of the torus semicircularis in the bowfin, Amia calva: a horseradish peroxidase study. Soc Neurosci Abstr 5:139

Braford MR Jr, Northcutt RG (1974) Olfactory bulb projections in the bichir, Polypterus. J Comp Neurol 156:165–178

Braford MR Jr, Northcutt RG (1978) Correlation of telencephalic afferents and SDH distribution in the bony fish Polypterus. Brain Res 152:157–160

Braford MR Jr, Northcutt RG (1983) Organization of the diencephalon and pretectum in ray-finned fishes. In: Davis RE, Northcutt RG (eds) Fish neurobiology. 2. Higher brain areas and functions. University of Michigan Press, Ann Arbor, pp 117–164

Daget J (1958) Sous-classe des brachioptérygiens. Traite Zool 13:2500–2521

Denton TE, Howell WM (1973) Chromosomes of the African polypterid fishes, Polypterus palmas and Calamoichtys calabaricus (Pisces: Brachiopterygii). Experientia 29:122–124

Echteler SM (1974) Connections of the auditory midbrain in a teleost fish, Cyprinus carpio. J Comp Neurol 230:536–551

Fasolo A, Mazzi V, Franzoni MF (1978) A Golgi study of the hypothalamus of Actinopterygii. II. The posterior hypothalamus. Cell Tissue Res 191:433–447

Finger TE (1983) The gustatory system in teleost fish. In: Northcutt RG, Davis RE (eds) Fish neurobiology. 1. Brain stem and sense organs. University of Michigan Press, Ann Arbor, pp 285–310

Gardiner BG (1967) Further notes on paleoniscoid fishes with a classification of the Chondrostei. Bull Brit Mus Nat Hist (Geol) 14:143–206

Goodrich ES (1907) On the scales of fish, living and extinct, and their importance in classification. Proc Zool Soc Lond: 751–774

Goodrich ES (1928) Polypterus, a palaeoniscid? Palaeobiology 1:87–91

Hocke Hoogenboom KJ (1929) Das Gehirn von Polyodon folium Lacép. Z Mikrosk Anat Forsch 18:311–392

Holmes RL, Ball JN (1974) The pituitary gland: a comparative account. Cambridge University Press, Cambridge

Holmes PH, Northcutt RG (1995) Afferent and efferent connections of the nonolfactory pallium in the Senegal bichir, Polypterus senegalus (Ostheichtyes: Cladistia). Soc Neurosci Abstr 21:432

Holmgren N (1922) Points of view concerning forebrain morphology in lower vertebrates. J Comp Neurol 34:391–440

Huxley TH (1861) Preliminary essay upon the systematic arrangement of the fishes of the devonian epoch. Mem Geol Surv UK Decade 10:1–40

Jarvik E (1947) Notes on the pit-lines and dermal bones of the head in Polypterus. Zool Bidrag Uppsala 25:60–78

Jarvik E (1968) Aspects of vertebrate physiology. In: Ørvig T (ed) Current problems of lower vertebrate phylogeny. Fourth Nobel Symposium. Almqvist and Wiksell, Stockholm, pp 497–527

Jeener R (1930) Evolution des centres diencéphaliques périventriculaires des Téléostomes. Proc Kon Ned Akad B 33:1–16

Johnston JB (1911) The telencephalon of ganoids and teleosts. J Comp Neurol 21:489–591

Kalmijn AJ (1978) Electric and magnetic sensory world of sharks, skates and rays. In: Hodgson ES, Mathewson RF (eds) Sensory biology of sharks, skates and rays. US Government Printing Office, Washington DC, pp 507–528

Kenemans P (1980) On the structural plan of the brain stem. Thesis, Nijmegen

Kerr T (1968) The pituitary in Polypterines and its relationship to other fish pituitaries. J Morphol 124:23–36

Lagios MD (1968) Tetrapod-like organization of the pituitary gland of the polypteriformid fishes, Calamoichthys calabaricus and Polypterus palmas. Gen Comp Endocrinol 11:300–315

Larsell O (1967) The comparative anatomy and histology of the cerebellum from myxinoids through birds. University of Minnesota Press, Minneapolis

Lauder GV, Liem KF (1983a) Patterns of diversity and evolution in ray-finned fishes. In: Northcutt RG, Davis RE (eds) Fish neurobiology. 1. Brainstem and sense organs. University of Michigan Press, Ann Arbor, pp 1–24

Lauder GV, Liem KF (1983b) The evolution and interrelationships of the actinopterygian fishes. Bull Mus Comp Zool 150:95–197

Le HLV, Lecointre G, Perasso R (1993) A 28S rRNA-based phylogeny of the Gnathostomes: first steps in the analysis of conflict and congruence with morphologically based cladograms. Mol Phylogen Evol 2:31–51

Masino T, Grobstein P (1990) Tectal connectivity in the frog Rana pipiens: tectotegmental projections and a general analysis of topographic organization. J Comp Neurol 291:103–127

Matthieu J-M, Eschmann N, Bürgisser P, Malotka J, Waehneldt TV (1986) Expression of the myelin proteins characteristic of fish and tetrapods by Polypterus revitalizes long discredited phylogenetic links. Brain Res 379:137–142

Mazzi V, Fasolo A, Franzoni MF (1977) The optic tectum of Calamoichthys calabaricus Smithi. A Golgi study. Cell Tissue Res 182:491–503

Mazzi V, Franzoni MF, Fasolo A (1978) A Golgi study of the hypothalamus of Actinopterygii. I. The preoptic area. Cell Tissue Res 186:475–490

McCormick CA (1978) Central projections of the lateralis and eighth nerves in the bowfin, Amia calva. Thesis, University of Michigan

McCormick CA (1981) Central projections of the lateral line and eighth nerves in the bowfin, Amia calva. J Comp Neurol 197:1–15

McCormick CA (1982) The organization of the octavolateralis area in actinopterygian fishes: a new interpretation. J Morphol 171:159–181

Morita Y, Ito H, Masai H (1980) Central gustatory paths in the crucian carp, Carassius carassius. J Comp Neurol 191:119–132

Morita Y, Murakami T, Ito H (1983) Cytoarchitecture and topographic projections of the gustatory centers in a teleost, Carassius carassius. J Comp Neurol 218:378–394

Moy-Thomas JA, Miles RS (1971) Palaeozoic fishes. Chapman and Hall, London

Neary TJ (1974) Diencephalic efferents of the torus semicircularis in the bullfrog, Rana catesbeiana. Anat Rec 178:425

Nelson GJ (1969) Origin and diversification of teleostean fishes. In: Petras JM, Noback CR (eds) Comparative and evolutionary aspects of the vertebrate central nervous system. Ann NY Acad Sci 167:18–30

Nieuwenhuys R (1962) Trends in the evolution of the Actinopterygian forebrain. J Morphol 111:69–88

Nieuwenhuys R (1963) The comparative anatomy of the Actinopterygian forebrain. J Hirnforsch 6:171–192

Nieuwenhuys R (1966) The interpretation of the cell masses in the teleostean forebrain. In: Hassler R, Stephan H (eds) Evolution of the forebrain. Thieme, Stuttgart, pp 32–39

Nieuwenhuys R (1967a) Comparative anatomy of the cerebellum. Prog Brain Res 25:1–93

Nieuwenhuys R (1967b) Comparative anatomy of olfactory centres and tracts. Prog Brain Res 23:1–64

Nieuwenhuys R (1969) A survey of the structure of the forebrain in higher bony fishes. Ann NY Acad Sci 167:31–64

Nieuwenhuys R (1983) The central nervous system of the Brachiopterygian fish Erpetoichthys calabaricus. J Hirnforsch 24:501–533

Nieuwenhuys R, Bodenheimer TS (1966) The diencephalon of the primitive bony fish Polypterus in the light of the problem of homology. J Morphol 118:415–450

Nieuwenhuys R, Oey PL (1983) Topological analysis of the brain stem of the reedfish, Erpetoichthys calabaricus. J Comp Neurol 213:220–232

Nieuwenhuys R, Bauchot R, Arnoult J (1969) Le dévelopement de télencéphale d'un poisson osseux primitif, Polypterus senegalus Cuvier. Acta Zool 50:101–125

Northcutt RG (1981) Localization of neurons afferent to the telencephalon in a primitive bony fish, Polypterus palmas. Neurosci Lett 22:219–222

Northcutt RG (1983) Evolution of the optic tectum in ray-finned fishes. In: Davis RE, Northcutt RG (eds) Fish neurobiology. 2. Higher brain areas and functions. University of Michigan Press, Ann Arbor, pp 1–42

Northcutt RG (1995) The forebrain of Gnathostomes: In search of a morphotype. Brain Behav Evol 46:275–318

Northcutt RG, Braford MR Jr (1980) New observations on the organization and evolution of the telencephalon of actinopterygian fishes. In: Ebbesson SOE (ed) Comparative neurology of the telencephalon. Plenum, New York, pp 41–98

Northcutt RG, Butler AB (1980) Projections of the optic tectum in the longnose gar, Lepisosteus osseus. Brain Res 190:333–346

Northcutt RG, Reiner A, Karten HJ (1988) An immunohistochemical study of the telencephalon of the spiny dogfish, Squalus acanthias. J Comp Neurol 227:250–267

Onstott D, Elde R (1986) Immunohistochemical localization of urotensin I/corticotropin-releasing factor, urotensin II, and serotonin immunoreactivities in the caudal spinal cord of nonteleost fishes. J Comp Neurol 249:205–225

Patterson C (1982) Morphology and interrelationships of primitive actinopterygian fishes. Am Zool 22:241–259

Patterson C (1994) Bony fishes In: Prothero DR, Schoch RM (eds) Short courses in paleontology, no 7. Paleontological Society, Knoxville, Tennessee, pp 57–84

Piñuela C, Northcutt RG (1995) Dopamine distribution in the forebrain of the Senegal bichir. Soc Neurosci Abstr 21:432

Platel R, Ridet J-M, Bauchot R, Diagne M (1977) L'organisation encéphalique chez Amia, Lepisosteus et Polypterus: Morphologie et analyse quantitative comparées. J Hirnforsch 18:69–73

Poll M (1965) Anatomie et systematique des Polypteres. Bull Acad R Belg Cl (5 ser) 51:553–569

Reiner A, Northcutt RG (1987) An immunohistochemical study of the telencephalon of the African lungfish. J Comp Neurol 256:463–481

Reiner A, Northcutt RG (1992) An immunohistochemical study of the telencephalon of the Senegal bichir (Polypterus senegalus). J Comp Neurol 319:359–386

Repérant J, Rio J-P, Amouzou M (1979) Analyse radioautographique des projections rétiniennes chez le Poisson osseux primitif Polypterus senegalus. CR Acad Sci [Paris] 289:D947–D950

Repérant J, Rio J-P, Miceli D, Amouzou M, Peyrichoux J (1981) The retinofugal pathways in the primitive african bony fish Polypterus senegalus (Cuvier 1829). Brain Res 217:225–243

Roth A (1973) Electroreceptors in Brachiopterygii and Dipnoi. Naturwissenschaften 2:S106

Roth A, Tscharntke H (1976) Ultrastructure of the ampullary electroreceptors in lungfish and brachiopterygii. Cell Tissue Res 173:95–108

Sawyer WH (1969) The active neurohypophysial principles of two primitive bony fishes, the bichir (Polypterus senegalis) and the African lungfish (Protopterus aethiopicus). J Endocrinol 44:421–435

Schnitzlein HN (1962) The habenula and the dorsal thalamus of some teleosts. J Comp Neurol 118:225–268

Senn DG (1976a) Brain structure in Calamoichthys calabaricus Smith 1865 (Polypteridae, Brachiopterygii). Acta Zool 57:121–128

Senn DG (1976b) Notes on the midbrain and forebrain of Calamoichthys calabaricus Smith 1865 (Polypteridae, Brachiopterygii). Acta Zool 57:129–135

Sheldon RE (1912) The olfactory tracts and centres in teleosts. J Comp Neurol 22:178–337

Smeets WJAJ (1981) Efferent tectal pathways in two chondrichthyans, the shark Scyliorhinus canicula and the ray Raja clavata. J Comp Neurol 195:13–23

Smeets WJAJ, Boord RL (1985) Connections of the lobus inferior hypothalami of the clearnose skate Raja eglanteria (Chondrichthyes). J Comp Neurol 234:380–392

Stensiö EA (1921) Triassic Fishes from Spitzbergen, part I. Holzhausen, Vienna

Striedter GF (1991) Auditory, electrosensory and mechanosensory lateral line pathways through the forebrain of channel catfishes. J Comp Neurol 312:311–331

van der Horst CJ (1919) Das Kleinhirn der Crossopterygii. Bijdr Dierk Kon Zool Gen Nat Artis Mag 21:113–118

van der Horst CJ (1925) The cerebellum of fishes. I. General morphology of the cerebellum. Proc R Neth Acad Sci [Amsterdam] 28:735–746

von Bartheld CS, Meyer DL (1986a) Central projections of the nervus terminalis in the bichir, Polypterus palmas. Cell Tissue Res 244:181–186

von Bartheld CS, Meyer DL (1986b) Central connections of the olfactory bulb in the bichir, Polypterus palmas, reexamined. Cell Tissue Res 244:527–535

von Bartheld CS, Meyer DL (1988) Central projections of the nervus terminalis in lampreys, lungfishes, and bichirs. Brain Behav Evol 32:151–159

Waldschmidt J (1887) Beiträge zur Anatomie des Zentralnervensystems und des Geruchsorgans von Polypterus bichir. Anat Anz 2:308–322

Wullimann MF (1988) The tertiary gustatory center in sunfishes is not nucleus glomerulosus. Neurosci Lett 86:6–10

CHAPTER 14

Chondrostean Fishes

R. NIEUWENHUYS

14.1 Introduction

The Actinopterygii or ray-finned fishes are usually subdivided into three superorders: the Chondrostei, the Holostei and the Teleostei. The present chapter is devoted to the Chondrostei, which are generally considered the most primitive and the most ancient of these three groups. This superorder contains 25 extant species arranged in two families: the sturgeons (or Acipenseridae) and the paddlefishes (or Polyodontidae).

The sturgeons are large, spindle-shaped fishes with a flattened, extended snout and a strongly heterocercal tail (Fig. 14.1). The skeleton, although highly ossified in the palaeozoic chondrostean ancestors, is almost wholly cartilaginous in recent sturgeons. Their skin has a partial armour, consisting of rows of large, plate-like scales. Their mouth, which is on the lower side of the head, is surrounded by tentacular fringes. Just in front of the mouth, four barbels are found. The sturgeons are bottom-dwelling fishes that feed mainly on invertebrates. Their eyes are small, but their olfactory organs are well developed. Taste buds are not confined to the mouth, but also occur on the barbels and perioral fringes. These external taste organs play an important role in locating food. In addition to mechanoreceptive canal organs, the lateral line system comprises numerous ampullary organs, the sensory epithelium of which bears a striking resemblance to that observed in the ampullae of Lorenzini, electroreceptive organs found in cartilaginous fishes (Norris 1925; Jørgensen et al. 1972; Teeter et al. 1980). Behavioral and electrophysiological evidence suggests that the ampullary organs of chondrosteans also subserve an electroreceptive function. Thus Jørgensen et al. (1972) observed that the introduction of an iron rod into a tank in which a *Polyodon* specimen was kept caused the fish to give a clear avoidance response. This response was similar to that observed in the dogfish *Scyliorhinus* and the teleosts *Ameiurus* and *Gymnarchus*, species which are all known to possess electroreceptors. When a wooden rod was introduced, the fish showed no response whatsoever. Teeter et al. (1980), who recorded single nerve fibres innervating ampullary receptor cells in *Scaphirhynchus*,

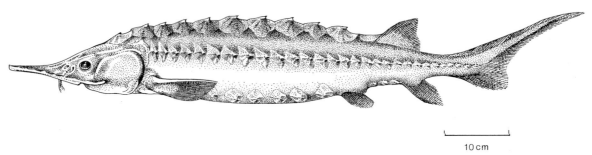

10 cm

Fig. 14.1. The shovelnose sturgeon *Scaphirhynchus platorynchus*

demonstrated that these cells are very sensitive to weak electric fields.

The family Acipenseridae contains four genera: *Acipenser* (16 species), *Huso* (two species), *Scaphirhynchus* (two species) and *Pseudoscaphirhynchus* (two species). They are all fishes of temperate waters of the Northern hemisphere. Some species live only in fresh water. Others spend part of their life cycle in the ocean, but return to fresh water to spawn.

The family Polyodontidae comprises only two living species, one in the Yangtse River basin of China, *Psephurus gladius*, and the other in the Mississippi drainage, the paddlefish *Polyodon spathula*. The skin of these animals is naked. Their mouth is wide and can be protruded to feed on small, floating organisms. Their snout has developed into a large, flattened spoon- or sword-shaped structure. Snout, head and opercula bear an enormous number (50 000–75 000) of ampullary organs (Nachtrieb 1910; Norris 1925; Jørgensen et al. 1972).

The literature on the chondrostean central nervous system is limited. Goronowitsch (1888) described the brain of *Acipenser ruthenus*. The brain of *Acipenser rubicundus* was thoroughly investigated by Johnston (1898a,b, 1901) using the Golgi technique, and Hocke Hoogenboom (1929) used the Weigert-Pal and Nissl techniques to analyse the structure of the brain of *Polyodon spathula*. New (1981) and McCormick (1982) described the area octavolateralis. The cerebellum was studied by Palmgren (1921), Larsell (1967) and Nieuwenhuys (1967), and in several papers attention has been paid to the telencephalon (Johnston 1898b, 1911; Nieuwenhuys 1963, 1966; Northcutt and Braford 1980). The hypothalamo-hypophysial system was extensively studied by Polenov and collaborators (Polenov and Garlov 1971a-c; Polenov and Pavlovic 1978; Polenov et al. 1972, 1976a,b, 1979, 1983).

As far as I am aware, experimental neuroanatomical studies are confined to the analysis by Northcutt and Braford (1980) of the secondary olfactory connections in the shovelnose sturgeon

Scaphirhynchus platorynchus, studies by New and Northcutt (1984a,b) on the central projections of the lateral line and trigeminal nerves, respectively, in the same species and the study by Repérant et al. (1982) on the retinofugal pathways in *Acipenser güldenstädti*. Finally, the site of origin of diencephalic retinopetal projections was experimentally established by Hofmann et al. (1993) in the Sterlet *Acipenser ruthenus*.

Histochemical and immunohistochemical studies on the chondrostean central nervous system are also limited in number. Kotrschal et al. (1985a) described the distribution of monoaminergic neurons in the central nervous system of *Acipenser ruthenus*, using the formaldehyde-induced fluorescence (FIF) technique. Immunohistochemical techniques were used for the localisation of the following neuroactive principles: urotensin I and II (Onstott and Elde 1986: *Polyodon spathula*; Oka et al. 1989: *Acipenser transmontanus*), corticotropin and corticotropin-releasing factor-like peptides (González et al. 1992: *Acipenser ruthenus*), gonadotropin-releasing hormone (Leprêtre et al. 1993: *Acipenser baeri*), neuropeptide Y (Chiba and Honma 1994: *Acipenser transmontanus*), and leu-enkephalin, substance P, tyrosine hydroxylase, serotonin and dopamine in the forebrain (Piñuela and Northcutt 1994: *Acipenser transmontanus*).

The development of the central nervous system of *Acipenser sturio* was described by von Kupffer (1893, 1906). Data on the ontogenesis of the diencephalon and telencephalon are found in the studies by Bergquist (1932) and Nieuwenhuys (1964), respectively.

The present chapter is based on the literature cited above and on our own histological material of the shovelnose sturgeon *Scaphirhynchus platorynchus*, a species which attains a length of about 1 m, distributed in the Mississippi delta and its tributaries (Fig. 14.1). Some material of the paddlefish *Polyodon spathula*, belonging to the collection of the Institute for Brain Research, Amsterdam, has also been consulted.

14.2
Gross Morphology

In chondrosteans, the five main divisions of the brain – rhombencephalon, cerebellum, mesencephalon, diencephalon and telencephalon – can be readily distinguished (Figs. 14.2, 14.3).

The *rhombencephalon* is relatively very large and can be clearly divided bilaterally into a basal plate and an alar plate. The ventricular surface of this brain part shows a number of intraventricular protrusions. Prominent among these are an elongated paramedian strip, which contains the large fasciculus longitudinalis medialis, a more laterally situated ridge, i.e. the lobus vagi, and a rostrodorsally situated eminence, which harbours the nucleus dorsalis, a lateral line centre. The latter is extraordinarily large in *Polyodon*, due to which feature the fourth ventricle in this species is much narrower than in sturgeons. Rostrolaterally, the fourth ventricle widens on either side to form a large recessus lateralis.

The *cerebellum* comprises a pair of large, lateral expansions, the auriculae, an unpaired central division protruding into the ventricular cavity. The auriculae partly surround the lateral recesses of the fourth ventricle. The central division has a caudal part, the corpus cerebelli, and a rostral part, which expands forward under the tectum as the valvula cerebelli. The boundary between the corpus and the valvula is marked dorsally by a transverse external groove, the plica valvulae (Fig. 14.3d).

The *mesencephalon* is rather poorly developed. It consists of the ventral tegmentum mesencephali and the dorsal tectum mesencephali. The tegmentum is continuous caudally with the rhombencephalic basal plate. The lateral part of its ventricular surface shows an eminence, the torus semicircularis. The lateral border of the tectum is marked by a slight external groove. Caudally, the tectum passes into the valvula cerebelli via the relatively thin velum medullare anterius.

The *diencephalon* is partly covered by the midbrain. On the dorsal side, only the habenular ganglia are externally visible. The thalamic portions are relatively small, but the hypothalamus is strongly developed. The walls of this brain part are largely evaginated into a pair of lobi inferiores, structures which, particularly in sturgeons, reach a very large size (Fig. 14.3b,c).

Rostrally to the habenular ganglia, the diencephalon is covered by a membranous tela diencephali. The latter forms a large saccus dorsalis, which covers most of the telencephalon. The slender, elongated epiphysial stalk rests on the dorsal surface of the saccus dorsalis. The epiphysis is

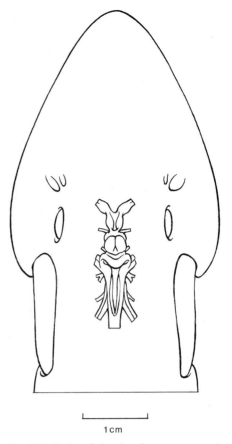

Fig. 14.2. Brain of the shovelnose sturgeon *Scaphirhynchus platorynchus* (juvenile specimen) in position

formed by a small evagination at the rostral end of the stalk.

The *telencephalon* comprises a pair of very large, hollow olfactory bulbs and the telencephalon proper. The latter is of the everted type, i.e. its walls have recurved laterally during ontogenesis. Due to this eversion, the dorsal closure of the telencephalic ventricle is formed by a membranous tela telencephali, which is attached at the lateral side of the hemisphere walls. Together with the ventral wall of the saccus dorsalis, the caudal part of the telencephalic roof constitutes a large, caudoventrally directed fold, the velum transversum (Fig. 14.3d).

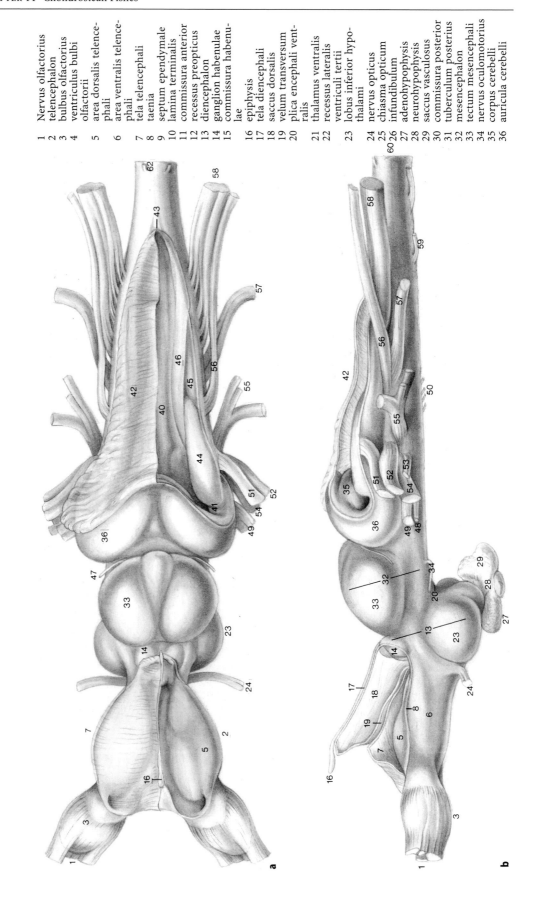

1   Nervus olfactorius
2   telencephalon
3   bulbus olfactorius
4   ventriculus bulbi olfactorii
5   area dorsalis telencephali
6   area ventralis telencephali
7   tela telencephali
8   taenia
9   septum ependymale
10 lamina terminalis
11 commissura anterior
12 recessus preopticus
13 diencephalon
14 ganglion habenulae
15 commissura habenulae
16 epiphysis
17 tela diencephali
18 saccus dorsalis
19 velum transversum
20 plica encephali ventralis
21 thalamus ventralis
22 recessus lateralis ventriculi tertii
23 lobus inferior hypothalami
24 nervus opticus
25 chiasma opticum
26 infundibulum
27 adenohypophysis
28 neurohypophysis
29 saccus vasculosus
30 commissura posterior
31 tuberculum posterius
32 mesencephalon
33 tectum mesencephali
34 nervus oculomotorius
35 corpus cerebelli
36 auricula cerebelli

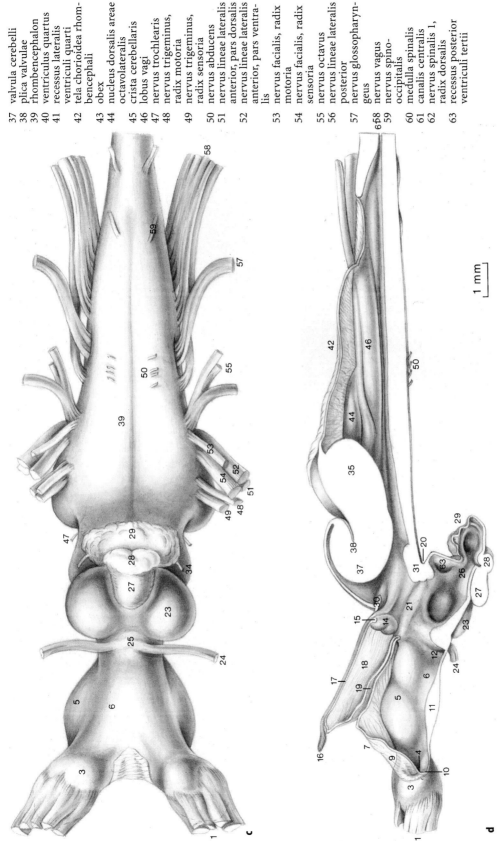

37 valvula cerebelli
38 plica valvulae
39 rhombencephalon
40 ventriculus quartus
41 recessus lateralis
    ventriculi quarti
42 tela chorioidea rhom-
    bencephali
43 obex
44 nucleus dorsalis areae
    octavolateralis
45 crista cerebellaris
46 lobus vagi
47 nervus trochlearis
48 nervus trigeminus,
    radix motoria
49 nervus trigeminus,
    radix sensoria
50 nervus abducens
51 nervus lineae lateralis
    anterior, pars dorsalis
52 nervus lineae lateralis
    anterior, pars ventra-
    lis
53 nervus facialis, radix
    motoria
54 nervus facialis, radix
    sensoria
55 nervus octavus
56 nervus lineae lateralis
    posterior
57 nervus glossopharyn-
    geus
58 nervus vagus
59 nervus spino-
    occipitalis
60 medulla spinalis
61 canalis centralis
62 nervus spinalis 1,
    radix dorsalis
63 recessus posterior
    ventriculi tertii

Fig. 14.3a–d. The brain of the shovelnose sturgeon *Scaphirhynchus platorynchus*; the telae choroideae are removed on the left side. **a** Dorsal view. **b** Lateral view. **c** Ventral view. **d** Medial view of the bisected brain.

14.3
Development and Overall Histological Pattern

14.3.1
Development

The following data on the ontogenesis of the central nervous system of *Acipenser* are largely based on the studies carried out by von Kupffer (1893, 1906).

In early embryonic stages, the neural tube has a flattened shape and a horizontally oriented cavity, but somewhat later the neural tube becomes circular in cross-section. In embryos approximately 2 days old, the anlage of the spinal cord has a thin roof-plate and a somewhat wider floor-plate, both consisting of simple cubic cells, whereas the lateral plates are thickened and made up of tall, cylindrical elements showing a pseudostratified arrangement. The central canal has the shape of a vertically oriented cleft. The series of cross-sections shown in Fig. 14.4 gives an impression of the further development of the cord. It can be seen that both the floorplates and roof-plates gradually thicken with consequent relative diminution of the central canal. Conspicuous large dorsal or Rohon-Beard cells develop, but these elements disappear soon after hatching. On the ninth larval day, they are in full regression. In 4-week-old larvae, large ventral horn cells can be clearly recognised. The peripherally situated fibre layer increases steadily in width. At the end of the larval period, a number of coarse, well-myelinated fibres can be distinguished within this layer. As in the adult stage, these fibres are concentrated in the ventral funiculi and in the deepest portions of the lateral funiculi.

The morphogenesis of the brain (Fig. 14.5) is characterised by the early appearance of the plica encephali ventralis in the midbrain region. The tuberculum posterius and the anlage of the epiphysis can also be recognised early. Due to the formation and the increase in height of the plica encephali ventralis, the floor of the rhombencephalon is strongly curved, a curvature which later in development disappears again. Dorsally, the walls of the rhombencephalon diverge, a process which leads to the transformation of the roof-plate into a wide tela ependymalis. The cerebellum originates from a paired anlage, but due to an extensive fusion of the bilateral cerebellar primordia, the medial portion of this brain is transformed into a solid body which expands into the ventricular cavity of the brain stem.

The mesencephalon essentially maintains the early embryonic tube-like shape. It is noteworthy that the ventral, tegmental portion of this brain part differentiates much earlier than the dorsal, tec-tal portion. As regards the diencephalon, it has already been mentioned that the epiphysis differentiates early in development. More rostrally, a second dorsally directed outpouching of the diencephalic roof leads to the formation of the saccus dorsalis. It is remarkable that in the ventral part of the diencephalon the caudal expansion of the hypothalamus occurs relatively late and that the same holds true for the outgrowth of the lobi inferiores hypothalami.

In the anlage of the telencephalon, the processus neuroporicus, i.e. the rostral site of closure of the neural tube, can be distinguished over a long period. The lateral walls of the telencephalon evert during ontogenesis. Nieuwenhuys (1964) studied this developmental process in *Acipenser* and noted that the eversion first manifests itself in 10-day-old larvae as a slight diverging of the lateral walls on the dorsal side. Four days later, this process has proceeded further and has led to a considerable widening of the initially narrow roof-plate (Fig. 14.6a). During the next 50 days of development, the shape of the forebrain anlage changes comparatively little. The dorsal parts of both side-plates thicken somewhat and are about to protrude into the common telencephalic ventricle. As a result of this intraventricular growth, lateral recesses of the ventricular space arise (Fig. 14.6b). A notable difference between the development of the chondrostean forebrain and that of the Holostei and Teleostei is that the intraventricular expansion of the dorsal parts of the side-plates in the latter groups starts much earlier. In the holostean and teleostean forebrain, the lateral shift of the taeniae (i.e. the lines of attachment of the roof-plate) and the dorsomedial growth of the dorsal portions of the side-plates occur simultaneously. Comparison of the 64-day stage of *Acipenser* (Fig. 14.6b) with the adult (Fig. 14.6c) shows that considerable changes of shape in the chondrostean forebrain occur late in development. It can be seen that the intraventricular expanse of the side-plates greatly increases and that the taeniae move ventrolaterally. In other words, the side-plates of the forebrain, which are still straight at the 64-day stage, show a clearly recurved or everted condition in the adult. Nieuwenhuys (1962) observed that a bending of the forebrain wall is always clearly reflected in the pattern of the blood vessels which traverse it. Figure 14.6b shows that, in the telencephalic side-plates of a 64-day-old sterlet, the blood vessels converge from the larger, meningeal surface towards the smaller, ependymal surface. In the adult, however, the vessels of the now everted telencephalic walls diverge from each other radially to the larger, ependymal surface (Fig. 14.6c).

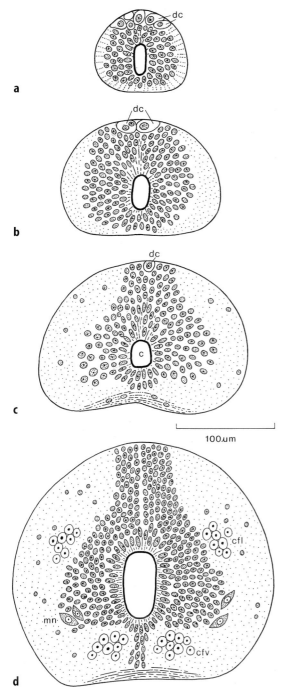

Fig. 14.4a–d. Transverse sections through the spinal cord of larvae of the sturgeon *Acipenser sturio*. All sections are taken at the level of the pectoral fin. **a** Hatching specimen. **b-d** Larvae 3, 9 and 28 days old, respectively. *c*, canalis centralis; *cfl*, coarse fibres in funiculus lateralis, *cfv*, coarse fibres in funiculus ventralis, *dc*, dorsal cells; *mn*, motoneurons. (Modified from von Kupffer 1906)

development in this region does not differ from that in the remaining portions of the forebrain anlage. In a 10-day-old larva, the olfactory nerve penetrates into the forebrain, and in the (rostrodorsally situated) area of entrance a number of cells are detached from the central grey to form a separate stratum cellulare externum. At the 14-day stage, the area of entrance of the olfactory nerve is slightly thickened and shows a local bending-inward or inversion (Fig. 14.7a). During further development, the inversion increases and gradually assumes the character of a rostrocaudally directed evagination. In 44-day-old specimens, the layer of olfactory glomeruli appeared to be clearly differentiated in the zone directly peripherally to the stratum cellulare externum. Figure 14.7b shows that, at the 52-day stage, the external cell layer has become somewhat less distinct. At the 64-day stage, the texture of the olfactory bulb appears to conform to that of the adult. The stratum cellulare is now represented by a rather diffuse layer, consisting partly of large mitral cells and partly of smaller elements.

14.3.2
Overall Histological Pattern

In the adult, the embryonic mantle layer appears to be largely transformed into a rather narrow and moderately dense zone of periventricular grey. Only in a limited number of regions have neurons migrated away over a larger distance from the ventricular surface. Thus, in the cerebellum and in the dorsal parts of the telencephalon, the neurons have spread over the entire width of the wall, and in all of the main regions of the brain one or a few groups of migrated cells occur which have clearly detached from the central grey. Prominent among these are the very elongated oliva inferior (see Figs. 14.11, 14.27) and the nucleus tori lateralis (see Figs. 14.17–14.20). The Golgi studies carried out by Johnston (1898a,b, 1901; Fig. 14.8) showed that most of the larger neurons are provided with large dendritic trees that spread in the wide, peripheral fibre zone. As in all gnathostomes, the latter consists of both myelinated and unmyelinated fibres. The fibre paths consist mostly of diffusely arranged fibres, a feature which, according to Hocke Hoogenboom (1929), is even more true of *Polyodon* than of *Acipenser*.

Finally, turning our attention to the rostralmost parts of the chondrostean forebrain, it may be stated that, during the first days after hatching, the

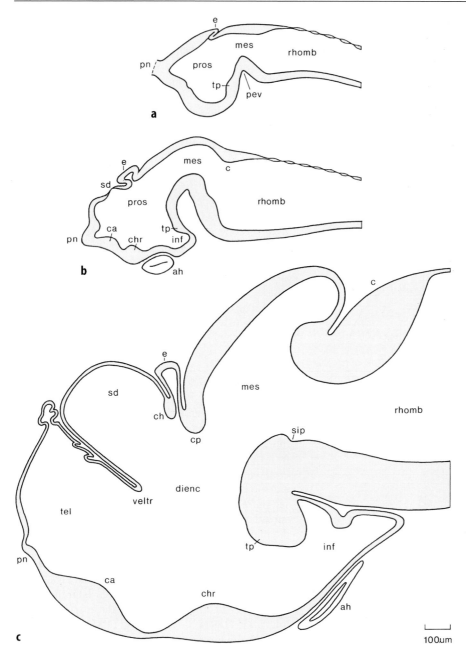

Fig. 14.5a–c. Median sections through the brain of the sturgeon *Acipenser sturio*. **a** Embryo 57 h old. **b** Embryo 85–90 h old (hatching). **c** Larva 4 weeks old. *ah*, adenohypophysis; *c*, cerebellum; *ch*, commissura habenulae; *chr*, chiasmatic ridge; *dienc*, diencephalon; *e*, epiphysis; *mes*, mesencephalon; *pev*, plica encephali ventralis; *pn*, processus neuroporicus; *pros*, prosencephalon; *rn*, recessus neuroporicus; *rhomb*, rhombencephalon; *sd*, saccus dorsalis; *sip*, sulcus intraencephalicus posterior; *tel*, telencephalon; *vt*, velum transversum. (Modified from von Kupffer 1906)

100µm

14.4
Spinal Cord and Spinorhombencephalic Transition Zone

Our knowledge of the histological structure of the chondrostean spinal cord is very limited. Goronowitsch (1888) mentioned some observations on the middle part of the cord of *Acipenser*, and the present author studied its rostralmost portion in *Scaphirhynchus*. The spinal grey matter can be subdi-

vided into a cornu ventrale, a zona intermedia and a cornu dorsale (Fig. 14.9). The ventrolateral parts of the cornu ventrale contain large neurons of different shapes. Although it is by no means certain that all of these large elements actually represent motoneurons, the group as a whole is usually denoted as the spinal motor column. The dorsomedial part of the cornu ventrale and the central part of the zona intermedia are mainly composed of small cells; however, in the superficial part of the intermediate zone, large, fusiform neurons are fre-

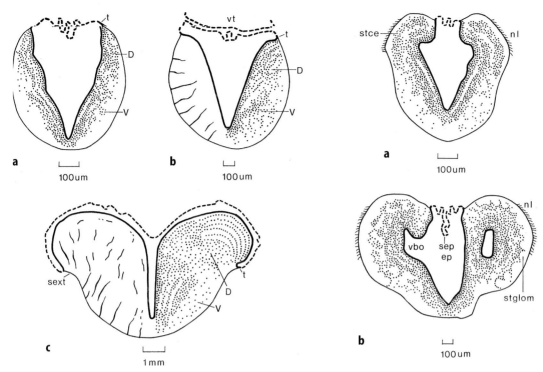

Fig. 14.6a–c. Transverse sections through the telencephalon of the sterlet *Acipenser ruthenus*. **a** Larva 14 days old. **b** Larva 64 days old. **c** Adult specimen. The *left halves* of Fig. 14.6B,C show the pattern of blood vessels. *sext*, sulcus externus; *vt*, velum transversum. (Modified from Nieuwenhuys 1964)

Fig. 14.7a,b. Transverse sections through the rostral part of the telencephalon of the sterlet *Acipenser ruthenus*. **a** Larva 14 days old. **b** Larva 52 days old.; *sep ep*, septum ependymale; *stce*, stratum cellulare externum; *stglom*, stratum glomerulosum; *vbo*, ventriculus bulbi olfactorii. (Modified from Nieuwenhuys 1964)

quently observed. The cornu dorsale is small and consists of small, rather diffusely arranged cells. Large neurons are entirely lacking in the dorsal horns, a feature which is in keeping with the observation of made by von Kupffer (1906) that the large embryonic dorsal cells degenerate and disappear during the early larval period (Fig. 14.4).

The three zones of spinal grey discussed above can be traced rostrally well beyond the level of the obex (see Fig. 14.27). The part of the spinal motor column situated rostrally to that landmark gives rise to two large spino-occipital nerve roots (Fig. 14.3). The zona intermedia is found to divide into two parts that become separated from each other by the opening out of the central canal into the fourth ventricle (Figs. 14.9, 14.10). The dorsal horn finally passes imperceptibly into the nucleus tractus spinalis nervi trigemini. Dorsal funicular nuclei could not be delineated.

As in all gnathostomes, the dorsal parts of the spinal ventral funiculi contain numerous large fibres which together form a direct caudal continuation of the fasciculus longitudinalis medialis. Conspicuous among these fibres is the coarse fibre of

Mauthner (Figs. 14.9, 14.10). Nothing can be said with certainty on the polarity of the fibres which constitute the lateral funiculi. However, I consider it likely that the more superficially situated portions of these funiculi contain numerous spinorhombencephalic and spinocerebellar fibres and that the coarse-fibred deepest part of these funiculi are equivalent to the chondrichthyan fasciculus medianus, i.e. a bundle which descends from the vestibular region to the cord. The small dorsal funiculi are composed of thin fibres. Whether these funiculi contain any long ascending primary afferent fibres is unknown.

The caudalmost part of the spinal cord of many groups of fishes contains neurosecretory cells, the axons of which project to a capillary network in which their secretory products are supposed to be released (Fridberg and Bern 1968; Kobayashi et al. 1986). In teleosts, the processes of the neurosecretory cells and the capillary network together constitute a macroscopically visible neurohaemal organ, the urophysis, which is located ventral to the cord (see Sect. 15.2.7). It is known that, in teleosts, the neurosecretory cells in the caudal cord produce two neuropeptides, urotensin I and II.

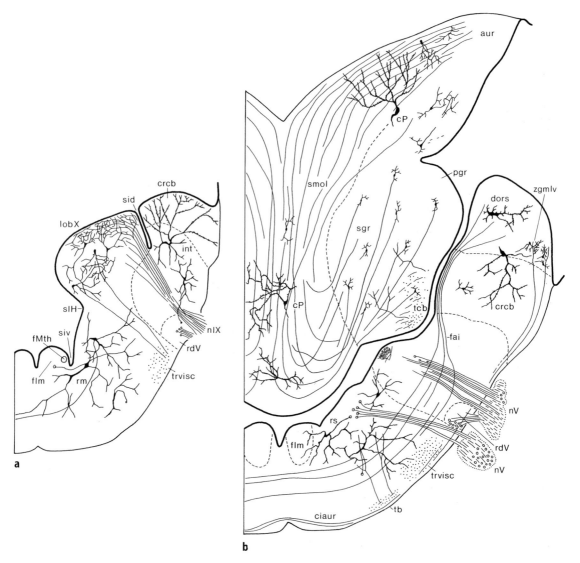

Fig. 14.8a,b. Transverse sections through the rhombencephalon of the sturgeon *Acipenser rubicundus*, showing elements impregnated using the Golgi technique. **a** Level of entrance of nerve XI. **b** Level of entrance of nerve V. (Based on Johnston 1898a)

Neurosecretory cells producing urotensin I and urotensin II have been observed in the caudal cord of both *Polyodon* (Onstott and Elde 1986) and *Acipenser* (Oka et al. 1989), but a distinct urophysis is lacking in these animals.

14.5
Rhombencephalon

14.5.1
Introductory Nnotes

Throughout most of the chondrostean rhombencephalon, three distinct ventricular grooves bilaterally mark the boundaries between four longitudinal cell zones or columns (see Fig. 14.27). These sulci are the sulcus limitans of His, which demarcates the ventromedial basal plate from the dorsolateral alar plate, the sulcus intermedius ventralis, which subdivides the basal plate into an area ventralis and an

Abbreviations

amtp	area magnocellularis tuberculi posterioris
aopta	area optica accessoria
ath	nucleus anterior thalami
aur	auricula cerebelli
brc	brachium conjunctivum
ca	commissura anterior
cans	commissura ansulata
caud	nucleus caudalis areae octavolateralis
cc	canalis centralis
ccb	corpus cerebelli
chab	commissura habenulae
chopt	chiasma opticum
ciaur	commissura interauricularis
cmsp	columna motoria spinalis
cnd	cornu dorsale
cnv	cornu ventrale
cP	cell of Purkinje
cp	commissura posterior
cpth	nucleus centralis posterior thalami
crcb	crista cerebellaris
ctm	commissura tecti mesencephali
D	area dorsalis telencephali
Dc	central zone of D
Dd	dorsal zone of D
dfll	nucleus diffusus lobi lateralis
dfmt	decussation of fasciculus medialis telencephali
Dl	lateral zone of D
Dm	medial zone of D
dors	nucleus dorsalis areae octavolateralis
Dp	posterior zone of D
dpth	nucleus dorsalis posterior thalami
emgrd	eminentia granularis, pars dorsalis
emgrl	eminentia granularis, pars lateralis
emgrld	eminentia granularis, pars laterodorsalis
emgrli	eminentia granularis, pars latero-intermedia
emgrlv	eminentia granularis, pars lateroventralis
emgrr	eminentia granularis, pars rostralis
emmed	eminentia mediana
emth	eminentia thalami
en	efferent neurons of lateral line system
ent	nucleus entopeduncularis
ep	epiphysis
fae	fibrae arcuatae externae
fai	fibrae arcuatae internae
fd	funiculus dorsalis
fdmopt	fasciculus dorsomedialis tractus optici
fhyt	fibrae hypothalamotegmentales
fl	funiculus lateralis
flm	fasciculus longitudinalis medialis
flt	fasciculus lateralis telencephali
fmt	fasciculus medialis telencephali
fMth	fibre of Mauthner
fpopt	fasciculus preopticus tractus optici
fr	fasciculus retroflexus
ft	fasciculi tegmentales
fthy	fibrae torohypothalamicae
fttd	fibrae tectotegmentales dorsales
fttv	fibrae tectotegmentales ventrales
fv	funiculus ventralis
fvisc	fibrae viscerales secundariae
gcm	griseum centrale mesencephali
gll	glomerular layer
hab	ganglion habenulae
habd	nucleus habenularis dorsalis
habv	nucleus habenularis ventralis
inf	infundibulum
int	nucleus intermedius areae octavolateralis
ip	nucleus interpeduncularis
is	nucleus isthmi
ith	nucleus intermedius thalami
lamt	lamina terminalis
lamv	nucleus laminaris ventralis
ll	lemniscus lateralis
lobinf	lobus inferior hypothalami
lobtub	lobus tuberis hypothalami
lobX	lobus vagi
lrz	lateral reticular zone
lv	nucleus lateralis valvulae
meV	nucleus mesencephalicus nervi trigemini
Mth	cell of Mauthner
nil	neurointermediate lobe of pituitary
nllad	nervus lineae lateralis anterior, pars dorsalis
nllav	nervus lineae lateralis anterior, pars ventralis
nllp	nervus lineae lateralis posterior
nspoc	nervus spino-occipitalis
nucb	nucleus cerebelli
nufl	nucleus funiculi lateralis
nuflm	nucleus of the fasciculus longitudinalis medialis
nufs	nucleus fasciculi solitarii
null	nucleus lemnisci lateralis
numtp	nucleus medianus tuberculi posterioris
nupc	nucleus postcommissuralis
nuptpd	nucleus periventricularis tuberculi posterioris, pars dorsalis
nuptpv	nucleus periventricularis tuberculi posterioris, pars ventralis
nurdV	nucleus of the radix descendens nervi trigemini
nutegm	nucleus tegmentalis medialis
nutld	nucleus tuberis lateralis, pars dorsalis
nutlv	nucleus tuberis lateralis, pars ventralis
nutp	nucleus tuberis posterior
nutsc	nucleus tori semicircularis
nI	nervus olfactorius

nII	nervus opticus	rpost	recessus posterior ventriculi tertii
nIII	nervus oculomotorius	rs	nucleus reticularis superior
nV	nervus trigeminus	rsV	radix sensoria nervi trigemini
nVIII	nervus octavus	rsX	radix sensoria nervi vagi
nIX	nervus glossopharyngeus	rub	nucleus ruber
oli	oliva inferior	sa	sulcus a
oll	tractus olfactorius lateralis	sac	stratum album centrale
olm	tractus olfactorius medialis	sce	stratum cellulare externum
opta	tractus opticus accessorius	sch	nucleus suprachiasmatis
optdm	tractus opticus dorsomedialis	sci	stratum cellulare internum
optl	tractus opticus lateralis	sdd	sulcus diencephalicus dorsalis
optm	tractus opticus medialis	sdm 1, 2	sulcus diencephalicus medius 1 and 2
optma	tractus opticus marginalis	secolf	secondary olfactory fibres
optvl	tractus opticus ventrolateralis	sfgs	stratum fibrosum et griseum superficiale
osc	organon subcommissurale	sgc	stratum griseum centrale
P	nucleus P	sgp	stratum griseum periventriculare
pd	pars distalis of pituitary	sgr	stratum granulare
pdhy	pars dorsalis hypothalami	sid	sulcus intermedius dorsalis
pdhyc	caudal portion of pars dorsalis hypothalami	sih	sulcus intrahabenularis
		sis	sulcus isthmi
pdhyr	rostral portion of pars dorsalis hypothalami	siv	sulcus intermedius ventralis
		slH	sulcus limitans of His
pgr	prominentia granularis	slmes	sulcus lateralis mesencephali
plencv	plica encephali ventralis	slt	sulcus limitans telencephali
pm	nucleus preopticus magnocellularis	sm	stria medullaris
pol	layer of primary olfactory fibres	smarg	stratum marginale
pp	nucleus preopticus parvocellularis	smi	sulcus medianus inferior
proh	tractus preopticohypophyseus	smol	stratum moleculare
prom	nucleus profundus mesencephali	sms	sulcus medianus superior
prtc	nucleus pretectalis centralis	sol	layer of secondary olfactory fibres
prtp	nucleus pretectalis periventricularis	sop	stratum opticum
prts	nucleus pretectalis superficialis	spcb	tractus spinocerebellaris
pv	portal vessels	ssh	sulcus subhabenularis = s. diencephalicus dorsalis
pvhy	pars ventralis hypothalami		
Q	nucleus Q	sv	saccus vasculosus
rai	nucleus raphes inferior	t	taenia
ranllav	radix ascendens nervi lineae lateralis anterioris, pars ventralis	tb	tractus tectobulbaris
		tcb	tractus tectocerebellaris
ranllp	radix ascendens nervi lineae lateralis posterioris	tect	tectum mesencephali
		tl	nucleus tori lateralis
ras	nucleus raphes superior	tla	nucleus tori lateralis, pars anterior
rb	area retrobulbaris	tldc	nucleus tori lateralis, pars dorsocaudalis
rdnllav	radix descendens nervi lineae lateralis anterioris, pars ventralis	tlvc	nucleus tori lateralis, pars ventrocaudalis
		tlong	torus longitudinalis
rdnllp	radix descendens nervi lineae lateralis posterioris	tp	tuberculum posterius
		trvisc	tractus visceralis secundarius
rdV	radix descendens nervi trigemini	tt	tela telencephali
ri	nucleus reticularis inferior	V	area ventralis telencephali
rism	nucleus reticularis isthmi et mesencephali	valvcb	valvula cerebelli
		vbol	ventriculus bulbi olfactori
rlat	recessus lateralis ventriculi tertii	Vd	dorsal nucleus of V
rm	nucleus reticularis medius	veltr	velum transversum
rmeV	radix mesencephalicus nervi trigemini	vem	nucleus vestibularis magnocellularis
rmV	radix motoria nervi trigemini	visc	nucleus visceralis secundarius
rmVII	radix motoria nervi facialis	Vl	lateral nucleus of V
rmX	radix motoria nervi vagi	vmth	nucleus ventromedialis thalami

Vn	nucleus situated in the dorsalmost part of V
Vp	postcommissural nucleus of V
Vs	supracommissural nucleus of V
Vv	ventral nucleus of V
zgmi	zona granularis marginalis, pars intermedia
zgmld	zona granularis marginalis, pars laterodorsalis
zgmli	zona granularis marginalis, pars laterointermedia
zgmlv	zona granularis marginalis, pars lateroventralis
zgmm	zona granularis marginalis, pars medialis
zgmr	zona granularis marginalis, pars rostralis
zi	zona intermedia
zld	zona limitans diencephali
zlt	zona limitans telencephali
III	nucleus nervi oculomotorii
IIIl	nucleus nervi oculomotorii, pars lateralis
IIIm	nucleus nervi oculomotorii, pars medialis
IV	nucleus nervi trochlearis
Vm	nucleus motorius nervi trigemini
Vpr	nucleus sensorius principalis nervi trigemini
VI	nucleus nervi abducentis
VIIm	nucleus motorius nervi facialis
VIIIa	nucleus anterior nervi octavi
VIIId	nucleus descendens nervi octavi
VIIIp	nucleus posterior nervi octavi
IXm	nucleus motorius nervi glossopharyngei
Xm	nucleus motorius nervi vagi

area intermedioventralis, and the sulcus intermedius dorsalis, which roughly coincides with the boundary of two zones present in the alar plate, i.e. the area intermediodorsalis and the area dorsalis. Johnston (1898a, 1901) thoroughly analysed the brain of *Acipenser rubicundus* and recognised the four cell zones mentioned above. He held that each of these zones is specifically related to one of the following fibre categories or fibre components of the cranial nerves: somatosensory, viscerosensory, visceromotor or somatomotor. Hence he used these terms to designate the dorsal, intermediodorsal, intermedioventral and ventral areas, respectively, of the present account. In several general publications, Johnston (1902, 1903, 1906) employed the sturgeon as a paradigm to elucidate the concept that the brain stem fundamentally consists of four functional zones (see Sect. 4.6.2).

In the following survey, the various rhombencephalic centres and their connections will be dis-

cussed. It will become apparent that the functional subdivision, as presented by Johnston (1901, 1902), is largely, but not entirely correct. Following a discussion of the primary afferent centres present in the alar plate, the cell masses contained within the intermedioventral and ventral zones will be dealt with. Brief consideration will then be given to the structural relations in the isthmus region; finally, our as yet very limited knowledge of the rhombencephalic fibre systems will be summarised.

14.5.2
Alar Plate

The rhombencephalic primary afferent centres are all situated in the alar plate and fall into three categories: special somatosensory, general somatosensory and viscerosensory.

Special Somatosensory Centres. The *special somatosensory* centres are related to the octavus and lateral line nerves. They were thoroughly studied by Larsell (1967) and, more recently, by New (1981) and McCormick (1982). According to these studies, the chondrostean area octavolateralis can be subdivided into three zones: dorsal, intermediate and ventral. The dorsal and intermediate zones receive first-order input from the lateral line nerves, whereas the ventral zone constitutes the area of termination of the fibres of the eighth nerve.

The dorsal zone is represented by the nucleus dorsalis areae octavolateralis, a large cell mass which forms a distinct eminence upon the dorsal surface of the alar plate. In sturgeons, this eminence extends from the cerebellar auricle to the level of entrance of the posterior lateral line nerve (Fig. 14.3a), whereas in the paddlefish, in which the dorsal nucleus is much larger, it extends caudally almost to the level of the obex (New and Bodznick 1985). The nucleus dorsalis overlies the crista cerebellaris (Figs. 14.14, 14.15). It is composed of cells of different types. Prominent among these are large elements, polygonal or fusiform in shape, and small granule cells. Many of the fusiform cells are arranged along the dorsal surface of the cerebellar crest and extend branching dendrites into that formation. Some also send dendrites dorsalward into the body of the nucleus. Johnston (1898a, 1901; Fig. 14.8b) was the first to describe these elements and referred to them as primitive Purkinje cells. Other, more deeply situated large neurons distribute their dendrites entirely within the dorsal nucleus (Fig. 14.8b).

The intermediate zone of the area octavolateralis is formed by two cell masses, the nucleus intermedius and the nucleus caudalis. The very elongated

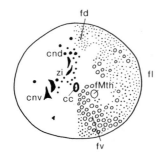

Fig. 14.9

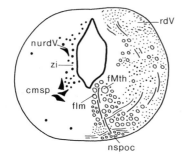

Fig. 14.10

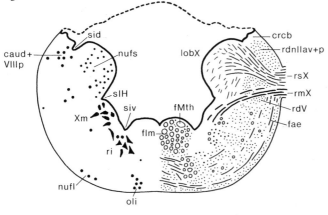

1mm

Fig. 14.11

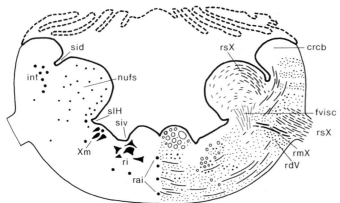

Fig. 14.12

Figs. 14.9–14.24. Series of transverse sections through the spinal cord and brain of the shovelnose sturgeon *Scaphirhynchus platorynchus*. The *left half* of each figure shows the cell picture based on Nissl-stained sections; the *right half* shows the fibre systems based on adjacent Bodian-stained sections. **Fig. 14.9.** Section through the second spinal segment. **Fig. 14.10.** Section through the obex region. **Fig. 14.11.** Section through the rhombencephalon at the level of entrance of one of the caudalmost vagal roots. **Fig. 14.12.** Section through the rhombencephalon at the level of entrance of the rostralmost vagal root. **Fig. 14.13.** Section through the rhombencephalon, just caudal to the level of entrance of the nervus lineae lateralis posterior. **Fig. 14.14.** Section through the rhombencephalon at the level of entrance of the octavus nerve. **Fig. 14.15.** Section through the rhombencephalon at the level of entrance of the trigeminal nerve. **Fig. 14.16.** Section through the rostralmost part of the rhombencephalon. **Fig. 14.17.** Section through the middle of the mesencephalon. **Fig. 14.18.** Section through the rostral mesencephalon and the caudoventral part of the diencephalon. **Fig. 14.19.** Section through the rostralmost part of the tectum mesencephali, the commissura posterior and the middle of the diencephalon. **Fig. 14.20.** Section through the rostral diencephalon. **Fig. 14.21.** Section through the caudalmost part of the telencephalon. **Fig. 14.22.** Section through the telencephalon at the level of the anterior commissure. **Fig. 14.23.** Section through the middle of the telencephalon. **Fig. 14.24.** Section through the caudal parts of the olfactory bulbs

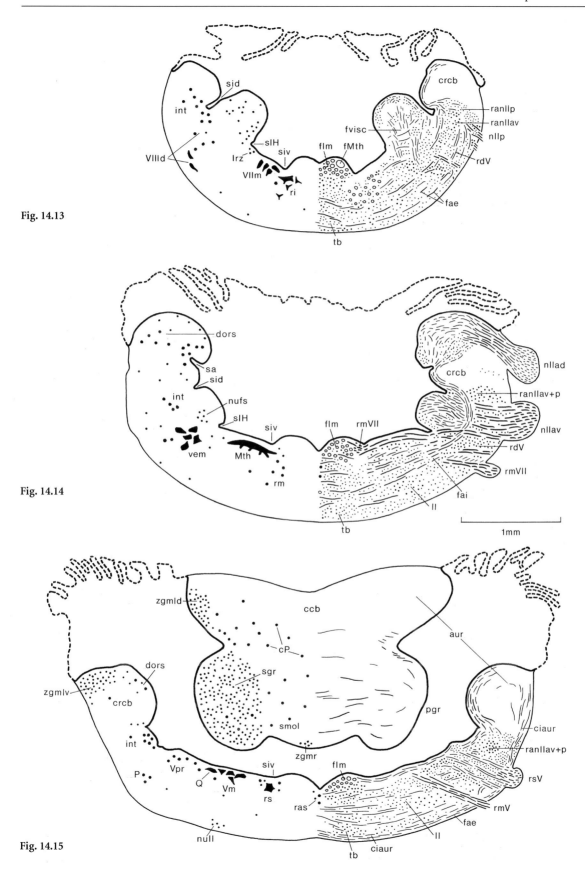

Fig. 14.13

Fig. 14.14

Fig. 14.15

1mm

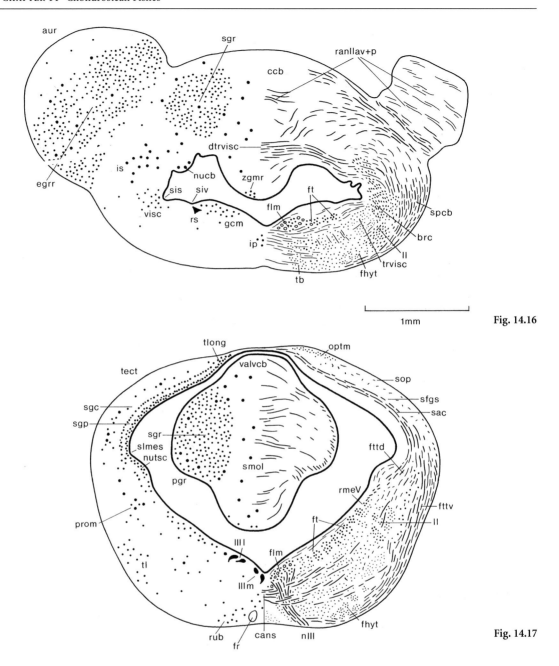

Fig. 14.16

Fig. 14.17

nucleus intermedius is situated close to the sulcus intermedius dorsalis (Figs. 14.12–14.15). The rather small nucleus caudalis forms a direct caudal continuation of the intermediate nucleus (Fig. 14.11; see also Fig. 14.27). Both nuclei are covered by the crista cerebellaris. The cells of the intermediate nucleus are very similar to those of the dorsal nucleus. Rostrally, the intermediate nucleus grades into the lateral part of the eminentia granularis (see Fig. 14.27). The cells of the caudal nucleus are mostly spindle-shaped and tend to be smaller than those of the intermediate nucleus.

According to New (1981) and McCormick (1982), New and Northcutt (1984a) and New and Bodznick (1985), the ventral zone of the area octavolateralis comprises four first-order octavus nuclei: the nucleus anterior, the nucleus magnocellularis, the nucleus descendens and the nucleus posterior.

According to these descriptions, the nucleus anterior nervi octavi consists of granular, multipolar and fusiform cells and extends from the level of entrance of the trigeminal nerve to a level slightly rostral to the level of entrance of the anterior lateral line nerve (see Fig. 14.25). Identification of this

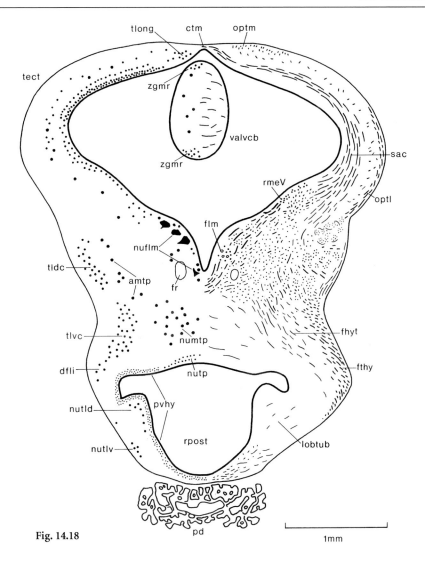

Fig. 14.18

nucleus was difficult in my *Scaphirhynchus* material. However, I consider it likely that a cell mass consisting of rather large, fusiform cells, provisionally designated by us as nucleus P (Fig. 14.15; see also Fig. 14.27), corresponds to the centre in question.

The nucleus magnocellularis, or nucleus vestibularis magnocellularis as it is termed here, is found at the level of entrance of the eighth nerve, caudal to the anterior nucleus. It consists of very large, multipolar cells situated close to the sulcus limitans at the ventrolateral border of the vagal lobe (Fig. 14.14; see also Figs. 14.25, 14.27).

The nucleus descendens is the largest component of the ventral octavolateral zone. It extends from a level just caudal to the entering eighth nerve to the region of the entrance of the glossopharyngeal nerve. Its dorsomedial part, which is sited directly ventral to the nucleus intermedius, consists mainly

of medium-sized, round and triangular elements. Its ventral part is constituted by large and medium-sized, fusiform cells. Together, these elements form a superficially situated zone that extends ventrally into the most lateral part of the basal plate (Fig. 14.13; see also Fig. 14.27).

The nucleus posterior lies directly ventral to and is co-extensive with the nucleus caudalis. A sharp boundary between these two nuclei cannot be drawn on cytoarchitectonic grounds; hence I have labelled them as one entity in Figs. 14.11 and 14.27.

The lateral line nerves of sturgeons and paddle fishes comprise a nervus lineae lateralis posterior and a nervus lineae lateralis anterior, the latter of which possesses separate dorsal and ventral roots. New and Northcutt (1984a) traced the central projections of these nerves in *Scaphirhynchus* using the horseradish peroxidase (HRP) and Fink-Heimer techniques (Fig. 14.25). These studies showed that,

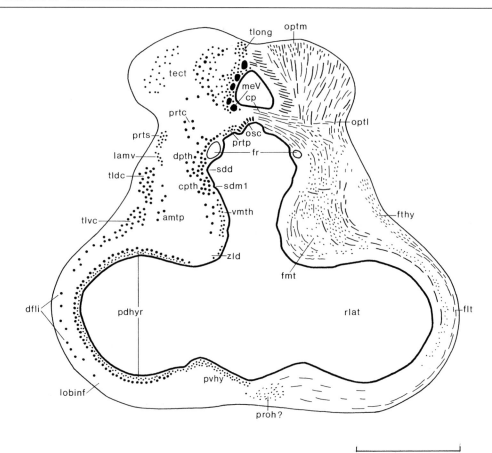

1mm

Fig. 14.19

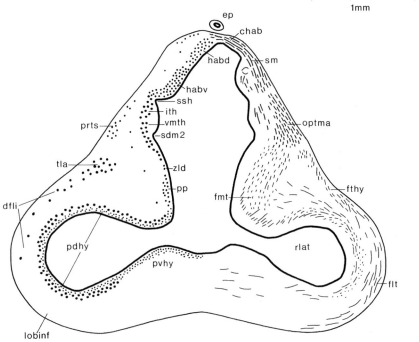

Fig. 14.20

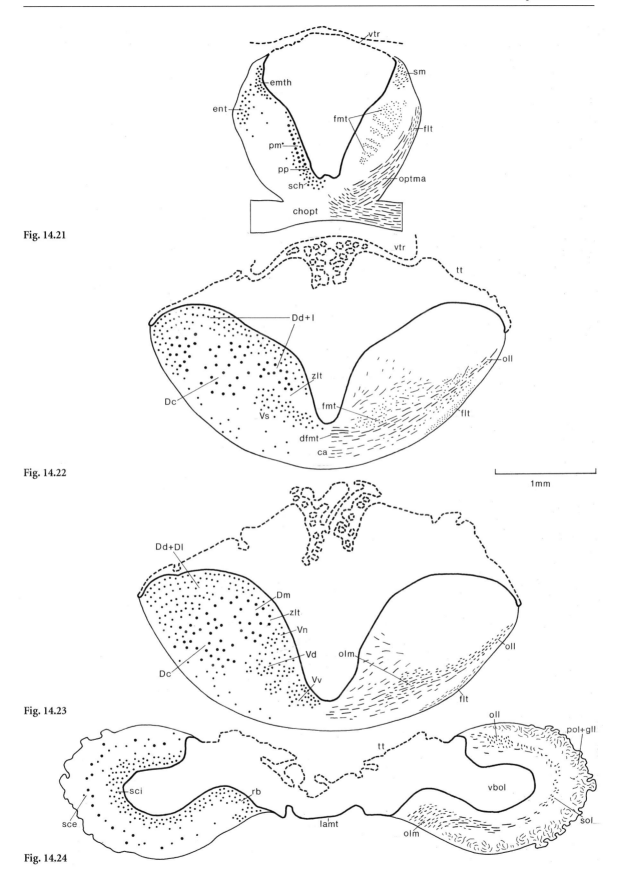

Fig. 14.21

Fig. 14.22

1mm

Fig. 14.23

Fig. 14.24

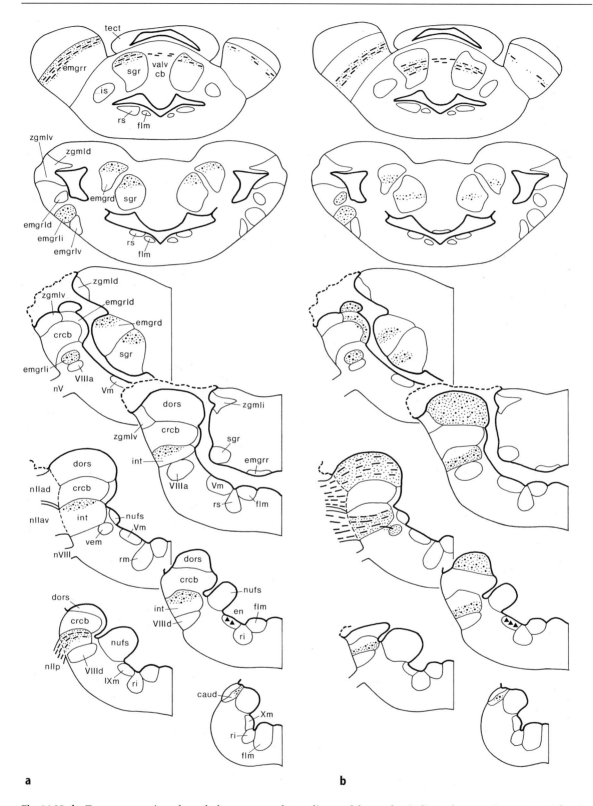

Fig. 14.25a,b. Transverse sections through the area octavolateralis of the shovelnose sturgeon *Scaphirhynchus platorynchus*, showing the projections of **a** the nervus lineae lateralis posterior and **b** the nervus lineae lateralis anterior. *Dashed lines* and *heavy dots* indicate degenerating axons. *Light stippling* indicates degenerating terminals. *Filled triangles* indicate retrogradely filled cell bodies. Fur further explanations, see text. (Redrawn from New and Northcutt 1984a)

upon entering the rhombencephalon, fibres of the dorsal root of the anterior lateral line nerve form ascending and descending branches that terminate within the ipsilateral nucleus dorsalis areae octavolateralis. Fibres of the ventral root of the anterior lateral line nerve and fibres of the posterior lateral line nerve enter the rhombencephalon ventral to the dorsal root of the anterior lateral line nerve, where some of the fibres terminate among the dendrites of the nucleus vestibularis magnocellularis. The bulk of the fibres form ascending and descending branches that terminate within the ipsilateral nucleus intermedius areae octavolateralis. Some of the descending fibres continue more caudally to terminate in the ipsilateral nucleus caudalis areae octavolateralis. Similarly, some of the ascending fibres continue more rostrally and terminate bilaterally in the cerebellum (Fig. 14.25). The latter will be dealt with in a later section.

The experimental studies carried out by New and Northcutt (1984a) also showed that both lateral line nerves carry efferent fibres, the cells of origin of which are positioned in the intermedioventral area, directly caudal to the large cells of Mauthner. Cells projecting efferent fibres to the anterior lateral line nerve appeared to be located rostral to those projecting efferent fibres to the posterior lateral line nerve.

The primary octavus afferents and the efferent projections of the cell masses contained within the octavolateral area have not been studied experimentally so far in chondrosteans. However, non-experimental material suggests that the octavus fibres bifurcate after having entered the rhombencephalon and distribute their fibres to the four nuclei which constitute the ventral zone of the octavolateral area.

As regards the efferents of the octavolateral area, non-experimental material suggests that the axons of the larger cells of the dorsal and intermediate nuclei form arcuate fibres which descend toward the basal plate (Figs. 14.8, 14.13, 14.14), where they form a large ascending bundle, the lemniscus lateralis (Figs. 14.14, 14.17). Some of the fibres join the ipsilateral lateral lemniscus, but most cross the median plane and ascend in the contralateral lemniscus. This pathway ascends to the tegmentum mesencephali and terminates in the torus semicircularis (Fig. 14.17). It is noteworthy that, whereas the rhombencephalic lateral line centres are large and well developed, the torus semicircularis is only small.

In the rostral part of the rhombencephalon of *Scaphirhynchus*, a superficially situated cell mass is found. Because this cells mass is clearly associated with the lateral lemniscus, it has been termed the nucleus lemnisci lateralis (see Figs. 14.15, 14.27). Hocke Hoogenboom (1929) observed a comparable cell mass in *Polyodon*.

The large cells in the octavus nuclei give rise to internal arcuate fibres that enter the ipsilateral and contralateral fasciculus longitudinalis medialis. Fibres descending with the latter bundle to the spinal cord constitute a tractus vestibulospinalis medialis. I consider it likely that other fibres, emanating from the large cells in the octavus nuclei, descend laterally to the medial longitudinal fasciculus to the spinal cord. Finally, it should be mentioned that, according to Larsell (1967), the smaller cells in the octavus area give rise to fibres that pass rostrally and that some of these probably reach the auricula cerebelli.

It was pointed out in the introduction to this chapter that, in addition to neuromasts, the lateral line system of chondrostean fishes comprises organs which closely resemble the chondrichthyan ampullae of Lorenzini and that some physiological and behavioural evidence is available suggesting that the latter subserve an electroreceptive function (Jørgensen et al. 1972; Teeter et al. 1980). It may be added that New and Bodznick (1985) presented experimental neuroanatomical and electrophysiological evidence indicating that, in the shovelnose *Scaphirhynchus platorynchus*, the atlantic sturgeon *Acipenser oxyrhynchus* and the paddlefish *Polyodon spathula*, the electroreceptive and mechanoreceptive lateral line afferents remain segregated in the rhombencephalon and that in these species the fibres innervating the electroreceptive ampullary organs project to the dorsal octavolateral nucleus via the dorsal root of the anterior lateral line nerve. New and Bodznick (1985) also mention preliminary experiments in chondrosteans in which retrograde labelling of Purkinje-like cells in the dorsal octavolateral nucleus was observed following HRP injections in midbrain electrosensory areas.

General Somatosensory System. The *general somatosensory* system is formed essentially by fibres of the trigeminal nerve and their centres of termination. After having entered the rostral rhombencephalon, the afferents fibres of the trigeminal nerve split up into three components:

1. The radix mesencephalicus nervi trigemini
2. A number of fibres which pass medially and terminate at the level of entrance of the nerve
3. The radix descendens nervi trigemini

The fibres constituting the mesencephalic trigeminal root originate in a collection of very large spherical and ellipsoidal cells, situated in the rostralmost part of the tectum mesencephali (see

Figs. 14.18, 14.27). The coarse processes arising from the nucleus mesencephalicus nervi trigemini course caudally through the tegmentum mesencephali (Figs. 14.16, 14.17) and the tegmentum isthmi and enter the trigeminal nerve after passing laterally along the motor trigeminal nucleus.

The trigeminal fibres passing medially after entrance terminate in a rather large, periventricular cell mass, the nucleus sensorius principalis nervi trigemini. This nucleus, which is situated directly lateral to the rostralmost part of the sulcus limitans, consists of small, rather loosely arranged cells (see Figs. 14.15, 14.27).

The radix descendens nervi trigemini is a conspicuous, superficially situated bundle which can be traced caudally into the rostralmost part of the spinal cord (Figs. 14.8, 14.12, 14.14). Hocke Hoogenboom (1929) reported that, in *Polyodon*, the descending trigeminal root occupies a much deeper position than in *Acipenser*. During its caudal course, the bundle is reinforced by small contingents of the glossopharyngeal and vagus nerves (Johnston 1901, 1903). The nucleus of termination of the radix descendens of nerve V consists of scattered cells, distributed throughout its length; however, in the caudalmost part of the rhombencephalon, these cells become more concentrated, earning the designation nucleus tractus descendens nervi trigemini. This nucleus caudally grades into the dorsal horn of the spinal cord. According to Hocke Hoogenboom (1929), in *Polyodon* most fibres of the radix descendens of nerve V terminate in a spinal trigeminal nucleus, a centre situated a short distance behind the obex.

The above description of the primary projections of the trigeminal nerve, which was primarily based on non-experimental silver material of the shovelnose *Scaphirhynchus platorynchus* (see also Nieuwenhuys and Pouwels 1983), has been confirmed and extended by the experimental study carried out by New and Northcutt (1984b). These authors studied both the shovelnose and the Atlantic sturgeon *Acipenser oxyrhynchus*, using HRP as a tracer, and observed that the descending trigeminal projections consist of two fascicles: (1) the fibres projecting to and terminating in the descending trigeminal nucleus and (2) the fibres of the trigeminal spinal tract. The descending trigeminal nucleus itself was found to consist of separate dorsal and ventral parts. The trigeminal spinal tract appeared to descend as a compact bundle to the cervical spinal cord. Along its length, fibres were given off to the ventral part of the descending trigeminal nucleus. Immediately caudal to the obex, fibres appeared to leave the spinal tract to terminate densely in the funicular nucleus, a region of dense neuropil and neurons in the dorsal part of the spinal cord. The existence of fibres of cranial nerves IX and X projecting to the descending trigeminal tract was experimentally confirmed by New and Northcutt (1984b).

Nothing is known at present about the efferent projections of the chondrostean sensory trigeminal centres.

Viscerosensory Centres. It has already been mentioned that small fibre contingents of the facial, glossopharyngeal and vagus nerves contribute to the radix descendens of the trigeminal nerve. However, the bulk of the afferent fibres of these nerves are viscerosensory in nature and pass to one common mass of grey, the nucleus fasciculi solitarii. This centre extends from the level of entrance of the sensory facial root caudalward to the level of the obex and forms a large, elongated projection into the fourth ventricle, known as the lobus vagi (see Figs. 14.3, 14.11–14.14, 14.27). It consists of small cells, the density of which varies from place to place. Many of the primary viscerosensory fibres end at their level of entrance, but others descend in the lobe to constitute a diffuse fasciculus solitarius. Bifurcating axons giving rise to ascending and descending branches have also been observed (Johnston 1901).

Efferent fibres passing from the nucleus fasciculi solitarii to the adjacent visceromotor nuclei could not be clearly distinguished in our non-experimental *Scaphirhynchus* material. However, it may be assumed that such connections exist. They have been observed in several teleosts and are known to form part of the reflex mechanisms for the intake of food and for swallowing (see Sect. 15.3.3).

Throughout the vagal lobe, numerous fibres were observed to leave this structure and to pass ventrolaterally (Figs. 14.12, 14.13). The Golgi studies carried out by Johnston (1898a, 1901) revealed that these fibres assemble directly ventral to the radix descendens of nerve V, forming a superficially situated ascending pathway, the tractus visceralis secundarius (Fig. 14.8). Upon reaching the isthmus region, the fibres turn dorsomedially and terminate in a centre known as the nucleus visceralis secundarius (Fig. 14.16).

There is some uncertainty regarding the exact location of the nucleus visceralis secundarius. Johnston (Fig. 10 in Johnston 1898a; Figs. 15, 16 in Johnston 1901) depicts this nucleus in *Acipenser rubicundus* as a large centre, situated laterally in the isthmus region, at some distance from the ventricular surface. According to Hocke Hoogenboom (1929), in *Polyodon* the centre in question is situated ventrolaterally to the nucleus cerebelli. In her Fig. 23, she depicts a nucleus isthmi which clearly occupies the position of Johnston's secondary visceral nucleus. I consider it likely that, in *Scaphirhynchus*, the nucleus visceralis secundarius is represented by a relatively small, periventricular cell mass, situated ventrolaterally to the nucleus cerebelli, close to a distinct ventricular groove, the sulcus isthmi. The centre designated by Johnston as the secondary visceral nucleus is also present in *Scaphirhynchus*. However, it appears to occupy the position of Hocke Hoogenboom's nucleus isthmi and has been labelled accordingly (see Figs. 14.16, 14.27). Unfortunately, this nucleus isthmi was misinterpreted in a previous study (Fig. 11 in Nieuwenhuys and Pouwels 1983) as the nucleus cerebelli. According to my current interpretation, the latter is situated medially to the nucleus isthmi, close to the ventricular surface (Fig. 14.16).

Data concerning the efferents of the chondrostean nucleus visceralis secundarius are not available at present.

Having discussed the various centres present in the alar plate and their connections as far as known, brief consideration may now be given to the functional significance of the two longitudinal zones in which this part of the rhombencephalon can be divided, i.e. the area dorsalis and the area intermediodorsalis. Johnston (1901, 1902, 1903) termed the area dorsalis of the present account the somatosensory zone. Because this area is entirely occupied by special somatosensory lateral line centres, there is no objection to make to this functional designation. However, it appears that Johnston's designation of the area intermediodorsalis as the viscerosensory zone is untenable. The viscerosensory nucleus fasciculi solitarii is, admittedly, by far the largest centre situated within this zone, but reference to Fig. 14.27 clearly shows that two general somatosensory centres – the nucleus sensorius principalis and the nucleus of the tractus descendens of nerve V – and three special somatosensory centres – the nucleus anterior, the nucleus magnocellularis and the nucleus descendens – are also located within its confines. The nucleus descendens even encroaches upon the lateralmost part of the basal plate.

14.5.3
Area Intermedioventralis

The area intermedioventralis is a longitudinal strip, bounded laterally by the sulcus limitans and medially by the sulcus intermedius ventralis (see Fig. 14.27). Johnston (1901, 1902, 1903) referred to this strip as the visceromotor zone, a term which is

correct in so far as centres belonging to that functional category prevail within it. However, this area appears to contain several centres that cannot be subsumed under that label.

From a general comparative point of view, it might be expected that in chondrosteans, as in other vertebrates, cranial nerves VII, IX and X contain general visceral efferent or parasympathetic preganglionic fibres, the cells of origin of which are situated in roughly the same area as those of the special visceromotor or branchiomotor fibres. In my non-experimental material of *Scaphirhynchus*, this subdivision could not be made; the motor nuclei of nerves VII, IX and X (and that of nerve V) have therefore been designated as visceromotor, without further specification. However, I consider it likely that the bulk of the cells contained within these centres belong to the branchiomotor category.

The nucleus motorius nervi vagi and the nucleus motorius nervi glossopharyngei together constitute one elongated cell column (Figs. 14.11, 14.12).

According to Theunissen (1914: *Acipenser*) and Hocke Hoogenboom (1929: *Polyodon*), the nucleus motorius nervi facialis is caudally continuous with the cell column representing the nuclei of nerves IX and X. However, in my *Scaphirhynchus* material, this nucleus could be clearly delineated as a separate entity (see Figs. 14.13, 14.27). It is noteworthy that the motor facial nucleus is situated caudal to the level of entrance of the corresponding nerve. Before issuing, the efferents from the nucleus form a conspicuous bundle which passes rostrally, just lateral to the medial longitudinal fasciculus (Fig. 14.14).

The large nucleus motorius nervi trigemini represents the rostral part of the visceromotor column. As in many other fishes, this nucleus is separated from the motor nucleus of nerve VII by a rather wide gap (see Figs. 14.15, 14.27).

In chondrosteans all visceromotor nuclei occupy a periventricular position.

The 'non-visceromotor' nuclei in the intermedioventral zone include the following: (a) a group of neurons efferent to the lateral line organs, (b) the nucleus funiculi lateralis, (c) the nucleus lemnisci lateralis, (d) the lateral reticular zone and (e) the large cell of Mauthner.

As has already been mentioned, the retrograde tracer studies carried out by New and Northcutt (1984a) showed that the intermedioventral area contains a number of neurons projecting efferents to the anterior and posterior lateral line nerves. These cells are situated in the area just caudal to the cell of Mauthner (Fig. 14.25).

The nucleus funiculi lateralis is a superficial cell mass situated in the caudal rhombencephalon (see

Figs. 14.12, 14.27). A centre occupying a corresponding topological position has been observed in most groups of vertebrates. In mammals, this cell mass, which is also known as the nucleus reticularis lateralis, receives afferents from the spinal cord and projects to the cerebellum. The latter projection has also been demonstrated in teleosts (Libouban and Szabo 1977; Finger 1978). In chondrosteans, its connections are entirely unknown.

The nucleus lemnisci lateralis also occupies a superficial position, but is situated in the rostralmost part of the rhombencephalon, peripheral to the motor trigeminal nucleus (see Figs. 14.15, 14.27). Its close relation to the lateral lemniscus, the main efferent projection of the first-order lateral line centres, has already been mentioned.

The lateral reticular zone is in most places a rather ill-defined zone of cells, situated just medial to the sulcus limitans. I consider the cell group designated provisionally with the letter Q (see Figs. 14.15, 14.27) as the more compact, rostralmost part of that zone. The cells of this zone are possibly intercalated in reflex arcs linking both somatosensory and viscerosensory centres with the visceromotor nuclei.

Finally, the cells of Mauthner appear bilaterally at the level of entrance of the octavus nerve. These very large neurons are transversely oriented and lie immediately under the ventricular surface (see Figs. 14.14, 14.27). It is of historical interest that the connection of the giant nerve fibres described by Mauthner (1859) with a pair of large neurons situated in the rhombencephalon was first observed by Goronowitsch (1888) in the chondrostean *Acipenser ruthenus*. Goronowitsch (1888, p. 499) wrote as follows: "In der Querschnittsebene des Austrittes des Acusticus findet eine Kreuzung der Mauthner'schen Fasern statt. Nach der Kreuzung verlaufen sie bogenförmig zum Unterhorne, wo sie in sehr grosse Nervenzellen enden" ("At the cross-sectional level of the issue of the acoustic nerve, the fibres of Mauthner decussate. Thereafter, they take an arcuate course to the inferior horn, where they terminate in very large nerve cells"). Curiously enough, Mauthner never described the cells which bear his name. The fibres of Mauthner are very coarse and well myelinated in chondrosteans. As shown in Fig. 14.9–14.13, they descend within the fasciculus longitudinalis medialis to the spinal cord.

14.5.4
Area Ventralis

The area ventralis, i.e. the medialmost of the four rhombencephalic columns, was designated by Johnston (1901, 1902, 1903) as the somatomotor zone. Although true somatic efferent centres occupy only a minor portion of this zone, the name is appropriate in that its chief constituents, the median and medial reticular formation and the fasciculus longitudinalis medialis, presumably form part of the somatomotor coordinating apparatus. However, there are a few 'non-somatomotor' centres in the area ventralis. These include the oliva inferior, the griseum centrale and the nucleus interpeduncularis.

There are only two rhombencephalic somatic efferent centres: the columna motoria spinalis and the nucleus nervi abducentis. The former extends over some distance rostral to the obex and in *Scaphirhynchus* gives rise to two spino-occipital nerve roots (see Figs. 14.3, 14.27). The nucleus nervi abducentis consists of rather diffusely arranged elements which are hard to distinguish from those of the medial reticular formation. These elements are situated medially to the motor facial nucleus, at the ventrolateral border of the medial longitudinal fasciculus (see Fig. 14.27).

The median reticular formation, which in chondrosteans is composed of small and medium-sized cells, consists of two nuclei: the nucleus raphes superior (Fig. 14.15) and inferior (Fig. 14.12).

The medial reticular formation is constituted by a continuous strand of large and medium-sized cells, extending throughout the rhombencephalon. Local differences in the density of the large elements allow a subdivision into a nucleus reticularis inferior, medius and superior (see Figs. 14.11–14.16, 14.27). The latter continues rostrally into another reticular centre, the nucleus reticularis isthmi. The main distinguishing feature of this nucleus is its lack of large neurons (see Fig. 14.27).

The oliva inferior is an elongated, small-celled nucleus, occupying a superficial position in the caudal rhombencephalon (see Figs. 14.11, 14.27). The axons of its cells decussate to the opposite side and pass as fibrae arcuatae externae to the region in which the cerebellum joins the rhombencephalon (Theunissen 1914).

The griseum centrale and the nucleus interpeduncularis are both situated in the rostral part of the rhombencephalon. Both consist of small neurons. The interpeduncular nucleus takes up the fasciculus retroflexus, but for the rest the fibre connections of these centres are unknown.

14.5.5
Isthmus Rhombencephali

Although the zonal pattern discussed above is present throughout most of the rhombencephalon, this pattern becomes less distinct in the isthmus, i.e. the tapered, rostralmost portion of that brain part. The ventral zone, flanked by the intermedioventral sulcus, is the only one of the four morphological zones which clearly continues into the isthmus region. The lateral wall of the latter is sculptured by a distinct sulcus isthmi, but this ventricular groove cannot be traced as being continuous with any of the more caudally situated ventricular sulci. Three cell masses – the nucleus visceralis secundarius, the nucleus cerebelli and the nucleus isthmi – are present in the lateral wall of the isthmus region. These cell masses have already been briefly discussed. It should be emphasised that their interpretation, as given here, largely rests upon their positional relations. Experimental neuroanatomical studies are needed to further clarify their identity. However, such studies are entirely lacking at present. Suggestions concerning the connections of these nuclei, based on non-experimental material, may be summarised as follows: The nucleus visceralis secundarius represents the end-point of a large, mainly ipsilateral pathway, originating from the nucleus fasciculi solitarii (see above). According to Hocke Hoogenboom (1929), the nucleus cerebelli receives efferents from the auricula cerebelli and contributes efferent fibres to the brachium conjunctivum. The same author considered it likely that the nucleus isthmi receives fibres from undetermined rhombencephalic sources and is reciprocally connected with the cerebellum. It is, however, important to note that experimental neuroanatomical studies have shown that, in holosteans and teleosts, the nucleus isthmi is reciprocally connected with the tectum mesencephali (see Sect. 15.3.6).

14.5.6
Summary of Rhombencephalic Fibre Systems

Because experimental neuroanatomical studies are lacking, our knowledge of the rhombencephalic fibre systems in chondrosteans is very limited indeed. All that can be done here is to summarise pertinent data based on non-experimental histological material and to add some comparative neuroanatomical comments. The primary afferent and efferent fibres have been sufficiently dealt with above; the cerebellar connections will be discussed in the next section.

Our knowledge of the rhombencephalic fibre systems in chondrosteans can be summarised as follows:

1. The lateral funiculi of the spinal cord pass rostrally into the white matter of the rhombencephalon. This fibre continuum most probably contains several ascending systems, but so far none of these has been demonstrated with sufficient clarity.

2. It is likewise probable that the bulbar general somatosensory centres, i.e. the nucleus of the radix descendens of nerve V and the nucleus sensorius principalis of nerve V, give rise to fibres ascending to higher levels, but again such fibres have not been described so far.

3. The large viscerosensory nucleus of the tractus solitarius gives rise to a distinct ipsilateral tractus visceralis secundarius, which terminates in the nucleus of the same name.

4. The lateral line nerve centres give rise to a distinct, mainly crossed lemniscus lateralis, which can be traced to the torus semicircularis in the midbrain.

5. As in all vertebrates, the ganglion habenulae situated in the dorsal diencephalon projects to the nucleus interpeduncularis.

6. Johnston (1901) reported that, in *Acipenser*, fibres emanating from various parts of the hypothalamus reach the upper rhombencephalon. According to his observations, these fibres constitute two bundles, which he termed the lobobulbar and mamillobulbar tracts. Corresponding fibres have been observed by Hocke Hoogenboom (1929) in *Polyodon* and by the present author in *Scaphirhynchus*. As far as I could ascertain, in the latter species these fibres connect the large lobus interiou hypothalami (Fig. 14.3) with the tegmentum rhombencephali. The polarity of these fibres could not be established in my non-experimental material. However, I consider it likely that most of them are descending; hence they are designated in Figs. 14.16–14.18 as fibrae hypothalamotegmentales. These fibres might conceivably play an important role in the mediation of feeding reactions.

7. A mainly crossed tractus tectobulbaris passes from the tectum mesencephali to the rhombencephalon. Within the latter, the tract descends in the superficial part of the area ventralis (Figs. 14.13–14.16). Its exact caudal extent cannot be determined in non-experimental material.

8. In the periventricular zone of the tegmentum mesencephali and the adjacent tegmentum isthmi, a number of well-myelinated, longitudinally oriented fasciculi tegmentales, with unknown origins and terminations, are present (Figs. 14.16, 14.17).

9. The fasciculus longitudinalis medialis is very large in chondrosteans. It can be easily traced from upper mesencephalic levels to the caudal rhombencephalon, where it continues into the spinal ventral funiculus (Figs. 14.9–14.18). Our

non-experimental material of *Scaphirhynchus* strongly suggests that the bundle receives (a) fibres from the nucleus of the fasciculus longitudinalis medialis, a large-celled, reticular centre situated in the rostralmost part of the tegmentum mesencephali (Fig. 14.18), (b) large numbers of fibres from the rhombencephalic medial reticular formation and (c) a substantial contingent of fibres from the larger neurons situated in the ventral (octavus) zone of the area octavolateralis. Finally, there can be no doubt that the large fibre of Mauthner descends with the fasciculus longitudinalis medialis to the spinal cord.

10. I consider it likely that the larger cells in the octavus area also give rise to an ipsilateral vestibulospinal tract, descending laterally to the fasciculus longitudinalis medialis. The very coarse fibres present in the deepest part of the spinal lateral funiculus (Figs. 14.4d, 14.9) may well form part of this tract.

14.6
Cerebellum

14.6.1
Gross Relations

The cerebellum of chondrosteans is strongly developed. Its central part consists of a massive body which protrudes into the ventricular cavity. Laterally, this central body is continuous on either side with a large structure which, after having surrounded a lateral recess of the fourth ventricle, continues into the rhombencephalic alar plate (Fig. 14.3). The caudal part of the central body represents the corpus cerebelli, whereas its rostral part represents the valvula cerebelli. The latter projects rostrally into the mesencephalic ventricle, where it turns dorsally and caudally to become continuous with the tectum mesencephali via a short velum medullare anterius (Fig. 14.3d). As in polypterids, the central body of the chondrostean cerebellum surrounds an extension of the extra-encephalic cavity, but in the latter group this extension is very slight and only represented by a rostroventrally directed groove. According to Hocke Hoogenboom (1929), this groove marks the boundary between the corpus and the valvula. More ventrally, there is no landmark which clearly distinguishes these two structures from each other.

The lateral parts of the central body are occupied by large masses of granule cells that protrude into the ventricular cavity, thus forming a prominentia granularis on either side (Nieuwenhuys 1967; Figs. 14.15–14.17). In between these two granular

masses, but extending further rostrally, ventrally and caudally, there is an unpaired zone of molecular substance which passes dorsolaterally over into the molecular layer of the lateral portions of the cerebellum (Fig. 14.15). The peculiar structural relations found in the central part of the chondrostean cerebellum can be explained by the fact that, during ontogenesis, the walls of the bilateral halves of the cerebellar anlage fold inward and that the medial surfaces of these halves, thus brought into close apposition, fuse in the median plane.

The lateral parts of the chondrostean cerebellum were called lateral lobes by Johnston (1901) and lateral lobules by Larsell (1967). Following Palmgren (1921) and Hocke Hoogenboom (1929), they will be termed the auriculae cerebelli here. It is unfortunate that the latter term, which clearly belongs in the realm of gross morphology, has been employed by some authors (e.g. Pearson 1936) to designate microscopic structures. Larsell (1967) was of the opinion that the lateral parts of the chondrostean cerebellum, together with a commissure and a band of grey to be discussed below, forms a unit that is comparable to the cerebellar vestibulolateral lobe of cartilaginous fishes and other vertebrates.

14.6.2
Histological Structure

The granule cells in the chondrostean cerebellum constitute three different formations, the stratum granulare of corpus and valvula, the zona granularis marginalis and the eminentia granularis. These formations, the rather intricate spatial relations of which are represented in Fig. 14.26, will now be discussed.

The *stratum granulare of corpus and valvula* encompasses most of the large cell masses which are found bilaterally in the prominentiae granulares. The Golgi studies carried out by Johnston (1898a, 1901) showed that the cells in these formations are provided with a few short dendrites and with long, non-bifurcating axons, many of which traverse the unpaired, medial molecular zone, after which they turn dorsolaterally into the molecular layer of the auricle (Fig. 14.8b).

In addition to the small granule cells described above, the granular layer of corpus and valvula contains several varieties of larger, Golgi-type II cells (see also Johnston 1901; Larsell 1967). The elements of one of these types is particularly found in the lateralmost part of the prominentia granularis; their dendrites are specifically related to the tecto-cerebellar tract, the fibres of which are concentrated in that area (Fig. 14.8b).

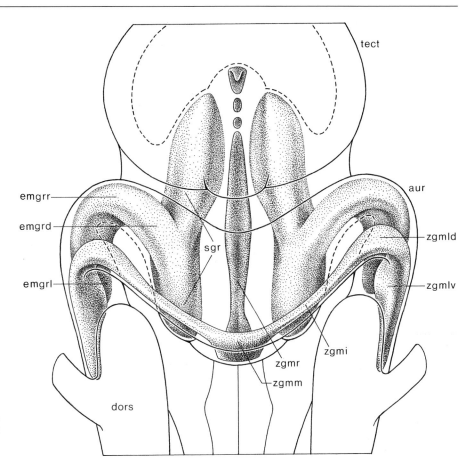

Fig. 14.26. Stereogram showing the granular cell masses in the cerebellum of the shovelnose sturgeon *Scaphirhynchus platorynchus* in dorsal view

The *zona granularis marginalis* is, as its name implies, largely located along the caudal margin of the cerebellum. Somewhat schematically, it may be subdivided into five different portions, which I have designated as the medial, rostral, intermediate, laterodorsal and lateroventral parts of the zone (Fig. 14.26).

The pars medialis is represented by an accumulation of granule cells situated in the caudalmost part of the cerebellum. Palmgren termed this structure the pars medialis auriculi. According to Larsell (1967), it corresponds to the granular ridge of the pars medialis of the vestibulolateral lobe of cartilaginous fishes.

The pars rostralis starts from the medial part and extends rostrally along the ventral surface of the corpus and the valvula (Figs. 14.15, 14.16). In the ros tralmost part of the latter, this structure is represented by isolated clusters of cells. It corresponds to the medial string of granular cells referred to by Palmgren (1921).

The pars intermedia is medially continuous with the pars medialis and extends rostrolaterally along the taenia of the cerebellum. Upon reaching the auricula, the zone enlarges somewhat and forms the

pars laterodorsalis (Fig. 14.15). It then arches ven trocaudally and enters the lateralmost part of the rhombencephalic alar plate. This pars lateroventralis of the zona granularis marginalis flanks the nucleus dorsalis of the area octavolateralis (see Figs. 14.15, 14.27).

The structures described above have been given different names in the literature. Thus the pars intermedia plus the pars laterodorsalis correspond to the lateral string of granular cells referred to by Palmgren (1921), to the interauricular granular band plus the superior fold of the auricle referred to by Larsell (1967) and to the upper leaf of the marginal granular cell group referred to by New (1981). Larsell denoted the pars lateroventralis of the present description as the external granular layer of the dorsal nucleus, while New (1981) termed it the lower leaf of the marginal granular cell group.

The *eminentia granularis*, like the zona granularis marginalis, describes an arch extending into the rhombencephalic alar plate (Fig. 14.26). Following New (1981), I subdivide this granular formation into a pars dorsalis, a pars rostralis and a pars lateralis.

The pars dorsalis of the eminentia granularis becomes visible in the caudal portion of the corpus cerebelli, where it gradually dissociates itself from the dorsolateral part of the stratum granulare of this structure (Fig. 14.25). Rostrolaterally, it is continuous with the pars rostralis of the eminentia granularis, a large cell mass which forms a wide band across the auricle (Figs. 14.16, 14.25). The pars lateralis of the eminentia granularis extends caudally from the ventrolateral portion of the pars rostralis into the rhombencephalic alar plate (see Fig. 14.27). New (1981) observed that the caudal portion of the pars lateralis of the granular eminence branches into three components, a dorsal, an intermediate and a ventral one. These have been termed the pars laterodorsalis, the pars laterointermedia and the pars lateroventralis here (Fig. 14.25). According to New (1981), the lateroventral part extends caudally to contact the anterior octavus nucleus, the latero-intermediate part is continuous caudally with the nucleus intermedius of the octavolateral area, and the laterodorsal part extends caudally slightly dorsomedial to the other two components and then courses dorsomedially to cover the medial surface of the cerebellar crest.

In the auricle, there are scattered cells in many places between the zona granularis marginalis and the eminentia granularis, but a true fusion does not occur.

In addition to numerous granule elements, the granular eminences include two varieties of Golgi type II cells (Larsell 1967).

The molecular layer of the cerebellum consists largely of fine axons, among which the dendrites of the Purkinje cells spread widely. The medial molecular zone of the corpus cerebelli is continuous dorsolaterally with the molecular layer of the auricula (Fig. 14.15), which in turn continues into the cerebellar crest.

Only in the transitional area of corpus and auricula do the Purkinje cells show a tendency toward layer formation (Fig. 14.15). In the corpus and valvula, these elements lie scattered in the medial molecular zone (Figs. 14.15–14.17). In the eminentiae granulares, irregularly arranged Purkinje cells occur at or near the borders of the molecular layer. We have not observed Purkinje elements specifically related to the zona granularis marginalis.

Golgi preparations show that the Purkinje cells send richly branching and widely spreading dendrites into the molecular zone (Johnston 1898a, 1901; Larsell 1967; Fig. 14.8b). Their axons could only be traced over short distances.

The ventralmost part of the cerebellum of Scaphirhynchus contains a nucleus cerebelli. This cell mass, which has also been described for Acipenser (Larsell 1967) and Polyodon (Hocke Hoogenboom 1929), forms a small protrusion into the lateral part of the fourth ventricle (Fig. 14.16). It should be emphasised that the nucleus cerebelli, as described by Nieuwenhuys and Pouwels (1983), represents the nucleus isthmi according to my present interpretation.

In the teleostean cerebellum, the Purkinje cells impinge on large, so-called eurydendroid cells, and the axons of the latter constitute the cerebellar output system. The somata of the Purkinje and eurydendroid cells together form a ganglionic layer, intermediate between the molecular and granular layers. A deep nucleus in which the cerebellar efferent neurons are concentrated is probably lacking (Nieuwenhuys and Nicholson 1969; Pouwels 1978b; Meek and Nieuwenhuys 1991). In view of these findings, the following questions arise:

1. Do some of the large stellate cells observed in Golgi material of the chondrostean cerebellum actually represent eurydendroid cells?
2. What are the connections of the cell group designated in chondrosteans as the nucleus cerebelli?

Only experimental tracer studies can provide answers to these questions.

14.6.3
Fibre Connections

Our knowledge of the fibre connections of the chondrostean cerebellum is mainly based on the work of Johnston (1901), Hocke Hoogenboom (1929) and Larsell (1967), who all studied nonexperimental material. So far, only the lateral line nerve projections to the cerebellum in *Scaphirhynchus* have been explored using experimental techniques (New 1981; New and Northcutt 1984b).

Afferent Projections. The following afferent projections have been described:

1. *Lateral line nerve fibres.* The experimental studies performed by New and Northcutt (1984a) showed that numerous fibres of the ventral root of the anterior lateral line nerve enter the auricula cerebelli. Within this structure, these fibres pass rostrally and then dorsomedially and caudally, terminating in the latero-intermediate, rostral and dorsal parts of the eminentia granularis. The experiments also revealed fibres extending medially to the stratum granulare, where they bifurcate into descending and ascending branches, the former reaching the valvula. A number of fibres were observed to decussate across the molecular layer of the corpus cerebelli to the contralateral stratum granulare and the

contralateral dorsal and rostral parts of the eminentia granularis, terminating within these cell masses (Fig. 14.25b). Some fibres of the dorsal root of the anterior lateral line nerve could be traced to the laterodorsal part of the eminentia granularis. The fibres of the posterior lateral line nerve terminate in the same cerebellar grisea as those of the ventral root of the anterior lateral line nerve (Fig. 14.25a). However, the fibres of these two nerves appeared to show a definite somatotopic organisation. Within the ipsilateral and contralateral eminentia granularis and stratum granulare of corpus and valvula, the fibres of the posterior lateral line nerve remain strictly confined to the dorsal portions of these cell groups, while those of the anterior lateral line nerve are confined to the intermediate portions of the same cell groups, directly ventral to the projections of the posterior lateral line nerve (Fig. 14.25).

2. *Octavus nerve fibres.* Johnston (1901) and Larsell (1967) reported that primary octavus fibres pass to the cerebellum. Preliminary experiments in *Scaphirhynchus* (J.G. New and R.G. Northcutt, unpublished observations) indicate that the ventralmost portion of the eminentia granularis, i.e. that part of the structure which is not in receipt of lateral line nerve fibres (see Fig. 14.25), is occupied by ascending octavus fibres. Corresponding observations have been made in *Amia* (McCormick 1981).

3. *Fibres ascending from the area octavolateralis.* Larsell (1967) reported that fibres originating from the octavolateral area ascend to the eminentia granularis and to the valvula and that a small fascicle of these fibres follows the zona granularis marginalis.

4. *Trigeminocerebellar fibres.* Primary trigeminocerebellar fibres reaching the corpus cerebelli were described by Johnston (1901) and Larsell (1967). According to the latter author, these trigeminal root fibres are augmented by fibres from the sensory trigeminal nucleus.

5. A largely uncrossed *spinocerebellar system*, corresponding to the ventral spinocerebellar tract of teleosts, was described by Hocke Hoogenboom (1929). Larsell, who also observed this bundle, mentioned that it reaches the lateral part of the corpus cerebelli.

6. As has already been mentioned, an *oliva inferior*, giving rise to decussating fibres, is present in chondrosteans. Theunissen (1914) reported that these decussating efferents reach the cerebellum, thus forming an olivocerebellar projection.

7. A *tractus tegmentocerebellaris*, originating partly from the nucleus lateralis valvulae and partly from other centres in the tegmentum of the midbrain and terminating, largely or entirely decussating, in the valvula cerebelli, was observed in *Polyodon* by Hocke Hoogenboom (1929) and Larsell (1967). A corresponding fibre system was described by Johnston (1901) in *Acipenser* as bundle y.

8. A *tractus tectocerebellaris*, which, unlike its teleostean homologue, passes directly from the lateral margin of the tectum into the valvula, was observed by Johnston (1901), Hocke Hoogenboom (1929) and Larsell (1967). In fact, Johnston described two separate tectocerebellar tracts in *Acipenser*, but Hocke Hoogenboom and Larsell considered it likely that one of these actually represents the spinocerebellar system mentioned in this list (see item 5).

9. The presence of crossed and direct *lobocerebellar tracts*, originating from the inferior lobes of the diencephalon, was reported by Johnston (1901) and Larsell (1967) in *Acipenser*. According to the latter author, many of the fibres of these systems terminate in the valvula. Hocke Hoogenboom (1929) considered it likely that a corresponding tract exists in *Polyodon*.

Cerebellar Commissural Connections. Hocke Hoogenboom (1929) noted that both auriculae are interconnected by a large commissural system, the fibres of which, beneath the ventral surface of the rhombencephalon, form a superficial band-like layer which crosses the median plane. A corresponding commissure also exists in *Acipenser* (Fig. 14.8b) and *Scaphirhynchus* (Fig. 14.15). According to our observations, it originates mainly from the pars lateroventralis of the zona granularis marginalis.

Efferent Projections. According to the literature, the cerebellar efferents of the chondrostean cerebellum constitute three fibre systems, the tractus cerebellomotorius, the tractus cerebellobulbaris and the brachium conjunctivum.

According to Hocke Hoogenboom's observations, the *tractus cerebellomotorius* originates in the auricula, probably directly from the Purkinje cells situated within this structure. Its axons pass as internal arcuate fibres along the ventricular wall. Many cross the median plane and continue caudally in the fasciculus longitudinalis medialis. A smaller number of these fibres ascend, after decussating, with the fasciculus longitudinalis medialis to the midbrain.

According to Larsell (1967), the *cerebellobulbar tract* consists of fibres that originate in the eminentia granularis and terminate diffusely in the rhombencephalon.

Finally, the *brachium conjunctivum* reportedly originates partly from Purkinje cells situated in the auricula and in the corpus (Larsell 1967) and partly from the nucleus cerebelli. The bundle passes ventrolaterally and then ventromedially to decussate beneath the fasciculus longitudinalis medialis in the caudalmost part of the midbrain. After decussation, the fibres pass rostrally in the ventromedial part of the tegmentum mesencephali. Larsell (1967) considered it likely that the brachium conjunctivum reaches the diencephalon.

14.7
Mesencephalon

14.7.1
Introductory Notes

The mesencephalon of chondrostean fishes is rather poorly developed (Fig. 14.3). It consists of two main parts, the dorsal tectum mesencephali and the ventral tegmentum mesencephali. The latter can be subdivided into a lateral and a medial tegmental zone. The sulcus limitans of His does not extend into the midbrain, and thus a morphological landmark on the basis of which this brain part could be subdivided into a alar plate and a basal plate is lacking.

The valvula cerebelli protrudes into the mesencephalic ventricular cavity (Figs. 14.17, 14.18). Over a certain distance, the ventrolateral part of the valvula is continuous with the caudolateral part of the tegmentum mesencephali (Fig. 14.27).

14.7.2
Tectum Mesencephali

The tectum mesencephali is not clearly differentiated into bilateral lobes (Fig. 14.3). Dorsally, the two tectal halves are connected by a commissure (Fig. 14.18). The rostralmost part of the tectum surrounds a recess of the mesencephalic ventricle, which extends dorsally to the posterior commissure (Fig. 14.19). The caudomedial part of the tectum consists of a thin lamella (Fig. 14.17), but throughout most of its remainder, this structure is differentiated into six layers (Figs. 14.17, 14.18, 14.28). From external to internal, these are as follows:

1. The very thin stratum marginale (smarg).
2. The stratum opticum (sop), consisting of both myelinated and unmyelinated fibres.
3. The stratum fibrosum et griseum superficiale (sfgs), which, in addition to loosely arranged small and medium-sized cells, contains a dense feltwork of fibres of various diameters.
4. The stratum griseum centrale (sgc), which,

unlike layer 3, mainly consists of neuropil, in which scattered neurons are embedded.

5. The stratum album centrale (sac), which, as its name implies, contains numerous myelinated fibres. Medially, such fibres constitute the commissura tecti mesencephali; laterally, they can be observed to contribute to the tectal efferent system. The very coarse axons of the cells in the nucleus mesencephalicus nervi trigemini pass, by way of the deepest zone of these layers, toward the tegmentum mesencephali. The cells within these two layers vary in size. Among them, scattered large, triangular and multipolar neurons are found. It is noteworthy that, within these layers, the cell density in the lateralmost part of the tectum is considerably higher than in the remainder of this structure (Figs. 14.17, 14.18).
6. The stratum griseum periventriculare (sgp), which consists largely of densely packed, medium-sized cells. Throughout most of the tectum, these elements are arranged in two or four sublaminae. In the external zone of this layer, large, ellipsoid and multipolar cells occur. The rostralmost part of its internal zone contains the very large neurons which together form the nucleus mesencephalicus nervi trigemini (Figs. 14.19, 14.27).

The lamination pattern described here for *Scaphirhynchus* corresponds largely with that observed by Repérant et al. (1982) in *Acipenser güldenstädti*. It is, however, worthy of note that these authors referred to the stratum opticum and the stratum fibrosum et griseum superficiale of the present description as a single layer, which they termed the stratum fibrosum et griseum externum, and subdivided the stratum griseum periventriculare into four alternating cellular and plexiform sublayers.

In *Polyodon*, the medialmost parts of the tectal halves extend into the ventricle to form bilateral, elongated eminences, known as the tori longitudinalis. These structures contain densely packed granule cells. Homologues of the tori longitudinalis are present in *Acipenser rubicundus* (Johnston 1901) and *Scaphirhynchus*; however, in these forms they are merely represented by a small accumulation of cells in the dorsomedial part of the tectum that does not protrude into the ventricular cavity (Figs. 14.17, 14.18). The torus longitudinalis is a unique feature of the brain of actinopterygian fishes. It has been experimentally established that, in the tectum of teleosts, the efferents emanating from the torus longitudinalis constitute a separate stratum marginale, situated superficially to the stratum opticum (see Sect. 15.4.4). I identified only an extremely narrow marginal layer in our non-

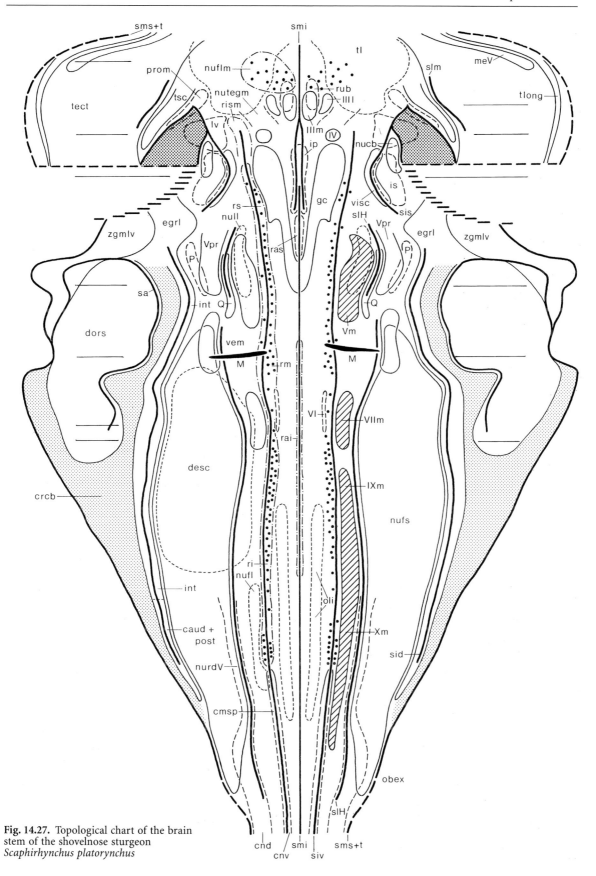

Fig. 14.27. Topological chart of the brain stem of the shovelnose sturgeon *Scaphirhynchus platorynchus*

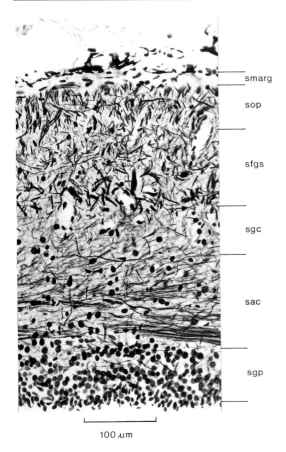

smarg
sop
sfgs
sgc
sac
sgp

100 μm

Fig. 14.28. Transverse section through the tectum mesence-phali of the shovelnose sturgeon *Scaphirhynchus platoryn-chus*. Bodian preparation

experimental material of *Scaphirhynchus*. One would expect this layer to be well developed in *Poly-odon* with its large torus longitudinalis, but perti-nent data are lacking.

The tectum receives afferents from the retina by way of the optic tract. At the rostral pole of the tec-tum, this tract splits up in the usual manner into medial and lateral optic tracts (Figs. 14.17–14.19). The experimental studies performed by Repérant et al. (1982) showed that, in *Acipenser güldenstädti*, the retinotectal fibres terminate in the stratum opti-cum and in the stratum fibrosum et griseum super-ficiale. In addition to a strong contralateral retino-tectal projection, a smaller ipsilateral projection, confined to the rostral part of the tectum, was demonstrated.

Our knowledge of the non-retinal tectal afferents in chondrosteans is scant. In his Golgi material of *Acipenser*, Johnston (1901) traced axons from neu-rons situated in the dorsal part of the telencephalon to the tectum; this is an interesting finding, because experimental studies using the HRP technique have

revealed that such a direct telencephalotectal pro-jection exists in holosteans (Northcutt 1982) and teleosts (see Sect. 15.6.4.5).

Numerous axons descend from the deeper layers of the tectum into the tegmentum, where they can be observed to form two fibre streams, the smaller *fibrae tectotegmentales dorsales* and the much larg-er *fibrae tectotegmentales ventrales* (Fig. 14.17). The former terminate in the torus semicircularis and presumably also in other centres in the lateral teg-mentum. The fibrae tectotegmentales ventrales swing ventromedially through the tegmentum mesencephali. Many of these fibres decussate in the commissura ansulata (Fig. 14.17) and descend into the rhombencephalon as the tractus tectobulbaris cruciatus. A smaller contingent remains ipsilateral and forms a tractus tectobulbaris rectus. Both of these fibre contingents pass caudally in the medio-ventral part of the tegmentum rhombencephali (Figs. 14.13–14.15). I consider it likely that, in chondrosteans, the tectum also sends fibres to the pretectal area and to the thalamus, but such projec-tions could not be observed with sufficient clarity in our non-experimental material.

14.7.3
Lateral Tegmental Zone

The lateral tegmental zone contains three centres: the nucleus lateralis valvulae, the nucleus profun-dus mesencephali and the nucleus tori semicircu-laris.

The *nucleus lateralis valvulae* is a compact cell mass situated directly ventrolateral to the area of continuity between the valvula cerebelli and the tegmentum mesencephali (Fig. 14.27). The nucleus lateralis valvulae projects by way of the tractus teg-mentocerebellaris to the valvula cerebelli. Hocke Hoogenboom (1929) suggested that the centre under discussion receives afferents from the rhombencephalic lateral line centres.

The *nucleus profundus mesencephali* is a rather small, elongated centre which lies embedded in the dense fibre zone of the lateral tegmentum. It con-sists of loosely arranged, fairly large cells (Figs. 14.17, 14.27). Nothing can be said with cer-tainty about its fibre connections.

The *nucleus tori semicircularis* occupies a peri-ventricular position in the lateral tegmentum. Its inner zone is formed by a lamina of cells, which is laterally continuous with the stratum griseum peri-ventriculare of the tectum. Its outer zone consists of scattered elements (Fig. 14.17). The nucleus tori semicircularis is the centre of termination of the lemniscus lateralis, which originates in the rhomb-encephalic lateral line (and perhaps the octavus) centres and also receives input from the tectum. Its efferents are unknown. In *Polyodon*, the nucleus tori semicircularis is considerably larger than in sturgeons such as *Acipenser* and *Scaphirhynchus*. This difference is most probably related to the fact that, in the former, the rhombencephalic lateral line centres are much more strongly developed than in the latter.

14.7.4
Medial Tegmental Zone

The medial tegmental zone may be considered as a rostral continuation of the rhombencephalic area ventralis. It contains three primary efferent centres – the trochlear nucleus and the medial and lateral oculomotor nuclei – and four relay centres – the nucleus tegmentalis medialis, the nucleus of the fasciculus longitudinalis medialis, the nucleus tori lateralis and the nucleus ruber.

The *nucleus nervi trochlearis* and the *nuclei oculomotorii medialis* and *lateralis* consist of large neurons and occupy a periventricular position (Figs. 14.17, 14.27). Norris (1925) reported that, in *Acipenser* and *Scaphirhynchus*, a ciliary ganglion is present and that the oculomotor nerve between the brain and this ganglion contains (in addition to numerous myelinated fibres) a distinct, nonmedullated, apparently preganglionic contingent. However, nothing is known concerning the localisation of the parent neurons of this fibre contingent.

The *nucleus tegmentalis medialis* is a local condensation of rather small cells situated centrally in the medial tegmentum. Both rostrally and caudally, it grades into the diffuse tegmental grey (Fig. 14.27). This centre probably corresponds to parts of the ventral tegmental column observed by Bergquist (1932) in *Acipenser* and to parts of the lateral tegmental nucleus delineated by Northcutt and Butler (1980) in *Lepisosteus*.

The *nucleus of the fasciculus longitudinalis medialis* is strongly developed and consists of very large multipolar neurons (Figs. 14.18, 14.27).

The *nucleus tori lateralis* is an aggregation of rather diffusely arranged, medium-sized cells situated superficially in the tegmentum mesencephali. Rostrally, it extends into the diencephalon (Figs. 14.17, 14.18). This nucleus has been given different names in the literature, e.g. nucleus ruber tegmenti (Johnston 1901) and nucleus entopeduncularis dorsalis (Hocke Hoogenboom 1929). Following Bergquist (1932), I have termed it the nucleus tori lateralis. The fibre connections of this cell mass are unknown. Johnston (1901) believed that its efferents join fibres which pass from the hypothalamus to the cerebellum and to the rhomb-

encephalon. However, Northcutt (1981) demonstrated experimentally that, in *Polypterus*, the homologue of this nucleus projects to the telencephalon.

The ventromedial part of the superficial tegmental zone contains a small cell mass, which I have tentatively labelled as the *nucleus ruber*. Although experimental hodological data in support of this interpretation are lacking, I feel that the position of this cell mass and the fact that it lies in the area of passage or termination of the brachium conjunctivum justify our tentative label.

As regards the fibre connections, the periventricular area of the medial tegmental zone contains, from medial to lateral, the initial part of the fasciculus longitudinalis medialis, the fasciculi tegmentales and the mesencephalic root of the trigeminal nerve. In the superficial area of the medial tegmental zone, the fasciculus retroflexus and, more laterally, the fibrae hypothalamo-tegmentales can be clearly recognised (Figs. 14.17, 14.18). Repérant et al. (1982) experimentally established that primary optic fibres, travelling with a tractus opticus accessorius, reach the rostralmost part of the tegmentum mesencephali. These fibres terminate in a zone of neuropil, mixed with a few cell bodies, which Repérant et al. (1982) designated as the area optica accessoria.

14.8
Diencephalon

14.8.1
Introductory Notes

Braford and Northcutt (1983) published a survey of the organisation of the diencephalon and pretectum of actinopterygian fishes. This survey is primarily based on a comprehensive cytoarchitectonic analysis in three species: the bichir *Polypterus palmas*, the holostean *Lepisosteus osseus* and the teleost *Carassius auratus*. Braford and Northcutt placed the rostral boundary of the diencephalon directly caudal to the commissura anterior, and they considered the caudal boundary of that brain part to pass between the posterior and tectal commissures dorsally and between the nucleus of the medial longitudinal fasciculus and the oculomotor nucleus more ventrally, reaching the ventral surface of the brain at the plica encephali ventralis. On the basis of their personal observations and taking into account suggestions made in a thorough embryological study on the diencephalon of anamniotes (Bergquist 1932), Braford and Northcutt divided the actinopterygian diencephalon and pretectum into the following regions: area preoptica, epithalamus, dorsal thalamus, ventral thalamus, hypothalamus, tuber-

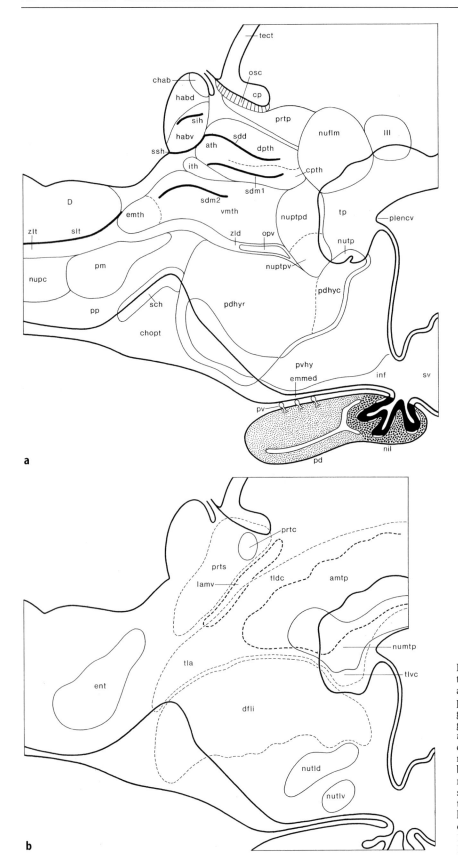

Fig. 14.29a–d. Reconstructions of the diencephalon and the adjacent mesencephalic and telencephalic regions of the shovelnose sturgeon *Scaphirhynchus platorynchus.* **a** Periventricular cell masses. **b** Migrated cell masses. **c** Fibre connections, based on non-experimental material. **d** Retinofugal and secondary olfactory projections toward the diencephalon. (Based on experimental data from Repérant et al. 1982; Northcutt and Braford 1980)

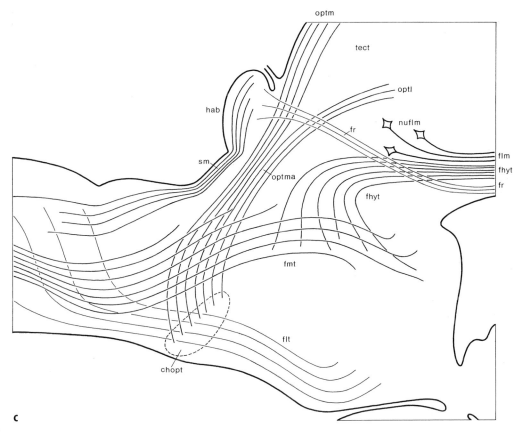

Fig. 14.29c

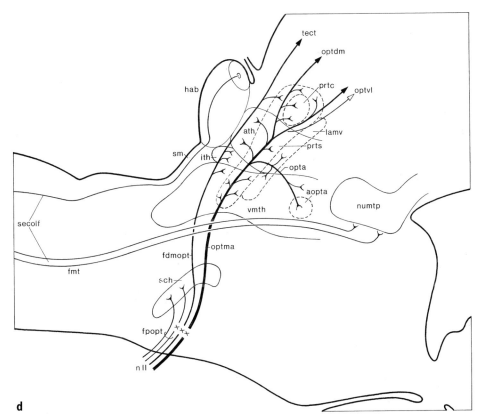

Fig. 14.29d

culum posterius, synencephalon (including part of the pretectum) and the remainder of the pretectum. Within each of these regions, Braford and Northcutt delineated two or more nuclei or areas and provided a consistent nomenclature for all of these grisea.

Although several authors (Johnston 1901; Hocke Hoogenboom 1929; Bergquist 1932; Repérant et al. 1982) have devoted attention to the structure of the diencephalon of chondrostean fishes, a systematic cytoarchitectonic analysis of the entire region in this group has not been made. In order to fill this gap, the present author analysed the diencephalon of the shovelnose sturgeon *Scaphirhynchus platorynchus*. It appeared that the subdivision of the polypteriform, holostean and teleostean diencephalon presented by Braford and Northcutt (1983) is to a very large extent also applicable to this chondrostean species. Hence, in the following survey, this subdivision and the nomenclature attached to it have been followed wherever possible. Before discussing the general relations and the cell masses in the diencephalon of *Scaphirhynchus* and other chondrosteans, the following general prefatory remarks seem appropriate:

1. Following Johnston (1911), I define the rostral boundary of the diencephalon as running from the lateral attachment of the velum transversum dorsally to the chiasma opticum ventrally. By doing so, I assign the entire regio preoptica to the telencephalon.
2. In accordance with views on the neuromeric organisation of the diencephalon (see Sect. 4.5.5 and Figs. 4.20 and 4.37), the present author considers it likely that the parencephalon anterius gave rise to the ventral thalamus plus most of the tuberculum posterior, that the parencephalon posterius gave rise to the epithalamus, the dorsal thalamus and a small part of the tuberculum posterius and that the synencephalon, i.e. the caudalmost diencephalic neuromere, gave rise to the pretectum and the nucleus of the fasciculus longitudinalis medialis. In spite of its presumed diencephalic origin, the nucleus of the fasciculus longitudinalis medialis has been dealt with in the section on the tegmentum of the midbrain (Sect. 14.7.4).
3. In the diencephalon of *Scaphirhynchus*, the grey matter is unusually diffusely arranged (a similar observation was made by Repérant et al. 1982 in *Acipenser güldenstädti*). This means that in several places the boundaries drawn are arbitrary in character. The boundaries between the dorsal and ventral thalamus and between the latter and the nucleus periventricularis tuberculi posterioris (Fig. 14.29a) are illustrative cases in point.

4. The diencephalon of *Scaphirhynchus* contains several distinct, longitudinal sulci, but in contrast to what is seen in other anamniotes (e.g. *Latimeria*, urodeles), these sulci mostly do not coincide with cytoarchitectonic boundaries (Fig. 14.29a).
5. Throughout the diencephalon, a cell-poor zona limitans marks the ventral boundary of the ventral thalamus. Rostrally, this structure is continuous with a similar telencephalic zone (Fig. 14.29a).

14.8.2
General Relations and Cell Masses

The *epithalamus* comprises the ganglia habenula and the epiphysis. The epiphysis consists of a tube-shaped stalk, which originates in the diencephalic roof between the habenular and posterior commissures, and a rather small terminal vesicle. In *Polyodon*, the epiphysis extends rostrodorsally into a canal of the cartilaginous cranium, and its terminal vesicle corresponds with an elliptic foramen in the dermal skeleton, situated between the eyes. Subjacent to this foramen, the cartilage of the skull is more translucent than elsewhere. In *Acipenser*, the terminal organ is less closely apposed to the surface than in *Polyodon*. Moreover, a foramen in the dermal skeleton is lacking here (Garman 1896). Nerve fibres have been observed in the chondrostean epiphysial stalk. According to Johnston (1901), part of these fibres terminate in the habenula.

The ganglia habenulae show a marked asymmetry. In *Acipenser* (Johnston 1901) and *Scaphirhynchus*, the right ganglion is much larger than the left, but in *Polyodon* (Hocke Hoogenboom 1929) this relation is reversed. Ventrally, the boundary of the habenular ganglion is marked by a distinct sulcus diencephalicus dorsalis; caudally, it is bounded by the dorsal thalamus. The dorsal parts of both ganglia are connected by a well-developed commissura habenulae (Fig. 14.20). The habenulae contain a grey layer of variable thickness which borders the ventricular cavity. Both are composed of dorsal and ventral habenular nuclei. The cells of the ventral habenular nucleus are slightly smaller and more densely packed than those of the dorsal habenular nucleus (Fig. 14.20). The surface of the ganglion habenulae contains a rather slight sulcus intrahabenularis. As shown by Fig. 14.29a, this groove does not correspond with the boundary between the dorsal and ventral habenular nuclei.

The *thalamus dorsalis* consists of a zone of periventricular cells that extends from the habenular ganglion rostrally to the nucleus of the fasciculus longitudinalis medialis caudally. It is bounded dorsally by the nucleus pretectalis periventricularis and ventrally by the thalamus ventralis and by the nucleus periventricularis tuberculi posterioris (Fig. 14.29a). Within the confines of the dorsal thalamus, three cell groups can be recognised: the nucleus anterior, nucleus dorsalis posterior and nucleus centralis posterior. The nucleus anterior occupies approximately the rostral third of the dorsal thalamus. It is characterised by two somewhat irregular, parallel laminae medially and some scattered cells laterally. The nucleus anterior is replaced dorsally by the nucleus dorsalis posterior and ventrally by the nucleus centralis posterior. The cells of both nuclei last mentioned form rostrocaudal columns along the ventricular surface (Fig. 14.19).

Hofmann et al. (1993) demonstrated that, in the Sterlet *Acipenser ruthenus*, the periventricular zone of the dorsal thalamus contains retinopetal cells (two to three ipsilaterally and five to seven contralaterally).

The *pretectum* of *Scaphirhynchus* comprises three grisea: a periventricular zone, a small cell group situated at some distance of the periventricular grey and a large superficial cell mass. Following Braford and Northcutt (1983), I have termed these three grisea the nucleus pretectalis periventricularis, nucleus pretectalis centralis and nucleus pretectalis superficialis.

The *nucleus pretectalis periventricularis* (Figs. 14.19, 14.29a) is bordered ventrally by the fasciculus retroflexus, which separates it from the dorsal thalamus. Its rostral portion lies immediately ventral to the subcommissural organ, whereas its caudal portion abuts on the nucleus of the fasciculus longitudinalis medialis. It consists of a layer of small, densely packed cells that lies directly against the ependyma.

The *nucleus pretectalis centralis* (Fig. 14.29b) is a poorly differentiated, small cell group, consisting of scattered, medium-sized neurons. It is situated laterally to the nucleus pretectalis periventricularis and the nucleus dorsalis posterior. This nucleus probably corresponds to the nucleus pretectalis of Hocke Hoogenboom (1929) and to the nucleus pretectalis dorsalis of Repérant et al. (1982).

The *nucleus pretectalis superficialis* (Figs. 14.19, 14.20, 14.29b) is a large, elongated cell mass that for the most part lies embedded in the tractus opticus marginalis. It consists primarily of small neurons, although scattered, somewhat larger elements are present throughout its extent. Its cell density changes from site to site. This nucleus corresponds to the corpus geniculatum laterale of Hocke Hoogenboom (1929) and Bergquist (1932) and includes the nucleus thalamicus tractus optici marginalis

and the nucleus pretectalis ventralis of Repérant et al. (1982).

A fourth nucleus which may well form part of the pretectum is the nucleus laminaris ventralis. This nucleus, which was first described by Repérant et al. (1982) for *Acipenser güldenstädti*, appeared to also be present in *Scaphirhynchus*. It occupies a superficial position and is located directly ventral to the caudal quarter of the nucleus pretectalis superficialis (Figs. 14.19, 14.29b). Its small, fusiform cells form a thin lamina which accompanies the fibres of the tractus opticus accessorius over some distance. The nucleus laminaris ventralis has no obvious homologue in other actinopterygian groups.

In addition to the centres discussed above, Repérant et al. (1982), in their study of the primary optic projections of *Acipenser güldenstädti*, described several other pretectal groups, i.e. a nucleus commissurae posterioris, a nucleus intercalatus medialis and a nucleus intercalatus lateralis. I was unable to distinguish these nuclei with sufficient clarity in *Scaphirhynchus*.

The *thalamus ventralis* is bounded dorsally by the dorsal thalamus, ventrally by the zona limitans and caudally by the nucleus periventricularis tuberculi posterioris. The rostral pole of the ventral thalamus is formed by a zone of densely packed, small cells directly apposed to the ependyma (Fig. 14.21). This zone of cells, which represents the eminentia thalami, is situated just behind the caudal pole of the telencephalic area dorsalis (Fig. 14.29a). The dorsocaudal part of another telencephalic cell mass, the nucleus entopeduncularis (Figs. 14.21, 14.29b), merges with the eminentia thalami. Caudal to the eminentia thalami, two nuclei can be recognised in the ventral thalamus: the nucleus ventromedialis and the nucleus intermedius thalami.

The nucleus ventromedialis thalami is structurally heterogeneous. Its rostral part is formed by two or three laminae of medium-sized cells (Fig. 14.20), but more caudally it consists of a layer of small cells separated from the ventricular surface by a rather wide cell-free zone (Fig. 14.19).

The nucleus intermedius thalami is confined to the rostral part of the ventral thalamus (Fig. 14.29a). It consists of a group of cells that is not differentiated into layers and that is situated somewhat further away from the ventricular surface than the adjacent thalamic nuclei (Fig. 14.20).

Although the dorsal and ventral thalami of other groups of actinopterygian fishes contain groups of cells which have migrated away from the ventricular grey, such nuclei are lacking in chondrosteans.

Braford and Northcutt (1983) have pointed out

that the thalamus of actinopterygian fishes comprises, in addition to the dorsal and ventral thalamic regions already discussed, a third, caudal region, the *regio tuberculi posterioris*. They included in this region not only some periventricular grisea, but also a number of migrated nuclei, some of which extend far in front of the posterior tubercle. By doing so, they followed the interpretation, given by Bergquist (1932) in his extensive treatise on the ontogenesis of the diencephalic cell masses in anamniotes. In the following survey of the cell masses in the regio tuberculi posterioris of *Scaphirhynchus*, the nomenclature of Braford and Northcutt (1983) has been employed. However, it appeared to be impossible to homologise all of the cell masses observed with centres present in other actinopterygians. Thus some grisea have been labelled with provisional, purely descriptive terms.

The periventricular zone of the tuberculum posterius is bounded dorsally by the nucleus of the fasciculus longitudinalis medialis and by the dorsal thalamus, rostrally by the ventral thalamus and ventrally by the hypothalamus (Fig. 14.29a). This zone contains a large, rostral nucleus periventricularis tuberculi posterioris, which can be subdivided into a pars dorsalis and a pars ventralis, and a smaller, caudal cell mass, the nucleus tuberis posterior. The dorsal part of the nucleus periventricularis tuberculi posterioris consists of a zone of small, rather loosely arranged cells. Rostrally, this nucleus cannot be sharply delimited from the ventral thalamus. The ventral part of the nucleus periventricularis tuberculi posterioris is composed of larger elements forming a compact band, situated at some distance from the ventricular surface. It corresponds to the pars dorsalis hypothalami of Bergquist (1932). An immunohistochemical study (Piñuela and Northcutt 1994) has shown that, in *Acipenser transmontanus*, the nucleus periventricularis of the posterior tuberculum contains dopamine- and tyrosine hydroxylase-immunoreactive cells.

The nucleus tuberis posterior lines part of the ventral surface of the tuberculum posterius. Its medium-sized cells show in places a laminar arrangement (Fig. 14.18). Laterally, this nucleus grades into the pars dorsalis hypothalami. Johnston (1901), who first described this cell mass, noted that it receives axons originating from the epithelial cells of the saccus vasculosus. Accordingly, it has been designated by Hocke Hoogenboom (1929) and Bergquist (1932) as the ganglion or nucleus sacci vasculosi.

Braford and Northcutt (1983) also included the paraventricular organ (PVO), a strip of specialized cerebrospinal fluid-contacting neurons, in the periventricular zone of the tuberculum posterius. This organ is situated in the vicinity of the zona limitans diencephali (Fig. 14.29a).

The following three migrated cell masses may be assigned to the posterior tubercle: the nucleus medianus tuberculi posterioris, the nucleus tori lateralis and the area magnocellularis tuberculi posterioris (Fig. 14.29b).

The nucleus medianus tuberculi posterioris is a prominent cell group, which occupies a central position in the region under discussion (Fig. 14.18). The two sides of this nucleus fuse in the median plane of the posterior tubercle. A dorsocaudally directed process of this nucleus extends into the tegmentum mesencephali, where it grades into the cell mass tentatively designated as the nucleus ruber. The nucleus medianus has an obvious homologue in both polypterids (Nieuwenhuys and Bodenheimer 1966) and holosteans (Northcutt and Butler 1980; Braford and Northcutt 1983).

The nucleus tori lateralis is a very large, superficially situated cell mass which, extending rostrally from the tegmentum of the midbrain, attains the rostralmost part of the diencephalon. The entire cell mass is composed of medium-sized cells which in many places tend to form small clusters. Slight cytoarchitectonic differences and the presence of local cell-poor zones allow a subdivision of this cell mass into three moieties: the pars rostralis, pars dorsocaudalis and pars ventrocaudalis. The pars rostralis is situated between the rostral parts of the nucleus pretectalis superficialis and the nucleus diffusus lobi inferioris (Figs. 14.18, 14.19, 14.29b). The pars dorsocaudalis of the nucleus tori lateralis is, as its name implies, situated in the dorsocaudal part of the diencephalon and in the adjacent part of the tegmentum mesencephali (Figs. 14.17–14.19, 14.29b). The pars ventrocaudalis of the nucleus tori lateralis is separated from the pars dorsocaudalis by a cell-poor zone. Its ventral part encroaches upon the dorsocaudal part of the hypothalamus (Figs. 14.18, 14.29b).

A mesencephalic nucleus tori lateralis extending rostrally into the diencephalon is not only present in chondrosteans, but has also been observed in polypterids (see Figs. 13.14–13.16), holosteans (see Fig. 15.32) and teleosts (see Figs. 15.18, 15.19). However, in chondrosteans this nucleus extends far more rostrally than in the other groups, and the question arises whether the three moieties described above are merely integral parts of the nucleus or rather represent separate cell masses. Bergquist (1932), who studied *Acipenser ruthenus*, incorporated the pars dorsocaudalis of the nucleus tori lateralis of the present description also in the nucleus of the torus lateralis. However, he considered the pars rostralis and the pars ventrocaudalis as separate entities, which he designated the nucleus praerotundus-posterior thalami and nucleus subrotundus-supramamilaris, respectively. Comparison of the relations in *Scaphirhynchus* with those in the holostean *Lepisosteus* (Braford and Northcutt 1983) suggests that rostral and ventro-caudal parts of the nucleus of the torus lateralis of the present description positionally correspond to the part of the cluster formed by Braford and Northcutt's 'nucleus glomerulosus' and the preglo-merular nuclear complex.

The area magnocellularis tuberculi posterioris is formed by an irregular sheet of large, triangular and multipolar elements situated directly medial to the nucleus tori lateralis. It is in contact with all three parts of the nucleus tori lateralis; however, it extends less far dorsal, rostral and caudal than do the dorsocaudal, rostral and ventrocaudal parts of that nucleus, respectively (Figs. 14.18, 14.19). The area magnocellularis tuberculi posterioris has, to our knowledge, never been described before as a cytoarchitectonic unit. However, it probably includes the large-celled nucleus posterior thalami, which Braford and Northcutt (1983) delineated in *Lepisosteus*.

The *hypothalamus* forms the largest part of the chondrostean diencephalon. It comprises a large rostral part and a much smaller caudal part. The walls of the rostral part are evaginated into inferior lobes, which surround paired diverticula of the third ventricle, known as the recesses laterales hypothalami. In *Scaphirhynchus* (Figs. 14.3, 14.19, 14.20) and in sturgeons in general, the inferior hypothalamic lobes are much larger than in *Polyodon*. The caudal part of the hypothalamus is formed by the unpaired lobus tuberis. This portion of the hypothalamus surrounds the recessus posterior hypothalami. The latter displays small, laterally directed diverticula (Figs. 14.3, 14.18). Ventrocaudally, the lateral and tuberal lobes of the hypothalamus pass over into the infundibulum. Rostroventrally, the infundibular wall is differentiated into a median eminence, a structure which is situated just dorsal to the pars distalis of the pituitary gland. This part of the wall produces long, finger-like diverticula which penetrate deeply into the tissue of the pars intermedia. Caudally, the infundibulum passes over into the thin, strongly folded epithelial wall of the saccus vasculosus (see Figs. 14.3, 14.29a, 14.30). The well of this organ contains numerous *Krönchenzellen* or coronet cells (see Sect. 4.4). Kotrschal et al. (1983) observed that, in the Sterlet *Acipenser ruthenus*, coronet cells are not confined to the saccus vasculosus, but are rather distributed over the entire diencephalic floor and on the preoptic recess as well.

The hypothalamus is bordered dorsally by the cell-poor zona limitans diencephali. Rostrally, the boundary between the preoptic region and the hypothalamus is marked by an abrupt increase in the packing density of the cells in the periventricular zone (Fig. 14.20). I have divided the hypothalamus into two longitudinal zones: the pars dorsalis hypothalami and the pars ventralis hypothalami (Fig. 14.29a). Within the pars dorsalis hypothalami, a large rostral portion and a much smaller caudal portion can be distinguished. The rostral portion contains a compact periventricular layer which characteristically consists of an inner zone of small cells and an outer zone of medium-sized, darkly staining elements (Figs. 14.19, 14.20). In the caudal portion of the pars dorsalis hypothalami, the cells do not differ markedly in size, but show a distinct laminar pattern. Caudodorsally, this area grades into the nucleus tuberis posterior (Figs. 14.18, 14.29a). In addition to the periventricular grey discussed above, the pars dorsalis hypothalami contains a superficially situated nucleus diffusus lobi inferioris. This nucleus, which consists of loosely arranged, medium-sized cells, is in some places not sharply delimitable from the nucleus tori lateralis (Figs. 14.18, 14.20, 14.29b).

The pars ventralis hypothalami is rather small in the rostral part of the hypothalamus, but more caudally it increases in size, and in the lobus tuberis it extends over the entire wall. The periventricular zone of the pars ventralis contains a layer of small, compactly arranged cells (Figs. 14.18, 14.20, 14.29a).

In the hypothalamus of *Acipenser fulvescens* Sathyanesan and Chavin (1967) found a compact nucleus lateralis tuberis extending from the level of the middle of the pars distalis of the pituitary gland to the anterior margin of the pituitary gland. Moreover, these authors found a group of scattered cells in the infundibular floor adjacent to the pituitary gland, some of which stained positively with Gomori's aldehyde fuchsin. I found two potential nucleus tuberis lateralis candidates in *Scaphirhynchus*, which I have tentatively labelled as the nucleus tuberis lateralis dorsalis and ventralis (Figs. 14.18, 14.29b). These nuclei probably correspond to the cell groups observed by Sathyanesan and Chavin (1967). The nucleus tuberis lateralis dorsalis consists of an accumulation of fusiform cells, situated directly lateral to the hypothalamic periventricular grey. Its rostral portion is situated at the level of the border between the dorsal and ventral parts of the hypothalamus, but its caudal portion is entirely covered by the pars caudalis hypothalami on its inner side. The nucleus tuberis lateralis ventralis is represented by a small group of superficially situated, spherical and triangular cells. In both nuclei, numerous cells with a particularly darkly staining protoplasm can be observed.

14.8.3
Diencephalic Fibre Systems

Within the white matter of the chondrostean diencephalon, only a limited number of fibre systems are sufficiently individualised to be clearly traceable in non-experimental material, and the number of experimental studies on these fibre systems is as yet extremely small. For these two reasons, the following survey of the diencephalic fibre systems in chondrosteans will necessarily be very fragmentary.

The *retinofugal fibres* pass with the optic nerves to the rostroventral diencephalon, where most of them decussate in the chiasma opticum. The bulk of these fibres then pass via the marginal optic tract to the rostral pole of the tectum, where they split into a medial and a lateral optic tract (Figs. 14.3, 14.18–14.21, 14.29c). Repérant et al. (1982) have studied the course and sites of termination of the retinofugal fibres in the sturgeon *Acipenser güldenstädti* with the aid of three different experimental techniques: Fink-Heimer, amino acid autoradiography and HRP. Their main findings may be summarised as follows (see also Fig. 14.29d).

1. Fibres of the marginal optic tract terminate massively in the nucleus pretectalis superficialis, contribute to a large terminal field located immediately lateral to the anterior, intermediate and the rostral part of the ventromedial thalamic nuclei and ultimately reach the tectum mesencephali.
2. During its course through the diencephalon and pretectum, the tractus opticus marginalis gives rise to three fascicles: the tractus opticus accessorius, the tractus opticus dorsomedialis and the tractus opticus ventrolateralis.

The tractus opticus accessorius separates at the level of the thalamus from the marginal optic tract. After having innervated the nucleus laminaris ventralis, it passes caudally in a ventromedial direction and then subdivides into two branches, one projecting to the caudal part of the nucleus ventromedialis thalami and the other ending in a mesencephalic area located dorsal to the caudal part of the hypothalamic lateral lobes. This area, which consists of neuropil mixed with a few cell bodies, was designated by Repérant et al. (1982) as the area optica accessoria.

The tractus opticus dorsomedialis detaches at the pretectal level from the marginal optic tract. It innervates the nucleus pretectalis centralis, the cau-

dal part of the nucleus pretectalis superficialis and the intermediate and deep layers of the stratum fibrosum et griseum superficiale of the tectum.

The thin tractus opticus ventrolateralis separates from the marginal optic tract at the level of the caudal part of the posterior commissure. This fascicle projects essentially to the superficial part of the ventrolateral stratum fibrosum et griseum superficiale in the intermediate and caudal parts of the tectum.

3. Fibres stemming from the intermediate and dorsal parts of the chiasma opticum decussate to form the fasciculus dorsomedialis tractus optici and the fasciculus preopticus tractus optici, respectively. The fibres of the fasciculus dorsomedialis tractus optici gather in a compact bundle, which courses dorsocaudally, immediately adjacent to the periventricular cell plate of the diencephalon. After having supplied the nucleus intermedius thalami and, less densely, the rostral part of the nucleus ventromedialis thalami, the fibres of this tract reach the mediodorsal part of the tectum, where they arborise in the superficial part of the stratum fibrosum et griseum superficiale.
   The fibres of the fasciculus preopticus tractus optici terminate in a neuropil zone immediately adjacent to the perikarya of the nucleus suprachiasmaticus.
4. All of the optic fascicles described above contain, in addition to numerous crossed retinofugal fibres, modest numbers of uncrossed fibres from the same source, and all of the grisea mentioned receive not only a strong contralateral optic input, but also a much weaker, ipsilateral optic input. However, the contralateral and ipsilateral projections to the nucleus suprachiasmaticus are of equal size.

The *stria medullaris* is a distinct bundle, the fibres of which assemble in the caudal part of the telencephalon and pass along the rostrodorsal border of the thalamus ventralis to the ganglion habenulae (Figs. 14.20, 14.21, 14.29c). On the basis of non-experimental material, Johnston (1911) suggested that the area dorsalis telencephali (primordium hippocampi in his terminology), various parts of the area ventralis telencephali (medial and lateral olfactory nuclei in his terminology), the preoptic region and the entopeduncular nucleus (area somatica in his terminology) contribute fibres to the stria medullaris. He considered it likely that the fibres originating from the area dorsalis pass via the commissura habenularum to the contralateral area dorsalis and thus form a true commissure.

Northcutt and Braford (1980) established experimentally that, in *Scaphirhynchus*, secondary olfactory fibres enter the stria medullaris, decussate in the habenular commissure and distribute to the contralateral telencephalon (Fig. 14.29c).

The *fasciculus retroflexus*, which consists exclusively of thin, unmyelinated fibres, originates in the ganglion habenulae and passes caudoventrally along the ventricular surface to the basal region of the midbrain, where it decussates and terminates in the nucleus interpeduncularis. The asymmetry of the habenular ganglia is clearly reflected in the size of the two tracts (Figs. 14.18, 14.19, 14.29d).

The *fasciculus medialis telencephali* or medial forebrain bundle is a large and complex bundle which can be traced from the area ventralis telencephali toward the diencephalon, where it roughly follows the zona limitans. Gradually diminishing in size, the bundle attains the region of the tuberculum posterius, where it is lost (Figs. 14.19–14.21, 14.29c). I consider it likely that the fasciculus medialis telencephali primarily connects various parts of the area ventralis telencephali with the ventral thalamus, the dorsal part of the hypothalamus and the regio tuberculi posterioris. Johnston (1911) indicated that this bundle (i.e. the basal forebrain bundle in his terminology) contains the following: (a) fibres passing from the area ventralis telencephali to the lateral lobes of the hypothalamus, (b) fibres passing from the preoptic region to the hypothalamus and (c) crossed and uncrossed fibres ascending from the hypothalamus to the area dorsalis telencephali. Axon degeneration experiments (Northcutt and Braford 1980) revealed that, in *Scaphirhynchus*, secondary olfactory fibres pass with the medial forebrain bundle to the caudal part of the diencephalon, where they terminate close to the median plane in a neuropil zone situated just ventral to the nucleus medianus tuberculi posterioris (Fig. 14.29d).

The *fasciculus lateralis telencephali* or *lateral forebrain bundle* is the third large fibre system which interconnects telencephalon and diencephalon. Unlike the stria medullaris and the medial forebrain bundle, it consists exclusively of fine, unmyelinated fibres (Figs. 14.18–14.23, 14.29c). Johnston (1911), studying Golgi material of *Acipenser rubicundus*, observed that the fibres of this bundle (the tractus pallii in his terminology) ascend from the lateral lobes of the hypothalamus, pierce the optic tract in several fascicles and pass forward and upward along the lateral surface of the telencephalon. Within the latter brain part, the tract gives off many small bundles which ascend to the area dorsalis. In a few cases, Johnston was able to trace axons from pallial cells into the lateral forebrain

bundle; hence he considered it likely that this bundle, though largely composed of ascending axons, contains a descending component. Johnston, who also studied Bielschowsky material of *Polyodon*, reported that in this species the fibre system under discussion forms a compact bundle on the lateral surface of the telencephalon and that rostrally its fibres pierce the layer of olfactory tract fibres, fanning out in the area dorsalis (see Fig. 33).

From the dorsal part of the lobus inferior hypothalami, numerous fibres arise which form smaller and larger fascicles. These fascicles initially course dorsally, but then arch caudally, to enter the tegmentum mesencephali, where they are lost. I have termed these fibres collectively the fibrae hypothalamotegmentales (Figs. 14.17, 14.18, 14.29c), but this name is not intended to suggest that all of these fibres descend toward the brain stem. In fact, their polarity is unknown. The fibrae hypothalamotegmentales probably include the tractus mamillobulbaris of Johnston (1901), the tractus lobopeduncularis or mamillopeduncularis of Hocke Hoogenboom (1929) and the direct and crossed lobocerebellar tracts described by Johnston (1901), Hocke Hoogenboom (1929) and Larsell (1967).

I am under the impression that large numbers of fibres connect the torus lateralis complex with the lobi inferiores hypothalami (Figs. 14.18–14.20). Johnston (1901) observed corresponding fibres in *Acipenser*.

Finally, it should be mentioned that the diencephalon of *Scaphirhynchus*, in addition to the commissura habenulae and the chiasma opticum already discussed, contains three other decussations: the commissura posterior, the commissura tuberculi posterioris and the commissura postoptica. Although the fibres participating in the formation of these decussations can be traced over some distance, nothing can be said at present concerning their origins or sites of termination.

If we survey the data presented above it appears that very little is known on the specific connections of most of the grisea delineated in the diencephalon of chondrosteans. As in all vertebrates, the habenular ganglion represents a centre in a conduction route from the stria medullaris to the fasciculus retroflexus, relaying impulses from various telencephalic sources to the nucleus interpeduncularis in the upper brain stem. The dorsal thalamus is differentiated into three nuclei: the nucleus anterior, nucleus dorsalis posterior and nucleus centralis posterior. All we know at present about the connections of these centres is that the nucleus anterior receives a direct retinal projection. It is known that (a) in the holostean *Lepisosteus* (Northcutt and Butler 1980) and in teleosts (Striedter 1990), the nucleus dorsalis posterior receives a bilateral tectal input, (b) in the holostean *Amia* (Braford and McCormick 1979) and in teleosts (Echteler 1984; Striedter 1991), the the nucleus centralis posterior also receives a bilateral input from the torus semicircularis and (c) in both polypterids (Northcutt 1981) and teleosts (Echteler and Saidel 1981; Striedter 1990), the nucleus anterior, the nucleus dorsalis posterior and the nucleus centralis posterior all project to the telencephalon; thus it may be speculated that, in chondrosteans, these three dorsal thalamic nuclei also represent relay centres linking the retina, the tectum and the octavolateral system with the telencephalon. If this is indeed the case, these thalamotelencephalic projections are not sufficiently developed to be detected in non-experimental histological material. It would be of particular interest to know the part of the telencephalon in which these ascending thalamic connections terminate in chondrosteans, because, while in teleosts these connections terminate in the area dorsalis or pallium (Echteler and Saidel 1981; Striedter 1990), in polypterids they have been found to terminate mainly in the subpallial dorsal nucleus of the area ventralis telencephali (Holmes and Northcutt 1995).

We have seen that, in chondrosteans, most of the pretectal nuclei receive a retinal input, but the non-experimental material studied revealed nothing of the other connections of these nuclei. On comparative grounds, it may be expected that most of these nuclei are reciprocally connected with the tectum (Northcutt and Butler 1980; Northcutt 1982; see also Sect. 15.5.1).

As regards the ventral thalamus, non-experimental material suggests that, in chondrosteans, this complex receives a projection from and/or sends fibres toward the telencephalon via the medial forebrain bundle. In teleosts, the ventromedial thalamic nucleus is reciprocally connected with the telencephalic area dorsalis (Ito et al. 1986; Striedter 1990). Retinal projections to the intermediate and ventromedial thalamic nuclei, similar to those in *Acipenser*, have been shown to exist in the holosteans *Lepisosteus* (Northcutt and Butler 1976) and *Amia* (Butler and Northcutt 1992) and in teleosts (Braford and Northcutt 1983; Striedter 1990).

Our knowledge of the connections of the various centres assigned to the chondrostean regio tuberculi posterioris is extremely scant. A direct projection from the olfactory bulb to the neuropil immediately ventral to the nucleus medianus tuberculi posterioris, as observed in *Scaphirhynchus*, has also been described in the holostean *Lepisosteus* (North-

cutt and Braford 1980) and in some teleosts (Finger 1975; Bass 1981). Because the tuberculum posterius complex of chondrosteans closely resembles that of brachiopterygians, it may be of particular interest that in *Polypterus* the nucleus periventricularis tuberculi posterioris, the nucleus medianus tuberculi posterioris and the nucleus tori lateralis have been demonstrated to project to the telencephalon (Northcutt 1981). The nucleus periventricularis of the tuberculum posterius of *Acipenser transmontanus*, like its brachiopterygian (Reiner and Northcutt 1992) and teleostean homologue (see Sect. 15.5.4.5), contains dopaminergic cells and may well be the source of an ascending projection of a dopaminergic input to the subpallial dorsal nucleus of the area ventralis (Piñuela and Northcutt 1994). The nucleus medianus and the rostral part of the nucleus of the torus lateralis probably represent undifferentiated primordia of the holostean and teleostean preglomerulosus complex (Braford and Northcutt 1983; Northcutt and Butler 1993; see Sect. 15.5.4.1). In teleosts, this complex is known to receive ascending tectal, auditory, mechanosensory lateral line, electrosensory (where present) and gustatory inputs and to be reciprocally connected with the area dorsalis, i.e. the pallial part of the telencephalon (see also Sect. 15.5.4.1).

Experimental data concerning the fibre connections of the chondrostean hypothalamus are entirely lacking. However, observations in non-experimental material suggests that two fibre assemblies, the fibrae hypothalamotegmentales and the lateral forebrain bundle, connect the huge lobi inferiores hypothalami with the brain stem and the dorsal part of the telencephalon, respectively (Figs. 14.19, 14.20, 14.29c). Experimental studies have shown that, in teleosts, the lobus inferior hypothalami receives visual input from the pretectum and gustatory input from the nucleus visceralis secundarius, projects to the motor nuclei of nerves V and VII and is reciprocally connected with the area dorsalis of the telencephalon (see Sect. 15.5.5.3). It may be speculated that these connections also exist in chondrosteans and that in both chondrosteans and teleosts they subserve similar functions. By way of the ascending projections from the brain stem and the hypothalamic inferior lobe, gustatory information probably attains the telencephalic level, whereas the visual and gustatory information converging on the ipsilateral lobes and the branchiomotor output of the latter may well be involved in feeding.

14.8.4
Hypothalamo-hypophyseal Relationships

The pituitary gland lies against the caudoventral surface of the hypothalamus and consists of a rostral pars distalis and a caudal intermediate lobe (Figs. 14.3, 14.29a). It contains a large cleft-like cavity comprising a vertical and a horizontal part. The former almost completely separates the neurointermediate lobe from the pars distalis, while the latter splits the pars distalis into dorsal and ventral zones (Fig. 14.29a). This remarkable hypophyseal cavity arises as a schizocoel, i.e. a cleft in a solid anlage (Holmes and Ball 1974). The neurointermediate lobe is situated close to the transition of the infundibulum to the saccus vasculosus. Its neural component consists of a part of the infundibular wall which forms a system of hollow, slightly branching processes. These processes penetrate deeply among the cell cords that constitute the pars intermedia (Fig. 14.29a). The neurohypophyseal processes consist of tanycyte-like ependymal cells, among which a few scattered pituicytes and numerous unmyelinated neurosecretory fibres are found. Neural and intermedia tissues are separated by a connective tissue layer containing very wide sinusoidal capillaries. The part of the hypothalamic floor lying dorsal to the pars distalis is differentiated into a typical median eminence. The narrow space intervening between this eminence and the pars distalis contains blood vessels that form part of a hypothalamo-hypophyseal portal system.

Two types of neurosecretory fibres have been found to pass via the hypothalamic floor to the neurohypophysis. These fibres, which, in accordance with Knowles (1965), were designated by Polenov et al. (1972) as type-A and type-B fibres, will now be discussed.

The type-A fibres are peptidergic and originate in the nucleus preopticus magnocellularis. Sathyanesan and Chavin (1967) reported that the preoptico-hypophyseal fibre system can be divided into a ventral and a dorsal part. The ventral part is formed by a compact bundle lying close to the ventral wall of the hypothalamus. The dorsal part consists of sparsely distributed axons situated close to the ependymal lining of the third ventricle. Some peptidergic fibres, presumably originating in the preoptic nucleus, are present among the cells of the nucleus lateralis tuberis.

In the neurohypophysis, large neurosecretory, peptidergic fibres 1.5–3 µm in diameter are predominant. They can be divided into two types: the so-called A1 fibres, containing elementary granules measuring 140–180 nm in diameter, and the so-called A2 fibres, with elementary granules of 100–140 nm in diameter (Polenov et al. 1972). The A2 fibres appeared to be more numerous than the A1 fibres. There is evidence to suggest that the A2 fibres contain arginin-8-vasotocin, whereas the A1 fibres contain an oxytocin-like polypeptide (Polenov et al. 1979). Both types of fibres have been observed to terminate on the connective tissue layer separating the neural and intermedia elements and in the pericapillary space of the sinusoid capillaries situated within that layer.

The fibres of the preoptico-hypophyseal tract are beaded and provided with terminal swellings. The largest of these terminals, with a diameter of up to 50 µm, are known as Herring bodies. Most of these bodies are located between the ependymal cells lining the recessus hypophyseus, where they protrude into the lumen. Only rarely are such Herring bodies situated in the pericapillary space. Polenov and Garlov (1971a) carried out a detailed analysis of the ultrastructural organisation of Herring bodies. They observed that some of these large neurosecretory terminals contain numerous elementary granules, that others are entirely devoid of these granules and that all sorts of intermediary forms between these two extremes occur. Polenov and Garlov conjectured that the various types of large terminals observed reflect different functional states in the process of a secretory cycle. Evidence was presented suggesting that the Herring bodies protruding into the recessus hypophyseus release their neurosecretory product into the ventricular cavity through ruptures into their plasmalemma, i.e. a process known as the macro-apocrine type of secretion. It was suggested that the mechanism of release of the neurosecretory product from Herring bodies into the pericapillary space is similar to that in the recessus neurohypophyseus. However, the discharge of the neurosecretory product from smaller neurosecretory terminals and from preterminal parts of the fibres occurs mainly by exocytosis (see Polenov et al. 1979). It is worth noting that Polenov and collaborators (Polenov and Garlov 1974; Polenov and Pavlovic 1978; Polenov et al. 1976a, 1979) were able to demonstrate that, in *Acipenser güldenstädti*, the preoptico-hypophyseal system shows profound structural and ultrastructural changes related to spawning and osmotic stress.

Numerous thin, type-B fibres (0.5–1.5 µm in diameter) are present among the neurosecretory fibres in the neurohypophysis. In addition to numerous small, clear 'synaptic' vesicles, these fibres and their terminals also contain dense-core vesicles of 40–110 nm in diameter. Comparison of histofluorescence and electron microscopy data has led to the conclusion that these dense-core vesicles may be considered as carriers of catecholamines

(Polenov et al. 1972). The type-B fibres presumably arise from the nucleus lateralis tuberis, although direct evidence for this origin is lacking. Their terminal swellings in the neurohypophysis are smaller than those of the type-A fibres. The terminals of the type-B fibres, like those of the type-A fibres, make contact with the connective tissue layer separating the neurohypophysis from the intermediate lobe and with the pericapillary space of the numerous sinusoid capillaries situated within that layer. Some of the type-B fibre terminals make synaptoid contacts with ependymal cells in the neural lobe. Neither the pars intermedia nor the pars distalis is penetrated by nerve fibres (Hayashida and Lagios 1969; Polenov and Garlov 1971c). However, it is assumed that the contents of type-A and type-B fibre terminals diffuse through the connective tissue layer into the parenchyma of the intermediate lobe and that the glandular activity of the intermediate lobe is therefore under the dual control of catecholaminergic and peptidergic elements of the hypothalamus (Polenov et al. 1972).

The presence of a proximal neurosecretory contact region or eminentia mediana in chondrosteans was reported by Hayashida and Lagios (1969: *Acipenser transmontanus*, *Polyodon spathula*), Polenov et al. (1976b: *Acipenser güldenstädti*, *Acipenser stellatus*, *Acipenser ruthenus*) and Kotrschal et al. (1985b: *Acipenser ruthenus*). Within this region, the following four distinct zones can be recognised:

1. An *ependymal zone*, consisting of a single compact layer of tanycyte perikarya. Among these tanycytes, numerous coronet cells and intraventricular protrusions of monoaminergic and peptidergic cerebrospinal fluid-contacting neurons are found. Peripheral tanycyte processes pass toward the surface of the brain.
2. A *periventricular zone* of neuronal perikarya of monoaminergic and peptidergic neurons. Both cell types are most abundant in the caudal part of the median eminence.
3. A *fibre zone*, through the external part of which the axons of the preopticohypophyseal tract pass.
4. A *superficial zone* consisting of the end feet of the peripheral tanycyte processes and of the terminals of granule-containing nerve fibres. Immunohistochemical studies have shown the presence of fibres containing corticotropin-releasing factor (González et al. 1992), gonadotropin-releasing hormone (Leprêtre et al. 1993) and neuropeptide Y (Chiba and Honma 1994) in this zone.

The meningeal tissue separating the ventral surface of the median eminence from the pituitary gland contains a large number of widened primary portal capillaries, from which short vessels pass into the underlying pars distalis. The bulbous end feet of the tanycyte processes and synaptoid endings of peptidergic type-A1 and type-A2 fibres and of monoaminergic type-B fibres have been observed to abut on the perivascular space of the primary portal capillaries. Polenov et al. (1976b) believed that, as in other vertebrates, these endings discharge their products into the portal circulation and that along the way hypothalamic neurons affect the activity of the glandular cells in the pars distalis.

14.9
Telencephalon

14.9.1
Introductory Notes

The telencephalon of chondrostean fishes can be subdivided into a rostral, an intermediate and a caudal part. The rostral part consists of a pair of very large, diverging olfactory bulbs containing wide ventricular cavities. The intermediate and caudal parts together constitute the telencephalon proper. The large, intermediate part is formed bilaterally by a solid, everted wall, whereas the wall of the small, caudal part has remained uneverted (Figs. 14.3, 14.21–14.24). The entire telencephalon proper is covered by a wide, ependymal tela telencephali. This membranous telencephalic roof is attached to the dorsolateral side of the everted walls and to the dorsocaudal edge of the olfactory bulbs. Its caudal part forms the rostroventral leaf of the velum transversum (Fig. 14.3).

It is noteworthy that Johnston (1910, 1911) and Herrick (1921, 1922) held that, whereas in most groups of vertebrates the telencephalon is largely evaginated, this bulging-out process in chondrosteans and other actinopterygian fishes proceeds only so far as to form the olfactory bulbs which enclose lateral ventricles. These authors considered the remaining part of the actinopterygian telencephalon to represent a primitive end-brain or telencephalon medium. However, Niewenhuys (1964) has pointed out that, in the telencephalon of most vertebrates, two entirely different evagination processes occur, one leading to the formation of the olfactory bulbs and the other to the formation of the cerebral hemispheres. The rostral evaginated portions of the actinopterygian telencephalon represent the olfactory bulbs and have nothing to do with the evaginated cerebral hemispheres of other vertebrates. It is true that the walls of the actinopterygian telencephalon proper do not evaginate

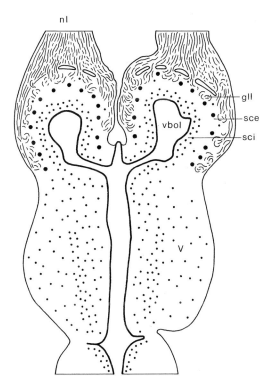

Fig. 14.30. Horizontal section through the telencephalon of the sturgeon *Acipenser ruthenus*. (Redrawn from Goronowitsch 1888)

and that the ventral parts of these walls essentially maintain the tube-like condition through which the forebrain anlage of all vertebrates passes (Fig. 14.30). However, the dorsal parts of the walls show profound changes of shape during development (Fig. 14.6). The dominant morphogenetic process occurring here, i.e. eversion, is the antithesis of the inversion and evagination processes found in the telencephalon of most other vertebrates. It is therefore incorrect to denote the actinopterygian telencephalon as a whole as 'primitive' or as a 'telencephalon medium'.

In the following survey, the olfactory bulb will be considered first. The large, everted, intermediate part of the telencephalon will then be discussed. Following Nieuwenhuys (1963) and Northcutt and Braford (1980), this part of the telencephalon will be divided into an area ventralis telencephali and an area dorsalis telencephali. Next, the regio preoptica, which is largely situated in the uneverted caudal telencephalic portion will be described; finally, attention will be devoted to the telencephalic fibre systems.

14.9.2
Olfactory Bulb

The chondrostean olfactory bulb displays a laminated structure (Figs. 14.24, 14.30). The layers are concentrically arranged around the bulbar ventricle and comprise the following (from superficial to deep): (a) a layer of primary olfactory fibres, (b) a glomerular layer, (c) a stratum cellulare externum, (d) a layer of secondary olfactory fibres and (e) a stratum cellulare internum. The structure of the olfactory bulb of *Acipenser rubicundus* was studied by Johnston (1898b, 1901) using the Golgi technique. His results may be summarised as follows (Fig. 14.31).

The primary olfactory fibres enter the bulb in small bundles; most of these fibres terminate in a single glomerulus. The stratum cellulare externum, which, according to Johnston, cannot be sharply delineated from the glomerular layer, contains four types of cells: large and small mitral cells, stellate cells and cells with short axons.

The large mitral cells possess two to five coarse dendrites, each of which supplies one or more glomeruli, and one or more slender dendrites, which end freely in the glomerular zone without participating in the formation of glomeruli. Their axons pass to the telencephalon proper. During their course through the bulb, these processes emit several collaterals that ascend toward the glomerular zone. The small mitral cells are provided with a single coarse dendrite, which breaks up into one or a few adjacent glomeruli. Their axons are usually directed centrally.

The stellate cells have two to five dendrites usually disposed parallel to the surface of the bulb; the cells thus have a stellate appearance in surface view. Their dendrites are richly branched and terminate in glomeruli. The axons of these cells could be traced centrally and, in a few cases, toward the telencephalon proper.

The cells with short axons are few in number. From one pole of these cells, a coarse dendrite arises, the branches of which end in glomeruli. From the opposite pole, an axon-like process arises, which breaks up into numerous smooth branches that are lost in the glomerular zone.

According to Johnston, the stratum cellulare internum is composed of cells of the following four types: granule cells, stellate cells, spindle-shaped cells and cells of Cajal.

The term *granule cell* was used by Johnston to denote rounded or pyramidal cells of various sizes, the dendrites of which terminate in glomeruli. He considered it likely that the axon of these cells pass to the telencephalon proper.

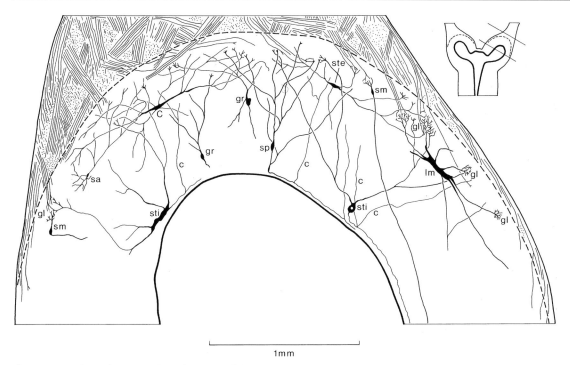

Fig. 14.31. Horizontal section through a part of the bulbus olfactorius of the sturgeon *Acipenser rubicundus*, showing elements impregnated using the Golgi technique. The position of the area depicted is shown at the *upper right*. *C*, cell of Cajal; *c*, axon collateral; *sa*, cell with a short axon; *gr*, granule cell; *gl*, glomerulosus; *lm*, large mitral cell; *sm*, small mitral cell; *sp*, spindle-shaped cell; *ste*, stellate cell of external cell layer; *sti*, stellate cell of internal cell layer. (Redrawn from Johnston 1898b)

The *stellate cells* are provided with three to five long, widely diverging dendrites, the branches of which end in the glomerular zone, where they enter the formation of olfactory glomeruli. Their coarse axons pass close to the ventricular cavity toward the telencephalon proper. They emit collaterals that branch into the superficial part of the stratum cellulare internum.

The *spindle-shaped cells* are bipolar elements that are oriented radially. From their peripheral end, a dendrite arises, the branches of which terminate in glomeruli. Their central end gives rise to an axon that passes along the ventricle toward the telencephalon. During their course through the bulb, these axons give off peripherally running collaterals which reach the glomerular layer.

The *cells of Cajal* are conspicuous, tangentially oriented neurons. Johnston observed that some dendrites of these cells terminate in glomeruli, but he considered it likely that other dendrites do not end in glomeruli and receive impulses from the many peripherally extending collaterals of the bulb. Their axons were observed to pass to the telencephalon. The name 'cell of Cajal' was chosen by Johnston because these elements resemble the conspicuous neurons present in the first layer of the

mammalian cortex, which were first observed by Cajal and later named after that author by Retzius.

From this brief survey, it appears that, according to Johnston, all cells in the olfactory bulb of the sturgeon, except those with short axons, receive impulses from primary olfactory fibres and transmit these impulses toward the telencephalon proper.

Story (1964) observed that, in *Polyodon*, the bulbar formation receives vomeronasal fibres and a nervus terminalis in addition to primary olfactory fibres. According to Story, the vomeronasal fibres terminate in a well-developed accessory olfactory bulb situated on the dorsomedial surface of the main bulb, whereas the fibres of the nervus terminalis have a distinct ganglion in the ventromedial part of the bulbar formation. The present author was unable to observe corresponding structures in *Scaphirhynchus*. It is noteworthy that, in the brain of the Siberian sturgeon *Acipenser baeri*, Leprêtre et al. (1993) observed a continuum of gonadotropin-releasing hormone-immunoreactive cells, extending from the olfactory nerves and bulbs, via the basal telencephalon and the preoptic region, to the mediobasal hypothalamus (see Fig. 14.34c). It seems likely that at least part of these gonadotropin-releasing hormone-immunoreactive cells in the telencephalon of this species represent ganglion cells of the terminal nerve, known to contain gonadotropin-releasing hormone in teleosts and other vertebrate groups (see Sect. 15.6.2). This is particularly true of a cluster of gonadotropin-releasing hormone-immunoreactive cells, situated in the transitional region of the olfactory bulb and the ventral telencephalic area (Leprêtre et al. 1993; see Fig. 14.34c: arrow).

In *Scaphirhynchus* and *Acipenser*, and probably in other chondrosteans as well, the caudomedial portion of the bulbar evagination is devoid of glomeruli and thus should be considered a part of the telencephalon proper. I denote the undifferentiated periventricular grey in this region as the area retrobulbaris (Figs. 14.24, 14.32b).

14.9.3
Area Ventralis Telencephali

In the rostral part of the area ventralis, three nuclei are present in a periventricular position; from ventral to dorsal, theses are as follows: Vv, a group of small cells packed against the ependyma, Vd, a triangular group of slightly larger cells in clusters migrated very slightly away from the ependyma, and Vn, a loosely organised circular mass lying between Vd and the area dorsalis (Figs. 14.23, 14.32c). Northcutt and Braford (1980) distinguished an additional cell mass in the area ventralis, contiguous with the lateral edge of Vn and the dorsolateral aspect of Vd. In view of its olfactory input (see below), these authors termed this cell mass Vl and suggested that it may be homologous to the lateral subpallial nucleus of polypteriforms, although it has not migrated as far from the ependyma in *Scaphirhynchus* (Fig. 14.32c). Nieuwenhuys (1963) included the above-mentioned cell group in *Acipenser* in Vd. The submeningeal zone of the area ventralis contains in most places only small, scattered neurons. However, these may locally condense so as to form a distinct cell group (nucleus postolfactorius lateralis of Johnston 1898b).

In the caudal part of the area ventralis, Vd is replaced by the supracommissural nucleus (Vs) at about the level of the height of the commissural ridge. As shown in Fig. 14.22, Vs consists of a group of cells oriented at an oblique angle to the ventricular surface. Vs is replaced caudally by the postcommissural nucleus (Vp), which caudally and ventrally adjoins the magnocellular and parvocellular preoptic nuclei, respectively (Figs. 14.29a, 14.32f).

The periventricular cell zone formed by the cell masses Vv, Vd and Vn corresponds to the nucleus olfactorius medialis of Johnston (1911), the area olfactoria medialis of Herrick (1922) and the septum of Hocke Hoogenboom (1929).

Piñuela and Northcutt (1994) studied the immunohistochemical organisation of the forebrain in *Acipenser transmontanus*. They observed that area Vd contains a very high concentration of substance P-, leu-enkephalin-, serotonin- and dopamine-positive fibres and interpreted this area on the basis of these findings as the homologue of the striatum in other vertebrates. They considered it likely that dopaminergic cells situated in the periventricular nucleus of the tuberculum posterius are the source of the dopaminergic input to Vd (see Fig. 14.34e).

14.9.4
Area Dorsalis Telencephali

Throughout most of the telencephalon, the area ventralis and area dorsalis are separated by a cell-poor zona limitans (Figs. 14.22, 14.23). In the caudal part of the telencephalon, the dorsal border of this zone is marked by a ventricular groove, the sulcus limitans telencephali (Fig. 14.29a). The greater part of the area dorsalis contains large pyramidal cells arranged in a number of layers, parallel to the ependymal surface. The region showing this arrangement of neurons occupies the dorsal and lateral parts of the area dorsalis and is designated accordingly as Dd+Dl (Fig. 14.22). Rostrally, in a medial position just dorsal to the area ventralis, lies a zone of cells which is somewhat less regularly laminated than Dd+Dl. This mass is Dm; however, in contrast to the situation in holosteans and teleosts, it does not extend the full length of the area dorsalis. Rather, it is gradually replaced from its medial aspect by Dd+Dl at successively more caudal levels. Consequently, Dd+Dl occupies in the caudal part of the telencephalon the entire ventricular zone of the area dorsalis. Northcutt and Braford (1980) found that the caudomedial part of the area dorsalis receives secondary olfactory fibres and thus interpreted this part as the posterior zone of the area dorsalis, Dp (Fig. 14.32). For the same reason, they also included the caudolateral part of Dd+Dl in Dp, despite the fact that the characteristic lamination pattern of Dd+Dl continues in this region (Fig. 14.32). In our opinion, the olfactory target area designated as Dp by Northcutt and Braford is not a cytoarchitectonic unit. It consists largely, if not entirely, of the caudal part of Dd+Dl.

At mid-telencephalic levels, a group of large, scattered cells in a migrated position constitutes the central zone, Dc (Figs. 14.22, 14.23, 14.32d). The submeningeal zone of the area dorsalis contains only small, sparsely scattered neurons.

The area dorsalis and its parts have been desig-

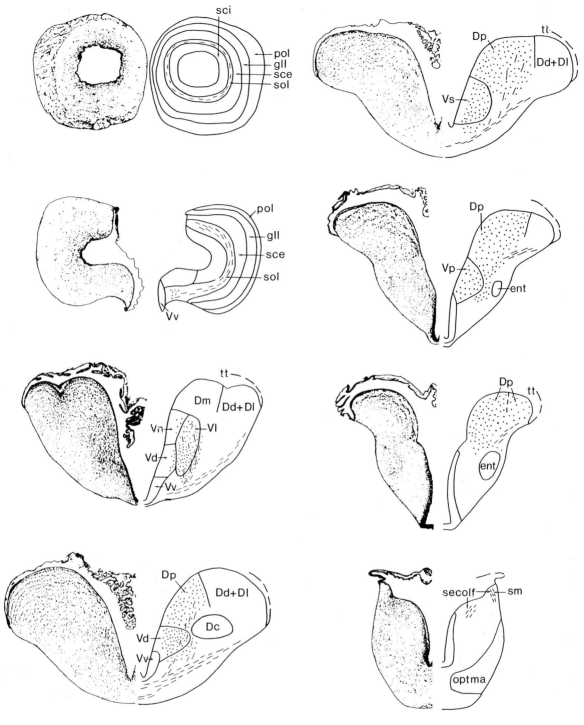

Fig. 14.32. Transverse sections through middle and caudal levels of the olfactory bulb and through the telencephalon proper in the shovelnose sturgeon *Scaphirhynchus platorynchus*, showing the pattern of degeneration following a unilateral aspiration lesion of the olfactory bulb. Note that the interpretation of the parts of this telencephalon, as presented by Northcutt and Braford, differs in several respects from our own. Thus, in our opinion, the area designated as Vl represents the lateral part of Vn and Vd, whereas Dp represents the caudal parts of Dd and Dl. For further explanations, see text. (From Northcutt and Braford 1980)

nated by previous authors with a variety of names. Johnston (1898b, 1901) initially termed Dc the striatum and Dm, Dd and Dl together the epistriatum, but he later labelled the entire area as the primordium hippocampi. Herrick (1922) denoted Dm, Dd and Dl together as the area olfactoria dorsalis and Dc as the area olfactosomatica, whereas Hocke Hoogenboom (1929), following the nomenclature introduced by Ariëns Kappers (1921), designated Dc as the striatum, Dm as the epistriatum and Dd and Dl together as the palaeopallium. In several previous papers, Nieuwenhuys (1962, 1966) expressed the opinion that the area ventralis and area dorsalis of the telencephalon of chondrosteans and other actinopterygian fishes represent the subpallium and pallium, respectively, of other vertebrates. Northcutt and Braford (1980) and Northcutt and Davis (1983) reported that the caudal part of Dm contains the highest concentrations of acetyl cholinesterase and high concentrations of catecholamines, features which in their opinion supported its homology with the caudatoputamen of land vertebrates. However, more recently, Piñuela and Northcutt (1994) rejected this homology for the following reasons: (a) Dm was poor in substance P- and leu-enkephalin-positive fibres, and (b) as already mentioned, the neuromediator profile of Vd appeared to be very similar to that of the striatum in other vertebrate groups.

Johnston (1898b, 1901) studied the area dorsalis using the Golgi technique. His main results may be summarised as follows: The laminated, periventricular zone of this area is composed of pyramidal cells, which direct their apices toward the meningeal surface. Their basal dendrites are oriented parallel to the ventricular surface, and their large apical dendrites, one to four in number, are directed peripherally. The axons of the pyramidal cells arise from the soma or from the initial part of one of the dendrites. They pass peripherally and issue numerous collaterals which ramify and terminate in Dc. The cells in the latter area are irregular in shape. Their axons, which give off few and short or no collaterals, pass toward the diencephalon.

14.9.5
Regio Preoptica

The regio preoptica is that portion of the telencephalic wall which surrounds the recessus preopticus of the telencephalic ventricular system. This recess is bounded rostrally by the commissural ridge of the anterior commissure and caudally by the chiasma opticum. Whereas in most fishes the anterior commissural ridge is high and well marked, in chondrosteans this structure is represented by an elongated thickening of the lamina terminalis (Fig. 14.3d). Thus the rostral boundary of the preoptic region in chondrosteans is less sharply defined than in other fish groups. Dorsally, it is bordered by a cell-poor zona limitans and caudally by the hypothalamus. The preoptic region contains three periventricular cell masses, the nucleus preopticus parvocellularis, the nucleus preopticus magnocellularis and the nucleus suprachiasmaticus, and one group of migrated cells, the nucleus entopeduncularis.

The *nucleus preopticus parvocellularis* extends throughout the length of the preoptic region, from the bed of the anterior commissure rostrally to the rostral border of the hypothalamus caudally. It is composed of small cells arranged in a few lamellae parallel to the ventricular surface (Figs. 14.21, 14.29a, 14.32f-h).

The *nucleus preopticus magnocellularis* occupies a central position in the preoptic region (Fig. 14.29a). It consists of a compact sheet of large, neurosecretory cells. As was mentioned in the previous section, the axons of these cells constitute the tractus preopticohypophyseus.

The *nucleus suprachiasmaticus*, which is composed of small, rather densely arranged cells, is situated directly dorsal to the rostral part of the chiasma opticum (Figs. 14.21, 14.29a). It was established experimentally by Repérant et al. (1982) that this nucleus receives a bilateral retinal input (Fig. 14.29d).

Finally, the *nucleus entopeduncularis*, is a large, submeningeally situated cell mass, consisting of small, rather loosely arranged elements (Figs. 14.21, 14.29b, 14.32 f,g). Its dorsocaudalmost part merges with the eminentia thalami, i.e. the rostralmost part of the thalamus ventralis.

14.9.6
Telencephalic Fibre Systems

The nerve fibres in the chondrostean telencephalon constitute a rather diffuse, superficially situated zone, within the confines of which more compact fibre systems can be distinguished (Figs. 14.22, 14.23, 14.33).

The tractus olfactorii arise gradually from the layer of secondary olfactory fibres of the olfactory bulb. The fibres of the large tractus olfactorius medialis assemble in the caudomedial part of the bulb and pass ventrally to the bulbar ventricle toward the area ventralis telencephali, whereas those of the smaller tractus olfactorius lateralis initially occupy a dorsal position in the caudal part of the bulb and from there enter the area dorsalis telencephali. Within the latter, they form part of the lateralmost portion of the telencephalic fibre zone (Figs. 14.22–14.24). Northcutt and Braford (1980) studied the efferent projections of the olfactory bulb in *Scaphirhynchus* using degeneration techniques. From this study, it appeared that secondary olfactory fibres distribute widely to the telencephalon (Fig. 14.32). The medial olfactory tract gives rise to a small terminal field lateral to Vv, but most of its fibres continue caudally, cross in the anterior commissure and distribute to the contralateral telencephalon. Dorsally directed fibres enter Vd and Vn. Scattered terminals are seen in the lateral parts of Vd and Vn rostrally, and considerably denser ones are seen in the lateral part of Vd more caudally. In the caudal area ventralis, both Vs and Vp are in receipt of olfactory input, the projection to Vs being the heavier of the two.

At rostral levels, the area dorsalis is free of degeneration, but at commissural levels Dm is replaced by a medial extension of Dd+Dl, which receives a large olfactory input. At successively more caudal levels, the olfactory terminal field in the area dorsalis extends more and more laterally and finally occupies the entire caudal pole of the area dorsalis (Fig. 14.32). Northcutt and Braford (1980) designated the olfactory projection field in the area dorsalis as a separate unit, Dp, but in our opinion this field represents the caudal part of Dd+Dl. A bundle of degenerating fibres passes via the medial olfactory tract and the medial forebrain bundle to the regio tuberculi posterioris, where it terminates just ventral to the nucleus medianus tuberculi posterioris (Fig. 14.29d).

The lateral olfactory tract appears to distribute fibres to the caudal part of Dd+Dl and to send fibres via the stria medullaris to the habenula. These fibres decussate in the habenular commissure and distribute to the contralateral telencephalon. In general, the olfactory projections to the contralateral telencephalon are symmetrical to the ipsilateral ones, but sparser. A large number of degenerating fibres were observed to enter the layer of secondary fibres of the contralateral olfactory bulb and to terminate on mitral cells and more sparsely on the most superficial cells of the stratum cellulare internum.

In addition to secondary olfactory fibres, the commissura anterior contains numerous axons originating from the telencephalon proper. The exact origin of these fibres is unknown; however, I am under the impression that numerous fibres originating in the various parts of the area ventralis decussate in the dorsal part of the anterior commissure and then join the medial forebrain bundle.

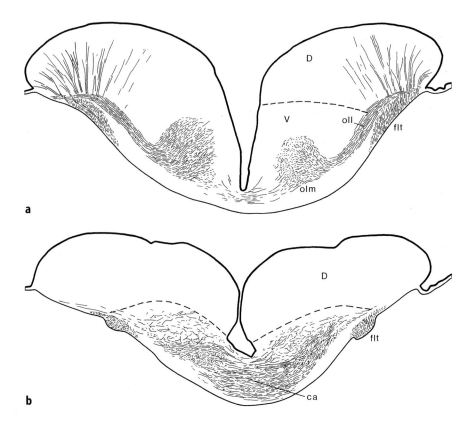

Fig. 14.33a,b. Transverse section through **a** the rostral and **b** the intermediate portions of the telencephalon of the paddlefish *Polyodon spathula*, drawn from Bielschowsky preparations. (Redrawn from Johnston 1911)

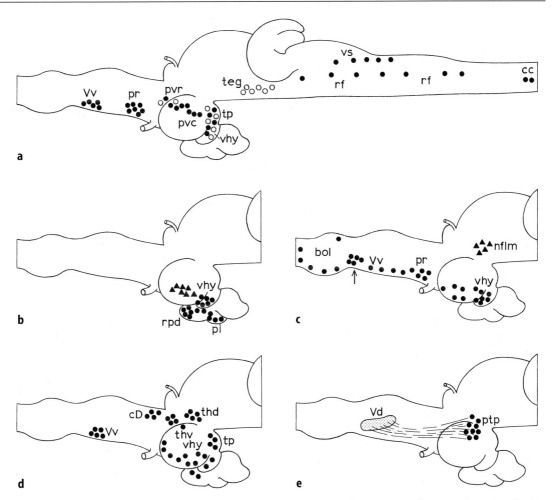

Fig. 14.34a–e. Distribution of cells containing particular neuromediators or neuromediator-related substances in the brain of various species belonging to the genus *Acipenser*. **a** Catecholamines (*filled circles*) and serotonin (*open circles*). **b** Corticotropin (*filled circles*) and corticotropin-releasing fac- tor (*closed triangles*). **c** Mammalian gonadotropin-releasing hormone (*closed circles*) and chicken gonadotropin-releasing hormone II (*closed triangles*). **d** Neuropeptide Y (*closed cir- cles*). **e** Dopamine (*closed circles*). For further information, see text

Fibres occupying a superficial position in the tel- encephalic fibre zone converge toward the anterior commissure and decussate in its ventralmost part. This system, which presumably interconnects both areae dorsales, was described by Johnston (1911) under the name commissura hippocampi. Johnston was of the opinion that this connection is inter- rupted in the ventrolateral area of the telencepha- lon, synapsing with the scattered, superficial situ- ated cells located in that area.

In summary, the following can be stated:

1. The major telencephalic olfactory targets in chondrosteans include at least parts of all of the cell groups present in the area ventralis and an extensive area Dp, which forms part of the area dorsalis (Fig. 14.32).
2. A considerable portion of the area dorsalis,

including Dm, Dc and the rostral parts of Dd and Dl, is devoid of a secondary olfactory input.

The fibre systems connecting the telencephalon with the diencephalon, i.e. the stria medullaris and the medial and lateral forebrain bundles, and their possible functional significance were discussed in Sect. 14.8.3. Suffice it here to mention that the stud- ies carried out by Johnston (1901, 1911; Fig. 14.33) on *Acipenser* and *Polyodon* strongly suggest that the non-olfactory part of the area dorsalis is connected with the hypothalamus by means of the lateral fore- brain bundle (i.e. the tractus pallii in his nomencla- ture).

Experimental neuroanatomical studies on the connections of the telencephalon with more cau- dally situated brain regions are warranted, and the results will be of considerable comparative interest.

14.10
Chemoarchitecture

Data on the distribution of neurotransmitters and other neuroactive principles in the chondrostean brain are scant. Kotrschal et al. (1985b) studied the distribution of monoaminergic neurons in the brain of the sterlet *Acipenser ruthenus* using the FIF technique. As shown in Fig. 14.34a, catecholamine-related (green) fluorescent cell groups appeared to be present in the ventral part of the telencephalic area ventralis (Vv), along the preoptic recess (pr), in the caudal part of the paraventricular organ (pvc), in the rhombencephalic reticular formation (rf), along the dorsal aspect of the rhombencephalic viscerosensory column (vs) and in the floor of the spinal central canal (cc). Serotoninergic (yellow) fluorescent cells were observed in the tegmentum of the midbrain (teg), but not in the raphe, whereas the rostral part of the paraventricular organ (pvr), the caudal part of the pars ventralis hypothalami (vhy) and the nucleus tuberis posterior (tp) contained an admixture of green and yellow fluorescent cells.

The distribution of some neuropeptides has been studied using immunohistochemical techniques. González et al. (1992) studied the localisation of corticotropin and corticotropin-releasing factor-like peptides in the brain and hypophysis of the sterlet *Acipenser ruthenus* (Fig. 14.34b) and found corticotropin-immunoreactive cells in the ventral part of the hypothalamus (vhy) and in both the rostral pars distalis (rpd) and the pars intermedia (pi) of the pituitary gland. The axons of the corticotropin-immunoreactive cells appeared to terminate in the ventral hypothalamus. Corticotropin-releasing factor-immunoreactive cells were found in the rostral part of the ventral hypothalamus (vhy). Their axons could be traced to the telencephalon, the dorsal and ventral parts of the hypothalamus, the median eminence, the pars nervosa of the pituitary gland and the tegmentum of the midbrain.

Leprêtre et al. (1993) described the distribution of two forms of gonadotropin-releasing hormone, i.e. mammalian gonadotropin-releasing hormone and chicken gonadotropin-releasing hormone II in the brain of the Siberian sturgeon *Acipenser baeri* (Fig. 14.34c). Neuronal perikarya-immunoreactive with anti-mammalian gonadotropin-releasing hormone appeared to be present in the olfactory nerve, the olfactory bulb and the forebrain. The latter were embedded in a continuum of mammalian gonadotropin-releasing hormone-immunoreactive fibres, extending from the mediobasal telencephalon via the preoptic region into the ventral hypothal-

amus. A cluster of large mammalian gonadotropin-releasing hormone-immunoreactive perikarya was consistently found directly caudal to the olfactory bulb (Fig. 14.34c). In teleosts with sessile olfactory bulbs, a group of cells containing gonadotropin-releasing hormone (or lactotrope hormone-releasing hormone) occupies a corresponding position. These cells are known to project to both the olfactory epithelium and the retina and to constitute the ganglion of the terminal nerve (see Sect. 15.6.2). In the Siberian sturgeon, Leprêtre et al. (1993) observed mammalian gonadotropin-releasing hormone-positive fibres close to the olfactory mucosa and in the optic tract and the retina. In view of the relations in teleosts, it seems likely that the retrobulbar cluster of mammalian gonadotropin-releasing hormone-immunoreactive cells in this species constitutes the origin of these fibres and represents the terminal nerve ganglion.

Chicken gonadotropin-releasing hormone II-immunoreactive cell bodies were observed by Leprêtre et al. (1993) in the nucleus of the medial longitudinal fasciculus of the Siberian sturgeon. Considering that in certain teleosts the nucleus of the medial longitudinal fasciculus projects to the caudal spinal cord, where it innervates the urotensin-producing caudal neurosecretory complex (see also Sect. 15.2.7), Leprêtre et al. (1993) suggested a similar function for this nucleus in the sturgeon. Fibres containing gonadotropin-releasing hormone were detected in many parts of the brain; mammalian gonadotropin-releasing hormone-immunoreactive fibres were particularly numerous in the telencephalon and the hypothalamus, whereas the chicken gonadotropin-releasing hormone innervation prevailed in the tectum, the cerebellum and the brain stem.

Chiba and Honma (1994) studied the distribution of neuropeptide Y in the telencephalon and diencephalon of the white sturgeon *Acipenser transmontanus* (Fig. 14.34d). They found groups of neuropeptide Y-immunoreactive cells in the ventral part of the telencephalic area ventralis (Vv), in the caudal part of the telencephalic area dorsalis (cD) and in the periventricular zones of the ventral thalamus (vth), dorsal thalamus (dth), ventral hypothalamus (vhy) and nucleus tuberis posterior (tp). Judging from their figures, the cell group designated as being located in the caudal part of the telencephalic dorsal area may well actually be located in the eminentia thalami, i.e. the rostralmost part of the ventral thalamus (see also Fig. 14.29a). Dense networks of neuropeptide Y-immunoreactive fibres were observed in the ventral part of the telencephalon and throughout most of the diencephalon.

Finally, Piñuela and Northcutt (1994) studied the

distribution of leu-enkephalin, substance P, tyrosine hydroxylase, serotonin and dopamine in the forebrain of the white sturgeon *Acipenser transmontanus* (Fig. 14.34e) and found a very high concentration of substance P-, leu-enkephalin-, serotonin- and dopamine-positive fibres in the dorsal part of the telencephalic area, Vd. They noticed that, in this respect, Vd resembles the striatum of other vertebrate groups and suggested that dopaminergic neurons in the periventricular nucleus of the tuberculum posterius (ptp) are the source of the dopaminergic input to Vd.

14.11
Concluding Remarks

In this final section, I will attempt to summarise the most salient data concerning the structure of the central nervous system of chondrostean fishes and to place these data in a comparative perspective.

The structure of the spinal cord does not deviate from that observed in most other groups of fishes.

The rhombencephalon is relatively very large. The grey matter in this brain part is arranged in four longitudinal zones, which appear as intraventricular protrusions. These zones, which have been designated as the area ventralis, area intermedioventralis, area intermediodorsalis and area dorsalis, largely, but not entirely, coincide with the so-called functional columns of Herrick and Johnston. The most important incongruity is that, in addition to the viscerosensory vagal lobe, the area intermediodorsalis contains several somatosensory centres. Corresponding observations have been made in other groups of fishes, including chondrichthyans, brachiopterygians, holosteans, teleosts and dipnoans and the crossopterygian *Latimeria*.

The area octavolateralis, which compriss both the area dorsalis and a part of the area intermediodorsalis of the rhombencephalon, can be subdivided into three zones, dorsal, intermediate and ventral, each of which is specifically related to a particular primary afferent input; thus the dorsal zone is associated with the electroreceptive lateral line system, the intermediate zone receives input from the mechanoreceptive lateral line system, whereas the ventral zone contains the centres of termination of primary afferents originating in the auditory and vestibular portions of the labyrinth. The chondrosteans share this organisational pattern of the octavolateral area with chondrichthyans, brachiopterygians, dipnoans, the coelacanth and even lampreys. However, the electroreceptive portion of the lateral line system and the associated dorsal zone is lacking in holosteans (McCormick 1981, 1982).

The cerebellum comprises a pair of large auriculae and an unpaired central body, protruding into the ventricular cavity. The caudal portion of this central body represents the corpus cerebelli, whereas its rostral part represents the valvula cerebelli. Distinct auriculae cerebelli are found in most groups of anamniotes, including chondrichthyans, brachiopterygians, dipnoans, the coelacanth and urodeles, and a homologue of the corpus cerebelli is present in all gnathostomes. However, a valvula cerebelli, i.e. a subtectal protrusion of cerebellar tissue, only occurs in brachiopterygians and actinopterygians. The configuration of the massive central body, consisting of the corpus and the valvula, is unique for chondrosteans. Larsell (1967) pointed out that this central body as a whole is similar to that of brachiopterygians, except for the disappearance of the fissure which separates the bilateral cerebellar halves in the latter group. This view is essentially correct; however, it should be added that, in the central cerebellar body of brachiopterygians, the Purkinje cells are concentrated in bilateral bands, whereas in the corresponding structure of chondrosteans these elements lie mostly scattered through the molecular layer. As in chondrosteans, the teleostean corpus cerebelli is represented by a solid body, but it should be emphasised that the apparent similarity of the ultimate configuration is ontogenetically attained in two different ways. In chondrosteans, it is the meningeal surfaces of both cellular halves which fuse in the median plane, but in teleosts the bilateral cerebellar anlagen form large intraventricular thickenings. These thickenings eventually fuse in the median plane, thereby obliterating the ventricular cavity (Schaper 1894a,b; Pouwels 1978a).

The mesencephalon is rather poorly developed and can be subdivided in the usual way into a dorsal tectum and a ventral tegmentum. The medial part of both tectal halves contains bilaterally an elongated mass of granule cells. These tori longitudinales are unique feature of the brain of actinopterygian fishes. Within the tegmentum mesencephali, two cell masses, the nucleus lateralis valvulae and the nucleus tori lateralis, deserve mention. The nucleus lateralis valvulae occupies a periventricular position in the caudal part of the tegmentum mesencephali and projects massively towards the valvula cerebelli. It occurs only in those groups of fishes which possess a valvula cerebelli, i.e. the brachiopterygians and the actinopterygians. The nucleus tori lateralis is an aggregation of cells situated superficially in the ventrolateral part of the tegmentum of the midbrain. This cell mass is present in chondrichthyans, brachiopterygians and actinopterygians, but is entirely lacking in cyclostomes, dipnoans and the coelacanth.

Although the regio pretectalis and the regio tuberculi posterioris may well both be composed of diencephalic and mesencephalic components, they have been dealt with here in the section on the diencephalon. As in most other groups of fishes, the pretectal region comprises a periventricular, a central and a superficial nucleus. The nucleus pretectalis superficialis (i.e. the corpus geniculatum laterale referred to by some previous authors) is extraordinarily large in chondrosteans, but the central pretectal nucleus is only poorly developed. It is our impression that, as far as the differentiation of the regio tuberculi posterioris is concerned, the chondrosteans are intermediate between the brachiopterygians and the actinopterygians.

The telencephalon can be subdivided into three principal parts: the olfactory bulbs, the cerebral hemispheres and the preoptic region. The olfactory bulbs are large and contain wide ventricular cavities. The cerebral hemispheres have thickened, everted walls and in this respect closely resemble the corresponding structures in holosteans and teleosts. The hemisphere walls can be subdivided into an area ventralis telencephali and an area dorsalis telencephali. These two areas correspond to the subpallium and the pallium, respectively, of other vertebrates. It has been experimentally established that a substantial portion of the area dorsalis telencephali is not in receipt of secondary olfactory fibres. Corresponding observations have been made in chondrichthyans (Ebbesson and Heimer 1970; Bodznick and Northcutt 1979), brachiopterygians (Braford and Northcutt 1974; von Bartheld and Meyer 1986), holosteans (Northcutt and Braford 1980) and teleosts (see Sect. 15.6).

A considerable number of fibre systems, including the lateral lemniscus, the fasciculus longitudinalis medialis, the tectobulbar tracts, the fasciculus retroflexus, the tractus opticus, the fibrae hypothalamotegmentales, the stria medullaris and the medial and lateral forebrain bundles, could be clearly observed in our non-experimental material. However, in order to unravel the central circuitry of the chondrostean brain, experimental studies will be necessary. To give an example, it would be of great comparative and functional interest to know which sources the non-olfactory portion of the area dorsalis telencephali receives its afferents from and which centres this area projects to.

Several previous authors have pointed out that the chondrostean brain has a number of characteristics in common with that of cartilaginous fishes. Thus Theunissen (1914) claimed that the motor nuclei show a 'selachian type of arrangement', and Norris (1925) stated that "in their cranial nerves the chondrosteans are plainly shark-like" and noted the presence of 'a distinct elasmobranch-like lobus lineae lateralis'. I believe that these similarities concern primitive characteristics that the chondrichthyans share with several other groups of fishes rather than derived characteristics that reveal a specific affinity between the chondrosteans and the chondrichthyans.

There can be no doubt that the two living chondrostean families, i.e. the Acipenseridae and the Polydontidae, form a monophyletic group. They share a number of typical derived characteristics, of which the structure of their central cerebellar body and the shape and histological differentiation of their telencephalon deserve special mention. However, the brain of sturgeons and paddlefishes show a number of differences in the development of particular sense organs:

1. The electrosensory lateral line system in paddlefishes is much more strongly developed than in sturgeons; consequently, the nucleus dorsalis of the rhombencephalic lateral line area and the mesencephalic torus semicircularis (to which the lateral line area projects) are much larger in the former.

2. Judging from the size and thickness of the tectum mesencephali, the visual system of paddlefishes is better developed than that of sturgeons.

3. In sturgeons, the gustatory system is particularly strongly developed. As has already been mentioned, in these animals taste buds are not confined to the mouth, but also occur on the barbels and perioral fringes. Such an external gustatory system is lacking in paddlefishes. Centrally, this difference in development of the taste system is reflected in the size of the rhombencephalic lobus vagi, this structure being considerably larger in sturgeons than in paddlefishes.

4. The most striking neuromorphological difference between sturgeons and paddlefishes concerns the lobus inferior hypothalami, which is much larger in the former. Johnston (1911) believed that this brain part receives a large tertiary gustatory projection; thus it may well be that the size of the lateral hypothalamic lobe is positively correlated with the development of the gustatory system.

In view of the differences in development of the various sense organs, it seems reasonable that in the bottom-dwelling sturgeons the olfactory and gustatory systems play a dominant role in the detection and selection of food, whereas the paddlefishes rely mainly on their olfactory, electroreceptive and visual systems for the procurement of food.

References

Ariëns Kappers CU (1921) Die vergleichende Anatomie des Nervensystems der Wirbeltiere und des Menschen, part 2. Bohn, Haarlem

Bass AH (1981) Telencephalic efferents in the channel catfish, Ictalurus punctatus: Projections to the olfactory bulb and optic tectum. Brain Behav Evol 19:1–16

Bergquist H (1932) Zur Morphologie des Zwischenhirns bei niederen Wirbeltieren. Acta Zool 13:57–304

Bodznick D, Northcutt RG (1979) Some connections of the lateral olfactory area of the horn shark (Heterodontus francisci). Soc Neurosci Abstr 5:139

Braford MR Jr, McCormick CA (1979) Some connections of the torus semicircularis in the bowfin, Amia calva: a horseradish peroxidase study. Soc Neurosci Abstr 5:139

Braford MR Jr, Northcutt RG (1974) Olfactory bulb projections in the bichir, Polypterus. J Comp Neurol 156:165–178

Braford MR Jr, Northcutt RG (1983) Organization of the diencephalon and pretectum in ray-finned fishes. In: Davis RE, Northcutt RG (eds) Fish neurobiology, vol 2. Higher brain areas and functions. University of Michigan Press, Ann Arbor, pp 117–164

Butler AB, Northcutt RG (1992) Retinal projections in the bowfin, Amia calva: cytoarchitectonic and experimental analysis. Brain Behav Evol 39:169–194

Chiba A, Honma Y (1994) Neuropeptide Y-immunoreactive structures in the telencephalon and diencephalon of the white sturgeon, Acipenser transmontanus, with special regard to the hypothalamo-hypophyseal system. Arch Histol Cytol 57:77–86

Ebbesson SOE, Heimer L (1970) Projections of the olfactory tract fibers in the nurse shark (Ginglymostoma cirratum). Brain Res 17:47–55

Echteler SM (1984) Connections of the auditory midbrain in a teleost fish, Cyprinus carpio. J Comp Neurol 230:536–551

Echteler SM, Saidel WM (1981) Forebrain connections in the goldfish support telencephalic homologies with land vertebrates. Science 121:683–685

Finger TE (1975) The distribution of the olfactory tracts in the bullhead catfish, Ictalurus nebulosus. J Comp Neurol 161:125–142

Finger TE (1978) Cerebellar afferents in teleost catfish (Ictaluridae). J Comp Neurol 181:173–182

Fridberg G, Bern HA (1968) The urophysis and the caudal neurosecretory system of fishes. Biol Rev 43:175–199

Garman H (1896) Some notes on the brain and pineal structures of Polyodon folium. Bull Illinois State Lab Nat Hist 4:298–309

González GC, Belenky MA, Polenov AL, Lederis K (1992) Comparative localization of corticotropin and corticotropin releasing factor-like peptides in the brain and hypophysis of a primitive vertebrate, the sturgeon Acipenser ruthenus L. J Neurocytol 21:885–896

Goronowitsch N (1888) Das Gehirn und die Cranialnerven von Acipenser ruthenus. Morphol Jahrb 30:427–574

Hayashida T, Lagios MD (1969) Fish growth hormone: a biological, immunochemical, and ultrastructural study of sturgeon and paddlefish pituitaries. Gen Comp Endocrinol 13:403–411

Herrick CJ (1921) A sketch of origin of the cerebral hemispheres. J Comp Neurol 32:429–454

Herrick CJ (1922) Functional factors in the morphology of the forebrain of fishes. Libro en honor de D Santiago Ramón y Cajal. Molina, Madrid, vol 1, pp 143–204

Hocke Hoogenboom KJ (1929) Das Gehirn von Polyodon folium Lacép. Z Mikrosk Anat Forsch 18:311–392

Hofmann MH, Piñuela C, Meyer DL (1993) Retinopetal projections from diencephalic neurons in a primitive actinopterygian fish, the sterlet Acipenser ruthenus. Neurosci Lett 161:30–32

Holmes RL, Ball JN (1974) The pituitary gland: a comparative account. Cambridge University Press, Cambridge

Holmes PH, Northcutt RG (1995) Afferent and efferent connections of the nonolfactory pallium in the Senegal bichir, Polypterus senegalus (Osteichthyes: Cladistia). Soc Neurosci Abstr 21:432

Ito H, Murakami T, Fukuota T, Kishida R (1986) Thalamic fiber connections in a teleost (Sebastiscus marmoratus): visual, somatosensory, octavolateral and cerebellar relay region to the telencephalon. J Comp Neurol 250:215–227

Johnston JB (1898a) Hind brain and cranial nerves of Acipenser. Anat Anz 14:580–602

Johnston JB (1898b) The olfactory lobes, fore-brain, and habenular tracts of Acipenser. Zool Bull 1:221–241

Johnston JB (1901) The brain of Acipenser. A contribution to the morphology of the vertebrate brain. Zool Jahrb Abt Anat Ontog 15:59–260

Johnston JB (1902) The brain of Petromyzon. J Comp Neurol 12:2–86

Johnston JB (1903) Das Gehirn und die Cranialnerven der Anamnier. Bergmann, Wiesbaden, pp 973–1112

Johnston JB (1906) The nervous system of vertebrates. Blakiston, Philadelphia

Johnston JB (1910) The evolution of the cerebral cortex. Anat Rec 4:143–166

Johnston JB (1911) The telencephalon of ganoids and teleosts. J Comp Neurol 21:489–591

Jørgensen JM, Flock A, Wersäll J (1972) The Lorenzinian ampullae of Polyodon spathula. Z Zellforsch 130:362–377

Knowles F (1965) Evidence for a dual control, by neurosecretion of hormone synthesis and hormone release in the pituitary of the dogfish, Scyliorhinus stellaris. Philos Trans R Soc Lond B 249:435–456

Kobayashi H, Owada K, Yamada C, Okawara Y (1986) The caudal neurosecretory system in fishes. In: Pang PKT, Schreibman MP (eds) Vertebrate endocrinology; fundamentals and biomedial implications. 1. Morphological considerations. Academic, New York, pp 147–174

Kotrschal K, Krautgartner W-D, Adam H (1983) Krönchenzellen im Zwischenhirn von Acipenser ruthenus (Acipenseridae, Chondrostei). J Hirnforsch 24:655–657

Kotrschal K, Krautgartner W-D, Adam H (1985a) Distribution of aminergic neurons in the brain of the sterlet, Acipenser ruthenus (Chondrostei, actinopterygii). J Hirnforsch 26:65–72

Kotrschal K, Lametschwandtner A, Adam H (1985b) Fine structure and vascular supply of the median eminence (ME) in Acipenser ruthenus (Chondrostei). J Hirnforsch 26:333–351

Larsell O (1967) The comparative anatomy and histology of the cerebellum from myxinoids through birds. University of Minnesota Press, Minneapolis

Leprêtre E, Anglade I, Williot P, Vandesande F, Tramu G, Kah O (1993) Comparative distribution of mammalian GnRH (gonadotrophin-releasing hormone) and chicken GnRH-II in the brain of the immature Siberian sturgeon (Acipenser baeri). J Comp Neurol 337:568–583

Libouban S, Szabo T (1977) An integration centre of the mormyrid fish brain: the auricula cerebelli. An HRP study. Neurosci Lett 6:115–119

Mauthner L (1859) Untersuchungen über den Bau des Rückenmarkes des Fische. Sitz Ber K Preuss Wiss 34:31–36

McCormick CA (1981) Central projections of the lateral line and eighth nerves in the bowfin, Amia calva. J Comp Neurol 197:1–15

McCormick CA (1982) The organization of the octavolateralis area in actinopterygian fishes: a new interpretation. J Morphol 171:159–181

Meek J, Nieuwenhuys R (1991) Palisade pattern of mormyrid Purkinje cells. A correlated light and electron microscopic study. J Comp Neurol 306:156–192

Nachtrieb HF (1910) The primitive pores of Polyodon spathula (Walbaum). J Exp Zool 9:455–468

New JG (1981) Central projections of the lateralis nerves in the shovelnose sturgeon, Scaphirhynchus platorynchus. MSc dissertation, Michigan

New JG, Bodznick D (1985) Segregation of electroreceptive and mechanoreceptive lateral line afferents in the hindbrain of chondrostean fishes. Brain Res 336:89–98

New JG, Northcutt RG (1984a) Primary projections of the trigeminal nerve in two species of sturgeon: acipenser oxyrhynchus and Scaphirhynchus platorynchus. J Morphol 182:125–136

New JG, Northcutt RG (1984b) Central projections of the lateral line nerves in the shovelnose sturgeon. J Comp Neurol 225:129–140

Nieuwenhuys R (1962) Trends in the evolution of the Actinopterygian forebrain. J Morphol 111:69–88

Nieuwenhuys R (1963) The comparative anatomy of the Actinopterygian forebrain. J Hirnforsch 6:171–192

Nieuwenhuys R (1964) Further studies on the general structure of the Actinopterygian forebrain. Acta Morphol Neerl Scand 6:65–79

Nieuwenhuys R (1966) The interpretation of the cell masses in the teleostean forebrain. In: Hassler R, Stephan H (eds) Evolution of the forebrain. Thieme, Stuttgart, pp 32–39

Nieuwenhuys R (1967) Comparative anatomy of the cerebellum. Prog Brain Res 25:1–93

Nieuwenhuys R, Bodenheimer TS (1966) The diencephalon of the primitive bony fish Polypterus in the light of the problem of homology. J Morphol 118:415–450

Nieuwenhuys R, Nicholson C (1969) Aspects of the histology of the cerebellum of Mormyrid fishes. In: Llinás R (ed) Neurobiology of cerebellar evolution and development. Am Med Ass Educ Res Fdn, Chicago, pp 135–169

Nieuwenhuys R, Pouwels E (1983) The brain stem of actinopterygian fishes. In: Northcutt RG, Davis RE (eds) Fish neurobiology. 1. Brain stem and sense organs. University of Michigan Press, Ann Arbor, pp 25–87

Norris HW (1925) Observations upon the peripheral distribution of the cranial nerves of certain ganoid fishes (Amia, Lepidosteus, Polyodon, Scpahirhynchus, and Acipenser). J Comp Neurol 39:345–432

Northcutt RG (1981) Localization of neurons afferent to the telencephalon in a primitive bony fish, Polypterus palmas. Neurosci Lett 22:219–222

Northcutt RG (1982) Localization of neurons afferent to the optic tectum in longnose gars. J Comp Neurol 204:325–335

Northcutt RG, Braford MR Jr (1980) New observations on the organization and evolution of the telencephalon of actinopterygian fishes. In: Ebbesson SOE (ed) Comparative neurology of the telencephalon. Plenum, New York, pp 41–98

Northcutt RG, Butler AB (1976) Retinofugal pathways in the longnose gar Lepisosteus osseus (Linnaeus). J Comp Neurol 166:1–16

Northcutt RG, Butler AB (1980) Projections of the optic tectum in the longnose gar, Lepisosteus osseus. Brain Res 190:333–346

Northcutt RG, Butler AB (1993) The diencephalon and optic tectum of the longnose Gar, Lepisosteus osseus (L): cytoarchitectonics and distribution of acetylcholinesterase. Brain Behav Evol 41:57–81

Northcutt RG, Davis RE (1983) Telencephalic organization in ray-finned fishes. In: Davis RE, Northcutt RG (eds) Higher brain areas and functions. The University of Michigan Press, Ann Arbor, pp 203–236 (Fish neurobiology, vol 2)

Oka S, Honma Y, Iwanaga T, Fujita T (1989) Immunohistochemical demonstration of urotensins I and II in the caudal neurosecretory system of the white sturgeon, Acipenser transmontanus Richardson. Biomed Res 3:329–340

Onstott D, Elde R (1986) Immunohistochemical localization of urotensin I/corticotropin-releasing factor, urotensin II, and serotonin immunoreactivities in the caudal spinal cord of nonteleost fishes. J Comp Neurol 249:205–225

Palmgren A (1921) Embryological and morphological studies on the mid-brain and cerebellum of vertebrates. Acta Zool 2:1–94

Pearson AA (1936) The acustico-lateral centers and the cerebellum, with fiber connections, of fishes. J Comp Neurol 65:201–294

Piñuela C, Northcutt RG (1994) Immunohistochemical organization of the forebrain in the white sturgeon. Soc Neurosci Abstr 20:997

Polenov AL, Garlov PE (1971a) The hypothalamo-hypohysial system in Acipenseridae. I. Ultrastructural organization of large neurosecretory terminals (Herring Bodies) and axoventricular contacts. Z Zellforsch 116:319–374

Polenov AL, Garlov PE (1971b) The hypothalamo-hypohysial system in Acipenseridae. III. The neurohypophysis of Acipenser güldenstädti Brandt and Acipenser stellatus Pallas. Z Zellforsch 136:461–477

Polenov AL, Garlov PE (1971c) The hypothalamo-hypohysial system in Acipenseridae. IV. The functional morphology of the neurohypophysis of Acipenser güldenstädti Brandt and Acipenser stellatus Pallas after exposure to different salinities. Cell Tissue Res 148:259–275

Polenov AL, Pavlovic M (1978) The hypothalamo-hypophysial system in Acipenseridae. VII. The functional morphology of the peptidergic neurosecretory cells in the preoptic nucleus of the sturgeon, Acipenser güldenstädti Brandt. A quantitative study. Cell Tissue Res 186:559–570

Polenov AL, Garlov PE, Konstantinova MS, Belenky MA (1972) The hypothalamo-hypohysial system in Acipenseridae. II. Adrenergic structures of the hypophysial neurointermediate complex. Z Zellforsch 128:470–481

Polenov AL, Garlov PE, Koryakina ED, Faleeva TI (1976a) The hypothalamo-hypohysial system in Acipenseridae. V. Ecological-histophysiological analysis of the neurohypophysis of the female sturgeon Acipenser güldenstädti Brandt during up-stream migration and after spawning. Cell Tissue Res 170:113–128

Polenov AL, Belenky MA, Garlov PE, Konstantinova MS (1976b) The hypothalamo-hypophysial system in Acipenseridae. VI. The proximal neurosecretory contact region. Cell Tissue Res 170:129–144

Polenov AL, Belenky MA, Garlov PE (1979) The hypothalamo-hypophysial system in Acipenseridae. VIII. Quantitative electron microscopic study of the functional state of neurosecretory terminals in the neurohypophysis of Acipenser güldenstädti Brandt during up-stream migration and after spawning. Cell Tissue Res 203:311–320

Polenov AL, Efimova NA, Kontantinova M, Senchik YI, Yakovleva IV (1983) The hypothalamo-hypophysial system in Acipenseridae. Cell Tissue Res 232:651–667

Pouwels E (1978a) On the development of the cerebellum of the trout, Salmo gairdneri. I. Patterns of cell migration. Anat Embryol (Berl) 152:291–308

Pouwels E (1978b) On the development of the cerebellum of the trout, Salmo gairdneri. IV. Development of the pattern of connectivity. Anat Embryol (Berl) 153:55–65

Reiner A, Northcutt RG (1992) An immunohistochemical study of the telencephalon of the Senegal bichir (Polypterus senegalus). J Comp Neurol 319:359–386

Repérant J, Vesselkin NP, Ermakova TV, Rustamov EK, Rio J-P, Palatnikov GK, Peyrichoux J, Kasimov RV (1982) The retinofugal pathways in a primitive actinopterygian, the chondrostean Acipenser güldenstädt. An experimental study using degeneration, radioautoradiographic and HRP methods. Brain Res 251:1–23

Sathyanesan AG, Chavin W (1967) Hypothalamo-hypophyseal neurosecretory system in the primitive actinopterygian fishes (Holostei and Chondrostei). Acta Anat 68:284–299

Schaper A (1894a) Die morphologische und histologische Entwickelung des Kleinhirns der Teleostier. Anat Anz 9:489–501

Schaper A (1894b) Die morphologische und histologische Entwickelung der Teleostier. Morphol Jahrb 21:625–708

Story RH (1964) The olfactory bulbar formation and related nuclei of the paddlefish (Polyodon spathula). J Comp Neurol 123:285–298

Striedter GF (1990) The diencephalon of the channel catfish, Ictalurus punctatus. II. Retinal, tectal, cerebellar and telencephalic connections. Brain Behav Evol 36:355–377

Striedter GF (1991) Auditory, electrosensory and mechano-sensory lateral line pathways through the forebrain in channel catfishes. J Comp Neurol 312:311–331

Teeter JH, Szamier RB, Bennett MVL (1980) Ampullary electroreceptors in the sturgeon Scaphirhynchus platorynchus (Rafinesque). J Comp Physiol [A] 138:213–233

Theunissen F (1914) The arrangement of the motor roots and nuclei in the brain of Acipenser ruthenus and Lepisosteus osseus. Proc Kon Ned Akad Wet (Amsterdam) 16:1032–1041

Von Bartheld CS, Meyer DL (1986) Central connections of the olfactory bulb in the bichir, Polypterus palmas, reexamined. Cell Tissue Res 244:527–535

Von Kupffer K (1893) Studien zur vergleichenden Entwicklungsgeschichte des Kopfes der Cranioten, vol 1. Lehman, Munich, pp 1–95

Von Kupffer K (1906) Die Morphogenie des Centralnervensystems. In: von Hertwig O (ed) Handbuch der vergleichenden und experimentellen Entwicklungsgeschichte der Wirbeltiere, vol 2, part 3. Fischer, Jena, pp 1–272